PRINCIPLES OF MARINE SCIENCE

PRINCIPLES OF MARINE SCIENCE

Editor
Richard M. Renneboog, M.Sc.
The University of Western Ontario

SALEM PRESS
A Division of EBSCO Information Services, Inc.
Ipswich, Massachusetts

GREY HOUSE PUBLISHING

Cover photo: iStock by Getty Images

Principles of Marine Science, published by Grey House Publishing, Inc., Amenia, NY, under exclusive license from EBSCO Information Services, Inc.

For information contact Grey House Publishing/Salem Press, 4919 Route 22, PO Box 56, Amenia, NY 12501.

∞ The paper used in these volumes conforms to the American National Standard for Permanence of Paper for Printed Library Materials, Z39.48 1992 (R2009).

Publisher's Cataloging-In-Publication Data
(Prepared by The Donohue Group, Inc.)

Names: Renneboog, Richard, editor.
Title: Principles of marine science / editor, Richard M. Renneboog, M.Sc., the University of Western Ontario.
Description: Ipswich, Massachusetts : Salem Press, a division of EBSCO Information Services, Inc.; Amenia, NY : Grey House Publishing, [2020] | Series: Principles of science ; [21] | Includes bibliographical references and index.
Identifiers: ISBN 9781642657661 | ISBN 9781642657678 (ebook)
Subjects: LCSH: Marine sciences. | LCGFT: Reference works.
Classification: LCC GC11.2 .P75 2020 (print) | LCC GC11.2 (ebook) | DDC 551.46—dc23

First Printing
Printed in the United States of America

Contents

Publisher's Note

Salem Press is pleased to add *Principles of Marine Science* to its growing list of *Principles of Science* titles. From *Information Technology* to *Sustainability*, this series introduces the researcher to the fundamentals of dozens of scientific topics. Using easy-to-understand language that provides both a solid background and a deeper understanding and appreciation of this important and evolving subject, *Marine Science* contains 135 entries which are arranged alphabetically, from "Acoustics" to "Yellow Sea Ecosystem," making it easy to find the topic of interest.

Each entry includes the following:

- A list of the *Fields of Study* covered by the article;
- An *Abstract* that provides a brief, concrete summary of the topic and its significance;
- A list of *Key Concepts* with definitions, important to a proper understanding of the topic;
- A detailed *Essay* that provides extensive background on the topic and explores its significance to the field of marine science;
- A list of *Further Reading* for those who wish to pursue the topic in more depth.

In addition, the book includes a number of helpful appendices, comprising the following:

- A *Glossary* that defines all the key specialized terms used throughout the book;
- A comprehensive list of *Further Reading*, comprising works that the authors drew upon in writing their articles as well as subjects for further study;
- A *Subject Index* that offers multiple points of entry for the reader.

Salem Press and Grey House Publishing extend their appreciation to all involved in the development and production of this work, especially volume editor Richard Renneboog who also contributed a comprehensive introduction. The entries have been written by experts in the field of marine science. Their names and affiliations follow this note.

Introduction

Seen from ground level, Earth's surface is rugged and craggy, leading it to be commonly classed among of the four so-called rocky planets. But to an observer from space, this brilliant blue-and-white marble that orbits the sun would most likely be called a water planet. Its skies are filled with masses of fluffy clouds composed of tiny droplets of condensed water vapor. Below them, the view of the surface is almost entirely one of liquid water and water ice interspersed with areas of "non-water." Even within those non-water areas, vast quantities of liquid water can be seen streaking across in thin ribbons and sometimes forming large bodies, water that seems somehow different from the world-wide mass of the oceans.

As the view of the surface of the planet gets closer, evidence of a multitude of different life forms comes into view. One sees vast green areas of innumerable kinds of plant species covering much of the solid areas of the planet, and detects an unimaginable richness of both plant and animal life in the water systems that punctuate the non-water areas and in the vast oceans of water that cover the rest of the globe. It quickly becomes obvious that this planet Earth is a dynamic global system in which the water controls all aspects of the environment and even of life itself.

Investigating further, something very interesting is discovered: all of the life forms are water-based! Some are almost entirely composed of water, and all of the rest, whether plant or animal, are at least two-thirds water internally. Their water content serves many purposes. It transports nutrients and waste products. It provides the medium in which hundreds of biochemical reactions take place in their cells to keep them alive. Another feature of interest comes to light as well. The water within all of the living organisms on the planet is essentially identical to the salty water that covers the surface of almost three-quarters of the planet. Yet, oddly enough, they all require the non-salty water typical of the non-water areas of the planet and that falls from the sky as precipitation all around the planet. Ingesting the salty water itself causes great physical distress to the life forms and may actually prove fatal.

Marine science is clearly a discipline that is central to our biological understanding of our planet. Water—the marine environment—is the major controlling factor of the entire planetary ecosystem. The interaction between the oceans and the atmosphere, affected as it is by the rotation of the planet on its axis, is by far the most noticeable mechanism at work on Earth. The atmosphere, composed entirely of gases, and the oceans, consisting entirely of liquid and solid water, may seem like independent bodies. But through the energy provided by the Sun, they behave as an integrated whole, each affecting the other in many ways.

The interplay of air and water begins when energy from the Sun strikes the surface of the planet directly along Earth's equatorial region. The energy is then absorbed and water evaporates from the surface to rise up into the atmosphere, carried aloft by air that becomes less dense as it gets warmer. Air is a fluid, and so cooler air at the surface is drawn toward the equatorial region from the north and the south to replace the air that has risen upward from the equatorial region. As this goes on, the rising air becomes cooler and moves to the north and south at high altitude, becoming denser as it cools and sinks toward the surface to replace the air that has been drawn towards the equator along the surface. This continual movement of air creates a continuous convection cell around the planet. This is a simple physical state of convection in which all air movement occurs strictly in a north-south direction. But it's not that simple, because within the atmospheric envelope, the planet spins on its axis.

This relative motion brings about something called "the Coriolis effect," so that the movement of the air has an east-to-west directional component to either side of the equator. This is what humans have termed "the trade winds." A similar convection system surrounds the planet farther to the north and south of the so-called "Hadley cell" system of the equatorial region. In these new convection cells, however, the air moves in the opposite direction. Beyond this cell system, to the north and south are the polar regions, where the cold, dense air is more or less held in place by northern and southern "jet streams"—high velocity wind currents like rivers of air flowing around the planet. As this colder air sinks away from the poles, it warms somewhat, becomes less dense and rises. This movement drives these northern and southern convection cells, and produces surface winds that now have a west-to-east directionality.

Wind moving over the surface of water causes water to move in the same direction. At the same time, thermal energy is transferred from the water to the air. The water becomes denser as it cools, and it sinks to lower levels to match its density. As water evaporates from the surface of the oceans, its salinity—salt dissolved content—increases, which also adds to the change in density. As a result, the water goes into what is called "thermohaline circulation," continually adjusting to differing conditions of temperature and salinity and changing density accordingly as it moves all around the planet. This is the engine that drives essentially all of the dynamic nature of the hydrosphere—the water part of planet Earth.

There is, in addition, another cycle at play, one that involves the areas of the planet that are not water. The water that rose into the atmosphere as vapor eventually cools to form clouds that are moved along by the various winds. Eventually, the clouds are cooled sufficiently so that the water can no longer be supported and it falls to the surface as liquid rain in some areas and as solid ice particles called snowflakes in others. In some areas it does not fall at all, or at least only very rarely. This precipitation that falls on water has completed its part in the cycle. But the precipitation that falls on areas that are not water is either absorbed into the soil or runs off across the surface to become streams, rivers and lakes, eventually returning to the oceans and completing the "hydrological cycle." As it goes through this cycle, it supports the multitude of plant and animal organisms on the land that depend on freshwater for life.

Many of the rocks of the planet's crust are composed partly of water. In the oceans, the crust moves slowly, great chunks of it grinding together inexorably, driven by convection currents in the layer of molten rock that is below the crust. In many places, there are mountain ridges on the deep ocean floor where tectonic plates are being pushed apart by the movement in the molten rock layer beneath them. As these plates slowly separate, new molten rock oozes up between them and freezes into solidity as it moves away from the rift. It stands to reason that if new seafloor is being pushed outward, then old seafloor must be going somewhere too and disappearing, otherwise the planet would be getting larger all the time as the crust expanded in size.

The answer is found near the continental land masses, where exceptionally deep waters can be found in trenches deep in the seafloor. In those places, the slow, inexorable movement of the seafloor is pushing downward beneath the less-dense continental rock masses. That means that the water trapped inside the seafloor rock as it solidified is being carried back into the molten layer below. The heat of the molten layer is sufficient to drive the water out of its mineral prison, enabling it to erupt as ultra-high pressure steam with enough power to blow apart mountains.

Near those places where the seafloor is spreading apart there are chimney-like structures where superheated mineral-heavy water shoots up into the cold water of the ocean depths. Black clouds of mineral particles precipitate from the superheated jets of some of the chimney structures as the water cools. From others, where the water shoots out at a lower temperature, white clouds of mineral particles precipitate out. Where does the superheated water come from?

Pressure forces water to infiltrate the rock of the seafloor driving it to move through cracks and fissures to depths where it becomes superheated, but not so far that it encounters the underlying molten layer—the mantle—of the planet. At some point, this superheated water reaches a location where there is a sudden opening up of the narrow cracks and fissures, and an equally sudden decrease of pressure as the water finds a new degree of freedom sufficient for it to be driven upward and out through the chimney above.

This superheated water is much too hot and mineral rich for any living organism to survive in, and no sunlight can reach to the depths of the seafloor. Yet, there are large colonies of living organisms near these hot mineral fountains. These life forms derive their energy from chemical reduction reactions rather than from sunlight, relying on chemosynthesis rather than photosynthesis. These colonies are one of the most unique ecosystems on the planet.

Earth's water then is the ultimate regulator and driving force behind all of the planet's various activities, second only to the Sun and the heat generated inside the planet by radioactive decay and friction due to the rotational motion of the interior of the planet. The water and the land areas absorb energy from the Sun and then return it to the atmosphere. The atmosphere moves against the water, making the water move and affecting its temperature and salinty, and thus its density, and the result is a global pattern of thermohaline circulation that functions like a "global conveyor belt" of energy. It is a wonderfully complex integrated system indeed, and home to a great multitude of living things.

Drawing back from a close inspection of the earth, we notice something startling: the water in different parts of the ocean expanse moves in a generally circular pattern, and in the center of these "gyres" there are tons of plastic. Some of the plastic looks very much like the natural food of the creatures that live in those waters. Other plastic things act like snares or traps, entangling creatures and preventing them from swimming and breathing. In other parts of the ocean, a sticky, black ooze insults the pristine surface of the water and sullies the clean shores of nearby land area, covering innumerable marine creatures in a thick tarry coating that proves fatal. All of these unnatural things happening in the oceans of the world are the result of human activities.

Marine science is the key to understanding the nature of planet Earth that we all call home. The articles in this volume have been carefully selected to provide a sound foundation and starting point for beginning the process of understanding Earth's marine nature. The waters that cover more than 70 percent of the planet, create a liquid realm that humans have known for as long as the human race has existed and yet understand so poorly.

—Richard M. Renneboog M.Sc.

Contributors

Arthur L. Alt
Independent Scholar

James T. Anderson
Independent Scholar

George K. Attwood
Independent Scholar

Michael P. Auerbach
Independent Scholar

Alexandra M. Avila
Independent Scholar

Victor R. Baker
University of Arizona

Kevin Bakker
University of Michigan School of Public Health

Hannah Bement
George Mason University

Harlan H. Bengtson
Southern Illinois University Edwardsville

John L. Berkley
Independent Scholar

Victoria M. Breting-García
Independent Scholar

Byron D. Cannon
University of Utah

Roger V. Carlson
Independent Scholar

Michael J. Caulfield
Gannon University

Zhiqiang Cheng
University of Hawaii at Manoa

Habte Giorgis Churnet
University of Tennessee at Chattanooga

John H. Corbet
University of Memphis

Robert L. Cullers
Kansas State University

Larry E. Davis
University of Pittsburgh

Albert B. Dickas
University of Wisconsin-Superior

Walter C. Dudley, Jr.
University of Hawaii at Hilo

Dean A. Dunn
American Chemical Society

David Elliott
Northern Arizona University

Joanna C. Ellison
University of Tasmania

Catherine G. Fontana
Independent Scholar

A. Kem Fronabarger
College of Charleston

Gillian Galford
The University of Vermont, Rubenstein School of Environment and Natural Resources

Joyce Gawell
Independent Scholar

Maria Jose González-Bernat
Interamerican Association for Environmental Defense

Lucia M. Gutierrez
Independent Scholar

Gina Hagler
Independent Scholar

Sarah H. Heller
Independent Scholar

Howard V. Hendrix
California State University, Fresno

Ruth M. Higgins
Institute of Marine Research

J.D. Ho
Independent Scholar

Robert M. Hordon
Rutgers University

Falk Huettmann
Institute of Arctic Biology

Samuel F. Huffman
University of Wisconsin, River Falls

Nasseer Idrisi
California Department of Fish and Wildlife

Micah L. Issitt
Independent Scholar

Zeehan Jaafar
National University of Singapore

Pamela Jansma
University of Colorado Denver

Sanghoon Kang
Eastern Illinois University

Christopher Keating
Independent Scholar

Narayanan M. Komerath
Georgia Institute of Technology

Jason Krumholz
McLaughlin Research Corporation

Bill Kte'pi
Independent Scholar

Gary G. Lash
SUNY Fredonia

David M. Lawrence
Independent Scholar

M. Lee
Independent Scholar

Manja Leyk
Independent Scholar

Yanna Liang
University at Albany

W. David Liddell
Utah State University

Donald W. Lovejoy
Palm Beach Atlantic University

R.C. Lutz
CII Group

David W. Maguire
Murdoch University

Susan Moegenburg
University of Vermont

Magdalena Ariadne Kim Muir
Arctic Institute of North America

Otto H. Muller
Alfred University

John E. Mylroie
Mississippi State University

Edward P. Nuhfer
California State University of the Channel Islands

George R. Plitnik
Frostburg State University

Elizabeth Rholetter Purdy
Independent Scholar

C. Nicholas Raphael
Eastern Michigan University

Richard M. Renneboog
Independent Scholar

Barbara J. Rich
Independent Scholar

Charlie Rosenberg
Independent Scholar

James L. Sadd
Occidental College

Santosh Kumar Sarkar
University of Calcutta

Panagiotis D. Scarlatos
Florida Atlantic University

Elizabeth D. Schafer
Independent Scholar

Rose Secrest
Independent Scholar

R. Baird Shuman
University of Pennsylvania

Robert J. Stern
University of Texas, Dallas

Frederick M. Surowiec
Independent Scholar

Lida Teneva
California Ocean Science Trust

Kenrick Vezina
Independent Scholar

Jennifer Rhode Ward
University of North Carolina, Ashville

Christine Watts
University of Sydney

Judith S. Weis
Rutgers University

Ken Whalen
Independent Scholar

Nan White
Independent Scholar

Thomas A. Wikle
Oklahoma State University

Grant R. Woodwell
University of Mary Washington

Lisa A. Wroble
Independent Scholar

Robin L. Wulffson
American Medical Writers Association

A

ACOUSTICS

Fields of Study: physics; marine science; marine biology; sonography

ABSTRACT

Acoustics is the science dealing with the production, transmission, and effects of vibration in material media. If the medium is air and the vibration frequency is between 18 and 18,000 hertz (Hz), the vibration is called "sound." Acoustics is also used in a broader context to describe sounds in solids, underwater, and structure-borne sounds. Mechanical vibrations, whether natural or human induced, have accompanied humans through the long course of human evolution, thus acoustics is among the most interdisciplinary of sciences.

KEY CONCEPTS

frequency: the characteristic of a sound, defined as the number of complete cycles of the sound frequency in one second, stated in hertz

hertz: the basic frequency unit, defined as one complete cycle of the frequency signal

infrasonic: the designation of sound frequencies that are below the lower limit of human hearing

piezoelectric: crystalline material that produces an electrical output in response to a physical stimulus

ultrasonic: the designation of sound frequencies that are beyond the upper limit of human hearing

DEFINITION AND BASIC PRINCIPLES

The words "acoustics" and "phonics" evolved from ancient Greek roots for hearing and speaking, respectively. Thus, acoustics began with human communication, making it one of the oldest if not the most basic of sciences. Because acoustics is ubiquitous in human endeavors, it is perhaps the broadest and most interdisciplinary of sciences, and its most profound contributions have occurred when it is comingled with an independent field. The interdisciplinary nature of acoustics has often consigned it to a subsidiary role as a minor subdivision of mechanics, hydrodynamics, or electrical engineering. Certainly, the various technical aspects of acoustics could be parceled out to larger and better-established divisions of science, but then acoustics would lose its unique strengths and its source of dynamic creativity. The main difference between acoustics and more self-sufficient branches of science is that acoustics depends on physical laws developed in and borrowed from other fields. Therefore, the primary task of acoustics is to take these divergent principles and integrate them into a coherent whole in order to understand, measure, and control vibration phenomena.

The Acoustical Society of America subdivides acoustics into fifteen main areas, the most important of which are:

- ultrasonics, which examines high-frequency waves not audible to humans;
- psychological acoustics, which studies how sound is perceived in the brain;
- physiological acoustics, which looks at human and animal hearing mechanisms;
- speech acoustics, which focuses on the human vocal apparatus and oral communication;
- musical acoustics, which involves the physics of musical instruments;
- underwater sound, which examines the production and propagation of sound in liquids; and
- noise, which concentrates on the control and suppression of unwanted sound.

Two other important areas of applied acoustics are architectural acoustics (including the acoustical design of concert halls and sound reinforcement systems) and audio engineering (recording and reproducing sound).

BACKGROUND AND HISTORY

Arguably, acoustics originated with human communication and music. The caves in which prehistoric humans displayed their most elaborate paintings have resonances easily excited by the human voice, and stalactites emit musical tones when struck or rubbed with a stick. Paleolithic societies constructed flutes of bird bone, used animal horns to produce drones, and employed rattles and scrapers to provide rhythm.

In the sixth century BCE, Pythagoras was the first person to correlate musical sounds and mathematics by relating consonant musical intervals to simple ratios of integers. In the fourth century BCE, Aristotle deduced that the medium that carries a sound must be compressed by the sounding body, and the third century BCE philosopher Chrysippus correctly depicted the propagation of sound waves with an expanding spherical pattern. In the first century BCE, the Roman architect and engineer Marcus Vitruvius Pollio explained the acoustical characteristics of Greek theaters, but when the Roman civilization declined in the fourth century, scientific inquiry in the West essentially ceased for the next millennium.

In the seventeenth century, modern experimental acoustics originated when the Italian mathematician Galileo explained resonance as well as musical consonance and dissonance, and theoretical acoustics got its start with Sir Isaac Newton's derivation of an expression for the velocity of sound. Although this yielded a value considerably lower than the experimental result, a more rigorous derivation by Pierre-Simon Laplace in 1816 obtained an equation yielding values in complete agreement with experimental results.

During the eighteenth century, many famous mathematicians studied vibration. In 1700, French mathematician Joseph Sauveur observed that strings vibrate in sections consisting of stationary nodes located between aggressively vibrating antinodes and that these vibrations have integer multiple frequencies, or harmonics, of the lowest frequency. He also noted that a vibrating string could simultaneously produce the sounds of several harmonics. In 1755, Daniel Bernoulli proved that this resultant vibration was the independent algebraic sum of the various harmonics. In 1750, Jean le Rond d'Alembert used calculus to obtain the wave equation for a vibrating string. By the end of the eighteenth century, the basic experimental results and theoretical underpinnings of acoustics were extant and in reasonable agreement, but it was not until the following century that theory and a concomitant advance of technology led to the evolution of the major divisions of acoustics.

Although mathematical theory is central to all acoustics, the two major divisions, physical and applied acoustics, evolved from the central theoretical core. In the late nineteenth century, Hermann von Helmholtz and Lord Rayleigh (John William Strutt), two polymaths, developed the theoretical aspects. Helmholtz's contributions to acoustics were primarily in explaining the physiological aspects of the ear. Rayleigh, a well-educated, wealthy English baron, synthesized virtually all previous knowledge of acoustics and also formulated an appreciable corpus of experiment and theory.

Experiments by Georg Simon Ohm indicated that all musical tones arise from simple harmonic vibrations of definite frequency, with the constituent components determining the sound quality. This gave birth to the field of musical acoustics. Helmholtz's studies of instruments and Rayleigh's work contributed to the nascent area of musical acoustics. Helmholtz's knowledge of ear physiology shaped the field that was to become physiological acoustics.

Underwater acoustics commenced with theories developed by the nineteenth-century mathematician Siméon-Denis Poisson, but further development had to await the invention of underwater transducers in the next century.

Two important nineteenth-century inventions, the telephone (patented 1876) and the mechanical phonograph (invented 1877), commingled and evolved into twentieth-century audio acoustics when united with electronics. Some products in which sound production and reception are combined are microphones, loudspeakers, radios, talking motion pictures, high-fidelity stereo systems, and public sound-reinforcement systems. Improved instrumentation for the study of speech and hearing has stimulated the areas of physiological and psychology acoustics, and ultrasonic devices are routinely used for medical diagnosis and therapy, as well as for burglar alarms and rodent repellants. Underwater transducers are employed to detect and measure moving objects in the water, while audio engineering technology has transformed music performance as well as sound reproduction. Virtually no area of human activity has remained unaffected by continually evolving technology based on acoustics.

HOW IT WORKS

Hydroacoustics (Underwater Sound). This field uses frequencies between 10 Hz and 1 megahertz (MHz).

Although the origin of hydroacoustics can be traced back to Rayleigh, the deployment of submarines in World War I provided the impetus for the rapid development of underwater listening devices (hydrophones) and sonar (sound navigation ranging), the acoustic equivalent of radar. Pulses of sound are emitted and the echoes are processed to extract information about submerged objects. When the speed of underwater sound is known, the reflection time for a pulse determines the distance to an object. If the object is moving, its speed of approach or recession is deduced from the frequency shift of the reflection, or the Doppler effect. Returning pulses have a higher frequency when the object approaches and lower frequency when it moves away. Hydroacoustics is a field that is routinely applied in marine research, for many purposes.

Ultrasonics. Dog whistles, which can be heard by dogs but not by humans, can generate ultrasonic frequencies of about 25 kilohertz (kHz). Two types of transducers, magnetostrictive and piezoelectric, are used to generate higher frequencies and greater power. Magnetostrictive devices convert magnetic energy into ultrasound by subjecting ferric material (iron or nickel) to a strong oscillating magnetic field. The field causes the material to alternately expand and contract, thus creating sound waves of the same frequency as that of the field. The resulting sound waves have frequencies between 20 Hz and 50 kHz and several thousand watts of power. Such transducers operate at the mechanical resonance frequency where the energy transfer is most efficient.

Piezoelectric transducers convert electric energy into ultrasound by applying an oscillating electric field to a piezoelectric crystal (such as quartz). These transducers, which work in liquids or air, can generate frequencies in the megahertz region with considerable power. In addition to natural crystals, ceramic piezoelectric materials, which can be fabricated into any desired shape, have been developed.

Physiological and Psychological Acoustics. Physiological acoustics studies auditory responses of the ear and its associated neural pathways, and psychological acoustics is the subjective perception of sounds through human auditory physiology. Mechanical, electrical, optical, radiological, or biochemical techniques are used to study neural responses to various aural stimuli. Because these techniques are typically invasive, studies are carried out with animals having auditory systems that are similar to the human system. In contrast, psychological acoustic studies are noninvasive and typically use human subjects.

A primary objective of psychological acoustics is to define the psychological correlates to the physical parameters of sound waves. Sound waves in air may be characterized by three physical parameters: frequency, intensity, and their spectrum. When a sound wave impinges on the ear, the pressure variations in the air are transformed by the middle ear to mechanical vibrations in the inner ear. The cochlea then decomposes the sound into its constituent frequencies and transforms these into neural action potentials, which travel to the brain where the sound is evidenced. Frequency is perceived as pitch, the intensity level as loudness, and the spectrum determines the timbre, or tone quality, of a note.

Another psychoacoustic effect is masking. When a person listens to a noisy version of recorded music, the noise virtually disappears if the music is being enjoyed. This ability of the brain to selectively listen has had important applications in digitally recorded music. When the sounds are digitally compressed, such as in MP3 (MPEG-1 audio layer 3) systems, the brain compensates for the information lost as the difference between analog and digital sound reproduction; thus one experiences higher fidelity sound than the stored content would imply. Also, the brain creates information when the incoming signal is masked or nonexistent, producing a psychoacoustic phantom effect. This phantom effect is particularly prevalent when heightened perceptions are imperative, as when danger is lurking.

Psychoacoustic studies have determined that the frequency range of hearing is from 20 to about 20,000 Hz for young people, and the upper limit progressively decreases with age. The rate at which hearing acuity declines depends on several factors, not the least of which is lifetime exposure to loud sounds, which progressively deteriorate the hair cells of the cochlea. Moderate hearing loss can be compensated for by a hearing aid; severe loss requires a cochlear implant.

Speech Acoustics. Also known as acoustic phonetics, speech acoustics deals with speech production and recognition. The scientific study of speech began with Thomas Alva Edison's phonograph, which allowed a speech signal to be recorded and stored for later analysis. Replaying the same short speech segment several times using consecutive filters passing through a limited range of frequencies creates a spectrogram, which visualizes the spectral properties of vowels and consonants. During the first half of the twentieth century, Bell Telephone Laboratories invested considerable time and resources to the systematic understanding of all aspects of speech, including vocal tract resonances, voice quality, and prosodic features of speech. For the first time, electric circuit theory was applied to speech acoustics, and analog electric circuits were used to investigate synthetic speech.

Musical Acoustics. A conjunction of music, craftsmanship, auditory science, and vibration physics, musical acoustics analyzes musical instruments to better understand how the instruments are crafted, the physical principles of their tone production, and why each instrument has a unique timbre. Musical instruments are studied by analyzing their tones and then creating computer models to synthesize these sounds. When the sounds can be recreated with minimal software complications, a synthesizer featuring realistic orchestral tones may be constructed. The second method of study is to assemble an instrument or modify an existing instrument to perform nondestructive (or on occasion destructive) testing so that the effects of various modifications may be gauged.

Noise. Physically, noise may be defined as an intermittent or random oscillation with multiple frequency components, but psychologically, noise is any unwanted sound. Noise can adversely affect human health and well-being by inducing stress, interfering with sleep, increasing heart rate, raising blood pressure, modifying hormone secretion, and even inducing depression. The environmental effects of noise are no less severe. The vibrations in irregular road surfaces caused by large rapid vehicles can cause adjacent buildings to vibrate to an extent that is intolerable to the buildings' inhabitants, even without structural damage. Machinery noise in industry is a serious problem because continuous exposure to loud sounds will induce hearing loss. In apartment buildings, noise transmitted through walls is always problematic; the goal is to obtain adequate sound insulation using lightweight construction materials.

Traffic noise, both external and internal, is ubiquitous in modern life. The first line of defense is to reduce noise at its source by improving engine enclosures, mufflers, and tires. The next method, used primarily when interstate highways are adjacent to residential areas, is to block the noise by the construction of concrete and other solid barriers or the planting of sound-absorbing vegetation. Internal automobile noise has been greatly abated by designing more aerodynamically efficient vehicles to reduce air turbulence, using better sound isolation materials, and improving vibration isolation.

Aircraft noise, particularly in the vicinity of airports, is a serious problem exacerbated by the fact that as modern airplanes have become more powerful, the noise they generate has risen concomitantly. The noise radiated by jet engines is reduced by two structural modifications. Acoustic linings are placed around the moving parts to absorb the high frequencies caused by jet whine and turbulence, but this modification is limited by size and weight constraints. The second modification is to reduce the number of rotor blades and stator vanes, but this is somewhat inhibited by the desired power output. Special noise problems occur when aircraft travel at supersonic speeds (faster than the speed of sound), as this propagates a large pressure wave toward the ground that is experienced as an explosion. The unexpected sonic boom startles people, can break windows, and may damage houses structurally. Sonic booms have been known to destroy rock structures in national parks. Because of these concerns, commercial aircraft are prohibited from flying at supersonic speeds over land areas.

Construction equipment (such as earthmoving machines) creates high noise levels both internally and externally. When the cabs of these machines are not closed, the only feasible manner of protecting operators' hearing is by using ear plugs. By carefully designing an enclosed cabin, structural vibration can be reduced and sound leaks made less significant, thus quieting the operator's environment. Although manufacturers are attempting to reduce the external noise, it is a daunting task because quieter components, such as the rubber tractor treads occasionally used to replace metal, are often not as durable.

APPLICATIONS AND PRODUCTS

Hydroacoustics. Applications for hydroacoustics include devices for underwater communication by acoustic means, remote control devices, underwater navigation and positioning systems, acoustic thermometers to measure ocean temperature, and echo sounders to locate schools of fish or other biota. Low-frequency devices can be used to explore the seabed for seismic research. Large creatures such as whales use very low frequency sound (infrasound) to communicate over long distances, while less massive aquatic creatures use audible and high-frequency sound both for communication and echolocation. Hydroacoustics is therefore used extensively in marine science for similar applications and the study of marine biota and underwater geological features.

Although primitive measuring devices were developed in the 1920s, it was during the 1930s that sonar systems began incorporating piezoelectric transducers to increase their accuracy. These improved systems and their increasingly more sophisticated progeny became essential for the submarine warfare of World War II. After the war, theoretical advances in underwater acoustics coupled with computer technology have raised sonar systems to ever more sophisticated levels, enabling their application in marine science for precise purpose with increasingly accurate observational results.

Ultrasonics. High-intensity ultrasonic applications include ultrasonic cleaning, mixing, welding, drilling, and various chemical processes. Ultrasonic cleaners use waves in the 150 to 400 kHz range on items (such as jewelry, watches, lenses, and surgical instruments) placed in an appropriate solution. Ultrasonic cleaners have proven to be particularly effective in cleaning surgical devices because they loosen contaminants by aggressive agitation irrespective of an instrument's size or shape, and disassembly is not required. Ultrasonic waves are effective in cleaning most metals and alloys, as well as wood, plastic, rubber, and cloth. Ultrasonic waves are used to emulsify two immiscible liquids, such as oil and water, by forming the liquids into finely dispersed particles that then remain in homogeneous suspension. Many paints, cosmetics, and foods are emulsions formed by this process.

Although aluminum cannot be soldered by conventional means, two surfaces subjected to intense ultrasonic vibration will bond—without the application of heat—in a strong and precise weld. Ultrasonic drilling is effective where conventional drilling is problematic, for instance, drilling square holes in glass. The drill bit, a transducer having the required shape and size, is used with an abrasive slurry that chips away the material when the suspended powder oscillates. Some of the chemical applications of ultrasonics are in the atomization of liquids, in electroplating, and as a catalyst in chemical reactions.

Low-intensity ultrasonic waves are used for nondestructive probing to locate flaws in materials for which complete reliability is mandatory, such as those used in spacecraft components and nuclear reactor vessels. When an ultrasonic transducer emits a pulse of energy into the test object, flaws reflect the wave and are detected. Because objects subjected to stress emit ultrasonic waves, these signals may be used to interpret the condition of the material as it is increasingly stressed. Another application is ultrasonic emission testing, which records the ultrasound emitted by porous rock when natural gas is pumped into cavities formed by the rock to determine the maximum pressure these natural holding tanks can withstand.

Low-intensity ultrasonics is used for medical diagnostics in two different applications. First, ultrasonic waves penetrate body tissues but are reflected by moving internal organs, such as the heart. The frequency of waves reflected from a moving structure is Doppler-shifted, thus causing beats with the original wave, which can be heard. This procedure is particularly useful for performing fetal examinations on pregnant women; because sound waves are not electromagnetic and of low energy, they will not harm the fetus. The second application is to create a sonogram image of the body's interior. A complete cross-sectional image may be produced by superimposing the images scanned by successive ultrasonic waves passing through different regions. This ultrasonography procedure, unlike an X-ray procedure, displays all the tissues in the cross section and also avoids any danger posed by the radiation involved in X-ray imaging.

Physiological and Psychological Acoustics. Because the ear is a nonlinear system, it produces beat tones that are the sum and difference of two frequencies. For example, if two sinusoidal frequencies of 100 and 150 Hz simultaneously arrive at the ear, the brain will, in addition to these two tones, create tones of 250 and 50 Hz (sum and difference, respectively). Thus, although a small speaker cannot reproduce the fundamental frequencies of bass tones, the difference between the harmonics of that pitch will recreate the missing fundamental in the listener's brain.

Another psychoacoustic effect is masking. When a person listens to a noisy version of recorded music, the noise virtually disappears if the individual is enjoying the music. This ability of the brain to selectively listen has had important applications in digitally recorded music. When sounds are digitally compressed, as in MP3 systems, the brain compensates for the loss of frequency information (the information lost when an analog signal is converted to a digital signal), thus creating a higher fidelity sound than that conveyed by the stored content alone.

As twentieth-century technology evolved, environmental noise increased concomitantly; lifetime exposure to loud sounds, commercial and recreational, has created an epidemic of hearing loss, most noticeable in the elderly because the effects are cumulative. Wearing a hearing aid, fitted adjacent to or inside the ear canal, is an effective means of counteracting this condition. The device consists of one or several microphones, which create electric signals that are amplified and transduced into sound waves redirected directly back into the ear. More sophisticated hearing aids incorporate an integrated circuit to control volume, either manually or automatically, or to switch to volume contours designed for various listening environments, such as conversations on the telephone or where excessive background noise is present.

Speech Acoustics. With the advent of the computer age, speech synthesis moved to digital processing, either by bandwidth compression of stored speech or by using a speech synthesizer. The synthesizer reads a text and then produces the appropriate phonemes on demand from their basic acoustic parameters, such as the vibration frequency of the vocal cords and the frequencies and amplitudes of the vowel formants. This method of generating speech is considerably more efficient in terms of data storage than archiving a dictionary of prerecorded phrases.

Another important, and probably the most difficult, area of speech acoustics is the machine recognition of spoken language. When machine speech recognition programs are sufficiently advanced, the computer is able to listen to a sentence in any reasonable dialect and produce a printed text of the utterance. Two basic recognition strategies exist, one dealing with words spoken in isolation and the other with continuous speech. In both cases, it is desirable to teach the computer to recognize the speech of different people through a training program. Because recognition of continuous speech is considerably more difficult than the identification of isolated words, very sophisticated pattern-matching models must be employed. Speech recognition technologies advanced greatly in the 2010s, with examples included on many smartphones and computers along with virtual personal assistants, such as Apple's Siri. Such systems are able to process full sentences, enact commands, and even respond with synthesized speech in a realistic manner.

Musical Acoustics. The importance of musical acoustics to manufacturers of quality instruments is apparent. During the last decades of the twentieth century, fundamental research led, for example, to vastly improved French horns, organ pipes, orchestral strings, and the creation of an entirely new family of violins. Acoustics opens up the possibilities of advanced technologies such as software instruments and other innovative forms of music making and recording.

Noise. One system for abating unwanted sound is active noise control. The first successful application of active noise control was noise-canceling headphones, which reduce unwanted sound by using microphones placed in proximity to the ear to record the incoming noise. Electronic circuitry then generates a signal, exactly opposite to the incoming sound, which is reproduced in the earphones, thus canceling the noise by destructive interference. This system enables listeners to enjoy music without having to use excessive volume levels to mask outside noise and allows people to sleep in noisy vehicles such as airplanes. Because active noise suppression is more effective with low frequencies, most commercial systems rely on soundproofing the earphone to attenuate high frequencies. To effectively cancel high frequencies, the microphone and emitter would have to be situated adjacent to the user's eardrum, but this is not technically feasible. Active noise control is also being considered as a means of controlling low-frequency airport noise, but because of its complexity and expense, this is not yet commercially feasible.

CAREERS AND COURSEWORK

Career opportunities occur in academia (teaching and research), industry, and national laboratories. Academic positions dedicated to acoustics are few, as are the numbers of qualified applicants. Most graduates of acoustics programs find employment in research-based industries in which acoustical aspects of products are important, and others work for government laboratories.

Although the subfields of acoustics are integrated into multiple disciplines, most aspects of acoustics can be learned by obtaining a broad background in a scientific or technological field, such as physics, engineering, meteorology, geology, or oceanography. Physics probably provides the best training for almost any area of acoustics. An electrical engineering major is useful for signal processing and synthetic speech research, and a mechanical engineering background is requisite for comprehending vibration. Training in biology is expedient for physiological acoustic research, and psychology course work provides essential background for psychological acoustics. Architects often employ acoustical consultants to advise on the proper acoustical design of concert halls, auditoriums, or conference rooms. Acoustical consultants also assist with noise reduction problems and help design soundproofing structures for rooms. Although a background in architecture is not a prerequisite for becoming this type of acoustical consultant, engineering or physics is.

Acoustics is not a university major; therefore, specialized knowledge is best acquired at the graduate level. Many electrical engineering departments have at least one undergraduate course in acoustics, but most physics departments do not. Nevertheless, a firm foundation in classical mechanics (through physics programs) or a mechanical engineering vibration course will provide, along with numerous courses in mathematics, sufficient underpinning for successful graduate study in acoustics.

SOCIAL CONTEXT AND FUTURE PROSPECTS

Acoustics affects virtually every aspect of modern life; its contributions to societal needs are incalculable. Ultrasonic waves clean objects, are routinely employed to probe matter, and are used in medical diagnosis. Cochlear implants restore people's ability to hear, and active noise control helps provide quieter listening environments. New concert halls are routinely designed with excellent acoustical properties, and vastly improved or entirely new musical instruments have made their debut. Infrasound from earthquakes is used to study the composition of Earth's mantle, and sonar is essential to locate submarines and aquatic life. Sound waves are used to explore the effects of structural vibrations. Automatic speech recognition devices and hearing aid technology are constantly improving.

Many societal problems related to acoustics remain to be tackled. The technological advances that support modern society have also resulted in more people with hearing loss. Environmental noise is ubiquitous and increasing despite efforts to design quieter machinery and pains taken to contain unwanted sound or to isolate it from people. Also, although medical technology has been able to help many hearing- and speech-impaired people, other individuals still lack appropriate treatments. For example, although voice generators exist, there is considerable room for improvement.

—George R. Plitnik

Further Reading

Bass, Henry E., and William J. Cavanaugh, editors. *ASA at Seventy-Five*. Acoustical Society of America, 2004.

Beyer, Robert T. *Sounds of Our Times: Two Hundred Years of Acoustics*. Springer, 1999

Bjorno, Leif. *Applied Underwater Acoustics*. Elsevier, 2017.

Crocker, Malcolm J., editor. *The Encyclopedia of Acoustics*. 4 vols. Wiley, 1997.

Everest, F. Alton, and Ken C. Pohlmann. *Master Handbook of Acoustics*. 6th ed., McGraw, 2015.

Rossing, Thomas, and Neville Fletcher. *Principles of Vibration and Sound*. 2nd ed., Springer, 2004.

Rumsey, Francis, and Tim McCormick. *Sound and Recording: An Introduction*. 7th ed., Elsevier, 2014.

Strong, William J., and George R. Plitnik. *Music, Speech, Audio*. 3rd ed., BYU Academic Press 2007.

Swift, Gregory. "Thermoacoustic Engines and Refrigerators." *Physics Today*, vol. 48, no. 7, pp. 22–28.

ADRIATIC SEA ECOSYSTEM

Fields of Study: oceanography; marine biology; marine geology

ABSTRACT

An arm of the Mediterranean Sea, the Adriatic Sea separates the Italian Peninsula from the Balkan Peninsula. The Adriatic Sea is bordered by Italy to the west and north; and by

Slovenia, Croatia, Bosnia and Herzegovina, Montenegro, and Albania to the east. The Adriatic stretches some 500 miles (800 kilometers) from the Gulf of Venice—the northernmost extent of the Mediterranean—to the Strait of Otranto, a 45-mi (72-km) channel separating Italy and Albania while connecting the Adriatic to the Ionian Sea area of the Mediterranean.

KEY CONCEPTS

bora: cold, dry winds that blow into Europe from the mountains ranges to the east

eutrophication: stagnation of a body of water by the intrusion of large quantities of organic matter, especially algae, which consumes the dissolved oxygen in the water, thereby rendering it incapable of supporting a healthy ecosystem

karstification: the formation of karst landscape through erosion of limestone beds by acidic rainfall and wind

scirocco: warm, moist winds that originate in northern Africa and blow across the Mediterranean Sea into Europe

sinkholes: essentially a vertical cave formation produced when acidified water seeps through an underlying limestone bed into an underground void, causing the roof of the void to collapse

DESCRIPTION

The Adriatic Sea is relatively narrow throughout, never spanning more than 120 mi (200 km) east to west. Its surface area is about 53,500 square km (138,600 square km); the sea is fed by freshwater flows from a catchment area nearly twice that size, or 90,750 sq mi (235,000 sq km). Indeed, the Adriatic supplies up to one-third of the freshwater flow received by the entire Mediterranean, mitigating the salinity of the greater body of water. From the shallows near the Venice lagoon and Bay of Trieste in the north, depths grow deeper moving southward, with a range from about 80 feet (25 meters) to 4,035 ft (1,240 m).

The western Adriatic coast tends to be alluvial or terraced,while the eastern coast is highly indented with pronounced karstification, the geologic term for chemical and mechanical erosion by water on soluble bodies of rock such as the limestone found here. This type of erosion results in sinkholes, towers, caves, and a complex subsurface drainage system, and facilitates significant groundwater contributions to the Adriatic Sea. More than 1,000 islands dot the eastern coast. Dominant circulation in the Adriatic is clockwise.

Map of the Adriatic Sea (Wikimedia Commons)

It is estimated that the Adriatic's entire volume is exchanged into the Mediterranean Sea through the Strait of Otranto every three to four years, a very short period and likely due to the combined contribution of rivers and submarine groundwater discharge. The Po River flowing west to east across Italy's north is the greatest single contributor of freshwater flows to the Adriatic, at 28 percent, and the largest deposition of sediments as well. Submarine springs along the Balkan, or Dalmatian, coast together contribute another 29 percent of freshwater flows.

The climate of the Adriatic Sea is Mediterranean, with hot dry summers, and mild wet winters. The predominant winter winds are the bora and scirocco. The bora brings cold, dry continental air from the mountains east of the Adriatic Sea. The scirocco brings humid and warm air from northern Africa.

BIOTA

The Adriatic Sea contains more than 7,000 species including many that are unique, rare, and endangered. The diversity and degree of specialization found here is due in part to the Dalmatian coastal islands, karst complexes, and submarine water flows, including some geothermal springs. Several marine protected areas have been established along this coast to protect the karst structures and their habitats.

The Kornati Islands National Park, for example, was established by Croatia in 1980 to protect 89 islands (since expanded to 109) and their marine environs. Among the signature protected species in the park is the noble pen shell (*Pinna nobilis*), a quite large, elongated clam. Kornati Park also hosts about half of the 682 known deepwater flora families of the Adriatic, such as red, brown, and green algae (*Rhodophyta, Phaeophyta,* and *Chlorophyta*). The park is home to an estimated 2,500 or more types of deepwater and tidewater species, including at least 177 species of mollusk; 127 species of bristle worms (*Polychaeta*); 64 species of sea stars, urchins, and cucumbers (*Echinodermata*); and 160 species of fish.

There are also more than 20 species of corals. Sponges are also found in relative diversity, although diminished from past human exploitation. Terrestrial species in the park include Dubrovnik knapweed (*Centaurea ragusina*), an endemic (evolved specifically and uniquely to a biome) plant that forms a centerpiece of the coastal scrub vegetation here; olive trees are by far the leading cultivar. Representative fauna include seagulls, cormorants, and owls; their prey in the form of snails and lizards; and some 60 or more species of moth and butterfly (*Lepidoptera*).

The Adriatic Sea features numerous species rated from declining to critically endangered, with representatives across the spectrum: various rockweeds and seagrasses including *Cystoseira zosteroides, Cystoseira spinosa, Zostera noltii,* and *Posidonia oceanica*; bivalves such as *Gibbula nivosa* and the ribbed Mediterranean limpet (*Patella ferruginea*); the European eel (*Anguilla anguilla*); Adriatic sturgeon (*Acipenser sturio*); the beluga or great sturgeon (*Huso huso*); the great white shark (*Carcharodon carcharias*); sea turtles such as the loggerhead (*Caretta caretta*) and leather- back (*Dermochelys coriacea*); and marine mammals including various cetacean species and the Mediterranean monk seal (*Monachus monachus* or *Monachus albiventer*).

Avian species under stress range from raptors like the white-tailed sea eagle (*Haliaeetus albicilla*), greater spotted eagle, (*Aquila clanga*), and golden eagle (*Aquila chrysaetos*); to typical seabirds such as the great white pelican (*Pelecanus onocrotaius*), Dalmatian pelican (*Pelecanus crispus*), little tern (*Sterna albifrons*), and Balearic shearwater (*Puffinus mauretanicus*); and wetlands-oriented species including purple heron (*Ardea purpurea*), European flamingo (*Phoenicopterus ruber roseus*), common crane (*Grus grus*), glossy ibis (*Plegadis falcinellus*), and northern pintail (*Anas acuta*).

More than two-thirds of the major commercially important fish species in the Adriatic are considered overfished. Those whose populations are thought to be in the "safe" range—even though they have in several cases seen marked declines in recent-year catch volume—include bogue (*Boops boops*), sprat (*Sprattus sprattus*), pilchard and sardinella (*Sardina pilchardus, Sardinella aurita*), jack mackerel and horse mackerel (*Trachurus picturatus, Trachurus trachurus*).

HUMAN INTERACTION

There is a long history of human settlement and use in the Adriatic Sea and adjacent coasts. Human activities have influenced marine ecosystems since Roman times, with this influence increasing in the nineteenth and twentieth centuries. Today, almost all of the original marine resources have been reduced to less than half of their former abundance, with large and midsized fish species being most affected. These changes may also make the overall ecosystem less resilient and more vulnerable in the future, as the regional effects from global warming gain in intensity, for example.

Fisheries, marine transport, and tourism are important throughout the Adriatic Sea. Technological improvements in the fishing fleet and increased activity driven in part by higher prices have resulted in a blanket decline in the fish catch rates per boat. Coastal pollution and eutrophication—due to excess nutrient inflow, typically from agricultural and municipal runoff—have been additional factors in declining fisheries yields. In many cases in the more urbanized northern Adriatic, fish kills have occurred due to algal blooms and low-oxygen conditions.

Shipping trade goods is another Adriatic Sea industry that dates from Roman times and before. Significant cargo port facilities ring the northern

Adriatic; today, tourism is an added generator of mechanized sea traffic and threats to coastal ecosystems. Seeking to encourage "green" practices, the Denmark-based Foundation for Environmental Education has found acceptance by the operating authorities of several hundred Adriatic Sea marinas and beaches of its Blue Flag certification program. These local businesses and governments seem to realize that the extra short-term expense to contain fuel spills, prevent debris dumping, and restrict coastal development will in the long run protect their investment in the healthy ecosystem qualities that attract customers to their facilities.

Venice illustrates a worst-case scenario for polluted coastal waters in the Adriatic, where shipping, transportation, farming, manufacturing, and wastewater combine to heavily pollute the sea. Venice also contains oil refineries, which fortunately have avoided a significant spill that would have extreme consequences on its marshes, species, commercial fisheries, and tourism. There are hydrocarbon resources underlying other sections of the Adriatic Sea, and various controversial proposals for developing them have been put forward. Just beyond the Strait of Otranto, Northern Petroleum, a United Kingdom concern, was poised in 2012 to commence seafloor drilling off the coast of the Puglia region, after receiving approval by the Italian Ministry of Environment—even though the 2,550-sq-mi (6,600-sq-km) area of the sea in question adjoins a host of variously conserved zones including a Marine Protected Area; a Specially Protected Area of Mediterranean Importance; a National Natural Reserve; and, under a European Union policy centerpiece known as Natura 2000, nine Sites of Community Importance.

—Magdalena Ariadne Kim Muir

Further Reading

Coll, Marta, et al. "Biodiversity of the Mediterranean Sea: Estimates, Patterns, and Threats." *PLoS ONE*, vol. 5, no. 8, 2010.

Cushman-Roisan, Benoit, Miroslav Gačić, Pierre-Marie Poulain, and Antonio Artegiani. *Physical Oceanography of the Adriatic Sea. Past, Present and Future.* Springer Science & Business Media, 2013.

Lotze, Heike K., Marta Coll, and Jennifer A. Dunne. "Historical Changes in Marine Resources, Food-Web Structure and Ecosystem Functioning in the Adriatic Sea, Mediterranean." *Ecosystems*, vol. 14, no. 2, 2011.

McKinney, Frank K. *The Northern Adriatic Ecosystem, Deep Time in a Shallow Sea.* Columbia UP, 2007.

Sala, Enric. "The Past and Present Topology and Structure of Mediterranean Subtidal Rocky-Shore Food Webs." *Ecosystems*, vol. 7, no. 4, 2004.

AEGEAN SEA ECOSYSTEM

Fields of Study: oceanography; marine biology; marine geology; marine ecology

ABSTRACT

Within the Mediterranean Sea basin, the Aegean Sea forms an embayment surrounded on the west and north by Greece, on the east by Turkey, and on the south by the island of Crete. The area of the Aegean is 82,625 square miles (214,000 square kilometers). It reaches a maximum depth, near Crete, of 11,624 feet (3,543 meters). The area has a complex topography marked by deep trenches and elevated structures; the eastern Mediterranean is a remnant of an ancient body of water known as the Tethys Sea. The waters are generally nutrient poor, sustaining an oligotrophic environment (an environment that is characterized by relatively slow growth and metabolic activity).

KEY CONCEPTS:

autochthonous: a synonym for "native"

climate forcing: the difference between energy received from sunlight and energy radiated back to space, the imbalance forcing the temperature of the atmosphere to rise or fall depending on whether the difference is positive or negative

column stratification: denotes the various temperature and salinity layers in a body of water; the water above any particular point of the seafloor is called the "water column"

mariculture: fish farming and other pursuits in which "crops" are raised in the sea rather than on land

pelagic: refers to the deep, open waters of the oceans

DESCRIPTION

To the northeast, the Aegean connects through Turkey to the Black Sea via the Sea of Marmara and the straits of the Dardanelles and the Bosphorus. The Aegean is dotted with thousands of mountain peaks peeping up from the seafloor. These island chains and the numerous craggy gulfs and bays surrounding them are unsurpassed in utility to seafarers. The islands include Crete and Euboea (the Mediterranean's fifth-largest and sixth-largest, respectively), the Sporades, the Cyclades, the Saronic, the Dodecanese, and the Northern and Eastern groups.

Map of the Agean Sea (Wikimedia Commons)

SALINITY AND TEMPERATURE

Like the Mediterranean as a whole, the Aegean Sea flows counterclockwise, its salty waters rising along its eastern boundary and then falling along its northern boundary, carrying the waters of the Black Sea down along its western Greek shoreline and out into the larger basin. The Aegean is influenced by water exchange with such regional sources as the Adriatic and Ionian Seas, as well as flow from the Nile River delta and the Red Sea via the Suez Canal. Changes in saltwater composition over time have periodically altered the circulation patterns and the fundamental ecology of the Aegean basin area.

Rising global temperatures now contribute to fluctuating carbon dioxide (CO_2) levels, in concert with CO_2 escaping from volcanic sources. One result is acidified marine zones that trigger losses in biodiversity and alter food-web functions. This situation is complicated by the Aegean basin being a somewhat enclosed province; pollutants such as heavy metals cannot be readily discharged and over time their levels become concentrated. These factors contribute to worrisome changes in ecology including the erosion of coral ecosystems and the extinction of species such as the sea turtle, the monk seal, and the Atlantic tuna, all of which spawn only in the Mediterranean.

The effects of climate change on pelagic (deep sea) marine ecologies have only begun to be documented. Some effects are unexpectedly beneficial to various life forms. Changes of interest include variations in the metabolic rates of marine organisms, and environmental conditions including changes in currents, column stratification, and nutrient production, all variables that affect food-web structure, population distribution, and community dynamics within an ecosystem.

Researchers marked a "transient event" in the area occurring in two phases during the 1990s. As a result of increased salinity by climate forcing, a mass of dense, warm water formed in the southern Aegean. This process was followed by a drop in temperature of 0.7 degrees Fahrenheit (0.4 degrees Celsius), creating another deeper layer of dense water. These changes triggered an upwelling of deep bottom water, creating an intermediate stratum of nutrient-rich water now known as the Transitional Mediterranean Water. This change, in turn, set into motion a threefold increase in biological activity along the continental shelf and upper slope.

PLANKTON AND HUMAN NEEDS

The total biomass of the Eastern Mediterranean is 46 percent lower than densities found at the western edge. Since the construction of the Aswan High Dam in Egypt and a major dam on the Ebro River in Spain, freshwater inputs into the Mediterranean Sea were significantly reduced. Continued flows of saltwater from the Atlantic Ocean and Red Sea contributed fractional increases in salinity. This, in turn, affects the quality of marine life throughout the region. The Atlantic Ocean waters that spill into the Mediterranean are poorer in nutrients, having served to feed the ocean's population of plankton.

As the waters move eastward toward the Aegean, bacteria and other organisms feed on the remaining nutrients, leaving little left for plankton and other marine life. Bacteria are thought to account for about 56 percent of all organisms found on the shallow shores of the Aegean Sea.

The Suez Canal is effectively a passageway for the migration of over 300 species of Erythraen marine biota from the Red Sea into the Mediterranean, an immigration that adds to environmental changes in the Aegean. This is known as the Lessepsian Migration, named after Ferdinand de Lesseps (1805–1894), the French mastermind of the nineteenth-century Suez Canal project. Exotic macrophytes, invertebrates, and fish include the bivalve *Brachiodontes pharaonis*, the gastropods *Cerithium scabridum* and *Strombu persicus*, the rock oyster *Spondylus spinosus*, the jellyfish *Rhopilema nomadica*, penaeid prawns, the American blue crab, mullids, the barracuda *Spyraena chrysotaenia*, thriving populations of clupeids, and pearl mollusks.

The competitive displacement of autochthonous species is also the result of coastal mariculture farming. Farm stocks include the Pacific oyster *Crassostrea gigas*, the soft-shell clam *Mya arenariea*, and the pearl oyster *Pinctada radiata*. Changes in bathymetric populations are a direct result of migration and displacement. Human population densities and subsequent coastal development have further decimated marine biota and altered fragile littoral (shoreside) ecosystems.

The general absence of phytoplankton contributes to the amazing clarity of the Aegean's waters. The most notable of the sea fauna include the blennies (Blenniidae), Cardinal Fish (Apogonidae), eels, picarels, electric rays, groupers (Serranidae), lizard fish, red mullets, and scorpion fish. The larger sea mammals include dolphins and seals. Divers and snorkelers enjoy touring the Aegean to study and to admire the habitats of sponges and sea urchins, octopus squids, and cuttlefish. Algae, kelps, and seagrasses are also common to the area.

COASTAL SPECIES

The Aegean Sea, its archipelago, Greece, and Turkey together form a distinct ecological region classified by the World Wildlife Fund (WWF) as a zone of sclerophyllous and mixed forests (designated PA1201). The eastern border of this ecoregion extends to plains and lowlands in Turkey, while Greece is replete with sparse, mountainous landscapes. Callabrian pine (*Pinus brutia*) and maquis are the dominant vegetation; north to south, the maquis are varied and include stands of *Arbutus andrachne, A. unedo, Spartium junceum,* and *Laurus nobilis*. Stands of sweetgum and Datça palms, endangered native species, are also found. Varied species of oak grow in areas where the pine stands have been removed.

Each of these types of forests provide habitat for European woodland birds; pine forests echo with the calls of wrens, blackbirds, tits, jays, and Krüper's nuthatch, a bird only found in southeast Turkey. Warblers, buntings, and partridges make their habitat in maquis scrublands. Other native wildlife includes loggerhead turtles, wolves, foxes, and boars. The entire basin is noted for its olive trees and viticulture.

—Victoria M. Breting-Garcia

Further Reading

Colasimone, Luisa, editor. *MedWaves: Focus on Biodiversity*. United Nations Environment Programme, Mediterranean Action Plan, 2010.

Danovaro, Roberto, Antonio Dell'Anno, Mauro Fabiano, Antonio Pusceddu, and Anastasios Tselepides. "Deep-Sea Ecosystem Response to Climate Changes: The Eastern Mediterranean Case Study." *Trends in Ecology & Evolution*, vol. 16, no. 9, 2001.

Gönülal, Onur. *Deep-sea Biodiversity in the Aegean Sea*. IntechOpen, 2017.

Gritti, E. S., B. Smith, and M. T. Sykes. "Vulnerability of Mediterranean Basin Ecosystems to Climate Change and Invasion by Exotic Plant Species." *Journal of Biogeography*, vol. 33, no. 1, 2006.

Şekercioglu, Çagan Hakki, Sean Anderson, Erol Akçay, and Raşit Bilgin. "Turkey's Rich Natural Heritage Under Assault." *Science*, vol. 23, no. 12, 2011.

Applied Physics

Fields of Study: physics; mathematics; marine sciences; physical sciences

ABSTRACT

Applied physics is the study and application of the behavior of condensed matter, such as solids, liquids, and gases, in bulk quantities. Applied physics is the basis of all engineering and design that requires the interaction of individual components and materials. The study of applied physics now extends into numerous other fields, including oceanography, physics, chemistry, biology, engineering, medicine, geology, and meteorology.

KEY CONCEPTS

condensed matter: matter that has definite quantifiable properties as a solid, liquid, or gas

parabolic: any two-dimensional (2D) motion or shape that can be described mathematically by an equation having the form $y = ax^2 + b$

quantify: to assign a standardized numerical value to a property in order to enable comparison to other values of that property and relation of the particular property to other properties

water cycle: the cycle of phase changes that water undergoes in the natural environment, from solid and liquid to gas and back to liquid and solid (snow, ice, and liquid water to water vapor and back to condensation, rain, and snow)

DEFINITION AND BASIC PRINCIPLES

Applied physics is the study of the behavior and interaction of condensed matter. Condensed matter is commonly recognized as matter in the phase forms of solids, liquids, and gases. Each phase has its own unique physical characteristics, and each material has its own unique physical suite of properties that derive from its chemical identity. The combination of these traits determines how something consisting of condensed matter interacts with something else that consists of condensed matter.

The interactions of constructs of condensed matter are governed by the interaction of forces and other vector properties that are applied through each construct. A lever, for example, functions as the medium to transmit a force applied in one direction so that it operates in the opposite direction at another location. Material properties and strengths are also an intimate part of the interaction of condensed matter. A lever that cannot withstand the force applied to it laterally will bend or break rather than function as a lever. Similarly, two equally hard objects, such as a train car's wheels and the steel tracks upon which they roll, rebound elastically with little or no generation of friction. If, however, one of the materials is less hard than the other, as in the case of a steel rotor and a composition brake pad, a great deal of friction characterizes their interaction.

BACKGROUND AND HISTORY

Applied physics is of necessity the oldest of all practical sciences, dating back to the first artificial use of an object by an early hominid. The basic practices have been in use by builders and designers for many thousands of years. With the development of mathematics and measurement, the practice of applied physics has grown apace, relying as it still does upon the application of basic concepts of vector properties (force, momentum, velocity, weight, moment of inertia) and the principles of simple machines (lever, ramp, pulley).

The modern concept of applied physics can be traced back to the Greek philosopher Aristotle (ca. 350 BCE), who identified several separate fields for systematic study. The basic mathematics of physics was formulated by Pythagoras about two hundred years before, but was certainly known to the Babylonians as early as 1600 BCE. Perhaps the greatest single impetus for the advance of applied physics was the development of calculus by Sir Isaac Newton (1642–1727) and the fundamental principles of Newtonian mechanics in the seventeenth century. Those principles describe well the behaviors of objects composed of condensed matter, only failing as the scale involved becomes very small (as below the scale of nanotechnology). Since the Industrial Revolution and into the twenty-first century, the advancing capabilities of technology and materials combine to enable even more advances and applications, yet all continue to follow the same basic rules of physics as the crudest of constructs.

HOW IT WORKS

As the name implies, applied physics means nothing more or less than the application of the principles of physics to material objects. The most basic principles of applied physics are those that describe and quantify matter in motion (speed, velocity, acceleration, momentum, inertia, mass, force). These principles lead to the design and implementation of devices and structures, all with the purpose of performing a physical function or action, either individually or in concert.

A second, and equally important, aspect of applied physics is the knowledge of the physical properties and characteristics of the materials being used. This includes such physical properties as melting and boiling points, malleability, thermal conductivity, electrical resistivity, magnetic susceptibility, density, hardness, sheer strength, tensile strength, compressibility, granularity, absorptivity, and a great many other quantifiable factors that determine the suitability of materials for given tasks. Understanding these factors also enables the identification of new applications for those materials.

Design. Applied physics is the essential basis for the design of each and every artificial object and construct, from the tiniest of nanoprobes and the simplest of levers to the largest and most complex of machines and devices. Given the idea of a task to be performed and a device to perform that task, the design process begins with an assessment of the physical environment in which the device must function, the forces that will be exerted against it, and the forces that it will be required to exert in order to accomplish the set task. The appropriate materials may then be selected, based on their physical properties and characteristics. Dimensional analyses determine the necessary size of the device, and also play a significant role in determining the materials that will be utilized. Accordingly, applied physics is of fundamental importance in designing devices and constructs that must survive and function in the pressure gradient environments of bodies of water such as rivers, lakes and oceans.

All of these factors affect the cost of any device. Cost is an especially important consideration, as is the feasibility of replacing a component of the designed structure. For example, a device operating in a local environment in which replacement of a component, such as an actuating lever, is easily carried out may be constructed using a simple metal rod. If, however, the device is something like the Mars Rover or a deep-sea submersible, for which replacement of worn parts *in situ* is not an option, it is far more reasonable and effective to use a less corrosive, stronger, but more costly material, such as titanium, to make movable parts. The old design tenet that form follows function is an appropriate rule of thumb in the field of applied physics.

APPLICATIONS AND PRODUCTS

In many ways, applied physics leads to the idea of creating devices that build other devices. At some point in time, an individual first used a handy rock to break up another rock or to pound a stick into the ground. This is a simple example of applied physics, when that individual then began to use a specific kind of rock for a specific purpose; this action advanced again when someone realized that attaching the rock to a stick provided a more effective tool for pounding. Applied physics has advanced far beyond the crude rock-on-a-stick hammer, yet the exact same physical principles apply now with the most elegant of impact devices. It would not be possible to itemize even a small percentage of the applications and products that have resulted from applied physics, and new physical devices are developed each day.

Marine Sciences. Applied physics is fundamental to all aspects of the different areas of marine science. Earth is divided into two different basic environments—the terrestrial, characterized by the medium of air, and the marine, characterized by the medium of water. The marine sciences are the efforts of air-breathing terrestrial humans to interact with and understand the "alien" marine environment. There is a great amount of overlap between the two environments through such phenomena as the water cycle, climate, erosion, plate tectonics, and many more. In addition, entering the marine environment from the terrestrial environment requires significant, and essential, planning with respect to the systems and materials necessary for survival in the aquatic medium. The primary considerations for this are, firstl, that a water environment does not support air-breathing organisms, and second, that the pressure exerted by the surrounding water increases with depth, adding the equivalent of the pressure at the surface with each increase of 33 feet (10 meters) in depth. Accordingly, an isolated air environment such as a submersible or scuba outfit must be used, and the "container" for that isolated air environment must be capable of withstanding the increased pressure to a safe limit. This requires attention to the materials used in the construction of that container, as well as to the chemical interaction of those materials with the water environment. For example, use of materials that dissolve in water, that absorb water or that do not have the required strength and toughness to resist pressure would be quite detrimental to the success of the endeavor.

Another major consideration of applied physics in marine sciences is the means of acquiring information and using that information successfully. Electronic developments such as sonar and its related technologies, hydroacoustics, lighting, videography, and photography are used to acquire information within the aquatic environment. The devices used for those purposes are necessarily adapted to be able to function in that environment. Being devices that operate on electricity, leakage of highly electrically conductive seawater into them would be the instantaneous end of the serviceability of the device, either requiring its removal and repair or more likely destroying it completely. Thus, such devices require the same degree of isolation from the aqueous environment as a human would require.

Movement within the aqueous environment is another major consideration. Air-breathing internal combustion motors are suitable for movement across the surface of the water, but below any depth at which acquiring air from the surface is feasible, such engines are not usable and must be replaced by electric motors. This requires storage of electrical energy in batteries, as well as protection of those batteries from any intrusion of water. This, of course, limits the amount of time that any incursion into the underwater environment can be carried out.

Given these major considerations, it should be apparent that marine sciences depend heavily on applied physics in any number of ways.

Civil Engineering. Applied physics underpins the design and construction of the human environment and its supporting infrastructure, the most obvious of which are roads and buildings, designed and engineered with consideration of the forces that they must withstand, both human-made and natural. In this, the highest consideration is given to the science and engineering of materials in regard to the desired end result, rather than to the machines and devices that are employed in the actual construction process. The principles of physics involved in the design and construction of supporting structures, such as bridges and high-rise towers, are of primary importance. No less important are the physical systems that support the function of a structure, including such things as electrical systems, internal movement systems, environmental control systems, monitoring systems, and emergency response systems, to name a few. All are relevant to a specific aspect of the overall physical construct, and all must function in coordination with the other systems. Nevertheless, they all also are based on specific applications of the same applied physics principles. On the other side of that coin, engineering design must also take into account the physical nature of the environment in which structures must function, and the forces that environment may be able to apply. For example, the interaction of a physical medium such as water with the materials and design of a structure presents a different set of variables that must be included in the design parameters. For example, water that is moving relative to a structure such as a road, a ship or a submersible affects that structure in an entirely different manner than does water that is not moving.

Transportation. Transportation is perhaps the single greatest expenditure of human effort, and it has produced the modern automotive, air, railway, and sea transportation industries. Applied physics in these areas focuses on the development of more effective and efficient means of controlling the movement of the machine while enhancing the safety of human and nonhuman occupants. While the physical principles by which the various forms of the internal combustion engine function are the same now as when the devices were first invented, the physical processes used in their operation have undergone a great deal of mechanical refinement.

A particular area of refinement is the manner in which fuel is delivered for combustion. In earlier gasoline-fueled piston engines, fuel was delivered via a mechanical fuel pump and carburetor system that functions on the Venturi principle. This has long since been replaced by constant-pressure fuel pumps and injector systems, and this continues to be enhanced as developers make more refinements based on the physical aspects of delivering a liquid fuel to a specific point at a specific time. Similarly, commercial jet engines, first developed for aircraft during World War II, now utilize the same basic physical principles as did their earliest counterparts. Alterations and developments that have been achieved in the interim have focused on improvement of the operational efficiency of the engines and on enhancement of the physical and combustion properties of the fuels themselves.

Sea transport, whether the largest of cruise ships, the little personal motorboat used by recreational fishers or the most modern of submarines and deep-sea research submersibles, uses the same basic screw principle of a motor-driven propeller. The physical requirements of the design vary drastically from surface craft to underwater craft. Accordingly, the materials must be appropriate for the demands those environments will place on the particular craft.

In rail transport, the basic structure of a railway train has not changed; it is still a heavy, massive tractor engine that tows a number of containers on very low friction wheel systems. The engines have changed from steam-powered behemoths made up of as many as one-quarter-million moving parts to the modern diesel-electric traction engines of much simpler design. Driven by the ever-increasing demand for limited resources, this process of refinement for physical efficiency progresses in every aspect of transportation on land, sea, and air. Paradoxically, it is in the area of railway transport that applied physics offers the greatest possibility of advancement with the development of nearly frictionless, magnetic levitation systems upon which modern high-speed bullet trains travel. Physicists continue to work toward the development of materials and systems that will be superconducting at ambient temperatures. It is believed that such materials will completely revolutionize not only the transportation sector but also the entire field of applied physics.

Medical. No other area of human endeavor demonstrates the effects of applied physics better than the medical field. The medical applications of applied physics are numerous, touching every aspect of medical diagnostics and treatment and crossing over into many other fields of physical science. The most obvious of these is medical imaging, in which X-ray diagnostics, magnetic resonance imaging (MRI), and other forms of spectroscopic analysis have become primary tools of medical diagnostics. Essentially, all of the devices that have become invaluable as medical tools began as devices designed to probe the physical nature of materials.

Spectrographs and spectrometers that are now routinely used to analyze the specific content of various human sera were developed initially to examine the specific wavelengths of light being absorbed or emitted by various materials. MRI developed from the standard technique of physical analytical chemistry called "nuclear magnetic resonance spectrometry," or NMR. In this methodology, magnetic signals are recorded as they pass through a material sample, their patterns revealing intimate details about the three-dimensional (3D) molecular structure of a given compound. The process of MRI that has developed from this simpler application now permits diagnosticians to view the internal structures and functioning of living systems in real time. The diagnostic X-ray that has become the single most common method of examining internal biological structures was first used as a way to physically probe the structure of crystals, even though the first practical demonstration of the existence of X-rays by German physicist Wilhelm Röntgen (1845–1923) in 1895 was as an image of the bones in his wife's hand.

Military. It is not possible to separate the field of applied physics from military applications and martial practice, and no other area of human endeavor so clearly illustrates the double-edged sword that is the nature of applied physics. The field of applied physics equally facilitates the most humanitarian of endeavors and the most violent of human actions. Ballistics is the area of physics that applies to the motion of bodies moving through a gravitational field. The mathematical equations that describe the parabolic motion of a baseball being thrown from center field equally describe the motion of an arrow or a bullet in flight, the motion of the International Space Station in its orbit, and the trajectory of a warhead-equipped rocket. The physical principles that have permitted the use of nuclear fission as a source of reliable energy are the same principles that define the functioning of a nuclear warhead.

In the modern age it is desired that warfare be carried out as quickly and efficiently as possible as the means to restore order and peace to an embattled area, with as little loss of life and destruction of property as possible. To that end, military research relies heavily on the application of physics in the development of weapons, communications, and surveillance systems.

Digital Electronics. Applied physics has also led to the development of the transistor, commonly attributed to William Shockley (1910–1989), in 1951. Since that time, the development of ever-smaller and more efficient systems based on the semiconductor junction transistor has been rapid and continuous. This has inexorably produced the modern technology of digital electronics and the digital computer. These technologies combine remarkably well to permit the real-time capture and storage of data from the many and various devices employed in medical research and diagnostics, improve the fine control of transportation systems and infrastructure, and advance a veritable host of products that are more efficient in their use of energy than are their nondigital counterparts.

SOCIAL CONTEXT AND FUTURE PROSPECTS

Applied physics has become such a fundamental underpinning of modern society that most people are unaware of its relevance. A case in point is the spin-off of technologies from the National Aeronautics and Space Administration (NASA) space program into the realm of everyday life. Such common household devices as microwave ovens and cell phones, medical devices such as pacemakers and heart monitors, and the digital technology that permeates modern society were developed through the efforts of applied physics in exploring space. For most laypersons, the term "applied physics" conjures images of high-energy particle accelerators or interplanetary telescopes and probes, without the realization that the physical aspect of society is based entirely on principles of applied physics. As these cutting-edge fields of physical research continue, they will continue to spawn new applications that are adapted continually into the overall fabric of modern society.

The areas of society in which applied physics has historically played a role, particularly those involving modern electronics, will continue to develop. Personal computers and embedded microcontrollers, for example, will continue to grow in power and application while decreasing in size, as research reveals new ways of constructing digital logic circuits through new materials and methods. The incorporation of the technology into consumer products, transportation, and other areas of the infrastructure of society is becoming more commonplace. As a result, the control of many human skills can be relegated to the autonomous functioning of the corresponding device. Consider, for example, the now-common built-in ability of some automobiles to parallel park without human intervention and the growing development of self-driving vehicles of all kinds.

The technologies that have been developed through applied physics permit the automation of many human actions, which, in turn, drives social change. Similarly, the ability for nearly instantaneous communication between persons in any part of the world, which has been provided through applied physics, has the potential to facilitate the peaceful and productive cooperation of large numbers of people toward resolving significant problems.

—Richard M. Renneboog

Further Reading

Askeland, Donald R. *The Science and Engineering of Materials*. 3rd ed., Chapman & Hall, 1998.
The Britannica Guide to the One Hundred Most Influential Scientists. Constable & Robinson, 2008.
Dera, Jerzy. *Marine Physics*. Elsevier, 1992.
Ewen, Dale, Neill Schurter, and P. Erik Gundersen. *Applied Physics*. 11th ed., Pearson, 2017.
Giancoli, Douglas C. *Physics for Scientists and Engineers with Modern Physics*. Pearson Prentice Hall, 2008.
Sen, P. C. *Principles of Electric Machines and Power Electronics*. 2nd ed., John Wiley & Sons, 1997.
Valencia, Raymond P., editor. *Applied Physics in the Twenty-First Century*. Nova Science, 2010.

Arabian Sea Ecosystem

Fields of Study: oceanography; marine biology; marine geology; climatology

ABSTRACT

The Arabian Sea composes the northwestern part of the Indian Ocean, lying west of India and covering a total area of approximately 1,491,000 square miles (3,862,000 square kilometers), and has formed part of the principal sea route between Europe and India for centuries. It is bounded to the west by the Horn of Africa and the Arabian Peninsula, with entrances to the Gulf of Aden/Red Sea and Gulf of

Oman/Persian Gulf respectively; to the north by Iran and Pakistan; to the east by India; and to the south by the remainder of the Indian Ocean. The largest river flowing into the Arabian Sea is the Indus River; others include the Netravathi, Sharavathi, Narmada, Tapti, Mahi, and the numerous rivers of the Indian state of Kerala.

KEY CONCEPTS

coy: a thermal gradient that develops in response to the different rates at which the land and the water absorb energy from the Sun

endemic: refers to species that have evolved and developed in adaptation to a specific area

eutrophication: the conversion of a viable aquatic environment to one that cannot support life due to the consumption of dissolved oxygen by excessive amounts of organic matter such as algae or rampant weed growth

nutrient loading: the introduction of higher-than-normal quantities of nutrients such as fertilizer or organic waste into a body of water

practical salinity: the salt content of water as determined by electrical conductivity, as opposed to absolute salinity which is determined as the weight of dry salts residue obtained from a specific volume of ocean water

DESCRIPTION

The Arabian Sea is a Large Marine Ecosystem (LME), as defined by the US National Oceanic and Atmospheric Administration (NOAA): "relatively large areas of ocean space of approximately 200,000 square kilometers [77,000 square miles] or greater, adjacent to the continents in coastal waters where primary productivity is generally higher than in open ocean areas." The Arabian Sea LME is considered a Class I, highly productive ecosystem based on SeaWiFS global primary productivity estimates. It has an average depth of 8,970 feet (2,734 meters).

The Arabian Sea has some of the most extreme climatic regimes, due to seasonal fluctuations in air and water temperatures. The sea's average temperature, which used to be 72 to 81 degrees Fahrenheit (22 to 27 degrees Celsius) until the 1980s, is now 81 to 90 degrees Fahrenheit (27 to 32 degrees Celsius), resulting in occurrence of much stronger cyclones. One example is Cyclone Gonu, which occurred in 2007 and was until then the strongest cyclone ever recorded in the northern Indian Ocean; it attained a peak wind intensity of 165 miles per hour (270 kilometers per hour). Recent increases in the average intensity of tropical cyclones in the Arabian Sea may be a side effect of increasing air pollution over the Indian subcontinent as well as the higher water temperatures in the sea itself.

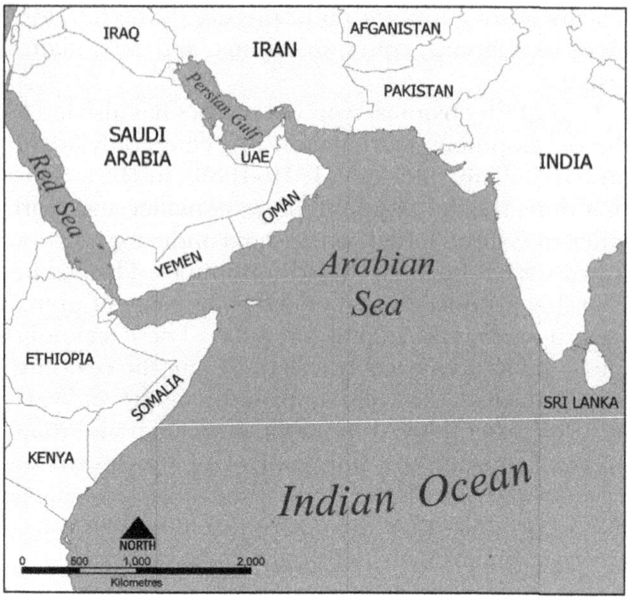

Map of the Arabian Sea (Wikimedia Commons)

The Arabian Sea contributes to a monsoon climate in the surrounding region by providing the water necessary for wet storms. In summer, strong winds blow from the southwest to the northeast, bringing rain to the Indian subcontinent. During the winter, the winds are milder and blow in the opposite direction, from the northeast to the southwest.

During the monsoon season, the generally southwest winds are particularly cold, and are so strong that they succeed in sweeping away some of the salt content of the upper levels of seawater. In monsoon season, the upper waters are less than 35 practical salinity units (psu), while in the nonmonsoon season salinity is over 36 psu. Monsoons are characteristic of the Arabian Sea and are responsible for the yearly cycling of its waters.

SPECIES AND ECONOMY

Scientists have found that because the Arabian Sea has naturally high productivity, it also has one of the thickest zones of oxygen-depleted water below a depth of 330 ft (100 m). Depending on monsoon winds, local topography, the width and depth of the continental shelf, and drainage of coastal areas, there are three coastal ecosystems in the Arabian Sea, each characterized by its own productivity and species distribution. The Arabian Sea's predictable, seasonally reversing monsoons drive one of the most energetic current systems in the world. The strength of the monsoon winds is regulated; during the monsoon season (May to October), the southwesterly winds in this region are from the southwest, inducing a great deal of evaporation from the warm waters of the Arabian Sea, and heavy rainfall along the coy, a thermal gradient that develops from differential heating of land and sea. The winds blow toward India and cause upwelling of low-oxygen waters. There is a concentration of fish in nearshore areas at that time. During the other half of the year, the winds blow in the opposite direction, and not as strongly.

The LMEs produce about 80 percent of the world's annual marine fisheries catch. The United Nations Food and Agriculture Organization (FAO) 10-year trend shows an increase in capture trends; from 1.9 million tons (1.72 million tonnes) in 1990 to 2.2 million tons (2 million tonnes) in 1999. There are catches of herrings, sardines, anchovies and crustaceans. India's southwest coast is fished for oil sardines (*Sardinella longiceps*), mackerels (*Rastrelliger kanagurta*), and tunas (*Euthynnus affinis* and *Auxis thazard*). Most of this catch comes from a narrow 6- to 10-mi (10- to 15-km) coastal belt, and accounts for 23.6 percent of India's fish catch.

The dominant fish species off India's central west coast ecosystem are Sciaenids (*Pseudosciaena diacanthus*), Carangidae (*Caranx* spp.), and Engraulidae (anchovies). The dominant species off India's northwest coast are prawn, Sciaenids (*Pseudosciaena diacanthus*), and Carangidae. Small tuna (*Euthynnus affinis*) migrates to this area to breed. There are also catches of herrings, sardines, anchovies, crustaceans and spiny lobster (*P. homarus megasculptus*), with zonal abundance in relation to surface circulation patterns.

A wide variety of invertebrates and algae exist, including *Sargassopsis zanardinii*, a marine species that is endemic to the Arabian coastline. The endemic (locally evolved and unique to a particular biome) fishes include certain members of the barracuda (*Sphyraena*), wrass (*Labridae*), and damselfish (*Pomacentridae*) families. Also found in the Arabian Sea are dugong (*Dugong dugon:* order *Sirenia* vulnerable International Union for Conservation of Nature [IUCN] 2.3) and several species of turtles, including the green turtle (*Chelonia mydas:* IUCN-listed endangered species), hawksbill turtle (*Eretmochelys imbricata,* critically endangered IUCN), and olive Ridley turtle (*Lepidochelys olivacea* vulnerable, IUCN 2.3).

Of the baleen whales, the following species have been recorded: Bryde's whales (*Balaenoptera edeni,* data deficient IUCN 3.1), minke whales (*B. acutorostrata,* least concern IUCN 3.1), fin whales (*B. physalus,* endangered IUCN 3.1), blue whales (*B. musculus,* endangered IUCN 3.1), and humpback whales (*Megaptera novaeangliae,* least concern IUCN 3.1). Toothed whales include sperm (*Physeter macrocephalus:* vulnerable A1d), orca (*Orcinus orca,* data deficient IUCN 3.1), and false killer (*Pseudorca crassidens,* data deficient IUCN 3.1). At least a dozen species of dolphins (family *Delphinidae*) as well as the finless porpoise (*Neophocaena phocaenoides,* vulnerable IUCN 3.1) exist in the Arabian Sea.

PRESSURES ON BIOTA

The Arabian Sea, due to its seasonal weather fluctuations, offers excellent examples of biological adaptation to environment. However, the diverse aquatic habitat is currently under threat from the oil industry, which uses the sea as a shipping lane. As a result, oil spills, anchor damage, sedimentation, and other pollution effects are severe threats and have long-standing effects. Mining operations, fishing pressures, destructive fish collecting practices (e.g., dynamiting), residential and commercial development, and effluent discharge have resulted in altered species composition in many areas. Recreation and tourism also contribute to eutrophication and reef degradation. War-related activities provide another source of environmental damage.

Oil from accidents and bilge washings can reach the coast and impact coastal ecosystems. On August 7, 2010, a spill of about 88,040 tons (79,869 tonnes) of lubricant oil was poured in the Arabian Sea due to a collision off two cargo ships. This was identified as a cause of tangible damage to mangrove plants and other pelagic organisms. Current fishing methods have resulted in the overexploitation of coastal fishery resources such as prawns, sardines, pomfrets and mackerel. However, most of the fish stocks breed in deeper offshore waters so there is an opportunity to rebuild stocks. The overexploitation is mostly because of large fishing vessels that fish illegally near the coast.

Human population expansion in India will continue to put enormous pressure on the living and nonliving coastal resources. This population pressure also creates major pollution problems, like untreated organic waste and sewage, which contribute to the nutrient loading of nearshore areas. Nutrient loading influences productivity cycles and depletes dissolved oxygen supplies. Rapid industrialization is also contributing pollution from industrial wastes and industrial effluents. Heavy metals such as cadmium, lead, manganese, and zinc have been found in core sediment samples to depths up to 18 inches (45 centimeters). Large amounts of pesticides are also deposited in the coastal areas, particularly off the Mumbai coast. This situation would require the dredging and removal of contaminated sediments for proper mitigation. Currently, plans are being considered that would protect the delicate wildlife of the Arabian Sea, particularly the turtle and coral populations.

Sixty-five percent of fish landings in the Arabian Sea derive from artisanal (small scale) fisheries. Coastal populations have traditionally relied on nonmotorized boats. Along its Arabian Sea coast, India operates about 180,000 country crafts, 26,000 motored traditional vessels, 34,000 mechanized boats, and a few large boats. The most important export is prawn. It is uneconomical to harvest the deepwater fisheries here because of low market price, except for tuna, which commands a good price.

Climate change is an interconnected event based on long-term change of regional weather that is thought to be caused by an increase in atmospheric carbon. It is thought that climate change is causing global warming. In the Arabian Sea, it has been recently suggested that climate change is also causing intensified monsoon winds and, in turn, intensified upwellings. This results in huge algal blooms in the western half of the Arabian Sea causing serious changes in productivity. This has detrimental effects by causing oxygen depletion at depth. In addition, incidence of fish mortalities is recorded in association with algal blooms.

—Santosh Kumar Sarkar

Further Reading

Barendse, R. J. *The Arabian Seas: The Indian Ocean World of the Seventeenth Century.* Routledge, 2016.

Food and Agriculture Organization (FAO) of the United Nations. *Trends in Oceanic Captures and Clustering of Large Marine Ecosystems—Two Studies Based on the FAO Capture Database.* FAO, 2003.

Goes, Joaquim I., Helga Gomes, Prasad Thoppil, Prabhu Matondkar, and Adnan Al Azri. *Eurasian Warming—Hydrography and Biological Productivity in the Arabian Sea.* Bigelow Laboratory for Ocean Sciences, 2007.

Hithin, N. K., V. Sanil Kumar, and P. R. Shanas. *Trends of Wave Height and Period in the Central Arabian Sea from 1996 to 2012: A Study Based on Satellite Altimeter Data.* Elsevier, 2015.

Jawad, Laith A. *The Arabian Seas: Biodiversity, Environmental Challenges and Conservation Measures.* Springer International Publishing, 2020.

ARCHAEOLOGY

Fields of Study: anthropology; oceanography; marine sciences

ABSTRACT

Archaeology is the field of applied science concerned with the techniques and practice of collecting, preserving, and analyzing the artifacts and physical remains left behind by human civilizations. The central purpose of archaeology is to study the lives and cultural practices of past societies, both ancient and modern. Archaeology has a deep intangible impact on human experience, as a means of understanding and maintaining the multifaceted cultural heritage produced by the human species. Of particular interest with regard to marine science is archaeology that must be carried out under water, as for example, the study of shipwrecks and former land sites that have disappeared beneath the waves.

KEY CONCEPTS

dendrochronology: a system of determining the absolute date of an artifact by relating it to the identifiable dates of growth rings in trees

multiscalar: involving study and analysis at scales ranging from the minutiae to the broad physical area in which the minutiae are found

radiocarbon dating: a technique that uses the ratio of the carbon isotope ^{14}C (carbon-14) in a nonliving material of organic origin to the ratio of ^{14}C that is maintained within the material while it was living

stratigraphy: a method of assigning dates to artifacts according to the relative age of buried strata in which they are found

DEFINITION AND BASIC PRINCIPLES

Archaeology is the study of past human cultures, both historic and prehistoric, through the systematic excavation, inspection, and interpretation of material remains such as tools, toys, clothing, bones, buildings, and other artifacts. In the United States, archaeology is considered a subfield of anthropology, which is the science concerned with the origin, evolution, behavior, beliefs, culture, and physical features of humankind.

The traditional image of archaeology held by many people (and reinforced by television programs and motion pictures featuring adventurous archaeologists such as Indiana Jones) is that it is a field largely concerned with examining ancient artifacts drawn from the extremely distant past. This is certainly true of many subfields of archaeology, such as Egyptology, Assyriology (study of ancient Assyria), classical archaeology, and prehistoric archaeology. However, archaeological tools and techniques can be and frequently are applied to the analysis of human cultures across all segments of time, from the very beginning of the species to much later epochs. For example, industrial archaeologists study the development and use of industrial methods, most of which did not truly come into being on a large scale until the eighteenth century, while urban archaeologists study the patterns of life revealed by the material past of metropolises—such as New York or Paris—that still exist and that have a long history of human habitation.

Archaeology as practiced in the modern world is characterized by several key principles. First, the field is not simply descriptive but rather explanatory. Archaeologists are concerned not just with the question of what happened in the past but also how and why it happened. To that end, they attempt to interpret the artifacts and features they uncover for clues about the belief systems, behaviors, traditions, and social, political, and economic lives of the cultures at hand. For example, archaeological excavations of the artifacts buried in the tombs of ancient Egyptian pharaohs have revealed a great deal of specific knowledge about how ancient Egyptians lived and what they believed to be true about the afterlife.

Second, it is multiscalar, meaning that the examination of a particular culture takes place simultaneously on many scales, each of which is intertwined with the others. For instance, an archaeologist may combine the small-scale analysis of individual pots, vases, and other clay artifacts—which give insight into the specific production processes used by the artisans who created them—with large-scale evidence about how those objects were handled on trade routes and marketplace procedures—which give insight into the overall historical trajectory of an entire civilization. By the same token, no matter which geographic region an archaeologist is studying, he or she is likely to examine collected data in the light of wider global trends.

BACKGROUND AND HISTORY

People have always been fascinated by the cultures that came before them. Even in ancient Babylon, for example, relics such as ruined temples were objects of interest. During the European Renaissance, an early form of archaeology arose that was mainly focused on classical antiquities and the investigation of large prehistoric sites and monuments, such as Stonehenge. The eighteenth century witnessed the first formal archaeological excavations, one of which was conducted by future president Thomas Jefferson (1743–1826) on an Indian burial mound located on his Virginia estate. In the nineteenth century, the site of the ancient city Pompeii, destroyed by the sudden volcanic eruption of Mount Vesuvius on August 24, 79 CE, became the focus of intensive archaeological investigations. Innovative techniques, such as the use of plaster of Paris to create molds of bones and other artifacts, were developed by early practitioners such as Italian scholar Giuseppe Fiorelli.

During the nineteenth century, archaeology began to develop the formality of a scientific discipline. The simultaneous maturation of the field of geology, with its theory of the stratification of rocks, helped establish similar ideas in archaeology. In addition, archaeologists began to work with the principle of uniformitarianism for the first time; this was the idea that past and present societies had more commonalities with each other than differences. Finally, the nineteenth century was the age in which naturalist

Charles Darwin's theory of evolution by natural selection, combined with the accumulating weight of physical evidence, led archaeologists to search for material evidence of human activity in the very distant past, far beyond the age of the Earth as defined by biblical scholars.

Archaeology underwent a transformation during the twentieth century, when many useful field techniques were developed that helped transform the field into a true science. Detailed records were kept at every excavation, including the location and description of each artifact. Sites were also divided into grids and searched systematically, and drawings and models were made of every dig. Advances were also made in absolute dating; particularly significant was the invention of radiocarbon dating by chemist Willard Libby (1908–1980) in the United States. Partly as a result of these techniques, archaeology became far more quantitative, or driven by data and statistical analysis, than it had previously been. Hypotheses and theories were carefully tested before being accepted.

HOW IT WORKS

Archaeological Surveys. The first step in conducting any archaeological investigation is to identify and take stock of the location in which the study will be held. This is known as a survey. The choice of location can be determined in several ways. Documentary sources, such as old maps or other written materials, can sometimes be used to accurately discover the location of an archaeological site. For example, clues in the writings of the ancient Greek poet Homer helped Heinrich Schliemann (1822–1890) to identify the location of the ruined city of Troy (at Hisarlik, in northwest Turkey). In other cases, archaeologists attempt to survey sites on which new developments such as roads or buildings are planned to unearth important remains before they are destroyed. This is known as salvage archaeology. Perhaps the most celebrated piece of salvage archaeology was the location and identification of the remains of the English King Richard III (1452–1485) beneath a parking lot in present-day Leicester, in 2012. On occasion, developers have integrated remains found by salvage archaeologists into their construction plans, as in the case of an Aztec temple dug up in Mexico City and used as part of the subway station that was built on that site. Other archaeological sites of interest are discovered through ground reconnaissance, in which either a judgmental or probabilistic sampling method is used to search for physical evidence, or aerial reconnaissance, which can sometimes turn up traces of human activity that cannot be seen from the ground.

Marine archaeology is of special interest for marine science for several reasons. For shipwrecks and sunken vessels, the archaeological site may, and often does, lie beneath hundreds of feet (or meters) of water, making access to the site for study rather problematic. In many cases, archaeological investigation can only be carried out remotely using unmanned robot submersibles, although specialized machines have been designed that can take a human operator to the deepest parts of the oceans. The wreck of the ocean liner *Titanic* is a prime example of this situation. In shallower waters, archaeological investigation can be carried out carefully by trained and suitably equipped scuba divers. A prime example of this type of investigation is the so-called sunken city of Cleopatra that was lost for some 1,600 years until it was discovered beneath the sea near Alexandria, Egypt, in 1991. In all cases, the study depends entirely on materials and devices that are developed in the field of applied physics, and the on-site studies themselves must be carried out with great care and respect for the marine environment.

Once a site has been identified as being of archaeological interest and an excavation begins, a deeper survey is conducted of the selected area. The same kinds of sampling methods used in ground reconnaissance are available to archaeologists at this point. Judgmental sampling is based solely on an archaeologist's judgment about which locations within the site should be searched for evidence. For example, since humans tend to settle near sources of water, judgmental sampling might concentrate the excavation along a riverbed that runs through the site. Probabilistic sampling, in contrast, is subject to chance and variation, and may help archaeologists turn up unexpected discoveries or make more accurate predictions about areas not sampled. Probabilistic sampling can take the form of simple random, stratified random, or systematic sampling. Simple random sampling is where any given location within a site has an equal chance of being sampled. Stratified random sampling is where the site is divided into naturally occurring regions, such as forest versus cultivated land, and the number of searches conducted in each region is based on its size in relation to the entire site. Systematic sampling is where a site is divided into equal parts and then sampled at consistent intervals (e.g., a search is made every 328 feet [100 meters]).

Types of Archaeological Evidence. The evidence used by archaeologists to form a picture of an ancient or past society can be classified into four major categories: material, environmental, documentary, and oral evidence. Of the four, material evidence is perhaps the most fundamental. It can consist of either organic or inorganic remains that have been used by or constructed by people. Material evidence includes buildings, tools, pottery, toys, textiles, baskets, and food remains. Environmental evidence can take many forms, including soil samples, minerals, pollen, spores, animal bones, shells, and fossils.

Because Earth has undergone many significant geological and climate-related changes over the course of its history, environmental evidence can help reveal what the world looked like during the time of a particular group's existence in an area. Environmental evidence can also offer clues about the relationships of the contemporary people with the landscape around them. For example, if the materials in a lower layer of a dig consisted mostly of wood, charcoal, and pollen, while those in a higher layer of a dig consisted mostly of cattle bones, one hypothesis might be that a region that had once been wooded had at some point been cleared by the inhabitants of the area and used for the rearing of animals.

Documentary evidence, or written records, can be extremely useful to archaeologists whenever it is available. Depending on the epoch, or span of history, from which these documents arise, they may consist of text inscribed on stone slabs, clay tablets, papyrus, or other types of materials. They may record laws, serve as proof of legal contracts (such as marriages or business agreements), list births and deaths in a city, or inventory commodities such as grain or shells. Finally, oral and ethnographic evidence—interviews and oral histories with modern-day inhabitants of a particular area—can help supplement the physical remains collected in an excavation. Certain practices such as the techniques of rural architecture or traditions of marriage, for instance, may have been preserved without much variation over time. In such cases, living members of a culture that has been around for many generations may be able to provide key information about the habits of their ancestors.

Dating Archaeological Evidence. Archaeologists have a plethora of techniques for placing objects in time. Relative dating methods identify artifacts as being older or younger than others, thus placing them into a rough inferred sequence. One of the most straightforward techniques of relative dating is to note the depth at which a particular object was found at a site; that is, in which stratum, or deposit layer, it was contained. This method, known as stratigraphy, is an effective way to gauge the relative age of objects because layers build up on top of other layers over time. Other means of relative dating include pollen analysis, ice-core sampling, and seriation. Seriation uses the association of artifacts with known dates of use, along with knowledge about how the frequency of use of those artifacts changed over time, as markers for other items found in close proximity.

Absolute dating methods are more accurate than relative dating techniques (which can be confused by the effects of natural disasters or animal activity on the organization of strata). They also produce more specific results. The absolute dating technique that is used more than any other by archaeologists is radiocarbon dating, which is capable of assigning an accurate age to artifacts of biological origin (including bone, charcoal, leather, shell, hair, textiles, paper, and glues) that are 50,000 years old or younger. Radiocarbon, or carbon-14, dating relies on the fact that the element carbon has a particular isotope—carbon-14—which is radioactive. This radioactive carbon combines with oxygen in the atmosphere to form carbon dioxide, which is absorbed by plants as they photosynthesize. Because plants form the base of every food chain, all living things on Earth contain some amount of radiocarbon within their bodies. Scientists are able to use the half-life of radiocarbon (the time it takes for half the amount in a given sample to decay) to measure the amount of radiocarbon in any organic material in comparison to the base quantity that is maintained in living organisms and calculate how long it has been around. Besides radiocarbon dating, other absolute dating techniques include dendrochronology, which relies on the annual growth rings present in the trunks of long-lived trees, and thermoluminescent dating, in which trace amounts of radioactive atoms in rock, soil, and clay can be heated to produce light, the intensity of which varies depending on the age of the object. This method is primarily used to date pottery and other human-created artifacts.

Interpreting Evidence. Three basic concepts dominate the approach an archaeologist takes when he or she approaches an artifact, a feature, or a site: context, classification, and chronology. Context, or where an object is found and what other items surround it, can reveal more of a complete story than any single artifact. For example, a decorated container found within a tomb could serve one purpose; the same container found in the kitchen of a dwelling could have been intended for an entirely different purpose. Classification (also known as typology) also helps place unearthed artifacts within a particular frame of reference. For example, determining that all the items found at a particular location belong to the same type—such as cooking utensils, hunting supplies, or bathing vessels—could indicate how a location was used or who used it. Finally, chronology is essential to understanding the relationship between the elements of any archaeological investigation. For example, unless they have been disturbed, objects located at lower levels, or strata, of a particular excavation site will be older than those located at higher levels. This can help archaeologists trace the development of a particular tool over time or to see and identify important changes in the cultural practices of a given society.

APPLICATIONS AND PRODUCTS

Cultural Resource Management. The major practical application of archaeological tools and techniques can be found in the area of cultural resource management (CRM). In a sense, CRM is a form of institutionally legislated salvage archaeology. It involves the use of archaeological skills to identify, preserve, and maintain important features of historic and prehistoric culture to benefit the public interest. In most countries, including the United States and Canada, any planned development of any significance, including the building of new gas or oil pipelines, residences, highways, golf courses, or any other construction, requires the developer—whether private or public—to conduct a CRM survey to comply with legal requirements. The goal of CRM is to ensure that important pieces of a nation's past are conserved. In the United States, the major pieces of legislation mandating CRM include the National Historic Preservation Act, the National Environmental Policy Act, the Archaeological Resources Protection Act, and the Native American Graves Protection and Repatriation Act. CRM studies are generally performed by trained archaeologists serving as consultants and involve reconnaissance and sampling intended to scan the area for the presence of significant archaeological sites, features, or artifacts. If any are found, the developers are responsible for safely excavating and preserving them. If this is impossible or the site is identified as especially historically significant, the project may be terminated. In the case of marine sites, permission must be obtained to recover artifacts from the site, and often sites such as shipwrecks in which people have "gone down with the ship" the site may be declared sacred and given the same reverential status as any terrestrial gravesite.

Waste Management. The field of garbage archaeology, or garbology, was pioneered by American urban archaeologist William Rathje in the 1970s. By applying the tools of archaeology to the waste dumped in landfills, Rathje and other scientists have been able to provide many practical insights that are useful to waste management specialists and city, state, and federal administrators concerned with reducing waste and conserving energy. For instance, they showed that even supposedly biodegradable artifacts, such as paper, wood, and even food, are preserved almost indefinitely when packed tightly together in a landfill rather than decomposing to their component materials as had been traditionally believed. This is because the dense conditions reduce the amount of oxygen available for decomposition to take place. One of the implications of this discovery is that simply switching from nonbiodegradable to biodegradable materials in the production process—banning plastic disposable cups in favor of paper ones, for example—will not be enough to reduce the amount of effectively permanent waste that is generated by consumers and filling landfills to the point of overflowing.

Agriculture. In some parts of the world, environmental archaeologists have discovered evidence that areas of land that are marked by barren soil in fact used to be rich and fertile. For example, the Negev Desert in Israel, with its high temperatures and meager rainfall, used to be the site of an ancient urban society that cultivated crops such as wheat and grapes.

Archaeologists also established that the climate in the area had not changed over the 2,000-year period in between. Using aerial reconnaissance techniques, scientists were able to show that the ancient farmers had employed a water-delivery system composed of terraces and cisterns to collect and redirect rainwater from the infrequent flash floods that occur in the area. This insight has proved to be of immense importance to modern-day agricultural scientists and farmers in Israel and has played a role in the country's successful efforts to "green" the Negev. Similarly, local farmers in rural Peru have begun using prehistoric field technologies unearthed by environmental archaeologists—primarily elevated fields that improve drainage and help protect plants from chilly nights—to dramatically increase their crop yields.

Crime Investigation. Forensic archaeologists apply the methodologies of archaeology to the investigation of crimes, working closely with coroners and police officers to collect and analyze physical evidence from the scene. They investigate fragments of bone, teeth, soil, fabric, jewelry, or other artifacts, while more subtle clues such as disturbed soil or markings from tools or weapons may also provide important leads. Just as in other forms of archaeology, the forensic archaeologist first establishes the boundaries of the site to be surveyed, divides it into grids, and performs a thorough excavation of the material and environmental evidence found in the area. Stratigraphy is an important element of forensic archaeology, particularly so when a body has been buried before being found; the deeper into the soil digging proceeds, the older the remains, personal artifacts, or other evidence that is found will be.

CAREERS AND COURSEWORK

The typical archaeology career path begins with an undergraduate degree in anthropology, with an emphasis on coursework covering the fields of biological, linguistic, and cultural anthropology. In addition, the aspiring archaeologist should be sure to take a set of courses in related sciences—such as geology, chemistry, ecology, environmental studies, evolutionary science, geophysics, anatomy, and paleontology. Mathematics and statistics are equally important subjects, since much archaeological analysis requires a keen understanding of these areas. In the humanities, history courses—preferably focused on a particular geographic area or epoch in time—are essential, as is an advanced knowledge of at least one foreign language. In addition, business management and technical writing skills will assist any archaeologist who chooses to go into the field of CRM. Finally, if at all possible, students should attempt to pursue archaeological internships, fieldwork placements, or other forms of practical work experience during their undergraduate careers. Such experiences not only are a means of gaining hands-on knowledge of the field techniques used in archaeology but also provide students with professional contacts.

A bachelor's degree will serve as sufficient minimum qualification for many professional archaeological positions in the private sector, particularly in consulting firms hired by developers to perform CRM surveys—in such settings, practical work experience is more important. However, additional graduate study at the master's or doctoral level is necessary to obtain a role as a crew supervisor and enter higher pay brackets. For archaeologists who wish to become faculty in academic institutions, a doctoral degree in archaeology and evidence of original research is required. Other organizations where archaeologists work include museums, city and state governments, and the federal government. Typical roles for archaeologists include conducting field investigations, performing analyses of found artifacts, curating museum archaeology collections, teaching courses, and publishing research papers.

SOCIAL CONTEXT AND FUTURE PROSPECTS

Archaeology is one of the most important and influential lenses through which scientists, historians, and other scholars view pieces of the past that otherwise would be lost. Its findings often reveal stunning insights into the nature of human civilization and the growth of human culture. For example, archaeologists piecing together skeletons found in Africa have been able to uncover the point at which the early human species *Homo neandertalensis* seemed to have died out and *Homo sapiens* (anatomically identical to modern humans) became the prominent human species, while the ancient cave paintings discovered in Lascaux, an area in the southwest of France, demonstrate that human beings were creating art as long ago as 15,000 BCE. In other words, archaeology is a science dedicated to telling the story of humanity itself.

Although archaeology is at its core a science concerned with the investigation of material objects, the future of archaeology may be surprisingly metaphysical: a small but growing movement exists that is determined to treat the World Wide Web as a treasure trove of archaeological artifacts. Internet archaeology, as this nascent subfield of archaeology is called, seeks to archive and interpret websites, pages, and graphics that were created during the early years of the internet and are no longer being updated by their owners.

—*M. Lee*

Further Reading

Bailey, Geoff, Nina Galanidou, Hans Peeters, Hauke Jöns, and Moritz Mennenga. *The Archaeology of Europe's Drowned Landscapes*. Springer International Publishing, 2020.

Ballard, Robert D., editor. *Archaeological Oceanography*. Princeton UP, 2008.

Delino-Musgrave, Virginia E. *Marine Archaeology: A Handbook*. Council for British Archaeology, 2012.

Fagan, Brian. *Before California: An Archaeologist Looks at Our Earliest Inhabitants*. Rowman, 2004.

Gamble, Clive. *Archaeology: The Basics*. 2nd ed., Routledge, 2008.

Kelly, Robert L., and David Hurst Thomas. *Archaeology*. 6th ed., Wadsworth, 2013.

McCarthy, John K., Jonathan Benjamin, Trevor Winton, and Wendy van Duivenvoorde, editors. *3D Recording and Interpretation for Maritime Archaeology*. Springer, 2019.

Murray, Tim. *Milestones in Archaeology: A Chronological Encyclopedia*. ABC-CLIO, 2007.

Renfrew, Colin, and Paul G. Bahn, editors. *Archaeology: The Key Concepts*. Routledge, 2005.

ARCTIC OCEAN ECOSYSTEM

Fields of Study: oceanography; marine biology; marine geology; marine chemistry

ABSTRACT

The Arctic Ocean is the most extreme ocean in regard to the seasonality of light and the year-round presence of ice. Knowledge of what lives in the Arctic Ocean is limited due to the logistical challenges imposed by ice and its inhospitable climate. Covering about 3 percent of Earth's total surface area, the Arctic Ocean is the smallest of the world's five ocean regions. Most of this nearly landlocked ocean region is north of the Arctic Circle. The Arctic Ocean is connected to the Atlantic Ocean by the Greenland Sea, and to the Pacific Ocean via the Bering Strait. The Northwest Passage (in Canada) and the Northern Sea Route (between Norway and Russia) are two important seasonal waterways.

KEY CONCEPTS

brine channels: small tunnel-like openings found in sea ice, caused by the incomplete formation and erosion of ice as it forms from saline seawater

pack ice: the uniform layer of solid ice that covers the Arctic waters in the cold season

sea ice: ice that has formed from saline seawater, typically refers to the broken remnants of the pack ice in the warm season

seasonality of light: the variation in hours of daylight received each day

sympagic: indicates a close relationship with ice

DESCRIPTION

The Arctic Ocean is unique among the world's oceans for many reasons. It is the most extreme ocean in regard to the seasonality of light and the year-round presence of ice. Knowledge of what lives in the Arctic Ocean is limited due to the logistical challenges imposed by its many years of ice buildup and its inhospitable climate. Covering about 3 percent of Earth's total surface area, the Arctic Ocean is the smallest of the world's five ocean regions (after the Pacific, Atlantic, Indian, and the recently delimited Southern Ocean). Most of this nearly landlocked ocean region is north of the Arctic Circle. The Arctic Ocean is connected to the Atlantic Ocean by the Greenland Sea, and to the Pacific Ocean via the Bering Strait. The Northwest Passage (in Canada) and the Northern Sea Route (between Norway and Russia) are two important seasonal waterways.

The center of the Arctic Ocean is covered by a drifting, persistent ice pack that has an average thickness of about 10 feet (3 meters). During the winter months, this sea ice covers much of the Arctic Ocean's surface. Higher temperatures in the summer months cause the ice pack to shrink by about 50 percent in extent. In recent years, this polar ice pack has thinned, and its summer extent has contracted, allowing for increased navigation as well as inducing undersea resource claims by neighboring nations and shipping disputes among countries bordering the region, essentially all of which has historically always been mapped as sovereign Canadian territory.

The Arctic Ocean has the widest continental shelf of any ocean, extending 750 miles (1,207 kilometers) from the coast of Siberia, but also has areas that are quite deep. The shallowest ocean region overall, it has an average depth of 3,450 ft (1,052 m). The maximum depth is 17,850 ft (5,450 m). The Chukchi Sea provides a connection with the Pacific Ocean via the Bering Strait, but this channel is very narrow and shallow, so most water exchange is with the Atlantic Ocean via the Greenland Sea. The Arctic Ocean connects to or envelops numerous other smaller seas and bays, including Baffin Bay, Barents Sea, Beaufort Sea, Chukchi Sea, East Siberian Sea, Hudson Bay, Hudson Strait, Kara Sea, and the Laptev Sea.

The polar climate is characterized by persistent cold and relatively narrow annual temperature ranges. The winters are characterized by near-continuous darkness, cold and stable weather conditions, and clear skies; the summers by continuous daylight, damp and foggy weather, and weak cyclones with rain or snow.

SEAMOUNTS AND SUBMARINE RIDGES

The floor of the Arctic Ocean is divided by three submarine ridges—Alpha Ridge, Lomonosov Ridge, and Gakkel Ridge—of which the Lomonosov separates the two great basin deeps, the Amerasian and Eurasian. A further division of the Amerasian deep is the relatively isolated Canadian Basin; this area is particularly interesting to scientists because its isolation could mean that it contains life forms found nowhere else on Earth. The Lomonsov Ridge is an underwater mountain chain that averages 10,000 ft 3,048 m) above the seafloor plain and in places comes to within 3,000 ft (914 m) of the sea's surface.

BIODIVERSITY

Researchers have only begun to scratch the surface of biodiversity in the Arctic Ocean, which has at least three distinct biological communities. The Sea-Ice Realm includes plants and animals that live on, in, and just under the surface ice. Because only 50 percent of this ice melts in the summer, ice flows can exist for many years and can reach a thickness of more than 6 ft (2 m).

Sea ice usually is not solid like an ice cube, but is riddled with a network of tunnels called "brine channels" that range in size from microscopic to more than 1 inch (25 millimeters) in diameter. Diatoms and algae inhabit these channels and obtain energy from sunlight to produce biological material through photosynthesis.

Bacteria, viruses, and fungi also inhabit the channels, and together with diatoms and algae, they provide food for flatworms, crustaceans, and other animals. This community of organisms is called "sympagic," which means "ice-associated."

Partial melting of sea ice during the summer months produces ponds on the ice surface that contain their own communities of organisms. Melting ice also releases organisms and nutrients that interact with the ocean water below the ice. Also, the ice in the Arctic can contain natural-history information about what the world was like thousands of years ago. Much history must be hidden in the Arctic ice, including information regarding historical weather patterns, carbon dioxide levels from the Stone Age, and what organisms were alive thousands of years ago.

The Pelagic Realm includes organisms that live in the water column between the ocean surface and the bottom. Melting sea ice allows more light to enter the sea, and algae grow rapidly because they receive sunlight 24 hours a day during the summer. These algae and phytoplankton provide energy for a variety of floating animals (zooplankton), including crustaceans and jellyfish. Zooplankton in turn are the main energy source for larger pelagic animals, including fishes, squids, seals, and whales.

When pelagic organisms die, they settle to the ocean bottom and become food for inhabitants of the Benthic Realm. A variety of sponges, bivalves, crustaceans, polychaete worms, sea anemones, bryozoans, tunicates, and ascidians are common members of Arctic benthic (floor-dwelling) communities. These animals provide energy for bottom-feeding fishes, whales, and seals.

SEAWATER DENSITY AND SALINITY

Ocean currents are caused by winds and changes in seawater density. Density can be changed by evaporation or by the input of freshwater (which raises or lowers the salinity, respectively), as well as by temperature changes. Near the equator, for example, evaporation causes seawater salinity and temperature to increase. The density of seawater increases as salinity rises and decreases as temperature rises. So even though the water is saltier, the higher temperature keeps the surface water from sinking. But as the water flows toward the poles (driven in part by wind), it becomes cooler and sinks due to the increased density.

CURRENTS

The main ocean current flowing into the Arctic Ocean is the West Spitsbergen Current, a northward extension of the Norwegian Atlantic Current, passing through the Fram Strait and then following a deep trench leading to the Arctic Ocean. The access via the Barents Sea is partially obstructed by shallows. The water in this current is relatively warm, above 37 degrees Fahrenheit (3 degrees Celsius), and relatively salty, with a salinity greater than 0.00349 percent (34.9 parts per million [ppm]). The amount of water transported is uncertain, but it is possible that as much as half of it circulates in the vicinity of the Fram Strait without entering the Arctic Ocean.

Alongside the West Spitsbergen Current is the East Greenland Current. This is the main current out of the Arctic Ocean, carrying cold—below 32 degrees Fahrenheit (0 degrees Celsius)—and relatively fresh (less than 0.00344 percent [34.4 ppm salinity) water southward. The Bering Strait is narrow at 53 miles (85 kilometers) and shallow, reaching just 164 ft (50 m) in depth, allowing only a small northerly flow.

The Transpolar Drift, a surface current, flows from the Siberian to the Greenland side of the Arctic Ocean, where it feeds into the East Greenland Current. Explorer Fridtjof Nansen (1861–1930) trusted this current to carry his ship, the *Fram,* across the Arctic Ocean in the 1890s into the vicinity of the North Pole. A less heroic demonstration of the Transpolar Drift has been provided by a consignment of about 29,000 floating plastic bath toys (including ducks, frogs, and turtles) that were lost from a container ship in the North Pacific in October 1992. It was proposed that the floating toys would be transported by the pack ice across the Arctic Ocean into the North Atlantic.

The pack ice, up to 39 ft (12 m) thick, plays an important part in conserving heat. The seawater, with a mean depth of about 3,937 ft (1,200 m) and a volume of about 4 million cubic miles (17 million cubic kilometers), provides an immense heat reservoir. The ice cover reduces heat transfer to the atmosphere by one or two orders of magnitude compared with that from open water. Furthermore, largely because of the inflow of river waters and Bering Strait waters, a layer of low-salinity water floats on top of the denser water in the Arctic basin, producing a marked halocline (a vertical zone in the oceanic water column in which salinity changes rapidly with depth) at 98 to 197 ft (30 to 60 m), which limits the convection that would otherwise mix the whole water column and promote heat loss. Consequently, the deep water remains at minus 30 to 31 degrees Fahrenheit (0.5 to minus 0.9 Celsius), above its freezing point of minus 28 degrees Fahrenheit (2 degrees Celsius). These various factors contribute to the generally higher temperatures of the Arctic Ocean compared with the Antarctic Ocean.

Plans to divert the southward part of the flow of some Siberian rivers to alleviate water shortages further inland perhaps need not create too much alarm. On present evidence, it seems unlikely that such diversions would have major effects on circulation or sea-ice distribution in the Arctic Ocean.

This complex circulation system in the Arctic, which affects the entire food web, is however, in a delicate balance. In recent years scientists have documented changes in the Arctic systems, including a dramatic reduction in sea-ice cover and a weakening of some of the circulation systems that are attributed to—and perhaps accelerating—climate change. The Arctic Ocean affects not only the Arctic native peoples, but also those living farther south in Europe and North America. As a result, the Arctic Ocean and the effects of changes that are taking place here are the focus of intense study by many nations.

—Alexandra M. Avila

Further Reading

Berkman, Paul Arthur, and Alexander N. Vylegzhanin, editors. *Environmental Security in the Arctic Ocean.* Springer, 2012.

Coad, Bryan W., and James D. Reist. *Marine Fishes of Arctic Canada.* U of Toronto P, 2018.

Johannessen, Ola M., Leonid P. Bobylev, Elena V. Shalina, and Stein Sandven, editors. *Sea Ice in the Arctic: Past, Present and Future.* Springer, 2019.

Lund-Hansen, et al. *Arctic Sea Ice Ecology: Seasonal Dynamics in Algal and Bacterial Productivity.* Springer, 2020.

Pongrácz, Eva, Victor Pavlov, and Niko Hänninen, editors. *Arctic Marine Sustainability: Arctic Maritime Businesses and the Resilience of the Marine Environment.* Springer, 2020.
Rothwell, Donald R. *Arctic Ocean Shipping. Navigation, Security and Sovereignty in the North American Arctic.* Brill, 2008.
Thomas, D. N., G. E. Fogg, and P. Convey. *The Biology of Polar Regions.* Oxford UP, 2008.

ATLANTIC OCEAN

Fields of Study: geography; oceanography; marine biology

ABSTRACT
The Atlantic Ocean separates the North and South American continents from Europe and Africa. With an area, excluding its dependent seas, of about 32 square miles (83 million square kilometers), it is second in size only to the Pacific Ocean. The dependent seas add another 8.9 sq mi (23 million sq km) to the Atlantic's total size.

KEY CONCEPTS

abyssal plain: the ocean floor between the edges of the continental shelves

continental shelf: the extension underwater of the continental mass

gyre: the slow, roughly circular movement of ocean water over the expanse of the particular ocean

mid-ocean ridge: an undersea mountain range situated approximately down the middle of the Atlantic Ocean where tectonic activity causes molten magma to ooze up as the seafloor slowly spreads away to east and west

Pangaea: a supercontinent that existed some 250 million years ago, when all of the current continental masses were conjoined in a single, giant continent

seafloor spreading: the movement of the seafloor to the east and west of a mid-ocean ridge driven by convective currents in the underlying magma of Earth's mantle layer

tectonic plates: segments of Earth's crust somewhat analogous to the shell of an egg that has cracked in several places yet maintains the egg's shape

LOCATION AND ORIGIN

The Atlantic Ocean, with an average depth of 10,827 feet (3,300 meters) and a maximum depth of 29,493 ft (8,380 m), touches the eastern coastlines of North and South America and the western coastlines of Europe and Africa. The Atlantic Ocean is often divided into the North Atlantic and the South Atlantic, although the entire oceanic mass may properly be considered a single ocean. The North Atlantic extends from the equator north to the Arctic Ocean. The South Atlantic, that part of the ocean south of the equator that extends as far as Antarctica, meets the Pacific Ocean at Cape Horn, the southernmost tip of South America, and the Indian Ocean at the Cape of Good Hope, the southernmost tip of Africa.

More than 200 million years ago, Earth's land surface consisted of a single supercontinent, referred to as Pangaea, surrounded by an enormous sea, called "Panthalassa." A large, shallow bay known as the Tethys Sea also protruded into the supercontinent from Panthalassa. The landmass, like contemporary landmasses, was a large, solid rock plate that floated on the mantle layer of molten rock. Over the intervening millions of years, convection currents in the mantle drove Pangaea to break apart into several smaller segments and become the present-day continents. The magmatic currents have driven, and are still driving, the continental masses and tectonic plates across the face of the planet. Looking at a map of the modern world and regarding it as a huge jigsaw puzzle, it is easy to visualize how the western coast of Africa fits into the eastern coast of the Americas and how the coastlines of other geographical areas fit neatly into those facing them.

As Pangaea split apart, the Tethys Sea essentially became enclosed by land as continental drift continued over millions of years, forming part of the present-day Arctic Ocean and Mediterranean Sea. The Pacific Ocean is the remnant of Panthalassa that exists today. The main rift in Pangaea separated large tectonic plates that have since formed North and South America, Asia, Africa, Europe, India, and Australia. The ever-widening space between the separating continents filled with water to form what is now the Atlantic Ocean. Current data regarding the movement of the continents and tectonic plates reveal that the North and South American continental structure and the Europe-Africa-Asia-Australia chain continue to move away from each other as though pivoting about a point in the North Atlantic Ocean approximately at the location of Iceland, such that the greater the distance from that location, the faster the tectonic plate is moving. Thus Europe and the easternmost coast of Canada are separating at the slowest rate, while Australia continues to move in a northeasterly direction at a relatively fast rate.

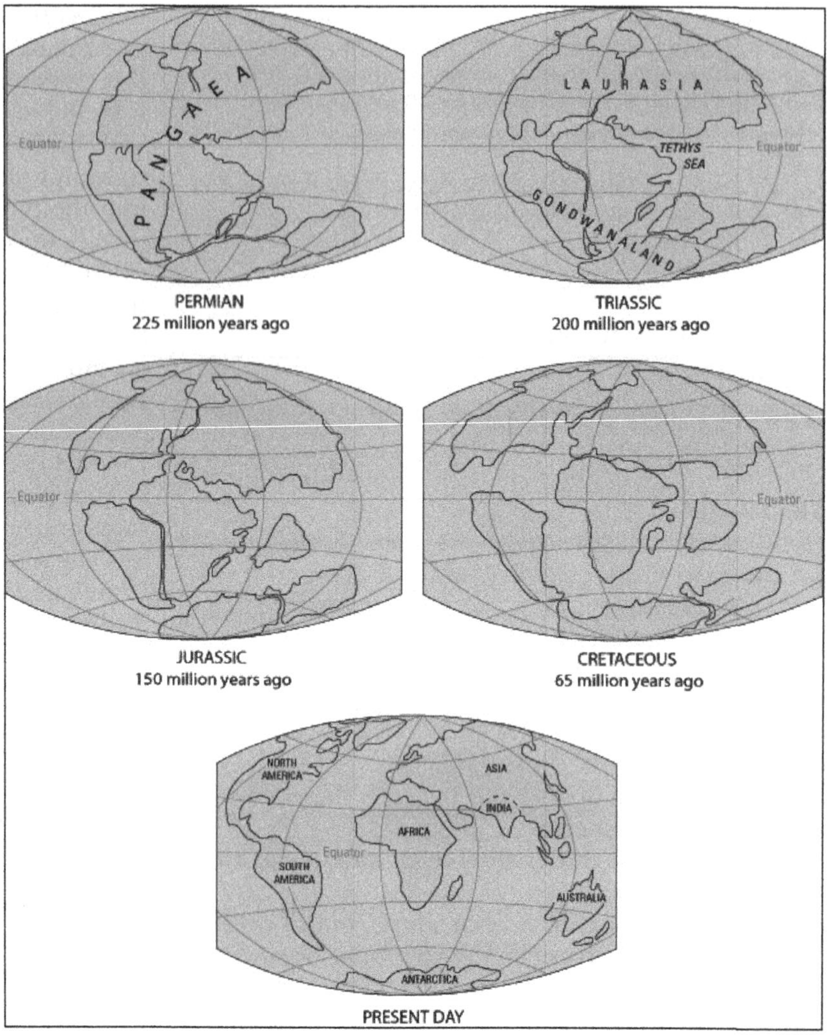

The breakup of Pangaea and motion of their continents to their present-day positions (United States Geological Survey via Wikimedia Commons)

Two major tectonic plates, each a segment of the Atlantic Ocean floor, are moving apart at the average rate of 0.5 inch (1.3 centimeters) annually from a volcanically active line that runs approximately down the middle of the Atlantic Ocean, called the "Mid-Atlantic Ridge." Where volcanic activity has been intense, molten rock has cooled into volcanic cones that sometimes protrude above the water's surface to create islands. However, in the Atlantic, unlike in the Pacific (where islands are scattered throughout the ocean), more islands are found close to shore than toward the middle of the ocean. The rate of movement of the Atlantic Ocean seafloor away from the Mid-Atlantic Ridge is such that the entire seafloor is replaced about every 100,000 years, precluding the formation of mid-Atlantic islands. The few exceptions are actually the exposed peaks of underwater volcanic mountains, some of which, such as Mount Tiede in the Canary Islands, extend more than 11,811 ft (3,600 m) above sea level.

Ancient people named the Atlantic Ocean after the mythological giant Atlas, who is said to have carried the world on his shoulders. Many ancients, believing that Earth was flat, feared that anyone sailing far enough into the ocean would eventually fall over the edge into an abyss. This belief inhibited early European explorers, although even in ancient Greece philosophers understood that Earth is round and later adventurers like the Vikings sailed far into the ocean and returned safely to shore.

CONTINENTAL SHELF AND COASTLINE

Close to shore, the Atlantic Ocean is generally shallow, seldom exceeding a depth of 1,788 ft (545 m). Waters cover a shelf of rock that, in most of the North Atlantic, extends between 40 and 50 mi (65 and 80 km) from the coastline into the ocean. At the end of this shelf, the ocean floor drops precipitously to a depth of about 12,926 ft (3,940 m), gradually leveling off into what is designated the abyssal plain. The continental shelf off the coast of Africa is much narrower than the continental shelf off the coast of North America.

The coastlines that border the Atlantic Ocean have been forming for millions of years, shaped largely by the motion and activity of the water that laps the shore. The rockiest areas are found in the northern parts of the North Atlantic and in the South Atlantic. The level of the ocean has risen considerably over the past 10,000 years as glaciers have melted. With the end of the Ice Age, huge quantities of water poured down rivers into the oceans. The point at which one of these rivers meet the ocean is called an "estuary." Among the deepwater estuaries are the Chesapeake Bay in the United States and the Falmouth Estuary in Great Britain, both of which are sufficiently deep to accommodate large ships.

Less rocky portions of coastline are often characterized by long stretches of sandy beaches that have been formed over the years as the moving waters of the ocean have pulverized rocks and shells, transforming them into sand. Sometimes sandbars form near the coast. These may be long expanses of sand, often permanently covered by water, while others are below the water's surface only at high tide. Some sandbars that are permanently above water may be broad enough to be inhabited, as are the Outer Banks of North Carolina, although at times devastating floods occur in communities built on sandbars. Sandbars endanger ships that venture too close to them, sometimes causing them to founder. Such ships may float off a sandbar as the water rises at high tide, but in many cases this does not happen, and the ship, trapped on the sandbar, must be hauled off. Most seaports employ pilots whose duty it is to guide incoming ships through deep channels away from dangerous sandbars or rock formations beneath the surface of the water. Lagoons sometimes form between the coast and sandbars that are not permanently covered with water. These are salty bodies of water in which marsh grasses and other vegetation grow. Many of them are rich in the microorganisms that fish need for their survival. Extensive fish and bird populations cluster around lagoons.

No ocean is fed by more rivers than the Atlantic. These rivers have, over hundreds of thousands of years, carried silt and sand toward the ocean they feed. These deposits have, in many places, built up to form deltas, roughly triangular areas whose broad base fronts the ocean. Some deltas, particularly those of the Amazon and Niger Rivers, are hundreds of miles wide with channels running through them. Often in tropical areas, mangrove swamps flourish in the deltas, the roots of the mangrove trees being covered with saltwater at high tide to provide them with the nutrients they require for their growth. Fish and birds flourish in mangrove swamps.

TIDES AND TRADE WINDS

Ocean tides are generally predictable. High tide and low tide each occur twice in any twenty-four-hour period. Tidal ranges—the depth variations between high tide and low tide—are most pronounced on continental shelves and in deepwater bays. In some parts of the world, the tidal range is dramatic. In the Canadian province of Nova Scotia, the Bay of Fundy has a tidal range of more than 39 ft (12 m), one of the largest in the world.

Gentler tides are observed as one approaches the equatorial regions, although there can be dramatic tidal activity in such areas during hurricanes and tropical storms, which have traditionally occurred in August through October in the Northern Hemisphere and February through April in the Southern Hemisphere. There is considerable concern at present that global climate change is causing more powerful hurricanes and extending the hurricane season. Hurricanes usually form in tropical areas but can move rapidly into more temperate zones where, if they strike land, they can result in injury, death, and substantial destruction. In some cases, such storms erode entire beaches and completely inundate waterside property. The Atlantic coastline in many places is shrinking rapidly as the result of water erosion from such storms.

Predictable wind currents, much like jet streams in the upper atmosphere, blow across the Atlantic Ocean. The trade winds near the equatorial areas blow in a westward direction. Such winds made it possible for early explorers to sail from European ports to the Americas. For their return trips, explorers depended upon winds to the north of the areas where the trade winds blow called the "westerlies," which blow from west to east. Along the equator is a narrow area where waters are calm and where there is virtually no wind. In this area, referred to as the doldrums, sailors can drift languidly for long periods of time with no wind to propel them.

ATLANTIC CURRENTS

Ocean water is never still. Much ocean water moves in one direction in predictable flows, in essence forming rivers through the ocean. Such rivers are called "ocean currents." The warm tropical currents near the equator push the waters of the Atlantic Ocean westward toward North and South America, moving clockwise in the Northern Hemisphere and counterclockwise in the Southern Hemisphere.

The northern equatorial current runs from just north of the equator along the northeastern coast of South America and on toward the southern coast of North America. There, in the ocean east of the United States, the Gulf Stream runs through the Atlantic, flowing in a northerly direction. The Labrador Current that flows from the Arctic Ocean meets the Gulf Stream off Newfoundland, causing temperature variations that result in high humidity and dense fogs. North of Labrador, the Gulf Stream divides into the North Atlantic Drift, which flows north toward Greenland, and the Canaries Current, which flows southeast through the Atlantic off the west coast of Africa. These clockwise currents flow in what is called the "North Atlantic Gyre."

South of the equator, the southern equatorial current moves counterclockwise off South America's east coast, proceeds south to the Brazilian Current, which then veers southeast to meet the Benguela Current that flows from just north of Antarctica to the Tropic of Capricorn. These counterclockwise currents are referred to as the South Atlantic Gyre.

The oceanic rivers that form the currents of the North and South American Gyres have a profound effect upon climate. The Gulf Stream, which brings warm, tropical waters north at the rate of some 81 mi (130 km) per day, makes far northern areas such as Greenland and Iceland warm enough for human habitation. Its southeasterly branch gives Great Britain a more temperate climate than one would expect at such latitudes. The northern regions of the United States are considerably warmer than they would be if the Gulf Stream did not flow along their coasts. Even areas within the Arctic Circle feel the effects of the Gulf Stream, which keeps Russia's Arctic port of Murmansk free of ice throughout the year.

The South Atlantic Ocean's Benguela Current propels cold waters from the Antarctic Ocean north along Africa's southwestern coast, keeping its temperatures much more temperate than would be expected in such latitudes. Cold air, however, does not hold moisture well, so when warm, humid winds from Africa's southwestern coast strike the cold air over the Benguela Current, they cause rain to fall over the ocean. The result is that the whole southwestern coast of Africa is arid. The South Atlantic Ocean there is bordered by desert.

FOOD CHAIN

The world's oceans are teeming with life. In a somewhat hierarchical arrangement called the "food chain," the sea's larger creatures feed on the smaller ones. At the base of the food chain are microscopic organisms called "plankton," which form the basis for life in oceans. Plankton, which exist in both animal and vegetable form, are the basic diet of many aquatic creatures and hence of all ocean life. Plankton cannot live without sunlight, so they are found close to the water's surface. Plankton depend upon nutrients borne in the water for sustenance, and such nutrients are usually found in coastal areas that are not very deep, such as mangrove swamps, where plankton often flourish. Some of the minerals the plankton require are washed into the ocean through estuaries that carry silt from inland areas. Other minerals come from deep in the sea, carried to the surface by ocean currents. In the Atlantic, the most abundant plankton are found in the extreme northern and southern reaches of the ocean, as well as off the west coast of Africa. These areas have a high concentration of the nutrients—abundant in deep, cool waters—on which plankton feed.

The presence of large amounts of plankton in these areas results in a wealth of other aquatic creatures, most of which live near the surface of the ocean where plankton are found, although exploration of the very deep oceans has revealed a remarkable amount of sea life in parts of the ocean so deep that it was previously thought that nothing could survive there.

The Atlantic has an intriguing population of creatures that range from among the smallest on Earth, such as plankton, to the largest, such as whales. Most of the fish caught by commercial fisherman for human consumption—notably sardines, smelt, cod, halibut, mackerel, hake, sole, and anchovies—swim in large schools near the ocean's surface to harvest the plankton and smaller fish that constitute their diets. Larger fish, such as sharks, whales, swordfish, and

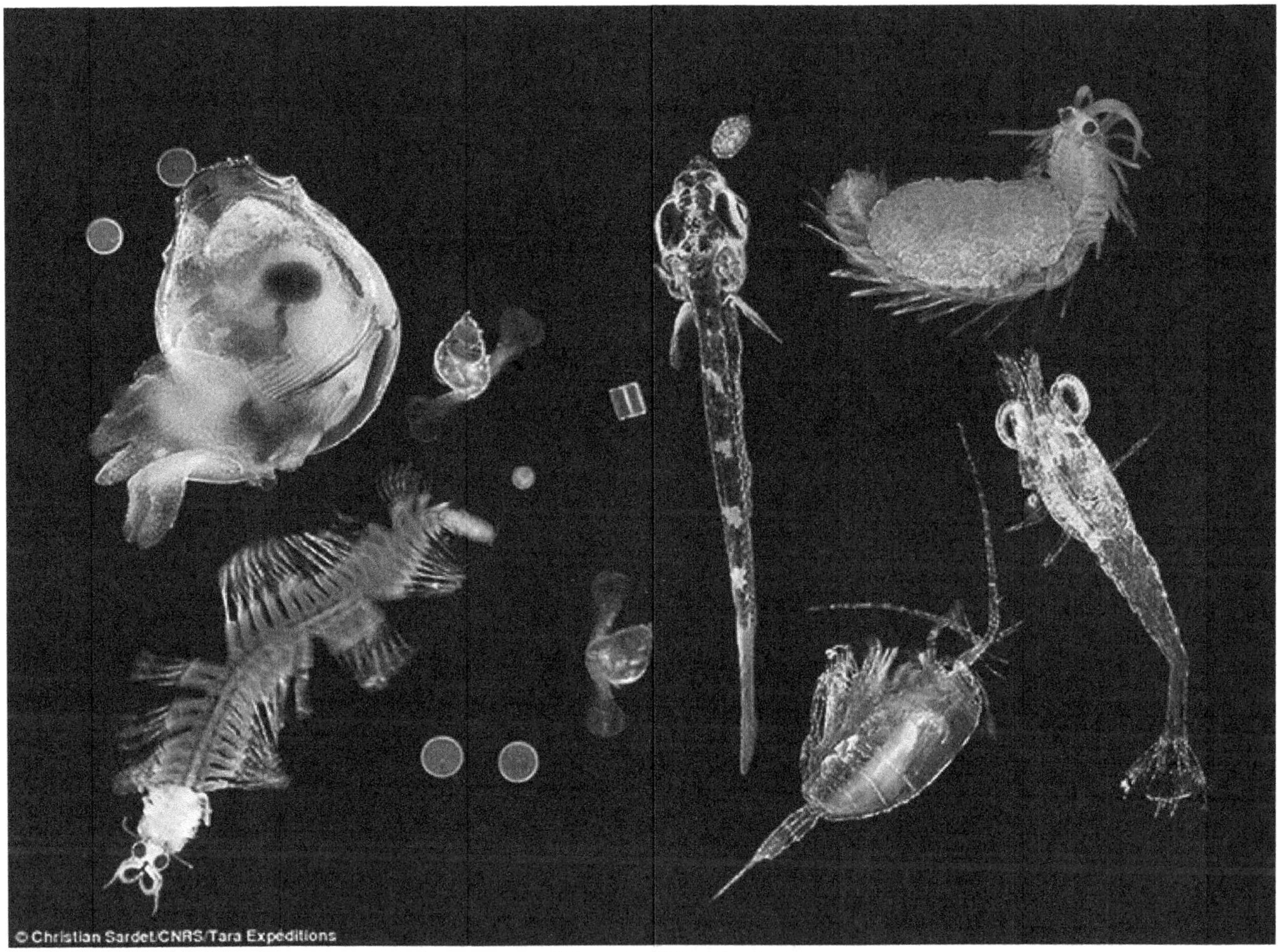

Plankton species diversity (Christian Sardet/CNRS/Tara expeditions via Wikimedia Commons)

sailfish, are fast-moving and live on a diet of smaller fish. They are often found in the cooler waters of the Atlantic, where plankton are abundant enough to nourish the small fish on which larger fish feed.

Some of the sea's creatures are extremely mobile. Whales travel from the polar regions to areas thousands of miles (or kilometers) away to give birth. Dolphins swim fast enough to cover 152 mi (245 km) in a single day. Conversely, some shellfish, such as mussels, attach themselves to formations in the sea and are immobile, protected from predators by their thick shells. Some forms of sea life are protected by camouflage that allows them to blend in with the surrounding area. Others are able to emit noxious jets of fluid into the waters around them when they are threatened.

Considerable bird life is associated with the Atlantic Ocean. Most of the birds that depend upon the Atlantic for their food supply can fly, but some of them, such as the gannet, the cormorant and the puffin, can swim as well. They dive into the water to catch their prey and can swim to catch it. Penguins, found in the Antarctic, cannot fly and can barely walk, but they swim as well as many fish. Cranes and other long-legged birds flourish in shallow waters, where they eat tiny sea creatures that they strain from the water and mud through their beaks.

In the middle of the North Atlantic is a large, relatively calm area of water known as the Sargasso Sea, whose surface is covered with huge fields of nutritious seaweed. The Sargasso Sea has become a spawning ground for eels that migrate to it during mating season from Europe and North America.

SEASIDE SETTLEMENTS

Throughout history, humans have tended to settle beside the sea or along rivers. The transportation opportunities provided by such locations still make them desirable places for settlement. Such areas usually offer a temperate climate as well. The ancient cultures of Rome and Greece grew up on the shores of the Mediterranean and the Aegean, two of the dependent seas of the Atlantic Ocean. The Iberian Peninsula borders on the Atlantic to the west, as do France and Great Britain. Advanced civilizations have flourished on the shores of the Atlantic, particularly in Europe and Africa, since long before recorded history.

The islands of the Atlantic—products of violent volcanic activity that created large, hilly outcroppings—were slow to develop. The exceptions are the Canary Islands and Madeira, both of which became necessary stopover points for explorers sailing from Europe to the Americas. In recent times, considerable development of the Atlantic islands has taken place in such areas as the Caribbean. The Falkland Islands off the southeastern coast of South America have a stable, permanent population, as do such island enclaves as Prince Edward Island and Newfoundland in the north.

Shipping and commerce have been the backbone of the economy in most of the areas that have developed along the Atlantic coastline. Raw materials are brought into Atlantic port cities such as New York, Philadelphia, Rio de Janeiro, and Lisbon to feed the manufacturing industries of those countries. The fishing industry has flourished for many centuries along the Atlantic coastline. Tourism has also become a major economic factor in the more temperate and scenic regions on the Atlantic coast.

Trade patterns were substantially altered with the opening in 1896 of the Suez Canal, which connects the Mediterranean Sea with the Indian Ocean, and with the opening in 1914 of the Panama Canal, which links the Atlantic and Pacific Oceans. These strategic canals opened up a great deal of trade worldwide and overcame the necessity of plying the turbulent and dangerous waters around Cape Horn and the Cape of Good Hope in order to deliver goods to distant markets.

—R. Baird Shuman

Further Reading

American Museum of Natural History. *Ocean*. Dorling Kindersley Limited, 2006.

Ballard, Robert D., and Malcolm McConnell. *Explorations: My Quest for Adventure and Discovery Under the Sea*. Hyperion, 1995.

Bertness, Mark D. *Atlantic Shorelines: Natural History and Ecology*. Princeton UP, 2006.

Broad, William J. *The Universe Below: Discovering the Secrets of the Deep Sea*. Simon & Schuster, 1997.

Brown, Joan. *Ocean Circulation: Prepared by an Open University Course Team*. Elsevier, 2016.

Drumond, Anita, Belen Rodrigues-Fonseca, Chris Reason, and Silvina A. Solman, editors. *Wet and Dry Periods in Regions Surrounding the Atlantic Ocean Basin*. Frontiers Media SA, 2016.

Goni, G. J., and Paola Malanotte-Rizzoli. *Interhemispheric Water Exchange in the Atlantic Ocean*. Elsevier B.V., 2003.

Morozov, Eugene G., Alexander N. Demidov, Roman Y. Tarakanov, and Walter Zenk. *Abyssal Channels in the Atlantic Ocean: Water Structure and Flows*. Springer, 2010.

Nicolas, Adolphe. *The Mid-Oceanic Ridges: Mountains Below Sea Level*. Springer Science & Business Media, 2013.

Oceanography Course Team. *Ocean Circulation*. 2nd ed., Butterworth-Heinemann, 2001.

Trujillo, Alan P., and Harold V. Thurman. *Essentials of Oceanography*. Prentice-Hall, 2010.

Ulanski, Stan L. *The Gulf Stream: Tiny Plankton, Giant Bluefin, and the Amazing Story of the Powerful River in the Atlantic*. U of North Carolina P, 2008.

Van der Voo, Rob. *Paleomagnetism of the Atlantic, Tethys and Iapetus Oceans*. Cambridge UP, 2004.

Waterlow, Julia. *The Atlantic Ocean*. Raintree Steck-Vaughn, 1997.

ATMOSPHERIC AND OCEANIC OSCILLATIONS

Fields of Study: oceanography; climatology; fluid physics; computer modeling

ABSTRACT

The oceans' effect on developing weather patterns has long been known and taken into consideration in weather prediction. Atmospheric and oceanic patterns fluctuate over a one- to twenty-year course. These fluctuations, or oscillations, create major climate change, such as the well-known El Niño Southern Oscillation (ENSO) in the tropical Pacific, which affects weather across the globe.

KEY CONCEPTS

Coriolis effect: the apparent deflection of the linear motion of air and water currents due to Earth's rotation

decadal: refers to phenomena that recur on an essentially regular ten-year cycle

Ekman spiral: a downward spiral movement of water from the ocean surface to depths under the influence of trade winds and the Coriolis effect

El Niño/La Niña: a complex system characterized by the movement of masses of warm surface water toward the west coast of South America in the former, and away from it in the latter

ocean conveyor belt: a global circulation system that ties all individual ocean currents into one continuous current system

upwelling: upward movement of relatively cold water from ocean deeps to the surface

AIR AND OCEAN INTERACTION

Currents in the water and the air push heated water toward cooler climates, where the heat is released, helping to regulate temperature on northern continents. As heat is released, currents in the ocean and atmosphere return the cooled water to warmer climates, and the process continues in an endless cycle. These exchanges are an important way of recycling energy through the ocean-atmosphere system, which tends to balance land temperatures and climate response. Ocean currents, which keep water in constant motion, are affected by Earth's rotation, the sun's energy, wind, and the salinity (salt content) and temperature of the ocean. Just as there are a series of air streams, pressures, and currents in the various levels of the atmosphere, the ocean also contains a similar network of circulation patterns and pressure zones.

Some deep-sea currents, such as the Gulf Stream, push through the waters much as rivers flow through land. By coupling the sciences of meteorology and oceanography, several major ocean currents have been detected using peak evaporation levels caused by warm, dry air in the subtropics. Strong evaporation throughout the year off the coast of South Africa characterizes the presence of the Agulhas Current, strong evaporation off the eastern United States in January characterizes the Gulf Stream, and strong evaporation throughout the year in the northeast Atlantic characterizes the North Atlantic Drift. These currents all have generally higher surface temperatures in relation to overlying air, especially in winter, and are

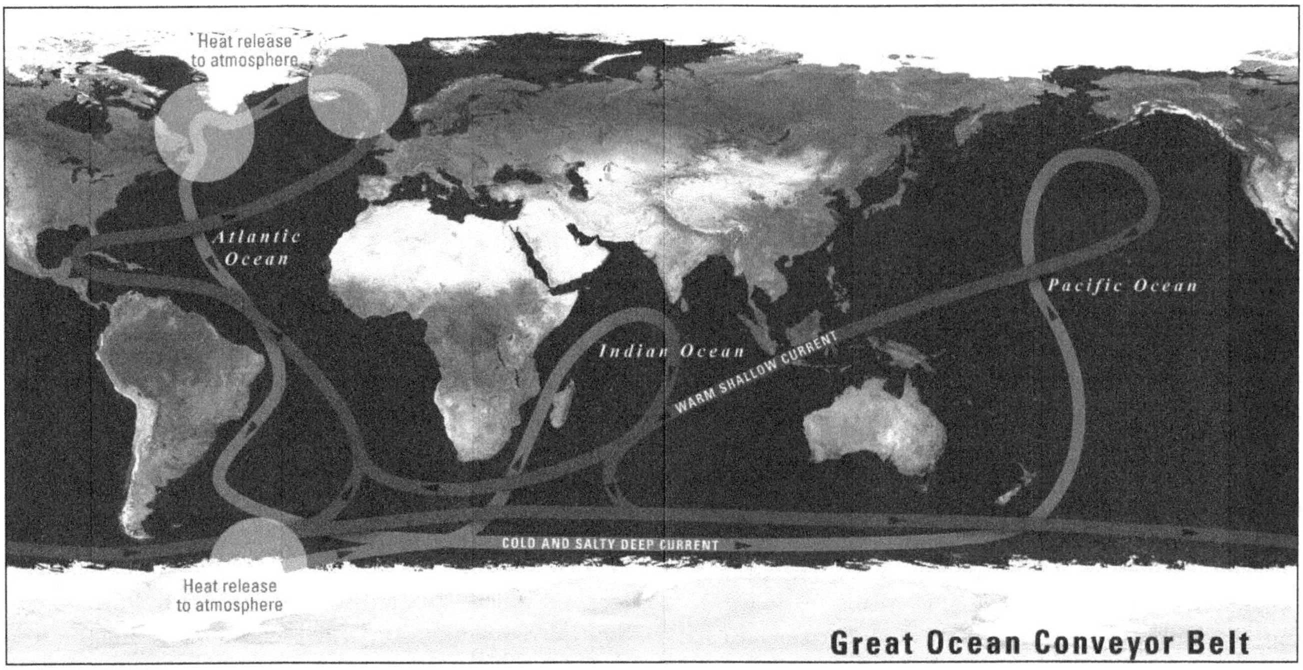

(United States Geological Survey)

frequently accompanied by strong winds. Approximately 40 percent of the total heat transported from the Southern to the Northern Hemisphere is through the action of surface ocean currents.

Another type of ocean current, called an "upwelling," brings deeper, colder waters to the surface. This colder water replaces warmer surface waters that are pushed away by the strong trade winds. Upwellings carry with them large amounts of nutrients that nourish the plankton, which make up the base of the food chain. Such currents are an excellent example of how atmospheric currents and ocean currents interact. Without the trade winds to push surface waters away, colder water could not surface.

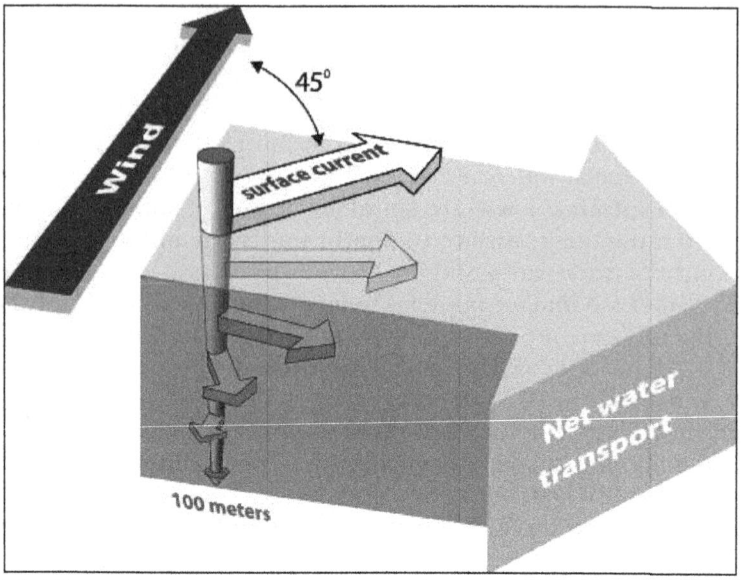

The Ekman spiral occurs as a consequence of the Coriolis effect (National Oceanic and Atmospheric Administration)

Ocean currents called "Ekman spirals" also demonstrate how the atmosphere affects ocean waters. Winds drive surface water in the same direction that the wind blows. Water just below the surface moves more slowly due to friction and the moving water is slightly deflected to the right by the Coriolis effect. Therefore, water at lower depths moves even more slowly, and in the opposite direction of the flow of surface water. This causes the water to flow in a downward motion. Because of the Ekman spiral, surface water actually flows at a 90-degree angle to the wind flowing out to sea along coastlines, thus creating the opportunity for upwellings.

CLIMATE SHIFTS AND GLOBAL IMPACT

Changes in ocean temperature and air currents influence weather, and weather changes over time determine climate. As warm-water masses move toward the coasts, they bring with them atmospheric moisture that becomes rainfall. When air currents shift, as when strong trade winds die down or reverse course, areas that normally get little rain may be flooded, and areas that normally get a lot of rain may have droughts.

The best-known climate shift has occurred for centuries. It was named *El Niño* (Spanish for "Christ child" or "the little boy") by Peruvian fisherman because it occurs during the Christmas season. Peru is known for its anchovy fisheries, and El Niño hampers this important harvest. Normally, strong trade winds push warmer surface waters away from Peru's coast so upwelling can occur. This upwelling brings rich nutrients from the lower ocean waters, and Peruvian fishermen reap strong harvests. During El Niño years, which cycle approximately every seven years and may last up to two years, trade winds weaken, and warmer surface waters remain along Peru's coast. Upwellings do not take place, surface waters heat up, and the anchovy harvest suffers.

The effects of El Niño do not end with the Peruvian anchovy fisheries, however. The phenomenon triggers a climatic ripple effect that disrupts weather patterns around the globe. Unusual numbers of storms may rage across North America. Drought conditions have been experienced in northeastern Brazil, southeastern Africa, and western Pacific islands. El Niño effects have also caused unusually wet springs in the eastern United States and elsewhere.

The 1982–1983 El Niño is estimated to have caused more than $8 billion in damage. The severe 1997-1998 El Niño, in turn, is thought to have caused more than $15 billion in damage. It was not until the 1982 El Niño, however, that meteorologists and oceanographers began to seriously study the phenomenon to learn ways to predict both its approach and the severe weather it often causes. Researchers learned that a fluctuating wind pattern known as the Southern Oscillation, which is the same fluctuation that causes the trade winds to weaken, triggers El Niño. Researchers also learned that the ocean-current pattern is dependent upon the wind pattern. Together, this climatic event is known as the El Niño Southern Oscillation (ENSO).

ENSO

Meteorologists and oceanographers continue to learn all they can about ENSO and other atmosphere-ocean oscillations around the globe. The ENSO pattern was the first to be studied, and many weather forecasting models have been created to aid the study of atmosphere-ocean oscillations. These computer models have also helped researchers study ENSO historical trends for the past five hundred years. Many have come to believe that ENSO events may have contributed to plagues and other similar disasters throughout history.

ENSO patterns fluctuate between the extremes of heated tropical waters associated with El Niño and a colder weather-front pattern known as *La Niña* ("the little girl"). This oscillation pattern takes about seven years to cycle, and each extreme (hot or cold) in the pattern may last up to two years. ENSO researchers have also linked the onset of El Niño with other climate events in surrounding ocean waters. Just before El Niño begins warming the Pacific Ocean, the tropical Indian Ocean warms. Warming then appears in the tropical Pacific, triggering ENSO, which causes warm winds to blow over South America. About nine months after this occurs, the circulation of the tropical Atlantic changes and the waters there begin warming.

The study of the ENSO pattern has led researchers to discover a major ocean circulation pattern called the "conveyor belt." This conveyor belt of water is thought to have existed for the several million years since the continents have occupied their current positions in the oceans. It serves to connect the major bodies of water making up the global ocean. Its significance was over-

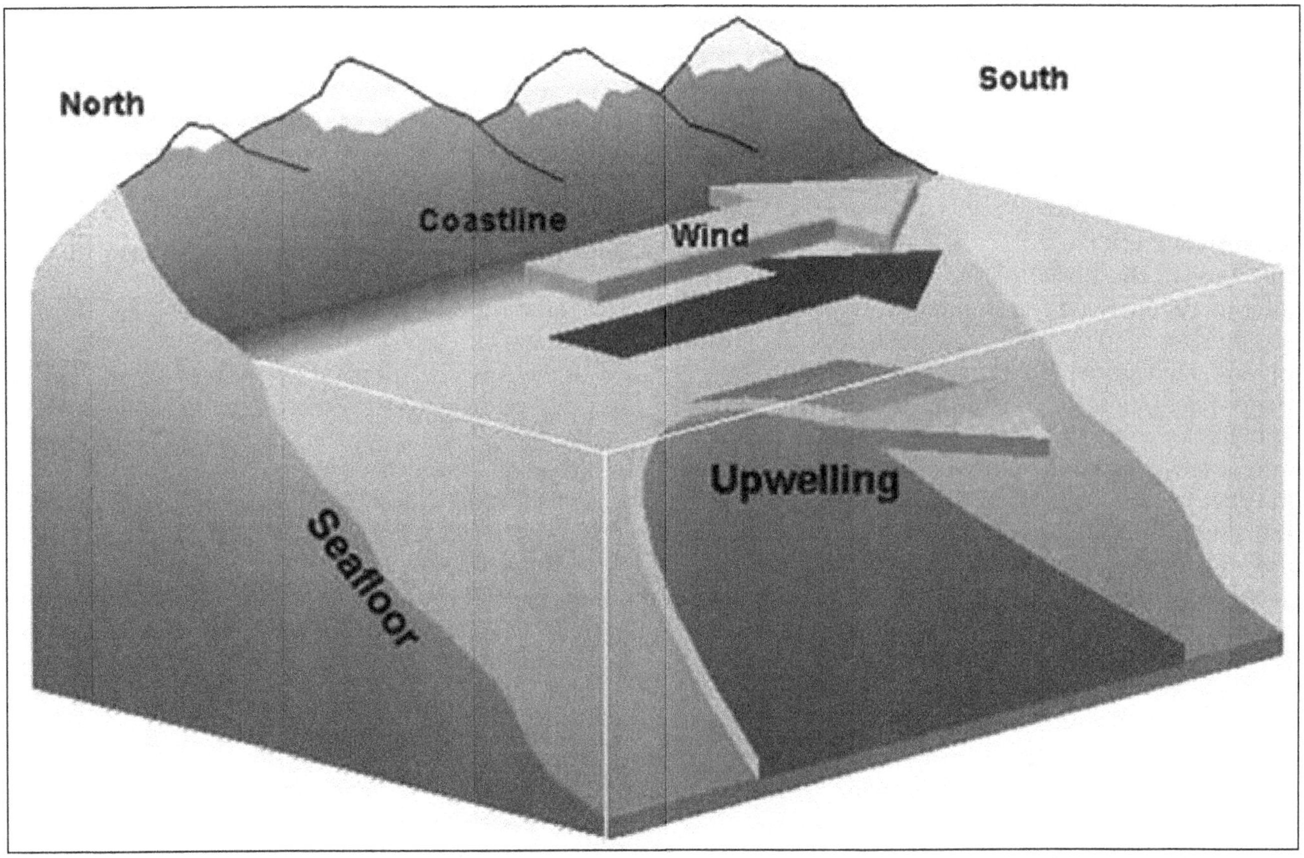

Upwelling occurs when winds push surface water away from the shore and deeper water rises to fill the gap (National Oceanic and Atmospheric Administration)

looked until oscillation patterns were studied. The conveyor belt current circulates heated ocean water from the tropical Pacific Ocean to the North Pacific Ocean, turning clockwise around the Pacific, to pass westward between Australia and Malaysia across the Indian Ocean, around the southern tip of Africa, and up into the North Atlantic Ocean. As the surface waters cool in the North Atlantic Ocean, they sink to a lower ocean level and return eastward across the Atlantic Ocean, around southern Africa, across the Indian Ocean, passing south of Australia, and back to the tropical waters of the Pacific Ocean.

Some researchers are concerned that if this conveyor belt current were to stop or slow down, heat would build up in the Southern Hemisphere, while the Northern Hemisphere would experience a severe drop in temperature. Such an occurrence may have caused the last ice age, and research findings continue to stress the importance of this conveyor belt current to maintaining global climate. Temperatures of surface water change in direct relation to changes in this conveyor belt current, the mass flow of which is estimated to equal that of the Amazon River a hundred times over, delivering a heat load to the upper North Atlantic equivalent to one-fourth of the solar energy that reaches the surface in that region.

NAO AND PADO

The study of the ENSO pattern and the use of computer modeling to forecast weather revealed other atmospheric-oceanic oscillations, such as the North Atlantic Oscillation (NAO), which fluctuates on a twenty-year timescale. The atmospheric behavior of NAO has long been known to meteorologists. It is typically a low-pressure, counterclockwise wind circulation that centers over Iceland. This weather pattern contrasts with a high-pressure, clockwise circulation near the Azores Islands off the coast of Portugal. Strong winds blow west to east between these two weather centers. NAO seesaws between these two centers, and strong winds drive heat from the Gulf Stream across Eurasia during high-index years, producing unusually mild winters there. At the other extreme, air pressure builds up over Iceland, weakening the warming winds and delivering bitter winters to Europe and Greenland.

Oceanographers have concluded the ocean must be the cause of the unique atmospheric patterns created by NAO. Because the ocean has a huge capacity for storing heat and reacts at a slower rate than the atmosphere, it must provide the input for atmospheric patterns operating in the same mode year after year. The source of the oceanic oscillator is a pipeline of warm water fed by the Gulf Stream. It takes twenty years to complete one cycle, thus setting the timing for the long-term swings of NAO. Researchers are unsure, however, how the temperature of the waters in the pipeline actually triggers NAO.

The discovery of the Pan-Atlantic Decadal Oscillation (PADO) resulted from a theory that tropical oscillations in the Atlantic actually extended beyond the tropical region. PADO covers an area of more than 6,835 miles (11,000 kilometers) extending from the southern Atlantic to Iceland. Meteorologists proposed the existence of PADO, claiming it fluctuates on a ten- to fifteen-year timescale. PADO consists of east-to-west bands of water spanning the Atlantic Ocean. The bands alternate between warmer and cooler water and are accompanied by changes in atmospheric circulation. An oscillation to the other extreme reverses the temperature variance. It is also believed that PADO is triggered by NAO.

The study of ENSO, NAO, and PADO has led many researchers to theorize that such oscillations are linked in a chain that circles the global ocean. The Arctic Oscillation, which reverberates from the far northern Atlantic Ocean to the far northern Pacific Ocean, is associated with the NAO, especially in winter. The two phases that characterize the Arctic oscillation correlate precisely to those of the NAO. Varying from day to day and decade to decade, the oscillation is a natural response of the atmosphere to the complex interactions of the ocean-atmosphere system.

The Pacific Decadal Oscillation (PDO) is believed to be another link in the chain of atmosphere-ocean oscillations. This high-latitude oscillation was first noted in 1977 when the northern Pacific Ocean cooled dramatically. An atmospheric low-pressure center off the Aleutian Islands intensified and shifted eastward. This brought more frequent storms to the West Coast of North America and warmed Alaska, while Florida experienced periodic winter freezes. A multitude of other environmental changes also resulted. Researchers have been able to pinpoint other shifts occurring in 1947 and 1925 and have also identified close ties between PDO and ENSO events.

SIGNIFICANCE

The greatest impact of the study of atmosphere-ocean oscillations has been the joining of two separate fields of science. As meteorologists and oceanographers research the manner in which atmospheric and oceanic patterns rely on each other and ultimately influence climate, they have become more dependent on coupled research.

Computer simulations, sometimes referred to as "oceans in a box," have become more sophisticated as researchers from both fields collaborate and combine data. Climate scientists began using computer simulations in the mid-1980s to study winds in the atmosphere to determine how they stir ocean currents and alter pressure patterns in the atmosphere that feed back on the ocean again. Joint research between oceanographers and climatologists extends knowledge about the complex interactions of atmosphere-ocean oscillations and their influence upon each other and the climate.

—Lisa A. Wroble

Further Reading

Bigg, Grant R. *The Oceans and Climate*. 2nd ed., Cambridge UP, 2003.

Christopherson, Robert W., and Mary-Louise Byrne. *Geosystems: An Introduction to Physical Geography*. Pearson Education Canada, 2006.

Garrison, Tom S. *Oceanography: An Invitation to Marine Science*. Brooks/Cole, Cengage Learning, 2010.

Gill, Adrian E. *Atmosphere-Ocean Dynamics*. Elsevier, 2016.

Marshal, John, and R. Alan Plumb. *Atmosphere, Ocean and Climate Dynamics: An Introductory Text*. Elsevier Academic Press, 2008.

Navarra, Antonio, editor. *Beyond El Niño: Decadal and Interdecadal Climate Variability*. Springer Science & Business Media, 2012.

Sarachik, Edward S., and Mark A Cane. *The El Niño-Southern Oscillation Phenomenon*. Cambridge UP, 2010.

TAO Project Office of NOAA/Pacific Marine Environmental Laboratory. *Upper Ocean Heat Content and ENSO*. Seattle: National Oceanic and Atmospheric Administration (NOAA) Website, www.pmel.noaa.gov/tao.

Vallis, Geoffrey K. *Atmospheric and Oceanic Fluid Dynamics: Fundamentals and Large-Scale Circulation*. Cambridge UP, 2006.

Woods Hole Oceanographic Institution (WHOI), www.whoi.edu/index.html.

Atmospheric Sciences

Fields of Study: climatology; oceanography; marine science

ABSTRACT

Atmospheric sciences include the fields of physics and chemistry and the study of the composition and dynamics of the layers of air that constitute the atmosphere. Related topics include climatic processes, oceanic temperatures, interaction between the atmosphere and the ocean, circulation patterns, precipitation patterns and amounts, chemical and particulate deposition, greenhouse gases, the ozone layer, climate change, air pollution, aerosol composition, atmospheric chemistry, modeling of pollutants both indoors and outdoors, and anthropogenic alteration of land surfaces that in turn affect conditions within the ever-changing atmosphere.

KEY CONCEPTS

anthropogenic: caused by or existing as a consequence of human activities

hydrosphere: that part of Earth that involves all forms of water

ozone: a highly reactive molecular form of oxygen consisting of three oxygen atoms rather the more stable diatomic form

scuba: acronym for self-contained underwater breathing apparatus, typically refers to the use of the equipment rather than just the equipment itself

DEFINITION AND BASIC PRINCIPLES

Atmospheric sciences is the study of various aspects of the nature of the atmosphere, including its origin, layered structure, density, and temperature variation with height; natural variations and alterations associated with anthropogenic impacts; and how it is similar to or different from other atmospheres within the solar system. The present-day atmosphere is in all likelihood quite dissimilar from the original atmosphere. The form and composition of the present-day atmosphere is believed to have developed about 400 million years ago in the late Devonian period of the Paleozoic era, when plant life developed on land. This vegetative cover allowed plants to take in carbon dioxide and release oxygen as part of the photosynthesis process.

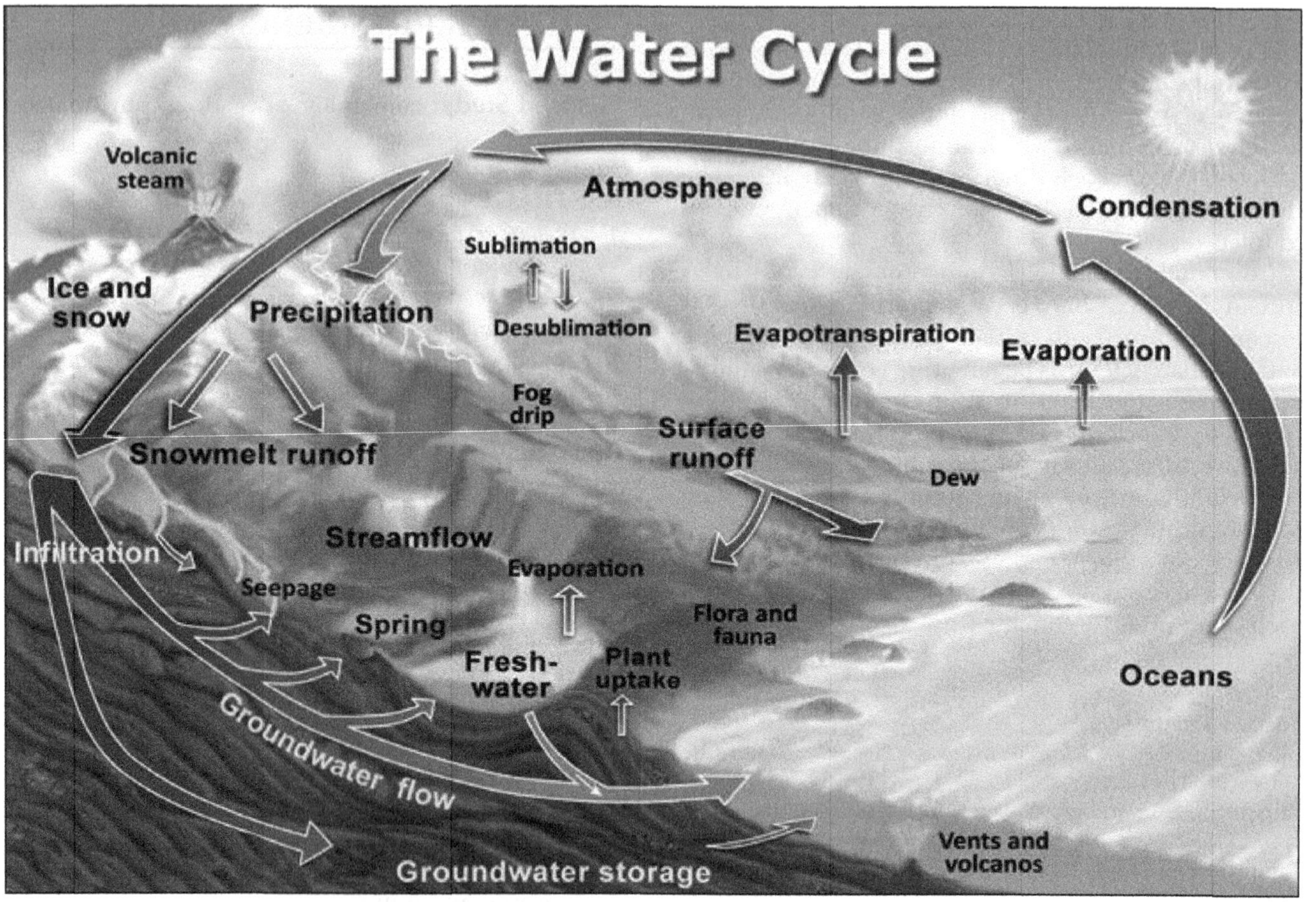

Diagram of the Water Cycle (Howard Perlman, USGS)

The atmosphere consists of a mixture of gases that remain in place because of the gravitational attraction of the Earth. Although the atmosphere extends about 6,000 miles (9,600 kilometers) above Earth's surface, the vast proportion of its gases (97 percent) are located in the lower 19 m (30 km). The bulk of the atmosphere consists of nitrogen (78 percent) and oxygen (21 percent). The last 1 percent of the atmosphere contains all the remaining gases, including an inert gas (argon), which accounts for 0.93 percent of the 1 percent, and carbon dioxide (CO_2), which currently makes up 0.0414 percent of Earth's atmosphere. Carbon dioxide has the ability to absorb longwave radiation leaving the Earth and shortwave radiation from the Sun; therefore, any increase in carbon dioxide in the atmosphere has profound implications for global warming.

BACKGROUND AND HISTORY

Evangelista Torricelli (1608–1647), an Italian physicist, mathematician, and secretary to Galileo, invented the barometer, which measures barometric pressure, in 1643. The first attempt to explain the circulation of the global atmosphere was made in 1686 by Edmond Halley (1656–1742), an English astronomer and mathematician. In 1735, George Hadley (1685–1768), an English optician, described a pattern of air circulation that became known as a Hadley cell. In 1835, Gustave-Gaspard Coriolis (1792–1843), a French engineer and mathematician, analyzed the movement of air on a rotating Earth, a pattern that became known as the Coriolis effect. In 1856, William Ferrel (1817–1891), an American meteorologist, developed a model of hemispheric circulation of the atmosphere that became known as a Ferrel cell.

Christophorus Buys Ballot (1817–1890), a Dutch meteorologist, explained the relationship between the distribution of pressure, wind speed, and direction in 1860.

Manned hot-air balloon flights beginning in the mid-nineteenth century facilitated high-level observations of the atmosphere. For example, in 1862, English meteorologist James Glaisher (1809–1903) and English pilot Henry Coxwell (1819–1900) reached 29,000 feet (12 kilometers), at which point Glaisher became unconscious and Coxwell was partially paralyzed so that he had to move the control valve with his teeth. In 1902, Léon Teisserenc de Bort (1885–1913) of France was able to determine that air temperatures begin to level out at 39,000 ft (12 km) and actually increase at higher elevations. In the twentieth century, additional information about the upper atmosphere became available through radio waves, rocket flights, and satellites.

RELATIONSHIP OF ATMOSPHERIC SCIENCE TO MARINE SCIENCE

In the strictest sense, atmospheric science deals only with the planet's gaseous envelope and its effects as a separate system of the Earth environment. However, the atmosphere is just one system of many that comprise our planet Earth. It affects, and is affected by, the other Earth systems. For people who would study the marine system in any way, transport goods and materials across it in ships, or make their livelihood by working on and in the seas, how the atmosphere affects the marine system is vitally important, as many a drowned sailor would tell you. One system of fundamental importance to all life on Earth is the "water cycle." In this system, vast quantities of water evaporate from oceans and other bodies, then are transported as water vapor and clouds in the atmosphere to condense and fall as rain, snow and ice in different areas of the continents. Farmers are happy to get the rainfall from water that has evaporated from the oceans, especially in times of drought. On the other hand, too much rain often results in flooding, property damage, erosion, and loss of lives. Large volcanic eruptions also affect the atmosphere, and as a secondary effect the oceans as well by the release of vast quantities of solid ejecta and materials such as sulfuric acid and carbon dioxide, which serve to affect the chemistry of ocean waters in ways that affect fish, coral reefs, and anything else in or on the seas. Accordingly, the study of marine science involves a significant understanding of the relationship between the hydrosphere and the atmosphere.

As might be expected, the greatest interaction between the marine environment and the atmosphere takes place at the surface, at the interface of air and liquid water. The water absorbs energy from sunlight and from heat sources within Earth, and the great currents that circulate throughout the world's oceans carry that energy with them. This is most noticeable at the surface as currents like the Gulf Stream deliver that energy, typically driving the formation of hurricanes, typhoons, and the El Niño Southern Oscillation. A significant area of marine science involves the study of such air-water interactions and the effects they have on the marine environment. Naturally, the effects of air-water interactions on local weather conditions also determine the ability of marine scientists to carry out studies in the field at any particular time.

LAYERS OF THE ATMOSPHERE

The heterosphere and the homosphere form the two major subdivisions of Earth's atmosphere. The uppermost subdivision (heterosphere) extends from about 50 mi (80 km) above Earth's surface to the outer limits of the atmosphere at about 6,000 mi (9,600 km). Nitrogen and oxygen, the heavier elements, are found in the lower layers of the heterosphere, and lighter elements such as hydrogen and helium are found at the uppermost layers of the atmosphere. The homosphere (or lowest layer) contains gases that are more uniformly mixed, although their density decreases with height. Some exceptions to this statement occur with the existence of an ozone layer at an altitude of 12 to 31 mi (19.2 to 50 km) and with variations in concentrations of carbon dioxide, water vapor, and air pollutants closer to Earth's surface.

The atmosphere can be divided into several zones based on decreasing or increasing temperatures as elevation increases. The lowest zone is the troposphere, where temperatures decrease from sea level up to an altitude of 10 mi (16 km) in equatorial and tropical regions and up to an altitude of 4 mi (6 km) at the poles. This lowermost zone holds substantial amounts of water vapor, aerosols that are very small, and light particles that originate from volcanic eruptions, desert surfaces, soot from forest and brush fires, and industrial emissions. Clouds, storms, and weather systems occur in the troposphere.

The tropopause marks the boundary between the troposphere and the next higher layer, the stratosphere, which reaches an altitude of 30 mi (48 km) above Earth's surface. Circulation in this layer occurs with strong winds that move from west to east. There is limited circulation between the troposphere and the stratosphere. However, manned balloons, certain types of aircraft (Concorde and the U-2), volcanic eruptions, and nuclear bomb tests are able to break through the tropopause and enter the stratosphere.

The gases in the stratosphere are generally uniformly mixed, with the major exception of the ozone layer, which is found at an altitude range of 12 to 31 mi (19 to 50 km) above the surface of Earth. This layer is extremely important because it shields life on Earth from the intense and harmful ultraviolet radiation from the Sun. The ozone layer has been diminishing because of the release of chlorofluorocarbons (CFCs), organic compounds containing chlorine, fluorine, and carbon, that were commonly used as propellants in aerosol sprays and as refrigerant liquids. In 1978, the use of CFCs in aerosol sprays was banned in the United States, but they are still used in some refrigeration systems. Other countries, however, continued to use CFCs, which eventually get into the ozone layer and result in ozone holes of considerable size. In 1987, members of the international community took steps to reduce CFC production through the Montreal protocol, and by 2003, the rate of ozone depletion had begun to slow down. Although the manufacture and use of CFCs is now strictly controlled, natural events that are detrimental to the ozone layer cannot be prevented. For example, the 1991 eruption of Mount Pinatubo in the Philippines reduced the ozone layer in the midlatitudes nearly 9 percent.

Temperatures decrease with elevation at the stratopause, where the mesosphere layer begins at about 30 miles and continues to an altitude of about 50 mi (80 km). The mesopause at about 50 mi (80 km) marks the beginning of the thermosphere, where the density of the air is very low and holds minimal amounts of heat. However, even though the atmospheric density is minimal at altitudes above 155 mi (249 km), there is enough atmosphere to have a drag effect on spaceships.

ATMOSPHERIC PRESSURE

The gas molecules in the atmosphere exert a pressure due to gravity that amounts to about 14.7 pounds per square inch (101 kilopascals) on all surfaces at sea level. As the distance from Earth's surface increases, in contrast to the various increases and decreases in atmospheric temperature, atmospheric pressure decreases at an exponential rate. For example, air pressure at sea level varies from about 13.8 to 15 psi (95.7 to 105 kPa) inches of mercury, averaging 14.7 psi (101.325 kPa). The pressure at the top of Mount Everest at 3.8 mi (6,069 m) can get as low as 4.37 psi (29.9 kPa). This means that each inhalation of air at this altitude is about one-third of the pressure at sea level, producing severe shortness of breath.

Earth's Global Energy Balance. Earth's elliptical orbit about the Sun ranges from 91.5 million mi (146.4 million km) at perihelion (closest point to the Sun) on January 3 to 94.5 million mi (151.2 million km) miles at aphelion (furthest from the Sun) on July 4, averaging 93 million mi (148.8 million km). The Earth intercepts only a tiny fraction of the total energy output of the Sun. Upon reaching Earth, part of the incoming radiation is reflected back into space, and part is absorbed by the atmosphere, land, or oceans. Over time, the incoming shortwave solar radiation is balanced by a return to outer space of longwave radiation.

Earth-Moon Differences. Scientists believe that the moon's surface has a large number of craters formed by the impact of meteorites. In contrast, there are relatively few meteorite craters on Earth, even though, based simply on its size, Earth is likely to have been hit by as many or even more meteorites than the Moon. This notable difference is attributed to Earth's atmosphere, which burns up incoming meteorites, particularly small ones (the Moon does not have an atmosphere). Larger meteorites can pass through Earth's atmosphere, but their impact craters may have been filled in or washed away over millions of years. Only the more recent ones, such as Meteor Crater in northern Arizona, with a diameter of 4,000 ft (1.2 km) and a depth of 600 ft (182 m), remain easily recognizable.

Air Masses. Different types of air masses within the troposphere, the lowest layer of the atmosphere, can be delineated on the basis of their similarity in temperature, moisture, and to a certain extent, air pressure. These air masses develop over continental and maritime locations that strongly determine their physical characteristics. For example, an air mass starting in the cold, dry interior portion of a continent develops thermal, moisture, and pressure differences that can be substantially different from an air mass that develops over water. Atmospheric dynamics also allow air masses to modify their characteristics as they move from land to water and vice versa.

Air mass and weather front terminology were developed in Norway during World War I. Norwegian meteorologists were unable to get weather reports from the Atlantic theater of operations; consequently, they developed a dense network of weather stations that led to impressive advances in atmospheric modeling that are still being used.

The Radiation Budget. The incoming solar energy that reaches Earth is primarily in the shortwave (or visible) portion of the electromagnetic spectrum. Earth's energy balance is attained by about one-third of this incoming energy being reflected back to space and the other two-thirds leaving Earth as outgoing longwave radiation. This balance between incoming and outgoing energy is known as the radiation budget. The decades-long, ongoing National Aeronautics and Space Administration (NASA) program known as Clouds and the Earth's Radiant Energy System (CERES) is designed to measure how much shortwave and longwave radiation leaves the Earth from the top of the atmosphere.

Clouds play a very important role in the global radiation balance. For one thing, they constantly change over time and in type. Some clouds, such as high cirrus clouds found near the top of the troposphere at 40,000 ft (12.1 km), can have a substantial impact on atmospheric warming. Accordingly, the value of CERES is based on its ability to observe if human or natural changes in the atmosphere can be measured even if they are smaller than large-scale energy variations.

Greenhouse Effect. Selected gases in the lower parts of the atmosphere trap heat and then radiate some of that heat back to Earth. If there was no natural greenhouse effect, Earth's overall average temperature would be close to 0 degrees Fahrenheit (–18 degrees Celsius), rather than the existing 57 degrees Fahrenheit (15 degrees Celsius).

The burning of coal, oil, and gas makes carbon dioxide (CO_2) the major greenhouse gas, accounting for nearly half of the total amount of heat-producing gases in the atmosphere. Before the Industrial Revolution in Great Britain in the mid-eighteenth century, the estimated level of carbon dioxide in the atmosphere was about 280 parts per million by volume (ppmv). Estimates for the natural range of carbon dioxide for the past 650,000 years range from 180 to 300 ppmv. In 2013, the Mauna Loa Observatory reported that the global concentration of carbon dioxide in the atmosphere had reached 400 ppmv for the first time in history, and that number was subsequently surpassed by 2016. All these values are less than the 414 ppmv recorded in February 20, 2020. Carbon dioxide levels have been increasing since 2000 at a rate of 1.9 ppmv each year. The radiative effect of carbon dioxide accounts for about one-half of all the factors that affect global warming. Estimates of carbon dioxide levels at the end of the twenty-first century range from 490 to 1,260 ppmv.

The second most important greenhouse gas is methane (CH_4), which accounts for about 14 percent of all global warming factors. The origin of this gas is attributed to the natural decay of organic matter in wetlands, but anthropogenic activity—rice paddies, manure from farm animals, the decay of bacteria in sewage and landfills, and biomass burning (both natural and human induced)—results in a doubling of the amount of this gas over what would be produced solely by wetland decay. Methane is many times more effective than carbon dioxide as a greenhouse gas, but there is much less methane than carbon dioxide in the atmosphere.

Chlorofluorocarbons (CFCs) absorb longwave energy (warming effect), but they also have the ability to destroy stratospheric ozone (cooling effect). The warming radiative effect is three times greater than the cooling effect. CFCs account for about 10 percent of all global warming factors. Tropospheric ozone from air pollution and nitrous oxide (N_2O) from motor vehicle exhaust and bacterial emissions from nitrogen fertilizers account for about 10 and 5 percent, respectively, of all global warming factors.

Several kinds of human actions lead to a cooling of Earth's climate. For example, the burning of fossil fuels results in the release of tropospheric aerosols, which acts to scatter incoming solar radiation back into space, thereby lowering the amount of solar energy that can reach Earth's surface. These aerosols also lead to the development of low and bright clouds that are quite effective in reflecting solar radiation back into space.

APPLICATIONS AND PRODUCTS

Atmospheric science is applied in many ways. It is used to help people better understand their global and interplanetary environment and to make it possible for them to live safely and comfortably within that environment. By using the principles of this field, researchers, engineers, and space scientists have developed a vast number of applications. Among the most important are those used to track and predict weather cycles and climate.

Remote Sensing Techniques. Oceans cover about 71 percent of Earth's surface, which means that large portions of the world do not have weather stations or places where precipitation can be measured with standard rain gauges. To provide more information about precipitation in the equatorial and tropical parts of the world, NASA and the Japan Aerospace Exploration Agency initiated the Tropical Rainfall Monitoring Mission (TRMM) in 1997. The orbit of the TRMM satellite allows monitoring of Earth between the 35 degrees north and 35 degrees south latitudes. The goal of the study is to obtain information about the extent of precipitation, along with its intensity and length of occurrence. The major instruments on the satellite are radar to detect rainfall, a passive microwave imager that can acquire data about precipitation intensity and the extent of water vapor, and a scanner that can examine objects in the visible and infrared portions of the electromagnetic spectrum. The goal of data collection is to obtain the necessary climatological information about atmospheric circulation in this portion of Earth in order to develop better mathematical models for determining large-scale energy movement and precipitation.

Geostationary Satellites. Geostationary operational environmental satellites (GOES) enable researchers to view images of the planet from what appears to be a fixed position above Earth. The satellites are actually circling the globe at a speed that matches Earth's rotation. This means that a satellite at an altitude of 22,200 mi (35,520 km) will make one complete revolution about Earth's axis of rotation above the equator in the same twenty-four-hour period and in the same direction that Earth is rotating about its axis. At this height, the satellite is in a position to view nearly one-half of the planet at any time. Onboard instruments can be activated to look for special weather conditions such as hurricanes, flash floods, and tornadoes. Onboard instruments are also used to make precipitation estimates during storm events.

Doppler Radar. Doppler radar was first used in England in 1953 to pick up the movement of small storms. The basic principle guiding this type of radar is that backscattered radiation frequency detected at a certain location changes over time as the target, such as a storm, moves. A transmitter is used to send short but powerful microwave pulses. When a foreign object (or target) is intercepted, some of the outgoing energy is returned to the transmitter, where a receiver can pick up the signal. An image (or echo) from the target can then be enlarged and shown on a screen. The target's distance is revealed by the time that elapses between transmission and return. The radar screen cannot only indicate where the precipitation is taking place but also reveal the intensity of the rain by the intensity of the echo's brightness. In short, Doppler radar has become a very useful device for determining the location of a storm and the intensity of its precipitation and for obtaining good estimates of the total amount of precipitation.

Responses to Climate Change. Since the 1970s, many scientists have pointed out the possibility that human activity is having more than a short-term impact on the atmosphere and therefore on weather and climate. Although much debate continues on the full impact of human activities and greenhouse gas emissions, the atmospheric sciences have led to conferences, United Nations conventions, and agreements among nations on ways that human beings can alter their behavior to halt or at least mitigate global climate change. The impact of these agreements, still in their infancy, remains unknown—as does the overall effect of human activity on weather and climate (the models for which are highly complex). However, the insights contributed by the atmospheric sciences to the overall debate on whether climate change is primarily anthropogenic (human caused)—and whether global warming is actually taking place—have caused many nations and individuals to modify their attitudes toward human relationships with the global environment, resulting in national and intergovernmental changes in policies concerning carbon emissions, as well as personal decisions ranging from the consumption of "green" building materials to the purchase of vehicles fueled by noncarbon sources of energy.

CAREERS AND COURSEWORK

The study of the physical characteristics of the atmosphere falls within the purview of atmospheric scientists. Those interested in a career in this technical area should recognize that there are several categories of specialization. The major group of specialists are operational meteorologists, who are responsible for weather forecasts. They have to carefully study the temperature, humidity, wind speed, and barometric pressure from a number of weather stations to make daily and long-range forecasts. They use data from weather satellites, radar, special sensors, and observation stations in other locations as input data for sophisticated computer modeling programs to make forecasts.

In contrast to meteorologists, who focus on short-term weather forecasts, the study of changes in weather over longer periods of time such as months, years, and in some cases centuries, is handled by climatologists. Other atmospheric scientists concentrate on research. For example, physical meteorologists are concerned with various aspects of the atmosphere, such as its chemical and physical properties, energy transfer, severe storm mechanics, and the spread of air pollutants over urbanized areas. The growing interest in air pollution and water shortages has led to another group of research scientists known as environmental meteorologists.

Given the importance of weather forecasting on a daily basis, operational meteorologists who work in weather stations may work on evenings, weekends, and holidays. Research scientists who are not engaged in weather forecasts may work regular hours.

In 2016, the American Meteorological Society estimated that there are more than one hundred undergraduate and graduate atmospheric science programs in the United States that offer courses in such departments as physics, earth science, environmental science, geography, and geophysics. Entry-level positions usually require a bachelor's degree with at least twenty-four semester hours in courses covering atmospheric science and meteorology. The acquisition of a master's degree enhances the chances of employment and usually means a higher salary and more opportunities for advancement. A doctorate is required only for those who want a research position at a university.

In 2014, excluding research positions in colleges and universities, about 11,800 atmospheric scientists were working, according to the Bureau of Labor Statistics. Above-average employment growth is projected until 2024, and the median annual average salary for atmospheric scientists in 2015 was $89,820.

SOCIAL CONTEXT AND FUTURE PROSPECTS

Climate change may be caused by both natural internal/external processes in the Earth-Sun system or by human-induced changes in land use and the atmosphere. Article 1 of the United Nations Framework Convention on Climate Change (entered into force March, 1994) states that the term "climate change" should refer to anthropogenic changes that affect the composition of the atmosphere rather than natural causes, which should be referred to "climate variability." An example of natural climate variability is the global cooling of about 0.5 degrees Fahrenheit (–17.5 degrees Centigrade) in 1992–1993 that was caused by the 1991 eruption of Mount Pinatubo in the Philippines. The 16.5 million to 22 million tons (15 million to 20 million tonnes) of sulfuric acid aerosols ejected into the stratosphere reflected incoming radiation from the sun, thereby creating a cooling effect. Many suggest that the above-normal temperatures experienced in the first decade of the twenty-first century provide evidence of climate change caused by human activity. Based on a variety of techniques that allow scientists to estimate the temperature in previous centuries, the years 2005, 2010, and 2014 were the warmest in the last thousand years. A 2009 article published by the American Geophysical Union suggests that human intervention in Earth systems has reached a point where the Holocene epoch of the past 12,000 years is becoming a new Anthropocene epoch in which human systems have become primary Earth systems rather than simply influencing natural systems.

Numerous observations strongly suggest a continuing warming trend. Snow and ice have retreated from areas such as Mount Kilimanjaro in Tanzania, which at 19,340 ft (5,895 m) is the highest mountain in Africa. Glaciated areas in Switzerland also provide evidence of this warming trend. Loss of sea ice cover in the Arctic Ocean during winter, rapid melting of the Greenland icecap, record temperatures and thawing of permafrost in Siberia and northern Canada, and the growing number of severe storms on land and sea are also indicators of this global warming trend. The Special Report on Emission Scenarios issued in 2001 by the Intergovernmental Panel on Climate Change (IPCC) examined the broad spectrum of possible concentrations of greenhouse gases by considering the growth of population and industry along with the efficiency of energy use. The IPCC computer climate models were used to estimate future trends.

—Robert M. Hordon

Further Reading

"Atmospheric Scientists, Including Meteorologists." *Occupational Outlook Handbook*. Bureau of Labor Statistics, 17 Dec. 2015, www.bls.gov/ooh/life-physical-and-social-science/atmospheric-scientists-including-meteorologists.htm.

Boorstein, M. and Richard M. Renneboog, editors. *Earth Science: Earth's Weather, Water and Atmosphere*. Salem Press, 2013

Christopherson, Robert W. *Geosystems: An Introduction to Physical Geography*. 8th ed., Prentice-Hall, 2012.

Coley, David A. *Energy and Climate Change: Creating a Sustainable Future*. John Wiley & Sons, 2008.

Ellis, Erle C., and Peter K. Haff. "Earth Science in the Anthropocene: New Epoch, New Paradigm, New Responsibilities." *EOS, Transactions, American Geophysical Union*, vol. 90, no. 49, 2009, p. 473.

Gautier, Catherine. *Oil, Water, and Climate: An Introduction*. Cambridge UP, 2008.

Lutgens, Frederick K., and Edward J. Tarbuck. *The Atmosphere: An Introduction to Meteorology*. 11th ed., Prentice-Hall, 2010.

Strahler, Alan. *Introducing Physical Geography*. 5th ed., John Wiley & Sons, 2011.

Wolfson, Richard. *Energy, Environment, and Climate*. W. W. Norton, 2008.

B

Barometry

Fields of Study: climatology; oceanography; marine science

ABSTRACT

The science and engineering of pressure measurement in gases take its practitioners far beyond its original realm of weather prediction. The sensors used in barometry range from those for the near vacuum of space and the small amplitudes of soft music to those for ocean depths and the shock waves of nuclear-fusion explosions.

KEY CONCEPTS

linear: describing the manner in which one property, such as pressure, changes with respect to another property, such as depth or altitude

manometer: a device used to measure differences in pressure between two systems

one atmosphere: the standard pressure experienced at Earth's surface

pressure: the force exerted uniformly by a fluid over all parts of a surface, commonly due to the weight of the overlying air or water column

DEFINITION AND BASIC PRINCIPLES

Barometry is the science of measuring pressure. Derived from the Greek words for "heavy" or "weight" (*baros*) and "measure" (*metron*), it refers generally to the measurement of gas pressure. In liquids, such as the marine environment, pressure tends to remain constant at any particular depth since it is determined by the mass of the water column above. At the surface, the pressure of "1 atmosphere" (14.7 pounds per square inch, or 101 kilopascals) is determined by the mass of the air column above. The pressure exerted by the water column increases essentially linearly with depth such that the equivalent of the pressure at the top of the water column is added gradually for each 33 feet (10 meters) of depth. Thus, at 33 ft (10 m) below the surface, the pressure is double that at the surface; at 66 ft (20 m) below the surface the water pressure is three times that at the surface, and at a depth of 3 miles (5000 meters), the pressure is about 500 atmospheres (7,350 psi, or 50,500 kPa). That is, about 3.85 tons (3.5 tonnes) on a space of 1 square inch (6.25 cm^2). Accordingly, water pressures are not subject to the wide and rapid variability that can affect gas pressures. However, all of the principles of gas barometry also apply to liquid barometry.

In gases, pressure is fundamentally ascribed to the momentum flowing across a given surface per unit time, per unit area of the surface. Pressure is expressed in units of force per unit area. Although pressure is a scalar quantity, the direction of the force due to pressure exerted on a surface is taken to be perpendicular and directed onto the surface. Therefore, methods to measure pressure often measure the force acting per unit area of a sensor or the effects of that force. Pressure is expressed in newtons per square meter (pascals), in pounds per square foot (psf), or in pounds per square inch (psi). The pressure of the atmosphere at standard sea level at a temperature of 59 degrees Fahrenheit (288.15 Kelvin) is or 14.7 psi (101,325 pascals). This is called 1 atmosphere. Mercury and water barometers have become such familiar devices that pressure is also expressed in inches of water, inches of mercury, or in torrs (1 torr equals about 0.019 psi, or 133.3 pascals).

The initial weather-forecasting barometer, the Torricelli barometer, measured the height of a liquid column that the pressure of air would support, with a vacuum at the closed top end of a vertical tube. This barometer is an absolute pressure instrument. Atmospheric pressure, or barometric pressure, is obtained as the product of the height, the density of the barometric liquid, and the acceleration because of gravity at the Earth's surface. The aneroid barometer uses a partially evacuated box the spring-loaded sides of which expand or contract depending on the atmospheric pressure, driving a clocklike mechanism to show the pressure on a circular dial. This portable instrument was convenient to carry on mountaineering, ballooning, and mining expeditions to measure altitude by the change in atmospheric pressure. A barograph is an aneroid barometer mechanism adapted to plot a graph of the variation of pressure with time, using a stylus moving on a continuous roll of paper. The rate of change of pressure helps weather forecasters to predict the strength of approaching storms.

The term "manometer" derives from the Greek word *manos*, meaning "sparse," and denotes an instrument used to measure the pressure relative to a known pressure. A U-tube manometer measures the pressure difference from a reference pressure by the difference between the height of a liquid column in the leg of the U-tube that is connected to a known pressure source and the height of the liquid in the other leg, exposed to the pressure of interest. Manometers of various types have been used extensively in aerospace engineering experimental-test facilities such as wind tunnels. The pitot-static tubes used to measure flow velocity in wind tunnels were initially connected to water or mercury manometers. Later, electronic equivalents were developed. Inclined tube manometers were used to increase the sensitivity of the instrument in measuring small pressure differences amounting to fractions of an inch of water.

BACKGROUND AND HISTORY

In 1643, Italian physicist and mathematician Evangelista Torricelli (1608–1647) proved that atmospheric pressure would support the weight of a thirty-five-foot water column leaving a vacuum above that in a closed tube and that this height would change with the weather. This is considered the first measurement of atmospheric pressure and the basis for all later barometry. Later Torricelli barometers used liquid mercury to reduce the size of the column and make such instruments more practical.

The technology of pressure measurement has evolved gradually since then, with the aneroid barometer demonstrating the reliability of deflecting a diaphragm. This led to electrical means of measuring the amount of deflection. The most obvious method was to place strain gauges on the diaphragm and directly measure the strain. Later electrical methods used the change in capacitance caused by the changing gap between two charged plates. Such sensors dominated the market until the 1990s at the low end of the measurement range. Piezoresistive materials expanded the ability of miniaturized strain gauge sensors to measure high pressures changing at high frequency. Microelectromechanical system (MEMS) technology enabled miniaturized solid-state sensors to challenge the market dominance of the diaphragm sensors. In the early twenty-first century, pressure-sensitive paints allowed increasingly sensitive and faster-responding measurements of varying pressure with very fine spatial resolution.

HOW IT WORKS

Barometry measures a broad variety of pressures using an equally broad variety of measurement techniques, including liquid column methods, elastic element methods, and electrical sensors. Electrical sensors include resistance strain gauges, capacitances, piezoresistive instruments, and piezoelectric devices. The technologies range from those developed by French mathematician Blaise Pascal (1623–1662), Greek mathematician Archimedes (287–212 BCE), and Torricelli to early twenty-first century MEMS sensors and those used to conduct nanoscale materials science.

Pressures can be measured in environments from the near vacuum of space to more than 203,052 megapascals (1,400 megapascals) and from steady state to frequencies greater than 100,000 cycles per second. Sensors that measure with respect to zero pressure or absolute vacuum are called "absolute pressure sensors," whereas those that measure with respect to some other reference pressure are called "gauge pressure sensors." Vented gauge sensors have the reference side open to the atmosphere so that the pressure reading is with respect to atmospheric pressure. Sealed gauges report pressure with respect to a constant reference pressure.

Where rapid changes in pressure must be measured, errors due to the variation of sensitivity with the rate of change must be considered. A good sensor is one whose frequency response is constant over the entire range of frequency of fluctuations that might occur. Condenser microphones with electromechanical diaphragms have long been used to measure acoustic pressure in demanding applications such as music recording, with flat frequency response from

0.1 cycles per second (hertz) to well over 20,000 Hz, covering the range of human hearing. Pressure-sensitive paint in certain special formulations has been shown to achieve excellent frequency response to more than 1,600 hertz but only when the fluctuation amplitude is quite large, near the upper limit of human tolerance. Using digital signal processing, inexpensive sensors can be corrected to produce signals with frequency response quality approaching that of much more expensive sensors.

In the 1970s, devices based on the aneroid barometer principle were developed, in which the deflection of a diaphragm caused changes in electrical capacitance that then were indicated as voltage changes in a circuit. In the 1980s, piezoelectric materials were developed, enabling electrical voltages to be created from changes in pressure. Micro devices based on these largely replaced the more expensive but accurate diaphragm-based electromechanical sensors. Digital signal processing enabled engineers using the new small, inexpensive devices to recover most of the accuracy possessed by the more expensive devices.

APPLICATIONS AND PRODUCTS

Barometry has ubiquitous applications, measured by a broad variety of sensors. It is key to control of industrial gas-phase chemical reaction methods, weather prediction and measuring the altitude of aircraft or the depth of submersibles, as well as to measuring blood pressure to monitor health. Pressure-sensitive paints enable measurement of surface pressure as it changes in space and time. The accuracy of measuring and controlling gas pressure is fundamental to manufacturing processes. The necessity of monitoring the pressure experienced by people and structures functioning below the surface of a body of water, especially as depth increases, should be self-evident.

Weather Forecasting. Scientists learned to relate the rate of change of atmospheric pressure to the possibility of strong winds, usually bringing rain or snow. For example, if the pressure drops by more than 0.0145 psi (3 millibar) per hour, winds of up to 31 miles (50 kilometers) per hour are likely to follow. Powerful storms may be preceded by drops of more than 0.35 psi (24 mb) in twenty-four hours. If the pressure starts rising, clear calm weather is expected. However, these rules change with regional conditions. For instance, in the Great Lakes region of Canada and the United States, rising pressure may indicate an Arctic cold front moving in, causing heavy snowfall. In other regions, a sharply dropping pressure indicates a cold front moving in, followed by a quick rise in pressure as the colder weather is established. As a warm front approaches, the pressure may level out and rise slowly after the front passes.

Modern forecasters construct detailed maps showing contours of pressure from sensors distributed over the countryside and use these to predict weather patterns. Aircraft pilots use such maps to identify safe routes and areas to avoid. Using Doppler radar wind measurements, infrared temperature maps and cloud images from satellites, and computational fluid dynamics, modern weather bureaus are able to issue warnings about severe weather several hours in advance for smaller local weather fronts and storms and several days ahead for major storms moving across continents or oceans. However, the number of weather-monitoring pressure sensors available to forecasters is quite inadequate to issue accurate predictions for minor weather changes, particularly when predicting rain or snow.

Electrical Gauges. Gauges operating on the electrical changes induced by deflection of a diaphragm are used in industrial process monitoring and control where computer interfacing is required. Unsteady pressure transducers come in many ranges of amplitude and frequency. Piezoresistive sensors are integrated into an electrical-resistance bridge and constructed as miniature self-contained, button-like sensors. These are suitable for high amplitudes and frequencies, such as those encountered in shock waves and explosions, and transonic or supersonic wind-tunnel tests. Condenser microphones are used in acoustic measurements. As computerized data-acquisition systems became common, but pressure sensors remained expensive, pressure switches enabled dozens of pressure-sensing lines connected through the switches to each sensor to be measured one at a time. This required a long time to collect data from all the sensors, spending enough time at each to capture all the fluctuations and construct stable averages, making it unsuitable for rapidly changing conditions. Inexpensive, miniaturized, and highly sensitive solid-state piezoelectric sensors and fast, multichannel analogue-digital converters have made it possible to connect each pressure port to an individual sensor, vastly reducing the time between individual measurements at each sensor.

Aircraft Testing. Water and mercury manometers were used extensively in aerospace test facilities such as wind tunnels, where banks of manometers indicated the distribution of pressure around the surfaces of models from pressure-sensing holes in the models. Pressure switches connecting numerous pressure-sensing ports to a single sensor became common in the 1970s. In the 1990s, inexpensive sensors based on microelectromechanical systems technology enabled numerous independent sensing channels to be monitored simultaneously.

Sphygmomanometers for Blood Pressure. Other than weather forecasting, the major common application of pressure measurement is in measuring blood pressure. The device used is called a "sphygmomanometer." The device consists of an inflatable cuff that is placed around the upper arm, a squeeze-bulb control to inflate and deflate the cuff, and a mercury-filled tube next to a graduated scale (the manometer). The high and low points of pressure reached in the heartbeat cycle are noted on the mercury manometer tube synchronized with the heartbeat sounds detected through a stethoscope. Electronic versions of this device use calibrated electronic sensors rather than a mercury manometer.

Bourdon Tubes for Household Barometry. The Bourdon tube is a pressure-measuring device in which a coiled tube stretches and uncoils depending on the difference between pressures inside and outside the tube and drives a levered mechanism connected to an indicator dial. Diaphragm-type pressure gauges and Bourdon-tube gauges are still used in the vast majority of household and urban plumbing. These instruments are highly reliable and robust, but they operate over fairly narrow ranges of pressure.

Pressure-Sensitive Paints to Map Pressure over Surfaces. So-called pressure-sensitive paints (PSPs) offer an indirect technique to measure pressure variations over an entire surface, using the fact that the amount of oxygen felt at a surface is proportional to the density and thus to the pressure if the temperature does not change. These paints are luminescent dyes dispersed in an oxygen-permeable binder. When illuminated at certain ultraviolet wavelengths, the dye molecules absorb the light and move up into higher energy levels. The molecules then release energy in the infrared wavelengths as they relax to equilibrium. If the molecule collides with an oxygen molecule, the energy gets transferred without emission of radiation. Therefore, the emission from a surface becomes less intense if the number of oxygen molecules being encountered increases. This occurs when the pressure of air increases. The observed intensity from a painted surface is inversely proportional to the pressure of oxygen-containing air. Light-intensity values at individual picture element (pixel) are converted to numbers, compared with values at some known reference pressure, and presented graphically as colors. Typically, an accurate pressure sensor using either piezoelectric or other technology is used for reference. By the 2010s, pressure-sensitive paints had reached the sensitivity required to quantify pressure distributions over passenger automobiles at moderate highway speeds, given expert signal processing and averaging a large number of images.

Smart Pressure Transmitters for Automatic Control Systems. Wireless pressure sensors are used in remote applications such as weather sensing. Modern automobiles incorporate tire pressure transmitters. Manifold pressure sensors send instantaneous readings of the pressure inside automobile engine manifolds so that a control computer can calculate the best rate of fuel flow to achieve the most efficient combustion. Smart pressure transmitters incorporating capacitance-type diaphragm pressure sensors and microprocessors can be configured to adjust their settings remotely, perform automatic temperature compensation of data, and transmit pressure data and self-diagnosis data in digital streams.

Nuclear Explosion Sensors. Piezoresistive transducers have been developed to report the extremely high overpressure, as high as 10,007 psi (69 MPa), of an air blast of a nuclear weapon, with the microsecond rise time required to measure the blast wave accurately. One design uses a silicon disk with integral diffused strain-sensitive regions and thermal barriers. Another design uses the principle of Fabry-Perot interferometry, in which laser light reflecting in a cavity changes intensity depending on the shape of the cavity when the diaphragm bounding the cavity flexes because of pressure changes. This sensor has the response speed and ruggedness required to operate in a hostile environment, where there may be very large electromagnetic pulses. In such environments a capacitance-based sensor or piezoelectric sensor may not survive.

Extreme Applications of Barometry. The basic origins of pressure can be used to explain the pressure due to radiation as the momentum flux of photons. At Earth's orbit around the Sun, the solar intensity of 1.38 kilowatts per square meter causes a radiation pressure of roughly 4.56 micropascals. Solar sails have been proposed for long-duration missions in space, driven by this pressure. Close to the center of the Earth, the pressure reaches 3.2 to 3.4 million bars. Inside the Sun, pressure as high as 250 billion bars is expected, while the explosion of a nuclear-fusion weapon may produce a quarter of that. Metallic solid hydrogen is projected to form at pressures of 250,000 to 500,000 bars.

CAREERS AND COURSEWORK

Because barometry is so important in so many industries and so many branches of scientific research, most students who are planning on a career in engineering, other technological jobs, and the sciences must understand it.

Modern pressure-measurement technology integrates ideas from many branches of science and engineering derived from physics and chemistry. The pressure-measurement industry includes experts in weather forecasting, plumbing, atmospheric sciences, aerospace wind-tunnel experimentation, automobile-engine development, the chemical industry, chemists developing paint formulations, electrical and electronics engineers developing microelectromechanical sensors, software engineers developing smart sensor logic, and the medical community interested in using barometry to monitor patients' health and vital signs.

Pressure measurement therefore comes up as a subject in courses offered in schools of mechanical, chemical, civil, and aerospace engineering. The numerous other related issues come up in specialized courses in materials science, electronics, atmospheric sciences, and computer science.

SOCIAL CONTEXT AND FUTURE PROSPECTS

Instrumentation for measuring pressure, normal and shear stresses, and flow rate from numerous sensors are becoming integrated into computerized measurement systems. In many applications, such sensors are mass-produced using facilities similar to those for making chips for computers. Very few ideas exist for directly measuring pressure, as it changes rapidly at a point in a flowing fluid, without intrusive probes of some kind. Such nonintrusive measurements, if they become possible, could help us to understand the nature of turbulence and assist in a major breakthrough in fluid dynamics.

Measurement of pressure is difficult to make inside flame environments, where density and temperature fluctuate rapidly. Better methods of measuring pressure in biological systems, such as inside blood vessels, would have major benefits in diagnosing heart disease and improving health.

As of the late 2010s, pressure-measurement systems were still too expensive to allow sufficient numbers to be deployed to report pressure with enough spatial and time resolution to permit development of a real-time three-dimensional (3D) representation. Research in this area will doubtless improve the resolution and response and hopefully bring down the cost. With more pressure sensors distributed over the world, weather prediction will become more accurate and reliable.

—Narayanan M. Komerath

Further Reading

American Society of Mechanical Engineers. *Pressure Measurement*. American Society of Mechanical Engineers, 2010.

Avallone, Eugene A., Theodore Baumeister III, and Ali M. Sadegh. *Marks' Standard Handbook for Mechanical Engineers*. 11th ed., McGraw-Hill, 2006.

Benedict, Robert P. *Fundamentals of Temperature, Pressure, and Flow Measurements*. 3rd ed., John Wiley & Sons, 1984.

Burch, David. *The Barometer Handbook: A Modern Look at Barometers and Applications of Barometric Pressure*. Starpath Publications, 2009.

Gillum, Donald R. *Industrial Pressure, Level, and Density Measurement*. 2nd ed., International Society for Automation, 2009.

Green, Don W., and Robert H. Perry. *Perry's Chemical Engineers' Handbook*. 8th ed., McGraw-Hill, 2008.

Ryans, J. L. "Pressure Measurement." *Kirk-Othmer Encyclopedia of Chemical Technology*. 5th ed., John Wiley & Sons, 2000.

Taylor, George Frederic. *Elementary Meteorology*. Prentice-Hall, 1954.

"What Is Atmospheric Pressure?" *American Geosciences Institute*, 2018, www.americangeosciences.org/education/k5geosource/content/weather/what-is-atmospheric-pressure. Accessed 19 Mar. 2018.

Basaltic Rocks

Fields of Study: geology; volcanology; oceanography

ABSTRACT

"Basalt" is the term applied to dark, iron-rich volcanic rocks that occur everywhere on the ocean floors, as oceanic islands, and in certain areas on continents. It is the parent material from which nearly any other igneous rock can be generated by various natural processes.

KEY CONCEPTS:

fissure flow: typically refers to lava that flows relatively smoothly from a long fissure vent rather than being ejected from a volcano

lithosphere: Earth's solid crust

magma: molten rock that exists in the mantle layer below the lithosphere

plume: a rising column of superheated magma within Earth's mantle

seafloor spreading: the movement of seafloor on either side of a mid-ocean ridge such that basaltic magma wells up to fill the space that opens up between the separating sections of seafloor

subduction: a consequence of seafloor spreading, in which seafloor being displaced from the mid-ocean ridge eventually encounters a less dense continental mass and is forced beneath it and back into the mantle layer

BASALT COMPOSITION

Basalt is a dark, commonly black, volcanic rock. It is sometimes called "trap" or "traprock," from the Swedish term *trapp*, which means "steplike." On cooling, basalt tends to form hexagonal columns, which in turn form steplike structures after erosion. Excellent examples can be seen at Devils Postpile, in the Sierra Nevada in central California, and the Giant's Causeway, in County Antrim, Ireland. The term "trap," however, is used mostly by miners and nonspecialists; scientists prefer the word "basalt" for the fine-grained, volcanic rock that forms by the solidification of lava flows. Basaltic magma that crystallizes more slowly below the earth's surface, thus making larger mineral grains, is called "gabbro."

The importance of basalt to the evolution of planets like Earth cannot be overemphasized. Basalt is considered to be a "primary silicate liquid," in that the first liquids to form by the melting of the original minerals that made up all the so-called rocky planets (those composed mostly of silicates) were basaltic in composition. In turn, basalt contains all the necessary ingredients to make all the other rocks that may eventually form in a planet's crust. Furthermore, many meteorites (which are believed to represent fragments of planetoids or asteroids) are basaltic or contain basalt fragments, and the surfaces of the moon and the planets Mercury, Venus, and Mars are known to be covered to various degrees by basaltic lava flows.

Like most rocks in Earth's crust, basalt is composed of silicate minerals, substances whose principal component is a silicon-oxygen complex, hence the term "silicate." Compared with other silicate rocks, basalts contain large amounts of iron and magnesium and small amounts of silicon. This characteristic is reflected in the minerals in basalts, which are mostly pyroxene minerals (dark-colored, calcium-iron-magnesium silicates such as augite) and certain feldspar minerals called "plagioclase" (light-colored, calcium-sodium-aluminum silicates). Pyroxenes and plagioclase are essential components of basalts, but some types of basalt also contain the mineral olivine (a green, iron-magnesium silicate). The high abundance of dark green or black pyroxene and, in some cases, olivine gives basalts their characteristic dark color. This color can be mainly attributed to the high iron content of pyroxene and olivine.

OCEANIC BASALT DISTRIBUTION

Basalt is the most common type of igneous rock (that is, rock formed by the crystallization of magma) on or near the surface of Earth. It is the principal rock in the ocean floors and is common, though less abundant, on the continents as well. Drilling into the seafloor by specially designed oceanographic research ships reveals that basalt invariably lies just below a thin cover of fine, sedimentary mud. Ocean floor basalt flows out of mid-ocean ridges, a system of underwater mountain ranges that spans the globe. These ridges commonly trend roughly down the middle of ocean basins, and they represent places where Earth's lithospheric plates are being split apart by the force of convective currents in the underlying magma. In this process, basalt magma is generated below what is referred to as rift mountains; it flows onto the cold ocean floor and solidifies. Although oceanic ridges are normally hidden from view under the oceans, a segment of the Mid-Atlantic Ridge emerges above the waves as the island of Iceland.

Oceanic islands are also composed of basalt. The islands of Hawaii, Fiji, Mariana, Tonga, and Samoa, among others, are large volcanoes or groups of volcanoes that rise above sea level from the ocean floor. Unlike the basalts that cover the ocean floors, however, these volcanoes do not occur at oceanic ridges but instead rise directly from the seafloor, driven either by a stationary magma 'plume' (Hawaii) or seismic forces near the edges of tectonic plates. Basalt is also a fairly common rock type on island arcs, volcanic islands that occur near continental margins. These curvilinear island chains arise from melting along subduction zones, areas where Earth's lithospheric plates are colliding. This process generally involves dense seafloor crustal material from the ocean basins being forced under the less dense material of the continents; andesite (a light-colored rock) volcanoes are the main result, but some basalt erupts there as well. The Japanese, Philippine, and Aleutian island chains are examples of island arcs, as are the island countries of New Zealand and Indonesia.

CONTINENTAL BASALT DISTRIBUTION

Basaltic lava flows are not nearly as common on continents as in oceanic areas. Andesites, rhyolites, and related igneous rocks are far more abundant than basalt in continental settings. In North America, basaltic lava flows and volcanoes are most exposed to view in

Kilauea's Fissure 8 cone erupting in the early morning of June 28, 2018, in the Lower East Rift Zone (United States Geological Survey via Wikimedia Commons)

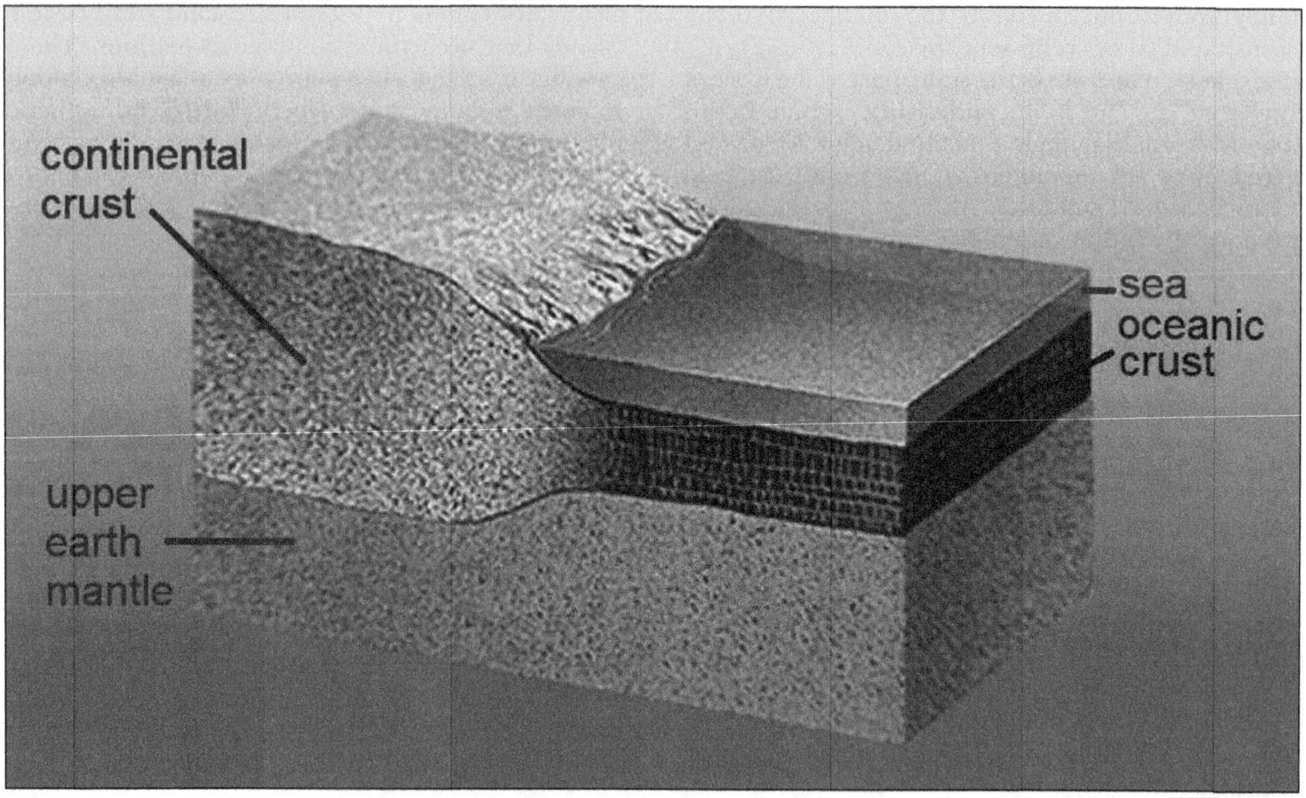

Lithosphere: continental and oceanic crust on the upper earth mantle (United States Geological Survey and Sciencia58 via Wikimedia Commons)

the western United States and Canadian provinces, western Mexico, Central America, and western South America. The greatest accumulations of basaltic lava flows in the United States are in the Columbia Plateau of Washington, Oregon, and Idaho. In this large area, a series of basaltic lava flows have built up hundreds of feet of nearly flat-lying basaltic flows over a few million years. These "fissure flows" result from lava pouring out of long cracks in the crust, or fissures. They are similar in many respects to the basalt flows produced at oceanic ridges, because no actual volcanic cones are produced—only layer after layer of black basalt. The Snake River Plain in southern Idaho has a similar origin, and other extensive basalt plateaus occur in the area of southwestern India called the "Deccan Traps," the Karroo area of South Africa, and Paraná State in Brazil.

Not all basalt is erupted as lava flows. If the lava is particularly rich in volatiles such as water and carbon dioxide, it will be explosively ejected from the volcano as glowing fountains of incandescent particles that rain down on the surrounding area. Conical volcanoes composed almost exclusively of basalt ejecta particles are called "cinder cones." Good examples of cinder cones are Mt. Fuji in Japan, Sunset Crater in northern Arizona, and the numerous cinder cones in Hawaii and Iceland.

BASALT CLASSIFICATION

Although basaltic rocks may all look alike to the nonspecialist, there are actually many different kinds of basalt. They are arranged by scientists into a generally accepted classification scheme based on chemistry and, to some extent, mineralogy. To begin with, basalt can be distinguished from the other major silicate igneous rocks by its relatively low silica content (about 50 percent). Within the basalt clan itself, however, other means of classification are used. Basalts are divided into two major groups: the alkaline basalts and the subalkaline basalts. Alkaline basalts contain large amounts of the alkali metal ions potassium and sodium but relatively small amounts of silica. In contrast, subalkaline basalts contain less potassium and sodium and more silica. As might be expected, this chemical difference translates into differences in the mineral content of the basalt types as well. For example, all alkaline basalts contain one or more minerals called "feldspathoids" in addition to plagioclase feldspar. They also commonly contain significant olivine. Subalkaline basalts, in contrast, do not contain feldspathoids, although some contain olivine, and they may be capable of crystallizing very tiny amounts of the mineral quartz. The presence of this very silica-rich mineral reflects the relatively high silica content of subalkaline basalt magmas versus alkaline basalt magmas. Within these two major groups are many subtypes, too numerous to discuss here.

STUDY TECHNIQUES

Like other igneous rocks, basalts are analyzed and studied by way of many techniques. Individual studies may include extensive field mapping, in which the distribution of various types of basalt is plotted on maps. By correlating, the history of magma generation and its relation to tectonic history (earth movements) can be reconstructed for a particular area. Geologic maps of basalt types with absolute ages are determined through radiometric dating techniques. Good examples of such studies are those conducted in recent years on the Hawaiian Islands. These studies indicate that the alkaline basalts on any given island are generally older than the subalkaline basalts, showing that magma production has moved upward, to lower-pressure areas, in the mantle with time. This finding supports the idea that oceanic island basalts such as those in Hawaii are generated within so-called mantle plumes—roughly balloon-shaped, slowly rising masses of mantle material made buoyant by localized "hot spots."

Samples of basalt are also analyzed in the laboratory. The age of crystallization of basalt is obtained by means of radiometric dating techniques that involve the use of mass spectrometers to determine the abundances of critical isotopes, such as potassium-40 and argon-40, or rubidium-87 and strontium-86 and -87. To obtain information on how basalt magma is generated and how it subsequently changes in composition before extrusion as a lava flow, scientists place finely powdered samples in metallic, graphite, or ceramic capsules and subject them to heating and cooling un-

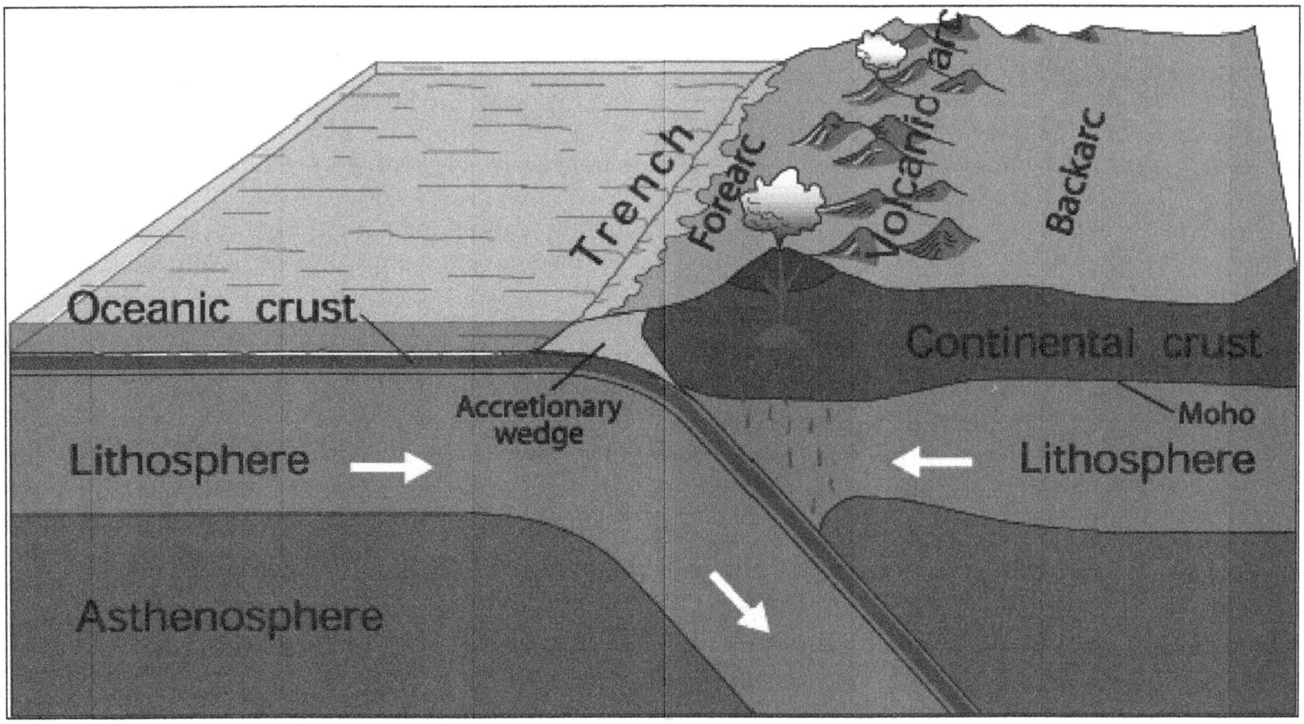

Subduction fault zone (United States Geological Survey)

der various conditions of pressure. Such procedures are known as experimental petrology. Such studies prove that nearly any other igneous magma composition can be derived from basalt magma by the process of crystal fractionation. Widely believed to be the major factor influencing chemical variation among igneous rocks, this process results in ever-changing liquid compositions, as the various silicate minerals crystallize and are thus removed from the liquid over time. Basalt's parental role gives it enormous importance in the discipline of igneous petrology.

TRACE ELEMENT ANALYSIS

Another useful avenue of research is the trace element analysis of basalts. Trace elements occur in such low abundances in rocks that their concentrations must usually be expressed in terms of parts per million or even parts per billion. Among the most useful substances for tracing the history of basalt are the rare-earth elements. The elements chromium, vanadium, nickel, phosphorus, strontium, zirconium, scandium, and hafnium are also used. There are many methods for measuring these elements, but the most common, and most accurate, is neutron activation. This method involves irradiating samples in a small nuclear reactor and then electronically counting the gamma-ray pulses generated by the samples. Since different elements tend to emit gamma rays at characteristic energies, these specific energies can be measured and the intensity of gamma pulses translated into elemental concentrations.

Once determined for a particular basalt sample, trace element abundances are sensitive indicators of events that have transpired during the evolution of the basalt. There are two reasons for this sensitivity. First, trace elements are present in such low concentrations (as compared with major elements—iron, aluminum, calcium, silicon, and the like) that any small change in abundance caused by changes in the environment of basalt production will be readily noticed. Second, different minerals, including those crystallizing in the magma and those in the source peridotite, incorporate a given trace element into their structures or reject it to the surrounding liquid to widely varying degrees. Therefore, trace element concentrations can be used to show the minerals that were involved in producing certain observed chemical signatures in basalts and those that were likely not involved.

For example, it is well known that the mineral garnet readily accepts the rare-earth element lutetium into its structure but tends to reject most lanthanum to any adjacent liquid, even though these elements are very similar chemically. Basalts with very little lutetium but much lanthanum were therefore probably derived by the melting of garnet-bearing mantle rocks. Since garnet-bearing mantle rocks can exist only at great depths, basalts with such trace element patterns must have originated by melting at these depths in the mantle. In fact, that is one of the most important lines of evidence to support that alkaline basalts originate at high-pressure regions in the mantle.

ECONOMIC APPLICATIONS

Basaltic islands, particularly in the Pacific basin, are some of the most popular tourist stops in the world. More importantly, however, basalt magma contains low concentrations of valuable metals that, when concentrated by various natural processes, provide the source for many important ores. Copper, nickel, lead, zinc, gold, silver, and other metals have been recovered from ore bodies centered in basaltic terrains. Some of the richest mines of metallic ores in the world are located in Canada, where ores are found associated with extremely old basaltic rocks, called "greenstones," from long-vanished oceans. The richest of these mines is Kidd Creek in northern Ontario. These ore-bearing basalts were first extruded more than 2 billion years ago, during what geologists call Precambrian times (the period from 4.6 billion to about 600 million years ago). Other notable ore deposits include the native, or metallic, copper in late Precambrian basalts that were mined for many years in the Keweenaw Peninsula of northern Michigan. The island of Crete in the Mediterranean Sea has copper mines that were mined thousands of years ago during the "copper" and "bronze" ages of human history. The basalt enclosing these ores is believed to have erupted from an ancient mid-ocean ridge formerly located between Africa and Europe.

Basalt can also be used as a building stone or raw material for sculptures, but its high iron content makes it susceptible to rust stains. It is also ground up to make road gravel, especially in the western United States, and it is used as decorative stone in yards and gardens.

—John L. Berkley

Further Reading
Decker, Robert, and Barbara Decker. *Volcanoes*. 4th ed., W. H. Freeman, 2005.
Faure, Gunter. *Origin of Igneous Rocks: The Isotopic Evidence*. Springer-Verlag, 2010.
Floyd, Peter A., editor. *Oceanic Basalts*. Springer Science + Business Media, 2012.
Gill, Robin. *Igneous Rocks and Processes: A Practical Guide*. Wiley-Blackwell, 2011.
Lutgens, Frederick K., Edward J. Tarbuck, and Dennis Tasa. *Earth: An Introduction to Physical Geology*. 10th ed., Prentice-Hall, 2010.
Molnar, Peter. *Plate Tectonics: A Very Short Introduction*. Oxford UP, 2015.
Philpotts, Anthony, and Jay Aque. *Principles of Igneous and Metamorphic Petrology*. 2nd ed., Cambridge UP, 2009.
Ross, Pierre-Simon, et al. "Basaltic to Andesitic Volcaniclastic Rocks in the Blake River Group, Abitibi Greenstone Belt: 2. Origin, Geochemistry, and Geochronology." *Canadian Journal of Earth Sciences*, vol. 48, 2011, pp. 757–77.
Sharma, S. R. *Thermo-Physical Properties of Rocks: Special Reference to Deccan Trap Basalts*. Allied Publishers, 2015.
Sobolev, A. V., et al. "The Amount of Recycled Crust in Sources of Mantle-Derived Melts." *Science*, vol. 316, 2007, pp. 412–17.
Tarbuck, Edward J., Frederick K. Lutgens, and Dennis Tasa. *Earth: An Introduction to Physical Geology*. 10th ed., Prentice-Hall, 2010.

Beaches and Coastal Processes

Fields of Study: geology; oceanography; marine engineering; marine ecology

ABSTRACT
The shoreline is the meeting place for the interaction of land, water, and atmosphere, and rapid changes are the rule rather than the exception.

KEY CONCEPTS
buffer zone: the area and structure of a shoreline that tends to absorb and mitigate the energy of incoming waves
downcurrent: an abbreviation of "downstream current," refers to locations farther downstream of a current from any arbitrary reference point
geomorphology: the study of how and why the shapes of geologic features change over time
longshore current: a current that is produced by wave action on a shoreline and that flows parallel to that shoreline
luminophors: particles that have been treated artificially so that they can be detected in the environment
tracers: the chemical materials used to treat particles to create luminophors

BEACH COMPOSITION
The processes that create, erode, and modify beaches are many. Marine processes, such as waves, wave refraction, currents, and tides, work concurrently. This suggests that a beach is a sensitive landform, and indeed it is. Generally a beach is a sedimentary deposit made by waves and related processes. Beaches are often regarded as sandy deposits created by wave action; however, beaches may be composed of broken fragments of lava, sea shells, coral reef fragments, gravel, or even large stones. A beach is composed of whatever sediment is available. Beaches have remarkably resilient characteristics. They are landforms made of loose sediment and are constantly exposed to wave and current action. On occasion, the coastal processes may be very intensive, such as during a hurricane or tropical storm. Yet, in spite of the intensity of wave processes, these rather thin and narrow landforms, although perhaps displaced, restore themselves within a matter of days. Coastal scientists are inclined to believe that the occurrence and maintenance of beaches are related to their flexibility and rapid readjustment to the varying intensity of persistent processes.

Sediment deposited on beaches is often derived from the continents. Rivers are one source of beach sediment, and the coast is another. As rivers erode the land, the sediment they carry ultimately finds its way to the shore. Because the sediment may be transported several tens or hundreds of miles, or kilometers, it is refined and broken down even further to finer-sized particles. Once the river reaches the sea, the sediment is distributed by longshore currents along the shoreline as a beach. Beaches also occur where the shoreline is composed of cliffs, such as along the Pacific coast of North America. Here, waves erode the sea cliffs, and the sediment is deposited locally as a beach. Under these conditions, the beach deposit is most often gravelly because the sediment is transported only a short distance and has not had an opportunity to break up into finer-sized sediment such as sand. Some island beaches are made up entirely of shell fragments.

WAVES

The most obvious mechanical energy source working on beaches is waves. Waves approaching a beach are generally created by winds at sea. As wind velocities increase, a wave form develops and radiates out from the storm. An oscillatory motion of the water occurs as the wave form moves across the water surface. It is important to note that the water movement within a wave is not the same as the movement of the wave form. In a wave created in deep water that is approaching a beach, the water particles move in a circular orbit toward the beach and back again, and little forward movement of the water occurs. The water at the surface moves from the top of the orbit (the wave crest) to the base of the orbit (the wave trough) and then back up. Thus, the water particles move in an oscillatory wave; this motion continues down into the water. Although the size of the orbits in the water column decreases, motion occurs to a depth referred to as the wave base. At this point, the depth of the wave base is less than the depth of the water. As wave crests and troughs move into shallow water, the water depth decreases to a point where it is equal to the wave base. From this point, the orbital motion is confined because of the shallowness of the water and takes on an elliptical path. The ellipse becomes more confined as the waves enter ever shallower water and eventually becomes a horizontal line. At this point, there is a net forward movement of water in the form of a breaker.

As a wave enters shallow water, many adjustments occur, such as the change in the orbital path of water particles already described and a change in the velocity of the wave form. Since the wave base is "feeling the bottom" in shallower water, friction occurs, slowing the wave down. As seen from an airplane, waves

Baker Beach, San Francisco (Daniel Schwen via Wikimedia Commons)

entering shallow water do so at an angle, not parallel to the shoreline. Therefore, one part of the wave enters shallow water and slows down relative to the rest of the wave. Thus a part of the wave crest is feeling bottom sooner than the rest of the wave, causing the breaking action of the wave to appear to peel itself along the beachfront. Because the wave crests and troughs have different velocities, the wave refracts or bends. In so doing, the wave crests and troughs try to parallel the shallow bottom topography that they have encountered.

The wave refraction is seldom completed, and the breaking wave surges obliquely up the slope of the beach and then returns perpendicular to the shoreline. The result is a current that basically moves water in one direction parallel to the beach in a zigzag pattern (a longshore, or littoral, current). It is a slow-moving current that is located between the breaking wave out in the water and the beach. Because longshore-current movement operates in shallow water and along the beach, it is capable of transporting sediment along the shoreline. "Longshore drift" refers to the movement of sediment along beaches. In a sense, the longshore current is like a river moving sand and other material parallel to the shore. Beaches are always in a state of flux. Although they appear to be somewhat permanent, they are constantly being moved in the direction of the longshore current. A prime example of this phenomenon is Ontario, Canada's Long Point peninsula in Lake Erie, which has slowly been extending towards the opposite shore of that lake for tens of thousands of years and continues to do in the present day. Along any shoreline, several thousand cubic yards (or cubic meters) of sediment are constantly in motion as longshore drift. Along the beaches of the eastern United States, about 26,159 cu yds (200,000 cu m) of sediment is transported annually within a longshore drift system.

A different type of ocean wave is one that is generated on the ocean floor rather than by the wind. Such waves are properly termed "tsunamis." Although they popularly have been coined "tidal waves," they are completely unrelated to tides or the movement of planets. Some type of submarine displacement, such as the creation of a volcano, a landslide on the seafloor, or an earthquake beneath the ocean bottom, displaces a volume of water, which triggers waves. Tsunami waves are low, subdued forms traveling thousands of miles or kilometers over the ocean surface at extremely high velocities, often exceeding 497 miles (800 kilometers) per hour. As a tsunami approaches shallow water or a confined bay, its height increases. There is no method for direct measurement of the heights of tsunami waves. However, in 1946, a lighthouse at Scotch Cap, Alaska, located on a headland 102 feet (31 meters) above the Pacific Ocean, was destroyed by waves caused by a landslide-generated tsunami. Video records of the 2011 tsunami that struck Japan following a massive subduction earthquake beneath the sea showed the incoming inundation to appear as though the sea level had been suddenly raised as the mass of dark, black water rolled over the tops of seawalls, rather than the classic seashore wave and nothing like the depiction of the clear water seen in such waves in popular movies.

The changing character of a beach is very dynamic because the properties of waves are variable. During storm conditions or when more powerful waves strike a beach during the winter season, for example, the beach commonly is eroded, has a steeper slope, and is composed of a residue of coarser sediment, such as gravel, that is more difficult to remove. During fair-weather or summer conditions, however, beaches are redeposited and built up.

STUDY OF BEACHES AND COASTAL PROCESSES

The study of beaches and coastal processes is not particularly easy because of constant wave motion and changes in the beach shape. Scientists have, however, developed field methods as well as laboratory techniques to study these phenomena. To study wave motion and related current action, several techniques have been devised that include tracers, current meters, aerial photography, and satellite-based monitoring. Two types of tracers are commonly used to determine the direction and velocity of sand movement: radioactive isotopes and fluorescent coatings to produce luminophors. (Luminophors are sediment particles coated with selected organic or inorganic substances that glow under certain light conditions.) In the former type, a radioactive isotope such as gold, chromium, or iridium is placed on the surface of the grains of sediment. Alternatively, grains of glass may be coated with a radioactive element. The radioactive sediment can then be detected relatively easily and quickly using just a Geiger counter. Both techniques can trace the direction and abundance of sediment along the seashore.

Bright-colored dyes or current meters can also be used to document the direction of longshore currents. Surveying instruments are used to determine the high and low topography of beaches and adjacent sandbars. By measuring beach topography before and after a storm, for example, scientists can record immediate changes. In this way, they can document the volume and dimensions of beach erosion or deposition that has taken place during a storm. Measurement of wave height and other wave characteristics can be done with varying degrees of sophistication. Holding a graduated pole in the water and visually observing wave crest and troughs is both the simplest and the cheapest method. To achieve more refined measurements, however, scientists place pressure transducers or ultrasonic

devices on the shallow seafloor to record pressure differences or fluctuations of the sea surface. These more precise instruments also record other wave data for later study and analysis.

All these methods are detailed field techniques. By comparing aerial photographs, detailed maps, satellite pictures, and in some cases government studies in a coastal sector, changes over long periods of time may be detected. Finally, because wave motion cannot be controlled on a shoreline, wave tank studies are used to derive wave theories. Normally, an elongated glass-lined tank containing water from 0.98 to 3 ft (0.3 to 1 m) in depth is used. A mechanical device called a "wave machine" displaces water at one end of the wave tank to create waves and sediment is introduced at the other end. The heights and other characteristics of the waves can be varied, as can the type of sediment, to form the beach. In this controlled way, the various beach and wave relationships can be studied.

SIGNIFICANCE

More than 70 percent of the population of the United States lives in a shoreline setting, and thus an understanding of how waves interact with beaches is important. Shoreline property is prized for residential use and hence highly valuable. On many beaches, the great investments made in hotels and condominiums suggest that beaches are the most sought-after environment on Earth, and shoreline frontage is sold by the foot or meter, not by the acre or hectare. Currently, however, such demand and investment are threatened with rising sea levels and continued coastal erosion.

Beaches represent the buffer zone that mitigates wave erosion. Waves generated in the open sea slow down in shallow water, and the beach deposit absorbs the impact of waves. Beaches are therefore constantly changing and are one of the most ephemeral environments on Earth. Thus, a sound knowledge of longshore currents and beach development is necessary prior to nearshore or marine construction. Sea walls and groins, for example, interfere with waves and longshore currents and may cause considerable erosion in selected areas, while inducing abnormal sedimentation in others. Sea walls are often constructed perpendicular to a shoreline to slow longshore currents so that a beach can be deposited. Downcurrent, beyond the area of beach deposition, erosion will take place. Similarly, rivers, which are a major source of the sediment that creates beaches, are sometimes dammed, thus depriving beaches of sediment and resulting in their erosion. Unless planners, developers, and builders understand these processes, major damage can result: Failure to understand coastal processes has, on occasion, caused a property owner to sue a neighbor who caused beach erosion to take place.

—C. Nicholas Raphael

Further Reading

Blumberg, Alan F., and Michael S. Bruno. *The Urban Ocean: The Interaction of Cities with Water*. Cambridge UP, 2018.

Ciavola, Paolo, and Giovanni Coco. *Coastal Storms: Processes and Impacts*. Wiley-Blackwell, 2017.

Davis, Richard Jr., and Duncan Fitzgerald. *Beaches and Coasts*. 2nd ed., John Wiley & Sons, 2020.

Davidson-Arnott, Robin, Bernard Bauer, and Chris Hauser. *Introduction to Coastal Processes and Geomorphology*. 2nd ed., Cambridge UP, 2019.

Fall, J. A., et al. *Long-Term Consequences of the Exxon Valdez Oil Spill for Coastal Communities of South-central Alaska*. Alaska Department of Fish and Game, Division of Subsistence, Anchorage, Alaska. Technical Report 163, 2001.

Haslett, Simon K. *Coastal Systems*. 2nd ed., Routledge, 2009.

Hobbs, Carl H. *The Beach Book: Science of the Shore*. Columbia UP, 2012.
Jackson, Derek, and Andrew D. Short, editors. *Sandy Beach Morphodynamics*. Elsevier, 2020.
Kaufman, Wallace, and Orrin Pilkey. *The Beaches Are Moving: The Drowning of America's Shoreline*. Duke UP, 1983.
Komar, Paul D., and J. Robert Moore, editors. *Handbook of Coastal Processes and Erosion*. CRC Press, 2018.
Micallef, Anton, and Allan Williams. *Beach Management: Principles and Practice*. Earthscan, 2009.
Schwartz, Maurice L. *Encyclopedia of Coastal Science*. Springer, 2005.

BERING SEA ECOSYSTEM

Fields of Study: oceanography; marine biology; marine geology

ABSTRACT

The Bering Sea is located in the North Pacific, joining that ocean with the Arctic Ocean by way of the Bering Strait that spans the gap between easternmost Asia and westernmost North America. This Large Marine Ecosystem encompasses the deepwater Aleutian Basin, with depths well in excess of 10,000 feet (3,048 meters), as well as a large area of shallow, broad continental shelf where depths seldom exceed about 200 ft (60 m). The rapid change in bathymetry (depths) between these two regions brings cold, nutrient-rich bottom water onto the shallow shelf, which thus supports a vast food web.

KEY CONCEPTS

doughnut hole: a relatively small area of international waters between the exclusive economic zones of the nations bordering the Bering Sea

exclusive economic zone (EEZ): the area of ocean waters extending from the shore of a nation to a distance of 200 miles (322 kilometers), which can only be harvested by fishers and other industries from the corresponding nation

individual transferable quota: an allotment of a portion of the allowable annual catch from the Bering Sea, which can be purchased by individual fishers who exceed their allotted quota from those who have not fulfilled their quota

seawater stratification: the formation of discernible layers in the ocean water as characterized by differences of density due to salinity or temperature

The Bering Sea is located in the North Pacific, joining that ocean with the Arctic Ocean by way of the Bering Strait that spans the gap between easternmost Asia and westernmost North America. This Large Marine Ecosystem (a designation of the US National Oceanic and Atmospheric Administration, or NOAA) encompasses in its western areas the deepwater Aleutian Basin, with depths well in excess of 10,000 ft (3,048 m), as well as a large area of shallow, broad continental shelf in the east, where depths seldom exceed about 200 ft (60 m). The rapid change in bathymetry (depths) between these two regions brings cold, nutrient-rich bottom water onto the shallow shelf, which thus supports a vast food web.

The Bering Sea is host to a rich and diverse assemblage of marine life, with a large variety of marine mammals, fish, invertebrates, and seabirds living within its bounds. The region also supports very large and productive fisheries for groundfish, salmon, and crabs. Many of these species were severely overfished with the onset of mechanized fishing in the second half of the twentieth century, but improved international treaties and domestic management regimes have contributed to stabilization of these high-value fisheries over the past decade.

The borders of the Bering Sea are defined by mainland Alaska on the east, the Kamchatka Peninsula of Russia on the west, Alaska's Aleutian Island chain on the south, and the Bering Strait, which separates the Bering Sea from the Arctic Ocean, on the north. The sea's total area is roughly 849,425 square miles (2.2 million square kilometers). Its borders lie mostly within the 200-mi (322-km) Exclusive Economic Zone (EEZ) of either the United States (in the eastern Bering Sea) or Russia (in the western Bering Sea), largely because of the large number of islands dotting the continental shelf and Aleutian chain, each of which commands a 200-mi (322-km) EEZ radius. A small area of international waters in the center is commonly referred to as the "doughnut hole."

OCEANOGRAPHY AND MARINE LIFE

The physical oceanography of this region is particularly important. Sharp changes in depth from the

Aleutian Basin to the North American continental shelf create localized upwelling events, which bring nutrient-rich bottom water to the surface and also generate strong currents throughout the shelf portion of the system. Strong winds and frequent storms create hazardous conditions for seagoing vessels, with waves frequently exceeding 30 ft (10 m) and winds often exceeding 50 miles per hour (80 kilometers per hour), but also create a great deal of mixing of deep ocean nutrients into the well-lit surface layer. These conditions are ideal for the growth of phytoplankton —the tiny marine plants that are the basis of the Bering Sea food web. Seawater stratification from freshwater formation along the edge of the ice front, which advances and retreats seasonally, also creates areas of high primary productivity.

This productivity supports a range of marine life from varied pelagic (surface-layer-dwelling) and demersal (near-bottom-dwelling) fish species, to large populations of benthic (bottom-dwelling) invertebrates such as crabs, which are scavengers that feed on the large flux of organic matter raining down from the richly productive waters above. The western Bering Sea has the largest population of gadoid (cod-like) fishes in the world; the eastern Bering Sea is home to an array of fish species from cod and haddock to flatfish (flounder, sole, and halibut), to pelagic species like salmon. The system also supports an abundance of marine-mammal species, ranging from seals, walruses, and toothed whales (which feed on the abundant fish stocks) to baleen whales (which feed on krill), as well as pelagic invertebrates.

Coastal regions of the Bering Sea also support large populations (up to 20 million individuals) of predatory fish-eating birds, such as puffins, albatross, kittiwakes, and eagles. Polar bears, too, prey along these shores. All told, more than 15 mammal, 30 bird, and 400 fish species call the Bering Sea home—almost as many fish species as can be found in the Caribbean Sea.

FISHERIES AND CHALLENGES

With the onset of industrialized fishing in the post–World War II era, large mechanized fishing fleets from various nations (chiefly the United States, Russia, Canada, China, and Japan) rapidly increased fishing pressure in the Bering Sea. A broad range of techniques and target species are employed in this fishery, although the principal fisheries are groundfish (cod, haddock, pollock, and halibut), salmon, and crabs (king, tanner, bairdi, and opilio). Of these species, the most valuable—and therefore most overexploited—is the red king crab, the population of which was reduced to less than 10 percent of historical levels by the mid-1980s.

Early management efforts by the National Marine Fisheries Service were predominantly based on single-species catches, setting size limits and total allowable catches. In large part, these measures created a derby-type fishery, in which large numbers of entrants fished target species extremely heavily for a very short time. In some cases, the season for red king crab was as short as four days. This situation created additional unnecessary environmental pressure. It also endangered fishing crews, put excessive pressure on the canning industry, and reduced the total value of the catch by flooding the market.

Recently, management efforts have shifted to a system dominated by individual transferable quotas (ITQs), in which each entrant to the fishery is assigned a specific percentage of the total quota and can fish that quota over a longer period. This management strategy has worked very well in this system, improving the biomass of fisheries species and the total value of the catch while dramatically reducing the fatality rate among fishermen. The system has also dramatically reduced the number of entrants into the fishery, however, creating some contention due to the high unemployment rates among local fishers.

Although the Bering Sea is in many respects a "poster child" example of the effectiveness of ITQ fisheries, it is not without major environmental concerns. Availability of food in this system is highly correlated to temperature, with cold years being more favorable for high production than warm ones. Climate change presents a particular concern to the Bering Sea system for this reason, and also because the specialized physical oceanographic features of the system prevent species from simply moving north to remain in their preferred temperature range.

Some scientists estimate that primary productivity in this system has already dropped by 30 to 40 percent, and they have shown that high-energy phytoplankton species (diatoms) are being outcompeted by lower-energy species (coccolithophores) during warm years, which results in less energy moving up the food chain. Loss of sea ice is also problematic to species such as polar bears, which hunt along the ice edge. Although cycles of warm and cold years are part of natural climate variability, a continued warming trend may significantly affect diversity and productivity in this system throughout the twenty-first century.

—*Jason Krumholz*

Further Reading

Berger, Joel. *Extreme Conservation: Life at the Edges of the World.* U of Chicago P, 2018.
Demuth, Bathsheba. *The Floating Coast: An Environmental History of the Bering Strait.* W. W. Norton and Company, 2019.
Gates, Nancy. *The Alaska Almanac: Facts About Alaska.* 30th anniversary ed., Alaska Northwest Books, 2006.
Herman, Yvonne, editor. *The Arctic Seas: Climatology, Oceanography, Geology and Biology.* Springer Science & Business Media, 2012.
Hilborn, Ray, and Ulrike Hilborn. *Overfishing: What Everyone Needs to Know.* Oxford UP, 2012.
MacLeish, Summer. *Seven Words for Wind: Essays and Field Notes from Alaska's Pribilof Islands.* U of Alaska P, 2008.
Tang, Q., and K. Sherman, editors. *Large Marine Ecosystems of the Pacific Rim.* Blackwell Science, 1999.

BIOGENIC SEDIMENTARY ROCKS

Fields of Study: marine biology; oceanography; environmental chemistry; geology

ABSTRACT

Biogenic sedimentary rocks represent the accumulation of skeletal material, produced by the biochemical action of organisms, to form limestone, chert, and phosphorites, or the accumulation of plant material to form coal. Biogenic sedimentary rocks are economically important as sources of energy, fertilizers, and certain chemicals, and they contain evidence of former life on Earth.

KEY CONCEPTS

biogenic: produced as a product of the biological processes of living organisms

cathodoluminescence: the emission of light following irradiation by a beam of electrons

cryptocrystalline: consisting of submicroscopic crystals

diagenesis: the process by which an accumulation of sediment is converted to sedimentary rock through the combination of compaction, cementation and the development of crystal growths between the sediment grains

interbedding: formation of a layer, or bed, of one kind of sedimentary rock between two beds of sedimentary rock of a different nature

luminescence: the property of emitting light

LITHIFICATION AND COALIFICATION

Numerous organisms, particularly those living in the oceans, produce a shell or some type of skeletal material composed of minerals calcite or aragonite (polymorphs of calcium carbonate), silica (silicon dioxide), or phosphate minerals, principally apatite. After the organism dies and the soft tissue decays, the skeletal material accumulates as sediment. This sediment then undergoes lithification, a process involving compaction and chemical cementation. The time necessary for lithification is highly variable and ranges from a few decades to millions of years. Biogenic sedimentary rocks composed of calcite or aragonite minerals are commonly referred to as limestones; those composed of silica are cherts, and those composed of phosphate minerals are known as phosphorites or sedimentary phosphate deposits. Limestones and cherts may form as the result of direct chemical precipitation of calcium carbonate or silica from a saturated solution, but the primary mode of formation is biochemical. Limestones are the most abundant of the biogenic sedimentary rocks and, economically, the most important. Iron formations, our principal source of iron ore, were also precipitated biogenically early in the earth's history. They are not discussed in this article.

Coal also is a biogenic sedimentary rock. Most coal forms from the accumulation of woody plant material in anoxic (oxygen-lacking) environments, such as stagnant lakes, swamps, and bogs. The formation of coal is referred to as coalification and represents microbiological, physical, and chemical processes whereby the percentage of carbon is increased and the volatile content decreased. Limestones, cherts, phosphorites, and coal are often closely associated and may occur as alternating beds in a sequence of sedimentary rocks.

PHOSPHORITES

Because of their biological affinities, biogenic sedimentary rocks contain important evidence of former life on Earth, a record of organic evolution (changes in organisms through geologic time), and clues to the past environments in which the rocks were formed. Limestones may serve as reservoir rocks for oil and gas, which form from the chemical breakdown of microscopic organisms that accumulate with the sediment at the bottom of the ocean and in some lakes. Phosphates, which are contained in phosphorites, are one of the chief constituents of fertilizers and are widely used in the chemical industry. Important evidence of past swamp communities, including both plants and animals, has been preserved in coal beds.

Phosphorus is one of the essential elements of life, and when chemically linked with other elements to form phosphate minerals (especially apatite) it becomes one of the major constituents of all vertebrate skeletons and some invertebrate hard parts. Phosphate is a primary nutrient in marine waters and, therefore, controls organic productivity. Phosphorites often occur as nodules, which are highly variable in size and shape; they may be several inches (or centimeters) in diameter and up to a few feet (or meters) in length. These nodules are usually rich in vertebrate skeletal debris, especially that of fish, and fecal material. Bedded phosphorites are commonly interbedded with limestones. Thin beds of phosphorite are rich in bones, fish scales, and fecal material. Modern phosphorites are forming in areas where cold, nutrient-rich waters rise from the ocean depths toward the surface, such as off the west coasts of Africa and North and South America.

CHERTS

Cherts, which are occasionally referred to as flint or novaculite, are not nearly as abundant as limestones; however, chert has two properties that have made it

Black coal (bjdlzx via iStock)

one of the most important rocks to humans. Chert has the same chemical composition as quartz but is cryptocrystalline; that is, its crystals are submicroscopic in size. Like quartz, chert is very hard and it tends to break along smooth, curved surfaces. With a little practice, one can produce a sharp edge by chipping a piece of chert with another hard object. Although early humans used numerous rock types for tools and weapons, chert was the most important for the production of arrowheads, knives, and scrapers. Consequently, chert might be considered to have contributed as much to the rise of civilization as did the development of the steam engine. In addition to chert's utility, Precambrian cherts (1.8 to 3.4 billion years old) are interbedded with important iron deposits known as banded iron formations. These cherts contain evidence of some of the earliest life to evolve on Earth.

Chert, which may be any color depending on the presence of impurities, occurs principally as spherical to oblong nodules 0.8 to 9.8 inches (2 to 25 centimeters) in diameter in limestones and dolostones, or as thin beds 0.8 to 9.8 in (2 to 25 cm) deep. Nodular cherts are commonly parallel to bedding or stratification. Bedded cherts may extend laterally for great distances and are commonly interbedded with limestones and dolostones. Most geologists believe that nodular chert forms from the replacement of limestone and that bedded chert forms either by the complete replacement of carbonate-rich beds or by the diagenetic alteration of siliceous ooze.

The source of the silica may be the chemical precipitation of silicon dioxide, volcanic ash, or skeletal material. Biogenic silica is opaline, which means that it may contain up to 10 percent water. Only skeletal silica will be considered in this article. The principal biogenic sources of silica for chert are sponges, diatoms, and radiolarians. Skeletal materials from these organisms are common constituents of oceanic sediments. As this siliceous skeletal material accumulates, the opaline silica is diagenetically converted to crystalline opal and reprecipitated as bladed crystals. Continued diagenesis results in the formation of quartz chert, a mosaic of microscopic quartz crystals. The final diagenetic change often obliterates the shape or structure of the original skeletal material.

Sponges, which are abundant in most marine environments, contain as part of their supportive structure microscopic rods of opaline silica called "spicules." When the sponge dies, these spicules accumulate and become part of the sediment. During burial, the opaline silica undergoes diagenesis and precipitation of crystalline opal occurs within pores in the sediment. As crystalline opal is converted into quartz chert, chemical replacement of the surrounding sediment (usually calcite) occurs, resulting in the formation of chert nodules.

Radiolarians and diatoms are microscopic organisms with disk-shaped, elongate, or spherical tests (shells) with spines and surface ornamentation composed of opaline silica. Radiolarians occur as part of the marine zooplankton, and diatoms are part of the marine and nonmarine phytoplankton. Radiolarian and diatom oozes accumulate on the deep ocean floor. In time, these beds of silica-rich ooze are diagenetically converted into thin, bedded cherts. Diatoms, in contrast, may occur in great numbers in lakes and accumulate to form diatomaceous earths, or diatomites. Diatomaceous earths have a wide variety of uses, such as in filtering agents, absorbents, and abrasives.

LIMESTONES

Limestones are the most abundant of the chemical sedimentary rocks. A minor amount of limestone forms from the inorganic precipitation of calcite from seawater or the deposition of calcite in caves and around hot springs. The majority of limestones form as the result of biological and biochemical processes that produce aragonite or calcite. Later, this material becomes part of the carbonate sediment. Once deposited, the carbonate sediment is often modified by the chemical and physical processes of diagenesis.

Numerous animal phyla, such as mollusks, brachiopods, echinoderms, bryozoans, coelenterates, and certain protozoans, produce aragonite or calcite as part of their skeletal structure. The skeletal remains of these organisms are important constituents of carbonate sediments and, eventually, of most limestones. Some marine, bottom-dwelling algae secrete aragonite or calcite. These calcareous algae represent a significant contribution to carbonate sediment. Calcareous algae called "coccoliths" and calcareous foraminifera are found in great numbers in the plankton of the open ocean. Accumulation of this calcareous material at the bottom of the ocean contributes greatly to the formation of chalk, a type of biogenic

Limestone Conch Bar Caves on the island of Middle Caicos in the Turks and Caicos Islands (krw14 via iStock)

limestone. Modern marine organisms principally secrete aragonite; consequently, modern calcareous sediments are initially composed predominantly of aragonite. Aragonite, however, is unstable at low pressures and quickly undergoes diagenesis to calcite.

It is not uncommon to find that ancient limestones have been partially or completely transformed into dolostones; this process is known as dolomitization. Dolomitization occurs when calcium carbonate minerals are diagenetically converted into dolomite. Diagenesis may take place soon after the calcite or aragonite has been deposited or a long time after the deposition. The diagenesis is the result of the magnesium-bearing waters (seawater or percolating meteoric water) moving through the carbonate sediment or limestone. One important aspect of dolomitization is that it often leads to the formation of pores, cavities, and fissures, which enable the rock to serve as a reservoir for oil, gas, and water. Recent research suggests that much dolomitization is caused by the circulation of deep, warm fluids long after burial.

Under certain conditions, limestones are relatively easy to dissolve. In areas such as Florida or Kentucky, where limestones are abundant and there is adequate rainfall, cave systems develop, and failure of the roof of a cave system results in the formation of a sinkhole. Regions with extensive cave systems and related features are referred to as karst, and in places, particularly in China, have produced breathtaking formations and landscapes.

STUDY OF ROCK SAMPLES

Initially, the study of biogenic sedimentary rocks involves field investigations. Where outcrops of these rocks occur, geologists plot and map their distribution and thicknesses; note changes in the rocks' character, both vertically and horizontally; observe their association with other types of rocks; and collect samples of both the rock and any fossils that may be present. When drilling for oil, gas, or water, engineers and geologists study the same phenomena.

Samples, which are brought back to the laboratory, are analyzed with a vast array of techniques, including X-ray analysis, electron microprobe analysis, scanning electron microscopy, cathodoluminescence, and thin-section examination with a polarized microscope. Samples containing large fossils are observed with a magnifying glass or dissecting microscope. Some samples, particularly limestones, are dissolved in acid to recover both megafossils and microfossils. When samples have been ground thin enough to allow light to pass through the rock, microfossils can be observed with the help of microscopes. Scanning electron microscopy is particularly useful for observing extremely small fossils, especially radiolarians, diatoms, and coccoliths.

AREAS OF RESEARCH

In general, the chemistry of biogenic sedimentary rocks is simple; however, diagenesis can alter the structure, texture, and mineralogy of a sediment during its deposition, lithification, and burial. The analysis of diagenetic changes, particularly in limestones, is one of the most important avenues of investigation. Limestones are greatly influenced by diagenetic modifications and undergo changes in sediment size, porosity (the spaces between sediment grains), and mineralogy. Early diagenetic changes include the conversion of unstable aragonite—which is produced by most calcium-carbonate-producing organisms—into calcite and changes in the calcite's magnesium content. (Since the atoms of calcium and magnesium have similar properties, they can often substitute for each other in the crystal structure.)

Dolomitization may also occur. The replacement of carbonate minerals by silica, so that chert is formed, is another important aspect of diagenesis. Stages of diagenesis can best be observed by using cathodoluminescence, which bombards a thin slice of rock with an electron beam. This process causes minerals within the rock to luminesce (emit light energy for a short interval after the energy source has been removed). The luminescences indicate various stages of diagenesis and are particularly important in studying the details of cements and crystal growth. X-ray analysis and electron microprobe techniques allow scientists to detect subtle differences in chemistries and the presences of trace elements or rare-earth elements.

Another major area of research is the classification of biogenic sedimentary rocks. The classification of limestones is of primary importance. Limestones have been classified using several criteria, but in general, classifications have focused on chemical and mineralogical composition; fabric features, such as fossils and cements; and special physical parameters, such as porosity. When examining chert, the geoscientist must ask: What was the source of the silica (volcanic ash, skeletal grains, chemical precipitate)? What was the time and rate of conversion of siliceous ooze to chert? What was the environment of deposition?

ECONOMIC AND EVOLUTIONARY SIGNIFICANCE

Biogenic sedimentary rocks are among the most important of all sedimentary rocks. Because of their biological affinities, they have preserved an important record of past life on Earth. In addition, some of these rocks were used for toolmaking by early humans; others are important resources for construction materials, fertilizers, and chemicals. Oil and natural gas, which also have biological affinities, are inseparably linked to biogenic sedimentary rocks.

Although not widespread, biogenic phosphorites are economically important, particularly to the fertilizer and chemical industries. Biogenic chert was of major importance in the past. Early humans used chert, or flint, to make tools and weapons. Biogenic limestone is the most common and widespread biogenic sedimentary rock. Economically, limestones are very important. Hydrocarbons (oil and gas) are commonly recovered from porous limestones and dolostones, and in some areas, limestones are reservoirs for groundwater supplies. Limestones often serve as host rocks for important mineral deposits, such as lead. Limestones are quarried as building stone; crushed to form construction materials, such as gravel; or processed into lime and cement. Biogenic limestones are the most important sedimentary rock containing fossils and, therefore, are the most important record of the evolution of life.

—*Larry E. Davis*

Further Reading

Blatt, Harvey, Robert J. Tracy, and Brent Owens. *Petrology: Igneous, Sedimentary, and Metamorphic*. 3rd ed., W. H. Freeman, 2005.

Boggs, Sam, Jr. *Petrology of Sedimentary Rocks*. Cambridge UP, 2009.

Bottjer, David J. *Paleoecology: Past, Present and Future*. Wiley-Blackwell, 2016.

Chernicoff, Stanley. *Geology: An Introduction to Physical Geology*. 4th ed., Prentice-Hall, 2006.

James, Noel P., and Brian Jones. *Origin of Carbonate Sedimentary Rocks*. John Wiley & Sons, 2015.

Middleton, Gerard V., editor. *Encyclopedia of Sediments and Sedimentary Rocks*. Springer, 2003.

Nichols, Gary. *Sedimentology and Stratigraphy*. John Wiley & Sons, 2013.

Riding, Robert E., and Stanley M. Awramik. *Microbial Sediments*. Springer-Verlag, 2010.

Stow, Dorrik A. V. *Sedimentary Rocks in the Field: A Color Guide*. Academic Press, 2006.

Tucker, Maurice E. *Sedimentary Rocks in the Field*. 4th ed., John Wiley & Sons, 2011.

BLACK SEA

Fields of Study: oceanography; geology; marine ecology

ABSTRACT

The Black Sea is a nearly landlocked body of water that was formerly linked to its eastern neighbor, the Caspian Sea. Geological changes separated it from the Caspian and created two narrow waterways, the Bosphorus Straits and the Dardanelles Straits, connecting it with the Mediterranean. Some of its most interesting physical features make it unique among the world's small- to medium-sized seas.

KEY CONCEPTS

cold zone: a thermally isolated band of water at a depth below the surface that is considerably colder than the water above and below it

hydrogen sulfide: having the molecular formula H_2S; as a gas, familiar as the compound responsible for the odor of rotten eggs; dissolved in water, acts as a weak acid, but can also react with dissolved oxygen to form the much stronger sulfuric acid molecule H_2SO_4; this reaction acts to deplete the amount of dissolved oxygen in the water

salinity count: the concentration of dissolved salt stated relative to the amount of water, as for example 17 parts per thousand, corresponding to a salt solution of 0.017 percent

GENERAL FEATURES

The Black Sea, with a surface area of more than 160.000 square miles (or 400,000 square kilometers) of brackish water, is located on the southern margin of the Eurasian landmass. Its shores include territories in several countries. From the southern and southwestern coast in Turkey, the shoreline proceeds in a clockwise direction through Bulgaria, Romania, Ukraine, Russia, and Georgia.

The name "Black Sea" is a literal translation of the Turkish *Karadeniz*. This name, suggesting a somewhat ominous image (particularly in comparison to the Turkish name White Sea, or *Akdeniz*, referring to the Mediterranean), dates from the Turkish occupation of the southern shores of the Black Sea between the thirteenth and the fifteenth centuries. The sense of blackness refers not to any impression of the color of its waters, but rather to a presumed inhospitable environment. When the Turks captured the Greek-named *Pontus Euxinus* (meaning "hospitable sea") from the Byzantine Greeks, they adopted a much earlier Greek denomination for the same body of water, namely *Pontus Axeinus* (or "inhospitable sea").

The major ports of the Black Sea littoral states are Trabzon, Samsun, Sinop, and Zonguldak (Turkey); Burgas and Varna (Bulgaria); Odessa (Ukraine); Sevastopol (Crimea); and Batumi (Georgia).

GEOLOGICAL ORIGINS

Geologically, the Black Sea appears to be a basin left behind by the retreating ancient Tethys Sea, a process that took place over about 200 million years. This long geological evolution reached a significant stage about 50 million years ago as the upward thrust of the Anatolian landmass (Turkey) and Western Iran split the Caspian Sea basin off from the Black Sea. Further mountain-forming activity, including the upheavals that created the Pontic, Caucasus, Carpathian, and Crimean landmasses, also affected the basin that eventually held the waters of the Black Sea. Finally, the geology associated with the Turkish Straits, which connect the Black Sea to the Mediterranean via the straits of the Bosphorus and the Dardanelles, with the Sea of Marmara in between, appears to have been the most recent stage contributing to the current configuration of the Black Sea, occurring less than 10,000 years ago.

The total area of the Black Sea is about 162,935 sq mi (422,000 sq km), although some sources put it at over 168,340 sq mi (436,000 sq km). At its widest, the sea extends more than 730 mi (1,175 km) from west to east. Its narrowest north-to-south point, running from the tip of the Crimean Peninsula to the Turkish coast at the Kerempe Burmi Cape, is about 162 mi (260 km).

The sea is fed by a number of important freshwater rivers. The most famous of these, the Danube, enters the sea along the coast of Romania. The Dniester, which flows parallel to the northern fringes of the Carpathian Mountain range, ends at the delta that feeds the farthest northwest area of the Black Sea. The Dniester Delta forms part of the major port complex of Odessa in Ukraine. To the east of the Dniester is the Dnieper, the last major river to enter the north coast of the Black Sea proper before its coastline is broken by the Crimean landmass and the waters and rivers associated with the Sea of Azov.

The Danube spills into the Black Sea (NASA via Wikimedia Commons)

The depth of the sea varies; the deepest area, in the southern part of the basin about midway from east to west, measures more than 1.3 mi (2,120 m) deep.

The fact that the deeper zones of the Black Sea receive very low levels of ventilation (circulation of oxygen) means that these areas do not support any significant plant growth or fish populations. In the deep waters, a high concentration of hydrogen sulfide causes whatever oxygen that exists to ionize. These conditions mean that only certain species of bacteria can survive at lower depths. The beginnings of low oxygen levels, with consequential decreases in marine life, occur at depths of around 495 ft (150 m). Where waters are sufficiently ventilated, there are substantial numbers of fish, although only a few species are sought by commercial fisheries.

The most outstanding irregularity at surface level along the coast is the Crimean Peninsula, which juts southward into the Black Sea from the mainland just east of the Dnieper River Delta. The Crimean Peninsula is the site not only of the traditional seaside vacation town of Yalta (where the Western allies met with the Soviet Union's Joseph Stalin (1879–1953) near the end of World War II) but also of the strategic port of Sevastopol, founded by Russian czars as a strategic maritime outpost for Russian access to the Mediterranean. Part of Russia since the eighteenth century, the Crimea, as the peninsula is also known, was made part of Ukraine in the 1950s but annexed again by Russia in 2014, resulting in armed conflict that is ongoing.

There are no major islands in the Black Sea. Most geographical accounts overlook the few smaller islands, mentioning only Zmiyini, located east of the Danube River delta, and Berezan, near the mouth of the Dniester.

The Kerch Strait, located to the east of the Crimean Peninsula where a city by that name stands, connects the Black Sea to the Sea of Azov. The latter, which forms part of the same residual basin complex left behind by the receding Tethys Sea, encompasses approximately 15,058 sq mi (39,000 sq km). The Sea of Azov receives the waters of two major rivers, the Don and the Kuban.

With the one major exception of the Crimean Peninsula, the coastline of the Black Sea is quite regular, reflecting its geological origins as a residual basin. The shallow coastal shelf is seldom deeper than 295 ft (90 m). It can be very wide but generally extends out from the shore about 6.8 mi (11 km). If one takes the average of 6.8 mi (11 km) and adds areas where the shelf is considerably wider, these shallow waters make up almost one-fourth of the total surface of the Black Sea.

TIDAL PHENOMENA AND CURRENTS

There is little regularly predictable tidal movement in the Black Sea. When moderate tidal movements do occur, their levels can be affected by a sudden rise or fall of atmospheric pressures in different regions either over the sea or on the landmasses around it. Only rarely do such tides cause dangers for boats or pier structures along the shorelines.

Satellite view of the Black Sea, Mediterranean Sea, and Caspian Sea (Norman Kuring/NASA via Wikimedia Commons)

Currents are strongest near the shores of the sea, where the water is not very deep. Such currents run an average of between 15.7 to 19.7 inches (40 and 50 cm) per second. In areas of deeper water, very slow currents are difficult to measure but generally do not run faster than 0.98 to 1.97 in (2.5 to 5 cm) per second. A particular phenomenon of currents exists in the Bosphorus Straits zone leading from the Black Sea to the Sea of Marmara. Here, a current near the surface carries water from the Black Sea past the city of Istanbul into the Sea of Marmara and eventually into the Mediterranean. At deeper levels, the current moves in the opposite direction, carrying water with higher saline content that originates in the Mediterranean into the Black Sea.

The salinity level of the Black Sea is roughly one-half that of the world's major oceans. Salinity counts (relative proportions of salt to water) near the surface are between seventeen and eighteen parts of salt per thousand parts of water. In the cold band that lies beneath the first layer of Black Sea surface water, the salinity count increases rapidly, reading twenty-one parts per thousand. The proportion of salt to water continues to increase as one descends to deeper zones, but the rate of increase is much more gradual than the apparently sudden increase that occurs in the intermediate-level cold band. The highest salinity count in the deep water is about thirty parts salt to one thousand parts water. Measurements are higher in the Bosphorus Straits, but this phenomenon is linked to the transfer of Mediterranean water into the Black Sea via subsurface currents.

The surface temperature of the Black Sea varies considerably from season to season. Winter temperatures in the northern zones hover just below 32 degrees Fahrenheit (0 degrees Celsius). Where the inflow of freshwater from major rivers is substantial, this means some freezing can occur near the coast. The saline content of most zones is sufficiently high, however, to prevent widespread freezing of Black Sea waters. In the southern half of the sea, winter temperatures are more moderate, reaching about 50 degrees Fahrenheit (10 degrees Celsius) in the southeast corner near the Turkish border with Georgia. Summer brings surface temperatures up to nearly 69.8 degres Fahrenheit (21 degrees Celsius). The warming effect, however, is limited to depths down to about 148 feet (45 meters). Below this level, down to about 295 ft (90 m), there is a more or less constant cold band, where temperatures remain at about 80.6 degrees Fahrenheit (7 degrees Celsius) throughout the year. In fact, deeper waters may even be a bit warmer than the cold band, but this phenomenon stems from an insulating effect and is not tied to rising temperatures at the surface.

NATURAL RESOURCES AND ECONOMIC POTENTIALS

Until the late twentieth century, perhaps the best-known natural resource exploited on an industrial scale in the Black Sea region was coal, shipped in large quantities from the Turkish port of Zonguldak. Caspian Sea petroleum transit shipments from the Black Sea port of Batumi were always a factor in the economic history of the Black Sea during the period of the Soviet Union, but their importance grew rapidly in the last decade of the twentieth century, as newly opened free international markets for Caspian Sea oil provided incentives for expanded construction of pipelines and port facilities in the southeast coastal region of the Black Sea.

Fishing is the most traditional form of economic activity in the Black Sea. It appears to have a limited capacity for expansion beyond levels reached by the middle of the twentieth century. The most common species of fish in the Black Sea are Mediterranean varieties, some of which actually move into and out of the Black Sea via the Sea of Marmara from season to season. Specialists have noted nearly two hundred different species of fish, but only about forty varieties are exploited commercially. The most important locally processed fish include horse mackerel and anchovies.

SIGNIFICANCE

Study of the Black Sea region can be particularly useful for comparison with small to moderately large inland seas that exist around the world. The scale of ecological considerations in such circumstances is drastically different from what would apply in the study of large oceans. In this case, two comparable bodies of water, the Black Sea and the Caspian Sea, are located much in the same region but demonstrate quite different characteristics. On the one hand, for example, the Black Sea has not been seriously affected by large-scale petroleum development, something that has had major effects on the ecology of the Caspian Sea. On the other hand, the Black Sea has considerably more large urban port cities around its coasts. Examination of pollution levels and marine biology in the Black Sea can lead to a better understanding of its own environmental prospects as well as those of similar bodies of water in other areas of the world.

—*Byron D. Cannon*

Further Reading

Dumont, Henri, Tamara A. Shiganova, and Ulrich Niermann, editors. *Aquatic Invasions in the Black, Caspian, and Mediterranean Seas*. Kluwer Academic Publishers, 2004.

FAO-UN. *The State of Mediterranean and Black Sea Fisheries 2018*. Food and Agriculture Organization of the United Nations, 2019.

Izdar, E., and James W. Murray. *Black Sea Oceanography*. Springer Science & Business Media, 2012.

King, Charles. *The Black Sea: A History*. Oxford UP, 2005.

Kostianoy, Andrey G., and Alexey N. Kosarev. *The Black Sea Environment*. Springer, 2007.

Ryann, Amy L., and Nathan J. Perkins. *The Black Sea: Dynamics, Ecology and Conservation*. NOVA Science Publishers, 2011.

Schwartzstein, Peter. "The Black Sea Is Dying, and War Might Push It over the Edge." *Smithsonian*, 11 May 2016, www.smithsonianmag.com/science-nature/black-sea-dying-and-war-might-push-it-over-edge-180959053.

Sorokin, I. O. U. *The Black Sea: Ecology and Oceanography*. Backhuys Publishing, 2002.

Sosson, M, et al., editors. *Sedimentary Basin Tectonics from the Black Sea and Caucasus to the Arabian Platform*. Special Publication 340. Geological Society of London, 2010.

Vespremeanu, Emil, and Mariana Golumbeanu. *The Black Sea: Physical, Environmental and Historical Perspectives*. Springer, 2018.

Yanko-Hombach, Valentina, et al., editors. *The Black Sea Flood Question: Changes in Coastline, Climate, and Human Settlement*. Springer, 2007.

BLEACHING OF CORAL REEFS

Fields of Study: marine biology; marine ecology; marine conservation; oceanography

ABSTRACT

Coral bleaching is a phenomenon in which normally brightly colored, hard, tropical corals lose their pigments and appear bright white. Bleached corals are starved and weakened, and will eventually die if they do not recover. Coral bleaching occurs when zooxanthellae, symbiotic algae that normally live in the tissues of the transparent corals, are expelled by the corals. Coral bleaching has many causes, both natural and anthropogenic, but large-scale mass bleaching events are primarily linked with rising ocean temperatures.

KEY CONCEPTS

anthropogenic: caused by or existing as a consequence of human activities

bleaching: the loss of colored symbiotes from coral polyps, leaving the innate white appearance of the coral structures

coelenteron: the central tubelike structure found in many invertebrate species, which is essentially their digestive tract

predation: the exploitation of "prey" pecies for food by "predator" species

salinity: the content of dissolved salt(s) in water

CORAL BIOLOGY AND CORAL-ALGAL SYMBIOSIS

Corals, like hydras, jellyfish, and sea anemones, are cnidarians: living organisms that belong to the phylum *Cnidaria*. All cnidarians have a single body cavity, known as a coelenteron, through which they ingest food and release waste.

Members of this phylum are also characterized by radial symmetry and the presence of stinging cells on their surfaces known as nematocysts. These nematocysts are used to poison and capture prey. The term "coral" is most often applied to hermatypic corals, which are corals that build hard skeletons around their soft bodies by secreting calcium carbonate (limestone); they also grow and reproduce in large colonies, forming impressively intricate branching structures called reefs. Reefs typically comprise thousands of individual coral polyps, each of which is a single invertebrate marine animal.

Most hermatypic corals share another critical attribute: Some of their cells contain vast numbers of a particular type of single-celled organism. These organisms are tiny algae known as zooxanthellae. A single square centimeter of tissue from the gastrodermis (the lining of the digestive cavity) of a coral may contain millions of zooxanthellae. The relationship between the corals and the zooxanthellae is generally described as being symbiotic, or mutually beneficial. The coral provides the algae that inhabit its cells with a protected environment, and the waste products of its digestion also provide the algae with beneficial inorganic nutrients, such as nitrates and phosphates.

In return, the zooxanthellae remove waste products from the coral's tissues, serve as a crucial source of oxygen, and produce carbohydrates that the coral can use to manufacture fats and its protective calcium carbonate skeleton. As a result of their photosynthesis, the zooxanthellae also provide the coral with many useful organic products, including nitrogen compounds, vitamins, amino acids, and nutrients that are essential for supporting and speeding the coral's growth.

Color is another byproduct of the symbiotic relationship between corals and zooxanthellae. Although the skeletons of hard tropical corals appear to be beautifully and brightly colored, they are actually white. Their apparent colors come from vast quantities of fluorescent pigments produced by the algae; these striking pigments can be seen through the clear tissues that make up each coral polyp's soft body.

CAUSES OF CORAL BLEACHING

Coral bleaching is the phenomenon in which a coral expels most or all of the zooxanthellae that normally inhabit its tissue. When this happens, the coral is left appearing bright white. This is a generalized reaction that occurs when corals, and therefore the algae they contain, are subject to stress.

Stressed zooxanthellae produce a form of oxygen that is toxic to the corals. Rather than suffer the cellular damage that would result from this toxic compound, corals are forced to expel the zooxanthellae. Most coral colonies go through natural, temporary periods of mild bleaching, often in response to seasonal alterations in water temperature, but they recover when conditions return to normal.

Other natural environmental stresses that can lead to bleaching include increased predation, changes in the salinity or oxygen levels of ocean water, or an outbreak of a coral disease, such as a bacterial infection. Two pathogens that have been shown to cause bleaching in corals are the bacteria *Vibro shiloi* and *V. patagonica*.

Normally, coral reefs display a high level of resilience to such localized environmental stresses. Mass bleaching events (when bleaching occurs to vast swathes of reef at the same time) rarely occur in nature, and in time corals that have expelled their zooxanthellae will typically regain them and recover their health. Since about 1980, however, human activities have reduced the resilience of coral reefs and have introduced new and much more powerful causes of bleaching. These harmful activities include pollution, oil spills, overfishing, and stress caused by reef tourism.

The most significant cause of mass coral bleaching in recent years, however, has been increased sea surface temperatures (SST). Elevations in SST of as little as 33.8 to 35.6 degrees Fahrenheit (1 to 2 degrees Celsius) above the long-term average, if they persist for between one to two months, can lead to bleaching. Widespread increases in SST are occurring as a result of global climate change.

The Great Barrier Reef, off of the coast of Australia, has experienced some of the greatest damage linked to increased water temperatures caused by global warming in the twenty-first century. For the first time in history, in 2017, the Great Barrier Reef was reported to have undergone a severe coral bleaching event for the second year in a row; in 2016, 90 percent of a portion of the reef 1,430 miles (2,300 kilometers) long had been damaged. In early 2018, in a study published by the journal *Science* in which researchers analyzed one hundred tropical reefs worldwide, it was reported that the average amount of time between severe bleaching events had decreased, from 27 years to 5.9 years. Therefore, as global warming patterns continued, scientists have grown increasingly concerned that coral reefs will, at some point, not have enough time to recover after a bleaching event.

NOAA scuba diver surveying bleached corals (National Oceanic and Atmospheric Administration via Wikimedia Commons)

According to the 2007 Intergovernmental Panel on Climate Change, unequivocal evidence now exists that both the atmosphere and oceans are warming, primarily as a result of anthropogenic greenhouse gas emissions, especially carbon dioxide. Temperature records indicate that the past twenty years with the warmest sea temperatures on record have all occurred since about 1980.

Rising global temperatures also are being linked with an increase in the frequency of El Niño Southern Oscillation (ENSO) events. ENSO events are fluctuations in the surface temperature and pressure of the Indian and Pacific oceans; the events are named for a warm-water current that flows along the equator toward the coast of Ecuador and Peru every few years. ENSO events can lead to such extreme weather events as floods and droughts and also may cause mass coral bleaching.

EFFECTS OF CORAL BLEACHING ON OCEAN ECOSYSTEMS AND HUMANS

In addition to losing their color, corals that lack zooxanthellae are deprived of up to 90 percent of the nutrients and oxygen they require for their survival, reproduction, and growth. The stress caused by zooxanthellae expulsion also leaves corals susceptible to disease and predation. Corals that have lost all their algae and do not recover them can typically survive a few months only before starving.

Research has shown that branching corals and plate corals, which grow quickly, are more susceptible to bleaching and less likely to survive it. Massive corals, which grow more slowly, are more resilient. Thus, even when coral reefs do eventually recover from bleaching events, their overall genetic and species diversity is likely to be reduced.

Bleaching also has a profound negative impact on the numerous other organisms that live in and around the coral ecosystem. For example, coral reef degradation has been shown to cause declines in fish communities. A variety of studies of coral reefs around the world indicate that bleaching events result in marked declines in the numbers of species that depend heavily on coral reefs for food and shelter. Some coral species also have been shown to disappear completely from particular areas, a phenomenon known as a local extinction, when reefs undergo mass bleaching. These types of negative effects often persist for many years after the bleaching event itself.

Finally, bleaching has a negative impact on human communities. Weakened reefs provide less protection from waves, which can lead to faster and more damaging erosion of the shorelines near coral reefs. In addition, colorless reefs are less likely to attract dive tourism, which may make it more difficult for local communities to sustain their economic activities.

—*M. Lee*

Further Reading

Camp, Emma F., Verena Schoepf, Peter J. Mumby, and David J. Suggett, editors. *The Future of Coral Reefs Subject to Rapid Climate Change: Lessons from Natural Extreme Environments*. Frontiers, 2019.

Crabbe, M. James C., et al. "The Impact of Weather and Climate Extremes on Coral Growth." *Climate Extremes and Society*, edited by Henry F. Diaz and Richard J. Murnane, Cambridge UP, 2011.

Davidson, Helen. "Coral Reef Bleaching 'the New Normal' and a Fatal Threat to Ecosystems." *The Guardian*, 4 Jan. 2018, www.theguardian.com/environment/2018/jan/04/coral-reef-bleaching-the-new-normal-and-a-fatal-threat-to-ecosystems.

Grimsditch, Gabriel D., and Rodney V. Salm. *Coral Reef Resilience and Resistance to Bleaching*. International Union for Conservation of Nature, 2006.

Hutchings, P. A., et al., editors. *The Great Barrier Reef: Biology, Environment, and Management*. Springer, 2008.

Kennedy, Merrit. "Great Barrier Reef Hit By Bleaching for the Second Year in a Row." *NPR*, 10 Apr. 2017, www.npr.org/sections/thetwo-way/2017/04/10/523254085/great-barrier-reef-hit-by-bleaching-for-the-second-year-in-a-row.

Lesser, Michael P. "Coral Bleaching: Causes and Mechanisms." *Coral Reefs: An Ecosystem in Transition*, edited by Zvy Dubinsky and Noga Stambler, Springer, 2011.

———. "Coral Reef Bleaching and Global Climate Change: Can Corals Survive the Next Century?" *PNAS*, vol. 104, no. 13, 2007, pp. 5259–60.

National Academy of Sciences, Engineering and Medicine, 2019. *A Research Review of Interventions to Increase the Persistence and Resilience of Coral Reefs*. National Academies Press, 2019.

Rosenberg, Eugene, and Yossi Loya, editors. *Coral Health and Disease*. Springer, 2004.

Van Oppen, Madeleine J. H., and J. M. Lough, editors. *Coral Bleaching: Patterns, Processes, Causes, and Consequences*. Springer, 2009.

C

CALCULUS

Fields of Study: mathematics; marine science

ABSTRACT

Calculus is the study of functions and change. It is the bridge between the elementary mathematics of algebra, geometry, and trigonometry, and advanced mathematics. Knowledge of calculus is essential for those pursuing study in chemistry, engineering, medicine, physics, and all related fields. Computer models of fluids in motion, such as ocean currents and rivers, rely heavily on the calculus of fluid dynamics. Generally, calculus is employed to solve a large variety of optimization problems; one example is the so-called least squares solution method commonly used in statistics and elsewhere. The least squares function best fits a set of data points and can then be used to generalize or predict results based on that set.

KEY CONCEPTS

definite integral: the value of the parent function of a derivative determined as the difference between two specific values of the parent function, e.g., the definite integral of $y = 2x$ between $x = 4$ and $x = 2$ is $4^2 - 2^2 = 16 - 4 = 12$. The integral notation specifies the starting and ending values of the variable of integration.

derivative function: a function that describes the change of another "parent" function with respect to a particular variable, e.g., the derivative of $y = x^2$ is $y' = 2x$

integral function: generally, the parent function of a derivative function, e.g., the integral function of $y = 2x$ is $y = x^2 + c$, where c is a constant value with no derivative function

modeling: the use of computer programs that employ data inputs, calculations and algorithms to describe conditions and predict the behaviors of a particular phenomenon that changes with respect to one or more variables

DEFINITION AND BASIC PRINCIPLES

Calculus is the study of functions and their properties. Calculus takes a function and investigates it according to two essential ideas: rate of change and total change. These concepts are linked by their common use of calculus's most important tool, the limit concept. It is the use of this tool that distinguishes calculus from elementary branches of mathematics such as algebra, geometry, and trigonometry. In elementary mathematics, one studies problems such as "What is the slope of a line?" or "What is the area of a parallelogram?" or "What is the average speed of a trip that covers three hundred miles in five and a half hours?" Elementary mathematics provides methods or formulas that can be applied to find the answer to these and many other problems. However, if the line becomes a curve, how is the slope calculated? What if the parallelogram becomes a shape with an irregularly curved boundary? What if one needs to know the speed at an instant, and not as an average over a longer time period?

Calculus answers these harder questions by using the concept of limit. The limit is found by making an approximation to the answer and then refining that approximation by improving it more and more. If there is a pattern leading to a single value in those improved approximations, the result of that pattern is called the "limit." Note that the limit may not exist in some cases. The limit process is used throughout calculus to provide answers to questions that elementary mathematics cannot handle.

The "derivative" of a function is the limit of average slope values within an interval as the length of the interval approaches zero. The integral calculates the total change in a function based on its rate of change function.

BACKGROUND AND HISTORY

Calculus is usually considered to have been developed in the seventeenth century, but its roots were formed much earlier. In the sixteenth century, Pierre de

Fermat (1601–1665) did work that was very closely related to calculus's differentiation (the taking of derivatives) and integration. In the seventeenth century, René Descartes (1596–1650) founded analytic geometry, a key tool for developing calculus.

However, it is Sir Isaac Newton (1642–1727) and Gottfried Wilhelm Leibniz (1646–1716) who share the credit as the (independent) creators of calculus. Newton's work came first but was not published until 1736, nine years after his death. Leibniz's work came second but was published first, in 1684. Some accused him of plagiarizing Newton's work, although Leibniz arrived at his results by using different, more formal methods than Newton employed.

Both men found common rules for differentiation, but Leibniz's notation for both the derivative and the integral are still in use. In the eighteenth century, the work of Jean le Rond d'Alembert (1717–1783) and Leonhard Euler (1707–1783) on functions and limits helped place the methods of Newton and Leibniz on a firm foundation. In the nineteenth century, Augustin-Louis Cauchy (1789–1857) used a definition of limit to express calculus concepts in a form still familiar more than two hundred years later. German mathematician Bernhard Riemann (1826–1866) defined the integral as a limit of a sum, the same definition learned by calculus students in the twenty-first century. At this point, calculus as it is taught in the first two years of college reached its finished form.

HOW IT WORKS

Calculus is used to solve a wide variety of problems using a common approach. First, one recognizes that the problem at hand is one that cannot be solved using elementary mathematics alone. This recognition is followed by an acknowledgment: there are some things known about this situation, even if they do not provide a complete basis for solution. Those known properties are then used to approximate a solution to the problem. This approximation may not be very good, so it is refined by taking a succession of better and better approximations. Finally, the limit is taken, and if the limit exists, it provides the exact answer to the original problem.

One speaks of taking a limit of a function $f(x)$ as x approaches a particular value, for example, $x = a$. This means that the function is examined on an interval around, but not including $x = a$. Values of $f(x)$ are taken on that interval as the varying x values get closer and closer to the target value of $x = a$. There is no requirement that $f(a)$ exists, and many times it does not. Instead the pattern of functional values is examined as x approaches a. If those values continue to approach a single target value, it is that value that is said to be equal to the limit of $f(x)$ as x approaches a. Otherwise, the limit is said not to exist. This method is used in both differential calculus and integral calculus.

Differential Calculus. Differentiation is a term used to mean the process of finding the derivative of a function $f(x)$. This new function, denoted $f'(x)$, is said to be "derived" from $f(x)$. If it exists, $f'(x)$ provides the instantaneous rate of change of $f(x)$ at x. For curves (any line other than a straight line), the calculation of this rate of change is not possible with elementary mathematics. Algebra is used to calculate that rate between two points on the graph, then those two points are brought closer and closer together until the limit determines the final value.

Shortcut methods were discovered that could speed up this limit process for functions of certain types, including products, quotients, powers, and trigonometric functions. Many of these methods go back as far as Newton and Leibniz. Using these formulas allows one to avoid the more tedious limit calculations. For example, the derivative function of sine x is proven to be cosine x. If the slope of sine x is needed at $x = 4$, the answer is known to be cosine 4, and much time is saved.

Integral Calculus. A natural question arises: If $f'(x)$ can be derived from $f(x)$, can this process be reversed? In other words, suppose an $f(x)$ is given. Can an $F(x)$ be determined whose derivative is equal to $f(x)$? If so, the $F(x)$ is called an antiderivative of $f(x)$, and the process of finding $F(x)$ is called integration. In general, finding antiderivatives is a harder task than finding derivatives. One difficulty is that constant functions all have derivatives equal to zero, which means that without further information, it is impossible to determine which constant is the correct one. A bigger problem is that there are functions, such as sine (x^2), whose derivatives are reasonably easy to calculate but for which no elementary function serves as an antiderivative.

The definite integral can be thought of as an attempt to determine the amount of area between the graph of $f(x)$ and the x-axis, usually between a left and right end point. This cannot typically be answered using elementary mathematics because the shape of the graph can vary widely. Riemann proposed approximating the area with rectangles and then improving the approximation by having the width of the rectangles used in the approximation get smaller and smaller. The limit of the total area of all rectangles would equal the area being sought. It is this notion that gives integral calculus its name. By summing the areas of many rectangles, the many small areas are integrated into one whole area.

As with derivatives, these limit calculations can be quite tedious. Methods have been discovered and proven that allow the limit process to be bypassed. The crowning achievement of the development of calculus is its fundamental theorem: The derivative of a definite integral with respect to its upper limit is the integrand evaluated at the upper limit; the value of a definite integral is the difference between the values of an antiderivative evaluated at the limits. If one is looking for the definite integral of a continuous $f(x)$ between $x = a$ and $x = b$, one need only find any antiderivative $F(x)$ and calculate $F(b) - F(a)$.

APPLICATIONS AND PRODUCTS

Optimization. A prominent application of the field of differential calculus is in the area of optimization, either maximization or minimization. Examples of optimization problems include: What is the surface area of a can that minimizes cost while containing a specified volume? What is the closest that a passing asteroid will come to Earth? What is the optimal height at which paintings should be hung in an art gallery? (This corresponds to maximizing the viewing angle of the patrons.) How shall a business minimize its costs or maximize its profits?

All of these can be answered by means of the derivative of the function in question. Fermat proved that if $f(x)$ has a maximum or minimum value within some interval, and if the derivative function exists on that interval, then the derivative value must be zero. This is because the graph must be hitting either a peak or the bottom of a valley and has a slope of zero at its highest or lowest points. The search for optimal values then becomes the process of finding the correct function modeling the situation in question, finding its derivative, setting that derivative equal to zero, and solving. Those solutions are the only candidates for optimal values. However, they are only candidates because derivatives can sometimes equal zero even if no optimal value exists. What is certain is that if the derivative value is not zero, the value is not optimal.

The procedure discussed here can be applied in two dimensions (where there is one input variable) or three dimensions (where there are two input variables).

Surface Area and Volume. If a three-dimensional (3D) object can be expressed as a curve that has been rotated about an axis, then the surface area and volume of the object can be calculated using integrals. For example, both Newton and Johannes Kepler (1571–1630) studied the problem of calculating the volume of a wine barrel. If a function can be found that represents the curvature of the outside of the barrel, that curve can be rotated about an axis and pi (p) times the function squared can be integrated over the length of the barrel to find its volume.

Hydrostatic Pressure and Force. The pressure exerted on, for example, the bottom of a swimming pool of uniform depth is easily calculated. The force on a dam due to hydrostatic pressure is not so easily computed because the water pushes against it at varying depths. Calculus discovers the answer by integrating a function found as a product of the pressure at any depth of the water and the area of the dam at that depth. Because the depth varies, this function involves a variable representing that depth.

Fluid Dynamics. One of the most complex uses of calculus is in the description of moving fluids such as ocean currents, rivers and gases or liquids moving through a conduit. Two situations apply. One considers the motion of the particles making up the fluid as it moves relative to a "stationary" solid such as a building, the hull of a ship, an island or the interior walls of the surrounding conduit. The other considers the motion of a solid, such as a submarine or a propeller blade, as it moves through the fluid. There is a point at which both situations become identical, since the motions involved are relative motions with respect to each other. This can be envisioned by standing on a bridge over a river and peering down at the water flowing around the base of the bridge support, and realizing that it looks just like the bow of a ship plowing through the water. Precise solutions to fluid dynamics problems are not satisfied by the general description of the event. For example, determining that a current moves a certain amount of water in a certain amount of time does not describe the actual movement of the current. A precise solution requires the accurate descriptions of the motion of the individual component parts of the current that then combine to yield the actual precise description of the motion of the current. In the extreme example of this, one would have to solve the equation of motion of each individual molecule of water in a current like the Gulf Stream in order to determine the exact description of the motion of the Gulf Stream. Calculus that describes particle motion, as well as other properties such as pressure variations, is essential to any computer modeling application that is used to analyze and predict fluid motion. However, the same basic principles are applicable to any phenomenon that can be described as "flowing," such as shoals of fish, blood cells in veins and arteries, the growth of populations and the movement of traffic on highways.

Arc Length. Algebra is able to determine the length of a line segment. If that path is curved, whether in two or three dimensions, calculus is applied to determine its length. This is typically done by expressing the path in parametric form and integrating the function representing the length of the vector that is tangent to the path. The length of a path winding through 3D space, for example, can be determined by first expressing the path in the parametric form $x = f(t)$, $y = g(t)$, and $z = h(t)$, in which f, g, and h are continuous functions defined for some interval of values of t. Then the square root of the sum of the squares of the three derivatives is integrated to find the length.

Kepler's Laws. In the early seventeenth century, Kepler formulated his three laws of planetary motion based on his analysis of the observations kept by Tycho Brahe. Later, calculus was used to prove that these laws are correct. Kepler's laws state that any planet's orbit around the sun is elliptical, with the sun at one focus of the ellipse; that the line joining the sun to the planet sweeps out equal areas in equal times; and that the square of the period of revolution is proportional to the cube of the length of the major axis of the orbit.

Probability. Accurate counting methods can be sufficient to determine many probabilities of a discrete random variable. This would be a variable whose values could be, for example, counting numbers such as 1, 2, 3, and so on, but not numbers in between, such as 2.4571. If the random variable is continuous, so that it can take on any real number within an interval, then its probability density function must be integrated over the relevant interval to determine the probability. This can occur in two or three dimensions.

One common example is determining the likelihood that a customer's wait time is longer than a specified target, such as ten minutes. If the manager knows the average wait time that a customer experiences at an establishment is, for example, six minutes, then this time can be used to determine a probability density function. This function is integrated to determine the probability that a person's wait time will be longer than ten minutes, less than three minutes, between five and thirteen minutes, or within any range of times that is desired. A considerably more complex example is the description of atomic orbitals in atoms. These are determined in three dimensions as the probability density function for the probability that an electron will be found within a certain range of energy values about the atomic nucleus.

CAREERS AND COURSEWORK

A person preparing for a career involving the use of calculus will most likely graduate from a university with a degree in mathematics, physics, actuarial science, statistics, or engineering. In most cases, engineers and actuaries are able to join the profession after earning their bachelor's degree. For actuaries, passing one or more of the exams given by the Society of Actuaries or the Casualty Actuarial Society is also expected, which requires a thorough understanding of calculus.

In statistics, a master's degree is typically preferred, and to work as a physicist or mathematician, a doctorate is the standard. In terms of calculus-related coursework, in addition to the calculus sequence, students will almost always take a course in differential equations and perhaps one or two in advanced calculus or mathematical analysis.

In 2014, about 627,700 people were working as either chemical, electrical and electronics, or mechanical engineers in the United States, most in either the manufacturing or service industries. Private industry and the government employed 30,000 statisticians.

Insurance carriers, brokers, agents, or other offices provided jobs for 24,600 actuaries. The Bureau of Labor Statistics counted only 3,500 individuals as simply mathematicians but tallied nearly 107,000 math and computer teachers at the postsecondary level.

SOCIAL CONTEXT AND FUTURE PROSPECTS

Calculus itself is not an industry, but it forms the foundation of industries. In this role, it continues to power research and development in diverse fields, including those that depend on physics. Physics derives its results by way of calculus techniques. These results in turn enable developments in small- and large-scale areas. An example of a small-scale application is the ongoing development of semiconductor chips in the field of electronics. Large-scale applications are in the solar and space physics critical for ongoing efforts to explore the solar system. These are just two examples of calculus-based fields that will continue to have significant impact in the twenty-first century.

—Michael J. Caulfield

Further Reading

Banner, Adrian. *The Calculus Lifesaver: All the Tools You Need to Excel at Calculus*. Princeton UP, 2007.

Bardi, Jason Socrates. *The Calculus Wars: Newton, Leibniz, and the Greatest Mathematical Clash of All Time*. Thunder's Mouth Press, 2006.

Bittinger, Marvin L., et al. *Calculus and Its Applications*. 11th ed., Pearson Education, 2016.

Comette, James L., and Ralph A. Ackerman. *Calculus for the Life Sciences: A Modeling Approach*. MAA Press, 2015.

Dunham, William. *The Calculus Gallery: Masterpieces from Newton to Lebesgue*. Princeton UP, 2005.

Dyke, Phil. *Modeling Coastal and Offshore Processes*. Imperial College Press, 2007.

Glover, David M., William J. Jenkins, and Scott C. Doney. *Modeling Methods for Marine Science*. Cambridge UP, 2011.

Simmons, George F. *Calculus Gems: Brief Lives and Memorable Mathematics*. 1992. Mathematical Association of America, 2007.

CARBON-OXYGEN CYCLE

Fields of Study: biogeochemistry; biology; oceanography; environmental chemistry

ABSTRACT

The carbon-oxygen cycle is the process by which oxygen and carbon are cycled through Earth's environment. The cycle includes phenomena such as photosynthesis, which makes oxygen by producing sugar from carbon dioxide, and respiration, which uses oxygen to break sugar down into carbon dioxide. The cycle is critical for the homeostasis of the environment.

KEY CONCEPTS

aerobic: requiring or in the presence of oxygen

anaerobic: requiring or in the absence of oxygen

Calvin cycle: a metabolic pathway of photosynthesis occurring in the stroma of chlorophyll-containing chloroplasts, during which carbon dioxide becomes fixed to an enzyme and is reduced to form carbohydrates

eukaryotic cells: cells in which the genetic material deoxyribonucleic acid (DNA) and ribonucleic acid (RNA) is carried within a distinct nucleus

oxygen catastrophe: a mass die-off of anaerobic life forms, essentially the only life on Earth at the time, due to poisoning by the rapidly forming abundance of oxygen in the atmosphere as photosynthesis became prominent

positive feedback loop: any process in which the presence of its product acts to increase the effect or output of the process

THE CARBON-OXYGEN CYCLE

The carbon-oxygen cycle involves the flow of carbon and oxygen through the environment. Carbon is recycled between carbon dioxide and carbon compounds in life-forms. The cycle includes the transfer mechanisms of the carbon compounds and oxygen in the biosphere and in the atmosphere, hydrosphere, and pedosphere. Climatological research is focusing on the rates of processing and the homeostasis of the system to determine carbon dioxide release and capture.

The carbon-oxygen cycle is the result of the storing of carbon from atmospheric (or in the case of algae and other aquatic organisms, dissolved) carbon dioxide as organic material through photosynthesis, and its release through respiration, combustion, or decomposition, all of which combine the carbon in the organic material with atmospheric or dissolved oxygen as CO_2. This CO_2 is then released into the atmosphere, where it is reabsorbed by photosynthetic

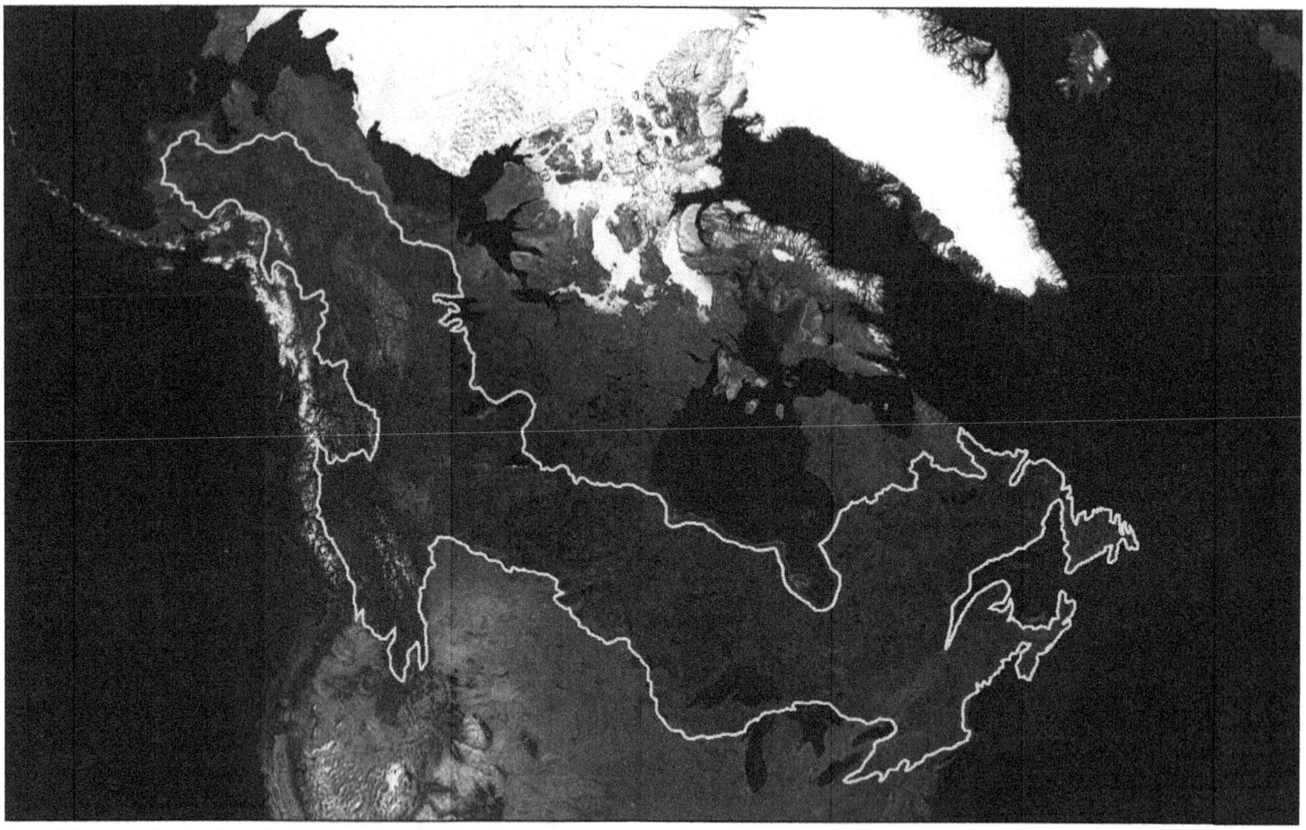

Boreal forest region in Canada (ATDoron via Wikimedia Commons)

organisms such as cyanobacteria, some protists, and plants, starting the process anew.

The cycle shows how different parts of Earth's environment and climate are interrelated. For example, analysis of the flow of material from rivers into seas allows scientists to consider the impact of biologic materials, and the release of the gases into the atmosphere and the capturing of carbon by plants allow scientists to see the effect of land use on climate.

By examining the environmental factors in a given area—factors such as land use or propensity for fire—one can get an idea of how a given region will impact other areas and can estimate the total amount of gas it might contribute to the atmosphere. Additionally, environmental factors provide hints as to effective methods of managing or sequestering carbon. A prime example of carbon sequestration can be seen in the trees of a forest, all of which are composed of materials derived from atmospheric carbon dioxide; in a sense, trees literally appear out of thin air.

Based on the carbon-oxygen cycle, it is known that trees take in a certain amount of carbon in time, that the trees will produce living space for some number of animals, and that those animals and trees will release a certain amount of carbon dioxide back into the environment throughout their lives and when they die and decompose. Another factor of consideration is whether a forest is prone to fires. By looking at such factors, one can get an idea of how much carbon dioxide a forest can remove from the atmosphere, for how long, and if there is anything humans can do to make that forest more efficient.

Most of the actual oxygen released into Earth's atmosphere is produced by oceanic plankton through photosynthesis, but rainforests also produce great quantities of photosynthetic oxygen as they absorb huge amounts of carbon dioxide. Northern (boreal) forests are another major site of such oxygen turnover.

PROCESSES

Photosynthesis is a set of chemical reactions that occurs in green plants and in some bacteria and some protists. Photosynthesis uses energy from the sun to turn carbon dioxide and water into sugar and oxygen.

The process itself is the interaction of several sets of reactions: the light-dependent reactions and the Calvin or dark cycle reactions. The light-dependent reactions separate the hydrogen from the oxygen in water and are thus the oxygen-producing portion of the cycle; they prepare reactants needed for the other cycles. The Calvin or dark cycle turns the carbon dioxide and light cycle products into sugar. The light cycle runs directly on captured photons, whereas the Calvin cycle does not depend directly upon sunlight.

Along with rates of release from combustion, which is the conversion of organic material to carbon dioxide and water vapor, also needing to be factored are rates of respiration and decomposition. Respiration is how organisms get energy from organic compounds, such as sugars. Plants and animals use aerobic respiration, which evolved after oxygen became prevalent in the atmosphere. Aerobic respiration occurs in special organelles of eukaryotic cells called "ribosomes." Here sugars are broken down into carbon dioxide and water in a process known as the Krebs or citric acid cycle.

Decomposition releases various greenhouse gases, as decomposing organisms are often excellent habitats for anaerobic life-forms; anaerobic respiration releases such gases as carbon dioxide, hydrogen sulfide, and methane. Thus the total impact of the disruption of an environment, such as the burning of a forest, is not only in the combustion of the trees but also in the drop in photosynthesis and decomposition of any detritus.

HISTORY OF THE CYCLE

The understanding of the carbon-oxygen cycle has existed since the eighteenth century, after the discoveries of Antoine Lavoisier (1743–1794), who found that respiration was a process similar to burning, and of Joseph Priestley (1733–1804), who outlined the idea of the relationship between blood and air and who also discovered oxygen. The full mechanics of the carbon-oxygen cycle and its implications, however, were discovered by scientists in subsequent centuries.

The workings of the chemical reactions of photosynthesis were discovered in the 1940s, and research continues into the role of quantum physics in the light-harvesting stage of photosynthesis. With the growing awareness of climate change, more work is being done to examine the cycle. Current research often takes samples from the area in question, be they soil samples to estimate bacterial carbon production or air samples to measure carbon levels at a given site. These samples are then analyzed together to get a picture of how each part is affecting the collective carbon-oxygen balance for that area. The carbon data are often mixed with data on other climatological factors, such as albedo, humidity, and temperature, to build a picture of the effects on climate.

ATMOSPHERIC COMPOSITION AND THE GREENHOUSE EFFECT

The carbon-oxygen cycle affects climate because it affects Earth's atmosphere. Ignoring water vapor, which is variable, it is known that the standard dry atmosphere is 78.09 percent nitrogen, 20.95 percent oxygen, 0.93 percent argon, and 0.00039 percent carbon dioxide; the rest of the atmosphere comprises various trace gases. Water vapor composes, on average, about 1 percent of the atmosphere. These percentages change slightly as the quantity of any of the component gases change. In 2020, for example, the carbon dioxide content was measured as 0.000415 percent (415 parts per million.

The carbon-oxygen cycle plays an important role in maintaining homeostasis in the atmosphere. The cycle "cycles" oxygen and carbon to maintain equilibrium. The gases in the atmosphere and their relative amounts have an important impact on climate. Some gases absorb infrared rays reflected or emitted from Earth's surface and send them back toward the surface and into the atmosphere, thus trapping the energy and making Earth warmer. This process is known as the greenhouse effect.

Atmospheric gases that are good at emitting energy in the infrared range are known as greenhouse gases and include water vapor, carbon dioxide, methane, nitrous oxide, and ozone. Water vapor and carbon dioxide make up the largest contribution to the greenhouse effect, with methane and ozone playing a smaller role. Many of these gases are produced in industrial reactions such as combustion. Combustion from cars and

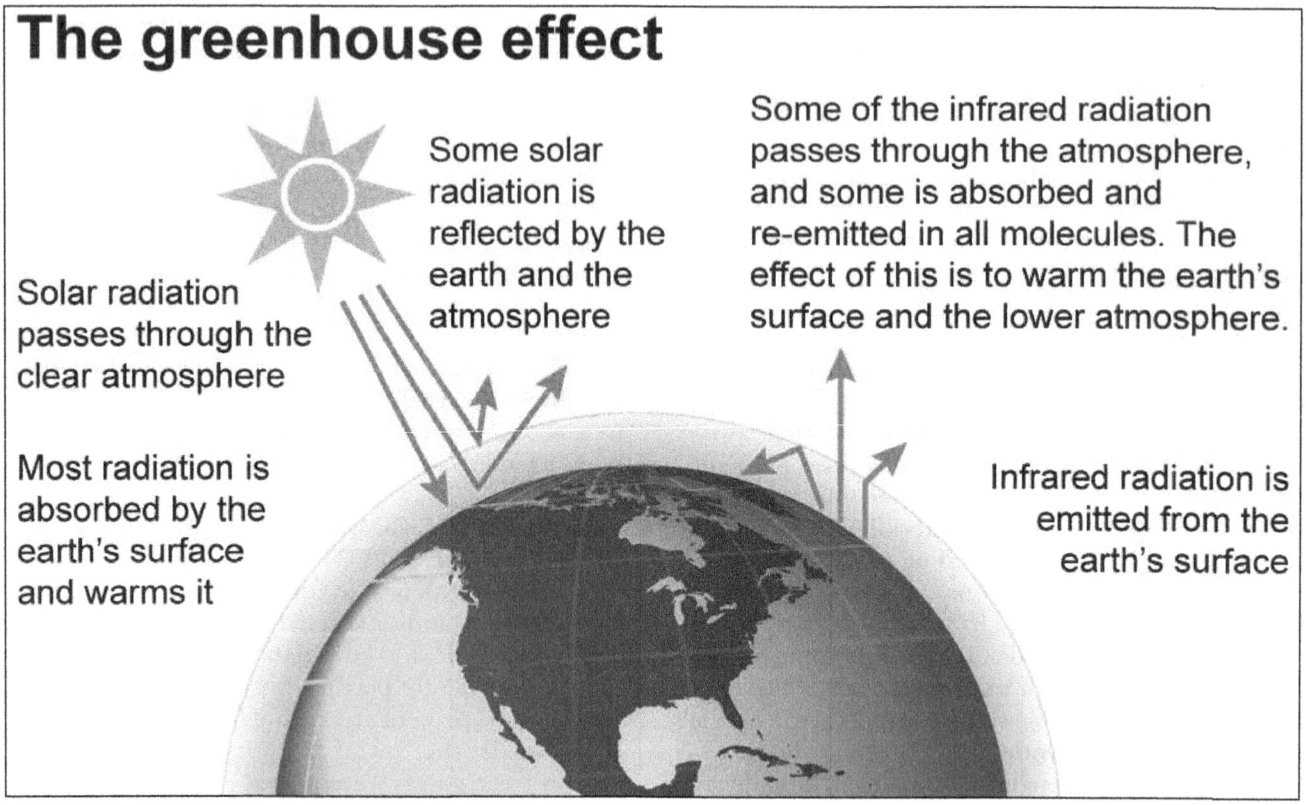

Adapted from U.S. Environmental Protection Agency

factories and from other human activities (since the start of the Industrial Revolution) have been major factors in anthropogenic climate change. Here the carbon-oxygen cycle comes into play. There is a maximum rate at which a particular environment can cycle the carbon dioxide into oxygen, meaning that excess carbon dioxide builds up in the atmosphere, increasing the greenhouse effect.

By studying the components of the cycle (that is, the plants and animals that are part of the process), scientists can find the effects of increased temperature on plants and animals. The environmental impact (increased environmental stress) leads to less capacity to handle the increased carbon dioxide. This creates a positive feedback loop that further increases the temperature. Increased global temperature could cause increased cloud cover, which would increase albedo. Increased temperature also could spawn algal blooms in the oceans, which would absorb carbon dioxide and thus reach a new equilibrium. Much environmental damage will occur before such an event happens.

The greenhouse effect is not wholly negative, however. It is actually essential for life to exist on this planet. In the absence of greenhouse gases, the planet would be significantly colder than it is now, essentially a frozen ball of ice with a mean temperature of –0.4 degrees Fahrenheit (–18 degrees Celsius). Additionally, Earth's atmospheric composition has varied throughout its history. Earth's early atmosphere was different from today's atmosphere. It consisted mostly of water vapor, carbon dioxide, hydrogen sulfide, and ammonia. It did not take long, in geologic time, for the oceans to form from rainwater. All of the greenhouse gases of the time kept the average Earth temperature fairly high.

Once photosynthetic life evolved about 3.5 billion years ago, that life began to break down the methane and carbon dioxide in the atmosphere, and the level of oxygen began to increase. This resulted in what is called the "oxygen catastrophe," in which many of the anaerobic organisms that existed at the time became extinct, leaving those that adapted to the new situation by developing aerobic respiration. This development had a great impact upon later life, as aerobic respiration became far more efficient.

The oxygen catastrophe also may have caused the Huronian glaciation event, which occurred most likely as a result of greenhouse gas depletion by photosynthetic life-forms. The event was one of the longest glaciation periods in Earth's history (300 million years). It and the later "Snowball Earth" periods that likely caused the evolution of multicellular life and the Cambrian explosion were ended only by the buildup of greenhouse gases from volcanic eruptions and by the weathering of rocks.

Equally, after this period, greenhouse gas levels have oscillated by period; for example, during the Devonian period, carbon dioxide was higher and oxygen was lower than present levels. In the Cretaceous period, both carbon dioxide and oxygen levels were higher. During both periods, the average temperature was higher than it is today.

A dramatic example of how the carbon-oxygen cycle figures into climate would be the Permian-Triassic mass extinction event (250 million years ago), in which a shutdown of the carbon-oxygen cycle may have been responsible for the greatest mass extinction in Earth's history. While the exact causes remain unknown, the mass extinction is thought to have begun with the eruption of the Siberian Traps, a massive set of volcanoes that released carbon dioxide and debris into the atmosphere.

The largest eruptions in known history probably caused a small period of global cooling, which stressed many environments. As a result, many trees died before they could process the carbon dioxide that was added to the atmosphere. Evidence seems to show that global warming set in around the peak of the extinction, and it is thought that the oceans became saturated with carbon dioxide, to the point in which there was a degassing event. The seas released their carbon in huge clouds that swept over the land, suffocating all nearby creatures. Massive erosion suggests that the climate became arid and the interior of continents were more or less lifeless. While another event on this scale is unlikely, it provides a strong example of the kind of phenomena that can arise from the disruption of the carbon-oxygen cycle.

GREENHOUSE GAS EMISSIONS

More than 3 million tons (2.72 tonnes) of carbon dioxide is released each year into the atmosphere. Compounding that are ongoing releases of other gases such as methane, which has twenty-five to sixty times the effect of carbon dioxide (as a greenhouse gas) in a one-hundred-year period.

The largest sources of anthropogenic emissions are deforestation and the impact of industrial technology, including factories, cars, and cement making (which releases carbon dioxide). However, agriculture also releases greenhouse gases; for example, cows release 16 percent of anthropogenic methane.

Humans have always released carbon dioxide into the environment. Various indigenous North American cultures burned forests annually to encourage habitats for large game animals. Also, much of Europe has been deforested for agricultural space and shipbuilding. However, the emissions from industrial and postindustrial technology are on a wholly greater scale. Of great importance is the fact that the carbon being released now has not been part of the carbon-oxygen cycle since it was sequestered as coal or petroleum deposits: in many cases, carbon dioxide released now was stored more than 250 million years ago in ancient swamps. Thus, with the combination of environmental disruption and greenhouse gas emission, recent activities are more akin to a sustained volcanic eruption than to any other Earth event.

As the system moves further from homeostasis, the compensatory mechanisms, such as increased plant growth, are overloaded or are not given the chance to function (as is the case with heavy deforestation). This forms positive feedback, pushing the environment further from equilibrium. Something as simple as planting additional trees can stall the harmful trend, however. An important point is that the carbon in a living tree has not been expelled; it remains stored as long as the tree lives. Once the tree dies the carbon dioxide is released as the tree decomposes over time.

CARBON SEQUESTRATION

Carbon sequestration involves storing carbon in a stable state to negate carbon's harmful effects on the environment. As discussed, trees serve as reservoirs for carbon.

Many efforts focus on manipulating the carbon-oxygen cycle to increase the amount of carbon stored.

Means of sequestering carbon range from genetically modifying plants to employ a more efficient form of photosynthesis to timing forest harvesting and replanting to ensure peak carbon storage, as in the Canadian lumber industry. Other forms of carbon sequestration under consideration include injecting carbon into exhausted coal seams; sinking carbon to the bottom of the ocean through algae growth; producing certain carbon-absorbing compounds; and conducting controlled burns that increase the health of an ecosystem, as in the Australian savanna. With proper manipulation of the carbon-oxygen cycle through methods such as carbon sequestration, many of the worst effects of climate change can be mitigated.

—Gina Hagler

Further Reading

Archer, David. *The Global Carbon Cycle*. Princeton UP, 2010.

Bashkin, V. N., and Robert W. Howarth. *Modern Biogeochemistry*. Kluwer Academic, 2002.

Glikson, Andrew Y. *Evolution of the Atmosphere, Fire and the Anthropocene Climate Event Horizon*. Springer, 2014.

Kirby, Richard R. *Ocean Drifters: A Secret World Beneath the Waves*. Studio Cactus, 2010.

Kondratyev, K. Y., V. F. Krapivin, and Costas A. Varotsos. *Global Carbon Cycle and Climate Change*. Springer, 2003.

Le'vêque, C. *Ecology from Ecosystem to Biosphere*. Science, 2003.

Vallero, Daniel A. *Environmental Biotechnology: A Biosystems Approach*. Academic Press, 2010.

Walker, Sharon, and David McMahon. *Biochemistry Demystified*. McGraw-Hill, 2008.

Williams, Richard G., and Michael J. Follows *Ocean Dynamics and the Carbon Cycle*. Cambridge UP, 2011.

CARBONATE COMPENSATION DEPTHS

Fields of Study: oceanography; marine chemistry; biogeochemistry; paleogeology

ABSTRACT

Carbonate compensation depths (CCDs) are the depths in ocean basins that separate calcium carbonate-containing seafloor sediments from carbonate-free sediments. The CCD may be different in different ocean basins, and it may rise or fall at different times in the same ocean basin as a result of the balance between surface production of carbonate and deep-water carbonate dissolution. Carbonate solubility increases with increasing pressure and decreasing temperature.

KEY CONCEPTS

biogenic: produced by living organisms rather than from mineral sources

carbonate compensation depth: the depth range at which the rate of carbonate introduction is balanced by the rate of carbonate dissolution

carbonic acid: H_2CO_3, a weak acid produced by the reaction of carbon dioxide (CO_2) and water (H_2O)

clastic: produced from mineral sources rather than by living organisms, the opposite of biogenic

lysocline: the physical region of ocean water at the carbonate compensation depth in which carbonate is actually dissolved

MARINE SEDIMENTS

Marine sediments are composed of a variety of materials of biological or mineral origin. Individual sediment grains in oceanic deposits may be either clastic or biogenic particles. Clastic sediments are materials derived from the weathering, erosion, and transportation of exposed continental rocks, and these grains are classified by particle diameter into gravels, sands, silts, and clays. Clastic sediment particles become less common with increasing distance from the continental landmasses and are nearly absent from deepwater sediments on the abyssal plains.

Biogenic sediment particles are composed of skeletons and tests precipitated by planktonic plants and animals living in the shallowest waters of the ocean. Biogenic particles composed of calcium carbonate or of opaline silica make up the majority of oceanic sediments, which are deposited at great distances from land. Deposition of biogenic sediments is controlled by two factors: the biological productivity of surface waters and the dissolution of biogenic particles by corrosive bottom waters. Increased levels of atmospheric carbon dioxide lead to an increase in the corrosiveness of bottom waters.

Biogenic sediment particles are produced in shallow, well-lighted surface waters as a result of the biochemical activity of microscopic plants and animals, which precipitate solid shells and tests formed from minerals dissolved in seawater. Biological productivity is a measure of the number of the organisms present and their rate of reproduction. The productivity of planktonic organisms is directly related to chemical and physical conditions in the surface waters. High-productivity waters have abundant supplies of oxygen, with dissolved nutrients and chemicals needed for the formation of shells and tests, and enough light for photosynthesis by plants. Generally, high productivity is found in warm-water areas with abundant dissolved oxygen and nutrients.

When planktonic plants and animals die, their shells and tests fall through the water column to be deposited on the bottom of the ocean. This "planktonic rain" causes deposition of seafloor sediments as the biogenic particles produced in the surface waters sink and accumulate on the seafloor, and it moves chemicals from surface waters to deep waters. The higher the productivity values in surface waters, the greater the supply of biogenic sediment particles to the seafloor. As biogenic particles sink through the water column to the seafloor, they may be dissolved by corrosive seawater. Surface waters are saturated with dissolved calcium carbonate, so most dissolution takes place in deep waters, which are undersaturated with carbonate.

Calcareous sediments are produced by the accumulation of biogenic particles of calcium carbonate that survive the fall through the water and are deposited and buried on the seafloor. Accumulation rates of biogenic sediments are controlled by the balance between surface productivity and deepwater dissolution: The higher the biological productivity in the surface, the greater the number of shells and tests that will sink to the ocean bottom. In certain high-productivity areas associated with upwelling of cold, nutrient-loaded water masses to the surface, seafloor sediment accumulation rates may be as high as 1.18 to

Calcareous open-cast mine in Gornozavodsk in Perm Krai, Russia (Мастер Chob via Wikimedia Commons)

1.97 inches (3 to 5 centimeters) per thousand years. In areas with higher dissolution rates, fewer calcareous particles will survive to be deposited on the ocean floor. In these low-productivity areas, all the carbonate produced at the surface may be dissolved, and sediment accumulation rates may be as low as 0.039 in (1 millimeter) per million years.

LEVEL OF CARBONATE COMPENSATION DEPTH

The CCD marks the boundary between carbonate-rich sediments (calcareous oozes and chalks) and carbonate-free sediments (red clays) in the oceans. It is the result of deepwater dissolution rates exceeding the rate of supply of calcium carbonate to the deep sea by surface productivity. The depth of the CCD marks the zone in which the supply of carbonate sinking in the planktonic rain from surface waters is exactly balanced by the rate of removal of carbonate dissolution in deep waters. Calcareous sediments will be deposited on the seafloor in water depths shallower than the CCD, because in these areas, the rate of supply of calcium carbonate is higher than the dissolution rate. In these areas, individual particles of calcium carbonate will survive the trip through the water column and be deposited as biogenic sediments on the seafloor.

Below the compensation depth, dissolution exceeds the rate of supply, so all carbonate particles supplied from the surface become dissolved, and seafloor sediments are carbonate-free. Surface sediments in water depths below the CCD tend to be red clays, or combinations of fine-grained materials derived from continental sources and carried to the deep sea by wind, mixed with micrometeorites and other particles from extraterrestrial sources. Red clays generally lack fossils, as a result of complete dissolution of carbonate and opaline silica, so only a few dissolution-resistant fossils, such as phosphatic fish teeth and whale ear bones, are found in these sediments. The reddish-brown color of deep-sea clay deposits is a result of the presence of iron particles, which have reacted with oxygen in seawater to form the rust-brown color.

Much information on the past history of the compensation depth in different ocean basins has been provided by ocean-floor drilling programs. The level of the CCD has changed dramatically throughout geologic history, with fluctuations of up to 6,562 feet (2,000 meters) being recorded in deep-sea sediments. Changes in the CCD are believed to be caused by changes in either the rate of supply of carbonate to the oceans or the rate of carbonate dissolution in the deep sea, which may be caused by changes in the shape of ocean basins, by changes in the location of carbonate deposition within different ocean basins, or by changes in the concentration of carbon dioxide in the atmosphere. During the last 100,000 years, carbonate sediments formed at greater depths during glacial intervals than during interglacials, indicating a less corrosive deepwater environment resulting from lower concentrations of atmospheric carbon dioxide.

CALCIUM CARBONATE DEPOSITION

Calcium carbonate is delivered to the oceans by rivers carrying particles eroded from continental rocks. Ocean sediments are the primary geochemical reservoir for calcium carbonate, so most of the calcium carbonate on Earth remains dissolved in ocean water or in the form of calcareous sediments on the seafloor. The amount of carbonate deposition on the seafloor depends on the input of calcium carbonate derived from continental weathering and delivered to the oceans by rivers. Because oceanic plankton can precipitate solid calcium carbonate at a much faster rate than the rate of input of dissolved carbonate from rivers, most of the calcium carbonate that is deposited in the oceans must dissolve in order to maintain the chemical balance between dissolved carbonate and solid carbonate in oceanic sediments.

Any changes in the locations of carbonate deposition may cause corresponding changes in the level of the CCD as the compensation depths in different oceans change as a result of bathymetric fractionation or basin-basin fractionation. In bathymetric fractionation, a balance is established between the rates of carbonate deposition in shallow-water and deepwater sedimentary basins. Greater deposition of calcium carbonate in shallow waters atop the continental shelves will cause a shallowing of the CCD as more deepwater carbonate deposits are dissolved so as to balance the shallow-water deposition. Similarly, the level of the compensation depth may vary by basin-basin fractionation of carbonate, which establishes a balance between the compensation depths in different ocean basins. For example, greater deposition of calcium carbonate in the Pacific Ocean will cause the Pacific CCD to become deeper, while at the same time the Atlantic compensation depth must become shallower, because greater deposition of carbonate in Pacific sediments will leave less dissolved carbonate for precipitation in the Atlantic Ocean.

Even within an ocean basin, the level of the CCD may vary, depending on the balance between carbonate productivity and dissolution in a local area. For example, in the equatorial Pacific Ocean, the compensation depth is 1,640 to 2,625 ft (500 to 800 m) deeper than in areas immediately to the north and south of this high-productivity area as a result of the greater supply of carbonate in the planktonic rain below these high-productivity surface waters. Also, the compensation depth tends to shoal near the edges of ocean basins, because higher biological productivity in shallow water near the continents causes rapid sinking of large amounts of organic carbon produced by planktonic plants and animals. Breakdown of this organic carbon by seafloor bacteria produces increased amounts of dissolved carbon dioxide gas, which reacts with water molecules to form carbonic acid; carbonic acid is corrosive to solid calcium carbonate. Greater carbonic acid concentrations lead to increased carbonate dissolution in bottom waters and cause upward migration of the CCD into shallower waters.

STUDY OF CARBONATE COMPENSATION DEPTHS

Carbonate compensation depths may be studied by obtaining a series of deep-sea sediment samples from different depths within an ocean basin to determine the relationship between sediment type and water depth. In 1891, a study was published describing the global patterns of seafloor sediment type in each ocean basin, based on sediment samples obtained on the HMS *Challenger* oceanographic expedition. It was discovered that virtually no calcium carbonate was present below depths of 14,764 ft (4,500 m) as a result of dissolution of carbonate.

The CCD in the modern ocean was first described in detail in 1935 based on sediment core transects taken across the South Atlantic Ocean by the 1925–1927 German *Meteor* expedition. Similar studies of sediment cores from different depths in the Pacific Ocean revealed that calcareous oozes are common seafloor sediments in water depths to 14,436 ft (4,400 m), with noncalcareous red clays being found in surface sediments deeper than 14,436 ft (4,400 m). Cores with calcareous oozes underlying red clays were obtained, however, in depths well below the present CCD, indicating that this geochemical boundary has migrated vertically throughout geologic history.

One innovative experiment to measure the rate of calcium carbonate dissolution with increasing water depth involved the placement of a stationary mooring for a period of months in deep water in the Pacific Ocean. Calcite spheres and calcareous microfossils were hung in permeable nylon bags at different water depths on the mooring; the nylon bags allowed seawater to come in contact with the calcium carbonate and thus permitted carbonate dissolution to occur. By measuring the weight loss of spheres suspended for a few weeks to months on the mooring, the rate of dissolution was determined for different water depths. In this experiment, little carbonate dissolution was observed at water depths shallower than 12,140 ft (3,700 m), while a rapid transition from minimal dissolution to extreme dissolution was seen at lower depths. Rapid loss of carbonate by dissolution occurred in carbonate spheres suspended between the lysocline (the depth at which carbonate dissolution first begins to occur) and the compensation depth. Below 14,764 ft (4,500 m), the carbonate compensation depth, all carbonate was removed within a matter of weeks, demonstrating the ability of bottom waters to dissolve calcium carbonate.

Information on compensation depths may also be provided from microfossils preserved in seafloor sediments. Calcite dissolution can be measured by enrichment of dissolution-resistant forms of planktonic organisms, by benthic-planktonic foraminiferal ratios (foraminifera are one-celled animals that secrete a calcium carbonate internal test), by fragmentation indices (the percentage of broken planktonic tests compared to whole tests), or by the coarse-fraction ratios of seafloor sediments. Different microfossils will have differing susceptibilities to dissolution, depending on the thickness of the walls of the microfossil tests. Thin-walled plankton will be more dissolution-susceptible, while thicker-walled tests will resist dissolution. Thin-walled planktonic tests in seafloor sediments deposited in water depths between the lysocline and the CCD will be removed by dissolution more rapidly than thicker-walled tests, leading to greater enrichment of dissolution-resistant fossils with greater carbonate dissolution. Relative dissolution rates may be measured by the ratio between dissolution-susceptible and dissolution-resistant planktonic shells in seafloor sediments.

Similarly, the deeper-water bottom-dwelling benthic foraminifera tend to have thicker walls than the tests of planktonic foraminifera, which live floating in the shallow surface waters. Seafloor sediments may contain both benthic and planktonic varieties of foraminifera. Dissolution of calcium carbonate will preferentially remove the thinner-walled planktonic foraminifera, thus leading to enrichment in the proportion of benthic forms remaining in the sediment. The relative amount of carbonate dissolution in sediments may be determined by measuring the proportion between benthic and planktonic foraminifera in sediment samples.

Finally, fragmentation of planktonic tests may provide an indication of dissolution of carbonate from marine sediments. The percentage of broken tests to whole tests will increase with greater dissolution of carbonate. Also, coarse-fraction percentages of sediments provide information on carbonate dissolution, because unbroken foraminiferal tests are sand-sized, with particle diameters greater than 63 microns in size (1 millimeter is equal to 1,000 microns). As foraminifera are dissolved, they tend to break into smaller fragments, so dissolution tends to break down sand-sized particles into smaller silt-sized fragments. By measuring the proportion between sand-sized and silt-sized particles (the coarse-fraction percentage) in calcareous sediments, it is possible to obtain an indicator of the relative amount of dissolution that has affected those sediment deposits. The more dissolution that has occurred, the smaller will be the percentage of coarse (sand-sized) sediment particles.

DEPTH ESTIMATES

All these methods provide similar depth estimates for the CCD at approximately 14,764 ft (4,500 m) below the sea surface, about halfway between the crests of the mid-ocean ridges and the abyssal plains. Individual compensation depths may vary in the different ocean basins of the world, however, as a result of basin-basin fractionation of carbonate. For example, in the Pacific Ocean, the CCD is typically between 13,780 and 14,764 ft (4,200 and 4,500 m), while in the Atlantic and Indian Oceans the CCD is deeper, being found near a depth of 16,404 ft (5,000 m). Even within an ocean basin, the level of the CCD may vary, depending on the balance between carbonate productivity and dissolution in a local area. In the centers of the North and South Pacific Oceans, the average compensation depth is between 13,780 and 14,764 ft (4,200 and 4,500 m), while near the equator it is found near 16,404 ft (5,000 m) because of the higher biological productivity of equatorial surface waters.

Analyses of deep-sea cores drilled by the *Glomar Challenger* have revealed that the level of the carbonate compensation depth has changed by up to 6,562 ft (2,000 m) in the South Atlantic, Indian, and Pacific Ocean basins. For example, one of the results of Deep Sea Drilling Project Leg 2 was the discovery of significant vertical excursions in the compensation depth of the Atlantic Ocean. In seafloor boreholes drilled in the North Atlantic, 2- to 5-million-year-old calcareous ooze sediments were found atop older (5- to 23-million-year-old) red clays, which in turn were deposited atop carbonate deposits older than 23 million years. In order for these sediments to have accumulated in that order, large vertical changes must have occurred in the compensation depth, starting when the seafloor was shallower than the CCD more than 23 million years ago. Red clays were deposited between 23 and 5 million years ago, when the CCD became shallower. After these red clays accumulated, deepening of the compensation depth allowed the deposition of younger calcareous sediments atop the carbonate-free red clays.

In order to study the past history of the compensation depth within an ocean basin, it is necessary to obtain a series of cores that were deposited at different paleodepths at the same time in the past. (The paleodepth is the depth at which ancient seafloor sediments were deposited.) Paleodepth estimates for seafloor sediments are calculated by studying the cooling history of seafloor basement rocks. After new seafloor is produced by volcanic activity at the mid-ocean ridge system, these rocks cool and contract, thus sinking to greater water depths as they move away from the ridge system. The older the seafloor, the greater its water depth will be, so sediments deposited atop volcanic seafloor will accumulate in progressively deeper water. Once a series of sediment cores deposited at the same time in the past has been obtained, it is possible for the oceanographer to determine the paleodepth of the compensation depth by finding the paleodepth below which no calcium carbonate is present in ancient sediment deposits.

—*Dean A. Dunn*

Further Reading

Ahr, Wayne M. *Geology of Carbonate Reservoirs*. John Wiley & Sons, 2008.

BouDagher-Fadel, Marcelle K. *Biostratigraphic and Geological Significance of Planktonic Foraminifera*. UCL Press, 2015.

Burdige, David J. *Geochemistry of Marine Sediments*. Princeton UP, 2020.

Emerson, Steven, and John Hedges. *Chemical Oceanography and Marine Carbon Cycle*. Cambridge UP, 2008.

Garrison, Tom S. *Oceanography: An Invitation to Marine Science*. Brooks/Cole, Cengage Learning, 2010.

Middelburg, Jack J. *Marine Carbon Biogeochemistry: A Primer for Earth System Scientists*. Springer Open, 2019.

Rigo, M., et al. "A Rise in the Carbonate Compensation Depth of Western Tethys in the Carnian (Late Triassic): Deep-Water Evidence for the Carnian Pluvial Event." *Palaeogeography, Palaeoclimatology, Palaeoecology*, vol. 246, 2007, pp. 188–205.

Segar, Douglas. *An Introduction to Ocean Sciences*. 2nd ed., W. W. Norton, 2007.

Swart, Peter K., Gregor Eberli, and Judith A. McKenzie, editors. *Perspectives in Carbonate Geology*. Wiley-Blackwell, 2009.

Talley, Lynne D., George L. Pickard, William J. Emery, and James H. Swift. *Descriptive Physical Oceanography: An Introduction*. 6th ed., Elsevier, 2011.

Williams, Richard G., and Michael J. Follows. *Ocean Dynamics and the Carbon Cycle: Principles and Mechanisms*. Cambridge UP, 2011.

CARBONATES

Fields of Study: geology; marine geochemistry; oceanography; environmental chemistry

ABSTRACT

Carbonate minerals are characterized by a composition that features the carbonate ion. Common carbonate minerals are divisible into calcite, aragonite, and dolomite groups. While natural precipitation of insoluble carbonate salts occurs, other carbonates are formed by living marine organisms as shells and tests. Both become rock through the process of diagenesis.

KEY CONCEPTS

diagenesis: the process by which an accumulation of sediment is converted to sedimentary rock through the combination of compaction, cementation and the development of crystal growths between the sediment grains

goniometer: a device used in the study of crystal structures for the measurement of angles between the faces of a crystal

isostructural: describes the unit cell structures of two crystalline materials having the same angular orientation of crystal faces in the unit cells

karst: a landscape of limestone, characterized by surface structures and the presence of cave formations that have been deeply eroded into the landscape by the action of mildly acidic water

plane-polarized light: light for which the normal radially symmetric electromagnetic vibration perpendicular to the direction of motion has been restricted to a single plane by reflection from a surface or by passing through a polarizing filter

polarizer: a filter that polarizes light that passes through it

polymorphism: a phenomenon in which a single compound or material may have different intrinsic forms in nature

CARBONATE MINERAL FORMATION

Carbonates are one among several classes of mineral, which are the materials that make up rocks. They are natural substances with a definite chemical composition and an ordered internal arrangement. Minerals can be divided into chemical groups based on their atoms and into structural groups based on the ordered arrangement of the atoms.

All carbonates contain the carbonate ion as their defining anionic group. An ion is an atom that has lost or gained electrons and so has become chemically reactive. An ion that has lost electrons has a positive charge and is called a "cation"; an ion that has gained electrons has a negative charge and is called an "anion." The number of electrons lost or gained represents the charge of the ion. The charge and the radius (size) of an ion determine how it can bond chemically to other ions, how strong the bonding is, and the number of ions that can be coordinated to it or surround it.

Carbonate travertines the natural pools during sunset, Pamukkale, Turkey (THANAN via iStock)

The carbonate ion contains one carbon ion in the middle of a triangle formed by three oxygen ions, which occupy the corners of the triangle. The central carbon ion has a formal charge of +4 and the oxygen ions have a charge of –2. As a result, the carbonate ion group acts as if it were a single ion with an overall –2 charge. To form a carbonate mineral, the negative charge on the carbonate ion must be balanced by cations such as calcium, magnesium, strontium, manganese, and barium.

CARBONATE MINERAL STRUCTURE

Atoms in a mineral can be pictured as lying on imaginary planes, which are called "atomic planes." These planes cut across other planes, forming definite and measurable intersection angles. The atomic planes that contain many atoms tend to develop into crystal faces. Minerals are crystalline, and the crystal faces of a mineral are related to the internal arrangement of atoms. The bonds between atoms of the same atomic plane tend to be stronger than the bonds across the atomic planes. Consequently, minerals commonly break or cleave along preferred planes, or cleavage planes. A set of parallel cleavage planes yields one cleavage direction. Some minerals have more than one cleavage direction; others have none, since the bond strength between atoms is the same in all directions. Common carbonate minerals have three cleavage directions at oblique angles and cleave into skewed boxlike shapes.

The atoms of a mineral are symmetrically related to one another. Since crystal faces are related to the internal arrangement of the atoms, these crystal faces are also symmetrically related. Scientists have found that all minerals can be grouped into thirty-two crystal classes based on their symmetry relations. A small group of atoms forms the basic building block of crystals. This building block is called a "unit cell"; it is an arrangement of the smallest number of atoms which, as a unit, may be repeated over and over again to form a visible crystal. The volume of a unit cell can be determined from the lengths of its lines and the angles subtended by them. Depending on the cell's shape, these imaginary lines may be parallel to its edges or may pass through opposite corners, sides, or edges. The lines are called "crystallographic axes." The thirty-two crystal classes of minerals can be grouped into six crystal systems based on the shapes of their unit cells.

The common carbonates belong to the two classes termed the orthorhombic and hexagonal systems. The orthorhombic crystal system is characterized by having three crystallographic axes that are unequal in length and perpendicular to one another. The unit cell is a rectangular box with unequal sides. The hexagonal system has four axes, all of which pass through a common center. Three of these axes lie on the same plane, are of equal length, and are separated from one another by 120 degrees. The fourth axis is different in length from the others and is perpendicular to them. The hexagonal unit cell is a box with 60-, 90-, and 120-degree angles.

The three common carbonates—calcite, aragonite, and dolomite—belong to three structural types. Calcite and aragonite are polymorphs of the same compound; that is, the same chemical compound occurs in different structures. Aragonite is orthorhombic and less symmetrical than calcite, which is hexagonal. Calcite and dolomite are both hexagonal, but they do not have identical structures. In calcite, calcium atomic planes lie between carbonate ion planes. In dolomite, alternating calcium planes are occupied by magnesium atoms. As a result, calcite has a higher symmetry content than dolomite, although both belong to the same crystal system. Also, dolomite shows more internal order than calcite because it requires the positioning of more different atoms in specific atomic sites.

Different chemical compounds that have identical structures are said to be isostructural. All carbonates that contain divalent cations (+2 charge) whose ionic radii are less than or equal to that of the calcium ion are isostructural to calcite, which has a hexagonal structure. Siderite (iron carbonate), magnesite (magnesium carbonate), and rhodochrosite (manganese carbonate) are isostructural to calcite.

CALCITE GROUP

Calcite is by far the most common of all carbonate minerals. It is commonly off-white, colorless, or transparent. It may exhibit different crystal shapes, but in all cases, a careful examination will reveal the hexagonal crystal structure. It is fairly resistant to abrasion, although it can be scratched by a steel knife. Calcite fizzes and dissolves in acid because the acid breaks down the carbonate ion and releases carbon dioxide. Calcite is the main component of a rock called "limestone." Most limestone is at least partly biological in origin, or biogenic, made of calcium carbonate secreted by marine organisms. Limestone is soluble in acid, and most surface water is slightly acidic. Many caves and some sinkholes are found in regions where the rocks are limestone, and formed because calcite is dissolved over time by groundwater seepage. This process is accelerated by acid rain that enters the seepage system. Sinkholes are formed when the roofs of caves collapse.

DOLOMITE GROUP

In the dolomite group, alternating calcium atomic planes are occupied by other divalent ions, such as magnesium in dolomite, iron in ankerite, and manganese in kutnahorite. Since the magnesium, iron, and manganese ions are smaller than the calcium ion, the dolomite group structure is not as highly symmetrical as the calcite structure. Consequently, although both the calcite and the dolomite group minerals are hexagonal, they belong to different classes among the thirty-two classes of crystals. Of the dolomite group minerals, the mineral dolomite is by far the most common. It is fairly common in ancient carbonate rocks called "dolostones." Dolomite is white to pinkish and is similar to calcite in many ways; however, it does not fizz readily when diluted acid is dropped on it.

For a long time it was a puzzle why dolostone was so abundant, as dolomite is not forming in many places today. We now know that most dolomite begins as limestone and is converted to dolomite later through diagenetic alteration. In some cases the conversion happens when fresh water mixes with salt water as the rocks are forming, but more recent research suggests that most dolomite forms when warm fluids circulating through the rocks replace calcium with magnesium.

ARAGONITE GROUP

In the aragonite group, carbonate ions do not lie on simple atomic planes. Adjacent carbonate ions are slightly out of line; also, adjacent carbonate ions face in opposite directions. Divalent cations whose ionic radii are larger than or equal to the calcium ion form carbonate minerals that are isostructural to aragonite. Of the minerals in the aragonite group, aragonite is the most common. It is generally white and elongate. It is the carbonate mineral that readily precipitates from marine water, but it is not a stable mineral, and in time it changes to the more symmetrical and hexagonal structure of calcite. Aragonite is the least symmetrical polymorph of calcium carbonate, and is found in metamorphic rocks that were formed under high pressure and comparatively low temperature.

In its most abundant form, aragonite is synthesized by aquatic organisms that are common in shallow marine environments of warm latitudes, such as the Gulf Coast. As marine organisms die, their shells, which are made of calcite, settle at the bottom. These shells may be broken into smaller fragments by browsing organisms or wave action. When buried to form limestone, the aragonite quickly recrystallizes to calcite.

IDENTIFICATION OF CARBONATE MINERALS

Carbonates are generally light-colored, soft, and easy to scratch with a knife, as compared with most other common rock-forming minerals. They form a white powder when scratched. They are harder than fingernails, however, and cannot be scratched by them. When acid is poured on carbonate powder, it fizzes, liberating carbon dioxide. Carbonates such as calcite do not even have to be powdered for the acid test, because they fizz readily. The cleavage planes and the angles between cleavages are another physical method by which carbonate minerals can be distinguished. Clear crystals such as those of calcite can produce double refraction of objects, another property that identifies carbonates without the aid of instruments.

Better mineral identification is done by scientists after a rock is cut to a small size, mounted on glass, and ground to a very thin section of rock (0.0012 inches, or 0.03 millimeters thick), which is then capable of transmitting light. The thin section is placed on a stage of a transmitted-light polarizing microscope. A lens below the stage polarizes light; it allows the transmission only of light that vibrates in one direction—east to west, for example. Another lens above the stage allows the passage only of light that vibrates in the other direction—north to south. When glass is placed on a stage and the lower polarizer is inserted across the light source, the color of the glass can be seen. When the upper polarizer is also inserted, however, the glass appears dark, because no light is transmitted. The optical properties of most minerals, including the carbonates, are different from those of glass and how the mineral permits the passage of plane-polarized light is an identifiable characteristic. Other accessories are used in addition to the polarizing lenses in order to determine the optical properties of minerals. Magnification by microscope permits the better determination of minerals' physical properties, such as their shape and cleavage. Special dyes can also be used to distinguish calcite from dolomite.

DETERMINATION OF CRYSTAL STRUCTURE

The crystal structure of carbonate crystals can be determined with the aid of a contact goniometer. Its simplest version is a protractor with a straight edge fastened in the middle. The goniometer is used to measure the angles between crystal faces, from which scientists can determine the crystal structure.

X-ray diffraction can ascertain the crystal structure of any substance, including carbonates. Diffraction peaks characteristic of each mineral can be displayed on a chart recorder when X-rays bombard a sample. Each diffraction peak results from the reinforcement of X-ray reflections from mutually parallel atomic planes within the sample. Several diffraction peaks from one mineral indicate equivalent numbers of sets of atomic planes within the minerals. The difference in the peak heights corresponds to the density of atoms in the pertinent atomic planes. The detection device does not have to be a chart recorder; it can be a photographic paper or a digital recorder that can be appropriately interfaced to a computer for the quick identification of minerals.

PRACTICAL APPLICATIONS

Carbonates are a common group of minerals that form in environments that range from arid lands to shallow seas. After silicates, they are the most abundant minerals on the surface of Earth. Hot springs are one place where calcite precipitates. Travertine, or tufa, is a banded rock that precipitates at the mouth of springs. Caliche, deposits of carbonate that precipitate from groundwater in arid climates, is a source of serious problems to irrigation farmers. Sinkholes that form through the collapse of roof rocks of near-surface caves in the limestone of warm and humid regions are a problem not only to farmers but also to homeowners. It is not unusual for part of a house, or the whole of it, to sink suddenly into a depression caused by the collapse of an underground cave. Whole lakes of surface water have also been seen to drain into a sinkhole that has opened beneath it.

Carbonates are important for their regulation of the pH, or acidity content, of ocean waters. Carbonate minerals dissolve when the acid content of water is raised and precipitate when the acid content is reduced. In this way, the pH of ocean water is regulated to a steady value of 8.1.

Carbonates are also known for their industrial applications, which range from dolomite tablets to building materials such as cement and mortar. One of the finest building rocks is marble, which is composed of carbonate minerals. Marble often is delicately banded with different colors. The banding arises because the minerals are lined up in directions perpendicular to the natural pressure under which an impure limestone was metamorphosed and converted to marble. If the original limestone was pure and composed entirely of calcite, the marble that is metamorphosed from it would be white and not banded. Polished marble is used as a building material, or often as decorative stone for doors or exteriors, and has been the most prized material for stonecarvers and sculptors. Polished travertine is also used as building stone, but it is placed in the interiors of buildings because of banded porous zones that can accumulate rainwater. Regular limestone is used in buildings and, most commonly, in retaining walls alongside houses and roads. Most limestone is used for cement. Cement is up to 75 percent lime or calcium oxide, created by heating limestone to drive off carbon dioxide; the rest is silica and aluminum. Recently, there has been significant criticism of the cement industry for the amount of anthropogenic carbon dioxide it releases into the atmosphere, and alternative materials have been developed that do not release carbon dioxide during their manufacture.

The most important natural application of carbonates is in regulating Earth's climate. Carbonate rocks lock up carbon dioxide that would otherwise trap solar heat and warm the planet. Weathering and other processes also break down calcite and release carbon dioxide. The balance between carbonate formation and breakdown is critical in maintaining Earth's natural climate balance.

—Habte Giorgis Churnet

Further Reading

Ahr, Wayne M. *Geology of Carbonate Reservoirs*. John Wiley & Sons, 2008.

Burdige, David J. *Geochemistry of Marine Sediments*. Princeton UP, 2020.

James, Noel P., and Brian Jones. *Origin of Carbonate Sedimentary Rocks*. John Wiley & Sons, 2015.

Klein, Cornelis, and C. S. Hurlbut, Jr. *Manual of Mineralogy*. 23rd ed., John Wiley & Sons, 2008.

Miglorie, Catherine. *Vermont's Marble Industry*. Arcadia, 2013.

Milliman, J. D. *Recent Sedimentary Carbonates. Part 1: Marine Carbonates*. Springer Science & Business Media, 2012.

Pastero, Linda, editor. *Carbonates*. MDPI, 2019.

Shanov, Stefan, and Konstantin Kostov. *Dynamic Tectonics and Karst*. Springer, 2015.

Swart, Peter K., Gregor Eberli, and Judith A. McKenzie, editors. *Perspectives in Carbonate Geology*. Wiley-Blackwell, 2009.

Tucker, Maurice E., and Robin G. C. Bathurst, editors. *Carbonate Diagenesis*. John Wiley & Sons, 2009.

CARIBBEAN SEA

Fields of Study: oceanography; marine geology; marine biology; marine conservation

ABSTRACT

The Caribbean Sea is an enclosed basin of the western Atlantic Ocean between the Gulf of Mexico and the Atlantic Ocean proper. The coastal areas around the Caribbean Sea include twenty-six continental and island countries as well as nineteen territories of the United States, France, Britain, and the Netherlands. The coast is made up of coral reefs, mangroves, expanses of seagrasses, sand beaches, and rocky shores.

KEY CONCEPTS

atoll reef: a circular coral reef structure that typically builds around the edges of a sunken volcanic crater, although some atoll reefs form naturally without the presence of a volcanic substrate on which to grow

basin: an area of deep-sea bottom defined by rough boundaries presented by continental shelves and island structures

biome: the flora, fauna and mineral structures existing within a particular defined environment

fringing reef: a coral reef that builds parallel to a shore rather than as an isolated structure such as an atoll

trough: a long, deep depression in the sea bottom, formed as sea floor tectonic plate sections move away from each other over time as is the case for the Cayman Trough which also has a developing mid-trough ridge

LOCATION AND DESCRIPTION

An enclosed basin of the western Atlantic Ocean, the western and southern sides of the Caribbean Sea form the boundaries of Central and South America, while it encompasses the Antilles Islands chain to the east and northeast. In the northwest, the Yucatan Channel links the Caribbean to the Gulf of Mexico. Stretching for approximately 8,389 miles (13,501 kilometers), the coastal areas around the Caribbean Sea include twenty-six continental and island countries as well as nineteen territories of the United States, France, Britain, and the Netherlands. The coast is made up of coral reefs, mangroves, expanses of seagrasses, sand beaches, and rocky shores.

Millions of years ago, the Greater Caribbean was separated from the Pacific Ocean by the closing of the Isthmus of Panama. Separated by water ridges and sills, the floor of the Caribbean Sea today is comprised of the Basins of Grenada, Venezuela, Colombia, Yucatan, and the Cayman Trough. The depth is recorded at more than 7,874 feet (2,400 meters) through three-quarters of the Caribbean, but it reaches 21,326 ft (6,500 m) in the Cayman Trough. The area of the Caribbean Sea is approximately 1.05 million square miles (2.72 million square kilometers).

BIODIVERSITY

The Caribbean has been identified as one of the world's five hot spots for marine biodiversity. While many species remain unknown, a team of experts led by marine biologist Patricia Miloslavich in 2010 identified 12,046 species of marine life in the Caribbean.

Among the species found in the Caribbean's ecosystem is the tiny bee hummingbird, among the smallest known birds in the world, which is found only in Cuba and on the Isle of Youth. Weighing only 0.064 ounces (1.8 grams), the bee hummingbird is often taken for an insect. In 2008, the smallest snake in the world was discovered on the Caribbean island of Barbados by Penn State biology professor Blair Hedges. Less than 4 inches (102 millimeters) long, this snake, named *Leptotyphlops carae*, looks like a strand of spaghetti. Blair determined that the snake's genetic makeup was new to science, with a unique color pattern and scales. At the other end of the size spectrum, the Caribbean Sea is home to at least two dozen species of marine mammals, from spotted dolphin (*Stenella attenuata*) to the humpback whale (*Megaptera novaeangliae*).

A total of 6,550 plant species are known in the Caribbean region. At least seven species of seagrass are present in the Caribbean Sea, with turtle grass (*Thalassia testudinum*) and manatee grass (*Syringodium filiforme*) the most widely distributed. While many animals tend to remain in a single habitat type for most of their activity cycles, a number split their time between seagrass beds, for instance, and coral reefs. Among fish, the bluestriped grunt (*Haemulon sciurus*) is known as a denizen of the reef, where it shelters during the day. However, the bluestripe migrates by night to seagrass beds for feeding. This behavior has been documented in both juveniles and adults; it is seen as evidence for habitat diversity in the support of maintaining strong populations of native fauna in protected areas. The interaction of mangrove-based species and reef-based species is another area of similar research.

The Atlantic hawksbill sea turtle (*Eretmochelys imbricata*), an endangered omnivore, depends on isolated beaches for nesting, sometimes migrating to the Caribbean from as far north as Labrador, where it feeds during summer. While enjoying some protection, the hawksbill continues to be a target of tortoiseshell poachers who trade in this illicit product.

Resident birds of the Caribbean Sea range from the common pelican (*Pelecanus occidentalis*) and magnificent frigatebird (*Fregata magnificens*) to snowy plover (*Charadrius alexandrinus*) and clapper rail (*Rallus longirostris*). Migratory species rely on many of the islands and wetlands surrounding the sea to make stopovers on their North American routes. These include the peregrine falcon and least tern.

REEFS AND ATOLLS

The Caribbean region is home to some sixty-five to seventy-five species of hard coral and at least 500 to 700 reef-associated fish species. The center of reef biodiversity in the Caribbean region is the area around the Belize Barrier Reef, which hosts forty to fifty species of hard coral and 80 percent of reef-associated species found in the biome. In addition to providing a home to a vast number of creatures and their ecosystems as well as offering a wealth of economic resources for local inhabitants and tourists, the reefs of the Caribbean offer crucial protection from the destructive hurricanes that frequent the region in the summer and fall months.

More than one-fourth of all fish species in the world are found in coral reef ecosystems. Across the planet, these reefs are under threat from climate change and direct human activity. That behavior generally involves such practices as overfishing, fishing techniques that employ explosives and cyanide, and coastal development. As climate change causes temperatures to rise in oceans and seas, associated coral bleaching is a challenge to coral growth and health. In addition, increased carbon dioxide concentrations are causing seawater acidity levels to rise, a related threat to coral sustainability.

Researchers at the Global Coral Reef Monitoring Network estimate that 20 percent of coral reefs are

Undersea scene with a starfish on massive starlet coral with tropical fish and seagrass, Caribbean sea (Damocean via iStock)

threatened or already harshly impacted. Another 35 percent of coral reefs could be lost in the first few decades of the twenty-first century. Only 45 percent of all reefs are considered to be at low risk. In the Caribbean Sea, about 38 percent of coral reefs have been identified as lost or critically threatened; another 24 percent are considered moderately threatened.

The region around the Caribbean Sea has two true barrier reefs. The largest of these is the Belize Barrier Reef, which stretches for some 137 mi (220 km) from the Yucatan in southern Mexico to the Gulf of Honduras. In the southwest Caribbean, the second major barrier reef is located north of Providencia Island, Colombia. In the Caribbean, the most dominant reefs are those identified as fringing reefs, which means that they are adjacent to mainland areas or to continental islands. Fringing reefs also encircle the majority of the smaller Caribbean islands, including Aruba, Bonaire, Antigua, and the Caymans.

There are only about ten to twenty true atoll reefs in the region of the Caribbean; most of these are in the offshore area that extends from the Yucatan to Nicaragua. Unlike most of the atoll islands of the Pacific, these atolls do not appear to be volcanic. Along the coast of southern Belize lies Glover's Reef, the best-developed atoll in the region. The reefs of the Caribbean region also are distinct from those in the Indo-Pacific region in other ways, including the fact that Caribbean reefs generally do not have prominent algal ridges on reef crests.

In the Indo-Pacific region, reef surfaces tend to be made up of hard corals and microalgae, while those in the Caribbean instead produce an abundant mixture of sponges and octocorals. Because of the differences present here, biodiversity is enriched and reef topography is more complex here.

In 2000, the Seaflower Biosphere Reserve was established, sponsored by the government of Colombia. This reef-based sanctuary includes about one-tenth of the Caribbean Sea. In 2005, a core area of the reserve was designated as a Marine Protected Area.

HUMAN IMPACT

Environmental damage and species disruption in the Caribbean Sea have come in the form of habitat destruction resulting from the development of farm runoff, aquaculture, petrochemical exploitation, and other threats. When the habitat of an endemic species is encroached upon, by definition there is nowhere else for the species to go. At least thirty-eight species are known to have become extinct recently in the Caribbean. Many others, such as two species of giant shrews and the Cuban crocodile, are considered highly threatened.

Additionally, forty-eight bird species, eighteen mammal species, and 143 amphibian species that are endemic to the Caribbean are now threatened. On the positive side, nearly 18,640 sq mi (30,000 km) have been classified as protected areas by various governments within the Caribbean region.

Since the late twentieth century, the reefs of the Caribbean area have become increasingly more vulnerable to overdevelopment along the coast and to overuse for recreational purposes. As the quality of the waters of the Caribbean Sea have been overtaxed by human uses, algae have begun to smother coral in some locales. To deal with reef degradation, Caribbean nations have identified Marine Protected Areas that are organized around protection of coral reefs and related habitats. However, environmentalists warn that much more needs to be done to adequately protect the ecosystems here. They are calling for better management of protected areas, improved monitoring of protected species and sites, stricter enforcement of environmental laws, and significant expansion of existing protected areas.

Climate change impacts in the Caribbean Sea are already particularly evident in sea-level rise, seawater temperature increase, heavier rains, and more violent hurricane events. Such effects are expected to multiply in the decades ahead. Scientific predictions include a doubling of hurricane damage for every 39-in (1-m) rise in sea level, for example. Beach erosion, intertidal zone habitat degradation, coral reef stress, and associated undercutting of biological diversity are among the unfortunate, but likely, outcomes.

—Elizabeth Rholetter Purdy

Further Reading

Beets, J., L. Muehlstein, K. Haught, and H. Schmitges. "Habitat Connectivity in Coastal Environments: Patterns and Movements of Caribbean Coral Reef Fishes with Emphasis on Bluestriped Grunt, *Haemulon sciurus*." *Gulf and Caribbean Research*, vol. 14, no. 2, 2003.

Ettenger, Kreg, and Samantha Hogenson. *Marine Tourism, Climate Change and Resilience in the Caribbean. Volume 1:*

Ocean Health, Fisheries and Marine Protected Areas. Business Expert Press, 2017.
Kappelle, Maarten. *Costa Rican Ecosystems.* U of Chicago P, 2016.
McKeown, Alice. "Coral Reefs Under Threat." *World Watch*, vol. 23, no. 1, 2010.
Miloslavich, Patricia, et al., editors. *Caribbean Marine Biodiversity: The Known and Unknown.* DEStech Publications, 2005.
Norton, Robert. L. *An Inventory of Sea Birds of the Caribbean.* UP of Florida, 2009.
Schrope, Mark. "Conservation: Providential Outcome." *Nature*, vol. 451, no. 1, 2008.
Sokolow, Susanne Helene. *Ecology of Caribbean Coral Disease.* U of California P, 2007.
Spalding, Mark. *A Guide to the Coral Reefs of the Caribbean.* U of California P, 2004.
Venkataraman, K., C. Sivareuman, and C. Raghunathan, editors. *Ecology and Conservation of Tropical Marine Faunal Communities.* Springer, 2013.

Caspian Sea

Fields of Study: limnology; marine ecology; marine biology; geology

ABSTRACT
The Caspian Sea, the largest inland body of water on Earth, holds petroleum and wildlife resources that are crucial for Eurasia's economic and political stability. It is a unique water body that exhibits some traits of a sea and in other ways more resembles a lake.

KEY CONCEPTS
caviar: preserved and processed sturgeon roe (eggs) considered a great delicacy; having very high commercial value, this has driven Caspian Sea sturgeon almost to extinction due to poaching and overfishing
Kara-Bogaz-Gol: one of the two bays in the Caspian Sea
landlocked: any body of water or land region that does not have an ocean shoreline or shipping ports
mud volcano: volcanic vents that emit mud rather than lava

CASPIAN SEA GEOGRAPHY
The Caspian Sea, the world's largest landlocked body of water, is located in central Eurasia. During the Miocene era 26 million years ago, the Caspian Sea was part of a larger sea that included the Black Sea and Aral Sea before tectonic uplift eventually transformed them into distinct basins. Actually, a saltwater lake, the Caspian Sea is bordered by five countries. Kazakhstan lies to the north and east of the sea, while Turkmenistan is also on the sea's east coast. Azerbaijan and Russia (containing the two autonomous republics of Daghestan and Kalmykia) are on the Caspian Sea's western border, and Iran is on the southern coast. Known to humans for centuries, the Caspian Sea was called by the ancient Roman name of "Caspium Mare." Its proximity to the trade route known as the Silk Road assured its importance as a center of trade and commerce until ships came to dominate trade routes.

Unsymmetrically shaped and somewhat resembling the silhouette of a seahorse, the Caspian Sea's size varies because it undergoes cycles of contraction and expansion. On average, the sea stretches about 746 miles (1,200 kilometers) from north to south and from 130 to 271 mi (210 to 436 km) from east to west, encompassing an approximate area of 143,244 square miles (371,000 square kilometers). Six rivers feed freshwater into the Caspian Sea, diluting its brine. The Volga and Emba Rivers enter from the north. The Ural River, considered the geographical division line between Europe and Asia, also ends at the sea's northern edge. The Gorgan and Atrek Rivers flow from the east, and the Kura River is the sea's western tributary. The Elburz and Greater Caucasus mountain ranges are located south and southwest of the Caspian Sea.

Deepest in its southern part, the Caspian Sea has an average depth of 558 feet (170 meters). There are two gulfs along the eastern coast, the Krasnovodsk Gulf and the shallow Kara-Bogaz-Gol, which evaporates and leaves salt deposits. Although the Caspian Sea's level fluctuates every year, it measures an average of 92 ft (28 m) below mean sea level. During the 1960s and 1970s, the Caspian Sea's level was substantially lowered because water that usually flowed into the sea from its tributaries was diverted for agricultural irrigation. In an attempt to stop water loss, engineers built a dike across the mouth of the Kara-Bogaz-Gol in 1980. Instead of creating a long-lasting lake, the dammed-up water evaporated within three years. Subsequently, an aqueduct was necessary to deliver water to the Kara-Bogaz-Gol. Other river and sea dams block both water and fish from moving naturally while protecting oil refineries on land.

Lacking a natural outlet to other large bodies of water, the Caspian Sea is connected by tributaries to adjacent waterways (such as the Black Sea) that can be navigated to reach distant markets. The Caspian Sea's primary ports are Krasnovodsk in Turkmenistan, Makhachkala in Russia, and Baku on the Apsheron Peninsula in Azerbaijan. The frigid winter climate causes ice to form in the northern parts of the Caspian Sea, hindering travel and fishing. Weather patterns produce dangerous storms, mostly moving southeast across the sea, damaging vessels and oil-drilling platforms. These storms are also hazardous to humans, plants and animals living in the water, and reefs and other geological structures.

PETROLEUM AND GEOPOLITICS

Humans have known about the Caspian Sea's valuable natural resources for hundreds of years. Thirteenth-century explorer Marco Polo (1254–1324) commented that Caspian oil was so plentiful that it gushed like fountains. By the 1870s, speculators equipped with drilling technology became wealthy selling oil, and the Caspian Sea region enjoyed prosperity until the 1917 Russian Revolution, when oil barons were arrested and their wells seized. Caspian Sea oil was considered a strategic raw material during both world wars. Soviet leaders abused Caspian oil production, negligently removing oil with inadequate tools and polluting adjacent areas in the process. Because the Caspian Sea contains substantial petroleum deposits, international competition to secure this valuable fuel resource was the catalyst for an oil rush in the late twentieth century that filled the sea with oil derricks.

The largest known oil deposits in the Caucasus region and Central Asia are located in the Caspian states of Kazakhstan and Azerbaijan. Geologists surveyed three major areas of petroleum lying in deposits that began about 11,811 ft (3,600 m) beneath the Caspian Sea's surface. Although the size of these deposits was not immediately determined, their potential seemed to rival the measured deposits of other oil-producing countries. One of the primary deposits is underneath the Caspian Sea. The others are onshore. The second deposit reaches eastward from Baku to Turkmenistan. The third area spreads west of Kazakhstan underneath the sea's northern tip. The Caspian Sea's oil offered an alternative source of petroleum to countries that had been relying on Persian Gulf reserves.

As oil exploration increased in the region in the twentieth century, experts suggested that almost 200 billion barrels (a barrel equals 42 US gallons, or about 159 liters) of oil remained in the Caspian Sea area. On average, 1.1 million barrels was removed daily, in addition to byproducts of sulfur. The area is also rich in natural gas, with an estimated 279 million cubic feet (7.89 million cubic meters) in the late twentieth century, and gas extraction grew rapidly along with petroleum operations. During the 1990s, European and American oil companies such as Amoco (later part of BP) and Chevron secured agreements with the governments of several of the countries bordering the Caspian Sea to invest money, equipment, and labor to build pipelines and extract petroleum. Local institutions such as the Azerbaijan State Oil Academy prepared workers for the technical demands of oil extraction. Geologists rely on seismic imaging to analyze the Caspian Sea for petroleum. Drillers encountered problems with currents and mud volcanoes on the seafloor, and scientists developed refined technology and geophysical extraction methods to locate and remove oil from deep reservoirs of hydrocarbons existing at high temperatures and pressures.

Investors were aware of political and social tensions in the Caucasus region and carefully planned to place pipelines so as to avoid territorial claims on oil. Competing commercially for the oil, foreign nations strove to achieve strategic alliances and to consider diplomatic consequences. They were aware of the vulnerability and instability of many Caspian Sea countries. Several treaties were signed by the littoral Caspian states to legally divide the sea into zones for each state to claim resources. These nations encouraged competition to boost prices because access to Caspian Sea energy resources became an urgent post–Cold War geopolitical objective. Such considerations continued to shape economic and political realities in the region into the twenty-first century, as fossil fuel extraction remained a key economic driver.

FLORA, FAUNA, AND OVERFISHING

The Caspian Sea and its surrounding lands provide habitats for an eclectic assortment of wildlife and plants that enhance the area's environment and economy. The sea's primary water source, the Volga River, forms a swampy delta near the confluence with the sea. This marsh is home to waterfowl, including ducks, swans, flamingos, and herons, as well as eagles that rely on the wetlands for shelter and food. Along parts of the uninhabited eastern shore, camels thrive and wildflowers grow among reeds on the dunes. Fish living in the Caspian Sea include salmon, perch, herring, and carp. The sea is also home to seals, tortoises, and crustaceans unique to the region.

Among the most significant wildlife are the several species of sturgeon found in the Caspian's salty water. Among these, the Beluga sturgeon stretches as long as 13 ft (4 m) and weighs 2,204 pounds (1,000 kilograms), the Russian sturgeon (Ossetra or Asetra) is about 6.6 ft (2 m) long and weighs 441 lbs (200 kg), and the starry sturgeon (sevruga) is just over 3 ft (1 m) in length and weighs approximately 55 lbs (25 kg). Approximately 90 percent of the global sturgeon population lives in the Caspian Sea, producing what gourmets consider the world's tastiest caviar. Eggs compose as much as 15 percent of a sturgeon's weight.

Russian czars generously served Caspian caviar, and Soviet and Iranian entrepreneurs became wealthy exporting the processed sturgeon eggs for hundreds of dollars per ounce (or per gram). In an effort to maintain a monopoly and keep prices high, these countries carefully controlled fishing and cultivated sturgeon in fisheries to replenish the population. After the collapse of the Soviet Union in 1991, sturgeon poaching became a serious problem that threatened the Caspian Sea ecosystem. Nineteenth-century Russians reported yearly catches of 50,000 tons (45,359 tonnes) of sturgeon. In the 1980s, the Soviet Union caught as much as 26,000 tons (23,587 tonnes) of sturgeon annually according to statistics maintained by the Caspian Fishery Research Institute. By the late 1990s, only 3,000 tons (2,721 tonnes)of sturgeon was reported to be harvested by commercial fishermen from Caspian countries each year due to sharp declines in the population and legal restrictions on fishing.

Officially reducing the allowable sturgeon harvest did not solve the overfishing problem, however. Motivated by potential profits and desperate economic conditions, Caspian fishermen poached sturgeon to survive financially by selling fish eggs and meat. One ounce of caviar often sold for more than ten times the amount of an individual's monthly salary. Poachers also saved portions of fish to feed their families. Placing nylon line traps beneath the surface of the Cas-

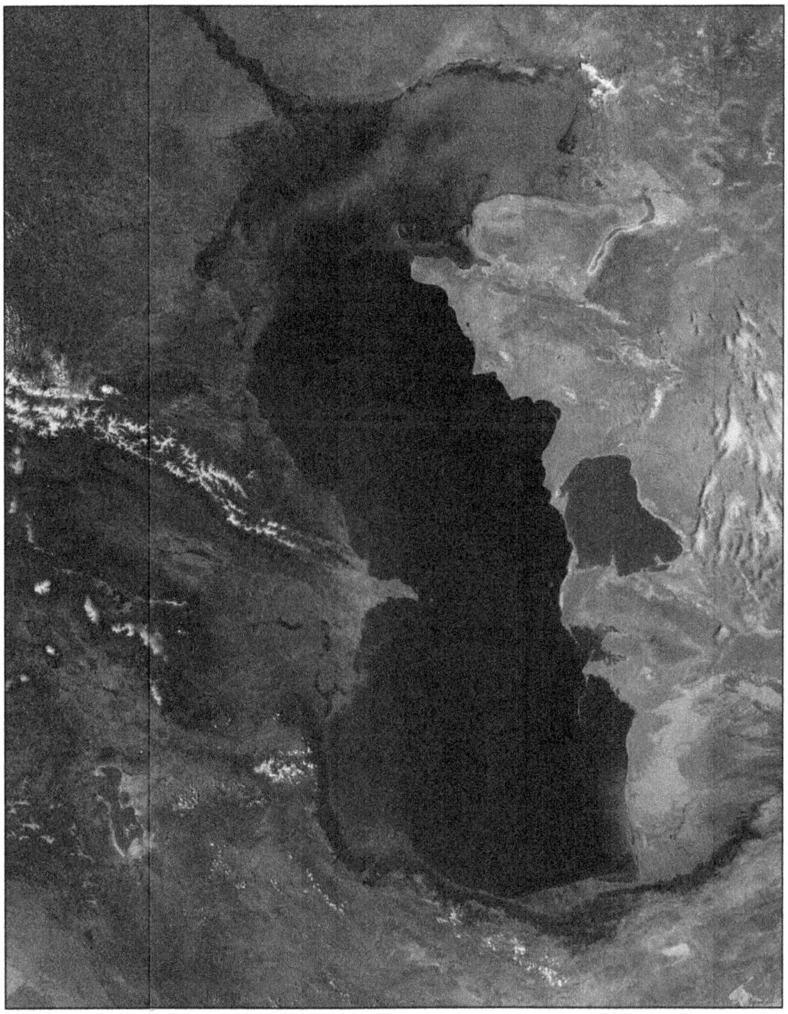

View from orbit of the Caspian Sea (Jeff Schmaltz, MODIS Rapid Response Team, NASA/GSFC via Wikimedia Commons)

pian's tributaries, poachers waited for spawning sturgeon to swim upstream. Such uncontrolled snaring reduced the sturgeon population to the point that it was eventually considered an endangered species. Experts stressed that overfishing resulted in the deaths of approximately 90 percent of sturgeons too immature to spawn (most sturgeons require an estimated eighteen years to develop reproductive capabilities), which detrimentally reduced the number of fish returning to the Caspian's rivers every year. Fewer Beluga sturgeon were captured annually, and not many reached their potential to be as heavy as 1.10 ton (1 tonne) and survive for a natural one-hundred-year life span. Despite ongoing efforts to reduce fishing, increase aquaculture, and otherwise regulate the fisheries industry, these environmental threats remained an important issue into the twenty-first century.

ENVIRONMENTAL ISSUES

Pollution also poses substantial dangers for sturgeon and other wildlife. The Caspian Sea has been contaminated by oils and petrochemicals from drilling platforms, as well as by pollutants and industrial residues dumped into rivers by factories. Careless drilling methods caused oil films as thick as 0.24 inches (6 millimeters) to float on the Caspian Sea, creating an unhealthy environment for fish. Raw sewage was also discharged from drains into the sea's tributaries. Scientists in the late twentieth century found that the waters near the cities of Baku and Sumqayit were so saturated with poisonous wastes and were so depleted of oxygen and nutrients that they could not harbor aquatic life.

The Caspian Sea benefits from the marshes in the Volga River's delta, which strain out some of the debris and contaminants before the water reaches the sea. Near the river's mouth, which consists of thousands of slender streams moving through the stalks of indigenous water plants, the Caspian Sea's shallow pools of sparkling blue water seem clean. Moving toward the sea's interior, however, the water becomes murky and appears gray in color.

Experts warn that the Caspian Sea is experiencing an environmental crisis that will only worsen as extraction of natural resources increases and industrialization expands in bordering countries. Authorities have blamed these pollutants for contributing to the growing rates of miscarriages and stillbirths in Caspian nations, and industries are now encouraged to reduce pollution and to cease using rivers for effluent disposal. Petroleum companies operating onshore oil fields and platforms in the Caspian Sea have been advised to adapt their operations to rigid standards regarding drilling to prevent tainting the water with contaminating fluids and detritus.

Ecologists want extractors of natural resources to become more aware of their impact and accountability to maintain the Caspian Sea's environmental quality. Scientists demand that bans on poaching and illegal caviar processors be enforced more diligently, with police monitoring the sea and arresting offenders. They also seek to prohibit open-sea fishing, which kills young sturgeon, and recommend that Caspian Sea nations agree on caviar quotas so that fair prices are set to discourage black-market profiteering and poaching. Public relations campaigns continue to work to promote public knowledge of the Caspian Sea's environmental crises and provoke more responsible treatment of its resources.

CHANGING SEA LEVEL

The Caspian Sea is well known for its periodical variation in water level, which is affected by various factors as the water body is a self-contained system subject to rainfall and evaporation. Biologists monitoring the sea noted that after a lengthy period of contraction, its average level rose 6.6 ft (2 m) from the late 1970s to the 1990s. Some researchers optimistically believed that the increased amount of water might weaken the effect of pollutants on the sea and on its inhabitants unless the amount of toxic material and oil spills entering the sea drastically increased. Many scientists speculated that global warming and the region's climate might have contributed to the rise in sea level.

Other scholars hypothesized that the water level increases might have been caused by natural processes, possibly initiated by climatic deviations that affect the atmosphere and sea over periods of years, decades, or greater periods of time. Scientists in the early twenty-first century have suggested that such trends, together with the broader effects of anthropogenic climate change, are likely to contribute to a marked decline in the sea level moving forward. System modeling pointed to evaporation as the driver of sea level decline observed from 1996 into the 2010s, leading to fears of a similar situation to the Aral Sea desiccation.

The erratic changes in the sea's composition and contents that were documented in the past and that periodically recur will crucially affect future development of the Caspian Sea's resources. The quantity and quality of the Caspian Sea's water are vital to ensure Eurasian economic and political stability and to maintain international technological and business interests in the region.

—*Elizabeth D. Schafer*

Further Reading

Amirahmadi, Hooshang, editor. *The Caspian Region at a Crossroad: Challenges of a New Frontier of Energy and Development*. St. Martin's Press, 2000.

Brunet, M.-F., M. Wilmsen, and J. W. Granath, editors. *South Caspian to Central Iran Basins*. Special Publication 312. The Geological Society of London, 2009.

Chen, J. L., et al. "Long-Term Caspian Sea Level Change." *Geophysical Research Letters*, vol. 44, no. 13, 2017, doi:10.1002/2017GL073958.

Cullen, Robert. "The Rise and Fall of the Caspian Sea." *National Geographic*, vol. 195, May 1999, pp. 2–35.

Dumont, Henri, Tamara A. Shiganova, and Ulrich Niermann, editors. *Aquatic Invasions in the Black, Caspian, and Mediterranean Seas*. Kluwer Academic Publishers, 2004.

Khodorevskaya, R. P., G. I. Ruban, and D. S. Pavlov. *Behaviour, Migrations, Distribution and Stocks of Sturgeons in the Volga-Caspian Basin*. World Sturgeon Conservation Society Special Publication No. 3, 2009.

Kostianoy, Andrey G., and A. Kosarev. *The Caspian Sea Environment*. Springer-Verlag, 2005.

Nihoul, Jacques C. J., Peter O. Zavialov, and Philip P. Micklin, NATO Scientific Affairs Division. *Dying and Dead Seas: Climatic versus Anthropic Causes*. Kluwer Academic Publishing, 2004.

Stolberg, F., United Nations Environment Program. *Caspian Sea*. United Nations, 2006.

Szalay, Jessie. "Caspian Sea: Largest Inland Body of Water." *LiveScience*, 24 Feb. 2017, www.livescience.com/57999-caspian-sea-facts.html.

Zonn, Igor S., et al. *The Caspian Sea Encyclopedia*. Springer-Verlag, 2010.

CHEMICAL PRECIPITATES

Fields of Study: chemistry; geochemistry; biogeochemistry; oceanography; geology

ABSTRACT

Chemical precipitates, and precipitates in general, are useful in nearly all human endeavors. Precipitates are materials formed by precipitation—the formation of solids from solution either as the insoluble products of chemical reactions or by the settling of suspended solid particles. Snow and rain are also precipitates, having formed from water vapor that is "dissolved" in air. Uses are numerous, including pigments and water treatment. Additionally, precipitates occur in nature to form rocks such as shale and limestone.

KEY CONCEPTS

carbonate compensation depth: the ocean depths at which carbonate materials become soluble

cementation: the binding of relatively large particles into a solid whole by the formation of crystalline structures in the spaces between the particles

diagenesis: the process by which an accumulation of sediment is converted to sedimentary rock through the combination of compaction, cementation, and the development of crystal growths between the sediment grains

flocculation: the process of binding suspended particles in a colloid

pH: formally defined in value as the negative logarithm of the hydrogen ion concentration of a solution, according to the equation: $pH = -\log[H^+]$

volatile: able to enter the gas phase by evaporation or sublimation

THE SOLVATION PROCESS

A proper understanding of how things dissolve is needed to discuss chemical precipitates and their uses. Chemical bonds are, in general, either ionic or covalent. Covalent bonds occur where atoms share electrons. These bonds are far stronger than ionic bonds, which occur when electrons are transferred. A covalent bond occurs when there is a large difference in electronegativity.

Electronegativity is the measure of how tightly an atom holds its electrons. Atoms with high electronegativity, such as fluorine and oxygen, hold their electrons tightly, whereas atoms of low electronegativity, such as sodium and boron, hold their electrons loosely. This is the result of properties of electron orbitals.

Atoms with eight electrons in their outermost orbital are the most stable. As a result, atoms with many

Water treatment plant process at sunset (tuachanwatthana via iStock)

electrons in their outermost orbital, such as fluorine (with seven electrons) and oxygen (with six electrons), easily gain more electrons; atoms with few electrons, such as sodium (with one electron) and calcium (with two electrons), easily lose electrons. When atoms of different electronegativity bond, the less electronegative atom transfers its electrons to the more electronegative atom. Once the electrons are transferred, the atoms become ions and are then held together by electrostatic attraction.

Equally covalent bonds result in sharing; and when one is more electronegative than the other, as in water, the result is a polar molecule. In such a case, the oxygen atoms hold the electrons more tightly than the hydrogen atoms, meaning that the oxygen portion has a net negative charge. The hydrogen portion is positive. Water's shape is also of note, as it is a bent molecule. However, regardless of the shape of the molecule, the electronegative principles are the same.

Taking the example of salt dissolving in water, one can see the effect of these properties. When table salt (sodium chloride, NaCl) is placed in water, it dissolves. Sodium chloride is an ionic compound. Because its bonds are weaker than the electrostatic attraction of the dipolar structure of water molecules, the sodium atoms are separated from the chlorine atoms and align to the water molecule's poles. The water molecules surround the sodium and chlorine ions, and they become distributed throughout the water.

Solution systems are not static; they exhibit dynamic equilibrium. The solute continuously reforms and redissolves, coming in and out of solution. When the rate of reformation exceeds the rate of dissolution, the solute comes out of solution as a precipitate; if the converse is true, the solute simple redissolves.

Many factors affect solubility. The behavior is particularly complex because it depends on the interactions within the crystal lattice of the solute material.

Only recently have computer models allowed for the prediction of solubility. The solubility of any given substance is related to temperature and pressure.

Also important to note is that not all soluble substances are soluble in the same solutes. The major division is between organic and inorganic solvents. In general, organic solvents dissolve organic solutes best, and inorganic solvents work best on inorganic solutes.

ROCK FORMATION

Some sedimentary rocks are formed by the process of precipitation as well. Water is a polar molecule and a good solvent. Seawater contains many kinds of solutes, and the average gallon (or liter) has about 35 grams of dissolved salts.

The oceans exhibit a great deal of variance in temperature and pressure across their expanse and depths, so it is only natural that they should also display a wide variance in what and how much they can dissolve. The varying nature of seawater leads to different rock formations.

A good example of this is the formation of limestone, which is made of calcium carbonate (a soluble). In equatorial surface seawater, however, there is so much limestone dissolved in the seawater that it is practically insoluble. As a result, it tends to come out of solution to eventually form rock in the process of diagenesis. The vast majority of limestone laid down, however, comes from sea-life shells and tests, which are composed of calcium carbonate as well. When these creatures die, they fall to the seabed and form layers of calcium carbonate. A different scenario is possible, however.

Should a creature's shell fall below a certain depth, it will dissolve. This depth is known as the carbonate (or calcite) compensation level. Under conditions that exist below a certain depth, calcium carbonate's solubility increases dramatically. The calcium compensation level depends upon many factors but is about 138 to 1,640 feet (42 to 500 meters) in the Pacific Ocean, except in the equatorial upwelling zone, where it is about 16,404 ft (5,000 m). Below this level, the seabed is mainly made up of silicates; above this level is carbonates.

Additionally, under certain conditions, dissolved substances precipitate from water. Warm water often releases many precipitates, causing the mineral deposits in many salt lakes. Additionally, the deposition of certain minerals reveals to climatologists the temperature of the seas.

Another result of precipitates is manganese nodules. These are metal-rich spheroids that form on the ocean bottom. They may prove important in future metal production, particularly for manganese and certain other metallic elements.

CEMENTATION

Sedimentary rocks are formed by minerals coming out of solution and settling on the bottom of a body of water. As this happens over time, the layers begin to compress. Water containing ions that flows between the grains and precipitates them out to form crystalline material between the grains, joining them. This can occur through groundwater or in seafloor sediments.

This process is used by humans in the making of cement. Various chemicals are mixed in an aqueous solution and then are left to dry. As the moisture content changes, the chemicals are left to precipitate out and crystallize. These crystals give concrete its strength.

COLLOIDS AND SUSPENSION

Solids can be contained in water by means other than through dissolving. Solids also can be contained as suspensions and as colloids. A suspension is a system in which the particles are larger than single molecules and ions but are still small enough to remain in the solvent. A colloid is a substance in which the particles are distributed homogeneously and stably; without changes to the system, the colloid will not separate. Suspensions, in contrast, will naturally separate over time. This process is called "flocculation."

Flocculation is the process wherein colloids come out of suspension in the form of clumps of particles called "floc." Flocculation differs from precipitation in that the particles are suspended in a colloid rather than dissolved in a solution. The flocs can then come out of suspension.

WATER TREATMENT

Water treatment is the process by which water is cleaned for human use. The use of precipitates is critical for treatment, but much water treatment does not rely on precipitates.

First, in water treatment, waste is screened out using filters. The chunks of waste are usually disposed of in a landfill after drying. Then, after water filtration, the resulting waste is directed to a settlement tank, where the waste particles are allowed to settle. The result is called "sludge," which can be refined into fertilizer. The water is then disinfected by chlorination.

The properties of precipitates can remove solutes from water. This is important because many solutes have negative effects upon water. Some solutes, such as iron, can stain objects, such as water fixtures, and can change the color of the water. Sometimes solutes are clear when drawn from a tap but turn yellow upon standing (in place without movement). Other solutes have unpleasant odors, such as hydrogen sulfide, which smells like rotten eggs. Another common concern is the solute pH (potential hydrogen), which can affect corrosion and other factors of water quality. These chemical properties are caused by things dissolved in water; precipitating removes the properties from the water.

Aeration, the placing of a solute in contact with air, is a common means of precipitation. In this method, a substance dissolved in water is removed from the water by tiny bubbles passing through the solution. The increase of surface area between the solution and the air bubbles creates a space for dissolved gases to react with the gases in the air bubbles, in this manner the air bubbles "scrub" the water clean of the volatile solutes.

Another common precipitation process is water softening. Hard water is water with a high dissolved mineral content. Most processes aim to remove calcium, magnesium, and other metal ions from the water. One way of doing this is to add compounds to the hard water that will react with the dissolved metals to form an insoluble compound. Another process known as ion exchange works by exchanging one type of ion in a substance for another. Because the goal in water softening is to remove certain kinds of ions, the exchange of ions works fine. Typically, the process works by passing water through a mass of specially treated resin beads.

Another method, often used in the American Midwest, Florida, and Texas, is lime softening, which involves adding lime to water. The water then undergoes several reactions that lead to the removal of ions through precipitation. This process is notable because the total dissolved particles decreases (with ion exchange, little to no change occurs).

Reverse osmosis is another water treatment method. Osmosis is a natural process that involves moving water (or any other liquid with different concentrations) through a semipermeable membrane. The liquids tend to move from regions of low solute concentrations to high concentrations to create chemical potential equilibrium. Reverse osmosis involves forcing a liquid with dissolved solute through the semipermeable membrane by applying a pressure greater than the osmotic pressure. This filters out the solutes. The pores in the filters are small enough to prevent the ions from passing through them.

Chelation is often used as well. In chelation, metal ions are dissolved by organic compounds. By dissolving them in the organic compound, they are removed from the water.

By all of these processes, salts and metal ions can be removed. Salts are ionic compounds that result from the neutralization of an acid with a base. Acids are acidic because they produce H^+ ions, whereas bases produce OH^- ions. When acids and bases come into contact, they undergo a special reaction called "neutralization." In neutralization, the H^+ and OH^- ions form water, leaving the anions (negative ions) of the acid and the cations (positive ions) of the base to form an electrically neutral compound. Because it is made of ions in this manner, it is an ionic compound.

To change the pH of the water, one can add either an acid or a base to neutralize the liquid's excess. (pH is a measure of how acidic or basic something is.) According to the system, pure water is neutral with a pH of 7. Solutions with pH less than 7 are acidic and those with pH greater than 7 are basic or alkaline. pH is a logarithmic scale based on hydrogen ion activity. Improper pH can cause problems, such as pipe corrosion and harm to plants and other vegetation, so managing pH is an important part of water treatment.

PIGMENTS

A pigment is a material that provides color to something as a result of absorbing and reflecting certain wavelengths of light. Pigments are used to make many things, ranging from clothing to paint. Pigments are suspensions, but precipitation is important for manufacturing a specific kind of pigment called a "lake pigment."

A lake pigment is made when a dye is precipitated with an inert binder, called a "mordant." A dye is a colored solution. Dyes do not always hold well, so they need to be changed chemically to supply color. The mordant is a metallic salt. Metallic salts are compounds like sodium chloride. The colored precipitant of a reaction of a dye and a mordant, such as alumina, is usually more stable and will remain in the object being colored better than the dye alone would.

This process is ancient. In earlier times chalk, white clay, and bones were used for their calcium carbonate and calcium phosphate. Now, however, the most common salts include aluminum hydroxide, alumina (aluminum oxide), calcium sulfate, and barium sulfate. No matter the era, these mordants were all insoluble in dyes and had neutral colors. The neutral colors are needed so that the color of the precipitate is dependent upon the dye.

—Gina Hagler

Further Reading

Burdige, David J. *Geochemistry of Marine Sediments.* Princeton UP, 2020.
Chang, Raymond. *Chemistry.* McGraw-Hill Higher Education, 2007.
Crittenden, John C. *Water Treatment Principles and Design.* Wiley, 2005.
Delamare, Francois, and Bernard Guineau. *Colors: The Story of Dyes and Pigments.* H. N. Abrams, 2000.
Edzwald, James K. *Water Quality and Treatment: A Handbook on Drinking Water.* McGraw-Hill, 2011.
Faulkner, Edwin B., and Russell J. Schwartz. *High Performance Pigments.* Wiley, 2009.
Lutgens, Frederick K., and Edward J. Tarbuck. *Essentials of Geology.* Prentice-Hall, 2012.
Middelburg, Jack J. *Marine Carbon Biogeochemistry: A Primer for Earth System Scientists.* Springer Open, 2019.
Williams, Richard G., and Michael J. Follows. *Ocean Dynamics and the Carbon Cycle: Principles and Mechanisms.* Cambridge UP, 2011.

CHESAPEAKE BAY

Fields of Study: oceanography; limnology; geology; paleogeology; marine biology

ABSTRACT

Chesapeake Bay is the largest of the 130 estuaries occurring in the conterminous United States and the third-largest in the world. The bay is 180 miles (290 kilometers) long and 3 to 35 mi (5 to 56 km) wide. The main bay averages less than 30 feet (9 meters) deep, and the entire bay, including tributaries, averages about 20 ft (6 m).

KEY CONCEPTS

bolide: a stony object from space, whether meteor or comet, that strikes Earth's surface

estuary: locations where a stream or river of freshwater meets and flows into an ocean or other large body of saline water

hydrology: study of the type and behavior of water

intertidal flat: a broad, flat area of land that is under water at high tide and exposed at low tide

oyster bed, oyster reef: an area of shallow sea bottom that supports colonies of oysters

DESCRIPTION

Chesapeake Bay is the largest of the 130 estuaries occurring in the conterminous United States and the third-largest in the world. Bordering the Atlantic Ocean, it extends from its mouth near Norfolk, Virginia, north to Havre de Grace, Maryland. The bay is 180 mi (290 km) long and 3 to 35 mi (5 to 56 km) wide. The main bay averages less than 30 ft (9 m) deep, and the entire bay, including tributaries, averages about 20 ft (6 m). Several deep holes indicating the remnants of the ancient Susquehanna River exceed 171 ft (52 m). Substrate is comprised primarily of gray silt and clay in the bay's center, with fine and medium sands near the shore.

More than 11,000 mi (17,000 km) of shoreline, including mainland and islands, are associated with the bay. The bay's shorelines are a mixture of developed and conserved lands. The Baltimore, Maryland, metropolitan area, with 2.7 million people, is the most populous city directly on the bay's shore. Two major roadways cross the bay: the 4-m (7-km) Chesapeake Bay Bridge near Annapolis to Kent Island, Maryland; and the 17-m (28-km) Chesapeake Bay Bridge Tunnel from Virginia Beach to Cape Charles, Virginia. Substantial revenues are generated from trade and commercial fishing in Chesapeake Bay waterways. In addition, tourism and recreation contribute billions of dollars to the regional economy.

HYDROLOGY

Chesapeake Bay started to form its present shape over 3,000 years ago, following repeated glacial melting that began about 18,000 years ago in the ancient Susquehanna River watershed. Evidence suggests that a bolide (meteor or comet) struck the region 35 million years ago, creating a 55-m-wide (89-km-wide) crater. These conditions eventually led to the formation and placement of Chesapeake Bay and its watershed, an area that stretches across 64,000 sq mi (165,000 sq km) of land.

The bay is fed by nineteen large rivers and over 150 smaller rivers, which in turn are fed by more than 100,000 streams in the watershed. The Susquehanna River provides the largest—nearly half—of freshwater inflow. Rivers on the western shore generally are larger than those on the eastern shore. The Chesapeake Bay watershed encompasses portions of Delaware, Maryland, New York, Pennsylvania, Virginia, and West Virginia. This land area is primarily forested (57 percent) although much reduced from historic levels due to agriculture and development. In 2016, an estimated 18.1 million people lived in the watershed.

Tides rise and fall twice daily, with tidal range varying from 1 to 3 ft (0.3 to 0.9 m) as of 2017. Salinity varies based on location, season, and depth, averaging 1.5 percent (15 parts per thousand) and varying from 0 to 3.6 percent (0 to 36 parts per thousand). Salinity is generally greater near the bottom, due to the lighter freshwater emptying into the bay from the streams and stratifying above the heavier ocean-derived saltwater. The mouth of Chesapeake Bay is saltier and salinity generally decreases farther north due to the greater influence of streams. The bay occurs in the humid subtropical climate zone, which exhibits hot, humid summers and mild to cool winters. Ice often builds up during the winter along the Delmarva

Eastern oysters are central to Chesapeake Bay's commercial fishery (NOAA Fisheries)

Peninsula, located on the eastern shore, due to the prevailing northwest winds. During extreme cold periods, ice may even accumulate on the western shore.

BIODIVERSITY

The Chesapeake Bay estuary supports more than 3,600 species of plants and animals. Typical habitat features of the bay include sand beaches, intertidal flats, shallow subtidal zones, wetlands, oyster reefs, and subtidal deep water. Sand beaches are most common in the southern part of the bay but overall are less abundant than on the Atlantic Coast proper.

Beach "fleas" (genus *Emerita*), ghost crabs (*Ocypode quadrata*), and horseshoe crabs (*Limulus polyphemus*) are common inhabitants of sandy beaches. Horseshoe crabs, with their characteristic domed bodies and spiked tails, are primitive crustaceans dating back 360 million years. These crabs lay their eggs in the sand, particularly in the southern and middle portions of the bay, during high spring tides. Although not as abundant as in Delaware Bay, horseshoe crabs are still extremely common here. Each crab lays 10,000 to 100,000 eggs, attracting shorebirds that consume many of the eggs. Humans harvest the crabs for use in the biomedical industry and as bait for commercial fishing.

Intertidal flats composed of mud and sand are valuable areas for a variety of invertebrates, including fiddler crabs (*Uca* spp.), marsh crabs (*Sesarma reticulatum*), and snails. Shallow subtidal zones are extremely productive areas, with an abundance of submerged aquatic vegetation (SAV) and a suite of fish, invertebrates, and waterbirds. SAV is a common and integral component of Chesapeake Bay; the twenty-six species occurring in the bay contribute to clean, productive water. Areas of higher salinity are dominated by widgeon grass (*Ruppia maritima*) and eelgrass (*Zostera marina*), which harbor a less diverse assemblage of species than freshwater areas of the bay. SAV is important for filtering pollutants and nutrients, including nitrogen and phosphorous, from the water and for providing food for fish, crabs, and waterfowl. If pollution levels become too high, SAV declines. In 1978, Chesapeake Bay had a recorded 41,000 acres (16,400 hectares) of SAV; as of 2016, this figure increased to about 97,433 ac (39,000 ha).

Chesapeake Bay is well known for native American or Eastern oyster (*Crassostrea virginica*) beds. These dense oyster reefs attract diverse assemblages of micro- and macroorganisms that live in and among the oyster beds. Eastern oysters are ecologically important inhabitants of the bay and are central to the region's commercial fishery. Their abundance has declined since the mid-1800s, but at their peak, they filtered the entire volume of water in the bay every three to four days, removing algae and particulates from the water.

Oyster beds also provide habitat for numerous other species. In 2011, the oysters were at less than 5 percent, and harvests were less than 2 percent of their historic densities. The decline is attributed to increased efficiency of harvest techniques, degradation of water quality and habitat, and increased prevalence of diseases and parasites. Historic harvest records are rare, but in 1980, 3.1 million bushels (109,000 cubic meters) were harvested, compared with only 370,000 bsh (13,000 cu m) in 2001. As of 2018, man-made oyster beds in Chesapeake Bay were helping to restore the Eastern oyster population.

The forested, shrub, and emergent wetlands, including both tidal and nontidal areas, span the gradient from freshwater to saltwater, providing the plant material that breaks down into detritus, forming one of the primary foundations of the bay's complex food web. Indeed, consumption of detritus by organisms is about ten times greater than herbivorous grazing in the bay. The eastern shores have larger expanses of marshes, providing habitat for migratory waterfowl during the spring and fall migrations, as well as during the winter. About 40 percent of the waterfowl in the Atlantic Flyway overwinter on Chesapeake Bay. Avian types range from ducks, terns, and gulls to apex predators including osprey (*Pandion haliaetus*) and bald eagles (*Halieaeetus leucocephalus*).

Large fish, including bluefish (*Pomatomus saltatrix*) and red drum (*Sciaenops ocellatus*), are among the most impressive inhabitants of the subtidal deepwater zones, but phytoplankton and zooplankton are dominant, integral contributors to the bay's productivity. Several anadromous fish species, including striped bass (*Morone saxatilis*), American shad (*Alosa sapidissima*), and American eel (*Anguilla rostrata*), live in the deepwater zones and enter the streams feeding Chesapeake Bay to spawn.

Great Blue Heron standing on a Chesapeake Bay beach at sunset (flownaksala via iStock)

HUMAN IMPACT AND CONSERVATION

Chesapeake Bay has a long human history; Indigenous peoples first inhabited the Chesapeake Bay watershed about 13,000 years ago, following the retreat of the glaciers. Oyster shells were used extensively by these early inhabitants as tools and jewelry, but the extent to which they were consumed for food is unclear. Certainly, mastodons (*Mammut* spp.), mammoths (*Mammuthus* spp.), and other large mammals were pursued for their food value. As the climate warmed in the region, these paleo-Indians were replaced by the archaic Indians about 9,000 years ago; they also primarily subsisted on hunting but apparently had a stronger connection to the aquatic resources, including fish and shellfish, and the migratory birds provided by the Chesapeake Bay. Woodland Indians moved into the Chesapeake Bay region some 3,000 years ago, remaining until after European settlers arrived; evidence of many villages and stronger relation to bay resources survives.

Chesapeake Bay has been negatively affected and will continue to be so, both by direct and indirect human impact According to US Fish and Wildlife reports, there is a long list of threatened and endangered species of flora and fauna in the bay area. Climate change is predicted to lower salinity due to increased flows from major rivers such as the Susquehanna, which could potentially affect important bay species that are intolerant of low salinity conditions, such as oysters and crabs. Moreover, sea-level rise from melting glaciers, due to global warming trends, will submerge some salt marshes. Models predict greater future increases in mean temperatures, and lower dissolved oxygen levels, which may affect numerous organisms.

The bay's productivity is limited by degradation of water quality caused by excessive chemical, nutrient, and sediment input. Sediment accretion rates in the bay have increased one and a half to seven times over the pre-European settlement rate. Europeans started settling the region in the early seventeenth and eighteenth centuries. Between 1830 and 1890, 80 percent of the land was cleared of timber, and much of it was plowed and put into agriculture.

Clearing of the land led to increased flow of freshwater into the bay, resulting in greater sediment deposition. With the advent of commercial fertilizers since the mid-1800s—most notably those containing nitrogen and phosphate compounds—fertilizer discharge has increased, as have sewage discharge and industrial waste. The 1940s brought about common agricultural practices implementing a variety of pesticides. Starting in the 1960s, efforts have been made to reduce the effects of contaminated runoff. Water quality continues to fluctuate, as do population numbers of various species in the region.

In 2009, Chesapeake Bay was declared a national treasure, and then-President Barack Obama issued an executive order to promote and restore its health, heritage, and natural resources. The US Environmental Protection Agency (EPA) mandated goals in the form of Total Maximum Daily Load for the amount of pollutants that can enter the bay. These goals and related solutions to reduce sediment, nitrogen, and phosphorous runoff were implemented for each state in the watershed. In addition to federal and state agencies, numerous local and regional watershed groups, such as the Chesapeake Bay Foundation, work to improve water quality in the bay. In 2017, the EPA recorded that, as a result of the Total Maximum Daily Load as well as other restoration efforts, more than 40 percent of bay tidal waters met water quality standards.

—*James T. Anderson*

Further Reading

Chesapeake Bay Program, 2018, www.chesapeakebay.net/. Accessed 29 June 2018.

Committee on Nonnative Oysters in the Chesapeake Bay. *Nonnative Oysters in the Chesapeake Bay*. National Academies Press, 2004.

"EPA Outlines Final Stage of Chesapeake Bay Restoration Effort." *Storm Water Solutions*, 28 June 2018, www.estormwater.com/epa-outlines-final-stage-chesapeake-bay-restoration-effort.

Gohn, Gregory S., Christian Koebert, Kenneth G. Miller, and Wolf Uwe Reimold, editors. *The ICDP-USGS Deep Drilling Project in the Chesapeake Bay Impact Structure: Results from the Eyreville Core Holes*. Geological Society of America, 2009.

Kennedy, Victor S. *Shifting Baselines in the Chesapeake Bay: An Environmental History*. Johns Hopkins UP, 2018.

Murdy, Edward O., and John A. Musick. *Field Guide to Fishes of the Chesapeake Bay*. Johns Hopkins UP, 2013.

National Oceanic and Atmospheric Administration. "Chesapeake Bay Office." chesapeakebay.noaa.gov.

National Research Council. *Achieving Nutrient and Sediment Reduction Goals in the Chesapeake Bay: An Evaluation of Program Strategies and Implementation*. National Academies Press, 2011.

United States Geologic Survey. "The Chesapeake Bay Bolide: Modern Consequences of an Ancient Cataclysm." woodshole.er.usgs.gov/epubs/bolide.

CHILE INTERTIDAL ZONES

Fields of Study: oceanography; marine ecology; marine biology

ABSTRACT

Chile is a long, narrow country that is defined by the Andes mountain range on the east and by the Pacific Ocean on the west. Chile is 2,672 miles (4,300 kilometers) long and an average of 109 mi (175 km) wide. This long strip of coastal land runs parallel to the Andes and extends from midcontinent to Cape Horn, the southernmost point of South America. The coastal climate is strongly influenced by the slow, cold Humboldt Current.

KEY CONCEPTS

intertidal zonation: the distribution of intertidal species according to the range of low and high tide water levels

DESCRIPTION

Chile is a long, narrow country that is defined by the Andes mountain range on the east and by the Pacific Ocean on the west. Chile is 2,672 mi (4,300 km) long and an average of 109 mi (175 km) wide. This long strip of coastal land runs parallel to the Andes and extends from midcontinent to Cape Horn, the southernmost point of South America.

The coastal climate is strongly influenced by the Humboldt Current. This slow, cold Pacific current sustains one of the world's richest fisheries along the

Chilean coast. Many forms of wildlife are also found in abundance along the adjacent land areas. The Humboldt Current begins north of the Antarctic and runs the full length of the Chilean coast, bringing nutrient-rich waters that nourish the intertidal zones. Intertidal zones along shorelines are the areas between the high tide and the low tide and can include many types of habitat: steep rocky cliffs, sandy beaches, wetlands, and marshes.

The organisms that live here have become specially adapted to survive the significant changes in temperature, salinity, moisture, acidity, dissolved oxygen, and food supply that occur on a daily basis due to the movement of the tides. Different plants and animals that have excelled at dealing with different sets of conditions dominate zones at different tidal heights. This distribution pattern of different species along the shore in vertical bands is referred to as intertidal zonation.

Varied climatic conditions along Chile's long coast also contribute to the diversity of intertidal species: An extremely dry climate is typical of the north, but the coast exhibits a rainforest microclimate, as moisture from the sea is trapped by cliffs and valleys; the central region of the country is characterized by the hot, dry summers and mild, wet winters of a Mediterranean climate; and the southern region is cooler and wetter, classified as an oceanic temperate climate.

The complex relationship between the Humboldt Current and the intertidal zones supports a wealth of oceanic and land organisms. Near the coast are found some of the world's greatest populations of shrimp, anchovies, squid, sardines, salmon, and mackerel. Marine birds and mammals—such as penguins, pelicans, flamingos, bottle-nosed dolphins, whales, and sea lions—live off of the smaller of these fish, mollusks, and crustaceans. Intertidal species such as sea urchins, starfish, and coral also contribute to this diverse, yet fragile ecocommunity, as do seaweeds, kelp, and shoreline grasses.

Many of these organisms are adapted to extremes, as the local water may be fresh, brackish, or salty. They may be subject to varied wave action, and must be able to live exposed to air and sun, or submerged in water. Many types of shellfish thrive here, most notably the Chilean blue mussel, which is harvested by humans regularly. Wild fisheries and salmon fish farms have become very economically important to Chile in the past 30 years.

HUMAN IMPACT AND CONSERVATION

Because there has been an increasing worldwide demand for fish over the past several decades, wild fisheries are being overexploited. Salmon farms consume many of these wild fish, and also have caused problems on some coastal areas due to pollution runoff, low oxygen zones, algal blooms, and the spread of viruses into the marine environment. Inland mining practices have also had a negative impact on shoreline species. Forestry, pulp, and paper industries are present in the south, releasing pollutants that harm the intertidal zones.

The Chilean coast, including the intertidal zone, is administered by the Ministry of National Defense through the Marine Subsecretary, which addresses issues related to national and state public goods in the intertidal zones—including harvesting of marine species. Under this Marine Subsecretary is another division that regulates such diverse activities as aquaculture and harbors. State ownership of the coast tends to occur in the extreme north or south of Chile, as the other coastal lands became private during the long colonial period.

Since the 1990s, there has been extensive Chilean research on intertidal zones, and two coastal marine research stations have been created: Estación Mehuín in southern Chile of the Universidad Austral de Chile, and Estación Costera de Investigaciones Marinas in Las Cruces in Central Chile of Pontificia Universidad Católica de Chile.

In 2001, Oceana was founded. It is the largest international organization focused solely on ocean conservation. The Oceana offices in North America, Central America, South America, and Europe work together on a limited number of strategic, directed campaigns to achieve measurable outcomes that will help return the oceans to former levels of abundance. (Oceana Chile is located in Santiago, Chile.) In its first decade, Oceana has achieved some policy victories for marine life and habitats. As oceans play a key role in regulating Earth's temperatures, this organization is currently considering the impacts of global warming and is moving forward with public education, legislation advocacy, and ecosystem protection.

—Magdalena Ariadne Kim Muir

Further Reading

Castilla, J. C. "Roles of Experimental Marine Ecology in Coastal Management and Conservation." *Journal of Experimental Marine Biology and Ecology*, vol. 250, nos. 1–2, 2000.

Fariña, José Miguel, and Andrés Camaño, editors. *The Ecology and Natural History of Chilean Salt Marshes.* Springer, 2017.

Heileman, S., R. Guevara, F. Chavez, A. Bertrand, and H. Soldi. "XVII-56 Humboldt Current: Large Marine Ecosystems #13." 2009, www.lme.noaa.gov/lmeweb/LME_Report/lme_13.pdf.

Kämpf, Johan, and Piers Chapman. *Upwelling Systems of the World: A Scientific Journey to the Most Productive Marine Ecosystems.* Springer, 2016.

Siegfried, W. Roy. *Rocky Shores Exploitation in Chile and South Africa.* Springer Science & Business Media, 2013.

Chukchi Sea

Fields of Study: oceanography; marine ecology; marine geology; marine biology

ABSTRACT

The Chukchi Sea, located off the coast of northwestern Alaska, connects the Arctic and Pacific Oceans. Its shallow seafloor and stratified water column are populated by a vast array of organisms due to the influx of nutrients and warmth from warmer Pacific waters.

KEY CONCEPTS

benthic organisms: organisms that live on or very near to the sea bottom

frazil ice: a loose formation of small, plate-like ice crystals that forms on the surface of freshwater under suitable conditions; in Arctic seawaters becomes compacted into pack ice, but can also be seen in terrestrial areas like Yosemite National Park where it can flow haphazardly over flooded ground like thick slush

pack ice: ice that forms in Arctic and Antarctic seawater as the seawater freezes and expels dissolved salts, first forming frazil ice that becomes compacted and bound by continuing ice formation into a solid mass up to several feet (or meters) thick

polynya: an area of open water that forms in pack ice when relatively warmer water upwells from below and relatively warmer air circulates above

DESCRIPTION

The Chukchi Sea, located off the coast of northwestern Alaska, is approximately 232,000 square miles (600,000 square kilometers), with most depths not exceeding 64 feet (50 meters). North of the Bering Sea, east of the East Siberian Sea, and west of the Beaufort Sea, it connects the Arctic and Pacific Oceans. Its shallow seafloor and stratified water column are populated by a vast array of organisms due to the influx of nutrients and warmth from warmer Pacific waters. This water flows from the Bering Sea through the Bering Strait before breaking into three different currents under the ice of the Chukchi Sea.

As with other areas of the Arctic, the Chukchi basin is overlain by pack ice for most of the year. However, indicators of global climate change are greatly in evidence here: Sea ice is disappearing at a rate of roughly 4 percent per decade, and alarmingly, this trend may be accelerating; sea temperatures are rising; shorelines are dramatically eroding; and summer sea ice is receding to historic lows. In addition to a changing climate, overfishing and commercial interest in natural resources such as gas and oil threaten the region. Thus, while rich in nutrients, marine species, and minerals, the Chukchi Sea is threatened and its future remains uncertain.

HYDROLOGY AND SEA ICE

The Chukchi Sea gets its name from the indigenous Chukchi people who inhabit the Chukchi Peninsula within the Russian Federation. Despite being 193 mi (500 km) wide and 309 mi (800 km) long, one of the defining physical characteristics of the sea is its shallow continental shelf.

The water column is highly stratified, with melting sea ice and runoff providing for a fresher surface layer and denser, saltier water near the bottom. The Arctic and Pacific Oceans are bonded by the Chukchi, whose waters are an important source of nutrients, heat, and freshwater for the Arctic Ocean.

Nutrients here support Arctic ecosystems; heat influences ice-melt; freshwater can stratify the Arctic Ocean and help protect the surface ice from the warmer waters below. Freshwater flow-through is an important part of global water cycles, and the pathways of Pacific waters in the Arctic is thought to affect the entire Arctic system.

Sea ice in the Chukchi Sea (Jannelle Trowbridge, USGS)

Melting sea ice constantly floating above the ocean currents is another source of nutrients, providing life support to a vast array of species. Sea ice in the Arctic grows and recedes seasonally, with a peak around March and a trough typically arriving in late September that leaves the Chukchi at its sea-ice minimum. Chunks of ice, or floes, are prone to converge and form thick ridges, creating a unique landscape of extremely dense sea ice. Because of these ridge formations, sea ice in the Arctic tends to have a longer life cycle than its flatter, thinner counterpart in the Antarctic. Sea ice in the Arctic, which receives very little snow due to the region's being surrounded by land, is on average about 7 ft (2 m) thick but can grow up to 16 ft (5 m). Sea ice is formed by cold temperatures—a minimum of 29 degrees Fahrenheit (-2 degrees Celsius)—acting on ocean saltwater, causing tiny ice platelets, or frazil ice, to eventually form pack ice. As it freezes, the crystals of the ice expel salt into the water, yielding crystals that consist of nearly pure freshwater.

Up until very recently, the average mean Arctic sea-ice extent (or the average annual maximum ice area) has been around 6 million sq mi (15 million sq km), and the average minimum sea-ice extent has been around 3 million sq mi (8 million sq km), which means that more than half of the ice melts every year. During the summer of 2012, however, Arctic sea-ice extent fell to just over 1.5 million sq mi (4 million sq km), surpassing the previous low set in September

2007, as reported by the US National Snow and Ice Data Center. The comprehensive outcome of these significant ecosystem changes will unlikely be positive for many of the organisms now sustained in the region.

BIODIVERSITY

When sea ice starts to retreat in the spring, polynyas, which are bodies of open water in the middle of ice floes, can be formed. Polynyas are unique to the polar oceans and formed via ocean currents, wind, or upwelling. When the sunlight hits one of the approximately twenty polynyas in the Chukchi Sea every year, a microalgae bloom occurs, creating the foundation of the marine food web. The combination of sea ice and the influx of nutrient-rich Pacific water provides for one of the most productive ecosystems in the world, with a multitude of seals, seabirds, walruses, polar bears, whales, fish, and invertebrates inhabiting its realm.

Beneath the sea ice, various species of fish such as the capelin, Arctic cod, and sculpin live in the open water, while anemones, mollusks, and numerous other benthic organisms are common sights on the seafloor. Meanwhile, rivers and inlets along the Chukchi Sea shores provide more than adequate breeding, nesting, and feeding spots for dozens of bird species. Spectacled eiders, another endangered species, feed on mollusks in shallow polynyas before migrating south for breeding.

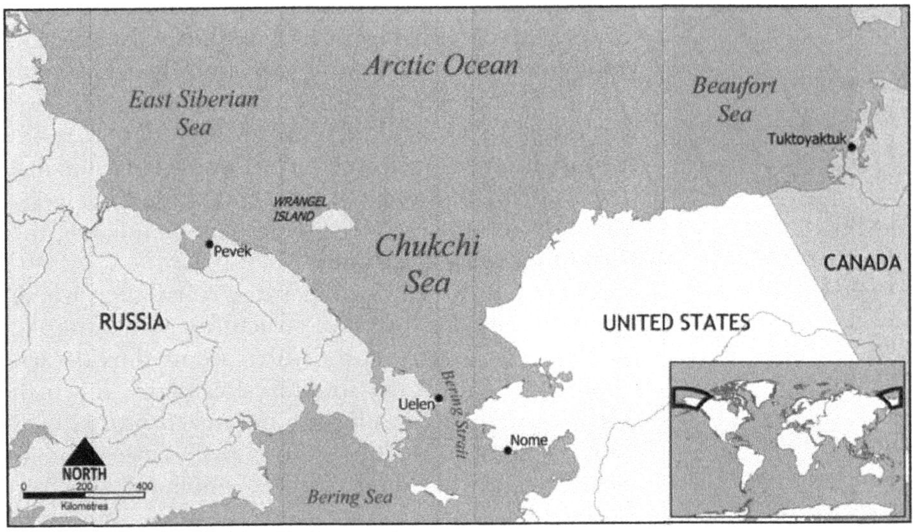

Map of Chukchi Sea (Wikimedia Commons)

Endangered fin, bowhead, and humpback whales, as well as the formerly endangered gray whale, feed and migrate throughout the Chukchi Sea. Beluga whales numbering about 3,500 mate, feed, and calve along the Kasegulak Lagoon off the coast of Alaska. Many other marine mammals call the Chukchi Sea home, with bearded, ribbon, spotted, and ringed seals as well as walrus living their lives both on and below the sea ice. Walruses, which are long-lived social animals, are considered to be a keystone species of the Arctic ecosystem, with Pacific populations spending their summers and falls in the Chukchi Sea region before migrating south to the Bering Strait for the cold, harsh winters.

Ringed seals are the most abundant, wide-ranging seals in the Northern Hemisphere, identified by the small head, plump body, and silver rings on their back, from which the seals get their name. They are the primary food source for the icon of the Arctic: the polar bear. The polar bear, now on the Endangered Species List, is a symbol of wildlife loss due to global warming.

ENVIRONMENTAL THREATS

With an estimated population of about 2,000, more than half of America's polar bears inhabit the Chukchi ecoregion. They depend on the sea ice for all aspects of life, from breeding and denning to hunting. In 2007, it was reported that surveys were showing that numerous polar bears were swimming far from land and risking drowning due to record loss of sea ice. Because of this increase of ice-melt due to climate change, the US Geological Survey predicts a comprehensive loss of Alaska's polar bears by 2050.

Like most of the Arctic, the Chukchi Sea is also environmentally threatened due to commercial interests in its natural reserves. Commercial fishing and whaling are putting pressure on the fragile ecosystem, and as of August 2012, Shell Oil was seeking extended time for drilling purposes. The area is projected to contain more than 30 billion barrels (4.8 billion cubic meters) of oil.

For now, the Chukchi Sea remains a diverse, pristine, and balanced ecosystem, providing a habitat for over a dozen threatened or endangered species, among others. However, with global climate change causing the extent of sea ice to retreat, a rise in sea level, and erosion along coasts due to thawing permafrost, legislative advocacy is in the forefront of all discussions regarding this biome. More research needs to be conducted; potential impacts and mitigation measures need to be adequately addressed.

—Kevin Bakker

Further Reading

Audubon Alaska. "Chukchi Sea." ak.audubon.org/chukchi-sea.

Danson, Ted, and Michael Orso *Oceana: Our Endangered Oceans and What We Can Do to Save Them.* Potter/Ten Speed/Harmony/Rodale, 2011.

Demuth, Bathsheba. *Floating Coast: An Environmental History of the Bering Strait.* W.W. Norton & Company, 2019.

Grebmeier, Jacqueline M., and Wieslaw Maslowski, editors. *The Pacific Arctic Region: Ecosystem Status and Trends in a Rapidly Changing Environment.* Springer, 2014.

Herman, Yvonne, editor. *The Arctic Seas. Climatology, Oceanography, Geology and Biology.* Springer Science & Business Media, 2012.

Horner, Rita A., editor. *Sea Ice Biota.* CRC Press, 2018.

Moore, Sue E., Douglas P. DeMaster, and Paul K. Dayton. "Cetacean Habitat Selection in the Alaskan Arctic During Summer and Autumn." *Arctic*, vol. 53, no. 4, 2000.

National Snow and Ice Data Center. "All About Sea Ice." nsidc.org/seaice/characteristics/difference.html.

Clathrates and Methane Hydrates

Fields of Study: oceanography; biogeochemistry; marine engineering

ABSTRACT

Clathrates (also known as gas hydrates) are solid crystals that occur when water molecules form a dodecahedral lattice around gas molecules. A wide range of gas hydrates occur naturally, but methane hydrates are the most common. Increased scientific attention has been paid to methane clathrates, both as an important factor in climate change and as a potential energy source.

KEY CONCEPTS

clathrate: a nonbonded compound that forms when water molecules freeze into a cagelike structure about a gas molecule such as methane

dodecahedron: a geometric shape having twenty (i.e., dodeca) individual faces, as a soccer ball does

hydrate: any molecule or compound that has the component elements of water in its structure

permafrost: a contraction of permanent frost, the name given to soils that remain well frozen throughout the entire year

BASIC PRINCIPLES

Methane hydrates (also known as clathrates) are lattices of solid ice enveloping (but not bonding with) methane molecules. In conditions in which temperatures are low and pressure is high, water crystallizes in a twelve-sided, three-dimensional (3D) lattice (a cagelike structure) known as a dodecahedron. Hydrates are commonly found in areas in which both water and methane are available (such as the bottom of the ocean and in polar and glacial regions).

Hydrates are in and of themselves structurally unstable, but when they envelop a guest molecule (such as that of methane and other gases and nongases), they become more stable. Still, the stability of clathrates is highly dependent on the environmental conditions in which they are located. If they are exposed to lower levels of pressure and to higher temperatures, they quickly destabilize. If brought to room temperature, for example, a clathrate will, as it melts, pop and sizzle as the methane contained therein escapes. If a clathrate is exposed to an open flame, the ice of the hydrate will burn.

There are three basic types of gas hydrates, which are differentiated by their structures. For example, methane hydrates (along with carbon dioxide hydrates) are part of the structure I category, in which forty-six water molecules are organized into lattices that contain two small cavities and six large cavities. Structure II hydrates are larger, containing multiple and smaller gas molecules such as propane and isobutane. Structure H hydrates also contain multiple gas

Gas hydrate under carbonate rock (United States Geological Survey via Wikimedia Commons)

molecules, although those molecules may be larger. Both structure II and structure H hydrates are rare, however, while structure I clathrates are found with greater frequency and in larger volumes around the world.

HISTORY

In 1810, scientists Humphrey Davy (1778–1829) and Michael Faraday (1791–1867), while experimenting with chlorine and water mixtures, noticed the presence of solid material forming when the liquids were just above freezing levels. For the remainder of the nineteenth century, clathrates were simply viewed as a scientific curiosity, as researchers attempted to understand the process by which an unstable water frame could form around, but not bond with, a guest molecule.

In the early twentieth century, however, prevailing attitudes toward clathrates evolved from curiosity to irritation. By the 1930s, modern civilization had developed the technology to tap into and extract natural gas and transport it through a system of pipes. E. G. Hammerschmidt, who at the time was working for the Texoma Natural Gas Company, noticed that in cold-weather environments, the company's pipelines were frequently clogged by frozen ice. However, the ice seemed to freeze at higher temperatures, which led Hammerschmidt to conclude that the frozen blockages were clathrates. Shortly thereafter, scientists began to study clathrates more closely in an effort to find ways to prevent them from forming in natural gas pipelines.

In the decades that followed, it became apparent that clathrates were naturally occurring compounds that could form not only in pipelines but also in the permafrost of polar regions and on the ocean floor. Scientists from around the world became interested in locating deposits of the substance, which had been dubbed solid natural gas. The United States, the former Soviet Union, India, and other countries all launched efforts to find and access what were believed to be large deposits of clathrates around the world. By the end of the twentieth century, those countries, along with other nations, coordinated their efforts to study methane clathrates with two general perspectives in mind: the role clathrates play in Earth's natural processes and the potential that methane hydrates represent as an energy source for the future.

DISTRIBUTION OF CLATHRATES

Methane hydrates are found wherever conditions are right for their stability. They are found on the ocean floor, in regions in which temperatures are just above freezing and in which pressure is high and consistent. Clathrates also are found in permafrost, the layer of soil beneath the snow and ice in polar and sub-polar regions like the Siberian region of Russia and in northern Canada. In both types of environment, the methane hydrates are found between 492 to 6.560 feet (150 and 2,000 meters) below the surface.

The volume of methane hydrates throughout the world is significant. The total amount of clathrates in the world far surpasses the volume of conventional natural gas. According to the most recent US Geological Survey review of the world's clathrate inventory, approximately 10,000 gigatonnes (a gigaton is the equivalent of 1 billion tons [0.90 billion tonnes]), or 11 trillion tons (10 trillion tonnes), of methane hydrates exist in 44 regions in which samples have been uncovered and in 113 other regions in which evidence suggests the presence of such clathrates.

A POTENTIAL ALTERNATIVE FUEL SOURCE

Scientists have become intrigued by the existence of methane hydrates in such large volume. It is therefore understandable that governments and private industry have developed an interest in pursuing clathrates as an alternative source of energy.

The challenge is extracting clathrates in a manner that is both safe and cost-effective. Governments, observing that the energy contained in methane hydrates exceeds the total volume of oil, conventional natural gas, and coal available in the world, have launched individual and intergovernmental efforts to

Thawing permafrost in Herschel Island, Yukon Territory, Canada (Boris Radosavljevic via Wikimedia Commons)

launch the clathrate market. Japan, for example, developed plans to tap into clathrates as an energy source by 2016. The United States has for decades invested research dollars toward this pursuit, and a growing number of other nations are also developing clathrate energy programs.

The dangers involved in "harvesting" methane hydrates at their remote sources are both immediate and long term. For example, drilling for clathrates on the ocean floor can destabilize the surface, creating safety risks for workers and potentially disrupting nearby fossil-fuel drilling operations. Furthermore, scientists must find ways to store and transfer methane from these clathrates once they have been harvested. When it is released naturally, methane can oxidize and release carbon dioxide, a greenhouse gas, into the atmosphere. Methane is itself a much more potent greenhouse gas than carbon dioxide. Therefore, using computer modeling and laboratory experiments, scientists are investigating ways to utilize the energy benefits of the methane clathrates while at the same time isolating carbon (a process known as carbon sequestration). This particular aspect of research on clathrates and methane hydrates remains a major challenge and, at the same time, an important key toward the successful utilization of clathrates as a fuel source.

METHANE HYDRATES IN THE CARBON CYCLE

Clathrates have undergone an evolution in terms of scientific pursuits. The concept was at first a curiosity and thereafter developed into a problem to be solved. Today, methane hydrates are considered extremely important, warranting significant investment of money and research from both the private and public sectors.

Methane hydrates are also seen as key to the carbon cycle. The carbon cycle is a concept whereby one of the most critical elements for life on Earth is shared among the planet's many different forms of life and systems. Within this framework, carbon dioxide is absorbed by plants through the air and the soil. The plants are eaten by animals, which in turn are eaten by others (including humans). The animals also exhale carbon dioxide and, when they die, decompose and release carbon back into the soil.

Clathrates add another dimension to the carbon cycle. By capturing methane, a carbon compound, clathrates have a role in maintaining a balance of Earth's carbon supply. However, as stated earlier, methane hydrates are only stable under certain temperature and pressure conditions, particularly those found in permafrost and at the bottom of the ocean. If temperatures rise sufficiently, the clathrate lattice can quickly destabilize, releasing its methane guest molecule. On a large scale, such events mean a rapid release of carbon, as methane and ultimately as carbon dioxide, into the atmosphere. Given that each molecule of methane, CH_4, that is oxidized releases one molecule of carbon dioxide, CO_2, into the atmosphere, this would be a "one-two-punch" in terms of the greenhouse effect.

Such events have happened before in Earth's history. For example, scientists have found plentiful evidence of clathrate involvement in changes that occurred during the late Precambrian era, hundreds of millions of years ago. Studies of Precambrian rocks show signs of rapid carbon releases in areas in which clathrates had existed. These samples suggest that in light of the significant volume of carbon released during these events, clathrates played a major role in the large-scale development of plant and animal life during that era.

CLATHRATES AND CLIMATE CHANGE

Methane hydrates represent a major enticement for alternative energy enthusiasts, particularly because they exist in much greater volume than other energy fossil sources and natural gas fuel sources. Additionally, if the myriad government and privately sponsored efforts to develop ways to harvest and contain methane from clathrates succeed, the resulting energy may be considerably cleaner than fossil fuels that produce polluting byproducts.

Separate from the issue of using clathrates as a clean energy source is the potentially significant danger that methane hydrates pose for Earth's atmosphere. As stated earlier, when destabilized naturally, methane hydrates simply release methane and carbon dioxide (under certain conditions, those gases are released into the air). However, in their present locations (on the ocean floor and deep within the permafrost), they remain stable under precarious temperature and pressure conditions. If these conditions are changed even slightly, the potential exists for a major release of methane into the air and the atmosphere. For example, clathrate breakdowns that are currently occurring in the permafrost could trigger landslides, sending sediment into the sea, contributing to rising sea levels. Geological evidence from the end of the Pleistocene epoch (nearly 12,000 years ago) shows that major changes to the slopes of the earth's continents were caused by the release of clathrate gases.

Additionally, scientists believe that clathrate breakdowns in both the permafrost and on the ocean floor could significantly hasten global warming. This disastrous scenario stems from the notion that such a release, caused by rising temperatures in the ocean, would abruptly set off a chain reaction among the planet's methane hydrates and, in turn, quickly raise the planet's temperature by several degrees. Some scientists argue that such events have happened in Earth's history, most dramatically at the end of the Permian period (about 251 million years ago). Adherents to the clathrate gun hypothesis believe that a major release of methane hydrate gas, the cause of which is unknown, triggered a mass extinction of nearly 90 percent of life on Earth.

The debate concerning the validity of the clathrate gun hypothesis continues. Many scientists believe that clathrate breakdowns were not necessarily the primary cause of the Permian extinction. Nevertheless, most researchers agree that clathrates likely contributed to some degree to that event, an event from which Earth took 30 million years to recover. In light of the severity of that event, the notion that a simple elevation of a degree or two in temperature could trigger a massive breakdown in sea- and land-based gas clathrates remains a concern among climatologists and global warming experts.

IMPLICATIONS AND FUTURE PROSPECTS

Clathrates, particularly methane hydrates, have evolved from a scientific curiosity to a major component in both the pursuit of alternative energy sources and the understanding of climate change. Scientists continually seek to understand the dynamics of these unusual geochemical configurations. Advances in relevant technologies, including the application of computer models and advances in thermal imaging systems, are contributing to the evolution of clathrate studies.

Although debate over the clathrate gun hypothesis continues, scientists are in agreement that methane hydrates would contribute to global warming and climate changes. However, researchers are also excited at the prospects of tapping into clathrates as an alternative energy source. To this end, scientists are pursuing technological advances to explore the safe extraction and use of methane from hydrates (including carbon sequestration practices). Their hope is to present to the modern world the use of clathrate-based energy that reduces greenhouse gas emissions and protects the planet from further warming trends.

—Michael P. Auerbach

Further Reading

Demirbas, Ayhan. *Methane Gas Hydrates*. Springer, 2010.

Dorton, Peter, Saman Alavi, and T. K. Woo. "Free Energies of Carbon Dioxide Sequestration and Methane Recovery in Clathrate Hydrates." *Journal of Chemical Physics*, vol. 127, no. 12, 2007.

Giavarini, Carlo, and Keith Hester. *Gas Hydrates: Immense Energy Potential and Environmental Challenges*. Springer, 2011.

Huo, Hu, et al. "Mechanical and Thermal Properties of Methane Clathrate Hydrates as an Alternative Energy Resource." *Journal of Renewable and Sustainable Energy*, vol. 3, no. 6, 2011.

Kurten, T., et al. "Large Methane Releases Lead to Strong Aerosol Forcing and Reduced Cloudiness." *Atmospheric Chemistry and Physics Discussions*, vol. 11, no. 3, 2011, pp. 9057–81.

Max, Michael D., editor. *Natural Gas Hydrate in Oceanic and Permafrost Environments*. Springer Science & Business Media, 2012.

Partain, Roy Andrew. *Environmental Hazards from Offshore Methane Hydrate Operations*. Wolters Kluwer, 2017.

Rajput, Sanjeev, and Naresh Kumar Thakur. *Geological Controls for Gas Hydrates and Unconventionals*. Elsevier, 2016.

Rogers, Rudy. *Offshore Gas Hydrates: Origins, Development and Production*. Elsevier, 2015.

Sloan, E. Dendy, Jr., and Carolyn Koh. *Clathrate Hydrates of Natural Gases*. 3rd ed., CRC Press, 2007.

Climate and Climatology

Fields of Study: atmospheric science; oceanography; climatology; physics; computer modeling

ABSTRACT

Climate is the long-term combined effects of atmospheric variations. Over shorter periods of time, climate is referred to as weather. Climate always refers to a specific geographical location or region and is determined by many factors, including wind belts, topography, elevation, barometric pressure, the movement of air masses, the amount of solar radiation available, proximity to oceanic influences, and planetary and solar cycles.

KEY CONCEPTS

empirical: describes any system based on direct observation

pibal: a pilot weather balloon that is typically followed by direct observation to determine wind speeds and directions at altitude

rawin: a weather balloon carrying a radar target to allow its progress to be monitored by radar rather than by direct observation

taiga: the evergreen forests typical of northern regions

tundra: treeless expanses that extend to the north from the taiga regions, typically overlying permafrost

SCIENCE OF CLIMATOLOGY

From early times, humans have attempted to classify climates. The ancient Greeks in the sixth century BCE visualized Earth as having three temperature zones, based on the sun's elevation above the horizon. (The Greek word *klima*, meaning "inclination," is the origin of the word "climate.") They called these three zones: torrid, temperate, and frigid. This system did not consider the differences between climates over land and over water, the effects of topography, or other atmospheric elements such as precipitation.

People tend to confuse weather with climate. The two phenomena are related, but weather is concerned with day-to-day atmospheric conditions, such as air temperature, precipitation, and wind. Climatology is concerned with the mean (average) physical state of the atmosphere, along with its statistical variations in both time and location over a period of many years. In addition to the description of climate, climatology includes the study of a wide range of practical matters determined by climate and the effects and consequences of climatic change. As a result, it has become an interdisciplinary science, relevant to and affected by a wide range of other fields, including geophysics biology, oceanography geography, geology, engineering, economics, statistics, solar-system astronomy, and political and social sciences.

For centuries, the content of the atmosphere and the amount of solar radiation reaching Earth's surface have been fairly constant. Until the middle of the 1950s, it was assumed that the climates of Earth were also relatively unchanging, but climate scientists now understand that climate is never constant. The global climate consists of effects generated by the oceans, the atmosphere, the cryosphere (areas of permanent ice), the land surface, and biomass. The various components of climate are linked to one another in a nonlinear feedback process, so that a change in any one of these factors produces a change or feedback effect in all of the other factors. Therefore, all or any of these processes can change the statistical state of the system called "climate."

The science of climatology has developed along two main lines. Regional climatology studies the discrete and characteristic qualities of a particular region of the globe. The second approach of climatology is a physical analysis of the basic relationships among various atmospheric elements such as temperature, precipitation, air pressure, and wind speed. A third branch is dynamical climatology, which originated in the 1960s; this branch uses models to simulate climate and climatic change based on the averaged forms of the basic equations of dynamic meteorology.

CLASSIFICATION OF CLIMATES

There are two main classification systems for climates: genetic and empirical. A genetic system is based on air masses and global wind belts, which control climates. The empirical classification system is based on the observed elements of climate, such as temperature.

The Köppen system is an empirical classification system with five designated general categories: tropical forest climates; dry climates; warm, temperate, rainy climates with mild winters; cold forest climates with severe winters; and polar climates.

Another empirical classification system, which may be more useful in describing world climates, is based on decreasing temperature and increasing precipitation. This system of classification is convenient for use in many other sciences as well. The first category is the desert climate, which has the highest temperatures and the lowest precipitation. The second category is the savanna, which may be either temperate or tropical and is characterized in either case by nearly treeless grassland. The steppes of eastern Europe and the prairie regions of Canada and the United States are examples of temperate savannas, with plants adapted to very hot conditions and extremely limited water supply. The grassy plains of western Africa are a typical tropical savanna. The third category or climate type is temperate and tropical Mediterranean climate. An example of temperate Mediterranean climate is the coast of Southern California; tropical Mediterranean climate is found typically in the northern region of Africa. The fourth climate type is the temperate and tropical rainforests. An example of a temperate rainforest is found on the Olympia Peninsula in Washington State, while the Amazon basin in Brazil is the standard example of a tropical rainforest.

The fifth and six climate types in this system are the coldest and have a great amount of precipitation in the form of snow. The first of these two, the taiga is characterized by great forests of evergreen coniferous trees. Taigas occur only in the temperate zone or at great elevation. Large portions of Canada and Russia are typical of this climate type. The last region is the tundra the coldest climate type, which may have small amounts of precipitation. Again, this region is found in either the temperate zone or the highest elevations on Earth. The best examples are the northern territories of Canada and much of eastern Alaska.

FACTORS OF CLIMATE CHANGE

Climate is not a steady, unvarying cycle of weather; it fluctuates and varies over time. One of the major factors causing these variations is the role of atmospheric circulation. The atmosphere extends outward from Earth's surface for hundreds of miles (or kilometers) and consists of five clearly defined regions. The troposphere the region closest to Earth, extends about 6.8 mi (11 km) above the surface. It is in this region that most weather phenomena affecting climate take place.

Filling pibal weather balloon (David Hercher/U.S. Forest Service, Southwestern Region, Kaibab National Forest via Wikimedia Commons)

In the tropics there is intense solar heating, whereas the polar regions have little solar heating. These differences of heating and cooling of the atmosphere result in a large-scale global circulation of air that carries excess heat and moisture from the tropical areas of intense heating into the higher latitudes, where there is little excess heat and moisture. The great movements of air masses that are produced create and alter various local climates.

Global cloud patterns are also extremely important factors in regulating worldwide climate. Heavy cloud cover reduces the amount of solar radiation reaching Earth's surface, producing a cooling effect. Yet cloud cover also acts as a kind of blanket, reducing the amount of heat lost from the ground by absorbing surface-emitted infrared radiation and directing about half of it back toward the surface and into the atmosphere. A clear sky in winter always results in colder temperatures at the surface than when there is cloud cover. Even so, only the solar heat that has already reached Earth's surface can be kept under this protective cover. Extended cloud cover will result in less solar heat reaching the surface, leading to a cooler climatic condition.

The activity of the climate in various parts of the globe is also strongly influenced by the movement of ocean currents and global wind belts. An oceanic effect termed "El Niño" is periodically responsible for dramatic climate changes, with effects felt on a global scale. Scientists are not completely certain of the initial cause of this phenomenon, but it appears to be dependent on the accumulation of warmer surface water driven by prevailing surface winds.

It is important to remember that climate has never been stable and unchanging. Many factors can cause changes in local and even global climate; when one factor changes, it inevitably modifies or influences other factors affecting climate.

FACTORS RESULTING FROM HUMAN ACTIVITY

Some factors causing climate change are of natural origin, while many others are anthropogenic, result-

Taiga in Quebec (peupleloup via Wikimedia Commons)

ing from human activity. The globe's tropical rainforests are vital to maintaining the equilibrium of carbon dioxide in the global atmosphere. Heavy deforestation in South America and Asia has endangered this balancing mechanism by removing great numbers of the trees that remove vast quantities of carbon dioxide from the atmosphere. At the same time, industrial activity and automobile emissions release massive quantities of carbon dioxide, contributing to the buildup of carbon dioxide in the atmosphere and resulting in measurable increases in global temperatures. Carbon dioxide absorbs heat emitted from Earth's surface and reradiates much of it back toward the surface, creating a greenhouse effect that normally raises the average global temperature to a level that supports the existence of life. Even a small increase of 35.6 degrees Fahrenheit (2 degrees Celsius) in the average global temperature, however, would have a potentially disastrous effect on climatic conditions. Global warming would cause the oceans to expand appreciably as their temperatures increased, as well as by adding massive amounts of water from melting glaciers and the miles- (or kilometers-) thick icecaps of Greenland and Antarctica. In addition, ocean temperature and the direction of upper-level winds are the main factors in the development of hurricanes. Some scientists argue that the warming of the oceans has already increased the frequency as well as the power and destructive force of hurricanes.

One of the main dangers of a warming climate is flooding due to rising sea levels. Scientists have calculated that if the average temperature increases by about 38.66 Fahrenheit (3.7 degrees Celsius), the sea level could rise approximately 32 inches (81 centimeters)—enough to flood huge areas of unprotected coastal land. Nearly 30 percent of all human beings live within about 37.3 mi (60 km) of a coastline. A rise in sea level of only 20 in (50 cm) would have a profound effect, as it would result in the flooding of many of the world's most important cities and ports.

Large urban areas alter the local climate through the heat island effect by which the average temperature of an urban area is raised sufficiently to form rising air columns that would not otherwise exist, which in turn influence the activity of the larger air mass. Various forms of pollution over large urban areas such as Los Angeles, Tokyo, and London have also caused local climate changes in these areas. Concentrations of ozone that build up at ground level from industrial and automobile exhausts cause extreme eye and lung irritation. Nitrogen oxides, carbon dioxide, sulfur dioxide, and particle-laden air, which can cause a depletion in the amount of total solar radiation, also accumulate and alter climate conditions over local urban areas. Changes in surface characteristics, such as converting forests and prairies to agricultural fields, building cities, damming rivers to create lakes, and spilling oil at sea, all affect local climate conditions. Usually these surface changes appear to have only a minor global effect, though they do alter local conditions significantly, but the cumulative effects of these minor changes are not well understood and should, therefore, not be overlooked.

There is widespread national and international interest in how human activity is fundamentally altering the global climate. Organizations are seeking to inform the general public regarding society's role in maintaining a healthy and healthful climate. The Climate Institute in Washington, DC, sponsors international conferences where scientists and government leaders from many nations gather, providing a forum for information exchange among climate researchers and analysts, policymakers, planners, and opinion makers.

DEPARTURES FROM THE "NORM"

For centuries, the content of the atmosphere and the amount of solar radiation reaching Earth's surface have been fairly constant. Scientists now accept that climate is never constant and that departures from the expected "norm" provide the greatest insight into climatic processes and have the greatest human impact.

The ice ages were a decided departure from the norm. Drought in central North America during the 1920s and 1930s resulted in the Dust Bowl disaster which disrupted human life and crop production in the fertile middle region of the continent. In the early 1970s, the impact of the devastating drought of the Sahel region in Africa led to increased desertification. Loss of the anchovy fisheries along the Peruvian coast in the 1970s (probably as a result of as El Niño), severe droughts in the United States' farm belt in the 1980s, and large-scale flooding in many parts of the world all attest to the fact that dramatic climatic fluc-

High altitude tundra at Rocky Mountain Park (MichaelKirsh via Wikimedia Commons)

tuations and changes are the true norm. It is now understood by climatologists that climate modification may even result in microclimates as small as portions of a backyard garden.

DATA-GATHERING TOOLS

Because the science of climatology is the study of weather patterns for a geographic area over a long span of time—generally a minimum of thirty years—records of scientific data on the weather are valuable for interpreting climate trends. Because of the international agreements and standardized procedures concerning climate that were adopted in 1853, about one hundred million observations have been taken from ships since then. The quantities recorded were sea-surface temperature, wind direction and speed, atmospheric pressure and temperature, the state of the sea, and cloudiness.

A great variety of instruments are used by meteorologists in gathering weather information. The net radiometer, pyranometer, and Campbell-Stokes sunshine recorder are used for measuring radiation and sunshine. The sling psychrometer is used to determine relative humidity. For remote sensing, a humidity gauge is used in which the variations of electrical transmission by the device depend on the passage of an electrical current across a chemically coated strip of plastic that is proportional to the amount of moisture absorbed at its surface. This type of gauge is used in radiosondes for upper-air observations.

Measurement of wind velocity, its speed and direction, involves a simple device called an "anemometer." Anemometers have four-bladed propellers that are driven by the wind to record wind speed. When the propellers are mounted at right angles, an anemometer responds to air movement in three dimensions, the sum of which reveals its true direction. Weather pilot balloons, or "pibals," are released and then tracked visually using a theodolite, or a right-angled telescopic transit that is mounted in a way to make possible the reading of both azimuthal (horizontal) and vertical angles. The progress of the balloon is then plotted minute by minute, and the direction and speed are computed for each altitude. This technique does not work, however, for upper-air wind observations when there is a low cloud cover.

Under these conditions, "rawin" observation is used. Here, a metal radar target is attached to the balloon and tracked by a radar transceiver or radio theodolite until the balloon breaks or is out of range.

Precipitation is measured in several ways. Rain measurements are typically made using a rain gauge, a simple, graduated cylinder calibrated to rain depth according to the cross-sectional area of the device. An improved version is the tipping-bucket gauge, which has a small divided metal bucket mounted so that it will automatically tip and empty measured quantities of rainfall. This tipping closes electrical contacts that activate a recording pen. Another type of improved rain gauge is a weighing-type gauge. Here, as precipitation falls on a spring scale, a pen arm attached to the scale makes a continuous record on a clock chart. Snow depth is determined by averaging three or more typical measured depths. Where snow depths become very great, graduated rods may be installed in representative places so that the depth can be read directly from the snow surface. Heat and air pressure are measured by thermometers and barometers, most of which record and transmit data automatically using digital electronics.

AERIAL PHOTOGRAPHS AND CORE SAMPLES

Specially equipped airplanes and satellites can obtain previously inaccessible data of immeasurable value in determining climate changes. The National Aeronautics and Space Administration (NASA) has supplied a modified U-2 plane to carry instruments to an altitude of 12 mi (20 km) for seven-hour flights up to 80 degrees north latitude. A DC-8 is also used at the same time because of its extended range. The Nimbus 7 satellite is also employed. Landsat satellites continually photograph Earth's surface, then transmit the images back to Earth stations, where they are computer enhanced to provide visual information about such things as the amount and type of vegetation, precipitation, and underground waterways. The enhanced photographs also provide information about changing conditions that may cause changes in local and global climates.

In order to gain information about the relationship between carbon dioxide and global temperature, scientists have drilled deep into Arctic and Antarctic ice to obtain core samples of the polar ice pack. By studying the amount of carbon dioxide trapped in air bubbles in very old layers of ice pack, scientists have been able to chart the ebb and flow of carbon dioxide in the atmosphere. Fossilized plant tissues indicate how warm the air was during the same period as the bubbles in the core samples. This clue helps provide a picture of the warming and cooling trends that have occurred in the past, making comparisons with current climate conditions possible.

GENERAL CIRCULATION MODELS

Climatologists have found mathematical and computer-modeling methods very useful, primarily with regard to general circulation models (GCMs). The most important part of a GCM is the parameterization of a wide range of physical processes too small to be resolved by the model. These may include all of the turbulent fluxes that occur in the surface boundary layer as well as the occurrence of convection, cloudiness, and precipitation. These relatively small processes must be parameterized in terms of the variables that are capable of being resolved by GCMs. It is primarily this parameterization that makes GCMs different from one another and leads to variations between the results obtained from different GCM models. The value of the models, however, depends mostly on the accuracy with which the model actually simulates natural conditions. Assumptions sometimes must be made in modeling, and these assumptions can give a false picture under certain conditions. Models, therefore, are intensely tested against actual weather data as newer methods of observation and more data become available in order to ensure the accuracy of the models.

—George K. Attwood

Further Reading

Ackerman, Steven A., and John A. Knox. *Meteorology: Understanding the Atmosphere*. 4th ed., Jones, 2015.

Ahrens, C. Donald. *Meteorology Today: An Introduction to Weather, Climate, and the Environment*. 10th ed., Brooks, 2013.

Aquado, Edward, and James E. Burt. *Understanding Weather and Climate*. 7th ed., Prentice, 2015.

Bolius, David. *Paleo Climate Reconstructions Based on Ice Cores: Results from the Andes and the Alps*. SVH, 2010.

Bonan, Gordon. *Climate Change and Terrestrial Ecosystem Modeling*. Cambridge UP, 2019.

Fagan, Brian M., editor. *The Complete Ice Age: How Climate Change Shapes the World*. Thames, 2009.

Hobbs, John E. *Applied Climatology*. Elsevier, 2016.
Lalic, Branislava, et al. *Agricultural Meteorology and Climatology*. Firenze UP, 2018.
Lutgens, Frederick K., and Edward J. Tarbuck. *The Atmosphere: An Introduction to Meteorology*. 12th ed., Prentice, 2013.
Marshall, John, and R. Alan Plumb. *Atmosphere, Ocean and Climate Dynamics: An Introductory Text*. Elsevier, 2008.
Oke, Timothy R., Gerald Mills, Andreas Christen, and James A. Voogt. *Urban Climates*. Cambridge UP, 2017.
Pierrehumbert, Raymond T. *Principles of Planetary Climate*. Cambridge UP, 2011.
Shen, Samuel S. P., and Richard C. J. Somerville. *Climate Mathematics Theory and Applications*. Cambridge UP, 2019.

Coastal Engineering

Fields of Study: civil engineering; fluid dynamics; marine biology

ABSTRACT
Coastal engineering is a branch of civil engineering that aims to solve coastal zone problems such as shoreline erosion and the destruction caused by tsunamis through the development, construction, and preservation of mechanisms and structures. Coastal engineering also entails understanding the theories and processes of wave actions and forces, ocean currents, and wave-structure interactions. Given that coastal areas are often highly populated and especially vulnerable to human impact, coastal engineering attempts to blend conservation and management of the world's coastal zones with the development requirements of humans.

KEY CONCEPTS
groin: solid structures that are perpendicular to the shoreline and extend into the water below normal wave heights as a means of controlling longshore drift

hard coastal engineering: engineering and construction of solid, man-made structures such as docks and ports, walkways, fences, etc., structures that are not part of the natural structure of coastlines

revetment: a solid structure consisting of stones or other materials that acts to prevent erosion of beach sands as waves break on and recede from the shoreline

soft coastal engineering: engineering and construction that seeks to protect or restore the natural structures of a coastline, such as beach "nourishment" and sand dune stabilization

DEFINITION AND BASIC PRINCIPLES
Coastal engineering is a branch of civil engineering involved in the development, design, construction, and preservation of structures in coastal zone areas. Its primary function is to monitor and control shoreline erosion, design and develop harbors and transport channels, and protect low-lying areas from tidal flooding and tsunamis. To achieve such objectives, coastal engineers must have a strong understanding of the sciences of engineering, oceanography, meteorology, hydrodynamics, geomorphology, and geology. They must also have a strong understanding of the theories and processes of wave motion and action, wave-structure interaction, wave-force forecasting, and ocean current prediction.

Coastal areas are socially, economically, and ecologically crucial. They are home to a significant proportion of the world's human population, serve as important breeding grounds for many animal species, and are important for tourism, aquaculture, fishing, transport, and trade. These multiple uses of a relatively small area have led to rapid environmental degradation and have created conflict and the need to achieve a sustainable balance.

BACKGROUND AND HISTORY
Although the specialty of coastal engineering emerged only in the latter half of the twentieth century, the ideas of coastal engineering and management have been practiced for many centuries. Human beings have long lived in coastal zones and, for just as long, have attempted to control and manage the effects of tidal flooding and wave action and to engage in land reclamation. People have developed ports to make transportation and trading easier for thousands of years. The first river port—port of Aur on the Nile—was built before 3000 BCE, and the first coastal port—the port on Pharos near Alexandria—about 2000 BCE. For centuries, many countries wanting to protect strategic coastal areas used structures such as embankments as sea defenses.

Early engineering projects mainly involved hard coastal engineering, which is the design and building of docks, ports, defenses, and walkways. Although coastal engineering has long been practiced, the actual coastal processes and their driving forces were little understood. Therefore, formal engineering development required much more comprehensive knowledge of the ocean and the processes and principles involved.

Traditionally, engineering projects in coastal zones were the province of civil and military engineers. However, because of the growing need for specific research in coastal engineering, the first International Conference on Coastal Engineering was held in 1950 in California, establishing this branch of applied science and introducing the terminology. The conference proceedings stated that coastal engineering was not a novel science or unconnected to civil engineering but rather was a branch of civil engineering that was strongly reliant on and influenced by sciences such as oceanography, meteorology, electronics, and fluid and structural mechanics.

By the 1960s, significant inroads had been made into understanding coastal processes and this, coupled with advances in modeling, highlighted the importance of incorporating such processes into design and construction. By the 1970s, greater understanding of coastal processes led to the introduction of soft coastal engineering, which included beach nourishment and artificial dunes. In the following decades, significant progress was made in the development of coastal engineering techniques, such as modeling techniques for deep-water wave prediction, and the understanding of processes, including wave-structure interaction and coastal sediment erosion.

Essentially, coastal engineering has transformed from a science that simply responds to anthropocentric demands (docks, ports, and harbors) to one that aims to balance human requirements and environmental protection, with a particular focus on mitigating both natural and human-induced coastal erosion.

HOW IT WORKS

It is the job of a coastal engineer to recognize the characteristics and natural processes of the shoreline environment and to apply fundamental engineering theories and philosophies to solving ecological problems. To be successful, a coastal engineer must have a strong understanding of other sciences such as oceanography, meteorology, geomorphology, structural and fluid mechanics, and hydrodynamics. In particular, this means clarifying and using theories of wave action, wave-structure interaction, wave-force forecasting, ocean current prediction, and beach profile modification.

Coastlines around the world are the natural division between the terrestrial landscape and the seas and oceans. The geological composition of these areas is distinctive, as are the natural and anthropogenic processes that affect them. Coastal engineering techniques must take into account the morphological development of coasts, such as erosion and accretion, and relate this to shore protection through engineering approaches and structures.

The action of waves on the shoreline forms one of the fundamental areas of coastal engineering research. Waves contain a very large amount of energy formed by the action of wind over vast tracts of open water, the gravitational attraction of the Sun and Moon, and occasionally seismic activity (as with tsunamis). Although this energy is collected over very large areas, it is released along relatively small areas of coastline. This release of energy, in the form of breaking waves, has a strong influence on the currents, sediment, and geological structures of coastal areas. The shoreline or beaches, which are frequently composed of sediment such as sand, shingle, or gravel, are constantly battered and reshaped by the actions of wind and water.

Erosion is a natural process, and the movement of sediment is achieved by wave uprush, which transports sand onshore, and backwash, which transports sand offshore. The natural processes of erosion in coastal areas are very complex and involve flow fields created through not only the action of breaking waves but also erratic turbulent sediment transport in the water column and a shifting shoreline. Worldwide research aims to develop predictive models of this erosion process. Over time and under normal conditions, erosion is also a relatively cyclical process. Sediment can be carried offshore during one season and dumped back onshore during the next season, as well as moved obliquely along the shore over time. Erosion processes can, however, be significantly disrupted by human activity and structures or can be considered undesirable because of human requirements and desired aesthetics. Coastal engineering structures can be, in fact, both detrimental and beneficial in regard to coast erosion, and most coastal engineers believe that only an integrated and holistic approach to planning and design can generate long-term sustainability.

Within this energetic and process-driven natural boundary between land and sea, people have constructed coastal engineering structures such as ports and harbors, which have had both positive and negative effects on the environment. The design of such structures must predict wave dynamics and their impact on individual coastal zones and beach environments. Since the emergence of coastal engineering science, much greater emphasis has been placed on understanding coastal processes, such as wave action, and designing and developing policies and structures to protect coastal areas from erosion.

Although the study of coastal erosion is a fundamental concept for coastal engineering, coastal engineers must also recognize and study coastal protection measures and applications such as hard and soft shoreline protection structures, understand the effects of these structures on the morphology of the coastal areas, and develop effective coastal zone management plans and policies.

APPLICATIONS AND PRODUCTS

Coastal engineering structures and activities vary significantly from country to country. They depend on social aspects, such as the country's history and development and its people's relationship to the ocean, and environmental and structural aspects, such as the nature of the ocean along the country's coast, geomorphological and geological conditions, the ecosystem, climate, extent of tectonic activity, currents, and wave action.

Generally speaking, however, there are two types of coastal engineering applications: hard and soft coastal engineering. Although hard stabilization techniques have been used for many years and are considered to be appropriate under certain circumstances, they can also be expensive and, in trying to rectify human impact, can actually disrupt and destroy natural shoreline processes and habitats, intertidal areas, and wetlands. In some cases, they can even increase erosion. As coastal engineering has evolved and become more sophisticated, engineers have moved away from constructing hard structures with little understanding of their impact on the environment and toward incorporating soft engineering and minimizing the ecological impact.

Hard Structures and Applications. Groins are solid structures, usually constructed from wood, concrete, or rocks, running perpendicular from the foreshore into the water (under normal wave levels). They aim to disrupt water flow and reduce the transportation of sediment from longshore drift. They trap sediment to extend or preserve the beach area on the up-drift side and reduce erosion on the down-drift side. The size and length of these structures is usually determined based on specific local conditions, including wave dynamics, beach slope, and environmental factors. Seawalls are very large, rigid, and usually vertical structures constructed from concrete. They are found at the transition between the low-lying beach and the higher mainland and run parallel to the shoreline. The main function of a seawall is to preserve the shoreline by preventing additional shoreline erosion and recession during direct wave-energy impact and flooding. Seawalls are, however, not effective in preventing longshore erosion.

Revetments are a concrete, rock, or stone veneer, or facing, constructed on a beach slope with the aim of preventing erosion caused by wave action and storm surges. Unlike seawalls, which can assist in flood prevention, revetments are not usually constructed for this function. Revetments are always built as sloping structures, and although they may sometimes be completely solid and rigid, they are often built with interlocking slabs or stone and designed to be permeable. This permeability tends to enhance the strength of the revetment and the absorption of wave energy while reducing erosion and wave run-up. Revetments can also be made from sand-filled bags, interlocking tires, concrete-filled bags, and wire-mesh stone-filled gabions (sunken cylinders filled with earth or stones). As revetments tend to be passive structures, their application is often limited to areas that are already protected in some other way or by some other engineering structure.

Breakwaters are fixed or floating structures built parallel or at an angle to the shoreline. The main function of a breakwater is to protect the shore and activities along the coastline by reducing the impact of wave energy. The ability and impact of a breakwater depend on whether it is submerged or floating, its distance from the shoreline, its length and orientation, and if it is solid or segmented. The four main types of floating breakwaters are box, pontoon, mat, and tethered float; the three main types of detached breakwaters are offshore, coastal, and beach.

Soft Structures and Applications. Shore nourishment has become one of the most common soft coastal engineering applications. The three main types are backshore, beach, and shoreface nourishment. As the name suggests, nourishment is the action of artificially adding sand to the backshore (upper part of the beach), beach, or shoreface (usually the seaside of the bar) in an attempt to modify the effects of erosion. Although nourishment replaces sand in an eroded area and is considered to be a rather natural form of coastal engineering, it does not address the causes or processes of erosion nor does it reduce the impact of wave energy.

Sand dune stabilization is a relatively basic and common form of soft coastal engineering. It involves the planting of vegetation to stabilize and protect sand dunes along the coastline. The creation of artificial dunes is also considered to be an effective form of soft coast engineering protection, particularly in conjunction with shore nourishment.

Beach drainage, or beach dewatering, is a system of shore protection based on the concept of physically draining water from a beach. Drainpipes are installed and buried below the beach and parallel to the shoreline. These pipes collect seawater and transport it to a pumping station, where it is collected and either returned to the ocean or used. Beach drainage helps reduce erosion by lowering the watertable in the uprush zone and reducing the force of the backwash by increasing the volume of water that seeps into the beach.

CAREERS AND COURSEWORK

Many universities offer undergraduate and graduate degrees in coastal engineering. Most commonly, students who wish to pursue a career in coastal engineering will study engineering or marine science as an undergraduate. By graduation, students should have a solid understanding of coastal engineering concepts, theories, processes, and practices, such as wave action, wave transformation, statistical and numerical analysis, tides and currents, beach dynamics and coastal structures, and the impact of natural and anthropogenic factors on the coastal environment. In addition, as coastal engineers need to provide the scientific base data used for coastal zone management, they are likely to require knowledge and experience in the use of modeling.

Students who study coastal engineering can enter such careers as consulting engineers, project managers, environmental consultants, and construction contractors and such fields as construction management, coastal and oceanographic engineering, water resource management, and hydrologic engineering. They may find employment in the private sector, with nongovernmental organizations, with specialized government organizations and agencies, or in universities undertaking teaching and research.

SOCIAL CONTEXT AND FUTURE PROSPECTS

The social, economic, and ecological consequences of coastal erosion are significant. Almost 60 percent of the world's population lives within 62 miles (100 kilometers) of a coast. Coastal areas not only provide significant economic benefits—fish and other maritime products, a means of transportation, and easier access to trading partners—but also offer numerous recreational opportunities. However, the economic and aesthetic appeal of coastal areas has created population pressure that has greatly increased the need for functional coastal structures, such as ports, wharfs, and jetties, and infrastructure, such as roads and sewerage facilities, which increase erosion and pollution. The erosion caused by human activities and structures can change beach dynamics and profiles, which in turn can have ecological effects, such as the loss of animal breeding grounds, and social effects, such as making the area unsafe for swimming and other recreational activities.

Research and climate change modeling has indicated that sea levels may rise significantly in the twenty-first century. This, coupled with an associated increase in coastal storm frequency and strength, will produce more floods and greater erosion, creating serious ecological repercussions for the coastal zones of the world, which are often densely populated and popular areas for tourism. The construction of such engineering structures as ports, harbors, recreational facilities, and resorts are considered necessary for continued economic and social development. Such engineering has, however, a very significant ecological impact on the coastal zone areas, particularly an increase in erosion. One of the primary goals in coastal engineering is to rectify erosion issues.

The difficulty for the future of coastal engineering is that some of the solutions to such environmental or aesthetic issues can be ineffective or ecologically unsound, thereby exacerbating the problem. As such, the future of coastal engineering requires integrated coastal zone management (ICZM), a holistic and integrated strategy and approach incorporating hard and soft engineering for the management of all aspects of coastal areas. It incorporates advances in modeling that are helping to develop a framework for predicting coastal erosion hazards and processes. Many coastal engineers are advocating an integrated approach, stating that the future ecological and economic sustainability of the world's fragile and vulnerable coastal areas depends on its adoption. Such a strategy is optimal, as it considers such issues as climate change, the rise in sea level, navigational needs, the impact on plants and animals, and aesthetic considerations.

—*Christine Watts*

Further Reading

Dean, Robert G., and Robert A. Dalrymple. *Coastal Processes with Engineering Applications*. 2002. Reprint. Cambridge UP, 2004.

Kamphuis, J. William. *Introduction to Coastal Engineering and Management*. 3rd ed., World Scientific Publishing, 2020.

Kraus, Nicholas, editor. *History and Heritage of Coastal Engineering: A Collection of Papers on the History of Coastal Engineering in Countries Hosting the International Coastal Engineering Conference 1950–1996*. American Society of Civil Engineers, 1996.

Mani, J. S. *Coastal Engineering*. 2nd ed., PHI Learning, 2018.

Reeve, Dominic, Andrew Chadwick, and Christopher Fleming. *Coastal Engineering: Processes, Theory and Design Practice*. 2nd ed., CRC Press, 2015.

Shibayama, Tomoya. *Coastal Processes: Concepts in Coastal Engineering and Their Application to Multifarious Environment*. World Scientific Publishing, 2009.

Sorensen, Robert. *Basic Coastal Engineering*. 3rd ed., Springer, 2006.

Young, Kim, editor. *Handbook of Coastal and Ocean Engineering*. World Scientific Publishing, 2010.

Coastal Seas

Fields of Study: oceanography; limnology; marine biology; marine geology

ABSTRACT

Coastal seas are extremely important for biodiversity and ecosystems. Coastal seas can be defined in terms of the continental shelves, as understood geologically, and by reference to international conventions. In geological terms, the continental shelf is the extended perimeter of the continent and the associated coastal plain. The continental margin, which is located between the continental shelf and the abyssal plain, consists of a steep continental slope, followed by a flatter continental rise. Sediment from the continent descends down the slope, and accumulates at the continental rise located at the base of the slope.

KEY CONCEPTS

anoxic: containing no dissolved oxygen

dead zone: a particular local volume within a body of water, in which life is not supported

desertification: the conversion of normally fertile and arable land into desert, occurring by the expansion of an existing desert, or by prolonged dryness and lack of precipitation resulting in desert formation

eutrophication: the conversion of a viable aquatic environment to one that cannot support life excessive amounts of organic matter such as algae or rampant weed growth encouraged by an influx of fertilizers consumes the dissolved oxygen in the water

hypoxic: containing low quantities of dissolved oxygen relative to normal oxygenation levels

polynya: a limited area of open water that does not freeze over within an unbroken ice pack

red tide: a type of algal bloom consisting of dynoflagellate organisms that produce a toxic red pigment rather than the green chlorophyll of other algae

UNDERSTANDING TYPES OF COASTAL SEAS

Some coastal seas are pelagic seas, located along an open ocean. Examples of pelagic seas include the North Sea and the South China Sea. Others are enclosed seas or semienclosed seas. Examples of enclosed seas include the Black Sea, the Baltic Sea, and the Mediterranean Sea, while semienclosed seas include the

Gulf of Mexico, the Gulf of Carpentaria, and Hudson Bay. Some seas may have multiple characteristics over time, such as the Wadden Sea, which was a pelagic sea that became enclosed over time through human action. Lastly, polar seas that are adjacent to the Arctic and Antarctic coasts can be considered a unique type of coastal sea because of their extreme environmental conditions and predominantly pristine nature.

Coastal seas, particularly estuaries and adjacent coasts, have played a critical role in human development for most of human history, and are the most affected by human activities. Historically, coastal seas were primary commercial fishing grounds. Due to their depth and seafloor, they can provide protection from moderate storms and extreme events. Coastal seas contain abundant life because of the way sunlight diffuses through the shallow waters over the continental shelves, in contrast to the sparser life of the deeper abyssal plains. Most commercial exploitation from the sea, such as mining and hydrocarbon development, also takes place on continental shelves.

Increasing numbers of the world's people live in coastal regions adjacent to coastal seas. Many of the world's major cities have been built near natural harbors and have port facilities. The coast is also a frontier for military invaders, smugglers, and migrants. Coastal beaches and warm waters can be important tourist attractions. As a result of these increasing populations and uses, coastal seas face many environmental challenges from human-induced impacts. While human activity and the associated adverse impacts have been affecting coastal seas since early times, the decline of coastal seas has significantly accelerated in the last few centuries. For example, records from all coastal seas illustrate that seagrass habitats have been destroyed, water quality has decreased significantly, and marine species and diversity have lessened.

A perfect storm of overexploitation, nutrient pollution, and climate change is creating an uncertain future for coastal seas. Land-based and marine pollution includes petrochemicals and biological waste, while climate change contributes to sea-level rise, which threatens coastal ecosystems. Jeremy Jackson of the Scripps Institution of Oceanography has warned that "synergistic effects of habitat destruction, overfishing, introduced species, warming, acidification, toxins, and massive runoff of nutrients are transforming once complex ecosystems like coral reefs and kelp forests into monotonous level bottoms, transforming clear and productive coastal seas into anoxic dead zones, and transforming complex food webs topped by big animals into simplified, microbially dominated ecosystems with boom and bust cycles of toxic dinoflagellate blooms, jellyfish, and disease."

Knowledge of coastal seas and their original ecosystems is important for understanding the current and possible future ecosystems. Managing and conserving the world's coastal seas and their ecosystems now and in the future will require science-based approaches, commitment, and resources for managing these seas and human activities.

PELAGIC SEAS

The North Sea is located on the European continental shelf and connects to the Atlantic Ocean through the English Channel in the south and the Norwegian Sea in the north. The North Sea is an important shipping lane and supports major fisheries. The sea is popular for tourism and recreation and is the site of significant hydrocarbon production and wind- and wave-energy generation. The coasts adjacent to the North Sea are subject to high populations, industrialization, and intense uses of the sea and coasts. Environmental impacts on the sea include overfishing, industrial and agricultural pollution, dredging, and dumping. Climate change is affecting its food web and encouraging its fisheries to migrate northward, while new species enter from southern waters.

The European Union has been implementing coastal zone management and marine spatial planning in the North Sea, and there is extensive regional cooperation. The Wadden Sea is adjacent to the North Sea and borders Denmark, Germany, and the Netherlands. It is a pelagic sea that has been modified through a system of dikes and causeways to become a semienclosed sea. The Wadden Sea has long been an important source of food and transport for adjacent communities.

Eutrophication and nutrient pollution have extended for over one hundred years. Given its significance, the Joint Declaration on the Protection of the Wadden Sea was agreed to by the three adjacent countries in 1982, and the Trilateral Wadden Sea Plan was adopted in 1997. The United Nations Educational, Scientific and Cultural Organization (UNESCO) declared the Wadden Sea Area a biosphere reserve in

1986 and added the Dutch and German conservation areas to its list of World Heritage sites in 2009. Most of the Danish conservation area was added five years later.

The South China Sea is bordered by China to the north; the Philippines to the east; Malaysia, Singapore, Indonesia, and Brunei to the south; and Thailand, Cambodia, and Vietnam to the west. The sea lies above a drowned continental shelf that was above the sea during the prior ice ages, when the global sea level was hundreds of feet (or meters) lower. The South China Sea contains many small islands, atolls, cays, shoals, reefs, and sandbars, many of which are uninhabited and below water at high tide. One of these archipelagos is the Spratly Islands. Jurisdiction over the South China Sea is disputed by the adjacent countries, which advocate different management regimes. The sea includes the second-most-used sea lane in the world, extensive proven oil reserves, and significant fishing reserves.

The South China Sea contains an estimated one-third of the world's marine biodiversity, making it an important ecosystem that includes mangrove forests as well as coral reefs and seagrass beds. Environmental threats include land-based pollution, overfishing, destructive fishing practices, the loss of mangrove forests, coral-reef degradation, and damage to seagrasses and wetlands. For example, most mangrove forests have been lost to shrimp farms, industrialization, or tourism.

ENCLOSED SEAS

The Baltic Sea is a brackish inland sea. The sea resembles a riverbed with two tributaries, the Gulf of Finland and the Gulf of Bothnia, and can be understood as the estuary of all rivers flowing into it. As a result of these inflows, its salinity is lower than that of the Atlantic Ocean. Baltic Sea water flows are complex: brackish surface waters discharge into the North Sea, while heavier saline waters enter from the North Sea and move in the opposite direction. The Baltic Sea is stratified, with the denser saline waters remaining near the bottom, resulting in a shifting dead zone along up to a quarter of the seafloor, in which only bacteria flourishes. Freshwater and marine species reside in the Baltic Sea, but there is limited diversity.

Algal blooms have occurred every summer for decades. Agricultural runoff from adjacent areas has encouraged these blooms and increased eutrophication. Climate change is likely to increase temperature and vertical stratification of the Baltic Sea, and the extent and duration of dead zones and algal blooms. Since the demise of the Soviet Union, all countries around the Baltic Sea are collaborating more extensively in the management of the sea and its watershed and the use of its resources.

The Mediterranean Sea is connected to the Atlantic Ocean and almost entirely enclosed by the Mediterranean region, which is characterized by hot, dry summers and cool, wet winters. Being enclosed affects many aspects of the Mediterranean Sea. For example, tides are very limited as a result of the narrow connection with the Atlantic Ocean. Evaporation exceeds precipitation and river flows, driving water flows. Evaporation is especially high in the eastern half, causing the water level to decrease and salinity to increase eastward.

The Mediterranean Sea has been very important throughout human history, with extensive agriculture and industrial development. The sea and the adjacent region are very vulnerable to climate change. Under the changing climate regime, sea-surface temperatures and salinity will increase. Depending on local characteristics, erosion, sediment deposition, drought, desertification, and flooding may intensify or shift. Coastal and beach tourism is an important source of income in the Mediterranean and south Atlantic regions, and the ongoing economic viability of these regions and local communities may hinge on the maintenance of the coastal and marine ecosystems on which tourism, as well as other activities such as commercial fishing, depend.

SEMIENCLOSED SEAS

The Gulf of Mexico is a coastal sea, largely surrounded by the United States, Mexico, and Cuba. It is connected to the Atlantic Ocean through the Straits of Florida between Florida, Cuba, and the Bahamas, and to the Caribbean Sea through the Yucatán Channel between Mexico and Cuba. The outer margins of the continental shelves of Yucatán and Florida receive cooler, nutrient-rich waters from the deep by a process known as upwelling, which stimulates plankton growth and attracts fish, shrimp, and squid. The Gulf Stream is a strong, warm Atlantic current that originates in the Gulf of Mexico and warms the coasts of eastern North America and western Europe. A number of rivers flow into the gulf, including the Mississippi River and the Rio Grande in the north and the Grijalva and Usumacinta Rivers in the south. The land around the gulf is mostly low lying and includes barrier islands, marshes, swamps, and sandy beaches.

Human activities along the coast include fishing, hydrocarbon development, shipping, petrochemical processing and storage, military use, paper manufacture, and tourism. Major environmental threats are agricultural pollution and hydrocarbon-related activities. There are frequent red-tide algal blooms that kill fish and marine mammals and cause health problems in humans. The gulf also contains a hypoxic dead zone that is aggravated by agricultural pollutants. In July 2015, the Associated Press reported that there were more than 30,000 abandoned oil and gas wells (4,081 temporarily abandoned, 25,928 permanently sealed) beneath the gulf that were not being consistently monitored.

The 2010 Deepwater Horizon oil spill in the Gulf of Mexico was the largest accidental marine oil spill to date and caused extensive damage to marine and wildlife species, ecosystems, fishing, and tourism throughout the gulf. The oil spill illustrated the vulnerability of coastal seas to offshore activities when those activities are not properly regulated and implemented. Subsequently, US President Barack Obama issued a directive to improve the regulation of offshore drilling, but his successor, Donald Trump, signed an executive order in June 2018 repealing those regulations.

Hudson Bay is a coastal sea in northern Canada that drains more than one-third of North America. The bay is shallow, with an average depth of 0.06 mile (100 meters), being 851 miles (1,370 kilometers) long and 656 mi (1,050 km) wide. It connects to the Atlantic Ocean through the Hudson Strait, and to the Arctic Ocean's Foxe Basin through the Foxe Channel. Hudson Bay is part of the north Atlantic Ocean, but it is also viewed as the southern extension of the Arctic Ocean. It has a lower average salinity than ocean water because of its low rate of evaporation, the large volume of freshwater entering the bay, and the limited circulation of this brackish water with the Atlantic Ocean. This lower salinity layer also decreases the overall time that the bay is free of ice.

Because of its size, the Hudson Bay marine ecosystem has many coastal ecozones and varied habitats that are used year-round by local peoples and Arctic and sub-Arctic species, and seasonally by migratory fishes, marine mammals, and birds. There are few settlements and limited industrial activity in the bay, so it is mostly affected by global environmental and climate change.

The Gulf of Carpentaria is a warm, shallow sea in northern Australia that measures 115,830 square miles (300,000 square kilometers) and is fed by the flows of over twenty rivers. The gulf is enclosed on the west by Arnhem Land, in Australia's Northern Territory, and on the east by Cape York Peninsula, Queensland. The gulf floor is the continental shelf common to Australia and New Guinea, with a very low gradient. Prawn fishing developed rapidly in the gulf in the late 1960s. Sport fishing is also prevalent.

Manganese and bauxite deposits are exploited around the gulf. As a result of these economic developments, settlement on the shores and islands of the gulf has increased from traditional indigenous populations to several thousand people. The Gulf of Carpentaria Commonwealth Marine Reserve, now the Gulf of Carpentaria Marine Park, was established over an area of 9,178 sq mi (23,771 sq km) in 2012 to protect important turtle nesting and bird breeding areas, submerged coral reefs, and high concentrations of marine life.

POLAR SEAS

The Beaufort Sea is a complex marine ecosystem within the Arctic Ocean. The nearshore productivity of the sea has been an important resource, supporting human occupation for thousands of years. The circulation system is dominated by the Beaufort Gyre, a clockwise rotating surface current that results in large-scale movements of sea ice and surface waters. Ice usually begins to form in October. Most of the marine system is covered by ice until spring breakup, limiting biological production. The Mackenzie River significantly influences the Beaufort Sea; its freshwater mixes with marine waters, and a relatively fresh, mixed layer forms along the coastal areas and stratifies the ocean waters, allowing numerous species to use the area. Marine mammals, fish, and birds aggregate and use the adjacent area of the sea as migratory routes and seasonal breeding and feeding areas.

Open-water areas in the ice known as polynyas, such as the Cape Bathurst polynya, are most important habitats and attract high densities of birds, benthic organisms, and marine mammals.

The Beaufort Sea is relatively pristine. There are large hydrocarbon reserves underlying the sea, but they have not yet been extensively developed. Renewable resources, such as fish and whales, are important for both subsistence uses by local peoples and tourism, but there are no commercial fisheries. Ecotourism is increasing in popularity. In Canada, the Beaufort Sea Large Ocean Management Area is one of five priority areas that have been identified for integrated ocean management planning. This area is located in the northwestern corner of Canada and includes the marine portion of the Inuvialuit Settlement Region, established under an indigenous land claim agreement.

The Ross Sea is one the most pristine marine ecosystems on Earth. Located adjacent to Antarctica and without human occupation, it has not been subject to pollution, invasive species, mining, or overfishing. The Ross Sea is the most productive stretch of water in the Southern Ocean and has high concentrations of wildlife, including large predatory fish, whales, seals, and penguins. Because of its pristine nature and abundance, the Ross Sea provides a unique opportunity to understand the functioning of healthy marine ecosystems. In 2016 the Commission for the Conservation of Antarctic Marine Living Resources (CCAMLR), the international body that manages living marine resources around Antarctica, established a Ross Sea marine park of more than 579,000 million sq mi (1.5 million sq km)—the world's largest marine park—of which 425 million sq mi (1.1 million sq km) is a "general protection zone" in which no fishing is allowed. Because its protections expire after thirty-five years, the Ross Sea marine park does not qualify as a "marine protected area" under International Union for Conservation of Nature and Natural Resources (IUCNNR) rules, which require permanent protection.

—*Magdalena Ariadne Kim Muir*

Further Reading

"Beaufort Sea Integrated Management Area." *Fisheries and Oceans Canada*, Government of Canada, 28 Oct. 2017, www.dfo-mpo.gc.ca/oceans/management-gestion/beaufort-eng.html.

Clark, John R. *Coastal Seas: The Conservation Challenge.* John Wiley & Sons, 2009.

Donn, Jeff. "AP Exclusive: Gulf Oil Wells Lingering with Temporary Seals." *AP News*, 18 July 2015, apnews.com/e68c094dbe7d4d1b91e3657c2314d364/ap-exclusive-share-aging-temporarily-sealed-wells-grows.

"Gulf of Carpenteria Marine Park." *Australian Marine Parks*, Parks Australia, parksaustralia.gov.au/marine/parks/north/gulf-of-carpentaria/. Accessed 9 July 2018.

Jackson, Jeremy B. C. "Ecological Extinction and Evolution in the Brave New Ocean." *PNAS*, vol. 105, supp. 1, pp. 11458–65, doi:10.1073/pnas.0802812105. Accessed 2 July 2018.

Kenchington, Richard, Laura Stocker, and David Wood, editors. *Sustainable Coastal Management and Climate Adaptation: Global Lessons from Regional Approaches in Australia.* CRC Press, 2012.

Milman, Oliver. "Deepwater Horizon Disaster Altered Building Blocks of Ocean Life." *The Guardian*, 28 June 2018, www.theguardian.com/environment/2018/jun/28/bp-deepwater-horizon-oil-spill-report.

Slezak, Michael. "World's Largest Marine Park Created in Ross Sea in Antarctica in Landmark Deal." *The Guardian*, 27 Oct. 2016, www.theguardian.com/world/2016/oct/28/worlds-largest-marine-park-created-in-ross-sea-in-antarctica-in-landmark-deal.

"The Value of Coastal Ecosystems." *UN Environment*, United Nations Environment Programme, 1 Feb. 2016, web.unep.org/coastal-eba/it/node/63.

Walsh, John J., editor. *Wind-Borne Illness from Coastal Seas: Present and Future Consequences of Marine Aerosols.* Academic Press, 2019.

COASTAL WETLANDS

Fields of Study: oceanography; limnology; marine biogeochemistry; marine biology

ABSTRACT

Coastal wetlands are flooded periodically by tides. Some of the plants and animals that live in these areas have origins as land species, others come from the sea. The plants must be able to live in saltwater and in soil that tends to be waterlogged and low in oxygen. Many of the marine animals must be able to survive exposure to the air, sun, wind, and possibly rain at low tide. Bays, inlets, harbors, and sounds are all places where freshwater from rivers and streams mixes with saltwater from the ocean. Salt marshes are found on protected shorelines and on the edges of estuaries. In the United States, they span the entire East Coast, and are extensive along the Gulf of Mexico, but are less common on the West Coast.

KEY CONCEPTS

estuary: locations where a stream or river of freshwater meets and flows into an ocean or other large body of saline water

intertidal area: the strip of terrain that becomes flooded during a high tide and then bared during a low tide

marsh accretion: the building up of sediments over time that gradually raises the bottom of the marsh to higher levels

waterlogged: indicates that the soil contains more water than it is capable of absorbing

zonation: the graduation of plant species from the most water-tolerant in the wettest part of a marsh to dryland plants outside of the marsh environs, according to the nature of the terrains in the intervening area

DESCRIPTION

Coastal wetlands are flooded periodically by tides, so that resident plants and animals spend part of the time in the water and part of the time in the air. Some of the plants and animals that live in these areas have origins as land species, such as grasses, insects, birds, and mammals. Others, such as mollusks, crabs, and fishes, come from the sea. The plants must be able to live in saltwater and in soil that tends to be waterlogged and low in oxygen. Twice a day, many of the marine animals must be able to survive exposure to the air, sun, wind, and possibly rain at low tide. These are harsh conditions, in which drastic changes occur regularly. Bays, inlets, harbors, and sounds are all places where freshwater from rivers and streams mixes with saltwater from the ocean; these areas are n "estuaries." Salt marshes are found on protected shorelines and on the edges of estuaries. In the United States, they span the entire East Coast, and are extensive along the Gulf of Mexico, but are less common on the West Coast.

On the Pacific Coast, most of the shoreline tends to be rocky, so salt marshes are relatively scarce, except in the major estuary/marsh systems of California's San Francisco Bay and Washington's Puget Sound. They are also found in other coastal regions, from the Arctic tundra to the tropics, and on every continent, except Antarctica. Salt marshes are extensive between New Jersey and northern Florida, particularly along the coasts of the Carolinas and Georgia. Farther south, in Florida, they are replaced by mangrove swamps, which are the tropical equivalents of salt marshes. Mangroves are not grasses, but trees, and the aerial prop roots of red mangroves provide special habitat for juvenile fishes, some of which, as they mature, move to coral reefs.

The lower portions of marshes on the East Coast of the United States are alternately flooded and drained twice a day by the tides, caused by the gravitational pull of the Sun and the Moon on the water. The term "flood tide" refers to the incoming tide, while "ebb tide" indicates the falling, receding tide. The exact height and timing of the tides at any given place is determined by the phase of the Moon and many other factors, such as the shape of the shoreline and the strength and direction of the wind. One extreme case is the Bay of Fundy in Nova Scotia, where there is a difference of almost 15 feet (16 meters) between high and low tides. In some places in the Caribbean, the difference between high and low water is less than 1 ft (3 m).

The environmental conditions in a salt marsh are highly variable because the incoming freshwater and ocean water are always mixing in the estuary. If a marsh is located right on the ocean, its water will be more saline than the water in a marsh located further upstream in an estuary that receives much freshwater. The ratio of saltwater to fresh predicts the habitat and resident species. A brackish marsh, for example, occurs in the portion of an estuary where the saltwater is more diluted with freshwater. Further up the estuary, there is a transition to freshwater marshes that are still affected by the tides, although the concentration of salt is low. Still further upstream are freshwater marshes that are untouched by the tides.

There is a parallel transition to less salt-tolerant plants and animals as the water in the marsh becomes fresher. Salinity also fluctuates, depending on the phase in the tidal cycle and the amount of recent rainfall. The same salt marsh may experience salinities ranging from almost freshwater to full strength seawater, and anything that lives there must be able to tolerate these wide swings in environmental conditions. The amount of dissolved oxygen undergoes similar swings as water alternately covers and uncovers the marsh. Temperature also varies widely across the seasons. The air in summer is much warmer than the water, but in winter, the water is warmer than the air. Plants and animals must cope with these variations on a daily basis, but despite these inhospitable and inconsistent conditions, salt marshes are thriving ecosystems.

COASTAL WETLAND PLANT AND ANIMAL SPECIES

Certain species of grasses found only in the shallow intertidal areas are highly specialized to deal with saltwater and salty soil. They are called "halophytes" (Greek *hals*, meaning salt, and *phyton*, meaning plant). Because they have to deal with being submerged in water part of the time, they are also considered hydrophytes (*hydro*, meaning water). In addition to having to tolerate salt and being immersed in water, the lack of oxygen in the waterlogged marsh soil also makes the environment stressful to plants. In East Coast marshes, this zone is occupied by cord grass, *Spartina alterniflora*. The low marsh zone in some other regions frequently lacks vegetation altogether; these areas are mudflats.

The diversity of plant species increases as one moves higher in the marsh to areas that are only affected by the saltwater during very high tides. In the high marsh, there are grasses, rushes, herbs, and shrubs, with different species adapted to different zones. Zonation is a typical feature of salt marshes, with different plant species found at different elevations, depending on their ability to tolerate immersion and salt, and their ability to compete for space.

Marine animals must find a way to keep moist during low tide, and when the tide is out, there is no food for those that obtain their food directly from the water. Most marsh animals are inactive during low tide, with a few exceptions. Fiddler crabs (genus *Uca*) forage at low tide and remain in their burrows at high tide, safe from predatory fish. Smaller crustaceans, called "amphipods" and "isopods," may also remain active on the marsh surface during low tide, and the salt-marsh mussel (*Geukensia demissa*) frequently "air gapes" (gape their shells open for air) at low tide. Virtually all residents of the salt marsh time their lives and their reproductive cycles according to the tides and the phases of the Moon. On the highest tides (at the full and new moon), fiddler crabs release their young where tidal currents can carry them out to the ocean. Horseshoe crabs come up onto the beach to lay their eggs at the high tides before the full moon.

MARSH ACCRETION

Salt marshes develop gradually when tides deposit sediments (soil, fine-grained clay, sand, and silt from rivers and shorelines) across low-lying land, creating wet mudflats or sand flats. Salt marshes develop in sheltered areas that are flat and drain slowly, where mats of microscopic algae and bacteria form and stabilize the sediment so that its elevation can increase. This process speeds up when plants take root as the marsh ecosystem develops. Salt-tolerant grasses arrive when plants raft in on the tides and seeds are carried in by the air or water. The plants slowly take hold and spread, usually by means of underground horizontal stems or rhizomes, which further stabilize the sediment through the growth of dense root systems and add volume, helping to raise elevations.

Once the grasses are in place, the force of wave action is reduced, thus permitting additional layering and settling of the sediment to occur. As the plants decay, fine sediment builds up, causing marsh accretion. As sea levels rise, the marsh pushes landward and seaward, and freshwater plants on the land are replaced by marsh grasses that can tolerate the salt. Underneath the marsh, peat accumulates from the belowground (root and rhizome) plant debris. Peat formation eventually raises the elevation of the sediment surface so that it is flooded less of the time. At that point, other somewhat less salt- and flood- tolerant plant species are able to survive. Eventually, this leads to the development of high marsh areas that are flooded less often by the incoming tides.

The bulk of the roots and rhizomes contribute to both sediment deposition and marsh accretion. The rate that matches marshland growth to changing sea level has been about 4 to 10 inches (10.16 to 25.4 centimeters) per century. With climate change causing thermal expansion of water and melting of glaciers, sea level is rising faster than the sedimentation rate in many areas. Unless marsh accretion can keep pace with sea-level rise, marshes may be replaced by open water. The result of the natural sedimentation is a general spreading of the marsh system and development of the pattern of zonation of plants, in which particular species are located in bands at different elevations.

A fully developed marsh includes creeks that form a network over the marsh and are routes used by tidal water as it enters and leaves the marsh, the primary link of the estuary to the land. The general ecology of salt marshes and the roles played by different species of animals and plants are seen worldwide. Salt marshes are important for many reasons. Many species of migratory shorebirds depend on tidal wetlands as stopover points during their migration between summer and winter habitats, and some birds overwinter in the marsh. Wading birds such as egrets and great blue herons feed in the productive salt marshes during summer months.

About 85 percent of waterfowl and migratory birds use coastal areas for resting, feeding, and breeding habitat. Estuarine animals in various stages of life can be found among the salt-marsh plants. At high tide, juveniles of small fishes move up and swim among the stems of the marsh grasses, where they are protected from larger predators that cannot penetrate the shallow waters of the creeks. It is estimated that over two-thirds of the commercially harvested fish on the East Coast spend some portion of their life cycle in the marsh-estuary ecosystem. Salt marshes are thus considered nurseries, where many species of fish and shellfish live during the early stages of their lives, depending on the marsh for food and shelter. The plants provide shelter for spawning and protect the juveniles of many different species of commercial and ecological importance, including blue crabs, croakers, flounders, and spot. Bluefish spawn out in the ocean, for example, but the juveniles (snappers) move into estuaries in the spring and remain there through the summer, feasting on the smaller fish that live in the marsh/estuarine system. Some fish, like the mummichog or common killifish (*Fundulus heteroclitus*), spend their entire lives in the marsh/estuary system.

FUNCTIONAL NATURE OF MARSHES

Salt marshes also serve to protect coastal areas from storms and floods. They can stabilize shorelines because they can take the brunt of storm waves, buffering the shore from flood and storm damage. One common source of flood damage occurs following storms when runoff from the land hits the coastal plain. Marshes are natural sponges that can absorb much of this runoff, reducing its impact on coastal environments and real estate. Established marsh grasses are also very effective against erosion. The roots and rhizomes of marsh plants help the sediments cohere and consolidate, resulting in less erosion in vegetated areas.

When marshes are removed, the effects of storms on coastal communities are much more severe and devastating, and therefore more costly. In places where marshes are strong, functional systems, storm damage to the land and coastal communities is much less severe. In Sri Lanka, for example, the Indian Ocean tsunami in December 2004 caused much less devastation in areas where mangroves were present than in areas where they had been removed. Hurricane Katrina in 2005 produced a catastrophe in Louisiana in part because of the loss of tidal wetlands.

Marshes also purify water, filtering out pollutants, sediments, and nutrients from the water that washes down from the land. Marshes along the shore slow the movement of runoff from land into the estuary while they increase sedimentation and the removal of wastes. Salt-marsh sediments retain or sequester many kinds of toxic contaminants, helping to reduce the degree of contamination of coastal waters, and intercept nutrients that arrive in runoff from the land, protecting adjacent estuarine areas from excess nutrients that can be harmful. Resident microbes in marshes can remove wastes from the water. They break down and process pollutants in their normal metabolism. In some areas, engineers have constructed wetlands specifically to treat wastewater, a process that capitalizes on this natural ability.

Marshes are highly productive, rivaling tropical rainforests in producing the most basic food energy per acre (or hectare). They are highly productive because they are rich in nutrients and cycle the nutrients very efficiently. They also have many different types of plants that perform photosynthesis: rooted grasses, seaweeds, and microscopic algae that are found in the water (plankton) and living on the surface of the mud. There have been some attempts to attach a dollar value to the services provided by salt marshes. For example, it has been estimated that the amount of annual storm surge protection services provided to areas most vulnerable to hurricane and tropical storms is $23 billion. However, monetizing their benefits and then making decisions about the conservation or destruction of that region in purely economic terms is an overly simplistic approach. Many of the important functions of a marsh are intangible and its benefits accrue to everyone, so fixing its dollar value is unrealistic, if not impossible. Unfortunately, US coastal wetlands are in decline, and are further threatened by increasing coastal development and rising sea levels.

COASTAL DISTURBANCES

Marshes can be disturbed by both natural factors and human activities. Natural disturbance by ice and floating debris can damage temperate marshes, particularly in the colder northern areas. Tidal action can lift up pieces of marsh and deposit them on higher areas, where they literally suffocate the underlying vegetation. Floating masses of dead plant material (wrack) are produced in the fall, when some plants die and collapse, and then are swept away by the tides. Winter ice can amass and destroy large chunks of low marsh vegetation. Fire and overgrazing by animals like geese can also destroy large areas of marshes. However, these effects are trivial compared with the damage that results from human activities. Early settlers in America found marshes important for raising livestock, but when agriculture moved inland because of expanded settlement, salt marshes were regarded as useless wastelands that were extensively filled for expanding towns and cities.

Since the 1950s, an average of over 65 square kilometers (25 square miles) per year of Louisiana's coastal wetlands has been lost because of numerous factors, including development and oil and gas exploration and production. The engineering of channels and the system of levees along the Mississippi River prevents regular tidal flooding, stopping the flow and process of salt-marsh sedimentation and vertical growth. The normal accumulation of sediments washing down into the marsh cannot occur, and damage from hurricanes is exacerbated. If the area had been undisturbed, accumulated sediments would have elevated the marshes and reduced the effects of the storm surge.

The 2005 hurricanes caused the additional loss of 215 sq mi (557 sq km) of salt marsh, an order of magnitude greater than the average annual loss. According to the Army Corps of Engineers, for every 3 sq mi (7.68 sq km) of wetlands, storm surge can be reduced by 1 ft (30 cm). Among other lessons, Hurricane Katrina and its aftermath demonstrate the importance that wetlands play in protecting communities from floods.

As more and more people choose to live in coastal areas, impacts on tidal marshes have increased. Today, larger coastal cities and towns hug the shores of estuaries, continuing to encroach on these vital habitats. Cities developed in these locations because they were ideal for trade when goods were carried by ships. Forest and agricultural products could be easily moved downriver to a coastal port. Ports were often dredged out of marshes, and networks of roadways and railroads were built that crossed the marshes in every direction. This construction disrupted the flow of water, impeded tidal exchange, and slowly strangled the marshes.

Marshes have been filled, drained, diked, ditched, grazed, and harvested. They have become depositories for pollutants that wash into streams and rivers from the land, as well as for contaminants introduced from coastal waters. They have been sprayed with insecticides for mosquito control, chemicals that have a wide range of negative effects on desirable estuarine animals and plants. A range of nonnative species have been introduced or arrived accidentally and have altered their ecology. East Coast marshes have been affected by a strain of the common reed, *Phragmites australis*, which has taken over many lower salinity marshes, reducing the diversity of native plant species, while the native *Spartina alterniflora* of East Coast marshes has become an undesired invasive species on the West Coast and in China.

Salt marshes have been treated as something devalued and "reclaimed"—that is, converted to a use that is considered to have more value. There was previously a widespread perception that wetlands were wastelands and pestilential swamps. They were drained, dredged, and filled with dredge materials for urban development, fenced in for livestock grazing, or polluted through use as garbage dumps. In the past several decades, however, people have begun to understand wetlands, to appreciate the "ecosystem services" they provide, free of charge.

Mudflats and salt marshes lack the public appeal and mystique of tropical rainforests and coral reefs, although they are interesting and ecologically valuable. While protective legislation in the 1970s slowed the rate of loss, marshes are still degraded by the invasion of exotic species, poor management practices, development of urban and suburban areas, and the rise in sea level concomitant with climate change. It is estimated that 65 percent of all coastal marshlands and swamps in the contiguous United States would be inundated with 3.1 ft (1 m) of sea-level rise. As marshes are destroyed and altered, even if at a slower rate than in the past, the vulnerability to floods and storms, the purity of water systems, the vitality of communities, and the health of populations are all at risk. A myriad of plant and animal species face dire consequences if coastal marshlands continue to be lost or exploited.

In the late 1980s, it was estimated that about 60,000 acres (24,000 hectares) corresponding to 97.3 sq mi (249 sq km) of wetlands were lost per year. Since President George H. W. Bush established a national policy of no-net-loss-of-wetlands in 1989, the overall rate of loss has slowed down. However, from 1996 to 2006, 431 sq km (1,116 sq km) of coastal marshes were lost from the United States. Numerous conservation and restoration projects have taken place, with varying degrees of success.

The practice of creating new salt-marsh habitats to replace ones that have been destroyed elsewhere is widespread. But in many cases, there has been little, if any, follow-up to measure the success of the project. Despite increased public awareness of the value of wetlands, they are still losing ground to development, and some people still view them merely as obstacles in the way of urban expansion. Salt marshes and mangroves also provide unique opportunities for fishing, kayaking, and educational field trips. They are fascinating places to study, and should be valued for their beauty, wilderness, and the sense of peace and tranquility that they provide.

Further Reading

Batzer, Darold P., and Rebecca R. Sharitz, editors. *Ecology of Freshwater and Estuarine Wetlands.* 2nd ed., U of California P, 2014.

Doody, J. Patrick. *Saltmarsh Conservation, Management and Restoration.* Springer, 2008.

Finkl, Charles W., and Christopher Makowski, editors. *Coastal Wetlands: Alteration and Remediation.* Springer, 2017.

Perillo, Gerardo M. E., Eric Wolanski, Donald R. Cahoon, and Mark M. Brinson, editors. *Coastal Wetlands: An Integrated Ecosystem Approach.* Elsevier, 2009.

Reddy, K. Ramesh, and Ronald D. DeLaune. *Biogeochemistry of Wetlands: Science and Applications* CRC Press, 2008.

Simon, Thomas P., and Paul M. Stewart. *Coastal Wetlands of the Laurentian Great Lakes: Health, Habitat and Indicators.* AuthorHouse, 2007.

Windham-Myers, Lisamarie, Stephen Crooks, and Tiffany G. Troxler, editors. *A Blue Carbon Primer: The State of Coastal Wetland Carbon Science, Practice and Policy.* CRC Press, 2019.

CONTINENTAL SHELVES ECOLOGY

Fields of Study: oceanography; marine ecology; marine biology; marine geology

ABSTRACT

Continental shelves are the submerged margins of land around landmasses, extending outward from the intertidal zone. These shelves eventually slope away into much deeper waters, the true ocean, known as the abyssal plain. At their widest, continental shelves can underlie entire seas, such as the North Sea. While continental shelves compose just 8 percent of the entire seabed, they are extremely important marine biomes because of their plant and animal life, as well as nonliving resources.

KEY CONCEPTS

benthic: designates an association with the seafloor

ecotope: a particular ecological zone

fjordic: resembling high, steep-walled ravines like the fjords of Norway

riverine: indicates that water and other materials that flow into the oceans are carried there by freshwater rivers

thermocline: a range of water depth that acts as a boundary layer between an upper mass of relatively warm water and a lower mass of relatively cold water

DESCRIPTION

For most humans, the experience of the marine realm is completely limited to continental shelf habitats, and the value attributed to the ocean as a whole is based on knowledge and perception of these regions. While the geological mineral composition of continental shelves varies considerably throughout their worldwide distribution, usually similar to that of the exposed terrain, they share several common features. Continental shelves are relatively shallow, less than 300 to 650 feet (100 to 200 meters), whereas the open ocean is typically greater than 1,312 ft (400 m) deep. Waters overlying continental shelves are often termed 'shelf seas' to distinguish them from the "high seas," or the ocean proper. The uses made of continental shelves are also similar the world over. Not only are most marine recreational activities focused in these areas, so too are aquaculture, mining, oil extraction, and other important economic activities. Natural influences on shelf waters are also similar and include deposition, pollution, ice scour (particularly in shallower areas), currents, stratification, plumes, turbulence, and upwelling.

Earth's landmasses are generally aligned north–south, meaning that major coastlines tend to be eastward or westward facing. There are distinct differences that arise as a result of these individual aspects, caused in part by the rotation of Earth. Eastern continental shelves, for example, are exposed to some of Earth's strongest currents, taking the form of gyres, such as the Pacific Gyre and the North Atlantic Gyre.

Further, sea and air interactions contrive in a complex manner, resulting in a greater variation in sea temperature, which in turn has repercussions for the communities of animals and plants that live there. Western continental shelves are notably productive, with four of the world's major upwellings occurring on western shelves. They are generally narrow in their extent, and adjacent landmasses are typically high and mountainous or fjordic (higher latitudes). Desert conditions are common on these shorelines, and riverine inputs are fewer than on eastern shelves. Western shelves of the Americas are particularly vulnerable to El Niño Southern Oscillation (ENSO) effects. These can manifest in periodically higher sea level, elevated water temperatures, and a depressed thermocline.

HABITATS

Continental shelf habitats are diverse and heterogeneous, with benthic (seafloor) habitats ranging from silty sands, muds, and clays to coarser gravel, stones, and boulders, and even bare rock. Different substrata support particular types of plants and animals, with organisms specifically adapted to a certain set of conditions. Creatures that live on or in sediment thrive in muddy and sandy areas, and typical animals that might be found in such environments include worms, clams, prawns, lobsters, and flatfish. Rocky areas provide more complex physical habitats and support organisms, including lush algae, anemones, sponges, bryozoans, urchins, starfish, octopus, and more agile predatory fish. The substratum is not the only habitat in continental seas; the water column is also an important habitat for a plethora of pelagic species of fish and invertebrates such as squid, jellyfish, sardines, haddock, whiting, cod, and even larger animals such as basking shark, porpoises, dolphins, and whales.

The marine habitat that humans are most familiar with is the intertidal zone, because it is accessible and readily witnessed. Even this single marine ecotope can offer extreme ambient conditions. At their simplest, intertidal shores may be exposed or sheltered, with levels of exposure rated along a gradient. Substrates may be rocky, stony, gravelly, sandy, or muddy. Salinity changes drastically according to proximity to constant freshwater inputs and the effects of meteorological conditions. Exposure to solar radiation, wind, and predation by terrestrial organisms is periodical, according to the tides, and can change drastically in its intensity according to the season. Fringing systems range from tidal mudflats and sand flats to mangroves, wetlands, marshes, and barrier reefs.

At its oceanward extreme, where the continental shelf becomes the continental slope, the seabed declines steeply. The seafloor of the abyssal plain curves upward at the continental shelf edge at a gentler gradient, to form the deeper continental rise. During this transition from shelf to seafloor, the geology of the benthos also changes from continental crust to oceanic crust. As a habitat, continental slopes and rises tend to be rich in mineral deposits dumped there by oceanic currents, or terrestrial runoff and glacial deposition over geological time when sea level was periodically low. Conditions are much colder in these waters compared to shelf seas, light is limited, and barometric pressure is much higher. Species of the deeper slope and rise area tend to be long-lived, with slower metabolism and growth rates.

CONTINENTAL SHELF COMMUNITIES

The relatively shallow waters of the world's continental shelves are some of the richest regions of the ocean with respect to marine life. This is partly because of the nutrient-rich runoff from rivers and streams on land. Shallow waters also allow sunlight to penetrate to the seafloor, which makes photosynthesis possible. Marine flora then sustain invertebrates, such as crustaceans, mollusks, cnidarians (urchins, starfish, and sand dollars), and worms, which in turn sustain fish, sharks, marine mammals, and seabirds. Nature is rarely simple, however, and in the marine realm, food webs are probably more complex than anywhere else on the planet.

Phytoplankton (plant) and zooplankton (animal) form the basis of life on the continental shelf. The world of plankton consists of minute, drifting, single-celled plants and tiny animals, fish larvae and eggs, foraminifera, copepods, radiolaria, and a whole host of other strange and astonishing creatures. Among these, perhaps krill is one of the best known: Although one of the smallest animals on the planet, this crustacean is the main food source of a number of marine mammals and predatory fish by virtue of its abundance, particularly in upwelling zones. Some organisms spend their entire life as plankton (holoplankton), while others are planktonic for just the early stages of their life cycle (meroplankton). Plankton, however, is not limited to continental shelf waters. Seagrass meadows are blankets of lush green foliage that are found on continental shelves throughout the world. These delicate, habitat-forming plants are hugely important in the marine biome, both as a source of food for larger animals such as sea turtles and dugongs, and as a refuge for very rare and endangered species such as seahorses. Seagrass meadows also provide ecological services through the stabilization of sediments and the absorption of excess nutrients in eutrophic runoffs.

Kelp forests are perhaps one of the lesser known assets in coastal waters. Not the most beautiful plant, kelp is rubbery-brown seaweed with a thick stalk that attaches to the seabed, topped with thick, wavy, herbaceous fronds. Despite their lack of aesthetic appeal, kelp stands are one of the most important living habitats in higher latitude waters, both a source of shelter and nutrition for everything from marine invertebrates to otters and seals. The jewel in the crown of coastal waters is coral reefs. These living ecosystems abound with diverse and prolific species, and are typical in the cleanest, warmest waters of the world. It has been suggested that coral reefs are home to about a third of all described fish species in the world. Coral reefs are abundant along western Pacific continental shelves, with Australia's Great Barrier Reef the largest-known coral reef in the world, which is in fact a collection of close to 3,000 reefs, rather than one continuous structure. Toward its northern extreme, the Great Barrier Reef lies only a few miles (or kilometers) from the coast, reaching several hundred miles offshore in its southern stretches.

As well as resident species, the transient vertebrate community is important on continental shelves. Marine vertebrates are numerous and diverse, ranging from fish and sharks to turtles, porpoises, whales, and dolphins to seabirds. Fish are among the most prominent life forms in continental shelf waters. They are important for a number of reasons, as prey and predators within ecosystems, as well as a valuable natural resource to humans. Continental shelf habitats are crucial to fish in a number of ways: as nursery habitats, feeding grounds and spawning grounds, and as a highway for diadromous species that spend part of their life cycle in freshwater and part in the sea. Continental shelves are also home to a plethora of marine mammal species, from sea otters to seals, sea lions, walruses, porpoises, dolphins, and whales.

The nutrient-rich waters of the shelf attract both the plankton-feeding baleen whales, as well as toothed predators such as sperm whales and orcas. Larger marine mammals are found throughout a wide range of pelagic conditions, from the shelf to slope and deeper oceanic waters, many will utilize shelf areas at least temporarily. In the northeastern region of the US Continental Shelf, for example, right, humpback, and minke whale species prefer shelf environments, whereas sei and blue whales show preference for the outer slope and oceanic depths.

LIVING RESOURCES

Most of the world's fisheries are concentrated along continental shelves. Fishing comes in many forms:

subsistence, artisanal, commercial, and large-scale commercial. For many thousands of years, marine fish and invertebrates have been exploited for subsistence and trade. Nowadays, fish products are collected not only for human sustenance, but also in the production of several secondary products such as animal feed and fertilizer, alternative medical treatments, and for the aquarium trade. Fishing is carried out with an array of gear and technologies. Gear used in continental shelf fisheries ranges from a handheld rod or line, to pelagic and bottom longlines, gillnets, purse seines, trawls, pots, traps, and dredges. Spear fishing and hand collection through free diving are also common practices in shallower, warmer waters. In US coastal waters, groundfish have long been the most heavily exploited stocks, including fish such as cod, haddock, flounder, and sole.

For many decades, cod was the currency behind North American prosperity; it fed millions and found a fond place at the heart of Western culture. The crash of cod stocks in the northeastern United States and Canadian continental shelves was unexpected. For at least 500 years, cod fishing along the Newfoundland and Labrador shelves was highly profitable. It is believed that Portuguese fishermen were crossing the Atlantic to catch these fish, drying and salting them, and shipping them back to Europe even before the long-held discovery of the New World in 1492. From the Georges Bank offshore of New England to the Grand Banks off Newfoundland and further still to the Gulf of St. Lawrence, cod aggregated in huge numbers to spawn.

Fishing these shelves remained sustainable for hundreds of years, but the post–World War II arrival of hi-tech factory ships marked the end of these cod stocks. These new ships could trawl and haul huge nets in virtually any weather conditions, 24 hours a day, remaining at sea for weeks at a time, as well as process and freeze fish onboard, and were equipped with sonar, radar, and "fish-finders." To make matters worse, at the time, these banks were considered international waters, and therefore did not fall under the jurisdiction of any one nation. Although the creation of the 200 miles (320 kilometers) exclusive economic zones (EEZs) in the early 1980s diminished the power of these large vessels, the damage to the stocks was already done. Catches continued to rise until 1968, and then took a turn for the worse, falling slowly until by the early 1990s, the great northeastern cod stocks were depleted.

OIL EXTRACTION AND MINING

The outer continental shelf (OCS) is the region of most interest with regard to extracting mineral resources. OCS regions are the seaward extents of the shelf that in the United States are managed under federal jurisdiction. Four regions have been demarcated in US shelf waters: the Atlantic OCS, Pacific OCS, Alaskan OCS, and Gulf of Mexico OCS. The most recent data from the Bureau of Ocean Energy Management reported that there were 3,409 functioning extraction facilities in the Gulf of Mexico OCS in 2010, and 23 facilities on the Pacific OCS, with no exploitation of the Alaskan or Atlantic shelf regions. Following a flurry of activity in the early 1980s, the overall trend in recent years has been downward, with fewer and fewer installations established. In 2010, approximately 30 percent of US oil production was from OCS reserves, a contribution that has increased steadily over time.

Natural gas accounted for around 10 percent of the national total. Marine sands and minerals are also highly prized. Since the 1990s, the United States has seen mining of marine sediments along the Atlantic and Gulf coasts for the replenishment of beaches where erosion has taken a toll. Mining activities in the United States within 3 nautical mi (5.56 nautical km) of the shoreline are generally regulated by individual states, while those beyond fall under federal jurisdiction and are regulated by the Bureau of Ocean Energy Management (formerly the Marine Mineral Service [MMS]).

PLATE TECTONICS AND SEA-LEVEL CHANGE

Evidence suggests that over geological time, sea levels have risen and fallen, both exposing and concealing areas of the continental shelf during this process. Sea level can change by two principal mechanisms, eustatic and isostatic change. Eustatic change is climate-driven and linked with ice formation and melt, while isostatic change is driven by the rise and fall of tectonic plates. As little as 21,000 years ago, sea levels were 410 ft (125 m) lower than at present, and in places the sea would have receded to the edge of the continental slope, exposing the entire continental shelf. Evidence of such changes in sea level and the geological history of continental shelves has been demonstrated by the discovery of fossilized tree remains in deep shelf waters.

Continental crust is less dense than oceanic crust, therefore, where these plates meet, the oceanic crust tends to be pushed under, or subducted by, the continental plate. Often, this causes an oceanic trench to form along the edge of the continental shelf plate edge. The impact causes the terrestrial plate to buckle and lift up, eventually forming mountain ranges, often associated with very narrow continental shelves with steep continental slopes. Although sediment runoff can be considerable from mountain streams, these are usually carried directly offshore to quite deep water. This type of continental shelf is typical along the Pacific coast of North and South America.

Earthquakes are common where plates meet, as is changing sea level. This is not because of changes in the water mass, but rather because of movements of the tectonic plates. These movements are characterized by gradual uplifting, interspersed with sudden dropping of the coastline when energy is released during seismic events. Rifts form where plates move apart; these trailing coastlines are typical of much of the African continent, as well as the eastern seaboard of North America. Here, continental shelves tend to be more extensive and more prone to sedimentation.

INTERNATIONAL LAW OF THE SEA

Throughout history, the sea was seen as a common resource that could be navigated and exploited by all. In the early seventeenth century, however, the principle of Freedom of the Seas, *mare liberum*, was born. This was a doctrine wherein sovereign nations were given jurisdiction over continental shelf waters within 3 nautical mi (6 nautical km) of shore, while the rest of the sea remained a common resource. By the mid-twentieth century, concerns about exploitation of fish stocks, pollution, and the potential profits available from both living and nonliving marine resources incited a number of coastal nations to unilaterally extend their claim to govern and manage adjacent waters. In 1945, US President Harry Truman (1884–1972) declared national jurisdiction over all continental shelf waters, mineral resources, and underlying geology, in a move to secure valuable oil, gas, and mineral reserves, and in doing so began a revolution in the way that shelf resources are managed.

Several South American countries soon laid claim to waters within 200 nautical mi (370 nautical km) of shore, while north African and Arabian Gulf nations extended their territorial claims to 12 nautical mi (22 nautical km). The world over, nations were disputing claims to marine resources, from tin to diamonds, gas and oil. The 1958 United Nations Convention on the Law of the Sea (UNCLOS I), which entered into force in 1964, defined "continental shelf" in terms of its use, rather than its geological extent. Here, the continental shelf was considered the seabed and subsoil, beyond territorial seas, extending to a depth of 656 ft (200 m), and coastal states were given exclusive sovereign rights over its exploitation.

Responding to continued international unrest, the United Nations convened the Third United Nations Conference on the Law of the Sea in 1973, a process that continued for nine years, finally ending in 1982 with the publication of the UN Convention on the Law of the Sea (UNCLOS III). The convention provided for the setting of limits, resolution of disputes, inauguration of EEZs, as well as exploitation, environmental protection, and research. Under UNCLOS III, territorial waters extended to 12 nautical mi (22 nautical km) from the mean low water mark, with a further contiguous zone extending to 24 nautical mi (44 nautical km). EEZs were waters extended up to 200 nautical mi (370 nautical km). The degree and nature of regulation that coastal nations can impose in each of these jurisdictions varies, but EEZs demarcate the waters within which a state has sole rights to exploitation of all-natural resources, both living and nonliving. In the United States, management of the continental shelf is shared by state and federal governments. In most cases, these limits are delineated by the 3-mi (5-km) limit (9-mile [14.5 km] limit in Florida), in nautical miles.

The legal definition of a continental shelf was modified from that of UNCLOS I, and now refers to the natural prolongation of territory to the continental margin or 200 nautical mi (370 nautical km), whichever is greater, satisfying nations with narrow natural continental shelves. To appease nations with a naturally wider shelf, the convention allows the extension of continental shelf designation of as much as 350 nautical mi (648 nautical km) or more where specific geological criteria are met. A further development in UNCLOS III was the creation of the International Seabed Authority, which governs mineral exploitation beyond EEZ limits. The Commission on the Limits of the Continental Shelf was set up to resolve disputes about the extent of shelf regions and their sovereignty.

—Ruth M. Higgins

Further Reading

Burk, C. A., and C. A. Drake, editors. *The Geology of Continental Margins.* Springer, 2013.

Chiocci, F. L., and A. R. Chivas, editors. *Continental Shelves of the World: Their Evolution During the Last Glacioeustatic Cycle.* Geological Society, 2014.

Flemming, Nicholas C. et al., editors. *Submerged Landscapes of the European Continental Shelf: Quaternary Paleoenvironments.* Wiley Blackwell, 2017.

Mossop, Joanna. *The Continental Shelf Beyond 200 Nautical Miles: Rights and Responsibilities.* Oxford UP, 2016.

Rose, George A. *Cod: The Ecological History of the North Atlantic Fisheries.* Breakwater Books, 2007.

Walsh, John J. *On the Nature of Continental Shelves.* Elsevier, 2013.

CORAL REEFS AND SEAGRASS BEDS

Fields of Study: oceanography; marine ecology; marine biology; marine biogeochemistry

ABSTRACT

Coral reefs are found on continental shelves at depths up to 164 feet (50 meters), in water with a mean temperature averaging above 68 degrees Fahrenheit (20 degrees Celsius), and with a salinity ranging from 32 to 42 practical salinity units. Coral reefs are home to approximately 25 percent of all marine life, comparable to tropical rain forests in terms of the biodiversity. Exact species counts on coral reefs are unknown, and many coral reef species remain undescribed. Coral reef ecosystems cover about 0.17 percent of Earth's surface, and are extremely productive, with gross primary productivity about 100 times greater than primary productivity in the ocean's pelagic zones.

KEY CONCEPTS

bioerosion: erosion or structural breakdown caused by the activity of living organisms

biogenic: produced by or the result of activity by living organisms

eutrophic: oxygen-poor water that contributes to a reduced ability to support life, such as cold water upwelling into a body of warmer water

photosynthetic compensation point: the depth to which sunlight reaches in ocean water where the rate of formation of products of photosynthesis is balanced by the rate of consumption of those products in respiration

practical salinity: the salt content of water as determined by electrical conductivity, as opposed to absolute salinity, which is determined as the weight of dry salts residue obtained from a specific volume of ocean water

trophic cascade: the result of a single change in the trophic chain (the food web and species distribution) of an environment, often as the result of a single change to one small part of the trophic chain

zooxanthellae: colored symbiont organisms that exist within the interstitial spaces between the cells of coral polyps; expulsion of these organisms due to stress results in coral bleaching

DESCRIPTION

Coral reefs are found in the euphotic zone of continental shelves at depths up to 164 ft (50 m), in water with a mean temperature averaging above 68 degrees Fahrenheit (20 degrees Celsius), and with a salinity ranging from 32 to 42 practical salinity units. They are found in both tropical and subtropical waters, in latitudes from 25 degrees north to 25 degrees south. Coral reefs have greater taxonomic diversity per area than any other marine environment and are home to approximately 25 percent of all marine life. They have been compared to tropical rain forests in terms of the diversity of life. Exact species counts on coral reefs are unknown, and many coral reef species remain undescribed. Some estimates state that up to 2 million different species live in and around coral reefs. Coral reef ecosystems cover about 0.17 percent of Earth's surface, and are extremely productive, with gross primary productivity estimates of 1.1 to 3.3 pounds (500 to 1,500 grams) of coral per square foot (or meter) per year; this is about 100 times greater than primary productivity in the ocean's pelagic zones.

Coral reefs are most common in the western Pacific, western Atlantic, and Indian Oceans, where waters are oligotrophic. More eutrophic waters, a consequence of upwelling, mean that coral reefs in the eastern Pacific and Atlantic have reduced species richness and are smaller. Mangrove forests and seagrass beds are often adjacent to coral reefs, and the former communities benefit coral reef ecosystems by absorbing nutrients and slowing rates of sedimentation.

Coral reefs are biogenic structures, composed primarily of calcium carbonate. Coral reefs take up over half of all the calcium ions that enter the world's oceans. Reefs are surrounded by biogenous sediments derived, in part, from the breakdown of this skeleton by bioerosion (particularly by boring sponges) and abiotic processes (particularly through wave action). The main taxonomic groups that create reef structures are hermatypic (reef-building) anthozoans, including stony corals and stony hydrozoans, though crustose coralline algae and mollusks are also important contributors to the reef's infrastructure. Other common organisms found in coral reef communities include sponges, gorgonians (sea fans), echinoderms, and fishes.

Hermatypic corals are cnidarians that grow clonally; about 75 percent of them can also reproduce sexually, and release of sperm and eggs is triggered in response to changes in temperature, day length, or lunar cycles. Corals can feed on plankton using their

Green Sea Turtle grazing seagrass at Akumal Bay (P. Lindgren via Wikimedia Commons)

tentacles, which contain pneumatocysts. They can also acquire planktonic food with their mesenterial filaments, which release digestive enzymes and can be everted to absorb food particles that have not entered the gastrovascular cavity. All hermatypic coral polyps also contain endosymbiotic, mutualistic dinoflagellates, termed "zooxanthellae," which live between the cells of the polyp's gastrodermis.

The highly colored zooxanthellae, which are members of multiple taxonomic groups, are autotrophs that exchange photosynthate and amino acids with their animal hosts in return for protection and nitrogenous waste. Zooxanthellae provide up to 90 percent of their primary productivity to the polyp; this allows the polyp to produce and deposit calcium carbonate, which it embeds in a carbohydrate matrix to form a calyx into which it anchors. Polyps continue to deposit calcium carbonate throughout their lifetimes, allowing the reef's skeleton to continue to grow in size.

Unfavorable environmental conditions can trigger polyps to expel their endosymbionts, a process called "bleaching." Such conditions include water temperatures in excess of 95 degrees Fahrenheit (35 degrees Celsius), high light conditions (particularly ultraviolet radiation), or eutrophication. As the mutualism is obligate for many coral species, the polyps often die in the days or months after a bleaching event. Some polyps are able to regain zooxanthellae, and recent studies have shown that zooxanthellae species and genotypes differ in their resistance to stress and likelihood to abandon their host polyps.

Since coral polyps are small, usually less than 0.12 inch (3 millimeters), in diameter), and calcium carbonate deposition rates are relatively low, a coral colony grows less than 3.9 in (10 cm) per year. Thus, formation of an entire reef structure takes up to 10,000 years. There are three main types of coral reefs, each of which has its topographic features determined by abiotic factors including light, temperature, the direction and strength of waves and currents, and water depth. Coral reef classification is based on both reef morphology and reef origin.

TYPES OF REEFS

The most common reef type worldwide is the fringing reef, which is found in narrow bands parallel to tropical shorelines. Fringing reefs are found in nearshore habitats with existing hard substrate to support the settlement of a coral's planula larvae, which can metamorphose into polyps after settlement. Regions of a fringing reef include the reef flat (inshore), which extends to the reef crest and drops off to the reef slope, which interfaces with the open ocean. The world's longest coral reef, in the Red Sea, is a fringing reef.

Barrier reefs like Australia's Great Barrier Reef are farther offshore, and a lagoon separates the reef from the shoreline. The reef begins with a back-reef slope, which transitions to a reef flat, then reef crest, which absorbs much of the energy from wave impacts. The reef crest ends in the forereef slopes, which trails off to the deep ocean.

Atolls, found mostly in the Pacific and Indian Oceans, consist of a lagoon ringed by a coral reef. The edge of the reef is the reef flat, and the outer slope moves into the open ocean. Atolls are formed around sunken volcanoes, a theory first proposed by Charles Darwin. Fringing reefs first form around islands. As islands slowly subside, reefs gradually gain height through calcium carbonate accretion by hermatypic corals and others. Over long periods of geologic time, the island is replaced by a lagoon.

Interspecific interactions like competition are the primary factors determining community assemblages in coral reefs.

Space is at a premium on reefs, and is the primary resource for which organisms struggle. Some scleractinian species actually digest neighboring species with their mesenterial filaments to acquire more space. Bottom-down interactions mediated by herbivory also affect the species composition of reefs. Mutualisms are also common in coral reef systems; these include cleaner symbioses and mutualisms between zooxanthellae and sponges, mollusks, or other cnidarians. Herbivores are important grazers of macroalgae, and reductions in macroalgae densities allow coral polyps and their symbionts to acquire sufficient light.

Coral reefs contribute to local economics by attracting tourism dollars, and they are important sources of organisms for both commercial and sustenance fisheries. Coral reefs can yield an average of about 33,000 pounds (15,000 kilograms) of fish and other seafood per approximately 247 acres (1 square kilometer) per year. Coral reefs also prevent shoreline erosion and dampen the energy of waves associated with storm activity.

Healthy coral reef in St. Croix, U.S. Virgin Islands (National Oceanic and Atmospheric Administration)

THREATS TO CORAL REEFS

Natural threats to coral reefs include damage from wind and waves associated with hurricanes and cyclones. The slow growth rates of the reef skeleton means that recovery times for reefs can be long. Predation from fishes and invertebrates (polychaetes, mollusks, crustaceans, and echinoderms) is also a threat to coral polyps. Extreme low tides or drops in water level can cause physiological stress to polyps and trigger bleaching. Though coral polyps secrete mucus, this mucus provides only limited protection from desiccation.

In 2012, nearly 85 percent of the world's coral reefs were at immediate risk because of human activity. When combined with rising ocean temperatures and coral bleaching, the risk rises to over 90 percent. The most threatened ecosystems are found in southeast Asia. Anthropogenic threats to coral reef ecosystems include overfishing, which can cause trophic cascade effects that allow macroalgae to overgrow coral reefs and trigger bleaching. Fishing practices that use toxins like cyanide can kill the corals' sessile polyps, and dynamite-based fishing or trawling can destroy reef structures. Pollutants can cause eutrophication, can be toxic to the polyps, can trigger bleaching, can reduce water clarity so that zooxanthellae are below their photosynthetic compensation points, or can interfere with episodic sexual reproduction events.

Reef harvesting, also called "coral mining," damages reefs long-term. Coral is removed and sold as souvenirs and is also used as material for road-fill, bricks, and cement. Approximately 25 percent of coral reefs worldwide are considered damaged beyond repair.

Sedimentation or siltation is particularly harmful to coral polyps, because it can reduce rates of photosynthesis or actually suffocate polyps. Although viral, bacterial, and fungal diseases occur naturally in coral reef ecosystems, the incidence of disease is heightened by other stressors to corals, including increased ultraviolet radiation or temperature. Finally, the effects of global climate change are particularly problematic for sensitive coral reef communities.

Increased oceanic temperatures can contribute to increased bleaching rates and resultant coral reef mortality. Rising levels of atmospheric carbon dioxide diffuse into oceans, where they decrease water pH in a process known as ocean acidification. Carbonate ions combine with carbon dioxide and water, producing bicarbonate while decreasing the relative availability of carbonate. Since hermatypic corals deposit carbonate in their exoskeletons, the ability of reefs to maintain or grow their structures has been impeded.

Sunscreen, or more specifically a common chemical found in the majority of sunscreens—oxybenzone—is significantly contributing to the deterioration and bleaching of coral reefs by breaking the coral down and robbing it of nutrients. Not only are swimmers at beaches to blame for providing increased oxybenzone in the world's oceans, but any person who applies sunscreen then washes it off in the shower or bath contributes to the chemical being added to the ecosystem.

When threats of climate change and direct human impacts are considered together, the World Resources Institute estimates that over 85 percent of the world's coral reef ecosystems are threatened. Although 33 percent of the world's coral reef ecosystems are in Marine Protected Areas, enforcement of protection laws is uneven, and many of the world's most threatened reefs completely lack protection.

SEAGRASS BEDS

Seagrasses are submerged aquatic angiosperms that live in relatively low-current regions within the euphotic zone of estuarine or saline waters. They have a broad thermal tolerance, living in waters with average temperatures ranging from minus 21 to over 104 degrees Fahrenheit (6 to over 40 degrees Celsius), and can tolerate salinities from 0 to 50 practical salinity units. All reproduce asexually via rhizomatous growth, and this process contributes to most of the growth rate of seagrass populations. Seagrasses, which comprise less than 0.02 percent of the described flowering plant flora, are distributed across two families (Potamogetonaceae and Hydrocharitaceae) and a total of 12 genera. Of these genera, seven are tropical and five are temperate. All but one seagrass species (*Enhalus acoroides*, the most recent to evolve) utilize hydrophilous pollination from simplified and reduced flowers. The number of seagrass species is estimated to range from 50 to 60. The precise number is still in dispute, as molecular and morphological taxonomists struggle to resolve their data. Other factors contributing to the confusion surrounding seagrass taxonomy is the difficulty in finding reproductive organs, most useful in anatomical classification schemes, for many species, and the fact that many vegetative traits used in classification are highly plastic.

Seagrasses evolved from either terrestrial (marsh) species or freshwater macrophytes, and the sparse fossil record indicates that they invaded coastal marine habitats about 100 million years ago. Biochemical and molecular evidence supports the polyphyletic origin of this evolutionary group. Current levels of population genetic diversity vary widely among species and regions and are affected greatly by seagrass species' extensive rhizomatous (clonal) reproduction.

Habitat requirements restrict the global distribution of seagrass-dominated biomes. Although some abiotic needs are species-specific, all seagrasses require saline waters, substrate in or on which to root, and high levels of light. Salinity levels ideal for seagrass growth range from 10 to 45 practical salinity units, with values above or below this resulting in solute stress, necrosis, and/or greater disease susceptibility, particularly to fungal pathogens. Most species root in substrates with high sand or organic contents where they absorb nutrients primarily through their root systems. Areas where particulate matter is fine are less likely to be colonized due to increased sediment turnover rates and decreased light penetration.

Rocky intertidal and subtidal regions are frequently inhabited by members of the genera

Phyllospadix, *Posidonia*, and *Thalassodendron*, which can attach their rhizomes and roots to rocks in intertidal or subtidal regions. Seagrasses require from 4 to 29 percent of incident radiation to stay above their compensation point, the light level at which the glucose generated via photosynthesis and that used by cellular respiration are in balance. These light levels are higher than those estimated for many shade-adapted terrestrial plants and macroalgae.

Because of their high light requirements, seagrass beds are restricted to the shallow, oligotrophic, littoral portions of continental shelves. They are found along the margins of all Earth's land masses, except Antarctica. Seagrasses are often separated into nine different flora based on their distribution: temperate east Pacific, temperate west Pacific, temperate north Atlantic, temperate south Atlantic, Caribbean, Indo-Pacific, Mediterranean, New Zealand, and South Australia. The species diversity of seagrasses is highest in the waters surrounding the Indo-Pacific region. Biogeographic boundaries between such flora are often sudden, and such shifts are usually determined by changes in water density or nutrient availability. Within a single region, species evenness is usually depressed compared to even the most depauperate of temperate regions, and many stands of seagrasses are composed of only a single species. As in temperate flora and other marine communities (including mangals and coral reef ecosystems), species diversity and evenness declines along a latitudinal gradient from the tropics to the poles.

SEAGRASS ECOSYSTEMS

Seagrass ecosystems are rich in macro- and microalgae, which can be benthic, or use the seagrass organs as a substrate. Epiphytic diversity within seagrass communities can be quite rich, composing up to 45 percent of all algae in that ecosystem, and algal primary productivity can meet or exceed that of the submerged aquatic macrophytes. All major phyla of animals can be found residing in seagrass beds, in infaunal, epifaunal, or motile forms. Experimental manipulations as well as mensurative studies have shown that community composition of seagrasses can be controlled by resource limitation, bottom-down forces, or a combination of these.

Although many animals use seagrass blades as a substrate, or graze on the seagrasses' epifloral, epifaunal, and periphyton communities, few consume seagrasses directly as a major component of their diets. This is because of the high cellulose content and high carbon to nitrogen ratio in seagrasses and the fact that many are chemically defended. Exceptions to the rule of direct consumption are found in communities which lack keystone predators, where large populations of isopods, amphipods, parrotfishes, or sea urchins may overuse their preferred algal food source and prey switch onto seagrasses. Some invertebrates selectively graze the inflorescences or seeds of seagrasses, with potential impacts on their sexual fitness. Grazing of periphyton, which lack structural or biochemical deterrents to herbivory, actually increases seagrass fitness, as it increases the amount of incident light available for photosynthesis and reduced hydrodynamic drag on seagrass leaves. Larger herbivores, like West Indian manatees (*Trichechus manatus*), have diets consisting mostly of seagrass species, and other Sirenia like dugongs (*Dugong dugon*) eat only seagrasses.

Aquatic birds like American widgeons or brant geese might rely on seagrasses for a major part of their diet during migration. After spending their first year of life in the pelagic zone, where isotopic studies have shown them to be omnivorous, green sea turtles (*Chelonia mydas*) have diets that consist exclusively of seagrass species. Grazing at intermediate levels is also beneficial to seagrass productivity and population growth, because the physiological integration among seagrass clones allows the plant to produce increased standing crops of shoots. Isotopic analyses have revealed that seagrasses are also an important component of detrital food chains in both nearshore (adjacent) and deeper oceanic environments. In fact, more than half of the primary productivity of a seagrass bed is connected to other trophic levels in the food web via detritus. Seagrass leaves themselves are rich in aerenchyma, can float over long distances, and can directly export carbon-rich leaves to adjacent ecosystems.

Seagrass beds are both ecologically and economically important. The extensive root and rhizome systems of seagrasses, as well as their baffle-like blades, allow them to stabilize sediments and improve water quality and clarity. They also do significant carbon and nutrient cycling, sequestering up to 12 percent of the ocean's total carbon and supporting primary productivity levels of as much as 0.236 ounces of coral per square yard (8 grams of coral per square meter) per day. Seagrass beds are vital nursery habitats for many organisms, though the number of species that they support and the commercial importance of those species varies by latitude. The hydrodynamic properties of seagrass beds help them entrap planktonic larvae and small nekton.

Seagrass beds also provide shelter from predation for a number of animals, particularly small invertebrates, and are higher than most ecosystems in their levels of secondary productivity. In addition, they are important feeding grounds and nursery habitats for coral reef and mangrove-dwelling fishes; these species often migrate diurnally from reefs to seagrass beds to feed, or they may recruit their offspring into the more quiescent seagrass habitat. Seagrass beds house more species than habitats with similar abiotic conditions that lack seagrasses, even those with extensive macroalgal communities. Rates of predation on both fish and invertebrate species are lower in seagrass beds than on bare substrate. Common taxonomic groups found as juveniles or adults as members of seagrass communities include fishes, crustaceans, mollusks, and polychaetes.

In 2009, the National Academy of Sciences (NAS) published a study that found that seagrass meadows are disappearing at a rate similar to that of coral reefs and tropical rainforests. The NAS also found that while the annual loss of seagrass meadows was at a rate of 1 percent per year before 1940, since 1990 the loss had accelerated to 7 percent per year.

Seagrass species are sensitive to disturbance, especially those that reduce light availability, such as sediment loading and dock building. Eutrophication, in addition to creating resource regimes more conducive to the growth of macroalgae, increases the abundance of epiphytes and phytoplankton, which can reduce light availability. Declines in seagrass can occur in response to ice scouring, storm activity, grazing by waterfowl, or infection (particularly wasting disease, a fungal infection). Recent decades have seen dramatic and worldwide declines in seagrass cover, and Frederick T. Short and Sandy Wyllie-Echeverria estimated that most of these declines can be attributed directly to anthropogenic disturbances.

Oil spills can reduce seagrass production, as can changes in salinity because of changes in freshwater input. One example of this occurs because of modifications of water flow through the Everglades, leading to hypersaline conditions in Florida Bay.

Mechanical damage from prop scarring or dredging activities also harms seagrass beds. Climate change and associated warming of waters might harm temperate species near the tropical edges of their biogeographic distributions. Sea-level rise might also harm seagrasses by increasing water depth and thus decreasing light availability. Current legislation in the United States requires that seagrass losses be mitigated, under provisions of the Clean Water Act. However, effective restoration of complete seagrass communities that mimic the function of undisturbed communities has proven elusive.

—Jennifer Rhode Ward

Further Reading

Bowen, James, and Mavourna Collits. *The Coral Reef Era: From Discovery to Decline, a History of Scientific Investigation from 1600 to the Anthropocene Epoch.* Springer, 2015.

Camp, Emma F., Verena Schoepf, Peter J. Mumby, and David J. Suggett, editors. *The Future of Coral Reefs Subject to Rapid Climate Change: Lessons from Natural Extreme Environments.* Frontiers Media, 2019.

"Coral Reefs." *WWF Global.* WWF—World Wide Fund for Nature, 2016. Accessed 8 Aug. 2016.

Fears, Darryl. "How We Are All Contributing to the Destruction of Coral Reefs: Sunscreen." *The Washington Post*, 20 Oct. 2015.

Goreau, Thomas J., and Robert Kent Trench, editors. *Innovative Methods of Marine Ecosystem Restoration.* CRC Press, 2013.

Hogarth, Peter J. *The Biology of Mangroves and Seagrass.* 3rd ed., Oxford UP, 2015.

Larkum, Anthony W. D., Robert J. Orth, and Carlos M. Duarte, editors. *Seagrasses: Biology, Ecology and Conservation.* Springer, 2006.

Monaco, André, and Patrick Prouzet. *Marine Ecosystems: Diversity and Functions.* Wiley, 2015.

Reynolds, Pamela L. "Seagrass and Seagrass Beds." *Ocean Portal.* Smithsonian Institution, 2015. Accessed 8 Aug. 2016.

Sheppard, Charles. *Coral Reefs: A Very Short Introduction.* Oxford UP, 2014.

Coral Triangle

Fields of Study: oceanography; marine biology; marine biogeochemistry; marine ecology

ABSTRACT

The Coral Triangle is a large marine habitat area in southeast Asia. The roughly triangular area of 2.3 million square miles (6 million square kilometers) covered by the Coral Triangle encompasses the national waters of six countries: Philippines, Indonesia, Malaysia, Papua New Guinea, Solomon Islands, and Timor Leste.

KEY CONCEPTS

coral bleaching: expulsion of the highly colorful zooxanthellae symbionts that inhabit intracellular spaces in coral polyps, leaving them colorless against the white calcium carbonate structure of the reef

live-fish trade: capture of desirable and colorful reef fish in the wild for sale to the aquarium market

ocean acidification: lowering of the pH of seawater to more acidic values as carbon dioxide and other acidic gases are absorbed and react with water molecules to produce carbonic acid and other acids

ocean warming: increase in the average temperature of ocean waters, locally or globally, as a reflection of the corresponding increase in average atmospheric temperature due to climate change or local environmental influences

pH: a relative measure of the acidity or alkalinity of a water solution, defined as the negative logarithmic value of the hydrogen ion concentration in the solution

AN UNDERSEA WONDERLAND

The Coral Triangle is a true nursery of the seas, containing more than 600 coral species, or about 75 percent of all coral species in the world, as well as six of the seven sea-turtle species found in the oceans and more than 2,200 species of fish, or about one-third of the world's known coral-reef fish species. At least fifteen of the corals in this biome have been identified as endemic, meaning they exist nowhere else on Earth. The epicenter of coral diversity here, with 553 species, is the Raja Ampat region near Bird's Head Peninsula in the West Papua province of Indonesia.

The reefs of the Coral Triangle sustain more than 120 million people, who depend directly on the reef's resources for food and income. More than 100 million people beyond the Coral Triangle also benefit.

ENVIRONMENTAL THREATS

The local threats are mainly pollution from coastal development and poor sewage treatment infrastructure; overfishing, especially of species that play critical roles in the ecosystem; destructive fishing methods such as cyanide poisoning and dynamite blast fishing, which kill corals and destroy habitat, lowering the productivity of the reef and its resources over time; and bycatch, which is the unintended catching of sea turtles, dolphins, sharks, seabirds, and the like in the context of long-line fishing for tuna. Many of the local threats persist due to high poverty levels, poor resource-management practices, and lack of political will, combined with high market demand outside the region for many of the area's reef and open-ocean fish.

The growing live-fish trade for restaurants in Asia and aquariums around the world is driving a dramatic increase in fishing efforts in the Coral Triangle, which has led to declines in many ecologically important fish species. These are the same species of fish that the local subsistence, artisanal fishers also consume for food. In addition, the Coral Triangle has immense areas for spawning and juvenile tuna, so the region harbors what is perhaps the largest tuna fishery in the world—one of the single most significant economic drivers in the area.

Heavily fished tuna species in the region include southern bluefin, big-eye, albacore, yellowfin, and skipjack. While a tremendous economy has been fueled by commercial fishing and tourism, the Coral Triangle's fish stocks are declining dangerously and are in need of effective management.

Population growth and continued reef resource exploitation can significantly weaken not only the health of the marine ecosystem, but also the food security in the region. Well over 100 million people depend on the reef for food. The resources need to be cared for to preserve them for future generations in order that a food crisis may be averted.

EFFECTS OF CLIMATE CHANGE

The reefs of the Coral Triangle also are subject to global threats, just like all other reefs around the world. The rising global demand for reef-derived fish is accompanied by rising ocean temperatures and acidity levels. The unprecedented rates of the global rise of carbon dioxide in the atmosphere is leading to fast rates of ocean warming and acidification. These climate-change factors can severely undermine the ability of coral reefs to grow, reproduce, and continue to act as hubs of marine life.

When sea-surface temperatures warm the local corals above usual highs and remain at that level for several months, severe coral bleaching and die-offs can occur. During coral bleaching, the corals expel from their tissues tiny photosynthetic organisms, called "zooxanthellate algae," which contribute significantly to bringing food and energy to the coral animal. These algae, living inside the coral, also impart colors to the outside of the corals. The reefs in the Coral Triangle are already living close to the thermal edge of what scientists have measured. If the waters continue to warm, the corals—the engineers of the whole bountiful ecosystem—will continue to be threatened.

Temperatures are rising just as levels of ocean acidity are increasing. Unfortunately, acidification adds stress to corals and other organisms on the reefs that make their shells out of calcium carbonate. The lower the pH (higher the acidity) of the seawater, the weaker corals and other calcifying organisms are in building their skeletons. Eventually, flimsy skeletons make for a more fragile overall habitat and a less productive ecosystem, which in turn could lead to fewer and smaller fish, lower fish catches, and less food, trade, and tourism.

By the end of the twenty-first century, the Coral Triangle is expected to experience some of the fastest rates and greatest amounts of acidification in the world. Continued scientific studies of the area exposed to the hottest temperatures and highest acidity levels will shed light on which communities are most vulnerable to deterioration and those that could be the most resilient. Such information can be used to design conservation strategies and efforts that are more robust and likely to resist the effects of climate change.

CONSERVATION EFFORTS

To inspire environmental stewardship of the local and regional reefs, many conservation organizations have developed strong, active programs that include working with local communities to sustain healthy reefs without compromising the human needs of food, tourism, and recreation. For example, environmental education programs aimed at raising awareness for ocean stewardship have been an important focus in Raja Ampat, in West Papua, Indonesia. There, Conservation International maintains an educational sailing ship, *Kalabia,* that travels to approximately 100 coastal villages in the region every year to educate children about the importance of the reefs on which their communities depend.

On another level, the World Wildlife Fund and The Nature Conservancy work with the six nations of the Coral Triangle to eliminate illegal and destructive fishing and to develop successful joint programs that focus on protection of the shared waters and reef resources. Also, many small community conservation organizations continue to work tirelessly with governmental bodies toward greater protection of these unique and valuable marine resources.

On land, there has been significant success for conservation, with wilderness areas set aside to preserve and to safeguard the resources. The Coral Triangle is now a top priority for marine conservationists because it is one of the most beautiful, biologically productive—and threatened—areas on the planet.

—Lida Teneva

Further Reading

Carpenter, Kent E., et al. "One-Third of Reef-Building Corals Face Elevated Extinction Risk from Climate Change and Local Impacts." *Science*, vol. 321, no. 1, 2008.

"Conservation International." Coral Triangle Initiative." www.conservation.org/global/marine/initiatives/oceanscapes/cti/pages/overview.aspx.

Kassem, Ken, and Eric Madeja. *The Coral Triangle.* John Beaufoy Publishing, 2014.

Leidy, Chris. *The Coral Triangle.* Assouline, 2020.

Mora, Camilo, editor. *Ecology of Fishes on Coral Reefs.* Cambridge UP, 2015.

Powderham, Alan J., and Sancia Van Der Meij. *At the Heart of the Coral Triangle: Celebrating Biodiversity.* Taylor & Frances, 2020.

Unsworth, Richard K.F., Julian Clifton, and Adrian J. Smith, editors. *Marine Research and Conservation in the Coral Triangle: The Wakatobi National Park.* Nova Science, 2010.

Veron, J., et al. "Delineating the Coral Triangle." *Galaxea, Journal of Coral Reef Studies*, vol. 11, no. 1, 2009.

D

Dead Zones

Fields of Study: oceanography; marine biogeochemistry; marine biology; marine geology

ABSTRACT

A dead zone is, as the name implies, a space within a body of water where no living, aerobic creature can be found, usually due to a lack of oxygen in the water. Dead zones can be found in both marine and freshwater environments. There are numerous and similar causes of dead zones in both environments. Remediation of a dead is problematic.

KEY CONCEPTS

eutrophication: depletion of the dissolved oxygen content in a body of water, typically due to an overabundance of algae, water plants, and other organic matter fueled by an influx of fertilizers and other nutrients that encourage the growth of plants

gyre: a general, slow circulation of water that spans entire ocean ranges, formed by the meeting of different ocean currents; generally, rotation is clockwise in the northern hemisphere and counterclockwise in the southern hemisphere

hydrologic cycle: also called the "water cycle," in which Earth's water changes state from liquid and solid to atmospheric gas, and condenses back to liquid or solid and falls as precipitation

oxygen: O_2, a biatomic molecule that aerobic biological systems use for production of biochemical energy

AQUATIC ENVIRONMENTS

There are two primary aquatic environments on planet Earth. Both are bulk water environments: one freshwater and the other saline water. Both environments are connected through the planetary hydrologic cycle, or water cycle. In that cycle, liquid water in the saline oceans absorbs energy from the Sun and evaporates. The water vapor is carried by air currents in the atmosphere until it condenses and falls back to the surface as freshwater in the form of rain and snow. This freshwater collects on the ground and flows into streams, rivers, and lakes as surface runoff, and is eventually returned to the saline oceans. In the process, the freshwater runoff acquires a load of dissolved materials, sediments, organic matter, and larger debris that it transports to whatever body of water it flows into, or until it exceeds the ability of the water to carry it along. Freshwater flowing into the oceans also brings about changes in salinity and temperature as it mixes with the saline water, and the density difference in density between freshwater and saline water plays a significant role in current flows in the oceans. The dynamic nature of this system of circulating water is integral to the formation pf aquatic dead zones in both freshwater and saline environments.

CAUSES OF DEAD ZONES

A number of causes of dead zones can be enumerated. A dead zone may be a natural result of a local circulation pattern in the water, or may form due to a persistent or ongoing natural feature of water movement. Others, however, tend to have anthropogenic causes. The common feature of all dead zones, however, is that they contain insufficient dissolved oxygen to maintain aerobic life. Thus, fish that enter a dead zone quickly suffocate, as do any fish eggs that may have been deposited in that location. By and large, however, a dead zone will be eerily vacant of anything but anaerobic bacteria and floating debris.

Natural Circulation Effect: Rarely, water currents persist in a configuration that precludes absorption of oxygen from the atmosphere. Not strictly speaking dead zones, the ability to support aerobic life within the area is limited. An upwelling current rising from depths at which dissolved oxygen is essentially nonexistent may produce a dead zone at some depth below the surface before reaching the level at which atmo-

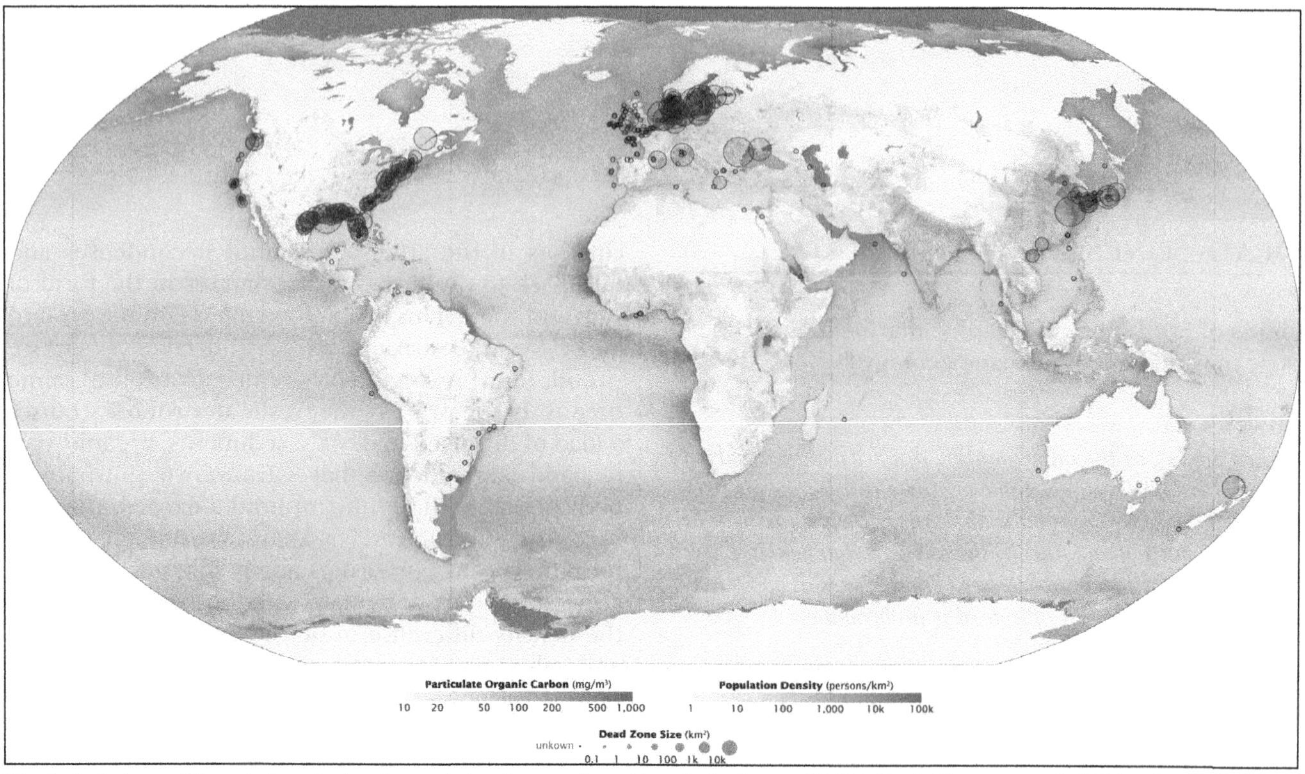

The dark circles on this map show the location and size of many of our planet's dead zones (NASA Earth Observatory via Wikimedia Commons)

spheric oxygen would begin to be absorbed. Fortunately, such deep currents bring large amounts of nutrients with them from the depths, and the aerated regions near the surface tend to be very productive in terms of life support. The cold Benguela Current near the western coast of South Africa, for example, is one of the world's great fisheries, as is the Humboldt Current near the west coast of Peru and Chile. The deeps from which these currents rise, however, have little to offer in regard to life support, and the vast majority of sea life there is found within just a few hundred feet (or meters) of the ocean surface.

The Sargasso Sea in the Atlantic Ocean provides another example of a circulation-driven dead zone. Again, though, it is not strictly speaking a dead zone. Indeed, there are large numbers of aerobic creatures living in and about the floating masses of sargassum weed that characterize the Sargasso Sea. It is the nursery area for European eels and is home to a number of unique small organisms. There are no large fish, however, and the space below the sargassum weed tends to be very oxygen-poor due to the weed itself consuming what oxygen there is as it falls away and rots.

Eutrophication: Eutrophication is a process by which excess organic matter consumes the available oxygen in a body of water as it grows, dies, and rots. The most readily apparent process occurs in both freshwater and marine bodies when there is a great algal bloom. Large lakes in central Canada—Lakes Winnipeg and Winnipegosis—have exhibited massive algal blooms that greatly exceed the ability of water exchange to remove them. Images from the International Space Station have clearly shown them as bright green rather than the brilliant blue normal of bodies of clean water. During such algal blooms, oxygen in the water is inhibited from being absorbed and is rapidly consumed by rotting algae. Organisms that are normally safe to eat by humans and animals become highly toxic, leading to accounts of human and animal death simply from having been in the water.

The broad definition of eutrophication as lack of oxygen in the water is associated with other causes of dead zones, as it is those causes that result in eutrophication of the body of water, whether marine or fresh. Starting in the 1980s, a large dead zone—with insufficient oxygen to support marine life—was discovered some 60 miles (97 kilometers) off the coast of Louisiana and Texas. By 2011, scientists said it was approaching 9,500 square miles (25,000 square kilometers) in size. It is thought to be the result of nitrate- and phosphate-rich agricultural runoff, particularly from fertilizers used far up the Mississippi River basin in the Corn Belt. There is evidence that some species of fish exhibit physical abnormalities because of oxygen depletion.

Sediment Flows: The infusion of such excess quantities of nutrients in the water from the freshwater flow of a river is associated with the river's sediment load, which can amount to millions of tons (or tonnes) annually in places. The sediment load entering the sea may build up a steep-sided structure over time, as it has in the Mississippi delta and the Nile delta. At some point, the mass of sediment may exceed the ability of the structure to support it, and there follows a collapse—essentially an underwater landslide of massive quantities of silt and other matter. Downslope movement of this mass is diffused and effectively sorted according to particle size as it moves through the water, with the heaviest and densest particles traveling the greatest distance away from the point of rupture and the light silt forming a thick, drifting cloud that smothers anything in its path. In such a case, while the water may be well oxygenated, it becomes fatal for many of the fish and other creatures as they become unable to obtain the oxygen through silt-impacted gills. This, in effect, creates a dead zone.

Chemical Pollution: Chemical pollution may be included in the sediment load of a freshwater river, acquired in agricultural runoff that takes place throughout the basin drained by a particular river system. Pesticides—both herbicides and pesticides—that are used in agricultural operations can be carried into the natural drainage systems of the watercourse and from there into lakes and eventually to the oceans. The chemicals used in these applications are also effective in water systems, especially those that do not undergo hydrolysis. In other cases, the products of hydrolysis may be even more harmful to the aquatic environment than the nonhydrolyzed compounds. Whereas excess nutrients contribute to the rampant growth of organic matter leading to eutrophication, herbicides that find their way into the water system can increase the rate of dying of aquatic plants, thus assisting the process of eutrophication.

Chemical pollutants from industry are another matter entirely. These include everything from plastic resins to heavy metals and human waste, and it is one of the great shames of the human race that rivers and oceans have been used as a ready dumping ground for such refuse for centuries. The case of Minamata Bay in Japan is a prime example. The population in the Minimata region are traditionally high consumers of fish and seafoods caught locally by Minamata Bay fishers. In the 1960s, the world became aware of a neurological disease that was affecting a large segment of the population. The cause was eventually traced to a large paper mill that had been using mercury in its production processes and releasing it as waste into the bay. Mercury is a highly potent neurotoxin that bioaccumulates up the food chain so that larger fish that eat the smaller fish that were ingesting the benthic organisms that also consumed mercury in the detritus of the seafloor were storing biologically active mercury in their fats and other organs. People who then ate the fish also acquired the full load of mercury the fish were carrying within their tissues, and thus acquiring what came to be known as Minamata Disease. While pollution such as this creates a "death zone" rather than a dead zone, the environment is rendered at least partially unable to support life, even though the waters may be well oxygenated. Other chemicals released from industries, such as dioxin and polychlorinated biphenyls (PCBs), also bioaccumulate in fatty tissues and may have a more direct role in the consumption of dissolved oxygen.

Plastic items that have been discarded after use further add to the problem. There exists in the oceans and seas surface circulation systems called "gyres." Plastic and other items that are released into the oceans in massive quantities each day are captured by the gyre currents as they float about, and are drawn into the area at the center of a gyre, where they become trapped in the low-energy, relatively still water. This has been likened to what happens when one flushes a toilet. In their respective journeys to the "garbage patch" in the gyre center, the plastics tend to absorb pollutants that they come into contact with.

Sea creatures that ingest the plastic items as they mistake them for food not only become poisoned by the toxins coating and absorbed into the plastic, but the plastic itself becomes an indigestible plug that eventually kills the creatures. Investigations of the Great Eastern Garbage Patch, in the North Pacific Ocean, have revealed that the plastic outnumbers the plankton in some areas by a factor of six-to-one. At current rates of disposal, it has been suggested that plastic in the oceans will outnumber the fish in the oceans by the year 2050.

Oil Spills: The petroleum exploration and recovery industry is prone to accidents that release massive quantities of crude oil directly into the oceans and seas of the world. The *Exxon Valez* spilled some 1.9 million gallons (7.25 million liters) of crude oil when it ran aground and ruptured, and no one knows how much crude oil actually spewed into the Gulf of Mexico from the *Deepwater Horizon* well blowout. In some regions of the world, freshwater lakes are also the scene of oil exploration, and despite the more stringent regulations and practices of the industry in those waters, accidents can also happen there. Crude oil is a very complex mixture of compounds, mainly hydrocarbons of all sizes. It also contains significant amounts of sulfur-containing compounds and water-soluble compounds that can permeate the area of a spill, becoming acids by reacting with water and dissolved oxygen in the process. The immediate area of any significant oil spill very quickly becomes an oxy-

The Sydney Tar Ponds in Nova Scotia, seen here in 2005, were one example of an anthropogenic dead zone, caused by runoff from the coke ovens of a now-decomissioned steel mill from the 19th century (David Smith via Wikimedia Commons)

The Sydney Tar Ponds and Coke Ovens Remediation Project was a ten year, $400 million endeavour to clean up one of the worst man-made environmental disasters in history (sshaw75 via iStock)

gen-deprived dead zone as the oil spreading over the surface of the water blocks absorption of oxygen from the atmosphere while it suffocates and poisons the area.

Atmospheric Contamination: In northern Ontario, Canada, the city of Sudbury is host to a long-standing nickel mining and refining operation of Inco, the International Nickel Company. A very significant byproduct of the smelting operation is the gas sulfur dioxide. To prevent, or at least limit contamination of the surrounding area, a giant exhaust stack was built so that the sulfurous emissions from the refinery would be ejected high into the air, ostensibly to be carried away safely by the prevailing winds a hundred feet (or meters) and more in the atmosphere. At least, that was the thinking of the time. What happened instead is that the sulfur dioxide reacted with water vapor and oxygen in the air to produce a series of sulfur acids culminating in sulfuric acid. The sulfuric acid aerosol became acid rain and snow as it dissolved in liquid water droplets in the clouds, falling to ground with a pH as low as 1. Chlorine-containing emissions similarly produced atmospheric hydrochloric acid, adding to the problem of acid precipitation. After several years of such precipitation, the vegetation in the immediate area of the refinery had withered, died and all but disappeared. Everything nearby from grass to trees in the rocky Canadian Shield landscape was affected.

Of much greater concern, however, was the observation of "dead" freshwater lakes that were devoid of any living thing, whether plant or animal, because of the acidic precipitation and runoff flowing into them. In the decades before the nickel refinery was established there, the area was known as a "fisherman's paradise" where numerous big, healthy fish could be caught. Fortunately, stringent emissions treatment and control measures and environmental remediation are now allowing the area to recapture some of its former glory. It is a work in progress, but life is returning to these once-dead waters.

Similar environmental effects have been observed in similar situations around the world, and one would hope that remediation efforts in those places will also be effective.

DEAD ZONE REMEDIATION

The is effectively only one way in which a dead zone can be remediated, and that is to eliminate the cause of the dead zone to allow Nature to restore itself. This process can be aided artificially, of course, but full remediation will only be achieved in Nature's own good time after the cause has been eliminated. In some cases, remediation will never be possible because the particular dead zone is itself an authentic part of Nature. Dead zones that exist as a result of natural current flows, for example, will always exist, until the current flows change and water exchange in the space of the dead zone can happen naturally. In all other cases, the dead zone is anthropogenic, caused by human activities, and can only be remedied by changing the environmental inputs that are the cause. Severe restrictions on emissions of pollutants, reliance on natural fertilizers and natural means of plant and animal pest control in agriculture, taking extra care to ensure that accidents involving oil exploration, recovery and transport do not happen, and ensuring that the aquatic environment is always actively and meaningfully protected in such activities will go a long way toward the prevention of dead zone formation. In the end, the best form of dead zone remediation is simply just prevention.

—Richard M. Renneboog

Further Reading

Hand, Carol. *Dead Zones: Why Earth's Waters Are Losing Oxygen.* Twenty-First Century Books, 2016.

Lymbery, Philip. *Dead Zone: Where the Wild Things Were.* Bloomsbury Publishing, 2017.

Maser, Chris. *Interactions of Land, Ocean and Humans: A Global Perspective.* CRC Press, 2015.

United Nations Environment Programme Staff. *One Planet, Many People: Atlas of Our Changing Environment.* UNEP/Earthprint, 2005.

Weis, Judith S. *Marine Pollution: What Everyone Needs to Know.* Oxford UP, 2015.

DEEP OCEAN CURRENTS

Fields of Study: oceanography; climatology; physics

ABSTRACT

Deep ocean currents, the dynamics of which are not yet well understood, involve significant vertical and horizontal movements of seawater. They distribute oxygen- and nutrient-rich waters throughout the world's oceans, thereby enhancing biological productivity. They also transport vast amounts of energy that influence climate all around the planet.

KEY CONCEPTS

benthic, benthonic: indicating deep sea bottom lifeforms and phenomena

Coriolis effect: the apparent curvilinear motion of air and water currents due to Earth's rotation

echo-sounding: the use of sonography, or sonar, for the specific purpose of measuring the vertical depths in a body of water to produce a topographical map of the bottom

gravity corer: a device, essentially a length of pipe, that is dropped vertically to the bottom of a body of water in order to recover a vertical sample of bottom sediment

piston corer: a modified type of gravity corer that can recover bottom sediment samples of much greater depth

saltation: the motion of particles under the influence of wind or water, by which individual particles bump into particles that lie ahead of them and essentially "leapfrog" over them

thermohaline circulation: circulation driven by differences in temperature and salt content

EVIDENCE FOR EXISTENCE

Deep-sea currents—ocean currents that involve vertical as well as horizontal movements of seawater—are generated by density differences in water masses that result in the sinking of colder, denser water to the bottom of the ocean. For many years, however, most oceanographers refused to accept the presence of these currents. Even when the Deep Sea Drilling Pro-

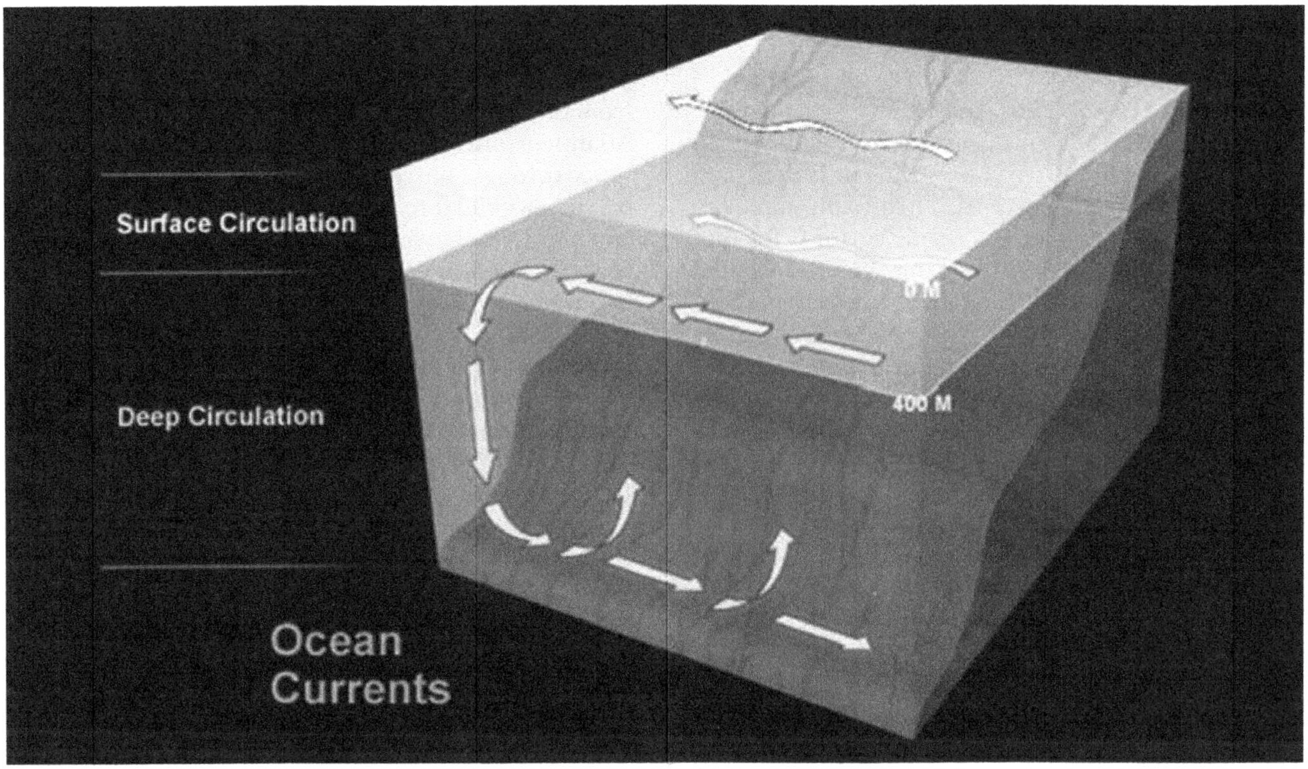

Diagram showing the different levels of ocean currents (National Oceanic and Atmospheric Administration)

ject, an international effort to drill numerous holes into the ocean floor, was initiated, most researchers envisioned the deep sea as a tranquil environment characterized by sluggish or even stationary water. More recently, oceanographers and marine geologists have accumulated abundant evidence to suggest the opposite: that the deep sea can be a very active area in which currents sweep parts of the ocean floor to the extent that they affect the indigenous marine life and even physically modify the seafloor.

In the 1930s, Georg Wust argued for the likelihood that the ocean floor is swept by currents. Furthermore, he suggested that these currents play an important role in the transport of deep-sea sediment. Wust's ideas were not widely accepted at first, but in the 1960s, strong evidence for the existence of deep-sea currents began to accumulate. In 1961, for example, oceanographers detected deep-sea currents moving from 2 to 4 inches (5 to 10 centimeters) per second in the western North Atlantic Ocean. These researchers also determined that the currents changed direction over a period of one month.

In 1962, Charles Hollister, while examining cores of deep-sea sediment drilled from the continental margin off Greenland and Labrador, noted numerous sand beds that showed evidence of transport by currents. The nature of these deposits suggested to Hollister that they did not accumulate from turbidity currents, dense sediment-water clouds that periodically flow downslope from nearshore areas. Moreover, it appeared to Hollister that the sand was transported parallel to the continental margin rather than perpendicular to it, as might be expected of sediment transported by a turbidity current. He argued that the sand beds in the cores were transported by, and deposited from, deep-sea currents moving along the bottom of the ocean parallel to the continental margin. Since then, extensive photography of the ocean floor has provided direct evidence for the existence of deep-sea currents. Such evidence includes smoothing of the seafloor; gentle deflection, or bending, of marine organisms attached to the seafloor, as though they were standing in the wind; sediment piled into small ripples by saltation; and local scouring of the seafloor.

THERMOHALINE CIRCULATION

Essentially all earth scientists now agree that the deep seafloor is swept by rather slow-moving (less than 0.79 in [2 cm] per second) currents. The driving force behind these currents, and all oceanic currents for that matter, is energy derived from the sun. Differential heating of the air drives global wind circulation, which ultimately induces surface ocean currents. The vertical circulation of seawater, and thus the generation of deep-sea currents, is controlled by the amount of solar radiation received at a point on Earth's surface. This value is greatest in equatorial regions; there, the radiation heats the surface water, the seawater that lies within the upper 984 to 3,280 feet (300 to 1,000 meters) of the ocean. As this water is heated, it begins to move toward the poles along paths of wind-generated surface circulation, such as the Gulf Stream current of the northwestern Atlantic Ocean.

The cold waters that compose the deep-sea currents originate in polar regions. There, minimal solar-radiation levels produce cold, dense surface waters. The density of this water may also be increased by the seasonal formation of sea ice, ice formed by the freezing of surface water in polar regions. When sea ice forms, only about 30 percent of the salt in the freezing water becomes incorporated into the ice. The salinity and density of the nearly freezing water beneath the ice are therefore elevated. This cold, saline seawater eventually sinks under the influence of gravity to the bottom of the ocean, where it moves slowly toward the equator. Deep-sea circulation driven by temperature and salinity variations in seawater is termed "thermohaline circulation" and is much slower than surface circulation; the cold, dense water generated at the poles moves only a few miles (or kilometers) per year. After moving along the bottom of the ocean for anywhere from 750 to 1,500 years, the cold seawater rises to the surface in low-latitude regions to replace the warm surface water, which, as noted above, moves as part of the global surface circulation system back to the polar regions.

Thermohaline circulation and related deep-sea currents are commonly affected by the shape of the ocean floor. Although sinking cold seawater seeks the deepest route along the seafloor, deep-sea currents may be blocked by barriers. The Mid-Atlantic Ridge the large volcanic ridge effectively bisecting the Atlantic Ocean basin, may prevent the movement of water from the bottom of the western Atlantic to the eastern Atlantic. Conversely, the funneling of deep-sea currents through narrow passages or gaps in seafloor barriers will lead to an increase in the velocity of the current. Once beyond the passage, however, the current spreads and velocity is reduced. Because both air and water act like a fluid, these effects and behaviors are entirely analogous to those of winds produced in the atmosphere by air-density differences resulting from the uneven distribution of solar energy over Earth's surface.

CORIOLIS EFFECT

The circulation pattern of deep-sea currents is controlled to a large extent by Earth's rotation. The Coriolis effect—the frictional force achieved by Earth's rotation that causes particles in motion to be deflected to the right in the Northern Hemisphere and to the left in the Southern Hemisphere—induces deep-sea currents to trend along the western margins of the major oceans. Thus, water sinking from sources in the North Atlantic Ocean and moving south toward the equator will be deflected to the right, causing it to run along the western side of the North Atlantic. Similarly, north-directed deep-sea currents generated by the sinking of cold water from the Antarctic region will also be deflected to the western margin of the Atlantic.

The Coriolis effect guides deep-sea currents along bathymetric contour lines, lines on a map of the ocean floor that connect points of equal depth. Deep-sea currents that have a tendency to move parallel to the bathymetric contours are known as bottom currents. Barriers to flow may locally deflect deep-sea currents from the bathymetric contours; nevertheless, bottom currents are most conspicuous along the western margins of the major oceans.

SHORT- AND LONG-TERM CONTROLS

The formation of the cold seawater required to set deep-sea currents in motion can itself be considered in terms of short- and long-term controls. Seasonal sea-ice formation is probably the most important process in the production of the north-flowing water generated at the south polar region, or the Antarctic bottom water (AABW). Velocities of the AABW are highest in March and April, that period of the year when sea-ice production in the ocean surrounding Antarctica is greatest. During Southern Hemisphere summers, however, the sea ice melts, and there is an increase in the freshwater flux to the ocean from the continent. Both of these factors reduce the salinity and therefore the density of the seawater, thereby decreasing AABW production.

Many oceanographers and marine geologists have argued that long-term variations in the production of the cold, dense bottom water required to generate deep-sea currents may be related to global climatic changes. More specifically, deep-sea currents appear to be most vigorous during glacial periods, when sea-ice production is enhanced and the sea ice remains on the ocean surface for a greater proportion of the year. Nevertheless, there is also evidence to suggest that the velocities of deep-sea currents in the North Atlantic Ocean were much lower during the most recent glacial periods than they were during the times between glacial phases. Much more work is required to gain a more complete understanding of long-term controls on deep-sea currents.

MEASUREMENT TOOLS AND TECHNIQUES

The most common methods for the study of deep-sea currents include direct measurement of current velocities, bottom photography, echo sounding, and the sampling of ocean-floor sediment. The speed and direction of deep-sea currents have been determined by the use of free-fall instruments, such as the free-instrument Savonics rotor current meter. This device is dropped unattached into the ocean, where it records current velocities and directions over a period of several days. It then returns automatically to the surface of the ocean, at which time a radio transmitter directs a ship to its position. Other current-measuring devices can be suspended at various depths in the ocean from fixed objects, such as buoys or light ships, to monitor currents for long periods. One such anchored meter measures the flow of water past a fixed point. Flowing water causes impeller blades, similar to the blades of a fan, to rotate at a rate proportional to the current's speed. In addition, the blades cause the meter to align with the current's direction. Electrical signals indicating the direction and speed of the current are transmitted by radio or cable to a recording vessel. Current velocities of less than 0.4 in (1 cm) per second can be detected by this meter.

Diagram of a gravity corer, with example of a sediment echosounder graph at the bottom (Hannes Grobe/AWI via Wikimedia Commons)

To get the most complete picture of the variability of the ocean, a combination of various measurement techniques with remote sensing may be employed. Such a multidimensional approach may involve the measurement of current velocity, pressure (a measure of depth), water temperature, and water conductivity (a measure of salinity). These data can be transmitted via satellite to a land station or even directly to a computer.

ADDITIONAL STUDY METHODS

Perhaps the most persuasive evidence for the existence of deep-sea currents and their influence on the ocean bottom has been gained through bottom photography. Sediment waves, or ripples, apparently formed by the saltation of sediment carried by deep-sea currents, along with evidence of current-induced scour of the ocean floor, were first photographed in the Atlantic Ocean in the late 1940s. Since then, the technology of bottom photography has advanced greatly. Bottom photography permits detailed study of some of the smaller features on the ocean floor apparently formed by deep-sea currents. The bending of benthonic, or benthic, organisms (marine organisms that live attached to the ocean floor) in the flow of the current is a particularly intriguing example of the phenomena recorded by this technique.

Echo-sounding studies of the seafloor have yielded abundant information on ocean-floor features that are either formed or modified by deep-sea currents. Notable among these are very long ridges in the North Atlantic evidently constructed from sediment carried by deep-sea currents. In echo sounding, a narrow sound beam is directed from a ship vertically to the sea bottom, where it is reflected back to a recorder on the ship. The depth to the seafloor is determined by multiplying the velocity of the sound pulse by one-half the amount of time it takes for the sound to return to the ship. The depths to the ocean floor are recorded on a chart by a precision depth recorder, which produces a continuous profile of the shape of the seafloor as the ship moves across the ocean.

Sediment transported and deposited by deep-sea currents can be studied directly by actually sampling the ocean floor. Sampling of these deposits is best accomplished by the use of various coring devices capable of recovering long vertical sections, or cores, of seafloor sediment. Sediment recovery is achieved by forcing the corer, a long pipe usually with an inner plastic liner, vertically into the sediment. The simplest coring device, the gravity corer, consists of a pipe with a heavy weight at one end. This type of corer will penetrate only 6.6 to 9.8 ft (2 to 3 m) into the seafloor. The piston corer, used to obtain longer cores, is fitted with a piston inside the core tube that reduces friction during coring, thereby permitting the recovery of 59 ft (18 m) or longer cores. Analysis of the sediment recovered from the ocean floor by these and other coring devices reveals much information about small-scale features formed by deep-sea currents.

IMPORTANCE TO LIFE ON EARTH

Because cold bottom-water masses often are nutrient-rich and contain elevated abundances of dissolved oxygen, deep-sea currents are extremely important to biological productivity. There are areas of Earth's surface where nutrient-rich cold bottom waters rise to the ocean surface. These locations, known as upwelling ecosystems or areas of upwelling, are generally biologically productive and therefore are important food sources. Especially pronounced upwelling occurs around Antarctica. Bottom waters from the North Atlantic upwell near Antarctica and replace the cold, dense, sinking waters of the Antarctic.

The great amount of time required for seawater to circulate from the surface of the ocean to the bottom and back again to the surface has become an important practical matter. If pollutants are introduced into high-latitude surface waters, they will not resurface in the low latitudes for hundreds of years. This delay is particularly important if the material is rapidly decaying radioactive waste that may lose much of its dangerous radiation by the time it resurfaces with the current. The introduction of toxic pollutants into a system as sluggish as the deep-sea circulation system, however, means that they will remain in that system for prolonged periods. Nations must, therefore, be concerned with the rate at which material is added to this system relative to the rate at which it might be redistributed at the surface of the ocean by wind-induced surface circulation.

The multinational Geochemical Ocean Sections (GEOSECS) program, introduced as part of the International Decade of Ocean Exploration and carried out from 1970 to 1980, attempted to better assess the problem of how natural and synthetic chemical substances are distributed throughout the world's oceans. The program yielded abundant information regarding the movement of various water masses and, among other things, the distribution of radioactive material in the oceans. For example, GEOSECS demonstrated that tritium produced in the late 1950s and early 1960s by atmospheric testing of nuclear weapons had been carried to depths approaching 3.1 mi (5 km) in the North Atlantic Ocean by 1973.

—*Gary G. Lash*

Further Reading

Broecker, Wally. *The Great Ocean Conveyor: Discovering the Trigger for Abrupt Climate Change*. Princeton UP, 2010.

Brown, Joan. *Ocean Circulation: Prepared by an Open University Course Team*. Elsevier, 2016.

Garrison, Tom S. *Oceanography: An Invitation to Marine Science*. 8th ed., Brooks, 2013.

Goni, G. J., and Paola Malanotte-Rizzoli, editors. *Interhemispheric Water Exchange in the Atlantic Ocean*. Elsevier, 2003.

Joseph, Antony. *Measuring Ocean Currents. Tools, Technologies, and Data*. Elsevier, 2014.

Schmittner, Andreas, John C .H. Chiang, and Sidney R. Hemmings, editors. *Ocean Circulation: Mechanisms and Impacts–Past and Future Changes of Meridional Overturning*. John Wiley & Sons, 2013.

Siedler, Gerold, John Church, and John Gould. *Ocean Circulation and Climate: Observing and Modelling the Global Ocean*. Academic, 2001.

Smith, F. G. Walton. "Measuring Trujillo, Alan P., and Harold V. Thurman. *Essentials of Oceanography*. 11th ed., Prentice, 2014.

Steele, John H., Steve A. Thorpe, and Karl K. Turekian, editors. *Ocean Currents*. Academic Press, 2010.

Vallis, Geoffrey K. *Atmospheric and Oceanic Fluid Dynamics: Fundamentals and Large-Scale Circulation*. Cambridge UP, 2006.

Deep-Sea Sedimentation

Fields of Study: marine geology; oceanography; stratigraphy; sedimentology

ABSTRACT

Deep-sea sedimentation occurs by the settling of particles to the ocean floor and by the transport of material from shallow to deep water. Knowledge of the distribution of sediments and the processes of sedimentation will help to evaluate the potential of the world's oceans for the mining of natural resources and the storage of waste.

KEY CONCEPTS

biogenic: derived from living creatures

calcareous ooze: seafloor sediment consisting primarily of planktonic carbonate remains (at least 30 percent)

pelagic: associated with the ocean

piston corer: a tubular device consisting of a length of heavy steel pipe with an internal piston, typically dropped from the surface to plunge into the floor sediment of a body of water to extract a core sample of up to 59 feet (18 meters) in length

sedimentation: the accumulation over time of layers of particles under the influence of gravity

siliceous ooze: seafloor sediment consisting primarily of silicate particles (greater than 70 percent)

INVESTIGATION OF DEEP-SEA SEDIMENTS

Sedimentation in the deep sea differs substantially from that in any other environment. Deep-sea sedimentation is pelagic. It typically occurs at great distances from continents and at depths in excess of 3,280 ft (1,000 m), primarily by the settling of particles through the overlying water. Every 26,417 gallons (100,000 liters) of seawater contains about 1 gram of extremely small particles. Their settling is extremely slow, and sediments accumulate at rates on the order of a few millimeters per thousand years. Transport of material by turbidity currents along the bottom of the ocean from shallow to deep regions is also a mechanism by which deep-sea sediments are deposited, but this is far less important than the settling of particles from the surface. In contrast, current transport dominates sedimentation in all other environments, where sedimentation rates typically exceed several inches (or centimeters) per year.

The oceans cover 70 percent of Earth, and, as a result, deep-sea sedimentation may be the most common sedimentary process. Early investigations of the seas were limited, however, to coastal processes and oceanic circulation because of the lack of techniques available to study the deep ocean bottom. Knowledge of deep-sea sedimentation was profoundly increased by the voyage of HMS *Challenger* from 1872 to 1876, which made the first systematic study of the ocean floor. Prior to the *Challenger* expedition, practical investigations of the oceans concentrated on winds, tides, and water depths in harbors to provide information for commercial navigation and for the laying of submarine telegraph cables. The *Challenger* crew collected data on the temperature and depth of the oceans, marine plants and animals, and sediments on the ocean bottom. Deep-sea sediments were described by John Murray (1841–1914), naturalist to the expedition. He noted that sediments in the deepest parts of the ocean are fine-grained and are composed of particles that settled from the surface. Comparison of rocks exposed on land with those dredged from the ocean led to the assertion that deep-sea sediments could not be found on land. This conclusion caused great debate, and the interpretation of some rocks as deep-sea deposits by a few scientists persisted despite opposition by most of the scientific community.

Investigation of deep-sea sedimentation flourished again during and immediately after World War II. Despite the voluminous amount of data available by 1950, the prevailing doctrine in the first half of the twentieth century was similar to that in the previous century: The continents and oceans were generally believed to be permanent features of Earth's surface that had developed to their present form near the beginning of geologic time. Thus, the ocean basins were billions of years old, ocean-bottom sediments were tens of miles (or kilometers) thick, having accumulated from early in the geological history of Earth to the present, and deep-sea sediments were not exposed on land, a point that was still severely contested.

Then, during the 1950s two discoveries shocked the oceanographic community. First, fossils taken from seamounts (submarine mountains) in the Pacific Ocean at depths of 1.24 miles (2 kilometers) were only 100 million years old, suggesting that subsidence of the ocean floor was recent. Second, deep-sea sediments were found to be less than 656 ft (200 m) thick. The advent of the theory of plate tectonics in the 1960s profoundly changed ideas about the ocean bas-

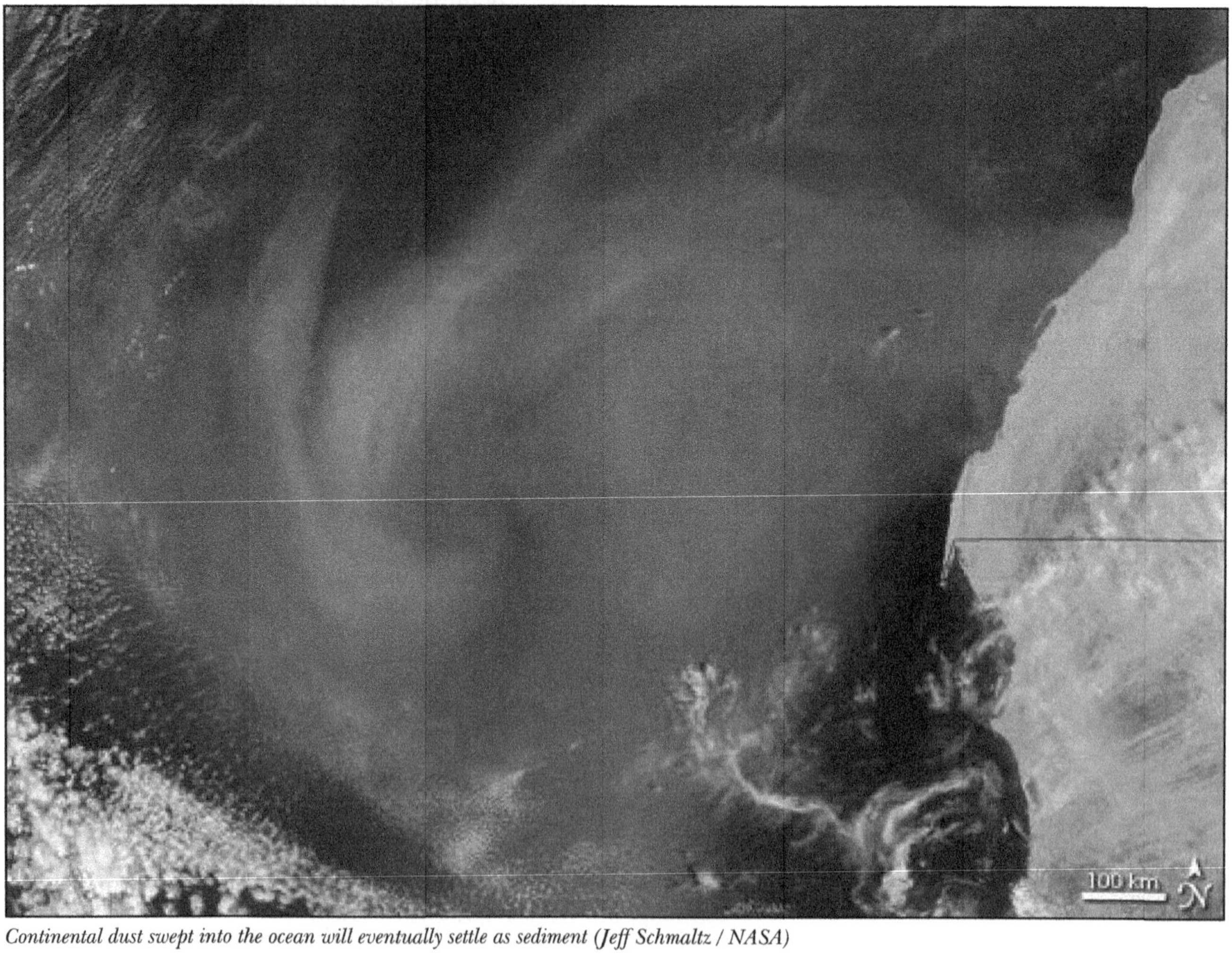

Continental dust swept into the ocean will eventually settle as sediment (Jeff Schmaltz / NASA)

ins by providing both an explanation for the two observations of the previous decade and a mechanism for the emplacement of deep-sea sediments onto continents. In addition, the motion and cooling of the oceanic plates predicted by plate tectonics explain the global distribution of deep-sea deposits.

PELAGIC SEDIMENTS

Pelagic sediments are characterized by their fine grain size. Few particles are larger than 0.025 millimeter, and most are smaller than 0.001 millimeter. Particles include terrigenous debris (derived from the continents), such as clay minerals quartz, and feldspar; volcanogenic grains (derived from volcanoes), such as volcanic glass pumice, and ash; biogenic material (derived from living organisms), such as fecal pellets (waste produced by organisms inhabiting the surface waters) and skeletons of planktonic plants and animals; and cosmogenic matter (derived from space or the cosmos), such as "space dust" and pieces of meteorites.

Pelagic sediments that contain more than 30 percent biogenic material are called "oozes." If the biogenic debris is composed of calcium carbonate, the sediment is a calcareous ooze. If the biogenic debris is composed of silica, the sediment is a siliceous ooze. Most calcareous oozes consist of foraminiferans, which are single-celled animals that secrete calcium carbonate and live in the surface waters. In contrast, siliceous oozes contain diatoms (green, unicellular algae) in cold water and radiolarians (silica-secreting single-celled animals) in warm water. Pelagic clays are sediments that contain less than 30 percent biogenic debris.

The two major controls on pelagic sedimentation are the calcium compensation depth (CCD) and the fertility of the surface waters. The CCD reflects the interplay between the release of carbon dioxide during the decay of surface organisms and the dissolution of the calcium carbonate skeletons as they descend through the ocean. Above the CCD, calcareous oozes are dominant, whereas below the CCD, siliceous oozes and pelagic clays are common. For example, in one region of the ocean, sediments decreased from 90 percent calcium carbonate at a depth of 3,280 ft (1,000 m) to 20 percent calcium carbonate at a depth of 16,404 ft (5,000 m).

Oozes accumulate below areas of high fertility where upwelling brings nutrient-rich bottom water to the surface. Because upwelling creates increased productivity of both calcareous and siliceous organisms, siliceous oozes form only where the water depth exceeds the CCD. Calcareous oozes, the most abundant biogenic sediments in the ocean, cover only 15 percent of the Pacific Ocean bottom but blanket 60 percent of the Atlantic Ocean floor. This difference reflects the combination of a shallower CCD and greater water depth in the Pacific Ocean than in the Atlantic Ocean. Siliceous oozes are in areas of elevated CCDs such as the equatorial regions, the subarctic and subantarctic zones, and the continental margins of northwestern South America and eastern Africa.

Sedimentation of pelagic clays is restricted to the central portions of the oceans where the fertility is low. This reflects the difference in settling rates between nonbiogenic particles and fecal pellets. Because of its small size, a particle may take many years to sink through the ocean. The concentration of many particles into fecal pellets by organisms feeding at the surface, therefore, is thought to be an important mechanism by which debris settles to the bottom. The greater size of the pellets allows them to fall much faster than unconsolidated particles. It is only in areas of low productivity, therefore, where biogenic material does not overwhelm the sediment. Wind is the primary means by which nonbiogenic particles reach the sea. Dust is carried in the atmosphere at great distances from continents and constitutes as much as 10 percent of pelagic sediment.

OTHER PROCESSES

Additional processes in deep-sea sedimentation include the rafting of glacial debris by ice, the transport of terrigenous and shallow-water material by turbidity currents along the ocean floor, and the precipitation of nodules on the ocean bottom. Pieces of ice calve off glaciers in the polar regions and float toward warmer waters where they eventually melt, releasing glacial sediment into the ocean. Although this process is most common in the south polar region today, glacial sediments greater than ten thousand years old elsewhere on the ocean floor suggest the process extended over a much greater area in the past.

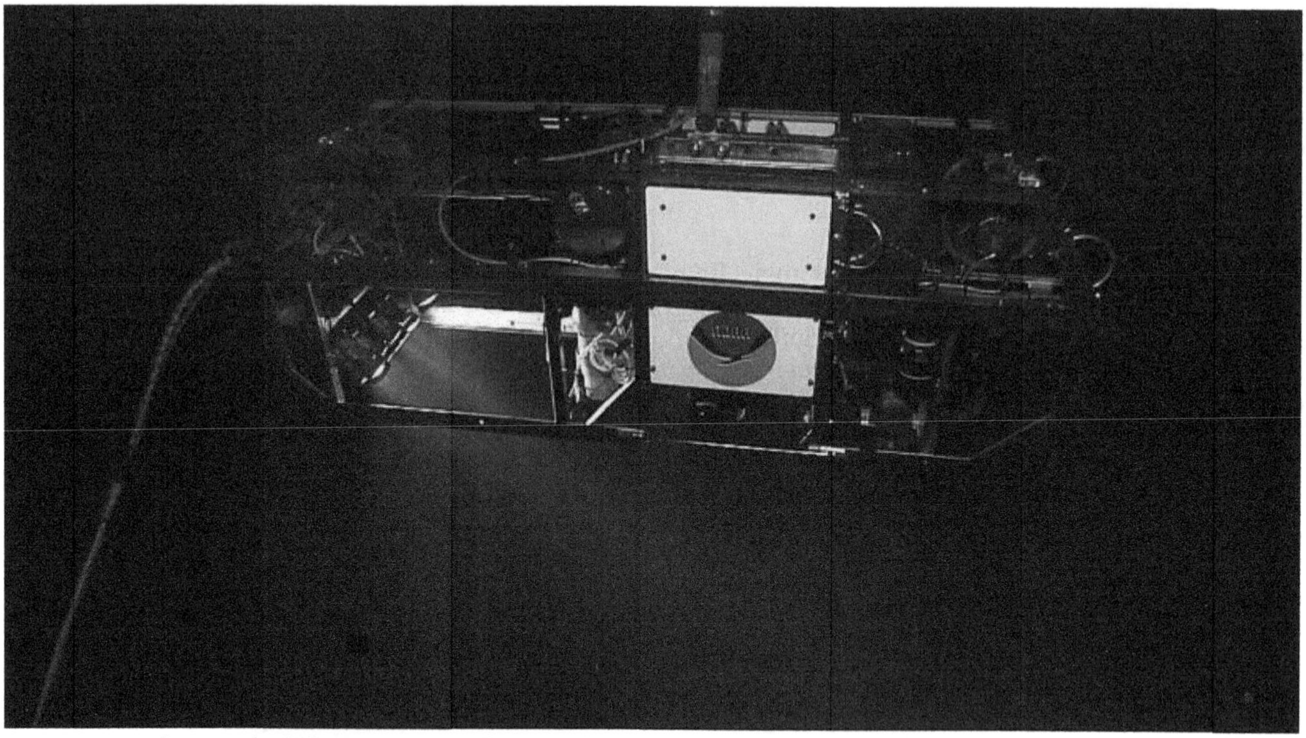

Deepwater remotely operated vehicle (ROV) (NOAA Okeanos Explorer Program via Wikimedia Commons)

Turbidity currents generated along continental slopes may traverse hundreds of miles (or kilometers) of ocean basin where ridges and depressions do not occur. Far away from continents, the currents commonly contain only silt but may deposit the grains over great distances. These currents also may locally erode the seafloor, creating breaks in the continuity of the sediment pile by removing the most recent sediments and exposing older sediments at the abraded surface.

Chemical reactions of seawater with the surface of the sediment cause the precipitation of ferromanganese (composed of iron and manganese) nodules in areas where sediments accumulate slowly. The nodules form by nucleating around a manganese source in the sediment and by accreting minerals from the seawater. Such nodules may cover more than 40 percent of the seafloor.

Sedimentation in the deep sea is a complex interaction between water depth, fertility of the surface waters, and current transport. To illustrate how the depth dependence of sedimentation affects the global distribution of sediments, one must understand the relationship between plate tectonics and the topography of the ocean floor. At a midocean ridge, which rises above the surrounding ocean basin, molten rock is extruded to form new seafloor. Calcareous oozes accumulate on the new seafloor because of its depth above the CCD. As more seafloor forms, the older seafloor moves away from the midocean ridge toward the deep ocean basin (seafloor spreading), sinking to lower depths as it does so. The old seafloor gradually descends below the CCD, where siliceous oozes accumulate above the calcareous oozes. Eventually the motion of the seafloor carries it below a region of low surface productivity such that pelagic clays are deposited above the siliceous oozes. This vertical sequence has been documented in several localities throughout the world's oceans.

STUDY OF DEEP-SEA SEDIMENTATION

Techniques used to study sedimentation in the deep ocean are diverse and include piston coring, seismic reflection profiling the use of deep-diving submarines, and experiments performed both at sea and in the laboratory. Piston coring is the most important technique because it enables direct sampling of the ocean floor. The device consists of a long, metal tube that is dropped from the surface to plunge into the seafloor sediment. An internal piston allows the capture of cylindrical core samples up to 59 ft (18 m) in length. The cores are 2.8 in (7 cm) in diameter and provide samples that are large enough to preserve features that yield insights into the processes by which the sediment accumulated. For example, current-produced features in some samples were the first direct evidence of currents in the deep seas. In addition, piston coring confirmed the vertical sequence of sediments predicted by plate tectonics.

To map the ocean floor using a 2.8-in (7-cm) core is time-consuming and inefficient. Seismic reflection profiling has the advantage of providing information about the thickness and distribution of layers in the sediments on the seafloor relatively quickly. The two major disadvantages are that it cannot determine the composition of the sediment and that it identifies only layers that are tens of feet (or meters) thick. In seismic reflection profiling, a low-frequency sound source is towed over the bottom. The sound waves penetrate and are reflected by the various layers within the seafloor to several miles (or kilometers) below the boundary between the sediment and the water. The returning signals are displayed on strip charts that reveal the water depth and the thickness and pattern of layers in the sediments. One of the most significant early contributions of seismic reflection profiling was the evidence that sediments on the ocean floor were too thin to have been accumulating since early in Earth's history, an observation explained by plate tectonics.

The use of deep-diving submarines has allowed scientists to observe the ocean floor directly. Previously the ocean floor was only photographed. Deep-diving submarines allowed scientists to confirm or reject hypotheses based on other indirect techniques. The ranges and endurance times of the submersible vessels, however, are limited. In addition, they frequently cannot be used in areas where current velocities are greater than 3 ft (1 m) per second.

Simple experiments either at sea or in the laboratory also can be illuminating. For example, the existence of the CCD was verified by lowering spheres of calcium carbonate to various water depths in the Pacific Ocean. When the spheres were examined, it was discovered that the spheres in the deepest water had dissolved the most, and the rate of dissolving exhibited a sharp increase at certain depths that are the definition of the CCD. Scientists in the laboratory conduct experiments on settling rates by dropping particles of various sizes in large beakers and measuring the time necessary for the particles to reach the bottom. Thus, the amount of time represented by the thickness of sediment on the ocean floor was roughly estimated at a few hundred million years.

One of the most important developments in deep-sea sedimentation research was the creation in 1965 of the Deep Sea Drilling Project (DSDP), now known as the Ocean Drilling Project (ODP), to drill and investigate the ocean floor. The project began in earnest in 1968 with the maiden voyage of the American ship the *Glomar Challenger*, which was outfitted with state-of-the-art navigational, positioning, and drilling equipment. International teams of scientists boarded the ship in ports all over the world and remained at sea for two months. Since its inception, the ODP has undertaken more than sixty-five voyages and contributed significantly to scientists' understanding of the ocean.

SIGNIFICANCE

The oceans cover nearly three-fourths of the surface of Earth. To study deep-sea sedimentation, therefore, is to understand one of the most globally prevalent processes. Understanding deep-sea sedimentation is important for several other reasons, however. First, sediments on the ocean floor yield insight into bottom currents and water depths, which affect the navigation of submarines and the installation of underwater communications cables. Second, the sediments preserve the climatic record of the past few hundred thousands of years. For example, glacial deposits approximately ten thousand years old are found over a much greater area of the bottom of the ocean than recent glacial deposits, suggesting that transport of glacial debris by ice was very common ten thousand years ago. This implies that the temperature of Earth's environment was lower in the past, allowing the formation of significant quantities of ice. Scientists can use this information to gauge present-day climate patterns and predict future fluctuations in the global climate. Third, the seas receive a significant amount of the garbage produced by modern society. Knowledge of sedimentation rates and currents in the deep sea helps to establish the length of time necessary to bury the waste, the direction the waste will travel prior to its burial, and the effect of dumping waste on the overall health of the ocean.

Fourth, and most important, understanding deep-sea sedimentation and the global distribution of deep-sea sediments provides a framework within which one can estimate the economic potential of the deep sea and minimize the damage from ocean exploitation. As the reserve of natural resources on the continents dwindles because of expanding demand from high-technology industries, exploding population growth and increasing consumer appetite, attention focuses on the oceans as a possible source of raw materials. Important resources that may be found locally in the deep sea include oil and gas. The decay and accumulation of plants and animals are essential to the generation of hydrocarbons and are processes that occur over a large area of the seafloor. Emphasis also is placed on mining the deep sea for important materials such as manganese, gold, cadmium, copper, and nickel. Similar to most new industries, large-scale exploitation of the natural resources of the deep sea is limited at present by the current state of technology. The incentive to mine the oceans, however, will grow as cheaper and more sophisticated technologies are created.

—Pamela Jansma

Further Reading

Blatt, Harvey, Robert Tracy, and Brent Owens. *Petrology: Igneous, Sedimentary, and Metamorphic*. 3rd ed., W. H. Freeman, 2005.

Burdige, David J. *Geochemistry of Marine Sediments*. Princeton UP, 2020.

Coe, Angela L., editor. *The Sedimentary Record of Sea-Level Change*. Cambridge UP, 2003.

Grotzinger, John, et al. *Understanding Earth*. 5th ed., W. H. Freeman, 2006.

Gurvich, Evgeny G. *Metalliferous Sediments of the World Ocean*. Springer-Verlag, 2010.

Huneke, H., and T. Mulder. *Deep-Sea Sediments*. Elsevier, 2011.

Joseph, P. *Deep Water Sedimentation in the Alpine Basin of SE France*. Special Publication 221. Geological Society of London, 2004.

McKinney, Frank. *The Northern Adriatic Ecosystem: Deep Time in a Shallow Sea*. Columbia UP, 2007.

Murray, John, and Alphonse J. Renard. *Deep Sea Deposits*. Neill, 1891.

Pickering, Kevin T., and Richard N. Hiscott. *Deep Marine Systems: Processes, Deposits, Environments, Tectonics and Sedimentation*. John Wiley & Sons, 2015.

Stein, Ruediger. *Arctic Ocean Sediments: Processes, Proxies and Paleoenvironment*. Elsevier, 2008.

Trujillo, Alan P., and Harold V. Thurman. *Essentials of Oceanography*. 10th ed., Prentice-Hall, 2010.

E

Earth Science and the Environment

Fields of Study: oceanography; ecology; geology; biology

ABSTRACT
An understanding of the systems of Earth, how they interact, and how they are affected by anthropogenic disturbances is essential to any study of the environment. As the model of Earth is refined and improved, it will be possible to predict the planet's response to anticipated perturbations with greater precision and confidence and adopt better public policy.

KEY CONCEPTS

Earth system science: an approach to the many disciplines of science, in which Earth is viewed as a single entity composed of many interrelated systems

El Niño/La Niña: a complex system characterized by the movement of masses of warm surface water toward the west coast of South America in the former, and away from it in the latter

greenhouse effect: increase of the average temperature of the global atmosphere due to the action of so-called greenhouse gases

greenhouse gases: atmospheric gases that absorb infrared radiation from Earth's surface and release a significant portion of the absorbed energy back toward the surface and into the atmosphere

isostasy: the downward deformation of Earth's crust under the weight of glacial ice during an ice age, thus forcing the basal rock of the crust back into the mantle

GLOBAL AND SOLID EARTH SYSTEMS

The planet Earth has a large number of interconnected and interacting systems. Although the rates with which these systems respond to perturbations vary greatly, over time they all influence one another. Some of these systems are global in scale, while others are local. Many provide feedback that can either amplify or suppress a change.

The systems that have the slowest response times involve the solid earth. The surface of the earth is divided into several tectonic plates. These plates show little interior deformation over hundreds of millions of years, and yet their boundaries are constantly changing. They move across Earth at rates on the order of inches (or centimeters) per year and have been doing so for at least several hundred million years. The environmental issues most closely related to this motion are the hazards of earthquakes and volcanoes; however, less obvious connections exist.

Large areas of uplift, such as the Himalayan Mountains and the Tibetan Plateau, dramatically change atmospheric circulation patterns, which in turn affect precipitation and evaporation. Rates at which tectonic plates are formed along mid-ocean ridges vary, with far-reaching effects. The new material is hot and expanded. The time it takes to cool and contract does not depend on the rate at which it is forming, so faster growth results in larger regions underlain by expanded rock. The sea level rises as a result of this process. Some scientists have proposed a connection between volcanic processes acting along the East Pacific Rise and the occurrence of El Niño events.

Although the plate tectonic system interacts with other Earth systems and has long-term effects, it operates at such a slow rate that it is unlikely to cause significant changes to the environment during the course of a few human generations. In one hundred years, a plate may move a few meters. This is more than enough to cause tremendous earthquakes, but it is not likely to produce noticeable climate change.

Continental glaciation causes vertical movements of Earth's crust, which occur much more rapidly than plate motions. Approximately 20,000 years ago, a mass of ice about 1.86 miles (3 kilometers) thick covered much of Canada, the northern United States, and Scandinavia. The weight of this ice forced the crust to sink into the mantle, a process called "isostasy"; when the ice melted away, the crust began to rise again. One result of this uplifting motion is that some ancient Viking fishing villages that were once on the seashore are now well above sea level.

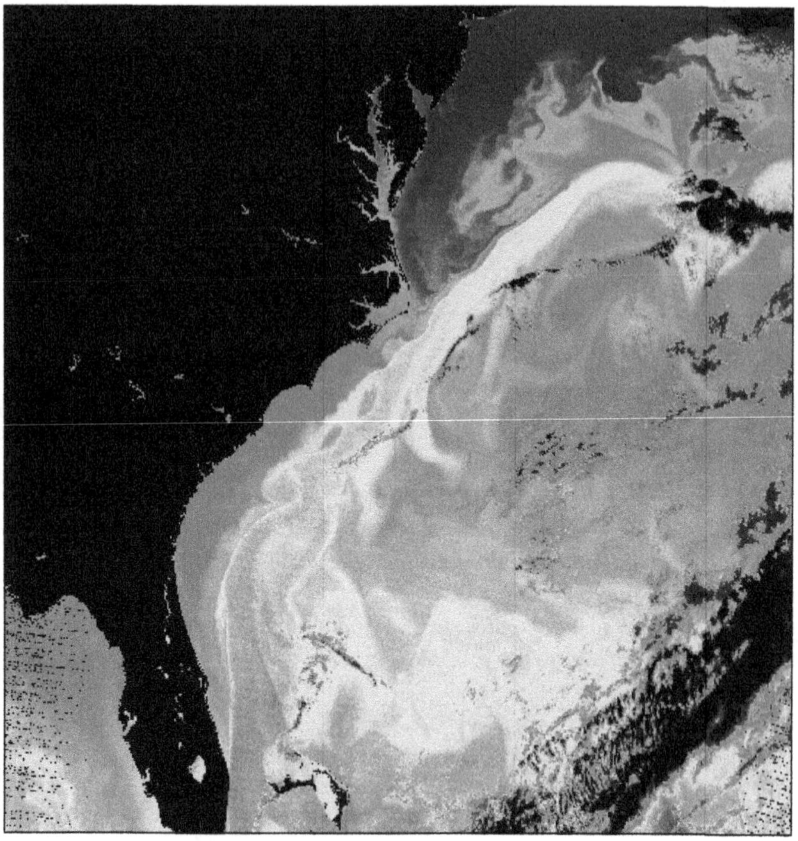
Image of the temperature of the Gulf Stream (Donna Thomas/MODIS Ocean Group NASA via Wikimedia Commons)

OCEAN SYSTEMS

Much faster than solid rock systems, the circulation of the oceans is one of the most significant processes affecting the environment. It is convenient to divide this circulation into surface current circulation and thermohaline circulation (involving the effects of both temperature and salinity).

Heat from the sun warms the surface water, which then expands and floats on the cooler water beneath it. This warm body of water has a lens shape, with the center a few meters higher than the edges. As water tries to flow down the slope of the surface, Coriolis force—a result of Earth's rotation—imparts some sideways motion as well. In the Northern Hemisphere, this produces a clockwise movement; in the Southern Hemisphere, it produces a counterclockwise movement. This ocean-scale circular current is called a "gyre." The Gulf Stream is the northwest segment of the North Atlantic gyre.

The Gulf Stream moves 46 million cubic yards (35 million cubic meters) of water past Chesapeake Bay every second. The edge of the Gulf Stream forms an abrupt boundary between cool, nutrient-rich waters near the continents and the warm, nearly lifeless waters of the Sargasso Sea. Although interconnected with weather and climate in a complex way, the operation of the Gulf Stream is not the principal source of heat responsible for the mild climate of Europe. The principal source of warmth for Europe is heat transported northward by atmospheric circulation, plus heat stored in Atlantic surface waters during the summer.

This heat transfer drives the thermohaline circulation. Cold, salty Antarctic water sinks to the bottom and moves northward, eventually rising to join the surface circulation. The largest amount of bottom water is in the Pacific, warming as it circulates southward and westward. Warm water from the Indian Ocean rounds Africa into the Atlantic, moves north to join the Gulf Stream, then cools and sinks in the Arctic. This circulation transports heat from regions around the equator to those around the pole and keeps the oceans mixed uniformly.

ATMOSPHERIC SYSTEMS

Atmospheric systems constitute the other major heat redistribution scheme on Earth. Because of Earth's spherical shape, the equatorial regions receive and absorb far more solar energy than upper latitudes.

As solar radiation reaches Earth, much of it is absorbed, raising the temperature of the surface. This heats the air above, which expands and rises into the atmosphere, just like a hot air balloon. The air cools as it rises, and the moisture therein condenses and precipitates, producing most of Earth's rain forests.

The cooler, drier air moves to the north and south until it eventually descends over belts of latitudes between 20 degrees north and 30 degrees north, and between 20 degrees south and 30 degrees south. Descending, this air warms up and gains moisture. At the surface, this air evaporates water effectively. Most of Earth's major deserts are in these latitudinal belts. The convection cell is completed by a return flow across the surface toward the equator. Because of the Coriolis force produced by the earth's rotation, this flow is deflected to come out of the northeast in the Northern Hemisphere and the southeast in the Southern Hemisphere. Regardless of direction, these winds are referred to as trade winds.

Another convection cell forms at somewhat higher latitudes, causing the surface winds to come out of the southwest in the Northern Hemisphere and out of the northwest in the Southern Hemisphere. These winds, called the "westerlies," blow across most of the United States. As they travel over the mountains in the western states, they are forced up into higher elevations, where the air loses its water. Deserts lie to the east of these mountains, in what is called the mountains' "rain shadow."

The trade winds of both hemispheres meet along the equator and result in surface winds that tend to push the surface water of the oceans to the west. This surface water is warmer and less dense than the water beneath it, and the boundary between the two is a region of rapidly changing temperature called the "thermocline." Because only warm water is pushed to the west, it piles up there, pushing the thermocline lower and lower. Under these conditions, the air just above the surface of the ocean is heated, and moisture is added to it as it travels from east to west. When it reaches the western edge of the Pacific, it rises and its water condenses. Every few years, however, the situation reverses itself. Surface winds weaken, the accumulated warm surface water flows eastward, rain comes to the deserts of Peru, and weather patterns over the entire planet are severely disrupted. The atmospheric effects of such a reversal are called the "Southern Oscillation," whereas the oceanic effects are called "El Niño." As they are linked and occur together, it has become common practice to refer to the phenomenon as the El Niño Southern Oscillation (ENSO).

GREENHOUSE GASES

Most of the energy that reaches Earth from the sun is electromagnetic in the visible light portion of the spectrum. A cloudless atmosphere lets this pass through easily. Upon reaching Earth's surface, some of this energy is absorbed by surface material, which then radiates the energy back, but at a much lower frequency in the infrared region, which humans detect as heat. On its way back to outer space, this energy may be absorbed by a greenhouse gas molecule. When this molecule reradiates the energy, the chances are about equal that it will be directed toward or away from the surface. In this way, a significant fraction of the infrared radiation is redirected back toward the surface of Earth.

The most important greenhouse gas is water vapor. The distribution of water vapor is uneven and subject to wide variations. It is generally assumed that the average relative humidity of the atmosphere remains constant. Because warm air contains more moisture at the same level of relative humidity than cooler air, the amount of water in the atmosphere will increase if global surface temperatures rise. Thus, water vapor is expected to enhance any effects produced by other greenhouse gases.

The greenhouse gas of greatest concern is carbon dioxide. It is well known that the trace amounts of carbon dioxide present in the atmosphere have increased have increased from about 0.025 percent (250 parts per million) to the present amount of 0.0425 percent (425 ppm) since the beginning of the Industrial Revolution, rising by 12 percent between 1958 and 1998 alone. The amount of coal, oil, and natural gas burned in just that 40-year time period is also known, and the 12 percent increase represents only about one-half of the carbon dioxide produced by this burning. Where the other half has gone remains a subject of research. Scientists are trying to determine if the observed increases in atmospheric carbon dioxide have led to changes in global surface temperatures. It appears that such temperatures rose approximately 32.9 degrees Fahrenheit (0.5 degree Celsius) between 1880 and 1980, a change well within historical rates of temperature fluctuations, leaving the significance of the increase in atmospheric carbon dioxide open to question.

Earth scientists studying gas bubbles trapped in the ice of Greenland and Antarctica have observed that carbon dioxide concentrations in the atmosphere and global temperatures changed together over the past 200,000 years. A causal relationship has not been proved, however, and the situation is complicated by biologic activity, the solubility of carbon dioxide in seawater, and many other factors that vary with temperature.

WATER AND THE SURFACE OF EARTH

Water evaporates from the oceans, is transported by weather systems, and falls on the land as rain and snow. Some of this water soaks into the soil, where it contributes to the chemical and physical breakdown of rocks, a process called "weathering." Some water can soak deep into the ground, where it is stored as groundwater. Much of it runs across the surface, transporting loose grains of sand and silt and eroding the surface. It takes only a few tens of millions of years to erode a mountain range flat.

In addition to being transported by water, loosened surface materials can move downhill by gravity—a process called "mass wasting." Slow mass wasting is called "soil creep." One common sign of soil creep is trees with bent bases; young trees are tipped by creep and keep trying to grow upright. Eventually, their roots become firmly enough established to resist creep, and the trees grow straight from then on. Rapid forms of mass wasting can be extremely destructive to life and property. Slump is a slope failure where a large block of material drops but remains largely intact. Mudflows and avalanches are fast-moving quantities of debris that can cause great losses of life and property.

LOCAL SYSTEMS

The environment of Earth is maintained by the interactions of global systems. Local areas—such as a floodplain, a watershed, or an earthquake-prone re-

Himalayas as seen from the International Space Station (NASA via Wikimedia Commons)

gion—also have interconnected systems of environmental significance. Here, too, Earth science provides essential background for policymakers.

Flooding occurs when too much water tries to move through a channel at once. Earth scientists can determine how much water a channel can hold and how long it will take various volumes of water to move down tributaries. Therefore, they also can predict how high a particular flood will be and when the crest will occur. By studying geological evidence of previous floods, scientists can establish the historical frequency of floods of various sizes. Taking into account the effects of agriculture and development, they can then estimate how frequently floods of different magnitudes are likely to occur in the future. This information can be used to determine how much money to invest in flood control or how restrictive to make zoning within the floodplain.

During floods there is an overabundance of water, but at other times water can be in short supply. In many places, underground aquifers are tapped for drinking water. Earth scientists can study a watershed and evaluate the quantity, quality, and flow patterns of its underground water. They can determine how much water can be safely withdrawn without long-term detrimental effects. If pollution occurs, scientists can determine the likely paths of the pollutants, trace them using remote sensing techniques, and suggest strategies to contain or eliminate their threat.

Threats posed by earthquakes are more difficult to perceive than those posed by pollution or flooding. Large earthquakes may, but are unlikely to strike a region twice within the span of one person's life. They may leave a record in the soils and structures of a region, and Earth scientists have been able to interpret these to extend knowledge of major earthquakes back several centuries. To detect small earthquakes, which recur frequently, Earth scientists have developed instruments called "seismographs." The behavior of geologic units subjected to prolonged periods of violent shaking can be very different from their behavior under normal conditions. Earth scientists have developed theories to explain such behavior, as well as criteria that can be used to identify those materials that pose the greatest risk. By combining results from these and other areas of research, researchers have been able to estimate the risks posed by earthquakes at various locations. This information can be used to develop building codes and zoning laws to minimize the risk of death and destruction posed by major earthquakes in the future.

SIGNIFICANCE

The environmentalists' slogan "Think globally; act locally" is an acknowledgment of the interconnectedness of many planetary processes and the importance of making local decisions based on some understanding of potential global consequences. Earth science endeavors to develop that understanding.

Earth scientists, by quantifying the many feedback loops with which systems interact with themselves and one another, can identify those natural processes that human influences are most likely to disturb. By studying the past behavior of these systems, scientists can make reasonable estimates of how they might respond to future disturbances. By informing public policy, legislation and law enforcement, and civic leaders in general, Earth scientists can use their knowledge and insight to influence the course taken by the governments of the world.

—Otto H. Muller

Further Reading

Chiras, Daniel D., and John P. Reganold. *Natural Resource Conservation: Management for a Sustainable Future*. 10th ed., Addison Wesley, 2009.

Dodds, Walter K., and Matt R. Whiles. *Freshwater Ecology: Concepts and Environmental Applications of Limnology*. 2nd ed., Academic Press, 2010.

Keller, Edward A. *Environmental Geology*. 9th ed., Prentice-Hall 2010.

Lenton, Tim. *Earth System Science: A Very Short Introduction*. Oxford UP, 2016.

Lundgren, Lawrence W. *Environmental Geology*. 2nd ed., Prentice-Hall, 2010.

Lutgens, Frederick K., Edward J. Tarbuck, and Dennis Tasa. *Essentials of Geology*. 11th ed., Prentice-Hall, 2011.

Middelburg, Jack J. *Marine Carbon Biogeochemistry: A Primer for Earth System Scientists*. Springer Open, 2019.

Sinding-Larsen, Richard, and Friedrich-W. Wellmer, editors. *Non-renewable Resource Issues: Geoscientific and Societal Challenges*. Springer, 2012.

Turner, David P. *The Green Marble. Earth System Science and Global Sustainability*. Columbia UP, 2018.

Zektser, Igor S., et al., editors. *Geology and Ecosystems*. Springer, 2010.

EARTH SYSTEM SCIENCE

Fields of Study: geology; biology; atmospheric science; oceanography; limnology

ABSTRACT

Earth system science is the study of four distinct but interconnected spheres within the earth's natural system. This field of study focuses on the lithosphere, the hydrosphere, the biosphere, and the atmosphere. Earth system science studies the interactions among these systems, including the effects of natural and human-made events on these spheres.

KEY CONCEPTS

asthenosphere: the layer of mobile liquid rock in the mantle directly beneath the lithosphere
atmosphere: the gaseous layer that envelopes the planet
biosphere: all living things that exist on and in the planet
cryosphere: the portion of Earth consisting of ice and snow
hydrosphere: the portion of Earth consisting of liquid water
lithosphere: generally called the "crust," consisting of the solid outermost rock layer and the very viscous molten layer between the crust and the asthenosphere
pedosphere: the layer of fertile earth or nonrock overlying the lithosphere

BASIC PRINCIPLES

The field of Earth system science (ESS) is based on the notion that the lithosphere, the hydrosphere, the biosphere, and the atmosphere are interconnected systems. ESS involves the combination of a number of natural sciences, including chemistry, physics, biology, mathematics, and applied sciences, to frame and analyze Earth's various processes and systems.

ESS developed relatively recently (it was first applied in the mid-1980s) to catalog the various elements of the planet and explain how these concepts related to one another. ESS also has demonstrated relevance in the study of natural disasters, such as research on hurricanes and earthquakes. ESS has become a useful tool for examining the effects of humans on the natural environment.

BACKGROUND AND HISTORY

In 1983, the advisory council of the National Aeronautics and Space Administration (NASA) established the Earth System Sciences Committee. This committee was charged with researching the earth's many natural systems and processes and examining how they interact. Scientists on the committee compared the committee's endeavor to the study of the solar system, wherein such individual concepts as gravity, solar wind, magnetism, and the planets themselves are often examined in relation to one another. A member of the committee, Moustafa Chahine of NASA's Jet Propulsion Laboratory, in a conversation with committee chair Francis Bretherton, argued that scientists had long focused on the solar system and that it was time to examine Earth's system, too.

The purpose of the development of ESS was, in part, to better understand the scope and the contributing factors of natural disasters and events. For example, volcanic activity in the lithosphere can send soot into the atmosphere and thus dramatically influence the habitats of nearby organisms. ESS has proved useful in understanding how the earth has evolved through millions of years.

The Earth System Sciences Committee also came into being as part of an effort to understand the concept of climate change and the effects that humanity was having on the planet. In 1988, the committee released a groundbreaking report, "Earth System Science: A Program for Global Change." This report included a schematic diagram that demonstrated how atmospheric, oceanic, biologic, geologic, and human factors interact, in some cases complementing one another and in other cases causing imbalances (such as climate change). The report became an important resource to which governmental and nongovernmental organizations pointed as they began to study global warming and climate change.

ELEMENTAL INTERACTION

ESS does not consider the four main elements of Earth's system as independent. Rather, it considers these spheres to be overlapping and interdependent. Each sphere plays a role in a cycle of biogeochemical processes.

For example, a volcano releases water vapor through the lithosphere and into the atmosphere. The water returns to Earth in the form of rain and snow.

The water is collected in the hydrosphere, from which plants and animals within the biosphere consume it. The organisms of the biosphere, in turn, return the water to the lithosphere in the form of waste, where it is broken down and recycled. The process then restarts.

Because ESS encompasses such a broad range of components, the study of this type of science requires the application of many different disciplines. Scientists must take and classify biological samples, conduct chemical analyses, utilize models of physics, develop and apply mathematical models, and conduct sociological surveys to better understand the processes that occur between each of the four main areas of ESS.

THE LITHOSPHERE

One of the main components on which ESS focuses is the lithosphere, which is the layer of solid and semisolid rock that forms Earth's outer crust and uppermost mantle. The lithosphere is fractured into giant tectonic plates, which are in constant motion atop the fluid asthenosphere (the softer and hotter layer just beneath the lithosphere). When tectonic plates come into contact or release themselves from contact, the energy released along the edges of the plates (faults) causes earthquakes at the surface. Similarly, molten rock that is pushed outward through the lithosphere typically flows through fissures in the lithosphere and out through volcanoes.

Lithospheric activity implicates other systems within the ESS framework. For example, volcanic and seismic activity can significantly alter the topography and the flow of river water in the surrounding area. In India, for example, a study of seismic and volcanic activity that occurred along a river basin through millions of years showed related changes in the terrain,

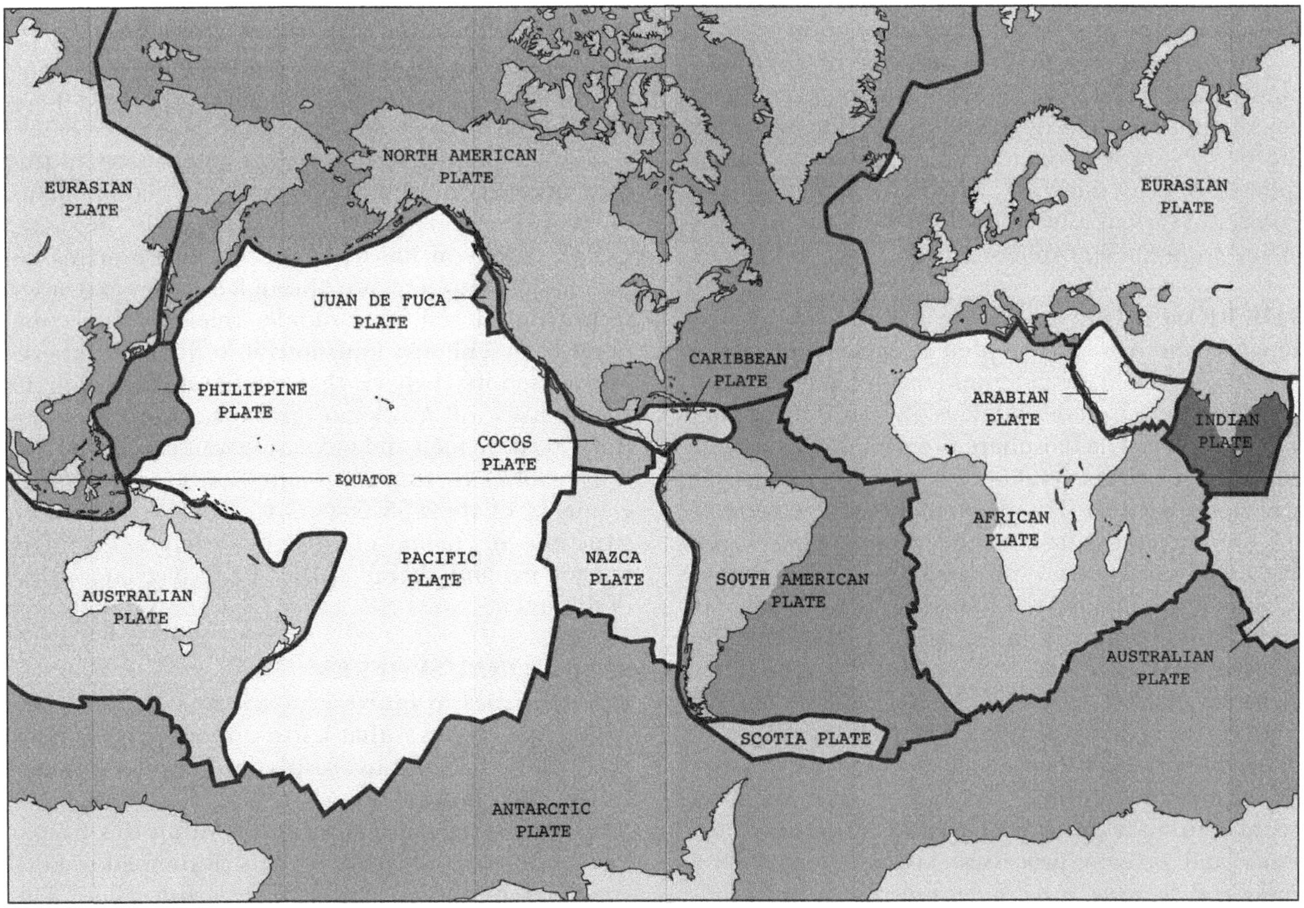

Tectonic plates of the Earth (United States Geological Survey via Wikimedia Commons)

forming hills and valleys and redirecting a river repeatedly.

In more dramatic fashion, a severe volcanic eruption can both level forests and kill countless people and animals, as was the case in 1980, when Mount St. Helens erupted in Washington State. However, volcanic activity can also create opportunities for life. The formerly devastated region has, since the eruption, seen more than 130 new ponds and lakes form, creating new ecosystems and habitats for wildlife.

THE HYDROSPHERE

The hydrosphere, which contains all of the planet's liquid water in oceans, lakes, rivers, ponds, and streams, plays a pivotal role in ecosystems and provides habitats for a wide range of organisms. It is central to the hydrologic cycle, whereby water collects as surface water from groundwater sources and ice, and is evaporated and transferred into the atmosphere, ultimately to return to Earth as precipitation.

For the purposes of ESS, the hydrosphere also plays a major role in the interaction among Earth's elemental spheres. The hydrosphere provides water for the biosphere, releases water vapor into the atmosphere, returns minerals, and is a vehicle for the transfer of various minerals and chemical compounds to and from the lithosphere.

THE BIOSPHERE

The biosphere is the component of the ESS model that houses the many different forms of life on Earth. The biosphere includes humans, as well as all animal and plant life. The biosphere even contains the most basic and simple forms of microscopic life, such as microbes and bacteria. The effects that other elements of the ESS framework have on the biosphere have been well explored. The study of ecosystems, the extinction of animal species, and the damage caused to habitats and forests by natural disasters are all examples of the study of how the biosphere is affected by the other elements.

Scientists are increasingly exploring the effects of the biosphere on the other elements of Earth's system. As technology improves, for example, scientists are able to study the role microbes play in the weathering and erosion processes, breaking down base rocks and creating and cycling minerals in the lithosphere. Furthermore, biologists are examining how the gases emitted by biologic species affect the atmosphere and, through weather conditions, the hydrosphere (where precipitation is deposited).

THE ATMOSPHERE

Since the development of ESS, increased attention has been paid to the study of global warming and how human-generated emissions are contributing to this problem. Millions of years before human life began, volcanic eruptions and even an occasional asteroid impact sent large volumes of carbon dioxide and other greenhouse gases high into Earth's atmosphere, causing Earth's average temperature to rise, melting ice caps and raising water levels. (Greenhouse gases are so named because they capture some of the energy radiating out from Earth's surface and return it to the atmosphere, thereby keeping the planet warm; this is a process known as the greenhouse effect.) However, when carbon dioxide and other greenhouse gases are retained at the surface level, a cooling effect takes place, eventually leading to the formation of glaciers and ushering in a period known as an ice age.

The ESS model recognizes the interaction, in this case, between the atmosphere and other components of the overall terrestrial system. Recently, however, greater attention has been paid to how humans (as part of the biosphere) are, through industry and other technological activities, rapidly releasing high volumes of greenhouse emissions into the atmosphere. Most scientists believe that continued emission of such gasses will accelerate global warming, causing the ice caps to melt and increasing water levels. Such a flood could have disastrous implications for the other elements of the ESS framework, including the destruction of many of the biosphere's countless ecosystems and changing the chemical composition of the soil.

ADDITIONAL SPHERES

ESS continues to evolve as a science. Although the general consensus is that Earth's intertwined systems are localized to the hydrosphere, biosphere, atmosphere, and lithosphere, some scientists believe that other spheres should be included within this framework. For example, some argue that the pedosphere must be considered a part of this discipline. The pedosphere is the layer of soil and organic matter that

View of the crescent moon through the top of the earth's atmosphere (NASA Earth Observatory via Wikimedia Commons)

rests on the surface of the earth. The pedosphere contains a vast number of minerals and chemical compounds essential for life on Earth. For this reason, it is often referred to as the skin of the planet, and like the skin of a living organism, it is influenced heavily by its relationship with the lithosphere, atmosphere, hydrosphere, and biosphere.

Others argue that another sphere should be considered a part of ESS: the cryosphere, which encompasses Earth's frozen water (its glaciers, ice packs, icebergs, and arctic regions). The cryosphere is an important resource for the world's water supply. It also is a critical indicator of global change. When icebergs, glaciers, and other frozen bodies melt under increasing temperatures, water levels rise. For this reason, scientists concerned with global warming carefully monitor the cryosphere for evidence of this trend.

—*Michael P. Auerbach*

Further Reading

Bockheim, J. G., and A. N. Gannadiyev. "Soil-Factorial Models and Earth-System Science: A Review." *Geoderma*, vol. 159, nos. 3–4, 2010, pp. 243–51.

Finley, Fred N., Younkeong Nam, and John Oughton. "Earth Systems Science: An Analytic Framework." *Science Education*, vol. 95, no. 6, 2011, pp. 1066–85.

Jacobson, Michael, et al. *From Biogeochemical Cycles to Global Changes*. Vol. 7 in *Earth System Science*. Academic Press, 2000.

Lenton, Tim. *Earth System Science: A Very Short Introduction*. Oxford UP, 2016.

Merritts, Dorothy, Andrew De Wet, and Kirsten Menking. *Environmental Geology: An Earth System Science Approach*. W. H. Freeman, 1998.

Middelburg, Jack J. *Marine Carbon Biogeochemistry. A Primer for Earth System Scientists*. Springer Open, 2019.

Schlager, Wolfgang. "Earth's Layers, Their Cycles, and Earth System Science." *Austrian Journal of Earth Sciences*, vol. 101, 2008, pp. 4–16.

Skinner, Brian J., and Barbara W. Murck. *The Blue Planet: An Introduction to Earth System Science*. 3rd ed., Wiley, 2011.
Turner, David P. *The Green Marble. Earth System Science and Global Sustainability*. Columbia UP, 2018.

EARTH TIDES

Fields of Study: geophysics; oceanography; astrophysics

ABSTRACT
Earth tides are deformations of Earth's crust as a result of gravitational interaction with the Moon and the Sun. Knowledge of the effects of these tidal forces is important to earth scientists who search for natural resources.

KEY CONCEPTS

angular momentum: the principle of physics by which a mass rotating about an axis tends to maintain its rotation in the same angular direction (*cf*: Newton's law of inertia, which states that an object in motion tends to stay in motion unless acted upon by an external force)

extensometer: a device that measures the strength of an Earth tide by measuring the tension it causes in a taut wire

gravimeter: a device that measures the strength of Earth tides by measuring the small change in Earth's gravity

oscillation: movement of a property or a structure between two values or positions that are determined by the the parameters of the physical system

spring constant: a unique, constant value, inherent to each particular spring, that determines the input force needed to deflect the spring; formulated in Hooke's Law equation as $F = -kx$ where F is the force required, x is the extent of deflection and k is the spring constant

synchronized rotation-revolution: the situation in which a body in orbit rotates about is axis at the same rate as its orbital rotation about its orbital center, as the Moon rotates on its axis at the same rate that it orbits Earth

GRAVITATIONAL ATTRACTION

Earth tides are the deformation of the solid portion of Earth's crust by the combined gravitational forces of the Moon and the Sun. Although other bodies within and beyond the solar system gravitationally attract Earth, the distances are great enough to make their tidal effect upon Earth negligible. Consider the Earth-Moon system. According to Sir Isaac Newton's law of gravity, every particle of mass in the universe is attracted to every other particle of mass by a force that is directly proportional to the product of the masses and inversely proportional to the square of the distance between them. This means that gravity is always an attractive force, but its magnitude depends to a great extent upon the distance between the two bodies in question.

Since gravity is an inverse square law, the following relationship holds true: If the distance between two bodies is doubled, the attraction of gravity becomes one-fourth as great. If the distance between the bodies is tripled, the attraction becomes one-ninth as great, and so on. According to this law, each particle of the Moon attracts each particle of Earth. Because these particles are not all equidistant from one another, the force of gravity varies in intensity. Gravitational attraction is greatest between the particles that are closest. Therefore, the surface of Earth nearest the position of the Moon is subjected to more attraction than is the surface of Earth opposite the Moon. It is this difference in relative position that causes the tidal force and thus the deformation of Earth.

Albert Michelson (1852–1931) measured Earth tides in 1913 by observing water tides in long horizontal pipes. He had assumed that Earth was rigid, but he did not observe the tidal values that the theory indicated he should. He used two 500-foot (15-meter) pipes at right angles, with 6-inch (15-centimeter) diameters and half-filled with water. They were buried in trenches 6 ft (2 m) deep with concrete-lined viewing pits at the ends. As expected, there was more deformation due to tides in the north-south pipe than in the east-west pipe. The difference could be accounted for when Earth was assigned a rigidity so that it was able to respond to lunar gravitational forces by raising crustal tides to a height of several centimeters.

OCEAN TIDES

The ocean tides may be considered as being analogous to Earth tides. Like Earth tides, ocean tides are caused by the gravitational forces of both the sun and the moon. Because of its relative closeness, the Moon is the greater factor. Its gravitation causes the water in the oceans to bulge outward a distance of one meter or so. There are two water bulges on the surface of Earth: one in the direction of the Moon and one in the opposite direction of the Moon. This latter bulge forms because of the reduced amount of gravity at that position on Earth's surface. Another way of looking at it might be as follows: Earth is being attracted toward the Moon or, in a sense, is falling toward the Moon. Therefore, the water on the lunar side is falling toward the Moon and is actually ahead of Earth's surface. The water on the opposite side of Earth is also falling toward the Moon but cannot quite keep up with Earth's surface and so forms a bulge. Theoretically, as Earth rotates with respect to the Moon, the water level rises and falls as these bulges of water are swept around Earth. In reality, the height and timing of tides may vary considerably. In some bays, the tidal water may accumulate to heights of 33 ft (10 m) and greater. Because there are two tidal bulges, there are two high tides per day.

The sun also exerts a tidal force on Earth, but because of its greater distance, its influence is only about one-half as great as the moon's. Extremely large high tides are generated when the Sun, the Moon, and Earth lie along a straight line. The tidal forces of the Sun and the Moon then act in the same direction. These tides are known as spring tides, though they have nothing to do with the season. The nature of the ocean tides provides an immediate observation and a fairly simple observation of the nature of tidal forces.

EARTH'S SHAPE

Tidal forces also have an effect on gravity, as does the shape of the planet. The ancient Greeks taught that Earth is a sphere. The philosopher Plato (ca. 428–347 BCE) reasoned that all heavenly bodies are perfect and therefore must be spherical; because Earth was a heavenly body, its shape was thus spherical. In about the year 230 BCE, Eratosthenes (ca. 276–194 BCE) calculated the circumference of Earth to be 250,000 stadia. If his calculations were based on the Egyptian *stadion* as the standard unit of distance measurement where his measurements were made, his calculation for the circumference of Earth corresponds to 24,662 miles (39,690 kilometers), which is within 1 percent of current measurements. During the seventeenth century, several measurements were made on Earth's surface. The size of one degree of arc in the Northern Hemisphere proved to be somewhat smaller than a degree of arc farther south. It was concluded from these studies that

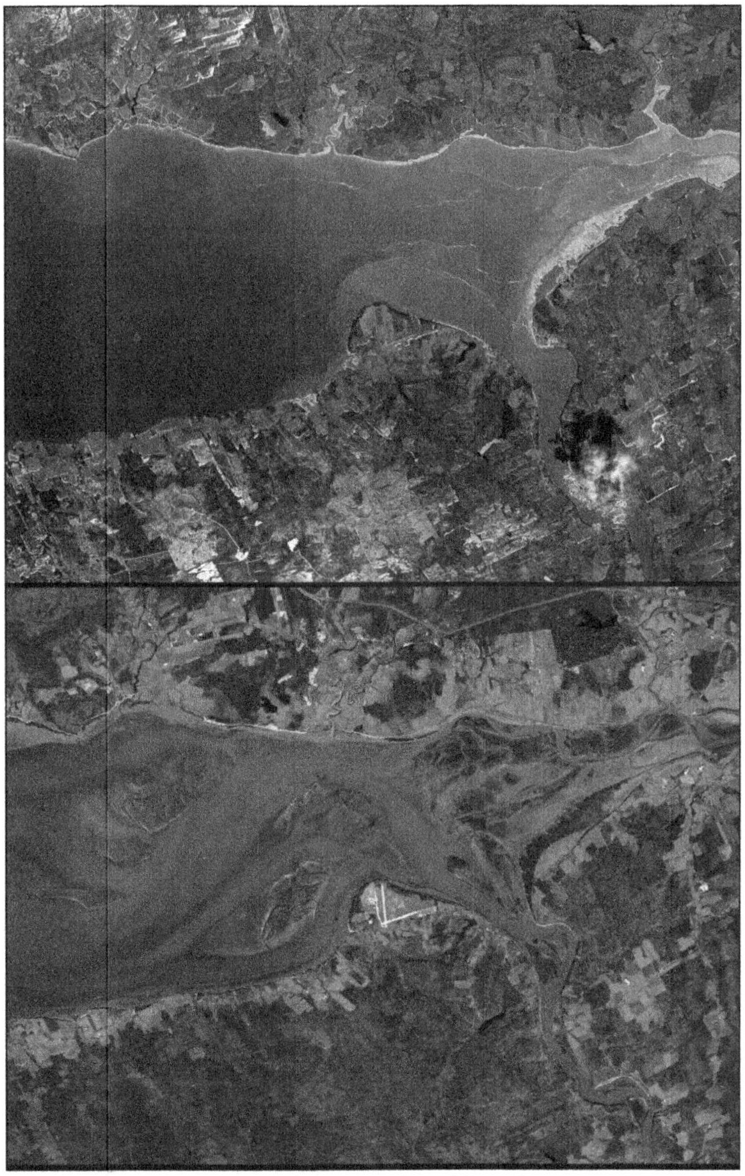

Thermal image of The Bay of Fundy at high and low tides (NASA/GSFC/METI/ERSDAC/JAROS, and U.S./Japan ASTER Science Team)

Earth is flattened somewhat toward the poles and thus is not quite spherical. The shape of Earth is rather an oblate spheroid, as explained by Sir Isaac Newton (1642–1727) in his famous work of 1687, *Principia*.

If Earth were a perfect sphere and homogeneous in composition, the gravity measurements at all points on the surface would be identical and the orbits of Earth satellites would be perfectly circular or elliptical. Because Earth's gravitational field is uneven, resulting from the fact that Earth is neither perfectly spherical nor homogeneous, the orbits of satellites are somewhat perturbed. The paths of satellites can be observed and plotted with a high degree of precision. The data indicate that Earth is an oblate spheroid, its radius 13 mi (21 km) longer at the equator than at the poles. It behaves as though it were a fluid balanced between gravitational forces, which tend to make it spherical, and centrifugal forces resulting from its rotation, which tend to flatten it.

ACCELERATION OF GRAVITY

The acceleration of gravity near Earth's surface is measured in gals, in honor of Galileo. A gal is the amount of force that will accelerate a mass 0.4 inch, or 1 centimeter per second squared. The total value for the acceleration of gravity is 980 gals, which is equivalent to the more familiar value of 32.15 ft per second sq (9.8 m per second sq). It is known that when the moon is directly overhead, at a position known as the zenith, the value for the acceleration of gravity at that point on Earth's surface is slightly less than if the Moon were in any other position. This phenomenon is a result of the gravitational influence or tidal force that the Moon exerts on Earth. The attraction of the Moon's gravity causes a point on Earth's surface to be distended slightly. Values for the amount of distension have been found to be about 0.24 ft (0.073 m). The fact that this point on Earth's surface has been gravitationally pulled away from the center of Earth will result in a slightly reduced value in the acceleration of gravity toward the center of Earth. These values have been found to be in the vicinity of 0.2 milligal. (A milligal is one thousandth of a gal.)

SYNCHRONIZED ROTATION-REVOLUTION

Subtle effects of tidal forces on Earth exist. When Earth and the oceans are subjected to tide-raising forces, energy caused by friction is dissipated. The result of this friction is the reduction in the period of Earth's rotation. In the case of a binary system such as Earth and the Moon, the result of tidal forces produces a synchronized state of rotation-revolution. In other words, the rate that the Moon rotates on its axis is the same as the rate at which it revolves around Earth in its orbit—which is the reason that the same face of the moon always points toward Earth. This particular phenomenon occurs elsewhere in the solar system; for example, the Sun and Mercury, as well as Pluto and its moon Charon, form other such binary systems.

There is a law in physics that states that angular momentum is conserved. If the rotation rates of Earth and the Moon are slowing but their masses stay the same, the distance between them must be increasing. Evidence from paleontological studies indicates that at one time, Earth had a faster rotation rate and the Moon was much closer than it is today. It is now known that the Moon is moving away from Earth at a rate of 1.3 in (3.2 cm) per year.

VERTICAL AND LINEAR DEFORMATION STUDIES

At the beginning of the nineteenth century, the concept that Earth was not perfectly rigid but in fact was somewhat deformable began to be accepted. The first studies of the deformation of Earth's crust were conducted in France in the early 1830s. These early studies were accomplished by using containers of mercury and comparing the motion of the liquid metal with the rise and fall of the ocean tides. The horizontal pendulum was the first instrument to record the effect of Earth tides with scientific precision. It consisted of a rigid bracket whose base contained three leveling screws. At both the top and the bottom of the bracket (which resembled a C-clamp), two metal wires were attached. These wires were all attached to a metal arm in such a way as to suspend it in position. At the end of the arm was attached a small mass. The slightest vibration caused by changes of the ground would cause the pendulum arm to begin oscillating back and forth. This instrument was but the first of many types and variations of the pendulum.

In the 1900s, the gravimeter came into use in the field of exploration geophysics and was later used to detect the minute changes in gravity brought about by Earth tides. Gravimeters are designed to measure the differences in the acceleration of gravity. There are several different types of these instruments, most of which consist of a mass suspended by springs. The greater the force, such as gravity, pulling on the mass, the more the spring stretches. The upward force is a function of the strength of the spring, or the spring constant. When the mass is in balance (not oscillating) the spring constant is equal to the force of gravity. Any change in gravity will then produce a corresponding change in the stretch of the spring. During a period of a maximum earth tide, gravity will be slightly reduced, resulting in a slight upward drift of the mass.

The pendulums and the gravimeter are used to study the vertical deformation of Earth's surface. The linear deformation may be measured by means of a device called an "extensometer." The first results from the use of this device were reported in the early 1950s. The extensometer consists of a wire 0.063 inch (1.6 millimeters) in diameter that is held nearly horizontal between two fixed supports about 66 ft (20 m) apart. A mass of 1.23 ounces (350 grams) is suspended from the center of the wire by a smaller wire with a diameter of 0.0079 in (0.2 mm). Variations in the 66-ft (20-m) distance between the two fixed supports as a result of linear deformations of Earth's surface can cause variations in the tension of the main wire. These variations cause the suspended mass to oscillate vertically. By methods of calibration, the oscillation can be translated into values of linear deformation.

ECONOMIC AND GEOLOGIC APPLICATIONS

The knowledge of how Earth tides function is necessary for an understanding of the deformable nature of Earth and of Earth's gravitational interaction with the Moon and the Sun. This knowledge is important to those who explore for the oil, gas, groundwater, and minerals that are necessary for life in the modern world. To the geophysicists who use the technique of gravity surveying, it is necessary to know whether the change in the value of gravity indicated by their instruments is caused by a subsurface geological structure or by the gravity of the Moon or the Sun.

For this reason, gravity surveyors must make what is known as a tide correction, which accounts for the time-varying gravitational attraction of the Sun and the Moon. The attraction is cyclic because the positions of the Sun and Moon are constantly changing with regard to a fixed position on the surface of Earth. To those Earth scientists who use the technique of searching for magnetic anomalies, or areas where Earth's magnetism is greater or less than expected, the Sun's effect on Earth's magnetic field is very important. The Sun's tidal force produces wind currents in Earth's ionosphere in the same way that it produces ocean tides. Since these winds in the ionosphere consist of waves of charged particles, there is an associated electric current. With this current comes a fluctuating magnetic field. The geophysicist, therefore, needs to know if the Sun's tidal force is causing deviations in the equipment being used.

—David W. Maguire

Further Reading

Braitenberg, Carla, Giuliana Rossi et al., editors. *Geodynamics and Earth Tides Observations from Global to Micro Scale*. Birkhauser, 2019.

Brosche, P., and Jürgen Sünderman. *Tidal Friction and the Earth's Rotation*. Springer Science & Business Media, 2012.

Hamblin, William K., and Eric H. Christiansen. *Earth's Dynamic Systems*. 10th ed., Prentice-Hall, 2003.

Jacoby, Wolfgang, and Peter L. Smilde. *Gravity Interpretation: Fundamentals and Application of Gravity Inversion and Geological Interpretation*. Springer-Verlag, 2009.

McCully, James Greig. *Beyond the Moon: A Conversational, Common Sense Guide to Understanding the Tides*. World Scientific Publishing, 2006.

Souchay, Jean, Stéphane Mathis, and Tadashi Tokieda, editors. *Tides in Astronomy and Astrophysics*. Springer, 2013.

Stephenson, F. Richard. *Historical Eclipses and Earth's Rotation*. Cambridge UP, 2008.

Tarbuck, Edward J., Frederick K. Lutgens, and Dennis Tasa. *Earth: An Introduction to Physical Geology*. 10th ed., Prentice-Hall, 2010.

Xu, Guochang, editor. *Sciences of Geodesy–I. Advances and Future Directions*. Springer, 2010.

Earthquakes

Fields of Study: geology; seismology; oceanography; marine geology; marine biogeochemistry

ABSTRACT

An earthquake is the sudden movement of the ground caused by the rapid release of energy that has accumulated along fault zones in Earth's crust. Earth's fundamental structure and composition are revealed by earthquakes through the study of waves that are both reflected and refracted from the interior of Earth.

KEY CONCEPTS

continental drift: the theory that the continents move about as they essentially float on the denser molten material of Earth's interior

paleomagnetism: the magnetic "signature" of Earth's magnetic field that is acquired by seafloor basaltic rock as it cools from its molten state and solidifies

tectonic plates: sections that form Earth's solid crust

triangulation: a method of finding the location of the epicenter of an earthquake by calculating its distance from three separate seismic detection sites, using the formula $d = (T_P - T_S)(V_P V_S) / (V_P - V_S)$ where d is distance, T is the time of detection of P and S waves, and V is the velocities of the P and S waves

TRANSVERSE SLIPPAGE AND SUBDUCTION EARTHQUAKES

In the early twentieth century, German meteorologist Alfred Wegener (1880–1930) proposed a theory of continental drift to account for the apparent manner in which the shelves of the various continents seemed to fit together as though they were parts of a planetary jigsaw puzzle. At the time, he could not describe a mechanism whereby continental drift could be justified. In the 1950s, a clue was found in the paleomagnetism of ocean floor rock, but it was still insufficient evidence to support the theory. Exploration of the Atlantic Ocean floor in the 1960s, however, revealed a feature termed the "Mid-Atlantic Ridge," a line of active volcanic peaks extending roughly along the ocean floor in a north-south orientation and including such volcanic and seismic centers as the Canary Islands, the Azores and Iceland. Eventually, observation of the activity of the ridge revealed the mechanism that had eluded Wegener: seafloor spreading.

In light of what had been revealed by the discovery, it became apparent that Wegener's theory of plate tectonics was indeed true. According to tectonics theory, Earth's crust is composed of a number of large tectonic "plates" that form the solid crust of the planet surrounding a molten core, much as the shell of an egg surrounds the egg's liquid interior. It has become very apparent from geological studies since then that the tectonic plates are in constant motion, however slow, and are propelled by convective currents in the molten rock of the mantle that underlies the crust of the planet. Seafloor spreading occurs where the convective currents act to push plates apart, allowing molten material to well up between them and create new seafloor rock. Where those plates encounter another plate or a continent, the force driving their movement causes two types of interaction to occur, both of which produce earthquakes of varying intensity.

Alfred Wegener, 1910 (Wikimedia Commons)

The first type is called a "transverse slippage earthquake," and is caused by sideways movement of one plate against another. This is the most common type of earthquake, resulting from the jerky movement of the irregularly shaped plates. They are generally not terribly severe, and indeed the vast majority of such earthquakes are not strong enough to be felt. The intensity of such earthquakes can be very great at times, however, depending on the amount of built-up stress that is released in the event. This is the type of earthquake produced by the San Andreas Fault in California and many others like it.

The second type is called a "subduction earthquake," and is caused when the motion of one plate forces it against a continental mass. Because the base rock of a continental mass is less dense than the basalt rock of the seafloor, the continental mass must go on top of the basaltic seafloor rock, and the seafloor rock is forced downward beneath the continental mass, where it re-enters Earth's molten mantle and is 'recycled'. Subduction earthquakes are considerably rarer than transverse slippage earthquakes, and much more powerful. Because they occur on the seafloor rather than on land, subduction earthquakes can have much more serious consequences as they are capable of generating highly destructive tsunamis, such as the one that struck the coast of Japan in 2011. The massive build-up of stress as a plate and a continental mass are forced together is only relieved by the sudden "dive" of the seafloor plate to go below the continental mass, shifting an equally massive amount of water above it. Video images of the 2011 tsunami coming ashore in Japan demonstrated that it did not arrive as a typical wave, but as a sudden raising of the sea level that subsequently flooded into the land.

STRESS

Earthquakes are sudden vibrational movements of Earth's crust and are caused by a rapid release of energy within Earth. They are of critical importance to humans because they reveal much about the interior of Earth and because they are one of the most destructive naturally occurring forces on the planet.

The outermost layer of Earth, called the "crust," is in constant motion as a result of large convection cells within the upper mantle that circulate heat from the interior of Earth toward the surface. The crust of Earth is about 3.1 miles (5 kilometers) thick in the oceanic basins (oceanic crust) and about 25 mi (40 km) thick in the continental masses (continental crust), while the upper mantle is about 435 mi (700 km) thick. Because the crust is relatively thin compared to the upper mantle, the crust is broken up into several plates that float along the top of each convection cell in the upper mantle. Most earthquakes occur along the boundaries separating the individual plates and are represented by faults that may be thousands of miles (or kilometers) long and tens of miles (or kilometers) deep.

Although the vast majority of earthquakes occur along these plate boundaries, some also occur within the plate interior. The rocks on either side of the fault fit tightly together, resulting in great resistance to movement. As the blocks of rock attempt to move against one another, the resistance to movement causes stress, which is a force per unit area, to build up along the fault. As the stress continues to build, the rocks in the immediate vicinity slowly deform, or bend, until the strength of the rock is exceeded at some point along the fault. Suddenly, the rocks break violently and return to their undeformed state, much as a rubber band snaps to its original shape when it breaks. This rapid release of stress is called "elastic rebound." The point at which the stress is released is called the "focus" of an earthquake, and that point at Earth's surface directly above the focus is called the "epicenter."

SEISMIC WAVE MOTION

The release of energy associated with elastic rebound manifests itself as seismic waves propagating away from the focus. When these waves of energy reach the surface of Earth, the land will oscillate, causing an earthquake. These waves move through Earth in two ways.

Primary (P) waves move in a back-and-forth motion in which the motion of the rock is in the same direction as the direction of energy propagation. This type of wave motion is analogous to placing a spring in a tube and pushing on one end of the spring. The motion of the spring in the tube is in the same direction as is the motion of the energy. These waves are called "primary" because they move through Earth faster than do other waves—up to about 15.5 mi per second (25 km per second). Thus, P waves are the first waves to be received at a seismic recording station. Because the individual atoms in a rock move back and forth along the direction of energy movement, P waves can move through solids and liquids and, for this reason, do not tell geologists much about the state (solid or liquid) of a given rock structure at depth.

For secondary (S) waves, in contrast to P waves, the rock motion is perpendicular to the direction of energy propagation. Guitar strings vibrate in a similar manner. Each part of the guitar string moves back and forth while the energy moves along the string to the ends. S waves are the second waves to be received at a seismic recording station and derive their name from this fact. Unlike P waves, S waves cannot move through liquids but can move through solids. Thus, when a P wave is received by a seismic station but is not followed by an S wave, seismologists know that a liquid layer is between the focus of the earthquake and the receiving seismic station.

Both S and P waves are bent, or refracted, as they move in Earth's interior. This refraction occurs as the result of the increase in density of rocks at greater depths. Furthermore, both types of waves are reflected off sharp boundaries, representing a change in rock type located within Earth. Thus, by using these properties of S and P waves, geologists have mapped the interior of Earth and know whether a given region is solid or liquid.

Although S and P waves represent the way seismic energy moves through Earth, once this energy reaches Earth's surface, much of it is converted to another type of wave. Love (L) waves move in the same manner as do S waves, but they are restricted to surface propagation of energy. L waves have a longer wavelength and are usually restricted to within a few miles (or kilometers) of the epicenter of an earthquake. These waves cause more damage to structures

The Mexico City Earthquake of 1985 resulted in serious damage and death (Wikimedia Commons)

than do P and S waves because the longer wavelength causes larger vibrations of Earth's surface.

EARTHQUAKE INTENSITY

The amount of energy released by an earthquake is of vital importance to humans. Many active fault zones, such as the famous San Andreas Fault in California, produce earthquakes on an almost daily basis, although most of these earthquakes are not felt and cause no damage to human-made structures. These minor earthquakes indicate that the stress that is accumulating along some portion of a fault is continuously being released. It is only when the stresses accumulate without continual release that large, devastating earthquakes occur. The intensity of an earthquake is dependent not only on the energy released by the earthquake but also on the nature of rocks or sediments at Earth's surface. Softer sediments such as the thick muds that underlie Mexico City will vibrate with a greater magnitude than will the very rigid rocks, such as granites, found in other parts of the world. Thus, the great earthquake that devastated Mexico City in 1985 was in part the result of the nature of the sediments upon which the city is built.

For a given locality, earthquakes occur in cycles. Stress accumulates over a period of time until the forces exceed the strength of the rocks, causing an increase in minor earthquake activity. Shortly thereafter, several foreshocks, or small earthquakes, occur immediately before a large earthquake. When a large earthquake occurs, it is usually followed by many aftershocks, which may also be rather intense. These aftershocks occur as the surrounding rocks along the fault plane readjust to the release of stress by the major earthquake. The cycle then repeats itself with a renewed increase in stress along the fault. Although seismologists can usually tell which part of the seismic cycle a region is experiencing, it is difficult to predict the duration of each of these cycles; thus, precise earthquake predictions are nearly impossible.

SEISMOGRAPHS

Seismographs are the primary instruments used to study earthquakes. All seismographs consist of five fundamental elements: a support structure, a pivot, an inertial mass, a recording device, and a clock. The support structure for a seismograph is always solidly attached to the ground in such a fashion that it will oscillate with Earth during an earthquake. A pivot, consisting of a bar attached to the support structure via a low-friction hinge, separates a large mass from the rest of the seismograph. This pivot allows the inertial mass to remain stationary during an earthquake while the rest of the instrument moves with the ground. The recording device consists of a pen attached to the inertial mass and a roll of paper that is attached to the support structure. Finally, the clock records the exact time on the paper so that the time of arrival of each wave type is noted.

When an earthquake wave arrives at a seismic station, the support structure moves with the ground. The inertial mass and the pen, however, remain stationary. As the paper is unrolled, usually by a very accurate motor, the wave is recorded on the paper by the stationary pen. Modern seismographs, however complex in design, always contain these basic elements. The clock, which each minute places a small tick mark on the recording, is calibrated on a daily basis by a technician using international time signals from atomic clocks. The recording pen often consists of an electromagnet that converts movement of the inertial mass relative to the support structure to an electrical current that drives a light pen. The light pen emits a narrow beam of light onto long strips of photographic film that are developed at a later date. Digital seismographs record measurements electronically using computers. A global network of these machines is overseen by the FDSN, the International Federation of Digital Seismograph Networks.

RICHTER AND MERCALLI SCALES

Seismologists have adopted two widely used scales, which are called the "Richter scale" and "Mercalli scale," to measure the energy released by an earthquake. The Richter magnitude scale is based on the amplitude of seismic waves that are recorded at seismic stations. Because seismic stations are rarely located at the epicenter of earthquakes, the amplitude of the seismic wave must be corrected for the amount of energy lost over the distance that the wave traveled. Thus, the Richter magnitude reported by any seismic station for a given earthquake will be approximately the same. Richter magnitudes are open-ended, meaning that any amount of seismic energy can be calculated. The weakest earthquakes have Richter magnitudes less than 3.0 and release energy less than

Seismographs at the U.S. Geological Survey record (1) north-south horizontal, (2) east-west horizontal, and (3) vertical components of an earthquake (J.K. Nakata, U.S. Geological Survey)

10^{14} ergs. These earthquakes are not usually felt but are recorded by seismic stations. Earthquakes between magnitudes 4.0 and 5.5 are felt but usually cause no damage to structures; they release energy between 10^{15} and 10^{16} ergs. Earthquakes that have magnitudes between 5.5 and 7.0 cause slight to considerable damage to buildings and release energy between 10^{18} and 10^{24} ergs. Earthquakes that are greater than 7.5 on the Richter scale generate energy up to 10^{25} ergs—as much as a small nuclear bomb.

The Mercalli intensity scale is based not on the energy released by an earthquake but rather on the amount of shaking that is felt on the ground; it rates earthquakes from Roman numerals I to XII. Unlike the Richter scale, the Mercalli scale provides descriptions of sensations felt by observers and of the amount of damage that results from an earthquake. Thus, an earthquake of Mercalli intensity I is felt by only a very few persons, while an earthquake of intensity XII causes total destruction of virtually all buildings.

Both the Mercalli and Richter scales have advantages and disadvantages. The Mercalli scale provides the public with a more descriptive understanding of the intensity of an earthquake than does the Richter scale. The damage caused by an earthquake is a function not only of the energy released by such an event but also of the nature of the sediments or rocks upon which the buildings in the vicinity are constructed. The Richter scale is best used to study specifically the amount of energy released by an earthquake. Finally, the Richter scale, which is purely quantitative, does not rely on subjective observations such as those required by the Mercalli scale.

TRIANGULATION TECHNIQUES

The exact location of an earthquake epicenter can be deduced from three seismographic stations using triangulation techniques. Because the P and S waves travel at different velocities in Earth, seismologists can determine the distance from the station to the epicenter. They calculate the difference in time between the first arrival of the P and S waves, respectively, at the station. They then multiply this time difference by the product of the P and S velocities and divide by the difference in wave velocities to obtain the distance to the epicenter. The earthquake must have occurred along a circle whose radius is the distance so calculated and whose center is the seismographic station; any three stations that record the event can be used to draw three such circles, which will intersect at a single point. This point is the epicenter.

EARTHQUAKE PREDICTION

Earthquakes are one of the most important processes that occur within Earth because they have such a profound effect on how and where people should develop cities. Geologists understand how and where earthquakes occur, yet despite their best efforts, they still cannot accurately determine when an earthquake will happen. They are merely able to predict that a large earthquake will occur in a particular region "in the near future." Very great earthquakes of magnitude 8 or greater, such as the San Francisco earthquake of 1906, occur about every five to ten years throughout the world. Industrialized societies, such as Japan, the United States, Chile, and many European countries, have developed buildings that are capable of withstanding devastating seismic catastrophes, but other countries are not as fortunate. Furthermore, some great earthquakes occur in regions that are not considered seismically active. The great Charleston, South Carolina, earthquake of 1886 and the Tangshan, China, earthquake of 1976 are examples of seismic events that could not have been easily predicted using modern technology. In such regions, buildings are not designed to withstand devastating earthquakes. Finally, many regions of the world do not experience earthquakes on a daily basis and, therefore, their governments lack the motivation to plan adequately for such potentially catastrophic events.

—A. Kem Fronabarger

Further Reading

Abaimov, S. G., et al. "Earthquakes: Recurrence and Interoccurrence Times." *Pure and Applied Geophysics*, vol. 165, 2008, pp. 777–95.

Emergency Management BC. *A Simple Explanation of Earthquake Magnitude and Intensity*. Ministry of Public Safety and Solicitor General, Provincial Emergency Program. 2007.

Grotzinger, John, et al. *Understanding Earth*. 5th ed., W. H. Freeman, 2006.

Hough, Susan. *Predicting the Unpredictable: The Tumultuous Science of Earthquake Prediction*. Princeton UP, 2010.

Kagan, Yan Y. *Earthquakes: Models, Statistics, Testable Forecasts*. John Wiley & Sons, 2013.

Moores, Eldridge M., and Robert J. Twiss. *Tectonics*. Waveland, 2014.

Prothero, Donald R. *Catastrophes!: Earthquakes, Tsunamis, Tornadoes, and Other Earth-Shattering Disasters*. Johns Hopkins UP, 2011.

Talwani, Pradeep, editor. *Intraplate Earthquakes*. Cambridge UP, 2014.

Tarbuck, Edward J., Frederick K. Lutgens, and Dennis Tasa. *Earth: An Introduction to Physical Geology, Global Edition*. Pearson Education, 2016.

Udías, Agustin, Raúl Madariaga, and Elisa Buforn. Source Mechanisms of Earthquakes: Theory and Practice. Cambridge UP, 2014.

East China Sea Ecosystem

Fields of Study: oceanography; marine ecology; marine biology; marine geology

ABSTRACT

Located directly offshore of East China, the East China Sea is a marginal sea in the West Pacific Ocean which covers an area of about 1,249,000 square kilometers (482,000 square miles.)

KEY CONCEPTS

average salinity: the mathematical average of salinities found in different parts of the sea

biodiversity: the different species of plants and animals found in a specific region

biome: a defined region and type of environment in which an associated range of biodiversity is found

endemic: describes species that are found in one region of Earth and nowhere else

trough: a deep depression in the seafloor, caused by the slow separation of two local tectonic plates

upwelling: the rise to the surface of relatively cold, nutrient-rich water from depths

DESCRIPTION

The East China Sea is the major sea off the eastern coast of China; it is a marginal sea of the North Pacific Ocean. The East China Sea is bound on the west by mainland China, where the large estuary of the Yangtze River flows in at Shanghai, via Hangzhou Bay. To the south, it is bound by the Taiwan Strait (the channel to the South China Sea) and by Taiwan. To the east, the East China Sea is separated from the Philippine Sea and greater North Pacific Ocean by the arc of the Ryukyu Islands chain, and to the northeast by Kyushu, southernmost of Japan's "big four" main islands. Directly north lies the Korea Strait (channel to the Sea of Japan), the South Korean island of Jeju, and the Yellow Sea.

The East China Sea is most often considered as including the Yellow Sea, which is tantamount to a large gulf opening off its northwest, between mainland China and the Korean Peninsula. The total area of the sea is about 482,000 square miles (1.25 million square kilometers). In addition to the Yangtze, about forty other large rivers flow into the sea, including the Qiantang, Ou, and Minjiang. Most of the sea lies over the continental shelf, which extends out as much as 300 mi (480 km) to the east. The East China Sea therefore is relatively shallow, seldom more than 656 feet (200 meters) deep. The Okinawa Trough, parallel to the Ryukyu Islands, is the deepest area, with depths to about 8,858 ft (2,700 m).

Because the East China Sea is in a subtropical zone, the average annual temperature is 68 to 75 degrees Fahrenheit (20 to 24 degrees Celsius); the annual temperature differential is 45 to 48 degrees Fahrenheit (7 to 9 degrees Celsius). The average salinity of the East China Sea is 31 to 32 percent, and rises to 34 percent in the eastern areas. This is somewhat less salty than most global oceans, but not out of the ordinary for coastal seas.

One of the vital characteristics of this biome is the constant cycling of water in the East China Sea, which benefits from river inflow and from exchange with other oceanic regions. Colder waters enter the sea from the rivers and from upwelling along coasts. The Kuroshio Current, a major warming influence, moves north from near Taiwan. A flow known as the Taiwan Warm Current enters the Taiwan Strait from the Philippine Sea, through the Bashi Channel. Another flow, the Yellow Sea Warm Current, enters the northern part of the East China Sea to create an ideal habitat for fish and shellfish breeding, feeding, and winter survival.

BIODIVERSITY

With its temperate climate, warm and shallow but well-stratified waters, seasonal flux in water exchange with other seas, freshwater supply from its rivers and multiple areas of strong tidal action, the East China Sea offers a very robust, differentiated, and broad base for a multitude of plant communities and for many types of invertebrates, fish, marine mammals, and birds. It has been estimated that some 13,000 species exist here, of which nearly 48 percent are endemic, that is, found nowhere else on Earth.

The food web here benefits from the widespread propagation and growth of phytoplankton and zooplankton. In turn, these creatures form the basis of life for great populations of crustaceans and mollusks, which provide sustenance to all types of fish here.

As such, the East China Sea is the most productive marine area around China for seafood; it is especially rich in both big and small yellow fish, beltfish, flatfish, croaker, chub and spotted mackerel, Pacific cod, bluefin and bigeye tuna, a full range of sharks, squid, broadclub cuttlefish, and more. Most of the fishing is done by small local boats, but larger trawlers are a used in some cases. The Zhoushan Islands fishing grounds within the East China Sea are known as the treasure trove of China seafood production.

Spotted seal, finless porpoise, minke whale, and the critically endangered gray whale are among key marine mammals in these waters, roaming seasonally from the deeps to preferred grounds along the coasts.

The Yellow Sea-East China Sea region is an absolutely vital component of the East Asian-Australasian Flyway, with a reported two million shore birds stopping over on their northward migrations. Two globally threatened species are among these: spotted greenshank and spoon-billed sandpiper. Fulmars, storm petrels, alcids, geese, mallards, ducks, and other birds make up some of the 160 distinct marine bird species either resident in or that migrate through the East China Sea.

The short-tailed albatross, a vulnerable species that favors coastal upwelling zones for feeding areas, has important colonies in the East China Sea. The eastern curlew and Asian dowitcher are two additional vulnerable migrants here.

ENVIRONMENTAL THREATS

Environmental pollution caused by fast economic growth in mainland China as well as other nearby industrialized nations has led to conditions such as eutrophication (excess nutrients), low-oxygen zones, and frequent red tides along the coast of the East China Sea. Similarly, there have been repeated events of giant jellyfish swarms.

Overfishing is a perennial challenge. Most of these conditions tend to be exacerbated by increasing water and air temperatures, with the onset of global warming. Sea-level rise, too, is a growing concern as it threatens coastal habitats with accelerated erosion and saltwater intrusion.

These factors have significantly affected the marine ecological web. In addition, air pollution from southeastern China is thought to form acid rain when passing above the East China Sea, thus precipitating heavy metal and toxic chemical fallout into the sea and along the north coast of Taiwan.

The East China Sea's continental shelf bears rich petroleum and natural gas deposits that have led to disputes among the bordering countries in the region, including China, Japan, South Korea, and Taiwan. The tension is focused on the control of the areas with potentially exploitable hydrocarbon reserves, particularly regions near deep trenches, straits, rocks, and uninhabited islands. Such disputes are incremental to already-existing conflicts over fishing rights in many of the same areas.

The short-tailed albatross, a vulnerable species, has important colonies in the East China Sea (Wikimedia Commons)

While diplomacy is required in the struggle over fossil-fuel ownership, there are single-nation efforts to protect some of the East China Sea biome. Within the areas of the sea and coastal zones that it legally controls, the government of China has established 17 nature reserves, seven special marine reserves, and a handful of fishery conservation areas. The total area of these marine protected areas (MPAs) is approximately 40,000 sq mi (103,000 sq km). The intent of setting up and enforcing MPAs is to provide a future environment that preserves and restores biodiversity, habitats, and fishery resources.

—*Zhiqiang Cheng*

Further Reading

Ding, H., et al. "An Overview of Spatial Management and Marine Protected Areas in the East China Sea." *Coastal Management*, vol. 36, no. 5, 2008.

Ishimatsu, Atsushi, and Heung-Jae Lie, editors. *Coastal Environmental and Ecosystem Issues of the East China Sea*. Terrapub and Nagasaki U, 2010.

Kämpf, Johan, and Piers Chapman. *Upwelling Systems of the World: A Scientific Journey to the Most Productive Marine Ecosystems*. Springer, 2016.

Nagai, Takeyoshi, Hiroaki Saito, Koji Suzuki, and Motomitsu Takahashi, editors. *Kuroshio Current. Physical, Biogeochemical, and Ecosystem Dynamics*. Wiley, 2019.

Park, Chul. *Yellow Sea and East China Sea Reported by PICES and Korean Monitoring Program*. Chungnam National U, 2012.

Qu, J., Z. Xu, Q. Long, L. Wang, X. Shen, J. Zhang, and Y. Cai. *East China Sea—Global International Waters Assessment*. Linnaeus U and United Nations Environment Programme, 2005.

Wang, Qian-Wen, Gui-Peng Yang, Ze-Ming Zhang, and Shan Jian. *Perfluoralkyl Acids in Surface Sediments of the East China Sea*. Elsevier Science, 2017.

Zhu, Qian. *Biological Assessment of the Ecologically Important Areas for Marine Mammals in the Yellow Sea Ecoregion*. Shandong U, 2008.

EAST SIBERIAN SEA

Fields of Study: oceanography; marine biology; marine geology; marine ecology

ABSTRACT

A marginal sea of the Arctic Ocean, located off the northern coast of Siberia, the East Siberian Sea is bordered to the east and west by the Chukchi and Laptev Seas; it is noted for its numerous ice fields, most of which are fully melted for less than two months a year.

KEY CONCEPTS

heavy metals: generally toxic metallic elements including cadmium, mercury, lead, arsenic, copper, and several others

hydrocarbons: organic compounds consisting only of carbon and hydrogen atoms, the products of the petroleum industry

phytoplankton: planktonic organisms that are plant in nature, as opposed to zooplankton, which are animal in nature

radionuclides: atoms that are radioactive isotopes of various elements that are used in different nuclear industries

DESCRIPTION

An obscure area in many respects, the East Siberian Sea has been less studied than much of the Arctic area, and commercial fishing has been little developed within it. The sea did not even receive a distinct name until 1935, when the government of the former Soviet Union gave the sea its present designation. Before that, the East Siberian Sea was variously called Kolymskoe, Sibirskoe, Ledovitoe, and other names, and sometimes went without specific designation on maps.

Because the shallow sea is frozen for most of the year, it is navigable only in August and September, limiting human presence. Even the coast is the site of several of Siberia's ghost towns, including the former trading post of Logashkino and the abandoned mining town of Valkumey.

Rivers flowing into the sea include the Indigirka, Alazeya, Ujandina, Chukochya, Kolyma, Rauchua, Chaun, and Pegtymel. There are two distinct hydrological zones in the sea: the west, which is influenced most strongly by the water of the Lena River (and marine- and terrestrial-derived sediments), and the east, influenced by Pacific Ocean waters. The main gulfs of the East Siberian Sea are in the south, with the north opening into the Arctic Ocean. It is home to few islands, many of which consist only of gradually eroding sand and ice.

Most of the remaining islands are grouped along the coastal limits, including the tundra island of Ayon off the coast of Chukotka, used by the Chukchi people to pasture herds of reindeer; the small adjacent islands Ryyanranot, Chengkuul, and Mosey; and the uninhabited Medvezhyi Islands in the west of the Kolyma Gulf. The *Medvezhyis*, the name of which means "bears" in Russian, are both covered by ice, the proximate sea generally staying frozen even in the summer. Though home to minor seasonal commercial fishing, the islands serve mainly as a breeding ground for polar bears.

CLIMATE AND ICE COVER

One reason the ice is so tenacious is because the sea is so shallow. Seventy percent of the area is less than 164 feet (50 meters) deep, and the deepest point is 509 ft (155 m) deep. The summer air temperature averages just above freezing, 32 degrees Fahrenheit (0 degrees Celsius), and in winter plummets to -20 degrees Fahrenheit (29 degrees Celsius). The ice cover prevents sunlight from penetrating far into the sea, and along with the seasonal change of the seawater salinity and chemical content due to the gradual freezing and thawing of the ice, this situation causes considerable complications for the ecosystem.

The entire phytoplankton life cycle takes place during the brief summer, lying dormant the rest of the year. Seasonal coastal erosion and river discharges provide key nutrients, especially in the Kolyma River, which contributes millions of tonnes to the sea every year. This can be expected to increase significantly as the rivers flow through permafrost lands that are now thawing at an accelerating rate due to climate change.

Climate change appears to have increased the summer period in the East Siberian Sea. In 2007, the National Aeronautics and Space Administration (NASA) satellite imagery confirmed that as early as June 16, melting was well underway, and patches of ocean water had become visible surrounding small slabs of ice in the northeast. By the end of July, the area between the Siberian coast and the New Siberian Islands had become almost entirely ice-free, and throughout the sea, ice was sparse.

While sea-ice loss was detectable throughout the Arctic during summer 2007, the majority of it occurred in the East Siberian Sea—but even this event was just a prelude. The summer melting in 2012 exceeded anything previously seen in the record of Arctic satellite imagery, with the East Siberian Sea especially void of surface ice; scientists have issued more urgent warnings about the need for environmental protection measures, especially the curtailment of burning fossil fuels.

The water in the sea is fairly clean by twenty-first-century standards. While the snow and sediment can be found to contain hazardous contaminants, including heavy metals, radionuclides, and hydrocarbons, their concentrations are very low except in areas immediately surrounding mineral-resource extraction zones. The Siberian rivers contribute industrial and agricultural runoff, and coastal currents carry pollutants along the continental shelf, but the levels of these substances are low compared with other parts of the ocean.

BIODIVERSITY

The sea is home to a limited number of gray whales in the summer months, populations of which may migrate between the East Siberian Sea and the western Chukchi Sea, wherever conditions are most favorable and ice conditions permit movement. Gray whales have been spotted in the early fall, before ice coverage is complete. Some scientists have speculated that the whales may be reoccupying habitats that had been abandoned after commercial whaling decimated the whale population.

Other fauna living on the coast and the drift ice include Pacific walruses, polar bears, Arctic foxes, seals, and seabirds. Prominent fish include salmon, pollock, halibut, and crab, mainly in the subestuary zones.

—Bill Kte'pi

Further Reading

Akulichev, Viktor Anatol'evich. *Far Eastern Seas of Russia.* Nauka Press, 2007.

Barry, Roger G., and Eileen A. Hall-McKim. *Polar Environments and Global Change.* Cambridge UP, 2018.

Herman, Yvonne. *Marine Geology and Oceanography of the Arctic Seas.* Springer Science & Business Media, 2012.

Kosyan, Ruben, editor. *The Diversity of Russian Estuaries and Lagoons Exposed to Human Influence.* Springer, 2017.

Robinson, Allan R., and Kenneth H. Brink. *The Global Coastal Ocean: Regional Studies and Syntheses.* Harvard UP, 2005.

Tedesco, M., editor. *Remote Sensing of the Cryosphere.* John Wiley & Sons, 2015.

EL NIÑO-SOUTHERN OSCILLATION (ENSO)

Fields of Study: climatology; oceanography; fluid dynamics; computer modeling

ABSTRACT

The El Niño-Southern Oscillation (ENSO) is a series of linked atmospheric and oceanic events in which a reduction or reversal of the normal large-scale pressure systems acting in the equatorial regions of the Pacific Ocean results in dramatic changes in precipitation, currents, and wind patterns. Repercussions occur over wide areas from the interior of North America to India. Although not regularly periodic, El Niño phenomena generally occur every two to seven years.

KEY CONCEPTS

Coriolis effect: a perceived force due to Earth's rotation that seems to cause a moving air mass to change its north-south direction of movement to an east-west direction in the northern and southern hemispheres

thermocline: a density boundary between warmer water above and colder water below, characterized by a marked difference in the water temperatures

trade winds: winds to the north and south of the equator that typically blow in an east-to-west direction, so named because sailing ships on trade missions used these winds to sail to the west

GLOBAL WIND PATTERNS

El Niño is the term used for the oceanic aspects of one phase of a climatological phenomenon more generally referred to as an ENSO event. *El Niño* is a Spanish term meaning "the little boy," familiarly interpreted to mean the Christ Child, that was used by Ecuadoran and Peruvian fishermen to refer to unusually warm ocean surface currents that arrived around Christmastime at irregular intervals. "Southern Oscillation" refers to the accompanying effects in the atmosphere. ENSO events typically recur every two to seven years and involve dramatic changes in sea surface temperatures (SST) precipitation, and wind patterns, accompanied by nutrient upwellings. Such environmental fluctuations influence, and in some cases may control, the climate and ecology of large areas of Earth where the effects of ENSO events are most pronounced.

The large-scale wind patterns of Earth result from the different amounts of solar energy received at the equator and at the poles. Generally, a low-pressure region of ascending air can be found at the equator. This air rises, cools, and releases its water vapor as rain, creating equatorial rainforests over land. High in the atmosphere, the dry air spreads out, eventually returning to Earth near the latitudes of the tropics, between 23 degrees and 30 degrees north and south. Most of Earth's large deserts are located in this latitude band. To finish its circuit, this air travels across the surface of Earth toward the equator. Because of Earth's rotation, however, its path is deflected, a phenomenon called the "Coriolis effect." In the Northern Hemisphere, the observed deflection is to the right, and in the Southern Hemisphere, it is to the left. Such a deflection causes the air to travel obliquely toward the west in both hemispheres, resulting in the system of winds that are known as the trade winds. (Most of the United States lies north of this area and is under the influence of a similar set of winds blowing in the opposite direction, known as the prevailing westerlies.)

In one pattern called the "Walker circulation warm," wet air rises over Indonesia, spreads out, and then falls as cool, dry air on the eastern Pacific Ocean and the coast of Peru. The existence of this phenomenon became apparent to British scientist Sir Gilbert

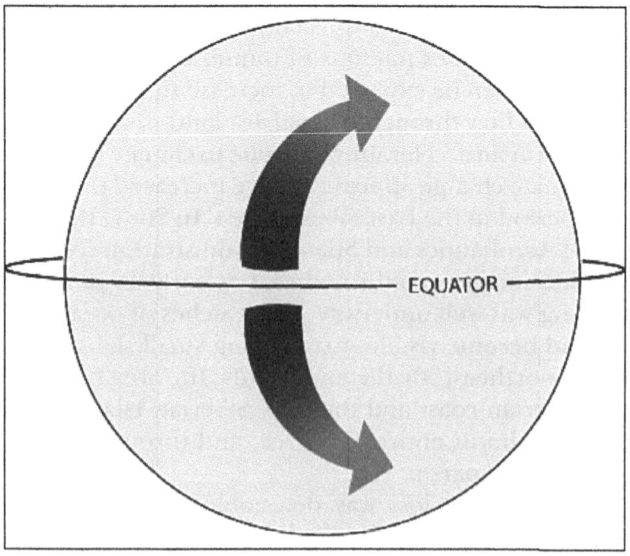

Coriolis effect (National Oceanic and Atmospheric Administration)

Walker (1868–1958) as he collected data from Tahiti and Darwin, Australia, during the early part of the twentieth century. He found that the atmospheric pressures in these two locations varied together but out of phase with each other. That is, if the pressure rose in Darwin, it fell in Tahiti, and vice versa. Sometimes conditions were normal, with high-pressure zones over Tahiti and lows over Darwin, while at other times they were reversed. Walker called this phenomenon the "Southern Oscillation" and showed that many other meteorological phenomena from around the world seemed to be tied to it, though his particular goal was predicting when the monsoons would fall in India.

As the trade winds move across the Pacific Ocean, they drag along great quantities of water. This water is at the sea surface, where it is warmed by incoming solar radiation, and builds up in the western Pacific in the vicinity of Indonesia. The elevation of the sea there can be 15.7 inches (40 centimeters) higher than in the eastern equatorial Pacific. The temperature of the sea changes rather abruptly between the water that has been warmed by the sun and the water beneath it. The depth at which this change occurs is called the "thermocline." Because the density of water varies with temperature, this abrupt temperature change corresponds to a density contrast that effectively acts as a barrier, and mixing of water across the thermocline does not readily occur. The buildup of warm surface water in the area east of Indonesia causes the thermocline to settle to as much as 656 feet (200 meters) deep. Off the coast of Peru, the thermocline is typically about 164 ft (50 m) deep. The difference of 15.7 in (40 cm) of water at the surface is balanced by a difference of 492 ft (150 m) on the thermocline because the density contrast across the thermocline, between the cold water and warm water, is much less than the density contrast between the water and air at the surface.

OCCURRENCE OF EL NIÑO AND LA NIÑA

El Niño and La Niña ("the little girl") are opposing phases of the ENSO cycle. Every few years, for reasons that are not well understood and are apparently the result of a complex atmospheric system, the strength of the trade winds decreases. Often this is accompanied by an eastward spreading of the surface pool of warm water. These two phenomena are linked: Each can and probably does, to some extent, cause the other. As the trade winds collapse, a major readjustment of the thermocline occurs. As the small surface slope dwindles, the much larger slope on the thermocline also disappears. As the thermocline rises, it generates Kelvin waves, which are large-scale gravity waves. These waves race across the Pacific Ocean, deepening the thermocline to the east as they go.

The deeper thermocline and warmer surface water reduce the effectiveness of upwelling in supplying nutrients to the surface water. This causes dramatic decreases in the primary productivity of the oceans off the coast of Peru and often spectacular die-offs in many species of fish and the seabirds and other animals that rely on them. During the 1972 El Niño, die-offs resulted in the collapse of the anchovy fishing industry, although its decline was exacerbated by questionable fishery management policies.

As the thermocline deepens and warmer surface waters move east, the area in the Pacific equatorial region where the ascent of warm, moisture-laden air produces intense precipitation shifts to the east. This area, called the "Intertropical Convergence Zone (ITCZ)," is generally found over the warmest water. During strong ENSO events, the Walker circulation may be completely reversed, with warm, moist air rising over coastal Peru, where cool, dry air normally falls. When this happens, regions that have not received a drop of rainfall for years are suddenly inundated with precipitation.

The paths and severity of many storms tracking across the continental United States are affected by the location of the ICTZ. During ENSO events, the jet stream is often moved or disrupted by the heavy storm activity above the zone. Sometimes this results in more precipitation across certain sections of the United States and sometimes less, but typically the result is abnormal weather.

Eventually, El Niño weakens, and the trade winds return to their usual strength. Cold water wells up from the depths to replace the warm, extra water as it dissipates, bringing nutrients to reestablish the high productivity of the waters. As warm surface water builds up in Indonesia, the thermocline there deepens again, and the ICTZ moves toward its earlier location. If, as sometimes happens, the return of the trade winds cools the surface waters beyond their normal temperatures, a phase called "La Niña" begins.

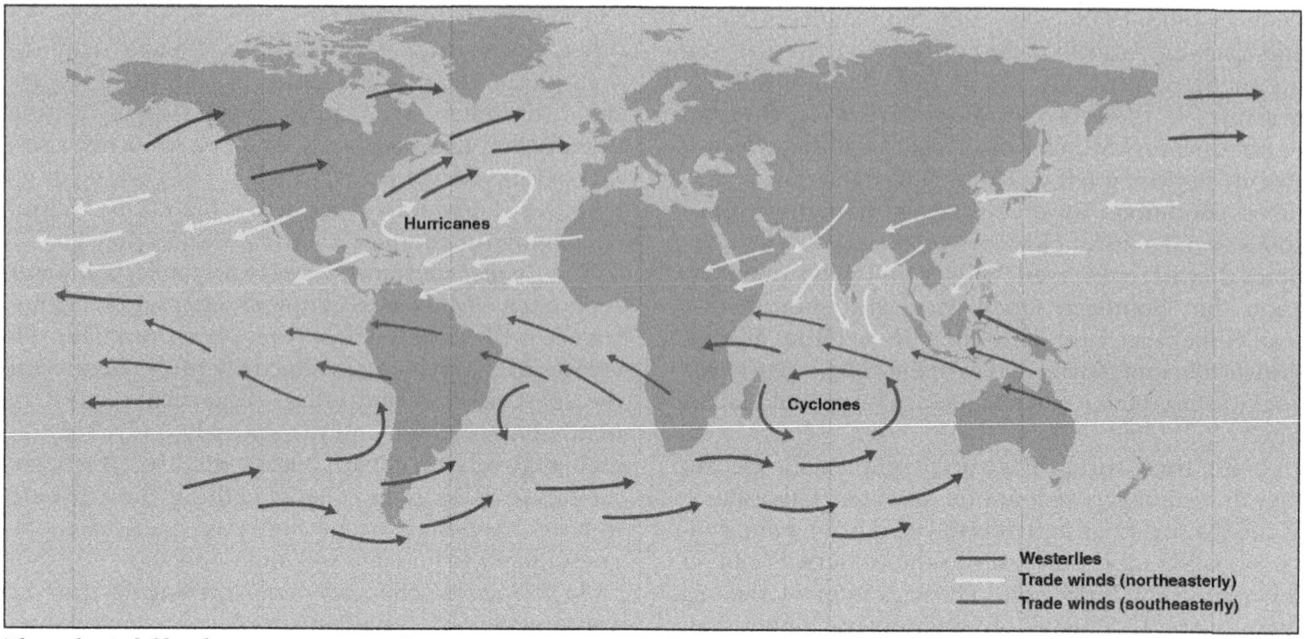

The trade winds blow from east to west near the equator (NASA/JPL-Caltech)

Apparently, the ocean/atmosphere system oscillates between the two states of La Niña and El Niño. What drives it from one state to the other remains a matter of considerable speculation. Although the complexities of the ocean/atmosphere system would seem more than adequate to produce almost any sequence of events, some researchers have suggested the involvement of yet another complex system, Earth's tectonic engine. ENSO events may also be correlated with eruptions of large equatorial volcanoes, which inject dust into the air and can have dramatic impact on the atmospheric components of the cycle.

There is a great deal of interest, both scientific and general, in learning more about ENSO events. Computational capabilities and theoretical developments have progressed rapidly, and although scientists do not yet understand this complex interaction of air and water, they have begun to deploy water temperature sensing buoys across the Pacific to study it.

STUDY OF EL NIÑO

El Niño-Southern Oscillation events are large-scale phenomena. They last longer than one year, involve massive amounts of air and water, and affect a substantial percentage of Earth's surface. During the course of an ENSO event, the changes that have already occurred in the system influence the changes that will occur as it progresses. Winds affect sea surface temperatures, and sea surface temperatures affect the winds. Such feedback interactions make it difficult to develop models with predictive capability. In the past, computer models used for atmospheric studies often oversimplified the ocean, and those used for oceanic studies oversimplified the atmosphere. ENSO events involve a tight coupling of ocean and atmosphere, and any successful model will need to treat both with sophistication and detail.

Large computer models are used to study ENSO events. As they evolve, deficiencies in the available data set are identified, and additional data are acquired. Much of the necessary data are of the type normally gathered during almost any routine oceanographic study: temperatures, salinities, and surface wind velocities and directions. Researchers would like to have such data from before, during, and after an ENSO event. Because these events recur so frequently, almost any data must be collected within a few years of an event. The difficulty lies in figuring out which data are important. Although there is agreement on what typically happens during an ENSO event, the specifics vary widely from event to event. Just when scientists think they understand it well enough to venture forth with a predictive model, nature comes along and shows them that their model is inadequate.

Still, models, whether sophisticated computer models or simple physical models, can further the understanding of ENSO events. One simple physical model uses a clear Pyrex baking pan as the ocean basin; water as the deep, cool, ocean water; vegetable oil as the warmer, less dense, surface water; and a hair dryer (set to blow unheated air) as the trade winds. If the flow of air produced by the hair dryer is directed across the oil, the oil piles up at the far side of the baking pan. The surface of the oil will form a slight slope, with the higher elevation at the far side of the baking pan. A more pronounced, readily visible slope will form where the oil meets the water. Friction from the air coming from the hair dryer pushes surface oil toward the far side of the pan, but as it builds up, it increases the pressure on the water beneath it. This causes the water beneath the oil to move back toward the near side of the baking pan. Eventually a wedge of oil sits over the water, its upper slope maintained by the hair dryer "wind," and its lower slope positioned where the pressures at depth are the same. (The fraction of an inch, or centimeter, of additional oil instead of air at the top surface is compensated by a considerably greater deflection of the oil-water boundary because the density contrast between oil and water is much less than that between oil and air.) This represents the normal state in the equatorial Pacific: a shallow thermocline off the coast of Peru; winds moving off the South American continent at Peru, preventing rainfall; a deep thermocline in Indonesia; and air moving from east to west across the ocean, becoming warm and moist, rising over Indonesia, and subsequently drenching it with rain.

In the model, the equivalent of an El Niño can be produced by turning off the hair dryer. The forces that maintained the slope on the upper surface of the oil are gone, so the oil sloshes down and its surface becomes horizontal. A wave forms where the oil meets the water, a boundary that corresponds to the thermocline. Such a wave is called an "internal wave" because it occurs between two layers of water with contrasting densities but in general is not visible at the surface. The wave races across the baking pan just as the waves race across the Pacific during an El Niño.

Although this physical model demonstrates many of the basics of an ENSO event, no physical model is capable of capturing the entire natural event because its scale is so enormous. The rotation of Earth plays a major role, and the nature of currents and waves as they move directly along the equator is also quite significant. Computer models can incorporate these effects, however, and developing such models is an active area of research.

Additional data regarding the ocean are also required. The Pacific Ocean between Indonesia and Peru has not been studied in enough detail to enable scientists to refine their models as much as they would like. The processes that occur during an ENSO event involve large areas and take place over a considerable period of time. Monitoring the sea to detect gradual changes in sea surface temperatures, depth of the thermocline, and surface wind velocity and direction is a complex, expensive project. As more data are acquired, understanding of ENSO events is certain to improve.

SIGNIFICANCE

The El Niño Southern Oscillation cycle is as much a part of Earth's weather and climate as the cycle of seasons. Just as people spend about one-fourth of the year in winter, people spend about one-fourth of their lives in an El Niño phase. Therefore, El Niño events are not truly anomalous. Unlike winter, however, the timing of the ENSO cycle is not regular. For example, from 1991 through 1994, three El Niños developed. Analysis of historic records shows that El Niños have occurred at nearly the same frequency since at least 1525, when record-keeping began in Peru. In addition to historic records, evidence for prehistoric ENSO events is found in tree rings, flood deposits, ice cores, and other geological and biological record keepers. These, too, suggest that the frequency and severity of ENSO events has been fairly constant for at least several thousand years.

The severity of El Niños can vary remarkably. Substantial effort has been made to identify predictors that would permit an early estimate of the occurrence and severity of ENSO events. Several promising predictors of the timing or intensity have emerged but have been abandoned as additional events transpired in which the predictor was present but the expected development did not occur.

Sir Gilbert Walker's suggestion that local weather in widely separated parts of the planet might be closely related was ridiculed by some of his contemporaries. However, his belief has been borne out by the data and understanding acquired subsequent to his early work. ENSO-related changes can also be seen in the Indian and Atlantic Oceans. Some droughts in Africa and India and most droughts in Australia seem connected to ENSO events. Additionally, weak floods on the Nile River, records of which have been kept since 622 CE, seem directly related to El Niño events.

Droughts and floods can probably never be prevented. However, many of the ensuing hardships could be avoided or reduced if they could be predicted. People and governments in affected areas could take preventive measures: Drought-resistant or flood-resistant varieties of crops could be planted, dikes could be reinforced, deforestation could be halted, and water could be temporarily impounded. Before these measures are taken, however, considerable confidence in any prediction is required. Preparing for a flood may worsen the effects of a drought. Someday, however, it may be possible to predict ENSO events with the same degree of accuracy with which meteorologists predict the arrival of a significant winter storm.

ENSO is also frequently discussed in the context of climate change. Climate change skeptics often claim that the existence of natural, unpredictable shifts in warming and cooling cycles can explain the trend of increasing average temperatures referred to as global warming, therefore denying the theory that humans have contributed to such warming (anthropogenic climate change). However, most scientists reject this belief. For example, 2015 set a new record for global average high temperatures, beating out the previous record set in 2014. Scientists around the world, including with National Aeronautics and Space Administration (NASA) and the US National Oceanic and Atmospheric Administration (NOAA), acknowledged that the El Niño experienced in 2015 (considered the strongest on record to that point) helped boost temperatures but determined that the year would have set the record even without the ENSO effect. Virtually all mainstream researchers agree that while ENSO can have an effect on local and even global temperatures, it is a relatively short-scale process, and climate change is responsible for the general trend of increasing temperatures. Scientists continue to study the connections between ENSO and global warming with many suggesting that the latter causes more extreme versions of the former. This holds critical implications for ecosystems around the world including humans.

—*Otto H. Muller*

Further Reading

Ahrens, C. Donald. *Meteorology Today: An Introduction to Weather, Climate and the Environment*. 8th ed., Thomson Brooks-Cole, 2007.

Brown, Joan. *Ocean Circulation: Prepared by an Open University Course Team*. Elsevier, 2016.

Clarke, Allan J. *An Introduction to the Dynamics of El Niño and the Southern Oscillation*. Academic Press, 2008.

Cropper, Thomas. "Did El Niño Drive the Record Heat of 2015?" *Niskanen Center*. Niskanen Center, 25 Jan. 2016.

Hatfield, Jerry L., et al., editors. *Impacts of El Niño and Climate Variability on Agriculture*. ASA Special Publication Number 63. American Society of Agronomy, 2001.

"NASA, NOAA Analyses Reveal Record-Shattering Global Warm Temperatures in 2015." *Goddard Institute for Space Studies*. NASA, 20 Jan. 2016.

Nash, J. Madeleine. *El Niño: Unlocking the Secrets of the Master Weather-Maker*. Warner Books, 2002.

Perry, Sarah J., Shayne McGregor, Alex Sen Gupta, and Matthew H. England. *Future Changes to El Niño-Southern Oscillation Temperature and Precipitation Teleconnections*. John Wiley & Sons, 2017.

Sarachik, Edward S., and Mark A. Cane. *The El Niño-Southern Oscillation Phenomenon*. Cambridge UP, 2010.

Wang, S.-Y. Simon, Jin-Ho Yoon, Christopher C. Funk, and Robert R. Gillies, editors. *Climate Extremes: Patterns and Mechanisms*. Wiley, 2017.

ENVIRONMENTAL CHEMISTRY

Fields of Study: environmental chemistry; marine chemistry; biogeochemistry; ecology

ABSTRACT

Environmental chemistry is the study of the chemical interactions that take place in a natural environment, how to control them when necessary, and what happens when unnatural chemicals are introduced into a natural environment. Air, water, and soil chemistry are all affected by materials injected into the corresponding environmental systems by way of human activity. All natural environments function according to the same chemical principles that apply in a chemistry laboratory. Regular testing and monitoring of environmental systems determines how those materials affect and are affected by environmental system processes.

KEY CONCEPTS

bioaccumulation: increasing amounts of toxic materials stored in body fats and organs progressing upwards in a food chain

chromatography: a means of physically separating components of a solution by using the differences in their abilities to stick to interact temporarily with a suitable solid phase while being eluted through it by a suitable fluid phase

greenhouse gas: an atmospheric gas capable of absorbing infrared radiation (heat) coming from Earth's surface and reemitting a significant portion of it back into the atmosphere rather than out into space

leachate: the solution of soluble materials that is produced as water percolates through a solid layer such as soil or landfill

microplastics: tiny fragments of plastics that find their way into the environment from normal human activities such as washing clothes made with synthetic polymer fibers

nanoparticulates: solid particles less than 1 micron in size, such as particles of lead oxide produced in the combustion of leaded gasoline and soot from burning wood or coal

THE CHEMICAL NATURE OF THE ENVIRONMENT

Modern chemical science is based on the quantum mechanical model of the atom, the underlying tenet being that all material in the universe is matter composed of atoms. Accordingly, anything and everything in the physical universe is chemical in nature. Matter interacts with matter and energy, as the atoms of matter interact with each other and with energy. The quantum mechanical model of the atom requires that the atoms interact in specific ways, however, and according to strict mathematical rules that are neither entirely known nor entirely understood.

The quantum mechanical model describes atoms of matter as consisting of a small, dense nucleus. This nucleus contains almost the entire mass of the atom in a specific number of discrete particles called "protons" and "neutrons." The protons each bear a single positive charge, while the electrically neutral neutrons fulfill the role of a sort of "nuclear glue," countering the mutual repulsion of the protonic charges. For the atom itself to be electrically neutral, it must also contain a number of negatively charged electrons equal to the number of protons. The electrons occupy a space around the nucleus that is about 100,000 times larger in diameter than the nucleus.

According to the mathematical rules of quantum mechanics, electrons are allowed only to have specific energies depending on their position in the space around the nucleus; each specific energy-position combination is referred to as an orbital. The lowest-energy electrons, and hence the most tightly bound electrons, are those situated closest to the nucleus, while any other electrons in the atom range outward in order of energy; therefore, the outermost electrons are the most energetic and the least tightly bound by the nucleus. Interaction between atoms takes place only at the level of the outermost electrons of the respective atoms. Within that restriction, any interaction that can take place will take place. This means that atoms of different elements can combine in an infinite number of ways, producing stable combinations of atoms with specific spatial arrangements called "molecules." Electrons in molecules also must adhere to stringent energy levels.

The interaction of matter, as atoms, with energy is somewhat more complicated, but must nevertheless obey the same rule system of quantum mechanics as applies to atomic structure. The most common interaction of atoms occurs through the direct action of light energy on the electrons within the atoms and molecules. Absorption of light energy by an electron in one atomic or molecular orbital raises the level of the orbital that it must occupy. This is often sufficient to break the molecule apart and, in so doing, drive uncontrolled reactions that are usually undesirable. The fading of color and the embrittlement of plastics left in bright sunlight are examples of matter interacting with light energy. In other cases, the interaction of light energy with specific molecules also may be highly desirable, as is the case when chlorophyll is activated by sunlight to produce glucose and oxygen from carbon dioxide and water.

THE PHYSICAL INTERFACE

Substances exist normally in one of three states: solid, liquid, or gas. Earth is an agglomeration of substances that are found in any and all of these states. As determined by ancient philosophy, the four basic elemental forms of water, air, earth, and fire represent the three basic states of matter plus energy, all of which are the material medium of environmental chemistry. The physical environment of Earth can thus also be considered in the context of physical interaction between matter in all of these states, and in the context of chemical interactions.

It is an unavoidable reality that the physical and chemical contexts affect each other directly in various ways. The rates of chemical reactions, for example, are affected by the extent to which different chemical materials come into contact with each other at a physical interface, such that the larger the interface area, the faster an overall chemical reaction occurs. Similarly, the more intimately mixed two gases become, such as ozone and fluorocarbon compounds, the more rapidly and completely they react.

The concept of the physical interface between materials is complex. In the case of gases, no definitive interfacial surface can be defined as existing between two masses. Diffusion occurs where the two masses come into contact; the only real definable interface for interaction between them is the "surface" of the individual atoms or molecules as they interact. At the other end of the interface surface spectrum are the massive interfaces between air and water, air and land surface, and water and the solid earth. In the overall structure of Earth, one could also include the more diffuse divisions between the crust and the underlying mantle of the planet. As the source of volcanic and seismic activity, these also have significant roles in the environmental chemistry of the planet.

ENVIRONMENTAL CHEMISTRY AND HUMAN ACTIVITIES

While there is nothing humans can do to control or even affect the natural geochemical processes associated with volcanic or seismic activity, the practical field of environmental chemistry focuses much more closely on those aspects of human society that enact change gradually. It has become increasingly apparent that human activities since the Industrial Revolution have had a significant impact on the natural workings of the planetary environment, especially with regard to air and water quality. Agriculture, industry, and transportation are the components of human activity that affect the environment most significantly. In all cases, the principal means of these changes is the injection of waste and unnatural materials into the environment.

Since the early nineteenth century, the main gaseous component has been carbon dioxide gas from the combustion of fossil fuels. This, in and of itself, represents a major detriment to the livability of Earth for different reasons. Most notably, carbon dioxide is a "greenhouse gas" that contributes to the retention of heat in the atmosphere in such a way that the more of it there is, the warmer the atmosphere becomes. The effect has no limit, and at a certain level it becomes a "runaway" effect, driving the average temperature of the planet ever higher. This is what is believed to have happened on the planet Venus, where the skies are filled with clouds of sulfuric acid, the ground-level temperature is hot enough to melt lead, and the rocks glow red with heat. Here on Earth, carbon dioxide can also dissolve in rain water to produce a weak solution of carbonic acid, a mild acid but nonetheless strong enough to have eroded limestone rock over centuries to produce karst landscapes above ground and vast caverns below ground.

Additionally, significant environmental effects have been observed as a result of other materials produced through combustion, unnatural quantities of naturally occurring materials (such as methane), and the indiscriminate use and release of chlorofluorocarbon (CFC) compounds as propellants and refrigerants. Unnatural materials and unnatural quantities of normal materials can now be found in the environment everywhere in the world because of the release of liquids and dissolved materials into the environment—as effluents from production and processing facilities, mining operations, industrial and transportation accidents, warfare, and many other human activities.

Solid materials produced as a result of human activities are also problematic with regard to environmental chemistry. Not the least of these materials is the mass quantity of solids placed in landfills every year. Foremost of these are the billions of tons (or tonnes) of plastics produced and discarded each year. In the beginning years of the twentieth century some 881,849 tons (800,000 tonnes) of discarded plastics find their way into the oceans of the world every day, and have given rise to floating "islands" of plastic items gathered in by ocean gyres. Plastic items floating in the oceans pose a significant risk to aquatic lifeforms that identify the items with food, and die as a result of developing an internal blockage composed of indigestible material. Fragments of plastic, called "microplastics," have been found contaminating the environment globally, from the Arctic to the Antarctic, by the year 2020, with environmental effects that no one can predict. Even more insidious environmental effects have been wrought through the nanoparticulate matter deposited in the environment by combustion fuels—namely, the millions of tons (or tonnes) of lead and manganese oxides that have been produced by internal combustion engines.

Human energy use and production also plays a role in environmental chemistry. Excess city light and heat have affected air and water quality, while the production of electrical energy through nuclear processes has led to an entirely separate environmental chemistry problem of its own: the management of waste and spent materials and the release of radioactive materials in quantity into the environment.

ENVIRONMENTAL CHEMISTRY OF THE ATMOSPHERE

The potential of atmospheric carbon dioxide has been mentioned above, but its effects are not readily visible. Perhaps the most visible and recognizable of atmospheric chemical effects is the generation of smog: an accumulation of noxious gases, smoke, and nanoparticulate matter that forms throughout cities. Gas-phase chemical reactions occurring between components of smog and normal atmospheric gases are known to contribute to an assortment of undesirable effects, including acid rain and asthma.

Nitrogen oxides and sulfur dioxide react with water and oxygen to produce nitrous, nitric, sulfurous, and sulfuric acids, all of which become dissolved in raindrops as water vapor condenses in the atmosphere. The resulting acidic solutions, often appearing long distances from their sources, react chemically with carbonate stone such as limestone, marble, and concrete. This results in the slow chemical erosion of infrastructure, directly bringing about destruction and occasioning structural upkeep that costs billions of dollars.

Acidic solutions also dissolve solid materials on the surface, carrying them along as they percolate through soils. These materials are then carried into waterways and into other sources of drinking water. In more extreme or prolonged cases, acidic precipitation has been identified as the agent responsible for the ecological deaths of otherwise pristine bodies of water, rendering them unable to support an aquatic ecology.

Other causes of smog include chlorine, bromine, and ground-level ozone, all of which react with oxygen and with each other in a multitude of ways by radical reaction mechanisms. These interactions produce various oxyacids and other chemical compounds that have inimical environmental effects. Breathing air that contains these reactive materials causes damage to lung tissue and can lead to emphysema, asthma, and other respiratory difficulties. In some cities around the world, smog has become a serious health issue, requiring constant monitoring of air quality. Government environmental regulatory agencies now issue daily statements of air quality as a regular feature of official meteorological reports.

While ground-level ozone is considered a health hazard, upper atmosphere ozone is essential to the survival of life on Earth. The diffuse layer of ozone at very high altitudes acts as a shield for the planet against the continuous influx of ultraviolet radiation from the Sun and stars. Absorption of the energy of ultraviolet light by the ozone molecule results in the formation of an electronically excited species that re-emits that energy at a different wavelength that is harmless with regard to living tissue.

In the twentieth century, the chemical compounds known as CFCs (chlorofluorocarbons) was commonly used as a propellant and refrigerant. The low boiling point and ready compressibility in the gaseous state of CFCs made them ideal for that application, but at the same time also made them easy to dispose of as they dissipated readily into the atmosphere. Ignorance of their environmental effects allowed them to be used for many years, until satellite-based monitoring of the upper atmosphere revealed that the accumulation of CFCs there had seriously depleted the ozone shield by way of chemical reactions that produced oxygen dichloride and other exotic chemical species that do not interact with ultraviolet radiation. A global ban on their production and use was subsequently agreed to, but in the second decade of the twenty-first century it was discovered that the quantities of the materials detectable in the atmosphere were once more on the rise.

Air quality and gaseous effluent monitoring is carried out through a number of methods. Direct sampling for instrumental analysis, using tandem gas chromatography-mass spectrometry (GC-MS), can detect airborne materials in the range of parts per billion. This is especially important when the materials of interest are highly carcinogenic materials (such as dioxins) that do not exist naturally in the environment. Other monitoring methods include ground-based and satellite-based infrared detection of composition and temperature and the use of monitoring tags for exposure to specific chemical compounds. Such tags use the occurrence of a specific chemical reaction to bring about a color change in the surface of the tag.

ENVIRONMENTAL CHEMISTRY OF WATER

Airborne materials can become dissolved in condensate water to produce acidic precipitation. Both acidic and nonacidic precipitation, including agricultural irrigation, dissolve materials in the soil and carry that material into the nearest waterways or into the water table as the liquid percolates through the soil. The leachates so formed can include every compound and element in the soil. For naturally occurring materials in their normal natural proportions, this is not problematic. However, the injection into the environment, through human activities, of many materials makes it essential that the chemical effects of those materials in the environment be monitored and understood, ideally before they are used.

For example, the use of tetraethyl lead in gasoline over a period of fifty years (before its use in North America was banned) had deposited some 19 billion pounds (9 billion kilograms) of lead (imagine a solid block of lead metal 100 yards (93 meters) in length, width, and height) into the environment as nanoparticulate lead oxides from vehicle exhaust gases. This resulted in a drastic increase in the amount of lead detectable in drinking water. In Europe, where estate wineries keep meticulous records of production, the amount of lead detectable in wine could be traced through analysis of vintage wine samples to the exact year that the use of leaded gasoline began in that region. In North America, the uptick in lead accumulation coincided with an increase in the levels of birth defects and disorders, and in mental health issues in children.

Lead is a known neurotoxin, as is the manganese that was used to replace the lead in gasoline fuels. Another neurotoxin is mercury, a common waterborne effluent of the mining and pulp and paper industries. Mercury has the insidious property of accumulating in the fatty tissue of organisms, as do many other heavy metals such as cadmium. The process of bioaccumulation works to concentrate those materials in organisms as it progresses up the food chain. By this process, creatures that consume smaller creatures as food, such as fish, may contain quantities of bioactive heavy metals that greatly exceed the baseline amount of those materials in the environment. Chemical compounds such as pesticides (e.g., dichloro-diphenyl-trichloroethane, or DDT), industrial chemicals (such as polychlorinated biphenyls, or PCBs) and carcinogenic chemicals (such as dioxin) also can be concentrated by bioaccumulation.

In agriculture, both organic and mineral fertilizers are used to augment the nutritive value of soils to increase or maintain crop yields. The primary components of mineral fertilizers, and to a lesser extent the organic fertilizers, are nitrogen (as nitrate) and phosphorus (as phosphate). Without this addition, mineral nutrients essential for plant growth in soils can become depleted as successive crops remove them faster than natural processes can replenish them.

Irrigation also works toward the depletion of minerals and other nutrients from the soil by dissolving and carrying them away as the water percolates down to the water table. Coupled with the overuse of inorganic fertilizers, this can and has led to elevated amounts of nitrates and phosphates in waterways and water tables. The presence of these extra nutrients contributes to the excess growth of algae and aquatic plants, and to the eutrophication of bodies of water. To an extent, this problem is alleviated by the practice of crop rotation, in which different crops are grown on the same plots of land in successive years.

Typically, a crop that draws nutrients from the soil in one year will be succeeded in the following year by a crop that incorporates nutrients naturally by nitrogen fixation or deep-root mineral extraction. For example, corn is often rotated with beans on heavy soils with a high clay component, and tobacco is generally rotated with rye grain on light, sandy soils. Additionally, the likelihood of a specific crop disease such as blue mold and tomato blight increases when the same crop is grown on the same fields year after year without rotation through other crops.

WATER TESTING

Water testing is carried out both regularly and upon request using a much broader variety of techniques. Because materials are dissolved in water, numerous instrumental analytical techniques are available to measure contents and properties of samples brought in for testing.

Spectrophotometric methods detect specific ions and organic compounds both qualitatively and quantitatively. Samples are prepared for analysis in different ways. In some cases, specific compounds are added that combine with dissolved ions or organic compounds to form specific complexes. Such complexes are often colored, whereas the dissolved ions and organic compounds are not. Colorimetric analysis can then be used to determine the concentration of the complexed material in the prepared sample, and hence the original concentration of the ions and organic compounds of interest.

Spectrophotometric analysis can be used to determine the presence and concentration of specific materials directly. In each case, analysis depends on the measurement of the absorption of light at characteristic wavelengths that are uniquely associated with the specific materials for which testing is being carried out. The Beer-Lambert law describes the relationship between the absorption of light energy and the concentration of the material that is absorbing the light.

Methods such as high-performance liquid chromatography (HPLC) and tandem GC-MS can determine concentrations as low as only a few parts per billion. Chromatographic methods are extremely versatile. They are based on the equilibrium between the differential adsorption of dissolved materials on a solid surface and the solubility of those materials in a fluid phase (either gas or liquid).

In chromatographic separations, the fluid phase containing the dissolved materials passes through a second, solid phase. Components of the gas or liquid solution equilibrate between being dissolved in the fluid medium and being adsorbed on the solid medium. As they progress through the column of the solid medium, they become increasingly separated from each other. Effective separation brings each component of the solution out of the column at different times.

When chromatographic methods such as gas chromatography or HPLC are connected in tandem with a second procedure, such as mass spectrometry, it becomes possible to identify and measure the components of complex mixtures such as environmental water samples in a single step. In tandem processes the efflux from the first stage becomes the input material for the second stage without any intermediate isolation or purification step. In some cases, such as the analysis of organic compounds in aqueous solutions, it is expedient to extract the organic compounds by partitioning to eliminate physical interference by water in the analysis.

As a general rule, the presence of water makes organic analysis very difficult simply because of the physical properties of that material. Organic solvents, being much more volatile than water, are also much more easily removed than water and so make analysis much easier.

—Richard M. Renneboog

Further Reading

Ahluwalia, V. K. *Advanced Environmental Chemistry.* TERI, 2017.

Ayers, Jonathan. *Introduction to Environmental Chemistry.* Callisto Reference, 2019.

Baird, Colin, and Michael Cann. *Environmental Chemistry.* W. H. Freeman, 2008.

Beard, James M. *Environmental Chemistry in Society.* Taylor & Frances, 2017.

Crawford, Christopher Blair, and Brian Quinn. *Microplastic Pollutants.* Elsevier, 2016.

Girard, James E. *Principles of Environmental Chemistry.* 2nd ed., Jones and Bartlett, 2010.

Karapanagioti, Hrissi K., and Ioannis K. Kalavrouziotis. *Microplastics in Water and Wastewater.* IWA Publishing, 2019.

Sposito, Garrison, editor. *The Environmental Chemistry of Aluminum.* 2nd ed., CRC Press, 2020.

Trimm, Harold H., and William Hunter, III. *Environmental Chemistry: New Techniques and Data.* Apple Academic Press, 2011.

United Nations Environmental Programme (UNEP). *Marine Plastic Debris and Microplastics: Global Lessons and Research to Inspire Action and Guide Policy Change.* United Nations, 2016.

Van Loon, Gary W., and Stephen J. Duffy. *Environmental Chemistry: A Global Perspective.* Oxford UP, 2010.
Weiner, Eugene. *Applications of Environmental Aquatic Chemistry: A Practical Guide.* 2nd ed., CRC Press, 2008.

ENVIRONMENTAL ENGINEERING

Fields of Study: civil engineering; marine biology

ABSTRACT

Environmental engineering is a field of engineering involving the planning, design, construction, and operation of equipment, systems, and structures to protect and enhance the environment. Major areas of application within the field of environmental engineering are wastewater treatment, water-pollution control, water treatment, air-pollution control, solid-waste management, and hazardous-waste management. Water-pollution control deals with physical, chemical, biological, radioactive, and thermal contaminants. Water treatment may be for the drinking water supply or for industrial water use. Air-pollution control is needed for stationary and moving sources. The management of solid and hazardous wastes includes landfill and incinerators for disposal of solid waste and identification and management of hazardous wastes.

KEY CONCEPTS

aerobic digestion: breaking down of organic materials by bacteria that require the presence of oxygen to function

anaerobic digestion: breaking down of organic materials by bacteria that do not function if oxygen is present or that do not require oxygen to function

stoichiometry: determination of the proportions of the materials with which complete chemical reactions take place

wastewater: all of the liquid waste, including the solid waste within it, that is produced by a community

DEFINITION AND BASIC PRINCIPLES

Environmental engineering is a field of engineering that split off from civil engineering as the importance of the treatment of drinking water and wastewater was recognized. This field of engineering was first known as sanitary engineering and dealt almost exclusively with the treatment of water and wastewater. As awareness of other environmental concerns and the need to do something about them grew, this field of engineering became known as environmental engineering, with the expanded scope of dealing with air pollution, solid wastes, and hazardous wastes, in addition to water and wastewater treatment.

While environmental engineering is primarily concerned with the terrestrial uses of water, it is important to note that how water is used on land affects the marine environment as well. The discharge of various effluents into rivers, for example, eventually results in the transport of contaminants into larger bodies of water and ultimately into the oceans. In the oceans, those contaminants affect the environment in a multitude of ways. Transported materials ranging from dissolved fertilizers to discarded plastics and other materials end up where they were never intended to be in nature. In a very few cases, transported solids may fall to the seafloor and provide a beneficial structure for aquatic creatures, but in most cases they simply float about and become hazardous to fish and other marine creatures. Dissolved materials such as fertilizers and biosolids may also contribute to the formation of algal blooms that deplete the oxygen from the water or be themselves toxic to marine organisms. Marine scientists study and monitor the effects of terrestrial pollutants on the marine environment, and play a very important role in calling attention to situations that require physical amendment and policy changes by governments.

Environmental engineering is an interdisciplinary field that makes use of principles of chemistry, biology, mathematics, and physics, along with engineering sciences (such as soil mechanics, fluid mechanics, and hydrology) and empirical engineering correlations and knowledge to plan for, design, construct, maintain, and operate facilities for treatment of liquid and gaseous waste streams, for prevention of air pollution, and for management of solid and hazardous wastes.

The field also includes investigation of sites with contaminated soil and/or groundwater and the planning and design of remediation strategies. Environmental engineers also provide environmental impact analyses, in which they assess how a proposed project will affect the environment.

BACKGROUND AND HISTORY

When environmental engineering, once a branch of civil engineering, first became a separate field in the mid-1800s, it was known as sanitary engineering. Initially, the field involved the water supply, water treatment, and wastewater collection and treatment.

In the middle of the twentieth century, people began to become concerned about environmental quality issues such as water and air pollution. As a consequence, the field of sanitary engineering began to change to environmental engineering, expanding its scope to include air pollution, solid- and hazardous-waste management, and industrial hygiene.

Several pieces of legislation have affected and helped define the work of environmental engineers. Some of the major laws include the Clean Air Act of 1970, the Safe Drinking Water Act of 1974, the Toxic Substances Control Act of 1976, the Resource Recovery and Conservation Act (RCRA) of 1976, and the Clean Water Act of 1977.

HOW IT WORKS

Environmental engineering uses chemical, physical, and biological processes for the treatment of water, wastewater, and air, as well as in-site remediation processes. Therefore, knowledge of the basic sciences—chemistry, biology, and physics—is important along with knowledge of engineering sciences and applied engineering.

Chemistry. Chemical processes are used to treat water and wastewater, to control air pollution, and for site remediation. These chemical treatments include chlorination for disinfection of both water and wastewater, chemical oxidation for iron and manganese removal in water-treatment plants, chemical oxidation for odor control, chemical precipitation for removal of metals or phosphorus from wastewater, water softening by the lime-soda process, and chemical neutralization for pH (acidity) control and for scaling control.

The chemistry principles and knowledge that are needed for these treatment processes include the ability to understand and work with chemical equations, to make stoichiometric calculations for dosages, and to determine size and configuration requirements for chemical reactors to carry out the various processes.

Biology. The major biological treatment processes used in wastewater treatment are biological oxidation of dissolved and fine suspended organic matter in wastewater (secondary treatment) and stabilization of biological wastewater biosolids (sludge) by anaerobic digestion or aerobic digestion.

Biological principles and knowledge that are useful in designing and operating biological wastewater treatment and biosolids digestion processes include the kinetics of the biological reactions and knowledge of the environmental conditions required for the microorganisms. The required environmental conditions include the presence or absence of oxygen and the appropriate temperature and pH.

Physics. Physical treatment processes used in environmental engineering include screening, grinding, comminuting, mixing, flow equalization, flocculation, sedimentation, flotation, and granular filtration. These processes are used to remove materials that can be screened, settled, or filtered out of water or wastewater and to assist in managing some of the processes. Many of these physical treatment processes are designed on the basis of empirical loading factors, although some use theoretical relationships such as the use of estimated particle settling velocities for design of sedimentation equipment.

Soil Mechanics. Topics covered in soil mechanics include the physical properties of soil, the distribution of stress within the soil, soil compaction, and water flow through soil. Knowledge of soil mechanics is used by environmental engineers in connection with design and operation of sanitary landfills for solid waste, in storm water management, and in the investigation and remediation of contaminated soil and groundwater.

Fluid Mechanics. Principles of fluid mechanics are used by environmental engineers in connection with the transport of water and wastewater through pipes and open channels. Such transport takes place in water distribution systems, in sanitary sewer collection systems, in storm water sewers, and in wastewater-treatment and water-treatment plants. Design and sizing of the pipes and open channels make use of empirical relationships such as the Manning equation for open channel flow and the Darcy-Weisbach equation for frictional head loss in pipe flow. Environmental engineers also design and select pumps and flow measuring devices.

Hydrology. The principles of hydrology (the science of water) are used to determine flow rates for storm water management when designing storm sewers or storm water detention or retention facilities. Knowledge of hydrology is also helpful in planning and developing surface water or groundwater as sources of water, and in protecting those sources from contamination.

Practical Knowledge. Environmental engineers make use of accumulated knowledge from their work in the field. Theoretical equations, empirical equations, graphs, nomographs, guidelines, and rules of thumb have been developed based on experience. Empirical loading factors are used to size and design many treatment processes for water and wastewater. For example, the design of rapid sand filters to treat drinking water was based on a specified loading rate in gallons per minute of water per square foot of sand filter. Also, the size required for a rotating biological contactor to provide secondary treatment of wastewater was determined based on a loading rate in pounds of biochemical oxygen demand (BOD) per day per 1,000 square feet (93 square meters) of contactor area.

Engineering Tools. Tools such as engineering graphics, computer-aided drafting (CAD), geographic information systems (GIS), and surveying are available for use by environmental engineers. These tools are used for working with plans and drawings and for laying out treatment facilities or landfills.

Codes and Design Criteria. Much environmental engineering work makes use of codes or design criteria specified by local, state, or federal government agencies. Examples of such design criteria are the storm return period to be used in designing storm sewers or storm water detention facilities and the loading factor for rapid sand filters. Design and operation of treatment facilities for water and wastewater are also based on mandated requirements for the finished water or the treated effluent.

APPLICATIONS AND PRODUCTS

Environmental engineers design, build, operate, and maintain treatment facilities and equipment for the treatment of drinking water and wastewater, air-pollution control, and the management of solid and hazardous wastes.

Air-Pollution Control. Increasing air pollution from industries and power plants as well as automobiles led to passage of the Clean Air Act of 1970. This law led to greater efforts to control air pollution. The two major ways to control air pollution are the treatment of emissions from fixed sources and from moving sources (primarily automobiles).

The fixed sources of air pollution are mainly the smokestacks of industrial facilities and power plants. Devices used to reduce the number of particulates emitted include settling chambers, baghouses, cyclones, wet scrubbers, and electrostatic precipitators. Electrostatic precipitators impart an electric charge the particles to aid in their removal. They are often used in power plants, at least in part because of the readily available electric power to run them. Water-soluble gaseous pollutants can be removed by wet scrubbers. Other options for gaseous pollutants are adsorption on activated carbon or incineration of combustible pollutants. Because sulfur is contained in the coal used as fuel, coal-fired power plants produce sulfur oxides, particularly troublesome pollutants. The main options for reducing these sulfur oxides are desulfurizing the coal or desulfurizing the flue gas, most typically with a wet scrubber using lime to precipitate the sulfur oxides.

Legislation has greatly reduced the amount of automobile emissions, the main moving source of air pollution. The reduction in emissions has been accomplished through catalytic converters to treat exhaust gases and improvements in the efficiency of automobile engines.

Water Treatment. The two main sources for the water supply are surface water (river, lake, or reservoir) and groundwater. The treatment requirements for these two sources are somewhat different.

For surface water, treatment is aimed primarily at removal of turbidity (fine suspended matter) and perhaps softening the water. The typical treatment processes for removal of turbidity involve the addition of chemicals such as alum or ferric chloride. The chemicals are rapidly mixed into the water so that they react with alkalinity in the water, then slowly mixed to form a precipitate that is able to settle out from the water (flocculation). After sedimentation, the water passes through a sand filter and finally is disinfected with chlorine. If the water is to be softened as part of the treatment, lime ($Ca(OH)_2$) and soda ash (Na_2CO_3)—the lime-soda process—are used in place of alum or ferric chloride, and the water hardness due calcium and magnesium ions is removed along with its turbidity.

Groundwater is typically not turbid (cloudy), so it does not require the type of treatment used for surface water. At minimum, it requires disinfection. Removal of iron and manganese by aeration may be needed, and if the water is very hard, it may be softened by ion exchange.

Wastewater Treatment. The Clean Water Act of 1977 brought wastewater treatment to a new level by requiring that all wastewater discharged from municipal treatment plants must first undergo at least secondary treatment. Before the passage of the legislation, many large cities located on a river or along the ocean provided only primary treatment in their wastewater-treatment plants and discharged effluent with only settleable solids removed. All dissolved and fine suspended organic matter remained in the effluent. Upgrading treatment plants involved added a biological treatment to remove dissolved and fine suspended organic matter that would otherwise exert an oxygen demand on the receiving stream, perhaps depleting the oxygen enough to cause problems for fish and other aquatic life.

Solid-Waste Management. The main options for solid-waste management are incineration, which reduces the volume for disposal to that of the ash that is produced, and disposal in a sanitary landfill. Some efforts have been made to reuse and recycle materials to reduce the amount of waste sent to incinerators or landfills. A sanitary landfill is a big improvement over the traditional garbage dump, which was simply an open dumping ground. A sanitary landfill uses liners to prevent groundwater contamination, and each day, the solid waste is covered with soil.

Hazardous-Waste Management. The Resource Conservation and Recovery Act (RCRA) of 1976 provides the framework for regulating hazardous-waste handling and disposal in the United States. One very useful component of RCRA is that it specifies a very clear and organized procedure for determining if a particular material is a hazardous waste and therefore subject to RCRA regulations. If the material of interest is indeed a waste, then it is defined to be a hazardous waste if it appears on one of RCRA's lists of hazardous wastes, if it contains one or more hazardous chemicals that appear on an RCRA list, or if it has one or more of the four RCRA hazardous waste characteristics as defined by laboratory tests. The four RCRA hazardous waste characteristics are flammability, reactivity, corrosivity, and toxicity. The RCRA regulations set standards for secure landfills and treatment processes for disposal of hazardous waste.

Much work has been done in investigating and cleaning up sites that have been contaminated by hazardous wastes in the past. In some cases, funding is available for cleanup of such sites through the Comprehensive Environmental Response, Compensation, and Liability Act of 1980 (known as CERCLA or Superfund) or its amendment, the Superfund Amendments and Reauthorization Act (SARA) of 1986.

CAREERS AND COURSEWORK

An entry-level environmental engineering position can be obtained with a bachelor's degree in environmental engineering or in civil or chemical engineering with an environmental specialization. However, because many positions require registration as an engineer in training or as a professional engineer, it is important that the bachelor's degree program is accredited by the Accreditation Board for Engineering and Technology (ABET). Students must first graduate from an accredited program before taking the exam to become a registered engineer in training. After four years of experience, the engineer in training can take another exam for registration as a professional engineer.

A typical program of study for an environmental engineering degree at the undergraduate level includes the chemistry, calculus-based physics, and mathematics that is typical of almost all engineering programs in the first two years of study. It also may include biology, additional chemistry, and engineering geology. The last two years of study will typically include hydrology, soil mechanics, an introductory course in environmental engineering, and courses in specialized areas of environmental engineering such as water treatment, wastewater treatment, air-pollution control, and solid- and hazardous-waste management.

Master's degree programs in environmental engineering fall into two categories: those designed primarily for people with an undergraduate degree in environmental engineering and those for people with an undergraduate degree in another type of engineering. Some environmental engineering positions require a master's degree. A doctoral degree in environmental engineering is necessary for a position in research or teaching at a college or university.

SOCIAL CONTEXT AND FUTURE PROSPECTS

Many major areas of concern in the United States and around the world are related to the environment. Issues such as water-pollution control, air-pollution control, global warming, and climate change all need the work of environmental engineers. These issues, as well as the need for environmental engineers, are likely to remain concerns for much of the twenty-first century.

Water supply, wastewater treatment, and solid-waste management all involve infrastructure, needing repair, maintenance, and upgrading, which are all likely to need the help of environmental engineers.

—Harlan H. Bengtson

Further Reading

Anderson, William C. "A History of Environmental Engineering in the United States." *Environmental and Water Resources History*, edited by Jerry R. Rogers and Augustine J. Fredrich, American Society of Civil Engineers, 2003.

Boulding, Russell J., editor. *EPA Environmental Engineering Sourcebook*. Routledge, 2019.

Cooper, C. David. *Introduction to Environmental Engineering*. Waveland Press, 2015.

Davis, Mackenzie L., and Susan J. Masten. *Principles of Environmental Engineering and Science*. 3rd ed., McGraw, 2013.

Juuti, Petri S., Tapio S. Katko, and Heikki Vuorinen. *Environmental History of Water: Global Views on Community Water Supply and Sanitation*. IWA, 2007.

Leonard, Kathleen M. "Brief History of Environmental Engineering: 'The World's Second Oldest Profession.'" *ASCE Conference Proceedings*, vol. 265, no. 47, 2001, pp. 389–93.

Mines Jr., Richard O. *Environmental Engineering Principles and Practices*. Wiley Blackwell, 2014.

Spellman, Frank R., and Nancy E. Whiting. *Environmental Science and Technology: Concepts and Applications*. 2nd ed., Government Institutes, 2006.

EROSION CONTROL

Fields of Study: ecology; marine biology

ABSTRACT

Preventing or eliminating the erosion of soil and rock protects water and air quality and the integrity and usability of public and private lands. As a result, erosion control requires both engineering and land-use management techniques. Erosion control is often seen as primarily an agricultural problem because the land area affected by farming practices is large. However, erosion control is also necessary in coastal regions and areas used for forestry, transportation, development, and recreational purposes because natural or human actions can change landscapes or soil covers and result in increased erosion.

KEY CONCEPTS

erosion: the movement or displacement of soil and larger particles by the action of a fluid in motion, such as by wind or water in motion

gabion: a large, usually rectangular, metal cage filled with rocks, used to construct erosion control structures

infiltration capacity: the ability of a soil to absorb water rather than allow it to flow away as surface runoff

saltation: the motion of particles under the influence of wind or water, by which individual particles bump into particles that lie ahead of them and essentially "leapfrog" over them

DEFINITION AND BASIC PRINCIPLES

Erosion control is the practice of preventing the movement of soil or rock by the action of wind or water. Uncontrolled erosion by either natural or human actions can cause water or air pollution or damage to property. Erosion is a natural result of the action of water or wind; however, human activity can accelerate this process by removing protective vegetation or creating instabilities in existing soil and rock structures.

Wind erosion acts by selectively transporting soil particles. In other words, the higher the wind velocity, the larger the soil particle that can be transported. Water erosion works both as precipitation hits the ground surface and as water flows over the land. Erosion-control techniques and structures work to reduce the potential for soil or rock transport either by reducing the exposure of land surfaces to the effects of wind and water or by modifying the landscape or installing structures that increase the stability of the landform. For example, mulching, revegetation, and the application of geotextiles are all approaches that work to reduce the vulnerability of soil surfaces to erosion. Gabions, retaining walls, and terracing are examples of structures or modifications to landforms.

Lack of erosion control can result in declines in soil productivity for agricultural purposes and loss of land stability for land development and uses. Examples of the large-scale effects of erosion include landslides and desertification.

BACKGROUND AND HISTORY

Before 1920, erosion was noted during farming practices but not necessarily considered reparable, and eroded lands were abandoned in favor of new fields. This mindset had its roots in the colonial days of the United States, where potential new farmland stretched to the horizon and was available to those willing to clear the land. Decades of intensive farming led to the loss of soil productivity and then the soil itself to water and wind erosion, with dramatic examples of wind erosion occurring during the Dust Bowl era of the 1930s.

In 1935, the US Congress established the Soil Conservation Service (now the Natural Resources Conservation Service) to provide for ongoing work to conserve the nation's soils. This act established soil conservation as a national priority independent of agricultural programs, which had far-reaching advantages borne out by more recent concerns about erosion related to stormwater runoff, coastal wave and storm action, and river movement.

Erosion-control practices are required in many different circumstances. Agricultural practices for animal husbandry or to produce food crops can have significant adverse impacts on soil health if soil conservation techniques are not applied. Land development for residential housing or commercial and industrial facilities can have short-term erosion impacts during construction or long-term effects from rainwater or snowmelt on the sites. Recreational uses near lakes, streams, or oceans can affect the stability of the shoreline. Transportation systems may create unstable slopes with increased potential for landslides from cuts through hills for roads and rail lines. Failure of such unstable structures due to gravity-driven collapse can potentially plunge large quantities of rock, earth and above-ground structures into any water system that may be below, and where this occurs the marine ecosystem below may be severely damaged or even destroyed.

HOW IT WORKS

Agriculture. Erosion-control practices in agriculture focus primarily on nonstructural techniques to retain and improve soil productivity. The intent of erosion-control measures is not only to protect the soil from raindrop impacts or wind transport but also to increase the infiltration capacity of the soil to lower the amount of water that runs off over land. One way that infiltration capacity is increased is to reduce the length of the slope that is available for water to travel by using terraces or contour plowing. Another way to increase infiltration is to slow the movement of water and physically protect the soil surface with mulch or vegetation, practices that also increase soil fertility by increasing organic content and biological activity.

Coastal Zone Management. Wave action and storm surges in coastal environments cause beach and headland erosion that is typically controlled through revegetation of coastal dunes and barrier islands or hardened structures like riprap, gabions, and retaining walls. An understanding of beach dynamics is needed, particularly of the seasonal transport of sand on- and offshore or the transverse migration of sediments along shorelines. Hard structural erosion-control features must be evaluated and designed to ensure that erosion is not increased elsewhere as an unintended consequence.

Transportation. The development and maintenance of transportation systems can increase erosion potential in landscape by altering the stability of landforms through either physically changing the landscape or by increasing water runoff. For example, road cuts can increase the potential for erosion and mass wasting by creating new steep slopes in previously stable hillsides. Unless control measures, such as gabions, geotextiles, or terraces, are used, erosional forces will work to reduce the steep slope to a more stable form. This erosive action can take the form of catastrophic landslides. Increased water runoff is also generated by road systems because of the impermeable nature of most road surfaces. The runoff, if not addressed, can increase erosion of road beds or nearby stream drainages, which can increase road-maintenance requirements or property damage.

Stormwater Management. Residential, commercial, industrial, and public-facility uses generate erosion-control needs during both the development phase and long-term occupation of the site. Site preparation includes activities such as tree and vegetation clearing and grading that expose soils and subsoils to erosion and alter the topography of the site. Erosion-control features during construction include phasing the work to minimize the amount of land exposed at any particular time, rapid revegetation of the site, and temporary approaches such as mulching.

Long-term land uses can also increase erosion potential if rainfall or snowmelt is not adequately accommodated on the site. In a similar manner to road systems, the impermeable surfaces created by roofs, driveways, parking areas, and walkways means that overland flow is increased and the infiltrative capacity of the site is reduced. Increased off-site erosion can result if excess overland flow is discharged to adjacent properties or the road system.

Recreation Management. The effects of recreational activities on landscapes may be dealt with through other disciplines and programs mentioned above. Effects of recreation often include soil compaction and vegetation removal through foot or vehicle travel. When the effects are limited in area, the overall effect on erosion potential can be slight. However, when the effects are widespread or occur on steep slopes or near bodies of water, the erosion potential can be high. Mitigating approaches may be to limit the area of travel, for example, designated trails or sites for activities, which will allow revegetation to occur naturally or through replanting. In some instances, the volume of traffic may be so great that structural approaches are needed. For example, a shoreline hardened with a retaining wall can allow increased access for fishing when foot traffic would heavily impact a naturally vegetated shoreline.

APPLICATIONS AND PRODUCTS

Agriculture. Many agricultural erosion-control measures emphasize changing farming practices rather than necessarily installing a structure or feature on the site. For example, conservation tillage works to minimize soil tilling to leave as much crop residue in place as possible on the surface of the soil. This approach requires that the crop residue is not plowed into the soil at the end of the growing season and the new crop is planted in rows plowed through the residue at the beginning of the next season. The crop residue acts as a mulch to help stabilize the soil, retain soil moisture, and increase the soil's organic content. In this approach, chemical weed control is often used to eliminate undesirable plants or volunteer plants from the previous year's crop. Alternative methods of controlling weeds are available if a reduction in use of herbicides is desired.

Other nonstructural approaches to erosion control on agricultural land include contour plowing and crop or pasture rotation. Crop rotation is a practice where a sequence of crops are cultivated on the land over a series of growing seasons or years. Rotating crops improves soil structure and makes it more resistant to erosion, particularly when paired with conservation-tillage techniques. In addition, varying the crops grown can reduce reliance on herbicides and pesticides by, for example, changing the crop to one that creates field conditions that are less favorable to the weeds or pests of concern.

Structural erosion-control measures for agriculture include terraces and large windbreaks. Although windbreaks are not constructed, they usually are composed of trees planted in rows perpendicular to the prevailing wind direction. They do require planning for location and long-term maintenance to ensure continued effectiveness. Terraces are constructed to create areas of flat land in an otherwise sloping terrain. Terraces retain water behind a dike or berm at the edge of the terrace and can help manage water on a site after heavy rains. Terraces cost more than other erosion-control measures in terms of labor and equipment to construct but can be effectively farmed with large equipment when properly designed, located, and constructed.

Coastal Zone Management. Erosion control in coastal zones can be necessary because the action of wind and waves erodes coastal lands so that it threatens structures or features such as roads, buildings, or navigation channels. Erosion control can also be needed when structures encroach into areas that are vulnerable to erosion or are otherwise unstable landforms. Examples include the construction of buildings on barrier islands or on exposed headlands.

Nonstructural erosion-control methods in coastal environments can be modeled after natural features such as coastal-dune environments or saltwater marshes to create buffers that absorb the energy of waves or storm surges. These approaches include limiting access for foot and vehicle traffic in areas vulnerable to erosion and revegetating with native plants. The plants trap wind-borne sediment and sand and anchor coastal soils with their root systems. Revegetation measures may include the use of geotextiles or mulches to protect the soil surface while the vegetation becomes established.

Structural erosion-control measures in coastal environments include riprap, gabions, and retaining walls. Riprap is rock placed to protect the toe of a slope or the base of a vertical wall. The rock absorbs the energy of wave action and prevents the undermining of existing structures. Gabions are large cages filled with rock, often the same rock that would be used for riprap, that armor the shoreline. Gabions can be stacked to create a retaining wall and can be used in place of riprap where space is limited. Gabions have an advantage over other armoring approaches in that the rock-filled cages drain water freely and are flexible enough to accommodate ground movement. In general, retaining walls need to be properly designed and installed to ensure that the pressure of the ground and subsurface water behind the wall can be withstood for the long term by creating weep holes for moisture to escape and using devices to prevent the soil load from transferring to the wall. Retaining walls can be constructed using either nonproprietary methods or proprietary products such as interlocking concrete blocks and grid materials.

Transportation. Transportation facilities, in particular roads and railways, are often constructed through hilly terrain using cuts into steep slopes. The resulting wall often creates an instability in the landscape that needs to be addressed to avoid damage to facilities and delays in shipping and travel. Slope-stabilization techniques are commonly used in road cuts to prevent rock fall and landslides. These techniques include structural solutions (such as gabions, retaining walls, and spray-on concrete) and less structural approaches incorporating geotextiles and revegetation.

In addition, transportation facilities need to be designed and constructed to ensure that unintended erosion impacts are not created by runoff of stormwater from impermeable surfaces. In these circumstances, facility designs need to include consideration of how stormwater can be infiltrated.

Stormwater Management. Construction and other site-development activities can significantly increase erosion potential on sites by removing vegetation and changing the topography by grading. Erosion control on construction sites can be achieved either temporarily during construction or for the long-term operation of the site after construction is complete. Temporary measures include phasing construction activities to minimize how much of the soil on the site is exposed at any one time and use of stabilizing materials such as mulches or geotextiles. Products such as sedimentation fences or storm-drain filters are not erosion-control devices. Instead, they control the movement of sediments that have already eroded to ensure they do not create off-site impacts to roads or adjacent lands.

Long-term stormwater management on developed sites includes the use of structural approaches, such as retaining walls, or nonstructural approaches, such as revegetation, to stabilize sites and prevent erosion as the site is used.

Recreation Management. Recreation activities can be either dispersed or focused in terms of impacts to a site's erosion potential. Activities such as hiking or canoeing can have small impacts on sites if traffic is light. However, heavily used trails or boat launches can become significantly eroded. Other activities, such as the use of off-road recreational vehicles, can significantly reduce vegetation and produce bare compacted soils vulnerable to wind and water erosion even with light usage. Erosion-control measures related to recreation areas will vary by need and facility. For example, boat launches for light craft in a rural area may consist of a graveled surface along the bank, whereas heavily used launches for heavy craft may consist of a concrete ramp extending into the water.

Erosion control along river channels historically emphasized engineering solutions such as riprap, retaining walls, and channelization where the stream channel was lined with concrete. Because of the unintended erosion and flooding effects experienced as a result of these approaches, more recent stream-erosion projects have mimicked natural systems by reestablishing riparian vegetation and features such as floodplains and meanders. These approaches include removing barriers to river movement and using products such as coconut fiber bales and geotextiles to help revegetate areas and improve infiltration.

CAREERS AND COURSEWORK

Coursework that supports careers in erosion control or soil conservation are based in the physical and applied sciences. Degree programs include soil science, agronomy, hydrology, hydrogeology, and geology. Civil engineering is another alternative degree program, although coursework from the other disciplines is required in order to create a focus on erosion control. A solid foundation of coursework in soil science, geology, and hydrogeology is required to understand the basic physical processes involved in this subject in addition to coursework in mathematics, physics, and engineering.

Students may focus on the engineering, land-use management, or regulatory approach to erosion control. A career oriented toward engineering or structural solutions to erosion-control requires an engineering degree, including coursework in physics, mathematics, and engineering dynamics. This career may also require professional certifications, such as being certified as a professional engineer, which potentially requires additional study for the examination that must be passed to achieve the certification. A career focusing on land-use management or a regulatory approach to erosion control may be founded on any of the degree programs listed above and includes additional coursework in land use and public-policy analysis.

A bachelor or master of science degree is typically required for career opportunities designing and implementing erosion-control practices. A degree is typically not required for a career as a construction contractor installing or building erosion-control structures or features, such as retaining walls or regrading and revegetation projects. However, certain professional certifications or licenses may be required for construction contractors depending on the jurisdiction.

SOCIAL CONTEXT AND FUTURE PROSPECTS

Erosion-control requirements affect many sectors of society and have become integrated into a wide variety of occupational disciplines. Given the potential for significant damage if erosion-control measures fail—for example, in areas with steep slopes or coastal zone environments—the emphasis on and expectations for high levels of success increase with increasing development pressures. Increasing development and market pressures can also affect food production in that farmers seek to maximize the productivity of their land while reducing costs. Soil-conservation measures can increase productivity while reducing the need for some herbicides and pesticides. Soil-conservation measures that increase the moisture-holding capacity of sites and make those sites less dependent on irrigation are also important for agriculture and urban or suburban areas where water supplies may be stressed by demands for domestic use.

—Barbara J. Rich

Further Reading

Baxter, Roberta. "Maintaining Vertical: Techniques for Slope Stabilization." *Erosion Control*, vol. 11, no. 2, Mar.–Apr. 2004.

Bell, F. G. *Basic Environmental and Engineering Geology*. Whittles Publishing, 2007.

Coetzee, Ken. *Caring for Natural Rangelands*. U of KwaZulu-Natal P, 2005.

England, Gordon. "Implementing LID for New Development." *Stormwater*, vol. 11, no. 5, July–Aug. 2010.

Garcia, Carmelo Consesa, and Mario Aristide Lenzi. *Check Dams, Morphological Adjustments and Erosion Control in Torrential Streams*. Nova Science, 2013.

Leposky, Rosalie E. "Retaining Walls: What You See and What You Don't." *Erosion Control*, vol. 11, no. 1, Jan.–Feb. 2004.

Lundgren, Lawrence W. *Environmental Geology*. 2nd ed., Prentice-Hall, 1999.

National Research Council. *Restoring and Protecting Marine Habitat: The Role of Engineering and Technology*. National Academy Press, 1994.

Ray, G. Carleton, and Jerry McCormick-Ray. *Marine Conservation Science Policy and Management*. Wiley Blackwell, 2014.

Thompson, J. William, and Kim Sorvig. *Sustainable Landscape Construction: A Guide to Green Building Outdoors*. Island Press, 2008.

Estuaries, Bays, Lagoons, and Coastal Rivers

Fields of Study: oceanography; limnology; physical geography; marine biology; marine ecology

ABSTRACT

Estuaries are partially enclosed bodies of water with a connection to open ocean in which seawater enters according to the rhythm of the tides. Seawater entering the estuary is diluted with freshwater flowing from rivers and streams, and the pattern of dilution depends on several factors. These ecosystems are known to thrive where the inflow of both seawater and freshwater provide high levels of nutrients in both the water column and sediments. Classification into general groups is made according to the geological processes that formed their embayment or according to their water circulation and mixing characteristics.

KEY CONCEPTS

dead zone: a volume within a body of water where there is insufficient oxygen or an amount of contamination that denies the support of a living ecosystem

eutrophication: rendering of a body of water to become less capable of supporting aerobic life due to the consumption of dissolved oxygen in the water

halocline: the intermediate layer or interface between a lower layer of water of higher salinity and density, and an upper layer of water of lower salinity and density

heavy metals: metals, such as cadmium, lead, mercury and others from the same region of the periodic table, that are readily absorbed and bioaccumulated within a trophic chain where it can cause neurological and other health issues

DESCRIPTION

Estuaries are partially enclosed bodies of water with a connection to open ocean in which seawater enters according to the rhythm of the tides. Seawater entering the estuary is diluted with freshwater flowing from rivers and streams, and the pattern of dilution depends on the volume of freshwater, tidal amplitude range, and evaporation rate. These ecosystems are known to thrive where the inflow of both seawater and freshwater provide high levels of nutrients in both the water column and sediments. The high concentration of nutrients supports phytoplankton, which is the primary producer in the ecosystem. Each estuary is unique with respect to its physical, chemical, and biological characteristics.

Classification into general groups is made according to the geological processes that formed their embayment or, alternatively, according to their water circulation and mixing characteristics. Dynamics inside an estuary depend on its geographic location, the shape of the coastline and ocean floor, the depth of the water, tidal patterns and currents, local winds, freshwater input, and any restrictions to water flow. Additionally, there is a constant fluctuation in dissolved oxygen, salinity, temperature, and sediment load. At any point, salinity and temperature can vary considerably over time and season. Sediment often settles in intertidal mudflats, and dissolved oxygen variations can limit livability for many organisms. The wide range of salinity, temperature, and other chemical conditions, coupled with high primary productivity, creates environments that support many species of fishes and invertebrates.

ESTUARY CLASSIFICATION BASED ON GEOMORPHOLOGY

Four types of estuaries are recognized: coastal-plain estuaries, bar-built estuaries, tectonic estuaries, and fjords. In coastal-plain estuaries, formation occurred at the end of the last ice age, between 10,000 and 18,000 years ago as sea level rose to flood river valleys. These estuaries, often called "drowned river valleys," are especially abundant on passive margins, such as the east coast of the United States. Their width-to-depth ratio is typically large, appearing wedge-shaped in the inner part, broadening and deepening seaward. Water depths rarely exceed 98 feet (30 meters). Chesapeake Bay, Delaware Bay, and New York Harbor are examples of coastal-plain estuaries. The Thames River in England, the Ems River in Germany, the Seine River in France, the Si-Kiang River in Hong Kong, and the Murray River in Australia are other examples. The Mississippi River delta is an example of a former coastal-plain estuary that has been filled with river-borne sediment.

Bar-built estuaries are formed when a sandbar is constructed parallel to the coastline by wave action and long-shore drift, and the bar separates the ocean from a shallow lagoon. Bar-built estuaries typically develop on gently sloping plains located along tectonically stable edges of continents and marginal sea coasts. They are extensive along the Atlantic and Gulf coasts of the United States in areas with active coastal deposition of sediments and where tidal ranges are less than 13 ft (4 m). The average water depth is usually less than 16 ft (5 m) and rarely exceeds 33 ft (10 m). Examples include Albemarle Sound and Pamlico Sound in North Carolina, Barnegat Bay in New Jersey, and Laguna Madre in Texas. Some estuaries have mixed characteristics. For example, the Hudson River estuary that passes through New York Harbor and Raritan Bay is primarily a drowned river valley, but the Sandy Hoop spit gives Raritan Bay certain bar-built estuary characteristics.

The formation of tectonic estuaries occurs when a section of land drops or tilts below sea level as a result of vertical movement along a fault, volcanoes, or landslides. Such estuaries are found on coasts along a subduction zone or on a transform fault plate boundary. There are only a small number of tectonically produced estuaries in the world. One example is the San Francisco Bay, which was formed by the movements of the San Andreas Fault system, causing the inundation of the lower reaches of the Sacramento and San Joaquin Rivers.

Fjord-type estuaries are found in drowned valleys that were cut by glaciers when sea level was lower. The U-shaped estuaries are generally steep-sided, both above and below sea level, and are often deep. Depth can exceed 980 ft (300 m). Many fjords have a shallow sill near the mouth that was formed by sediment deposited at the lower end of the glacier when it flowed through the valley. Fjords are common along the coast of Alaska, the Puget Sound region of western Washington State, eastern Canada, Greenland, Iceland, New Zealand, and Norway.

ESTUARY CLASSIFICATION BASED ON WATER CIRCULATION

Movement and mixing of freshwater and seawater in estuaries are affected by many factors, including tidal currents and mixing; wind-driven wave mixing; shape and depth of the estuary; rate of freshwater discharge; friction between moving freshwater and seawater layers, and between the water and the seafloor; and the Coriolis effect. For this, estuary circulation is extremely important because major cities are located on estuaries. These cities discharge large quantities of waste, particularly sewage treatment plant effluents and storm water runoff, into the estuaries. Additionally, circulation in many estuaries is altered by piers and other port structures, dredging, filling of wetlands, and construction of levees and other structures. These alterations can affect the life cycles of marine species that inhabit and transit estuaries. Based on the major characteristics in water circulation, five types of estuaries are recognized: salt wedge estuaries, partially-mixed estuaries, well-mixed estuaries, fjord estuaries, and inverse estuaries.

In salt wedge estuaries, river output exceeds marine input, and freshwater flows down the estuary as a surface layer separated by a steep density interface (halocline) from seawater flowing up the estuary. The differences in density form a wedge-shaped intrusion. Vertical mixing is slow, but a velocity difference between seaward-flowing river water and the underlying seawater generates internal waves mixing the seawater upward with the freshwater. Almost no freshwater is transferred into the lower seawater layer. Examples of salt wedge estuaries are the Mississippi, Columbia, and Hudson Rivers.

In partially mixed estuaries, freshwater and seawater layers are separated by a relatively weak halocline, and therefore the density gradient between the two layers is much less pronounced. Partially mixed estuaries are most common where tidal currents are relatively fast, river flow rate is moderate, and river current speed does not greatly exceed tidal current speed. Examples include the Chesapeake Bay and Narragansett Bay, the Puget Sound, and San Francisco Bay.

In well-mixed estuaries, tidal mixing forces exceed river output, and no halocline is present. Salinity decreases progressively toward the head of the estuary, although it is uniform from surface to bottom. Some variation may occur during the tidal cycle, increasing as the tide floods and decreases as the tide ebbs. This type of estuary tends to be wide and relatively shallow, with limited freshwater inputs and strong tidal currents. Some examples include the Delaware Bay and the Raritan River in New Jersey.

Most fjords have a deep, elongated basin where vertical mixing does not reach the bottom waters, even if tidal currents are strong. Consequently, almost all fjords have a halocline separating high-salinity bottom water from lower-salinity surface water. Fjords are almost never well-mixed. Fjord-type estuaries are found along glaciated coasts such as British Columbia, Alaska, Chile, New Zealand, and the Scandinavian countries.

In arid regions, shallow estuaries can sustain high evaporation rates where salinity is higher than the ocean water outside. High-salinity estuarine water flows seaward in the bottom layer, and ocean water flows landward as a relatively lower-salinity surface layer. Thus, estuarine circulation is inverted. Inverse estuaries generally occur in subtropical regions (30 degrees north and south), where evaporation is strong and rainfall is low. Some examples include the Red Sea, San Diego Bay in California, and Laguna Madre in Texas.

Estuaries are stressful environments for organisms because of variable temperature, salinity, turbidity, currents, and other environmental factors. Salinity can induce stressful changes and organisms adapted to live in an estuary must be able to cope with widely varying osmotic pressures. Because of the high-stress environment, estuaries support fewer species than the adjacent ocean or freshwater environment. However, there is an abundant supply of nutrients and sunlight, resulting in large biomass.

Despite being an ecosystem with a high stress environment, estuaries are ideal habitats for many larval and juvenile stages of marine animals because they provide abundant food and substantial protection. Many marine fish and invertebrates use estuaries as nursery rounds and other species briefly visit the estuary to spawn. Along the southeast coast of the United States, more than half of the commercially important marine fish species are known to use estuarine wetlands as nursery areas or for breeding. Because of the abundant supply of detritus, estuaries also support huge populations of commercially important shellfish, including clams, oysters, mussels, and crabs.

Estuaries are also among the most heavily populated areas throughout the world, with about 60 percent of the world's population living along estuaries and the coast. As a result, these ecosystems are suffering degradation by many factors, including sedimentation from soil erosion from deforestation, overgrazing, and poor farming practices; overfishing; drainage and filling of wetlands; eutrophication; and pollution with heavy metals and polychlorinated biphenyls (PCBs). These impacts in turn negatively affect not only the people living around the estuaries, but the degraded environment reduces habitable space for animal and plant species and can make those populations sick.

BAYS

A bay is a small area of water or broad inlet set aside from a larger body of water where the land curves inward. Bays have calmer waters than the surrounding sea, since land is blocking waves and often reducing winds. A large bay may be called a "gulf," a "sea," a "sound," or a "bight." A cove is a circular or oval coastal inlet with a narrow entrance, and some coves may be referred to as bays. Bays are essential for many bird, fish, and marine mammal species. For example, the Bay of Fundy located on the Atlantic (east) coast of North America, on the northeast end of the Gulf of Maine, has been compared in marine biodiversity to the Amazon Rainforest. Other examples include San Francisco Bay, off the coast in northern California. Other bays include the Bay of Pigs in Cuba, Hudson Bay in Canada, Chesapeake Bay in Maryland and Virginia, and the Bay of Bengal, near India. When large and deep enough, bays become natural harbors that are often of great economic and strategic importance.

Many of the great cities around the world are located on natural harbors. Shark Bay is located in Gascoyne, the most western point of Australia, in the federal territory of Western Australia, 497 miles (800 kilometers) north of Perth. Shark Bay is currently composed of two bays, several peninsulas, and islands with a coastline that extends around 932 mi (1,500 km) and covers an area of 3,861 square mi (10,000 square km).

Shark Bay gathers three of the most important climatic regions of the world, and has great ecological importance. This bay is home to more than 10,000 dugongs, a large community of bottlenose dolphins (*Turciops truncatus*), and is an important reproduction zone for hundreds of species of fishes, cnidarios, and crustacean, including sharks and rays that give its name to the bay.

Additionally, Shark Bay has also the largest and most varied area of seagrass in the world, with a cover of about 1,544 sq mi (4,000 sq km). The bay also hosts a unique community of microbes (located in Hamelin Pool) that are building colonies of algae around 3,000 years old (stromatolites).

The Bay of Fundy is located on the Atlantic coast of North America, on the northeast end of the Gulf of Maine between the Canadian provinces of New Brunswick and Nova Scotia, with a small portion in the state of Maine. It extends 170 mi (270 km) and it is known for having the highest tides in the world (54 ft, or 16 meters), with the potential of becoming one of the world's greatest producers of tidal energy. The marine biodiversity in the Bay of Fundy is vast; at least 12 species of whales can be found here. An abundant number of dolphins, porpoises, fish, seals, and seabirds populate its waters.

The Bay of Islands is one of the most popular holiday destinations in New Zealand, located in the Northland Region of the North Island, 37 mi (60 km) northwest of Whangarei. The bay is an irregular, 10-mi (16-km) wide inlet and contains 144 islands, many secluded bays, and sandy beaches. This bay has abundant marine life including whales, penguins, dolphins, and marlins. It is also a popular gathering place for sailing yachts and international sport fishermen. The bay is also historically significant because it was the first part of New Zealand settled by Europeans.

The Bay of Kotor is a winding bay on the Adriatic Sea, located in southwestern Montenegro. This bay is often mistaken for a fjord, but is in fact a submerged river canyon of the disintegrated Bokelj River. Because of its location and peculiar topography, this bay is the second-wettest place after Japan's Kii Peninsula, north of the Himalayas. The Bay of Kotor is considered a picturesque place to visit, with numerous beaches.

Phang Nga Bay is located over 60 mi (95 km) from the island of Phuket and the Malay Peninsula in the Andaman Sea. It is an extensive bay, and since 1981 an extensive section was protected as the Ao Phang Nga National Park. In 2002, this bay was declared a Ramsar Site of international ecological significance, comprising shallow marine waters and intertidal forested wetlands, with 28 species of mangroves, seagrass beds, and coral reefs. Species like the dugong and the black finless porpoise (*Neophocaena phocaenoides*) are also found in the bay.

San Francisco Bay is situated on the Californian coast, and it includes an entire group of interconnected bays. The area around it is considered the second-largest urban area, with approximately 8 million residents. Water drains from the Sacramento and San Joaquin Rivers from the Sierra Nevada Mountains, where both rivers then flow into the Suisun Bay. This water then flows through the Carquinez Strait, which meets with the Napa River at the entrance of the San Pablo Bay. This group of rivers finally drains into the Pacific Ocean through the San Francisco Bay. The bay occupies between 400 and 1,600 sq mi (1,040 and 4,160 sq km), and it is considered the largest estuary in the Americas. The San Francisco Bay and the Sacramento–San Joaquin Delta are considered California's most important ecological habitats, where organisms like the Dungeness crab, California halibut, and the Pacific salmon fisheries rely on the bay as a nursery. This bay is protected by the California Bays and Estuarine Policy.

LAGOONS

Coastal lagoons are among the most common coastal environments, occupying around 13 percent of the world's coasts. A lagoon can be defined as a shallow body of water, usually oriented parallel to the coast and separated from the ocean by a barrier, connected to the ocean by one or more restricted inlets. These ecosystems are found on all continents, from arctic to equatorial, and from arid to humid, but are less common on emergent high-latitude coasts. They may have one or more natural entrances from the ocean, which can be permanent or intermittent gaps through the enclosing barriers. Lagoons are characteristically shallow, typically 3 to 10 ft (1 to 3 m), and almost always less than 16 ft (5 m), with the exception of inlet channels and isolated relict holes or channels, and are subject to tidal mixing. They are strongly influenced by precipitation and evaporation, which results in fluctuating water temperature and salinity.

Salinity can range from completely fresh to hypersaline conditions. Size varies substantially, with surface areas ranging up to 3,938 sq mi (10,200 sq km), as in the case of Lagoa dos Patos in Brazil. Coastal lagoons serve as material sinks or marine filters because they trap inorganic sediment and organic matter. They often exhibit high primary and secondary productivity, and are valuable for fisheries and aquaculture, and sometimes for salt extraction. These are fragile ecosystems that experience forcing from river input, wind stress, tides, precipitation to evaporation balance, and surface heat balance. Lagoons are classified into three geomorphic types, according to water exchange with the coastal ocean. They are classified as: choked lagoons, restricted lagoons, and leaky lagoons.

Choked lagoons occur along high-energy coastlines, and have one or more long narrow channels that restrict water exchange with the ocean. Tidal oscillations are often reduced to 5 percent or less, compared to adjacent coastal tide. Circulation within choked coastal lagoons is characterized by long flushing times, dominated by wind patterns, and intermittent stratification events from intense solar radiation or runoff events. Examples of chocked coastal lagoons include Lagoa dos Patos and Lagoa de Araruama in Brazil, Lake St. Lucia in South Africa, the Coorong in Australia, and Lake Songkla in Thailand.

Restricted lagoons consist of a large and wide water body with two or more channels or inlets, and have a well-defined exchange with the ocean. Wind patterns influence circulation, and lagoons exhibit a variety of salinities, ranging from brackish to hypersaline. Examples include Indian River Lagoon and Pontchartrain in the United States, and Laguna de Terminos in Mexico.

Leaky lagoons are elongated shore-parallel water bodies with many ocean channels along the coast. Leaky lagoons are characterized by numerous wide tidal passes; unimpaired exchange of water with the ocean on wave, tidal, and longer timescales; strong tidal currents; and salinities close to that of the coastal ocean. Examples include Mississippi Sound in the United States and Wadden Zee in the Netherlands.

Coastal lagoons are extensive in low-lying coasts and are poorly developed on coasts dominated by high retreating cliffs, such as the Great Australian Bight, and the steep and rocky fjord coasts of Norway, Chile, and southern New Zealand. They are also rare on macro-tidal coasts such as the Bay of Fundy between Canada and the United States, and the Baie de Mont Saint Michel in France. Other examples include Albemarle Sound in North Carolina; Great Sound Bay between Long Island and the barrier beaches of Fire Island in New York; Isle of Wight Bay, which separates Ocean City, Maryland, from the rest of Worcester County, Maryland; Banana River in Florida; and Lake Illawarra in New South Wales. In India, Chilika Lake in Orissa, near Puri, and Vembanad Lake in Kerala are also considered lagoons. Some other well-known lagoons include Lagos Lagoon and Keta Lagoon in Ghana, Africa.

COASTAL RIVERS

Rivers are the result of precipitation falling across the land, coalescing into even larger streams and rivers. These ecosystems are one of the most dramatic features of a continent, delivering sediment and nutrients downstream, eroding valleys, and eventually reaching the sea or an inland lake. This water movement shapes terrain, creates a variety of freshwater environments, and allows the evolution of several species of plants, animals, and microbes. Rivers vary in latitude, topography, and size, contributing to a great variation in biodiversity and ecological characteristics. Climate, including temperature and precipitation patterns, further characterizes the landscape and is influenced greatly by global air circulation patterns, ocean currents, shape of the continent, and mountain ranges.

Coriolis forces have a predominant effect in river discharge patterns, and significantly influence the living marine resources of coastal areas. Gravity causes the higher-density oceanic water to sink beneath the less-dense surface water, and mixing processes produce water of intermediate density. Buoyancy-driven flows that are supplied by this mixed water formation cause positively buoyant materials and vertically migrating planktonic organisms to accumulate in the slowly sinking waters of the convergent frontal zone. In this scenario, the distribution of food particles becomes highly concentrated, and feeding organisms may dramatically increase caloric intake per unit. This process tends to work its way upward through the trophic levels.

For this reason, coastal rivers are extremely important to biological population dynamics. Major factors that threaten river biodiversity, and consequently the marine environment, include agriculture, deforestation, freshwater extraction, and human and industrial waste discharge, plus the effects of artificial changes to river hydraulic characteristics. Nutrient pollution, also known as overenrichment of nitrogen and phosphorous from coastal rivers, can lead to phytoplankton blooms, causing estuaries to receive more nutrients per surface area than any other ecosystem. In the United States, for example, more than 60 percent of coastal rivers and bays are moderately to severely degraded by nutrient pollution. Although such problems occur in all coastal states, the situation is particularly acute in the mid-Atlantic states, southeast, and Gulf of Mexico.

High sediment loads and stagnant nutrient-rich organic matter flows, plus excessive extraction of freshwater in coastal rivers, can also lead to high turbidity rates and bottom anoxia in many coastal waters. These changes in nutrients, light, and oxygen can favor some species over others and cause shifts in the structure of phytoplankton, zooplankton, and bottom-dwelling (benthic) communities. For example, blooms of harmful algae such as red and brown tide organisms can become more frequent and extensive, sometimes resulting in human shellfish poisonings and even marine mammal deaths. Other factors, such as how often a bay or estuary is flushed and its nutrients diluted by open ocean water, make some coastal ecosystems more susceptible to nutrient overenrichment than others. Ecosystems like coral reefs and seagrass beds are particularly susceptible to environmental changes in coastal river flows.

The largest coastal system affected by eutrophication in the United States is the dead zone in the Gulf of Mexico, an extensive area of reduced oxygen levels. In the early 1990s, the zone covered an estimated 3,668 sq mi (9,500 sq km) of the gulf, extending out from the mouth of the Mississippi River. By the summer of 1999, this hypoxic area had doubled to 7,722 sq mi (20,000 sq km), an area the size of Lake Ontario or New Jersey. Other severely impacted coastal systems in the United States include Chesapeake Bay, Long Island Sound, and the Florida Keys. In Europe, the Baltic, North, Adriatic, and Black Seas have all experienced problems from nutrient overenrichment. In the Black Sea, eutrophication was partially reversed during the early 1990s as nutrient inputs decreased following the collapse of the Soviet Union and fertilizer use in Eastern Europe dropped sharply. This decrease was temporary, and the Black Sea has recently reached high levels of nutrient overenrichment.

—Maria Jose González-Bernat

Further Reading

Anthony, Abigail, et. al. "Coastal Lagoons and Climate Change: Ecological and Social Ramifications in US Atlantic and Gulf Coast Ecosystems." *Ecology and Society*, vol. 14, no. 1, 2009.

Banyopadhyay, Jayanta. *Water, Ecosystems, and Society*. Sage Publishing, 2009.

Cyronak, Tyler et al. *Spatial and Temporal Variability of Seawater Chemistry in Coastal Ecosystems in the Context of Global Change*. Frontiers Media, 2019.

Davis, Richard A., and Duncan M. FitzGerald. *Beaches and Coasts*. 2nd ed., Wiley, 2020.

"Human Disturbances to Estuaries." *NOAA Ocean Service Education*, 25 Mar. 2008. http://oceanservice.noaa.gov/education/kits/estuaries/estuaries09_humandisturb.html.

Kosyan, Ruben, editor. *The Diversity of Russian Estuaries and Lagoons Exposed to Human Influence*. Springer, 2017.

Livingston, Robert J. *Climate Change and Coastal Ecosystems*. CRC Press, 2015.

McCabe, Declan. "Rivers and Streams: Life in Flowing Water." *Nature Education*, 2011. http://www.nature.com/scitable/knowledge/library/rivers-and-streams-life-in-flowing-water-23587918. Accessed on 2 Nov. 2016.

Wolanski, Eric, John W. Day, Michael Elliott, and Ramesh Ramachandran, editors. *Coasts and Estuaries: The Future*. Elsevier, 2019.

F

Fisheries

Fields of Study: oceanography; marine ecology; environmental studies; fisheries management

ABSTRACT

"Fisheries" is the term used to define industries involved in the harvesting, processing, and marketing of aquatic products. In a broad sense, fisheries is used to describe areas of water bodies, species of organisms being harvested, methods of harvesting, and the people involved in any of these processes. Fisheries products are not limited to fish; rather they include the entire span of aquatic organisms, both marine and freshwater. There is a great variety of such resources—and great challenges in terms of the sustainability of major fisheries in the face of regional and global ecosystem changes.

KEY CONCEPTS

bottom trawling: a method of capturing and removing large numbers of fish by indiscriminately dragging nets over the sea bottom

bycatch: the fish and other species that are captured that are not the desired species, but are removed from the aquatic environment anyway, usually dying in the process and being discarded as worthless

capture fisheries: any fishery in which fish are captured by some means and removed from their aquatic environment for use as food and other purposes

carnivorous fish: fish that eat other fish and aquatic animal species

fishing down the food web: the almost stepwise shift of capture fishing from species at the top of the food web to species at increasingly lower trophic levels

lotic: refers to water that is moving, that has a current flow

trophic level: a step in the ordering of a food chain or food web; at the highest level are the top predator species, and at the lowest level are the "bottom dweller" prey species, while in between are various species that are both predator and prey species

FISHERIES PRODUCTS

Aquatic organisms such as fish are among the most readily available protein sources in many countries. According to the United Nations Conference on Trade and Development (UNCTAD), fish accounted for approximately 17 percent of global animal protein consumption as of 2018, and fisheries provide the general population with access to these protein sources. Aquatic organisms included in the fisheries designation include seaweeds, cnidarians (jellyfish), echinoderms (sea cucumbers), crustaceans (crabs and shrimp), mollusks (snails, limpets, bivalves, squid, cuttlefish, and octopuses), marine reptiles (sea turtles), fish (cartilaginous and bony), marine mammals (manatees, dugongs, porpoises, sea lions, and whales), and even the marine worms (peanut worms) that are harvested in parts of Asia.

Aquatic organisms that are sold alive or chilled command the highest prices. Live seafood is extremely popular in countries in east Asia, such as China (especially in Hong Kong), Singapore, and Taiwan, where they fetch prices multiple times higher than those of frozen fish. Fresh fisheries products are those that have been chilled over short periods; these also cater to the people of Japan and Japanese restaurants all over the world as sashimi. To extend shelf life, aquatic organisms are processed in various ways, often via freezing, canning, fermenting, or drying, which dramatically reduces their commercial value as compared to fresh goods.

Fisheries products also refer to parts, by-products, or processed aquatic organisms. Caviar, caviar substitutes (roe from nonsturgeon fish), and roe from other marine organisms such as shrimp are included. Mollusks are harvested for nacre, a lustrous substance from which pearls and mother-of-pearl are made in nature. Shark and ray skins are byproducts that can be made into leather. Other by-products of fisheries include trash fish or unwanted fish parts that are ground up to make fish meal, which is used in feed for pets, cultured fish, and livestock.

CAPTURE FISHERIES

"Capture fisheries" is a term used for exploiting resources from the wild. Capture fisheries in marine environs occur at near-shore areas such as estuaries, mangroves, and coral reefs; or at offshore areas such as open oceans and seamounts. Capture fisheries also occur in lotic (moving water) habitats like freshwater streams and rivers, or in lotic freshwater lakes. There are several scales at which capture fisheries occur: commercial, artisanal, and recreational. Commercial or industrial fisheries are typically large-scale, and meet the demands of a broader national or international market. Artisanal or subsistence fisheries cater to local demands or are self-sufficient. Recreational fisheries are exclusively for pleasure and not for commercial purposes.

FISHING DOWN THE FOOD WEB

The post-World War II period saw an increase in commercial fishing that has resulted in greater (by tonnage) yields. The ability to capture more fish through better-equipped commercial vessels and more effective means of capture allowed for increased exploitation of aquatic resources. The sharp increase in the number, size, and engine power of vessels allowed for extended periods at sea. The spatial and temporal intensity of exploitation, along with evolved fishing

Catch from a deepwater tow featuring scarlet king crabs, pacific grenadier, and longspine (NOAA Fisheries)

gear, enhanced skills of fishers, and high catch efforts contributed to increased landings. Computers installed in fishing vessels to locate positions of shoals allowed for precision in harvesting. Intensive commercial fishing from the 1960s through the 1990s thereby saw the rapid depletion of global fisheries resources.

The compositions of captured fisheries transitioned from organisms high in the trophic level, such as carnivorous fishes, to those low in the trophic level, such as invertebrates and planktivorous fishes. The term "fishing down the food web" has been coined to describe this phenomenon.

Initially, global landings of organisms lower in the trophic levels were high, but soon stagnated or declined as overharvesting persisted. Countries where fisheries biomass decreased moved their efforts to deeper waters. Scientists have arrived at a consensus that the propensity for slow growth rates and high longevity of deep-water organisms render this resource type unsustainable.

Some researchers argue that collapses in various fisheries occurred before the twentieth century. This conclusion was reached via paleoecological records from marine sediments; archaeological records from ancient human settlements; historical written records, journals, and documents; and ecological records from scientific literature for more than a century. A group of scientists led by Jeremy Jackson (Scripps Institution of Oceanography at the University of San Diego), for example, studied patterns of population abundances and crashes of harvested marine organisms.

Scientists postulated that larger marine organisms were already trending toward population crashes before the twentieth century. Citing, among others, the example of the Steller's sea cows that became extinct in 1768, the scientists stated their belief that aboriginal hunting was responsible for decimating populations throughout the northern Pacific Rim.

Similar patterns were observed for organisms such as whales, dugongs, fur seals, and turtles, the difference being that small isolated populations of those species still exist. Global colonial infiltrations and the advent of vessels capable of long-haul cruises sealed the fate of these organisms. The dramatic reduction in populations of these organisms altered coastal ecosystems significantly and eventually led to the crashes presently observed. These authors did not discount the fact that the intensity and persistence of current exploitation of fisheries resources are unsustainable and, if continued unchecked, will lead to systemic ecosystem collapses in the future.

Current trends show that global fishing efforts are estimated to exceed the optimum factor by three- to fourfold. Excess fishing efforts contribute to billions of dollars in yearly losses through four main avenues: increased fishing, declining fishing stocks, stagnating or declining resource prices, and fishing subsidies that support overexploitation. The Food and Agriculture Organization of the United Nations (FAO) report based on data from 2007 showed that global fisheries were facing inevitable crashes. A total 52 percent of all fish stocks were already overexploited, 28 percent were overexploited or depleted, 20 percent were moderately exploited, and only 1 percent showed signs of recovery. According to UNCTAD, by 2018, approximately 90 percent of the world's fish stocks were fully exploited, overexploited, or depleted.

In 2014, capture fisheries' production was approximately 103 million tons (93.4 million tonnes), according to the FAO. If the current levels of capture fisheries continues, some scientists estimate that global fisheries will collapse by the mid-twenty-first century, with only smaller fish surviving.

FISHERIES COLLAPSE

There are two major persistent problems related to fisheries: bycatch and bottom trawling.

Bycatch is the incidental capture of nontarget fish that are often of no economic concern due to their low commercial value. At times, these are nontargeted species or species that are discarded due to their size. Often, organisms such as echinoderms, crustaceans, turtles, marine mammals, and birds are entangled in the nets set out for commercially targeted fish. Marine cetaceans often get entangled and consequently drown in tuna fisheries. Marine birds, and large marine organisms such as turtle and sharks, often are entangled in or caught on longlines when trying to obtain baited hooks targeting tuna, swordfish, and halibut.

Bycatches have been identified as the cause of the dramatic decline in many marine species. The unintentional yet systematic removal of nontarget species from the environment has strong ecological implications; it is considered one of the greatest threats to the marine environment. The highest incidence of bycatches is associated with shrimp capture fisheries, with shrimp capture responsible for a significant portion of total global bycatch.

Trawling is a fishing method involving the use of large nets and mechanized arms on fishing vessels. The nets are set into the water at specified depths and pulled through the water column over a period of time. Bottom trawling occurs when the net is dragged over the seabed, targeting marine organisms living on the seabed or just above it, such as bottom fish, shrimp, and squid. This method of fishing has been associated with heavy, negative effects on seabed communities. Nets dragged over the seabed not only cause seabed sediments to be suspended in the water column, but also physically damage reefs and deep-water corals. In addition, catches made by this method are nondiscriminatory in nature, resulting in large volumes of bycatches.

GLOBAL CLIMATE CHANGE AND FISHERIES

Scientists believe that in the chronological chain of events, factors such as pollution, physical destruction of habitats, and anthropogenic climate change are factors affecting fisheries. However, human-induced climate change has been a particularly confounding factor. Coupled with overexploitation, the effects of climate change on fisheries are only now beginning to be understood.

The oceans, being open systems, are governed by many interacting factors, which are often difficult to elucidate in isolation. It is generally agreed, however, that changes in the climate will cause distinct changes in the patterns of distribution of important fisheries resources. Marine organisms are likely to respond to global warming through shifts in latitudinal ranges, for instance.

Consequently, local extirpations, invasions by other organisms, and changes in the population structure of remaining or newly invaded species are likely to occur. In short, resultant community interactions for such ecosystems also are likely to change.

MITIGATION MEASURES

Farmed fisheries for both freshwater and marine organisms have been touted as the solution to meet increased demand for a waning natural supply. Economically important aquatic organisms are farmed either ex-situ in facility tanks; in-situ at enclosed inlets or bays; or in sea cages and pens. In 2014 alone, production from farmed fisheries made up 44.1 percent of total production by farmed fisheries and capture fisheries combined, according to the FAO.

However, several concerns about this industry have arisen. Large tracts of mangrove forests, particularly in southeast Asia, have been cleared to make way for unsustainable coastal shrimp and fish farms. With increasing bioaccumulation and toxin buildup over time, these areas are abandoned, and new mangrove areas are then deforested. Many aquafarms also are fundamentally unsustainable, requiring that fingerlings and hatchlings of aquatic organisms are still harvested from the wild and are only grown and fattened in farmed areas.

The introduction of antibiotics to farmed organisms, especially if these areas are connected to natural water bodies or other natural ecosystems, is a concern among many environmentalists who fear that resistance to parasites and disease is undermined by the overuse of antibiotics. The biological discharge from aquatic farms to the natural environment raises similar concerns. Yet another issue surfaced when it was found that fish farms consumed more wild fish than their final output. This situation has been monitored, and scientists are still refining ways to produce farmed aquatic resources with minimal environmental effect.

A major challenge faced by managers of aquatic ecosystems is illegal fishing. For example, in some reefs around southeast Asia, dynamite fishing was rife in past decades. Increased efforts by managers of marine ecosystems to curtail bombing of coral reefs have caused dynamite fishing to subside somewhat. Both the live-seafood and marine-aquarium trades are booming, with much of the resource base centering on southeast Asian seas. The use of low doses of cyanide to stun and capture live marine organisms is common in parts of southeast Asia.

The effects of this method of marine-organism collection are continuously being studied. In addition to the obvious removal of selected organisms, cyanide remains in the ecosystem and affects the remaining sea life. Catch efforts in intertidal pools are especially deleterious for all creatures left in the pools post-collection. Corals are known to be affected by cyanide, with reports of decreased productivity by symbiotic zooxanthallae. Because of such negative effects, as of 2018, researchers were working on perfecting a method for breeding those fish species of the most value to collectors (particularly tropical) in tanks. Managers of aquatic ecosystems are aware that these activities persist but face constraints in the implementation of maritime laws, due largely to enforcement agencies' insufficient capacities to patrol the areas under their purview.

THE WAY FORWARD

Stakeholders in the industry are making some progress in dealing with bycatches. Trawl nets have been redesigned to aid in excluding several species of organisms, notably turtles, for which turtle excluder devices have been created. These nets are equipped with a device above or below the nets so that large organisms can escape unharmed. Efforts such as those by World Wildlife Fund (WWF) in 2005–2009 to reward innovative ideas to reduce bycatches have been seen as a positive step toward solving global problems as a global community. With WWF and interested stakeholders such as the US National Oceanic and Atmospheric Administration (NOAA) and the US Department of Fisheries, some of these novel ideas are being refined and tested to assess their suitability for broad use by the industry. In the United States specifically, according to the NOAA, by June 2018, efforts to minimize overfishing had led to the rebuilding of forty-four fish stocks.

According to the Database of World Protected Areas, as of 2018 there are more than fifteen thousand marine protected areas, and there are different scientific recommendations for their use. In some areas, partial exploitation is allowed. In others, there are strict no-take zones where no fisheries activities are allowed. There also exists direct evidence of positive spillover effects, where larvae borne from organisms in protected areas are able to seed neighboring reefs.

The constant efforts by aquatic ecosystem managers and scientists to understand the processes behind the ecosystems under their care serve to improve the management of these imperiled ecosystems and, by extension, enhance the management of fisheries resources. These efforts are especially pertinent in open systems such as the oceans, where connectivity and common resources call for the participation of all stakeholders. Global initiatives led by organizations such as FAO are beneficial in elucidating ways in which fisheries-dependent countries can work together to achieve sustainable resources for the future.

—*Zeehan Jaafar*

Further Reading

Barange, M., et al., editors. *Marine Ecosystems and Global Change*. Oxford UP, 2010.

Bjorndal, Trond, and Gordon R. Munro. *The Economics and Management of World Fisheries*. Oxford UP, 2012.

Bland, Alistair. "Ending the Ecologically Harmful Capture of Tropical Fish." *Pacific Standard*, 9 Jan. 2018, psmag.com/environment/tropical-fish-collection-kills-coral-reefs.

'Fishery Stock Status Updates." *NOAA Fisheries*, 2018, www.fisheries.noaa.gov/national/population-assessments/fishery-stock-status-updates. Accessed 26 July 2018.

Food and Agriculture Organization of the United Nations. "Fisheries and Aquaculture Department Website." 2012. www.fao.org/fishery/en.

Jackson, B. C., et al., editors. *Shifting Baselines: The Past and the Future of Ocean Fisheries*. Island Press, 2011.

Kituyi, Mukhisa. "90% of Fish Stocks Are Used Up—Fisheries Subsidies Must Stop." *United Nations Conference on Trade and Development*, 13 July 2018, unctad.org/en/pages/newsdetails.aspx?OriginalVersionID=1812.

Longhurst, Alan. *Mismanagement of Marine Fisheries*. Cambridge UP, 2010.

Pauly, D. "Global Fisheries: A Brief Review." *Journal of Biological Research—Thessaloniki*, vol. 9, no. 3–9, 2008.

The State of World Fisheries and Aquaculture. Food and Agriculture Organization of the United Nations, 2016, www.fao.org/3/a-i5555e.pdf. Accessed 26 July 2018.

Williams, Rick. *A Future for the Fishery: Crisis and Renewal in Canada's Neglected Fishing Industry*. Nimbus Publishing, 2020.

Worm, B., et al. "Impacts of Biodiversity Loss on Ocean Ecosystem Services." *Science*, vol. 314, no. 1, 2006.

Fisheries Science

Fields of Study: biology; marine biology; oceanography; mathematics

ABSTRACT

Fisheries science is an interdisciplinary study concerned with the hunting and farming of aquatic organisms in oceans and

bodies of fresh- or saltwater as part of the ongoing effort to feed the world's population. Until recently, most aquatic organisms from fisheries were wild creatures. Farming and the use of genetic-engineering techniques to produce desirable domesticated aquatic food organisms are widespread, but the majority of fisheries science still involves the management of fish stocks for sustainability of the resource.

KEY CONCEPTS

aquaculture: farming of plants and animals that inhabit marine environments, as opposed to agriculture, the farming of plants and animals that inhabit terrestrial environments

fish emulsion: a fertilizer produced from the fluid remainder of fish already processed for fish oil and fish meal

fish meal: a high-protein byproduct of rendering and processing fish for fish oil, used as a food supplement in aquaculture.

fish oil: natural oils obtained from fish by various means of processing

public trust: the principle by which natural resources are deemed to belong to the people and are administered as a trust by governments

DEFINITION AND BASIC PRINCIPLES

Fisheries science is concerned with the continued extraction of aquatic organisms from marine, brackish, and freshwater environments for subsistence, commercial, or recreational purposes. As such, fisheries science necessarily involves issues of yield (the numbers of fish or other targeted organisms harvested from a given stock) and sustainability (the numbers of fish, etc. that must not be harvested if the particular stock is to demonstrate continued productivity). This seemingly simple biological situation, however, is made much more complex by ecological, economic, and social considerations.

No aquatic species—not even farmed fish—exists in isolation, unconnected to the food web of its surrounding ecosystem. To the extent that fishing or aquaculture alters the population of one species, it also inevitably alters the populations of other species in the same environment. The health of a fishery is dependent as much on the health of the entire ecosystem as it is on any particular species. Sustainability, then, must ultimately take into account not just numbers of a particular stock harvested or left behind but also the health of that stock's entire ecosystem.

Fisheries science is further complicated by the fact that fish, particularly in the sea, have long been assumed to be a limitless resource, available to all for the taking. Depleted stocks and increasing competition for declining numbers of fish have led to an understanding that even fish in the sea are also a limited resource. This has led to the search for new ways to manage marine fisheries and to produce more fish aquaculturally through fish farming of various types. Basic, applied, and developmental research in fisheries science is deeply involved with aquaculture and fisheries management. These are fields in which biological sustainability and productivity must be balanced with the economic productivity of those employed in the fishing and fish-farming industries and with the social productivity of communities who have long been dependent on the harvest of aquatic organisms.

BACKGROUND AND HISTORY

Evidence of aquaculture (particularly the raising of fish in ponds or estuaries) goes back at least 5,000 years, and evidence of the hunting and gathering of aquatic organisms goes back significantly more than 100,000 years. However, it was only in the twentieth century that fisheries science developed out of the biological study of aquatic organisms. The development of the discipline can be traced to the fact that, although there had been occasional spot depletion of fish stocks (particularly in freshwater environments or among large sea mammals such as whales), widespread depletion of fish stocks did not become a concern until the twentieth century.

The depletion and eventual collapses of fisheries resulted from technological advances in the fishing industry that increased both the efficiency in how a given fish stock was harvested and the rate at which that stock was exploited. Through the first half of the twentieth century these technological advances were primarily mechanical: More powerful steam and then diesel engines allowed larger and sturdier fishing vessels, hauling larger and stronger fishing gear, to travel farther and, along with better refrigeration, stay at sea longer. During the second half of the twentieth century, the technological advances were primarily electronic: sonar and the global positioning system (GPS), along with many other informational technologies, made it possible to find the fish more accurately and extract them.

Over the course of the twentieth century, each new technological advance allowed fishers to eat deeper into the natural capital of a given fish stock. Short-term-oriented market forces, driving an extractive technological arms race within the context of open-access fisheries, often reached the point that the fish stocks in question collapsed entirely. The need to manage fish stocks more rationally became increasingly obvious, and fisheries science has begun to serve this need.

HOW IT WORKS

The core of fisheries science is biological. It is concerned with understanding the life cycles of individual aquatic organisms, including growth rates, ages of sexual maturity, longevity, predation, and mortality. How these factors affect estimates of fish-stock population sizes and long-term population management of fish stocks are also all of key importance in fisheries science. Fisheries science also requires practical understanding of fish tagging and fish marking, the particular fishing gear used in specific fisheries, habitat improvement and bioremediation, fishways, screens, and guiding devices, the role of hatchery-raised fish and stocking, and small-pond, floating-enclosure, or net-pen management (these last being particularly important in aquacultural contexts).

Whether in aquacultural or traditional fishing approaches, however, fisheries scientists find they must go beyond a simply biological understanding of fish stocks. Fisheries science is profoundly influenced by the context of the cultural and legal framework within which fisheries management takes place. An important part of this framework is the common-law doctrine of the "public trust," particularly the idea that the resources of the rivers and seas within a nation's jurisdiction belong to the people, and the government holds them in trust for the public.

Fisheries as a Public Trust. The idea of the public trust has its roots in the principle of ancient Roman law that things such as the air, running water, the sea, and the shores of the sea are incapable of private ownership. In English law, and later in US law, the fish and wild beasts were added to the list of "common property" that the government holds in trust for the public. Resources relating to the public good cannot be given away to private interests, because public trust resources are inalienable—that is, they cannot be given over or transferred by the government to other entities or individuals. The government also must, by law, exercise continuing stewardship responsibility and authority over public trust resources, including fisheries.

In conjunction with improved extractive technologies and largely unregulated market forces, the open-access nature of fisheries in many cases resulted in the degradation of those fisheries. It thus became increasingly clear that government, as trustee of the fisheries, ought to exercise its stewardship authority over these public trust resources. In the United States, this stewardship responsibility is seen most clearly in the Magnuson-Stevens Fishery Conservation and Management Act of 1976 (which extended to 200 miles (322 kilometers) offshore the nation's jurisdiction over the sea as an "exclusive economic zone" for fishing) and the Sustainable Fisheries Act of 1996 (which called for further analysis of sustainability and closed to fishing many depleted fisheries, in hopes of their recovery).

The places where aquatic organisms (whether wild or farmed) live and from which they are to be harvested or protected from harvest will likely remain waters held primarily in public trust. As a result, fisheries scientists will continue to serve most often in the role of knowledge expert, assessing fish stocks and advising government, industry, and the public how best to use the fish stocks that are the common property of a nation's citizens.

Assessing Estimated Population Size and Ecological Risk: Traditional Fisheries. To estimate yield, to understand basic changes in population number and composition, and as a basis for sound management, the fisheries scientist must be able to estimate fish population reliably. Like much of the rest of fisheries science, population assessment is based in mathematical and systems-analysis approaches. Although opportunities do exist for direct counting of fish, most statistics on fish populations are estimates. Some examples of methods for estimating population include area density, mark-recapture or "single census," and catch-effort.

The area-density approach to estimating population involves counting the number of animals in a series of sample strips or plots distributed randomly or systematically throughout the total environmental area in which the fish stock population is to be determined. The sample count, once taken, is expanded to an estimate of the entire population by multiplying the aggregate sample count by a particular fraction (total area divided by the sum of sample areas). Because a subarea can also be sampled for time instead of space, the same equations applied to area-density can also be applied to partial-time coverage.

In the mark-recapture method of estimating fish population, a sample of fish is collected, marked, and released, and then at a later time a second or recapture sample is taken, which includes both marked and unmarked fish. This approach is based on the assumption that the proportion of marked fish recovered is to the total catch in the second sample as the total number of marked fish released is to the total fish population.

The catch-effort method of estimating fish population depends on the premise that all individual fish in a sample have the same chance of being caught and that the effort made by fishermen to catch the fish is constant. This approach works best when the population is closed, when chance of capture and constancy of effort remains unchanged from sample to sample, and when there are enough fish removed so that the population shows a decline.

Assessing Estimated Ecological Risk: Aquaculture. In aquaculture, population is much more controlled, so assessment emphasizes questions of ecological risk rather than determination of population. Because of the intensive nature of fish-farming practices, these questions of ecological risk include near-field and far-field effects of increased organic loading (mainly from uneaten fish feed, fish fecal material, and decomposing dead farm fish), increased inorganic loading (mainly nitrogen, phosphorus, trace elements, and vitamins in fish excretory products and uneaten feed), residual heavy metals (mainly zinc compounds in uneaten feed and fish feces), the transmission of disease organisms (often increased because of the high density of fish per volume of water in net-pen enclosures), and residual therapeutants (from biomedical treatments of the fish performed in response to the increased transmission of disease organisms resulting from crowded net-pen conditions).

Other ecological risks to be assessed for aquaculture include biological interaction of farmed fish that have escaped into wild populations and vice versa (with the potential not only for breeding but also for cross-infection with parasites and pathogens), the impact on marine habitat of fish-farm enclosures (including entanglements involving nets, anchors, and moorings), control of natural predators in the fish-farm environment, and increasing pressure on shoaling small pelagic fish populations to be fed to the farmed fish.

APPLICATIONS AND PRODUCTS

The primary role of most fisheries scientists continues to be consultative. That consultative role usually falls into three categories: assessor (the person responsible for making reliable estimates of fish populations, predicting what harvest levels those populations can support, and the environmental impacts of both fishing and fish-farming approaches), adviser (the person responsible for communicating to government, industry, and public all findings relevant to the health of a fishery and its environment), and educator (the person responsible for increasing governmental, industrial, and popular awareness of both the economic and ecological importance of aquatic organisms).

Some fisheries scientists, however, already find themselves involved in efforts with direct applications and products. These include new ways of growing or harvesting aquatic organisms, some as revolutionary as genetic modification of farmed fish, some as evolutionary as the development of more efficient fishing gear.

Aquatic and Marine Food. Fisheries scientists, as experts in the health of individual fish species and overall fishery populations, help industry provide high-quality seafood products for the consumer—an important activity, considering that seafoods are one of the world's primary sources of high-quality protein. Fisheries scientists are involved in product development, physicochemical principles, and process technology for aquatic food and marine bioproduct utilization, as well as in examining and improving aquatic and marine products and manufacturing processes.

In the aquacultural context, fisheries science helps to increase the aquacultural contribution to the food supply while also developing new methods of production and improved cultural practices for selective species. These production and cultural concerns include environmental, ecological, and disease considerations, selective breeding, feeding, processing, and marketing.

Aquatic and Marine Nonfood Products. In addition to foodstuffs, however, many other products are derived from aquatic and marine organisms. Fish oil, which contains omega-3 fatty acids and anti-inflammatory eicosanoids, are important for a healthy diet. Fish meal, a high-protein food supplement used in aquaculture, is a byproduct of rendering and processing fish for fish oil. Fish emulsion, a fertilizer, is produced from the fluid remainder of fish already processed for fish oil and fish meal. Fish skins, swim bladders, and bones are boiled to produce fish glue for specialized uses, while isinglass, a form of collagen obtained from dried swim bladders, is used for clarifying and refining wine and beer, for preserving eggs, and for conserving parchment. Kelp is steadily growing in popularity as a fertilizer and has long been a major source of iodine. Traditional royal or Tyrian purple is derived from sea snails of the *Murex* genus. Pearls and mother-of-pearl are key components of lustrous jewelry.

Fisheries scientists continue to explore aquatic and marine bioresources for pharmaceuticals, nutraceuticals, and novel biomaterials as well as investigate distribution and biodiversity of marine organisms important to industrial utilization. In finding innovative uses for aquatic and marine products and thereby increasing the value of specific fish stocks, fisheries scientists also make more likely the prospect that a given stock will be more sustainably harvested over the long term.

Aquaculture. In the aquaculture context, fisheries scientists not only examine the impacts of aquaculture practices on the environment (including habitat alteration, release of drugs and chemicals, interaction of cultured and wild organisms, and related environmental and regulatory issues) but also propose and evaluate methods for reducing or eliminating those impacts, including modeling, siting, and monitoring of aquaculture facilities and the use of polyculture and water-reuse systems. Fisheries scientists propose the design criteria, provide operational analysis, and develop management strategies for selected species in water-reuse systems.

Fisheries scientists provide in-depth understanding of the natural and social ecology of aquaculture ecosystems, applying principles of systems ecology to the management of the world's aquaculture ecosystems. They also evaluate the nature, causes, and spread of diseases limiting the success of freshwater and marine aquaculture projects. They provide diagnoses of diseases affecting hatchery management and other aquacultural contexts, as well as define appropriate prevention, control, and treatment strategies for those diseases.

Traditional Fisheries. For traditional or wild fisheries, much fisheries science involves advising and educating government, industry, and the public on the biology of aquatic-resource animals, as well as assessing fisheries' populations, stock abundance, and overall ecological health. Fisheries scientists also help frame the debate concerning aquatic resource management, conservation legislation, rehabilitation of depleted fisheries, and socioeconomic considerations involved in national and international fishery issues, practices, patterns, and public policy.

Fisheries scientists also are involved in the development of new fish-catching methods and technologies, including the development and assessment of electronic enhancements to fishing and fishing-vessel operation that have increased fishing power. Fisheries science contributes to the application of these new methods to scientific sampling, to commercial harvesting, to recreational and subsistence fishing. Fisheries scientists not only contribute to advancements and innovations in fishing-gear construction, maintenance, and operation, but also to the evaluation—through empirical, theoretical, model scaling, and statistical-analysis techniques—of the behavior and performance of fish-capture systems.

CAREERS AND COURSEWORK

Career titles in fisheries science involve both traditional wild fisheries and aquaculture. They include aquarium director, biologist, conservation ecologist, conservation officer, environmental consultant, fish culturist, fish-processing manager, fisheries manager, fisheries technician, fisheries biologist, fisherman, hatchery manager, lab technician, museum or aquarium curator, marine biologist, natural-resource specialist, nature interpreter, lake or pond manager, researcher, vessel captain, and wildlife manager.

Courses in ecology, biology, chemistry, and mathematics are foundational for students wishing to pursue careers in fisheries science. A fisheries technician, fish-hatchery manager, fish-processing manager, lab technician, or nature interpreter may need little more than this background, together with some basic skills in mechanical, electrical, or systems engineering.

The pursuit of the master's and doctoral degrees—which are often the necessary minimum qualification for more advanced academic, governmental, or industrial career in fisheries science—generally requires more specialized coursework beginning at the upper division of undergraduate studies or the beginning of graduate studies. More specialized courses may include genetics, marine biology, limnology, ichthyology, oceanography, and veterinary medicine.

Although fisheries science is primarily biological at its root, it is also strongly interdisciplinary and advisory, so background in fields such as hydrology, anthropology, law, economics, and political science can also prove very helpful.

SOCIAL CONTEXT AND FUTURE PROSPECTS

As more and more of the world's fisheries become overfished or fished out, and wild fish stocks are threatened globally, there is increased pressure to make the changeover from the mechanized hunting-gathering approach of commercial fishing to aquaculture. This must be done in decades, rather than the millennia involved in the analogous changeover to agriculture. Concerns about potential ecological risks, ranging from aquaculture's inherently dense and intense practices to genetic modification of fish species, have also increased as the rapidity of this changeover has increased.

One of the most pivotal roles of fisheries science in the future will be the assessment of how much of the near-shore oceans should remain wild and how much should be intensively farmed. This will involve not only the assessment of ecological risk associated with the growing aquaculture industry but also the determination of how fully and how rapidly depleted or damaged fisheries can recover. Marine reserves, aquatic refuges, and bioremediation all have roles to play here, as do concerns associated with reef die-offs, pollution and climate change, and economic and political pressures on the remaining 20 percent of fisheries worldwide that are not yet fully exploited.

During the twentieth century, the survival or extinction of many populations of aquatic organisms will depend on how the growing population of a terrestrial organism with a fondness for seafood—the human species—interacts with its global environment. Fisheries scientists, in their roles as assessors and advisers, are key to helping shape that interaction.

—Howard V. Hendrix

Further Reading

Beamish, Richard J., and Brian J. Rothschild, editors. *The Future of Fisheries Science in North America*. Springer, 2009.

Clover, Charles. *The End of the Line: How Overfishing Is Changing the World and What We Eat*. U of California P, 2008.

Committee to Review Individual Fishing Quotas, et al. *Sharing the Fish: Toward a National Policy on Individual Fishing Quotas*. National Academy Press, 1999.

Edwards, Charles T. T., and Dorothy J. Dankel, editors. *Management Science in Fisheries: An Introduction to Simulation-based Methods*. Routledge, 2016.

Everhart, W. Harry, and William D. Youngs. *Principles of Fisheries Science*. 2nd ed., Cornell UP, 1981.

Iudicello, Suzanne, Michael Weber, and Robert Wieland. *Fish, Markets, and Fishermen: The Economics of Overfishing*. Island Press, 1999.

McGoodwin, James R. *Crisis in the World's Fisheries: People, Problems, and Policies*. Stanford UP, 1990.

Molyneaux, Paul. *Swimming in Circles: Aquaculture and the End of Wild Oceans*. Thunder's Mouth Press, 2007.

Santhanam, R. *Fisheries Science*. Astral International, 2013.

Walters, Carl J., and Steven J. D. Martell. *Fisheries Ecology and Management*. Princeton UP, 2004.

FLOOD-CONTROL TECHNOLOGY

Fields of Study: hydrology; engineering; ecology; marine biology

ABSTRACT

Flood-control technology deals with the myriad techniques that can be employed to deal with water that overflows stream banks, thereby leading to deaths and injuries, property and crop damage, and severe erosion. The magnitude of the flood damage varies considerably, as it depends on the duration of the storm and the amount of precipitation. Each watershed has different physical features, such as size, slope, basin re-

lief, impoundments, flood history, soil types, and drainage characteristics. While flooding, and severe flooding, are natural events, the interaction of floodwaters with areas of human activity introduces the nonnatural byproducts of human activity into marine environments, where they may have many unforeseen consequences.

KEY CONCEPTS

flooding: the condition in which water has surpassed its normal environmental containment features, ranging from stream banks to the ocean's verge

hydrologic cycle: also called the water cycle, in which Earth's water changes state from liquid and solid to atmospheric gas, and condenses back to liquid or solid and falls as precipitation

hundred-year flood: the magnitude of a flood that has a one-in-one-hundred chance of occurring in any given year

tsunami: popularly, but erroneously, called a "tidal wave," the sudden uplift of local sea level caused by the displacement of ocean water due to earthquakes, volcanic activity or landslides

DEFINITION AND BASIC PRINCIPLES

Flood-control technology includes various structural and nonstructural measures that play a substantial role in mitigating flood damages. The methods vary from place to place, given the enormous heterogeneity of stream discharge, precipitation frequency and amount, and watershed factors that include slope, soil permeability, infiltration characteristics, degree of urbanization, varying land-management practices, governmental interest, and land-ownership practices.

While flooding, and severe flooding, are natural events, the interaction of floodwaters with areas of human activity introduces the nonnatural byproducts of human activity into marine environments, where they may have many unforeseen or even disastrous consequences. Human activities often contribute to increasing the severity of flooding. For example, areas in which flood-retarding natural growth has been removed for development present opportunities for large quantities of earth, and its contents, that would otherwise have been stabilized, to be introduced into marine environments. Industrial byproducts, chemical and biological wastes and many other types of debris may be carried away by floodwaters coursing through cities, to larger bodies of water and ultimately into the oceans. Tsunamis in particular represent a worst-case scenario of such flooding. The purpose of flood control technology is to mitigate or prevent damage from such occurrences as much as possible.

The structural or physical measures that are employed include levees or flood walls made of earth or concrete, channel modifications such as deepening or widening the stream itself, building dams or reservoirs to hold additional water, watershed-improvement techniques that include reforestation and planting of vegetative cover in bare areas, flood proofing in lower-risk floodplain areas that include dikes, elevating the buildings, and waterproofing.

The nonstructural or preventative measures that could be adopted include floodplain regulations such as zoning laws that stipulate where development may or may not occur, building codes that put restrictions on basements and the permissible amount of impervious cover allowed on the site, open-space requirements for large-scale developments, and designing flood warning systems to provide advance notice about impending problems.

BACKGROUND AND HISTORY

In ancient Egypt, which depended heavily on the regular flood cycle of the Nile River for agriculture, a relatively complex system of monitoring stations was devised for use by priests, who would use their observations to deliver predictions of when the floods would arrive and what they would be like. The beginnings of more scientific speculation about water movement (the hydrologic cycle) began in ancient times by such renowned philosophers as Homer, Thales, Plato, and Aristotle in Greece, and later on by Lucretius, Seneca, and Pliny in ancient Rome. The field of hydrometry, which pertains to stream-flow measurement, began in the seventeenth century and improved techniques increased with each century. For example, flow formulas, measuring instruments, and stream-gauging procedures became better established in the eighteenth and later centuries. Some of the many advances include civil engineer Theodore G. Ellis's and surveyor William Gunn Price's stream current meters in 1870 and 1885, respectively, Irish engineer Robert Manning's flow formula in 1891, civil engineer Allen Hazen's comments on increasing the use of statistics in flood studies in 1930, and

German statistician E. J. Gumbel's suggestion for using frequency analysis for floods in 1941.

Major changes at the governmental level in the United States assumed increasing importance as several hydrologic agencies were created in the nineteenth century: the Army Corps of Engineers in 1802, the Weather Bureau in 1870 (now called the National Weather Service [NWS]), the US Geological Survey (USGS), and the Mississippi River Commission in 1879.

HOW IT WORKS

Floods. In the hydrologic cycle, atmospheric precipitation falls to Earth and is either evapotranspired back to the atmosphere, absorbed by vegetation, or moved downslope as overland flow that eventually becomes large enough to form stream channels. Water generally stays within the channel for most of the year. However, if the precipitation is heavy enough, the channel cannot transport all of this water and the stream overflows its banks, thereby creating floods.

Tsunami's represent a special class of flooding in that, while caused by natural events, they are not the result of excess rainfall or too-rapid snowmelt. Rather, tsunamis are caused by the sudden displacement of a large mass of water either as the result of an undersea earthquake, volcanic activity or a landslide that suddenly deposits a large mass of earth and rock into a body of water. The sudden rise of sea level as the water displacement reaches shore overflows the normal confines of the water and drives inland with a great deal of force. The inundation of land surfaces by

Floodplains of the Waal river (Bemmell, Lingewaard, Netherlands) (Lendskaip via Wikimedia Commons)

such an event is almost always catastrophic, resulting in massive property damage and loss of life. Similarly, storm surge due to tropical storms and hurricanes pushes massive quantities of seawater onto land. The flooding by storm surge is caused by the dual effect of wind pushing surface water in large waves, and the rising of the water column under the low atmospheric pressure of the storm. The atmosphere surrounding the storm tends to press down on the water, pushing it up under the low pressure center, much as the atmosphere pushes soda up a straw when the pressure inside the straw is reduced.

Floodplains are low areas that can be found on one or both sides of a stream and are common in humid regions. Small watersheds with steep slopes often have flash floods that move very quickly with enough turbulence to damage buildings and vehicles easily. Since the flash floods occur so rapidly, people cannot be readily warned and drownings can occur.

Stream Discharge. In the United States, stream flow is measured in cubic feet per second at most gauging stations operated by the USGS. Most of the larger streams have higher discharges as they flow downstream. There are some exceptions, such as the Colorado River in the arid southwestern United States, where the demand for irrigation water and public supply literally dries up the stream at its mouth.

Flood Frequency. It would be wonderful if flood prediction were reasonably exact. However, the probability of knowing when floods would occur turns out to be a markedly difficult task as it involves frequency analysis, probability theory, data that need to be homogeneous and independent, and selected assumptions that existing stream-flow data would be similar to future flows. This fallacious assumption implies that there will be no changes in future land use and climate.

The longer the period of record, the better the estimation of future flood flows. Accordingly, a "one hundred-year flood" does not mean that the next one of that magnitude will occur one hundred years in the future. It does mean that there is a 1 percent chance that a one hundred-year flood could occur in any year. Accordingly, a one hundred-year flood could occur this year and again the following year, although the odds are against it. In hindsight, it would have been better if the term "one hundred-year flood" would have been expressed as a "one-in-one hundred chance flood" in order to reduce public confusion.

Floodplain Development. The stream channel itself occupies a relatively narrow area within a floodplain. The channel is bordered by low-lying land that is called a floodway. As the distance away from the river increases, the slope of the land gradually increases and a slightly higher landscape called a floodway fringe is created. A cross-sectional profile of this floodplain would then show a channel with a floodway on one or both sides of the channel and a floodway fringe farther away from the stream.

In a nonregulated development scenario that was common in earlier years and still exists in some areas, residential and commercial buildings were constructed in the floodway zone, which lead to higher flood levels, structural damage, and potential loss of life. Low bridges that cross the stream in this zone could partially block the flow and thereby increase flood levels.

The situation would be quite different in a regulated floodplain. Ideally, the floodway would be zoned for as much open space as possible, such as parks, picnic areas, and golf courses. This would facilitate easier passage of floodwaters and restrict buildings to higher ground wherever possible.

Floodplain Extent. It has been estimated that one hundred-year floodplains make up from 7 to 10 percent of the total land area of the United States. The floodplains with the largest areas are located in the southern portions of the country; those with large populations are located along the north Atlantic coast, the Great Lakes, and California.

Types of Floods. Damaging floods occur in varying locations. Flash floods are associated with quickly developing thunderstorms in mountainous areas. The rapid downslope movement of water, even if relatively shallow, can quickly move vehicles and their occupants to the extent that 50 percent of the flood deaths in the United States are caused by flash floods. If floodwater is flowing at a depth of only 2 feet (0.6 meters), a trapped vehicle experiences a lateral force of 1,000 pounds (454 kilograms) and a buoyant force of 1,500 pounds (680 kg), more than enough to flip the vehicle over and drown the passengers.

Regional floods are associated with level land that experiences long periods of heavy rain. The resulting damage to homes and properties in the floodplain can be enormous, as evidenced in the August 1993 flood in the lower Missouri and upper Mississippi rivers. About 69 percent of the levees built along the upper Mississippi River to protect the floodplain occupants were overtopped, drowning forty-eight persons, submerging seventy-five towns, destroying fifty thousand homes, and even excavating more than seven hundred coffins from a cemetery in Missouri.

Storm surges powered by tropical storms and hurricanes can wreak havoc along low-lying coasts. The settlers who founded New Orleans in 1718 along the lower Mississippi River experienced their first major flood the same year. Over time, many other floods have occurred in New Orleans and the lower reaches of the Mississippi, aided by the slow sinking of the floodplain as river sediments compact over time.

New Orleans, Louisiana in the aftermath of Hurricane Katrina (AP Photo/U.S. Coast Guard, Petty Officer 2nd Class Kyle Niemi via Wikimedia Commons)

APPLICATIONS AND PRODUCTS

Structural Flood-Control Objectives. Most structural flood-control measures are physical and expensive. They include construction of large reservoirs, diversion structures, levees and flood walls, channel alterations, modifications to bridges and culverts, and tidal barriers.

Flood-control reservoirs are built so that excess water from storms can be stored and released at a later time, thereby reducing peak discharge. The economic value of protecting property that may be damaged in floods can justify the construction costs of a reservoir. In addition, the stored water can also be used for hydropower, water-supply purposes, and recreation. Diversion structures are designed to reduce peak flows by forcing floods to go to another location via channels or tunnels. Levees and flood walls can physically keep water away from floodplains, thereby providing good potential for damage reduction. Levees are usually made of earth and flood walls of concrete. Both structures are usually set parallel to the stream.

Channel alterations to the stream can lower the height of water so that peak discharges are reduced at one point, but they can be increased downstream. The cost of these measures is substantial, but they can be justified in certain situations by their potential protection for valuable property. Possible long-term negative effects of these structural measures include aggradation or degradation of the downstream channel and sediment deposition in downstream bypass channels or tunnels. Bridge and culvert modifications are useful when the structures have inadequate capacity to handle flood flows. Repairing or raising the structures over the channel can lower damages by allowing more water to flow downstream. Tidal barriers along the coast can prevent high tides from moving upstream and damage developed areas. They are expensive to build and are generally used if large urbanized areas need protection. Note that the potential for tidal flooding is expected to increase as sea level rises because of climate change.

Nonstructural Flood-Control Objectives. These measures are employed in order to lower flooding susceptibility, thereby decreasing possible damage. They include flood warning, flood proofing, and a variety of land-use control procedures. Flood warning can reduce possible loss of life, or even better, eliminate that possibility. Flood-proofing procedures include rearranging the working space in buildings, waterproofing outside walls, and raising the height of buildings that occupy particularly susceptible locations. Land-use controls include many kinds of action within the floodplain that can reduce flood-hazard potential. These controls include proper building codes, purchasing flood insurance, zoning restrictions, or purchase of land and property in the less vulnerable portions of the floodplain.

Climate Change. There is growing concern that global climate change could substantially impact flooding in coastal locations. For example, since 1870 sea level has increased about 5.9 inches (15 centimeters). The overall sea-ice cover in the Arctic has been decreasing about 3 percent per decade and by 7 percent per decade in the summer. Estimates indicate that average global surface temperatures for the period 1990–2100 will increase between 3.1 and 10.5 degrees Fahrenheit (–16 and –12 degrees Centrigrade). As a result, sea level could increase between 8.3 to 18.5 in (21 to 47 cm) by the end of the twenty-first century. In addition, future peak precipitation events are expected to increase in intensity. To make matters worse, some of the climatic models employed suggest that rates of sea-level rise will double along the shorelines of certain portions of the eastern United States and the western North American and Arctic coasts.

Flood Mapping. It would be very useful to be able to generate flood maps in advance of approaching storms. Experimental work has been developed by the USGS and the National Weather Service (NWS) to make forecasting and mapping of potential floods readily available to local officials. The goal is to have NWS provide storm forecasts for a particular area, thereby enabling the USGS to generate maps showing potential flooded areas on the floodplain, arrival times, and the depth of the flood itself. The combination of new methodologies and techniques, such as light detection and ranging (LIDAR), advanced computer programs, and geographic information systems (GIS) have led to the strong possibility of creating useful flood-forecast maps.

Flood Warning Systems. A useful addition to the regulation of floodplain development occurs when flood-warning systems are employed by local governments. These systems use radar, rainfall, and stream-flow gauging that are connected by satellite transmitters to relay real-time data to computers at some central site, which in turn make the data available to interested parties via the internet. Automatic warnings are then dispatched to emergency-management officials who can then institute procedures, based on the predicted flood levels, that range from selective road closures to complete evacuation.

CAREERS AND COURSEWORK

There are a surprising number of varied courses and future jobs that are included in the field of flood-control technology. The most obvious ones include civil engineering for construction of dams and levees, hydraulic engineering to handle channel deepening where appropriate, and surveying. However, there are many other skills that are needed, and it would benefit those interested in this field to study geology, physical geography, hydrology, meteorology (storms), and water-resources management. Other useful courses of study include environmental planning, computer science, economics, and environmental law.

Many governmental organizations, including the Army Corps of Engineers, American Water Resources Association, and FEMA (Federal Emergency Management Agency), have job postings on their Web sites that encompass a range of employment opportunities. These agencies employ civil engineers, scientists, natural-resource specialists, administrators, hydrologic technicians, and ecologists, among others.

SOCIAL CONTEXT AND FUTURE PROSPECTS

Floods have occurred on numerous occasions over many millennia. Societies have designed ways to store some of this water in special collecting basins or small reservoirs, and in selected cases by creating diversion canals. In other cases, the floods have been very damaging to settlements and farms. What makes the situation even worse is the population growth on floodplains, the inhabitants of which have the mistaken notion that floods can be contained by structural techniques. Too many damaging floods have occurred in too many countries, and that renders naive the presumption that one is safe behind the levee.

One step in the right direction has been the growing recognition that floodplain development carries a substantial risk: predicting flood heights is fraught with uncertainty because of a complicated mix of storm tracks, soil-moisture conditions, available upstream reservoir storage, and the elevations of commercial facilities and homes. One can now recognize the benefits of nonstructural measures, such as zoning lands that are in the one hundred-year floodplain as not suitable for development. To make matters more interesting, and more complicated, how does one factor in shifts that could occur over many decades, such as climate change, river behavior, and sea-level rise that would impact coastal areas as well as the existing homeowners in the floodplain who simply enjoy living close to rivers?

Two things will certainly help out in this situation. One would simply be better explanation of the flooding danger as given by the probability of certain events occurring in any year. These odds could, of course, change as floodplain development continues or decreases. The second element is simply better discussion of future weather patterns, as unpredictable as they are, that could occur on a global scale.

—Robert M. Hordon

Further Reading

Bedient, Philip B., Wayne C. Huber, and Baxter E. Vieux. *Hydrology and Floodplain Analysis*. 5th ed., Prentice-Hall, 2013.

Cech, Thomas V. *Principles of Water Resources: History, Development, Management, and Policy*. 3rd ed., Wiley, 2010.

Chang, Fi-John, Kuolin Hsu, and Li-Chiu Chang, editors. *Flood Forecasting Using Machine Learning Methods*. MDPI, 2019.

Dunne, Thomas, and Luna B. Leopold. *Water in Environmental Planning*. Freeman, 1978.

Dzurik, Andrew A. *Water Resources Planning*. 3rd ed., Rowman, 2003.

Mays, Larry W. *Water Resources Engineering*. 2nd ed., Wiley, 2010.

Strahler, Alan. *Introducing Physical Geography*. 5th ed., Wiley, 2010.

Viessman, Warren, Jr., and Gary L. Lewis. *Introduction to Hydrology*. 5th ed., Prentice-Hall, 2003.

Wohl, Ellen. *Sustaining River Ecosystems and Water Resources*. Springer, 2018.

Wylson, Anthony. *Aquatecture: Architecture and Water*. Elsevier, 2013.

FLOODS

Fields of Study: hydrology; geography; geology; meteorology; climatology; oceanography

ABSTRACT

Floods are extreme conditions of flowing water. They generally occur because of inordinate amounts of rainfall or snowmelt, but floods may also result from other causes, such as dam failures or volcanic eruptions. Floods play a major role in shaping river systems, and their occurrence is critical to the human use of riparian lands.

KEY CONCEPTS

alluvial: describes a river or watercourse that carries and deposits its silt load in its associated floodplain

floodplain: the area bordering a river or watercourse that will become inundated by flood waters

headwater: the collective term for the area that begins the watershed of a particular watercourse, typically the farthest points from the exit point of the watercourse

hydrology: the study of natural water movement in the environment

infiltration capacity: the ability of a soil to absorb water before it can become surface runoff

riparian lands: the lands beside freshwater watercourses

CAUSES OF FLOODS

Floods involve flows of water in rivers and streams that exceed the ability of those watercourses to contain them. Some technical hydrological definitions of floods refer to stages, or heights, of water above some reference level, such as the banks of a river channel. In practice, however, floods can be thought of as any extreme flow of water that exceeds usual experience, often damaging or threatening life and property.

Floods may be caused by a variety of physical factors, including dam failures and the subsidence of land. The most common kind of flood occurs when excessive precipitation and physical factors of the land combine to produce maximum runoff of water. Precipitation can yield runoff directly, or water from melting snow can produce the flow. Very intense flash floods are usually associated with heavy, short-duration rainfall from thunderstorms. Such rainfall easily overwhelms the infiltration capacity of the ground, and water rapidly runs off from the local drainage area or watershed into adjacent stream channels. If enough water concentrates in the channel to exceed the carrying capacity of the channel, it will constitute a flood.

Physical factors on the land surface also determine the rate of concentration of floodwater in stream channels. For example, less permeable soils will allow water to run off faster, as will a lack of vegetation. Artificial enhancement of runoff occurs when slopes are covered by impermeable materials. This situation commonly occurs in construction, when the natural ground surface is replaced by buildings and pavement, resulting in extreme enhancement of runoff. In cities, the yield of floodwater from paved surfaces may be ten times greater than the same storm's yield under natural conditions. In this way, human construction tends to exacerbate flooding problems in urban areas.

SEDIMENT AND BEDROCK

As the flow of water increases in a stream channel, it is usually associated with considerable sediment. If such waterborne sediment also composes extensive deposits adjacent to the stream channel, the river is termed "alluvial." An alluvial river commonly has a floodplain of deposited sediment adjacent to the channel. The floodplain is really an intimate part of the river system, as sediment is added to it every time the river rises above its banks. As the water rises, the river's depth rapidly increases, causing an increased ability to transport sediment that is eroded from the bed and banks. If the stream is appropriately loaded with sediment, it will be deposited when the banks are overtopped, and the width of the flow greatly increases. If not enough sediment is supplied by erosion, however, the increasingly energetic flood flows will be erosive, attacking the banks, widening the channel, and restoring the appropriate sediment load to the stream.

When the bed and banks of a river are composed of bedrock these adjustments of sediment load to flow energy cannot occur. Because bedrock is very resistant to erosion, the energy level in the flow can rise spectacularly without being damped by sediment. The excess energy of such sediment-impoverished floods goes into the development of turbulence. Turbulence at high energy levels takes on an organized structure of powerful vortices that produce immense pressure changes. These pressure effects may be sufficient to erode the bedrock boundary by a "plucking" action.

A famous geological controversy once surrounded the problem of bedrock erosion by great floods. In the 1920s, University of Chicago geologist J. Harlen Bretz (1882–1981) proposed that immense tracts of eroded basalt bedrock in Washington State had been created by a catastrophic glacial flood. Bretz subsequently showed that the fascinating landforms of that region, known as the Channeled Scablands, were created when a great lake impounded by glacial ice burst. The lake, glacial Lake Missoula, had been more than 1,969 feet (600 meters) deep at its ice dam. It took Bretz nearly fifty years to convince his many critics that catastrophic flooding could explain all the bizarre features of the Channeled Scablands. Geologists now know that the physics of catastrophic flooding is completely consistent with Bretz's observations. Missoula flood flows moved at depths of 328 to 656 ft (100 to 200 m) and velocities of 66 to 98 ft (20 to 30 m) per second. The power (rate of energy expenditure) per unit area for such flows is thirty thousand times that for a normal river, such as the Mississippi, in flood. The reason for such immense flow power is that the Missoula flooding occurred on very steep slopes, giving the water great potential energy when the dam burst.

The Missoula floods occurred during the last ice age, more than twelve thousand years ago. There are, however, modern examples of glacial floods called

Geologist J Harlen Bretz (Dr. Julian Goldsmith via Wikimedia Commons)

Gígjökull glacier outlet of Eyjafjallajökull, Iceland, showing evidence of jökulhlaups (Richard Gould via Wikimedia Commons)

jökulhlaups. In Iceland, jökulhlaups occur where glaciers overlie active volcanoes. Volcanic heat releases water that is stored in subsurface reservoirs by melting the overlying ice. Lakes may also form adjacent to the ice masses. Because ice is less dense than water, such juxtapositions are inherently unstable. When the pressure is high enough, the water may lift the ice dam and burst out from beneath the glacier. The jökulhlaups move house-sized boulders and transport immense quantities of sediment.

DISCHARGE AND RECURRENCE INTERVAL

The volume of water released by a flood per unit of time is termed its "discharge." This quantity is the magnitude of the flow that will potentially inundate an area. The chances of experiencing floods of different magnitudes are expressed in terms of frequency. Large, catastrophic floods have a low frequency, or probability of occurrence; smaller floods occur more often. The probability of occurrence for a flood of a given magnitude can be expressed as the odds, or percent chance, of the recurrence of one or more similar or bigger floods in a certain number of years. Analyses of flood magnitude and frequency are achieved by measuring floods and statistically analyzing the data. Results are expressed in terms of the probability of a given discharge being equaled or exceeded in any one year. The reciprocal of this probability is the return period, or recurrence interval, of the flooding, expressed as a number of years.

The concept of a recurrence interval is sometimes confusing; it is simply a statistical measure, based on historical measures of the flood frequency of a river flow of the likelihood of a flood of a given magnitude occurring over a one-year period. A flood that has a probability of occurrence in one year of 0.1, or 10 percent, is called a "ten-year flood," because the one-year probability (0.1) multiplied by 10 is equal to 1. By extension, a larger flood that has a probability of occurrence of 0.01, or 1 percent, in a single year is called a "hundred-year flood," because 0.01 multiplied by 100 is equal to 1. Note that these numbers do not preclude several such floods occurring in a given period; for example, two or more floods of the one-hundred-year magnitude could occur in the same year. Such an event, while unlikely, does have a small probability of occurring. In fact, in August 2017, in advance of the flooding caused by Hurricane Harvey in Houston, Texas—later calculated to be a "thousand-year flood," the first such recorded event in US history, with a 0.001 probability (0.1 percent) of occurring in a single year—meteorologist Eric Holthaus reported that the Houston area had seen four hundred-year floods since May 2015.

Flood magnitude-frequency relationships vary immensely with climatic regions. In the humid-temperate regions of the globe, such as the northern and eastern United States, streamflow is relatively continuous. Even rare floods are not appreciably larger than more common floods. A result of this relationship is that stream channels have developed in size to convey the relatively common, moderate-sized floods with maximum efficiency. In contrast, the more arid regions of the southwestern United States have immensely variable flood responses. Stream channels may be dry most of the time, filling with water only after rare thunderstorms. The flash floods that characterize these streams may also be highly charged with

Flooding in Central Texas, near Lake Travis (RoschetzkyIstockPhoto via iStock)

sediment; indeed, the sediment may so dominate the flow that the phenomenon is called "mudflow or debris flow" rather than streamflow. Because extreme events dominate in these environments, the stream channels have developed in size according to these rare, great floods.

MONSOONS AND ANCIENT FLOODS

In tropical areas, some of the greatest known rainfalls are produced by tropical storms and monsoons. Monsoons are seasonal wet-to-dry weather patterns driven by atmospheric pressure changes over the oceans and continents, which result in alternations between dry periods that inhibit vegetation growth and immensely wet periods that facilitate runoff. Tropical rivers thus have a pronounced seasonal cycle of flooding, and some floods may be immense. These rivers also show channel-size development in accordance with rare, great flows. Some of the most populous places on Earth are situated on seasonal tropical rivers, such as the Ganges and Brahmaputra Rivers in India and Bangladesh. Immense tragedies have occurred when a particularly severe monsoon or tropical storm has produced especially great floods. In 1974, monsoon-related flooding in Bangladesh killed 2,500 people. This pales in comparison to a flood in 1970 that took 500,000 lives and another in 1991 that killed more than 100,000 people. India's National Disaster Management Authority has estimated that an average of 1,600 lives are lost annually due to flooding, with the greatest loss of life—11,316 people—having occurred in 1977.

Studies of ancient floods (paleofloods) through geological reconstruction of past discharges show that monsoons and other flood-generating systems have varied in the past. Between about ten thousand and five thousand years ago, floods were apparently much more intense in many world areas on the boundaries between the tropics and the midlatitude deserts. These intense floods may have been related to long-term glacial-to-interglacial cycles. Because of modern increases in atmospheric carbon dioxide and other greenhouse gases it appears quite likely that tropical floods may again become more intense, if this process has not already begun. Such a situation could have grave consequences for flood-prone tropical countries.

STUDY OF FLOODS

Hydrologists study floods by measuring flow in streams. Measurements are taken at stream gauges, locations at which mechanical devices are used to record the water level, or stage, of the river. To transform these stage measurements into discharge values, the hydrologist must perform a rating of the stream gauge. This is accomplished by measuring velocities in the stream channel during various flow events. A velocity meter with a rotor blade calibrated to the flow rate is used for this purpose. When the average measured velocity of the stream channel is multiplied by the cross-sectional area of the channel, the result is the discharge for that flow event. When several flows at different stages are measured, the data are used to generate a rating curve for the gauge. This curve shows discharges corresponding to any stage measured at the gauge.

The discharge values obtained at a stream gauge are collected over many years, constituting a record that can then be used in flood-frequency analysis. Several statistical procedures can be employed to plot the flood experience and to extrapolate to ideal values of the ten-year flood, the one-hundred-year flood, and so on. The discharges associated with these recurrence intervals are then used as design values for hazard assessment, dam construction, and other flood controls. "Flood control" is probably a misnomer, however, because flows larger than the design floods are always possible. Flood control really involves providing various degrees of flood protection.

Another approach to evaluating extreme flood magnitudes involves careful study of the precipitation values that generate the greatest known floods. By transposing the patterns of known extreme storms to other areas, scientists can, in theory, calculate what the runoff would be from hypothetical great storms. Such calculations involve the use of a rainfall-runoff model. These models are prevalent in hydrology because they can be easily programmed. The models give idealized predictions of how water from a given storm would concentrate. The flood discharge modeled from the assumed maximum rainfall is called the "probable maximum flood."

Unfortunately, there are problems inherent to both of these traditional hydrological approaches to flood studies. Both make assumptions in calculating potential flood flows. Another procedure is to study the natural records of ancient floods, or paleofloods, that are preserved in geological deposits or in erosional features on the land. It has been found for some sections of bedrock that nonalluvial rivers act as natural recorders of extremely large flood events. These natural flood gauges can be interpreted only by detailed studies that combine geological analysis with hydraulic calculations of the ancient discharges.

Paleoflood hydrology generates real data on the largest floods to occur in various drainages over several millennia. The data include the floods' ages, or the periods in which they occurred, and their discharges. The information can be used directly in a flood-frequency analysis, or it can be used to assess the probable validity of extrapolations from conventional data on smaller floods. Paleoflood data can also be compared with probable maximum flood estimates. In this way, the expense of overdesign and the danger of underdesign can be avoided in flood-related engineering projects.

SIGNIFICANCE

People have lived with floods since the beginning of civilization. The first great civilizations on Earth developed along the fertile but flood-prone valleys of rivers such as the Nile, the Tigris and the Euphrates, the Indus, and the Yangtze and the Huang He. Various means of coping with floods have been documented since the biblical accounts of Noah. Early societies merely avoided zones that tradition told them were hazardous. It has only been in the modern era that large cities have systematically developed on immense tracts of flood-prone lands. Thus the natural process of flooding has become an unnatural hazard to humans.

Floods can have immense consequences for both human life and infrastructure. In the 1930s, a national US program began to respond to such problems by constructing large dams. Despite, or perhaps because of, this expensive effort, flood damage to life and property is much greater today than it was in the 1930s. A single tropical storm system, Hurricane Agnes, killed 128 people overall and generated nearly $3 billion in flood damage to the northeastern United States in 1972; in 2011, the flooding of the Ohio, Mississippi and Missouri Rivers in April and May, plus the flooding caused by Hurricane Irene and Tropical Storm Lee in August and September, caused a total of 108 deaths and more than $8 billion in damages in the United States; and New Orleans may never fully recover from the flooding caused by Hurricane Katrina in 2005, which resulted in as many as 1,464 deaths and $70 billion in damages in that city alone. The flooding caused by Hurricane Harvey in 2017 was estimated to have caused between $65 billion and $190 billion worth of damage in Texas, the upper range of which would make it the costliest natural disaster in US history, while Hurricane Irma, which followed just days after, inflicted an additional $50 billion to $100 billion in damages in the southeastern United States.

Elsewhere, flooding is a particularly serious issue in South Asia, where monsoon season is an annual occurrence and flood-control measures are largely limited to Dutch-style embankments that do not take into account the greater amount of sediment present in the major rivers. In 2007, unusually heavy rains during monsoon season caused thirty-seven major floods throughout several South Asian countries, including India, Bangladesh, Pakistan, and Nepal, killing more than 4,000 people and displacing or otherwise affecting more than 45 million. In summer 2017, flooding in India, Bangladesh, and Nepal caused more than 1,200 deaths and destroyed close to a million houses, affecting more than 41 million people.

In addition to the dangers posed by flooding itself, the aftermath of a major flood can pose serious health risks to residents. Remaining floodwater may be contaminated with sewage and chemicals, and that contamination can be transferred to flood-damaged homes. Mold, which grows quickly in warm, damp conditions, can cause or trigger respiratory issues. Standing water attracts mosquitoes, increasing the risk of mosquito-borne illnesses, and in some areas the flooded environment may become temporarily inhabited by displaced dangerous creatures such as poisonous snakes and alligators or crocodiles. In addition, the stress, anxiety, and depression caused by the event can lead to serious mental health issues for survivors.

One trend in flood management has been to manage flood-hazard zones with multiple approaches that respond to the nature of the flood risk. The river is treated as a whole integrated system, rather than as individual segments, for engineering design. Management alternatives for this system are not limited solely to structural controls, such as dams and levees. Instead, options are considered for land-use adjustment. Flood-prone lands can be used for parks, greenbelts, and bikeways instead of industrial warehouses, stores, and housing. Even when construction must be done on floodplains, it may be possible to make provision for flood risks. Warehouses, for example, can be organized for rapid transfer of materials to second stories or to temporary, safe storage sites. Such adjustments require accurate and timely warning systems that involve measuring rainfall in headwater areas and rapidly predicting the flood consequences to downstream sites at risk.

There is a general need to educate the public that floodplains are a natural part of rivers. Living on a floodplain is really choosing to play a game of "floodplain roulette," in that it is known with certainty that a flood will happen, but it is not known when that flood will happen. For this reason, the most accurate and reliable methods of evaluating flood magnitudes and frequencies are necessary. The choices made for land use on floodplains cannot be based merely on idealized theories of how floods behave.

—Victor R. Baker

Further Reading

Adams, Thomas E., and Thomas C. Pagano, editors. *Flood Forecasting; A Global Perspective*. Academic Press, 2016.

Aldrete, Gregory S. *Floods of the Tiber in Ancient Rome*. Johns Hopkins UP, 2007.

American Society of Civil Engineers, Hurricane Katrina External Review Panel. *The New Orleans Hurricane Protection System: What Went Wrong and Why*. ASCE, 2007. *ASCE Library*, doi:10.1061/9780784408933. Accessed 6 Sept. 2017.

Baker, Victor R., editor. *Catastrophic Flooding: The Origin of the Channeled Scabland*. Dowden, Hutchinson & Ross, 1981.

Freitag, Bob, et al. *Floodplain Management: A New Approach for a New Era*. Island Press, 2009.

Leonard, Barry, editor. *Floods: The Awesome Power*. DIANE Publishing, 2009.

Martini, I. Peter, Victor R. Baker, and Guillermina Garzón, editors. Flood and Megaflood Processes and Deposits: Recent and Ancient Examples. John Wiley & Sons, 2009.

O'Loughlin, Karen Fay, and James F. Lander. *Caribbean Tsunamis: A 500-Year History from 1498–1998*. Kluwer Academic Publishers, 2003.

Prothero, Donald R. *Catastrophes! Earthquakes, Tsunamis, Tornadoes, and Other Earth-Shattering Disasters*. Johns Hopkins UP, 2011.

Sene, Kevin. *Flash Floods: Forecasting and Warning*. Springer, 2013.

Sene, Kevin. *Flood Warning, Forecasting and Emergency Response*. Springer, 2008.

Soennichsen, John. *Bretz's Flood: The Remarkable Story of a Rebel Geologist and the World's Greatest Flood*. Sasquatch Books, 2010.

Warner, Jeroen Frank, et al., editors. *Making Space for the River: Governance Experiences with Multifunctional River Flood Management in the US and Europe*. IWA Publishing, 2013.

FLUID DYNAMICS

Fields of Study: mathematics; engineering; marine engineering; oceanography

ABSTRACT

Fluid dynamics is an interdisciplinary field concerned with the behavior of gases, air, and water in motion. An understanding of fluid dynamic principles is essential to the work done in aerodynamics. It informs the design of air and spacecraft. An understanding of fluid dynamic principles is also essential to the field of hydromechanics and the design of oceangoing vessels and submersibles. Knowledge of fluid dynamics is essential for understanding the motion of ocean currents and circulation. Any system with air, gases, or water in motion incorporates the principles of fluid dynamics.

KEY CONCEPTS

compressible fluids: fluids that can change volume by the application of external pressure, particularly gases

incompressible fluids: fluids that do not change volume with the application of external pressure, particularly liquids

laminar flow: flow in which there is no turbulence in the fluid as it flows, as all particles making up the fluid flow uniformly in parallel paths, or "lamina"

turbulence: nonlaminar flow, characterized by chaotic motion within the fluid, as the flowing motion of particles making up the fluid is disrupted and forced to flow in nonparallel paths, or "lamina"

viscosity: the temperature-dependent property relating the ability of different fluids to flow

DEFINITION AND BASIC PRINCIPLES

Fluid dynamics is the study of fluids in motion. Air, gases, and water are all considered to be fluids. When the fluid is air, this branch of science is called aerodynamics. When the fluid is water, it is called hydrodynamics.

The basic principles of fluid dynamics state that fluids are a state of matter in which a substance cannot maintain an independent shape. A fluid will take the shape of its container, forming an observable surface at the highest level of the fluid when it does not completely fill the container. Fluids flow in a continuum, with no breaks or gaps in the flow. They are said to flow in a streamline, with a series of particles following one another in an orderly fashion in parallel with other streamlines. Real fluids have some amount of internal friction, known as viscosity. Viscosity is the comparative temperature-dependent property that describes a fluid's ability to flow. Fluids with lower viscosity flow more readily than fluids with higher viscosity. Thus, molasses flows more slowly than water at room temperature.

Fluids are said to be compressible or incompressible. Water is an incompressible fluid because its density does not change when pressure is applied. Incompressible fluids are subject to the law of continuity, which states that fluid flows in a pipe are constant. This theory explains why the rate of flow increases when the cross-sectional area of a pipe is reduced and vice versa.

The viscosity of a fluid is an important consideration when calculating the total resistance on an object. The point where the fluid flows at the surface of an object is called the boundary layer. The fluid "sticks" to the object, not moving at all at the point of contact. The streamlines further from the surface are moving, but each is impeded by the streamline between it and the wall until the effect of the streamline closest to the wall is no longer a factor. The boundary layer is not obvious to the casual observer, but it is an important consideration in any calculations of fluid dynamics.

Most fluids are Newtonian fluids. Newtonian fluids have a stress-strain relationship that is linear. This means that a fluid will flow around an object in its path and "come together" on the other side without a delay in time. Non-Newtonian fluids do not have a linear stress-strain relationship. When they encounter shear stress, their recovery varies with the type of non-Newtonian fluid.

A main consideration in fluid dynamics is the amount of resistance encountered by an object moving through a fluid. Resistance, also known as drag, is made up of several components, all of which have one thing in common: they occur at the point where the object meets the fluid. The area can be quite large, as in the wetted surface of a ship, the portion of a ship that is below the waterline. For an airplane, the equivalent is the body of the plane as it moves through the air. The goal for those who work in the field of fluid dynamics is to understand the effects of fluid flows and minimize their effect on the object in question.

BACKGROUND AND HISTORY

Swiss mathematician Daniel Bernoulli (1700–1782) introduced the term "hydrodynamics" with the publication of his book *Hydrodynamica* in 1738. The name referred to water in motion and gave the field of fluid dynamics its first name, but it was not the first time water in action had been noted and studied. Leonardo da Vinci (1452–1519) made observations of water flows in a river and was the one who realized that water is an incompressible flow and that for an incompressible flow, V = constant. This law of continuity states that fluid flow in a pipe is constant. In the late 1600s, French physicist Edme Mariotte (1620–1684) and Dutch mathematician Christiaan Huygens (1629–1695) contributed the velocity-squared law to the science of fluid dynamics. They did not work together, but they both reached the conclusion that resistance is proportional not to velocity itself but to the square of the velocity.

Sir Isaac Newton (1642–1727) put forth his three laws of motion in the 1700s. These laws play a fundamental part in many branches of science, including fluid dynamics. In addition to the term hydrodynamics, Bernoulli's contribution to fluid dynamics was the realization that pressure decreases as velocity increases. This understanding is essential to the understanding of lift.

Leonhard Euler (1707–1783), the father of fluid dynamics, is considered by many to be the preeminent mathematician of the eighteenth century. He is the one who derived what is today known as the Bernoulli equation from the work of Daniel Bernoulli. Euler also developed equations for inviscid flows. These equations were based on his own work and are still used for compressible and incompressible fluids.

The Navier-Stokes equations result from the work of French engineer Claude-Louis Navier (1785–1836) and British physicist George Gabriel Stokes (1819–1903) in the mid-nineteenth century. They did not work together, but their equations apply to incompressible flows. The Navier-Stokes equations are still used. At the end of the nineteenth century, Scottish engineer William John Macquorn Rankine (1820–1872) changed the understanding of the way fluids flow with his streamline theory, which states that water flows in a steady current of parallel flows unless disrupted. This theory caused a fundamental shift in the field of ship design because it changed the popular understanding of resistance in oceangoing vessels.

Laminar flow is measured today by use of the Reynolds number, developed by British engineer and physicist Osborne Reynolds (1842–1912) in 1883. When the number is low, viscous forces dominate. When the number is high, turbulent flows are dominant.

American naval architect David Watson Taylor designed and operated the first experimental model basin in the United States at the start of the twentieth century. His seminal work, *The Speed and Power of Ships* (1910), is still read. Taylor played a role in the use of bulbous bows on vessels of the navy. He also championed the use of airplanes that would be launched from naval craft underway in the ocean.

The principles of fluid dynamics took to the air in the eighteenth century with the work done by aviators such as the Montgolfier brothers and their hot-air balloons and French physicist Louis-Sébastien Lenormand's parachute. It was not until 1799, when English inventor Sir George Cayley (1771–1857) designed the first airplane with an understanding of the roles of lift, drag, and propulsion, that aerodynamics came under scrutiny. Cayley's work was soon followed by the work of American engineer Octave Chanute (1832–1910). In 1875, he designed several biplane gliders, and with the publication of his book *Progress in Flying Machines* (1894), he became internationally recognized as an aeronautics expert.

The Wright brothers are rightfully called the first aeronautical engineers because of the testing they did in their wind tunnel. By using balances to test a variety of different airfoil shapes, they were able to correctly predict the lift and drag of different wing shapes. This work enabled them to fly successfully at Kitty Hawk, North Carolina, on December 17, 1903.

German physicist Ludwig Prandtl (1875–1953) identified the boundary layer in 1904. His work led him to be known as the father of modern aerodynamics. Russian scientist Konstantin Tsiolkovsky (1857–1935) and American physicist Robert Goddard (1882–1945) followed, and Goddard's first successful liquid propellant rocket launch in 1926 earned him the title of the father of modern rocketry.

All of the principles that applied to hydrodynamics—the study of water in motion—applied to aerodynamics: the study of air in motion. Together these principles constitute the field of fluid dynamics.

HOW IT WORKS

When an object moves through a fluid such as gas or water, it encounters resistance. How much resistance depends upon the amount of internal friction in the fluid (the viscosity) as well as the shape of the object. A torpedo, with its streamlined shape, will encounter less resistance than a two-by-four that is neither sanded nor varnished. A ship with a square bow will encounter more resistance than one with a bulbous bow and V shape. All of this is important because with greater resistance comes the need for greater power to cover a given distance. Since power requires a fuel source and a way to carry that fuel, a vessel that can travel with a lighter fuel load will be more efficient. Whether the design under consideration is for a submarine or submersible, a tractor-trailer, an automobile, an ocean liner, an airplane, a rocket, or a space shuttle, these basic considerations are of paramount importance in their design.

APPLICATIONS AND PRODUCTS

Fluid dynamics plays a part in the design of everything from automobiles to the space shuttle. Fluid dynamic principles are also used in medical research by bioengineers who want to know how a pacemaker will perform or what effect an implant or shunt will have on blood flow. Fire flows are also being studied to aid in the science of wildfire management. Until now the models have focused on heat transfer, but new studies are looking at fire systems and their fluid dynamic properties. Sophisticated models are used to predict fluid flows before model testing is done. This lowers the cost of new designs and allows the people involved to gain a thorough understanding of the trade-off between size and power, given a certain design and level of resistance.

CAREERS AND COURSEWORK

Fluid dynamics plays a part in a host of careers. Naval architects use fluid dynamic principles to design vessels. Aeronautical engineers use the principles to design aircraft. Astronautical engineers use fluid dynamic principles to design spacecraft. Weapons are constructed with and understanding of fluids in motion. Automotive engineers must understand fluid dynamics to design fuel-efficient cars. Architects must take the motion of air into their design of skyscrapers and other large buildings. Bioengineers use fluid dynamic principles to their advantage in the design of components that will interact with blood flow in the human body. Land-management professionals can use their understanding of fluid flows to develop plans for protecting the areas under their care from catastrophic loss due to fires. Civil engineers take the principles of fluid dynamics into consideration when designing bridges and highways, as the "flow" of traffic exhibits fluid dynamic behavior. Fluid dynamics also plays a role in sports: from pitchers who want to improve their curveballs to quarterbacks who are determined to increase the accuracy of their passes.

Students should take substantial coursework in more than one of the primary fields of study related to fluid dynamics (physics, mathematics, computer science, and engineering), because the fields that depend upon knowledge of fluid dynamic principles draw from multiple disciplines. In addition, anyone desiring to work in fluid dynamics should possess skills that go beyond the academic, including an aptitude for mechanical details and the ability to envision a problem in more than one dimension. A collaborative mindset is also an asset, as fluid dynamic applications tend to be created by teams.

SOCIAL CONTEXT AND FUTURE PROSPECTS

The science of fluid dynamics touches upon a number of career fields that range from sports to bioengineering. Anything that moves through fluids or that is stationary within moving fluids such as air, water, or gases is subject to the principles of fluid dynamics. The more thorough the understanding, the more efficient vessel and other designs will be. This will result in the use of fewer resources in the form of power for inefficient designs and help create more efficient products.

—Gina Hagler

Further Reading

Anderson, John D., Jr. *A History of Aerodynamics and Its Impact on Flying Machines*. Cambridge UP, 1997.

Carlisle, Rodney P. *Where the Fleet Begins: A History of the David Taylor Research Center*. Naval Historical Center, 1998.

Çengel, Yunus A., and John M. Cimbala. *Fluid Mechanics: Fundamentals and Applications*. McGraw-Hill, 2010.

Darrigol, Olivier. *Worlds of Flow: A History of Hydrodynamics from the Bernoullis to Prandtl*. Oxford UP, 2005.

Eckert, Michael. *The Dawn of Fluid Dynamics: A Discipline Between Science and Technology*. Wiley-VCH, 2006.

Ferreiro, Larrie D. *Ships and Science: The Birth of Naval Architecture in the Scientific Revolution, 1600–1800*. MIT Press, 2007.

Johnson, Richard W., editor. *The Handbook of Fluid Dynamics*. CRC Press, 1998.

Mahan, A. T. *The Influence of Sea Power upon History, 1660–1783*. 1890. Reprint. Barnes & Noble Books, 2004.

Nazarenko, Sergey. *Fluid Dynamics via Examples and Solutions*. CRC Press, 2015.

Rieutord, Michel. *Fluid Dynamics. An Introduction*. Springer, 2015.

FRESHWATER CHEMISTRY

Fields of Study: limnology; hydrology; geochemistry; biogeochemistry

ABSTRACT

The unique properties of water that are so important to geological processes depend upon the distinctive polar structure of the water molecule. The chemistry of freshwaters is highly variable. Throughout the world, it is influenced by the atmosphere and locally influenced by reactions with the local soils and bedrocks. Changes through time in the chemistry of nat-

ural freshwaters are one of the best indicators of changes in Earth's surface environment.

KEY CONCEPTS

covalent bond: a bond that forms between two appropriate atoms through the sharing of two electrons between them

hydrogen bonding: not a true bond, but a pronounced attractive force between the unshielded positive charge on the nucleus of a covalently bonded hydrogen atom of one molecule and the exposed negative charge of the outermost electron pair(s) on a covalently bonded atom (typically a nitrogen or oxygen atom) of another molecule

ionic bond: a strong attraction between two ions carrying opposite charges, forming a rigid connection that does not involve sharing of electrons

ions: atoms and partial molecules that have either lost one or more electrons to carry the corresponding net positive charge, or that have gained one or more electrons to carry the corresponding net negative charge

magmatic differentiation: decomposition of mineral rock through melting in Earth's mantle layer, resulting in the release of water molecules that have been incorporated into the molecular structures of the minerals or that have been entrained and trapped within the rock as the mineral formed

pH: formally defined in value as the negative logarithm of the hydrogen ion concentration of a solution, according to the equation $pH = -\log[H^+]$

WATER AS A POLAR MOLECULE

The temperature of Earth lies within the so-called Goldilocks range—not too hot, not too cold, just right—that permits water to exist as a liquid, a gas, and a solid. Earth is sometimes called the "blue planet" because of the brilliant blue color, seen from outer space, that results from about three-quarters of the planet's surface being covered by water. Ultimately, this water is thought to have come from within the planet during magmatic differentiation and formation of the crust. The size of the planet provides a gravitational field that is sufficient to have kept water vapor from escaping into space throughout geologic time. Steam, the gaseous form of water above its boiling point, is the dominant constituent of volcanic gases, and pressure generated by steam accounts for the power of most explosive eruptions. Ice in glaciers has been a major force in influencing the topography of vast areas of the continent.

The chemical symbol for water, H_2O, refers to a molecule composed of one hydrogen ion (H^+) and one hydroxyl ion (OH^-). When combined to form a water molecule, the two hydrogen ions are arranged to one side and have an angle of about 105 degrees between them. Because the positively charged hydrogen ions lie on one side of the molecule, they impart a positive charge to that side of the molecule while leaving a negative charge of the oxygen exposed at the other end. Therefore, each water molecule is like a tiny magnet, with a positive pole and a negative pole. The molecule has polar charge distributions and thus is termed a "polar molecule."

Because each water molecule is like a magnet, as a substance, water will behave somewhat like a box full of very tiny magnets. The tendency for the magnets would be to line up positive pole to negative pole, and a similar tendency occurs in water, where the ionic attraction of the hydrogen of one water molecule for the oxygen of another water molecule actually draws water molecule to water molecule and creates a type of ordering. The cohesive ordering that results from this attraction is called "hydrogen bonding," and it accounts for some remarkable physical properties that belong only to water.

HYDROGEN BONDING

Water is one of the few liquid substances that do not simply become denser as they cool. The hydrogen bond allows liquid water molecules to pack themselves into a tighter pattern than would be possible if the molecules were not polar. Water achieves its maximum density of 62.43 pounds per cubic foot (1.0 gram per cubic centimeter) at 39 degrees Fahrenheit (3.94 degrees Celsius). Liquid water molecules vibrate rapidly (about 10^{12} vibrations per second), but when the water is cooled to its freezing temperature 32 degrees Fahrenheit (0 degrees Celsius) these vibrations decrease to about 10^6 vibrations per second. At that point, the strength of the hydrogen bonds is able to overcome the energy of the vibrations and stabilize the water molecules. The slowing of vibrations as water cools permits some rearrangement and some creation of open spaces as water molecules begin to organize into cage-like structures just below 39 degrees Fahrenheit (3.94 degrees Celsius), even though ice crystals do not form until a temperature of 32 degrees Fahrenheit (0 degrees Celsius) is reached. This accounts for the maximum density of water occurring at 39 degrees Fahrenheit (3.94 degrees Celsius) rather than at 32 degrees Fahrenheit (0 degrees Celsius).

When actual freezing occurs, water molecules arrange themselves into a stable hydrogen-bonded three-dimensional (3D) crystal structure that has open space between the molecules that is not present in liquid water. Water goes from a density of 62.42 lbs per cu ft (0.99987 g per cu cm) as liquid to 57 lbs per cu ft (0.917 gm per cu cm) as solid ice. The pronounced decrease in density causes ice cover to form on the surface of lakes and rivers rather than at the bottom. Without this property, aquatic life would probably not be possible, since ice on the surface insulates the water below, keeping it above 32 degrees Fahrenheit (0 degrees Celsius) and preventing complete freezing from bottom to top. The decrease in density is accompanied by an increase in volume; water expands as it freezes. When water seeps into cracks and pores of rock and soil and then freezes, the force generated by the expansion can cause rocks to break, soils to heave and swell, and small grains and crystals to spall away. Therefore, it is the hydrogen bond of water that ultimately permits water to become a powerful mechanical agent in weathering, as exemplified through frost heave in soils or frost wedging in rock.

Above 39 degrees Fahrenheit (3.94 degrees Celsius), the density of water decreases with increasing temperature. At temperatures likely to be encountered in lakes and streams, the density of pure water can be closely approximated with a formula. The small changes in density that occur during heating, cooling, and freezing are responsible for seasonal circulation of water in freshwater lakes.

Water is unlike other substances in that its freezing point is lowered, rather than raised, by pressure, a result of a tendency at near-freezing temperatures for the stabilized hydrogen bonds of ice to collapse back to the freely mobile hydrogen bonds of liquid water. Simple pressure is sufficient to enact the transition. When an ice skater places pressure on the blades of the skate, the pressure melts the ice at the sharp edges of the blade, and the skater glides on a thin layer of water that is created momentarily by the pressure. Such skating is not possible on other substances such as solid carbon dioxide (dry ice). This same liquefaction under pressure permits glacial ice to flow over a thin layer of water created at the base of the glacier and for this water to flow into joints and cracks, refreeze, and "pluck" out sections of the bedrock as the glacier moves.

ELECTROSTATIC INTERACTIONS

Water is a powerful solvent because the water molecule is polar. The magnet-like quality of the water molecule strongly influences substances that are bonded primarily by electrostatic (ionic) bonds. Ions bound into many solid minerals by virtue of the ionic bond may therefore be dissolved into water. In this way, water picks up dissolved materials from the soils and rock. The dissolving power provided by the polar nature of the molecules is further assisted by their vibration. As the molecules vibrate faster, they act like high-speed jackhammers against solid particles, and those solids held together with ionic bonds are particularly affected. Heating is one way to increase the vibration of molecules. Therefore, freshwater tends to be a more effective solvent in warm climates.

Freshwaters contain dissolved solids and gases. The substances that are likely to occur in waters of this planet are those substances that are abundant and that occur in forms that are easily soluble. More than 99 percent of Earth's crust can be accounted for by only twelve elements (oxygen, silicon, aluminum, iron, calcium, magnesium, sodium, potassium, titanium, hydrogen, manganese, and phosphorus—in order of abundance from greatest to least), and nearly 100 percent of the atmosphere by just five elements (nitrogen, oxygen, argon, hydrogen as water vapor, and carbon as carbon dioxide).

The most common dissolved substances, as ions, in freshwater are bicarbonate, calcium, sodium, magnesium, potassium, fluorine, iron, phosphate, sulfate, chloride, and dissolved silica. These substances enter natural waters both from the atmosphere and from the rocks and soils contacted by the waters. Even rainwater is not completely pure and contains small amounts of silica, sulfate, calcium, chlorine, nitrate, carbonic acid, sodium, potassium, iron, and aluminum.

Some elements are abundant in Earth's crust but are not abundant as dissolved species in freshwaters because they form insoluble compounds. For example, aluminum and iron are abundant elements, but in most water they form nearly insoluble hydroxides—aluminum hydroxide and iron hydroxide. Silicon is extremely abundant, but it forms strong covalent bonds with oxygen that are not possible for water to break. In nature, silicon forms silicate minerals that include silicon-oxygen structures with very low solubility in water.

ACIDIC AND ALKALINE WATERS

Although water is a good solvent because of the polar nature of the water molecule, it can be made an even more effective solvent if it can be rendered acidic. The pH scale expresses the degree to which a solution is acidic or alkaline. The scale runs normally from 0 (most acid, least alkaline) to 14 (least acid, most alkaline), with a pH value of 7 termed "neutral." PH values less than 0 and more than 14 are also possible, but are not usual. Waters with a pH of less than 7 are termed "acidic," and those with pH values greater than 7 are termed "alkaline." Natural rainwater is acidic and has a pH of 5.6, because carbon dioxide in the atmosphere dissolves in the rainwater to make carbonic acid when a molecule of carbon dioxide reacts with a molecule of water to produce H_2CO_3. (The atmosphere has a powerful effect on the composition of freshwater; the character of freshwater on this planet has changed through geologic time in accord with changes in the atmosphere.) Once rainwater strikes the ground, the weak acid is neutralized as it dissolves minerals in the rocks and soils. Therefore, even though natural rainwater is acidic, most fresh groundwater and surface waters are near neutral, with pH values between 6 and 8.5.

Unusually acidic waters occur around volcanic vents, geysers, and fumaroles, where gases such as sulfur dioxide react with water to produce sulfuric acid. Acidic waters also occur in bogs and marshes (pH 3.3 to 4.5), where carbon dioxide released during the decay of organic matter reacts with water to produce carbonic acid. The strongest acids occur where sulfide minerals such as pyrite and marcasite oxidize in the presence of air and water to produce sulfuric acid. Sulfuric acid is a strong acid, and its production through the weathering of sulfide minerals is so common around coal and metal mines that the waters released are termed "acid mine drainage." Streams flowing from these sites often have pH values below 2. Thousands of miles of streams have become polluted by acid mine drainage. Alkaline waters occur in limestone terrains, where pH values between 8 and 9 are common. Extremely alkaline waters occur in playa lakes in contact with sodium carbonate or sodium borate, where pH values above 12 have been noted.

Extremely acidic and extremely alkaline waters permit substances that remain immobile in normal waters to be dissolved. For example, aluminum is normally insoluble but does dissolve in both very acidic and very alkaline waters. Acidic waters dissolve and transport metals such as iron and manganese and nonmetals such as phosphorus that would normally be present in most natural waters at low or undetectable levels.

PH VERSUS ACID STRENGTH

Acid materials come in a variety of strengths, and are generally classed as either strong acids or weak acids. It is sometimes the case that the terms "pH" and "acid strength" are used interchangeably, but this is an erroneous usage of the terms. In aqueous, that is, water solutions, strong acids are those that dissociate completely to release H^+ ions and the corresponding negatively charged counterions. There is none of the undissociated compound remaining in solution. Weak acids, on the other hand, are those that do not dissociate completely in solution, leaving a significant amount of the undissociated material in solution as a neutral compound. Hydrogen chloride (HCl), hydrogen nitrate (HNO_3) and dihydrogen sulfate (H_2SO_4) are examples of strong acids, becoming hydrochloric acid, nitric acid and sulfuric acid when dissolved in water. Dihydrogen carbonate (H_2CO_3) is an example of a weak acid, forming a solution of carbonic acid in water. Carbonic acid is the most ubiquitous acid found to occur in nature. In a solution of carbonic acid, however, most of the H_2CO_3 remains undissociated and does not dissociate until some of the H^+ ions have been consumed in a neutralization reaction, as for example by contact with alkaline rock.

The most important difference between strong and weak acids is the chemical activity of the acidic species. For example, both strong and weak acid solutions having the same pH must by definition have the same hydrogen ion concentration. However, the strong acid solution will be far more potent in its interactions with susceptible materials than the weak acid solution will be. Reaction will be faster with the strong acid than with the weak acid, and the strong acid will be neutralized more abruptly than will the weak acid. This difference lies in the presence of undissociated molecules of the weak acid. As neutralization proceeds, these molecules continue to dissociate and provide more acid to be neutralized. Thus, even though the two acids may have the same pH, and therefore the same hydrogen ion concentration, how they interact and what they interact with will be very different. Consider, as an example, their interaction with aluminum metal. A piece of aluminum foil dropped into even a very weak solution of hydrochloric acid immediately begins to decompose, releasing bubbles of hydrogen gas and producing a yellow solution of aluminum trichloride. However, an identical piece of aluminum foil dropped into a "strong" solution of carbonic acid with the exact same pH, typically just carbonated water such as Club Soda, will remain unaffected for a very long time. The difference is not in the pH, but in the acid strength of the compounds.

AQUEOUS GEOCHEMISTRY

Freshwater chemistry is a part of the field of aqueous geochemistry, or low-temperature geochemistry, which is a specialty field of many geologists. Proper collection, analyses, and interpretation of data from natural waters require more than knowledge of chemistry. Success also requires knowledge about the natural environmental system, and this is why hydrology and aqueous geochemistry fall more properly within the province of geology than that of pure chemistry.

A typical water analysis includes determination of the levels of silica, aluminum, iron, manganese, copper, calcium, magnesium, strontium, sodium, potassium, dissolved oxygen, carbonate, nitrate, bicarbonate, sulfate, chloride, fluoride, phosphate, arsenic, selenium, boron, total dissolved solids, pH, acidity or alkalinity, temperature, and conductivity. If waters are polluted with unusual substances such as pesticides, sewage, or industrial wastes, specialized tests must be undertaken for these.

As soon as a sample is taken, the water is removed from its actual environment, and changes start to occur. Dissolved gases may leave, precipitates may form, and bacteria and microscopic algae may metabolize substances or die and release substances that were formerly solids. Because some parameters change so quickly and easily, some tests must be done immediately in the field. The temperature, conductance, pH, dissolved oxygen, and sometimes acidity or alkalinity is measured in the field. Small battery-operated instruments such as specific ion meters (for nitrate, sulfate, carbonate, and a number of elements), colorimeters, and conductivity meters are made by several manufacturers to be used in field analyses.

Pure water is a poor electrical conductor, but water's ability to conduct electricity increases markedly with the amount of dissolved solids. Therefore, a conductance test in the field is a rapid method to use to make a rough estimate of the amount of total dissolved solids. Sometimes rapid colorimetric tests for sulfate, phosphate, and some metals are done in the field, but more often the analyses of those substances are done in the laboratory.

Samples to be taken to the laboratory are filtered in the field to remove any microscopic suspended solids, because it is important that the water analyses show only dissolved substances. After the water is filtered, it is acidified with a few milliliters of high-purity acid to ensure that all dissolved species remain dissolved. The water is then put into a plastic container and filled to the top so as not to admit any air. The samples are usually placed in a dark cooler and kept refrigerated until they are analyzed in the laboratory.

DANGER OF CONTAMINANTS AND POLLUTANTS

Of the vast amount of water present on the planet's surface, only about 2.5 percent is freshwater; the remainder is in oceans and inland seas. Of that 2.5 percent, 2.15 percent is locked away in glaciers and polar ice caps. About another 0.3 percent of the total is accounted for by fresh groundwater. All the surface water commonly seen in freshwater lakes and streams amounts to only 0.009 percent of Earth's total water.

The small percentages do not reveal the true importance of freshwater. Freshwater is the major sculpting agent of the planet's land surface, and it is an essential substance for all terrestrial life-forms. Changes in the chemical composition of freshwater constitute one of the most sensitive indicators of changes in the environment. Changes may be local, as exemplified by the eutrophication of a small lake that receives phosphorus from a few local septic tanks, or they may be global, as exemplified by the acidification of sensitive lakes in northern Europe and Canada as a result of the burning of fossil fuels, smelting and other industrial operations, all of which release carbon dioxide and oxides of nitrogen and sulfur into the atmosphere.

The category "freshwater" includes a group of waters from wells, lakes, and streams with diverse chemistry. Even natural substances in excessive amounts pose problems for water consumption or industrial use. In addition to the toxic content of acid mine drainage, excessive amounts of nitrate, sodium, and fluoride present health hazards. Excessive amounts of calcium and magnesium promote buildup of scale in water tanks and boilers, and waters that are high in these elements (hard waters) typically have these constituents chemically removed with a water softener prior to domestic and industrial use. High phosphate contents of waters promote algal blooms and eutrophication. High boron content is important in irrigation, because boron is toxic to many crops. Of all-natural resources, freshwater ranks as one of the most essential to human survival. Continual monitoring of the chemistry of freshwaters is essential to ensure that these valuable resources are not rendered unsuitable by human-made contaminants and pollutants.

—Edward B. Nuhfer

Further Reading

Ahuja, Satinder, editor. *Chemistry and Water: The Science Behind Sustaining the World's Most Crucial Resource.* Elsevier, 2017.

Craig, John F., editor. *Freshwater Fisheries Ecology.* Wiley Blackwell, 2016.

Dodds, Walter K., and Matt R. Whiles. *Freshwater Ecology: Concepts and Environmental Applications of Limnology.* 2nd ed., Academic Press, 2010.

Finlayson, C. Max, Angela H. Arthington, and Jamie Pittock, editors. *Freshwater Ecosystems in Protected Areas: Conservation and Management.* Routledge, 2018.

Gobas, Frank A. P. C. *Chemical Dynamics in Freshwater Ecosystems.* CRC Press, 2018.

Greenberg, Arnold, et al., editors. *Standard Methods for the Examination of Water and Wastewater.* 21st ed., American Public Health Association, 2005.

Hem, J. D. *Study and Interpretation of the Chemical Characteristics of Natural Water.* US Geological Survey Water Supply Paper 2254. UP of the Pacific, 2005.

Kernan, Martin, Richard W. Battarbee, and Brian Moss, editors. *Climate Change Impacts of Freshwater Ecosystems.* Wiley Blackwell, 2010.

Krauskopf, Konrad B. *Introduction to Geochemistry.* 3rd ed., McGraw-Hill, 2003.

Lam, Buuan, et al. "Major Structural Components in Freshwater Dissolved Organic Matter." *Environmental Science & Technology*, vol. 41, 2007, pp. 8240–47.

Wagner, Martin, and Scott Lambert, editors. *Freshwater Microplastics: Emerging Environmental Contaminants?* Springer Open, 2018.

G

Geographic Information Systems (GIS)

Fields of Study: geography; computer science; ecology; oceanography

ABSTRACT

Geographic information systems (GIS) capture, manage, manipulate, and present geographical data. GIS involves a combination of hardware, software, and collated data to analyze conditions, trends, and events in the natural environment. It may be used to map and track natural resource, seismological, volcanic, meteorological, and climatological conditions around the world and is an invaluable tool for the analysis of regional and global systems.

KEY CONCEPTS

algorithm: a computing routine designed to identify and coordinate data points according to user-defined criteria

GPS: the global positioning system, an array of satellites in geosynchronous orbit for the purpose of identifying the position of any point on Earth's surface by triangulation

raster mapping: portraying map data as an array of rectangular sections, with the sections having equal dimensions but different data values, like an image made by using different background colors in the cells of a spreadsheet

surficial: in relation to Earth's surface

topography: the three-dimensional (3D) features of the surface of Earth's crust

vector mapping: portraying map data as an array of points, each having a unique value and directon derived from the data itself, such as elevation relative to a set base level

BASIC PRINCIPLES

Geographic information systems (GIS) are integrated devices used to map natural events, trends, and conditions. GIS employs a number of different and separate hardware and software tools, such as satellite and aerial sensors and cameras, along with specialized computer databases and other systems.

GIS software employs a geographic reference, such as a digitized map, to study a given region. Depending on the type and complexity of the scientific pursuit, the GIS then applies one or more layers, applying the compiled data to that map. The resulting image (or rendering) provides a composite of the area being studied according to the respective scientist's research.

Three general areas make up an integrated GIS program. The first is a series of numerical algorithms, which enable researchers to assign data to values on the geographic reference. The second is a table of statistics. In GIS research, this table is frequently extensive, comprising a large volume of data. The third is optimization: In the application of GIS to a specific area of focus, the user or users must be able to modify the software appropriately to conform to the research.

GIS is utilized in a number of scientific disciplines and fields, including the social sciences (such as sociology, political science, and history), engineering, epidemiology, the natural sciences, and geoinformatics. For the purposes of earth science, GIS has a wide range of applications of value. Scientists using this technology can map changes in the earth's topography, follow climate shifts, analyze water resources, and study the impact of volcanic eruptions.

GIS focuses on the collection of data from a specific geographical location. This information, when combined with data from other locations, can help scientists create a composite of a particular concept. For example, GIS is used in the long-term study of the melting of glaciers. It also can be used to map the destruction caused by volcanic eruption or to monitor pollution in a given area.

HISTORY

GIS has its roots in cartography. Indeed, scientists maintain that GIS could not exist without knowledge of the way maps are (and have been) created. One of the earliest examples of GIS comes from the middle of the nineteenth century, when British physician John Snow (1813–1858) began investigating the spread of cholera in London. Snow hypothesized that an outbreak in 1854 was caused by a bacterium in contaminated water. He drew a map of London and, using data from the hospitals in the area, plotted on the map the locations in which cholera deaths had occurred. His map revealed that most victims lived near a particular water pump. Upon his recommendation, that pump was shut down, leading to an end to the outbreak.

GIS development continued much in the same manner as it did in the mid-nineteenth century. Before the advent of computer technology, one of the more popular methods employed by scientists when studying natural events and trends in a geographic location was to use an existing map of a region. When analyzing data, scientists would simply lay a clear plastic sheet over the map, showing the points of interest. GIS continued to develop along these lines; even today, many GIS systems require the use of computer-aided design (CAD) software, a system that scans high-resolution hard copies of existing maps. The images derived from these scans are used as the basis for overlapping images suited for a given study.

As a computer-based system, GIS was largely developed for governments. However, by the start of the twenty-first century, the private market for GIS software and hardware had developed quickly. By 2000, the private sector spent between $15 billion and $20 billion on GIS. Furthermore, the use of GIS programs has grown worldwide, with software and technology available and affordable for even the most austere budgets.

(Autotype from a Presentation Portrait, 1856, and Autograph facsimile.—B. W. R.)

John Snow, 1856 (Wellcome Collection gallery via Wikimedia Commons)

RASTER AND VECTOR DATA

Two types of reference data are utilized on a GIS rendering, in addition to a separate data table. The first of these reference data is a set of indicators known as vectors. Vector data appear in three general manifestations: polygon, line, and point. On the image, a polygon (a multisided geometric shape) represents a city or other boundary (such as a body of water or forest region). A line, or arc, depicts linear features, such as streets, rivers, railways, and trails. The third form of vector is the point. This type of data identifies singular points of interest, such as buildings and bridges. Vectors simplify locations on a given rendering, reducing clutter on the image. The user can obtain an enhanced view with the vectors, which can show more detail. Adjacent structures or physical features may come into view with the zoom.

The second type of reference data in a GIS image is the raster. A raster focuses on surfaces rather than features. They are, in essence, high-resolution images (such as aerial and satellite photographs) comprised of a grid of pixels. Rasters appear in two forms: continuous and discrete. Continuous rasters vary consistently with no clear boundaries. Examples of this type of cell include surface temperature and geographical elevation. Discrete rasters, however, have boundaries with attached categories, classes, and descriptions. Some examples of discrete rasters are data for population density and political boundaries.

Advantages and disadvantages come with the use of rasters and vectors. Vectors, for example, use less of a computer system's memory because of their relative simplicity and low volume of data. However, because of their simplicity, vectors can be easily adjusted for scale. Rasters, on the other hand, use a complex system of values, allowing for a number of different operations, including mathematical calculations. Then again, the high resolution and complexity of rasters require the user's computer system to contain a significantly larger memory to perform any spatial analysis.

Graphic showing the layers of GIS (Bureau of Ocean Energy Management)

The ease with which vector data are managed in GIS analyses has made vector analysis a preferred choice among many researchers. Software developers are, as a result, introducing GIS programs that can convert reference data from raster to vector formats. Still, a large number of GIS programs have various raster-vector conversion programs, enabling the user to access and manipulate a range of maps and images.

SPATIAL ANALYSIS

GIS accesses and analyzes spatial (or geographic) data. In spatial analysis, the data for each map layer are provided in an adjoining table. Each data point is assigned a value, which is typically performed through numerical algorithms and statistics. Analysis considers the relationships among each point. In many cases, these points lie within a larger feature (such as a building or geographic area). Spatial analysis also may require assessing the relationships among two or more points in different geographic locations.

The data provided on a given area may not automatically correspond to a predesigned map, or it may be anticipated that the map will change in light of certain conditions. For example, satellite-based thermal images may not delineate a mountain or a structure. Additionally, a scan of a shrinking rainforest may require multiple images and points of reference to demonstrate a chronological change. In such cases, it is important to generate an appropriate map. In these situations, GIS software typically calls upon modeling and forecasting programs to create a base image for spatial analysis.

GIS-generated spatial analysis helps scientists collate and utilize the attributes of a given reference point (or series thereof). Such data help researchers develop clear illustrations of a given geographic area in a far more comprehensive manner than that which is uncovered simply by flying over or driving through a target area.

For example, scientists in Israel conducted research on the availability of drinking water in nearby Gaza Strip. The project entailed the acquisition of data on rainfall, population increases, and the number of reservoirs available in Gaza, and the generation of a useful map to serve as the basis for the spatial analysis. Using GIS, this project created an analytical framework that identified the areas of highest consumption from groundwater reservoirs. The models produced by this GIS application also provided accurate short-term forecasts for groundwater usage in this area.

MAPPING TOPOGRAPHY

Until the mid- to late twentieth century, scientists were limited in their ability to study Earth's topography. Although aerial photographs and ground-based analysis prove effective in some areas, scientists were until this period unable to study a wide range of locations and trends in the natural environment.

One such location is the ocean floor. Although hidden from plain sight, the ocean floor contains fish habitats and provides indications of the impact of humanity on the natural environment. To better survey the suboceanic environment, scientists are increasingly calling upon GIS to map the ocean floor.

For example, scientists at Oregon State University utilized GIS to create a surficial geologic habitat map. This map is seen as critical for the region and its fishing industry, which has been in crisis because of concerns of depleted fish populations. Using a series of different mapping software that employs sonar data and density information, and seismic imaging and other databases and systems, researchers can effectively map the floor off the coast of Oregon with 3D images and other relevant applications.

In another example, GIS technology was employed to study the bedrock of Oklahoma. Here, researchers used a digital elevation model to map the sandstone and shale that compose that state's bedrock. Understanding the geology of this area is important for that state, with implications in its agricultural and civil engineering sectors.

NATURAL RESOURCES

Applying GIS to the study of Earth's natural resources has been evolving since the late 1970s. However, this endeavor has seen increased success with the development of computer modeling and satellite-based technologies. In rural areas of North America and Pakistan, for example, GIS has proved effective in

Flock of gulls near Tuz Lake, Turkey (Zeynel Cebeci via Wikimedia Commons)

mapping forest resources. The use of such technology is particularly useful for studying the decline of forest cover due to overdevelopment and unsustainable logging practices.

In an era in which sustainable development and protection of natural resources, such as drinking water and agriculture, are of increasing importance, GIS is a highly effective vehicle for analyzing trends and, ultimately, creating environmentally friendly policies. In developing countries and regions, this issue is particularly critical. For example, the area around Turkey's Tuz Lake (the second-largest lake in Turkey and one of the largest salt lakes in the world) is considered an important region for agriculture and for many different forms of animal life. Scientists began to develop a strategy for land use in the area, considering the area's delicate ecological balance and such elements as soil and water quality, climate, and land suitability. Using GIS mapping, researchers created an overlapping series of maps of the region accounting for each of these ecological factors. The result of the study was a comprehensive land-use framework that outlined sustainable agricultural use, effective waste management, and the protection of the lake area's wildlife.

GIS SOFTWARE

GIS involves integrated and open-source software technologies. Among these systems is computer assisted design (CAD) software. CAD is used to design a high-resolution image that can be manipulated by zooming in and out, adding overlapping images, and viewing segments of an image from multiple angles. Although many CAD systems are built into appropriate hardware, a growing number of CAD software programs may be obtained independently, including through the internet.

In addition to CAD, GIS requires the integration of spatial indexing software. This type of system enables spatial data to be stored, modified, and analyzed within the database. Spatial indexing software is used to collate statistics and data and assign them to an appropriate vector and raster. Spatial indexing and development software must be compatible with the CAD program in use, as the resulting reference data (vectors and rasters) will be overlain on the main composite image. Like CAD, this software is now available on the internet. Furthermore, because GIS-based studies commonly feature a large amount of data, it is important that the integrated system in use includes a high-volume database application.

In addition to web-based software, another innovation useful to GIS researchers is mobile GIS applications. Because the memory storage of smartphones, tablets, and other mobile systems is ever-increasing, researchers can now access and utilize GIS software from their mobile devices.

IMPLICATIONS AND FUTURE PROSPECTS

GIS has evolved so extensively that it is being applied across a broad spectrum of scientific disciplines. Social scientists use GIS to map and analyze demographics and residential populations. Civil engineers employ GIS to assess the viability of constructing roads, bridges, and other structures. Public health professionals use it not only to track diseases but also to analyze potential health threats and benefits for populations.

The evolution of GIS also has been aided by the parallel evolution of sensory hardware. In terms of earth science, advances in thermal imaging, radar, seismography, and other technologies add a number of additional resources to spatial analysis. Furthermore, the accessibility of the global positioning system and other satellite-borne hardware systems now enables scientists to study a geographic location with ever-greater detail. Many of these advanced systems can "see" through weather, foliage, and even topographical features, all of which were previously considered obstacles to spatial analysis.

As with other scientific disciplines, it is likely that GIS will continue to have relevance for researchers in the field of earth science. The private sector market for GIS continues to grow because of its applicability to so many fields. Government and nongovernmental research on such issues as sustainable development, climate change, and natural resource management will only continue to drive this interest.

—Michael P. Auerbach

Further Reading

Brewer, Cynthia A. *Designing Better Maps: A Guide for GIS Users*. 2nd ed., ESRI Press, 2015.

Bolstad, Paul. *GIS Fundamentals: A First Text on Geographic Information Systems*. 5th ed., XanEdu, 2016.

Fotheringham, Stewart, and Peter Rogerson. *Spatial Analysis and GIS*. CRC Press, 2013.

Harder, Christian, and Clint Brown. *The ArcGIS Book: 10 Big Ideas About Applying the Science of Where*. ESRI Press, 2017.
Knowles, Anne Kelly, and Amy Hillier, editors. *Placing History. How Maps, Spatial Data and GIS Are Changing Historical Scholarship*. ESRI Press, 2008.
Law, Michael, and Amy Collins. *Getting to Know ArcGIS*. 4th ed., ESRI Press, 2015.
Longley, Paul A., et al. *Geographic Information Systems and Science*. 4th ed., Wiley, 2015.
Pick, James B. "Geographic Information Systems: A Tutorial and Introduction." *Communications of AIS*, vol. 14, 2004, pp. 307–31.
Sei-Ichi, Saitoh, et al. "Some Operational Uses of Satellite Remote Sensing and Marine GIS for Sustainable Fisheries and Aquaculture." *ICES Journal of Marine Science/Journal du Conseil*, vol. 68, no. 4, 2011, pp. 687–95.
Thrall, Grant Ian. "MapInfo Professional 8.5: Internet-Enabled GIS." *Geospatial Solutions*, vol. 16, no. 9, 2006, pp. 36–39.
Wade, Timothy G., et al. "A Comparison of Vector and Raster GIS Methods for Calculating Landscape Metrics Used in Environmental Assessments." *Photogrammetric Engineering and Remote Sensing*, vol. 69, no. 12, pp. 1399–405.

Geoinformatics

Fields of Study: computer science; oceanography; geography

ABSTRACT

The term "geoinformatics" refers to a collection of information systems and technologies used to create, collect, organize, analyze, display, and store geographic information for specific end-user applications. The field represents a paradigm shift from traditional discipline-based systems such as cartography, geodesy, surveying, photogrammetry, and remote sensing to a data systems management protocol that includes all earlier technologies and combines them to create new models of spatial information. Computation is an essential foundation of all geoinformatics systems.

KEY CONCEPTS

coordinate system: a means of identifying location relative to a designated starting point by means of standardized directions and distances from that point; for example, latitude and longitude relative to the equator and the meridian of Greenwich, respectively

geodesy: measurement of the shortest distance between two points on Earth's surface

map: a symbolic representation of anything that may be described as a terrain, such as the surface of the planet, a range of connected ideas, etc.

photogrammetry: the practice of reproducing a three-dimensional (3D) representation or mapping of an area from two-dimensional (2D) photographic images

DEFINITION AND BASIC PRINCIPLES

Geoinformatics is a complex, multidisciplinary field of knowledge specializing in the creation, collection, storage, classification, manipulation, comparison, and evaluation of spatially referenced information for use in a variety of public and private practices. Its technologies are rooted in mapping, land surveying, and communication technologies that are thousands of years old. The exponential growth of science-based knowledge and mathematical expertise during the nineteenth and twentieth centuries greatly assisted the accumulation of verifiable geographic information describing Earth and its position among the myriad celestial entities occupying the known universe. Detailed geometric descriptions and photographic materials make it possible to translate, measure, and order the surface of Earth into multidimensional coordinate systems that provide a rich and detailed visual language for understanding the relationships and features of locations too vast to be easily comprehended by the human senses. Geographic communication systems are greatly enhanced by the proliferation of computation technologies including database management systems, laser-based surveys, digital satellite photo technologies, and computer-aided design (CAD). Given that fully three-quarters of Earth's surface is beneath the seas and oceans, the field of geoinformatics is also of vital importance with regard to oceanography and the marine sciences.

BACKGROUND AND HISTORY

The components and design structures of maps form the fundamental language of geographic information systems (GIS). Maps illustrate a variety of environments, both real and imagined, and the structures and life forms that fill them. The geographies of Earth's land masses and the relationships of land and water to the Moon and stars are the foundations of advanced mathematics and the physical sciences, subjects that are continually modified by new measurements of spatial-temporal coordinates. Aerial and nautical photography introduced a new and vital reality to cartographic representation in the twentieth century. Advances in sonar and radar technologies, the telephone and telegraph, radio broadcast, mass transportation, manned space flight, spectrometry, the telescope, nuclear physics, biomedical engineering, and cosmology all rely on the power of the map to convey important information about the position, structure, and movement of key variables.

Cartography. Mapping is an innate cognitive ability; it is common in a variety of animate species. In human practice, mapping represents an evolutionary process of symbolic human communication. Its grammar, syntax, and elements of style are composed of highly refined systems of notation, projections, grids, symbols, aesthetics, and scales. These time-honored features and practices of cartographic representation demand careful study and practice. Maps are the most ancient documents of human culture and civilization and, as such, many are carefully preserved by private and public institutions all over the world. The advent of the World Wide Web (WWW) revolutionized the field by making it possible to share, via the internet, facsimiles of these precious cultural artifacts; the originals are protected for posterity. Scholars study the intellectual and technical processes used to create and document maps. These provide valuable clues about the beliefs and assumptions of the cultures, groups, and individuals that contract and prepare them.

Geodesy. The cartographer artfully translates the spatial features of conceptual landscapes into two- and 3D documents using data and symbols selected to illustrate specific relationships or physical features of a particular place for the benefit or education of a group or individual enterprise. Geodesy (or geodetics) is an earth science the practitioners of which provide timely geographical measurements used by cartographers to create accurate maps. Since earliest recorded times, geodesists have utilized the most current astronomical and geographical knowledge to measure the surface of Earth and its geometric relationship to the Sun and Moon. The roots of geodetic measurement systems are buried in the ancient cultures and civilizations of Egypt, China, Mesopotamia, India, and the Mediterranean. Enlightened scholars and astute merchants of land and sea traveled the known world and shared manuscripts, instruments, personal observation, and practical know-how for comprehending the natural world. Hellenic scholars brilliantly advanced the study of astronomy and Earth's geography. Their works formed the canon of cartographical and astronomical theory in the Western world from the time of the great voyages of discovery in the fifteenth and sixteenth centuries.

Alexandria was the intellectual center of the lives of the Greek polymath Eratosthenes of Cyrene (ca. 276–ca. 194 BCE) and the Roman astronomer Ptolemy. Both wrote geographical and astronomical treatises that were honored and studied for centuries. Eratosthenes wrote *Peri tes anametreoeos tes ges* (On the Measurement of the Earth) and the three-volume *Geographika* (Geography), establishing mathematics and precise linear measurements as necessary prerequisites for the accurate geographical modeling of the known world. His works are considered singular among the achievements of Greek civilization. He is particularly noted for his calculations of Earth's circumference, measurements that were not disputed until the seventeenth century.

Ptolemy, following Eratosthenes, served as the librarian at Alexandria. He made astronomical observations from Alexandria between the years 127 and 141, and he was firm in his belief that accurate geographical maps were derived from the teachings of astronomy and mathematics. The *Almagest* is a treatise devoted to a scientific, methodical description of the movement of the stars and planets. The beauty and integrity of his geocentric model set a standard for inquiry that dominated astronomy for centuries. He also wrote the eight-volume *Geographia* (Geography), in which he established latitude and longitude coordinates for the major landmasses and mapped the world using a conic projection of prime and linear meridians and curved parallels. His instructions for creating a coordinate system of mapping, including sectional and regional maps to highlight individual countries, are still in practice.

At the end of the nineteenth century, geophysics became a distinct science. Its intellectual foundations provided entities such as petroleum corporations new technologies for identifying and classifying important land features and resources. In the United States, extensive surveys were conducted for administrative and military purposes. The US Coast Survey was established by President Thomas Jefferson (1743–1826) in 1807. The Army Corps of Engineers, the Army Corps of Topographical Engineers, and the United States Naval Observatory supported intensive geophysical research including studies of harbors and rivers, oceans and land topographies. In 1878, the US Coast Survey became the US Coast and Geodetic Survey. In 1965, the US Coast and Geodetic Survey was reincorporated under the Environmental Sciences Services Administration, and in 1970 it was reorganized as the National Oceanic and Atmospheric Administration (NOAA).

Underwater acoustics and geomagnetic topographies were critical to the success of naval engagements during World War I and World War II. The International Union of Geodesy and Geophysics (IUGG) was founded in 1919, and became one of scientific unions participating in the International Council for Science (ICSU). That same year the American National Committee of the International Union of Geodesy and Geophysics combined with the Committee on Geophysics of the National Research Council; in 1972 the committee was independently organized as the American Geophysical Union (AGU). Graduate geophysics programs became prominent in major universities after World War II.

Geodesy is a multidisciplinary effort to calculate and document precisely the measurements of Earth's shape and gravitational field in order to define accurate spatial-temporal locations for points on its surface. Land surveys and geomensuration, or the measurement of Earth as a whole, are essential practices. Geodetic data is used in GIS—these include materials from field surveys, satellites, and digital maps.

Earth's shape is represented as an ellipsoid in current mathematical models. Three-dimensional descriptors are applied to one quadrant of the whole in a series of calculations. All cartographic grid systems and subsequent measurements of Earth begin with a starting point called a datum. The first datum was established in North America in 1866 and its calculations were used for the 1927 North American Datum (NAD 27). In 1983, a new datum was established (NAD 83) and it is the basis of the standard geodetic reference system (GRS 80). It was again modified and became the World Geodetic System in 1984 (WGS 84). WGS 84 utilizes constant parameters for the definition of Earth's major and minor axes, semimajor and semiminor axes, and various ratios for calculating the flattening at the poles. The geoid is another geodetic representation of Earth, as is the sphere.

Photogrammetry. The word "photogrammetry" refers to the use of photographic techniques to produce accurate 3D information about the topology of a given area. Accurate measurements of spaces and structures are obtained through various applications including aerial photography, aerotriangulation, digital mapping, topographic surveys, and database and GIS management. Precise, detailed photographs make it possible to compare and analyze particular features of a given environment. Aerial photographs are particularly useful to engineers, designers, and planners who need visual information about the site of a project or habitat not easily accessible by other means. The International Society for Photogrammetry and Remote Sensing (ISPRS) advances the knowledge of these technologies in more than 100 countries worldwide.

Remote Sensing. Like photogrammetry, remote sensing is an art, science, and technology specializing in the noncontact representation of the Earth and the environment by measuring the wavelengths of different types of radiation. It includes passive and active technologies, both of which collect data from natural or emitted sources of electromagnetic energy. These include cosmic rays, X-rays, ultraviolet light, visible light, infrared and thermal radiation, microwaves, and radio waves. Aerial photography and digital imaging are traditional passive remote sensing technologies based on photographic techniques and applications. Later applications include manned space and space shuttle photography and Landsat satellite imagery. Radar interferometry and laser scanning are common examples of active remote sensing technologies. These products are used for the documentation of inaccessible or dangerous spaces. Examples include studies of particular environments at risk or in danger.

Global Navigation Satellite Systems (GNSS). Satellites and rapidly advancing computer technologies have transformed GIS products. Satellite systems of known distance from the Earth receive radio transmissions, which are translated and sent back to Earth as a signal giving coordinates for the position and elevation of a location. The Global Positioning System (GPS; originally called Navstar), a US government system originating from a World War II radio transmission system known as loran (long-range navigation), is the oldest and best-known GNSS system. Others include Russia's Global Navigation Satellite System (GLONASS), the European Union's Galileo, and India's NAVIC. GNSSs provide data for many systems including consumer navigation systems and tracking.

The National Aeronautics and Space Administration (NASA) maintains the Earth Observing System (EOS) program, made up of elements such as the Landsat and Terra satellites. These satellite systems are used to collect data including measurements of the Earth's ozone shield, cloud mappings, rainfall distributions, wind patterns, studies of ocean phenomena, vegetation, and land-use patterns.

HOW IT WORKS

Geographic information systems are built on geometric coordinate systems representing particular locations and terrestrial characteristics of Earth or other celestial bodies. The place and position of particular land features form the elementary and most regular forms of geographic data. Maps provide visual information about key places of interest and structural features that assist or impede their access. Land-survey technologies and instrumentation are ancient and are found in human artifacts and public records that are thousands of years old. They are essential documents of trade, land development, and warfare. Spatial coordinates describing the Moon and stars and essential information about particular human communities are inscribed into rocks, painted on the walls of caves, and hand-printed on graphic media all over the world. Digital photographs, satellite data, old maps, field data, and measurements provide new contexts for sharing geographic information and knowledge about the natural world and human networks of exchange.

GIS devices include high-resolution Landsat satellite imagery, light detecting and ranging (lidar) profiles, CAD data, and database management systems. Cross-referenced materials and intricately detailed maps and overlays create opportunities for custom-designed geographic materials with specific applications. Data that is created and stored in a GIS device provide a usable base for building complex multi-relational visualizations of landscapes, regions, and environments.

How a GIS component is used depends on the particular applications required. Service providers will first conduct a needs assessment to understand what information will be collected, by whom, and how it will be used by an individual or organization. The careful design of relationships connecting data sets to one another is an essential process of building an effective GIS system. Many data sets use different sets of coordinates to describe a geographical location and so algorithms need to be developed to adjust values that can have significant effects on the results of a study.

Depending on the needs of an organization, the rights to use some already-established data sets can be acquired. Other data can be collected by the user and stored in appropriate data files. This includes records already accumulated by an individual or organization. Converting and storing records in GIS format is particularly useful for creating the documents needed for a time-series analysis of particular land features and regional infrastructures. Digital records protect against the loss of valuable information and create flexible mapping models for communicating with internal and external parties.

APPLICATIONS AND PRODUCTS

Satellite and computer technologies have transformed the way spatial information is collected and communicated worldwide. With exponential increases in speed and detail, satellites provide continuous streams of information about Earth's life systems. More than fifty satellites provide free access to GPS coordinates. These are used on a daily basis by people from all walks of life, providing exceptional mapping applications for determining the location and elevation of roadways and waterways.

The Environmental Systems Research Institute (ESRI) is a pioneer in the development of GIS landscape-analysis systems. This includes the development of proprietary automated software products with web applications. Users can choose from a menu of services including community web-mapping tools; ArcGIS Desktop with ArcView, ArcEditor, and

ArcInfo applications; and applications for use in educational settings. ESRI software is an industry standard used worldwide by governments and industries. The ESRI International User Conference is attended by thousands of users and is one of the largest events of its kind. While ESRI continues to dominate the field, other GIS software has also been developed and become influential, including open-source systems that are free and customizable by users. Another product that had a significant influence in geoinformatics was Google Maps, which became a highly recognizable and widely used web-mapping service after its launch in 2005.

The Global Geodetic Observing System (GGOS) is a program developed by the International Association of Geodesy. It provides continuing observations of Earth's shape, gravity field, and rotational motion. These measurements are integrated into the Global Earth Observing System of Systems (GEOSS), an application that provides high-quality reference materials for use by groups such as the Group on Earth Observations and the United Nations (UN).

CAREERS AND COURSEWORK

Coursework and certification in geoinformatics at the undergraduate and graduate levels can be specialized or completed in tandem with coursework in other earth science and engineering fields. Courses include selections of introductory materials, computer programming, database management and design, statistics, bioinformatics, geostatistics, remote sensing, various foundational courses in mapping techniques and spatial analysis, and computer laboratory exercises to gain familiarity with a number of GIS software programs. Some coursework will emphasize applications in land development, transportation analysis, environmental science, public health, and a variety of engineering and architectural design schematics.

Careers specializing in geoinformatics technologies are directly involved in data-driven hardware and software applications. Practitioners must master the organizational and design skills necessary to produce highly detailed, error-free maps and related documents for use in private industry and government-related agencies. Familiarity with field research and land-survey technologies are also desirable.

SOCIAL CONTEXT AND FUTURE PROSPECTS

Due to the increasing prevalence of the internet, computer and telecommunications technologies continue to provide novel platforms for connecting individuals, groups, and communities worldwide. Location is an essential feature of such networks, and geographic information systems are needed to provide timely spatial information for individual and cooperative ventures. However, questions about the safety and integrity of global communications systems and data systems can pose challenges to political values in free and democratic societies. Nevertheless, geographic information systems will continue to be integrated into a variety of applications, contributing to the safety of individuals, the integrity of the world's natural resources, and the profitability of enterprises around the globe.

—Victoria M. Breting-García

Further Reading

Bartlett, Darius, and Louis Celliers. *Geoinformatics for Marine and Coastal Management*. Taylor & Francis, 2019.

Bender, Oliver, et al., editors. *Geoinformation Technologies for Geocultural Landscapes: European Perspectives*. CRC Press, 2009.

Chandra, A. M. *Geoinformatics*. New Academic Science, 2017.

DeMers, Michael N. *Fundamentals of Geographic Information Systems*. 4th ed., John Wiley & Sons, 2009.

Galati, Steven R. *Geographic Information Systems Demystified*. Artech House, 2006.

Harvey, Francis. *A Primer of GIS: Fundamental Geographic and Cartographic Concepts*. Guildford Press, 2008.

Karimi, Hassan A. *Big Data: Techniques and Technologies in Geoinformatics*. CRC Press, 2017.

Kolata, Gina Bari. "Geodesy: Dealing with an Enormous Computer Task." *Science*, vol. 200, no. 4340, 28 Apr. 1978, pp. 421–66.

Konecny, Gottfried. *Geoinformation: Remote Sensing, Photogrammetry and Geographic Information Systems*. Taylor & Francis, 2003.

Rana, Sanjay, and Jayant Sharma, editors. *Frontiers of Geographic Information Technology*. Springer, 2006.

Robinson, Ian S. *Measuring the Oceans from Space: The Principles and Methods of Satellite Oceanography*. Springer/Praxis, 2004.

Scholten, Henk J., Rob van de Velde, and Niels van Manen, editors. *Geospatial Technology and the Role of Location in Science*. Springer, 2009.

Thrower, Norman J. W. *Maps and Civilization: Cartography in Culture and Society*. 3rd ed., U of Chicago P, 2007.
Wright, Dawn J., editor. *Undersea with GIS*. ESRI Press, 2002.

Global Energy Transfer

Fields of Study: geoscience; oceanography; atmospheric physics; bioenergetics

ABSTRACT

Earth is best thought of as a system with energy inputs and outputs. Energy, mostly heat, comes from the sun and from Earth's core. As this energy transfers around and through the planet, it drives major processes such as plate tectonics, climate, ocean and air currents, and the distribution of life across the globe.

KEY CONCEPTS

climate change: the observed principle that Earth's average temperature is increasing over time and the global climate is adjusting to the increase in average temperature

geothermal: literally, "Earth heat," describing phenomena that are driven by the heat of Earth's interior

greenhouse effect: the capture of infrared energy by atmospheric gases, resulting in the maintenance of an elevated atmospheric temperature

infrared energy: electromagnetic radiation having wavelengths longer than red in the visible portion of the electromagnetic spectrum, experienced as "heat"

methane: the simplest of the hydrocarbon molecules, having the molecular formula CH_4, about sixty times more potent than carbon dioxide as a greenhouse gas

EARTH AS A SYSTEM

Earth is governed by many integrated processes, which are together called a "system." The study of global energy (heat) transfer aims to map the way these systems interact to move energy around the planet. The transfer of energy, in turn, drives everything from global weather patterns to the formation of mountains to the evolution of plants and animals.

The total amount of energy that moves through the Earth system amounts to hundreds of terawatts (10^{12} watts). For perspective, a typical automobile engine is rated at 50,000 to 100,000 watts, and incandescent lightbulbs at hundreds of watts; Earth receives and transfers hundreds of billions of watts.

The rate at which Earth absorbs and radiates energy is essential to the planet's ability to support life. The atmosphere, which slows the release of heat back into the universe, allows the planet to maintain a temperature conducive to liquid water and to the existence of life.

EARTH'S SUBSYSTEMS

Earth's system can be broken down into four basic subsystems: atmosphere, lithosphere, hydrosphere, and biosphere. The atmosphere is most notable for the way it acts as a shell around the planet.

The atmosphere reflects back into space much of the incoming radiation that strikes it, but what the atmosphere does let through, it also partially traps on Earth; the atmosphere captures and returns energy radiating from the surface back to the surface and into the atmosphere before that energy can leave the system.

The greenhouse gases, such as carbon dioxide (CO_2), get their name because they capture heat in the planetary system in a manner similar to a greenhouse, which traps heat. Earth's greenhouse gases trap heat by absorbing outgoing infrared radiation from the surface and by then reemit it in all directions. About one-third of the captured energy is radiated back toward the surface, another third is radiated laterally into the atmosphere, and the remaining third is radiated out toward space. The movement of air around the planet, as wind also transfers some energy. As the carrier for water vapor and pressure cells, the atmosphere can hold and transport massive amounts of energy. Air movement drives some significant ocean currents, such as the El Niño phenomenon, which move energy about and contribute to the "conveyor belt" of ocean currents.

The lithosphere is Earth's landmass. Land is better than the atmosphere at holding heat, but its most important role in global energy transfer comes from its relationship to the planet's core. At areas of high geothermal activity (volcanoes being perhaps the most dramatic example), energy from Earth's core escapes. The gases and chemicals produced by geothermal activity also can influence global energy transfer, such as when ash and gases spewed by volcanic eruptions lead to cooling because they reflect more incoming radiation than does the atmosphere.

The hydrosphere is arguably the most important element of Earth's energy system. The hydrosphere is the water that covers 70 percent of the planet's surface, largely the oceans, and the oceans serve as the major energy reservoir for the planet. This is because of water's high specific heat (water can store a large amount of energy before its temperature increases). The oceans move this energy around the planet, shaping the climate globally.

As an example, one can consider London's comparatively mild winters, despite that city's northerly latitude. London benefits from the Gulf Stream, a major warm-water current that carries energy from near the equator, up the east coast of North America, and then northeast across the Atlantic Ocean to Europe. The interaction of the hydrosphere and atmosphere is largely responsible for Earth's weather, as heat moves between the oceans and the air above them.

Finally, the biosphere comprises all the living things on the planet, from microbes to plankton to humans to whales. Life uses the energy in Earth's system to power itself; photosynthetic organisms turn the sun's radiated energy into a form (that is, food) usable by other forms of life. Life-forms are most important to global energy transfer because of the roles they play in the way gases and chemicals cycle around the planet. Plants help trap carbon, preventing it from building up in the atmosphere as carbon dioxide and adding to the greenhouse effect as a greenhouse gas. The processes of digestion in animals (cattle, for instance) can produce greenhouse gases such as methane. The reef-building of corals can help shape coastlines and the flow of energy in ocean currents.

Humans and industry alter the composition of the atmosphere and the way energy moves around the planet at an unprecedented rate. Burning fossil fuels, for example, releases stored energy as heat while simultaneously releasing vast quantities of carbon dioxide into the atmosphere, and the overall global effects of fuel-burning are significant.

EARTH'S ENERGY SOURCES: SOLAR FLUX AND GEOTHERMAL ENERGY

Across all four basic subsystems, energy comes from two major sources: the sun and Earth's core. Solar flux is the term for the total amount of incoming solar radiation that enters Earth's system. Because of the tilt in Earth's axis, sunlight impinges upon different parts of the planet with different intensities depending on the season. This uneven heating creates the three major climate zones: tropical, temperate, and polar, and also creates weather systems.

Warmer fluids are less dense than colder fluids; warm air and water in the skies and oceans will be pushed away and upward by heavier, colder air and water. This, coupled with the rotation of the planet, sets in motion the major ocean currents and wind patterns, which are important in moving energy around the planet.

The system's internal heat source, radiation from Earth's core, creates currents in the liquid rock that surrounds the core. These currents are what move the continents about on the surface of the earth. The motion of the crust sliding on the subsurface magma is what drives plate tectonics, which in turn gives the planet its mountains and basins and other physical features.

CLIMATE CHANGE

All efforts to mitigate the effects of global warming are predicated on a thorough grasp of how energy moves around the planet. Global warming is, after all, essentially a global energy surplus as more energy remains in Earth's system than leaves it, thus raising global temperatures.

Through the use of observational satellites scientists can estimate how much total radiation enters and exits the Earth system. However, the details of what happens to this energy once it enters the system remain poorly understood. Scientists know that Earth currently has a large surplus of unaccounted-for energy that has "disappeared" somewhere on the planet. Earth is not in a state of radiative balance. For this reason, a detailed accounting of Earth's energy budget and a comprehensive model of the processes that drive global energy transfer are fundamental requirements for success in the face of the challenges presented by global climate change.

Consider once again London's mild climate. Knowing how ocean and air currents move warm water (energy) around supports predictions of some of the seemingly counterintuitive effects of global warming. London, for instance, could experience a drop in temperature if the Gulf Stream, a major warm water current, shifts due to climate change. Without the warm water mediating the temperature of the air around London, the city could see much more severe winters.

—*Kenrick Vezina*

Further Reading

Bridgman, Howard A., and John E. Oliver. *The Global Climate System. Patterns, Processes and Teleconnections.* Cambridge UP, 2014.

Kandel, Robert. "Understanding and Measuring Earth's Energy Budget: From Fourier, Humboldt, and Tyndall to CERES and Beyond." Surveys in Geophysics, Jan. 2012, pp. 1-12.

Marsh, William M., and Martin M. Kaufman. "Physical Geography" Great Systems and Global Environments. Cambridge UP, 2012.

Smithson, Peter, Kenneth Addison, and Kenneth Atkinson. "Energy and Earth." Fundamentals of the Physical Environment. 3rd ed., Routledge, 2002.

———. "Energy Flows and Nutrient Cycles in Ecosystems." Fundamentals of the Physical Environment. 3rd ed., Routledge, 2002.

———. "Heat and Energy in the Atmosphere." Fundamentals of the Physical Environment. 3rd ed., Routledge, 2002.

Trenberth, K. E. "An Imperative for Climate Change Planning: Tracking Earth's Global Energy." Current Opinion in Environmental Sustainability, vol. 1, 2009, pp. 19-27.

Trenberth, K. E., and John T. Fasullo. "Tracking Earth's Energy." Science, vol. 328, Apr. 2010, pp. 316-17.

Wells, Neil C. *The Atmosphere and Ocean: A Physical Introduction.* John Wiley & Sons, 2012.

Global Warming

Fields of Study: atmospheric science; climatology; physics; oceanography; thermodynamics

ABSTRACT

"Global warming" is the term applied specifically to indicate rising average global air temperatures. This rise in temperature has the potential to cause drastic changes in climate and weather patterns worldwide by disrupting the equilibrium between incoming solar energy and the thermal energy that is reradiated away from the surface.

KEY CONCEPTS

El Niño: a quasi-regular phenomenon that brings abnormal quantities of warm surface water across the Pacific Ocean toward South America and subsequently affects climate globally

greenhouse effect: the maintenance of an unusually high atmospheric temperature due to the functioning of certain gases in the atmosphere, essential for life on Earth as the planet would have an average atmospheric temperature of -0.4 degrees Fahrenheit (-18 degrees Celsius) without the effect

greenhouse gas: an atmospheric gas capable of absorbing infrared radiation (heat) coming from Earth's surface and reemitting a significant portion of it back into the atmosphere rather than out into space

urban heat island effect: an elevated average temperature over the area covered by a city due to the generation and enhanced absorption of heat in that area

GREENHOUSE EFFECT

"Global warming" is the term for the rise in Earth's average atmospheric temperature. It is known that Earth's average global temperature has been rising slowly over time, a trend that has been particularly noted since the beginning of the Industrial Revolution in the second half of the eighteenth century. It is generally conceded that increases in air temperature could alter precipitation patterns, change growing seasons, result in coastal flooding, and turn some currently fertile areas into deserts. Although the global climate is an exceedingly complex system reliant on many factors, including long-term cycles and natural events, there is a strong scientific consensus that human activity is responsible at least in part for global warming and related climate change.

When the ground surface is heated by sunlight, it emits much of the solar energy it has absorbed as infrared radiation. Gases in the atmosphere absorb the reradiated infrared radiation and prevent it from escaping into space. This is called the "greenhouse effect" because it was once believed that the glass panes of greenhouses captured the infrared radiation given off by the soil inside the greenhouse. Although it has since been shown that greenhouses work by trapping heated air and not allowing it to blow away, the name has stuck.

The atmosphere eventually releases its contained heat into space and the amount of heat stored in the atmosphere remains fairly constant as long as the composition of gases in the atmosphere does not change. Some gases, including carbon dioxide, water vapor and methane, capture and store infrared radiation, or heat, more efficiently than others and are called "greenhouse gases." If the composition of the atmosphere changes to include more of these greenhouse gases, the atmosphere retains more heat and becomes warmer.

Global levels of greenhouse gases have been steadily increasing. In 1990 they were more than 14 percent higher than in 1960. According to the National Oceanic and Atmospheric Administration (NOAA) which measures long-lived greenhouse gases and publishes an Annual Greenhouse Gas Index (AGGI), greenhouse gases increased 43 percent from 1990 to 2018, and of that, carbon dioxide accounted for the largest percentage of the increase. Measurement of the carbon dioxide content of the atmosphere in 2020 pegged its concentration at 0.0425 percent (425 parts per million), compared to 0.027 percent (270 ppm) at the beginning of the Industrial Revolution. These numbers, according to the NOAA, are directly proportional to a similar rise in average global temperature. Meteorological records show that from 1975 to 2017, the average global temperature rose at a rate of 34.7 to 35.24 degrees Fahrenheit (1.5 to 1.7 degrees Celsius) per century. In comparison, the difference between the average global temperature in the 1990s and in the last ice age was approximatley 50 degrees Fahrenheit (10 degrees Celsius), and it is estimated that a drop of as little as 39.2 to 41 degrees Fahrenheit (4 to 5 degrees Celsius) could trigger the formation of continental glaciers. Therefore, the rise in average temperatures is significant and already appears to be causing changes in the

Sobradinho dam in Bahia, Brazil during the drought of 2015 (Marcello Casal Jr/Agência Brasil via Wikimedia Commons)

global climate. Documented changes include the net loss of ice due to melting glaciers and slowly rising sea levels as the ocean gets warmer and its waters expand. Also, measurements of plant activity indicate that the annual growing season has become approximately two weeks longer in the middle latitude regions.

Such observations led to increasing scientific attention to global warming and its potential disruptive effects. In the 2010s researchers warned that human-induced climate change was accelerating global warming to unprecedented levels. The year 2014 broke heat records but was quickly surpassed by 2015, which showed an exceptional average temperature increase over the previous year. In early 2016, climate scientists from organizations throughout the world (including National Aeronautics and Space Administration (NASA) and NOAA) confirmed that 2015 had been the hottest year on historical record to that point, and that almost all of the ten warmest years on record had occurred since 2000. One of the causes of this extreme rise in temperature was attributed to the El Niño weather pattern, which was one of the largest seen in one hundred years. El Niño's contribution to global warming continued into 2016, disrupting the circulation of the atmosphere. It was later announced that 2016 overtook 2015 as the hottest year on record, due to both an exceptionally strong El Niño and the continued buildup of greenhouse gas emissions from human activity.

The NOAA ranked 2017 as the third-warmest year on record worldwide and noted the continued shrinking of sea ice extent in both the Arctic and Antarctic regions. Another year of above-average precipitation, the year was also marked by an especially large amount of damaging severe weather events, including three major hurricanes (Harvey, Irma, and Maria). The warming trend continued to be observed, as 2018 became the fourth-warmest year on record.

EFFECTS OF GLOBAL WARMING

A common misunderstanding is that global warming simply means that winters will be less cold and summers will be hotter, while everything else will be basically the same. Actually, Earth's global weather system is very complex, and higher global temperatures will result in significant changes in climate and in weather patterns. Most changes will be observed in the middle and upper latitudes, particularly in North America, Europe, and most of Asia, with equatorial regions witnessing fewer changes. The Southern Hemisphere should experience less severe effects because it contains more water than the Northern Hemisphere does, and it takes more energy to heat water than land.

It is difficult to predict precisely what these changes will be, but observation of the changing climate and scientific studies allow researchers to make some estimates of the kinds of changes that will occur. Summers will be hotter, with more severe heat waves. Because hot air holds more moisture than cool air, rain will fall less frequently in the summer. Droughts can be expected to be more common and more severe. Through the late 1980s and into the early 1990s, as annual temperatures climbed higher each year, summer heat waves became more frequent. This is a particularly troubling problem in areas where homes are typically built without air conditioning. The air in a closed-up house during a heat wave can reach temperatures well over 104 degrees Fahrenheit (40 degrees Celsius). Because these temperatures exceed people's normal body temperature, which is about 97.7 degrees Fahrenheit (36.5 degrees Celsius), it becomes very difficult for the body to cool down. For this reason, heat waves are particularly dangerous for the

Temperature at the top of Hurricane Harvey (Contains modified Copernicus Sentinel data 2017, European Space Agency via Wikimedia Commons)

very young, the elderly, and the sick. More frequent heat waves will cause increases in the use of air conditioning which requires more energy consumption that will in all likelihood lead to the release of additional greenhouse gases. Many countries experienced extreme heat waves amidst the record-setting heat in 2015, with India suffering its second-most deadly heat wave.

Global warming also produces more severe autumn rains. Overheated summer air cools in the autumn and is no longer be able to hold all of the moisture it was storing. The moisture is then released as rains, which can cause flooding. This phenomenon has already been observed, but not for a period of time that is scientifically significant as the difference between long-term changes and short-term fluctuations are not differentiable with only a few years of observations.

Changing rain patterns, as droughts and severe autumn storms, will have a strong effect on Earth's landscapes. Some fertile areas may become deserts, while other areas may be transformed from plains to forests.

One of the stranger aspects of global warming is that it is predicted to result in not only hotter periods during the summers but also colder periods during the winters. With more atmospheric energy available to move large cold-air masses from the polar regions in winter, it is possible that large winter storms will be colder, more violent, and more frequent. This pattern has become more evident since about the mid-1970s. Winter storms have brought record low temperatures and enough snow to close cities for days. However, this time span is again too short to determine whether this is a temporary phenomenon or a trend. It is possible that some smaller, less permanent event than global warming is responsible for the more severe winter storms. Another consideration is that a shorter, though more severe, winter may produce a surge in pest populations and diseases that are normally controlled by longer winters.

Global warming will cause ocean levels to rise in part because water expands as it becomes warmer, and because of the additional water that will enter the oceans due to melting glaciers on Greenland and Antarctica. These terrestrial icecaps are as much as 2.5 miles (4 kilometers) thick and lie atop land rather than ocean water. The melting of the ice in the northern polar areas should not contribute to rising ocean levels because that ice, unlike the ice on Greenland and Antarctica, is already in the ocean. Just as the melting of an ice cube in a drink does not cause the level of the drink to rise, the melting of the northern oceanic ice sheets will not affect sea levels.

In 2019, the United Nations reported that climate change was one of the causes of an unprecedented decline of species population and an increasing rate of species extinction in the late 2010s. In addition to directly affecting animals and plants through temperature changes, global warming impacts species by changing the way humans interact with the natural environment. For example, increasing drought may cause humans to tap into new water sources, disrupting the species that already relied on that water. Similarly, the impact of invasive species is potentially heightened as global warming increases the range of certain animals and plants and shrinks that of others.

CAUSES OF GLOBAL WARMING

The vast majority of scientists are convinced, based on the abundant evidence, that global warming is occurring. While climate change skepticism is often a high-profile subject, serious researchers agree that the warming trend is real. However, there is still some dispute over which factors most drive the complex phenomenon. Some scientists point to gradual climatic cycles over extremely long periods of time, or other little-understood natural processes. Yet most experts believe anthropocentric climate change is a crucial component, especially in the unprecedented increase in temperatures observed since the beginning of the Industrial Revolution.

One significant factor in global warming, the rise in greenhouse gases, is strongly linked to the activities of humankind. Burning forests to clear land and operating factories and automobiles produce carbon dioxide and water vapor. Livestock herds and rotting vegetation release methane, and fertilizers used on farms also release greenhouse gases. Power plants that consume fossil fuels such as coal, oil, or natural gas release massive amounts of carbon dioxide into the air. The United Nations' Intergovernmental Panel on Climate Change (IPCC), made up of experts from around the world, determined that the probability of such human activity playing a role in global warming was over 95 percent certain. However, no one knows with certainty whether humankind's

Power plants that consume fossil fuels produce greenhouse gasses (Hramovnick via iStock)

activities are the only reason for the increase in global temperature.

To some degree the warming trend could also be linked to natural, cyclical changes in climate. The last ice age ended recently in geologic terms, and a number of changes are still taking place as the globe recovers from the presence of huge ice sheets on its surface. It is possible that the world's climate is still warming up from the last ice age. History provides examples of dramatic short-term changes in climate, including one period from 1617 to 1650 that was so unusually cold that it is called the "Little Ice Age." Therefore, Earth may be merely experiencing another cyclical change in its climate. Some researchers have also suggested that solar energy emissions may be responsible for global warming, although many experts note that this theory does not fit with modern observed evidence, as solar energy has remained relatively constant since 1750 and all layers of the atmosphere have not warmed as would be expected if the sun were responsible.

STUDY OF GLOBAL WARMING

The greenhouse gas theory first emerged in the late nineteenth century, and by the mid-twentieth century increasing scientific attention was paid to the global climate. By the 1990s the scientific community widely recognized the occurrence and potential impact of global warming, and the study of the phenomenon grew steadily. However, skepticism prevailed in part due to the innate challenge in studying something as complex as the global climate system, which operates on many different scales of time. It may seem straightforward to record temperatures for a number of years and then compare them. However, detailed records on the weather have not been kept for more than a few decades in many areas, and even the oldest, most reliable written accounts provide only a glimpse at larger-scale cycles and trends. Scientists are therefore forced to rely on interpretations of historical accounts and the clues left in fossil records. Tree rings, old ice from deep within glaciers, and sedimentary deposits are examined to provide data about the global climate in the past.

Another complication is that existing records must be reviewed carefully to identify local changes that may not reflect global ones. For example, as towns grow into cities, the temperature climbs simply because larger cities are warmer than smaller ones, a phenomenon known as the urban heat island effect. Measurements taken years ago in a more rural environment should be lower than those taken after the population around the measuring station increased. This problem can be overcome with balloons. By sending instruments high in the atmosphere on weather balloons, air temperatures can be measured without being affected by urbanization. Data recorded this way show a consistent rise in global temperature.

Measurements of the level of greenhouse gases in the atmosphere are also affected by local factors such as urbanization. As a small town becomes a city, levels can be expected to rise. However, recording stations located in regions far removed from cities and factories also show an increase in the level of greenhouse gases. One station in Hawaii showed a steady rise in the amount of carbon dioxide present in the air since early 1958, with similar reports coming from stations in Point Barrow, Alaska, and Antarctica.

Another important variable in looking at global warming is sea surface temperature. Measurements can be skewed by local effects that have no impact on the global climate. One method used to make detailed measurements of seawater temperature is to broadcast a particular frequency of sound through the water and measure it at distant locations. The speed and frequency of the sound are affected by the temperature of the water.

However, more accurate and more global data became available through NASA's Earth Science program (formerly Mission to Planet Earth). The program placed satellites in orbit to study Earth and make a variety of detailed measurements, many of which are attributed to factors that contribute to global warming. These data have helped immensely in enabling scientists to understand the global climate and the changes it is undergoing.

Similarly, international efforts such as the IPCC (founded in 1988) have helped both to generate data on global warming and to study its potential effects. That organization's outlook grew more dire as research on the subject increased greatly in the twenty-first century. In 2018, the IPCC published a report stating that in order to avoid disastrous consequences including increased fires, floods, extreme heat, and drought, the increase in the global temperature had to be kept below 34.7 degrees Fahrenheit (1.5 degrees Celsius) relative to preindustrial levels. An increase of just half a degree more than that, the report said, would lead to 50 percent more people being exposed to water stress and 10 million more people being affected by sea level rise by 2100. At the time of the report's publication, the global average temperature was already 33.8 degrees Fahrenheit (1 degree Celsius) above preindustrial levels.

Also, in 2018, the US government published a report of its own saying that the country could lose as much as 10 percent of its gross domestic product (GDP) and see thousands more premature deaths per year by 2100 if climate change proceeded at then current rates. The report suggested an immediate decrease in fossil fuel use and greenhouse gas emissions.

SIGNIFICANCE

The global climatic environment is very complex because of many contributing factors, which together produce the climate in which people live. For this reason, it is impossible to say if changes are attributable simply to a particular natural event or to the activities of humankind. However, at the same time, the activities of humankind are so large-scale and widespread that it is safe to conclude that they have definitely changed the world. Scientists agree that global warming is among these changes, and that the ongoing warming trend holds enormous consequences not only for human civilization but for the entire natural world.

If the changes in the climate approach what is expected based on scientific models of global warming, the impact on the environment and humanity will be staggering. The changes are such that virtually every person on Earth would be affected. If rain patterns change drastically, it is possible that the food supplies of the world will be sharply reduced. In some areas, if the average temperature increases, certain diseases that are endemic to areas with warmer climates will begin to affect people in other areas. An increase of only 10 feet (3 meters) in the mean ocean level will inundate many of the world's coastal areas and the large cities found in those areas. In addition to these large-scale changes, changes in the weather will have a great impact on the day-to-day lives of the majority of the world's people.

Environmental and climate advocacy groups such as the IPCC argue that people must take action to preserve the environment, implementing large-scale behavioral changes to halt global warming and avoid catastrophic consequences. However, many politicians, businesses, and other groups oppose such changes due to their likely short-term impact on the economy and society. For this reason global warming (and climate change in general) became a major political issue in the 2010s. Along these lines, during this time some news outlets changed the terminology used to describe global warming, including phrases such as "climate crisis" and "climate emergency" in order to highlight the growing importance of the situation.

—*Christopher Keating*

Further Reading

Archer, David. *Global Warming: Understanding the Forecast.* 2nd ed., Wiley, 2012.

Black, Brian, and Gary J. Weisel. *Global Warming.* Greenwood, 2011.

Butler, James H., and Stephen A. Montzka. "The NOAA Annual Greenhouse Gas Index (AGGI)." *Earth System Research Laboratory Global Monitoring Division*, National Oceanic & Atmospheric Administration, Spring 2019, www.esrl.noaa.gov/gmd/aggi/aggi.html. Accessed 22 May. 2019.

"The Causes of Climate Change." *Global Climate Change: Vital Signs of the Planet*, NASA, 1 July 2019, climate.nasa.gov/causes/.

Fagan, Brian M., editor. *The Complete Ice Age: How Climate Change Shapes the World.* Thames, 2009.

Global Warming of 1.5 °C. UN Intergovernmental Panel on Climate Change, 2018.

Lucero, Olivia C., and Carter B. Keyes. *New Developments in Global Warming Research.* Nova Science, 2013.

Maslin, Mark. *Global Warming: Causes, Effects and the Future.* MBI, 2007.

Mooney, Chris. *Storm World: Hurricanes, Politics and the Battle Over Global Warming.* Harcourt, 2007.

Sommerville, Richard C. J. *The Forgiving Air: Understanding Environmental Change.* 2nd ed., American Meteorological Society, 2008.

Vaughn, Jacqueline. *Environmental Politics: Domestic and Global Dimensions.* 6th ed., Wadsworth, 2011.

Wallace, John M., and Peter V. Hobbs. *Atmospheric Science.* 2nd ed., Academic, 2006.

Great Barrier Reef Ecosystem

Fields of Study: oceanography; marine biogeochemistry; marine biology; marine ecology

ABSTRACT

The Great Barrier Reef is composed of over 3,000 individual reefs and islands. Located in the Coral Sea, off the coast of Queensland, Australia, and covering approximately 344,400 square kilometers (133,00 square miles), it is the largest coral reef system in the world.

KEY CONCEPTS

biodiversity: the variety of plant and animal organisms found within an ecosystem

coral bleaching: a stress response of corals that causes symbiotic zooxanthellae to be expelled from within the coral polyps, leaving them essentially colorless against the white calcium carbonate structure of the reef

crown-of-thorns starfish: an aggressive, voracious species of starfish that feeds exclusively on coral polyps; attempts to kill the starfish by catching it and chopping it to pieces have historically failed because each of the pieces subsequently grows and develops into a new starfish if they are simply tossed back into the sea

fringing corals: corals that build structures that develop around the perimeter of a small rocky island

platform reefs: reefs forming a relatively long, broad structure that is of a consistent height throughout its extent

AUSTRALIA AND ITS COASTLINES

Australia, an island continent with an extensive tropical coastline and 17 percent of the world's total area of reefs, has the world's largest total area of coral reefs after Indonesia. Conditions for reef development vary considerably along the island's coastline. The Great Barrier Reef runs along the northeastern coastline of Australia, extending out to the margins of the continental shelf, from the Warrior Reefs in the northern Torres Strait for well over 1,200 miles (2,000 kilometers) to the Capricorn-Bunker Group of reefs and islands in the south.

This reef system, extending to Papua New Guinea, actually is comprised of more than 3,000 individual reefs, coral shoals, and high island fringing reefs, which can range in size from scarcely visible to over 39 square miles (101 square kilometers). The shape and structure of the individual reefs are also of great variety. Two main categories have been defined: lying platform or patch reefs, resulting from radial growth; and wall reefs, resulting from elongated growth, often in areas of strong water currents. There are approximately 300 coral cays, including those with vegetation or not, and over forty low-wooded islands. There are also 618 continental islands that were once part of the mainland.

Continental drift brought the northern coastline of Australia into tropical latitudes about two million years ago, when the Great Barrier Reef originated with some minor development, but widespread reef development is thought to have begun about 500,000 years ago.

CLIMATE AND CURRENTS

Complex patterns of currents in Australian waters are key driving forces and contribute to the different ecological areas within the region. The Great Barrier Reef, bounded by the Coral Sea to the east and the Torres Strait to the north, is characterized by the counterclockwise gyre of the South Equatorial current.

Cyclones are seasonally common here in this tropical marine climate. Air temperatures vary between an average maximum of approximately 86 degrees Fahrenheit (30 degrees Celsius) in January and 73 degrees

An aerial view of the Great Barrier Reef in Austraila (mevans via iStock)

Fahrenheit (23 degrees Celsius) in July, and an average minimum of approximately 74 degrees Fahrenheit (24 degrees Celsius) in January and 64 degrees Fahrenheit (18 degrees Celsius) in July. Mean water surface temperature is at a maximum during February and at a minimum during July.

Current patterns across the continental shelf induce localized upwelling and are predominantly driven by prevailing southeast trade winds, northwest monsoons, and patterns of tidal flow.

The continental shelf of Australia's northern shores connects across the relatively shallow waters of the Torres Strait to Papua New Guinea, which loads considerable freshwater and sediments into the strait. The westernmost areas have the shallowest and most turbid waters, and soft mud dominates the shallow surface of the reefs in the area. The easternmost areas of the strait are characterized by fringing corals. There are very extensive platform reefs around Darnley Island, stretching out toward the edge of the continental shelf and a nearly continuous line of reefs. This particular section of reefs marks the northern edge of the outer Great Barrier Reef. The northern section of the Great Barrier Reef is unique, with well-developed ribbon-type barrier reefs on the outer edge and wide areas of shoals, banks, and fringing reefs.

The southernmost reefs of the Great Barrier Reef are known as the Capricorn-Bunker Group and are a little more than 30 mi (50 km) offshore south of the Swain Reefs Complex. At this relatively small reef complex, the continental shelf narrows rapidly, with steeply sloping reef edges, deep waters, shallow lagoons, and well-developed coral cays. The best-known reefs of the entire Great Barrier Reef system are at One Tree Island at the Tropic of Capricorn, whose cooler waters are largely responsible for the lower coral diversities in these islands.

BIODIVERSITY

The Great Barrier Reef has very high levels of biodiversity, with some 1,500 to 2,000 species of fish, more than 4,000 species of mollusks, and 350 hard coral species. The high number of coral species decreases toward higher latitudes, with the number decreasing to about 240 species in the southern sections. The differences in the higher levels of nutrient loads, sediments, and amounts of freshwater present near the mainland, and the low inputs on the outer reefs, have led to considerable variation in species assemblages in different locations. Fish populations tend to be relatively small, in that the waters of the Great Barrier Reef are generally nutrient-poor. Some nutrients are supplied by Coral Sea surface water and upwellings, and also by terrestrial runoff. Tidal mixing is a major contributor to the nutrient recycling dynamics of this ecoregion.

The Great Barrier Reef serves as a nesting ground for green, hawksbill, loggerhead, and flatback sea turtles. Raine Island and Pandora Cay are the largest breeding grounds for green turtles, whereas islands in the northern section of the area are nesting sites for hawksbill turtles. About twenty-six species of cetaceans are found within the region, with frequent reports of humpback, minke, and killer whales, and of bottlenose, spinner, and Irrawaddy river dolphins. Humpbacks breed along the east coast of Australia in winter, between July and October.

There are some twenty-three species of breeding birds on more than fifty-five major nesting islands in the northern and southern sections of the Great Barrier Reef, with most bird communities reported in the Capricorn-Bunker Group. Raine Island is a major seabird rookery. Saltwater crocodiles are found in mangrove swamps and river estuaries along the region's coastal fringe, but individuals are seen around offshore coral reefs. Giant clams are also common in the Great Barrier Reef.

Seagrass and mangrove communities are notably important in the Great Barrier Reef, as they are breeding and nursing grounds for many reef and marine species. Seagrass beds are important for marine turtles and provide breeding grounds for significant numbers of dugongs. Seagrass beds both shallow and deep—as much as 49 feet (15 meters) below the surface—are widespread, with an estimate of at least 1,930 sq mi (5,000 sq km) of cover. The largest known populations of dugongs occur in the northern and southern sections of the Great Barrier Reef region, but the southern groups are declining due to an increase in the number of boat collisions and entanglement in gill nets and shark nets placed near swimming beaches. Mangroves, for the most part, are located away from coral reef communities; some mangroves have been recorded on a few fringing reef systems, with the highest levels of diversity north of Cairns.

The Great Barrier Reef is the site of a mass spawning event that takes place once a year for a few nights after a particular full moon in the late austral spring, when the majority of stony corals, sponges, sea cucumbers, marine worms, and giant clams reproduce, releasing eggs and sperm in synchronicity. Occurring globally, mass spawning on coral reefs allows cross-fertilization between colonies and increases the chances of survival of individual larvae.

HUMAN INTERACTION

The types of fishing activities that are popular in the Great Barrier Reef are recreational fishing for groupers, emperors, lobsters, fish for the aquarium trade, sea cucumbers, and trochus; and trawling of fish, prawns, scallops, and crustaceans. Repetitive trawling and the size of bycatch have become major concerns associated with trawling activities, as 50 to 90 percent of hauls typically include unusable species of benthic (deep-water) organisms as well as fish, sea snakes, and turtles. Illegal trawling often occurs in some of the protected reef areas and seagrass communities. Other destructive fishing practices have threatened the health of the Great Barrier Reef, including cyanide fishing, blast fishing, and *muro-ami*, which involves banging on reefs with sticks or blocks of cement to scare fish from their hiding places into nets.

Reef health varies, depending on the type and location of reefs, local conditions, and the effect of anthropogenic activities. For a long time, the Great Barrier Reef was not heavily affected by human activities, as the majority of reefs are far offshore, but the effects of human activity can be seen on reefs closest to the mainland. Increases in human activities such as deforestation, poor agricultural practices, high concentrations of agricultural chemicals, nutrients in terrestrial runoff, land-based sediments, overgrazing, oil spills, intensive recreational use, and commercial fisheries are of major and immediate concern.

Cyclones, bleaching events, and coral predators such as the crown-of-thorns starfish have devastating effects on reef structure and development. Bleaching—which can be caused by environmental stress and seawater temperature rise—occurs when corals expel zooxanthellae, a unicellular flagellate (that is photosynthetic), with which the coral have a symbiotic relationship. This expulsion causes a bleached or whitened appearance, and can eventually lead to the decline in the health of the coral. In 2016, the Great Barrier Reef region experienced a widespread bleaching event due to substantially warmer-than-usual ocean temperatures, in large part due to the effects of climate change. In 2016, the Great Barrier Reef Marine Park Authority reported an average coral death rate of 50 percent on the reefs of the inner Cape York Region and 75 percent on the reefs north of Port Douglas. Following a study conducted to survey the amount of coral bleaching that had occurred in 2016, a team of scientists returned to determine how much of the Great Barrier Reef had died due to the extremely high oceanic temperatures caused by global warming that year. In early 2017, they announced that in a large section north of Port Douglas, 67 percent of the corals had died and may not recover.

Factors such as changes in salinity, decreases in zooplankton (caused by fishing), changes in water temperatures, bacterial infections, and climate change due to global warming are all associated with human activity. Both bleaching events and outbreaks of the crown-of-thorns starfish may be human-induced. The starfish predation of coral is of greatest concern in Australia because of the potential economic effect. The first outbreak of the crown-of-thorns starfish was observed on Green Island, off Cairns, in 1962, but most outbreaks have been recorded in the central sections of the Great Barrier Reef. Evidence suggests that increased soil runoff may aggravate outbreaks, and in some instances may cause them. Ocean acidification due to climate change and the increased uptake of carbon dioxide by the world's oceans will increasingly restrict coral growth and survival.

The great majority of Australia's reefs fall within protected areas. Most of the lagoons, and all offshore reefs from the Capricorn-Bunker Group to the northern section of Cape York Peninsula receive legal protection as the Great Barrier Reef Marine Park, managed since 1975 by the federal Great Barrier Reef Marine Park Authority (GBRMPA) and the Queensland Department of Environment and Heritage. The remaining coastal waters and offshore islands fall within other protected areas.

The marine park has a detailed zoning plan of its 133,205 sq mi (345,000 sq km), providing areas of strict protection alongside much larger areas of multiple use. About 80 percent of its total area is open for general use, including permitted commercial fishing and trawling, and a further 16 percent is open for general use but no trawling. Only about 5 percent is closed to fishing activities, but this area includes more than 120 reefs, and there are thirteen fisheries habitat reserves. The reefs of the Torres Strait are outside the jurisdiction of the Great Barrier Reef and thus do not fall under its legal protection, but a fisheries management agreement with Papua New Guinea offers these reefs some kind of loose protection. The Great Barrier Reef has been a World Heritage Site since 1981. In 2015, the United Nations considered placing the Great Barrier Reef on its list of heritage sites that are in danger but held off on the designation following commitments by the Australian government to address water pollution and coastal developments that threatened the health of the reef.

—*Lucia M. Gutierrez*

Further Reading

Arup, Tom. "Great Barrier Reef Coral Bleaching More Widespread Than First Thought, Surveys Reveal." *Sydney Morning Herald*, 31 Mar. 2016, www.smh.com.au/environment/great-barrier-reef-coral-bleaching-more-widespread-than-first-thought-surveys-reveal-20160331-gnux7x.html.

Australian Government. "About the Reef." *Great Barrier Reef Marine Park Authority*, 2016, www.gbrmpa.gov.au/about-the-reef. Accessed 22 Aug. 2016.

Beltran, Carmenza Duque, and Edisson Tello Camacho, editors. *Corals in a Changing World.* IntechOpen, 2018.

Cave, Damien, and Justin Gillis. "Large Sections of Australia's Great Reef Are Now Dead, Scientists Find." *The New York Times*, 15 Mar. 2017, www.nytimes.com/2017/03/15/science/great-barrier-reef-coral-climate-change-dieoff.html.

Claudino-Sales, Vanda. *Coastal World Heritage Sites.* Springer, 2019.

Great Barrier Reef Marine Park Authority. *Great Barrier Reef Outlook Report 2014: In Brief*, 2014, www.gbr.qld.gov.au/documents/gbr-outlook-report-2014-brief.pdf. Accessed 6 Apr. 2017.

Hutchings, Pat, Mike Kingsford, and Ove Hoegh-Guldberg, editors. *The Great Barrier Reef: Biology, Environment and Management.* 2nd ed., CRC Press, 2019.

Mathiesen, Karl. "World's Oceans Facing Biggest Coral Die-Off in History, Scientists Warn." *Guardian*, 8 Oct. 2015, www.theguardian.com/environment/2015/oct/08/worlds-oceans-facing-biggest-coral-die-off-in-history-scientists-warn.

Veron, J. E. N. *A Reef in Time: The Great Barrier Reef from Beginning to End.* Belknap Press/Harvard UP, 2008.

GREAT LAKES ECOSYSTEM

Fields of Study: limnology; marine ecology; marine biology; marine geology

ABSTRACT

The Great Lakes comprise the largest group of lakes on Earth's surface, covering over 94,000 square miles (244,000 square kilometers) as one continuous drainage basin spanning 750 mi (1,200 km) and containing one-fifth of the world's surface freshwater.

KEY CONCEPTS

alewife: a nonnative species of fish found in the Great Lakes

epilimnion: the warm upper layer of water that is heated by incoming sunlight

hypolimnion: the lower colder layer of water below the level that is heated by sunlight

isostatic rebound: the gradual uplifting of an area of Earth's crust that has been depressed by the weight of glacial ice that has been on top of it

thermocline: the region between the epilimnion and the hypolimnion

turnover: an exchange of level as surface water that has been cooled by cold winter air becomes denser and drops to a level below the warmer and less dense water beneath it

DESCRIPTION

The Great Lakes are five bodies of water spanning the eastern United States and the southern Canadian border. Together, Lake Superior, Lake Michigan, Lake Huron, Lake Erie, and Lake Ontario comprise the largest group of lakes on Earth's surface, covering over 94,000 sq mi (244,000 sq km), bordered by approximately 10,9000 mi (17,549 km) of shoreline, and containing an estimated 35,000 islands. This one continuous drainage basin spans 750 mi (1,200 km) and contains one-fifth of the world's surface freshwater, totaling 6 quadrillion gallons (23 quadrillion liters) of fresh water. Their combined surface area is larger than the combined surface area of the US states of New York, New Jersey, Connecticut, Rhode Island, Massachusetts, Vermont, and New Hampshire. The Great Lakes are framed by the states of Minnesota, Wisconsin, Illinois, Indiana, Michigan, Ohio, New York, and Pennsylvania, and the provinces of Ontario and Quebec. The United States has more shoreline on the Great Lakes than it has on the Atlantic Ocean and Gulf of Mexico combined. Michigan's 2,232 mi (3,592 km) of shoreline is second only to Alaska's.

The Great Lakes formed when the Laurentide, or Wisconsin, glaciation began to retreat 14,000 years ago. As the edge of the nearly 2 mi (3.29 km) thick, moving ice sheet melted, the giant basins carved by the glacier filled with water, forming huge Lake Agassiz, which held more water than all lakes on Earth today combined. The repeated drainage events of Lake Agassiz formed the first versions of the Great Lakes, which were initially much larger than they are today, and flowed east to west. The lakes continued to change shape and size as the climate shifted and the land recovered from the weight of the glaciers, a process known as isostatic rebound. Around 4,000 years ago, the Great Lakes took on their present configuration. They flow from west to east, starting at Lake Superior's southeastern corner through the St. Mary's River to Lakes Michigan and Huron, then down the St. Clair and Detroit Rivers to Lake Erie, then flowing north through the Niagara River, over Niagara Falls, and into Lake Ontario, and finally into the St. Lawrence River, joining the Atlantic Ocean over 2,000 mi (3,219 km) from its origins.

CLIMATE, WATER LEVELS, AND LAKE STRATIFICATION

Great Lakes weather is affected by three prominent air systems. There is a very dry and cold arctic system that comes from the north; another dry, but warm Pacific system that comes from the west; and finally, a warm, wet tropical system from the south and the Gulf of Mexico. Because of their large size and volume, the Great Lakes produce "lake effects" on regional weather. For example, the eastern shores and nearby land in Michigan, Ohio, Pennsylvania, New York, and Ontario can receive especially heavy snowfall. This happens when westerly winds blow across ice-free water in the middle of the lakes in the winter, picking up warmth and moisture. When these moisture-laden winds reach the colder landmasses on the eastern shores, heavy snowfall can occur. Another "lake effect" is the moderation of seasonal temperatures in a band around each lake. The latent summer heat in the water causes later autumn frosts, and the cold water keeps summer temperatures cooler lakeside than they are inland.

Like all lakes, the Great Lakes are dynamic bodies of water that undergo complex processes and have a variety of subsystems that change seasonally and on longer cycles. One such cycle is the stratification or layering of water in the lakes because of changes in water density caused by seasonal temperature changes. Colder water is denser, so it stays deep in the lake, forming a layer called the "hypolimnion during the summer months when the sun warms the surface water, forming a layer called the "epilimnion." A middle layer, called the "thermocline," is an area of rapid temperature change between the cold and warm waters. Scientists reported in 2015 that the summer stratification for Lake Superior is earlier each year by approximately one-half day each year, which in turn makes the fall mixing of warm and cooler waters later. Scientists from the American Society of Limnology and Oceanography noted in 2008 that the summer stratification season of Lake Superior increased from 145 days to 170 days during the previous one hundred years.

The warm waters of the epilimnion are the most productive, growing algae that form the base of the food chain. When temperatures fall in autumn, surface waters cool, becoming denser and eventually sinking, causing a mixing of layers, called "turnover." In winter, surface waters are often colder than those below. The layering and turnover of water annually are important for water quality. Turnover is the main way in which oxygen-poor water in the deeper areas of the lakes can be mixed with surface water containing more dissolved oxygen. This prevents anoxia, or complete oxygen depletion, of the lower levels of most of the lakes. However, the process of stratification during the summer also tends to restrict dilution of pollutants from effluents and land runoff.

Water levels also go through cycles in the Great Lakes. Seasonal fluctuations of about 12 to 18 inches (30 to 46 centimeters) result from regular changes in water amounts in the lakes. In the fall and early winter, when the air above the lakes is cold and dry and the lakes are relatively warm, evaporation from the lakes is greatest. This can be seen when fog or mist forms and appears to float over the lakes. Consequently, water levels decline to their seasonal lows. As the snow melts in the spring, runoff to the lakes increases. Evaporation from the lakes is also least in the spring and summer, when the air above the lakes is warm and moist and the lakes are cold. With more water entering the lakes than leaving, water levels rise to their peak in the summer.

Changes in water levels can also be short-term. For example, a strong sustained wind from one direction can push the water level up at one end of a lake, which makes the level drop by a corresponding amount at the opposite end. These events are most common on Lake Erie because of its east–west orientation in an area of prevailing westerly winds and its generally shallow western end. Long-term fluctuations occur over periods of consecutive years.

Continuous wet and cold years will cause water levels to rise. Likewise, consecutive warm and dry years will cause water levels to decline. Over the last century, the range from extreme high to extreme low water levels has been nearly 4 feet (1.2 meters) for Lake Superior, and between 6 and 7 ft (1.8 and 2.1 m) for the other Great Lakes.

AQUATIC LIFE

It is estimated that there were as many as 139 species of fish indigenous to the Great Lakes. Those inhabiting the nearshore areas included smallmouth (*Micropterus dolomieu*) and largemouth bass (*Micropterus salmoides*), muskellunge (*Esox masquinongy*), northern pike (*Esox lucius*), and channel catfish (*Ictalurus punctatus*). In the open water were lake herring (*Coregonus artedi*), lake whitefish (*Coregonus clupeaformis*), walleye (*Sander vitreus*, formerly *Stizostedion vitreum*), sauger (*Sander Canadensis*), freshwater drum (*Aplodinotus grunniens*), lake trout (*Salvelinus namaycush*) and white bass (*Morone chrysops*), blue pike (*Sander vitreus glaucus*), and Atlantic salmon (*Salmo salar*). The last two species are now considered extinct from the Great Lakes. Because of their different characteristics, the lakes are varied in their species composition.

There are at least 34 known nonnative species of fish in the Great Lakes that were introduced to the lakes through the discharge of ballast water from cargo ships, intentional introduction (such as three species of Pacific salmon), migration around barriers in canals, and escaping from aquaculture facilities. Several species of nonnative fish are considered invasive and harmful to the aquatic environment.

Warm, shallow Lake Erie is the most productive, while deep Lake Superior is the least productive. Currently, hatchery-reared coho salmon (*Oncorhynchus kisutch*) and chinook salmon (*Oncorhynchus tshawytscha*) are the most plentiful top predators in the open lakes, except in the western portion of Lake Erie, which is dominated by walleye. This diversity of fish formed part of a complex food web, with sunlight, dead organic matter, and bacteria at its base; plankton and small-bodied fish in the middle; and predaceous fish, birds, and humans at the top. Changes in the species composition of the Great Lakes basin in the last 200 years have been the result of human activities. Many native fish species have been lost by overfishing, habitat destruction, contaminants, or the arrival of exotic or nonindigenous species.

EARLY INHABITANTS AND CURRENT POPULATION

Because humans are thought to have been in North America as early as 14,000 years ago, they probably experienced some of the changes that the Great Lakes underwent. In particular, their shoreline settlements may have been affected by changes in drainage events. Over time, approximately 120 bands of native peoples have occupied the Great Lakes basin. Descendants of the first settlers were using copper from the south shore of Lake Superior 6,000 years ago. The population in the Great Lakes area is estimated to have been between 60,000 and 117,000 in the sixteenth century, when Europeans began their search for a passage to the Orient through the Great Lakes. The native people occupied widely scattered villages, fished and hunted, and grew corn, squash, beans, and tobacco. Present-day names for these tribes include the Chippewa, Fox, Huron, Iroquois, Ottawa, Potawatomi, and Sioux. Today, Native American fishing treaties protect the fishing rights of Great Lakes tribes.

The first written record of the Great Lakes is of Lake Huron, called *La Mer Douce*—the sweet sea—by French explorer Samuel de Champlain in 1615, and Lake Ontario appears shortly thereafter. Accounts of Lake Superior and Lake Michigan appeared in 1622 and 1634, respectively, and of Lake Erie in 1669. After European settlement, major cities grew up rapidly on the shores on the Great Lakes. The largest of these include Chicago, Milwaukee, and Detroit on Lake Michigan; Cleveland on Lake Erie; and Toronto on Lake Ontario. One-tenth of the population of the United States, and one-quarter of the population of Canada, currently live near the Great Lakes and depend on them for drinking water, industry, agriculture, recreation, food, transportation, and tourism.

COMMERCIAL FISHERY

Fishing has been a historically important industry of the Great Lakes, increasing about 20 percent per year after 1820. Commercially important species include lake trout, salmon, lake whitefish, and yellow perch. The greatest fishing harvests were recorded in 1889 and 1899, at about 147 million pounds (67,000 million tonnes). By 1929, tens of thousands of people were generating income from commercial fishing on the Great Lakes, with the first commercial fishing law in Michigan being instituted that same year. Over the course of the next several decades, however, commercial fishing and size of harvests decreased significantly due to overfishing, pollution, habitat degradation, and the introduction of exotic and invasive species. The value of the fishery also decreased when it shifted to smaller, low-value species. Catches of Lake Michigan whitefish declined from over 5,000 tons (4,536 tonnes) in 1930 to just 25 tons (23 tonnes) in 1957. Even more dramatic, lake trout catch declined from almost 7,000 tons (6,350 tonnes) in 1943 to a mere 3 tons (2.7 tonnes) in 1952. Populations of these species and others plummeted after the invasion of the lake by the sea lamprey.

The commercial and sports fishery on the Great Lakes is estimated to produce over $4 billion annually and is dependent on a mix of native species such as yellow perch, whitefish, walleye and introduced species such as smelt (*Osmerus*), alewife (*Alosa pseudoharengus*), splake, a hybrid between a lake trout and a brook trout (*Salvelinus fontinalis*). and Pacific salmon (*Salmo* and *Onchorynchus*). Commercial fishing of lake trout is allowed only in Lake Superior, where populations appear stable. In addition, researchers estimate that 25 to 40 percent of the salmon and trout populations in Lakes Michigan, Huron, and Ontario are now self-reproducing, citing improved habitat, water quality, and stream conditions. A number of species, including coho and chinook salmon, are raised in hatcheries and stocked in the lakes. The Great Lakes Fishery Commission was established in 1955 by the Canadian/US Convention on Great Lakes Fisheries. The commission coordinates fisheries research, works to control the invasive sea lamprey, and facilitates cooperative fishery management among the state, provincial, tribal, and federal management agencies.

NAVIGATION, SHIPPING, AND CANALS

Although the Great Lakes were naturally connected to the Atlantic Ocean via the St. Lawrence River, this passage was not available to vessels because of rapids and waterfalls, especially Niagara Falls. In 1959, the St. Lawrence Seaway, a series of canals, locks, and dams, was completed.

As part of this system, the 26-mi (42-km) Welland Canal connects Lake Ontario to Lake Erie, allowing ships to bypass Niagara Falls. The locks are not wide enough to allow passage of the widest freighters; much of the shipping moves materials among Great Lakes ports. The primary commodities are iron ore (more than 6.6 million tons [6 million tonnes] annually), coal (more than 4.4 million tons [4 million tonnes] annually) and limestone (more than 3.3 million tons [3 million tonnes] annually). Other cargoes include cement, salt, grain, and sand. The fleet of 300 ships from the last century has been reduced to about 140 in recent years. The largest boats are 1,000 ft (304.8 m) long.

The Great Lakes are also connected to the Mississippi River via the Chicago Sanitary and Ship Canal. Completed in 1900, the canal is 28 mi (45 km) long, 202 ft (62 m) wide, and 24 ft (7.3 m) deep. The canal was constructed to divert Chicago's treated sewage from entering Lake Michigan, the source of the city's drinking water. The canal reversed the flow of water in the Chicago River, which naturally flowed into Lake Michigan. Once the canal was opened, approximately 2,068 million gal (7,860 million L) per day from Lake Michigan began to flow into it, pushing Chicago's treated sewage toward the Mississippi River and the Gulf of Mexico. The Great Lakes have a long history of shipwreck, groundings, and collisions. From the 1679 sinking of *Le Griffon* with its cargo of furs to the 1975 loss of the *Edmund Fitzgerald*, thousands of ships and thousands of lives have been lost; many of these losses involved vessels in the cargo trade. The Great Lakes Shipwreck Museum estimates that over 6,000 ships and over 30,000 lives have been lost.

INVASIVE SPECIES

The Great Lakes Fishery Commission estimates that there are thirty-four nonnative fish species in the Great lakes, many of which are considered invasive and therefore harmful to the lake ecosystems. The most invasive species are sea lampreys, alewives, and zebra mussels, all of which are a major problem for Great Lakes ecosystems. Many nonnative species have entered the lakes via the St. Lawrence Seaway and the Chicago Sanitary and Shipping Canal. One of the most established invasive species is the sea lamprey (*Petromyzon marinus*), which is native to the Atlantic coast of North America, the Finger Lakes in New York, and Lake Champlain in New York and Vermont. It is thought to have invaded Lake Ontario via the Erie Canal. It spread to Lake Erie by way of the Welland Canal, and to Lakes Huron, Michigan, and Superior soon afterward. Sea lamprey attack other fish by attaching to them with a suction cup-like mouth and scraping away their flesh with sharp teeth and tongue. These fish usually die from blood loss and infection. Sea lamprey decimated populations of lake trout, lake whitefish, chub, and lake herring in the 1930s and 1940s, taking a huge toll on the commercial fishing of the Great Lakes region. Control efforts include applying electric currents, chemical lampricide, and physical barriers.

The alewife (*Alosa pseudoharengus*) is another invasive species whose population increased greatly when lampreys reduced populations of native fish. Alewives are native to the Atlantic coastal areas, but entered the Great Lakes through the Welland Canal in the 1940s. From time to time, large numbers of alewives die when they are exposed to rapid temperature changes, lack of food, or stress associated with living in freshwater or spawning.

Zebra mussels (*Dreissena polymorpha*) are believed to have come from the Caspian Sea in Europe. They first entered the Great Lakes in the late 1980s in the ballast of ships. Zebra mussels can be up to 2 inches (50 millimeters) long, have striped shells, and can live four to five years. They form dense colonies in sediment and drainage pipes, blocking water flow. They also consume plankton by filter-feeding, competing with fish for food. Annual cost to manage zebra mussel populations exceeds $250 million.

Other invasive species of concern include Asiatic clam (*Corbicula fluminea*), bighead carp (*Aristichthys nobilis*), bloody-red shrimp (*Hemimysis anomala*), European rudd (*Scardinius erythrophthalmus*), European ruffe (*Gymnocephalus cernuus*), New Zealand mud snail (*Potamopyrgus antipodarum*), quagga mussel (*Dreissena rostriformis bugensis*), round goby (*Apollonia melanostomus*), rusty crayfish (*Orconectes rusticus*), silver (or Asian) carp (*Hypophthalmichthys molitrix*), white perch (*Morone americana*), and the aquatic plants European frog-bit (*Hydrocharis morsus-ranae*), hydrilla (*Hydrilla verticillata*), and Eurasian watermilfoil (*Myriophyllum spicatum*).

WATER QUALITY, POLLUTION, AND WATER DIVERSIONS

Great Lakes water and sediments contain more than 800 toxic contaminants. The US Environmental Protection Agency has identified forty-three areas that have water impaired by pollutants such as mercury and polychlorinated biphenyls (PCBs). These invisible chemicals, many of which are no longer discharged, inhabit the sediment of the lake bottoms. Wave action and bottom-feeding organisms bring them back into the water column and the food chain, including fish. Pollutants enter Great Lakes waters from three main sources: point sources, nonpoint sources, and the atmosphere. Point sources of contaminants, such as heavy metals and human waste, include industrial discharge drainpipes and sewage treatment plants. Nonpoint source pollution, including agricultural pesticides and fertilizers, salts and oils from roads, sediments from construction sites and eroding stream banks, and animal waste from farm pastures, comes from many different sources and is transported to the Great Lakes when rain and snowmelt runoff pick up pollutants and drain them into waterways. Airborne contaminants, such as mercury, phosphorus, and PCBs originate from the burning of fossil fuels and waste and enter the lakes via rain and snow.

In March and April 1993, an outbreak of the protozoan Cryptosporidium occurred in the drinking water system of Milwaukee, Wisconsin. Approximately 403,000 people became sick with stomach cramps, diarrhea, and dehydration, and 104 people died, making this the largest documented disease outbreak caused by a waterborne pathogen in US history. The United States draws over 40 billion gal (152 billion L) of water from the Great Lakes every day, much of it for electricity generation. Some towns, such as Pleasant Prairie, Wisconsin, and Akron, Ohio, are allowed to divert Great Lakes water for public supply. Some large-scale schemes to divert Great Lakes water to, for example, Great Plains states to recharge the Ogallala aquifer, to Wyoming for mining, or to Asia, have been denied. In 2014, a major public health crisis occurred in the Toledo, Ohio, region when toxic algae in Lake Erie forced a shutdown of water supplies for over 500,000 people in the area. Unfortunately, Lake Erie has suffered from centuries of human mistreatment, which has resulted in high levels of pollution and increased eutrophication, or natural aging of the lake.

LEGISLATION, TREATIES, AND POLICIES

Several treaties, agreements, and laws govern use of water from the Great Lakes. The 1909 Boundary Waters Treaty between the United States and Canada deals with all matters concerning flow and quality of waters passing over the international border. It also established the International Joint Commission to resolve disputes and review any projects that would affect the lakes' flow and uses, including boating and navigation. In the 1972 Great Lakes Water Quality Agreement, the United States and Canada agreed to control pollution, research and monitor lake health, and work toward restoration with an ecosystem approach. The 1985 Great Lakes Charter is a nonbinding agreement between all Great Lake states and provinces to cooperatively manage the lakes.

Accordingly, before any new diversion or consumptive use of more than 5 million gal (19 million L) per day of any Great Lakes water is begun, all parties must agree. The 1986 Water Resources Development Act gives veto power to each of the eight Great Lakes states' governors to block diversion or export of water from the Great Lakes outside the region. This act was reauthorized in 2000.

The 2001 Great Lakes Charter Annex between the eight Great Lakes governors and two premiers set the intention to develop new protections to prevent harmful water withdrawals and diversions of Great Lakes water. The most recent agreement, the Great Lakes–St. Lawrence River Basin Compact and Agreement of 2008, seeks to ban the diversion of waters, with some limited exceptions, and to set responsible standards for water use and conservation within the basin. Each of the eight Great Lakes states must develop policies for water conservation, according to the compact. These accords have improved the management of the Great Lakes. In 2009, a record $475 million was authorized by the US government to improve the condition of the Great Lakes for future generations.

In 2012, the United States and Canada signed the updated Great Lakes Water Quality Agreement (GLWQA) that authorizes each country to act on potential and real threats to the water quality of any of the Great Lakes. Updated provisions of the agreement included measures to address invasive species, degradation of habitat, climate change, and threats to human health from toxic chemicals, algae, and ship discharges and spills.

—*Susan Moegenburg*

Further Reading

"About Our Great Lakes: Ecology." *Great Lakes Environmental Research Laboratory*. NOAA Great Lakes Environmental Research Laboratory, n.d. Accessed 22 Aug. 2016.

Annin, Peter. *The Great Lakes Water Wars*. Island Press, 2018.

Austin, Jay, and Steve Colman. "A Century of Temperature Variability in Lake Superior." *American Society of Limnology and Oceanography*, vol. 53, no. 6, 2008, pp. 2724–30.

Brant, Cory. *Great Lakes Sea Lamprey: The 70 Year War on a Biological Invader*. U of Michigan P, 2019.

Egan, Dan. "Commercial Fishing, Once a Great Lakes Way of Life, Slips Away." *LA Times*, 30 Aug. 2011.

Egan, Dan. *The Death and Life of the Great Lakes*. W. W. Norton & Company, 2017.

Goniea, Tom. "The Story of State-Licensed Commercial Fishing History on the Great Lakes." *Michigan Department of Natural Resources*. State of Michigan, n.d. Accessed 22 Aug. 2016.

McDonnell, Michael. *Masters of Empire: Great Lakes Indians and the Making of America*. Hill and Wang, 2015.
Rapai, William. *Lake Invaders: Invasive Species and the Battle for the Future of the Great Lakes*. Wayne State UP, 2016.
Riley, John L. *The Once and Future Great Lakes Country: An Ecological History*. McGill-Queens UP, 2013.

Gulf of Aden

Fields of Study: oceanography; marine biology; marine biogeochemistry; geology

ABSTRACT

The Gulf of Aden is located between the African and Arabian plates, extending the Great Rift Valley. This tropical zone between the southern coast of the Arabian Peninsula and the Horn of Africa possesses extraordinary biotic richness. In the northwest, it connects with the Red Sea and in the east opens into the Arabian Sea.

KEY CONCEPTS

anthropogenic: having been caused directly by or occurring as a result of human activities

diurnal tide: high and low tides that occur once per lunar day

drilling mud: an engineered mixture used in drilling of oil wells, designed to stabilize the piping in the bore hole against subterranean pressures while also cooling and lubricating the drillhead

organophosphorus compounds: organic molecules that include a phosphorus-based component in their molecular structure; highly toxic, and typically interact with biochemical systems such as the central nervous system

piracy: "seaway robbery," in which a group of heavily-armed individuals use high-speed boats to accost a commercial vessel in open ocean waters, then board the vessel by force to take control of the ship and its crew; crew quarters and cargo are typically ransacked for valuables, and at times the cargo, crewmembers and even the ship itself may be held for ransom

seagrass: collective term for a variety of plant species that grow on shallow sea bottoms in large colonies

DESCRIPTION

Tectonically, the funnel-shaped Gulf of Aden is located between the African and Arabian plates, extending the Great Rift Valley. This tropical zone between the southern coast of the Arabian Peninsula and the Horn of Africa, surrounded by the arid coastlines of Yemen, Somalia, and Djibouti, possesses extraordinary biotic richness. In the northwest, it connects with the Red Sea through the Bab-el-Mandeb Strait (Gate of Tears; Mandab Strait). In the east, it opens into the Arabian Sea, a northern part of the Indian Ocean. At its widest point, the Gulf of Aden is 217 miles (349 kilometers) from north to south and stretches about 783 mi (1,260 km) from east to west, covering a total surface area of 158,300 square miles (409,995 square kilometers). In the northwest, it is limited by the Red Sea, more precisely by a line joining Husn Murad and Ras Siyan.

The depth of the Gulf of Aden increases from 2,867 feet (874 m) in the west to 16,500 ft (5,029 m) in the east, where the gulf opens into the Indian Ocean. This markedly deep system so close to land is due to the area's conjunction with the regional geographically dominant feature of the Great Rift Valley, of which the Gulf of Aden can be considered an extension. A small part of its water flows underground, moving westward into the saline Lake Assal in Djibouti. A predominantly diurnal tide occurs, with an extreme range of about 10 ft (3 m) at Aden and Djibouti.

The mountains surrounding the Gulf of Aden reach elevations of more than 8,000 ft (2,438 m): Shimbiris, in Somalia, is 7,927 ft (2,416 m) above sea level, and Jabal Thamar, in Yemen, is 8,248 ft (2,514 m) above sea level. Coastal environments here range from rocky outcrops and sand beaches to muddy inlets and drowned river mouths. Intertidal features include seagrass beds, salt marshes, and mangroves. Hard shell and leatherback turtles are present in some abundance, with nesting grounds ranging from the Bab-el-Mandeb Strait in the west to the island chain of Socotra at the eastern end of the gulf, where it joins with the Arabian Sea. The beaches of Socotra are of particular importance as an egg-laying area for the loggerhead sea turtle (*Caretta caretta*) population in the northwestern Indian Ocean and Arabian Sea.

Phytoplankton bloom in the Gulf of Aden (NASA)

FISHERIES AND CORAL

Temperatures vary seasonally between 59 degrees Fahrenheit (15 degrees Celsius) and 82 degrees Fahrenheit (28 degrees Celsius). Additionally, monsoons influence wind patterns, which affects the water exchange between the Red Sea and the Gulf of Aden. During the winter months (October to May), the northeast monsoon winds blow into the Gulf of Aden and drive surface water into the Red Sea from the south-southeast. The water inflow into the Red Sea is affected by high evaporation rates and low precipitation, causing high salinity. During the summer months (June to September), the southwest monsoon winds cause intense upwelling of deep, cold, nutrient-rich ocean water along the southern Arabian coastline. Abundant available nutrients inhibit coral growth but allow kelp beds to thrive; hence, coastal fisheries are rich.

Despite upwelling water and sandy shorelines, the gulf's coral reefs are surprisingly diverse, especially around the Socotra archipelago, albeit not as splendid as the reef systems in the Red Sea. Altogether, 194 species of corals have been recorded along the Saudi Arabian coast. Other habitat types include seagrass beds, limited mangrove communities, salt marshes, and salt flats. Seagrass diversity is relatively high compared with that of surrounding areas, constituting productive ecosystems that provide feeding grounds for green turtles and many species of sea cucumbers. Turtles are caught and eaten opportunistically throughout the region, so all the marine turtles of the region are considered to be threatened, and both the green and hawksbill turtles are declared endangered species. Among the mollusks, cuttlefish and dried sea cucumbers are the main exports. Commercially important invertebrates include rock and deepsea lobster species. Due to temperature conditions and high salinity levels, the diversity of mangroves is low.

ECONOMIC AND MILITARY HISTORY

Periplus Maris Erythraei (Circumnavigation of the Red Sea) was written in the mid-first century by an Egyptian-Greek author. The text is an important source of information about ancient Roman trade in the Red Sea, including the Gulf of Aden. The port of Aden was an important shipping hub of the medieval world. Records from the tenth century note that the city was a busy port linking Red Sea ports with India during the ninth century. The opening of the Suez Canal in 1869 between Port Said and Port Tawfiq, Egypt, linked the Mediterranean Sea with the Gulf of Suez and the Red Sea, including the Gulf of Aden.

All shipping routes from the Mediterranean Sea to the Indian Ocean still use the Suez Canal and, thus, the Gulf of Aden. In 2008 the US Navy reported that approximately 21,000 vessels passed through the Gulf of Aden annually. Thousands of vessels pass through the gulf each year. About 11 percent of the world's petroleum passes through the Gulf of Aden each day. Main port cities are Aden, Yemen; Djibouti City, Djibouti; and Zeila, Berbera, and Boosaaso, Somalia.

The Gulf of Aden falls within the area patroled by the US Navy's Fifth Fleet. The US Navy considers the Strait of Bab-el-Mandeb to be one of three critical choke points in the area it patrols, along with the Strait of Hormuz and the Suez Canal. On October 12, 2000, the USS *Cole* was bombed by suicide terrorists affiliated with al-Qaeda. The attack ripped a hole in the side of the Navy Destroyer as it refueled at the port of Aden. The bombing killed seventeen US Navy personnel.

In the 2000s piracy in the Gulf of Aden increased. Pirates from Somalia perpetrated thousands of attacks on cargo ships passing through the Gulf of Aden and the Indian Ocean, and reduced international trade by about 1.9 percent each year between 2000 and 2010. NATO's efforts to combat piracy in the Gulf of Aden and other international waters in the Middle East include Operation Ocean Shield (OOS) and the NATO Shipping Centre, the organization's point of contact with merchant shipping. Increased international naval patrols and private security led to a decline in piracy. In 2015 the International Chamber of Commerce (ICC) Commercial Crime Services reported that worldwide incidents of maritime piracy totaled 190 by October of that year. The ICC further advised that although piracy in the Gulf of Aden had been reduced, the threat of piracy remained.

ENVIRONMENTAL STRESS AND RESPONSE

Most of the coastal areas and the waters of the region are considered to be biologically intact, but increasing anthropogenic activities introduce new threats. Power plants and desalination plants, refineries, and chemical plants along the coastline leak their effluents—including oil, organic pollutants, heavy metals, heated brine, and cooling water—into the gulf, often in uncontrolled fashion. Pressures on the marine ecosystems include increased salinity and particularly chlorine and thermal pollution.

Spilled hydrocarbons smother corals. Drilling mud from oil exploitation can suffocate reefs and disrupt their growth. In 1997–1998, reefs were affected by a worldwide coral bleaching event. Furthermore, corals have been mined for the manufacture of cement and lime, as well as for use as jewelry. Chlorinated and organophosphorus pesticides, insecticides, and herbicides, introduced into the marine environment by agricultural runoff, cause eutrophication. Further pressure on corals comes from a natural vector: the population of crown of thorns starfish (*Acanthaster planci*), for which hard corals are the main sustenance. The crown of thorns is known to thrive when its predators are overfished or displaced by habitat disruption. This has occurred periodically in the Gulf of Aden region.

The problems of pollution and sewage call for sustainable management measures in the industrial and agricultural sectors, but no regulations exist in the fishery sector to ensure sustainable reproduction. Currently, blast fishing with homemade explosives composed of fertilizer, fuel, and fuse caps inserted into empty beer bottles causes massive damage to corals. When the reef structure has been weakened, it is much more vulnerable to wave action, and the reef becomes unable to protect the coastline. Another destructive fishing method involves the use of cyanide to stun fish, allowing fishers to collect them and sell them to the live-fish trade. Monitoring could prevent overfishing of shark and ray, as well as numerous by-catch varieties. Declines in catches have been reported for diverse species, including Indian mackerel, shark, cuttlefish, shrimp, and Trochus.

To organize effective regional cooperation, the Programme for the Environment of the Red Sea and Gulf of Aden (PERSGA) was initiated in 1974 by the Arab League Educational, Cultural and Scientific Organization (ALECSO) and the United Nations Environment Programme (UNEP). In 1982, the Regional Convention for the Conservation of the Red Sea and Gulf of Aden Environment—the Jeddah Convention—was adopted by the neighboring nations of Egypt, Djibouti, Jordan, Eritrea, Palestine, Saudi Arabia, Somalia, Sudan, and Yemen. It was complemented by relevant provisions of various internationally recognized Law of the Seas bodies and covenants.

The Jeddah Convention and its associated protocols through 2006 include a range of features aimed at preventing and mitigating such threats as oil spill pollution and other features that promote conservation of biological diversity. This latter goal is supported by the ongoing development of protected areas in the Red Sea and Gulf of Aden. An overall Strategic Action Plan was implemented in 1999 as a coordinated effort by PERSGA, UNEP, and related regional and global bodies. This plan strives to strengthen administrative structures and regional cooperation; to reduce navigation risks and maritime pollution; and to support sustainable use of living marine resources, habitat and biodiversity conservation, development of a network of marine protected areas, and management of an integrated coastal zone.

—*Manja Leyk*

Further Reading

Feidi, Izzat H. "Mechanism for Fisheries Management in the Red Sea and Gulf of Aden." *Arab Agricultural Statistics Yearbook 2012*, no. 1, 2012.

Jeffreys, Elizabeth, and Ruth Gertwagen. *Shipping, Trade and Crusade in the Medieval Mediterranean*. Ashgate, 2012.

Margariti, Roxani Eleni. *Aden & the Indian Ocean Trade: 150 Years in the Life of a Medieval Arabian Port*. U of North Carolina P, 2007.

Morewood, Steve. "Suez: The Canal before the Crisis." *History Today*, vol. 56, no. 11, 2006, p. 38.

Schoff, Wilfred Harvey, and Aman Aman. *The Periplus of the Erythraean Sea: Travel and Trade in the Indian Ocean*, Creative Media Partners, 2018.

Ulrichsen, Kristian Coates. *Insecure Gulf: The End of Certainty and the Transition to the Post-Oil Era*. Oxford UP, 2015.

Vogt, H. "Coral Reefs and Coral Bleaching in the Region, Reefs in the Red Sea and Gulf of Aden." *Al Sanbouk*, vol. 2, no. 1, 2000.

Gulf of Mexico Ecosystem

Fields of Study: oceanography; marine ecology; marine biogeochemistry; marine conservation

ABSTRACT

The Gulf of Mexico is an arm of the Atlantic Ocean forming the marine border between the extreme southeastern section of the United States and eastern Mexico, and between the United States and northern Cuba. The Gulf of Mexico flows into the North Atlantic Ocean past northern Cuba, by way of the Straits of Florida, and along southern Cuba merges with the Caribbean Sea via the Yucatán Channel. The immense flow of warm water from the Gulf of Mexico becomes the Gulf Stream after it enters the Atlantic Ocean.

KEY CONCEPTS

Corexit: a chemical dispersant used to break up oil from oil spills and cause it to disperse into the water

dead zone: a region of the aquatic environment in which the ability of the region to support aerobic life has been depleted, usually due to eutrophication

eutrophication: depletion of the dissolved oxygen content in a body of water, typically due to an overabundance of algae, water plants and other organic matter fueled by an influx of fertilizers and other nutrients that encourage the growth of plants

DESCRIPTION

The Gulf of Mexico is an arm of the Atlantic Ocean that extends for about 700,000 square miles (1.8 million square km), forming the marine border between the extreme southeastern section of the United States and eastern Mexico, and between the United States and northern Cuba. The gulf flows for more than 1,100 mi (1,770 km) from west to east and for some 800 mi (1,287 km) from north to south. The Gulf of Mexico flows into the North Atlantic Ocean past northern Cuba, by way of the Straits of Florida. Along southern Cuba, the Gulf of Mexico merges with the Caribbean Sea via the Yucatán Channel.

As warm water from the Caribbean flows into the gulf, it creates a loop current along the coasts. In the Florida Straits, that loop current turns into the Florida Current as it exits and joins with the Gulf Stream. The largest lobes of the Gulf of Mexico are the Bay of Campeche in Mexico, and Apalachee Bay in Florida. At 12,714 feet (3,875 meters), the deepest part of the gulf is Sigsbee Deep, 875 mi (1,408 km) off the coast of Mexico.

The largest river, by far, that empties into the gulf is the Mississippi. Other major flows come from the Alabama, Brazos, and Rio Grande Rivers. The chief ports of the gulf are Tampa and Pensacola in Florida, Mobile in Alabama, New Orleans in Louisiana, Galveston and Corpus Christi in Texas, Tampico and Veracruz in Mexico, and Havana in Cuba. In the United States, the Intracoastal Waterway runs along the northern coast of the gulf. Along the shores of the Gulf of Mexico, the land tends to be low, sandy, and marshy.

More than half of all coastal wetlands in the United States are located in the Gulf of Mexico. Louisiana alone is home to 40 percent of all American coastal wetlands. According to a report issued by the National Oceanic and Atmospheric Administration (NOAA) in 2010, the loss of coastal wetlands in the United States has amounted to 25 sq mi (65 sq km) each year for the past five decades. More than 1 million sq mi (2.6 million sq km) have been lost over the past century. A total of 90 percent of all US coastal wetlands loss occurs in Louisiana.

BIODIVERSITY

The complex ecosystem of the Gulf of Mexico includes open water columns, floating sargassum mats, deep-sea soft coral, hard coral reefs, sandy bottom, muddy bottom, marshes, submerged aquatic vegetation, bays, lagoons, and sandy beaches where sea turtles lay their eggs. Shrimp is the major commercial fishery in the gulf; the diverse aquatic habitats here support brown shrimp, white shrimp, pink shrimp, royal red shrimp, seabobs, and rock shrimp (although neither seabobs nor rock shrimp are much prized by commercial fishers).

Whale species include Bryde whales and sperm whales, an endangered species. Other cetaceans that frequent the gulf are the dwarf sperm whale, pygmy sperm whale, Blainville's beaked whale, Gervais's beaked whale, short-finned pilot whale, orca (killer whale), false killer whale, and pygmy killer whale. Bottlenose dolphin, Atlantic spotted dolphin, Risso's dolphin, rough-toothed dolphin, Fraser's dolphin, pantropical spotted dolphin, striped dolphin, Clymene dolphin, and spinner dolphin are either permanent or seasonal residents here. All these species are protected under the Marine Mammal Protection Act. Five species of sea turtles also live in the gulf: the endangered Kemp's Ridley turtle, and the leatherback, loggerhead, green, and hawksbill. The olive Ridley, a threatened species, can sometimes, but only rarely, be found in the gulf.

DISASTERS: NATURAL AND HUMAN-MADE

The most recent major impact event threatening Gulf of Mexico habitats and species was the April 2010 BP *Deepwater Horizon* oil well blowout off the coast of Louisiana. By November 2010, the US Fish and Wildlife Service had issued a report on damage, declaring that 6,000 birds, 600 sea turtles, and 100 mammals had died within the spill area. Experts believed that the actual count was much higher because so many dead animals had not been found. Researchers also located dozens of dead and dying coral deposits.

An ecologist from the Lawrence Berkeley National Laboratory discovered a new species of bacteria that consumed oil and was able to survive up to 3,600 ft (1,097 m) below the surface. Great concern over the effects of the dispersant Corexit on the biota—BP pumped massive quantities of the toxic brew into Gulf waters—remains an open question.

Assessing damage to marine ecosystems in light of the disaster, the International Union for Conservation of Nature (IUCN) determined that there were 15,419 recorded species of marine life in the Gulf of Mexico, 10 percent of which were endemic, or found nowhere else on Earth. The IUCN Red List identified fifty-three threatened and twenty-nine near-threatened species, including five of twenty-eight mammals, all five species of sea turtles, forty-three of 131 species of sharks and rays, more than half of the twenty-two species of groupers, and three of forty species of seabirds. Also in jeopardy: two of nine seagrasses, and eleven of sixty reef-building corals.

Starting in the 1980s, a large dead zone—with insufficient oxygen to support marine life—was discovered some 60 mi (97 km) off the coast of Louisiana and Texas. By 2011, scientists said it was approaching 9,500 sq mi (25,000 sq km) in size. It is thought to be the result of nitrate- and phosphate-rich agricultural runoff, particularly from fertilizers used far up the Mississippi River basin in the Corn Belt. There is evidence that some species of fish, such as the female croaker, are already showing physical abnormalities because of oxygen depletion. Conservationists cite the 1990s dead-zone reversal in the Black Sea as an example that could be a goal for here.

Anticipated climate change impacts in the Gulf of Mexico include rising sea levels and increased storm activity. Coral reefs are at risk from rising seawater temperature, which tends to drive away their symbiotic caretaker species, leading to bleaching events. Excess greenhouse gas absorption into the seas is also linked to lower reproductivity among crustaceans and mollusks; putting such pressure on native species in the Gulf of Mexico allows faster and broader penetration by invasive species, further destabilizing habitats and food webs.

—Elizabeth Rholetter Purdy

Further Reading

Blomberg, Lindsey E. "Dead in the Gulf." *The Environmental Magazine*, vol. 22, no. 5, 2011.

Buster, Noreen A., and Charles W. Holmes, editors. *Gulf of Mexico: Origin, Waters, and Biota*. Texas A&M UP, 2011.

Dale, Virginia H., et al. *Hypoxia in the Northern Gulf of Mexico*. Springer, 2010.

Felder, Darryl L., and David K. Camp, editors. *Gulf of Mexico—Origin, Waters, and Biota: Vol. 1, Biodiversity*. Texas A&M UP, 2009.

Freudenburg, William R., and Robert Gramling. *Blowout in the Gulf: The BP Oil Spill Disaster and the Future of Energy in America*. MIT Press, 2011.

Murawski, Steven A., Cameron H. Ainsworth, Sherryl Gilbert, et al. *Deep Oil Spills. Facts, Fate, and Effects*. Springer, 2020.

Snedden, John W, and William E. Galloway. *The Gulf of Mexico Sedimentary Basin: Depositional Evolution and Petroleum Applications*. Cambridge UP, 2019.

Tunnell, John W. Jr., et al., editors. *Coral Reefs of the Southern Gulf of Mexico*. Texas A&M UP, 2007.

Ward, C. Herb, editor. *Habitats and Biota of the Gulf of Mexico: Before the Deepwater Horizon Oil Spill. Volume 1*. Springer Open, 2017.

GULF STREAM

Fields of Study: oceanography; fluid dynamics; thermodynamics; climatology

ABSTRACT

The Gulf Stream is a geostrophic surface current that constitutes the northwestern part of the North Atlantic Gyre. It moves huge quantities of water at remarkably fast velocities across vast distances with many geological, physical, and biological repercussions. By itself, however, it is not responsible for the mild climate of Western Europe.

KEY CONCEPTS

Coriolis effect: the apparent deflection of the linear motion of air and water currents due to Earth's rotation

geostrophic current: an oceanic current in which the force of downslope movement of an elevated mass of surface water is balanced by the Coriolis force

gyre: an immense, ocean-spanning circular movement of surface currents

meander: a current that branches away from the flow of the main current and may or may not rejoin the main current to form an oxbow if the current is terrestrial or an eddy circle if the current is oceanic

thermohaline circulation: vertical circulation of an ocean current due to differences in temperature and salinity of the water

WATER CIRCULATION

It is convenient to consider any systematic movement of water at sea as being part of either a wind-driven circulation system or a thermohaline circulation system. In the former, the linkage with atmospheric movement is direct, the currents are usually at or near the sea surface, and the velocities of the flows are often in the range of several knots (or centimeters) per second. In thermohaline circulation, the driving force is gravity, which causes denser water to sink and flow to the deepest parts of the sea, and the currents are usually much slower. In the North Atlantic Ocean thermohaline circulation occurs on a vast scale as saline surface water gives up its heat, sinks, and eventually flows across the bottom of both the North and South Atlantic basins.

Wind-driven circulation develops a gigantic clockwise circular motion called the "North Atlantic Gyre." The most intense flow of this gyre is along its northwestern boundary and is called the "Gulf Stream." Because the flow is circular, it does not actually have a beginning or end. As it is usually geographically defined, however, the Gulf Stream begins in the straits of Florida, where water leaving the Gulf of Mexico and the Caribbean joins with water continuing to go around the gyre. There the Gulf Stream moves about 1,059 million cubic feet (30 million cubic meters) per second past any point. The volume of water entrained in this flow continues to increase, and by the time it reaches Cape Hatteras, North Carolina, there are about 3,002 million cu ft (85 million cu m) moving by any point every second. When the Gulf Stream reaches longitude 65 degrees west, off the Grand Banks of Newfoundland, it moves 5,297 million cu m (150 million cu m) per second.

To put this in perspective, consider that 35 cu ft (1 cu m) of water has a mass of 2,204 pounds (1,000 kilograms). A 150-lb person has a mass of 68 kg. Therefore, 35 cu ft (1 cu m) of water has the mass of about fifteen people. The flow off Cape Hatteras would be equivalent to 1.2 billion people, roughly the population of China, streaming by every second. By the time it gets to the Grand Banks, the flow would be larger by 2,295 million cu ft (65 million cu m), or 975 million more people. The Gulf Stream truly dwarfs most of nature's other wonders. The total hydrologic cycle which includes all the rain, snow, sleet, and hail that falls on Earth (oceans plus continents), moves an average of only 353 million cu ft (10 million cu m) of water per second.

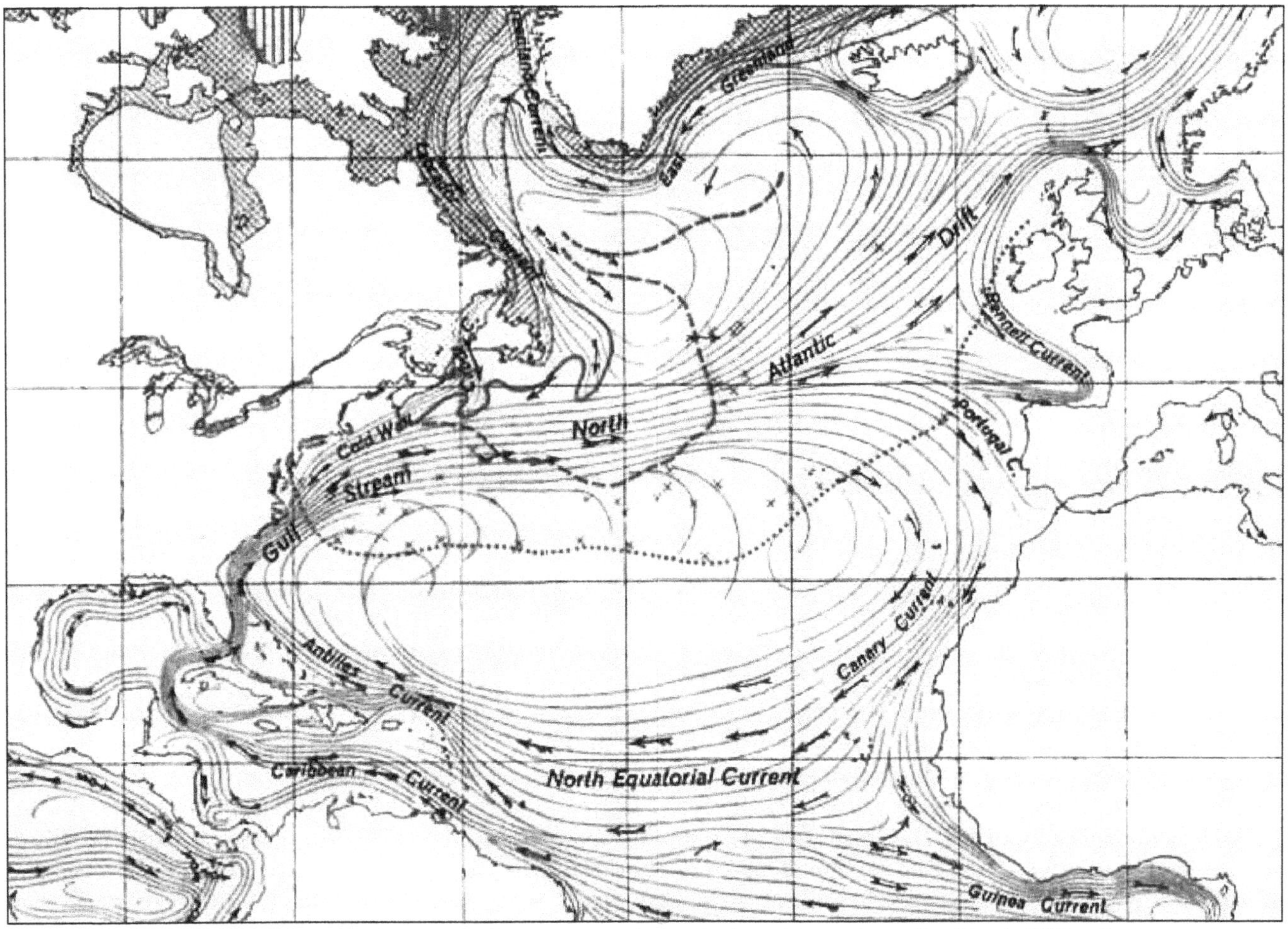

The North Atlantic Gyre, a clockwise-swirling vortex of ocean currents in the Northern Atlantic Ocean (Jrockley via Wikimedia Commons)

Fig. 173.—FRANKLIN'S CHART OF THE GULF STREAM.

Benjamin Franklin published this map of the Gulf Stream in 1769, 200 years before a submersible named after him studied this river in the ocean (National Oceanic and Atmospheric Administration)

CORIOLIS EFFECT

The secret to maintaining such huge flows of water lies in their circular nature. A "hill" of warm, low-density water sits over the core of the North Atlantic Gyre. Water on or within this hill is driven by gravity or a pressure gradient to move downslope or away from the center of this hill. This moving water is deflected by the Coriolis effect which is a consequence of the planet's rotation. In the Northern Hemisphere, the Coriolis effect causes things moving with a horizontal velocity to move to their right. Therefore, water continuously trying to move out from the center of this hill is deflected to move around the gyre instead. Eventually a balance is achieved between the Coriolis effect and the pressure gradient forces. As a result, the water does not move away from the center of the hill but instead circles it in a clockwise fashion. This kind of current is called a "geostrophic" current.

Earth's spherical shape means that the Coriolis effect varies with latitude. One result is that the flow within the North Atlantic Gyre varies with location. The hill is not symmetrical but has its steepest slopes on its northwest edge. Because currents must balance slopes, the gyre is most intense there. The center of the gyre is in the western Atlantic not far from Bermuda. The hill also slopes away to the east, but very gradually, so that the southern flows of the gyre are slow and spread out over a very large region.

Driving the gyre and maintaining the Bermuda High are the winds over the Atlantic Ocean. This subtropical atmospheric feature is a region of descending, dry air near the center of the North Atlantic Ocean. After descending, this air pushes out across the ocean. It, too, is deflected by the Coriolis effect, developing into somewhat circular, clockwise winds. Near the center of this system, winds are weak, and precipitation is uncommon. Early sailors, faced with long, dry periods of calm, sometimes made their thirsty horses "walk the plank." This is the origin of the term "horse latitudes," sometimes used to describe this region.

With little cloud cover and a subtropical latitude, this region receives intense solar radiation that warms the surface water and causes intense evaporation. This causes the water of this area, the Sargasso Sea to become extraordinarily saline and very warm. Just as oil floats on vinegar in a salad dressing, this warm, and correspondingly less dense, saline water floats on top of the cooler, denser water below.

It is this warm water that forms the hill driving the Gulf Stream. Gravity drives the warm water to spread out and flow over cooler surface waters far from the center of the gyre. The Coriolis effect makes it move around the hill, not down it, and the boundary between this warm, rotating mass of water and the cooler water it is trying to flow over is where the currents are most obvious. This is the Gulf Stream, a distinct boundary between the productive coastal waters, which are green and teeming with life, and the dark blue, nearly lifeless Sargasso Sea waters.

GULF STREAM AS A BOUNDARY CURRENT

For decades, schoolchildren have been taught that the mild climate of the British Isles and Western Europe gets its heat from the Gulf Stream. The Gulf Stream is often presented as a river of warm water moving north and then east, eventually to deliver its heat to the European continent. The significance of the Gulf Stream as a source of heat changes when it is recognized for what it is, just a boundary current. It separates the huge hill of warm, saline Sargasso Sea water from the colder surrounding waters.

The Gulf Stream is a very active boundary region with motions driven by the wind and geostrophic currents that surround the huge quantity of surface water, heated by the sun in subtropical high-pressure zones and made especially salty by accompanying evaporation. The Gulf Stream gets all the press, but it is actually this huge quantity of water that conveys heat to Europe.

In the North Atlantic, beyond the northern extent of the circulating gyre, cold winds remove heat from the surface waters. In the process, these winds are warmed and bring pleasant temperatures to Europe.

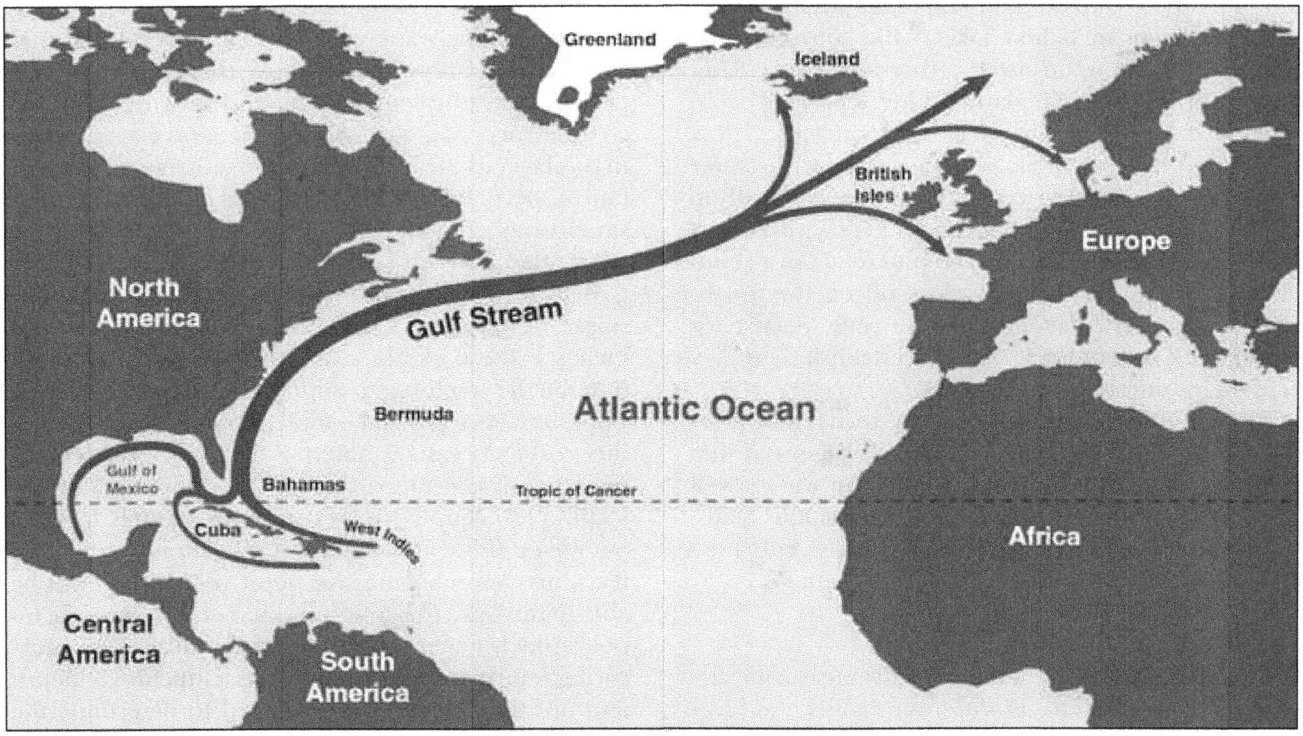

The Gulf Stream is an ocean current that carries warm water up the eastern coast of the United states and Canada and on to western Europe (National Oceanic and Atmospheric Administration)

However, by removing heat, these winds cool the saline surface waters until they become dense enough to sink to the bottom of the sea. This is called "thermohaline circulation." The sinking waters are replaced by a gradual northward flow of surface water. It is likely that most of this surface water spent time in the North Atlantic Gyre, but its transport northward was independent of that circular motion. If the gyre were to stop tomorrow, the thermohaline circulation would continue, and Europe would stay just as warm as it is today. In fact, some researchers have suggested that the strength of the Gulf Stream actually reduces the warming effects of this thermohaline circulation. If they are correct, were the Gulf Stream to stop, Europe might grow warmer.

Although its role as a heat-delivery system may have been overstated, the Gulf Stream is still an incredibly powerful element of oceanic circulation and, as such, greatly influences the biology and chemistry of the surface ocean. This significance is easily seen where meanders develop on the Gulf Stream, typically beyond Cape Hatteras. Just as meanders can grow and develop on a slow-moving river, they also develop on the Gulf Stream. Whereas a stream meander may form an oxbow lake if the course of the stream closes in upon itself, a meander in the Gulf Stream produces a circulating eddy separated from the rest of the stream when it closes upon itself. These eddies will have a core of warmer, less fertile water and rotate in a clockwise manner if they close off on the southeastern side of a meander. They will have a core of cooler, more fertile water and rotate in a counterclockwise manner if they close off on the northwestern side of a meander. These rings persist for months to a year or more and may establish their own ecosystems during their lifetimes.

The Gulf Stream disperses eggs, seeds, and juvenile and adult organisms. As a chemical agent, it stirs up the surface waters, keeping its warm waters well mixed. As a physical agent, it moves enormous quantities of water. It is clearly a remarkable current and a very important part of the global ecosystem.

STUDY OF THE GULF STREAM

The Gulf Stream is studied by directly measuring the strength of its currents at different depths and locations and by examining the effects of its currents through monitoring the position of floats released within it and designed to stay at particular depths.

Floating current meters can be moored to anchors at the bottom of the sea. Their depth is controlled by the length of the tether keeping them attached to the anchor. They can record data electronically, storing it in computer memory. When a vessel is at the surface, ready to retrieve the meter and its data, it transmits a special coded sound pulse. This instructs the meter to release itself from the tether and rise to the surface, where it transmits a radio signal allowing the vessel to home in on it for recovery. The data are incorporated in complex computer models that tie together the results obtained from hundreds of current meters deployed during overlapping time periods. Snapshots of the current system can then be obtained, and sequences of these snapshots reveal the behavior of the currents over time.

Floating objects can be released at sea and tracked by satellite. To ensure that these objects are being moved by ocean currents rather than surface winds, they usually have a large parachute or sail deployed in the water beneath them. Because the density of seawater increases with depth, floats are often designed to have neutral buoyancy at a particular depth (neither sinking below, nor floating above that depth). A layer of the ocean (the sound fixing and ranging [SOFAR] channel) acts as a wave guide for sound waves. Floats in this layer can transmit sounds over tremendous distances, permitting them to be tracked efficiently by a small number of surface ships with sonar receivers suspended into this layer.

Because geostrophic currents are driven by the slopes of the ocean's surface, any technique that can measure those slopes can provide valuable insight into the driving forces behind the Gulf Stream. These slopes are very gradual—total dynamic relief over all the world's oceans is about 7 ft (2 m)—and, consequently, direct measurement is difficult. Satellite techniques, coupled with computer models to filter out waves, tides, and dozens of other confounding effects, are approaching the point where they will be able to measure this topography. Yet this dynamic topography is generally determined indirectly by measuring temperature and salinity as a function of depth and position. These data are used to determine the density of the seawater as a function of depth and position. By assuming that at some depth the horizontal pressure gradients have disappeared, it is possible to reconstruct the differences in the height of the water column needed to accommodate these variations in density. Then the velocities and directions of the resulting currents can be calculated. When the theoretical models are compared with currents measured by moored meters or revealed by the paths of floating objects, the results are in agreement. This gives strong support to the theoretical concepts underlying the study of ocean currents.

Many of the approaches used to study of the Gulf Stream are exercises in applied mathematics. That computed and measured results agree so well is a triumph of geophysical fluid dynamics.

SIGNIFICANCE

The North Atlantic Gyre, which has the Gulf Stream as its northwest boundary current, dominates surface flow in the Atlantic Ocean. Voyages of discovery, exploration, conquest, and exploitation all were affected to some extent by this system and the winds that accompany it.

The Gulf Stream, studied since the eighteenth century, has provided the basis for much of what scientists know of ocean currents. Elaborate mathematical constructs, including the entire concept of geostrophic currents, have been developed to describe, analyze, and comprehend this powerful system.

As scientists have learned more about the Gulf Stream, they have discovered many new areas of study, including branches in the stream, countercurrents at the surface and at depth, and fluctuations in flow and velocity with time and place. Some researchers devote their entire careers to studying just the rings of the Gulf Stream, which contain entire ecosystems. Others investigate the location and strength of the Gulf Stream in the distant past, during and even before the ice ages. There is evidence that at times the current has taken different paths across the continental shelf perhaps scouring out valleys in the ocean floor in the process.

Scientists' understanding of the dynamics of the planet relies on comprehending the transfer of energy from the equator to the poles. The Gulf Stream is an important component of this transfer. As people have become more aware of the fragility of the environment and become more concerned about issues of global climate change, the role of the Gulf Stream and thermohaline circulation in influencing temperatures in Europe and elsewhere on the planet has taken on a new importance.

Drifting buoys help scientists study waves in the Gulf Stream (National Oceanic and Atmospheric Administration)

—*Otto H. Muller*

Further Reading

Broecker, Wally. *The Great Ocean Conveyor*. Princeton UP, 2010.

Brown, Joan. *Ocean Circulation: Prepared by an Open University Course Team*. Elsevier, 2016.

Colling, Angela. *Ocean Circulation*. 2nd ed., Butterworth-Heinemann, 2001.

Robinson, A. R. *Eddies in Marine Science*. Springer Science & Business Media, 2012.

Segar, Douglas. *Introduction to Ocean Sciences*. 2nd ed., W. W. Norton & Co., 2007.

Ulanski, Stan L. *The Gulf Stream: Tiny Plankton, Giant Bluefin, and the Amazing Story of the Powerful River in the Atlantic*. U of North Caroline P, 2008.

Vallis, Geoffrey K. *Atmospheric and Oceanic Fluid Dynamics: Fundamentals and Large-scale Circulation*. Cambridge UP, 2006.

Voituriez, Bruno. *The Gulf Stream*. UNESCO, 2006.

Wells, Neil C. *The Atmosphere and Ocean: A Physical Introduction*. John Wiley & Sons, 2012.

H

HUDSON BAY ECOSYSTEM

Fields of Study: oceanography; limnology; marine ecology; marine biology

ABSTRACT

Hudson Bay is the second-largest bay in the world after the Bay of Bengal. It is a marine bay and marginal sea in northern Canada that drains most of the central area of the country as well as parts of the upper Midwest of the United States, about 1.5 million square miles (3.9 million square kilometers) in all. Hudson Bay connects with the North Atlantic Ocean via the Hudson Strait, and to the Arctic Ocean via the Northwest Passages.

KEY CONCEPTS

anadromous fish: fish species that split their life cycle between freshwater and salt water, typically spawning in freshwater then developing to maturity in salt water before returning to freshwater to spawn

climate change: the observed principle that Earth's average temperature is increasing over time and the global climate is adjusting to the increase in average temperature

fauna: the animal life found in a particular region, whether terrestrial or marine

flora: the plant life found in a particular region, whether terrestrial or marine

heavy metals: metals, such as cadmium, lead, mercury and others from the same region of the periodic table, that are readily absorbed and bioaccumulated within a trophic chain where it can cause neurological and other health issues

hydrocarbons: organic compounds consisting only of carbon and hydrogen atoms, the products of the petroleum industry

seagrass: collective term for a variety of plant species that grow on shallow sea bottoms in large colonies

DESCRIPTION

Named for Henry Hudson (ca. 1565–1611), an English navigator exploring for the Dutch, Hudson Bay is the second-largest bay in the world after the Bay of Bengal. It is a marine bay and marginal sea in northern Canada that drains most of the central area of the country as well as parts of the upper Midwest of the United States, about 1.5 million sq mi (3.9 million sq km) in all. Hudson Bay connects with the North Atlantic Ocean via the Hudson Strait, and to the Arctic Ocean via the Northwest Passages.

HYDROLOGY AND GEOGRAPHY

Many streams and rivers contribute to Hudson Bay's waters, but not nearly enough to offset its brackish salinity. Hudson Bay is surrounded by vast expanses of wetlands, shrubland steppes, taiga or boreal forest, and tundra ecosystems.

Hudson Bay is a relatively shallow saltwater body, with an average depth of 330 feet (100 meters). It is about 900 mi (1,450 km) long north to south—including its southernmost lobe, James Bay—and about 650 mi (1,050 km) across at its widest. Total surface area is approximately 450,000 million sq mi (1.16 million sq km). Hudson Bay is ice-covered for about half the year; ice generally begins to form in early November, and the whole bay has very low year-round average temperatures. The ice cover historically begins to melt in mid-June, usually clearing from the eastern shores first. Water temperature rises to 46 to 48 degrees Fahrenheit (8 to 9 degrees Celsius) on the western side of the bay by late summer.

While it is a saltwater or brackish body, Hudson Bay receives freshwater inflow at an annual volume of some 170 cubic miles (700 cubic kilometers); even more comes from precipitation and ice melt. Consequently, and also due to its limited hydraulic exchange with the Atlantic Ocean, the bay is less salty than the ocean.

RICH BASE OF LIFE

Because of its vast dimensions, the Hudson Bay marine ecosystem has many aquatic and coastal ecozones with varied habitats that are used year-round by Arctic and sub-Arctic species, as well as seasonally by migratory fish, marine mammals, and birds. The sea ice, for instance, supports the seals upon which polar bears depend. Millions of geese and shorebirds feed and breed in vast salt marshes. Productive submarine eelgrass beds provide food for waterfowl migrating to and from breeding habitat in the Arctic Islands, and shelter for crustacean and fish populations. Numerous estuaries provide habitat corridors for anadromous fish like salmon and Arctic char, and support beluga whales.

Hudson Bay supports great numbers of large mammals, both aquatic and terrestrial. Migratory marine mammal species such as belugas, narwhals, and bowhead whales frequent the region as permitted by ice conditions across the northern channels. Other cetaceans, such as orca and minke whales, are rarer visitors. Large concentrations of up to 20,000 beluga whales are found in the estuaries of the Nelson and Churchill Rivers in July and August. Ringed seals, bearded seals, and harbor seals, as well as walruses, reside in the bay year-round, while harp and hooded seals visit seasonally. Subsistence harvesting of these marine mammals is important to the aboriginal peoples and the economy of the bay. Arctic foxes and polar bears reside on coasts in the summer, and on sea

A Beluga Whale in Churchill River (Hudson Bay). Large concentrations of Belugas are found in the river in July and August. (Ansgar Walk via Wikimedia Commons)

ice during the rest of the year; they, too, are hunters of the seals.

At least 130 species of birds migrate to Hudson Bay's waters each summer, and depart in late fall. Waders, water fowl, and raptors share pelagic, intertidal, and wetland habitats here. Most such species—at least 100—find breeding zones on shore. Cranes, plovers, herons, bitterns, loons, pelicans, ducks, geese, swans, osprey, falcons, eagles, and owls share the Hudson Bay ecosystem.

Less is known about the prevalence and range of marine fish species here, due to the near-absence of commercial fisheries; even traditional subsistence fishing has been somewhat limited. Some sixty or so fish species are known in these waters, however, and many of these are adapted to the relatively shallow, brackish environment here. Of the recognized species, about two dozen stay in the marine environment all their lives. Half that many species are marine but frequent the estuaries as seasonal nursery grounds; another nine feed in the brackish coastal realm, but spawn in freshwater zones. As many as sixteen freshwater species have evolved salinity tolerances sufficient to spend some time here in brackish coastal reaches or estuaries. One species—the Atlantic salmon—winters in saltwater in between the freshwater spawn and final act of its anadromous life cycle.

CONSERVATION AND CLIMATE CHANGE

The governance of the Hudson Bay area is largely a series of compromise decisions and shared responsibilities between federal, provincial, local, and First Nations bodies. As global warming proceeds—and faster here in its effects than in many if not most other parts of the world—such organizations will have to work more closely in concert to help determine the growing number of issues. These will range from plans for hydroelectric dams and other water diversion schemes; new deep-water port facilities and related marine transport strategies; mining, agriculture, grazing, and aquaculture rights; tourism pursuits such as hunting and ecotourism; and real estate development. Encompassing all of these areas is the issue of environmental conservation.

There has already been considerable focus on the impact of climate change and transboundary contaminants on Hudson Bay. Historic commercial whaling, particularly for bowheads and belugas, depleted these populations, some of which have not entirely recovered. The steady increase in regional temperatures over the last 100 years has caused a lengthening of the ice-free period, and continues to jeopardize the sustainability of ice-dependent species such as polar bears and seals. These ice-dependent marine mammals are vulnerable to both airborne contaminants and global warming. Because these carnivorous animals are high in the food chain, they accumulate contaminants such as heavy metals, carcinogenic hydrocarbons, and radioactive materials in their fatty tissues and organs.

Diminished sea-ice cover, decreasing flow from its tributary rivers, and increasing precipitation are some of the long-term trends already detected in Hudson Bay. Each of these vectors is projected to intensify, along with other effects such as sea-level rise, which will have as yet undetermined but certainly serious impacts on the habitats here.

—Magdalena Ariadne Kim Muir

Further Reading

Butts, Edward. *Henry Hudson: New World Voyageur.* Dundurn Press, 2009.

Derocher, Andrew E. *Polar Bears: A Complete Guide to Their Biology and Behavior.* Johns Hopkins UP, 2012.

"Ecosystems of the Hudson Bay Coast." *Churchill Wild*, 27 Apr. 2017, www.churchillwild.com/ecosystems-of-the-hudson-bay-coast/.

Ferguson, Steve H., Lisa L. Loseto, and Mark L. Mallory, editors. *A Little Less Arctic: Top Predators in the World's Largest Northern Inland Sea, Hudson Bay.* Springer, 2010.

Horner, Rita A. *Sea Ice Biota.* CRC Press, 2018.

March, James H. "Hudson Bay." *The Canadian Encyclopedia*, Historica Canada, 4 Mar. 2015, www.thecanadianencyclopedia.ca/en/article/hudson-bay/.

Riley, John L. *Wetlands of the Hudson Bay Lowland: An Ontario Overview.* Nature Conservancy of Canada, 2011.

Stewart, D. B., and W. L. Lockhart. "An Overview of the Hudson Bay Marine Ecosystem." Fisheries and Oceans Canada, 2005. www.dfo-mpo.gc.ca/libraries-biblio theques/toc-tdm/314704-eng.htm.

HYDROLOGIC CYCLE

Fields of Study: oceanography; limnology; geography; fluid dynamics; biogeography

ABSTRACT

Water circulates on Earth through a system called the "hydrologic cycle." This water cycle functions through the biosphere, the atmosphere, the lithosphere, and the hydrosphere. The sun and the force of gravity provide energy to drive the cycle from vegetation and animals, groundwater and surface water to atmospheric moisture, which then returns to the land and oceans as precipitation.

KEY CONCEPTS

aquifer: a mass of water contained within an underground soil or rock structure, maintained from a recharge point and flowing toward a discharge point

base flow: the minimum amount of water that normally flows in a watercourse or aquifer

capillary force: the effect of the intermolecular forces between water molecules that are in narrow channels and interparticle pores

convectional precipitation: precipitation resulting from the elevation of humid air in an atmospheric convection cycle

discharge: the amount of water flowing past a particular reference point in a specific amount of time

frontal precipitation: precipitation resulting from the interaction of warm and cold fronts

interception capacity: the ability of different types of foliage to capture rainfall

nomograph: typically, a one-page document that shows a graphical representation of the relationship between a cause and effect, such as rate of surface evaporation and ambient temperature

orographic precipitation: precipitation caused by the rising of humid air to regions of cooler temperatures as it approaches and surmounts a mountain range

EVAPORATION, CONDENSATION, AND PRECIPITATION

The unending circulation of water on Earth is called the "hydrologic cycle." This system is driven by the heat energy received from the sun. Gravity pulls water that falls on the surface back to the oceans to be recycled once again. The total amount of water on Earth is an estimated 0.85 billion cubic miles (1.36 billion cubic kilometers). Most of this vast amount of water, some 97.2 percent, is found in the oceans. Another 2.15 percent is contained in the Greenland and Antarctic ice caps. The remaining 0.65 percent is divided among rivers (0.0001 percent), freshwater and saline lakes (0.017 percent), groundwater (0.61 percent), soil moisture (0.005 percent), the atmosphere (0.001 percent), and the biosphere and groundwater below 13,123 miles (4,000 meters) (0.0169 percent). While these percentages seem small, the total volume of water contained in each of these reservoirs is immense.

A description of the hydrologic cycle must begin with the oceans, as most of Earth's water is located there. Each year, about 76,772 cu mi (320,000 cu km) of water evaporate from the world's oceans. Evaporation is the process whereby a liquid changes into a gas. Adding energy in the form of heat to the water causes the water molecules to become increasingly active and to move more rapidly, weakening the chemical and physical forces that bind them together. As the temperature of the water increases, water molecules tend to move from the ocean's surface into the overlying air. Factors that influence the rate of evaporation from free water surfaces are solar radiation temperature, humidity, and wind velocity. It is estimated that an additional 14,395 cu mi (60,000 cu km) of water enter the atmosphere from rivers, streams, and lakes or through transpiration from plants every year. Thus, a total of about 91,167 cu mi (380,000 cu km) of water are evapotranspired on Earth every year.

The amount of water vapor that can be present in the air depends largely on the temperature. More vapor can be present at higher temperatures. As vapor-laden air is lifted and cooled at higher altitudes, the vapor condenses to form droplets of water. Condensation is aided by small dust and salt particles or nuclei in the atmosphere. As droplets collide and coalesce, clouds begin to form and precipitation can begin. Wind may transport moisture-laden air long distances, and most precipitation events are the result of three causal factors: frontal precipitation, or the lifting of an air mass over a moving weather front; convectional precipitation related to the uneven heating of Earth's surface, causing warm air currents to rise and cool; and orographic precipitation, resulting from a moving air mass being forced to move upward over a mountain range, cooling the air as it rises.

Each year, about 68,135 cu mi (284,000 cu km) of precipitation fall on the world's oceans. This water has completed its cycle and is ready to begin a new cycle. Approximately 23,032 cu mi (96,000 cu km) of precipitation fall on the land surface each year. This precipitation follows a number of different pathways in the hydrologic cycle. It is estimated that some 14,395 cu mi (60,000 cu km) evaporate from the surface of lakes or streams or transpire directly back into the atmosphere. The remainder—about 8,637 cu mi (36,000 cu km)—is intercepted by human structures or vegetation, is infiltrated into the soil or bedrock, or becomes surface runoff.

INTERCEPTION, RUNOFF, AND INFILTRATION

Although the amount of water intercepted by and evaporated from human structures—the surfaces of buildings and other artificial surfaces—may approach 100 percent, much urban water is collected in storm sewers or drains that lead to a surface drainage system or that spread water over the land surface to infiltrate the subsoil. Interception loss from vegetation is dependent on interception capacity (the ability of the vegetation to collect and retain falling precipitation), wind speed (the higher the wind speed, the greater the rate of evaporation), and rainfall duration (the interception loss will decrease with the duration of rainfall, as the vegetative canopy will become saturated with water after a period of time). Broad-leaved forests may intercept 15 to 25 percent of annual precipitation, and a bluegrass lawn may intercept 15 to 20 percent of precipitation during a growing season.

When the duration and intensity of the rainfall are greater than the soil's ability to absorb it, the excess water begins to run off, a process termed "overland flow." Overland flow will begin only if the precipitation rate exceeds the infiltration capacity of the soil. Infiltration is the process whereby water sinks between soil particles directly into the soil surface or into fractures of rocks. It is dependent on the characteristics of the soil or rock type and the nature of the vegetative cover. Sandy soils have infiltration rates of 1.42 to 1.5 inches (3.6 to 3.8 centimeters) per hour, and clay rock soils average 0.79 to 0.905 in (2.0 to 2.3 cm) per hour. Nonporous rock would have an infiltration rate of zero, and all precipitation would become runoff. The presence of vegetation impedes surface runoff and increases the potential for infiltration to occur.

Water infiltrating into the soil or bedrock encounters two forces: capillary force and gravitational force. A capillary force is the tendency of water in the subsurface to adhere to the surface of soil or sediment particles, which may actually draw the water upward against the downward pull of gravity. Capillary forces are responsible for the soil moisture found a few inches below the land surface.

COMPLETION OF THE CYCLE

Growing plants are continuously extracting soil moisture and passing it into the atmosphere through a process called "transpiration." Soil moisture is drawn into the plant rootlet because of osmotic pressure. The water moves through the plant to the leaves,

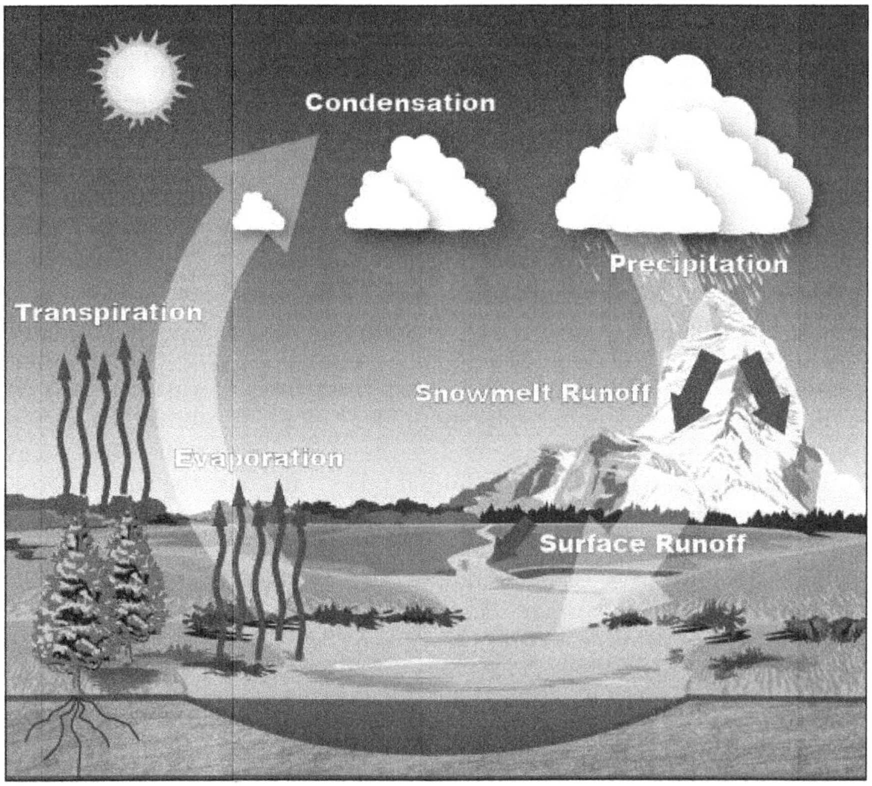

The basic hydrologic (water) cycle (National Oceanic and Atmospheric Administration)

where it is passed into the atmosphere through specialized leaf openings called "stomata." The plant uses less than 1 percent of the soil moisture in its metabolism with the remainder being used for temperature regulation and the transport of nutrients and metabolites between the roots and leaves. Thus, transpiration is responsible for most water-vapor loss from the land in the hydrologic cycle. An oak tree, for example, may transpire 39,943 gallons (151,200 liters) per year.

The water that continues to move downward under the force of gravity through the pores, cracks, and fissures of rocks or sediments will eventually enter a zone of water saturation. This source of underground water is called an "aquifer"—a rock or soil layer that is sufficiently porous and permeable to hold and transport water. The top of an aquifer, or saturated zone, is the water table. This water is slowly moving toward a point where it is normally discharged into a lake, spring, or stream. Groundwater that feeds and maintains the flow of a stream is called "base flow." Base flow is typical of so-called spring-fed streams, and it enables such streams to continue to flow during droughts and through cold winter months. Groundwater may flow directly into the oceans along coastlines.

When the infiltration capacity of the soil surface is exceeded, overland flow can begin as broad, thin sheets of water, no more than a few inches (or millimeters) thick, called "sheet flow." After flowing a short distance, the sheets break up into threads of current that flow in tiny channels called "rills." The rills can coalesce into gullies and, finally, into streams and rivers. While evaporation losses do occur from the stream surface, much of the stream's water is returned to the oceans in this way, completing the hydrologic cycle.

Scientists are interested in how long it takes water to move through the hydrologic cycle. The term "residence time" refers to how long a molecule of water would remain in the various components of the hydrologic cycle. The average length of time that a water molecule would stay in the atmosphere is about one week; in a river, about two weeks; and in a lake, ten years. It would take four thousand years for all the water molecules in the oceans to be recycled just once. Groundwater may require anywhere from a few weeks to thousands of years to move through the cycle. This time period may appear extremely long to humans, yet it suggests that over the course of geologic time every water molecule on Earth has been recycled millions of times.

STUDY OF THE HYDROLOGIC CYCLE

Scientists have developed a vast array of mathematical equations and data-collecting instruments with which to quantify the complexities of the hydrologic cycle. The geographic, secular, and seasonal variations in temperature, precipitation, evapotranspiration, solar radiation, vegetative cover, and soil and bedrock type, among other factors, must be evaluated to understand the local, regional, or global hydrologic cycle.

Precipitation, an extremely variable phenomenon, must be accurately measured to determine its input into the hydrologic cycle. The United States has more than thirteen thousand precipitation stations equipped with rain gauges placed strategically to compensate for wind and splash losses. Techniques have been developed to determine the average depth of precipitation falling on a given area or drainage basin. The effective uniform depth method uses a rain-gauge network of uniform density to determine the arithmetic mean for rainfall in the area. The isohyetal and polygonal methods are used to determine the arithmetic mean for an area or basin with a nonuniform distribution of rain gauges. The amount of water in a snowpack is estimated by snow surveys. The depth and water content of the snowpack are measured and the extent of the snow cover mapped using satellite photography.

The amount of precipitation lost by interception can be measured and evaluated. Interception is determined by the type of vegetation, the amount of evaporation that occurs during the storm, and the length of the storm. Most often, interception is determined by measuring the amount above the vegetative canopy and the amount received at the surface. The difference is deemed to be the amount lost to interception.

The volume of water flowing by a given point at a given time in an open stream channel, measured in cubic meters per second (CMS), is called "discharge." Discharge is determined by measuring the velocity of water in the stream channel with a current meter. The Price AA meter and the Price pygmy meter meet the specifications of the US Geological Survey and have

Groundwater flowing from a karstified limestone aquifer in Reimold quarry, Gemmingen, Southern Germany (Jonathan Trautmann & Randolf Rausch/Hydro-Berlin via Wikimedia Commons)

historically been the current meters most frequently used by the USGS. The cross-sectional area of the stream channel is determined at a specific point and multiplied by the stream velocity to determine discharge. Automated stream-gauging stations are located on most streams to supply data for various hydrologic investigations.

The National Weather Service maintains about five hundred stations using class A land pans to measure free-water evaporation. These pans are 48 in (122 cm) in diameter and 10 in (25.4 cm) deep, and they are made of unpainted galvanized metal. Water depths of 6.7 to 7.9 in (17 to 20 cm) are maintained. The wind velocity is also determined. Errors may result from splashing by raindrops or from birds. Because the metal pan will also heat and cool more rapidly than will a natural reservoir, a pan coefficient must be employed to compensate for this phenomenon. A lake-evaporation nomograph is employed to determine daily lake evaporation. The mean daily temperature, wind velocity in miles (or kilometers) per day, solar radiation, and mean daily dew point are the variables used to determine daily lake evaporation.

The amount of evapotranspiration can be measured using a lysimeter, which is a large container holding soil and living plants. The lysimeter is set outside, and the initial soil moisture is determined. All precipitation or irrigation is measured accurately.

Changes in the soil-moisture storage determine the amount of evapotranspiration.

All of these techniques are used to determine the water budget for different geographic areas. The information gathered is then used to estimate the total water budget of the hydrologic cycle.

—Samuel F. Huffman

Further Reading

Dodds, Walter Kennedy, and Matt R. Whiles. *Freshwater Ecology: Concepts and Environmental Applications of Limnology*. 2nd ed., Academic, 2010.

Gat, Joel R. *Isotope Hydrology: A Study of the Water Cycle*. Imperial College Press, 2010.

Greenwood, Gregory B., and J. F. Shroder Jr., editors. *Mountain Ice and Water: Investigations of the Hydrologic Cycle in Alpine Environments*. Elsevier, 2016.

Hamblin, Kenneth W., and Eric H. Christiansen. *Earth's Dynamic Systems*. 10th ed., Prentice, 2003.

Hunt, Constance Elizabeth. *Thirsty Planet: Strategies for Sustainable Water Management*. Room 400, 2004.

Ingebritsen, Steve, Ward Sanford, and Chris Neuzil. *Groundwater in Geologic Processes*. 2nd ed., Cambridge UP, 2006.

Lim, Wee Ho, and Michael L. Roderick. *An Atlas of the Global Water Cycle: Based on the IPCC AR4 Climate Models*. ANU E Press, 2009.

Lutgens, Frederick K., and Edward J. Tarbuck. *Essentials of Geology*. Illus. Dennis G. Tasa. 12th ed., Prentice, 2015.

Mays, Larry W. *Ground and Surface Water Hydrology*. Wiley, 2011.

Moore, John E. *Field Hydrogeology: A Guide for Site Investigations and Report Preparation*. 2nd ed., CRC, 2012.

Tang, Quihong, Qi Youcum, Zhihui Wang, and Yun Pan, editors. *Remote Sensing of the Terrestrial Hydrologhic Cycle*. MDPI, 2020.

Tarbuck, Edward J., and Frederick K. Lutgens. *Earth: An Introduction to Physical Geology*. Illus. Dennis G. Tasa. 11th ed., Prentice, 2014.

HYDROTHERMAL MINERALIZATION

Fields of Study: oceanography; geology; marine geology; marine biology

ABSTRACT

Hydrothermal mineralization occurs when mineral particulates precipitate from water. This process is closely tied to tectonics and volcanic activity, which create the heat and convection necessary for the transportation and deposition of minerals. Hydrothermal mineralization can occur in submarine or terrestrial environments. Hydrothermally created mineral deposits are economically important, and they provide data that help geologists to understand Earth processes.

KEY CONCEPTS

convergent boundary: the boundary where the edges of tectonic plates are pushed towards each other by underlying magma currents

divergent boundary: the boundary between two tectonic plates that are being driven away from each other by underlying magma currents

hydrothermal mineralization: the deposition of mineral deposits from hot water solutions from natural sources, such as hot springs and undersea thermal vents

isotopic tracing: using the relative abundances of specific isotopes in rocks to determine the timeline of their formation

magma: molten rock that exists in the mantle layer below the lithosphere

mineralization: the deposition of mineral deposits from natural aqueous solutions, whether hot or cold

HYDROTHERMAL MINERALIZATION: THE BIG PICTURE

Liquids frequently contain minerals, and when the liquids are heated, the minerals tend to fall from solution (precipitate). An everyday example of this is the mineral layer that sometimes forms on the bottom of tea kettles or in coffee makers.

In geologic terms, hydrothermal mineralization means that water is heated by magmas or hot rocks, and as the water travels, it leaves behind the minerals it was holding in solution. Other factors that affect precipitation are chemical reactions, pressure, oxygen levels, acidity, and rate of cooling.

To understand how water becomes heated and chemically active, it is necessary to understand wider geologic forces. Humans are familiar with Earth's crust because it is the layer of inhabitation, but beneath the crust are the lithosphere (composed of tectonic plates), the asthenosphere, and the mantle, which is hot. Convection causes the plates to move.

Sites where two tectonic plates are forced together are called "convergent boundaries." Sites where they are moving apart are called "divergent boundaries."

Movement of tectonic plates creates seismic activity and facilitates the release of heat and hot magma from Earth's interior. Evidence of this process comes in the form of volcanoes. Hydrothermal activity occurs at divergent boundaries (often midocean ridges) because new, hot material comes to the surface as the plates spread apart. Hydrothermal processes also occur at subduction zones (associated with convergent boundaries), where surface crust slides under another plate as they collide and descends to an area of increased heat and pressure. The rock in the crust then melts and is less dense than surrounding material, allowing the rock to again rise to the surface.

Because hydrothermal activity is associated with plate boundaries and hot spots (localized areas where a column of magma rises from the mantle), one can infer that hydrothermal mineralization occurs at divergent boundaries such as the East African Rift, the Mid-Atlantic Ridge, the East Pacific Rise, and the Red Sea Rift. Hydrothermal mineralization also occurs at subduction zones, such as the ones along the west coast of the Americas. Hydrothermal activity also occurs at oceanic hot spots such as the Hawaiian Islands and at continental hot spots such as the Yellowstone National Park region of the United States.

HYDROTHERMAL PROCESSES AND MINERAL DEPOSITION

Hydrothermal vents—fissures in which water heated by magmatic activity rises and escapes to the surface—can occur in submarine or terrestrial environments. The types of mineral deposits associated with these vents depend upon the composition of the nearby rock and magma. Magma from divergent plate boundaries is made of a high percentage of iron and magnesium, but at convergent plate boundaries, the magmas tend to be higher in silicon and oxygen because they are mixing with continental sediments rich in these elements.

At submarine vents, ocean water heated to temperatures of 662 degrees Fahrenheit (350 degrees Celsius) or more rises, leaching metals from the basaltic rocks it passes through. The metals later precipitate from the water, creating mineral deposits. These deposits can occur in the form of black smokers, which are tower-like structures built from iron-rich sulfides. White smokers, which have lower temperatures, contain calcium sulfate and silica deposits. Some silver, lead, zinc, and copper sulfides are produced near submarine hydrothermal vents, too.

Deep ocean deposits are neither large enough nor accessible enough to mine economically. However, in the future, resource scarcity and improved mining technology may make these deposits more attractive to exploit. Some exploitable ore deposits can be traced to ancient seafloor spreading centers that have moved great distances in time and have been lifted by plate collision. An example can be seen in the island nation Cyprus, where copper and other minerals, formed eons ago, are now near the surface.

Silicates, including feldspar, quartz, and mica, are not as economically valuable as metal deposits, but they are also formed by hydrothermal processes. Continental hydrothermal deposits are often rich in silica because igneous rocks that make up continental crust are high in silica. Silica-rich igneous rock that cools and solidifies underground is called "granite," but silica-rich igneous rock that cools at the surface is known as rhyolite. These rock compositions are important because hydrothermal fluids pass through the rock and pick up minerals from them.

Hydrothermal fluids seep into existing spaces and precipitate minerals called "cavity-filling deposits." Sometimes, however, the hydrothermal fluids can react with the rocks they pass through and alter the rock to form a deposit made from both hydrothermal precipitates and host-rock material. This process is called "hydrothermal replacement."

Terrestrial hydrothermal vents occur in several forms associated with volcanic activity. Examples can be seen in and around Yellowstone National Park, which is located at a hot spot. A fumarole is a type of gas- and steam-producing vent that occurs in volcanic craters or along their sides. Deposits form when the vent's gases cool. The Yellowstone area is rich in rhyolite, so the hydrothermal fluids there are silica-rich. Hot springs or geysers frequently contain silicon dioxide and calcium carbonate. Precipitation of these minerals can be aided by the biologic activity of organisms that live in and around hot springs.

Other continental hydrothermal deposits were formed in ancient subduction zones where metals were liquefied as they made their way down toward

Steam phase eruption of Castle Geyser, Yellowstone National Park (Brocken Inaglory via Wikimedia Commons)

the mantle then rose into fractures in the crust. As the fluids flowed through, such metals as copper, molybdenum, silver, gold, lead, and zinc were left behind. These metals are called "porphyry metal deposits" because they occur in porphyritic igneous rock associated with hydrothermal activity. Deposits of this type have been found in North America, and are being mined.

READING EARTH

Even when hydrothermal deposits have no economic value, they have scientific value. Because of the association of seafloor spreading centers with hydrothermal activity and the creation of mineral deposits, scientists can use deposits and rock samples from and near oceanic ridges to learn about cycles of spreading, subduction, and sediment transport. In addition, because humans cannot descend into the mantle to gather data about it, the material brought to the crustal layer through geologic events is a valuable extrapolative tool.

Hot hydrothermal fluids emerge from point sources (vents) and begin to spread out and cool as they mix with cold ocean water. Scientists can track the movement of these fluids by using a method called "isotopic tracing," which uses isotopes (forms of an element with the same number of protons but differing numbers of neutrons) to provide information about the vertical and horizontal movement of water. Isotopes also can be used to determine the age of particular rock samples or the conditions under which they formed. This is helpful in establishing patterns in tectonic plate movement or the rate of seafloor spreading.

Because many of the hydrothermal deposits and magmas at seafloor spreading centers contain iron, which can be magnetized, they also provide a record of Earth's magnetism. The needle of a compass will invariably indicate north because that is the direction in which Earth's north magnetic pole is located. This has not always been true. Earth's magnetic poles reverse approximately every few hundred thousand years. The process occurs gradually and does not follow strict cycles, but because seafloor spreading occurs with magmatic activity, and because iron-containing magma becomes magnetized at a specific point during cooling, scientists can construct a relatively precise timeline of the changes of Earth's magnetism based on the magnetism recorded in rocks and deposits near seafloor spreading centers.

One of the other benefits of studying mineral deposits at seafloor spreading centers is to compare magnetic and other qualities of the deposits on each side of the ridge. Minerals on one side of the ridge should have a matching counterpart on the other side of the ridge. By looking for similarities, scientists can see plate movement outward from a spreading center, but they also can track horizontal movement, such as the type of movement at transform fault boundaries.

—J. D. Ho

Further Reading

Chernicoff, Stanley, and Donna Whitney. *Geology: An Introduction to Physical Geology*. Prentice-Hall, 2007.

Demina, Liudmila L., and Sergey V. Valkin, editors. *Trace Metal Biogeochemistry and Ecology of Deep-Sea Hydrothermal Vent Systems*. Springer, 2016.

Ishibashi, Jun-ichiro, Kyoko Okino, and Michinari Sunamura, editors. *Subseafloor Biosphere Linked to Hydrothermal Systems*. Springer Open, 2015.

Jerram, Dougal, and Nick Petford. *The Field Description of Igneous Rocks*. Wiley-Blackwell, 2011.

Keller, Edward A. *Introduction to Environmental Geology*. Prentice-Hall, 2012.

Libes, Susan M. *Introduction to Marine Biogeochemistry*. Academic Press, 2009.

Lowell, Robert P., Jeffrey S. Seewald, Anna Metaxas, and Michael R. Perfit, editors. *Magma to Microbe: Modeling Hydrothermal Processes at Oceanic Spreading Centers*. John Wiley & Sons, 2013.

Morgan, Lisa A. *Integrated Geoscience Studies in the Greater Yellowstone Area: Volcanic, Tectonic, and Hydrothermal Processes in the Yellowstone Geoecosystem*. U.S. Geological Survey, 2007.

Rona, Peter A., Colin W. Devey, Jérôme Dyment and Bramley J. Murton, eds. *Diversity of Hydrothermal Systems on Slow Spreading Ocean Ridges*. John Wiley & Sons, 2013.

HYDROTHERMAL VENTS

Fields of Study: marine biology; marine geochemistry; marine chemistry; oceanography

ABSTRACT

Hydrothermal vents are openings on the seafloor where hot water is released. Hydrothermal water forms when cold seawater percolates down into the seabed rock and toward the underlying mantle, where it is heated by hot rocks. Both rocks and water are changed by this interaction. Deposits rich in metals form around these vents, and many ore deposits originally developed in similar environments. Numerous kinds of organisms live around hydrothermal vents, supported by bacteria that derive their energy from the reduction of sulfide to sulfate.

KEY CONCEPTS

biotic zones: areas within a particular environment where life is supported

chemosynthesis: a natural process of producing carbohydrates and other compounds from carbon dioxide and water that uses energy derived from chemical reduction reactions rather than sunlight

hydrothermal plume: a column of relatively warm water rising through the cold water column above a hydrothermal vent or other source of heated water

immiscible liquids: two liquids that do not dissolve in each other due to their different chemical natures, for example oil and water

millimole: simply, one one-thousandth of a mole, in which the mole is defined as the weight of a pure element or compound that corresponds to the number of identical particles equal to the Avogadro constant of 6.02259 X 10^{23}; accordingly, 1 mole of ^{12}C atoms weighs exactly 12 grams

CHARACTERISTICS

The late 1970s discovery of hydrothermal vents and the unique communities that live around them astounded scientists. Studies of these features have provided important and exciting research opportunities for biologists, geologists, and marine chemists. Almost all of the deep seafloor is a cold and relatively quiet place, but this is not true near hydrothermal vents. In contrast to the homogeneous and stable environment of the deep sea, hydrothermal vents are constantly changing, and scientists have observed many changes in hydrothermal vents and their associated communities in the relatively short period of time that they have been studied. Hydrothermal vents are easily the most spectacular features, both geologically and biologically, on the floor of the abyss. There are almost certainly many more hydrothermal vents remaining to be discovered along midocean ridges, where hydrothermal vents are concentrated.

Hydrothermal vents are hot springs on the ocean floor, the "exhaust pipes" of seafloor hydrothermal systems. These vents are the most accessible parts of seafloor hydrothermal systems, driven by heat sources that lie several hundred feet or miles (or meters or even kilometers) beneath the ocean floor. Hydrothermal systems form where near-freezing seawater from the bottom of the ocean penetrates deep into the crust along fissures until it comes into contact with hot rock structures. The water may be heated to as high as 752 degrees Fahrenheit (400 degrees Celsius) before it rises back up through the crust and jets back into the deep sea at the vents. Development of hydrothermal systems requires two things: hot rock structures and a way for seawater to percolate along cracks into the hot zone and return to the surface. How vigorous the resulting hydrothermal system is depends on both these factors. The hottest rock structures are found at

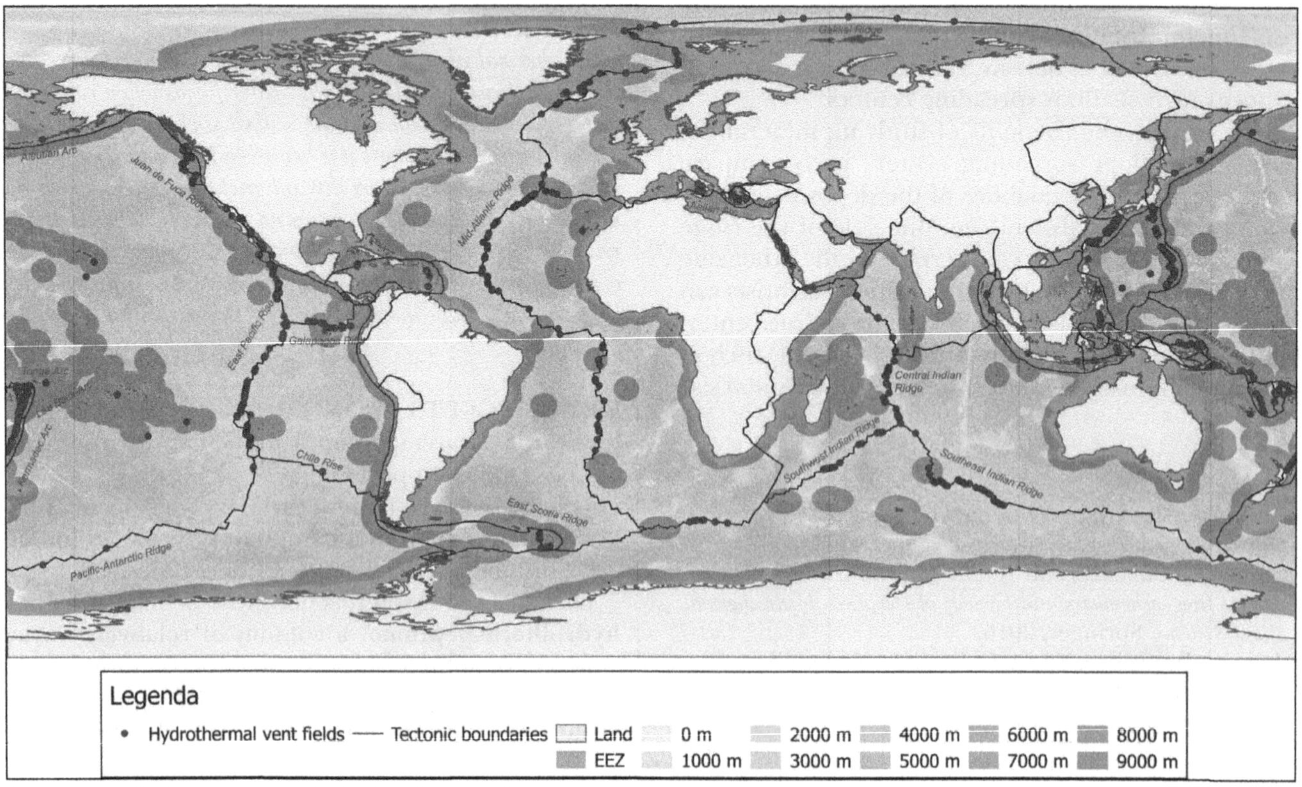

Distribution of hydrothermal vent fields (DeDuijn via Wikimedia Commons)

midocean ridges and other submarine volcanic centers such as the Loihi seamount off the coast of Hawaii where frequent eruption of basalt lavas maintains conditions that are most favorable for developing and maintaining hydrothermal systems.

Yet it is not enough for hot rock to lie beneath the surface; there must also be a way for seawater to be brought into contact with the hot rock and kept there until the water is sufficiently heated. This requires just the right configuration of fractured hot rock: If the cracks are too wide or the hot rock is not deep enough below the seafloor, the water will not be heated sufficiently before it returns to the surface. Similarly, if the cracks are too narrow, not enough water will flow through the system to develop a robust hydrothermal system, or the channels may be easily blocked by minor seismic or tectonic movements or by mineral deposits. The sensitivity of deep-sea hydrothermal systems to the sub-seafloor circulation geometry may be why hydrothermal events experience rapid changes and are generally short-lived features.

Seafloor hydrothermal systems are similar to hot springs or geysers found on the continents in that both form as hot rock heats and chemically modifies cold water. Seafloor hydrothermal systems differ, however, in their much higher temperatures. The boiling temperature of water increases as pressure increases, and pressures on the seafloor are several hundred times greater than at the surface. This relationship between pressure and the temperature at which a liquid boils is well known to most cooks, who are familiar with the fact that water boils at a lower temperature at high altitude, where the pressure is lower, than it does at sea level. Hydrothermal systems on land, such as those of Yellowstone National Park have a maximum water temperature of not much more than about 212 degrees Fahrenheit (100 degrees Celsius). At greater temperatures, the water is superheated and "flashes," or turns instantly to steam. This is what makes a geyser such as Old Faithful erupt, as the water in contact with the hot underlying rock turns to steam and violently forces out the colder, overlying water. The higher pressure of the seafloor prohibits flashing, however, and allows water to be heated to much higher temperatures.

Although most hydrothermal systems are found along the midocean ridges, where magma lies not more than a few miles (or kilometers) beneath the ocean floor, hydrothermal water is not heated by magma. Water heated by magma would be much closer to magmatic temperatures of 2,192 degrees Fahrenheit (1,200 degrees Celsius). Instead, the water is heated as it passes through hot but solid rock—the solid rock is heated by underlying magma. The upper limit of about 752 degrees Fahrenheit (400 degrees Celsius) for seafloor hydrothermal systems may reflect the maximum rock temperature at which fractures can form and remain open, or it may reflect separation of the fluid into immiscible fluids. It may be that at temperatures much higher than 752 degrees Fahrenheit (400 degrees Celsius), basaltic rock begins to slowly flow, closing any fractures that have formed. Another possibility is that, at pressures corresponding to typical depths of midocean ridges (about 300 bars [4,351 pounds force per square inch]), seawater separates into two fluids at about 752 degrees Fahrenheit (400 degrees Celsius). One phase is enriched in salt relative to seawater, while the other fluid contains a lower concentration of salt than seawater. The more concentrated phase will be the denser phase, and this may be why it is not found among waters issuing from hydrothermal vents. About one-third of the total heat lost from the planet's interior through the seafloor is lost as a result of seafloor hydrothermal systems.

METAL DEPOSITS

The seawater moving through a seafloor hydrothermal system is chemically changed as it is heated, and it alters the rocks through which it passes as well. All of the magnesium and sulfate in the seawater is absorbed by the hot rock, and large amounts of metals such as manganese, cobalt, copper, zinc, and iron are lost from the hot rock to the circulating water. In fact, magnesium is so completely removed from seawater that the depletion of this element in seawater can only be explained by seafloor hydrothermal activity. Oxygen-rich seawater is transformed into oxygen-poor hydrothermal water, and a high concentration of metals can be dissolved in such water. Seawater, for example, contains negligible magnesium and iron, whereas hydrothermal waters contain up to 1 millimole of magnesium and 6 millimoles of iron per liter.

When hot, chemically transformed water returns to the ocean at hydrothermal vents, the result is often spectacular. Near the vent, the hot, oxygen-poor, hydrogen-sulfide-rich hydrothermal water mixes with

Base of "black smoker" chimney, Pacific Ocean (W.R. Normark, Dudley Foster/United States Geological Survey via Wikimedia Commons)

cold, oxygen-rich, hydrogen-sulfide-poor seawater. This may cause the metals in the hydrothermal fluid to precipitate. Some of it may precipitate in fractures just beneath the seafloor. These "stockwork" deposits are most likely to be preserved and many ore deposits represent ancient stockwork deposits. Some metals precipitate around the vent itself, forming metal-rich chimneys. These chimneys are typically composed of an outer layer of anhydrite and inner deposits of copper-iron sulfides. These chimneys can have fantastic shapes, resembling spires, columns, cones, and beehives. Scientists visiting these chimneys give them fanciful names: "nail," "fir tree," and "moose" are names given to vent features from one hydrothermal field. Chimneys have been found that are up to 98 feet (30 m) in diameter and up to 148 ft (45 m) tall. Their growth rates can be on the order of 3 ft (1 m) per year, and vent chimneys are thus among the fastest-changing of all geologic phenomena. Fast-growing vent chimneys rapidly become unstable and collapse, then rise again. Cycles of growth and collapse continue as long as the hydrothermal fluids continue to issue from the vent.

Metals that do not precipitate as stockwork or vent chimneys rise with the hydrothermal water issuing from the vent. The hydrothermal water exits the vents in a jet stream due to the pressure exerted from below. Because this water is so hot, it is less dense than seawater and the jet forms the base of a hydrothermal plume. The plume becomes larger as it rises because cold seawater mixes turbulently with the hydrothermal water, and these waters quickly become well mixed. The hottest hydrothermal fluids (those heated to 572 degrees Fahrenheit [300 degrees Celsius] or more) are sufficiently metal-rich and acidic that they precipitate large quantities of microscopic grains of iron sulfides, zinc sulfides, and copper sulfides upon mixing with seawater. Much of this precipitates as the plume rises, producing a sulfide "cloud." The rapid precipitation of sulfides darkens the hydrothermal plume, giving it the appearance of smoke. Those vents releasing the hottest fluids have the greatest density of sulfide precipitation and are often referred to as "black smokers." In contrast, vents releasing cooler water (212 to 572 degrees Fahrenheit [100 to 300 degrees Celsius]) do not contain sufficient sulfide or metals in solution to cause this effect. Instead, mixing of these hydrothermal waters with seawater causes white particles of silica, anhydrite, and barite to form, forming a white cloud in the mixing plume and giving rise to the name "white smoker." White smokers may reflect mixing just below the surface, with the result that most of the metals associated with white smoker vents may be deposited just below the seafloor. Because the original hydrothermal vent waters are cooler and thus denser, plumes from white smokers do not rise as far as those from black smokers before obtaining neutral buoyancy and spreading laterally.

Mixing progressively dilutes the hydrothermal water until the mixed water cools to the point that it has the same density as ambient seawater and no longer rises. A tremendous amount of seawater must be mixed with the hydrothermal fluids before the plume attains neutral buoyancy. This may ultimately involve 10,000 or even 100,000 times as much seawater as hydrothermal water. During dilution, the mixture rises tens to hundreds of feet (or meters) to a level of neutral buoyancy, eventually spreading laterally as a distinct hydrographic and chemical layer, recognizable hundreds or even thousands of miles (or kilometers) away from the hydrothermal vent. Continued settling of hydrothermal iron and magnesium from these layers is the source of most of the metals in manganese nodules of the abyssal seafloor, far away from the midocean ridges.

Vast quantities of metals are deposited as sulfides around hydrothermal vents, especially as stockwork and in sediments around chimneys. These are particularly rich in iron, copper, zinc, and lead, and the exact composition of these deposits reflects several controlling factors, including temperature of the hydrothermal fluid, water depth, composition of the

Dense mass of anomuran crab Kiwa around deep-sea hydrothermal vent (A. D. Rogers et al. via Wikimedia Commons)

source rock, and flow regime in the portion of the hydrothermal system that lies beneath the seafloor. Many of the world's great ore deposits seem to be fossil seafloor hydrothermal systems, with the possible difference being that these may have mostly formed at convergent plate margins instead of divergent plate margins, where most modern hydrothermal systems are known. This may be part of the reason why ancient massive sulfide deposits are typically much larger than modern ones.

STUDY OF HYDROTHERMAL VENTS

To a marine biologist, hydrothermal vents are the oases of the deep seafloor. In contrast to most of the ocean floor, which supports few animals, hydrothermal vents teem with life. In fact, the analogy to a desert oasis falls short because there is much more life around a hydrothermal vent. Life in most of the deep sea is scarce because food is scarce. There is no sunlight, so plants, the basis of most food chains cannot grow. In contrast, food is abundant around hydrothermal vents; often, life is so crowded that the animals obscure the seafloor.

Vent fields can be divided into three biotic zones reflecting distance from the hydrothermal vents: the vent opening, the near-field, and the periphery. The bulk of the biomass is at the vent openings, where the density of life is so great that it appears to be limited by space, not food. Around hydrothermal vents on the East Pacific Rise and Galápagos Rift, vent openings are dominated by endosymbionts such as tube worms, clams, and mussels. These endosymbionts can grow to great size; some tube worms are 3 ft (1 m) long with 9-ft (3-m) tubes, while clams up to 12 in (30 cm) long are common. Other animals live with these, including limpets, polychaete worms, bresiliid shrimp, crabs, and fish. Vent chimneys at several Mid-Atlantic Ridge vents are almost entirely populated by bresiliid shrimp.

Autotrophic bacteria are the primary producers around hydrothermal vents. These use sulfur to convert carbon dioxide water, and nitrate into essential organic substances in a fashion similar to the way in which plants use sunlight during photosynthesis. This process is called "chemosynthesis." Autotrophic bacteria live within the subsurface plumbing system of

Tube worm colony at warm vent on Galapagos Rift (National Oceanic and Atmospheric Administration via Wikimedia Commons)

the vents, on the seafloor, and suspended in and around the hydrothermal plume itself, sometimes in such abundance that they color the water a milky blue or carpet the seafloor in white or bright yellow mats. Some of these bacteria can tolerate inordinately high temperatures, up to 230 degrees Fahrenheit (110 degrees Celsius) or more. All the other life around a hydrothermal vent ultimately feeds off the autotrophic bacteria. Some of the autotrophic bacteria are symbiotic, living within larger vent animals in a mutually beneficial relationship. The bacteria provide food, and the animals provide essential inorganic nutrients.

The flow of water around hydrothermal vents controls the distribution of life. Because cold water mixes with the hydrothermal plume, cold seawater flows in from all directions to converge on the rising plume. This means that the sulfur on which the autotrophic bacteria depend is not distributed around the vent. For a similar reason, the animals that feed or depend on the bacteria cannot survive away from the vent opening. The near-field is mostly populated by suspension feeders, animals that capture bacteria and other organisms that drift away from the vent opening. It is presumed that these animals live as close as possible to the vent but are forced to maintain a certain distance because of toxic effects resulting from very high concentrations of heavy metals. Animals living on the vent periphery include scavengers and other types that sustain themselves from bacteria that settle out of the hydrothermal plume.

Most of the animals that live around hydrothermal vents live nowhere else except in other sulfur-rich, reducing environments, such as in the rotting carcasses of whales or in "cold seeps," where cold, chemically altered seawater percolates up through the seafloor. Many vent animals may be ancient, originating in Mesozoic or earlier times. These animals may have been insulated from the effects of surface catastrophes, such as the meteor collision at Chicxulub believed to have killed off at least 75 percent of all life on the planet, including the dinosaurs about 65 million years ago.

There are a number of fascinating features about the life around hydrothermal vents, but one of the most intriguing is the suggestion that this environment is similar to what existed when life was just developing about 4 billion years ago. It is entirely possible that the first autotrophic life-forms were chemosynthetic, not photosynthetic.

—Robert J. Stern

Further Reading

Demina, Liudmila L., and Sergey V. Galkin, editors. *Trace Metal Biogeochemistry and Ecology of Deep-Sea Hydrothermal Vent Systems.* Springer, 2016.

Hopkins, Rae. *Hydrothermal Vents: Treasures of the Deep.* Heinemann, 2020.

Ishibashi, Jun-ichiro, Kyoko Okino, and Michinari Sunamura, editors. *Subseafloor Biosphere Linked to Hydrothermal Vents: TAIGA Concept.* Springer Open 2015.

Kallmeyer, Jens, editor. *Life at Vents and Seeps.* Walter de Gruyter, 2017.

Kennish, Michael J. *Practical Handbook of Marine Science.* 3rd ed., CRC Press, 2001.

McCarthy, Kevin T., Thomas Pichler, and Roy E. Price. "Geochemistry of Champagne Hot Springs Shallow Hydrothermal Vent Field and Associated Sediments, Dominica, Lesser Antilles." *Chemical Geology*, vol. 224, 2005, pp. 55–68.

Monroe, James S., Reed Wicander, and Richard Hazlett. *Physical Geology: Exploring the Earth*. 6th ed., Thomson Higher Education, 2007.

Parsons, Timothy R., et al. *Biological Oceanographic Processes*. Elsevier, 2013.

Pinet, Paul R. *Invitation to Oceanography*. 5th ed., Jones and Bartlett Publishers, 2009.

Rona, Peter A., Kurt Boström, Lucien Laubier, and Kenneth L. Smith Jr., editors. *Hydrothermal Processes at Seafloor Spreading Centers.* Springer Science & Business Media, 2013.

Talley, Lynne D., George L. Pickard, William J. Emery, and James H. Swift. *Descriptive Physical Oceanography: An Introduction*. 6th ed., Elsevier, 2011.

I

Indian Ocean

Fields of Study: oceanography; marine biology; marine geology; climatology

ABSTRACT

The Indian Ocean shares the broad ecological and oceanographic features of the other major oceans of the world, but it possesses many unique and interesting characteristics. In general terms, the Indian Ocean's size means that both its shores and its depths are so varied that it is difficult to describe it as a single geographical unit. Its westernmost tides arrive at the shores of a continent that bears little resemblance to the shores of its easternmost limits. Most of the truly unique characteristics of the Indian Ocean, however, lie beneath its surface in the form of extensive mountain ridges and, in one case at least, a very deep rift that represents an unparalleled underwater world of its own.

KEY CONCEPTS

current: a directional flow of water within a body of water such as a sea or ocean, driven by wind or as thermohaline circulation

deep: a relatively broad expanse like a deep valley in the ocean floor

Gondwonaland: one of two supercontinents formed by the sundering of the original supercontinent Pangaea about 200 million years ago

tide: the periodic rising and falling of ocean water under the influence of the gravitational force of the Moon and the Sun

trench: a relatively narrow and more sharply defined deep region of the ocean floor where the edge of a tectonic plate of the basaltic ocean floor meets a less dense continental mass and is forced downward into the mantle layer, often the site of massive earthquakes and volcanic activity

GEOLOGIC ORIGINS

The geologic origins of the Indian Ocean make it unique among the world's major oceans. Compared with the Pacific and Atlantic, the Indian Ocean is considerably smaller (covering an area of about 28 million square miles (73 million square kilometers), in contrast to the Atlantic's nearly 32 million (84 million) and the Pacific's nearly 64 million sq mi (166 million sq km). It is also of more recent origin geologically than the Atlantic, Pacific, or Arctic Oceans. Geologists specializing in the evolutionary history of Earth and plate tectonics estimate that about 150 million years ago, a giant southern continent called "Gondwanaland" began to break apart. The movement of segments both westward (to what became the African continent) and northeastward (to what became India a central section of which became known as Gondwana) took at least 100 million years. The collision of the Indian subcontinent with the Eurasian landmass about 50 million years ago brought about the violent upheaval of the Himalaya Mountains that continues in the present day. One of the effects of this phenomenon was to define new shorelines of the "youngest" of the world's major oceans.

The final product of these major geological upheavals was an oceanic body extending over the area between Australia in the east and Africa to the west. Its northernmost point corresponds to the Tropic of Cancer, where the Indian subcontinent joins the Eurasian landmass. From India's western coast to the southeastern tip of Arabia, the waters of the Indian Ocean form the Arabian Sea. On the opposite side of India, the Bay of Bengal and the Andaman Sea extend eastward to the coasts of Southeast Asia (Myanmar, Thailand, the Malay Peninsula and the Indonesian Archipelago). If one includes the two smaller subsidiary seas, the Persian Gulf and the Red Sea, the Indian Ocean extends even farther north, to 30 degrees north latitude. To the south, the Indian Ocean technically goes as far as Antarctica. Two features of the Indian Ocean's floor define the point at which it separates the Atlantic and Pacific Oceans: the Atlantic Indian Basin and the South Indian Basin.

ASSOCIATED SEAS AND MAJOR RIVERS

Although the formation of the Indian Ocean created several important seas and gulfs as distinct subsections of its total surface, there are fewer such bodies here than in the other oceans of the world. One should contrast general geographical denominations such as the Bay of Bengal (comprising most of the area east of India and touching the coasts of Southeast Asia) or the Arabian Sea (separating western India from the coasts of the Arabian Peninsula) with the geographical uniqueness of the Red Sea and the Persian Gulf. The latter two are, in fact, clearly separated from the main body of the Indian Ocean by narrow straits—the Mandab Straits and the Straits of Hormuz, respectively. Both the Red Sea and the Persian Gulf have ecologies that are very different from that of the main body of the Indian Ocean. This is not the case for the two other semicontained gulfs of the Arabian coast, the Gulf of Oman and the Gulf of Aden. The area known as the Great Australian Bight is simply the slightly curved central southern coast of Australia and is therefore even less circumscribed than the Bay of Bengal west of India. The Andaman Sea lies north of the Indonesian Archipelago and is enclosed geographically from the Bay of Bengal by a line of islands (actually an extension of the Indonesian islands) called the "Nicobar and Andaman Islands."

Several major rivers pour large amounts of freshwater into the Indian Ocean. Such rivers are probably much older than the Indian Ocean itself, even though their pattern of flow was different in earlier geological ages. The Zambezi in East Africa, the Indus in northwest India, and the Ganges in northeast India were all probably flowing from their respective continental freshwater sources toward what eventually became the Indian Ocean's coastline. Each of these has had, over the long period since the formation of the ocean, a notable effect on the configuration of the coast where it empties into the ocean. Freshwater currents have, for example, cut actual canyons into the continental shelf area adjacent to the coast. In the case of the Ganges, an immense zone of sediment has built up, affecting both marine life and local currents in its delta area in the Bay of Bengal.

OCEAN DEPTHS AND SUBMARINE GEOLOGICAL FEATURES

The average depth of the Indian Ocean is in the range of 2,26 to 2.45 miles (3,636 to 3,940 meters). Several extremely deep but limited areas, notably the Java Trench to the south of Indonesia are nearly twice as deep. The continental shelf along the coasts of the Indian Ocean is generally narrower than that of the other oceans, averaging 76 mi (122 km) before

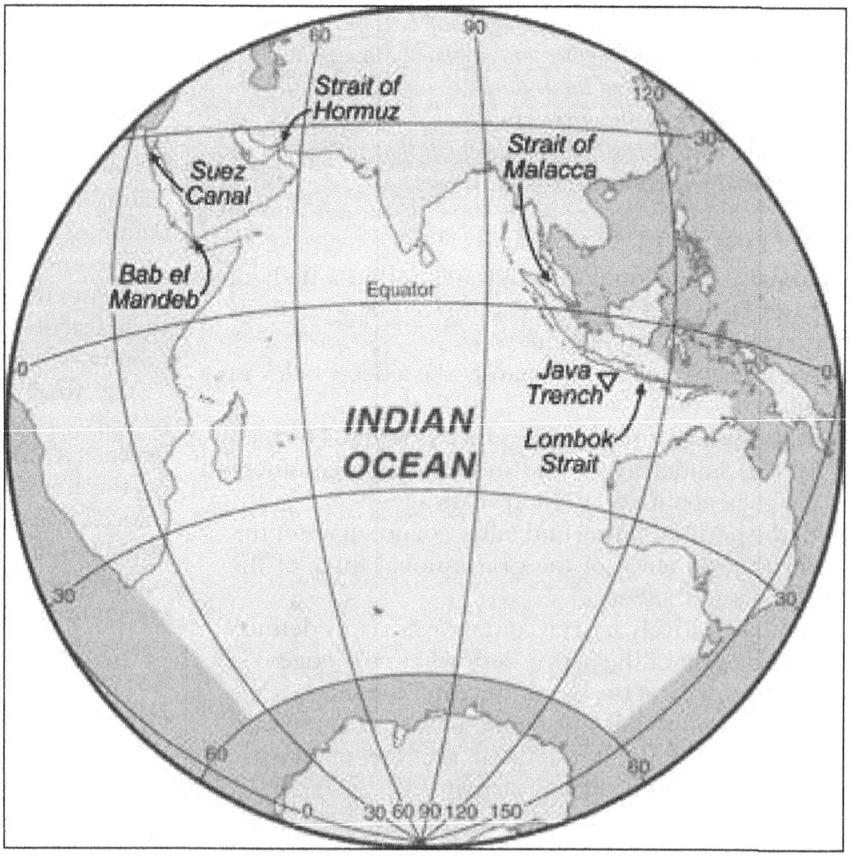

The Indian Ocean (CIA World Factbook)

deeper waters begin. The area west of the Indian coast, off the major city of Mumbai, is an exception. There, the continental shelf extends almost 202 mi (325 km) into the ocean.

The floor of the Indian Ocean is crisscrossed by a number of underwater mountain ranges. Although notable ridges exist, its underwater topography is certainly not as complex or as spectacular as the eastern half of the Pacific, where extensive archipelagos with many small islands and some very large island formations (Japan, the Philippines, New Guinea and New Zealand for example) predominate. The most concentrated area of subsurface mountains in the Indian Ocean is centered near 30 degrees longitude, which is about halfway between the west coast of India and the Gulf of Aden, south of Arabia.

A number of small but historically important islands mark high points along the mountain ridges between 30 and 60 degrees longitude. Mauritius and the Seychelles are examples of these. The huge island of Madagascar (227,028 sq mi [588,000 sq km]) is the most prominent surface example of the complex north-to-south submarine mountain systems located east of the African coast. The first of these is the Mauritius Ridge, marked on the surface at its southernmost point by Mascarene Island (due east from Madagascar) and by the Seychelles Islands to the north. The next range, the Carlsberg Ridge, is longer than the Mauritius Ridge, but none of its peaks emerge to form islands.

Finally, a long ridge extends due south off the southwest coast of India. This range is marked at the surface by the Laccadive and Maldive Islands. Again, like Mascarene Island and the Seychelles near Madagascar, the Maldives are dwarfed by the single major island just to the southeast of the tip of India, Sri Lanka (formerly Ceylon). Sri Lanka, however, is the tailing part of the Indian subcontinental landmass rather than the tip of a submarine mountain range.

Beginning with the 90-degree east longitude line and moving toward the eastern shores of the Indian Ocean, the topography of the ocean floor is quite different from that of the western half. First, the very name of one subsurface range, the Ninety East Ridge, suggests a very regular pattern extending from north to south. The Ninety East Ridge, discovered only as recently as the 1960s, has gained the distinction of being the longest and straightest underwater mountain range in the world. Unlike the other ridges of the eastern basin of the Indian Ocean, most notably those south of the Indonesian coast, the Ninety East Ridge appears to be seismically inactive.

Between 90 degrees east longitude and the western shores of Australia one finds the third-deepest point in the Indian Ocean, the Wharton Basin, measuring nearly 3.96 mi (6,364 m) deep. Farther south, off the southwestern tip of Australia, is the Diamantina Deep. Neither of these deep points is, however, associated with the rapid fall-off from mountain ridges to deep valleys that is characteristic of ocean trenches. Trenches are characteristic of the eastern rim of the Pacific Ocean but only one such phenomenon occurs in the Indian Ocean. The Java Trench off the southern coast of Indonesia is more than 3.77 mi (6,060 m) deep. Pioneer scientists examining flora and fauna in this area of what were then still undiscovered ocean trenches hypothesized that Indonesia was very close to the dividing line between tectonic plates. They observed that plants and animals to the east of Java appear to be biologically isolated in their evolution from species farther west.

The long, curved pattern of the ranges that constitute the Indonesian Archipelago actually extends far beyond the northern tip of Sumatra. Its peaks can be found in the chain of islands known as the Nicobar and Andaman Islands. The presence of these island chains west of the coasts of Thailand and Myanmar helps define the Andaman Sea area east of the main body of the Bay of Bengal.

TIDES AND CURRENTS

The immense size of the Indian Ocean means that tidal phenomena are variable, both in type and in the volume of tidal movement registered. The most common tides are semidiurnal (occurring twice daily). These are characteristic of the subequatorial eastern shores of Africa and, farther north and much farther east, in the Bay of Bengal. Australia's southwest coast, which is roughly opposite the subequatorial eastern coast of Africa, has an entirely different tidal pattern. Australia's Indian Ocean shores experience diurnal (once per day) tides that are extremely light by comparison to those of other coasts.

The Indian Ocean is the only ocean in the world with asymmetric reversing surface currents. Asymmetric conditions apply when currents in the northern half of the ocean are moving in a different direction from those of the southern half. The complexity of Indian Ocean currents close to the surface goes well beyond the relatively simple question of north-south asymmetry. One finds, for example, that wind conditions contribute to the creation of gyres, circular or spiral movements that break the broad pattern of the surface current into localized segments.

It is particularly important to note that broad patterns of currents in the Indian Ocean reverse according to the season. Currents, like so many other factors determining the overall ecology of the Indian Ocean, are affected in large part by the major monsoonal wind and weather conditions that are characteristic of this region of the world.

Probably the most famous current in the Indian Ocean is the so-called Somali Current, which moves, in certain seasons of the year, in a fairly rapid clockwise direction from the northeast coast of the Horn of Africa. In the summer season, this current goes as far as the coast of India. At its farthest point moving east in the summer months, it meets the southwesterly monsoon current off the Indian subcontinent. During the winter, the direction of the Somali Current reverses, a situation that created near-ideal seasonal sailing conditions for centuries for ships sailing between the Arabian peninsular zone, particularly the Persian Gulf, and the Horn of Africa.

CLIMATIC ZONES AND WIND PATTERNS

Because of the great north-to-south distance covered by the Indian Ocean and its associated seas, there are several quite distinct climatic zones according to geographical location. The most famous, and most important for sustaining seasonal agriculture in the entire Indian Ocean area, is the so-called monsoon zone, which runs from 10 degrees north of the equator to about 10 degrees south. The region between 10 and 30 degrees south of the equator is the zone of what has traditionally been called the "trade winds." It is the predictability of steadily blowing southeast winds in this wide region (as distinct from the area of

Grand Anse on the island of La Digue, Seychelles, on the Indian Ocean (Tobi 87 via Wikimedia Commons)

the Somali Current) that made maritime communication, and therefore trade, between the opposite shores of the Indian Ocean possible.

Farther south, near the global band of the Tropic of Capricorn (running through the island of Madagascar on the western shores and Australia on the east) lies the subtropical to temperate zone, between 30 and 45 degrees latitude. Some 20 degrees south of the Tropic of Capricorn, climatic conditions begin to show the temperate cooling influence of the extreme southern extent of Indian Ocean waters leading to the last climate zone. From 45 degrees latitude southward to the Antarctic ice cap, the beginnings of sharply cold antarctic waters mark the end of the gradual transition separating some of the world's hottest climates in the Indian Ocean proper from the extreme cold of the southern zone of the globe. Here, three of the world's four oceans almost literally fuse in the Antarctic ice cap.

MONSOON WIND CONDITIONS

Generally stated, monsoon conditions are semiannual reversing wind patterns. Extensive areas of high pressure "empty" their air in the direction of equally vast low-pressure zones. When this happens, winds moving across water carry the moisture they pick up, which is typically precipitated as rain before they reach their low-pressure destinations. In the case of Indian Ocean monsoons the widespread heating of the landmass in the Northern Hemisphere during summer creates conditions of low atmospheric pressure over Asia. This low-pressure zone becomes an attractive force for masses of air that are pressed downward by high-pressure conditions over Australia.

The resultant winds that move in a northwesterly direction across Southeast Asia and the Indian subcontinent bring with them a much-needed monsoon season of heavy rains that lasts until the particular atmospheric conditions cease to apply. Generally, the monsoon rainy season is predictable, but the arrival of torrents of rain to quench the dry agricultural fields of south Asia and Southeast Asia is not guaranteed. When the typical monsoon wind pattern develops but insufficient moisture is collected to bring rains, areas that depend on monsoon waters can face serious drought for at least one year, as there is no chance of humid air movements over the landmass once the directional wind pattern created by Asia's summer heating ends.

NATURAL RESOURCES

The Indian Ocean contains many key minerals that are extracted to supplement the local economies of several of the countries along its shores. Manganese is found in several areas off South Africa and Australia. Other minerals include tin and chromite. Mineral wealth in the water is, however, overshadowed by the vast petroleum reserves concentrated in one of its seas: the Persian Gulf. These are estimated to be perhaps the largest oil reserves in the world. Other locations where petroleum wealth is important are on the island of Sumatra and in its offshore waters. Similar intermediate potential for petroleum production is found in the Red Sea and off the shores of western India.

The fishing industries that depend on the Indian Ocean are varied. Many depend on the phenomenon of upwellings, movements of water from lower depths that carry phytoplankton—a basic food source for many fish species—close to the surface. The most common commercialized fisheries seek large schools of sardines, mackerel, and anchovies. The principal single species fished in many areas of the Indian Ocean is shrimp.

SIGNIFICANCE

The Indian Ocean and its associated seas represent one of the most diversified ecological marine environments in the world. This applies both to its coastlines and to the world beneath its surface. The African landmass along the ocean's western shores reveals a variety of natural characteristics that hardly resemble what one finds along its eastern shores. Dry coastal regions characterize both the southern and northern reaches of the African landmass, with some of the driest deserts in the world extending from the Somalian Horn of Africa into the Red Sea and around the southern shores of Arabia into the Persian Gulf. Although these conditions continue all the way along the northwestern coastline to the Gujarati coast of India, the Indian Ocean ecosystem is clearly distinct once one passes the Indian subcontinent.

From the Bay of Bengal southward along the eastern shores of the Indian Ocean, the environment becomes increasingly tropical, passing through some of the most extensive tropical rainforests in the world, especially in the Indonesian Archipelago. Indeed, parts of the west coast of Australia, generally thought to be nearly a desert environment, exhibit a tropical ecology that contrasts notably with the dry climates of the African coastline of the Indian Ocean. With this physical ecological diversification comes an enormous variation in plant and animal life along the ocean's shorelines and in the relatively shallow waters of its continental shelf. Studying the diversity of the Indian Ocean's biological ecology and submarine geology is like observing multiple different worlds on one segment of the planet.

—*Byron D. Cannon*

Further Reading

Aleshire, Peter. *Ocean Ridges and Trenches.* Chelsea House, 2007.

American Museum of Natural History. *Ocean.* Dorling Kindersley, 2006.

Charabe, Yassine, and Salim al-Hatrushi. *Indian Ocean Tropical Cyclones and Climate Change.* Springer, 2010.

Chatterjee, Piu, and Dona Ganguly. *India and Indian Ocean: Past, Present and Future.* Mittal Publications, 2018.

Clift, Peter D., and R. Alan Plumb. *The Asian Monsoon.* Cambridge UP, 2008.

Gupta, Rabin Sen, and Ehrlich Desa, editors. *The Indian Ocean–A Perspective. Volumes 1 & 2.* A. A. Balkema Publishers, 2001.

Lighthill, James, and Robert Pearce, editors. *Monsoon Dynamics.* Cambridge UP, 2009.

Mukhopadhyay, Ranadhir, Anil K. Ghosh, and Sridhar D. Iyer. *The Indian Ocean Nodule Field: Geology and Resource Potential,* edited by M. Hale, Elsevier, 2012.

Nairn, Alan E. M., and Francis G. Stehli. *The Ocean Basins and Margins Volume 6: The Indian Ocean.* Springer Science & Business Media, 2013.

Pearson, Michael Naylor. *The Indian Ocean.* Routledge, 2003.

Rasul, Najeeb M. A. and Ian C. F. Stewart, editors. *The Red Sea: The Formation, Morphology, Oceanography and Environment of a Young Ocean Basin.* Springer, 2015.

Wiggert, Jerry D. *Indian Ocean Biogeochemical Processes and Ecological Variability.* American Geophysical Union, 2009.

INTERNATIONAL SYSTEM OF UNITS (SI)

Fields of Study: mathematics; marine science

ABSTRACT

The metric system, the basis of the International System of Units (Le Système International d'Unités; known as SI), during the French Revolution in the mid-eighteenth century coincided with the beginning of the age of modern science. The value of physical measurements in the conduct of science became apparent and the common nature of science demanded an equally consistent system of units and measurements. The units in the SI have been defined by international accord to provide consistency in all fields of science and technology. The basic SI units are defined for only seven fundamental properties of matter. All other consistent units are derived as functions of these seven fundamental units.

KEY CONCEPTS

measurement: determination of the magnitude of a physical property relative to a fixed unit standard for that property

metrology: the study of methods, practices and applications of measurement

mole: defined as a quantity of individual particles such as atoms, molecules, subatomic particles or ions equal in number to the Avogadro constant

DEFINITION AND BASIC PRINCIPLES

The International System of Units (SI) is the internationally accepted standard system of measurement in use throughout the world. The units of the SI are ascribed to seven fundamental physical properties and two supplementary properties. These are length, mass, time, electric current, thermodynamic temperature, the amount of a substance, luminous intensity, and the magnitude of plane and solid angles.

Length is the extent of some physical structure or boundary in two dimensions, such as the distance from one point to another, how tall someone is, or the distance between nodes of a sinusoidal wave. The SI base unit associated with length is the meter (m).

Mass refers to the amount of material in a bulk quantity. The term is often used interchangeably with weight, although the mass of any object remains constant while its weight varies according to the strength of the gravitational field to which it is subject. The SI base unit for mass is the kilogram (kg), which is now defined in terms of the Planck constant (h).

Time is a rather more difficult property to define outside of itself, as it relates to the continuous progression of existence of some state through past, present, and future stages. The SI base unit of time is the second.

Electric current is the movement of electronic charge from one point to another, and it is described by the SI base unit of the ampere. One ampere is defined in relation to the elementary charge constant (*e*) as the movement of 1 coulomb of charge, as 1 mole of electrons, for a period of 1 second.

Thermodynamic temperature refers to the measurement of temperature in relation to absolute zero. The SI base unit used is the kelvin (K), which is expressed in relation to the Boltzmann constant (*k*) of $1.380\ 649 \times 10^{-23}$. It was previously based on the triple point of water, or the temperature and pressure in which water's three phases are in thermodynamic equilibrium.

The amount of a substance is defined by an SI base unit called the "mole." The term is essentially never used outside of the context of atoms, molecules, and certain subatomic particles such as electrons. A mole of any substance is the quantity of that substance containing $6.02214129 \times 10^{23}$ of these elementary entities; that number is known as the Avogadro constant.

The term "luminous intensity" refers to the brightness, or the quantity, of light or other forms of electromagnetic energy being emitted from a source. Initially luminous intensity was referred to by comparison to the light of a candle flame, but the variability of such a source is not conducive to standardization. The SI base unit for luminous intensity is the candela. This is about the amount of light emitted by a candle flame, but it has been standardized to mean an electromagnetic field strength of 0.00146 watts.

The derived unit known as the plane angle refers to the angular separation of two lines from a common point in a two-dimensional (2D) plane. The SI derived unit for plane angles is the radian. A complete rotation about a point origin is an angular displacement of 2π radians. The extension of this into three dimensions is known as the solid angle and is measured by the SI derived unit called the "steradian."

Measurement of any property or quantity is accomplished by comparing the particular amount of the property or quantity to the amount represented by the standard unit. For example, an object that is 3.62 times as long as the distance defined as 1 meter is said to be 3.62 meters long. Similarly, an object that is proportionately 6,486 times as massive as the quantity defined as 1 kilogram is said to weigh or to have a mass of 6,486 kilograms. The use of standard reference units such as the SI units ensures that measurements and quantities have the same meaning in all places where that system is used.

BACKGROUND AND HISTORY

Traditionally, systems of measurement have employed units that related to various parts or proportions of the human body. Ready and convenient measuring tools in earlier times included the hand, the foot, the thumb, and the pace. Because these human measures tend to vary somewhat from person to person, no work that depended on measurements could be repeated exactly, even if the same person carried it out. Specialization in tasks eventually led to the realization that standard units of measurement would be beneficial, and determination of certain units were established by royal decree as long ago as the signing of the Magna Carta.

The Industrial Revolution in Europe and the growth of science as a common international pursuit drove the need for a unified system of measurement that would be independent of human variability and consistent from place to place. The SI was developed and first put to use in about 1799. It represents the first real standardized system of measurements. Prior to this time, a broad variety of measurement systems were in use because many countries had developed their own measuring standards for use internally and in any territories that it held. For example, the British standard measurement of 1 yard, or 3 feet, compares to the Imperial Russian standard of the *arshin*, thought to be equivalent to one pace of a marching soldier of the Tsar's Imperial Russian Army.

The need for a standardized system of measurement had been recognized by various luminaries and was proposed in 1670 by French scientist Gabriel Mouton. The incompatibility and variability of the many measurement systems in use at that time often resulted in unfair trade practices and power struggles. In 1790, the French Academy of Sciences was charged with developing a system of measurement that would be fixed and independent of any inconsistencies arising from human intervention. On April 7, 1795, the French National Assembly decreed the use of the meter as the standard unit of length. It was defined as being equal to one ten-millionth part of one-quarter of the terrestrial meridian, specified by measurements undertaken between Dunkirk and Barcelona. The liter was later defined as being a unit of volume equal to the volume of a cube that is one-tenth of a meter on a side, and the kilogram as being equal to the mass of 1 liter of pure water. Larger and smaller quantities than these are indicated by the use of increasing and decreasing prefixes, all indicating units that were some power of ten times greater or lesser than the base units.

Standard reference models of the metric units were made and kept at the Palais-Royale, and it was then required that measuring devices being used for trade and commerce be regularly checked for accuracy against the official standard versions. Eventually, as methods of measurement became far more precise, it became clear that even these rigorously maintained reference models were susceptible to physical change over time, disrupting consistency. While adjustments were occasionally made, a movement to connect the definitions of the base units to more universally stable properties arose.

HOW IT WORKS

All measurement practices relate the actual properties and proportions of something to a defined standard that is relevant to that property or proportion. For example, 2D linear quantities are related to a standard of length, and the mass of an object is proportionately related to a standard unit of mass. In the measurement of length, for example, the standard SI unit is the meter. An object that is determined to be, for example, 6.3 times longer than the base unit of 1 meter is therefore said to be 6.3 meters long. Similarly, an object that is, say, 5.5 times heavier than the base unit of 1 kilogram is said to weigh 5.5 kilograms, and something that has a volume of 22.4 times that of the base unit of 1 liter is said to have a volume of 22.4 liters.

The modern metric system is simple to use, especially when compared with its most common predecessor systems, the British Imperial and the US customary systems. Whereas these predecessor systems use units such as inches, feet, yards, ounces, and pounds that relate to one another in irregular ways, the metric system has always used the simple relation of factors of ten for all of its units. For example, the decameter is equal to 10 meters (10 × 1 meter), while the decimeter is equal to one-tenth of a meter (0.1 × 1 meter). A complete series of prefixes indicates the size of the smaller and larger units that are being used in a measurement. Thus, a decimeter is ten times smaller than a meter, a centimeter is a hundred times smaller than a meter, a millimeter is a thousand times smaller, and so on. Correspondingly, there are 10 millimeters in a centimeter, 10 centimeters in a decimeter, and ten decimeters in a meter. There are 1,000 milliliters in a liter, 100 liters in a hectoliter, 1,000 grams in a kilogram, and 1,000 milligrams in a gram. The uniformity of the system readily allows one to visualize and estimate relative sizes and quantities.

The basic units of the SI have always been defined with the intention of relating to some universal and unalterable standard. In some cases, the definitions have been changed to relate the unit to a more permanent universal feature or a more accurately known property. For example, as the accuracy of time measurement greatly improved, the definition of the meter was changed from being a specific fraction of the distance between two fixed terrestrial points to "the length of the path traveled by light in vacuum during a time interval of 1/299792458 of a second."

A major turning point along these lines came in November, 2018, when sixty countries participating in the General Conference on Weights and Measures unanimously voted to revise the definitions of the SI base units so that all seven would be based on universal constants. While the meter, candela, and second already conformed to this requirement, the kilogram, kelvin, mole, and ampere were officially redefined in terms of constants (and the definitions of the other three units were made more rigorous), effective May 2019. Widely hailed as a major development in science despite little impact on daily measurements for most people, the decision meant that the seven base units would be tied to the most stable properties known.

The units of measurement of other properties and quantities reflect the fact that those properties and quantities can be viewed as combinations of the seven fundamental properties. That is, for any dimensional property *Q*, the dimensions of *Q* are derived from the expression:

$$\dim Q = L^{\alpha}M^{\beta}T^{\gamma}I^{\delta}\Theta^{\varepsilon}N^{\zeta}J^{\eta}$$

The exponents designated by the Greek letters α, β, γ, δ, ε, ζ, and η (or a, b, c, d, e, f, and g when the standard English letters are used) are integers representing the degree of involvement of the corresponding property of length (L), mass (M), time (T), electric current (I), thermodynamic temperature (θ), amount of material (N), and luminous intensity (J). (Remember that any quantity raised to the power 0 has the numerical value of 1.)

APPLICATIONS AND PRODUCTS

Applications and products related to the metric system essentially fall into two categories: educational devices and training to promote familiarity with the system, and metric versions of existing products and the tools required for their maintenance.

Educational Devices and Training. The long history of independent measurement systems has served to entrench those systems in common usage. Therefore, a very large body of materials, products, and devices have been constructed using nonmetric measurements. More significantly, those independent systems have been so deeply entrenched in the education systems of many nations, including the United States, that several generations of people have grown up using no other system. The familiarity gained through a lifetime of using a particular system generally gives individuals the ability to visualize and estimate quantities in terms of that measurement system. Making the change to a different measurement system (such as US efforts to introduce the metric system in the 1970s) represents a paradigm shift that leaves many individuals unable to associate the new measurement units with even the most familiar dimensions. Such a change, however, can be accommodated in a number of ways as the world continues to adopt the metric system of measurement as its universal standard.

The most basic method of replacing one system with another is to incorporate the new system into the basic education system, teaching it in such a manner that it becomes the entrenched measurement system as children progress through school. It is, therefore, important not only that teachers are educated in the use of the metric system but also that they actively replace their own reliance on any former system that they have used. This guarantees that, in time, the older system is completely displaced from the public lexicon and the metric system becomes the primary measurement system over a period of essentially one generation.

Those who have already left the school system can learn the metric system through training programs. Training in the metric system can be incorporated into professional development programs at the workplace or take the shape of formal training programs offered by local educational institutions or third-party providers.

Metric Versions of Existing Devices. The vast majority of goods and devices produced in countries that have not adopted the metric standard must be maintained using their original component dimensions because metric values and nonmetric values are not generally interchangeable. The fundamental difference between the two bases of measurement ensures that any coincidence of size from one system to the other is exactly that: coincidental. Therefore, switching to metric goods and devices requires that complete new lines of products be made with metric rather than nonmetric dimensions. This requirement places an odd sort of constraint on the situation because nonmetric parts and tools must still be produced in order to maintain existing nonmetric devices. At the same time, new devices to replace those that fail are produced in metric dimensions. This means that tradespeople, such as automobile mechanics, maintenance workers, and engineers, must obtain double sets of tools, and supply stores must maintain double sets of components and supplies. In addition, tools for taking measurements must be capable of using both metric and nonmetric dimensions, although the incorporation of electronic capabilities into many tools has greatly minimized the difficulties that arise from this dual requirement.

CAREERS AND COURSEWORK

By itself, the International System of Units is not broad enough to provide a field of study or advanced coursework and is unlikely to form the basis of a career. However, during the process of converting from a nonmetric measurement system to a metric system, opportunities will arise for those familiar with SI to create and provide training to facilitate understanding and use of the metric system.

Anyone entering a technical or scientific field must become familiar with the metric system, as this has been the standard system of measurement in those areas since the SI was developed. As the metric system becomes more and more widely adopted and accepted in the United States, students and others should expect to carry out measurements and calculations using the appropriate metric system units.

Careers that depend specifically on measurement include quality-control engineering and most branches of scientific research and physical engineering. Specific examples of measurement-based careers include civil engineering, medical and biochemical analysis, analytical chemistry, metrology, mechanical engineering, and industrial chemical engineering.

SOCIAL CONTEXT AND FUTURE PROSPECTS

The 2018 redefinition of the SI base units represented a significant moment in history for the scientific community. It marked the first time that all the measurements within the metric system would be based on fundamental natural properties rather than physical characteristics. While most individuals' use of the metric system was unchanged by the redefinition, it was hailed by scientists for its potential to spark new theoretical and technological innovations. In addition, it was expected to cut down on the complexity and cost of calibration for precision devices in laboratories and industrial applications. However, the changes were criticized by some members of the scientific community who questioned the specifics of the new definitions and even debated the legitimacy of some of the base units.

Efforts continue to make the metric system the standard system of measurement in the United States, as it already is in Canada, Mexico, and most of the rest of the world. As international trade and offshore manufacturing increase, the necessity for American industry to adopt the SI increases. In addition, educational systems have increased their focus on science and technology, making metric systems more familiar to children in American schools. However, resistance to abandoning the US customary system of measures remains widespread, and metric adoption is subject to changing political and social attitudes.

In the meantime, development of other measurement systems and hybrids of measuring systems progresses. For example, the manufacturing community adopted a standard unit called the "metric inch," which is the standard inch divided into hundredths, to be used on visual measuring devices. There is some logic to this adoption, as this measurement apparently corresponds with the limits of differentiation of which the human eye is capable to a better degree than does the millimeter division of the metric system. However, such developments are likely only to delay and otherwise interfere with the adoption of the metric system.

—*Richard M. Renneboog*

Further Reading

Butcher, Kenneth S., Linda D. Crown, and Elizabeth J. Gentry, editors. *The International System of Units (SI): Conversion Factors for General Use*. National Institute of Standards and Technology, 2006.

Cardarelli, François. *Encyclopaedia of Scientific Units, Weights and Measures: Their SI Equivalences and Origins*. Trans. M. J. Shields. Rev. and expanded ed. Springer, 2003.

Finucane, Edward W. *Concise Guide to Environmental Definitions, Conversions, and Formulae*. Lewis, 1999.

Göbel, Ernst O., and Uwe Seigner. *The New International System of Units (SI): Quantum Metrology and Quantum Standards*. Wiley-VCH, 2019.

Himbert, M. E. "A Brief History of Measurement." *European Physical Journal Special Topics*, vol. 172, 2009, pp. 25–35.

"Redefining the SI Units." *National Physical Laboratory*, www.npl.co.uk/si-units/redefining-the-si-units/. Accessed 12 Dec. 2018.

Shoemaker, Robert W. *Metric for Me! A Layperson's Guide to the Metric System for Everyday Use with Exercises, Problems and Estimations*. 2nd ed., Blackhawk, 1998.

"SI Redefinition." *National Institute of Standards and Technology*, US Department of Commerce, 2018, www.nist.gov/si-redefinition. Accessed 12 Dec. 2018.

Intertidal Zones

Fields of Study: marine ecology; marine biology; oceanography; marine biogeochemistry

ABSTRACT

Intertidal zones are common in coastal ecosystems worldwide and are subject to the influence of high and low tides. Many

factors that are either directly or indirectly related to the shoreline influence the type of habitat available for colonization by organisms. The heterogeneity in topography has a subtle influence on intertidal shores, with localized irregular features that can serve as refuges from harsh environmental conditions and predators.

KEY CONCEPTS

dessication: the loss of water from within an aquatic organism through exposure to the open air and heat, which can lead to physiological stress and death through dehydration and may ultimately result in rendering the body of the organism completely dry

emersion: removal of the aqueous medium that normally surrounds an aquatic species, as when the tide goes out; the opposite of submersion

intertidal zone: the area of a shoreline that is alternately covered by water at high tide and uncovered at low tide

submersion: the return of the aqueous medium that normally surrounds an aquatic organism, as when the tide comes in; the opposite of emersion

tidal zonation: also called "vertical zonation," identifiable zones of characteristic organisms corresponding to different water levels in the intertidal zone

tide pools: low spots or catchments in the intertidal zone that are refreshed and replenished with water during high tide and then retain that water during low tide

THE HETEROGENOUS ZONE

Intertidal zones, which are fundamentally marine, are common in coastal ecosystems worldwide and are subject to the influence of high and low tides on a daily basis, and organisms living on these shorelines are exposed to air during periods of low tides. Many factors that are either directly or indirectly related to the shoreline influence the type of habitat available for colonization by organisms. The effect of heterogeneity in topography has a subtle influence on intertidal shores, with localized irregular features such as cracks, crevices, undersides of boulders, gullies, overhangs, caves, tide pools, damp areas, and areas of freshwater drainage, all of which can serve as refuges from harsh environmental conditions and predators.

The type of substratum rock type, and rock texture determine how well water drains from the shore. Exposure to wave action varies greatly between intertidal shores because prevailing winds and the slope of the shore influence the type of waves and the point at which they break. Waves break far out from the shoreline on a relatively flat shoreline, then "spill" over the shore. On steep shorelines, waves come close before "surging" up the rock face.

Surging waves on steep shores may increase the effective intertidal zone to many times the height predicted purely from tidal rise and fall. Gentle, sloping shores will drain more slowly than steep slopes, allowing organisms that are characteristic of lower zones to expand their ranges into the upper intertidal area. Shores with heavy wave action increase the splash area on the upper intertidal sections that would normally be covered by the tides. Waves can also exert a destructive mechanical effect on the shoreline, encourage scour by sand and shingle, circulate the water, disturb or deposit sediment, renew oxygen, and reduce dissolved carbon dioxide.

They also affect the movement of organisms, limiting feeding raids or keeping away predators. Conditions during emersion are not only affected by the position of the shore with respect to water currents, climate, and latitude, but also by the direction in which the shore faces and the amount of sunlight that it receives. The time at which the greatest low tides occur may be a critical factor, as evaporation and dessication may be intense during the middle of the day, but in places where it occurs in early morning or late afternoon, there will be less physical stress.

Tidal and wave exposure, in addition to substratum type (among other factors), have a great influence on distributions of organisms on the intertidal zones. They play a role in the myriad of adaptations of form, function, and life history that are observed among the organisms of the shore to counteract the effects of the changing physical conditions of the environment in which they live. On rocky headlands, organisms must withstand the hydrodynamic force generated by the water and tremendous impact of tons of water breaking over them, and here tough or hard bodies and strong attachment mechanisms are an adaptation. Organisms abundant in relatively protected locations are simply not equipped to take the battering and the shearing force of heavy surf.

In addition to considering the physical conditions on the shore, vertical zonation is also controlled by the biological interactions that occur between species, such as larval settlement, competition between species and among individuals of the same species, predation, grazing, and behavior. Many mobile organisms have been shown to possess endogenous activity rhythms that trigger activity at each high tide, and may help to maintain the organisms in the appropriate zone, reducing the possibility of being stranded above the normal resting level on shore.

TIDES

Periodic emergence in air, or emersion, is an overwhelming feature affecting life on intertidal shores, and the rise and fall of tides over the shore can be referred to as the emersion-submersion cycle. The tidal cycle is produced by the interrelation of the gravitational and centrifugal forces between Earth, Moon, and Sun on the ocean waters of a rotating Earth. Earth and the Moon rotate about each other in space and are held in position relative to one another by the balance of the gravitational forces that tend to attract the two masses, and centrifugal forces that keep them apart. The centrifugal force from this rotation on Earth is everywhere equal in magnitude and direction, with the direction being away from the Moon. The Moon's gravitational pull is directed toward the Moon and is strongest on the side of Earth closest to the moon, but is directly opposing the centrifugal force of the Earth-Moon rotation. Depending on the location on Earth relative to the Moon's position, the net effect of these two opposing forces will vary. On the side of Earth closest to the Moon, strong gravitational forces override the centrifugal forces and result in a pull away from Earth and toward the Moon. On the opposite side of Earth, centrifugal forces override the weaker gravitational forces and again result in a net force directed away from Earth.

Tidal fluctuation is then related to the interplay of gravitational and centrifugal forces as these combined forces would create two bulges of water on opposite faces of the planet, one closest to the Moon and one farthest from it. Every 24 hours, Earth rotating on its axis would then seem to rotate under these bulges of water. A point on Earth's surface thus would experience alternately a high tide, a low tide, another high tide, and another low tide almost every 24 hours. During the time taken for Earth to complete one revolution on its axis, the Moon has advanced somewhat on its orbit, requiring an additional 50 minutes of Earth's rotation for a point to catch up to its previous position directly opposite the Moon. This means that the tides would occur approximately 50 minutes later each day as a complete tidal cycle of two high tides and two low tides takes about 24 hours and 50 minutes, or one lunar day. Tide tables show predictions for the time and height of tides in a number of standard ports around the world.

Differences in height between tides are observed in one day, but larger changes are usually seen over periods of weeks because of changes in the relative tide-raising forces generated by the Sun and the Moon. The influence of the sun on producing tides on Earth is only half that of the Moon because the Sun is much farther away, even though it is much larger. During a new or full moon, when the Sun and the Moon are nearly in a line with Earth, their forces combine to produce spring tides, which are of much greater range than average because the tide-raising forces are at a maximum. The word "spring" comes from the Anglo-Saxon word meaning "to rise," not from the season.

A few days after these occurrences, high tides are very high, while low tides are very low, thus tidal amplitude is very large. During a first or fourth quarter moon, when the Sun and Moon are at right angles to each other relative to Earth, the gravitational attraction of Sun and Moon act in opposition and the tidal forces of the Sun partially cancel those of the Moon. This produces high and low tides of minimum range in which high tides are not very high, and low tides are not very low. In these cases, tidal amplitude is very small and the tides are called "neap tides." The monthly cycle contains two neap tides and two spring tides. The highest spring tides occur at the equinoxes (September 21 and March 21), whereas the spring tides are at their lowest amplitude at the solstices (June 21 and December 21).

GLOBAL VARIATIONS

Tidal patterns and amplitudes vary immensely around the world, as well as times of tidal occurrence. A common one is the semidiurnal tidal pattern, when there are two high tides and two low tides a day. A simple pattern of two equal tides per day (with a delay of approximately 50 minutes from day to day) is also common, but not universal. These tides can be of different sizes. The overall variation of tidal patterns runs from fully semidiurnal through semidiurnal with unequal tides, to diurnal, where there is only one high and one low tide each day. Many factors are involved in determining the type and magnitude of the tidal changes' characteristic of a particular location at a particular time, like position of land masses, and local conditions such as weather, barometric pressure, and onshore winds.

As a result, organisms on shore in different parts of the world experience quite different patterns in the times when they are forced to tolerate life in air. Intertidal zones are characterized by many different organisms, which in one way or another have adapted to life in this zone. On shores with similar exposure to wave action, the distribution patterns of organisms are remarkably similar around the world. Almost everywhere, organisms are found in distinct vertical bands or zones at particular heights, and are not distributed randomly; this is known as vertical or tidal zonation. Organisms occur in certain vertical zones, but do not always occur uniformly in those zones. Patches of organisms, or even bare space, frequently occur. One boulder in a boulder field may be covered with mussels, and next to it, another may be covered with algae. The mosaic in the horizontal dimension is a vital contributor to the richness of intertidal life.

Shores that are neither extremely sheltered nor pounded by extreme wave action usually have three distinct vertical zones and are appropriately called the "littoral fringe, eulittoral zone, and the sublittoral zone." The high shore zone, or littoral fringe, often extends well above the levels reached by tidal cover, and is dominated by small snails and black lichen with blue-green algae. The midshore, or eulittoral zone, is characterized by barnacles and limpets, mussels, and algae.

The sublittoral zone at the bottom of the shore extends gradually downward to regions well below those that are ever uncovered by the tides, and contains red algae and kelps (e.g., laminarian algae) or large tunicates (as in parts of the Southern Hemisphere). Even though the zones have been defined entirely in biological terms as they are related to tidal levels, zonation patterns on the shore are caused by the relative influence of multiple biological and physical factors.

In many cases, intertidal species are controlled by some physical factor at their upper limit of distribution, while they tend to be controlled by sets of biological factors at their lower limit. A good example is the distribution of the barnacle *Chthamalus stellatus* because its upper limit is set by tolerance to dessication, while the lower limit is set by competition with the barnacle *Semibalanus balanoides* and predation by the dog whelk *Nucella lapillus*. Second, the zones are defined using conspicuous organisms as indicators, but these organisms are far from evenly distributed across the zone. Some eulittoral zones may have clumps of algae forming an irregular and changing mosaic among a background of barnacles. Third, the boundaries between zones are not necessarily sudden, but can occupy a transition belt. For example, barnacles and black lichens might overlap considerably.

INHABITANTS OF INTERTIDAL ZONES

Organisms on intertidal zones can be divided into four large subgroups: algae, grazers, suspension feeders, and predators. Algal growth is a major factor influencing shore ecology because they provide a primary energy source and thus form much of the basis for intertidal food webs. Great biomass of brown, green, and red algae, collectively known as macroalgae or macrophytes, can be abundantly present in some shorelines. Shores that do not have abundant macroalgae usually have a film of microalgae, which is a collective term that includes the blue-green algae, diatoms, and other protists such as the euglenoids, and the spores and sporelings of macroalgae. The microalgae are often more important than the macrophytes in providing a food supply for grazers because, in spite of their low biomass, they have very high rates of production.

On most exposed shores, there is a rich assemblage of algae at the bottom of the littoral and sublittoral zones, with brown kelps and often a belt of short red algal turf above them. On sheltered shores, the diversity of algae increases and the whole shore may show extensive algal cover, and in many cases this may be because of growth of brown algae. On all shores, the variety of species increases toward the sublittoral, and red algae tend to take over the browns at the bottom of the shore. Green algae do not usually occur in large masses, but some of their genera are found worldwide.

The grazers are herbivorous gastropods that are specialized for intertidal zones, such as limpets, winkles, topshells, neritids, chitons, and urchins. Some of these organisms can be found worldwide. Herbivores rasp microfilms off rock surfaces, taking in microalgae, sporelings, fungi, and probably algae detritus. Urchins that feed sublittorally may take both live and dead macroalgae. Suspension feeders, which are sessile, such as barnacles and mussels, usually outnumber mobile organisms on rocky shores. On top of rocks, barnacles and mussels may entirely cover rocks, while the underside of rocks and macroalgae are covered with bryozoan colonies, tunicates, sponges, hydrozoans, and polychaetes. Predators such as sea stars, crabs, whelks, and nemertine worms are common on all vertical zones of the intertidal.

The slope of the shore will affect the foraging ranges of the many predators that move up in the intertidal to capture prey during high tide, and then retreat to lower zones at low tide. On steep slopes, such predators can cover greater vertical distances in a given amount of time, and as result, the vertical distribution of both predator and prey may be affected. Despite their mobility, gastropods such as winkles and crustaceans such as crabs usually maintain a well-defined zonation pattern. When covered by high tide, many species roam widely. Some crabs move upshore as the tide rises, but retreat again as the tide falls. Some species of snails begin to move downshore as soon as the tide covers them, but return to their usual zone when the tide rises again. Some species, especially grazing gastropods, have been shown to exhibit movements that are random in direction and extent, at least in the short term. But since these animals usually remain within a relatively restricted zone, there must in the long term be some limits to movements in the vertical plane. Very mobile species, such as crabs, show directional movements. Grazing gastropods displaced from their normal position are often capable of returning to it.

Wave exposure may influence the distribution of organisms indirectly through the effect it can have on the foraging ability of predators and distribution of sessile organisms. On exposed areas to full wave impact, predators such as crabs or predatory snails may be largely restricted to protective crevices from which they can only make quick foraging raids on prey populations. On the other hand, for other organisms such as barnacles, the vertical distribution may become broader along a vertical gradient of increasing exposure to waves. This is often observed with sessile, upper intertidal organisms as the more frequent wetting of the upper zones by splash and surf in exposed areas lessens the restrictive influence of desiccation and allows these organisms to live higher on the shore than they would under conditions of quieter water.

One of the structural features of intertidal zones are tide pools, which are dynamic and often constantly submerged, undergoing changes in temperature of the water, carbon dioxide concentrations, and pH, depending on the time for which they are isolated from the sea. The amazing diversity of fauna and flora in tide pool systems varies greatly in relation to height on the shore, shelter, shading, size and depth of the tide pool, and also from random disturbance events. Pools high on the shore undergo the consequences of prolonged separation from the main body of the sea, and conditions within them fluctuate dramatically. For example, in tide pools on the Atlantic coast of Cape Peninsula, South Africa, the maximum temperature during the day increases from about 60.8 degrees Fahrenheit (16 degrees Celsius) in pools near the mean low water in spring tide, to near 86 degrees Fahrenheit (30 degrees Celsius) near mean high water in spring tides.

The situation with oxygen concentration, carbon dioxide concentrations, and pH is less simple because here organisms directly affect the variables, as well as being affected by them. Salinity may increase from evaporation, or decrease from rainfall and water seepage, and these changes can be overwhelming in high shore pools. Algae in tide pools produce oxygen during the day by photosynthesis, and oxygen concentrations may rise above saturation value, so that bubbles rise from dense beds of algae. Carbon dioxide, absorbed by algae, is followed by a rise in pH, which can be extreme. Tide pools can exhibit equally drastic changes in the opposite direction during the night, when photosynthesis stops.

—Lucia M. Gutierrez

Further Reading

Davis, Richard A., and Robert W. Dalrymple, editors. *Principles of Tidal Sedimentology.* Springer, 2012.

Hurd, Catriona, Paul J. Harrison, Kai Bischof, and Christopher S. Lobban. *Seaweed Ecology and Physiology.* 2nd ed., Cambridge UP, 2014.

Martin, Karen L. M. *Beach-Spawning Fishes. Reproduction in an Endangered Ecosystem.* CRC Press, 2015.

Murray, Steven N., Richard F. Ambrose, and Megan N. Dethier. *Monitoring Rocky Shores.* U of California P, 2006.

Nagarkar, Sanjay. *The Ecology of Intertidal Epilithic Biofilms with Special Reference to Cyanobacteria.* BiblioBazaar, 2017.

Raffaelli, David, and Stephen Hawkins, editors. *Intertidal Ecology.* Springer Science & Business Media, 2012.

K

Klamath Estuary

Fields of Study: marine biology; marine ecology; marine biogeochemistry; oceanography

ABSTRACT
The Klamath Estuary connects the Klamath River with the Pacific Ocean in northern California. The area at the mouth of the estuary is shared by the Yurok Indian Reservation and Redwood National Park. The estuary and river are important for fish migration; this was once the third-most-productive salmon river on the Pacific Coast, as more than one million salmon returned here to spawn each year. The estuary is home to coho salmon, a federally protected species.

KEY CONCEPTS

anadromous fish: fish species that split their life cycle between freshwater and salt water, typically spawning in freshwater then developing to maturity in salt water before returning to freshwater to spawn

flyway: the regular route by which flocks of migrating birds travel between seasonal habitats

hydrologic absorption: retention and retardation of the flow of quantities of water that would otherwise move as erosive and potentially dangerous surface runoff

riparian: refers to the ecological environment bordering freshwater streams and rivers

PLACE AND FUNCTIONALITY

The Klamath estuary area has been home to the Yurok Tribe of native Americans for 7,000 years. The Yurok depend on the estuary fishery for sustenance, and on culturally important plants including some of the willows (*Salix* spp.) and ferns (*Pteridophyta*), which are essential for basket-making and regalia. Many plants here also are important for their medicinal qualities, used for healing and in Yurok ceremonies.

The Klamath Estuary biome provides habitat support for a great many animal and plant species, as well as opportunities for hunting, fishing, hiking, and bird watching. It plays important overall ecological roles as well; the wetlands here control floods by hydrologic absorption and storage of large volumes of water. They collect sediment, adding to vital bank and bottomland structure, while filtering water so that channels within the estuary are clear for species requiring low-turbidity conditions, and water is clean for the many species that live in it.

FAUNA

The estuary is an important habitat to resident and migratory birds, particularly waterfowl. This is a vital part of the Pacific Flyway between the Klamath and Siskiyou mountains. The endangered willow flycatcher (*Empidonax traillii*) is one species that makes its summer home in these wetlands. The great blue heron and mallard ducks are common to the area, and the Yurok use their feathers in ceremonies.

The Klamath River has higher diversity of anadromous fish—which are born in the river, migrate to the sea, where they live, and then return to their birthplace to lay their eggs—than any other in California. The population of such fish has declined over the years, and the ongoing restoration programs require strong partnerships between government and local people, all dedicated to preserving their numbers. Coho (*Oncorhynchus kisutch*), chum (*Oncorhynchus keta*), and chinook salmon (*Oncorhynchus tshawytscha*), as well as steelhead species, are key anadromous fish found here.

Chinook salmon, the largest species of Pacific salmon, is often referred to as king salmon because it can grow to 100 pounds (45 kilograms) or more, and 22 to 56 inches (56 to 142 centimeters) long. Chum salmon—from the chinook word *tzum,* which means *spotted*—are the second-most-abundant salmon species of all the Pacific varieties, and are sometimes called "dog salmon."

Salmon populations are at risk when any part of their life cycle is disrupted. All species require adequate spawning habitats, such as a clean river location that is cool and well oxygenated. Without these conditions, salmon roe (eggs) may not develop properly. The salmon species of the Klamath estuary require such food as plantonic diatoms, copepods, kelps, seaweeds, jellyfish, and starfish that are found in these brackish waters. When young, they feed on insects, amphipods, and other crustaceans. They typically feed on other fish when older.

Today, there is great controversy over water use in the Klamath Basin. Dams have damaged the salmon migration, and much of the water from the upper basin is diverted for irrigation. Industrial pollution, mining, roads, and logging also have contributed to poor water quality. Most of the river basin is sparsely populated and rural. This may make it easier for restoration of the salmon populations, compared to rivers in more urban areas.

FLORA

Vegetation of the Klamath Estuary biome occurs in five layers: aquatic, short, medium, tall, and extra tall vegetation. The aquatic layer includes macrophytes such as ditchgrass (*Ruppia cirrhosa)*, water buttercup (*Ranunculus aquatilis*), and leafy pondweed (*Potamogeton foliosus*), which create floating or buoyant canopies at or near the water surface. This layer also includes nonrooted aquatic plants such as duckweed (*Lemna* spp.) and water hyacinth (*Eichhornia crassipes)*.

Short vegetation, plants taller than 20 in (50 cm), includes small emergent vegetation and plants such as watercress (*Rorippa nasturtium aquaticum*), saltgrass (*Distichlis spicata*), jaumea (*Jaumea carnosa*), creeping buttercup (*Ranunculus flamula*), and arrowhead (*Sagitaria* spp.), among others.

Medium vegetation never exceeds 29 in (75 cm) tall; this includes pickleweed (*Salicornia virginica*), saltbrush (*Atriplex* spp.), rushes (*Juncus* spp.), and curly dock (*Rumex crispus*).

Tall vegetation, plants up to 59 inches (1.5 meters) high, includes emergent vegetation and the larger shrubs such as broad-leaved cattail (*Typha latifolia*), bulrush (*Scirpus californicus*), California blackberry (*Rubus ursinus*), and coyote brush (*Baccharis piluaris*).

Very tall vegetation includes plants such as shrubs, vines, and trees that are taller than 59 in (1.5 m) high. These include trees like western sycamore (*Plantanus racemosa*), Fremont cottonwood (*Populus fremontii*), red alder (*Alnus rubra*), blue elderberry (*Sambucus mexicanus*), and hazelnut (*Corylus californicus*).

In addition to the native plants, the Klamath estuary is home to a federally endangered plant, the western lily (*Lilium occidentale*).

Because of habitat degradation and disturbances from development, the region's wetlands can be invaded by aggressive, highly salt-tolerant, nonnative vegetation, such as reed canary grass (*Philaris Urundinacea*), purple loosestrife (*Lythrum salicaria*), water hyacinth (*Eichornia crassipes*), and salvinia (*Salvinia molesta*). The most prevalent invasive species in the area, across several of the height layers, are reed canary grass (*Philaris urundinacea*), Himalayan blackberry (*Rubus procerns*), common reed (*Phragmites australis*), and the yellow pond lily (*Nuphar lutea*).

ENVIRONMENTAL RESTORATION

In 2010, restoration and recovery projects began in the Klamath River basin. In the estuary, efforts were initiated to improve in-stream and riparian river habitats in order to benefit threatened coho salmon, as well as chinook salmon and steelhead trout. These included tree planting and expansion of a local plant nursery by the Yurok Tribe that was started in 2009 with funding supplied by the American Recovery and Reinvestment Act.

The tribe also installed complex in-stream structures and stabilized more than 304 m (1,000 ft) of eroding stream bank. Ponds were built off these streams to provide winter habitat for the threatened coho salmon. Preliminary results show that 250 juvenile coho and 1,500 young chinook salmon were using this new habitat within the first year.

The Klamath Basin Restoration Agreement also started that year; it included a plan to remove four dams. Scientific studies and reports are underway to determine the impact of dam removal as a way to increase the chances of fish survival as they migrate upstream to spawn.

CLIMATE CHANGE

A recent report by the National Research Council suggested that the sea level offshore in this region will rise by 4 in (10 cm) by 2030, and as much as 2 ft (61 cm) higher by 2100, which may cause wetlands within the

Klamath Estuary biome to disappear. The plants in these areas now require more sediment to keep up with the rise in sea level. The deposition of sand and sediment is unlikely to change, however. Therefore, eventually these wetlands will be under water. Naturally, the river delivered sand to the estuary, but today much of this sediment is trapped behind upstream dams.

The wetlands also could also migrate inland if there is sufficient room. Removing upstream dams would increase sediment flow into the estuary and could offset sea-level rise in part. Further, sea-level rise in northern California is lessened slightly by upthrust in the land through plate tectonics and the ongoing uplift along the San Andreas Fault.

Climate change already is affecting the fragile estuary; scientists continue to prepare weather and climate models to help officials at all levels forecast and deal with the oncoming changes. Three such models agree that future summers will be drier than historical records, with precipitation declining 11 to 24 percent in forecasts. Vegetation is likely to shift to favor grasslands such as sagebrush and juniper in the upper basin; oaks and madrone will be favored over maritime conifer forest of redwood, Douglas fir, and Sitka spruce, which are all projected to decline. Projections also indicate that wildfires will increase 11 to 22 percent by late century. Heat waves, severe precipitation, and prolonged drought are all expected to increase. A foretaste of these projections may be seen in the massive forest wildfires that have struck California and the entire west coast in 2019 and 2020.

Scientists are working to find ways to help increase the resilience of local plant and animal communities, to blunt the negative impacts of climate change. This sense of environmental stewardship seems to be on the upswing in the Klamath River area.

—Gillian Galford

Further Reading

Barr, Brian R., Marni E. Kooperman, Cindy Deacon Williams, et al. *Preparing for Climate Change in the Klamath Basin.* National Center for Conservation Science & Policy and the Climate Leadership Initiative, 2010.

Doremus, Holly, and A. Dan Tarlock. *Water War in the Klamath Basin: Macho Law, Combat Biology and Dirty Politics.* Island Press, 2008.

National Research Council. *Hydrology, Ecology and Fishes of the Klamath River.* National Academies Press, 2008.

Patterson, William D. *Klamath River Estuary Wetlands Restoration Prioritization Plan v1.0.* Yurok Tribe Environmental Program, 2009.

Terence, Erica, et al. "Restoring the Shasta for Coho Salmon." *Klamath River News*, vol. 8, Summer 2012.

L

Landslides and Slope Stability

Fields of Study: geology; marine geology; oceanography; engineering

ABSTRACT

Landslides occur under specific geological conditions that are usually detectable. Site assessments done by qualified geologists are important to land-use planning and engineering design; much of the tragedy and expense of landslides is preventable. The same forces that cause terrestrial landslides also function under the sea, but are both exacerbated and mitigated by the water environment. Undersea landslides may behave very differently from terrestrial landslides.

KEY CONCEPTS

creep: gradual downslope movement of soil under the force of gravity, typically identified along gullies and other depressions by the presence of trees that lean toward the downslope side or that have grown with a distinct upward bend from the downslope direction

liquefaction: an effect of seismic activity that causes groundwater to separate from between the pores of soil, turning the solid soil into a mobile slurry

mass wasting: the downslope loss of material mass due to creep, slip, slump or landslide

slip: the sudden downslope movement of material that has sheared off from the face of a structure or that has otherwise become unstable, the typical landslide or avalanche

slump: a failure of slope integrity in which a large body of soil shears under gravitational stress and moves suddenly downslope with a rotational motion, characterized by an arc-shaped shear face

solifaction: melting of water in frozen soils such as permafrost, turning them from hard solids to muddy liquid slurries

FALLS, SLIDES, AND FLOWS

Slope failure, or mass wasting, is the gravity-induced downward and outward movement of Earth materials. Landslides involve the failure of the structural integrity of Earth materials under shear stress and/or flowage. When slope failures are rapid, they become serious hazards. Areas of the United States that are particularly susceptible to landslides include the West Coast, the Rocky Mountains of Colorado and Wyoming, the Mississippi Valley bluffs, the Appalachian Mountains, and the shorelines and bluffs around the Great Lakes. The downslope movement of soil and rock is a natural result of conditions on the planet's surface, both on land and under water. The constant stress of gravity and the gradual weakening of Earth materials through long-term chemical and physical weathering processes ensure that, through geologic time, downslope movement is inevitable.

Slope failures involve the soil, the underlying bedrock, or both. Several types of movements (falling, sliding, or flowing) can take place during the failures. Simple rockfalls, or topples, may occur when rock overhangs a vertical road cut or cliff face. Other failures are massive and include flows and slides. Slides involve failure along a discrete plane. The failure planes in soils are usually curved, as in the illustration of a rotational slide or slump. The failure planes in bedrock can be curved or straight. Failures often follow planes of weakness, such as thin clay seams, joints, or alignment of fabric in the rock. Slides may be slow or rapid but usually involve coherent blocks of dry material. Flows, in contrast, behave more like a fluid and move downslope much like running water. Earthflows, mudflows, sand flows, debris flows, and avalanches occur when soils or other unconsolidated materials move rapidly downslope in a fluidlike manner. The movement destroys the vegetative cover and leaves a scar of hummocky deposits where the flow occurred. Although flows usually involve wet materials, rare exceptions, such as the destructive flows in

Landslides in Kaliwadas, Mount Slamet, Central Java (Ganjar95 via Wikimedia Commons)

Kansu, China, occur in certain types of dry materials that are finely grained and loosely consolidated.

"Slides" and "flows" are terms applied to failures that produce rapid movement. Rock slides are those slides that involve mostly fresh bedrock; debris slides include those movements that are mostly rock particles larger than sand grains but with significant amounts of finer materials; mudslides involve even finer material and water, but the failure plane is straight. Earthflows involve mostly the soil overburden and move over a slope or into a valley rather than failing along a rock bedding plane; mudflows involve more water than earthflows and have a downslope movement much like flowing water.

CREEP

"Creep" is a term given to very slow movement of rock debris and soils. Creep in itself does not usually pose a life-threatening danger. When creep occurs beneath human-made structures, however, it leads to economic damage that requires repair or reconstruction in a new location. Examples include the gradual cracking and destruction of buildings, misalignment and breaking of power lines and fences, the filling of drains along highways, and the movement of soil into streams and reservoirs. Sometimes creep precedes a more rapid failure, and therefore new evidence of creep requires careful monitoring and an evaluation of the conditions that produce it.

Solifluction is a special type of creep that occurs in permafrost regions, in cold climates where the soil is frozen most of the year. In summer, the ice in the upper layer of soil melts, and the soil becomes waterlogged and susceptible to downslope movement. Solifluction is an important consideration for the design of structures in cold climates: For example, the Alaska pipeline, used for transporting petroleum, could not be buried but had to be set above ground on supports that were anchored at a depth below the seasonal depth of thaw to escape solifluction movement. Houses and other buildings in such areas must be set on supports and insulated in order to keep heat from the structures from melting the underlying soils.

ROLE OF WATER

Water is an important agent in promoting instability in slopes. Where soils are saturated, water in large pores will flow naturally in a downward direction. The resulting pressure of the water pushing against the soil grains is called "pore pressure." As pore pressures increase, the grains are forced apart and cohesion decreases. The saturated soils become easier to move downslope. Water flowing along a bedding plane or joint can also exert pressure on either side of the joint and decrease cohesion, causing the block above the discontinuity to move downslope.

Loose volcanic materials often can absorb so much water that they flow quickly down even gentle slopes. A mudflow buried the Roman city of Herculaneum at the base of Mount Vesuvius in 79 AD (although the city had already been hit by pyroclastic flows); and mudflows generated by the Mount St. Helens eruption of May 18, 1980, destroyed many properties. An especially destructive mudflow on the volcano Nevado Ruiz in Colombia in 1986 struck an unsuspecting town at night and killed 22,000 people.

In the underwater environment, water has a dual action in regard to landslides and slope stability. As may be readily imagined, rock and soil structures that are below the surface are saturated with water. This provides a lubricating effect for the downslope movement of material, making it significantly easier for material structures to lose their integrity and begin movement under the influence of gravity. At the same

Landslide in Pingdi Village, Jichang Town, Shuicheng County, Guizhou, China (Huangdan2060 via Wikimedia Commons)

time, the water medium through which the material must travel is considerably denser than the air through which terrestrial landslides must travel. As a result, underwater landslides quickly become less coherent in nature and move at an equally slower speed than their terrestrial counterparts, although nonetheless capable of reaching high speeds as they fall. Two factors seem to be involved in this, one being the overall density of the landslide and the size of the material particles making up the landslide. The movement of smaller particles, like sand, are more strongly hindered by the water and are driven apart from each other to form "sea fans," whereas larger particles, like rocks, are relatively unhindered as they move through the water.

Landslide during the 2001 El Salvador earthquake (United States Geological Survey via Wikimedia Commons)

ROLE OF HUMAN DEVELOPMENT

Designers may unwittingly assemble, in a human-made structure, conditions that produce slope failures. Dry materials such as mine wastes have sometimes been stacked into piles that are steeper than their angle of repose after saturation. Much later, a rainstorm or Earth tremor can send the piles into motion, destroying all structures around them, as happened at Aberfan, Wales, in 1964. When the supporting toe is removed from the base of a slope, as during such excavations that occur for a highway or building foundation, this action produces many landslides, evidence of which can be found on most highways constructed through hilly terrain. An unstable slope can be set into motion by loading the slope from above, which occurs when a structure such as a building, a storage tank, or a highway is built on materials that cannot remain stable under such a load. Catastrophic failures have occurred in which mudflows were produced when dams built from mine tailings burst as a result of slope failure. The mine wastes used for these dams were susceptible to swelling, absorption of water, and weakening over time.

Human development alters the natural drainage of an area and increases runoff. Occasionally, water from storm drains, roof gutters, septic tanks, or leaking water mains reaches a sensitive slope and generates movement. This instability is particularly likely to occur where intensive housing development takes place in several levels on a long slope.

REGIONAL STUDIES

The stability of slopes is evaluated over large regions from aerial photographs, satellite photographs, and images made by remote sensing techniques. Investigators look for telltale signs, such as hummocky topography and old scars left by slides, that may not be evident when viewed from the ground.

Regional studies involve an evaluation of the history of past landslides within the region. History often reveals particular geological formations that have an association with landslides. For example, the shales of the Pierre formation are well known by engineering geologists in the area of Denver, Colorado, as materials in which many slope failures occur. A geologic map that shows where this formation is exposed at the ground surface reveals potentially dangerous areas. Ignoring evidence of past landslides invites disaster.

The regional study also defines loose surficial materials that are likely to fail. Soils rich in clay minerals that swell and expand when wet are notorious for slumping and flowing. Usually, movement occurs in these soils in the spring, when the soil is very saturated from soil thaw and snowmelt. Other soils fail simply because they have low cohesion and large amounts of open space (pore space) between the tiny soil grains. Collapse of these soils requires no wetting; strong vibrations can trigger the movement. The most tragic example of this type took place in 1920, when thick deposits of fine loess (a type of soil deposited by the wind) settled rapidly during an earthquake in Kansu, China, and the resulting flows toppled and buried the many homes built upon them; more than 100,000 people died.

Regional studies also look at the earthquake history of an area, because a tremor, even a fairly mild one, can provide the coup de grâce to a slope that has been resting for decades in a state of marginal instability. Huge blocks of the shoreline slid beneath the ocean at Valdez, Alaska, during the 1964 earthquake when the rotational slumping of materials occurred below sea level. Landslides triggered by that same earthquake destroyed much of Anchorage, Alaska. Much of the failure was due to the sudden liquefaction of buried clay layers.

Finally, the regional study includes a history of weather events. When the right geological conditions exist, periods of intense rainfall can trigger the movement of unstable slopes. In the southern and central Appalachian Mountains, periods of increased slide frequency often coincide with severe local summer cloudbursts and thunderstorms. Intense rainfall

The only standing buildings left in the town of Armero in Tolima Department, Colombia, after the Nevado del Ruiz erruption and subsequent mudflow (United States Geological Survey via Wikimedia Commons)

events associated with hurricanes that have moved inland also trigger landslides over larger areas. Studies in the Canadian Rockies reveal a definite link between rainstorms and rockfalls, and landslides are particularly abundant during the rainy season along the West Coast of North America, as well as in Japan, China, and other Asian countries.

ON-SITE INVESTIGATION

Once data is collected on the region, more specific questions are considered about the local site itself. The investigator first looks at the steepness of the slopes and the earth materials present. In the case of loose materials and soils, the angle of repose is very important. Dry sand poured carefully onto a table to form an unsupported conical pile cannot achieve a cone with sides steeper than approximately 30 degrees, because the cohesion between loose dry sand grains is not strong enough to allow the material to support a steeper face. The 30-degree angle is the maximum angle of repose for dry sand. The angle of repose changes with water content, mineral content, compaction, grain shape, grain size, and sorting. Soils that contain clay may be tough and cohesive when dry and have natural repose angles greater than 30 degrees. When wet, their angle of repose may be only 10 degrees. This is particularly true if the soils contain clay minerals such as montmorillonite that absorb large amounts of water. Those soils, sometimes called "quick clays," can fail instantaneously and flow downslope almost as rapidly as pure water.

The orientation of discontinuities in rocks is as important in determining the stability of a slope as is the type of rock involved. Bedding planes that dip downslope serve as directions of weakness along which failure may occur. Other planar weaknesses may develop along joints and faults and parallel fabrics produced by the alignment of platy and rod-shaped minerals that are oriented downslope.

The investigator will check to see if natural processes are removing the supporting material at the bases of slopes. Landslides are particularly common along stream banks, reservoir shorelines, and large lake and seacoasts. The removal of supporting material by currents and waves at the base of a slope produces countless small slides each year. Particularly good examples are found in the soft glacial sediments along the shores of the Great Lakes of Canada and the United States.

Finally, the investigator will look for evidence of actual creep at the site. Damage to structures already on the site—such as curved tree trunks (where tilting is compensated for by the tree's tendency to resume vertical growth), the offset of fences and power lines, or the presence of hummocky topography on slopes—can demonstrate the presence of recent motion.

COSTS AND REMEDIATION OF STABILITY PROBLEMS

An annual economic loss of between $1 billion and $1.5 billion is a reasonable estimate for costs of landslides within the United States. Expenses include the loss of real estate around large lakes, rivers, and oceans; loss of productivity in agricultural and forest lands; depreciated real estate in areas of slide development; public aid for victims of large landslides; and the contribution of sediment to streams, which decreases water quality, injures aquatic life, and results in the loss of reservoir storage space. Small-scale damage from soil creep is not dramatic but very widespread all the same. In the United States, approximately twenty-five lives are lost each year from landslides. Elsewhere, in densely populated areas, single landslide events can cause death tolls in the thousands.

The remediation of slope stability problems involves contributions from both geologists in the investigation of the site and civil engineers in the design of the project. Geologists employed by state geological surveys and the US Geological Survey provide a tremendous service in several ways—using geographic information systems (GIS) to construct geological and slope stability maps based on knowledge of the soils and rock formations, use of remote-sensing methods such as satellite and high-altitude photography, and field study of suspect areas. These maps are made readily available to engineers, contractors, developers, and homeowners; they show color-coded areas of active and potentially active landslides. Such maps have been produced for many areas with a high population density. Residents in the United States may contact their local state's geological survey, which distributes these maps.

—Edward B. Nuhfer

Further Reading

Bromhead, Edward N., and Vernon R. Schaefer. *Stability of Slopes and Landslide Engineering*. 3rd ed., CRC Press, 2021.

Clague, John J., and Douglas Stead, editors. *Landslides: Types, Mechanisms and Modeling*. Cambridge UP, 2012.

Cornforth, Derek H. *Landslides in Practice*. John Wiley & Sons, 2005.

Duncan, J. Michael, Stephen G. Wright, and Thomas L. Brandon. *Soil Strength and Slope Stability*. 2nd ed., Wiley, 2014.

Glade, Thomas, Paola Albini, and Felix Francés. *The Use of Historical Data in Natural Hazard Assessments*. Springer Science & Business Media, 2013.

Keller, E. A. *Environmental Geology*. 9th ed., Prentice-Hall, 2010.

Norris, Joanne E., et al., editors. *Slope Stability and Erosion Control: Ecotechnological Solutions*. Springer, 2010.

Ogata, Kei, Andrea Festa, and Gian Andrea Pini, editors. *Submarine Landslides: Subaqueous Mass Transport Deposits from Outcrops to Seismic Profiles*. Wiley, 2020.

Sassa, Kyoji, and Paolo Canuti, editors. *Landslides: Disaster Risk Reduction*. Springer, 2009.

Yalçiner Ahmet C., Efim N. Pelinovsky, Emile Okal, and Costas E. Synolakis, editors. *Submarine Landslides and Tsunamis*. Springer Science & Business Media, 2012.

LAPTEV SEA ECOSYSTEM

Fields of Study: oceanography; marine biology; marine ecology

ABSTRACT

A marginal sea of the Arctic Ocean, the Laptev Sea lies between the Arctic Ocean and the northern coast of Siberia, with the Kara Sea to its west and the East Siberian Sea to its east. Like those seas, it is frozen for most of the year, leading to limited populations of flora, fauna, and human activities.

KEY CONCEPTS

continental shelf: the portion of the continental landmass that extends into ocean waters below the surface

sea ice: ice that has formed from saline seawater, typically refers to the broken remnants of the pack ice in the warm season

DESCRIPTION

A marginal sea of the Arctic Ocean, the Laptev Sea lies between the Arctic Ocean and the northern coast of Siberia, with the Kara Sea to its west and the East Siberian Sea to its east. Like those seas, it is frozen for most of the year, leading to limited populations of flora, fauna, and human activities.

The Lena River, which begins in the Baikal Mountains and is one of the longest rivers in the world, is the largest river flowing into the Laptev. The craggy shores are full of gulfs and bays, and the coast becomes mountainous in places. Half the sea is shallow, resting on the continental shelf, which keeps average depths to around 164 feet (50 meters). In the north, the sea bottom drops suddenly, and almost a quarter of the sea has a depth of 0.6 mile (1 kilometer). The large freshwater inputs of the Lena and other rivers have significant effects on the ecosystem. About 60 percent of the diatom flora consists of freshwater species transported here from Siberian lakes.

Because of its proximity to the North Pole, polar night on the Laptev lasts three months in the south and five months in the north, when the sun disappears entirely for the winter; the midnight sun of summer is as lengthy. As elsewhere in the Arctic region, this affects the life cycles and available sunlight for what local life there is. Even in the warmest months, temperatures are only slightly above 32 degrees Fahrenheit (0 degrees Celsius), except on the coasts. Winter's coldest months drop to -29 degrees Fahrenheit (34 degrees Celsius).

A ringed seal peeks its head above water in the Laptev Sea (Wikimedia Commons)

The Laptev Sea is frozen most of the year, leading to limited populations of flora, fauna, and human activities (Ansgar Walk via Wikimedia Commons)

The Laptev contributes more sea ice than the Barents, Kara, Chukchi, and East Siberian seas combined, with the outflow of ice varying considerably—from 96,912 square miles (251,000 square kilometers) in 1985 to 282,627 sq mi (732,000 sq km) in 1989, for example. Ice formation typically begins in October in the coastal south, and September in the Arctic north, creating a continuous sheet of ice covering about 30 percent of the sea.

BIODIVERSITY

Fish populations in the Laptev are scanty because of the ice and the poor availability of sunlight. The lack of commercial fishing in the area diminishes the body of available data, but most of the pelagic fish in the region are whitefish species of genus *Coregonus.*

Pollution in the sea is slight and results principally from chemical spills in coastal areas. River runoff contributes some industrial pollution because of the lack of local facilities in the coastal settlements for processing waste, but because the human population and industrial activity are so sparse in the first place, this is not a problem of nearly the same magnitude that it would be in a more urban or industrialized area. It could become a serious concern should the Siberian coast become more populous in the future.

HUMAN IMPACT

The Laptev shores were inhabited for millennia by the indigenous Yukaghirs, Evens, and Evenki, who herded reindeer and fished the frigid waters. It was previously known as the Tatar Sea, the Lena Sea, the Siberian Sea (from whence the East Siberian Sea was disambiguated), the Icy Sea, and the Nordenskjold Sea. Mammoth remains are well preserved on many of the sea's islands, and may have been hunted by the

Yukaghirs's prehistoric ancestors. Russian explorations began in the seventeenth century along rivers like the Lena, Anabar, Olenyok, and Khatanga that empty into the Laptev Sea.

Today, there is little commercial activity in the area other than navigation along the Northern Sea Route, which passes through, and mining on the shore. The lengthy winter prevents the feasibility of commercial fishing. Ice typically melts only in August and September in this remote part of the world.

—*Bill Kte'pi*

Further Reading

Alexandrov, Vitaly Y., Thomas Martin, Josef Kolatschek, et al. "Sea Ice Circulation in the Laptev Sea and Ice Export to the Arctic Ocean: Results from Satellite Remote Sensing and Numerical Modeling." *Journal of Geophysical Research*, vol. 105, no. C7, July 2000.

Kassens, Heidemarie, Henning A. Bauch, Igor A. Dmitrenko, et al. *Land-Ocean Systems in the Siberian Arctic: Dynamics and History*. Springer Science & Business Media, 2012.

LIMNOLOGY

Fields of Study: limnology; marine engineering; marine biology; marine biogeochemistry

ABSTRACT

Limnology is the study and science of natural freshwater systems. It is a part of marine science through the hydrologic cycle, and through the interface of freshwater and saline marine systems. Freshwater systems are essential to all life on Earth, and support a vast array of plant and animal species.

KEY CONCEPTS

aquifer: essentially an underground stream or river of water that flows through the interparticle spaces in the soil matrix

eutrophication: the conversion of a viable aquatic environment to one that cannot support life excessive amounts of organic matter such as algae or rampant weed growth encouraged by an influx of fertilizers consumes the dissolved oxygen in the water

fish ladder: a structure built in association with a dam that redirects a small amount of the river water to cascade down over a series of stages, allows fish to ascend to the higher level of the dam by leaping up stage by stage to the top

riparian: refers to the areas bordering freshwater systems

thermohaline circulation: vertical circulation of an ocean current due to differences in temperature and salinity of the water

THE HYDROLOGIC CYCLE

Earth's vast oceans contain an immense amount of water, but it is saline water containing about 3.6 percent (36 parts per thousand) of dissolved salts. That has been calculated to be enough salt to cover all of the land area of the planet to a depth of about 7 feet (2 meters). The saline water of the oceans is a unique environment.

The water on land areas is, with a very few exceptions, all freshwater and freshwater is essential for all life on the planet. Even the creatures and plants that inhabit the saline environments are dependent on freshwater to maintain life. Very few are able to ingest saline water without fatal effects. All others must extract freshwater from the saline water in order to maintain life functions and the biochemical reaction medium of those functions. The terrestrial freshwater systems and the saline oceans, however, are inextricably connected through the mechanism of the hydrologic cycle.

Energy from the Sun strikes Earth's surface and is absorbed by both land and water. The immense area of saline water in the oceans naturally captures the majority of the solar energy, and in so doing large quantities of water evaporate and rise into the atmosphere. Atmospheric air curre3nts move the water vapor all about the planet, until it has lost enough energy to cool, condense and fall as rain and snow. The precipitation that falls on land areas collects and flows into streams, rivers and lakes, eventually making its way back to the oceans to complete the cycle and begin a new one. The hydrologic cycle is much more complex than this simple description, but this is essentially how it works. It is Nature's equivalent of distillation that separates freshwater from the salts in saline water.

THE INTERFACE OF FRESHWATER AND SALTWATER

Inevitably, freshwater rivers flow into the saltwater of the oceans. This seemingly simple action is in fact remarkably complex, since freshwater and saltwater have different densities and do not automatically blend together. The freshwater flows over the saltwater as the two very slowly mix where they actually meet, but the mixing is not an automatic process. It is possible in some parts of the ocean, far from land, to scoop up freshwater from the surface flow. The freshwater outflow of the Amazon River, for example, is known to push a surface flow of freshwater some 124 miles (200 kilometers) and more out into the Atlantic Ocean.

The freshwater-saltwater interface is made more complicated by the ocean tides. When the tide rises in an estuary, a "wedge" of the denser saltwater flows in beneath the outward-flowing freshwater, both moving in opposite directions. It is conceivable that this phenomenon may have driven the evolution of anadromous fish that spend part of their life cycle in freshwater and the other part in the saltwater of the oceans.

The influx of freshwater into the saline oceans is an important feature for two reasons. One reason is that the freshwater affects the movement of ocean currents by the differences in density and temperature it provides. The greatest mover of water, and energy, is the thermohaline circulation of the "global conveyor belt." This system includes such currents as the warm Gulf Stream, the cold Benguela and Humboldt currents, and many others. The influx of freshwater into the oceans is a relatively small component of this system, but it is essential to maintaining the equilibrium of the system. The second reason is what the freshwa-

An example of a Fish Ladder, a structure often built in association with a dam that allows the fish to reach their spawning grounds (Scoo via Wikimedia Commons)

ter flow transports from the land areas and deposits them in the seas. This includes a vast quantity of mineral sediments, organic matter and nutrients that replenish the benthic supply for marine organisms. It is therefore essential for maintaining the viability of the marine environment.

FRESHWATER ORGANISMS

The freshwater flowing through land areas is home to a wide variety of fish and other organisms that cannot live in a saltwater environment. Rivers and lakes contain largemouth and smallmouth bass, trout species, pickerel or walleye, pike, muskellunge (thought of as the "shark of the north"), sturgeon, catfish and many other lesser species. In addition to those are the anadromous fish species that include several salmon and large trout species, Arctic char, eels and others. Anadromous fish species typically come into freshwater systems to spawn and live there until they attain a size that allows them to relocate to the oceans, where they spend the rest of their lives until the spawning urge drives them back to freshwater. After spawning, anadromous species such as salmon generally die, becoming food for land creatures such as bears, raccoons, birds and other carnivores. The carcasses of the dead also provide food for benthic creatures such as crabs, crayfish and some bottom-feeding fish.

Riparian swamps and wetlands are primary habitat for many aquatic or water-dwelling creatures, as well as for other large land-dwelling animals that enter the water to eat water plants. Beaver, otter, muskrat, moose, deer, several species of bear, large cats, alligators, crocodiles, caymans, turtles, frogs, toads, and many other kinds of reptiles and amphibians, as well as ducks, geese, flamingos and other waterfowl all either live in or depend on this habitat. Of course, all other living organisms of the land area, including humans, depend on the freshwater of the rivers and lakes for the water they require for life.

DIFFICULTIES, CHALLENGES AND OUTRIGHT MISTAKES

Historically, due to an incredible lack of foresight on the part of humans and the erroneous impression that the oceans are inexhaustible, rivers have been used as the "terrestrial sewer system" for millennia. In the twenty-first century, humanity has come much closer to understanding how wrong this view has been. Unfortunately, as the human population continues to grow and its demands on the environment continue to grow accordingly, changes to overcome the entrenched methods of the past are proving to be slow to happen and difficult to achieve. Industrial pollutants are still being injected into rivers and the water flow itself is being diverted to other purposes such as power generation and irrigation.

Agricultural practices and uses of freshwater represent perhaps the greatest single challenge to recovering the health of freshwater systems. Fertilizers and pesticides used in agriculture are transported into the freshwater drainage basin along with various sediments. This not only encourages eutrophication in susceptible bodies of water downstream, but also introduces toxic materials throughout the entire system. Both factors may also contribute to the formation of dead zones in both freshwater and saline bodies of water. Agricultural runoff also includes animal wastes from pasturage and use for soil amendment. Biological contaminants, including veterinary pharmaceuticals and their metabolites, are thus also introduced into freshwater systems, with the potential to have unpredictable serious adverse effects there.

In many parts of the world, dams have been constructed either for flood control or power generation, or both, and though well-intentioned, the construction of some of those dams can only be called an outright mistake. The main ecological problems with these dams have been the interruption of anadromous fish spawning ability, and the significant effects dams in general have on the sediment load of the corresponding river.

In some cases, it has been observed that riparian environments have been detrimentally altered by the changes in water flow and sediment distribution caused by the dam. While it might be thought that a dam simply holds back a quantity of water until the reservoir is filled to the desired level. The flow is then restored in a controlled manner, and the river downstream of the dam lives happily ever after. What has been historically overlooked in this is the role of the river's current velocity in the overall health of the river and its riparian environs. An indication of this can be seen in the history of the Nile River, in Egypt. Historically, the great annual flooding of the Nile River deposited large quantities of mineral rich silt and organic matter on the floodplains, enriching the

Prior to the construction of the Aswan High Dam in the 1960's, the downstream riparian agricultural lands along the Egyptian Nile suffered immensely (Wikimedia Commons)

soil and allowing farmers to grow abundant crops. In some years, however, drought upstream failed to supply the copious water flow needed in the Nile River to transport the necessary silt load, and the downstream riparian agricultural lands suffered immensely. In the 1960s, Egypt constructed the Aswan High Dam, at Abu Simbel, for both flood control and power generation. Since that time, Egyptian agriculture downstream has had to rely on irrigation and fertilizers for production of their crops. The difference between the before and the after of the Aswan High Dam is the ability of the river velocity to pick up and deliver the sediments along ther watercourse that would otherwise have been delivered to the floodplain. Obviously, the benefits to the downstream population outweigh the negative effect on the natural riparian environment.

The lesson in this is simply that construction of a dam has great ramifications for the riparian environment downstream. But there are also upstream ramifications. One is that as the upstream flow enters the reservoir, its velocity slows as the water spreads out and it loses its ability to carry the sediment load. The sediment drops out and settles into the basin of the reservoir. Some hydraulic engineers have calculated that the average viable lifetime of any dam is just 100 years, more or less, and after that length of time so much sediment will have become trapped in the reservoir that it is rendered ineffective for its original purposes of flood control and power generation.

The other ecological problem occurs upstream when construction of a dam blocks access of fish species to their specific spawning grounds. Salmon species especially always return to spawn in the exact same location where they were spawned. This may even be a very small minor tributary stream in the headwaters of the associated river. Until marine scientists began using tagging to study the anadromous life cycle of salmon, this return to the spawning grounds was thought to be just by random selection, with salmon going to whatever spawning stream they ended up in. It was an unexpected finding that salmon fry tagged in a specific stream before moving downstream and out to the ocean returned to that same location when they were due to spawn and die. How they do so remains a mystery, but the current theory is that they are somehow perfectly tuned to the "taste" of the water flowing from that spawning area, and following that taste is what leads them to travel unerringly from the oceans and up hundreds of miles (or kilometers) of watercourses to their specific spawning ground.

When a dam is constructed in such a river, access tops the spawning ground has historically been blocked off with obvious negative effects for the salmon population. The only way to amend this is by construction of "fish ladders" that will allow the fish to ascend past the dam. While this does allow fish to reach their spawning grounds, it circumvents the natural selection process of "survival of the fittest" by allowing fish to spawn that would otherwise have not survived the journey to spawn. As a result, the robustness of the overall salmon population suffers. Clearly, there are no easy answers.

GROUNDWATER

Another aspect of limnology is the study of what happens to freshwater that does not flow into a watercourse. A significant amount of freshwater precipitation is absorbed into the ground, where it enters an aquifer to flow slowly through the underground matrix until it emerges as a spring in some distant location. Some aquifers emerge offshore under a lake or an ocean. Many others remain shallow and emerge to feed streams and rivers with cold spring water.

Every aquifer has a recharge point as well as a discharge point such that the amount of water in the aquifer remains essentially constant. For aquifers that discharge into saltwater below sea level, this constancy is vitally important, especially if freshwater is being pumped out of the aquifer for some uses. Failure to maintain a positive flow out of such an aquifer's discharge point allows saltwater to intrude into the aquifer, and if depletion of the aquifer continues the aquifer will be changed from a freshwater aquifer to a saltwater aquifer, rendering it unusable. For landlocked aquifers, the danger of removing too much water from the aquifer will allow its recharge point to collapse, which marks the death of that aquifer. Another problem that comes with removing water from an aquifer is subsidence of the land above. For example, in California so much water has been pumped out of a major aquifer that the ground level above it has subsided by around 16 ft (5 m) or more, and severe restrictions on how much water can be removed have been put in place to protect the remainder of the aquifer.

As the study and science of freshwater systems, limnology and limnologists have an important task ahead in ensuring that the natural freshwater systems of the planet are understood and protected. It is a task that has historically been understaffed and neglected.

—Richard M. Renneboog

Further Reading

Dumont, Henri J. *The Nile: Origin, Environments, Limnology and Human Use.* Springer Science & Business Media, 2009.

Gharibreza, Mohammadreza, and Muhammad Aqeel Ashraf. *Applied Limnology: Comprehensive View from Watershed to Lake.* Springer, 2014.

Gökçe, Didem, editor. *Limnology: Some New Aspects of Inland Water Ecology.* IntechOpen, 2019.

Scudder, Thayer. *The Future of Large Dams: Dealing with Social, Environmental, Institutional and Political Costs.* Earthscan, 2005.

Tundisi, José Galizia, and Takako Matsumura-Tundisi. *Limnology.* CRC Press, 2011.

Vincent, Warwick F., and Johanna Leybourn-Parry, editors. *Polar Lakes and Rivers: Limnology of Arctic and Antarctic Aquatic Ecosystems.* Oxford UP, 2008.

M

Manganese Nodules

Fields of Study: oceanography; marine mining; marine engineering

ABSTRACT

Manganese nodules are rough spheres of manganese and other minerals deposited by seawater on the seabed. These nodules, about the size of a potato, are being considered as a potential source of manganese, which is essential for steel production. Manganese also is used to make pigments, alkaline batteries, and fertilizers.

KEY CONCEPTS

carbonate compensation depth: the depth in the ocean at which carbonate becomes highly soluble due to the increased pressure and cold temperature

crystallization: the formation of a solid material from a solution, either by combining dissolved ions to form an insoluble material, or by exceeding the solubility of the material in that solvent under the existing conditions

electric dipole: a condition that exists when the distribution of negative charge (electrons) in a molecule is not symmetrical

electrolysis: the process by which ions are extracted from solution by an electric field to be deposited on a collector as solid metals; because atoms have different electrical properties, different voltages can extract different materials

electronegativity: term for the propensity of a particular atom to gain or surrender electrons relative to another kind of atom

marine snow: the remains of dead planktonic and other creatures that constantly falls through the water to the seafloor

MANGANESE

Manganese nodules are vaguely spheroid rocks made of deposited sediments, primarily the eponymous manganese. These nodules also contain enough other metals that they are also referred to as polymetallic nodules. The nodules form through millions of years of deposition from sea water, and the highest concentrations are on the abyssal plains of the oceans, at about 2.5 to 3.7 miles (4,000 to 6,000 meters) deep. Manganese nodules have been suggested as potential sources of manganese for industry.

CHEMISTRY OF MANGANESE NODULES

Uses for manganese nodules come from their chemical properties. Manganese is a transition metal, like iron or nickel, and has several common oxidation states, with the most stable being +2. Manganese forms from a variety of minerals, though the majority of the minerals in a nodule consists of manganese hydroxides.

Seawater, an important element in the formation of manganese nodules, is not pure dihydrogen monox-

Marine snow is a shower of organic material falling from upper waters to the deep ocean (National Oceanic and Atmospheric Administration via Wikimedia Commons)

ide (H_2O). Instead, it includes many other dissolved substances, but mostly such minerals as sodium chloride (table salt, NaCl). Water is a good solvent because it is a polar molecule. Water is a polar molecule not simply because of its bent shape, but also because of its electric dipole resulting from the difference in electronegativity of the hydrogen and oxygen atoms in its molecular structure. The oxygen in water is more electronegative than is the hydrogen; thus, the electron density is greater about the oxygen, such that an electric dipole exists with more negative charge at the oxygen end and more positive charge at the hydrogen end.

Limits exist to the amount of any substance that can be dissolved, however. The maximum amount of solute depends on properties such as temperature and pressure but varies for every material combination. Some substances increase in solubility with temperature and others decrease, all due to the binding characteristics of the compounds. The bonds of compounds are either more ionic or more covalent. As a metal, manganese compounds tend to be ionic and to dissolve in water. When the solubility of the water changes, either through cooling or through the addition of more solutes, the dissolved manganese precipitates out. It then crystallizes around a core, such as a fragment of a nodule, a shark's tooth, microfossils, or basalt fragments.

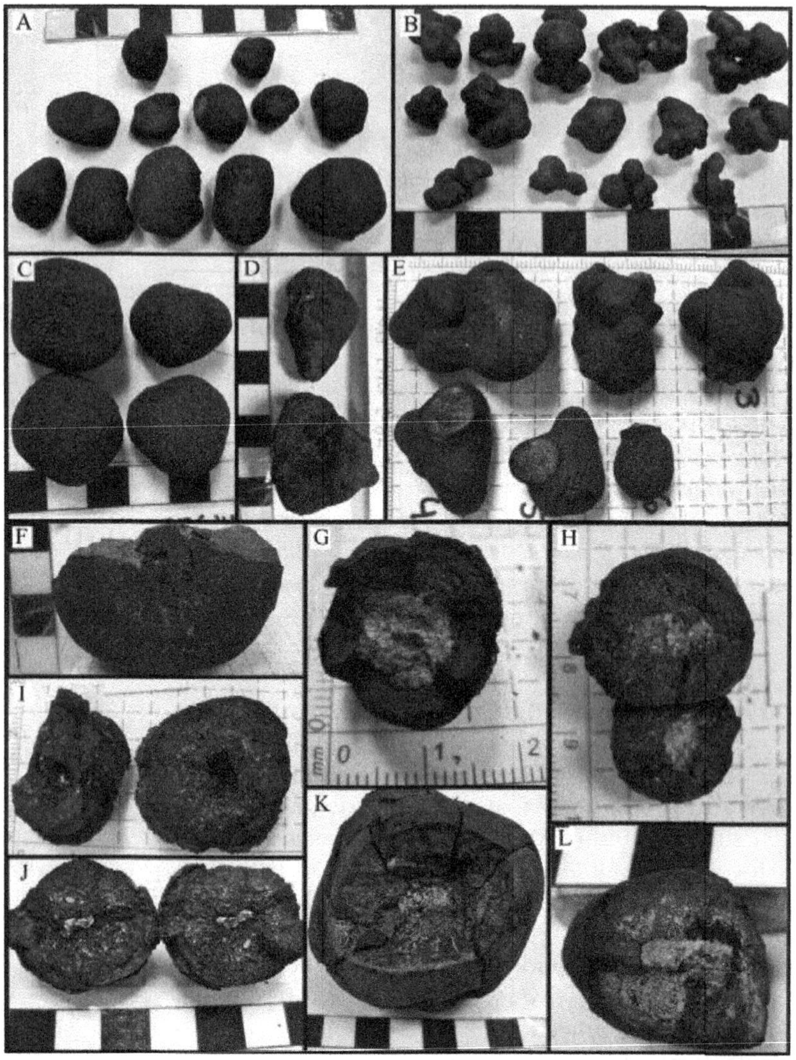

Manganese nodules from the deep seafloor off the Cook Islands in the southwest Pacific Ocean (United States Geological Survey)

Crystallization is the formation of crystals of a substance as they precipitate from solution. Crystals are ordered arrangements of atoms or ions. They are organized by the bonding properties of the atoms in question. However, the precise pattern is still difficult to predict (computers and algorithms are leading to progress in this field, however). At any rate, continued deposition of the component of the crystal will lead to crystal growth.

NODULE FORMATION AND LOCATION

Nodules can be found in any body of water suitable for their growth, including some lakes, but most nodules are on the abyssal plains of the oceans. In particular, the northern regions of the Pacific equatorial seabed and the Atlantic have been found to have high concentrations. The Indian Ocean also appears to have concentrations of the nodules north of the equator.

These patterns in Pacific distribution seem to be a result of seafloor composition. North of the equator, the tropical Pacific is mostly siliceous in content, made of the noncarbonate portions of the skeletons of plankton. Because the seafloor here is deeper than the carbonate compensation depth (calcium carbonate is what shells and tests are made of), the carbonate remnants dissolve. Calcium carbonate dissolves in water naturally, but because sufficient amounts exist in surface-level water, shells made of calcium carbonate are virtually insoluble.

Carbonate becomes more soluble with increasing cold and pressure. At the carbonate compensation depth, carbonate solubility increases dramatically. As a result, the seafloor at lower depths is made not of calcium carbonate but of other materials, such as silicates, which do not dissolve. This effect leads to the creation of siliceous ooze. Calcareous sediments are formed where the seabed is above the carbonate compensation level. The carbonate compensation level depends upon many factors, but it is at about 138 to 1,640 feet (or 42 to 500 meters), in the Pacific, except in the equatorial upwelling zone (where it is at 3 mi, or 5,000 m). Farther from the tropics, the seafloor tends to be made of red pelagic clay and has little biological material.

These differences in seafloor composition mean that the composition of nodules can be different, too. Where calcareous ooze, formed by sedimentation of calcareous shells, predominates, concentrations of nodules are lower. Equally, deposits in red clay regions have higher iron content than other deposits.

The sedimentation rate of siliceous ooze and pelagic clay regions is far faster than the rate of growth of nodules; therefore, the means by which nodules remain on the seabed and avoid getting buried are unknown. It has been suggested that deep-sea fauna might occasionally run into the nodules, pushing them over and thus preventing them from being buried. It also is possible that bottom currents push the nodules to the top or that buried nodules dissolve and are redeposited on the surface nodules.

THE ABYSSAL PLAIN

The abyssal plain is the seafloor beneath most of the ocean, typically at a depth of between 1.8 to 3.7 mi (3,000 and 6,000 m). The plain covers more than one-half of Earth's surface and is among the flattest landscapes on the planet. The plain also is one of the least explored surfaces on Earth. Despite the overall flatness, the plain also contains midocean ridges and oceanic trench subduction zones. These features are caused by plate tectonics.

In a phenomenon known as marine snow, organic detritus, such as dead organisms, fecal matter, land sediment, and inorganic sediment, continually fall from the upper levels of the ocean, making this detritus the main source of nutrients for the abyssal plain's ecosystem; the detritus also provides much of the sediment for magnesium nodules. Once thought lifeless, the abyssal plain is biologically diverse. Many species are bottom-dwelling species that subsist on the marine snow. Despite recent discoveries, however, much remains to be learned about these organisms.

Experts fear that manganese-nodule mining could negatively affect the seafloor environment. Disruptions to the seabed could increase the toxicity of the water near the seabed. Another possible problem is the formation of sediment plumes. A sediment plume is caused when tailings from mining are dropped back into the water.

DEEP SEA MINING

Deep-sea mining of manganese remains speculative. Modern prospecting methods tend to use remotely operated vehicles to collect samples from prospective mining areas. After a site has been found, a mining ship or platform is placed into position.

Two main mining methods are being considered: the continuous-line bucket (CLB) system and the hydraulic suction system. CLB is the primary method under consideration and is akin to a giant conveyer belt. It operates by using dredging buckets to pull material from the seabed. The material is processed on a platform and the waste materials, called "tailings," are returned to the seabed in the dredging buckets. The other method uses hydraulic suction to pull the materials through a large tube. Another tube returns the waste material to the seabed.

The CLB system was put into use by a syndicate of thirty companies in the 1970s using a 5-mi (8-m) cable launched from a former whaling vessel. The system worked but was prone to tangling. The hydraulic system was tried by an American consortium in the Pacific Ocean. The system used a dredge on the seafloor attached through a tube to a surface object, such as a ship or a platform. This allowed it to reach an area of seabed. The initial tests ran into problems because the cables to the dredge could not be fully waterproofed; later dredges had problems that included muck and hurricanes. However, the system was demonstrated successfully in 1979.

Many current designs work on variants of this system. The fully operational system would have a central mining platform with a fleet of ships to supply and transport ore back to shore. To be of economic interest, however, the abundance of manganese must be 22 pounds, or 10 kilograms, per square meter or more, with an average of 33 lb (15 kg) per square meter in several tenths of a square mile (or kilometer).

Sonar is used to map the seafloor for mining, too. Modern sonar systems can survey strips of more than 12.5 mi (20 km) at a time from the surface. This mapping can be supplemented with the use of arrays, which are towed above the seabed. Additionally, free-fall devices can take readings as they descend and return to the surface on their own. They even can take samples and photographs from the seafloor. Cable-operated apparatus also can perform this function.

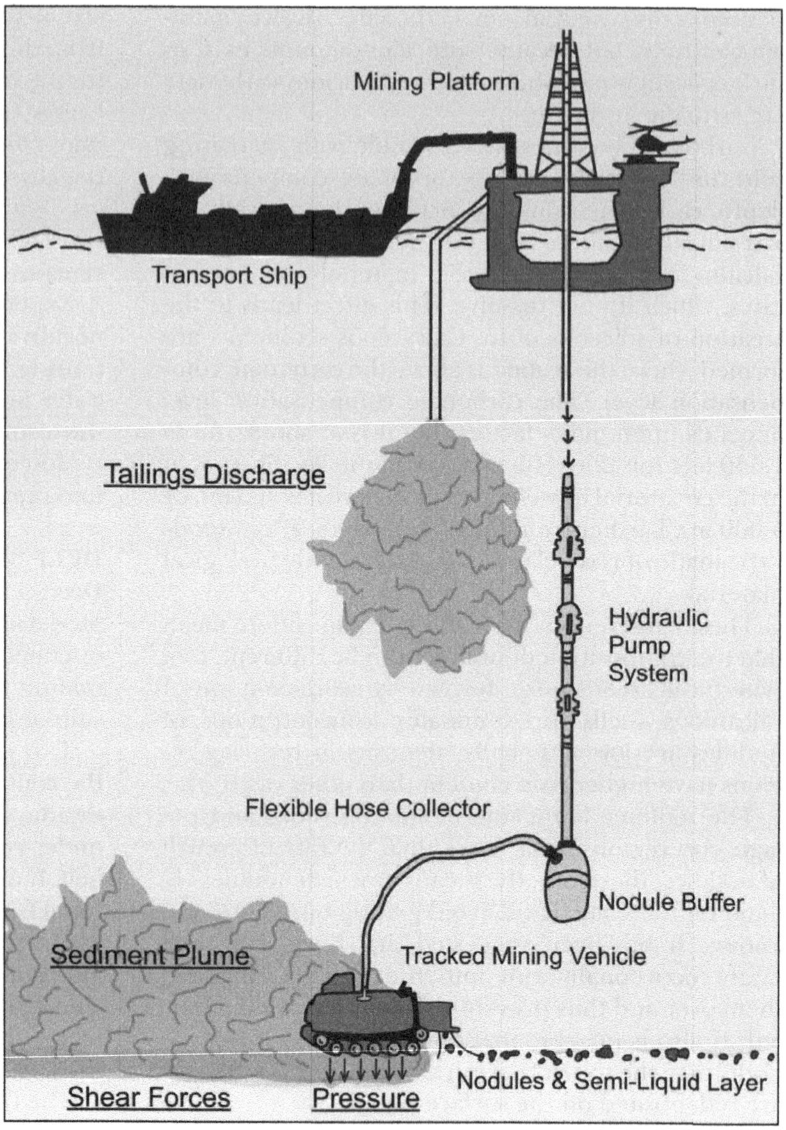

A schematic of manganese nodules mining on the deep sea floor, with environmental impacts underlined (MimiDeepSea via Wikimedia Commons)

MANGANESE NODULE PROCESSING

The idea for processing manganese nodules was first proposed in the 1960s by John Mero in his landmark paper on the subject, "Mineral Resources of the Sea." The paper received considerable attention, leading to several efforts to utilize the manganese resources.

Once on shore, several methods exist for refining. Some of the most popular methods include the cuprion process, sulfuric leaching, and smelting. The cuprion process, developed by Kennecott, an American mining company, involves grinding the nodules into slurry. The slurry is reduced by carbon monoxide in the presence of ammonia, in a low-temperature tank. The metals are made soluble and then are electrowinned, or separated by electrolysis. Recovery of manganese from the remaining ferromanganese residue remains problematic for the mining industry.

Sulfuric leaching works by dissolving the nodules in sulfuric acid at high temperature and pressure. Copper, nickel, and cobalt are precipitated with hydrogen sulfide. The resultant materials are then refined. These products are electrowinned, and the ferromanganese residue is smelted after drying.

CURRENT EFFORTS

Harvesting manganese nodules remains expensive, given available technology. Many countries are placing claims on regions of the seafloor, however, to use in the event manganese mining becomes viable.

—Gina Hagler

Further Reading

Bardi, Ugo. *Extracted. How the Quest for Mineral Wealth Is Plundering the Planet: The Past, Present and Future of Global Mineral Depletion. A Report to the Club of Rome*. Chelsea Green Publishing, 2014.

Baturin, G. N. *The Geochemistry of Manganese and Manganese Nodules in the Ocean*. Springer Science & Business Media, 2012.

Bischoff, James L., and David Z. Piper. *Marine Geology and Oceanography of the Pacific Manganese Nodule Province*. Springer Science & Business Media, 2013.

Cronan, D. S. *Handbook of Marine Mineral Deposits*. CRC Press, 2000.

Halfar, Jochen, and Rodney M. Fujita. *Science*, vol. 316, no. 5827, 2007, p. 987.

Koslow, J. A. *The Silent Deep: The Discovery, Ecology, and Conservation of the Deep Sea*. U of Chicago P, 2007.

Mukhopadhyay, Ranadhir, Anil K. Ghosh, and Sridhar D. Iyer. *The Indian Ocean Nodule Field: Geology and Resource Potential*. 2nd ed., Elsevier, 2018.

Rona, Peter A. "Resources of the Seafloor." *Science*, vol. 299, no. 5607, 2003, pp. 673–74.

Maps and Mapping

Fields of Study: geography; cartography; graphic design

ABSTRACT

Mapping encompasses methods for representing geospatial information in paper and digital forms as well as newer digital technologies such as remote sensing and geographic information systems. Maps are needed in areas ranging from civil engineering to regional planning. Computer technologies and the internet have brought profound changes to the ways maps are created and used, and a growing number of mapping applications have been developed for portable devices such as car navigation systems.

KEY CONCEPTS

coordinate system: a system of designating the position of a point in space relative to a fixed standard origin; for example, latitude, relative to the equator, and longitude, relative to the meridian of Greenwich

elevation isobar: a line marking a specific elevation about a feature on a topographical map

scale: the ratio of the depiction of a feature to the actual physical size of the feature

topographical map: a two-dimensional (2D) map that shows the relative elevations of three-dimensional (3D) features, typically as elevation isobars

DEFINITION AND BASIC PRINCIPLES

Maps assume many forms, including spherical globes, folded paper charts, wall-sized murals, floor plans in architectural drawings, genealogical "family trees," images displayed on tiny electronic screens, "heads up" displays in vehicles and even as holographic projections.

As symbolic representations at varying scales, maps show a selection of information including some component of space. The most well-known form of map is a geographic 2D depiction of a 3D area for means of orientation, as in the standard road map. However, maps may also be 3D themselves, be interactive, or present much more abstract spaces or concepts. In addition to describing features on the surface of Earth, maps can represent underwater areas, the interiors of caves, celestial bodies such as planets, the structure of deoxyribonucleic acid (DNA), or the relationships between the characters in a work of fiction, among many other examples. The term "cartography" refers to the process of creating maps and related geographic products. According to the International Cartographic Association, cartography is the art, science, and technology of making maps.

Maps can be divided into two general categories: general-purpose and thematic maps. General-purpose maps show a variety of information, commonly including lakes, rivers, roads, cities, and administrative or political boundaries. Most general-purpose maps, such as road and topographic maps, are designed to serve as reference tools. In contrast, the purpose of thematic maps is to show patterns and distributions corresponding to a single topic or theme, such as rainfall or corn production.

BACKGROUND AND HISTORY

For more than four thousand years, maps have been used to record and communicate information about Earth. The ancient Greeks created maps to characterize Earth's spherical shape, while legions of Roman soldiers used them as tools in the conquest of new territories. Drawn by hand, early maps showed landscape features using pictorial symbols to represent mountains, rivers, and other physical features. Later technologies such as printing eliminated the need for maps to be painstakingly copied and increased the speed at which maps could be reproduced and disseminated. Exploration, especially during the fifteenth and sixteenth centuries, created a demand for maps to assist in ocean navigation and to document newly discovered locations.

Modern maps play many roles in society, from guiding aircraft to providing reference information within school textbooks. Under most circumstances, maps are superior to charts and graphs in their ability to represent distances, directions, and the relative sizes of objects over space. In addition to showing locations of features, maps are able to efficiently represent geographic patterns and the spatial extent of physical entities such as rivers and mountains. They are also useful for showing cultural features such as property boundaries and political or administrative areas. As archives of spatial information, maps are used extensively for recording the locations of historical events as well as spatially referenced scientific data. In addition to their use in storing information, maps serve as important research tools for visualizing geographic data needed for evaluating hypotheses about spatial distributions. The development of computer technology allowed geographic information systems (GISs) and other mapmaking tools to drastically increase the power and potential applications of maps and mapping.

HOW IT WORKS

Scale and Projection. All maps of Earth's surface require some level of reduction. Map scale is an important concept to maps and mapping because it determines the level of map reduction necessary. Scale is the ratio of a distance measured on a map to the distance it represents in the real world, using the same units of measurement. For example, one inch (2.5 centimeters) measured on a map with a scale of 1:24,000 represents 24,000 in (60,960 cm) (or 2,000 feet [610 meters]) in the real world. The scale used on any given map determines the size of the area that can be represented as well as the level of detail that can be depicted. In addition to the representative fraction (e.g., 1:100,000), scale can be shown using a verbal statement such as "one inch equals one mile" or a graphic or bar scale printed directly on the map.

Cartographers draw a distinction between small- and large-scale maps. Small-scale maps represent very large areas of the world and are useful for representing entire, countries, continents, or the globe. In contrast, large-scale maps are capable of showing very small areas at higher levels of detail. City road maps and topographic maps are examples of large-scale maps. Given that geographic mapmaking involves reductions, the cartographer must make choices about the features to be shown and the symbols used to represent those features. This process, called "selection," must be closely tied to the purpose of the map. In addition, complex objects such as coastlines must be simplified to reduce unnecessary detail.

While most maps use a single scale, some may vary in scale throughout in order to achieve a desired effect. Others intentionally distort scale, whether to simplify use (as in many subway maps that stylize the train lines to improve readability) or to emphasize a point (such as a cartogram that displays countries' sizes based on population).

Another important consideration is the choice of projection for displaying mapped information. Projection refers to a mathematical transformation of geographic information from the spherical Earth to a flat or developable surface. Because spherical surfaces cannot be flattened without distortion, the cartographer must select the most important properties to maintain during this transformation. For example, on a world map, it may be desirable to maintain the relative sizes of countries for making comparisons. Although maintaining the property of equivalence, this type of projection distorts the shapes of countries near the poles. The well-known Mercator projection, first published in 1569, is useful for ocean navigation because it preserves shapes as well as lines of constant geographic direction as straight lines. However, a disadvantage of the Mercator projection is that it severely distorts the relative sizes of polar areas, making the island of Greenland in the North Atlantic look larger than the continent of South America, which is actually fifteen times larger than Greenland.

Geographic Grids. Cartographers use geographic grids, or coordinate systems, for the placement of points, lines, and areas to systematically organize geographic information. Most maps are created using either the latitude and longitude coordinate system or a rectangular coordinate system. Latitude-longitude and rectangular grids use Cartesian coordinates (x, y). The point where the prime meridian (0 degrees longitude) and the equator (0 degrees latitude) cross serves as an origin against which all other locations are referenced. Smaller geographic areas are often mapped with rectangular coordinates that use an origin located outside and to the southwest of the area of interest to make all coordinate values positive.

Map Design. Map design is a systematic process involving the selection and arrangement of map elements in a way that facilitates a user's correct interpretation of ideas and concepts. Cartographic design has benefited from developments in both graphic design and cognitive psychology. Because the principal objective of mapmaking is communication, it is important for cartographers to select visual elements and arrangements that present concepts and spatial relationships in ways that can be easily interpreted by the viewer. The design process can require several stages as alternatives are tested and modified. Basic graphic elements manipulated in the design process include points, lines, and area symbols. In addition, the selection and placement of text is an important design element. As suggested by French cartographer Jacques Bertin, graphic elements used in cartography can be made more or less prominent through changes in hue (color), value (darkness or lightness), texture, orientation, size, and shape. Design is affected not only by the choice and design of symbols but also by the arrangement of map elements. In the design process, cartographers must be careful to use space efficiently and to avoid symbols that create unwanted attention or noise. Other important concepts within map design include achieving balance in the arrangement of map elements and creating a visual hierarchy that draws the map reader's attention to features most important to the map's purpose.

Map Output. Developments in map reproduction have followed innovations in printing, beginning with the invention of the printing press by Johannes Gutenberg (1400–1468) in 1440. Early printing technology employed a process called "letterpress," in which ink was applied to raised portions of wooden blocks that were then pressed against paper. Introduced in the early 1700s, engraving involved the application of ink to depressions in a metal printing plate. Engraving improved map reproduction because it enabled the use of finer line widths and improved gray tones and patterns.

Lithography, a printing technique based on the incompatibility of grease and water, was introduced in the 1800s. Most modern paper maps are reproduced using lithographic plates, which are applied to a rotating drum to transfer an image to a moving sheet of paper. Four-color maps use lithographic plates corresponding to the subtractive primary colors of cyan, magenta, and yellow plus black. A wide variety of colors can be created by combining percentages of each primary color. Modern printing presses use computer files to create plates needed to produce color maps, with sheets of paper passing through the press once for each plate.

Maps can also be reproduced in smaller quantities using low-cost output devices such as laser printers and ink-jet plotters. Electrostatic (or xerographic) copying is a common method for reproduction. Laser and ink-jet printers have also become a popular method for reproducing small numbers of maps. The readability of map lettering and the sharpness of lines and gray tones are a function of the printer's resolution as represented in dots per inch (dpi). Low-end printers typically offer 300-dpi output, meaning that each square inch has 90,000 dots. Higher-end printers and plotters offer resolutions exceeding 2,000 dpi.

The introduction of personal communications devices with high-resolution screens has led to an increasing number of maps that are never printed. These virtual maps may be used for tasks such as checking weather conditions or finding a street address. Internet-based Google Maps has been an innovator in making virtual maps, aerial photographs, and satellite images accessible to the public. Google Maps and Google Earth enable the simultaneous viewing of aerial photographs and map information in the form of roads, contour lines, and political boundaries. Mapping tools enable the user to zoom in or out, pan in any direction, and select specific map features and points of interest to be displayed.

Computers and Mapmaking. The single most significant change to cartography during the twentieth century was the introduction of computers that enabled tasks previously requiring hours to be carried out in seconds. Computer cartography, or the use of computers in map production, has been particularly beneficial in enabling cartographers to experiment. Using mapping software, the cartographer can resize and reposition map elements such as text labels, scale bars, and legend boxes. Computers also facilitate mathematical and statistical transformations of tabular data needed for making thematic maps. Some programs enable map images to be combined with other spatial data such as aerial photographs or satellite images. Increasingly, digital databases are replacing paper maps as the principal method for storing and retrieving geographic information.

Before geographic information can be manipulated by a computer, it must be converted to a digital form. The process of capturing points, lines, and areas shown on a paper map in a digital format is called "digitizing." Analogue information can be digitized by tracing features of interest individually using a tablet or table-sized digitizer. Digitizing can also be accomplished by scanning an entire map sheet. Before scanned data are useful, they must be transformed into a system of coordinates such as latitude and longitude. This process is called "geocoding" or "georeferencing."

APPLICATIONS AND PRODUCTS

Since 1985, a significant amount of data used in computer mapping has been converted into digital format. For example, the US GeoData digital line graphs are computer files containing transportation and other line information that has been captured from topographic maps maintained by the United States Geological Survey (USGS). The set of digital files called "digital raster graphics" includes information scanned from USGS topographic maps. USGS digital elevation model data files contain grids of elevation values captured every 98 ft (30 m) and are useful for constructing contour maps, 3D terrain diagrams, and computer simulations such as aerial fly-bys.

Computer technology also makes it possible to create new types of cartographic visualizations such as animated maps. Maps that use a sequence of digital images are called "temporal animations." This type of map is useful for representing changes over time, such as the movement of military fronts during a battle.

Computer technologies have influenced maps and mapping through developments in closely related fields such as aerial photography, GIS, and remote-sensing satellites. Remote sensing involves the capture of information about the Earth using sensors and is commonly divided into photographic methods such as aerial photography and nonphotographic imaging that include sensors looking outside visible portions of the electromagnetic spectrum.

Digital orthophoto quadrangles (DOQs) have emerged as an important source of information for constructing maps. DOQs are aerial photographs in digital form that have been rectified (corrected) for camera distortion. In addition to serving as a source from which to extract the location of physical and cultural features, they can be used as background images for maps. Multispectral scanners used by Landsat and other satellites capture large amounts of data using scanners that are sensitive to narrow wavelengths of energy emitted from Earth's surface. Such data can be useful for revealing information that cannot be seen in photographic images, including stress or disease in vegetation.

Radio detection and ranging (radar) is another nonphotographic imaging system that can be used for mapping surface features, underground areas, and atmospheric conditions. Radar uses transmitted energy that is reflected by objects and then captured and converted into an image. Imaging radar used on space shuttle missions has become an important source of elevation data used in a variety of mapping applications. Light detection and ranging (lidar) is an innovation for mapping Earth by measuring the properties of scattered light as it comes into contact with distant objects. Lidar images are created by measuring the time needed for laser pulses to reach Earth's surface. Some applications for lidar include 3D mapping of urban areas and assessments following natural disasters.

CAREERS AND COURSEWORK

College and university courses in cartography are most often taught within departments of geography. Students seeking careers in the mapping sciences should complete college coursework in geography, mathematics, statistics, and technical writing. Depending on the type of maps they will be involved in producing, they may also take coursework in graphic arts, computer programming, physics, remote sensing, computer-aided drafting, database management, planning, or geology. A wide variety of agencies and organizations employ cartographers, GIS analysts, and remote-sensing specialists. The US Geological Survey (USGS) employs cartographers to maintain both paper and digital map products. Its National Imagery and Mapping Agency supports mapping and imagery needs for the military and policy making. The US Department of Agriculture's Natural Resources Conservation Service maintains the Soil Survey Geographic Database, an important spatial data set defining soil boundaries within the United States. The Central Intelligence Agency provides mapping products used internally and by other intelligence agencies. State and local government agencies and private organizations also employ geospatial experts. State agencies produce road maps used for promoting tourism, and local government departments maintain property and planning maps. Maps and proprietary data used in mapping are also created by private companies.

SOCIAL CONTEXT AND FUTURE PROSPECTS

Maps and other cartographic products are important within areas ranging from civil engineering and planning to the humanities. For example, maps are used in weekly magazines, in newspapers, and as part of websites. Journalistic maps typically show the countries, cities, or accident sites that are the focus of news events.

Transportation maps such as nautical and aviation charts are of critical importance to the safe movement of ships and aircraft. Within the United States, the National Ocean Service of the National Oceanic and Atmospheric Administration (NOAA) publishes four types of nautical charts: general, sailing, coast, and harbor charts. The Federal Aviation Administration publishes several aviation charts, including the world aeronautical charts at a scale of 1:1,000,000 and sectional charts that represent smaller areas in greater detail at a scale of 1:500,000. Nautical and aviation charts must be updated frequently to reflect changes that can affect safety. Other types of transportation maps include street and highway maps, road atlases, maps showing airline or train routes, and bicycle or walking trail maps.

Topographic maps show rivers, lakes, mountains, and other surface features, as well as cultural features such as roads, cities, levees, and towers. Elevations on topographic maps are shown using contours defined as lines that join points of equal elevation. Bathymetric or hydrographic charts are a related type of map that uses lines to represent underwater depths. The USGS maintains topographic maps as well as maps showing geological cross sections, groundwater areas, earthquake faults, and seismically active areas. Geologic maps have a wide range of uses, from mineral exploration to civil engineering. Most geologic maps use colors, gray tones, patterns, and sometimes letter codes to represent different rock types. In addition to government agencies, energy and mining companies also maintain maps showing fossil fuels, mineral reserves, and areas with the potential to generate power using solar or wind energy.

Maps are important for recording the location of property boundaries and for planning applications. Cadastral maps are maintained by local government agencies to show ownership boundaries as a basis for property taxes. This type of map is also important to insurance companies. Many local government planning agencies maintain maps showing the city or county infrastructure, such as water resources, sewers, electric lines, and gas lines. City and regional planners use maps for zoning changes and to maintain 911 emergency response systems.

The NOAA's National Weather Service is the principal federal agency responsible for disseminating information about weather conditions. Weather maps are used by forecasters and others to monitor or predict atmospheric conditions including temperature, precipitation, wind direction and speed, humidity, cloud cover, and anomalies such as thunderstorms, hurricanes, and tornadoes. Weather maps may show a single parameter such as surface pressure or multiple types of information. Climate maps include some of the same information shown on weather maps but for longer timeframes. Examples include maps depicting monthly temperature highs and lows or average yearly rainfall.

Widespread availability of the Internet has created possibilities for the publication of new types of mapping products such as electronic atlases that allow users to generate customized maps by selecting the data to be shown. Electronic atlases enable users to search on place-names and easily locate cities and find other information. An added benefit of the internet is its ability to incorporate updates as soon as new information becomes available.

—Thomas A. Wikle

Further Reading

Cartwright, William, Michael P. Peterson, and Georg F. Garner, editors. *Multimedia Cartography*. 2nd ed., Springer, 2007.
Dodge, Martin, Rob Kitchin, and Chris Perkins, editors. *Rethinking Maps: New Frontiers in Cartographic Theory*. Routledge, 2009.
Field, Kenneth. *Cartography: A Compendium of Design Thinking for Mapmakers*. ESRU, Press, 2018.
Hancock, James G., Hennie Haworth, Stuart Hill, and Sarah King. *The Art of Map Illustration*. Quarto Publishing, 2018.
Kent, Alexander, and Peter Vujakovic, editors. *The Routledge Handbook of Mapping and Cartography*. Routledge, 2017.
Kraak, Menno-Jan, and Ferjan Ormeling. *Cartography: Visualization of Spatial Data*. 3rd ed., Prentice-Hall, 2010.
Okada, Alexandra, Simon J. Buckingham Shum, and Tony Sherborne, editors. *Knowledge Cartography: Software Tools and Mapping Techniques*. Springer, 2008.
Robinson, Arthur H., et al. *Elements of Cartography*. 6th ed., Wiley, 1995.
Wilford, John N. *The Mapmakers: The Story of the Great Pioneers in Cartography from Antiquity to the Space Age*. Pimlico, 2002.

Marine Conservation

Fields of Study: oceanography; marine ecology; marine biology; marine biogeochemistry

ABSTRACT

Earth's vast oceans are surprisingly fragile ecosystems that are easily damaged, something that has been masked by their very size. Human activities have abused marine ecosystems for centuries as a result. As awareness of the sensitivity of the oceans and of the damage already done increases, efforts and practices to make amends are also increasing.

KEY CONCEPTS

benthos: the marine seafloor region

biota: the flora and fauna of a particular ecological niche or environment

by-catch: the undesired aquatic species that are collected along with the target species in fishing operations

ocean acidification: lowering of the pH of seawater as carbon dioxide and other gases are absorbed from the atmosphere and react with water to form carbonic acid and other acids

overfishing: harvesting species of fish in quantities that exceed the ability of the population of the target species to maintain its numbers

THE ECOSYSTEMS OF LAKES AND OCEANS

Large bodies of water offer a correspondingly large variety of ecological niches and ecosystems, whether freshwater or saline bodies. The variety of ecosystems in a large freshwater body is much simpler than that of the oceans. The movement of water through a large freshwater body is limited to just simple flow-through, with some relatively minor eddy currents in play. For example, there are no structures in freshwater bodies like the coral reefs of the oceans. Submerged tree skeletons and shallow-water stands of bottom growth are at best weak reflections of the extensive coral reefs and towering kelp and thick seagrass beds found in the oceans. Nevertheless, large freshwater bodies have unique ecologies and are more vulnerable to abuse than oceans due to their smaller size.

That vulnerability is masked by the rapid exchange of water that they exhibit as flow-through currents. The flowing water tends to carry contaminants and foreign materials away fairly quickly. This is a problem in the deep-water pools and the eddy currents, which are both affected by the geography of the particular lake basin. Here, currents may not be as fast, and a deep "well" in the lake bed may act as a trap for still, cold water that is effectively an oxygen-depleted dead zone. Evidence for this can be found in the many sunken shipwrecks that can be found in the Great Lakes of Canada and the United States. Those in warmer, shallow water with high levels of dissolved oxygen typically exhibit a great deal of substantial and ongoing decomposition, while those in deep, cold, poorly oxygenated water tend to be pristine, with very little decomposition.

Large freshwater bodies offer a very broad assortment of ecologic environments ranging from shallow, reed-choked marginal marshes and sand bottom flats to steep-sided rock walls reminiscent of fjords. Each different environment has its own unique biota, while also interacting with the general biota of the entire lake.

Ecological environments in the oceans are far more varied in nature than are those of large freshwater bodies, though there are basic similarities. Ocean environments range from shallow marginal salt marshes and sand bottom flats to the abyssal depths of the basaltic ocean floor that is the outermost and thinnest part of Earth's crust. They also include great coral reefs, hydrothermal vent systems, undersea mountains that rival the Himalaya Mountains in size, forests of towering kelp plants and other features not found in freshwater lakes. The undersea environment of the oceans is in many ways like a water-shrouded version of the terrestrial environment of the continents.

Water exchange by ocean currents is also a far more complex phenomenon than the simple gravity-driven flow-through exchange that takes place in freshwater bodies. Ocean circulation is a globe-spanning system driven by changes in temperature and salinity, and hence density, rather than just gravity. This circulation system moves truly massive quantities of ocean water throughout the global ocean of the planet in numerous identified currents. In so doing, the different currents are associated with specific large-scale ecological environments with their own biota. For example, an upwelling cold current brings nutrients to the surface from the ocean floor. When this happens, a great number of fish can be found in near surface waters, but only in that location. Similarly, warm currents promote and maintain coral reefs in the oceans, but only where the water is kept warm enough, but not too warm. Thousands of species unique to the coral reef system will be found there, but nowhere else in the oceans.

View of a shipwreck in the Big Tub Harbour of Georgian Bay. Wrecks in deep, cold, poorly oxygenated water tend to be pristine, with very little decomposition. (Shunrei via iStock)

PRESSURES AND THREATS

It is more than a little shocking to realize that all pressures on and threats to the health of the oceans are anthropogenic in nature. Humans have taken fish and other creatures without regard to the long-term effects of doing so for thousands of years as a sustainable and infinitely renewable food resource. In just the past two hundred years, however, technologies have been developed that have multiplied the annual catch many times over, exceeding the ability of fisheries to maintain their populations of targeted fish species. Fishing quickly became overfishing driven by greed for the extra monetary profits that could be had from broader distribution of the enhanced catches. In that same 200-year expanse of time, the "inexhaustible" cod fisheries of the North Atlantic Ocean have been driven to collapse, several species of great whales have been hunted to extinction, shark populations have been decimated in some areas, and fishing operations have at times been more akin to armed conflict and piracy as top-level fisheries became depleted and the fishers began more desperately to "fish down the food web" in order to maintain themselves. With the removal of top-level predator species—especially sharks, which are a keystone marine species—low-level species such as jellyfish bloom uncontrolled.

Another problem with enhanced fishing capabilities is the by-catch, the nontarget species that are scooped up by the methods of fishers along with the desirable species. The by-catch includes everything from inedible bottom-dwelling species and coral to sea turtles, sharks, dolphins, seals and whale species. While great whales are not taken in the by-catch, they also suffer losses from the fishing industry. Disposable nets that are miles (or kilometers) long get dumped at sea, ostensibly to sink to the seafloor. Instead, they tend to float about at various depths according to their buoyancy. A significant number of great whales are lost annually as they become entangled in the discarded nets, which then impair their ability to swim properly and the whales subsequently drown. Only a very few are saved by rescuers who happen to find them and work to cut away the nets and free the great creatures. Interestingly, most such whales immediately afterward put on a display of breaching that many believe to be an expression of gratitude to their rescuers.

The fishing industry affects ocean health in other ways, as does the shipping industry, simply by its very existence. The problem is noise pollution from the

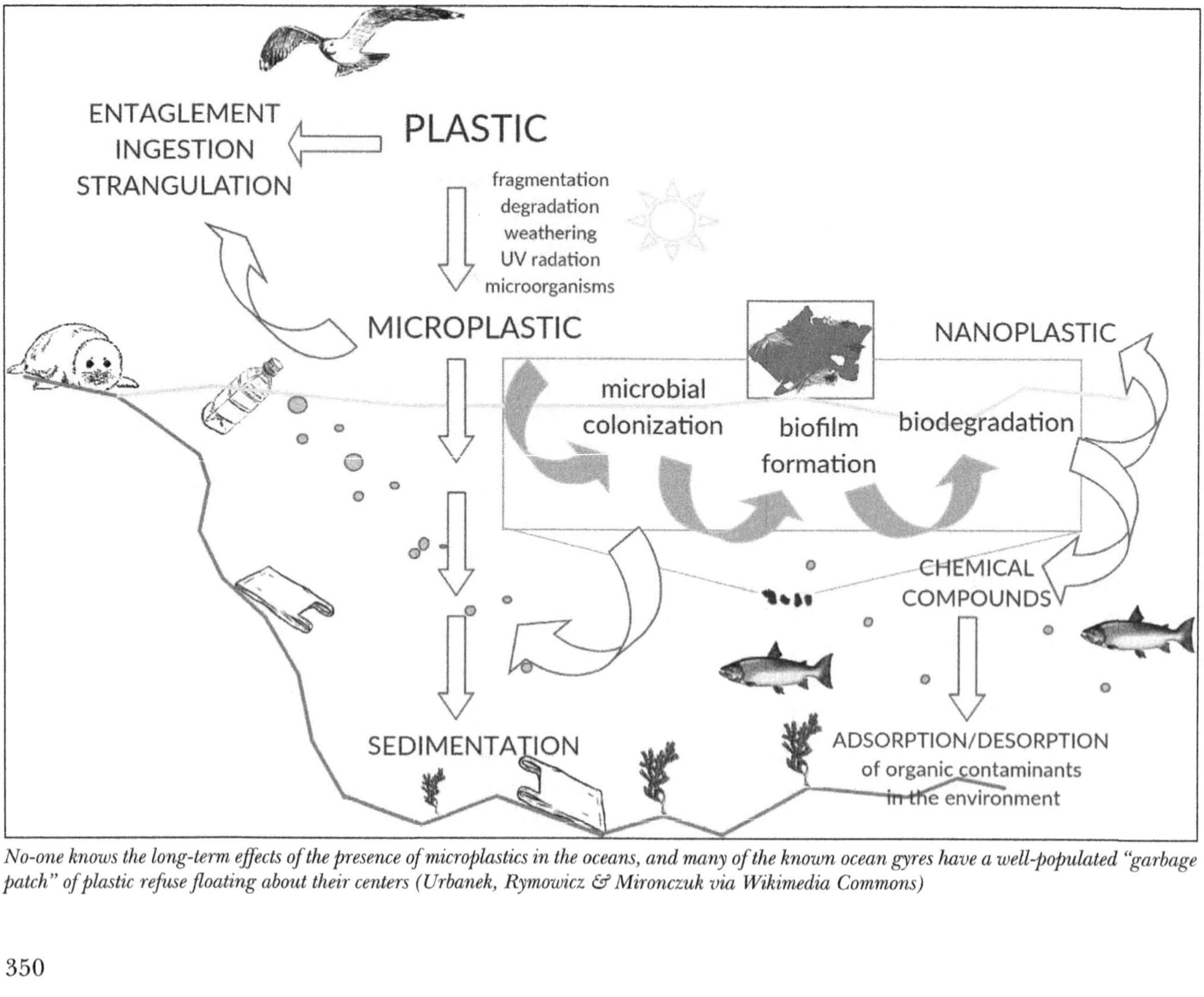

No-one knows the long-term effects of the presence of microplastics in the oceans, and many of the known ocean gyres have a well-populated "garbage patch" of plastic refuse floating about their centers (Urbanek, Rymowicz & Mironczuk via Wikimedia Commons)

powerful engines, propellers and onboard machinery of the ships. Water conducts sound about 700 times better than does air. It has been theorized that the unnatural sounds adversely affect schooling behavior of fish, disrupt echolocation by dolphins and other species, and confuse communication in pods of whales to the extent that they end up beached, where they suffer out of water and many eventually die. Evidence for this, however, is slim as no conclusive studies have been carried out to test the theory.

In addition, container ships carrying stacks of shipping containers traverse the oceans. On occasion, some containers or even the entire ship itself are lost at sea and sink rapidly to the ocean floor. In vulnerable areas, such losses can do severe damage on the benthos, and if the actual cargo in the containers is released, it may wander about the oceans at the whim of the currents for decades. In one such incident, a shipment of about 29,000 rubber bathtub toys was set adrift in the Pacific Ocean. No one knows how many were mistaken for normal food by any number of sea creatures that subsequently died after ingesting the plastic items and developing a fatal blockage.

Plastics floating about in the oceans have proved fatal to many creatures for that very reason, including great whales. The plastics come from offshore dumping and discarded materials that are washed into the oceans from rivers. It is very much just a case of "out of sight, out of mind" as these materials are disposed of in this way. Unfortunately, plastics do not biodegrade in the natural environment. But they can be physically broken down to smaller and smaller bits and pieces, eventually becoming microplastics. No one knows the long-term effects of the presence of microplastics in the oceans, and many of the known ocean gyres have a well-populated "garbage patch" of plastic refuse floating about their centers.

Tourism, a seemingly innocuous human activity, also has negative effects on ocean health. The many huge cruise ships that are used carry thousands of people, every one of whom generates waste. That waste is routinely dumped into the sea as ships are not required to have the facility to store the waste so it can be returned to shore for treatment before being reintroduced to the environment. Quantities of uneaten food waste also makes its way into the seas. Once at their destinations, travelers may go diving or boating, and these activities in coral-bearing regions are known to put stress on the reef ecosystems, sometimes resulting in physical damage to the coral itself.

The oceans also receive inputs of fertilizers, pesticides, herbicides and animal wastes from agriculture, and occasionally even untreated human waste from treatment plants that have been flooded. These pollutants are picked up from the land and carried into rivers by natural precipitation runoff, and the rivers then carry this burden into the oceans. This contributes to the formation of dead zones such as the one that occurs on an annual cycle in the Gulf of Mexico, south of the Mississippi delta.

The human population's seemingly endless desire for oil and natural gas represents another threat to the ocean environment. Large natural reservoirs of crude oil are located on the continental shelves around the world. Obtaining this petroleum requires drilling operations in water that is hundreds of feet (or meters) deep, and the oil that is obtained must then be transported to refining facilities. In both operations, accidents can and do happen that result in millions of liters of crude oil being spilled or spewed into the otherwise healthy environment. The effects of such accidents are not known in their entirety and affect ocean health for many years afterwards. The blowout of the *Deepwater Horizon* well in the Gulf of Mexico, and the wreck of the *Exxon Valdez* in Alaska are probably the most well-known of such disasters.

Perhaps the most insidious threat to the health of the oceans comes from the air, due to the climate change that is understood to be exacerbated by anthropogenic injection of vast quantities of carbon dioxide into the atmosphere from the use of carbon-based fossil fuels over the past 200 years and longer. Carbon dioxide is a so-called greenhouse gas that acts to trap a certain amount of heat energy in the atmosphere. This tends to raise the average temperature of the atmosphere, the process being termed the "greenhouse effect." Without a stable level of greenhouse effect, Earth would be a lifeless ball of ice floating in space. But with too much greenhouse gases, and therefore too much greenhouse effect, the temperature continues to climb. This leads to melting of the planet's ice caps and a corresponding rise in sea level all around the world. The influx of freshwater from the melting ice caps will alter the salinity of the oceans and severely affect the global thermohaline circulation that plays such an important role in maintaining the homeostasis of Earth's climate. As of 2020, the carbon dioxide content of the atmosphere stands at about double what it was when measurements were first taken; now 0.0475 percent (475 parts per million) as opposed to 0.0254 percent (254 ppm), and it continues to increase.

Yet another danger to ocean health that comes from increased carbon dioxide levels is an increase in the acidity of ocean waters. Carbon dioxide (CO_2) that is absorbed at the ocean surface reacts with water (H_2O) to produce carbonic acid (H_2CO_3). As the acidity of the water increases it becomes more destructive of carbonate structures such as coral reefs and the shells of many sea creatures, making it increasingly difficult for them to survive.

POSITIVE ACTIONS

It should go without saying that the first and foremost of positive actions with regard to marine conservation must be to study and understand as completely as possible marine ecosystems. How the marine environment interacts with and directs the atmosphere, and by extension life on Earth, is a very complex system that has taken many millions of years to attain its stable state. However, as with any dynamic system, altering any single factor can perturb the system in unpredictable ways. Only by fully understanding the dynamic systems of the oceans can humans have any hope of maintaining the oceans as a healthy, vibrant and infinitely renewable environment.

Equally important is reducing and eliminating the use of the oceans as a receptacle for mass quantities of terrestrial garbage by ending any and all oceanic dumping activities. Apart from physically collecting every piece of plastic and other anthropogenic litter already in the oceans, this is a process that is of necessity entirely land-based. Only by preventing the trash from getting into the oceans can this step be achieved. Enhanced and complete recycling coupled with the enforced use of recyclable plastics only would go a long way to fixing this situation. Unfortunately, it may not be possible to amend much of the damage already done over decades of dumping mixed garbage directly into the oceans.

There is a school of thought that argues new technological developments can and will solve all problems. It is conceivable that developing technologies specific to purposes of amending ocean pollution problems can be successful, but in present human society it is monetary return on the effort that will be the only determining factor.

In the short term, the most effective means of marine conservation is the establishment of marine sanctuaries and protected zones, with strict enforcement of the rules governing those areas. This necessarily requires an application of additional manpower and resources to police the areas effectively. It also demands timely and just prosecution by the host country of any charges that would result. The hydrosphere of this world is in reality, just a single ecological niche, a single highly varied and wonderfully complex environment, and as such deserves attention and protection by all nations.

Further Reading

Beckman, Daniel W. *Marine Environmental Biology and Conservation.* Jones & Bartlett, 2013.

Campbell, Maria Shauna. *Fisheries, Marine Conservation, Marine Renewable Energy and Displacement: A Fresh Approach.* U of Plymouth P, 2017.

Hoyt, Erich. *Marine Protected Areas for Whales, Dolphins and Porpoises: A World Handbook for Cetacean Habitat Conservation and Planning.* Earthscan, 2011.

Mackelworth, Peter, editor. *Marine Transboundary Conservation and Protected Areas.* Routledge, 2016.

Masud, Muhammad Mehedi. *Conservation of Marine Resources and Sustainable Coastal Development in Malaysia.* Palgrave Macmillan, 2019.

Probert, P. Keith. *Marine Conservation.* Cambridge UP, 2017.

Ray, G. Carleton, and Jerry McCormick-Ray. *Marine Conservation: Science, Policy and Management.* Wiley Blackwell, 2014.

Roff, John, and Mark Zacharias. *Marine Conservation Ecology.* Earthscan, 2011.

MARINE MINING

Fields of Study: geology; oceanography; engineering; marine ecology

ABSTRACT

Marine mining is the mining of minerals and other natural resources from the ocean. Valuable resources in the near-shore zone include sand and gravel for construction, salt, phosphate deposits needed for fertilizers, coral for marine aquariums and jewelry, carbonate sands, and diamonds, gold, tin, and other minerals in placers. Farther out are oil and natural gas deposits, sulfur for industry, and gas hydrates. On the seafloor are nodules and crusts containing strategically important metals such as manganese and cobalt and sulfide deposits around the black smokers in the rift zones, which contain gold, silver, and other useful metals.

KEY CONCEPTS

methane hydrate: a type of compound in which methane molecules have become associated with water molecules to form a solid complex under the influence of cold water temperatures and high pressure on the seafloor

placer: a deposit of sand or gravel containing particles of gold or other valuable minerals that can be washed out readily

Venturi principle: the phenomenon by which a fast-moving stream of a fluid draws more fluid along with it as it flows, as a flow of water under pressure will draw surrounding water along with it.

Venturi tube: tubing used in undersea mining and dredging in which an injected stream of water under pressure causes more water to flow into the open intake end of the tube, so that the tube acts to "suck up" material from the seafloor that is then ejected from the other end of the tube for collection

DEFINITION AND BASIC PRINCIPLES

Marine mining deals with the extraction of economically valuable substances from the ocean. These may be gases, liquids, minerals, rocks, base and precious metals, or ornamental stones such as pearls, diamonds, and coral. Near-shore mining has been done for years, and drilling for oil and gas at sea began in the 1890s, but the mineral deposits on the seafloor have been largely ignored because of the technical and technological difficulties in obtaining them. The substances obtained by marine mining fall into three categories: rocks and minerals found in shallow, near-shore waters; oil and gas deposits found in waters up to thousands of feet deep; and metal-rich deposits found on the seafloor, two miles or more down. Near-shore sand and gravel and the minerals in placers can be scooped or suctioned up by giant dredges, salt can be evaporated in shallow coastal ponds, and pearls and ornamental coral for aquariums and jewelry can be obtained by divers. Oil wells have been drilled in the ocean for more than one hundred years, and during that time, marine oil and gas production has grown so rapidly that by the twenty-first century, 34 percent of the world's oil and 28 percent of the world's natural gas come from below the seafloor. The gas hydrates, manganese nodules, and sulfide deposits may potentially be tapped as well.

BACKGROUND AND HISTORY

The ocean has provided salt for human dietary purposes since ancient times, so evaporation of salt from seawater probably represents one of the earliest examples of marine mining. Another early example was the collecting of minerals and gemstones washed into the sea by rivers to form placers. Shoreline tin was mined in Indonesia in the early 1900s using dredges, and diamonds carried into the sea by the Orange River were mined along the coast of South Africa at about the same time. Oil has been produced on land using dug or drilled wells for hundreds, if not thousands, of years, but the first offshore oil wells were not drilled until the early 1890s, using piers or pilings driven into shallow water.

Not until the 1930s did oil drillers venture more than a mile from shore, and the immense floating platforms used in the twenty-first century were not introduced until the 1960s. The first manganese nodules were discovered on the seafloor by the HMS *Challenger* expedition in the 1870s, but the techniques for mining them and the related sulfide deposits are still being developed.

HOW IT WORKS

Near-Shore Resources. Humans must have found early on that the salt crystals in dried-up lagoons near the ocean were useful as a seasoning and for the curing and preserving of foods. The construction of evaporating ponds was a logical next step, and these are common in countries that have access to the ocean and lack underground salt deposits. The mining of placers uses large, floating dredges that either suck the minerals up in long tubes using the Venturi principle, or scoop them up in a series of circulating buckets. Other valuable substances found in shallow water include phosphate for fertilizer, pearls for jewelry, live coral for aquariums, and pure carbonate sand. Although phosphate deposits are widely present in the ocean, they are not mined because phosphate deposits on land can be tapped at a lower cost. Pearls are primarily grown by "seeding" oysters in pearl farms, which consist of floating wooden rafts or posts driven into the bottom of the sea, though a few "wild" pearls are still brought up by divers. Divers also bring up the live coral so prized by marine aquarium hobbyists, and the highly colored, but dead, coral skeletons used to produce some fashion jewelry. In the Bahamas, shallow-water shoals of pure carbonate sand are mined as a source of lime for industry.

Oil and Natural Gas. To extract oil and natural gas, a fixed or floating platform is built so that a hole may be drilled into the seabed. The first wells used fixed platforms, either built on pilings at the ends of piers or on artificial islands. The first true offshore well in the Gulf of Mexico was spudded in 1937, one mile from shore in the then-intimidating water depth of 14 feet (4 meters). The first well located out of sight of land in the Gulf of Mexico was drilled in 1947; it was 10 miles (16 kilometers) from shore and in 18 ft (5 m) of water. In the 2010s, companies drill wells in the Gulf of Mexico that are 50 mi (80 km) out and in nearly 10,000 ft (3,048 m) of water. For wells in moderate water depths, the drilling platform rests on a concrete structure or a steel tower. Deeper waters necessitate the use of semisubmersible rigs in which the platform rests on underwater pontoons, moored to the bottom by steel cables or kept in place by dynamic positioning.

Related Products. Sulfur is commonly associated with oil deposits, especially when salt domes are present. As the salt moves upward because of its lesser density, it arches up the sediments, and chemical reactions between the salt and the sediments may form a cap rock that is rich in sulfur. The sulfur is mined by drilling into the cap rock, melting the sulfur with high-pressure steam, and bringing it to the surface as a liquid for processing. The methane gas hydrates in the ocean, which contains a large amount of natural gas, and has not been mined yet. In these hydrates, the gas is bonded with water in the form of ice instead of being associated with oil.

Deep-Sea Resources. Technology for mining the manganese nodules and sulfide deposits found on the seafloor is still in the developmental stages. The nodules are lumps about the size of hamburgers and are scattered about. Someday it may be possible to harvest them through methods developed as an entirely undersea operation, since they are found at depths below which methods such as Venturi hoses and bucket series can function. The metal sulfide deposits around the black smokers present a greater challenge. They are welded to the seafloor and would have to be physically broken loose on location.

APPLICATIONS AND PRODUCTS

Sand and Gravel. Sand and gravel for construction purposes are in short supply in many parts of the world. When available deposits on land have been exhausted, the only other supplies are in the ocean. This situation is particularly true for some of the older countries in Europe, such as England, where construction has gone on for hundreds of years and real estate values are so high that the few remaining deposits on land are prohibitively expensive. Deposits of sand and gravel in the ocean have the advantage of being well sorted, with grains all of the same size, so they are suitable for building purposes once the salt has been washed out of them. Sand and gravel are mined along the coast of the United States to rebuild beaches that have lost their sand because of erosion. The sand is pumped directly onto the beach from the offshore zone in a process known as beach nourishment. The cost of nourishing a beach may exceed several million dollars for each mile of beach treated, and a significant problem is that if the sand is fine-grained, it may wash away quickly in storms. In the Bahamas, shallow-water shoals of carbonate sand are mined as a source of lime for making cement and for separating iron from its ore in blast furnaces.

Placers. Gold, tin, and diamonds are the most sought-after substances found in placers, but platinum, iron minerals, and other gemstones may also be found in them. The gold, platinum, and gems are used in the manufacture of jewelry, with the gold also being needed for dentistry, coinage, and monetary purposes. Both gold and platinum have electrical properties that are useful for industry, and platinum is required in the manufacture of catalytic converters for automobiles. Tin was highly prized in ancient times because it was needed for the manufacture of bronze and pewter. In modern times, it is used in solder and for the tin plating of steel to prevent corrosion. The familiar so-called tin can used to package food is simply a steel container with a thin, protective coating of tin.

Salt. Because of its dietary importance and usefulness for food preservation, salt has been valued by humans throughout history. It is also widely used in the chemical industry as a water softener, for snow and ice removal, and by anyone who makes homemade ice cream. However, only 86 percent of the dissolved material in the ocean is table salt. The remainder includes bromine, which is used in medicines, chemistry, and antiknock gasoline; magnesium, an important lightweight metal used in fireworks and flares because it is flammable; potassium, which is used in fertilizers; and calcium sulfate (gypsum), an important component of wallboard.

Oil and Natural Gas. By far the most valuable of the resources obtained from the sea are oil and natural gas. In addition to being used for fuels, such as gasoline, diesel fuel, kerosene, fuel oil for heating homes, and jet fuel, oil is the raw material that is the basis of grease, motor oil and other lubricants, asphalt and tar, and a variety of chemical products, including nylon fibers for stockings. Even the paraffin wax used to seal jars of homemade jelly comes from oil. The natural gas found with oil is used domestically for cooking and heating, in the power industry to generate electricity, and in the production of ammonia for making fertilizer. The methane gas hydrates deposited on the seafloor, which have not yet been exploited commercially, may someday have similar uses.

Sulfur. Although sulfur is not derived from the seafloor in the twenty-first century, it has been in the past. Sulfuric acid, which is widely used in chemistry and industry, is its most important product. Because of its toxicity to bacteria, sulfur is an ingredient in medicines, insecticides, fungicides, and pesticides. Sulfur is also flammable and is used in the manufacture of matches and gunpowder.

Manganese Nodules. Although the manganese nodules and crusts found on the seafloor are not being mined commercially, the United States has an interest in them because of the metals they contain. Three of the metals—iron, needed for making steel; copper, used in wiring; and nickel, important in the manufacture of stainless steel—are already available on land. The other two metals, strategically important in wartime, are not available in the United States and must be imported. They are manganese, used in making armor plate, and cobalt, which is needed for the high-temperature steel used in tools and aircraft engines. Of even greater importance than these are the metallic and semimetallic elements used in cell phones, modern electronics and high-value batteries. The availability of these elements from land-based resources is very limited, and it is the value of these elements that drives the interest in harvesting the nodules from the ocean floor.

Sulfide Deposits. The sulfide deposits found on the seafloor have not been mined yet but are of interest because of their metal content. They contain not only the five metals found in the manganese nodules but also zinc, which is needed for galvanizing and for alloys such as brass; lead, which is used in typical lead-acid car batteries, scuba divers' weight belts, and as shielding for atomic reactors; silver, which is needed for coinage and in industry; and gold, with its monetary and industrial uses.

CAREERS AND COURSEWORK

The only marine mining done in the United States as of 2010 is the dredging of sand and gravel for beach nourishment and oil and gas production in the Gulf of Mexico, along the Southern California coast, and off the north slope of Alaska. Dredging jobs are limited because the work is intermittent and seasonal and takes place largely in resort areas where beach erosion is a problem. Advanced training and professional degrees are not required. A captain is needed to run the dredge, a hydrographic surveyor has to chart the bottom and locate the deposits of sand and gravel, a mechanic has to keep the equipment running, and deckhands are required to do routine tasks. By contrast, a wide variety of jobs are available in the oil and gas industry, both on land and at sea.

A good place for the beginner to start is with an undergraduate, college-level course in geology. This could lead to graduate work in geology and a career in oil and gas exploration, or to a degree in petroleum engineering and a job designing, constructing, and maintaining offshore drilling platforms. The oil industry also offers jobs doing seismic surveying, drilling and logging wells, and producing, refining, and marketing the oil that is produced. Management positions also are available, so courses in business management are also recommended.

SOCIAL CONTEXT AND FUTURE PROSPECTS

As glaciers and ice sheets melt because of global warming, the water added to the sea by this melting will result in rising sea levels. This, in turn, will increase the amount of beach erosion around the world and consequently the need to nourish beaches in resort areas. Jobs in the dredging industry should continue to be available as a result. Jobs in the oil and gas industry will continue to be available as well. More than three-quarters of world oil production goes into powering cars, trucks, ships, and planes, so the demand for oil is likely to grow as more and more people in the developing nations are able to afford to buy cars.

One problem that may lie ahead in marine mining involves the manganese nodules and sulfide crusts found on the seafloor. The developed nations are already investigating ways to mine these deposits. They are the only nations with the technological expertise to bring up these nodules and crusts. However, the law of the sea states that the profits from mining minerals in the area in which the nodules and crusts are found must be shared with the developing nations, and the developed nations may find it hard to agree to this condition.

—Donald W. Lovejoy

Further Reading

Drew, Lisa W. "The Promise and Peril of Seafloor Mining: Can Minerals Be Extracted from the Seafloor without Environmental Impact?" *Oceanus*, vol. 47, no. 3, 2009, pp. 8–14.

Drew, Lisa W. "Who Regulates Mining on the Seafloor?" *Oceanus*, vol. 47, no. 3, 2009, p. 15.

Fang, Huacan, and Menglan Duan. *Offshore Operation Facilities: Equipment and Procedures*. Gulf Professional Publishing, 2014.

Garrison, Tom. *Oceanography: An Invitation to Marine Science*. Thomson, 2007.

McLachlan, Anton, and Alec Brown. *The Ecology of Sandy Shores*. Elsevier, 2006.

Rice, Tony. *Deep Ocean*. Smithsonian Institution, 2009.

Sharma, Rahul, editor. *Deep Sea Mining: Resource Potential, Technical and Environmental Considerations*. Springer, 2017.

———. *Environmental Issues of Deep Sea Mining: Impacts, Consequences and Policy Perspectives*. Springer, 2019.

Welland, Michael. *Sand: The Never-Ending Story*. U of California P, 2009.

MARINE PALEONTOLOGY

Fields of Study: geology; oceanography; marine ecology; marine biology

ABSTRACT

Paleontology is an interdisciplinary field that is concerned with the study of the record of life through time and the application of that information to solve scientific problems. Fossils are particularly used in the relative dating of rocks, which is important information for the minerals and petroleum industries. However, paleontology also provides valuable data toward an understanding of past life and therefore an understanding of extinction events and the effects of climate change. Modern technology is increasingly used to help solve problems, including computed tomography (CT) scanning to reconstruct the anatomy of extinct organisms.

KEY CONCEPTS

biostratigraphy: the assignment of geological period and age to geologic strata through identification of their fossil content

cladistics: also called "phylogenetic analysis," a method that recognizes and orders characters in an objective fashion to show the most likely evolution of organisms through time

disarticulation: the separation of skeletal parts from each other by various means, typically due to predation and environmental activity such as flooding and landslides

taphonomy: the science of the processes that lead to fossilization

thermal maturation: essentially, cooking; the acquisition of certain characteristics according to the time and temperature to which something has been exposed

DEFINITION AND BASIC PRINCIPLES

Paleontology is the study of fossils, or once-living organisms preserved within sediment. The term "paleontology" is derived from the Greek *palaios* for "ancient," *ontos* for "being," and *logos*, for "study." Therefore, paleontology does not encompass the allied field of archaeology and overlaps only slightly with anthropology, both of which focus on humans. It is, however, an adjunct branch of oceanography and marine science as the field of paleoceanography. Traditionally, the study of fossils has been aimed at reconstructing the organisms and analyzing them to develop an understanding of their evolutionary relationships. This has been enhanced by the advent of cladistics, a computer-based analytical method that greatly increases the speed and rigor of such studies. In addition, the careful tabulation of fossils over several hundred years of study has enabled the development of a relative geological time scale that is the basic tool in all study of Earth's crust. Paleontology is an important industrial tool in that it helps extractive industries such as the petroleum and minerals industries understand where to find oil and minerals.

Increasingly, paleontology is being viewed as a potentially valuable predictive tool. The results of earlier climate changes are preserved in the fossil record and may help scientists understand the effects of global warming or other future climate changes.

BACKGROUND AND HISTORY

Fossils were probably picked up and admired long before they were studied, but the Ancient Greeks recognized that some of these objects represented marine organisms and could be used to determine where oceans had once been. In medieval times, fossils were often viewed as magical objects. For example, fossil shark teeth were called *glossopetrae*, or "tongue stones," and were thought to provide protection from snakebite. They were viewed as structures that had "grown" within the rocks rather than the remains of actual organisms. In the seventeenth century, Nicolaus Steno (1638–1686) demonstrated that *glossopetrae* were shark teeth by dissecting a modern shark and showing the similarity of its teeth to the fossils. At about the same time, Robert Hooke (1635–1703) used early microscope technology to show that fossils and modern wood and shells had the same internal structure.

Once fossils were accepted as representing past life, they could be studied as such. In some sequences, scientists found clear evidence of mass extinction events in which whole faunas were replaced over short periods of time. In the early nineteenth century, work by French naturalist Georges Cuvier (1769–1832) in the Paris basin showed that numerous mammals that lived when those rocks were laid down no longer existed. Studies of this type showed that long periods of time had been necessary for the evolution and extinction of groups of fossil organisms. At about the same time, English canal surveyor William Smith (1769–1839) determined that fossils could be used as time markers, representing the period they were present on the surface of Earth, and the theory became known as stratigraphy. His findings were developed into the geological time scale, which is still being refined but provides a basis for dating rocks in a relative sense.

HOW IT WORKS

Dating. Understanding the spatial relationships of rocks developed from the need to predict where materials of economic value could be found. Initially, distinct rock beds were used to correlate between places, but this method was imprecise over long distances or in areas where the rocks were folded and faulted. Rock units could not be dated effectively until Smith realized that the geological succession corresponded to a regular, nonrepeating succession of fossils. Fundamental to this process of biostratigraphy is the use of zone fossils that represent a period of time equal to the duration of the species before it became extinct and that are widespread enough to enable useful correlations to be made. Increasingly, microscopic fossils (such as the marine, planktonic, and unicellular foraminifera) are used as they have the advantage of being very abundant in sediments and of having short time ranges, which allows for accurate dating.

Taphonomy. The quality of the fossil record must be understood for it to be useful in applications such as dating or understanding mass extinctions or biodiversity loss. Taphonomy, which is the study of the processes that affect organisms from death to burial in sediment, is integral to such an understanding as it allows an evaluation of the extent to which the fossil record mirrors the original faunal situation. Most organisms, especially marine and oceanic organisms with cartilaginous skeletons, do not become fossils, and those that do will first undergo decay of the soft tissues, then transport and breakage of the hard tissues, and finally their burial and modification. For example, soft-bodied organisms can make up 50 to 60 percent of the organisms in present-day marine settings, and with very few exceptions, none of those organisms will be fossilized. Experimentation on modern forms shows that the presence of oxygen is vital to the initial breakdown of soft tissues, and its absence can lead to conditions of special preservation that result in unusually well-preserved faunas such as the Burgess Shale fauna (a Middle Cambrian age site in Canada) and the recently discovered, almost pristine *borealopelta* found in its entirety rather than as a skeleton in the Alberta Oil Sands (Canada). The breakage and disarticulation of organisms can be simulated in the laboratory or observed in nature. These studies show that organisms go through a number of stages as they progress toward fossilization. Factors that affect whether they are preserved include the presence of hard parts, how common the organisms are, and their life environment, as organisms living in areas of sedimentation (particularly aquatic environments) have a better chance of preservation. Processes within the rock, such as water filtering through, may result in replacement of the fossil or its total removal. Of course, before the information contained in the fossils can be used, the fossils must be found by a paleontologist and studied.

Taxonomy and Phylogeny. An understanding of the relationships of organisms enables scientists to appreciate and evaluate extinction events and the effects of climatic change, as well as to use fossils in dating. Naturalist Charles Darwin's views on natural selection led to an understanding not only of the evolutionary processes that result in the development of new species but also of phylogeny, the pattern of relationships of organisms, often shown as a branching diagram.

Phylogeny has been greatly aided by the method termed "cladistics," or phylogenetic analysis, developed by the German entomologist Willi Hennig in the mid-1900s. Hennig's method recognizes and orders characters in an objective fashion to show the most likely evolution of organisms through time. As it involves the processing of large amounts of data, this system lends itself well to the use of computer programs such as PAUP (Phylogenetic Analysis Using Parsimony). Results are presented as regularly branching diagrams, or cladograms, that show the relative sequence of speciation events.

APPLICATIONS AND PRODUCTS

Geological Dating. The ability of geological industries to accurately locate the materials they seek depends largely on an understanding of the structure of Earth's crust. It is important initially to realize that rocks are normally laid down horizontally and that they become younger toward the top of any sequence. Detailed understanding, however, requires the ability to relate rocks of the same time period to one another, and this was not possible until fossils were recognized as providing unique temporal markers. Once this was understood, the sequences of rocks that had been recognized in discrete outcrops could be related to one another. During the first half of the nineteenth century, most of the relative time scale was assembled, showing the age relationships of rocks and detailing the sequence of fossils that could be used. The oldest fossil-bearing rocks were described as from the Paleozoic (ancient life) era, followed by Mesozoic (middle life) and Cenozoic (new life) eras. Within these eras, the rocks were divided into periods, such as the Cambrian, Ordovician, and Silurian—which are part of the early Paleozoic era—and the Cretaceous period, the last in the Mesozoic era.

As the study of dating became more laboratory based—and in response to the requirements of the oil industry, whose samples were typically drilling chips, which were too small to contain recognizable fossils—paleontologists moved toward using microfossils. These can be the remains of very small organisms or small parts of larger organisms, as long as they are common and widespread and show evidence of having evolved rapidly.

The field of applied micropaleontology started in the late 1800s in Poland when paleontologist Józef Grzybowski described the foraminifera from the Carpathians in a series of monographs and then applied his findings to foraminifera from the Galician oil fields. He was soon able to demonstrate that it was possible to correlate subsurface strata in wells drilled for petroleum exploration using these fossils. The methods pioneered by Grzybowski continue to be used, and the study of microorganisms has been very much centered on the requirements of the petroleum industry.

Thermal Maturation. Organic microfossils such as spores and pollen can be used to indicate the temperature reached by the rocks (thermal maturity) that contain them because temperature increases as the rock is buried deeper. Organic microfossils are progressively altered by the loss of hydrogen and oxygen, and the resultant changes to physical properties such as color, reflectivity, and fluorescence can then be measured.

Reflectivity studies using vitrinite, an organic component, were initially carried out on coal to determine its rank, or thermal maturity. These studies were then applied to hydrocarbon generation, as hydrocarbons such as oil and gas are generated over time by the action of heat on fossil organic material. The reflectivity of vitrinite in the hydrocarbon source rocks reveals maturity and the likelihood of the presence of oil and gas in the sediments.

Spores and pollen change color from pale yellow to orange to dark brown as the temperature increases, and these changes can be used to show the level of organic maturity. A similar change is seen in conodonts, tiny jaw elements from primitive craniates that were abundant in the Paleozoic era, and an eight-point scale has been developed detailing a color change from pale yellow through black to colorless or clear. Below 140 degrees Fahrenheit (60 degrees Celsius), organic material in rock is converted to kerogens and bitumen. Rock generates and expels most of its oil at temperatures between 140 and 320 degrees Fahrenheit (60 and 160 degrees Celsius), and above 320 degrees Fahrenheit (160 degrees Celsius), natural gas is formed. It is important for industry to understand the temperature level the rock has reached, and the application of thermal indices based on color or reflectivity provides that information.

Environment and Climate Change. Organisms are intimately connected to environments and therefore can be used to deduce past environments. This is particularly true of plants, which, unlike marine organisms, are directly exposed to the atmosphere. The size of leaves is related to temperature as well as to humidity and light levels; leaves decrease in size as the temperature or humidity drops. Similarly, the ratio of serrated leaf margins to smooth margins varies according to temperature, and a 3 percent change in this ratio is equivalent to a change of 1 Fahrenheit degree (or 1 Celsius degree) in the mean annual temperature. In addition, the numbers of stomata (small openings on the surface through which plants lose oxygen and water and gain carbon dioxide) vary according to the carbon dioxide level. When carbon dioxide levels are high, plants need only a small number of stomata, but when the levels are low, plants will commonly have large numbers of stomata to enable them to get all the carbon dioxide they need.

Because carbon dioxide is a greenhouse gas and traps radiation from the sun, the higher the atmospheric carbon dioxide levels, the higher the temperature. Therefore, a correlation can be made between high global temperatures and low numbers of stomata on fossil leaves. If this technique is applied to the end of the Triassic period extinction, the large drop in the number of stomata on leaves that occurred at this time indicates an increase in atmospheric carbon dioxide and a consequent increase in temperature. Calculations suggest that the temperature would have increased by 41 degrees Fahrenheit (5 degrees Celsius) globally and possibly up to 61 degrees Fahrenheit (16 degrees Celsius) locally. In the twenty-first century, carbon dioxide levels are rising because of the burning of fossil fuels, and these studies of carbon dioxide in past eras can provide information that will help scientists understand what changes in biodiversity are likely to occur because of higher carbon dioxide levels.

To obtain information about the temperature of ancient ocean waters, it is possible to use oxygen isotopes extracted from calcareous marine fossils. The ratios of oxygen-16 and oxygen-18 vary, with oxygen-18 increasing as the temperature decreases. Using isotope ratios from foraminifera found in deep-sea sediments, it has been possible to show that over the last 100 million years, there has been a steady increase in the oxygen-18 level, indicating a decrease in ocean temperatures of up to 59 degrees Fahrenheit (15 degrees Celsius) over that period.

Evolution and Extinction. Studies on modern taxa have provided much information on how change in organisms takes place, but the fossil record is still the best evidence of evolution through time and also of its correlative, periods of extinction. Both of these areas are of interest because of their relevance to the changes in biodiversity that are taking place in the twenty-first century. To detail changes through time, it is necessary to develop a database of information on taxa and their longevity. That database is constantly gaining new entries, and therefore never complete; however, it has reached the point where information can be derived from it using statistical techniques aided by computers. Extinctions are always occurring, of course, but these studies show when the level of extinction exceeds that of the development of new species by a significant amount, a mass extinction has taken place. Although numerous extinction events have been recorded, there are seven major ones. The first occurred in the Middle Vendian period about 650 million years ago, followed by events at the end of the Cambrian, Ordovician, Devonian, Permian, Triassic, and Cretaceous periods, of which the end-Permian event was by far the most severe. Analysis of extinction rates of shallow marine organisms at the family level by University of Chicago paleontologists David M. Raup and Jack Sepkoski showed a 26-million-year periodicity to extinction events through the Mesozoic and Ceonozoic eras that was interpreted to indicate an extraterrestrial cause for the extinction events. Corroborative evidence is generally lacking for all but the Late Cretaceous event, however, leaving uncertainty as to the underlying cause of the periodicity.

CAREERS AND COURSEWORK

Paleontologists may be trained initially as biologists or as geologists, but in either case, a solid foundation of basic science courses is required. For jobs in industry, a master's degree or a doctorate is required, and the field of study would normally have to be in some aspect of micropaleontology, as that is the area of interest to the petroleum industry. However, as positions in the petroleum industry decreased, universities eliminated or reduced the size of their training programs. In the mid-1980s, more than thirty universities in North America had programs in stratigraphic palynology (the use of spores in dating), but by the late 1990s, only two universities had such programs. Partnerships between industry and university programs are being touted as the way to maintain a flow of adequately qualified paleontologists.

Paleontologists are employed by universities and museums and by the US Geological Survey (USGS). However, the number of university paleontologists in North America has not grown since the 1980s, and most of them are approaching retirement age. The trend is for these paleontologists to be replaced by faculty in other fields, particularly environmental geology, which became the top geology degree program in the 1980s and 1990s. The USGS has actually reduced the number of its paleontology staff since the 1990s, and although museums have consistently responded to the popular interest in dinosaurs, this has not resulted in a significant increase in jobs for paleontologists.

SOCIAL CONTEXT AND FUTURE PROSPECTS

In 2013, approximately 81 percent of world energy consumption came from fossil fuels, which includes oil, coal, and natural gas, according to the World Bank; this reflected a generally steady trend beginning around 1990. Although there is a growing movement toward greater use of renewable resources, it will be many years before renewable resources provide a majority of energy use. Therefore, paleontologists will still be needed to provide some of the basic data required in exploration to find fossil fuels. However, the numbers of paleontologists needed are likely to decline rather than to rise. Similarly, the need for paleontologists in academia and governmental agencies is declining.

One area in which paleontology could well have an effect in the future is in environmental studies and the study of climate change. As more emphasis is placed on the protection of natural resources, an understanding of past changes and an ability to monitor present-day changes using microorganisms will become increasingly important.

—*David Elliott*

Further Reading

Benton, Michael J. *Vertebrate Palaeontology*. 4th ed., Wiley, 2015.

Benton, Michael J., and David A. T. Harper. *Introduction to Paleobiology and the Fossil Record*. Wiley, 2009.

Foote, Michael, and Arnold L. Miller. *Principles of Paleontology*. 3rd ed., Freeman, 2007.

Haq, Bilal U., and Anne Boersma, editors. *Introduction to Marine Micropaleontology*. Elsevier, 1998.

Hu, Xiumian, Chengshan Wang, Robert W. Scott, Michael Wagreich, and Luba Jansa, editors. *Cretaceous Oceanic Red Beds: Stratigraphy, Composition, Origins and Paleoceanographic and Paleoclimatic Significance*. SEPM, 2009.

Lipps, Jere H. "What, if Anything, Is Micropaleontology?" *Paleobiology*, vol. 7, no. 2, 1981, pp. 167–99.

Martin, James E., and David C. Parris, editors. *The Geology and Paleontology of the Late Cretaceous Marine Deposits of the Dakotas*. Geological Society of America, 2009.

Prothero, Donald R. *Bringing Fossils to Life: An Introduction to Paleobiology*. 3rd ed., Columbia UP, 2013.

Tyler, Carrie L., and Chris L. Schneider, editors. *Marine Conservation Paleobiology*. Springer, 2018.

MEDITERRANEAN SEA ECOSYSTEM

Fields of Study: oceanography; marine geology; marine ecology; marine geology

ABSTRACT

The name Mediterranean originated in Roman times, when the sea was referred to as Mediterraneus, meaning "center of the world." The Mediterranean Sea is subdivided into various smaller lobes; these include: the Strait of Gibraltar, Alboran Sea, Balearic Sea, Ligurian Sea, Tyrrhenian Sea, Ionian Sea, Adriatic Sea, and the Aegean Sea. Smaller water-body divisions exist throughout, in the form of gulfs, bays, straits, and channels.

KEY CONCEPTS

chemosynthetic communities: communities of organisms that derive energy from chemosynthesis rather than from photosynthesis

endemism: the occurrence of endemic species unique to a particular environment or ecosystem

foundation species: species that the major food web of a particular ecosystem is dependent upon for its maintenance

hydrothermal vents: seafloor structures that emit volcanically superheated mineral-laden water, found at mid-seafloor ridges where two tectonic plates are being slowly driven apart

seamount: the mesa-like subsurface remnant of an undersea volcanic mountain

PLACE IN THE WORLD

The Mediterranean Sea is bordered by the three continents of Europe, Asia, and Africa, with twenty-one countries along the coastline. The sea can count among its national borders the countries of Spain, France, Monaco, Italy, Malta, Slovenia, Croatia, Bosnia, Montenegro, Albania, Greece, Turkey, Cyprus, Syria, Lebanon, Israel, Egypt, Libya, Tunisia, Algeria, and Morocco. Water boundaries for the Mediterranean include the Atlantic Ocean via the Strait of Gibraltar at the westernmost point, the Sea of Marmara by the Dardanelles and beyond it the Black Sea, by the Bosphorus Strait to the east. The Sea of Marmara is often considered to be part of the Mediterranean Sea, whereas the Black Sea generally is not. The 101-miles (163-kilometer) human-made Suez Canal in the southeast connects the Mediterranean to the Red Sea. Because of its central location, the Mediterranean Sea has become one of the most studied water bodies in the world.

BIODIVERSITY

This semi-enclosed sea is rich in islands and underwater beds. It is a critical area for wintering, reproduction, and migration. Although it covers less than 0.8 percent of the world's ocean area, this biome is one of the major reservoirs of marine and coastal biodiversity on the planet, with a great number of endemic (found nowhere else) species, as well as 7 percent of the world's marine fauna species and 18 percent of its marine flora represented here. The Mediterranean is a true crossroads of marine biodiversity.

The marine fauna and flora of the Mediterranean are derived from several biogeographical categories. More than 50 percent of Mediterranean species have their origins in the Atlantic Ocean; 17 percent originated in the Red Sea; and 4 percent are relict species, which are witnesses from earlier geologic periods, when the Mediterranean was subjected to a tropical climate. Recent estimates of Mediterranean marine species, taken from compilations of former biological studies, show that approximately 10,000 to 12,000 species are present in its waters. The presence of such a large number of distinct species qualifies the region as a biodiversity hot spot. The rate of endemism here is greater in the comparatively small Mediterranean than in the entirety of the Atlantic Ocean. About 19 percent of endemic Mediterranean species, however, are threatened with extinction; 7 percent are considered to be vulnerable, 7 percent are endangered, and 5 percent are critically endangered.

The marine biota of the Mediterranean Sea is mainly derived from origins in the Atlantic Ocean as a result of a flood more than 5 million years ago that introduced Atlantic biota to the Mediterranean through the Strait of Gibraltar. The marine life introduced to the Mediterranean Sea from the Atlantic Ocean has had to adapt to different climactic conditions. The North Atlantic was then and is now considerably colder and more nutrient-rich than the Mediterranean. Certain species form the foundations of the dense life in this area, and are consequently known as foundation species; they inhabit a multitude of habitats here.

HABITATS AND BIOTA

The Mediterranean Sea's diverse habitats together complete the landscapes and biological ranges of the Mediterranean coastal areas. Submarine meadows of endemic Neptune grass (*Posidonia oceanica*) and coral beds are prime examples. Many other sensitive habitats are found here, including deep-sea coral communities, underwater caves, dune areas, coastal forests, lagoons, and wetlands.

The Mediterranean Sea also has a vast array of deep-sea habitats, including hydrothermal vents, seamounts, and deep-sea coral reefs. Underwater canyons are of major importance for many species as places for reproduction and feeding for fish and marine mammals such as Risso's dolphin (*Grampus griseus*) and the sperm whale (*Physeter macrocephalus*). Such habitats also represent a remarkable reservoir of endemism among jellyfish and annelid worms. Chemosynthetic communities harboring essential microorganisms—hydrothermal vents, deep hypersaline habitats, and cold-water corals—are of great ecological value, but are threatened by deep-sea trawling by humans and by the direct seawater temperature-elevating effects of global warming.

The flora of the Mediterranean Sea has a high incidence of aquatic plant endemism, ranging as high as about 20 percent of the aquatic plants. More than 1,000 species of flora exist in the Mediterranean in all.

The Alboran Sea region of the Mediterranean is the most important feeding grounds in Europe for loggerhead sea turtle. In addition, the Alboran Sea hosts important commercial fisheries, including those for sardines and swordfish.

The most commonly seen marine mammals here include the fin whale (*Balaenoptera physalus*), sperm whale, striped dolphin (*Stenella coeruleoalba*), Risso's dolphin, long-finned pilot whale (*Globicephala melas*), bottlenose dolphin (*Tursiops truncates*), common dolphin (*Delphinus delphis*), and Cuvier's beaked whale (*Ziphius cavirostris*). Other, less-prevalent cetaceans here include: the minke whale (*Balaenoptera acutorostrata*), orca (killer whale) (*Orcinus orca*), false killer whale (*Pseudorca crassidens*), rough-toothed dolphin (*Steno bredanesis*), and humpback whale (*Megaptera novaeanglia*). There is also a small population of harbor porpoises (*Phocoena phocoena*) that range into the Black Sea.

The Mediterranean monk seal (*Monachus monachus*) is the only pinniped to be found within the Mediterranean Sea. It is now very rare and listed as an endangered species, with the only known colonies in the Alboran Basin and the Aegean Sea. The Alboran Sea, with the largest population of bottlenose dolphins in the western Mediterranean, is also home to the last population of the harbor porpoise in the Mediterranean.

ENVIRONMENTAL THREATS

Some of the greatest ecological concerns for the marine biodiversity of this area are human population growth that has come in tandem with habitat destruction, increased pollution, invasive species, and climate change. The coasts are under constant and growing human pressure caused by the activities of 150 million residents and the arrival of 200 million tourists every year, the consequences of which have for decades been nearly uncontrollable.

Unchecked urbanization and the overexploitation of resources, combined with the introduction of nonnative species; increased maritime transport; and increased runoff of sediments, nutrients, and heavy toxics have created devastating environmental consequences. Degradation of essential biodiversity, scarcity of the most sensitive species, and destruction of unique and irreplaceable coastal habitats, such as Port-Cros National Park and the Zembra Archipelago, are just a few of the most blatant effects of human activity across the region.

Worldwide, nonnative introduced species are of great ecological concern to regional habitats. Invasive species have become a major problem for the Mediterranean ecosystem in particular, and have endangered many local and endemic species. Analyzed populations of exotic species present in the Mediterranean Sea show that more than 70 percent of the nonindigenous decapods and about 63 percent of the exotic fish occurring in the Mediterranean are of Indo-Pacific origin, introduced into the Mediterranean through the Suez Canal. This pathway through the Red Sea is complicated by the differing salinity, nutrient, and climate regimes between the two—which has often allowed the somewhat hardier and more adaptable Red Sea species to colonize the Eastern Mediterranean basins, crowding out native taxa.

When the Suez Canal was initially constructed, the natural hypersaline condition of the Red Sea at first acted as a barrier, blocking the migration of Red Sea species into the Mediterranean for decades. Over time, however, the salinity in the channel roughly equalized with that of the Red Sea, opening the doors to invasion and colonization of the eastern area of the Mediterranean by species adapted to the lower-nutrient, higher-salinity, warmer-temperature conditions of the Red Sea, Arabian Sea, and Indian Ocean.

—Alexandra M. Avila

Further Reading

CIESM (International Commission for Scientific Exploration of the Mediterranean Sea). "The Mediterranean Science Commission." www.ciesm.org.

Fuerst-Bjelis, Borna, editor. *Mediterranean Identities: Environment, Society, Culture.* IntechOpen, 2017.

Hughes, Terrence B. *Mediterranean Sea: Ecosystems, Economic Importance and Environmental Threats.* NOVA Publishers, 2013.

Migon, Christophe, Paul Nival, and Antoine Sciandra, editors. *The Mediterranean Sea in the Era of Global Change 2: 30 Years of Interdisciplinary Study of the Ligurian Sea.* Wiley, 2020.

Pinet, Paul R. *Invitation to Oceanography.* Jones & Barlett Learning Publishers, 2009.

Stambler, Noga. *Life in the Mediterranean Sea: A Look at Habitat Changes.* NOVA Science, 2012.

Würtz, Maurizio. *Meditarranean Pelagic Habitat: Oceanographic and Biological Processes: An Overview.* IUCN, 2010.

MID-PACIFIC GYRE "GARBAGE PATCH"

Fields of Study: oceanography; marine ecology; environmental science

ABSTRACT

One of the five vast oceanic gyres, or roughly circular current systems, the Mid-Pacific Gyre is formed by four major currents. Together, they rotate through the entire North Pacific Ocean in a generally clockwise motion. The currents—North Equatorial, Kuroshio, North Pacific, and California—stretch for thousands of miles (or kilometers), influenced by and influencing air and water temperature, humidity, air pressure, and the biota that inhabit the Pacific, its islands, and continental shorelines.

KEY CONCEPTS

boundary currents: currents that border the eastern and western ranges of a gyre

gyre: a general, slow circulation of water that spans entire ocean ranges, formed by the meeting of different ocean currents; generally, rotation is clockwise in the northern hemisphere and counterclockwise in the southern hemisphere

microplastics: small bits and pieces of plastics, generally as microbeads from commercial products that have been disposed of down municipal drains, and as fragments produced as larger pieces are ground down against rocks by the action of waves and moving water

OCEANIC GYRES

There are five major ocean-wide gyres—the North Atlantic, South Atlantic, North Pacific, South Pacific, and Indian Ocean gyres. Each is flanked by a strong and narrow "western boundary current," and a weak and broad "eastern boundary current."

The California Current moves southerly toward Central America to meet the North Equatorial Current on the Mexican Coast before heading westerly to Asia as the North Equatorial Current. From these equatorial waters and along the coast of Japan, the system picks up warmth as it becomes the Kuroshio Current. This segment flows northeastward to confront the cold subarctic Oyashio current; their merger fuels the warm North Pacific Current, which flows easterly to its encounter with North America, where it splits into the northbound Alaska Current and the California Current, completing the gyre.

With few major island chains or continental masses, and with generally low winds, the surface water of the gyre tends to push floating matter toward its low-energy center. In recent decades, this trend has precipitated a phenomenon that has been variously dubbed the Trash Vortex, the Great Pacific Garbage Patch, the Eastern Garbage Patch, and other dubious monikers.

A SEA OF PLASTIC

This vast extent of mostly human-generated debris is composed of up to 90 percent polymer or plastic materials. Generally, the mundane products of consumer society, the leading recorded items include plastic bottles, cups, bags, and packing materials. Some estimates are that up to 11 million tons (10 million tonnes) of plastic materials enter the world's oceans each year; perhaps one-quarter to one-third of this flotsam finds its way ultimately to this vortex. While much of the material degrades into smaller particles, or microplastics, virtually none of it is naturally recyclable into the environment. Instead of decomposing, then, it remains a threat for marine organisms of all sizes and scales from tiny bottom-feeding creatures to birds and whales.

Although this diffuse Pacific Ocean garbage patch was first observed in the 1970s, it came to wide public attention in 1997 when Captain Charles Moore, founder of the Algalita Marine Research Foundation of Long Beach, California, observed the mass and began to take measurements. In many samples scooped out of the ocean, Moore discovered that there were six times more plastic in the water than plankton.

AQUATIC DANGER

The garbage gyre presents a dire threat—beyond the disturbing appearance of floating trash such as fishnets, clothing, bottles, plastic bags, and cigarette lighters. Believing that the larger items are prey or plants, many seabirds and sea turtles eat the plastic—which in many cases has absorbed toxins during its manufacture, its intended use, and/or after disposal. Digestive processes can release many such toxins. The chips and pieces of plastic also cause digestive, pulmonary, and/or respiratory blockage that eventually kills the creature that ingests it. Autopsies on dead whales, for example, have revealed masses of more than 100 pounds (45 kilograms) of ingested plastic blocking the gut. Similarly, carcasses of sea turtles that eat on jellyfish have revealed blockages composed of quantities of clear plastic bags that had been mistaken for jellyfish as they floated in the water.

In this way, dead fish, birds, and marine mammals also become caught up in the vortex. Scientists have estimated that around the world, more than 1 million seabirds and in excess of 100,000 mammals and sea turtles die each year from ingesting plastic that floats out to sea. Jellyfish, too, may eat the poisoned debris; they in turn are eaten by fish. When fish containing the concentrated and released toxins are caught by commercial fishers, those fish can then taint the human food chain.

In the summer of 2010, a team of scientists from Woods Hole, Massachusetts, recovered 20,000 bits of floating plastic per 1 square mile (2.6 square kilome-

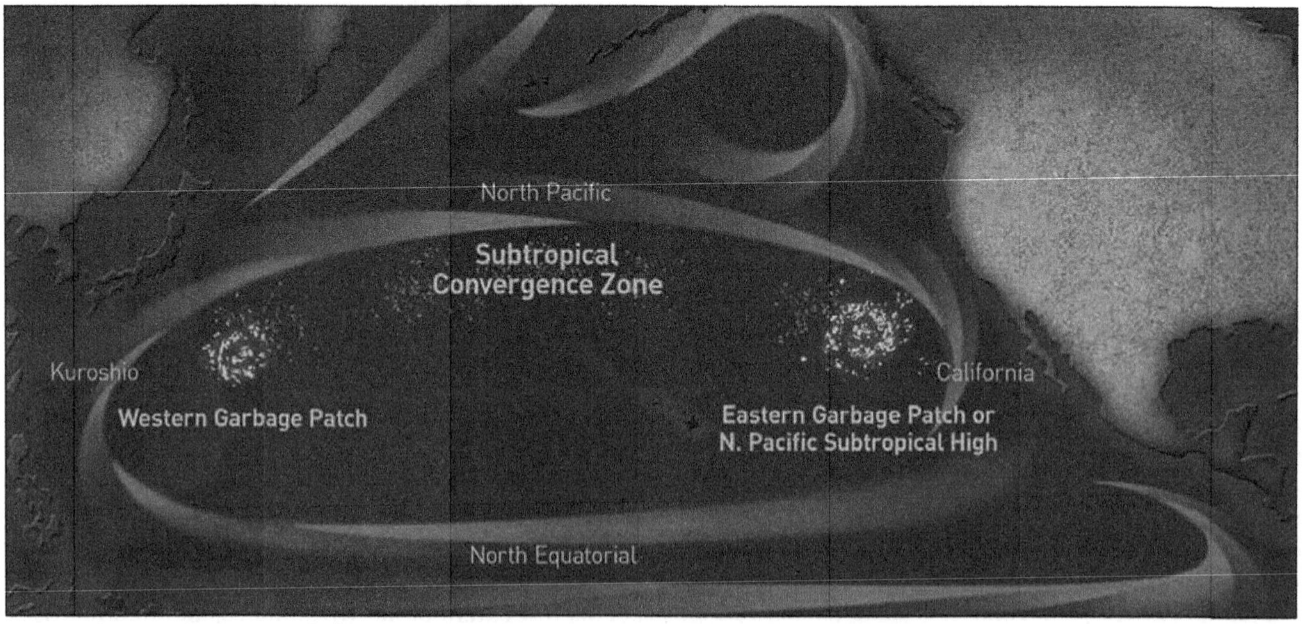

Diagram depicting the Great Pacific "Garbage Patch" (National Oceanic and Atmospheric Administration)

The garbage gyre presents a dire threat to the creatures that ingest its chips and pieces of plastic (RomoloTavani via iStock)

ters) from the Great Garbage Patch, giving some idea of the scope and intensity of the problem.

While plastics and polymers are the chief component of the Great Garbage Patch, this is by no means the only problem. Commercial fishers who use 20-mi (32-km) castaway fishnets snare everything in their nets from dolphins to sea turtles, which after being discarded as worthless by-catch, then in turn become debris heading toward the patch. There have been several incidents of whales becoming entangled in such discarded netting and drowning as it interfered with their ability to control swimming.

The problem of floating garbage also exists in the North Atlantic Ocean, to a somewhat lesser extent. Between 1986 and 2008, scientists collected 64,000 tiny particles of plastic from the Atlantic. Another large area known as the Western Garbage Patch lies off the coast of Japan. Other researchers believe that the biggest patch of all may be located in the South Pacific, but no dedicated scientific examination of that area has been undertaken.

In addition to the plastic and debris that end up in various surface garbage patches of the world's oceans, researchers believe up to 70 percent of ocean debris sinks to the bottom, winding up on the ocean floor. No one knows or can predict what havoc that debris may be creating for the ecosystems deep in the oceans.

MITIGATION AND PREVENTION

Many scientists believe that the only solution to the issue of floating garbage vortices is to prevent plastic and other debris from ending up in the ocean in the first place. Even if that task were accomplished, it is possible that current damage could persist for thousands of years, because plastic does not decompose.

Raising and activating public awareness will be crucial to get anything meaningful done about this threat. Environmentalist Moore, still active more than 15 years after his first discoveries, has concentrated his efforts on investigating the phenomenon, and informing both the scientific community and the public about the extent of the problem. Moore and his Algalita Foundation team, always seeking to better map the garbage gyre, think that it may be 100 feet (30 meters) deep or more.

Various suggestions for keeping plastic out of the ocean include recycling and reusing plastic, opting for items that use fewer or no plastic materials in either content or packaging, and initiating local cleanup activities on all the beaches, estuaries, lagoons, and marine coasts and waterways of the world. In one such local effort, for example, Mary Crowley, who owns a California-based boat-chartering business, in 2010 launched Project Kaisei to call attention to the Great Garbage Patch by assembling a team of volunteers to begin hauling debris away from the patch one boatload at a time.

—Elizabeth Rholetter Purdy

Further Reading

Bergmann, Melanie, Lars Gutow, and Michael Klages, editors. *Marine Anthropogenic Litter.* Springer Open, 2015.

Blomberg, Lindsey E. "The Great Pacific Garbage Patch." *Environmental Magazine*, vol. 22, no. 3, 2011.

Galgani, Luisa, Ricardo Beiras, Francois Galgani, et al. *Impacts of Marine Litter.* Frontiers Media, 2019.

Kostigen, Thomas M. "The World's Largest Dump: The Great Pacific Garbage Patch." *Discover*, July 2008, discovermagazine.com/2008/jul/10-the-worlds-largest-dump.

Moore, Charles, and Cassandra Phillips. *Plastic Ocean: How a Sea Captain's Chance Discovery Launched a Determined Quest to Save the Oceans.* Avery, 2011.

Perkins, Sid. "Oceans Yield Huge Haul of Plastic." *Science News*, vol. 177, no. 7, 2010.

Wagner, Martin, and Scott Lambert, editors. *Freshwater Microplastics: Emerging Environmental Contaminants?* Springer Open, 2017.

Mineralogy

Fields of Study: inorganic chemistry; geochemistry; geology; oceanography

ABSTRACT

Mineralogy is the study of the chemical composition and physical property of minerals, the arrangement of atoms in the minerals, and the use of the minerals. Minerals are naturally occurring elements or compounds. The composition and arrangement of the atoms that make up minerals is reflected in their physical characteristics. For example, gold is a naturally occurring mineral containing one element that has a definite density of 19.3 grams per milliliter and a yellow color and is chemically inactive. Sometimes mineral resources are broadened to refer to oil, natural gas, and coal, although those materials are not technically minerals.

KEY CONCEPTS

lava: molten rock that has been ejected from a volcano or other volcanic source

magma: molten rock contained within Earth's crust and that has not been ejected onto the surface

mass spectrometer: an apparatus used to identify the mass of atoms introduced into it, by providing them with an electrical charge and measuring their detection as they pass through a magnetic field

X-ray diffraction: the refraction of a beam of X-rays from the atoms within a crystal to produce a characteristic pattern of refracted beams that is then used to determine the relative orientations of the atoms making up that crystal

X-ray fluorescence: when X-rays of a specific energy impinge upon a mineral, the atoms of the mineral absorb the energy and then re-emit it as X-ray energies that are characteristic of the particular atoms

DEFINITION AND BASIC PRINCIPLES

Minerals are solid elements or compounds that have a definite but often not fixed chemical composition and a definite arrangement of atoms or ions. For instance, the mineral olivine is magnesium iron silicate, with the formula $(Mg, Fe)_2SiO_4$, meaning that it has oxygen (O) and silicon (Si) atoms in a ratio of 4:1 and a total of two ions of magnesium (Mg) and iron (Fe) in any ratio. The magnesium and iron component can vary from 100 percent iron with no magnesium to 100 percent magnesium with no iron, and all variations in between. Other minerals, however, such as gold, have nearly 100 percent gold atoms. Minerals are usually formed by inorganic means, but some organisms can form minerals such as calcite (calcium carbonate, with the formula $CaCO_3$).

Minerals make up most of the rocks in the earth, so they are studied in many fields. In geochemistry, for example, scientists determine the chemical composition of the minerals and rocks to derive hypotheses about how various rocks may have formed. Geochemists might study what rocks melt to form magmas or lavas of a certain composition. Environmental geologists attempt to solve problems regarding minerals that pollute the environment. For instance, they might try to minimize the effects of the mineral pyrite (iron sulfide, with the formula FeS_2) in natural bodies of water. Pyrite, which is found in coal, dissolves in water to form sulfuric acid and high-iron water, which can kill some organisms. Forensic mineralogists may determine the origin of minerals left at a crime scene.

Economic geologists discover and determine the distribution of minerals that can be mined, such as lead minerals (galena), salt, gypsum (for wall board), or granite (for kitchen countertops). Geophysicists may study the minerals below Earth's surface that cannot be directly sampled. They may, for example, study how the seismic waves given off by earthquakes pass through the ground to estimate the kinds of rocks present.

BACKGROUND AND HISTORY

Archaeological evidence suggests that humans have used minerals in a number of ways for tens of thousands of years. For instance, the rich possessed jewels, red and black minerals were used in the creation of prehistoric cave drawings in France, and minerals such as gold were used for barter. Metals were apparently extracted from ores for many years, but the methods of extraction were conveyed from person to person without being written down. In 1556, Georgius Agricola (1494–1555) wrote *De re metallica* (English translation, 1912) in which he described many of these mineral-processing methods. From the late seventeenth century into the nineteenth century, many people studied the minerals that occur in definite crystal forms such as cubes, often measuring the angles between faces.

In the nineteenth century, the fields of chemistry and physics developed rapidly. Jöns Jacob Berzelius (1779–1848) developed a chemical classification system for minerals. Another important development was the polarizing microscope, which was used to study the optical properties of minerals to aid in their identification.

During the late nineteenth century, scientists had theorized that the external crystal forms of minerals reflected the ordered internal arrangement of their atoms. In the early twentieth century, this theory was confirmed through the use of X-rays. Also, it became possible to chemically analyze minerals and rocks so that chemical mineral classifications could be further developed. Finally, in the 1960s, the use of many instruments such as the electron microprobe allowed geologists to determine variations in chemical composition of minerals across small portions of the minerals so that models for the formation of minerals could be further developed.

HOW IT WORKS

Mineral and Rock Identification. A geologist may tentatively identify the minerals in a rock using characteristic such as crystal form, hardness, color, cleavage, luster (metallic or nonmetallic), magnetic properties, and mineral association. The rock granite, for instance, is composed of quartz (often colorless, harder than other minerals, with rounded crystals, no cleavage, and nonmetallic luster) and feldspars (often tan, softer than quartz, with well-developed crystals, two good nearly right-angle cleavages, and nonmetallic luster), with lesser amounts of black minerals such as biotite (softer than quartz, with flat crystals, one excellent cleavage direction, and shiny nonmetallic luster).

The geologist then slices the rock into a section about 0.0012 inch (0.03 millimeters) thick (most minerals are transparent). The section is examined under a polarizing microscope to confirm the presence of the tentatively identified minerals and perhaps to find other minerals that could not be detected by eye because they were too small or present in very low quantities. The geologist can determine other relationships among the minerals, such as the sequence of crystallization of the minerals within an igneous rock.

Identification Using Instruments. The minerals in a rock can be analyzed a variety of other ways, depending on the goals of a given study. For instance, X-ray diffraction may be used to identify some minerals. One of the most useful applications of X-ray diffraction is to identify tiny minerals such as clay minerals that are hard to identify using a microscope. The wavelengths of the X-rays are similar to the spacing between atoms in the clay minerals, so when X-rays of a single wavelength are passed through a mineral, they are diffracted from the minerals at angles that are characteristic of a particular mineral. Thus, the mixture of clay minerals in a rock or soil may be identified.

The electron microprobe has enabled the analysis of tiny portions of minerals so that changes in composition across the mineral may be determined. The instrument accelerates masses of electrons into a mineral that releases X-rays with energies that are characteristic of a given element so that the elements present can be identified. The amount of energy given off is proportional to the amount of the element in the sample; therefore, the concentration of the element in the mineral can be determined when the results are compared with a standard of known concentration. The electron microprobe can also be used to analyze other materials such as alloys and ceramics.

The scanning electron microscope uses an electron beam that is scanned over a small portion of tiny minerals and essentially takes a photograph of the mineral grains in the sample. Some electron microscopes are set up to determine qualitatively what elements are present in the sample. This information is often enough to identify the mineral.

Other Analytical Techniques. Many instruments and techniques are used to analyze the major elements, trace elements, and the isotopic composition of minerals and rocks. Commonly used methods are X-ray fluorescence (XRF), inductively coupled plasma mass spectrometry (ICP-MS), and thermal ionization mass spectrometry. X-ray fluorescence is used to analyze bulk samples of minerals, rocks, and ceramics for major elements and many trace elements. A powdered sample or a glass of the sample is compressed and is bombarded by X-rays so that an energy spectrum that is distinctive for each element is emitted. The amount of radiation given off by the sample is compared with a standard of known concentration to determine how much of each element is present in the sample.

The inductively coupled plasma mass spectrometer is used to analyze many elements in concentrations as low as parts per trillion by passing vaporized samples into high-temperature plasma so that all elements have positive charges. A mass spectrometer sorts out the ions by their differing sizes and charges as they travel through a magnetic field, which permits a determination of the elemental concentrations when the results are compared with a standard of known concentration. Up to seventy-five elements can be rapidly analyzed in a sample with a precision of 2 percent or better.

Thermal ionization mass spectrometers can be used to analyze the isotopic ratios of higher mass elements such as rubidium, strontium, uranium, lead, samarium, and neodymium, which may be used to interpret the geologic age of a rock. The mineral or rock is placed on a heated filament so that the isotopes are ejected into a magnetic field in which the ions are deflected by varied amounts depending on the mass and charge of the isotope. The data may then be used to calculate the amount of a certain isotope in the sample and eventually the isotopic age of the sample.

Other specialized instruments are also available. For instance, gemologists use specialized instruments to study and cut gemstones.

APPLICATIONS AND PRODUCTS

Abundant Metals and Uses. The most abundant metals are iron, aluminum, magnesium, silicon, and titanium. Iron, which is mostly obtained from several minerals composed of iron and oxygen (hematite and magnetite), accounts for 95 percent by weight of the metals used in the United States. Much of the hematite and magnetite is obtained from large sedimentary rock deposits called "banded-iron formations" that are up to 2,297 feet (700 meters) thick and may extend for thousands of square miles . The banded-iron formations formed 1.8 billion to 2.6 billion years ago. They are abundant, for instance, around the Lake Superior region in northern Minnesota, northern Wisconsin, and northwestern Michigan. The banded-iron formations have produced billions of tons of iron ore deposits. The ores are made into pellets and mixed with limestone and coke to burn at 2,912 degrees Fahrenheit (1,600 degrees Centigrade) in a blast furnace. The iron produced in this process is molten and can be mixed with small amounts of other elements called "ferroalloys" to produce steel with useful properties. For instance, the addition of chromium gives steel strength at high temperatures, and it prevents corrosion. The addition of niobium produces strength, and the addition of copper increases the resistance of the steel to corrosion.

Much of the aluminum occurs in clay minerals in which the aluminum cannot be economically removed from the other elements. Thus, most of the aluminum used comes from bauxite, which is a mixture of several aluminum-rich minerals such as gibbsite ($Al(OH)_3$) and boehmite (AlOOH). Bauxite forms in the tropics to subtropics by intense chemical weathering from other aluminum-rich minerals such as the clay minerals. The production of aluminum from bauxite is very expensive because the ore is dissolved in molten material at 1,742 degrees Fahrenheit (950 degrees Celsius) and the aluminum is then concentrated by an electric current. Electricity represents about 20 percent of the total cost of producing aluminum. Recycling of aluminum is economical because the energy expended in making, for example, a new can from an old aluminum can is about 5 percent of the energy required to make a new aluminum can from bauxite.

Scarce Metals and Uses. At least thirty important trace metals occur in Earth's crust at concentrations of less than 0.1 percent. Therefore, these elements are trace elements in most minerals and rocks until geologic processes concentrate them sufficiently over time that they become significant portions of some minerals. For instance, gold is present in Earth's crust at only about 0.0000002 percent by weight. Gold often occurs uncombined with other elements and precipitated from hydrothermal solutions as veins. Gold may also combine with the element tellurium to form several kinds of gold-tellurium minerals. Also, gold does not react very well chemically during weathering, so it may concentrate in certain streams to form placer deposits. Gold has been used for thousands of years in jewelry, dental work, and coins. In the past, much world exploration has been motivated by the drive to find gold.

Gold is a precious metal. Other precious metals are silver and the platinum group elements (platinum, palladium, rhodium, ruthenium, osmium, and iridium). Silver, like gold, has been used for thousands of years in jewelry and coins. In modern times, silver also finds uses in batteries and in photographic film and papers because some silver compounds are sensitive to light. Silver occurs in nature as silver sulfide, and it substitutes for copper in copper minerals so that much of the silver produced is a side product from copper mining. The platinum metals occur together as native elements or combined with sulfur and tellurium. They concentrate in some dark-colored igneous rocks or as placers. The platinum group metals are useful as catalysts in chemical reactions; one common use is in catalytic converters in vehicles.

Another group of scarce metals are base metals, which are of relatively low economic value compared with the precious metals. The base metals include copper, lead, zinc, tin, mercury, and cadmium. Copper minerals are native copper and various copper minerals combined with sulfate, carbonate, and oxygen. Copper minerals occur in some igneous rocks, including some hydrothermal deposits, mostly as dilute copper-sulfur minerals. These minerals may be concentrated during weathering, often forming copper minerals combined with oxygen and carbonate. Copper conducts electricity very well, and much of it is used in electric appliances and wires. Lead and zinc tend to occur together in hydrothermal deposits in combination with sulfur. Lead is used in automobile batteries, ammunition, and solder. Lead is very harmful to organisms, which restricts its potential use. Zinc is used as a coating on steel to prevent it from rusting, in the production of brass, and in paint.

Fertilizer and Minerals for Industry. Fertilizers contain minerals that have nitrogen, phosphorus, potassium, and a few other elements necessary for the growth of plants. In Peru, deposits of guano (seabird excrement), which is rich in these elements, have been mined and used as fertilizer. Saltpeter (potassium nitrate) has also been mined in Peru. A calcium phosphate mineral, apatite, occurs in some sedimentary rocks, so some of these have been mined for the phosphorus.

Some minerals are a source for sodium and chlorine. Halite, commonly known as rock salt, is used as a flavoring and to melt ice on roads. Halite is produced in some sedimentary rocks from the slow evaporation of water in closed basins over long periods of time, a process that is occurring in the Great Salt Lake in Utah.

A variety of rocks or sediment—including granite, limestone, marble, sands, and gravels—are used for the exteriors of buildings, construction materials, or for countertops. Sands and gravels may be mixed with cement to make concrete.

CAREERS AND COURSEWORK

Anyone interested in pursuing a career in the minerals industry should study geology, chemistry, physics, biology, chemical engineering, and mining engineering. A bachelor's degree in one of these subjects is the minimum requirement for working in oil exploration or as a mining technician or water-quality technician. A master's degree or a doctorate is required for more challenging and responsible jobs in geochemistry, geophysics, environmental geology, geochronology, forensic mineralogy, and academics. Geochemists should have a background in geology and chemistry. Geophysicists use physics, mathematics, and geology to remotely tell what kinds of rocks are below the surface of the earth. Environmental geologists should have backgrounds in geology, chemistry, and biology. Geochronologists use natural radioactive isotopes to give estimates of the time when some kinds of rocks formed. They should have a background in geology, physics, and chemistry. Forensic mineralogists should have a good background in mineralogy and criminology. Economic geologists should major in geology, with a concentration in courses concerned with ore mineralization. Those interested in an academic career should have a PhD in geology, mining engineering, or chemical engineering. They will be expected to teach and do research in their subject area.

SOCIAL CONTEXT AND FUTURE PROSPECTS

Mineral resources are not evenly distributed throughout the world. In the nineteenth century, the United States did not import many mineral resources, but it has increasingly become an importer of mineral resources. The need to import is driven by many factors, including an increase in the minerals used. In 2016, the United States spent $122 billion on importing minerals; minerals mined within the country that year were valued at $74.6 billion. Other factors include the depletion of some US sources of minerals and the increased use of minerals that are not naturally available in the United States. In 2016, foreign sources supplied more than 50 percent each of fifty mineral commodities consumed in the United States. Twenty of those minerals were 100 percent foreign sourced. The United States must import much of the manganese, bauxite, platinum group minerals, tantalum, tungsten, cobalt, and petroleum that it uses. In contrast, the United States still has abundant resources of gold, copper, lead, iron, and salt. Manganese and several other high-value elements are found in nodules that are strewn about the ocean floors, and their recovery would do much to alleviate shortages of supply. However, the means to retrieve the nodules effectively and efficiently has not yet been developed.

The need to import and export mineral resources has forced countries to cooperate with one another to achieve their needs. The mineral resources traded in the largest quantities, in descending order, are iron-steel, gemstones, copper, coal, and aluminum. The total annual world trade in mineral resources is approaching $700 billion and is likely to rapidly increase as other countries use and produce more mineral resources.

As the supply of minerals decreases worldwide and environmental and safety concerns raised by mining gain in importance, it is likely that mineralogy research may turn to improving extraction and processing methods and looking for ways to recycle materials, capture minerals remaining in wastes, and restore mining areas after the minerals have been depleted. Earth's ocean floors represent a huge repository of many valuable minerals, but retrieving them is an extremely difficult and expensive operation due to the depths and high pressures of the water medium under which they are sequestered. Ultimately, it would become necessary and more feasible to harvest mineral resources from space, and a significant aspect of current efforts in space research is the eventual use of lunar and Martian resources in the study of what other resources may be found "out there."

—Robert L. Cullers

Further Reading

Chester, Roy. *Marine Geochemistry*. Springer Science and Business Media, 2012.

Craig, James R., David J. Vaughan, and Brian Skinner. *Earth Resources and the Environment*. Prentice, 2010.

Guastoni, Alessandro, and Roberto Appiani. *Minerals*. Firefly, 2005.

Kearny, Philip, editor. *The Encyclopedia of the Solid Earth Sciences*. Blackwell, 1994.

Klein, Cornelis, and Barbara Dutrow. *Manual of Mineral Science*. Wiley, 2008.

Pellant, Chris. *Smithsonian Handbook: Rocks and Minerals*. Dorling, 2002.

Perkins, Dexter. *Mineralogy*. Prentice-Hall, 2011.

Sinding-Larsen, Richard, and Friedrich-Wilhelm Wellmer, editors. *Non-renewable Resource Issues: Geoscientific and Societal Challenges*. Springer, 2010.

N

Navigation

Fields of Study: mathematics; oceanography; electronics engineering

ABSTRACT

Navigation is the process of plotting movement from one point to another. This is accomplished by determining one's exact position on Earth by visual or electronic means and determining the course to the exact position of the intended destination. Specific points are often defined as measurements of longitude and latitude.

KEY CONCEPTS

navigation: basically, the practice of finding one's way from point A to point B by relating direction to fixed points of reference

radar: an electronic method of detection at a distance by measuring the elapsed time between transmitting a radio frequency signal and receiving the reflected signal at a detector

sonar: similar to radar, but uses a sound pulse traveling through water rather than a radio frequency signal

triangulation: a method of locating a specific location based on the relationships of angles and the fact that the interior angles of a triangle always add up to a total of 180 degrees

trigonometry: the mathematics of angles and their measurement, particularly of triangles

DEFINITION AND BASIC PRINCIPLES

Location of one's position is based on trigonometry and the process of triangulation. Triangulation involves determining the location of a point by measuring angles to it from two or more known points. The intersection of the lines from the known points represents one's position.

Prior to the development of electronic instruments such as radio, radar, and the global positioning system (GPS), the accurate determination of one's position was accomplished using a sextant. The sextant was able to pinpoint one's location accurately by measuring the angular distance between the horizon and a celestial object. Repeated sightings were plotted on a nautical chart. This method was suitable for ships and propeller aircraft.

After the development of radio, a position could be determined by triangulating radio stations with known positions. With the development of radar, images of known landmasses or buoys could be used for triangulation.

GPS also involves triangulation. The device electronically determines one's position by triangulation from orbiting satellites and marks the position on a map overlay.

Navigation also involves correction for ocean currents and obstacles such as landmasses and other vessels. Experienced seamen past and present, aided by experience, can predict current with reasonable accuracy. Wind affects currents, which can be estimated by the sailor or measured with an anemometer (wind gauge). Airplanes are affected by currents known as jet streams. Spacecraft are affected by gravitational forces, which increase in power as the craft approaches. That force can be used to change the direction of the craft substantially.

BACKGROUND AND HISTORY

Human migration and the discovering of new lands by navigation of the oceans was accomplished by many ancient cultures, including the Phoenicians, ancient Greeks, the Norse, the Persians, the Chinese, and the Polynesians. Primitive navigation depended on knowledge of ocean currents and the position of celestial objects.

The magnetic compass was first used in China around 200 BCE. These ancient compasses used a piece of lodestone, a naturally magnetic iron-containing ore called "hematite," which pointed to the magnetic North Pole. In eighth-century China, the lodestone was replaced by a magnetized needle.

Navigation charts first appeared in Italy at the close of the thirteenth century. Called "portolan charts," they were rough maps based on accounts of sailors plying the coastlines of the Mediterranean and Black seas. The octant, precursor to the sextant and developed around 1730, made it possible to determine latitude accurately but not longitude. The sextant and accurate chronometers added the ability to determine longitude. At the close of the nineteenth century, radios that transmitted and received Morse code began to appear on oceangoing vessels.

The prototype of the modern radar was installed on the USS *Leary* in 1937. In 1942, the first long-range navigation (loran) system was installed at various points along the Atlantic coast of the United States. Loran is based on using the intersection of two radio waves to determine one's position. The subsequent proliferation of satellites, after the Soviet Union launched *Sputnik I*, the first Earth-orbiting satellite, in 1957, led to the development of satellite technology and the highly accurate GPS.

HOW IT WORKS

Compasses. The magnetic compass needle points to magnetic north and floats over a 360-degree compass face, which displays the cardinal points: north (0 degree), east (90 degrees), south (180 degrees), and west (270 degrees). The deviation (declination) of magnetic north is not a constant and is updated annually. Nautical charts display the amount of needed correction from the baseline, which is the declination value of the year the chart was published. The magnetic compass is a simple, reliable instrument; however, its readings can be affected by any magnetic field in the vicinity. Marine electronics located near the compass generate magnetic fields; ferrous (iron) material often has a magnetic field. To improve accuracy, the compass can be calibrated by small magnets or other devices to correct for extraneous magnetic fields. The compass must be recalibrated with the addition of any new electronics. A gyrocompass points to true north and is unaffected by stray magnetic fields. The gyrocompass, first invented in practical form in 1906, employs a rapidly spinning gyroscope, which is motor driven. A disadvantage of the gyrocompass is that electrical or mechanical failure can render it useless.

Loran. Loran is an acronym for long-range navigation. The system relies on land-based low-frequency radio transmitters. The device calculates a ship's position by the time difference between the receipt of signals from two radio transmitters. The device can display a line of position, which can be plotted on a nautical chart. Most lorans convert the data into longitude and latitude. Since GPS became available, the use of loran has markedly declined.

Radar. Radar consists of a transmitter and receiver. The transmitter emits radio waves, which are deflected from a fixed or moving object. The receiver, which can be a dish or an antenna, receives the wave. Radar circuitry then displays an image of the object in real time. The screen displays the distance of the object from the radar. If the object is moving, consecutive readings can calculate the speed and direction of the object. If the object is airborne, and the radio is so equipped, the altitude is displayed. Radar is invaluable in foggy weather when visibility can be severely reduced.

Sonar. Sonar, which is an acronym for sound navigation ranging, transmits and receives sound waves for underwater navigation by submarines. Passive sonar is a related technology in which the equipment merely listens for underwater sound made by vessels. Active sonar emits a pulse of sound (a "ping") then listens for a reflection or echo of the pulse. The distance of the object is determined by the time difference from transmission to reception of the ping (the speed of sound in water is constant). To measure the bearing, two or more separated transmitter/receivers are used for triangulation of the object. Sonar can also be used by both submarines and surface vessels to determine the depth and contour of the bottom of the ocean (or other body of water). By consulting a nautical chart, which is marked with depth gradients, the distance from the shoreline can be calculated.

Global Positioning System (GPS). GPS is a space-based global navigation satellite system that provides accurate location and time information for any point on Earth where there is an unobstructed line of sight to four or more GPS satellites. GPS can function under any weather condition anywhere on the planet, except underwater—a submarine must surface to use a GPS system. The technology depends on triangulation, just as a land-based system such as loran employs. GPS is composed of three segments: the space segment, the control segment, and the user segment. The US Air Force operates and maintains both the space and control segments. The space segment is made up of satellites, which are in medium-space orbit. The satellites broadcast signals from space, and a GPS receiver (user segment) uses these signals to calculate a three-dimensional (3D) location (latitude, longitude, and altitude). The signal transmits the time, accurate within nanoseconds.

Navigational Aids. According to the US Coast Guard, a navigational aid is a device external to a vessel or aircraft specifically intended to assist navigators in determining their position or safe course, or to warn them of dangers or obstructions to navigation. Navigational aids include buoys, lighthouses, fog signals, and day beacons. Buoys are used worldwide to mark nautical channels. They are color coded: Red buoys mark the right (starboard) side of the channel for a vessel returning to a port; green buoys mark the port (left) side of the channel; and red-and-green-striped buoys mark the junction of two channels. Buoys often contain gongs, which are activated by wave motion and solar-powered lights. Some buoys do not mark safe channels and are referred to as "nonlateral markers." They contain shapes identifying their purpose. Squares depict information, such as food, fuel, and repairs. Diamonds warn of dangers such as rocks. Circles mark areas of reduced speed. Crossed diamonds indicate off-limits areas such as dams and places where people may be swimming.

APPLICATIONS AND PRODUCTS

Navigational aids have many applications, which are tailored to underwater, water surface, terrestrial, atmospheric, and space use.

Water-Surface Navigation. Any vessel, which is capable of venturing offshore, must contain some basic navigation equipment. A magnetic compass is mandatory, as is a marine radio, which can be used to call for help. A knot meter, which consists of an underwater paddle wheel connected to a dial, can give the relative speed over water. The vessel should contain nautical charts of the area, which are prepared by the National Oceanic and Atmospheric Administration (NOAA). Another essential is an accurate timepiece (chronometer). By combining the readings of speed registered on the knot meter, the compass heading, and chronometer, one can plot the vessel's approximate position on the nautical chart. All inboard marine engines are equipped (or can be equipped) with a tachometer, which registers engine speed in revolutions per minute (rpm). A prudent skipper can run his boat over a known distance (between measured mile [or kilometer] markers) and construct a chart of the vessel's speed at various rpm. If the vessel does not have a knot meter, or it fails, the approximate boat speed can be determined via the tachometer. Another relatively inexpensive piece of equipment is a simple depth sounder, which displays depth (more expensive models display a graphic representation of the seabed). The depth sounder can aid in determining the vessel's position, and more importantly, alert the skipper to shallow water, which could damage the hull. Binoculars, which can help identify shoreline features, are another inexpensive necessity. The sextant, which can pinpoint one's position, is another inexpensive navigation aid; however, most recreational boaters are unfamiliar with its use.

Many recreational vessels of modest size (32 to 42 feet [10 to 13 meters] long) have an automatic pilot, radar, and GPS in addition to the aforementioned items. Some utilize a loran receiver, which may be less expensive than a GPS, and it can accurately display the ship's position. An automatic pilot adjusts the rudder(s) to maintain a desired heading. Radar can be adjusted for the distance it displays; ranges of less than one mile and more than thirty miles are commonplace. In many situations, the shortest range is the most valuable because it can clearly depict an approaching vessel or a small landmass (an islet or an offshore oil rig). Even in clear weather, radar can provide invaluable navigation information. Ship-mounted GPS devices can be purchased and installed relatively inexpensively and are becoming commonplace on recreational vessels.

Large commercial and naval vessels (and many luxury yachts) contain sophisticated navigation equipment. Large vessels also boast onboard computers, which interpret data from radar, GPS, and the tachometers to display detailed navigation information. These computers also replace the automatic pilot found on smaller vessels. The computer can adjust the rudders and throttles to maintain the ship's heading and time of arrival. Depth sounders with graphic displays and sonar are components of most large vessels as well as fishing boats, which use them for locating schools of fish.

Underwater Navigation. Unlike surface vessels, submarines must navigate in three dimensions. Active sonar is employed by distance to a target and depth. Transmission of sound from two separate sources can give the submarine's heading. Submarines are equipped with GPS, which can be used to determine position only when the vessel is on the surface. When below the surface, an inertial navigation system (INS) is employed. The INS is composed of precise accelerometers and gyroscopes to record every change in the submarine's speed and direction. The data from these instruments is fed to a computer, which determines the vessel's position. Over time, small errors accrue; these errors are corrected when the submarine surfaces and the GPS is activated.

Terrestrial Navigation. When traveling on a road, a motorist can navigate using street and highway signs as well as a map of the area. Many motor vehicles are equipped with a GPS. The GPS gives a visual display of the vehicle's position on a map overlay. Useful information such as distance to the next turn and the destination are given. The GPS in most automobiles has an audio system, which informs the motorist of upcoming turns. The GPS of some vehicles also gives additional information such as gasoline stations, charging stations, rest areas, local restaurants, and hospitals. GPS units for off-road use are also available. The motorist on a budget can purchase an inexpensive handheld GPS, devices that are also popular with campers and hikers. Foot travelers and off-roaders also rely on compasses for navigation.

Celestial navigation with a sextant can also be utilized. Being able to locate Polaris (the North Star), which is the brightest star in the constellation Ursa Minor (the Little Dipper or Little Bear), is a necessary skill for all off-road travelers. The rising and setting sun or other visual aids such as the glow from a city at night can also aid in navigation.

Atmospheric Navigation. Aircraft navigate in 3D space, so latitude, longitude, and elevation must be known. All aircraft contain basic instrumentation, which includes: a magnetic compass; an altimeter, which displays the altitude; an airspeed indicator; a heading indicator, which is similar to a gyrocompass; a turn indicator, which displays direction and rate of turn; a vertical-speed indicator, which displays the rate of climb or descent; and an attitude indicator. The attitude indicator, which is also known as an artificial horizon, is an invaluable navigation instrument. It contains a gyroscope that reflects both the horizontal and vertical alignment of the aircraft. When visibility is reduced, the pilot is unable to visualize the horizon. Many small aircraft and virtually all large aircraft contain a GPS. Other navigation equipment commonly found on larger aircraft include a course deviation indicator (CDI) and a radio magnetic indicator (RMI). However, a GPS is much superior to either a CDI or RMI. The CDI displays an aircraft's lateral position in relation to a track, which is provided by a very high frequency (VHF) omnidirectional range (VOR) or an instrument landing system. A VOR is a ground-based station that transmits a magnetic bearing of the ship from the station. An instrument landing system is a ground-based system that can provide precision guidance of the aircraft to a runway. It includes radio signals, and often, high-intensity lighting arrays. An RMI is usually coupled to an automatic direction finder (ADF); the RMI displays a bearing to a nondirectional beacon (NDB), which is a ground-based radio transmitter.

Unmanned aerial vehicles (drones) and guided missiles must also be navigated. In the case of an unmanned drone, a pilot sits at a console on the ground and directs the flight via radio. Guided missiles are either internally or externally guided, sometimes both. Some missiles contain inertial navigation systems similar to those found on submarines. The system is programmed to strike a specified target. External control can be via radio waves or laser. In some cases of laser guidance, the missile contains a laser homing device in which the projectile "rides" the laser beam to the target. More commonly, a technique referred to as semiactive laser homing (SALH) is employed. With SALH, the laser beam is kept pointed at the target after the projectile is launched. Laser energy is scattered from the target, and as the missile approaches the target, heat sensors home in on this energy.

Space Navigation. Space navigation is a complex and highly technical science. In addition to involving three dimensions it also requires plotting a course between two moving objects (Earth and a space station or an orbiting space shuttle and the Earth). Space navigation also entails calculating the gravitational force of celestial objects such as the Moon and planets. Space navigation requires the collaboration between computers (both ground and ship based) and complex instrumentation. It also requires collaborative effort between highly skilled personnel and either astronauts or navigation equipment within unmanned spacecraft.

CAREERS AND COURSEWORK

Any navigation course requires knowledge of trigonometry. Whether navigation is based on visual or electronic information, triangulation to fix a position is required. The military offers many careers and coursework involving navigation. Following military service, this training can be applied to many civilian job opportunities, up to and including piloting an aircraft or navigating a ship. The US Coast Guard Auxiliary and the US Power Squadrons offer free boating-safety courses, which include navigational skills. Courses are available throughout the United States as well as online. For individuals interested in more advanced courses, such as ones for preparation for the US Coast Guard captain's license, reasonably priced courses are available throughout the United States and online. For those not interested in a military career, navigation training is available at civilian institutions. These include California Maritime Academy in Vallejo; Great Lakes Maritime Academy in Traverse City, Michigan; Maine Maritime Academy in Castine; Massachusetts Maritime Academy in Buzzards Bay; State University of New York Maritime College in the Bronx; Texas Maritime Academy, which is part of Texas A&M University at Galveston; and the United States Merchant Marine Academy in Kings Point, New York.

Although the basic principles of navigation can be learned and applied by any high school graduate, more advanced topics require a college degree and often a postgraduate degree such as a master of arts (MA) or doctorate (PhD). An MA requires one year of study after four years of college; however, an academically aggressive student can earn an MA concurrently with a bachelor's degree. A PhD requires two to three additional years of study followed by submission of a thesis or dissertation. Coursework should include mathematics, engineering, computer science, and robotics. Positions are available for individuals with a degree in laser engineering in both the government and private sector. The ability to be a team player is often of value for these positions because ongoing research is often a collaborative effort.

SOCIAL CONTEXT AND FUTURE PROSPECTS

Although the basic concepts of navigation have remained unchanged for centuries, modern navigation technology is highly advanced and continues to evolve. The frontier of navigation research lies in military and extraterrestrial applications. The military is focused on guidance systems for missiles and drones. Extraterrestrial applications range from the navigation of Earth-orbiting shuttles to interplanetary (and beyond) navigation. Navigation is an essential component of daily life—both civilian and military.

—*Robin L. Wulffson*

Further Reading

Burch, David. *Emergency Navigation: Find Your Position and Shape Your Course at Sea Even if Your Instruments Fail*. 2nd ed., McGraw-Hill, 2008.

Burns, Bob, and Mike Burns. *Wilderness Navigation: Finding Your Way Using Map, Compass, Altimeter and GPS*. 2nd ed., Mountaineers, 2004.

Cutler, Thomas J. *Dutton's Nautical Navigation*. 15th ed., Naval Institute, 2004.

Launer, Donald. *Navigation Through the Ages*. Sheridan House, 2009.

Lele, Ajey. *Strategic Technologies for the Military: Breaking New Frontiers*. Sage, 2009.

Prinet, Dominique F. *Celestial Navigation*. Friesen Press, 2015.

Sonnenberg, G. J. *Radar and Electronic Navigation*. 6th ed., Elsevier, 2013.

Weintrit, Adam, editor. *Navigational Systems and Simulators*. CRC Press, 2011.

North Atlantic Ocean

Fields of Study: oceanography; marine biology; marine geology

ABSTRACT

The North Atlantic Ocean, the portion of the Atlantic that lies north of the equator, is a vast and growing expanse of water with complex ecology due to numerous connections with adjoining oceans, seas, gulfs, bays, and currents that help set the pace of the Earth's thermohaline circulation—a temperature- and salinity-driven system that exerts a strong influence on the planet's climate.

KEY CONCEPTS

aphotic zone: the lower part of the ocean water below the maximum depth to which any sunlight can reach

disphotic zone: the region between the euphotic and aphotic zones, where sunlight can still be detected but is insufficient to support photosynthesis

euphotic zone: the upper layer of ocean water to the maximum depth to which there is sufficient sunlight to support photosynthesis

littoral zone: the area of shoreline between high and low tide levels

nekton: term for all creatures that move through the water by their own self-propulsion, as opposed to merely floating

plankton: term for organisms that float in the water and are incapable of self-propulsion

thermohaline circulation: global ocean circulation system driven by differences in temperature and density

DESCRIPTION

As a whole, the Atlantic is Earth's second-largest ocean, but it is expanding while the largest ocean (the Pacific) is shrinking as a result of plate tectonics. On the west, the North Atlantic is bordered by the continents of South and North America and by the Caribbean Sea, Gulf of Mexico, Bay of Fundy, and Gulf of St. Lawrence. On the north, it is bounded by Earth's largest island, Greenland, as well as Iceland and the Greenland and Norwegian Seas. On the east, it is bounded by the continents of Europe and Africa; the British Isles; the North, Scottish, and Irish Seas; the Bristol and English Channels; the Bay of Biscay, Mediterranean Sea, and Gulf of Guinea.

The North Atlantic is slightly larger in area than the South Atlantic, 16,020,000 square miles (41,490,000 square kilometers) to 15,550,000 sq mi (40,270,000 sq km) as of 2010. But the North Atlantic has a lower average depth, 11,545 feet (3,519 meters) compared to 13,034 ft (3,973 m), and smaller volume at 35,027,263 cubic miles (146,000,000 cubic kilometers) compared to 38,386,041 cu mi (160,000,000 cu km).

ZONAL CHARACTERISTICS

There are numerous ways to classify the environmental zones of the oceans. One is by type of environment, whether pelagic (open water) or benthic (sea bottom). The dominant residents of the pelagic environment are either plankton, free-floating organisms that are usually microscopic in size; or nekton, free-swimming organisms that are typically fish, but include sea mammals such as whales, and reptiles such as sea turtles. The dominant residents of the benthic zone are invertebrates such as hard and soft corals, polychaetes, crustaceans, and mollusks.

The pelagic environment can be subdivided into two provinces: the neritic, which extends from the high tide line out to the edge of the continental shelves at about 660 ft (200 m) deep, and which is subject to influence from adjoining terrestrial environments; and the oceanic, which encompasses the deepwater environment beyond the edge of the continental shelves and is less influenced by the input of nutrients, sediment, and other pollutants from land.

The pelagic environment can be further subdivided into zones based on depth. The epipelagic zone is the surface layer down to 660 ft (200 m). By definition, all of the neritic province would fall into the epipelagic zone. The mesopelagic zone extends from 660 ft (200 m) down to 3,300 feet (1,000 m). The bathypelagic zone extends from 3,300 ft (1,000 m) down to 13,000 ft (4,000 m). Below 13,000 ft (4,000 m) lies the abyssopelagic zone.

The benthic or sea-bottom environment can likewise be subdivided into two provinces: the subneritic, which encompasses the continental shelves down to about 660 ft (200 m); and the suboceanic, which encompasses the sea bottom in the deeper waters beyond the edges of the continental shelves. As with the pelagic environment, the benthic environment can be divided into zones based on depth. The littoral zone corresponds with the edge of the sea—the area between the high- and low-tide lines. The sublittoral zone encompasses the rest of the seafloor of the subneritic province, from the low-tide line down to about 660 feet (200 m).

THE SUNLIGHT FACTOR

This littoral zone can be further subdivided into inner and outer sublittoral zones. The inner sublittoral is defined as the portion where sufficient light reaches the ocean floor to permit the growth of photosynthetic marine algae (such as kelp) attached to the bottom. The maximum depth of the inner sublittoral varies based on the amount of solar radiation reaching the ocean surface as well as the turbidity (cloudiness) of the water, but is usually about 160 ft (50 m).

The bathyal zone corresponds with the location of the continental slope—a drop-off from the edge of the continental shelf down to the abyssal plains of the deep ocean—from 660 ft (200 m) down to13,000 ft (4,000 m). The abyssal zone is the seafloor in waters between 13,000 ft (4,000 m) and 20,000 ft (6,000 m) in depth. Below 20,000 ft (6,000 m) lies the hadal zone.

Another classification system is based on what is arguably the single most important variable influencing the distribution of marine life—sunlight. The euphotic zone has sufficient light for photosynthesis to occur. It extends from the surface down to about 330 ft (100 m). The primary photosynthetic organisms are phytoplankton (microscopic organisms such as diatoms, coccolithophores, dinoflagellates, and cyanobacteria) along with fixed algae, attached to the bottom in shallower waters. Because of the incredible productivity made possible by the energy in sunlight, the euphotic zone is home to the vast majority of all marine life.

Below the euphotic zone is the disphotic zone, which extends to about 3,300 ft (1,000 m) in depth. While not sufficient to permit photosynthesis, the disphotic zone has measurable quantities of light which may have significant ecological effects, such as in concealing prey from or revealing prey to predators. The aphotic zone lies below the disphotic zone. No sunlight penetrates those depths. Any light that exists in the aphotic zone is the product of bioluminescence—light produced by the organisms themselves.

NORTH ATLANTIC GYRE

Centuries ago, oceanographers realized that a series of currents flowed through and subdivided the North Atlantic. The first of these currents to be discovered was the Gulf Stream, known to European explorers such as Juan Ponce de León (1460–1521) as early as the beginning of the 16th century and named and mapped by Benjamin Franklin (1706–1790) in 1770.

The Gulf Stream originates off the tip of Florida where the Caribbean Current and Antilles Current—both offshoots of the Atlantic North Equatorial Current—rejoin one another. Atmospheric circulation drives the current northward along the east coast of North America and east into the Atlantic south of Greenland. There it splits into two currents: the North Atlantic Drift, which flows northeastward along the coast of Scandinavia into the Arctic Ocean; and the Canary Current, which flows east and south along the western coasts of the Iberian Peninsula (Spain and Portugal) and North Africa. The Canary Current turns west off the coast of Africa, becoming the Atlantic North Equatorial Current and completing the final leg of a clockwise-rotating system known as the North Atlantic Gyre.

By bringing warm tropical water into the higher latitudes, the Gulf Stream helps make northern and western Europe warmer than they would otherwise be. It also permits the development of an extensive coral reef system around Bermuda, which at 32 degrees north latitude is the most northerly expanse of coral reefs in the entire Atlantic.

Where the Gulf Stream comes into contact with the colder Labrador Current, which flows southward between Canada and Greenland into the northwestern North Atlantic, it creates potentially stormy conditions, but it also creates some of the richest fisheries on the planet. One such contact is off the Grand Banks, an undersea rise southeast of Newfoundland; another is off the Outer Banks barrier island system of North Carolina.

The wind-driven surface currents of the North Atlantic Gyre are connected with the density-driven, deep currents of the thermohaline circulation—where colder and saltier, hence denser, water sinks to form strong currents that circulate through Earth's ocean basins—to ultimately form the global oceanic "conveyor belt." The conveyor begins in the North Atlantic off Greenland and Iceland where cold water sinks to form North Atlantic Deep Water, which bends southward along the continental slopes of the Americas to join with Antarctic waters that in turn spread into the Indian and Pacific Ocean basins.

By its interactions with the atmosphere as well as all Earth's seas, the global ocean conveyor helps act as a thermostat on Earth's climate, keeping global temperatures relatively stable. Disruptions in the North American portion of the conveyor—such as a massive influx of cold freshwater from rapidly melting glaciers in Greenland—could trigger sudden and devastating climate change.

FAUNA PROFILE

The rich fisheries off the east coast of North America have had significant impact on the history of Europe and the Americas. Arguably the most important was that of the North Atlantic cod, which drew European sailors across the Atlantic—hundreds of years before Columbus's 1492 voyage—and which fueled 16th-century Europe's ambitions for colonizing the northeastern United States and Canada. Cod survived intense fishing pressure for 400 years until technological innovations allowed commercial fishers to so thoroughly deplete the population that the fishery has collapsed throughout the species' range.

While abundant fisheries have historically been found off the east coast of North America and on the northwestern fringes of Europe, inside the great loop of the North Atlantic Gyre it is something of a biological wasteland. The nutrient-poor waters have relatively depauperate animal populations, except where upwelling currents bring cold, nutrient-laden water up from the depths.

What is found in abundance inside the gyre is floating plastic debris. Plastic refuse, discarded on land, washes into coastal waters and is picked up by the currents circulating around the edges of the gyre. As a result of the wind effect known as Ekman transport, surface waters—and the pollution they carry—are deflected to the right of the direction of the prevailing winds that drive the gyre's currents in the Northern Hemisphere (in the Southern Hemisphere, the deflection is to the left). This transport concentrates plastic debris in the center of the gyre, where the debris frequently enters marine food webs after ingestion by marine animals. The cloud of debris surrounding the gyre is known as the North Atlantic Garbage Patch; it was estimated in 2010 to have an overall concentration of 4,000 pieces of garbage per square mile, reaching as high as 250,000 pieces per square mile in places.

In addition to cod, the cooler waters of the western North Atlantic are home to Atlantic herring, whose populations have also suffered from overfishing in recent decades. Among notable mammalian species are beluga whale, a formerly heavily hunted species now suffering from the effects of water pollution and diseases; the endangered northern right whale; fin whale; humpback whale; and grey seal. The northern right whale, whose population was initially depleted by hunting, began to rebound in the late twentieth and early twenty-first centuries, but after 2010 went into decline once more, due to fatalities caused by entanglement in various types of commercial fishing gear. It was estimated in 2018 that the number of remaining northern right whales might be as low as 430 individuals. At the rate of decline that the species has experienced in the 2010s, it is estimated that the whales may be extinct by 2040.

The warmer waters of the western North Atlantic, under the influence of the Gulf Stream, support a rich temperate fish fauna—which includes snappers, groupers, grunts, and porgies—with a significant number of residents more typical of tropical water, such as wrasses and damselfish. Development, pollution, and introduced species and diseases are significant environmental concerns in this region.

The northern European waters offer largely shallow water environments with areas of upwelling that bring cool, nutrient-rich waters to the surface that have traditionally supported robust fisheries in species such as Arctic cod, Arctic char, and capelin. They are also home to porbeagle shark; several cetacean species, including the blue whale, minke, orca (killer), beluga, narwhal, bowhead, and northern bottlenose; and the grey and harbor seal. Overfishing and pollution are the primary environmental threats to the region.

TROPICAL POPULATIONS

While the coral reefs of the tropical North Atlantic are less diverse than their Indo-Pacific counterparts, they are diverse enough to support major feeding grounds that draw a wide range of species from the temperate North Atlantic. Among the more important residents are parrotfish, wrasses, Cherub fish, large grouper including Nassau grouper, drum, and conch. The region serves as a breeding haven for the humpback whale, offers crucial habitat for such littoral fauna as the American crocodile and manatees, and hosts a robust array of sea turtles including loggerhead, green, hawksbill, leatherback, olive ridley and Kemp's ridley. Loss of habitat from development and disturbance, epidemic disease, coral bleaching, and pollution threaten many reefs in the region.

The waters of the northeast Brazil shelf offer a diversity of habitats, including mangroves and coral reefs, that are home to a number of sea turtle species as well as breeding grounds for humpback whales and smalleye hammerhead sharks. Development, pollution, and destructive fishing practices are significant environmental concerns here.

Several areas of upwelling occur where the Canary Current approaches the coast of southern Europe and northwestern Africa. These areas support significant fisheries for deep-sea fish and lobsters. The Canary Current region is also home to several economically important sardine species (sardine, Spanish sardine, and pilchard), horse mackerel, hake, green and olive ridley sea turtles, and West African manatee; and serves as a breeding ground for humpback whale. As with other marine provinces near areas of dense human habitation, development and pollution are significant concerns.

—*David M. Lawrence*

Further Reading

Earle, Sylvia A., and Linda K. Glover. *Ocean: An Illustrated Atlas.* National Geographic Books, 2008.

Hutchinson, S., J. R. E. Lutjeharms, B. McMillan, J. Musick, B. Stonehouse, and M. Tomczak. *Illustrated Atlas of the Sea.* Weldon Owen, 2010.

Kurlansky, Mark. *Cod: A Biography of the Fish That Changed the World.* Walker, 2007.

Laist, David W. *North Atlantic Right Whales: From Hunted Leviathan to Conservation Icon.* Johns Hopkins UP, 2017.

Law, Kara Lavender, Skye Morét-Ferguson, Nikolai A. Maximenko, et al. "Plastic Accumulation in the North Atlantic Subtropical Gyre." *Science*, vol. 329, no. 5996, 2010.

Miles, Kathryn. "Can Anyone Save the North Atlantic Right Whale?" *The Boston Globe*, 11 Apr. 2018, www.bostonglobe.com/magazine/2018/04/11/can-anyone-save-north-atlantic-right-whale/iS76fBOUn9PdnL9LySgmhN/story.html.

Morozov, Eugene G., Alexander N. Demidov, Roman Y. Tarakanov, and Walter Zenk. *Abyssal Channels in the Atlantic Ocean: Water Structure and Flows.* Springer, 2010.

Roberts, Callum. *The Unnatural History of the Sea.* Island Press, 2007.

Siedler, Gerold, Stephen M. Griffies, John Gould, and John A. Church, editors. *Ocean Circulation and Climate: A 21st Century Perspective.* Academic Press, 2013.

Spalding, Mark D., et al. "Marine Ecoregions of the World: A Bioregionalization of Coastal and Shelf Areas." *BioScience*, vol. 57, no. 7, 2007.

North Pacific Ocean

Fields of Study: oceanography; marine biology; marine ecology

ABSTRACT

The Pacific Ocean is the world's largest, coldest, and deepest ocean, covering approximately one-third of Earth's total surface. It is commonly divided into North and South Pacific segments. The North Pacific Ocean is a major world habitat of marine organisms. Harvest of these organisms plays an important role in the economies and international relations of bordering countries—and of the planetary food web. The region itself extends from the near-tropical waters off southern Japan to the Arctic waters of the Bering Sea.

KEY CONCEPTS

gyre: a general, slow circulation of water that spans entire ocean ranges, formed by the meeting of different ocean currents; generally, rotation is clockwise in the northern hemisphere and counterclockwise in the southern hemisphere

phytoplankton: small marine organisms that rely on photosynthesis for life; essentially green plants composed of as little as one cell

semienclosed body of water: one that is bordered on three "sides" by land, thus connecting to the ocean by a single relatively narrow opening

subgyre: a smaller gyre adjacent to a main gyre

zooplankton: small marine organisms that rely on consumption of phytoplankton and other zooplankton for life; essentially animals composed of as little as one cell

DESCRIPTION

The North Pacific Ocean is characterized by strong latitudinal gradients in surface temperature and salinity (high in the south and low in the north), multiple water masses, major oceanic gyres, mesoscale eddies of up to 124 miles (200 kilometers) across, and a complex bathymetry of deep trenches and remote seamounts. Climate-system variability around the world is intimately linked to the natural variability in this oceanic region.

High biological production is typical in the area, and it is responsible for major commercial fisheries of tuna and Pacific salmon. The North Pacific Ocean biome is also an important sink region for atmospheric carbon dioxide (CO_2), and therefore plays an important role in the ultimate fate of CO_2 on Earth.

Seven physically complex subregions are identified in the North Pacific Ocean: Yellow Sea and East China Sea, Okhotsk Sea, Oyashio Current System, Kuroshio Current System, Bering Sea, Alaska Current System, and California Current System. These seven subregions are part of the main gyre, the North Pacific Gyre. This gyre occupies approximately 7.7 million square miles (20 million square kilometers), with a clockwise circular pattern. The Aleutian Island chain constricts the main gyre, causing recirculation with two subgyres: the Western Subarctic and Alaskan gyres. These gyres and seas have distinct characteristics, often supporting different species and patterns of production.

YELLOW SEA AND EAST CHINA SEA

The Yellow Sea is a shallow marginal sea with a surface area of 146,719 sq mi (380,000 sq km) and an average depth of 144 feet (44 meters). It is broadly connected with the East China Sea to the south, and contains a semienclosed gulf in the north, the Bohai Sea. The People's Republic of China, the Democratic People's Republic of Korea, and the Republic of Korea (South Korea) are populations located on the Yellow Sea, making it one of the most densely populated regions in the world. Three main currents are found in the Yellow Sea: Kuroshio, Tsushima, and the Yellow Sea Warm Current (YSWC).

Due to its jagged coastline and many islands scattered around the shallow waters, the Yellow Sea has diverse marine habitats. Intertidal flats are the most significant type of coastal habitat, but mudflats, salt marshes, sandflats with gravel beaches, sand dunes, eelgrass beds, and mixed flats are also found. These habitats provide feeding, wintering, and summering grounds for migratory birds, and support important food resources and ecological niches for many species. Approximately 1,600 species have been reported from marine and coastal habitats in the Korean part of the Yellow Sea, including 400 phytoplankton, 300 marine macroalgae, 50 halophytes (salt-tolerant species), 500 marine invertebrates, and some 389 vertebrate species. Among them, 166 zooplankton and 276 fish have been reported as resident species in the Yellow Sea. Additionally, 100 commercial species have been identified in the region, comprising demersal fish (66 percent), pelagic fish (18 percent), cephalopods (7 percent), and crustaceans (7 percent).

SEA OF OKHOTSK

The Sea of Okhotsk is a semienclosed marginal sea bounded by Russia to the north and west and Japan to the south, with an area of 590,000 sq mi (1,528,100 sq km), similar in magnitude to the Bering Sea. The bottom topography is rugged, featuring the deep Kuril Basin. The shelf zone occupies almost 40 percent of the total area, and is somewhat isolated from direct water exchange with the open ocean by the Kamchatka Peninsula. Seasonal changes are distinctive, and ice can be found from November (rarely, in October) to June. In March, the Sea of Okhotsk is mostly covered by ice, except near Kamchatka and the Kuril Islands.

OYASHIO AND KUROSHIO

The Oyashio is the western boundary current of the sub-Arctic North Pacific, and it is characterized by low temperature, low salinity, and high nutrient concentrations. The eastward-flowing Oyashio forms the Subarctic Front (Oyashio Front), which is a distinctive temperature front. The Oyashio Front mixes with the Kuroshio Extension Front, where cold and warm waters mix and many mesoscale features are formed. This region is called the Kuroshio-Oyashio-Transition Zone (KOTZ), or mixed water region. The Oyashio brings nutrient-rich water into the western North Pacific, resulting in high productivity in the confluence zone of the two currents.

Along the south side of the Japanese Archipelago flows the Kuroshio Current, a western-boundary warm current. The width of this current is about 62 mi (100 km). The high temperature and high salinity qualities of Kuroshio water bring a high diversity of tropical marine life northward to the coastal area of the Japanese archipelago and off eastern Japan.

Although many species of invertebrates (such as corals) and vertebrates (such as butterfly fishes) can be found near the south coast of Japan, most of them cannot survive and reproduce in the coastal area because of the cold winter water temperatures and the low nutrient concentrations of the Kuroshio. Nutrient concentrations and biological production in the Kuroshio are much lower than in the cold western boundary Oyashio current. Nevertheless, the Kuroshio is a major spawning area for many species of pelagic fishes, such as Japanese sardine (*Sardinops melanostictus*) and Pacific saury (*Cololabis saira*), that have their nursery and feeding grounds in the Oyashio area.

BERING SEA

The North Pacific and Arctic Oceans are connected by a semienclosed sub-Arctic sea, the Bering Sea. Bounded by the Bering Strait to the north and the Aleutian archipelago to the south, the Bering Sea consists of a deep central basin, a northwestern shelf in the Gulf of Anadyr that reaches south along the Kamchatka Peninsula, and a broad eastern shelf that stretches from the Alaska Peninsula to Russia and the Bering Strait.

High biological productivity in the Bering Sea supports more than 400 species of fish and at least fifteen species of squid. Of these, at least forty species are of some commercial importance, and catches are dominated by walleye pollock (*Theragra chalcogramma*), flatfish (*Pleuronectidae*), Pacific cod (*Gadus macrocephalus*), crab (*Paralithodes* spp. and *Chioenocetes* spp.), rockfish (*Sebastes* spp.), and five species of Pacific salmon (*Oncorhynchus* spp.). Rich benthic communities with a large biomass and production of flatfish, Pacific cod, crab, and cephalopods are found in the broad shelves along the eastern and western margins of the Bering Sea.

Forage fishes such as capelin (*Mallotus villosus*), eulachon (*Thalichthys acificus*), Pacific sand lance (*Ammodytes hexapterus*), and juvenile walleye pollock and cephalopods can be locally abundant and provide important food sources to upper-trophic-level species. The surface waters of the central basin of the Bering Sea comprise important feeding areas for abundant Pacific salmon. The midwater community in the basin is poorly known, but it is likely dominated by lantern fishes (*Myctophidae*, in particular *Stenobrachius leucopsarus*), and deep-sea smelts (*Bathylagidae*).

By the second decade of the twenty-first century, scientists have reported that the ice extent of the Bering Sea, which is heavily connected to the marine ecosystem around it, has been steadily decreasing. By April 2018, the National Aeronautics and Space Administration's (NASA) Earth Observatory found that the sea's covering of sea ice was drastically below the average for that point in the year. This unusual pattern of melting is likely to disrupt the blooms of phytoplankton, which will in turn negatively impact the marine life dependent upon them.

ALASKA CURRENTS

The Alaska Current System is comprised by the shoreward Alaska Coastal Current (ACC); the offshore Alaska Current (AC); and the Alaskan Stream, eddies, and meanders. Currents extend along the continental shelf and shelf break, reaching as far south as the mouth of the Columbia River and as far west as the Aleutian Islands. The eastern Gulf of Alaska reaches abyssal depths of 9,843 ft (3,000 m) and plummets to 22,966 ft (7,000 m) in the Aleutian Trench. The continental shelf has a total area of approximately 142,858 sq mi (370,000 sq km), and ranges from 3 to 124 mi (5 to 200 km) in width. Islands, banks, ridges, and numerous troughs and gullies cut across the shelf, resulting in a complex bathymetry that promotes exchange between shelf water and deeper waters.

The ACC distributes sub-Arctic plankton communities around the region and into protected inside waters. During the summer months, the ACC has local reversals and small eddies that concentrate plankton and small fishes in convergence zones for foraging fish, birds, and marine mammals. Species including the southeast Alaska Pacific herring (*Clupea pallasii*), pink salmon (*Oncorhynchus gorbuscha*), sablefish (*Anoplopoma fimbria*), Alaska rockfish (*Sebastes* spp.), Pacific Ocean perch (*Sebastes alutus*), northern rockfish (*Sebastes polyspinus*), tanner crab (*Chionoecetes bairdi*), and Dungeness crab (*Metacarcinus magister*) are of commercial and ecological importance.

Marine mammals such as the stellar sea lion (*Eumetopias jubatus*), beluga whale (*Delhinapterus leucas*), orca (killer whale) (*Orcinus orca*), humpback whale (*Megaptera noraengeliae*), and sea otter (*Enhydra lutris*) are also found in the gulf. However, according to the National Oceanic and Atmospheric Association, the once thriving population of beluga whales in the Cook Inlet had declined by 75 percent as of 2016.

CALIFORNIA CURRENT

The California Current is a year-round Equator-directed flow that extends 1,864 mi (3,000 km) from the northern tip of Vancouver Island to Baja California Sur.

Cool, fresh, and nutrient-rich water is carried through the California Current System. Fishery resources include invertebrate populations, especially in near-shore waters; groundfish populations along the continental shelf; and migratory pelagic species such as salmon (*Oncorhynchus* spp.), Pacific sardine (*Sardinops caeruleus*), Pacific hake (*Merluccius productus*), and Pacific herring (*Clupea harengus*).

At the southern end, the northern anchovy (*Engraulis mordax*) and market squid (*Loligo opalescens*) are important. This system also supports large and diverse seabird and marine mammal populations.

OCEANIC BIOME STATUS

Throughout the North Pacific, annual rates of primary and secondary productivity are generally similar among regions. In general, biomass has been decreasing in the deepwater areas of the eastern and western North Pacific over the past several years. Phytoplankton and zooplankton biomass has decreased in the northern part of the Oyashio Current and in the southern part of the California Current, but increased in the Yellow Sea and the coastal parts of the Gulf of Alaska.

Fish fauna of North Pacific Ocean coastal systems have declined. More than fifty species of marine mammals in all are found in the North Pacific. Although not all species are considered to be of commercial importance, the Steller's sea cow (*Hydroanalus gigas*) is known to have been exploited to extinction before 1800, and the Japanese fur seal (*Zalophus californianus japonicus*) is thought to be extinct.

The North Pacific Ocean is known to be one of the most productive and diverse marine ecosystems in the world. This region has witnessed a rapid increase in human population, with important technological, economic, and social transformations over the past several years. Due to human pressures and the chemical deterioration of many regions in the North Pacific Ocean, significant losses of marine biodiversity have already been caused.

Observing an area as large as the North Pacific Ocean is a significant task, resulting in a lack of information for many coastal regions. Scientific research on many fronts has addressed global warming and its potential effects on Pacific Ocean water-layer modalities, salinity, acidity, current structure and speed, atmospheric changes, seasonal "marching," and island and shoreline effects—to say nothing of observations and theories on the climate-change impacts sustained by many affected marine species. Long-term monitoring programs in this ecosystem that sample physical and biological conditions are necessary to understand the long-term effects of global environmental changes and human activity.

—Maria Jose González-Bernat

Further Reading

Bakun, A. "The California Current, Benguela Current, and Southwestern Atlantic Shelf Ecosystems: A Comparative Approach to Identifying Factors Regulating Biomass Yields." In Sherman, K., Alexander, L.M. and Gold, B.D. eds. *Large Marine Ecosystems: Stress, Mitigation, and Sustainability*. Washington, DC: Am. Ass. Advancement of Science, pp. 199–224.

Barange, M., et al. *Marine Ecosystems and Global Change*. Oxford UP, 2010.

Beamish, Richard J. *Impacts of Climate and Climate Change on the Key Species in the Fisheries in the North Pacific*. North Pacific Marine Science Organization, 2008.

"Beluga Whale." *NOAA Fisheries*, www.fisheries.noaa.gov/species/beluga-whale. Accessed 27 July 2018.

"Characterising Meso-Marine Ecosystems of the North Pacific," S. D. Batten, et al. *Deep-Sea Research II*, vol. 53, 2006.

Fache, Elodie, and Simmone Pauwels, editors. *Fisheries in the Pacific: Challenges of Governance and Sustainability*. Pacific-Credo Publications, 2018.

Hansen, Kathryn. "Historic Low Sea Ice in the Bering Sea." *NASA Earth Observatory*, earthobservatory.nasa.gov/images/92084/historic-low-sea-ice-in-the-bering-sea. Accessed 27 July 2018.

McKinnell, S. M., and M. J. Dagg, editors. *Marine Ecosystems of the North Pacific Ocean 2003–2008*. PICES, 2010.

Nagai, Takeyoshi, Hiroaki Saito, Koji Suzuki, and Motomitsu Takahashi, editors. *Kuroshio Current. Physical, Biogeochemical and Ecosystem Dynamics*. Wiley, 2019.

Polovina J. J., et al. "The Ocean's Least Productive Waters Are Expanding." *Geophysical Research Letters*, vol. 35, 2008.

Segar, D. A. *Introduction to Ocean Sciences*. 2nd ed., W. W. Norton & Company, 2007.

Wilson, Samuel T., and Matthew J. Church, editors. *Microbial Ecology in the North Pacific Subtropical Gyre*. Frontiers Media, 2018.

NORTH SEA ECOSYSTEM

Fields of Study: oceanography; marine biology; marine geology; marine resource management

ABSTRACT

The North Sea is a semienclosed, continental-fringe sea in the North Atlantic Ocean. Given its strategic position as a doorway to the Atlantic, it has had a major part in shaping European history—and this has heavily affected its ecology. Commercial productivity of the region lies mainly in its fisheries, hydrocarbon fuels, and renewable energy. Conservation is also of the utmost importance, with many of Europe's most valuable natural habitats occurring around its perimeter.

KEY CONCEPTS

bank: an extensive plateau-like region of seafloor that may be an extension of a continental shelf or a segment of continental rock that lies at some depth below the surface

fjord: a deep, narrow canyon-like coastal inlet with steep sides that was typically formed by glacial action

pelagic: refers to the open seas above the abyssal plain of an ocean

wind farm: an array of windmill-driven electric generators

A SEA OF IMPORTANCE

The North Sea lies between Great Britain and the coastline of continental Europe, with Norway at its northern extremity and France to the south. It is the only marine interface for Belgium and the Netherlands, and also a major component of the marine territory of Great Britain, Germany, Denmark, Sweden, and Norway. It spans approximately 289,577 square miles (750,000 square kilometers); and its gently sloping topology is shallowest in the south, typically around 164 feet (50 meters) deep, reaching 328 ft (100 m) in the center of the basin, and more than 2,297 ft (700 m) in the Norwegian Trench.

Scattered throughout the basin are several large sandbanks, including the Dogger Bank, the Broad Fourteens, and the Deep Forties. With an area of approximately 6,795 sq mi (17,600 sq km), the Dogger Bank is one of the most productive fishing grounds in the North Sea.

Current flow tends to be counterclockwise, with the greatest inflow entering via the Norwegian Sea. Atlantic water also enters via the English Channel, but this flow is somewhat more restricted. Brackish water also infiltrates the North Sea from the Baltic Sea via the Kattegat and Skagerrak. The salinity and temperature of North Sea waters fluctuate with the North Atlantic Oscillation.

SPECIES AND BIOMASS

The North Sea has an extremely diverse coastal extent, including fjords, sandbanks, mudflats, large estuaries, and deltas, but the seabed itself is mostly sandy or muddy. There is a general pattern of increased species richness around the circumference of the North Sea. A variety of invertebrates are found on the North Sea benthos, including polychaete worms, sea urchins, brittle stars, shrimp, and lobsters.

Extensive beds of coldwater coral *Lophelia* have been mapped along the shores of Norway, particularly along the Norwegian margin of the Skagerrak, where a unique yellow variety has been recorded. *Sabellaria* reefs are also known to exist, although their distribution has yet to be fully studied.

The pelagic fish community is dominated by herring, with summer peak abundances in mackerel and horse mackerel. Cod, haddock, whiting, and saithe are also important, in addition to flatfish such as sole and dab. The relative abundance and distribution of fish varies from year to year in accordance with climatic conditions. In recent decades, highly migratory species such as tuna and halibut that once were abundant have now disappeared or become rare. Spiny dogfish, the region's major shark species, is now considered to be depleted. Catches of rays and skates have also decreased significantly over time.

Although overfishing is often blamed for the disappearance of species, scientists also suspect that climate change has had major impacts on the food chain in the North Sea, reducing the food supply for larger sea creatures. Evidence is also at hand that the average size of individual fishes is shrinking, also based on increasing difficulty of finding suitable foods, among related pressures.

The basin is a key ecosystem for seabirds, and around 5 million birds of 28 species breed here each year. These birds occupy niches from surface feeders to divers and waders. Foraging expeditions can range over hundreds of miles (or kilometers) at a time. Many of these seabirds interact with fisheries in the area through competing for resources and consuming discards.

Marine mammals in the North Sea tend to be transient, with a few resident populations of harbor and gray seals, occurring particularly along the British coast. Seals compete for resources with fishing boats and commonly steal food from the catch in passive fishing gear. Three main marine mammal species frequent this sea: minke whales, white-beaked dolphins, and harbor seals. Minke whales tend to remain in the northern reaches of the sea, while dolphins are more abundant in the south, and it is thought that the North Sea may be the most important habitat in the world for these animals.

RESOURCES

The commercial and recreational value of the North Sea basin is considerable, especially in terms of fishing and boating. It is also a major conduit for freight ships, military vessels, and underwater cables. Considerable oil and gas deposits lie beneath the North Sea. For this reason, the area has attracted investment from the world's energy producers. Several European states now have well-established operations within the basin.

European nations have been harnessing wind power in the North Sea since the early 1990s; the world's largest offshore facility, the Thanet wind farm, was inaugurated in 2010 and is located close to the shore of southeast England. There are also test facilities established in the Scottish Orkney Isles to explore the potential of wave and tidal energy. Plans are under way to establish an electricity supergrid connecting renewable energy resources from across the North Sea area.

CONSERVATION EFFORTS

European waters are conserved through several international policies, including the Habitats Directive and the Birds Directive. Numerous locations have been designated as Natura 2000 sites to meet the objectives of these two directives, while the Ramsar Convention designates wetlands of international importance, and other conventions aim to establish a network of marine protected areas.

Perhaps the most important region of conservation interest is the Wadden Sea, which follows 311 mi (500 km) of coast from the northern reaches of the Netherlands, along the German shoreline, to the western seaboard of Denmark. Its extensive mudflats provide important habitat for seabirds and shorebirds, and rich feeding grounds for migratory species. It is an immensely valuable ecosystem because of its rich diversity, and part of the Wadden Sea is now listed as a World Heritage Site.

—Ruth M. Higgins

Further Reading

Calewaert, Jan-Bart, and Frank Maes. *Science and Sustainable Management of the North Sea: Belgian Case Studies.* Academia Press, 2007.

European Environment Agency. "Natura 2000 Data—The European Network of Protected Sites." www.eea.europa.eu/data-and-maps/data/natura-2000.

International Council for the Exploration of the Sea. "ICES Advice Book 6: North Sea." www.ices.dk/committe/acom/comwork/report/2008/2008/6.1-6.2 North Sea ecosystem overview.pdf.

Ludewig, Elke. *On the Effect of Offshore Wind Farms on the Atmosphere and Ocean Dynamics.* Springer, 2015.

Pedersen, Soren Anker, et al. "Natura 2000 Sites and Fisheries in German Offshore Waters." *ICES Journal of Marine Science*, vol. 6, 2009.

Quante, Marcus, and Franciscus Colijn, editors. *North Sea Region Climate Change Assessment.* Springer Open, 2016.

O

Ocean and Tidal Energy Technologies

Fields of Study: oceanography; engineering; marine ecology

ABSTRACT

Every continent on the planet is surrounded by a cleaner, safer, more efficient energy resource. As conventional energy supplies are depleted, means are being developed, and in some cases are in actual operation, to convert the energy found in waves, tidal currents, ocean and river currents, ocean thermal gradients, and offshore wind into usable electric power for utility-scale grids, independent power producers, and the public sector.

KEY CONCEPTS

anthropogenic: caused by or existing as a consequence of human activities

barrage: a structure constructed across a river mouth or estuary with openings to house generators to be driven by the influx and efflux of water as the rising tide drives water to flow into the estuary and the falling tide allows the water to flow out again

biofouling: fouling of marine devices by the buildup of algae, barnacles and other biological entities, leading to disruption of the optimum functioning of the particular device

corrosion: typically, the degradation of metal through oxidation processes that are chemical or biochemical in nature

tide mill: a mill constructed at the shore of the ocean and powered by the rise and fall of the tides

wind farm: an array of large windmills that "harvest" the energy of wind to drive the blades of the mill, causing them as they turn to drive a generator that converts the energy of the wind to electrical energy

DEFINITION AND BASIC PRINCIPLES

The tides were the earliest source of obtaining power from the ocean. The requirements were simple: a dam to contain a head of water that was brought in by high tide and a waterwheel to turn as the water was let out, thus generating the power. The obtaining of power from crashing waves was another dream of oceanographers, and it has finally been realized in a variety of modest projects that generate power in a number of ingenious ways. Another dream of oceanographers has been deriving power from fast-moving currents in estuaries on the ocean itself. Several so-called in-stream devices have been developed, but the ultimate goal of obtaining power from Florida's famous Gulf Stream, the fastest-moving ocean current in the world, is still elusive although study is under way. Another interesting project was the famous ocean thermal energy conversion (OTEC) project to obtain power from the temperature difference between warm- and cold-ocean waters. This project had a success in Hawaii during the 1980s and 1990s, and is being looked at further by several countries. And a spectacular newcomer for obtaining power from the ocean is the development of offshore wind turbines. These giant windmills, whirling high on stilts, are enjoying great success in the coastal waters of Europe, although development in the United States has been very slow.

BACKGROUND AND HISTORY

Power has been generated from the tides in Europe since at least the Middle Ages, and tide mills were common in Great Britain, France, Ireland, and along the east coast of the United States until the middle of the nineteenth century. By 2011, the world had four power stations generating electricity from the tides. The largest was located on the Rance River estuary in France and rated at 240 megawatts. Others were on the Annapolis River in Nova Scotia, Canada (rated at up to twenty megawatts), the Kislaya Guba, Russia, plant (rated at one to two megawatts), and the Xiamen,

China, plant (rated at up to three megawatts). As of 2017, more than 25 megawatts of ocean energy capacity had been installed worldwide, according to the International Energy Agency's Ocean Energy Systems (OES) committee. The generation of power from the waves has been a more recent development. Small-scale installations are generating power in Scotland, Spain, and at other locations, and projects involving various new methods are under way.

HOW IT WORKS

The model for a successful tidal barrage plant is on the Rance River estuary in Brittany, France. Here the tide range is about 40 feet (12 meters). As the rising tide passes through the circular openings in the barrage, which is located near the mouth of the estuary, rotors spaced at regular intervals in the circular openings generate power as the water level rises. After six hours, the tide will have turned and the flow of water is in the opposite direction. Then the rotors are turned 180 degrees so that the plant can generate power as the tide flows out to sea.

Wave Power. As many as a thousand ingenious devices have been designed to obtain electric power from waves. Several promising devices are in the developmental stage, but no single design has been judged the best. One type is the point absorber, a bottom-mounted or floating structure that can absorb wave energy coming from all directions. A second type is the terminator, which reflects or absorbs all the wave energy coming at it. Another type is the linear absorber, which is oriented parallel to the direction of the oncoming waves. It is composed of interlocking sections, and the pitching and yawing of these sections, because of the waves, pressurizes a hydraulic fluid that turns a turbine. A fourth device is an overtopping device, which, much like a hydroelectric dam, produces electricity by emptying a reservoir of seawater and thereby turning turbines. Factors that must be considered are the corrosive and occasionally violent marine environment; biofouling, which begins the moment any device is placed in the ocean; and competing uses and environmental degradation of potential sites. Advantages include consistent, predictable generation and the higher energy density of wave energy as compared to solar power.

Power from Tidal and Ocean Currents. The dream of obtaining power from the ocean's fast-moving currents may finally be nearing realization. Tidal generators are being tested in several estuaries and ocean channels, using either bottom-mounted rotors or rotors suspended from floating barges. Power generation is nearly continuous, with the rotors turning for both flood and ebb currents. The greatest challenge will be harnessing power from the ocean's famous Gulf Stream. The Southeast National Renewable Energy Center at Florida Atlantic University in Boca Raton, Florida, plans to mount one or more giant rotors in this current as it flows between Florida and the Bahamas to pilot prototypes. Water depths approach 2,500 ft (762 m) there, so attaching the rotors to the seabed will be a challenge. In addition, there will be the usual problems of biofouling, corrosion, and getting the power to shore. The rotors will constitute a potential hazard for passing ships, submarines, and large marine mammals, such as whales, so these concerns will also have to be addressed as well.

Ocean Thermal Energy Conversion (OTEC). An OTEC plant operated successful in Hawaii during the 1980s and 1990s and is being considered by several European countries, despite its high cost. This plant derives power from the differential between 40 degrees Fahrenheit (4.5 degrees Celsius) deep water and 80 degrees Fahrenheit (27 degrees Celsius) surface water. The cold water, drawn from thousands of feet down, is used to condense ammonia, which is then brought in contact with the warm surface water, causing it to vaporize explosively, driving a turbine. One problem is the disposal of the cold water after it has been warmed in the vaporization process. It cannot be returned to the ocean, where it would kill reefs and tropical fish, so the solution in Hawaii was to pipe it through the soil where it fooled cool-weather crops such as strawberries and asparagus into growing in a tropical climate.

Wind Power. The latest method for obtaining energy from the ocean is the giant turbines turning in the near-shore, shallow waters of Europe, where nearly 16,000 megawatts of wind capacity had been erected by the end of 2017, according to the Global Wind Energy Council (GWEC). These huge turbines function much like windmills on land, except that they are firmly anchored to the seafloor, often in plain sight of coastal residents. Environmental and aesthetic concerns have long held up the installation of such turbines along the east coast of the United States, and continue to do so.

APPLICATIONS AND PRODUCTS

The number of tidal barrage power plants in the world is extremely limited because of the large tidal range required—a minimum of a 10-ft (3-m) rise between low tide and high tide. Few coasts in the world have that great a rise in a body of water narrow enough to be dammed, and even fewer have it in an area that is sufficiently populated to provide a market for the power generated. One problem faced by all tidal barrage plants is that the power generated is intermittent. The time suitable for power generation shifts steadily as the Moon orbits around the Earth. This means that the supply of power and the demand for power will not always coincide. For several nights, customers will have ample power to cook their supper and enjoy evening activities, but the next few nights they will have no power at all if their power supply comes only from a tidal barrage power plant.

Power Generation from Waves. OES figures show that by 2017, installations generating a total of about 8 megawatts of power had been created worldwide, mostly on an experimental or demonstration project basis. One that has been generating power successfully since 2000 is the Limpet, the world's first commercial wave-power station, on the rockbound coast of the Island of Islay in Scotland. The plant is of the oscillating water column design with a fortresslike exterior fronting the waves just at sea level along the rocky shore. When a breaking wave enters the long, concrete tube, with its opening just below water level, it drives air in and out of a pressure chamber through a specially designed air turbine, generating electricity. The Limpet produces 0.5 megawatts of power, which is fed into the island's power grid. The design makes the Limpet easy to build and install, and its low visible profile does not intrude on the coastal landscape or the ocean views.

Power from Currents. Besides the proposed giant rotors in the Gulf Stream, which would be an open-ocean device, a number of projects have been designed for obtaining power from tidal currents in estuaries and other constricted passages using bottom- or surface-mounted turbines. These turbines do not require construction of an expensive barrage. The flow of water simply turns the turbine as the tide comes in, and, if the turbine is reversed 180 degrees, it can also generate power as the tide goes out. An experimental array of six turbines was installed in New York City's East River in 2007. Similar arrays have been installed in the United Kingdom, Italy, Korea, Canada, China, and France. By 2017, more than 17 megawatts of tidal current capacity had been installed worldwide, according to the OES. The turbines strongly resemble torpedoes, with the rotor at one end, and they stand on a stout pedestal firmly attached to a channel bed or they can be suspended from floating barges. Power is transmitted to shore by means of cables.

Wind Farms. The major components for a wind turbine are a tower, a rotor with hub and blades, a gear box, and a generator. Offshore systems are larger than those on land because of the greater cost to install and service them. Most are rated at three to five megawatts, compared with one and one-half to three megawatts for those on land. They are usually fixed in water depths of 15 to 75 ft (5 to 23 m) and require a foundation driven deep into the seabed to support the weight of the tower and the rotor. Frequently the turbines are arranged in arrays to reduce maintenance and cabling costs. The power generated is sent to shore through a high-voltage cable buried in the seabed. A service area is always required, with boats and a hoist crane for repairs and maintenance.

Wind farms, both onshore and offshore, are becoming more common worldwide. GWEC reported in 2018 that by the end of 2017, almost 540,000 megawatts of total wind power capacity had been installed worldwide, primarily in China, the United States, Germany, India, Spain, and the United Kingdom. In the United States, the leading states for installed wind power capacity in late 2017 were Texas, Oklahoma, Iowa, California, and Kansas, according to the US Department of Energy. GWEC figures showed that offshore wind farms were most common in the United Kingdom, China, Denmark, the Netherlands, and Belgium. The United States' first commercial offshore wind installation was the five-turbine, 30-megawatt Block Island Wind Farm, which went into production off the Rhode Island coast in May 2017.

CAREERS AND COURSEWORK

College-level students seeking careers in marine renewable energy are advised to take a general oceanography course in order to familiarize themselves with the characteristics of the environment in which they will be working. In addition, they should supplement this study with basic courses in math and physics so that they can understand the technology involved in the design of the various marine-energy projects. For those students planning to go on to graduate study, engineering courses are highly recommended, especially for those students interested in designing and building ocean energy generators. The jobs available in the United States for students seeking opportunities in the field of marine-renewable energy remain somewhat limited because most of the American marine energy projects operating are just getting under way or are still in developmental stages. Additional wind farm projects are planned but the permit process for them is a lengthy one. Several coastal wave energy projects are now operating, as well as a few tidal stream projects in rivers and estuaries with strong tidal flows, but these projects are all in developmental stages. Once they mature into full-time operations, job opportunities in these areas should begin to appear.

SOCIAL CONTEXT AND FUTURE PROSPECTS

For many years the United States has been totally dependent on traditional energy sources, such as oil, natural gas, and coal, as well as water power and nuclear energy. Attention is turning to the generation of electricity from marine renewable energy sources. Projects in this field are still in the developmental stages, with tidal stream and wind farm installations showing the most promise. The potential for them to make a significant contribution to US energy needs is great because nearly half of the states have access to the oceans along their borders. Although the United States has ample supplies of coal, natural gas, and oil, the burning of these fossil fuels harms the environment and contributes significantly to anthropogenic climate change. Attention to marine-energy sources cannot help but increase in the coming years. The deployment of an offshore wind farm for Rhode Island is an encouraging sign. Such wind farms are already making a significant contribution to energy needs in Europe, and coastal conditions along the Atlantic and Gulf coasts of the United States are equally favorable. The cost of these projects is large, but they will provide many jobs for the installation and maintenance of the equipment, as well as for the manufacture of the turbines.

—Donald W. Lovejoy

Further Reading

Annual Report 2017. Ocean Energy Systems, International Energy Agency, report2017.ocean-energy-systems.org. Accessed 29 Mar. 2018.

Bedard, Roger, et al. "An Overview of Ocean Renewable Energy Technologies." *Oceanography*, vol. 23, no. 2, June 2010, pp. 22–31.

Boon, John D. *Secrets of the Tide: Tide and Tidal Current Application and Prediction, Storm Surges and Sea Level Trends*. Horwood, 2004.

Carter, R. W. G. *Coastal Environments: An Introduction to the Physical, Ecological, and Cultural Systems of Coastlines*. Academic Press, 1988.

"Global Wind Statistics 2017." Global Wind Energy Council. *GWEC*, 14 Feb. 2018, gwec.net/wp-content/uploads/vip/GWEC_PRstats2017_EN-003_FINAL.pdf.

Greaves, Deborah, and Gregorio Iglesias, editors. *Wave and Tidal Energy*. Wiley, 2018.

Khare, Vikas, Cheshta Khare, Savita Nema, and Prashant Baredar. *Tidal Energy Systems: Design, Optimization and Control*. Elsevier, 2019.

Rusu, Eugen, and Vengatesan Venugopal. *Offshore Renewable Energy: Ocean Waves, Tides and Offshore Winds*. MDPI, 2019.

Segar, Douglas A. *Introduction to Ocean Sciences*. 2nd ed., W. W. Norton, 2007.

Sverdrup, Keith A., and E. Virginia Armbrust. *An Introduction to the World's Oceans*. 10th ed., McGraw-Hill, 2009.

Thresher, R., and W. Musial. "Ocean Renewable Energy's Potential Role in Supplying Future Electrical Needs." *Oceanography*, vol. 23, no. 2, June 2010, pp. 16–21.

Ulanski, Stan. *Gulf Stream: Tiny Plankton, Giant Bluefin, and the Amazing Story of the Powerful River in the Atlantic*. U of North Carolina P, 2010.

Ocean-Atmosphere Interactions

Fields of Study: oceanography; climatology; computer modeling; atmospheric physics

ABSTRACT

The complex interactions between the oceans and the atmosphere are basic to the understanding of oceanography and meteorology. The liquid and gaseous envelopes surrounding Earth have powerful effects on weather and climate on a global scale.

KEY CONCEPTS

Equatorial countercurrent: a current that exists between the northern and southern equatorial currents and moves in the opposite direction, returning surface water that has been moved by the equatorial currents

gyres: slow circular surface currents produced in oceans by the action of wind and the Coriolis effect

trade winds: winds to either side of the equator that blow in a westerly direction, so named because they were favored by sailing ships sent out on trade missions to western destinations

upwelling: the rise of nutrient-rich cold water from depths to the surface, typically the result of thermohaline circulation

wind stress: directional surface pressure exerted against surface water by wind causing the surface water to move in the same direction

ATMOSPHERE AND OCEAN

The largest fraction of the heat energy the atmosphere receives toward maintaining its circulation is derived from the condensation of water vapor originating mainly from marine evaporation. Therefore, fundamental to the understanding of atmospheric behavior and oceanic behavior is an understanding of the processes occurring at the air-sea boundary. The interactions between the marine and atmospheric environments involve constant exchanges of moisture, heat, momentum, and gases.

The study of oceanic and atmospheric interactions involves both a huge gaseous body and a massive liquid body, neither of which is ever homogeneous in content. The makeup of the atmosphere varies continually, depending on the areas over which it flows. The content of the oceans also varies in density, temperature, salinity, rate of movement by regular currents, and surface movement under the influence of winds. As a consequence, this interaction is very complex and not entirely understood. Certain conditions that are regularly met, however, can act as a general guide to understanding ocean-atmosphere interactions.

Atmospheric circulations depend on heat rising from Earth's surface. Because of the large area of Earth's surface that is covered by oceans, the main source of heat to the atmosphere is the sea surface. In the oceans, heat is supplied primarily from the sun, with some additional contribution from atmospheric sources. Heat supply processes are important in the development of convection currents in the surface

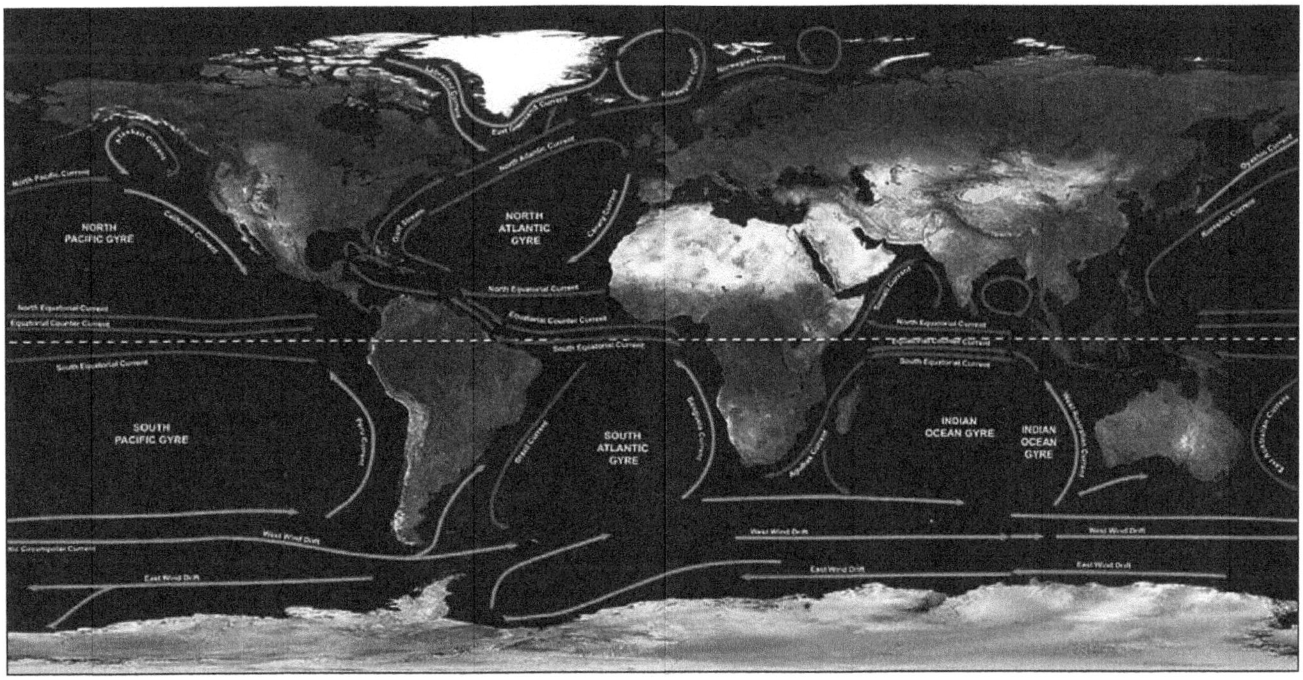

There are five major gyres, which are large systems of rotating ocean currents (National Oceanic and Atmospheric Administration)

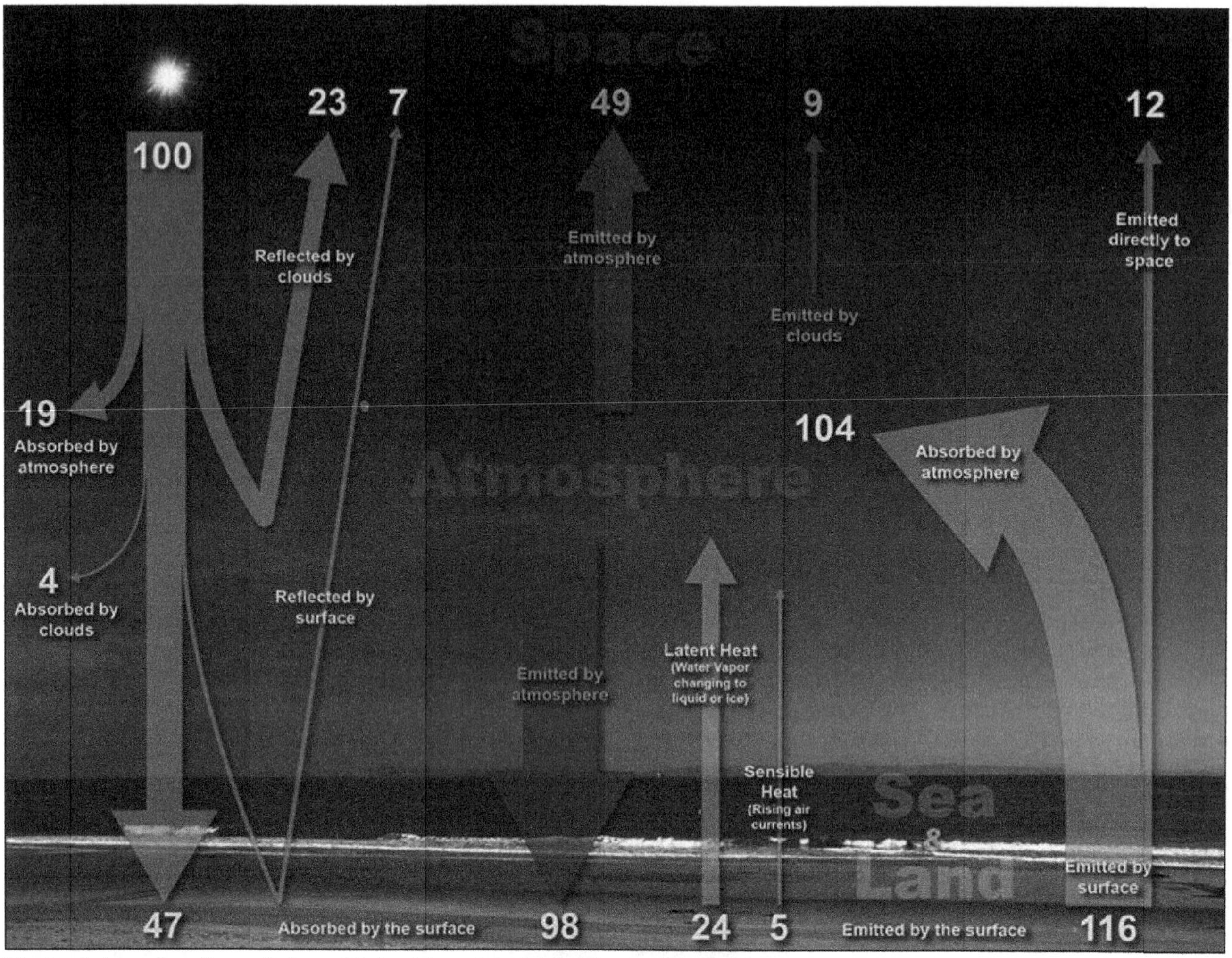

The Earth-Atmosphere Energy Balance (National Oceanic and Atmospheric Administration)

layer of the ocean, for the local exchange of energy with the atmosphere, and with slower, deepwater circulation currents. Heat exchange between the ocean and atmosphere has a pattern similar to evaporation. Wherever the surface of the ocean is warmer than the atmosphere, heat is transferred from the ocean to the air, usually as latent heat and is moved to great heights by eddies and convection currents in the air.

The range of weather extremes is smaller over the ocean than over land because of the enormous heat-storage capacity of the ocean, which tends to stabilize atmospheric conditions and properties. The upper layer of the oceans (to a depth of approximately 230 feet [70 meters]) can store some thirty times more heat than the atmosphere. Ocean climates, therefore, are largely determined by the atmospheric circulation and latitude. Ocean climate and atmospheric circulation are both affected by the solar distribution over Earth's surface, which is a function of the latitude and the season of the year. Northern and southern latitudes receive proportionately less solar radiation than latitudes at the equator; in winter, polar latitudes receive practically no solar radiation. Because of the greater capacity of the oceans to store heat, for a given change in heat content, the temperature change in the atmosphere will be around thirty times greater than in the ocean. Therefore, the ocean will lose its heat content by radiation much more slowly than the air. Land, which is intermediate with regard to heat storage and heat loss, can be modified by the effect of heat storage and heat loss of the ocean.

PROCESSES AT THE AIR-SEA BOUNDARY

The atmosphere adjacent to the oceans is constantly interacting with the oceans. Air does not simply flow along the surface of the sea but has a frictional effect or wind stress, which causes lateral displacement of the surface water. Wind stresses on the surface of the sea produce ocean waves storm surges, and shallow ocean currents. Pure wind-driven currents are the result of frictional wind-surface stresses and Earth's rotational motion. This rotational motion can be seen in the Coriolis effect, by which air currents and ocean currents are deflected from true linear motion to the left in the Northern Hemisphere and to the right in the Southern Hemisphere. Because the sea is continually in contact with the atmosphere, the gases that are present in the atmosphere are also found in seawater. The concentration of those gases depends on their solubilities and on the chemical reactions in which they become involved. Their concentrations are affected by temperature, which is determined by many factors, and by wind and wave actions.

The sea also has a large storage and regulating capacity with respect to processes involving carbon dioxide in the atmosphere and in the sea, including those processes relevant to photosynthesis. This whole group of reactions concerning carbon dioxide, generally referred to as the carbon cycle is extremely complex. Carbon dioxide, a "greenhouse gas" has a high capacity for absorbing infrared energy or heat. Solar energy that is absorbed at the ground surface is radiated back into the atmosphere and is absorbed by carbon dioxide. The heated carbon dioxide molecules reradiate that infrared energy as they cool, returning at least 50 percent of the heat they have absorbed into the atmosphere and back to the surface. This "greenhouse effect" acts to raise the functional temperature of the air and maintain it well above what it would be in the absence of the effect.

The largest fraction of radiant energy absorbed by the oceans is used in evaporation. The maximum evaporation and heat exchange between the sea and the atmosphere occur where relatively cold continental air flows over warm ocean currents moving toward the poles. Examples of where this phenomenon is most evident include the Kuroshio (Japan Current) and the Gulf Stream. The radiant energy that is absorbed and stored by the oceans at tropical latitudes may thus be moved and given off to the atmosphere elsewhere. This process is important to understand in terms of the Southern Oscillation and the effect of El Niño and La Niña.

Major wind systems are responsible for the formation and operation of the broadly symmetrical patterns of surface-water movement known as gyres, which rotate clockwise around the North Pacific and North Atlantic and counterclockwise around the South Pacific and South Atlantic oceans. Their tropical segments are the North and South Equatorial currents, which are driven westward by trade winds. The Equatorial Countercurrent, a compensating flow, travels west to east in the Pacific between the North and South Equatorial currents along a course that averages a few degrees north of the equator.

EL NIÑO AND LA NIÑA

Marked by warm water and high winds from the western Pacific, El Niño typically brings heavy winter rains to Peruvian deserts and warm weather to the West Coast of the United States. El Niño arises through interaction between the oceanic and atmospheric systems. During El Niño, the southeast trade winds over the equatorial Pacific collapse, allowing warm water from the western Pacific to flow eastward along the equator. This warm-water flow suppresses the normal upwelling of cold, nutrient-rich water and leads to the northward displacement of fish normally feeding in the nutrient-laden cold water.

El Niño is part of a gigantic meteorological system called the "Southern Oscillation" that links the ocean and atmosphere in the Pacific. This system normally functions as a kind of huge heat pump, distributing energy from the tropics at the equator to the higher latitudes through storms that develop over the warm western Pacific. Another part of the Southern Oscillation has been dubbed La Niña, which brings cold water to the central Pacific. La Niña exaggerates the normal conditions of the system. During this activity, easterly trade winds are stronger, the waters of the eastern Pacific off South America are colder, and ocean temperatures in the western equatorial Pacific are warmer. Atmospheric and oceanic conditions in the equatorial Pacific region can generate powerful effects on global weather. Therefore, the study of this interaction is essential.

STUDY OF OCEAN-ATMOSPHERE INTERACTIONS

Because the study of the ocean-atmosphere interactions is concerned with the boundary between marine and air masses, it is of necessity an interdisciplinary type of study. The data used in such studies are gathered by oceanographers and meteorologists who have made this interaction the focus of their research. The same instruments are used that are employed in oceanographic and meteorologic research, drawing heavily upon the same type of data collected, but with a shift in emphasis. Because gathering data is expensive and utilizes costly specialized equipment, nearly all studies are conducted by government scientists or are sponsored by government grants.

Scientists who study ocean-atmosphere interactions are interested in seeing how these huge bodies of matter affect each other and how these effects influence the weather and climate in the rest of the world. They are also interested in the ability to predict weather and climate changes more accurately. Each element researched, such as salinity, is compared with some other element, such as temperature, to see what relationship may exist. Air temperature and air movement are compared with wave movement, changing ocean currents, and water temperature. Each of a vast number of data points is examined for possible interrelations and interactions. When relationships are found, the data are fed into a computer model for correlation with other data. Computer models have been effective to some extent in predicting the results of interactions between marine and atmospheric environments. Separate computer models are used with the output aimed at the interactions between these forces.

Scientists do not fully understand all the mechanisms that link certain phenomena of the ocean and atmosphere interactions, such as El Niño and La Niña. It is their hope that in studying and under-

Flooding in Bangladesh (Badal Chandra Sarker via Wikimedia Commons)

standing the mechanisms involved, they may be able to make accurate long-range predictions of the amount and area of precipitation in specific regions.

SIGNIFICANCE

The interaction between the oceans and the atmosphere can cause immense problems on a global scale. An example of this interaction and its consequences is seen in the El Niño phenomenon. Roughly every three to five years, the surface waters of the central and eastern Pacific Ocean become unusually warm at the equator. Warm currents and torrential rains are brought to the normally dry desert area of central Peru and nutrient supplies for marine life from upwelling cold water along the west coast of South America are disrupted. Hardship can occur as a result, including widespread flooding in Peru.

La Niña brings an effect opposite to that of El Niño; easterly trade winds are stronger, the waters of the eastern Pacific off South America are colder, and ocean temperatures in the western equatorial Pacific are warmer than normal. As a result, the deserts in Peru and Chile become drier than normal, and the Indian subcontinent is inundated by heavier-than-usual rainfall and flooding as changes in the Southern Oscillation combine with the East Asian monsoon. In Bangladesh in late 1988, heavy rains and flooding killed more than 1,000 people, destroyed the homes of 25 million people, inundated 5 million acres of rice land, and damaged 43,496 miles (70,000 kilometers) of roads. While the storms were attributed to the La Niña phenomenon, much of the flooding was the result of massive deforestation in the Himalaya and foothills, which allowed water to rush down from the barren, eroded hills onto Bangladesh near sea level.

—George K. Attwood

Further Reading

Bigg, Grant. *The Oceans and Climate*. 2nd ed., Cambridge UP, 2003.

Brown, Joan. *Ocean Circulation: Prepared by an Open University Course Team*. Elsevier, 2016.

Chassignet, Eric P., and Jacques Verron, editors. *Ocean Weather Forecasting: An Integrated View of Oceanography*. Springer, 2006.

Clarke, Allan J. *An Introduction to the Dynamics of El Niño and the Southern Oscillation*. Academic Press, 2008.

Hamilton, Kevin, and Wataru Ohfuchi, editors. *High Resolution Numerical Modeling of the Atmosphere and Ocean*. Springer Science+Business Media, 2008.

Kagan, Boris A. *Ocean-Atmosphere Interaction and Climate Modeling*. Cambridge UP, 2006.

Le Quéré, Corinne, and Eric S. Saltzman, editors. *Surface Ocean–Lower Atmosphere Processes*. John Wiley & Sons, 2013.

Marshal, John, and R. Alan Plumb. *Atmosphere, Ocean and Climate Dynamics: An Introductory Text*. Elsevier Academic Press, 2008.

Sarachik, Edward S., and Mark A. Cane. *The El Niño-Southern Oscillation Phenomenon*. Cambridge UP, 2010.

Thurman, Harold V., and Alan P. Trujillo. *Introductory Oceanography*. 10th ed., Prentice-Hall, 2003.

Vallis, Geoffrey K. *Atmospheric and Oceanic Fluid Dynamics: Fundamentals and Large-Scale Circulation*. Cambridge UP, 2006.

Wells, Neil C. *The Atmosphere and Ocean: A Physical Introduction*. 3rd ed., John Wiley &Sons, 2012.

OCEAN COLONIZATION

Fields of Study: oceanography; marine ecology; marine biology

ABSTRACT

Ocean colonization is characterized by uniquely identifiable ecological zones, each with its own characteristic biota. The ocean being a dynamic place, ecological zones can change with changes in the oceans themselves. While the biota of a zone remains basically the same, the appearance of a new zone triggers colonization by the appropriate biota.

KEY CONCEPTS

barnacles: types of mollusks that attach themselves to submerged surfaces such as pier foundations, rocks, and the hulls of ships

biota: the living things, both plant and animal, that an ecosystem or ecological niche

intertidal zone: the area of a shoreline that is alternately covered by water at high tide and uncovered at low tide

pelagic: refers to the open water of the oceans above the abyssal plains

sea fans: an undersea avalanche, and the manner in which the debris is scattered over the ocean floor with the heaviest, densest particles traveling the furthest from the site of collapse and the fine silt being spread out in the water before settling; the name comes from the appearance of the materials after they have settled

THE ECOLOGICAL NICHE

An ecological niche is a unique, definable type of environment within the environment as a whole. Earth, as a living planet, can be broken down into a hierarchy of ecological niches, beginning with the whole environment, the entire planet. Within this master niche are three broadly defined ecological niches: the atmosphere, the lands and the water (the hydrosphere). Within each of these, there are smaller distinct niches. For example, the atmosphere can be subdivided into any number of altitude-related niches, the hydrosphere can be subdivided into any number of depth-related aquatic niches, and the lands can be subdivided into any number of terrestrial niches related to terrains. In addition, all of these can be further subdivided into more niches according to latitude and longitude.

The hydrosphere—the rivers, lakes and oceans—is a special case in that it consists of freshwater, saline water and frozen water (the cryosphere). In addition, all three of these water-based niches are intimately connected through the planet's hydrologic cycle and the globe-spanning thermohaline circulation in the world's oceans and seas.

There are a number of superficially similar niches common to both the saline and freshwater niches, such as shorelines, depths, and currents, but even where these features share the same space, as in riverine estuaries, they are still fundamentally separate.

COLONIZATION

In any environment, no matter how stable, changes occur that significantly alter the area within it. On land, for example, wildfire may eradicate the vegetation in an area. Similarly, clear-cutting of a forest lays bare the area that had been taken up by the large, mature trees of the forest, and usually the entire forest floor. In both cases, the area now becomes available for colonization by opportunistic plants and animals. The first requirement for colonization to occur is that the new species that would do the colonizing must be able to physically get to the new area. This can occur in different ways, of course, and what happens over time is that a succession of new species comes in to colonize the area in accord with its existing conditions, until eventually a mature forest stands there once more.

This process of opportunistic succession can be easily observed over time by watching a plowed field that is left unused next to a woodlot. First come grasses, followed by weeds that are taller than the grass, then more robust weeds and small bushes and shrubs slowly expanding the area they occupy toward the center of the field. Fast-growing trees will appear next here and there in the field, creating little groups where they have become established. Finally, the large softwood and hardwood trees will come. Each species comes into the field to colonize as the conditions become appropriate for that species. Natural competition also comes into play as the characteristics of the current dominant species in the field are such that they tend to discourage the establishment of other species that would subsequently be using the same resources. Accordingly, grasses growing thickly tend to prevent the seeds of other plants from reaching the ground and germinating, the spreading bushes and trees tend to discourage the growth of smaller plants by depriving them of sunlight, and so on.

While there are no wildfires in the marine environment, there are nevertheless events that can have the same effect. For example, a landslide or the collapse of a river bank may drop large quantities of sand and stones into a body of water, where it spreads over and smothers an existing niche area, essentially "wiping the slate clean." That area now becomes available for colonization by aquatic plants and creatures.

In areas where a pond may be scooped out to provide a groundwater supply for irrigation, the colonization process that follows is readily observed. At first the bottom of the pond can be seen through the crystal-clear water that infiltrates and fills the pond as a completely barren sandy expanse. Through the action of winds and rains, and with the assistance of passing waterfowl and local amphibians such as toads and salamanders, this isolated ecological niche is changed. Algae begins to grow, which will draw aquatic insects. An aquatic plant may be seen here and

In areas where a pond may be scooped out to provide a groundwater supply for irrigation, the colonization process that follows is readily observed. In spring, the pond will come alive with toads that have found it a suitable place to spawn. (crbellette via iStock)

there growing up from the bottom, especially if there are free-ranging animals such as muskrats nearby that may unknowingly transport a bit of water plant or a seed with them when they happen to visit this new water habitat. In spring, the pond will come alive with the toads that have found it a suitable place to spawn, followed shortly after by the appearance of a multitude of tadpoles. Deposits left by waterfowl such as cranes and herons may contain undamaged fish eggs from one of their last meals that later mature and hatch to provide a population of fish in the maturing pond. Eventually, the pond will become a thriving ecosystem in its own right.

The same sort of progressive colonization that occurred in the new pond will also occur in that new, barren sandy expanse that was created by that landslide or riverbank collapse. Sea fans occurring in the oceans have this same effect as they drop large quantities of silt, sand and stones cascading down an undersea slope to a lower level. Shoreline erosion, from whatever cause, also creates new niche terrains. Sea-level changes move the location of the intertidal zones accordingly, and over long periods of time this may have huge effects. Geological evidence has demonstrated that at one time the intertidal zones of the oceans were as much as 656 feet (200 meters) and 328 ft (100 m) higher than they are in the present time. In the ocean deeps, eruption of a new hydrothermal vent on a mid-ocean ridge creates a new ecological niche that will become inhabited by extremophile species that form the base of a food web supporting colonies of highly specialized creatures. Should that vent instead be a volcano that breaks the surface to form an entirely new island, like the island of Surtsey, that island begins life with an absolutely clean slate with regard to ecological niches and colonization, once it becomes stable, right from its base on the seafloor to its highest peak.

In essence, then, colonization occurs when a new environmental space becomes available that has suitable conditions to support the appropriate biota for that space.

OCEAN NICHES

Despite the vastness of the oceans, the number of different types of ecological niche is quite limited. These include intertidal zones, the continental shelves and banks, deep-sea colonies, seabird colonies, coral reefs and the pelagic ocean itself. As should be expected, within each of these there are numerous subdivision niches.

Intertidal Zones: The intertidal zone is the narrow expanse of shoreline that is bared at low tide and under water at high tide. Some include the ocean water at the base of this zone as a part of the intertidal zone as well. Intertidal zones are typically areas of high bioproductivity and are generally very active locales at low tide, as different organisms, including humans, forage over them for food species. Intertidal zones also include salt marshes and mangrove forests that receive tidal inundations of ocean water.

Continental Shelves and Banks: The continental shelves are integral parts of their respective continental land masses, rather than the deep ocean floor. Dense, basaltic rock forms the crust that is the seafloor of Earth's oceans. The continental land masses are composed primarily of less dense silicate rock, which "floats" on top of the denser basaltic rock of the crust. The whole acts as a giant, planet-sized "basin" for the waters of the oceans. The sea level of the water currently exceeds the level of the outer reaches of the continents by several hundred feet (or meters), and it is these water-covered outer reaches that are the continental shelves. They are home to a wide variety of ecosystems ranging from seagrass and kelp beds to coral reefs (in warmer water), up to the beginning of the intertidal zone.

Deep-Sea Colonies: These include the various organisms that inhabit the oceanic benthos (the floor of the ocean), as well as the hydrothermal vent environments, and the deeps and trenches of the different seas. For centuries, the deep-sea floors were believed to be just still, lifeless, water-filled voids untouched by light of any kind. It was not until deep-sea exploration revealed the mid-ocean ridge systems and the riotous life surrounding the hydrothermal vent systems that attention turned to the deep seafloor itself. Cameras and later deep-sea submersible craft revealed the presence of many previously-unknown lifeforms, many of which are bioluminescent, and plants moving gently in the currents that now obviously exist at those depths. These were findings that revolutionized marine science only about 60 years ago.

Seabird Colonies: These are included here for their relationship to the upwelling ocean currents that bring large quantities of nutrients from the seafloor to the surface. Large numbers of fish congregate in these areas, forming a food web based on the massive blooms of phytoplankton and zooplankton that use the nutrients from the cold upwelling currents. This also draws seabirds in great numbers that feed on the fish and colonize the area nearby.

Coral Reefs: Coral reefs are highly specialized ecological niches, in that they consist of coral polyps of many kinds and colors living on the calcium carbonate structure that they produce. The structure and its environs also provide a dynamic habitat for thousands of species of fish and other creatures. Coral reefs are easily damaged by human activities, and are presently in danger from increasing water temperatures causing coral bleaching and structural damage from physical contact. There are different kinds of coral reefs, including fringing reefs, plateau reefs and atolls. They are home to thousands of different reef-dwelling species. Coral reefs are found only in tropical waters to depths of no more than about 492 ft (150 m), although it was reported in 2020 that marine scientists had discovered a large forest of deep-sea coral that may one day develop into a deep-sea coral reef.

The Pelagic Ocean: This is the open ocean overlying the deep-sea floor. This niche is the "free range" niche for many species of oceanic creatures, including the great whales, squid of all kinds, ocean sunfish, sharks and many others. This is where banks and guyots can be found as well. A bank is a flat-topped piece of continental rock that is too small to break the surface and so is covered by water that is much shallower than the surrounding water above the abyssal plains. A guyot is an undersea volcanic mountain that has since either subsided into the seafloor, or that has had its peak eroded away, leaving an undersea mesa. Banks such as the Grand Banks, near Canada's province of Newfoundland and Labrador, in the North Atlantic Ocean. Are often the sites of large fisheries. The Grand Banks themselves were famous for the immense number of north Atlantic cod that could be found there. Overfishing has since caused that fishery to collapse.

ODD AND UNEXPECTED NICHES

Nature is nothing if not opportunistic, and colonization in the oceans can take some surprising and unexpected turns. Sailors have long fought the attacks on their ships by barnacles and shipworms. Before humans took ships into the oceans, barnacles were satisfied with submerged natural rock and wooden surfaces that could be found in the oceans, and shipworms could only be found in submerged wood such as trees and branches that had washed into the seas. The absolutely new and unnatural surfaces of the hulls of ships and pilings for wooden piers offered a new niche to be colonized, and they were. Barnacles continue to be problematic for oceangoing ships as they attach themselves to the hulls by secreting an ad-

Perhaps the most haunting of artificial niches are the cities that have been lost to the seas over past centuries, their pristine sculptures still on guard and buildings, though collapsed, in remarkably good condition (PlanctonVideo via iStock)

Man using a pressure washer to clean barnacles from the bottom of a boat while it is in dry dock. Barnacles continue to be problematic for ocean-going ships as they attach themselves to the hulls by secreting an adhesive material stronger than any known substance. (diatrezor via iStock)

hesive material stronger than any known synthetic bonding agent. The resistance their growths provide to movement of ships through the seas has cost billions of dollars in revenue over the years, and their elimination often requires physically removing the ship from the water, or at least employing the services of specially-trained maintenance divers. Hundreds of millions of dollars have been spent in developing antifouling agents to be applied to the hulls, and in finding other ways to defeat the barnacle *blitzkrieg* that continues to this day. Shipworms are not as much of a problem since the advent of metal and composite hulls, but there are still many wooden-hulled ships whose owners are not impressed when shipworms are found in the woodwork.

Another odd niche is the artificial reefs produced by sinking ships-of-war that have been defeated in wartime operations or decommissioned in peacetime. Each one provides a clean slate for colonization, and though fish are the first species to move into the new living space, it is not long before the wrecks are festooned with growths of sea plants, barnacles and other species. In the proper water temperatures, a sunken ship will even become encrusted with coral growths, beginning the formation of a new coral reef in that location.

Perhaps the most haunting of artificial niches are the cities that have been lost to the seas over past centuries, including those that have been purposely abandoned and flooded in the reservoir behind newly constructed dams. The port of Alexandria, in Egypt, was once an island that housed the palace of Queen Cleopatra some two thousand years ago. About 800 years ago, a massive earthquake brought about the subsidence of the island, drowning the palace and grounds entirely. Marine archaeologists discovered the site only a few short decades ago, with its pristine sculptures still on guard and buildings, though collapsed, in remarkably good condition. One can only imagine what their first impressions were on seeing an ancient city among the fishes, a once glorious site of human history made into just another ecological niche by impervious Nature.

THE DARK SIDE OF COLONIZATION

In accord with that opportunistic nature of Nature, not all colonization is good. The ease with which humans have been able to cross oceans and colonize lands that were not their usual "niche"— think of the many voyages of discovery and ocean transport that have taken place to facilitate that colonization—has seen the introduction of alien species into areas Nature had never positioned them to enter. Many plant and animal species today are recognized as "invasive species." What typically happens when these species are introduced into these new ecological niches where they have proper conditions for their growth and none of the natural predators that keep them in check is that they run rampant. The endemic species also have no experience of these invasive species and may be eradicated by their failure to compete successfully. A handful of rabbits introduced to Australia, where there had never been rabbits before, reproduced rapidly and quickly became a pest of unrivaled proportions. Lamprey eels, an anadromous species, quickly found their way into the Great Lakes between Canada and the United States following the construction of canals to circumvent the rapids in the St. Lawrence River. With no natural predators there, they found freshwater streams conducive to spawning. Lampreys ravaged, and continue to ravage, the native populations of lake trout and other large fish species. The Florida Everglades were never home to boa constrictors and pythons, until humans thought it a good place to discard the snakes they could not keep for whatever reasons. Now, these snakes have multiplied so prolifically and grown to such sizes that some people make a good living by hunting them down, yet after years of this their populations remain unchecked, to the detriment of the natural wildlife of the Everglades.

There are, in fact, hundreds and perhaps thousands of cases of aquatic animal and plant species that have been transplanted to an environment in which they are not native. This assisted colonization process has allowed them to colonize take over every ecological niche that has amenable conditions for them to prosper. This is an aspect of ecology that humans must take great care to amend.

—Richard M. Renneboog

Further Reading

Bellwood, Peter, editor. *The Global Prehistory of Human Migration.* Wiley Blackwell, 2013.

Fusetani, Nobuhiro, and Anthony S. Clare, editors. *Antifouling Compounds.* Springer, 2006.

Hawksworth, David L., and Alan T. Bull, editors. *Marine, Freshwater, and Wetlands Biodiversity Conversation.* Springer, 2006.

Johnson, Craig Richard. *Seaweed Invasions: A Synthesis of Ecological, Economic and Legal Imperatives.* Walter de Gruyter, 2007.

Nentwig, Wolfgang, editor. *Biological Invasions.* Springer, 2008.

Railkin, Alexander I. *Marine Biofouling: Colonization Processes and Defenses.* CRC Press, 2004.

Rilov, Gil, and Jeffrey A. Crooks, editors. *Biological Invasions in Marine Ecosystems.* Springer, 2006.

OCEAN DRILLING PROGRAM

Fields of Study: oceanography; marine geology; marine engineering; stratigraphy

ABSTRACT

The series of ocean-drilling efforts called the Ocean Drilling Program (ODP) revolutionized understanding of Earth's structure, climate, and available minerals. It has also allowed researchers to collect data about Earth's cosmological history.

KEY CONCEPTS

core sample: a cylindrical rod of material recovered either from a corer, a piston corer or a circular drill corer; the sample exhibits the strata of the material within the sediments or rock that has been sampled

dynamic positioning: the process of actively monitoring and continually adjusting the position of a drilling platform ship above the borehole by using stationary reference points, made much easier in the present day through computer-controlled thrusters and a global positioning system (GPS)

heave compensator: a system that functions to prevent the drillhead and piping from rising and falling in the borehole as waves raise and lower the drilling platform ship

marker fossils: fossils of specific types and species that can be coordinated to identifiable geologic times and terrains

piston corer: essentially a length of heavy steel pipe containing an internal piston that can be dropped directly into seafloor sediments to extract a core sample of those sediments up to 60 feet (18 meters) in length

MOHOLE PROJECT

Scientific drilling on the ocean floors began almost as a stunt, but it evolved into a long-term research program that revolutionized geology, contributed vital clues about climate change and about the extinction of the dinosaurs, discovered a major new type of subsea hydrocarbon deposit that may fuel the world, and aided in the study of rich metal ores that may also be tapped. Along the way, scientific drilling pioneered many techniques that have been used in offshore drilling for petroleum and natural gas.

Drilling samples have been a major part of geology since the second half of the nineteenth century. Drillers seeking water or oil could sample pieces of rock drilled from varying depths and log them into their drill records. For scientific purposes, drilling with circular drills allowed the cutting of a long cylinder or "core" that could be pulled up and carefully measured for position and composition. By the middle of the twentieth century, hard-rock miners were using cores to sample for ores that could be mined. Data from these drillings were correlated into three-tal (3D) maps of distinctive strata showing dips and faults.

In the late 1950s, a number of geologists envied the space program with its romantic goal of flight to the Moon. In 1959, they proposed a similarly dramatic program called the "Mohole project." In 1909, Andrija Mohorovičić (1857–1936) had analyzed seismic waves from earthquakes and concluded that the rock of the crust changed significantly about 10 to 25 miles (16 to 40 kilometers) below the surface as it changed to partially melted mantle. There was speculation that rocks might be different at this so-called Mohorovièiæ discontinuity, or "the Moho," and researchers proposed drilling all the way to the Moho for samples. Drilling such a tremendously deep hole would be expensive and maybe impossible. Several marine geologists suggested that oceanic drilling to the Moho would be cheaper because the crust is thinner under the ocean floor. Also, a drill core through the seafloor might yield a complete fossil record of tiny marine shells.

Drilling into the ocean floor required several major innovations. The drilling locations were in deep water, so the drilling platform had to be an oceangoing vessel rather than a tower resting on the bottom. Anchoring in those depths would be difficult, so the vessel had to actively maintain position. This "dynamic positioning" had to keep the drill ship within two ship lengths of straight, or the drill string would break. While out of sight of land, the drilling vessel crew had to navigate within this small area with no landmarks whatsoever. This was managed originally by taut, moored buoys and later by satellite position-finding and acoustic beacons on the seafloor. Since the drill bit would hang as much as several thousand feet (or meters) below the drill ship, bottom-hole assemblies were required for getting the drill started and for returning to the same hole. Finally, because the drill platform rose and fell with the waves, the rig needed a heave compensator.

Project Mohole started with engineering tests in 1961 in waters off California by *Cuss I*, a drilling barge developed for offshore oil drilling and built by Global Marine. (A number of famous drill ships from this company begin with the abbreviation "*Glomar*.") Initial tests off Baja California, Mexico, were promising, but the project eventually ended when technical difficulties and political management problems caused projected drilling costs to increase significantly. It was later discovered that the Moho outcrops at Earth's surface in certain areas and that most areas of the ocean are comparatively young geologically, so continuation of the Mohole project would have been of less use than its backers had hoped.

However, the possibility of deep-ocean drilling had been demonstrated. Oil drillers, who had previously worked in depths of fewer than 328 ft (100 m), saw new possibilities. Project Mohole stirred interest in dynamic positioning for oil and gas exploration; the offshore drilling industry subsequently developed many technologies that passed back to the scientific drillers. Oceangoing rigs allowed exploratory drilling in deep areas of the continental shelf before making multibillion-dollar investments in production platforms. Thus, the Mohole project was a major factor in opening many offshore oil fields.

Furthermore, the Mohole demonstration had obtained one major data point. A mysterious lower layer visible on sonar graphs was found to be not another layer of sedimentary rock but rather basalt, an igneous rock formed by recrystallization of molten rock and often associated with volcanoes. This discovery suggested that the oceans were younger than expected and that the Mohole project, as originally planned, would have been able to gather minimally useful data about the geological history of the Earth. Conversely, it gave support to the then-radical theory of continental drift.

Testing continental drift required not one deep hole to the Mohorovičić discontinuity but many shorter holes surveying many areas. These shorter holes were still a tremendous advance. Before oceanic drilling, the only data about the seafloor came from dredge hauls and piston cores. Dredges only pull a jumbled mass of material from the first few inches (or centimeters) of the seafloor. A piston corer is essentially a weighted pipe that is allowed to fall as fast as possible to the seafloor; its weight and momentum drive it into soft sediment. Its limitations are that it penetrates only a few feet (or meters) and that it cannot penetrate rock strata.

JOIDES

In 1964, the Joint Oceanographic Institutions for Deep Earth Sampling (JOIDES) was formed; it has become an international organization that includes universities and government research organizations. In April and May of 1965, JOIDES used the drill ship *Caldrill I* to test upgraded methods by drilling six holes on the Blake Plateau off the coast of Florida to sub-bottom depths of more than 3,280 ft (1,000 m). Based on that success, JOIDES proposed an eighteen-month program of scientific drilling in the Atlantic and Pacific Oceans to the US National Science Foundation. The resultant Deep Sea Drilling Project was operated by the Scripps Institution of Oceanography. The DSDP drill ship *Glomar Explorer* was capable of drilling 2,493 ft in 20,0013 ft (760 m in 6,100 m) of water. It began operations in July 1968, and ultimately made ninety-six voyages (or drilling legs) for JOIDES that focused on sites on or near the mid-ocean ridges. The cores retrieved revolutionized geology by proving the theory of continental drift.

In 1985, the *Glomar Explorer* was replaced by the *JOIDES Resolution* run by Texas A&M University as part of the one-third internationally funded Ocean Drilling Program (ODP). The ship was designed to drill in 5 mi (8 km) of water for a total drill-string length of 5.6 mi (9 km), and it can drill 1,25 mi (2 km) into the seafloor in shallower waters. Both the *Glomar Explorer* and the *JOIDES Resolution* worked in concert with remote instruments and piloted instruments in a number of revolutionary developments. By 2003 the *JOIDES Resolution* had conducted 110 ODP expeditions at 2,000 ocean-basin drill holes around the world

Building on the achievements of DSDP and ODP, the Integreated Ocean Drilling Program (IODP) started operations in 2003. Comprised of twenty-six participating nations, the IODP financed drilling platforms, including a revamped *JOIDES Resolution*, the Japanese deep-sea drilling vessel *Chikyu*, and specialized, mission-specific platforms. These platforms conducted fifty-two expeditions in previously undrilled areas of Earth's subsurface. In October 2013, the IODP changed its name to the International Ocean Discovery Program: Exploring the Earth Under the Sea, but kept its acronym, IODP, unchanged. As of 2017, twenty-four nations are funding the organization's Science Plan 2013–2023, *Illuminating Earth's Past, Present and Future*. The plan's scope of inquiry includes climate change, deep life, planetary dynamics, and geohazards.

PLATE TECTONICS

Drilling in marine strata has yielded tremendous advances in geological knowledge for several reasons. First, oceans cover more than 70 percent of Earth, so a proportionate number of discoveries should be expected from oceanic data. Second, tens of thousands of drill cores have been logged on land, so many of the land discoveries have already been made. Third, land is often subject to erosion, so there are more gaps than in marine strata. Fourth, land strata have been subjected to more compression, heat, and chemical attack—all of which could confuse possible geological data—than seafloor strata. Fifth, some of the mechanisms that happen under several miles (or kilometers) of water in the oceans happen under several miles (or kilometers) of rock on land, so these mechanisms are easier to study with marine drilling. Finally, drilling on land requires disassembly and reassembly of equipment, often in difficult terrain; a drill ship simply sails to the next drill site.

In 1620, Francis Bacon (1561–1626) noted similarities between the coastlines of South America and Africa and suggested that they had once been interconnected. In the early twentieth century, Alfred Wegener (1880–1930) proposed the former supercontinent of Pangaea to explain similar fossils from areas now widely separated. However, these theories lacked a provable mechanism for moving entire continents through the rock underlying the ocean floor.

It was eventually noted that the Mid-Atlantic Ridge had an identical, though reversed, pattern of bands of different magnetic orientation on either side. These bands corresponded to reversals of Earth's magnetic field that occur every several thousand years. It appeared that new melted rock was crystallizing (and thus freezing a weak magnetic field signature) at the ridge and was then shoved away from the ridge by new rock. The crystallizing rock that cooled and hardened out of magma would quite likely be basalt. The Mohole finding of basalt in its deeper cores was encouraging, but conclusive proof required correlating the age from a large number of drill samples. The initial *Glomar Explorer* drill cores allowed geologists to concentrate on that dating using marker fossils and radioisotopic dating. Data from the cores showed that the ocean floor approaching the Mid-Atlantic Ridge was progressively younger. Thus, it appeared that Europe, North America, Africa, and South America were joined until sometime between 200 and 170 million years ago, when the Atlantic Ocean began opening between them.

This revolutionary discovery allowed development of plate tectonics theory, which proposes that large plates of connected rock exist on Earth's surface. Plate tectonics has many implications. Because the matter that makes up the planet is finite, Earth cannot expand indefinitely. Thus, the idea that new rock accretes along mid-ocean ridges means that plate material must be disappearing elsewhere. That explains the existence of deep ocean trenches (such as the Mariana Trench) where oceanic strata are diving steeply down toward the mantle and some of the heated rock spurts up, forming island arcs. One plate riding over another provides the mechanism for raising up mountains, such as the Andes Mountains, which are riding up over the edge of the Pacific plate. A more extreme example is the Indian subcontinent, which is burrowing under the Tibetan Plateau and lifting the Himalaya.

HYDROTHERMAL VENTS

Continued study of plate tectonics along the mid-ocean ridges and other plate boundaries led to the unexpected discovery of hydrothermal vents, which are somewhat like geysers and hot springs but which are found on the ocean floor; however, the scale and results of the activity surprised researchers. In 1965, a sample of rocks dredged from the seafloor yielded rock samples of two types. One was depleted in certain chemicals that were present in the other type. Also, in the 1960s, it was noted that thermal readings in sediments near plate boundaries did not have as much heat as expected.

Hydrothermal vents were suggested as the mechanism for moving the large amounts of heat. Seawater percolating down to the molten rock beneath the seafloor would be heated and return to the ocean as hot springs. The heated water would be much less dense than the near-freezing seawater near the ocean floor, so convection would drive the process.

In 1977, researchers in a piloted submersible discovered an active field of hydrothermal vents with "black smokers," where plumes of superheated water meet colder waters and dark minerals begin to precipitate out and form "chimneys." This discovery revealed the importance of such vents in circulating heat from Earth's core, circulating minerals into and out of the ocean waters, and sustaining life driven by chemosynthesis (chemical energy) rather than photosynthesis (light energy).

However, only drilling could supply data on subsurface events. For example, bacteria living by chemosynthesis have been found several hundred feet (or meters) below the ocean floor, suggesting that such life-forms may be more biomass-based than photosynthesis-based life. Also, drilling in mounds built by successive chimneys revealed large amounts of carbonate salt that eventually dissolves when the area cools. Thus, old, inactive hydrothermal fields might have rich mining potential. Finally, the sulfide-enriched upper layer may be underlain by a copper-enriched layer, also with good mining potential. More importantly, comparable areas of former seabed have been lifted up and become land. These areas, such as the Klamath Mountains of northwestern North America and areas of Cyprus, have already served as mining areas, and better knowledge of the mineral-formation processes will probably improve land mining long before any ocean mining.

MARKER FOSSILS

Probably the most important dividends from the ODP are carefully logged and partially analyzed cores from every ocean except the Arctic. By the end of the twentieth century, there were more than 149 mi (240 km) of cores available for new studies and new instruments. These cores have helped provide increasingly fine chronological correlation among strata from various locations. Oceanic core data allow the identification of "marker fossils" that (ideally) lived for a short time but over a wide area. Those marker fossils can then be used to date match with other oceanic cores and land cores. This correlation is crucial for tracking individual layers of rock through folds and faults. The dated and mapped layers allow one to calculate what might have happened in the past, what might happen in the future, and where ores may exist at present.

The types of fossils in the cores are generally tiny microfossil remains of plankton that lived in the surface waters. The types of fossils indicate the climate in those waters when they were deposited. Likewise, percentages of carbon and oxygen isotopes vary with sea surface temperature. Data from isotopes and microfossils can be checked against one another and against climate data from land to calculate past climates (paleoclimatology). With better estimates of past climate, climatologists may better project possible climate change in the future.

Ocean drilling provides another climate indicator: sediments from land. Drilling confirmed massive catastrophic flows when ice dams broke at the end of the last ice age, releasing torrents of water down the Columbia and St. Lawrence River Basins. Drilling also shows the timing, direction, and quantity of windborne sand and dust from places such as North Africa and Central Asia. Cores around Antarctica have corroborated that Antarctica's continental ice sheet has existed for millions of years; however, continued studies are needed to detail fluctuations of ice level.

Drill cores also confirmed that turbidity currents (dense masses of sediment-laden water) are a major mode of deposition in the ocean. (The low relief of the abyssal plains in the Atlantic Ocean, which has a high amount of sediment compared to total area, is caused by turbidity flows filling in low areas.) Turbidities indicate past ocean currents, another paleoclimate factor.

Petroleum and natural gas traces in deep waters are another major discovery of ocean drilling. However, the *Glomar Challenger* and *JOIDES Resolution* had to stop drilling whenever they encountered such traces. These craft have room for laboratories because they were not fitted with a "riser," which collects anything coming out of the drill hole and pumps down drilling mud. The riser can be used to stop a flow of fluid from a borehole. Therefore, the riserless drill ships risked causing an uncontrolled leak into the ocean if they continued drilling in areas with any sign of petroleum or natural gas.

METHANE HYDRATES

Still another hydrocarbon deposit researched by ocean drillers is methane hydrate. Scattered reports of flammable ice and fizzing material brought up by piston cores led to a theory that methane (natural gas) might be frozen into the structure of ice in deposits beneath the deep ocean floors. These deposits could form because the waters are only slightly above the freezing point of fresh water and they have high pressure—the two requirements for methane hydrate. The methane source would probably consist mostly of the decay of biological material in sediments. Sonar readings even showed layers in seafloor sediments that might be such hydrates.

Drill cores from several sites confirmed extensive methane hydrate deposits, and their analysis led to several conclusions. Hydrates may contain several times as much energy as all other fossil fuel deposits combined. Hydrates might act as a cap rock to contain conventional gas and petroleum. It might be some time before hydrates are exploited because they are often spread thinly in sub-seafloor ocean strata. Hydrates might be a major factor in climate change because disturbances on the ocean floor or lowered sea levels might release methane, a greenhouse gas; conversely, raised sea levels would increase the area of high pressure, allowing more hydrates to form, thus decreasing the flow of greenhouse gases to the atmosphere. Finally, hydrates are stronger than unfrozen seafloor muds and oozes. Consequently, a hydrate-melting event could cause a major collapse of sediments on the continental slope and result in a major turbidite flow.

Drill cores also corroborated the asteroid-impact explanation for extinction of the dinosaurs. A thin dead zone marked the end of the Cretaceous period, the last great age of dinosaurs, and the beginning of the Tertiary. This Cretaceous-Tertiary (K-T) boundary zone was noted in a few rock outcrops in Europe, and rock from this layer had elevated levels of platinum-group metals, specifically iridium which is very rare on Earth (similar to elevated levels in asteroids) in that strata. However, this radical theory was met with doubt, until multiple drill cores closed in on the probable impact site near the Yucatán Peninsula of Mexico with thicker corresponding deposits off Florida and in the Caribbean Sea. Besides the platinum-group-enriched dead layer, those deposits had large amounts of shocked quartz, suggesting cosmic impact.

Sea levels over time have also been deduced from oceanic cores. One major sea-level surprise was massive evaporite deposits in the Mediterranean Sea from when sea levels were low enough to make the Strait of Gibraltar dry land and the Mediterranean a salty inland sea. Finally, remote stations have been implanted in several oceanic boreholes. These stations take seismic data to augment land seismic stations, and they also have temperature sensors.

—Roger V. Carlson

Further Reading

Exon, Neville. "Scientific Drilling Beneath the Oceans Solves Earthly Problems." *Australian Journal of Maritime and Ocean Affairs*, vol. 2, 2010, pp. 37–47.

"History." *International Ocean Discovery Program: Exploring the Earth Under the Sea*, 2017, www.iodp.org/about-iodp/history. Accessed 11 Sept. 2017.

Karson, Jeffrey A., Deborah S. Kelley, Daniel J. Farnari, et al. *Discovering the Deep: A Photographic Atlas of the Seafloor and Ocean Crust*. Cambridge UP, 2015.

National Research Council. *Scientific Ocean Drilling. Accomplishments and Challenges*. National Academies Press, 2011.

National Research Council. *Macondo Well* Deepwater Horizon *Blowout: Lessons for Improving Offshore Drilling Safety*. National Academies Press, 2012.

Normile, Dennis, and Richard A. Kerr. "A Sea Change in Ocean Drilling." *Science*, vol. 300, 2003, p. 410.

Oreskes, Naomi. *Plate Tectonics: An Insider's History of the Modern Theory*. CRC Press, 2018.

Stein, Ruediger, Donna K. Blackman, Fumio Inagaki, and Hans-Christian Larsen, editors. *Earth and Life Processes Discovered from Subseafloor Environments: A Decade of Science Achieved by the Integrated Ocean Drilling Program (IODP)*. Elsevier, 2014.

OCEAN-FLOOR DRILLING PROGRAMS

Fields of Study: oceanography; marine geology; stratigraphy; marine engineering

ABSTRACT

Ocean-floor drilling programs have allowed geologists and oceanographers to extend their knowledge of the earth's history by analyzing long marine sediment cores and basement rock cores recovered from the seafloor. Data from ocean-floor drilling have provided evidence supporting the theories of seafloor spreading and plate tectonics and have permitted the investigation of the paleoclimatic and paleoceanographic history of the earth.

KEY CONCEPTS

basalt: the rock type that is formed when molten magma breaks through the ocean floor and solidifies to form new ocean floor

basement rock: the most fundamental rock layer of Earth's crust, typically ocean floor basalt

black smoker/white smoker: seafloor hydrothermal vents; those spewing hotter water quickly produce clouds, like smoke, of inky black sulfide minerals, those spewing less hot water produce white clouds of mineral particles as their ejection contacts the cold water of the ocean deep

core barrel: the hollow cylindrical opening within the first section of a drill string, which also functions as the drill bit; typically contains a plastic liner that encases the core sample as it is produced

drill string: the long line of attached pipes that connects the drill head to the engine that drives the boring apparatus

igneous rock: rock such as granite and basalt that is formed by crystallization of molten rock

ORIGINS OF THE TECHNOLOGY

Most of our knowledge of the history of Earth comes from the study of sedimentary rocks, as sediments contain the preserved fossil remains of ancient plants and animals, while sedimentary structures record the processes of deposition. Sedimentary rocks exposed on land tend to have an incomplete record because they may be deformed by folding and faulting, which may destroy both the fossils and the sedimentary structures, and because they may be eroded by wind, water, and ice moving across the surface. In contrast, marine sedimentary sequences contain a more complete record of accumulation because they accumulate in a lower-energy environment, which is not as affected by folding, faulting, erosion, and post-depositional alteration as are terrestrial sediments. As a result, the deep-sea sediments tend to preserve a continuous record of sediment deposition in the ocean basins.

Ocean drilling techniques were originally developed in the 1950s by petroleum exploration companies searching for shallow-water hydrocarbon and petroleum deposits located on the continental shelves. These industrial exploration methods were adapted in the 1960s to obtain long sediment cores from the seafloor in deep-water areas on the continental slopes and abyssal plains. By drilling through the entire sediment record into the harder basement rocks below marine sediments, geologists hoped to acquire the complete history of sediment deposition within an ocean basin from the time that sediments were first deposited atop volcanic basalts.

GLOMAR CHALLENGER

Preliminary attempts to drill the ocean floor included engineering tests for Project Mohole by the drilling barge *Cuss I*, which in 1961 drilled marine sediments off La Jolla, California, and at the deep-water Experimental Mohole Site east of Guadalupe Island, off Baja California, Mexico, in a water depth of 11,700 feet (3,566 meters). Although further Project Mohole development was not undertaken because of a combination of political conflicts and increasing cost estimates for the project, in 1964 four American universities formed a consortium to initiate a program of scientific deep-sea drilling. JOIDES, the Joint Oceanographic Institutions for Deep Earth Sampling, successfully operated a drilling program in April- May, 1965, using the vessel *Caldrill I* to drill six holes on the Blake Plateau off Florida to sub-bottom depths of more than 1,000 meters, with continuous core recovery.

Following these successful trials, JOIDES proposed an eighteen-month program of scientific drilling in the Atlantic and Pacific oceans—to be called the Deep Sea Drilling Project (DSDP)—and operated by the Scripps Institution of Oceanography of La Jolla, California, using the drilling vessel *Glomar Challenger*. *Glomar Challenger* left Orange, Texas, on July 20, 1968, on Leg 1 of the Deep Sea Drilling Project. The results of DSDP drilling on the first nine cruises in the Atlantic and Pacific oceans caused the National Science Foundation to extend the drilling program beyond the initial eighteen-month period, with further drilling in the Indian Ocean and in the seas surrounding Antarctica.

When DSDP began operations, many other American oceanographic institutions joined JOIDES in support of the drilling program, and the success of DSDP also attracted scientific participation and financial support from foreign countries. The International Program of Ocean Drilling (IPOD) started in 1975 when the Soviet Union, the Federal Republic of Germany, France, the United Kingdom, and Japan joined JOIDES, with each country providing $1 million yearly to support drilling programs. DSDP/IPOD drilling activities continued through the early 1980s, leading to international scientific exchange of information between oceanographers.

SEDCO/BP 471

Because the initial JOIDES proposal was only for eighteen months, it was never expected that ocean drilling would continue for fifteen years. Because of demands for ocean drilling in deeper waters and in high-latitude polar areas, JOIDES proposed in the early 1980's that a larger drilling vessel be acquired for continued drilling. The last cruise of the *Glomar*

Challenger, DSDP Leg 96, ended in Mobile, Alabama, on November 8, 1983, with the retirement of the drilling vessel from service. In 1983, responsibility for scientific supervision of the international project, now called the Ocean Drilling Program (ODP), passed from the Scripps Institution to Texas A&M University, and the drilling vessel *Sedco/BP 471* replaced *Glomar Challenger*. The first cruise of a ten-year ODP drilling program began on March 20, 1985, when the *Sedco/BP 471*, informally called the *JOIDES Resolution*, left port to begin drilling on ODP Leg 101. Leg 186 was scheduled for completion in the year 2000, and planning continued for cruises up to Leg 201 in 2002.

The results of each cruise, or leg, have been published in a series of books, entitled *Initial Reports of the Deep Sea Drilling Project*, which are published by the US Government Printing Office. The cores recovered from the DSDP and ODP holes represent an invaluable record of the history of ocean sediment deposition around the globe. These recovered sediment cores are studied by a variety of scientists, who are interested in the sediment type, fossil content, geochemistry, magnetic orientation and strength, shear strength, and other sedimentary properties of the samples.

DRILLING PROCEDURES

In shallow water, drilling is accomplished either by building a drilling platform directly atop the seafloor or by firmly anchoring a drilling vessel to the bottom. In deep-water ocean drilling, however, it is not possible to anchor the drilling vessel to the bottom, so the technique of dynamic positioning is used to maintain the position of the vessel above the hole being drilled. In dynamic positioning, an acoustic beacon emitting sounds at either 12.5 kilohertz or 16 kilohertz is dropped to the seafloor. Four hydrophones, located at different points on the hull of the drillship, receive the signal from the acoustic beacon at slightly different times, depending on the position of the hull relative to the beacon. The position of the ship is maintained by a shipboard computer, which interprets the information from the hull hydrophones and controls the position of the ship by driving both the main propellers and two laterally oriented propellers, or hull thrusters. If the vessel is pushed off location by waves or surface currents, the shipboard computer attempts to compensate by using the propellers and hull thrusters to maintain the ship's location relative to the seafloor beacon.

To drill sediment and rock samples from the seafloor, a drill bit is attached to the bottom of a 31-foot-long (9.5-meter-long) piece of hollow cylindrical stainless-steel drill pipe. More individual lengths of pipe are connected on the rig floor of the drillship to make a "drill string," which extends from the vessel through the water down to the seafloor, where coring may begin. Usually, about 450 to 510 lengths of drill pipe are required simply to reach the bottom, and the assembly of this drill string may take twelve hours before bottom drilling may be started. Once the string reaches the seafloor, the drill pipe is rotated by hydraulic motors on the rig floor, and the rotary action combined with the weight of the drill string causes the drill bit to spiral down into the seafloor. Sharp iron carbide or diamond cutting teeth on the drill bit assist the penetration of the drill string into the sediment and the rock on the ocean bottom.

Samples of sediment and rock are retrieved from the seafloor by drilling a hole about 10 inches (25 centimeters) in diameter, using a drill bit with a 3-in (7.5-cm) hole in its center. In effect, sediment is cored by "drilling the doughnut and saving the doughnut hole": Rotating the drill string grinds the outer ring of sediment to small pieces against the diamond teeth of the drill bit, while the material in the center of the drill hole is saved as a core of drilled sediment 2.6 in (6.6 cm) in diameter. As the drill string is lowered deeper into the drilled hole, the core is pushed up into a plastic core liner in a steel "core barrel" within the lowest stand of drill pipe. After 31 ft (9.5 m) of the seafloor has been drilled, the core barrel is pulled up to the rig floor by a cable lowered through the drill string. By the time ODP ended in 2004, Joides Resolution had completed 110 expeditions, collecting about 2,000 deep sea cores from major geological features in the oceans of the world.

ATLANTIC OCEAN DRILLING

The first three cruises of the *Glomar Challenger* provided information proving that seafloor spreading had occurred in the Atlantic Ocean. A series of DSDP holes across the Mid-Atlantic Ridge showed that the age of bottom sediments increased with distance from the ridge crest and indicated that the ages of sediments with depth correlate from one hole to the next.

The total thickness of sediment atop basaltic basement also increased with greater distance from the ridge crest, on both the east and west sides of the Mid-Atlantic Ridge. Further DSDP and ODP drilling has provided evidence that seafloor spreading has occurred in all the earth's ocean basins. In addition, ocean drilling has confirmed the relative youth of the ocean basins, as predicted by plate tectonics; the oldest seafloor yet discovered is Early Jurassic in age (160 million years old), compared to continental rocks, which may be as old as 4.5 billion years.

PACIFIC OCEAN DRILLING

Glomar Challenger and *JOIDES Resolution* have operated from the Norwegian Sea to the Ross Sea off Antarctica and have drilled holes in water depths from 633 ft (193 m) on the Oregon continental shelf to 23,130 ft (7,050 m) in the Mariana Trench near Guam, in the western Pacific Ocean. The deepest hole through seafloor sediment deposits is more than 5,471 ft (1,750 m) below the seafloor, and one site in the equatorial Pacific Ocean west of South America (DSDP Hole 504B) has been drilled through 985 ft (300 m) of sediment and 4,291 ft (1,500 m) of volcanic basement rock.

GLOBAL STRATIGRAPHIC CORRELATION

Seafloor drilling has indicated that deep-sea sediments contain long sequences of well-preserved microfossils, which may be used for global stratigraphic correlation, in contrast to the fragmentary record preserved on land, where structural deformation of sediment deposits may complicate the problem of correlating different sedimentary sequences. Analysis of these sediments has revealed the history of deposition in the different ocean basins and has provided information on ancient climates and oceanographic conditions (such as the position, strength, and temperature of past ocean currents).

Sediment cores have indicated the presence of great shifts in oceanic climate conditions during the geologic past and have demonstrated that the Antarctic continent has been covered by glacial ice caps for at least 40 to 50 million years, rather than the 5 million years accepted prior to DSDP drilling near Antarctica. Another startling result of ocean drilling has been the discovery that the Mediterranean Sea dried up between 12 and 5 million years ago. Massive salt and evaporite mineral deposits below the Mediterranean basin indicate that the Strait of Gibraltar connection to the Atlantic Ocean was blocked during this time. Blockage of the Gibraltar connection allowed the water in the basin to evaporate, causing the deposition of vast salt and evaporite mineral deposits as the Mediterranean dried up.

STUDY OF SEAFLOOR BASEMENT ROCKS

Not all the information provided by ocean drilling has been concerned with the sediment column. Drilling into basement rocks has allowed geophysicists to compare the structure of seafloor basement to that of layered igneous-rock deposits that have been uplifted above sea level on the edges of certain continents. Similarly, direct drilling through these basalt and gabbro layers has allowed a comparison of the rock type to sound velocities measured by marine geophysicists. Some other results of hard-rock seafloor drilling have been the investigation of sediment and mineral deposition by hydrothermal processes at rapidly opening mid-ocean ridge segments, such as the sulfides deposited by high-temperature fluids emitted by "black smoker" and "white smoker" structures near the Galápagos Islands west of South America. Drilling of bare basement rocks along mid-ocean ridges in the Atlantic, Pacific, and Indian oceans has enabled geochemists and igneous petrologists to study the frequency at which seafloor volcanic rocks are produced at individual ridge segments and to determine whether temporal changes occur in the chemistry of basalts erupted from one location on the ridge. These studies of seafloor basement rocks may be applied to mineral exploration of marine rocks that have been uplifted above sea level and exposed on continents.

IMPROVEMENTS IN DRILLING TECHNOLOGY

In addition to their scientific results, DSDP and ODP operations have resulted in improvements in drilling technology by developing the ability to reenter seafloor boreholes, by devising techniques for "bare-rock" drilling on the seafloor, and through the development of new coring bits. During DSDP Leg 1, it was discovered that existing drill bits could not penetrate hard chert beds; thus, they also would not be able to penetrate through deeper igneous rocks below seafloor sediments. The drag bits were solid, consisting of a central opening and radial curved ridges of steel or tungsten carbide, capped with industrial diamonds and designed to churn through soft sediments. As a result, DSDP began a drill-bit design program, which led to the development of roller bits capable of penetrating both chert layers and seafloor basalts. These bits consist of four conical cutting heads studded with tungsten carbide or diamond cutting teeth, situated around the central core opening in the bit.

Another important technical development of DSDP, first successfully accomplished on Leg 15, was the ability to reenter a drilled borehole on the seafloor. Even with roller bit designs, drill bits wear out from the stresses of rotary coring through seafloor sediments and rocks. When a bit fails, the entire drill string has to be "tripped," or pulled up to the vessel to replace the bit at the lower end of the string, which in most deep-ocean drilling sites requires pulling the string up not only several hundred feet (or meters) from below the seafloor but also through 6,562 to 16,404 ft (2,000 to 5,000 m) of seawater. During early DSDP legs, bit failure forced the abandonment of a hole because after the fatigued bit was replaced, it was impossible to reenter the original borehole. Successful reentry techniques were facilitated by the development of a steel reentry cone 20 ft (6 m) in diameter, topped with three sonar reflectors and a rotating sonar scanner that can be lowered through the drill string. In areas where hardened sediment layers are anticipated, requiring bit replacement to complete drilling, the reentry cone is placed on the seafloor prior to drilling the initial borehole. As bits become worn, they may be replaced and the hole reentered by using the sonar scanner to locate the reentry cone (and thus the original hole).

DSDP and ODP drilling specialists have also devised methods to enable drilling in hard seafloor areas, such as mid-ocean ridges, which were not previously drillable by existing techniques. Development of a seafloor "guide base" for drilling has allowed successful drilling and reentry of boreholes in these areas and has permitted the implantation of seafloor sensing devices, such as earthquake-measuring seismometers, in these holes.

ROLE IN DEVELOPMENT OF PALEOCEANOGRAPHY

Ocean-floor drilling programs have enabled scientists to correlate apparently unconnected phenomena through the theory of plate tectonics, a global synthesis of geology and oceanography. Ocean drilling has provided verification of the seafloor spreading hypothesis as it applies to plate tectonics and has indicated that seafloor spreading has occurred in all of Earth's ocean basins.

Before long sediment cores could be acquired from the seafloor, scientists' knowledge of seafloor geology was sparse, based on limited samples available from dredging and shallow coring of the seafloor by oceanographic vessels. Prior to DSDP, global stratigraphic correlation was based on a fragmentary record preserved on land, where structural deformation of sediment deposits may complicate the problem of correlating different sedimentary sequences; DSDP drilling, however, has revealed that deep-sea sediments contain long sequences of well-preserved microfossils. Furthermore, seafloor sediment cores have revealed the history of the ocean basins and have provided information on ancient climates and oceanographic conditions (such as the position, strength, and temperature of past ocean currents). A new science, paleoceanography, has been developed based on this information from DSDP and ODP drilling. Analysis of the earth's ancient climates may provide information to predict future shifts in the biosphere.

INDUSTRIAL APPLICATIONS

Ocean drilling has also provided evidence for the existence of deep-water hydrocarbon accumulations, which has enabled petroleum exploration companies to drill petroleum deposits on the continental slopes and may eventually lead to the discovery of significant hydrocarbon deposits in ocean basins. If future technology is developed, humans may be able to exploit these deep-water petroleum resources. Furthermore, scientific ocean drilling has enabled geologists to understand the processes controlling the deposition of "black shale" deposits and other high-productivity seafloor sediments, which may be altered by burial into source beds for the generation of petroleum hydrocarbons. Understanding of the processes affecting the formation and distribution of these sediments may assist in future exploration for fossil fuel resources. In addition, studies of seafloor basement rocks may lead to a more complete understanding of the nature of mineral deposition at mid-ocean ridges, which may be applied to mineral exploration of similar marine rocks that have been uplifted above sea level and exposed on continents.

Finally, deep-ocean drilling has led to technological innovations in the tools and techniques used to sample the seafloor. These methods have been adapted by industrial companies exploring for hydrocarbons buried beneath marine sediments and for mineral deposits on the seafloor. In 2004 the Ocean Drilling Program was transformed into the Integrated Ocean Drilling Program (IODP). The final report of ODP was published as the "Ocean Drilling Program Final Technical Report 1983-2007." The IODP is an international program utilizing a variety of drilling platforms to investigate the earth deep below the seafloor.

—Dean A. Dunn

Further Reading

Davidson, Jon P., Walter E. Reed, and Paul M. Davis. *Exploring Earth: An Introduction to Physical Geology*. 2nd ed., Prentice-Hall, 2001.

Earle, Steven. *Physical Geology*. BC Open Textbook Project, 2019.

Exon, Neville. "Scientific Drilling Beneath the Oceans Solves Earthly Problems." *Australian Journal of Maritime and Ocean Affairs*, vol. 2, 2010, pp. 37–47.

Hamblin, William K., and Eric H. Christiansen. *Earth's Dynamic Systems*. 10th ed., Prentice-Hall, 2003.

"History." *International Ocean Discovery Program: Exploring the Earth Under the Sea*, 2017, www.iodp.org/about-iodp/history. Accessed 11 Sept. 2017.

Karson, Jeffrey A., Deborah S. Kelley, Daniel J. Farnari, et al. *Discovering the Deep: A Photographic Atlas of the Seafloor and Ocean Crust*. Cambridge UP, 2015.

National Research Council. *Scientific Ocean Drilling. Accomplishments and Challenges*. National Academies Press, 2011.

Normile, Dennis, and Richard A. Kerr. "A Sea Change in Ocean Drilling." *Science*, vol. 300, 2013, p. 410.

Oreskes, Naomi. *Plate Tectonics: An Insider's History of the Modern Theory*. CRC Press, 2018.

Segar, Douglas. *An Introduction to Ocean Sciences*. 2nd ed., Wadsworth, 2007.

Seibold, Eugen, and Wolfgang Berger. *The Sea Floor: An Introduction to Marine Geology*. 4th ed., Springer, 2017.

Stein, Ruediger, Donna K. Blackman, Fumio Inagaki, and Hans-Christian Larsen, editors. *Earth and Life Processes Discovered from Subseafloor Environments: A Decade of Science Achieved by the Integrated Ocean Drilling Program (IODP)*. Elsevier, 2014.

Ocean Pollution and Oil Spills

Fields of Study: engineering; marine chemistry; oceanography; marine biology

ABSTRACT

Oil spills resulting from human error often affect marine and coastal areas. Past oil spills in different areas of the world demonstrate that environmental damage depends on the toxicity and the persistence of the oil, both of which vary widely depending on a variety of factors.

KEY CONCEPTS

aliphatics: hydrocarbon compounds in which the molecular structures consist solely of single carbon-carbon and carbon-hydrogen bonds

aromatics: hydrocarbon compounds in which the molecular structures are cyclic and contain a series of alternating single and double carbon-carbon bonds, which imparts greater stability and chemical reactivity to them relative to their aliphatic and nonaromatic counterparts

biodegradation: the breaking down of out-of-place materials by living organisms and metabolic processes

condensate: a mixture of light hydrocarbon compounds that separate or "condense out" during the processing of natural gas

crude oil: as the name implies, the crude oil material, petroleum, that is taken from oil wells before any processing

dead zone: a region within an ocean or other body of water where life is not supported and so does not exist

mousse: a stable mixture of oil and water as a rough emulsion

volatile: describes materials that can readily enter the gas phase at ambient temperatures

WHY OIL SPILLS OCCUR

Oil has been the lifeblood of modern lifestyles in the twentieth century, and continues to be central in the twenty-first, especially in the most economically developed nations of the world, including North America. With only about 6 percent of the world's population, the United States consumes more than 25 percent of the world's petroleum supply, much of which is imported from other nations. According to the US Energy Information Administration, in 2016, the United States consumed a total of 7.21 billion barrels (a barrel equals 4 US gallons, or about 159 liters) of petroleum products. The percentage of imported oil that Americans use grew steadily after 1982, largely negating the lessons learned about cartel politics and energy conservation in the late 1970s. The amount of oil imported into the United States continued to grow even during periods when its merchandise trade deficit was in decline. Americans have also demonstrated that they are willing to go to war for oil. In the 1991 Gulf War with Iraq, the United States paid a huge cost to protect Kuwaiti oil fields and world oil markets in spite of the fact that very little of the oil produced in the Arabian Gulf is imported to the United States. Most of the nation's imported oil comes from Canada, Africa, Central and South America, and the Indo-Pacific region. Domestic production continues to be an important but dwindling source, with most coming from the north slope of Alaska via the Alaska Pipeline and Port Valdez in Prince William Sound.

Almost all imported and Alaskan oil is transported to US refineries and consumers by oceangoing bulk shipping vessels, or tankers, although dedicated pipelines have risen in importance. Most oil spills result from marine transportation accidents, with human error often being the cause or at least playing a major role. Navigation errors, equipment malfunctions, bad judgment, and even the inability of all crew members to speak a common language have all been major contributing factors in the largest and most environmentally damaging oil spills. Spilled oil can decimate plant and animal populations by a combination of mechanical and chemical toxicity effects resulting from an organism's physiological reaction to the chemicals present in oil.

ENVIRONMENTAL DAMAGE

The most common sight during an oil spill is dark, gelatinous masses of "mousse," an oil and water emulsion that floats on top of the water, sticking to everything with which it comes into contact. Mousse usually causes the majority of the environmental damage during an oil spill by processes of mechanical toxicity, as it suffocates and smothers organisms that ingest it or are covered by it. Seabirds and furry marine mammals are highly susceptible to this process, succumbing to exposure, dehydration, or starvation.

Crude oil is a complex mixture of thousands of different chemicals called "hydrocarbons," named after their molecular structures that consist solely of hydrogen and carbon atoms. Different hydrocarbons vary in their chemical properties, toxicity, and behavior during an oil spill. The major groups are classified by molecular geometry and weight.

The simplest and lowest-molecular-weight molecules (aliphatics) are generally single-bonded, chain-shaped molecules, such as those that make up gasoline. They are the most volatile, and are acutely toxic. Such compounds tend to evaporate or burn easily during an oil spill and therefore do not persist in the environment for long periods. The aromatics are

Bergan oil field fire during the Gulf War, Kuwait and Persian Gulf (National Oceanic and Atmospheric Administration via Wikimedia Commons)

Lingering oil slick from the Deepwater Horizon spill on the Mississippi Delta (NASA/GSFC, MODIS Rapid Response via Wikimedia Commons)

ring-shaped molecules, such as benzene. They also tend to be highly volatile and are more reactive than the simple hydrocarbons, and so can cause biological impacts because of both acute and chronic toxicity.

Aromatic hydrocarbon compounds and aliphatic hydrocarbons of higher molecular weight are more environmentally persistent than the simple aliphatics. Because many are carcinogenic, they can also cause different forms of biological damage, disease, and death even after a long time period and in low doses.

The highest-molecular-weight oil compounds include polycyclic aromatic hydrocarbons having structures composed of ring shapes bonded together to form larger molecules, and very large aliphatic and aromatic hydrocarbons whose molecular structures consist of long chains of carbon-hydrogen units. These typically have the consistency of heavy lubricating oils, greases, waxes, and even the tar used in road paving. Although they are generally not very chemically reactive under environmental conditions and do not dissolve well in water, many are carcinogenic, and they tend to be very environmentally persistent.

For hundreds of millions of years before human beings evolved, oil "spilled" naturally into the world's oceans from natural oil seeps, which are fractures that tap into deep, oil-bearing rocks. A variety of natural processes act to reduce the environmental impacts of this oil, and these same processes also take place during a human-caused oil spill. Oil is dispersed from the oil slick and into the larger environment by five basic processes. Evaporation of the low-molecular-weight hydrocarbon compounds removes much of the oil relatively quickly. Sunlight can degrade additional oil in a process called "photodegradation," if the oil is exposed for enough time. Because oil is an organic substance, additional oil is removed by natural biodegradation thanks to "oil-eating" microorganisms. Most of the rest of the oil either washes up onto a coastal area or breaks up into heavy "tar balls" rich in high-molecular-weight hydrocarbons that eventually sink.

Some oil spills put so much oil into the environment that these processes cannot respond quickly enough to prevent environmental damage. Other factors can also enhance environmental damage from oil spills. Some types of oil or refined petroleum products are more toxic than others. Oil spills in cold climates generally cause more damage because cold temperatures retard evaporation and the microbial metabolic rates necessary for rapid oil removal. Furthermore, sunlight is often of low intensity, which retards photo degradation. Wave conditions and tidal currents can affect how much oil washes up onto a coastal area and how rapidly it is moved elsewhere or removed. Finally, the amount of environmental damage from an oil spill is highly dependent on the type of coastal environment affected by the spill, as coastal environments vary in the density (or biomass) and varieties of wildlife.

Coastlines also vary in the degree to which they are sheltered from natural oil-removal processes. In general, rocky headlands, wave-cut rock platforms, and reefs exposed to high wave activity suffer far less damage during an oil spill than do sheltered marshes, tidal flats, and mangrove forests. The damage on beaches is related to the grain size of the beach sediment. Fine-sand beaches are relatively flat and hard-packed, and oil does not soak into the sediment or persist for long. Oil will soak deeply into coarse sand, gravel, and shell beaches, causing more damage over a longer period.

Most of what has been learned about oil spill behavior, environmental damage, and oil spill cleanup techniques comes from studying past spills. In most cases, spill prevention is far cheaper and infinitely more effective than spill response, and cleanup efforts usually capture very little of the spilled oil.

In 2018, an oil spill in the East China Sea evoked great concern from scientists regarding how different types of oil affect ocean life. On January 6, the Iranian tanker *Sanchi*, reportedly carrying about 1 million barrels of condensate, crashed into a Chinese freighter. After catching fire, the tanker burned for several days before ultimately sinking. While accidents involving ships carrying condensate had occurred previously, this incident was the first of such a great magnitude, and less is generally known about the impact of the chemicals that compose condensate on marine ecology than that of crude oil.

A worker cleans up oily waste on Elmer's Island, just west of Grand Isle, Louisiana, May 21, 2010 (DVIDSHUB via Wikimedia Commons)

Though condensate, which is a light petroleum byproduct of natural gas production, does not cause the same type of thick oil slicks that spread and linger in the spills of crude oil, it is believed that it causes a deep underwater plume to form that could be extremely harmful to marine life; therefore, scientists are particularly concerned about possibly severe effects occurring over a short-term period versus the long-term effects typical of a crude oil spill. In the weeks after the spill, Chinese officials reported that traces of petroleum were found in some fish taken from the area. Because condensate is lighter and highly flammable, the usual cleanup tactics for crude oil likely could not be used.

GULF OF MEXICO OIL SPILLS

The Ixtoc I spill of June 3, 1979, was the result of an explosion, or "blowout," of an offshore oil well that was drilling into a subsurface oil reservoir. Although human error was definitely a factor, the cause of the blowout remains unresolved. It has been blamed on the use of drilling mud that was not dense enough to counteract the pressure of the oil and gas at depth, as well as on the improper installation of the blowout preventer, a fail-safe device used on drilling rigs to prevent just this type of disaster. The result was a continuous 290-day oil spill, during which an estimated 523,598 tons (475,000 metric tonnes) of crude oil (1 metric ton equals approximately five barrels) was released into the environment. In addition to doing considerable environmental damage on the coast of Mexico, oil fouled much of the barrier island coast of Texas. However, most of the oil did not make it to shore, and the final accounting for this spill gives a good indication of the long-term fate of spilled oil in offshore areas: 1 percent burned at the spill site, 50 percent evaporated, 13 percent photodegraded or biodegraded, 7 percent washed up on the coast (6 percent in Mexico, 1 percent in Texas), 5 percent was mechanically removed by skimmers and booms, and 24 percent sank to the seafloor (assumed by mass balance).

A similar accident occurred in 2010 when the blowout preventer on BP's *Deepwater Horizon* well site failed, causing an explosion at the drilling platform. In the resulting spill, oil jetted continuously from the wellhead on the sea bottom for a prolonged period of time, as various efforts to recap the well failed. Massive amounts of crude oil expanded over a large portion of the water's surface and washed up along the coastline between Louisiana and Florida. The incident decimated the fisheries in the Gulf of Mexico creating an extensive "dead zone" where natural water currents had localized much of the toxic spill materials and created areas that could no longer support plant or animal life. The disruption of the beaches in the area also had a great negative impact on the tourist industry upon which the region depends.

THE *EXXON VALDEZ*

The *Exxon Valdez* oil spill—which occurred in Prince William Sound, Alaska, on March 24, 1989—is a good example of how environmental damage follows human error and inadequate response. After departing Port Valdez with a full cargo, the *Exxon Valdez* oil tanker struck a well-charted submerged rock reef located 1 mile (1.6 kilometers) outside the shipping lane. The ship was under the command of an unlicensed third mate in calm seas and left the shipping lane with permission from the Coast Guard to avoid ice. However, it strayed too close to the reef before evasive action was attempted. The captain, who had a history of drunk-driving convictions, was in his cabin under the influence of alcohol during events leading to the accident. His blood alcohol level nine hours after the grounding was measured at 0.06 percent; the estimate at the time of the accident was 0.19 percent. Convicted of negli-

Fire boat response crews battle the blazing remnants of the offshore oil rig Deepwater Horizon (US Coast Guard via Wikimedia Commons)

gence and stripped of his commander's license, he was subsequently employed as an instructor to teach others to operate supertankers.

Leaking oil was observed immediately. Oil-spill response crews funded by Exxon and the Alyeska Pipeline Consortium, oil companies that used the Port Valdez terminal, were poorly prepared and reacted too slowly and with inadequate equipment. The first response arrived ten hours after the accident with insufficient booms and skimmers. Chemical dispersants applied to break up the oil slick were ineffective in the calm seas and caused the oil slick to thin and spread more rapidly. Four days later, the weather changed: 71-mi- (114-km-) per-hour winds mixed the oil with seawater, creating a frothy mousse. More than 71,650 tons (65,000 metric tonnes) of oil spilled out of the stricken vessel over the next several weeks. About 6,023 square mi (15,600 square km) of ocean and 808 mi (1,300 km) of shoreline were affected.

Federal estimates of wildlife mortality include 3,500 to 5,500 otters; 580,000 seabirds; and 300 deer poisoned by eating oil-contaminated kelp. Economic damages totaled more than $5 billion. The long-term effects on commercial marine organisms, larval organisms, and bottom-dwelling life are not known.

Exxon promised to clean nearly 311 mi (500 km) of shoreline by September 1989, but cleaned only 1.24 mi (2 km) during the first month after the spill. Exxon and its contractors used a variety of cleanup techniques, including placing booms and skimmers, sopping up oil with absorbent materials, scraping oil by hand from rocks, stimulating the growth of oil-eating bacteria cultures, and washing coastal areas with cold water, hot water, and steam. The use of hot water and steam was effective at cosmetically removing surface oil, but it did not remove oil that had soaked into the sediment; the technique subsequently killed most of the organisms that had escaped the oil. The oil

The Exxon Valdez *in Prince William Sound, Alaska (National Oceanic and Atmospheric Administration via Wikimedia Commons)*

Oil clean-up efforts on Smith Island (US Navy via Wikimedia Commons)

washed from the beach was to be collected by booms and skimmers offshore, but this process was so inefficient that much of the oil migrated to tide pools that had not been affected by the spill directly. Ironically, only eighteen months after the spill, life had returned to oiled coasts that had received little or no cleanup, while beaches that had been cleaned with hot water were still relatively sterile and required several years to repopulate.

Exxon announced that it would not return to clean more shoreline in 1990 but relented under threat of a court order from the Coast Guard to enforce federal cleanup requirements. During the summer of 1990, shoreline cleanup resumed, including application of fertilizer to stimulate growth of naturally occurring oil-eating bacteria, a technique that is not very efficient in the cold waters of southern Alaska.

The tale of the *Exxon Valdez* is not complete without mentioning that the Port Valdez Coast Guard did not have state-of-the-art radar equipment for monitoring ship movement in this heavily used and environmentally sensitive area. In the early 1980s, federal and state funds for monitoring the Port Valdez oil companies' compliance with oil-spill preparedness legislation had been cut by more than 50 percent. The original environmental impact statement for oil-handling activity in Prince William Sound included an agreement that defines cleanup responsibility for oil spills. Exxon, as the company responsible for the spill, was to pay the first $14 million of cleanup costs, with $86 million in additional cleanup funds from the Alyeska contingency fund. Thus, the maximum financial responsibility to oil companies from a spill was $100 million unless the spill was judged to be caused by negligence. Cleanup activities ceased eighteen months after the spill with total expenditure of $2.2 billion—most of this at taxpayer expense. In 1994, a federal court unanimously awarded $5.3 billion in punitive and compensatory damages, the largest-ever jury award, to some 35,000 people impacted by the spill. By June 1999, Exxon had yet to pay a single dollar as the case continued through the legal process. Finally, it is interesting to note that Exxon's estimate of cleanup costs in late 1989 were $500 million, and it carried $400 million of oil spill liability insurance. Exxon saved $22 million by not building the *Exxon Valdez* with a double hull; its 1988 annual profits were $5,300 million.

According to National Oceanic and Atmospheric Administration (NOAA) estimates, less than 1 percent of *Exxon Valdez*'s oil burned at the site, 20 percent evaporated, 8 percent was mechanically removed, and nearly 72 percent was deposited on the seafloor. According to Exxon's estimates, 7 percent of the oil burned at the site, 32 percent evaporated, 9 percent photodegraded or biodegraded, 15 percent was mechanically removed, and 37 percent was assumed deposited on the seafloor.

THE GULF WAR OIL SPILL

In January 1991, the Gulf War oil spill, the largest oil spill in history at that time, occurred when the Iraqi military opened valves and pumps at Sea Island Terminal, a tanker loading dock located 9.94 mi (16 km) off the coast of Kuwait. This facility had a production capacity of 100,000 barrels per day, about three *Exxon Valdez* loads each week. The Iraqis also opened plugs on five Kuwaiti tankers, spilling an additional 60,000 barrels. The estimate for the entire spill is 6 million barrels, or roughly 30 times the volume of the *Exxon Valdez*. About 251 sq mi (650 sq km) of coast was heavily contaminated. Three days after starting the

Birds killed as a result of oil from the Exxon Valdez *spill (*Exxon Valdez *Oil Spill Trustee Council via Wikimedia Commons)*

spill, the Iraqis ignited the oil leaking from the terminal. This was the best thing to happen from an environmental perspective. During most spills, more oil is removed by natural evaporation than by any cleanup technique; igniting the oil merely speeds up this process.

Burning can be an important mechanism for removing oil from the sea and avoiding environmental damage, and tests have shown purposeful ignition in open water away from the coast to be an excellent oil slick-fighting strategy. However, this must be done within the first few hours of the spill. In order to maintain the fire, the slick must be more than 1 millimeter thick and must contain relatively little emulsified water. To maintain thickness, the slick is best surrounded with fireproof booms. However, at the time of the Gulf War oil spill, almost all the fireproof boom in the world was in Prince William Sound. Saudi Arabia also used dispersants on portions of the slick, but this effort was too late to be effective before a thick mousse had formed.

The prime objectives of causing the spill were to hamper an amphibious military landing by oiling the beaches and to disrupt desalinization of drinking water at Khafji and Jubail, the two primary sources of potable water for Saudi Arabia. The Saudis used booms to protect the plant intakes with great success. The retreating Iraqis also ignited more than seven hundred out of about one thousand inland wells, resulting in an additional 6 million barrels per day burned. This volume eventually made the marine spill insignificant, and the burning created 3 percent of total global carbon emissions during the time period of the event.

The Arabian Gulf is an unusual body of water. It is very shallow (average 108 ft [33 m]) and is nearly enclosed as a marine basin. Because it is also microtidal (the tidal range is less than 1.97 ft [0.6 m]), it flushes out slowly (once every two hundred years, compared

USAF aircraft of the 4th Fighter Wing fly over Kuwaiti oil fires, set by the retreating Iraqi army during Operation Desert Storm in 1991 (US Air Force via Wikimedia Commons)

with once every few days for Prince William Sound). It is also important to remember that this is not a pristine marine environment. Natural oil seeps are very common, there is a general lack of environmental standards and poor cooperation among Persian Gulf nations, and virtually no oil spill preparations or equipment were present in this part of the world. Earlier spills had occurred in the region, but they typically were associated with ongoing wars; the hostile environment made it difficult to utilize spill abatement specialists and equipment. For example, during the Iran-Iraq War, Iraq attacked an Iranian offshore platform (Nowruz) in 1982, spilling more than 2 million barrels of oil, a volume nearly half as large as the Gulf War spill. Losses of marine mammals and birds were great.

During the Gulf War oil spill, about 112 mi (180 km) of Saudi Arabian coastline was oiled (40 mi [65 km] was severely damaged), and oil reached south as far as the United Arab Emirates and Bahrain. Much of the southern Kuwaiti coast, made up of sea-grass beds, marsh, and mangroves, was severely damaged, and about 25 percent of the Saudi shrimp industry was lost. Although some twenty thousand wading birds were killed, no deaths of dolphins or dugongs were reported. However, these animals suffered greatly during the Iran-Iraq War. Estimates of the time required for ecological renewal of the Persian Gulf following the Gulf War spill (one to four years) were relatively short for two reasons: The high-water temperature results in high microbial activity and biodegradation of the oil, and much of the oil was burned.

—James L. Sadd

Further Reading

Curley, Robert. *New Thinking About Pollution*. Britannica Educational Publishing, 2011.

Fingas, Merv, editor. *Handbook of Oil Spill Science and Technology*. Wiley, 2015.

Graham, Bob, et al. *Deep Water: The Gulf Oil Disaster and the Future of Offshore Drilling*. National Commission on the BP Deepwater Horizon Oil Spill and Offshore Drilling, 2011.

Hilgenkamp, Kathryn. *Environmental Health: Ecological Perspectives*. Jones and Bartlett Publishers, 2005.

"How Much Oil Is Consumed in the United States?" *US Energy Information Administration*, 27 Sept. 2017, www.eia.gov/tools/faqs/faq.php?id=33&t=6.

Kaye, Catheryn Berger, and Philippe Cousteau. *Going Blue*. Free Spirit Publishing, 2009.

Letcher, Trevor, and Daniel Vallero, editors. *Waste: A Handbook for Management*. Academic Press/Elsevier, 2011.

Madrigal, Alexis C. "The World Has Never Seen an Oil Spill Like This." *The Atlantic*, 19 Jan. 2018, www.theatlantic.com/technology/archive/2018/01/the-oil-spill-that-wasnt/550820/.

McKinney, Michael L., et al. *Environmental Science: Systems and Solutions*. 4th ed., Jones and Bartlett Publishers, 2007.

Murawski, Steven A., Cameron H. Ainsworth, Sherryl Gilbert, et al, editors. *Deep Oil Spills. Facts, Fate and Effects*. Springer, 2020.

Novotny, Vladimir. *Water Quality: Diffuse Pollution and Watershed Management*. 2nd ed., John Wiley & Sons, 2003.

Wiens, John A. *Oil in the Environment. Legacies and Lessons of the Exxon Valdez Oil Spill*. Cambridge UP, 2015.

OCEAN POWER

Fields of Study: marine engineering; marine ecology; oceanography

ABSTRACT

Ocean power encompasses several distinctly different approaches to power generation, which, if developed properly, promise potentially large amounts of clean, renewable energy. The importance of developing such alternative energy sources for a world largely dependent on fossil fuels—subject to escalating costs and exhaustibility, and creating environmental pollution—cannot be overstated.

KEY CONCEPTS

head: the vertical difference between low and high tides at a certain location

osmosis: the passage of water molecules through a semipermeable membrane from the region of lower salinity to the region of higher salinity

semipermeable membrane: a porous membrane in which the pores are large enough to allow water molecules to pass through, but not large enough to allow dissolved salts to pass through

submarine topography: the shape and changes of seafloor depth in a body of water

thermocline: a virtual surface between warmer and colder water, where there is an abrupt change from one temperature to the other

wave-powered pump: a pump system in which the power to drive the pump mechanism is provided by the up-and-down motion of waves

TIDAL FLOW POWER

Attempts at harnessing ocean power involve the use of specialized technologies developed to exploit natural flows of energy within the marine environment. These energy flows are generated by the interaction of the ocean's waters with the effects of the sun's energy; the gravitational pull of other celestial bodies, such as the moon; and, to a much lesser extent, such influences as geothermal activity occurring on the sea bottom. Many engineering schemes have been devised to try to tap into these natural energy flows. These schemes recognize the fact that Earth's surface is mostly ocean—71 percent of it is covered by the sea—and that, due to an interplay of natural processes, this immense fluid environment is always in motion.

The most ambitious—and, as of 2012, most productive—ocean power schemes have been tidal power projects. The efficacy of these projects is dependent on how well engineered they are to take advantage of the key factors involved. Such factors include the character of a tide at a particular coastal locale as determined by local bottom topography, surface coastline geography, and the orientation of the coast to the open sea. Submarine topographic influences can accentuate the rise of an incoming tide, acting like a wedge to lift the oncoming bulge of tidal water. Thus, tides can reach up to 49 feet (15 meters) on coasts having the right tide-enhancing topography and orientation.

The maximum rise and fall of a tide experienced at a particular location is important in tidal power, as it represents the amount of usable head; "head" is a term used in hydraulic engineering to describe the

Aerial view of the tidal barrage on the Rance and of Saint Malo (Tswgb via Wikimedia Commons)

difference in elevation between the level at which water can flow by gravity down from an upper to a lower level, thus making itself available to do work. Unfortunately, only about one hundred coastal sites worldwide are classified as having significant head and qualify as optimal candidates for tidal power installations. Scientists conservatively estimate that the global, dissipated tidal power amounts to 3 terawatts per year, or 3 trillion watts. Of this amount, perhaps only 0.04 terawatt would ever be exploitable from feasible tidal power sites.

The French government has been the world pioneer in realizing tidal flow power engineering schemes, constructing a large, functioning tidal energy station at the estuary of the La Rance River in Brittany in northern France. At La Rance, a combination of factors produces a useful hydraulic head and has proved itself economically profitable for several decades. Twenty-four 10,000-kilowatt turbine generators operate within conduits inside the tidal dam, turning at the low speed of 94 revolutions per minute. The turbine blades are designed to operate bidirectionally, in response to either an incoming or an outgoing tide. Thus, the French plant exploits the free tidal water movement almost continuously.

OCEAN THERMAL ENERGY CONVERSION

Ranking alongside tidal flow power, in terms of economic feasibility and available technology, is the ocean thermal energy conversion (OTEC) approach. This method could ultimately be a large-scale global operation, as there are many more sites that can be used for OTEC than are available for tidal power. Projected, theoretical limits to this energy source are in the neighborhood of 1 terawatt per year. OTEC involves either the construction of floating, open-ocean plants or coastal, land-based plants that exploit the temperature differences existing between water masses at varying depths in the tropical seas. Only small pilot plants have been built thus far, and they have never been run for more than short periods.

The optimal conditions for the most efficient OTEC sites have been calculated to be those where an 64.6 degrees Fahrenheit (18 degrees Celsius) temperature difference exists between the surface and depths in the range of 1,969 to 3,280 ft (600 to 1,000 m). Among the regions of the ocean that meet or approach these thermal conditions are Puerto Rico and the West Indies, the Gulf of Guinea, the Coral Sea, many of the Polynesian island groups, and the northwest African island groups. Only in these warm seas is there a large enough temperature difference between surface and deep sea.

The actual process of converting the thermal difference of such areas involves a system in which a turbine is turned by heat from the warm, surface water layer. The heat is transferred through devices termed "heat exchangers," which introduce the thermal energy into a closed system. A working fluid such as ammonia, contained in sealed pipes, propels the system through the process of controlled convection. The warmed ammonia is heated to a vapor in an evaporator unit that drives the turbine. Then the used vapor is conducted through a condenser unit, where cold water drawn from below cools it down to a liquid state for another usage cycle. The cool water is brought into contact with the system by the deployment of a very

Ocean Thermal Energy Conversion (OTEC) plant built by Makai Ocean Engineering in Hawaii (Vitafougue via Wikimedia Commons)

long pipe, hundreds of feet (or meters) in length, which projects through the thermocline, or boundary, between the upper and lower water masses. Although the efficiency of the system is typically low, only 2 to 3 percent, the thermal reservoir is immense and constantly replenished by the sun. To make such projects attractive economically, they need only be built on a sufficiently large scale.

MARINE CURRENT ENERGY CONVERSION

Similar to OTEC in its use of the major oceanic flows of thermal energy are plans to exploit large-scale currents. One such current, the Gulf Stream—often called the "Florida Current"—constantly conveys many millions of liters past a given point in the ocean. One plan is a direct approach, involving large turbines placed within the main flow of the current. Ideally, the generating site would be close to major electrical consumers, such as coastal cities.

A good case can be made for implementing marine current energy conversion along the eastern coast of Florida. The city of Miami, a very large consumer of electrical power, is within ten miles of the Florida Current. This current is estimated to carry approximately 1,059 million cubic feet (30 million cubic meters) of water per second past the city at a rate sometimes reaching 8.2 ft (2.5 m) per second. One scheme would involve anchoring a large cluster of special, slow-speed turbines, or water windmills, to the bottom; the cluster would ride midwater in controlled buoyancy. This complex would function at a depth ranging from 98 to 427 ft (30 to 130 m) and stretch some 12 miles (20 kilometers) across the flow of the Florida Current directly adjacent to Miami. Estimates of the power output of this array are on the order of 1,000 megawatts, provided constantly on a twenty-four-hour, year-round basis. It would extract roughly 4 percent of the usable kinetic energy of the Florida Current at this point in its flow, which is calculated to be in the range of 25,000 megawatts. Other strong currents exist worldwide that also may be good potential sites for future turbine arrays.

OCEAN WAVE POWER

Another way to take advantage of oceanic energy is to utilize the kinetic energy of surface waves directly to power mechanical devices used to generate electricity. This approach capitalizes on the fact that waves raise and lower buoyant objects. If the floating object is also long and perpendicular to the ocean surface, with most of its mass below the waterline, it is inherently stable and less subject to damage. Such a device is embodied in several variations of wave-powered pumps that are beginning to see practical economic applications. In one form, a large, vertical cylinder floats in the waves. Inside, the lower end is open to the lifting and sinking of the water level in unison with wave motion. Because of its great length (tens of feet (meters) or more), it amplifies the wave motion in the column of air resting above the water. This motion propels air up and down through a double-flow electrical air turbine. As long as it remains in sufficiently deep water, it receives very little damage, no matter the magnitude of wave energy. A buoy-like device floats at the cylinder top, keeping out wave splash and snugly maintaining inside air pressure. The cylinder can be anchored by flexible moorings to the sea bottom, and electrical cables can feed the power output to shore.

Coastal-based variants also have been built solidly into rocky cliffs; they respond to wave-driven air pressure that enters from a conduit at its base. One example of the coastal-based type is the plant at Tostestallen in Norway. Numerous rocky coasts worldwide could be similarly utilized for wave power generation.

SALINITY GRADIENT ENERGY CONVERSION

One form of ocean power that has been envisioned but not implemented is the use of salinity gradients within the sea. This idea is still in the theoretical stage because of the lack of key materials necessary for the effective technologies to work. A major drawback at this point is the lack of appropriately tough and efficient synthetic membranes necessary for this energy to be economically practical. Salinity gradient energy works on the principle of extracting energy from osmotic pressure by the use of semipermeable membranes. Such membranes would most likely be fabricated from some type of plastic that possesses the correct, chemical properties and would take advantage of the osmotic pressure between water masses with different percentages of dissolved salts. Influenced by osmotic pressure, less salty water will naturally flow through a semipermeable membrane to the side of greater salinity. The membrane would therefore be designed to be permeable to the fresher water but impermeable to the saltier water. Osmotic pressure would propel a controlled, one-way flow that could be employed to propel water-driven turbines for electricity.

A PROMISING ENERGY SOURCE

Global human population growth is still explosively on the rise. Going hand in hand with population growth is the trend of increasing worldwide urbanization and industrialization. Because of these trends, consumption of energy to run manufacturing processes, transportation, food production, climate control, and communication systems has escalated dramatically. Coupled with this ravenous energy use are the byproducts of mounting energy consumption: widespread pollution and a general degradation of the world environment. The world industrial society already faces the specter of diminishing energy sources and the eventual, ultimate exhaustion of all fossil fuels. New energy sources, such as the nuclear option, seem to possess, so far, many drawbacks to widespread usage.

Renewable energy sources that are clean and have a low impact on the quality of the environment are ideal long-term solutions. Some alternative energy forms fitting this description are either technically feasible today or almost so. The various forms of ocean power that have been developed or are in theoretical stages are excellent candidates for helping to alleviate some of the world's more pressing energy-related problems. For at least some geographic areas, ocean power is not only an efficient and clean power source but is economically competitive with fossil fuels and nuclear power. Tidal flow power is an excellent example of the increasing viability of some forms of ocean power. By the end of the twentieth century, tidal power was producing less than 300 megawatts worldwide, although the largest site (producing 240 megawatts) has been in operation at La Rance River in France since 1967. As of 2011, fewer than a dozen small demonstration plants had been built. The most successful test OTEC system has been the one at Kailua-Kona, Hawaii, which has produced more than 50 kilowatts in net power. The ocean, however, remains Earth's largest solar collector, and supporters of ocean power remain optimistic. Forms of ocean power represent a means to live in harmony with the environment without compromising the use and demand for energy.

—Frederick M. Surowiec

Further Reading

Andrews, John, and Nick Jelley. *Energy Science: Principles, Technologies, and Impact*. Oxford UP, 2007.

Boyle, Godfrey, editor. *Renewable Energy*. 2nd ed., Oxford UP, 2004.

Charlier, Roger H., and Charles W. Finkl. *Ocean Energy. Tide and Tidal Power*. Springer, 2009.

Chen, Falin. *The Kuroshio Power Plant*. Springer Science & Business Media, 2013.

Hamblin, Kenneth W., and Eric H. Christiansen. *Earth's Dynamic Systems*. 10th ed., Prentice-Hall, 2003.

Kaligh, Alireza, and Omer C. Onar. *Energy Harvesting: Solar, Wind, and Ocean Energy Conversion Systems*. CRC Press, 2017.

Khare, Vikas, Savita Nema, and Prashant Baredar. *Ocean Energy Modeling and Simulation with Big Data*. Butterworth-Heinemann, 2020.

Letcher, Trevor M., editor. *Future Energy: Improved, Sustainable, and Clean*. Elsevier, 2008.

Neill, Simon P., and M. Reza Hashemi. *Fundamentals of Ocean Renewable Energy. Generating Electricity from the Sea*. Academic Press, 2018.

Pecher, Arthur, and Jens Peter Kofoed, editors. *Handbook of Ocean Wave Energy*. Springer Open, 2017.

Tarbuck, Edward J., Frederick K. Lutgens, and Dennis Tasa. *Earth: An Introduction to Physical Geology*. 10th ed., Prentice-Hall, 2010.

Wright, Glen, Sandy A. Kerr, and Kate Johnson. *Ocean Energy: Governance Challenges for Wave and Tidal Stream Technologies*. Taylor & Frances, 2018.

OCEAN STRUCTURE

Fields of Study: physical oceanography; acoustics; marine chemistry

ABSTRACT

The ocean has a complex structure, both at its surface and in the vertical dimension descending to the ocean floor. This internal structure results in layering with respect to temperature, salinity, density, and the way in which the ocean responds to the passage of light and sound waves.

KEY CONCEPTS

aphotic zone: the lower part of the ocean below the maximum depth to which sunlight can penetrate

convective overturning: an event in which colder, denser water forms above less dense, less cold water, resulting in an inversion of relative position as the colder water sinks and the warmer water rises

density stratification: the formation of identifiable layers in ocean water according to density

photic zone: the upper part of the ocean that can be penetrated by sunlight

Secchi disk: a round, white disk of standard size that is used to measure the depth to which it can still be seen from the surface

SOFAR (sound fixing and ranging) channel: an oceanic layer between two salinity layers that have the effect of trapping and propagating sound such that it travels long distances without significant loss of energy

TEMPERATURE LAYERING

One highly significant aspect of ocean structure is the layering of water based on temperature and salinity differences. In order to understand the reasons for the thermal layering of the ocean, one must bear in mind that the primary source of heating for the ocean is sunlight. About 60 percent of this entering radiation is absorbed within the first meter of seawater, and about 80 percent is absorbed within the first 33 feet (10 meters). As a result, the warmest waters in the ocean are found at its surface.

That does not mean, however, that surface temperatures in the ocean are the same everywhere. Because more heat is received at the equator than at the poles, ocean surface temperatures are closely related to latitude. As a result, they are distributed in bands of about equal temperature extending east and west, parallel to the equator. Temperatures are highest along the equator because of the near-vertical angles at which the sun's rays are received here. As latitude increases toward the poles, ocean temperatures gradually cool as a result of the decreasing angle of incidence of the incoming solar radiation.

Measurements of ocean surface temperature range from a high of 91.4 degrees Fahrenheit (33 degrees Celsius) in the Persian Gulf, a partly landlocked, shallow sea in a desert climate, to a low of 28.4 degrees Fahrenheit (–2 degrees Celsius) in close proximity to ice in polar regions. There, the presence of salt in the water lowers the water's freezing point below the normal 32 degrees Fahrenheit (0 degrees Celsius) level. Because the salinity of this cold water is so high, it is denser and sinks to the ocean floor to travel along it for substantial distances as part of the thermohaline circulation current in the ocean. Ocean surface temperatures may also vary with time of year, with warmer waters moving northward into the Northern Hemisphere in the summertime and southward into the Southern Hemisphere in the wintertime. These differences are most noticeable in mid-latitude waters. In equatorial regions, water and air temperatures change little seasonally, and in polar regions, water tends to be cold all year long because of the presence of sea ice.

Vertically downward from the equator, toward the ocean floor, water temperatures become colder. This results from the facts that solar heating affects the surface waters only and that cold water is denser than warm water. When waters at the surface of the ocean in the polar regions are chilled by extremely low winter temperatures, they become denser than the underlying waters and sink to the bottom. They then move slowly toward the equator along the seafloor, lowering the temperature of the entire ocean. As a result, deep ocean waters have much lower temperatures than might be expected by examination of the surface waters alone. Although the average ocean surface temperature is 63.5 degrees Fahrenheit (17.5 degrees Celsius), the average temperature of the entire ocean is a frigid 38.3 degrees Fahrenheit (3.5 degrees Celius).

Oceanographers recognize the following layers within the ocean, based on its temperature stratification: First, there is an upper, wind-mixed layer, consisting of warm surface water up to 1,640 ft (500 m) thick. This layer may not be present in polar regions. Next is an intermediate layer, below the surface layer, where the temperature decreases rapidly with depth; this transitional layer can be 1,640 to 3,281 ft (500 to 1,000 m) thick and is known as the main thermocline. Finally, there is a cold, deep layer extending to the ocean floor. In polar regions, this layer may reach the surface, and its water is relatively homogeneous, with temperature slowly decreasing with depth.

Because the upper surface layer is influenced by atmospheric conditions, such as weather and climate, it may contain weak thermoclines as a result of the daily cycle of heating and cooling or because of seasonal variations. These are temporary, however, and may be destroyed by severe storm activity. Nevertheless, the majority of ocean water lies below the main thermocline and is uniformly cold, the only exception being hot springs on the ocean floor that introduce water at temperatures of 572 degrees Fahrenheit (300 degrees Celsius) and higher. Plumes of warmer water emanating from these hot springs have been detected within the ocean.

SALINITY

A second phenomenon responsible for layering within the ocean is variation in the water's salinity. For the ocean as a whole, the salinity is 3.5 percent (35 parts per thousand). Considerable variation in the salinity of the surface waters from place to place results from processes that either add or subtract salt or water. For example, salinities of 4 percent (40 ppt) or higher are found in nearly landlocked seas located in desert climates, such as the Red Sea or the Persian Gulf, because high rates of evaporation remove the water but leave the salt behind. High salinity values are also found at the surface of the open ocean at the same latitudes where there are deserts on land (the so-called horse latitudes). There, salinities of 3.6 to 3.7 percent (36 to 37 ppt) are common.

At the equator, however, much lower salinity values are encountered, despite the high temperatures and nearly vertical rays of the sun. The reason is that the equatorial zone lies in the so-called doldrums, a re-

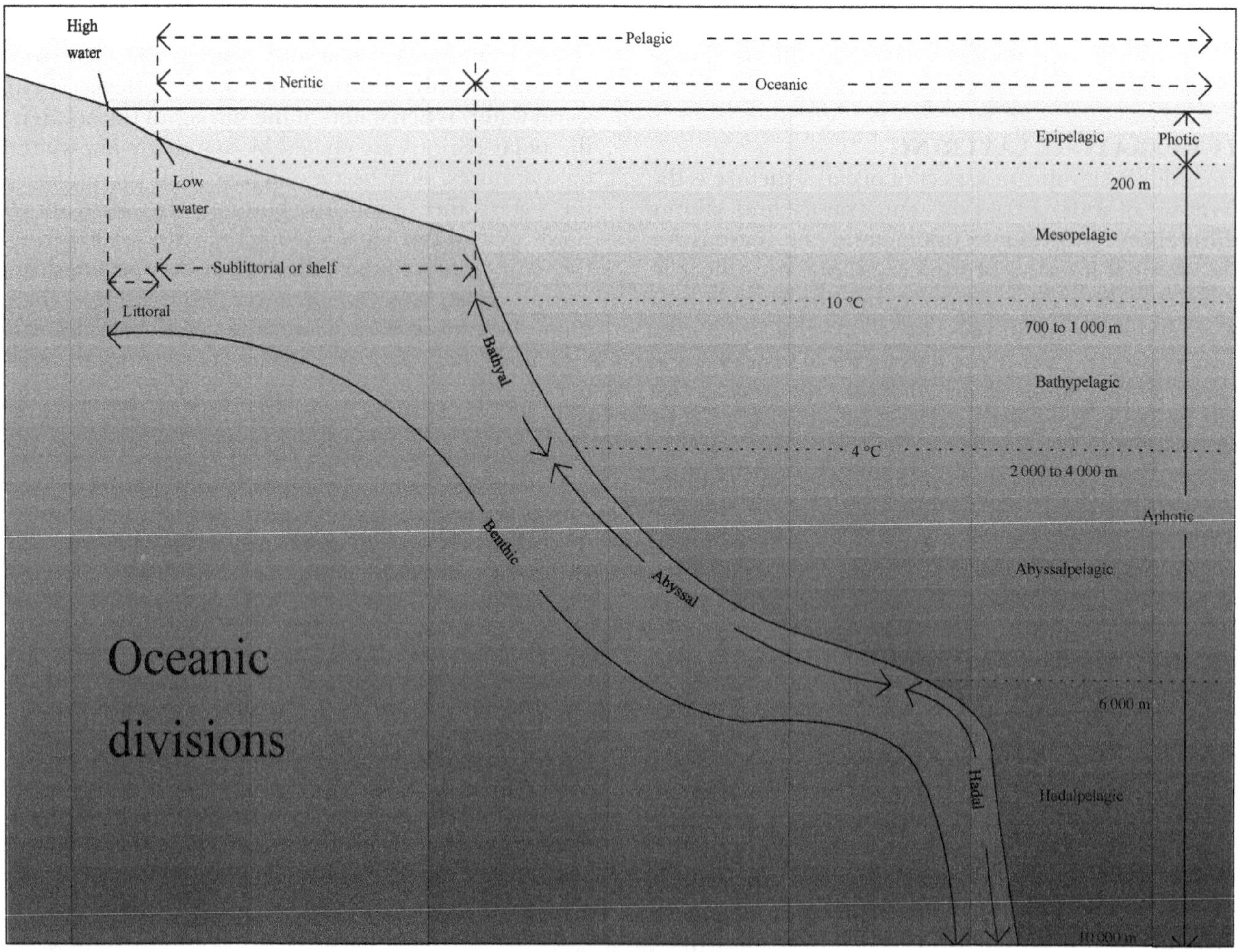

Diagram showing the divisions of the worlds oceans (Chris huh via Wikimedia Commons)

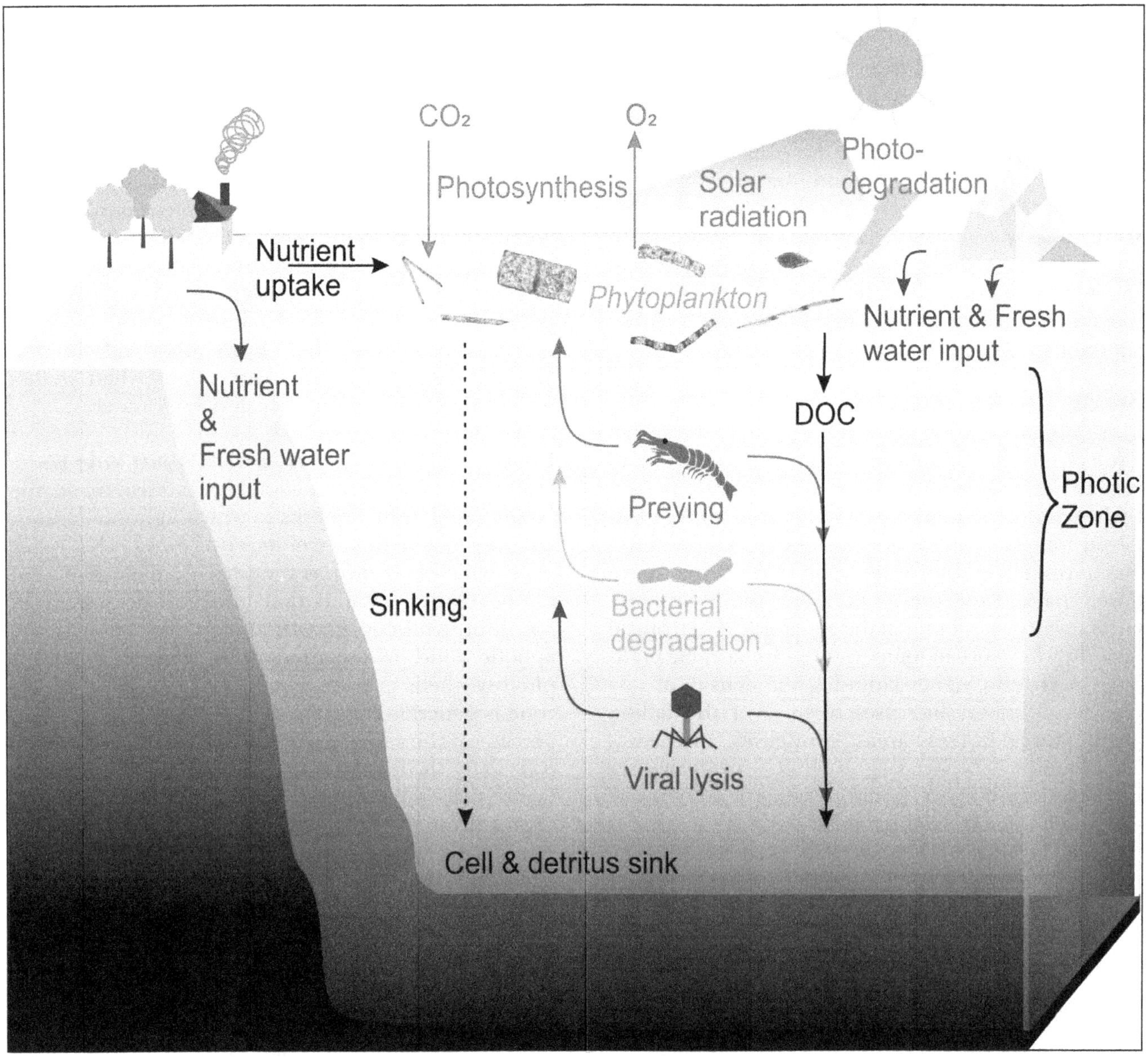

Cycling of marine phytoplankton, which live in the photic zone of the ocean (Käse L and Geuer JK via Wikimedia Commons)

region of heavy rainfall. The ocean's surface waters are therefore diluted, which keeps the salinity relatively low. Similarly, low salinities are also found in coastal areas, where rivers bring in large quantities of freshwater, and in higher latitudes, where rainfall is abundant because of numerous storms.

Despite the variation in salinity in the ocean's surface water, the deep waters are well mixed, with nearly uniform salinities ranging from 3.46 to 3.49 percent (34.6 to 34.9 ppt). Consequently, in some parts of the ocean, surface layers of low-salinity water overlie the uniformly saline deep waters and, in other parts of the ocean, surface layers of high-salinity water overlie the uniformly saline deep layer. Between these layers are zones of rapidly changing salinity known as haloclines. An important exception to this picture is a few deep pools of dense brine—such as are found at the bottom of the Red Sea, for example—where salinities of 27 percent (270 ppt) have been recorded.

Haloclines are very common in coastal areas. Off the mouth of the Amazon River, for example, a plume of low-salinity river water extends out to sea as far as 198 miles (320 kilometers), separated from the normally saline water below by a prominent halocline. In many tidal rivers and estuaries, a layer of heavier seawater will extend many miles (or kilometers) inland beneath the freshwater discharge as a conspicuous saltwater wedge.

DENSITY STRATIFICATION

A prominent density stratification within the ocean results from the variation in ocean temperatures and salinity just described. As noted, two factors make water heavier: increased salinity, which adds more dissolved mineral matter, and decreased temperature, which results in the water molecules being more closely packed together. Therefore, the least-dense surface waters are found in the equatorial and tropical regions, where ocean temperatures are at their highest. Toward the higher latitudes, the density of surface ocean waters increases because of the falling temperatures. In these areas, low-density surface water is found only where large quantities of freshwater are introduced, by river runoff, by high amounts of precipitation or by melting ice.

Vertical density changes are even more pronounced. As water temperatures decrease with depth, water densities increase accordingly. This increase in density, however, is not uniform throughout the ocean. At the poles, the surface waters are almost as cold as the coldest bottom waters, so there is only a slight increase in density as the ocean floor is approached. By contrast, the warm surface waters in the equatorial and tropical regions are underlain by markedly colder water. As a result, a warm upper layer of low-density water is underlain by an intermediate layer in which the density increases rapidly with depth. (This middle layer is known as the pycnocline.) Below it is a deep zone of nearly uniform high-density water.

Convective overturning takes place when this normal density stratification is upset. In a stable density-stratified system, the less dense surface water floats on top of the heavier, deeper water. Occasionally, however, unstable conditions will arise in which heavier water forms above lighter water. Then, convective overturning takes place as the mass of heavier water sinks to its appropriate place in the density-stratified water column. This overturning may occur gradually or quite abruptly. In lakes and ponds, it occurs annually in regions where winter temperatures are cold enough. In the ocean, convective overturning is primarily associated with the polar regions, where extremely low winter temperatures result in the sinking of vast quantities of cold water. In addition, convective overturning has been observed in the Mediterranean during the wintertime, when chilled surface waters sink to replenish deeper water.

LIGHT PENETRATION AND SOUND WAVES

Oceanographers also recognize stratification in the ocean based on the depths to which light penetrates, and they divide the ocean into two zones. The upper zone, which is known as the photic zone, consists of the near-surface waters that have sufficient sunlight for photosynthetic growth. Below this zone is the aphotic zone, where there is insufficient light for photosynthetic growth. The lower limit of the photic zone is generally taken as the depth at which only 1 percent of the surface intensity of sunlight still penetrates. In the extremely clear waters of the open ocean, this depth may be 656 ft (200 m) or more.

Stratification in the ocean based on the behavior of sound waves has also been observed. Because sound waves travel nearly five times faster under water than in the air, their transmission in the ocean has been extensively studied, beginning with the development of sonar, the echo sounder. So-called scattering layers have been recognized as regions that reflect sound, usually because of the presence of living organisms that migrate vertically, as layers within the water column, depending on light intensity. The SOFAR (sound fixing and ranging) channels are density layers within the ocean where sound waves can become trapped and can travel for thousands of miles (or kilometers) with extremely small energy losses. These channels have the potential to be used for long-distance communications. Shadow zones are also caused by density layers within the ocean. These layers trap the sound waves and prevent them from reaching the surface. One advantage of shadow zones is that submarines can travel in them undetected.

STUDY OF THE OCEAN

The measurement of water temperatures at the ocean's surface is quite simple. A thermometer placed in a bucket of water that has been scooped out of the ocean at the bow of a boat will suffice, provided the necessary precautions have been taken to prevent temperature changes caused by conduction and evaporation. For oceanwide or global studies, satellites provide near-simultaneous readings of ocean surface temperatures within an accuracy of 33.8 degrees Fahrenheit (1 degree Celsius). These satellites utilize infrared and other sensors, which are capable of measuring the amount of heat radiation emitted by the ocean's surface to within 32.36 degrees Fahrenheit (0.2 degrees Celsius).

Measuring the temperature of the deeper subsurface waters posed a problem, however, because a standard thermometer lowered over the side of a ship will "forget" a deep reading on its way back to the surface. As a result, the so-called reversing thermometer was developed in 1874. This thermometer has an S-bend in its glass tube. When the thermometer is inverted at the desired depth, the mercury column breaks at the S-bend, thus recording the temperature at that depth. Today, electronic instruments record subsurface water temperatures continuously. These devices can be either dropped from a plane or ship or moored to the ocean floor.

The measurement of the salinity of seawater is not as easy as one might think. An obvious way to determine water salinity would be to determine the amount of dried salts remaining after a weighed sample of seawater has been evaporated, but in actual practice, that is a messy and time-consuming procedure, hardly suitable for use on a rolling ship. A variety of other techniques have been used over the years, based on such water characteristics as buoyancy, density, or chloride content. By far the most popular relies on seawater's electrical conductivity. In this method, an electrical current is passed through the seawater sample; the higher the salt content of the water, the lower the electrical resistance of the solution and the faster this current is observed to pass. Using this method, oceanographers have been able to determine the salinity of seawater samples to the nearest 0.0003 percent (0.003 ppt). This is an important advantage, in view of the fact that the salinity differences between deep seawater masses are very minute.

Measurement of the densities of surface water samples can easily be accomplished by determining the water sample's buoyancy or weight, but the real difficulty comes with attempts to determine a subsurface water sample's density. If this water sample is brought to the surface, its temperature, and therefore its density, will change. Although sophisticated techniques are available for the determination of density at depth, in actual practice the density is not measured at all; instead, it is computed from the sample's known temperature, salinity, and depth. The density of the water is almost wholly dependent on these three factors.

A student lowers a secchi disk into the water to measure water clarity (Virginia State Parks via Wikimedia Commons)

Various methods are available for measuring the depth of light penetration in seawater. A crude estimate can be made using the Secchi disk, which was first introduced in 1865. This circular, white disk is slowly lowered into the water, and the depth at which the disk disappears from sight is noted visually. More sophisticated measurements can be made using photoelectric meters. Another good indicator of the maximum depth of light penetration in the sea is the lowest level at which photosynthetic growth can take place. For sound studies within the ocean, various methods are used to create the initial sound, including the use of explosives in seismic profiling. The returning echo is detected by means of a receiver known as a hydrophone.

—*Donald W. Lovejoy*

Further Reading

Broecker, Wally. *The Great Ocean Conveyor*. Princeton UP, 2010.

Charlier, R. H., and C. W. Finkl. *Ocean Energy: Tide and Tidal Power*. Springer-Verlag, 2009.

Emelyanov, Emelyan M. *The Barrier Zones in the Ocean*. Springer-Verlag, 2005.

Etter, Paul C. *Underwater Acoustic Modeling and Simulation* 4th ed., CRC Press, 2017.

Knauss, John A., and Newell Garfield. *Introduction to Physical Oceanography*. 3rd ed., Waveland Press, 2017.

Schwartz, M. *Encyclopedia of Coastal Science*. Springer, 2005.

Segar, Douglas. *Introduction to Ocean Sciences*. 2nd ed., W. W. Norton & Co., 2007.

Soloviev, Alexander, and Roger Lukas. *The Near-Surface Layer of the Ocean: Structure, Dynamics and Applications*. 2nd ed., Springer, 2014.

Talley, Lynne D., et al. *Descriptive Physical Oceanography: An Introduction*. 6th ed., Academic Press, 2011.

Tomczak, Matthias, and J. Stuart Godfrey. *Regional Oceanography: An Introduction*. Elsevier, 2013.

OCEAN TIDES

Fields of Study: oceanography; marine engineering; fluid dynamics; computer modeling

ABSTRACT

Tides are the displacements of particles on Earth caused by the differential attraction of the moon and the sun. There are atmospheric tides, land or crustal tides, and ocean tides. Of these, ocean tides are the most apparent because the ocean, as a fluid, is more easily stretched out of shape by the gravitational pull of the moon.

KEY CONCEPTS

conjunction: when the sun and the moon are on the same side of Earth and the three bodies are positioned in a straight line with each other

diurnal: high and low tides occurring once in a period of one lunar day

lunar day: a period of 24 hours and 50 minutes, the length of time for the moon to complete one orbit around Earth

microtidal: characterized by very low tide heights

opposition: when the sun and moon are on opposite sides of Earth and the three bodies are positioned in a straight line with each other

semidiurnal: high and low tides occurring twice in a period of one lunar day

CAUSES OF TIDES

Each particle in the ocean moves in response to the force of gravitational attraction exerted on it by both the sun and the moon. Although the sun is some 27 million times the size of the moon, the moon is the primary factor in the tidal movement of ocean waters. In fact, the moon's power is more than double the periodic tidal-stretching force exerted by the sun because it is much closer to Earth. This is described by Newton's universal law of gravity, which states that the gravitational force between two bodies is directly proportional to their masses, but inversely proportional to the square of the distance between their centers of mass. Thus, the proximity of the moon counts for more than the distant mass of the sun in solar and lunar relations with Earth.

In order to explain how tides are caused, tidal scientists use the concept of a theoretical "equilibrium tide." This concept is based on an ideal in which the ocean waters are always in static equilibrium and in which no continents obstruct the flow of water on Earth's surface. When the moon is directly above a particular location, its force of attraction causes the mass of water to bulge up directly under the moon and also on the opposite side of Earth. Meanwhile, in the opposite quadrants of the globe, two low-water troughs result from the movement of the water away from these areas. Thus, Earth rotates beneath tidal bulges and troughs, which results in high and low tides.

Tides generally follow the lunar day, which is twenty-four hours and fifty minutes, or the time it takes the moon to complete one full orbit of Earth. Some complex tidal cycles, however, result from the combined influences of the sun and moon. If there were no moon, the sun's influence alone would cause tides to occur at the same time each day. However, because the plane of the moon's orbit around Earth is in a different plane from that of Earth's orbit around the sun, combinations of full diurnal and semidiurnal tides result. In most areas, semidiurnal tides are the rule, with high and low tides occurring twice each lunar day, averaging twelve hours and twenty-five minutes apart. Diurnal tides occur when only one high tide and one low tide take place in one lunar day. They are found in parts of the China Sea and the Gulf of Mexico. Mixed tides result from a combination of both diurnal and semidiurnal tidal oscillations. Such tides are found along the Pacific coast of North America and in parts of Australia.

OTHER TIDE VARIABILITY FACTORS

Other factors that contribute to the variability of ocean tides are the phases of the moon, the position of the sun and moon relative to Earth, and the latitude and topography of the tide's location on Earth. When the sun, moon, and Earth are aligned (in "conjunction" or "opposition"), then the combined gravitational effects of these bodies will exert additional gravitational pull on Earth, resulting in increased tidal amplitude. This phenomenon, when lunar and solar tides reinforce each other, is called the "spring tide" and occurs around the full and new moons. In between the spring tides, a neap tide takes place when the sun, moon, and Earth are positioned at the apexes of a triangle. At this time, during the first and third quarters of the moon's phase, solar high tides are superimposed on lunar low tides, so resulting tides are the lowest in the month. In the open ocean, spring tides may be more than 3 feet (1 meter) high, while neap tides may be less than 3 ft (1 m). Tidal amplitude will also vary, depending on latitude and the declination of the moon.

Despite the complexity of the many variables that determine tidal behavior, tidal scientists can now predict the time and height of a tide anywhere, on any past or future date, given the one condition that they have sufficient information on how the local topography of the site modifies the tide. Local geographical conditions such as the width of a bay's mouth, the uneven slope of the bottom, and the depth of the body of water are the types of features that determine the range, amplitude, and time of the local tide. It is observed that the island of Nantucket, off the coast of the US state of Massachusetts, is microtidal and experiences a difference of no more than 1 ft (0.3 m) between high and low water, while only a few hundred miles (or kilometers) away, the Bay of Fundy, in the Canadian province of Nova Scotia, has the highest tides in the world, with a rise of 50 ft (15 m) during spring tides. To account for such phenomena, scientists have developed a model of tidal oscillation, in which the ocean structure is divided up among a great many basins of water, each with its own depth, length, and resulting period of oscillation. The boundaries of each basin are determined by the surrounding land forms, both above and below the ocean, and the influences of gravitational attraction in each are always changing, as are the currents that flow in. Ordinarily, when water rocks up and down in a basin, the water at the rim is most active, while the least amount of motion occurs in the center of the basin around a tideless node. Thus, the physical dimensions of these basins determine the period of oscillation of the waters throughout the basin.

OPPOSING TIDAL BULGES

When the pull of the moon creates a high tide on the side of Earth closest to it, a high tide occurs simultaneously on the opposite side of the planet. Logically, one would not expect this to be the case. To understand this, it is necessary to know that the moon and Earth revolve not only around each other but also around a common center of gravity located 994 mi (1,600 km) below the surface of Earth. As the Earth and moon revolve, the centrifugal force stretches the oceans outward against Earth's gravity. At its center, Earth is not still but is moving in a circle that is a small fraction of the size of the moon's orbit. This invisible revolution of Earth around the moon produces a centrifugal force throughout Earth, which varies as the moon revolves and pulls Earth's surface out of shape. It is the resulting "prolate," or lemon-shaped, elonga-

Kislaya Guba tidal power station, Russia (RusHydro via Wikimedia Commons)

tions of Earth that are observed as the tides. The tides nearest the moon are caused by gravitational attraction. On the side of Earth farthest from the moon, however, the centrifugal force is greater than the pull of the moon. To compound the complexity of the situation, the moon's gravitational force is also pulling the ocean floor of that area away from the waters there. Thus, high tides are produced on both sides of Earth in line with the moon.

Because tides are actually long waves, an observer on the moon might expect to see the two tidal bulges move around Earth at a speed consistent with the pace of the moon. Instead, tidal waves move out of step with the moon. Keeping pace with the moon would be possible under two conditions: if the oceans of the world were 14 mi (22 km) deep (they average a bit more than 2 mi [3 km] deep) and if there were no continents obstructing the movement of tidal waves. Thus, the speed of the movement of the tides is only 684 to 808 mi (1,100 to 1,300 km) per hour, and the tides do not keep up with the moon as it travels westward around the terrestrial globe.

EARTH'S ANGULAR MOMENTUM

The angular momentum of Earth is slowed by ocean tides. Because the momentum of the Earth-moon system is conserved, the slowing of Earth results in a speeding up of the moon's rotation around Earth. For unknown geophysical reasons, several abrupt increases in the length of Earth's day, some approaching 1 millisecond, have been observed over the past several decades. Ocean tides are not responsible for these abrupt changes but rather for small, steady increases over millennia. However, the US National Bureau of Standards must sometimes add a "leap second" to the Earth year to take into account the observed slowing of Earth. Assuming that the position of Earth's orbital distance with respect to the sun has not changed, the number of days in a year has changed from about 400 during the Devonian period to about 365 at the current time. Thus, the current length of a day is now quite a bit longer than it was during the Devonian, about 400 million years ago. Over this same time period, the moon's angular momentum is estimated to have increased by about 1.6 percent. As Earth's rotation slows and the moon revolves around Earth more quickly, the moon moves farther away from Earth. Accordingly, the moon has moved about 1 mi (1.6 km) away from Earth in the last 100,000 years.

STUDY OF OCEAN TIDES

Tides are not simple to predict. Qualitative prediction of the tides has been going on in harbors around the world for centuries, but quantitative prediction began in the nineteenth century, when people first designed tide-predicting machines to help forecast the tides. The first such machine was invented in 1872 by Lord Kelvin (1824–1907), who is often referred to as the first electrical engineer. Kelvin's machine was capable of drawing a line picture of the curve of the tide, and for this achievement he was knighted. Soon after this breakthrough, an employee of the United States Coast and Geodetic Survey invented a tide-predicting machine that showed the times and heights of the tides. The survey later designed a simpler machine that combines the capabilities of both previous inventions, giving the curve of the tides as well as the times and the heights of the tides. Unfortunately, these machines are not completely reliable because other factors, such as heavy storms, winds, or accumulation of sand as a result of wave action, can have dramatic impact on the water levels. Tide tables can only give the approximate high and low tides.

Today, tide-predicting machines have been replaced by faster digital computers. Tidal analysis still involves many complex computations, and predictions can be made only for places where a long series of observations are available. For each given spot on Earth, and for given time intervals, observations must be made that provide the value of the gravitational acceleration, the deflection of the vertical, and the measurements for the elevation of the water level. With this set of numbers in hand, the matter of prediction becomes one of extrapolating from the past into the future. Thus, around the coasts of the world, in inlets, in tidal rivers, and on islands, the water levels caused by tidal forces are carefully recorded. These measurements are analyzed locally. Once these measurements of the water level at a given place are taken within specified time intervals, the phases and amplitudes of the tide can be determined by a number of mathematical methods. Then, knowing the phases and amplitudes, scientists can reproduce the measurements according to a harmonic series. This harmonic method is a fairly reliable means of tidal prediction for deep-water ports, but in shallow-water areas, nonharmonic methods may need to be used. Finally, a determination of harmonic constants is made by national authorities, and the resulting data are sent to the International Hydrographic Bureau in Monaco. As a result, hydrographic offices in countries around the globe publish tide tables forecasting the high- and low-water times and water heights for the world's ports.

The problem of the measurement of tidal displacements in the open ocean has yet to be solved. This situation results from the fact that tidal sea-level records pertain primarily to coastal locations, and there are few or no measurements from the open ocean.

SIGNIFICANCE

Knowledge of ocean tides serves crucial purposes in the fields of navigation coastal engineering and tidal power generation. In addition, it holds a key position in relation to geophysics, marine geodesy, and astronomy.

Tides are of vital importance in navigation. Although the tides have become more predictable, and hence tamer, they still both help and hinder all mariners. The *Coast Pilots and Sailing Directions* for different parts of the world reveals the menacing possibilities which tides in various places are known to cause. Tidal currents often move violently when opposed by winds or confined in narrow channels. People have been swept off boats by an onslaught of giant waves when sailing in a flood tide through narrow straits. At certain stages of the tide, the waters can have dangerous eddies, whirlpools, or bores. A bore is created when a large portion of the flood tide enters a channel at once as one wave. Wherever they occur, they control the schedule of all shipping, as well as the rhythm of harbor life in the area. Even where there are no bores, the largest oceangoing liner must wait for slack water before entering a harbor where rushing tidal currents can fling it against piers. Since it is to the ship's advantage to sail in the direction of the tidal current flow, knowledge of tides is invaluable. All navigators approaching a coast rely on tide tables to supplement the information on depths in their nautical charts.

Coastal engineering work is dependent on knowledge of the tides for such undertakings as the management of tidal estuaries, construction of harbors, and damming of tidal rivers. Another practical aspect of tide information concerns the handling of problems that arise from the pollution of coastal waters and the ocean.

Tidal power generation is a new field that has gained increasing attention because of the shortage of available energy sources. Humans have long dreamed of harnessing the tidal forces for their energy needs. In 1966, the first tidal power station ever built was completed in the La Rance estuary in France. Construction of the project took place just after the Suez crisis when France felt uncertain about the future of its oil supply. The half-mile (805 m) Rance Dam was built to harness energy from the very large tides of the area—with a mean range of nearly 28 ft (8.5 m) and rising to more than 43 ft (13 m) at equinoctial spring tides. This power installation transmits electricity to Paris and the surrounding area, producing more than 580 billion watt-hours of energy a year but costing slightly more than the cost of operation of hydroelectric plants. In 1968, the Soviet Union finished construction on a 400-kilowatt tidal plant north of Murmansk, at a site where the maximum tide is less than 13 ft (4 m) in height. In 2004, China announced plans for a tidal plant that would be the world's largest, sited where a famous bore occurs on the mouth of the Yalu River. In the United States, at Passamaquoddy Bay in Maine a major tidal power plant project was abandoned because of the expense of maintaining the pipes and machinery in saltwater, and of transmitting the electricity generated to the nearest big users.

Additional areas of practical concern involving tides include the correlation of tides with earthquakes volcanic eruptions and geyser activity. Some scientists believe that the pressures exerted by water moving with the tides can trigger earthquakes.

—*Nan White*

Further Reading

Boyle, Godfrey, editor. *Renewable Energy*. 2nd ed., Oxford UP, 2004.

Christopherson, Robert W., and Mary-Louise Byrne. *Geosystems: An Introduction to Physical Geography*. Pearson Prentice Hall, 2006.

Freuchen, Peter. *Peter Freuchen's Book of the Seven Seas*. The Lyons Press, 2003.

Garrison, Tom S. *Essentials of Oceanography*. 6th ed., Brooks/Cole Cengage Learning, 2012.

______. *Oceanography: An Invitation to Marine Science*. Brooks/Cole, Cengage Learning, 2010.

Hamblin, Kenneth W., and Eric H. Christiansen. *Earth's Dynamic Systems*. 10th ed., Prentice-Hall, 2003.

Koppel, Tom. *Ebb and Flow: Tides and Life on Our Once and Future Planet*. Dundern Press, 2007.

McCully, James Greig. *Beyond the Moon: A Conversational, Common Sense Guide to Understanding the Tides*. World Scientific Publishing, 2006.

Morosov, Eugene G. *Oceanic Internal Tides: Observations, Analysis and Modeling. A Global View*. Springer, 2018.

Rusu, Eugen, and Vengatesan Venugopal. *Offshore Renewable Energy. Ocean Waves, Tides and Offshore Wind*. MDPI, 2019.

Schwartz, M. *Encyclopedia of Coastal Science*. Springer, 2005.

Talley, Lynne D., George L. Pickard, William J. Emery, and James H. Swift. *Descriptive Physical Oceanography: An Introduction*. 6th ed., Elsevier, 2011.

Thomson, William. *Tides and the Ocean: Water's Movement Around the World from Waves to Whirlpools*. Running Press, 2018.

OCEAN WAVES

Fields of Study: oceanography; wave physics: marine engineering; computer modeling

ABSTRACT

Waves continually shape and reshape beaches, and wave energy can be harnessed to generate power. Waves generated by the force of wind pose great dangers to life and human-made structures, and waves generated by seismic activity have killed thousands of people.

KEY CONCEPTS

capillary waves: the first and smallest identifiable waves to form when wind blows across still surface water

fetch: the extent of open water the wind pushes a surface wave before striking a beach

tidal wave: the long waveform of the tides following the gravitational "pull" of the moon around Earth along the equator**tsunami:** a large, dangerous wave produced by sudden, severe displacement of water caused by earthquake, landslide or volcanic eruption; often erroneously called a "tidal wave"

wave orbit: the circular motion of water in a wave

wave refraction: an effect of bottom topography that waves running parallel to the shoreline to break on an angle against the shore

CAUSES OF WAVES

The waves that agitate a lake or ocean are rhythmic, vertical disturbances of the water's surface. Their appearance may vary from a confused seascape of individual hillocks of water, each with a rounded or peaked top, to long, orderly swell waves with parallel, rounded crests. Waves involve a transfer of energy from place to place on the ocean's surface. An earthquake that jolts the California coast one evening may generate a tsunami that races across the Pacific and destroys a pier in Japan the next morning. The water itself, however, does not move; it is the wave form, or the energy impulse, that travels. The water stays where it is but oscillates as the wave form goes past.

Waves can originate in many ways. Tsunamis, erroneously called "tidal waves," are shock waves resulting from a sudden displacement of water by a submarine earthquake, landslide, or volcanic eruption. Shock waves are also generated when a pebble is tossed into a pond or a moving ship creates a wake. A second type of wave is the type produced by the gravitational pull of the sun and the moon, the tides that raise and lower the ocean's surface. These 'tidal' waves, which are not tsunamis, are the longest ocean waves of all, stretching halfway around the world as they travel along the equator at speeds of up to 994 miles (1,600 kilometers) per hour, following the gravitational influence of the moon.

The great majority of waves originate through the action of wind. Ordinary waves on the ocean or a lake form in this way. If there is no wind, the water surface is calm and smooth. If a slight breeze arises, the water surface instantly becomes roughened by patches of tiny capillary waves, the smallest waves of all. As the wind continues to blow steadily and in the same direction, ripples will appear because the surface roughness created by the capillary waves provides a vertical surface for the wind to push against. Soon the crests of adjacent ripples are pushed together to create larger and larger crests. This process continues as the intensity of the wind increases, with small waves steadily giving way to larger and larger ones.

Gulf Shores, United States (Caleb Jones via Wikimedia Commons)

Three factors determine the size of the waves ultimately produced: the wind speed, the duration or length of time that the wind blows in a constant direction, and the fetch or extent of open water over which the wind blows. A 23-mi- (37-km-) per-hour wind blowing for ten hours along a fetch of 75 mi (120 km) will generate waves 10 ft (3 m) high, but a 57-mi- (92-km-) per-hour wind blowing for three days along a 1,491-mi (2,400-km) fetch will generate waves 98 ft (30 m) high. Fortunately, such waves are very rare.

Rather than measuring the height of individual waves, scientists use a measure known as "significant wave height," which takes the average of the highest third of waves that occur during a particular storm. As of 2018, the largest recorded significant wave height is 62 ft (19 m), about the height of a six-story building, and was recorded in the North Atlantic between Iceland and the United Kingdom on February 4, 2013. It was certified as the largest significant wave height on record by the World Meteorological Organization (WMO) three years later. A record for the largest individual wave in the Southern Hemisphere was set in 2018 by a 78-ft (23.8-m) wave in the Southern Ocean off the coast of New Zealand. The significant wave height, 48.8 ft (14.9 m), was a record for the Southern Ocean but did not exceed the 2013 world record.

One way to understand the motion of water particles within a wave is to analyze the direction of water movement at various places in the wave. One can do so by sitting in a boat beyond the breakers. As the forward slope of a wave crest approaches, a lifting motion is experienced, followed by a forward push as the crest passes beneath the boat. This forward push is seen when waves break at the beach and their crests are thrown forward in a violent rush of water. Once the crest has passed, the boat is on the back slope of the wave, and now a downward motion is experienced. Next comes a backward motion, as the trough passes beneath the boat. This backward motion is also in a beach's breaker zone; after the crest has crashed forward on the beach, there follows a strong outward surge of water. This outward surge represents the backward water motion in the wave's trough. When all the preceding observations are combined, it can be seen that when a wave passes, the water particles move first up, then forward, then down, and finally back. This circular path is known as the wave orbit.

OTHER WAVE TYPES

The term "heavy sea" is often encountered in descriptions of the ocean. A heavy sea results from the prolonged action of strong winds over the open ocean. The waves are large, peaked, and confused, totally lacking in orderly arrangement by size. Frequently, there is much spray in the air as a result of the tops be-

Waves breaking with ships waiting to enter the porto of Sines, Porto Covo, Portugal (Alvesgaspar via Wikimedia Commons)

ing blown off the waves. A heavy sea is what one would expect to encounter in a hurricane or a violent storm. The term "swell," in contrast, refers to waves that have moved out of the wind-generating area. As these waves approach the beach, they appear as long rows of smoothly rounded wave crests, evenly spaced at wide intervals and of uniform height. These swells have been produced by a distant storm at sea and have then moved out of the wind-generating area. As they travel outward, their original irregularities are diminished. Very little energy is lost, however, because a wave traveling at the ocean's surface encounters very little friction.

Groups of larger swell waves will be interspersed with groups of smaller ones. Oceanographers believe that such variation is caused by two or more wave patterns traveling together across the ocean's surface. When the crest of one wave pattern is superimposed on the crests of the other wave pattern, larger swell waves will result. This is known as "constructive interference." When the trough of one pattern is superimposed on the crests of the other pattern, the swell waves will be smaller. This is known as "destructive interference."

Oceanographers also recognize two major categories of wind-generated waves: deep-water and shallow-water waves. Deep-water waves travel in water with depths greater than one-half of the wave length, and wind waves in the open ocean are generally in this category. Shallow-water waves, in contrast, travel in water so shallow that their wave orbits are affected by friction with the bottom. The shallower the water becomes, the more slowly the wave moves forward. This reduction in speed as waves approach the shoreline results in a process known as wave refraction, in which apparently straight wave crests approaching a shoreline from an angle are seen to be bent when viewed from above.

Close to the beach is the surf zone. There, the forward speed of waves is progressively slowed as the water depth decreases, and their crests become bunched more closely together. The shape of the crests changes from nearly flat to broadly arched, and there is a conspicuous increase in the height of the wave. In addition, the water in the crest of the wave begins moving faster than the water in the trough because of the friction created by the bottom, and this friction soon causes the crest to collapse in a torrent of water. The wave has "broken." Oceanographers recognize two types of breaker. The first is a plunging breaker, in which the wave crest curls smoothly forward and over, enclosing a tubular pocket of air below. The other type is known as a spilling breaker, in which foaming water spills down the forward slope of the crest as the breaker advances. This type of breaker has no air-filled tube.

The final zone, found between the breakers and the beach, is a narrow strip characterized by the rhythmic alteration of water rushing shoreward on the beach and water sliding back out to sea. The inward rush of water is known as the swash; it is a miniature wall of foaming water filled with air bubbles. The backward flow is a thin, glistening film of water termed the" backwash."

TSUNAMIS AND STORM SURGES

Two additional wave types require special mention. The first is a giant ocean wave known as a tsunami, which can be caused by submarine earthquakes volcanic eruptions or a landslide dumping massive amounts of debris into a body of water. In the open ocean, tsunamis behave just as any other ocean wave. They have crests and troughs that vary in height by 3 ft (1 m) or so while the wave is still at sea. The tsunami wave length is enormously long, however, averaging perhaps 149 mi (240 km) between crests. Tsunamis also have astonishingly high speeds, sometimes 403 mi (650 km) per hour or more. A tsunami does not come ashore as a plunging breaker; rather, the crest rushes in as a surge of foaming water whose motion is more indicative of a sudden rise in water level than of a typical wave.

The second special wave type is known as a storm surge. Storm surges are drastic rises in sea level accompanying hurricanes or other severe coastal storms. Several factors combine to create such a surge. One factor is the reduced atmospheric pressure that occurs in the eye of a hurricane. This reduction may allow the ocean's surface to rise 3 ft (1 m) or more, pushed upward in the low air pressure zone of the storm by the higher air pressure surrounding the storm. This is the same way liquid is pushed up a straw when the pressure inside the straw is reduced. If, in addition, the sun and moon are aligned in such a way as to produce unusually high tides, the storm surge may rise 3 ft (1 m) higher. A third contributing factor

"Ships in a Heavy Sea Running Before a Storm," Pieter Mulier the elder (c.1615-1670) (Art UK via Wikimedia Commons)

is the presence of strong onshore winds. In a major hurricane, these winds will push an accumulating mass of water toward the coast by preventing it from washing completely back into the sea as it normally would, increasing the impact of the storm surge. Finally, the nature of the offshore bottom plays a role. Shallow offshore bottoms permit wind to get a better "grip" on the water, raising its level higher. As a result, a hurricane that creates a13-ft (4-m) storm surge along Florida's east coast, with its deep offshore waters, would be able to raise a 33-ft (10-m) storm surge on Florida's west coast, where the bottom is flat and shallow for a distance of 99 mi (160 km) offshore.

STUDY OF OCEAN WAVES

Until the early 1940s, the principal method for studying waves was to observe the sea's surface and to record the length, height, speed, and period of individual waves. Based on an analysis of wave period, it was determined that the various types of ocean wave could be arranged in an increasing spectrum of size. Capillary waves were found to be the smallest ocean waves, with periods of less than 0.1 second. Ripples came next, with periods of 0.1 to 1 second. Ordinary wind waves followed, with periods ranging from 1 second to 1.4 minutes. Larger still were tsunamis, with periods averaging 17 minutes, and finally the tides, with periods of 12 or 24 hours.

One basic wave-measuring instrument is the tide gauge, which is used to study tsunamis. The tide gauge is usually mounted on a pier in the quiet waters of a harbor, where it will not be exposed to damaging surf. It consists of a float inside a vertical, hollow pipe. The float is free to rise and fall with the water level, and a continuous record is made of the float's movement. The pipe is sealed at the bottom in such a way that only the long-period waves associated with a tsunami can force the float to rise. In this way, the tide gauge can record the preliminary waves of a tsunami and serve as a warning of the larger waves that follow. After the destruction of Hilo,Hawaii by a tsunami on April 1, 1946, a seismic sea wave warning system was set up for the Pacific Ocean utilizing seismograph records and the type of tide gauge just described. Present-day tsunami warning systems linked to seismic observatories are able to provide warning of tsunami activity up to several hours in advance, depending on the location of the causal event.

Other instruments measure the impact of storm waves against pilings, piers, and deep-water structures. The measurements obtained from these instruments have enabled engineers to design structures that can better withstand the impact of storm waves. Before the 1950s, lighthouses and breakwaters were the structures most vulnerable to wave attack, but since that time, large oil drilling and production platforms have been built in the open ocean many miles (or kilometers) from shore. During severe storms a number of these platforms have been capsized by the impact or damage caused by oncoming waves, resulting in tragic losses of life.

In laboratory experiments, ocean waves can be simulated in wave tanks. These range from tabletop models with glass sides that look like aquariums to outdoor tanks that can hold large boats and generate breakers several feet (or meters) in height. The mathematical treatment of waves is facilitated by the regularity of their pattern, and computers are able to predict wave heights and other wave characteristics with a high degree of accuracy.

Stimulated by the energy crisis in the late 1970s, intensive consideration has been given to the possibility

Half a foot of storm surge, Pompano Beach, Florida (John Spade via Wikimedia Commons)

of harnessing wave energy to generate power. Many systems have been designed and tested, and some have been constructed and now produce millions of watt-hours of energy each year. The fundamental principle on which most wave generators work is that the motion of a passing wave exerts force that is similar to the rise and fall of a piston. This can then be mechanically harnessed and used to drive a generating system. Tidal systems such as the one in France in the La Rance estuary, however, use the inflow and outflow of tides to turn the blades of propeller-driven electrical generators

SIGNIFICANCE

Beaches owe their origin to wave action. The waves carry in the sand from which the beach is formed over time, and then smooth it daily, erasing imprints with their in-and-out motion. New sand is also formed as stones and rocks are made to grind against each other, chipping and eroding them slowly into sand grains. Storm waves, however, are capable of doing a great deal of damage in a much shorter span of time. The height of such waves in the open ocean can be dramatic; the USS *Ramapo* measured waves 116 ft (34 m) high in 1933, for example. When waves this height reach shore, the destruction can be enormous due to the energy and momentum of the moving mass of water. Concrete blocks weighing 65 tons (59 tonnes) or more have been torn loose from breakwaters by such waves. Sandy coastlines are also susceptible to attack. A longshore current develops when the waves encroach upon the shoreline at an angle, and this current can transport vast quantities of sand along a coast. Where this transported sand is trapped by an obstacle, such as a harbor jetty, excessive deposition will take place, perhaps requiring expensive dredging. At other points along the coastline, beach erosion may occur, also causing problems.

One of the most feared wave types is the tsunami, which is most frequently encountered on the shores of the Pacific due to the seismic activity associated with the so-called "Ring of Fire" of the Pacific Rim, but can occur in any body of water. Although tsunamis are almost imperceptible in the open ocean, their nature changes dramatically when they reach shore. Video records of incoming tsunamis amply demonstrate the immense, destructive force that such waves bring to bear against ships, buildings, and people. There are records of tsunamis rushing up mountain sides to elevations of 98 ft (30 m) and more. The tsunami generated by the explosion of Krakatau for example, is known to have transported an ocean-going vessel several miles (or kilometers) inland, depositing it in the jungle far up a mountainside. An added danger is that tsunamis have a series of crests that arrive approximately twenty minutes apart, and often the third or fourth to arrive is the largest.

Storm surges are another dangerous wave type. They are commonly encountered along the Atlantic and Gulf coasts of the United States and have been known to carry oceangoing ships 0.62 mi (1 km) or more inland. Six thousand people lost their lives in one such storm surge in Galveston, Texas, in 1900, and hundreds of thousands have drowned in a single storm surge on the shores of India's Bay of Bengal. Expensive storm surge barriers now guard Providence, Rhode Island; London; and the Netherlands.

Two other wave types that may present hazards for humans are rogue waves and seiches. Rogue waves are huge, solitary waves occasionally encountered in the ocean. They are particularly associated with the southward-flowing Agulhas current off South Africa and have been credited with sinking or severely damaging several large cargo ships. Seiche waves are oscillations in an enclosed water body such as a lake or bay. One such seiche, created by a hurricane, overflowed the dike surrounding Florida's Lake Okeechobee in 1928 and drowned two thousand people.

—Donald W. Lovejoy

Further Reading

Charlier, R. H., and C. W. Finkl. *Ocean Energy: Tide and Tidal Power*. Springer-Verlag, 2009.

Clarke, Allan J. *An Introduction to the Dynamics of El Niño and the Southern Oscillation*. Academic Press, 2008.

Cruz, Joan. *Ocean Wave Energy: Current Status and Future Perspectives*. Springer-Verlag, 2010.

Geggel, Laura. "78-Foot Wave Is the Largest Ever Recorded in Southern Hemisphere." *LiveScience*, 11 May 2018, www.livescience.com/62543-largest-wave-hits-southern-hemisphere.html.

Grossman, David. "Scientists Confirm the Largest Wave Ever Recorded." *Popular Mechanics*, 15 Dec. 2016, www.popularmechanics.com/science/environment/a24326/worlds-largest-wave-recorded-near-scotland.

Joseph, Antony. *Tsunamis: Detection, Monitoring and Early Warning Technologies*. Academic Press/Elsevier, 2011.

Pelinovsky, Efim, and Christian Kharif, editors. *Extreme Ocean Waves*. 2nd ed., Springer, 2016.
Schwartz, M. *Encyclopedia of Coastal Science*. Springer, 2005.
Sundar, V. *Ocean Wave Mechanics: Applications in Marine Structures*. John Wiley & Sons, 2016.
Talley, Lynne D., George L. Pickard, William J. Emery, and James H. Swift. *Descriptive Physical Oceanography: An Introduction*. 6th ed., Elsevier, 2011.
Trujillo, Alan P., and Harold V. Thurman. *Essentials of Oceanography*. Prentice-Hall, 2010.
Young, Ian, and Alexander Babanin. *Ocean Wave Dynamics*. World Scientific, 2020.

OCEANOGRAPHY

Fields of Study: oceanography; geology; marine biology; mathematics

ABSTRACT

More than 70 percent of the Earth's surface is covered by saltwater oceans and seas. Oceanographers study a wide range of subjects, including major underwater geological formations, such as very high mountain ranges and very deep canyons, as well as the tectonic and volcanic movements that continue to change the submarine environment. They also study the plant and animal life in the oceans of the world and analyze the chemical bodies—gases and minerals—in the waters.

KEY CONCEPTS

bathysphere: a spherical steel chamber devised by William Beebe and Otis Barton, capable of descending to a depth of 2,952 feet (900 meters)

constructive interference: a phenomenon of wave physics in which the coincidence of two or more waves adds together both their peak height and energy, resulting in the formation of a much larger wave; applies to waves of all types

echolocation: the identification of size and distance by the generation of sound waves and measurement of the time lag and frequency of their reflection from an object

paleoceanography: the study of oceanography of Earth's oceans in the distant past through the records left in geological formations and how that relates to Earth's oceans in the present

rogue wave: also called a "monster wave," a wave of unusual size formed as a result of constructive interference

DEFINITION AND BASIC PRINCIPLES

Oceanography is not a single applied science but rather a combination of fields of study. Because the oceans and seas of the world contain both living creatures and inanimate physical components, the basic sciences that contribute to oceanographic knowledge are marine biology, chemistry, and geology. However, the ecological effects caused by phenomena such as oceanic tides, currents, and temperature variations must be studied before the marine biologist, chemist, or geologist can obtain meaningful results from research.

BACKGROUND AND HISTORY

Since early historical times, people have been interested in the oceans and seas. For many centuries, attitudes toward the ocean were strongly influenced by superstition and mythology. Even as late as the publication of Jules Verne's (1828–1905) *Vingt mille lieues sous les mers* (1869–1870; *Twenty Thousand Leagues Under the Sea*, 1873), depictions of the ocean depths were dominated by fantastic and mysterious images rather than based on science. Exactly when the study of oceanography became a science is hard to say, but each time technology allowed scientists to study the ocean at a greater depth, knowledge advanced significantly. As scientific knowledge of oceans and seas increased, people began to realize that a knowledge of geology and, to some extent, astronomy, was valuable in understanding ocean phenomena.

The desire to develop submarines for use in warfare, particularly in World War I and World War II, spurred advances in ocean exploration. In the 1930s, Americans Otis Barton and William Beebe (1877 - 1962) developed the bathysphere. A pressure-resistant spherical steel apparatus weighing 4,464 pounds (2,025 kilograms), the bathysphere could descend more than 2,952 ft (900 m), six times the depth that could be reached by helmeted divers. The bathysphere was connected to a ship on the surface by heavy cables and a hose that held electric and communication wires.

In 1951, Rachel Carson (1907–1964) published *The Sea Around Us*, which popularized the study of oceanography. In her book, she focused on the vital interrelationships between submarine and surface phenomena and the delicate balance these relationships represent for the future of the world. Many of her concerns were adopted by later generations as the predominant issues confronting oceanography.

Another major figure who contributed to global recognition of the wonder of the oceans' secrets was the French oceanographer Jacques Cousteau (1910–1997), whose contribution to knowledge of the undersea world began with the 1943 film *épaves* (Shipwrecks), followed by his launching of the French navy's underwater research group. The navy group was the first of many to explore the ocean using methods developed by Cousteau and his team.

In 1953, Cousteau published *The Silent World*, which described his work as a diver and established the first scientific procedures for echolocation of large fish populations underwater. His accomplishments led to his appointment as director of the famous Oceanographic Institute and Museum in Monaco in 1957. Soon he developed the SP-350, known as the diving saucer, which became a prototype for descending—with significantly improved maneuverability—to ever-deeper reaches. The SP-350 reached a depth of 1,640 ft (500 m) in 1965. Although early developers of the bathysphere had descended to almost 820 ft (250 m) in 1930 and to more than 2,953 ft (900 m) in 1934, the limited maneuverability of their invention could not compare with that of Cousteau's diving vessel. Before Cousteau died in 1997, he and his associates witnessed major developments in fields of oceanography that they and others had pioneered.

HOW IT WORKS

Measuring Tides. Perhaps more than any other area of oceanography, the study of ocean tides and waves requires knowledge from several fields of science. Principles of physics and astronomy, for example, help explain timing lapses and the strength of coastal tides. The rising and falling of tides are the result of gravitational pull from both the Moon and the Sun, and the strength and direction of tides are affected by forces connected with Earth's rotation. Oceanographers may be involved in complicated mathematical calculations using the laws of physics to compare major tidal movements, while hundreds of relatively simple tidal gauge stations using float devices provide practical data for record keeping across the globe.

Systematic data gathering may not be enough, however, to predict potentially disastrous tsunamis or so-called monster waves in the middle of the ocean. Tsunamis are often caused by major earthquakes at fault points deep beneath the ocean surface. The earth movements unleash swells that race toward coasts. The task of oceanographers in such cases is to measure the speed and force in those waves in an attempt to limit the damage when they reach the shore.

Oceanographers continue to investigate monster or rogue waves that appear rarely but suddenly on the high seas and pose major threats to shipping. Rogue waves, which can be tall enough to sink ships, are thought to be the result of several tidal movements that come together at one location in mid-ocean and combine their forces, a phenomenon of wave physics called "constructive interference."

Mapping Ocean Currents. Ocean currents can be at the surface or much deeper. The best known of the surface currents is the Gulf Stream, which can reach speeds of 10 ft (300 centimeters) per second, followed by the Kuroshio off the east coast of Japan. Westward-flowing currents north and south of the equator (caused mainly by the trade winds over tropical regions of the Atlantic, Pacific, and Indian oceans) are separated by a 984-ft (300-m) wide countercurrent moving eastward. Most challenging for oceanographers are several major deep-ocean currents, which have both lateral and vertical patterns of movement. To study these currents, oceanographers use instruments to locate underwater strata that differ either in temperature, levels of salinity or pH. They use these data to attempt to explain why massive volumes of water either drop or rise, displacing other bodies and creating currents in the depths of the ocean.

Surveying the Ocean Floor. Two obvious and spectacular applications of geology to oceanography are the study of deep trenches and mid-ocean mountain ridges. The Mariana Trench north of New Guinea, for example, is more than 1,580 miles (2,542 kilometers) long and, in places, 43 mi (69 km) wide. At its deepest point, the Challenger Deep, the Mariana Trench plunges to more than 7 mi (10,924 m) below sea level. The Mid-Atlantic Ridge (discovered in 1872) rises from more than 1,243 mi (2,000 m) below sea level to form island mountains that are more than 1,243 mi

(2,000 m) high (notably in Iceland and the Azores). Less spectacular but of considerable practical importance is the study of the continental shelves. Oceanographic studies of coastal ocean zones contribute increasingly to key economic development schemes, most notably (but with increasingly controversial environmental implications) in seeking zones for commercial exploitation of undersea petroleum reserves.

How El Niño Works. Although oceanographers are still investigating the phenomenon of El Niño (part of the larger El Niño-Southern Oscillation (ENSO) pattern), certain facts have been established. When there is an area-wide warming (El Niño conditions) or cooling (La Niña conditions) of surface waters in the Equatorial Pacific occurring at the same time that atmospheric pressure causes surface waters of the western Pacific to rise or fall (the Southern Oscillation), major climatic repercussions occur in the subtropical eastern Pacific. These conditions may bring extreme drought or excessive rain. The reasons for the ocean-wide coupling of this cause-and-effect situation are not clear. Oceanographers estimate that the phenomenon recurs in cycles of about five to seven years. The phenomenon is known to have a pronounced effect on weather around the world.

Paleoceanography. Paleoceanography represents a special branch of the broader science of oceanography. It is interesting to note that the water present in the oceans of today is the same water, almost drop for drop, that has been in the oceans of Earth throughout the 4.5-billion-year history of the planet. The processes that are occurring in the oceans today are the same processes that were occurring millions and millions of years ago. But Earth is a dynamic planet, and the positions of the continental land masses have changed many times over that period. The dynamic behavior of the oceans has also changed accordingly, though not the processes taking place within them. The record of those changes is written in the geology of the planet. Paleoceanography is the science of studying the oceans of the past through the signs that have been left in the geological record of Earth. Paleontologists unearth the remains of creatures that inhabited the seas in the distant past, piecing together a history of life on this planet and discovering how the land masses have changed. Geologists use the geological structures left by ancient seas to find resources such as oil deposits and mineral ores that are the result of or that have been affected by biological activity in the ancient past. In examining the history of ancient seas, many valuable insights are gained by oceanographers with regard to the current and potential future conditions of Earth's oceans in the present day. In forensics it is said that every crime leaves a clue, no matter how small. It is equally true that everything that has affected Earth's oceans and the life within them in the past has left a clue in the geological record of the planet. Paleoceanographers, drawing on several scientific disciplines, seek to find those clues and apply what insights they glean from them to the oceans of today.

APPLICATIONS AND PRODUCTS

Harvesting Microscopic Organisms. Oceanographers have increasingly directed their attention to microscopic plant and animal organisms, most particularly to the smallest forms of plankton in the food chain and multiple varieties of algae. Extremely tiny and exponentially numerous holoplankton (as distinct from meroplankton, which are mainly tiny larvae that become larger independent organisms such as jellyfish, sea urchins, and starfish) never develop sufficient motor independence to propel themselves and instead drift like clouds in the world's seas and oceans. Concentrated masses of plankton provide nourishment to many varieties of fish, which in turn are eaten by larger fish, and ultimately by humans. Therefore, oceanographers—as potential sources of information for the fishing industry—focus on studying not only the temperatures and saline/chemical content that contribute to plankton propagation but also the currents that can transport them to different areas of the ocean, creating concentrated feeding grounds for larger, commercially desirable fish species.

Commercial Uses of Algae. Oceanographers have contributed information to industries wishing to control the growth of certain species of algae. Industrially, organized algae cultivation is already used in producing fertilizers and feed for livestock, dyes, some pharmaceuticals, and increasingly for the production of biofuels. In general terms, algae (larger forms of which are called "seaweed") are plant plankton that, like all plants, use photosynthesis to convert carbon dioxide into various organic compounds and, most important for the environment, into oxygen. Algae's ecologically important role as an oxygen producer, along with its potential as a source for fuel, pose challenges for oceanographers. They are likely to be called on to identify the most relevant species for various industrial uses and the optimal conditions for their propagation under controlled conditions.

CAREERS AND COURSEWORK

Individuals interested in preparing for a career in oceanography should not only plan their education carefully but also consider the employment paths likely to be open at various levels. It is generally recommended that undergraduates focus on the core sciences (such as biology and chemistry) and mathematics. For some entry-level jobs relating to oceanography, an undergraduate degree with a heavy emphasis on science may meet the basic requirements. Those seeking master's or doctoral degrees, however, must attend a university offering a graduate program in specialized subareas of oceanography, such as biological, chemical, geological, or physical oceanography. A master's degree or doctorate in a subarea of oceanography is required to work in academic institutions (as teachers or researchers), private companies with involvement in various maritime economic pursuits, international or national organizations involved in oceanography, a variety of governmental agencies involved in environmental issues, and certain branches of the military.

A number of specifically defined areas are likely to attract the professional expertise of oceanographers in the future. Several of these are connected with growing environmental concerns, while others are more clearly tied to changing economic markets. Environmental concerns include climate change and global warming issues and questions regarding the disposal of industrial and agricultural wastes. Increasingly, corporations are turning to oceanographers to develop ways to exploit the ocean waters. For example, fish farming based on scientifically grounded methods plays an ever-expanding role as world demand grows for common edible seafoods. Industrial extraction of minerals and the development of the oceans' biomedical and pharmaceutical resources will require a variety of oceanographic technical skills to guarantee proper responses to both ecological and health safety concerns.

The main US government agency concerned with oceanographic issues is the National Oceanic and Atmospheric Administration (NOAA). There are also numerous private-sector research organizations and professional associations of interest to oceanographers. Three of the most famous research institutes in the United States are the Scripps Institution of Oceanography in La Jolla, California; the Woods Hole Oceanographic Institution in Woods Hole, Massachusetts; and the Lawrence Berkeley National Laboratory in Berkeley, California. Professionals in these institutes (together with their fellows in many other private research organizations) continue to play significant roles in defining the new frontiers of oceanography.

A variety of private companies that employ oceanographers (such as the Woods Hole Group, General Oceanics, and Sound Ocean Systems) share contracts and research projects with Scripps and Woods Hole, as well as similar research institutions in other countries. Those interested in career possibilities in oceanography should familiarize themselves with a number of professional associations or interest groups whose members are oceanographers. These include the Consortium for Oceanographic Research and Education in Washington, DC; the Reef Environmental Education Foundation in Key Largo, Florida; the Atlantic Coastal Zone Information Steering Committee in Halifax, Nova Scotia; and the National Geographic Society in Washington, DC.

SOCIAL CONTEXT AND FUTURE PROSPECTS

A number of socially relevant issues are likely to involve professional oceanographers. Several of these are connected with growing environmental concerns, and others are tied to changing economic markets. In the first area, global climate change and questions of industrial-waste disposal rank very high. Rising concerns around the world over the effects of pollution on oceanic ecosystems suggest that very aggressive conservationist movements (such as Greenpeace) are likely to be joined by an increasingly broad spectrum of environmental organizations focused on the world's oceans and seas.

Meanwhile, economically attractive activities involving controlled exploitation of the ocean waters is already involving oceanographers in key areas that seemed to be of only marginal importance before the twenty-first century. Scientifically controlled fish farming, for example, is growing in importance as demand for seafood increases. Conservationists hope that fish farms will decrease pressures on the natural habitat of many commercially attractive species. Also, oceanographers are likely to need to create better ways to extract minerals from the world's oceans and to develop marine pharmaceutical resources without damaging the environment.

Because ecological and climate issues are being recognized as vital for the future of the planet, oceanographic research projects are being conducted in these areas. Such programs include NOAA's Climate Variability and Predictability program, which seeks to find the causes of climate variability; the Integrated Ocean Drilling Program, an international organization that studies the history of Earth through rocks and sediments beneath the ocean floor; and NOAA's Sustainable Seas Expeditions, which explores various marine ecosystems. High-tech environmental research projects include the Surface Heat Budget of the Arctic Ocean (SHEBA) of the National Center for Atmospheric Research's Earth Observing Laboratory, which is investigating the relationship between the Arctic ice pack and climate.

Development of remotely operated vehicles, greatly expanded since the 1980s, has made it possible to penetrate ocean depths that were unreachable by manned undersea devices. In May 2009, the Woods Hole Oceanographic Institution used a remotely operated vehicle called the *Nereus* to revisit the depths of the Mariana Trench, which had been partially explored by a Japanese remotely operated vehicle in 1998. The small dimensions of the *Nereus* 8 ft [2 m] wide and 14 ft [4 m] long) enabled it to maneuver in tight spaces, and the vehicle combined several advanced technologies, most notably a remarkably thin fiber-optic tether. This feat was eclipsed in 2012 when James Cameron piloted a free-moving one-man submersible to the bottom of the Mariana Trench.

—Byron D. Cannon

Further Reading

Ballard, Robert D., editor. *Archaeological Oceanography*. Princeton UP, 2008.

Carson, Rachel L. *The Sea Around Us*. 1951. Oxford UP, 2003.

Denny, Mark. *How the Ocean Works*. Princeton UP, 2008.

Earle, Sylvia A. *The World Is Blue: How Our Fate and the Ocean's Are One*. National Geographic Society, 2009.

McLellan, Hugh J. *Elements of Physical Oceanography*. Elsevier, 2016.

Millero, Frank J. *Chemical Oceanography*. 4th ed., CRC Press, 2013.

Nichols, C. Reid, et al. *Recent Advances in Oceanography*. Greenwood, 2003.

Rohling, Eelco. *The Oceans: A Deep History*. Princeton UP, 2017.

Schopf, Thomas, and J. M. Schopf. *Paleoceanography*. Harvard UP, 2013.

Seinfeld, J. H. *Atmospheric Chemistry and Physics of Air Pollution*. John Wiley & Sons, 1986.

Tartar, Luc. *An Introduction to Navier-Stokes Equation and Oceanography*. Springer, 2006.

Thurman, Harold V., and Alan P. Trujillo. *Essentials of Oceanography*. 10th ed., Pearson Education, 2009.

OCEANS' ORIGIN

Fields of Study: paleogeology; paleoceanography; astrophysics; marine geology

ABSTRACT

Oceanic waters are believed to be derived from the outgassing of hydrated minerals bound up during Earth's formation. Subsequent evolution of the waters primarily involved ions dissolving in the fluid medium by interactions with the continental and oceanic bottom sediments to produce the basic salinity of Earth's oceans.

KEY CONCEPTS

albedo: the percentage of incoming light that is reflected from Earth's atmosphere and surface

fumarole: a volcanic vent that emits steam and other gases, but not lava

hydrated minerals: minerals that contain water as part of their crystal structure or that is entrained within the overall structure of the mineral without being part of the crystal structure

infalling: objects from space that are captured by Earth's gravity and eventually fall to the surface of the planet

mass spectrometer: a sophisticated device used to determine the mass of particles by vaporizing them to impart an electrical charge and passing the resulting ions into a chamber under high vacuum where they interact with a tunable magnetic field and follow a unique curved path determined by their mass-to-charge ratio (m/e); a detector receives the ions and their numbers for each m/e are counted to produce a mass spectrum

outgassing: the release of trapped or entrained gases, either directly or by vaporization

pH: formally defined in value as the negative logarithm of the hydrogen ion concentration of a solution, according to the equation $pH = -\log[H^+]$

EARTH'S OCEANS

Of all the planets in the solar system, Earth stands out as a watery world, distinguished from the others by great quantities of liquid water. Encompassing more than 0.32 billion cubic miles (1.35 billion cubic kilometers), Earth's seas contain enough salt to cover all of Europe to a depth of 3.1 mi (5 km). The salt content of the oceans is composed primarily of sodium chloride, some 86 percent of the ions by weight, in association with other ions of magnesium, calcium, potassium, sulfate, and carbonate groups, providing a salinity of 3.5 percent (35,000 parts per million), with pH 8 for the hydrogen-ion concentration (slightly alkaline in nature). Of all the water on Earth, 97 percent of it is salty, the remainder being proportioned among ice (77 percent of total freshwater) and continental and atmospheric waters. The ice itself, principally in the Arctic-Greenland area (1.66 million sq mi [1.72 million sq km], 10,498 ft [3,200 m thick]) and the Antarctic area (4.6 million sq mi [12 million sq km], 13,123 ft [4,000 m] thick), provides effects ranging from climatic control to habitats for living organisms and sources of additional seawater that, in the past, have caused sea levels to rise more than 328 ft (100 m).

Composite image of southern Africa and the surrounding oceans (Ocean Biology Processing Group at NASA's Goddard Space Flight Center via Wikimedia Commons)

Consideration of the oceans' origin is twofold: the primordial origin of the water itself and the origin and rate of addition of the salt ions present in the past and contemporary waters. The data sources for consideration include the chemistry of water, the amounts and types of runoff delivered by rivers into the sea, and the composition of volcanic gases, geysers, and other vents opening to the surface. In addition, most researchers find the oceans and atmosphere to be linked in origin, providing even more data for analysis.

Numerous sources for the world's water have been proposed, although a definitive resolution has not been achieved. Possible sources include the primordial solar nebula the solar wind acting over time, water-containing bodies colliding with Earth, impact degassing, and outgassing from the planetary interior. A complete consideration of the origin of Earth's oceans necessarily involves the investigation of factors controlling water on Earth, particularly the rates, amounts, and types of outgassing, modes of planetary formation possible chemical reactions providing water, loss rates of gases to space, and, finally, internal feedback mechanisms such as changes in Earth's albedo (reflective power), temperature, alteration of mass, and other factors not clearly understood.

SOLAR WIND COLLIDING BODIES, AND IMPACT DEGASSING

The solar wind as a primary source of terrestrial water can be eliminated for several reasons. Its basic constituents, charged protons, can formulate water in the atmosphere by reactions with oxygen, but all evidence indicates that there was no free oxygen in the primordial atmosphere. The geologic record shows the presence of liquid water at least 4 billion years ago that was essentially devoid of free oxygen. Astrophysical evidence suggests that the solar flux of energy, associated with the solar wind, was such that, early in Earth's history, water on Earth should have been frozen, not liquid, if that was the primary source. The presence of liquid water through at least 80 percent of Earth's history, however, has been established.

Colliding bodies would include two primary sources: meteorites and comets. The occurrence of cometary bodies striking Earth has yet to be documented. However, the basic chemical makeup of comets consisting of various ions, metals, organic molecules, and dust grains in a mass of water ice, could theoretically have supplied enough water to account for the amount of water present on the planet, providing that large numbers of cometary objects struck the early Earth during the first half-billion years of its history. No evidence for such happenings is available at present, although a theory of Earth still being bombarded incessantly by small comets composed primarily of water ice has been fiercely debated.

Meteoritic impact, particularly during the earlier stages after final planetary accretion, would also have added water to the crust via two mechanisms. Through the study of carbonaceous chondrites (the oldest and most primitive meteorites), abundant volatiles, such as water, are found to be bound chemically to minerals such as serpentine. Additional waters, trapped in crustal and mantle rocks, would have been released during impact, particularly from large meteoritic rocks. It has been calculated that such impact degassing could have released 10^{22} kilograms of volatiles, quite close to the currently estimated value of 4×10^{21} kilograms for Earth as a whole. Remnants of such ancient astroblemes are lacking, however, because of subsequent erosion, filling in by molten magma, or shifting of the continental masses over 4 billion years.

OUTGASSING

The most widely accepted origin for the oceans and atmosphere combines the features of the primordial solar nebula and slow outgassing from within the solidifying Earth. Original water would have been combined, under gravitational collapse, with silicates and metallic materials during the planetary accretion process, with the hydration of minerals assisted by the heating of Earth due to infalling bodies and radioactive elemental decay. Such "wet silicates" appear to hold large quantities of bound water for indefinite periods of time. The primordial Earth is believed to have formed by cold accretion, and would have trapped the water molecules. If it had been too hot during the accretion, all the minerals would have been dehydrated, and if too cold, no water would have been released. A delicate balance of temperature must therefore have been achieved. Further, the volatiles forming the atmosphere must have outgassed first, because water must be insulated from solar radiation in order to form a liquid phase.

A secondary problem examines how swiftly the fluids would have outgassed, whether all at once, as individual events, or in a continuous fashion. Most data suggest the continuous mode of emission, with the greatest reliance on data from still-active sources—mainly volcanoes, undersea vents, and associated structures. Fumaroles, at temperatures of 932 to 1,112 degrees Fahrenheit (500 to 600 degrees Celsius), emit copious quantities of water, sulfur gases, and other molecules. These bodies grade gradually into hot spots and geysers, areas where water is moved upward through the crust from great depths. Magmatic melts rising in volcanoes release water and other gases directly to the surface.

In Hawaii for example, the Halemaumau Pit, the volcano Kilauea's most active vent, emits—in terms of material—68 percent water, 13 percent carbon dioxide and 8 percent nitrogen, with the rest mostly sulfurous gases. Similar types of values are found in ridge-axis black and white smokers, where hydrothermal accretions result in spectacular deposits of minerals falling out of solution from the emerging hot waters. Detailed studies show water trapped in the structures of altered minerals within the basaltic crustal rock of the oceanic plates, with 5 percent of the rocks, by weight, in the upper 1.24 to 1.86 mi (2 to 3 km) being water and hydroxide ions. Free water is known to be extremely buoyant, rising in the crust along shallow dipping faults. Bound water, subducted to great depths, would be expected to cook, moving upward as the rock density lessens, then acting as a further catalyst for melting the surrounding rocks.

In Earth's earliest stages, the primordial atmosphere probably escaped from Earth's gravitational pull because of overheating. In the second phase, gases released from molten rocks, with a surface temperature of 572 degrees Fahrenheit (300 degrees Celsius), provided 70 percent water and large quantities of carbon dioxide and nitrogen. In stage three, the atmosphere and oceans gradually changed, with gases and liquid water ejected from volcanoes resulting in more and more water deposited as liquid as the temperature fell. Oxygen was added to the atmosphere through thermal dissociation of water molecules, photochemical breakdown of high-altitude water, or photosynthetic alteration of carbon dioxide to oxygen in plants.

OCEAN SALT

The saltiness of the oceans can be accounted for by the inordinately large dielectric constant of water, a property that essentially ensures that it does not remain chemically pure. Geologic evidence shows the general composition to be similar over time, the stability of water content attributable to the continuous seawater-sediment interface. Geophysicist John Verhoogen determined that only 0.7 percent of the present ocean has been added since the Paleozoic era primarily from lava materials. The salty quality is a product of acidic gases from volcanoes (forming hydrochloric, sulfuric, and carbonic acids) that act to leach out the common silicate rocks. Paleontological evidence indicates that the change in ions must have been extremely slow, as demonstrated by the narrow tolerance of organisms then alive, such as corals, echinoderms brachiopods, and radiolarians. Ion concentrations in present-day river waters differ drastically from the ocean's values, however, indicating a different atmospheric environment in the past. Geochemists Robert M. Garrels and Fred T. Mackenzie have divided the oceans into three historical periods. In the earliest, water and volcanic acidic emissions actively attacked the crustal rock, leaching out ions and leaving residues of alumina and silicates. During the next period, from 3.5 to 1.5 billion years ago, slow continuous chemical action continued to attack sedimentary rocks, adding silica and ferrous ions. During the third period, from 1.5 billion years ago to the present, ions have accumulated to the point of modern concentrations, such that the composition is in apparent equilibrium with a mixture of calcite, potassium-feldspars illite-montmorillonite clays, and chlorite.

Fred T. Mackenzie (Wikimedia Commons)

Because it is known that, in an equilibrium state, the output of ions is equal to the input of ions, a new problem, that of geochemical "sinks," has been identified. Calcium carbonate (limestone) is removed from solution by living organisms to form skeletons, as is silica for opaline skeletons. Metals are dropped from seawater as newly formed mineral clays, oxides sulfides, and zeolites, and as alteration products at the hot-water basaltic ridges. Sulfur is removed as heavy-metal sulfides precipitating in anaerobic environments, while salts are moved in pore waters trapped in sediments. Residence times for many of the ions have been determined. For example, sodium cycles in 210 million years, magnesium in 22 million years, calcium in 1 million years, and silicon in 40,000 years. With such effective removal systems, it is truly a measure of the geochemical resistance of Earth's ocean waters to change that has allowed their composition to achieve stability over 4 billion years.

STUDY OF THE OCEAN

Numerous avenues of approach have been used to investigate the ocean and its ions, including geological, chemical, and physical means. Geology has supplied basic data on the types and makeup of rocks from the earliest solidified materials to present depositional formations. Using a petrographic microscope to view thin sections of rock under polarized light enables the identification of minerals and provides quantitative measurements of the water attached to the minerals themselves. Paleontological studies of fossil organisms and paleosols indicate the range of ions in the sea at diverse geologic periods, both by the ions themselves left in the deposited soils and rocks and through studies of the ion tolerance ranges for similar, twentieth-century organisms. Such studies, along with sedimentology investigations of the rates and types of river depositions, dissolved ion concentrations, and runoff rates for falling rain, provide determinants for comparing present-day ion concentrations with those of the past for continentally derived materials.

Chemical analysis reveals the various ions present in seawater and rocks via two principal methods. Use of the mass spectrometer identifies the types and quantities of ions present by use of a magnetic field to accelerate the charged ions along a curved path with a radius that is strictly determined by the weight and charge of the ions. Collection at the end of the path would provide a pure sample of the different ions present, although mass spectrometry is not used in this manner given the very small quantities of atoms in a typical analytical sample and the difficulty associated with collecting then under conditions of high vacuum. For solid samples, electron beam probe studies provide analysis from an area only one micron in diameter. The electrons, fired at the sample, cause characteristic X-rays to arise from the point. The energy of each type of X-ray is characteristic of a specific element or compound. By use of various optics to focus the X-rays, identification of even minute variations in concentration is possible.

Solubility studies provide residence times as a means of geochemical analysis for cyclical research. Similar laboratory projects, testing the ability of water to dissolve and hold ions in solution, argue for a primordial atmosphere that was essentially neutral, or mildly reducing in nature. Such reduction characteristics are based on the study of planetary composition for Earth and for other solar planets as supplemented by the various "lander missions" to Venus and Mars, and by modern astrophysical satellite observations. Chemical analysis from such missions in interplanetary space has also determined compositions for meteoritic gases, cometary tails and nuclei, and the mixing ratios for noble gases, all important for determining the origin of the solar system. The latter study, involving analysis of radioactive isotopes such as helium-3, an isotope of helium that is formed in the mantle, has prompted geophysicists to consider the mantle as a major source and sink for elements in various geochemical cycles.

Laboratory analysis reaches two other areas. Petrographic studies of returned lunar rocks suggest that the moon is essentially devoid of water, lacking even hydroxyl ions, at least on the surface. This discovery initially helps eliminate the solar wind and meteoritic impact as major factors in the initial formation of oceans. However, the recent identification of bound subsurface water on the moon is cause for reconsideration of that conclusion. Furthermore, high-temperature/high-pressure metallurgical and chemical studies indicate molten granite, at temperatures of 1,652 degrees Fahrenheit (900 degrees Celsius) and under 1,000 atmospheres of pressure, will hold 6 percent water by weight, while basalt holds 4 percent by weight. Based on geochemical calculations of the amounts of magma in the planet and lavas extruded over the first billion years, all the oceans' waters can be accounted for, particularly if parts of the fluid, as steam under pressure, are a result of oxidation of deep-seated hydrogen deposits trapped within or combined with mantle rocks. This supposition is thus considered an excellent likelihood from evidence gathered on radioactive decay in Earth's interior.

—Arthur L. Alt

Further Reading

Hamblin, Kenneth W., and Eric H. Christiansen. *Earth's Dynamic Systems*. 10th ed., Prentice-Hall, 2003.
Kandel, Robert. *Water from Heaven: The Story of Water from the Big Bang to the Rise of Civilization and Beyond*. Columbia UP, 2003.
Maxlow, James. *On the Origin of Continents and Oceans: A Paradigm Shift in Understanding*. Terrella Press, 2014.
Plummer, Charles C., Diane H. Carlson, and Lisa Hammersley. *Physical Geology*. 13th ed., McGraw-Hill, 2009.
Schwartz, M. *Encyclopedia of Coastal Science*. Springer, 2005.
Seibold, E., and W. Berger. *The Sea Floor*. 3rd ed., Springer-Verlag, 2010.
Shaw, George H. *Earth's Early Atmosphere and Oceans, and the Origin of Life*. Springer, 2016.
Zalasiewicz, Jan, and Mark Williams. *Ocean Worlds. The Story of Seas on Earth and Other Planets*. Oxford UP, 2014.

Offshore Wells

Fields of Study: geology; marine geology; marine ecology; oceanography

ABSTRACT

Offshore wells are drilled in favorable oil and gas areas that are inundated with oceanic (salt) or inland lake (fresh) waters. Most of the world's future hydrocarbon reserves may be discovered in offshore provinces. The ramifications for marine wildlife are manifold.

KEY CONCEPTS

directional drilling: a method of well-drilling that allows control of the direction in which the drill head will cut

gravimeter: a device that measures the apparent force of gravity in a small area, that force being dependent on the density of underlying rock structure in that area

magnetometer: a device that detects and measures the embedded magnetic "signature" embedded in underlying rock structures

seismic reflection: a system that uses sound impulses and their reflection from underlying rock structures to identify the stratigraphy of an area

stratigraphy: study of the rock types and ordering of different layers, or strata, of underlying rock structures

ESTABLISHING OFFSHORE DRILLING LOCATIONS

Offshore drilling is the extension of on-land oil and gas drilling techniques in waters that either cover or lie adjacent to Earth's landmasses, such as the Great Lakes of North America and the continental margins that surround each continent. The continental margins are the principal arena in which offshore drilling is conducted. They constitute approximately 21 percent of the surface area of the oceans and may contain a majority of the world's future reserves of oil and gas.

As of the last decade of the twentieth century, oil and gas production had been established off the coasts of more than forty countries in the continents of Africa, Australia, South America, North America, Asia, and Europe, and offshore drilling had been completed in the waters of more than half the nations on Earth. Drilling platforms capable of operating in water deeper than 5,905 feet (1,800 meters) probe the hydrocarbon potential of the outer continental margin. Even farther offshore, self-propelled drill ships capable of drilling more than 6,562 ft (2,000 m) into the seafloor within waters as deep as 19,685 ft (6,000 m) have been analyzing the deeper portions of the oceans since the early 1960s.

The specific offshore drilling location must be established at the outset. This is the responsibility of the geologist, who, by studying the various rock exposures and their structures, decides on the best surface site for the drilling equipment. Such exposures are submerged offshore, and geophysical methods of determining the geology of the seafloor are used.

Offshore geophysics is an indirect technique of studying the rocks composing the seafloor by measuring their physical properties. A magnetometer towed behind an aircraft flying low over the water can be used to measure the magnetic properties of underlying rocks. With a gravimeter mounted in a slow-moving boat or aircraft, the gravity field associated with the seafloor can be analyzed. A combination of these methods will help the geologist locate a site underlain by rocks that may contain hydrocarbons.

By far the most common offshore geophysical tool is the seismic reflection survey. Seismology depends on the d generation of an elastic sound wave and its transmission through the layered, sedimentary rocks underlying the seafloor within the continental margin zone. These waves reflect off sedimentary rocks and are transmitted back to the surface of the water, where a vessel will record the time difference between transmission and reflection. Millions of such combined reflection arrival times are interpreted as a cross-sectional view of the underlying seafloor. A series of intersecting seismic sectional views presents a simple three-dimensional (3D) portrayal of the best site for offshore drilling equipment.

DRILLING METHODOLOGY

The sole method used for drilling offshore wells is the rotary method. Rotary motion is supplied by diesel engines to a length of interconnected drill pipe (the drill string), to the bottom of which is attached the drill bit. Drilling bits come in a variety of styles designed to drill through differing types of sedimentary rock. Each type of drill bit contains an arrangement of high-strength alloy teeth that tear through the rock when rotated under pressure. As the hole deepens, new sections of drill pipe are added. Periodically, the inside of the borehole is lined with cemented casing, which prevents the hole from collapsing. During drilling, drilling fluid is circulated through the pipe and hole. This circulating fluid, a mixture of water, special clays, and other minerals and chemicals, is necessary to maintain a safe temperature and pressure in the borehole and to clean the hole of newly created rock chips. The drilling fluid can be formulated to stabilize chemically active rock layers. Since it is dense, the drilling fluid supports (partially floats) 10 to 20 percent of the weight of the drill pipe—a significant consideration when drilling a deep well. All drilling activities take place within the derrick, an open steel structure that often is 197 ft (60 m) tall. The derrick holds the draw works, whereby the drill string can be drawn out of the borehole and disassembled, one length at a time.

Should a drilling operation discover new reserves of oil or gas, production platforms must be constructed on-site after the movable drilling platform is deployed elsewhere. From these fixed production platforms, as many as fifty or more additional wells are drilled to determine the new hydrocarbon field's size and volume. Each of these boreholes is drilled from the same general location on the production platform; however, at a predetermined depth below the seafloor, individual boreholes veer away from one another, allowing different sectors of the field to be economically developed from one production platform. This process of deviation is termed "directional drilling"; it requires the assistance of a specialist, as the bottom positions of the boreholes must be very carefully controlled for effective hydrocarbon production. After the final directional hole has been drilled and the limits of the field fully defined, the production derrick is replaced with equipment used to gather the oil and gas from the many active wells flowing into the platform. An offshore pipeline is laid, connecting the producing platform to onshore pipeline and refining systems, and the new field begins production.

Offshore drilling rig (M L via Wikimedia Commons)

Drill Ship "Pacific Khamsin," owned by Pacific Drilling Services, Houston, Texas (Carlos Teixidor Cadenas via Wikimedia Commons)

DRILLING PLATFORMS

The type of platform that will be used to contain the drilling equipment is critical, considering the inhospitable weather periodically encountered offshore. There are four types of platforms: the submersible, semisubmersible, and jack-up platforms, and the drill ship. A submersible platform is stabilized by flooding the hollow legs and pontoons of the platform and establishing seabed moorings. Since these platforms are in contact with the seafloor, they cannot be used in excessively deep waters. In such deep waters, a semisubmersible design is employed; the hollow pontoons are only partially flooded, permitting the platform to settle below the surface of the water but not to rest on the bottom. Inherent in this design is buoyancy that, along with anchor moorings, is sufficient to maintain the platform safely over the drilling site. The ratio of semisubmersible to submersible units in use is approximately 4:1.

By far, the most popular offshore unit is the jack-up platform. These movable structures are towed to the drill location and stationed by lowering massive steel legs to the seabed. In essence, the platform is "jacked up" on the legs to a level sufficient to protect the unit from storm waves. The fourth design is the drill ship, a free-floating, usually self-propelled vessel that contains an open-water drilling unit in the ship's center. The drill ship is kept on location by computer-activated propellers that maintain the horizontal and vertical motion of the standing ship within safe limits.

Regardless of the basic design, all offshore platforms contain a drilling derrick, storage and machinery housing, living and eating quarters, recreation and basic health facilities, and a helicopter landing pad. The entire structure must be of a size sufficient to house and maintain a normal working crew of forty to sixty individuals.

MULTIPLE BOREHOLE PLATFORMS

Because of the expense associated with the siting of an offshore drilling platform, more than one borehole is commonly drilled from the same surface location. More than sixty-five boreholes have been drilled from a single platform. Most are drilled into the seabed at a predetermined angle, allowing many wells to "bottom" in an oil or gas reservoir rock over a distance as much as 1.86 miles (3 kilometers) laterally from the platform site. In this manner, the hydrocarbon reservoir can be exploited economically with a minimum risk to the aquatic environment.

Multiple borehole platforms are designed to extend a safe height above sea level. Platforms in the Gulf of Mexico, the North Sea, and the Santa Barbara Channel of California have been designed to endure the most severe storms likely to occur within a one-hundred-year cycle, including hurricane-force storms and earthquake tremors. Special submersible drilling platforms are designed for operation in polar waters, where ice-free waters exist only for a short time period during summer. These structures, which must be able to withstand strong currents and floating pack ice, are protected below the waterline by a steel or cement caisson. Often the drilling deck is circular, allowing easier passage of floating ice.

Floating platforms, semisubmersibles, and drill ships move with wind and water currents. Unless compensated for, these motions affect drilling efficiency. When boreholes are drilled in very deep waters, compliant platforms that yield to weather and water currents are employed.

AN INTERNATIONAL PRIORITY

Offshore drilling for oil and gas, even with the very high costs dictated by its architectural, meteorological, engineering, and safety requirements, is an international priority; it is driven by need, economic return, and national security. Except for difficult-to-explore and environmentally sensitive regions, such as the Arctic, the South American interior, and central Africa, many onshore regions of the world have entered a mature stage of hydrocarbon exploration: Most of the accessible large-volume oil and gas fields have already been discovered. The United States is an excellent example, for it imports almost 50 percent of its daily petroleum requirements, leaving it vulnerable to foreign military and political disturbances. Studies have consistently shown that the best opportunities for increasing the reserves of American oil and gas lie in the continued exploration of the continental margins. Although not all offshore areas contain oil and gas, analyses conducted by the US Geological Survey indicate that within the continental margin of the United States, as much as 40 billion barrels (a barrel equals 42 US gallons, or about 159 liters) of oil and more than 201 trillion cubic square feet (5.7 trillion cubic meters) of natural gas are yet to be discovered.

The degree and specific locations of offshore exploration and drilling activities depend on political, environmental, and economic factors, all of which affect petroleum products' availability and price. Certain offshore drilling regulations have already been established. In the United States, states' rights generally prevail to approximately 3 nautical miles (5,500 meters), offshore. From there out to designated international waters, federal government policies must be followed. The outer limits of American waters and those of other signatory nations are controlled by the 1982 Third United Nations Convention on the Law of the Sea, which allows nations to define Exclusive Economic Zones extending 370 km (200 nautical mi) offshore. Because of economic and engineering constraints, these outer limits have not yet been tested widely with actual drilling.

A common cause for the delay in exploration drilling is public concern for the marine environment. An example of a highly publicized and heavily studied American offshore oil spill is one that occurred in the Santa Barbara Channel, off California, in 1969; another is the 1989 spill from the tanker "*Exxon Valdez*" in Prince William Sound, in the Gulf of Alaska, the effects of which are still being discovered thirty years later. In 2010, the Deepwater Horizon drill rig, operated by BP, sank in the aftermath of an explosion. Attempts to activate shutoff devices on the seafloor failed, and it took approximately six months to fully seal the wells. About 4.9 million barrels (210,000,000 US gallons; 780,000 cubic meters) of oil were spilled—about 20 times the amount of the 1982 Alaska oil spill. The long-term effects of these spills are unknown. One study, however, has shown that between 1950 and 1980, the commercial fish-catch weight in the Gulf of Mexico increased fourfold, suggesting that offshore compatibility between the fishing and hydrocarbon industries is possible. Yet, as environmentally fragile continental margin areas, such as the shores of Alaska and Canada, continue to be evaluated, intensified concerns for the environment may delay exploration agendas. Finally, the costs of finding and producing oil and gas offshore will continue to climb as deeper waters are probed.

—*Albert B. Dickas*

Further Reading

Bar-Cohen, Yoseph, and Kris Zacny, editors. *Drilling in Extreme Environments: Penetration and Sampling on Earth and Other Planets*. Weinheim, Wiley-VCH. 2009.

Burgan, Michael. *Exxon Valdez. How a Massive Oil Spill Triggered an Environmental Catastrophe*. Compass Point, 2018.

Cavnar, Bob. *Disaster on the Horizon. High Stakes, High Risks, and the Story Behind the Deepwater Well Blowout*. Chelsea Green Publishing, 2010.

Graham, Bob, et al. *Deep Water: The Gulf Oil Disaster and the Future of Offshore Drilling*. National Commission on the BP Deepwater Horizon Oil Spill and Offshore Drilling, 2011.

Hyne, Norman J. *Nontechnical Guide to Petroleum Geology, Exploration, Drilling, and Production*. 2nd ed., PennWell Books, 2001.

Laik, Sukumar. *Offshore Petroleum Drilling and Production*. CRC Press, 2018.

Segar, Douglas. *An Introduction to Ocean Sciences*. 2nd ed., W. W. Norton & Company, 2007.

Speight, James G. *Handbook of Offshore Oil and Gas Operations*. Elsevier, 2015.

Sverdrup, Keith A., Alyn C. Duxbury, and Alison B. Duxbury. *An Introduction to the World's Oceans*. 8th ed. McGraw-Hill Science, 2004.

OXYGEN, HYDROGEN, AND CARBON RATIOS

Fields of Study: chemistry; geochemistry; geology; marine geology; climatology

ABSTRACT

Oxygen-18/oxygen-16, deuterium/hydrogen, and carbon-13/carbon-12 ratios in rocks are heavy- to light-isotope ratios of oxygen, hydrogen, and carbon, respectively. These ratios can give clues to the geologic conditions under which rocks were formed.

KEY CONCEPTS

covalent bond: a connection between two atoms in which each atom "shares" an electron from the other atom in order to form a more stable electron configuration around its nucleus

daughter isotope: in the process of radioactive decay, an unstable "parent" isotope releases part of its nucleus, producing a daughter isotope of a lower atomic number

diagenesis: the process whereby sediment particles are transformed into rock through pressure and the formation of opaline microcrystalline matter in the pores between the sediment particles

ion: an atom or molecule that contains either more or less electrons than it contains as a neutral atom or molecule and thus carries either a negative or a positive electric charge respectively

ionic bond: a connection between atomic or molecular ions of opposite charge

isotope: specific term for an atom based on the number of neutrons that it contains within its nucleus; almost all elements have atoms that contain different numbers of neutron but the same number of protons in their nucleus; for example, all carbon atoms have 6 protons, but ^{12}C has 6 neutrons, ^{13}C has 7 neutrons and ^{14}C has 8 neutrons

isotopic ratio: the relative amounts of different isotopes of an element in a sample; the ratio may be used to determine the age of the material in the sample according to the mathematics of radioactive decay processes

STABLE ISOTOPES

Rocks are divisible into three major types: igneous, sedimentary, and metamorphic. Igneous rocks are formed from hot, molten rock material called "magma." Sedimentary rocks are formed by diagenesis, the compaction and cementation of mineral grains and rock fragments, which collectively are called "sediments." Metamorphic rocks are formed by the alteration of rocks caused by increased heat and pressure and interaction with water in pore spaces and fractures. The temperature of formation and the types of water that interacted with the rocks are among the geologic conditions that scientists can determine from stable isotope ratios in rocks.

Rocks are aggregates of minerals. Minerals are natural compounds made of atoms that are arranged in an ordered fashion. The atoms in minerals exist as different isotopes. Some isotopes are unstable, or radioactive; others are stable. Radioactive isotopes, or parent isotopes, change into isotopes of other elements, or daughter isotopes. This change, which is called "radioactive decay," occurs at a specific rate and can be determined via experimental work. The amounts of parent and daughter isotopes in rocks and the decay constant of the isotope are used by scientists to determine rocks' ages. Unlike radioactive isotopes, stable isotopes do not decay into isotopes of other elements. Thus, stable isotopes are not used for age-dating purposes. Rather, the ratios of stable isotopes—particularly of low atomic number elements, such as hydrogen, carbon, and oxygen—are used to determine the geologic conditions under which rocks are formed.

IONIC AND COVALENT BONDS

The three elements oxygen, hydrogen, and carbon are key components of the minerals which make up rocks. Oxygen, the most common element in the crust, the outermost solid layer of Earth, is found in all kinds of rocks. Carbon is abundant in coal. Carbon and oxygen are important components of a fairly common group of sedimentary rocks called "limestones" and their metamorphosed product, marble. Hydrogen and oxygen are constituents of water and can be used to characterize the types of watery solution that interact with rocks.

Oxygen, hydrogen, and carbon form bonds that range from ionic to covalent. Ionic bonds are formed by the electrostatic attraction of adjacent atoms; covalent bonds are formed by the sharing of electrons and are stronger bonds. Scientists have found that bonding characteristics affect the behavior of isotopes in natural conditions; elements which form only one type of bond in all conditions are not useful for isotopic work.

ISOTOPIC SEPARATION

Oxygen, hydrogen, and carbon all have low atomic weights. The relative mass difference between the heavy and light isotopes is large for such elements, unlike the elements of high atomic weight. For example, deuterium, the heavy stable isotope of hydrogen, is heavier than the light isotope of hydrogen by about 100 percent. In contrast, the stable strontium isotopes differ from each other by only 1.2 percent. Some conditions and processes favor the incorporation of heavy isotopes into a material; others favor light isotopes. Thus, depending on the geologic conditions, there will be a difference between the heavy- to light-isotope ratio in one mineral and the ratio of the same isotopes in another mineral or in the source material. This syndrome is called "isotopic separation" or "fractionation." Isotopic separation is significant and detectable only for elements with low atomic numbers, such as oxygen, carbon, and hydrogen. The conditions and processes that cause separation include temperature, evaporation and diffusion, and oxidation reduction reactions.

A mineral is a compound formed by the chemical bonding of atoms. The internal energy of a mineral is controlled by factors including the vibration of atoms. The atoms of light isotopes vibrate with higher frequencies than do atoms of heavy isotopes. Since the bond strength between atoms depends on the vibration frequencies of atoms, light isotopes are bonded slightly less strongly than their heavier counterparts. Such bonds can be broken more readily upon dissolution or by bacterial action. Significantly, with increased temperature, the vibrational frequencies of all isotopes of the same element become nearly equal. Therefore, all other factors being equal, a mineral that is formed at low temperatures will contain a higher heavy- to low-isotope ratio than a similar mineral which is formed at high temperatures. A change in temperature during mineral formation can cause isotopic separation. Consequently, scientists can use isotopic separation to determine temperatures of mineral formation; in other words, isotope ratios can be used as geothermometers. Commonly, oxygen-isotope ratios in two different oxygen-bearing minerals which were formed at the same time and from the same source are used to determine the temperature at which a rock formed.

WATER-ROCK INTERACTIONS

Applications of oxygen isotope work have provided scientists with additional insights into the formation of rocks. It is now known that groundwater interacts with magmas. Part of this water is incorporated into the magma, and part of it circulates through rocks, changing their nature (metamorphosing them) in the process. Furthermore, it is now possible to determine whether a magma has been contaminated by the incorporation and subsequent melting of roof rocks.

Evaporation and diffusion also cause isotopic separation. Water is a compound of oxygen and hydrogen. During evaporation, the light isotopes of oxygen (^{16}O, or oxygen-16) and of hydrogen break through the water surface and escape to the atmosphere in the form of water vapor, while the heavy isotopes (^{18}O, or oxygen-18 and ^{2}H, or deuterium), concentrate in reservoirs. Similarly, molecules with lighter isotopes move across a boundary (diffuse) faster than the same molecules with heavier isotopes. Evaporation and diffusion lead to different waters' having different isotopic ratios. Scientists have determined $^{18}O/^{16}O$ and $^{2}H/^{1}H$ ratios in many kinds of water: lakes, rivers, rain, snow, oceans, and water that comes from molten rocks. When these waters enter into interconnected pore spaces and fractures of rocks and circulate through the rocks at depth, they become hot solutions. These solutions interact with the rocks, leading to the diffusion (movement) of atoms from rocks to the solutions and from the solutions to the rocks. Such water-rock interaction changes the nature of both the rocks and the solutions. From stable isotope ratios, scientists can determine the nature of the rock, the nature of solution, and the type of water-rock interaction.

OXIDATION AND REDUCTION

Oxidation reactions are another cause of isotopic separation. Atoms are said to be in an oxidized state if they have a lower number of electrons and in a reduced state if they have a higher number of electrons when compared with other atoms of the same element. For example, the element carbon can occur in the form of C^{4+}, C, and C^{4-}. C^{4+} is a highly oxidized state of carbon, formally a positively charged ion formed by the loss of four electrons. Such ions combine with negative ions of other elements, such as oxygen ions, to form compounds such as carbon dioxide. In a highly reduced state, carbon atoms gain four electrons, forming the negatively charged carbon ion, C^{4-}. These ions combine with positively charged ions to form compounds such as methane. (It should be noted that for elements such as carbon, such charge designations are a "bookkeeping" formality; the electrons are involved in covalent bonds rather than lost or gained as actual ions.)

Oxidized carbon is enriched in the heavy carbon isotope ^{13}C, or carbon-13, and reduced carbon is enriched in ^{12}C, or carbon-12. Thus, if the same source material were to permit the formation of two compounds, one with oxidized carbon and the other with reduced carbon, the one with the oxidized carbon will be enriched in ^{13}C relative to ^{12}C. The application of this simple principle is quite involved, however, because carbon can cycle through living organisms and other environments.

Green leaves of plants photosynthesize carbon dioxide, reducing the carbon and making it part of organic compounds. This reduction is done in enzyme-mediated stages, which are different in different plants. Thus, some plant types can be distinguished by their ^{13}C values. Generally, land plants have lower values of ^{13}C than marine plants; however, marine algae have values within the range of land plants. Evaporation leads to the enrichment of seawater in ^{13}C. Condensation in clouds leads to the enrichment of ^{13}C in raindrops as compared with water vapor. Consequently, repeated evaporation and rain cause seawater to be richer in ^{13}C; the atmosphere is lower. Carbon dioxide in soils and groundwater has an even lower ^{13}C value, because the carbon there has cycled through decaying plants.

DIAGENESIS

It appears that freshwater limestone should have lower ^{13}C values than marine limestone. Also, marine limestone should have variable ^{13}C values, depending on the amount of carbon inherited from algae (lower values indicating higher algal content). Most limestones, however, undergo a change called "diagenesis," which involves the reconstitution of carbonate minerals under conditions different from the original ones. That makes it difficult to determine precisely the original conditions under which the rock formed or the subsequent conditions that resulted in the diagenetic change. Commonly, low values of ^{13}C are obtained from the diagenesis of marine limestones.

ANALYTICAL TECHNIQUES

Scientists use many analytical techniques, including mass spectrometers, mass spectrographs, and ion microprobes, to determine isotopic ratios. In each of these techniques a sample containing the elements is bombarded by electrons or by (in the case of the microprobe) a beam of negatively charged oxygen atoms. All utilize the fact that different isotopes separate from each other and arrive at a detector at different times when they travel through a magnetic field from the respective ionization chamber.

The separated isotopes are detected electronically in mass spectrometers. In mass spectrographs, they are detected by nonelectronic methods, such as photographic devices. For these methods of analysis, the elements or molecules of interest are chemically separated and introduced into an ionization chamber in a gaseous form, or they are deposited as solids on filaments and are then vaporized in the ionization chamber. In the ion microprobe method, chemical separation of the sample is not necessary, and the original sample does not have to be destroyed. A small sample of a rock is polished and then coated with gold or carbon. A beam of negatively charged oxygen is focused on the sample, over an area of less than 0.01 millimeter. That causes ionization of the sample; the ions are accelerated through a magnetic field to a detector, and the isotopes are measured by a mass spectrometer or spectrograph.

MASS SPECTROMETERS AND SPECTROGRAPHS

There are many different spectrometers and spectrographs, but the principles involved can be understood by considering one type of spectrometer. In this device, an appropriate voltage applied across a filament, possibly of tungsten, produces a stream of electrons. These electrons bombard a sample and cause the removal of electrons from the atoms of the sample. The resulting positive ions of different isotopes are accelerated by a high-voltage field and are collimated into a beam. Since the kinetic energies of isotopes of the same element are identical, the lighter isotopes travel faster than the heavier isotopes.

The ion beam passes through a magnetic field, which is constructed in such a way that different isotopes are separated from each other as they exit the field and enter a collector cup. The accelerating voltage and the magnetic field can be adjusted so that an ion beam of one isotope can be focused through a collector slit to enter a detector cup. The focused ions are neutralized by electrons that flow through a resistor, and the voltage difference across the resistor can be measured with a voltmeter. The ensuing electrical signals can be digitized or, more commonly, displayed on a strip-chart record. A series of peaks and valleys—each peak representing an isotope, with the peak height being proportional to the abundance of the isotope—can be recorded by adjusting the accelerating voltage or the magnetic field, which would vary the ions being focused through the collector slits. In this way, the various isotopes and their relative abundances can be determined. In turn, the ratios of heavy to light isotopes can be calculated.

Modern commercial mass spectrometers, equipped with multiple collectors for the simultaneous detection of different isotopes and with digital computers, have improved both the speed of acquisition and the reliability of isotopic data.

APPLICATIONS OF STABLE ISOTOPIC GEOCHEMISTRY

Many kinds of chemical analysis can be used to determine the concentration of ions in groundwater. If different ions originate from different source areas, then these areas can be distinguished by such analysis. When the source areas produce the same ions, the isotopic ratios of the ions may be different because of different environmental conditions. In such instances, stable isotope geochemistry can be used to specify the source area, such as a groundwater-polluting factory. Stable isotope geochemistry, then, is useful in groundwater pollution studies. One of the uses of mass spectrometers is to identify elements by their isotopes, and by using a procedure called "isotopic dilution," scientists can identify elements whose amounts in a sample are extremely small. The isotope dilution method provides the added advantage of identifying the groundwater pollutants.

Scientists have used stable isotopic geochemistry to study the diet of prehistoric humans. The remains of wood in ancient campfires have been analyzed for their stable isotopic ratios. Such ratios can help the researcher to distinguish between land and marine plants and between groups of land plants. The ion microprobe, with its ability to analyze particles smaller than 0.01 millimeter, has a potential application in forensic science, where trace amounts of hair and clothes are used to identify criminals. The isotopic dilution method is another procedure that can be applied to such an effort.

—Habte Giorgis Churnet

Further Reading

Albarede, Francis. *Geochemistry: An Introduction*. 2nd ed., Cambridge UP, 2009.

Cazes, Jack. *Ewing's Analytical Instrumentation Handbook*. 3rd ed., Marcel Dekker, 2005.

Faure, Gunter. *Isotopes: Principles and Applications*. 3rd ed., John Wiley & Sons, 2004.

Hoefs, J. *Stable Isotope Geochemistry*. 6th ed., Springer-Verlag, 2009.

Krauskopf, Konrad B. *Introduction to Geochemistry*. 3rd ed., McGraw-Hill, 2003.

Krebs, Robert E. *The History and Use of Our Earth's Chemical Elements: A Reference Guide*. 2nd ed., Greenwood Press, 2006.

Richardson, S. M., H. Y. McSween, Jr., and Maria Uhle. *Geochemistry Pathways and Processes*. 2nd ed., Prentice-Hall, 2003.

Santamaria-Fernandez, Rebeca. "Precise and Traceable Carbon Isotope Ratio Measurements by Multicollector ICP-Ms: What Next?" *Analytical & Bioanalytical Chemistry*, vol. 397, 2010, pp. 973–78.

Szczepanek, MalGorzata, et al. "Hydrogen, Carbon, and Oxygen Isotopes in Pine and Oak Tree Rings from Southern Poland as Climatic Indicators in Years 1900–2003." *Geochronometria: Journal on Methods & Applications of Absolute Chronology*, vol. 25, 2006, pp. 67–76.

P

Pacific Ocean

Fields of Study: oceanography; marine biology; marine geology; volcanology

ABSTRACT

The Pacific Ocean is the world's largest ocean and has an overall area of 70 million square feet (182 million square kilometers). It is twice the size of the next largest ocean, the Atlantic. The Pacific Ocean covers approximately one-third of Earth's surface and is larger than Earth's entire landmass area.

KEY CONCEPTS

equatorial currents: currents flowing from east to west across oceans on either side of the equator, as part of their respective gyres

guyot: an underwater volcanic mountain whose top has been sheared off by the action of waves, leaving a submerged mesa

gyres: slow, circulations of the ocean water on either side of the equator

Ring of Fire: the name given to the coastal regime of the Pacific Ocean due to its volcanic character and many active volcanoes

subduction zone: the area at which a segment of the dense basaltic ocean floor is driven against a less dense continental mass by the action of seafloor spreading and is forced downward into the mantle; characterized by seafloor trenches, and capable of producing massive earthquakes and volcanic activity

LOCATION OF THE PACIFIC OCEAN

The Pacific Ocean, with a mean depth of 13,960 feet (4,255 meters) and a maximum depth of 35,990 ft (10,970 m), extends from its southern extreme at Antarctica some 9,662 mi (15,550 kilometers) north to the Bering Strait which separates North America from Asia. Its east-west area extends almost 12,079 mi (19,440 km) from the western coast of Colombia in South America to Asia's Malay Peninsula.

The Pacific and Arctic Oceans meet at the Bering Strait in the north. In the far south at Drake Passage, south of Cape Horn, the Atlantic and Pacific Oceans come together. Exactly where the Pacific meets the Indian Ocean is more difficult to define because the two bodies of water intermingle along a string of islands that extends to the east from Sumatra through Java to Timor and across the Timor Sea to Australia's Cape Londonderry. South of Australia, the Pacific runs across the Bass Strait to the Indian Ocean and continues from Tasmania to Antarctica.

The eastern Pacific generally follows the Cordilleran mountain system, which runs the entire length of North and South America from the Bering Strait to Drake Passage and includes both the Rocky Mountains and the Andes. The east coast of the Pacific is, comparatively, quite regular except for the Gulf of California and the fjord regions in its northern and southern extremes. The continental shelf of the eastern Pacific is narrow, and the continental slope at times quite steep, as is characteristic of subduction zones.

The Pacific's western extreme in Asia is, by contrast, quite irregular. As on the eastern coast of the Pacific, the western coast is bordered by mountain systems that run roughly parallel to the coast. The western Pacific has many dependent seas, notable among them the Bering Sea, the East China Sea, the Sea of Japan, the Sea of Okhotsk, the South China Sea and the Yellow Sea. The eastern extremes of these seas are characterized by peninsulas jutting out toward the south, by island arcs or by both.

Unlike the Atlantic, much of whose water flows into it from the rivers that feed it, only about one-seventh of the Pacific's water comes from direct river flow. Most of its water comes from the dependent seas that are nourished by such East Asian rivers as the Amur, the Xi, the Mekong, the Yangtze, and the Yellow Rivers.

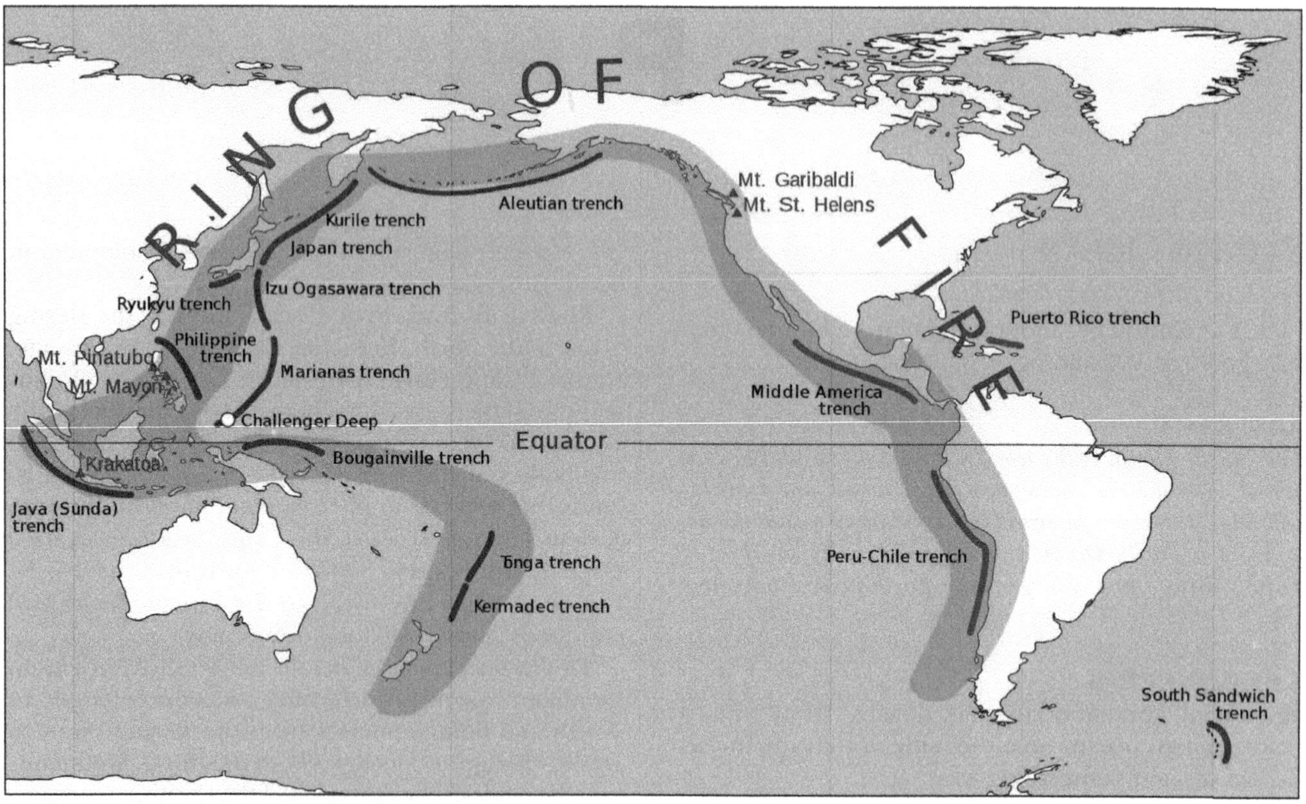

Pacific Ring of Fire (Gringer via Wikimedia Commons)

ORIGINS AND DIVISIONS

The Pacific Ocean is the present-day remnant of the world ocean called "Panthalassa" and has been evolving to its present state for some 200 million years, since the single landmass called "Pangaea" began to split apart through tectonic action. Through continental drift that has occurred over the intervening millions of years, the landmass divided into the several smaller fragments that are the present-day continents. As the spaces between the continental segments became larger, they filled with the water that surrounded them and, through eons, formed the various oceans and seas that encompass the surface of the planet.

The Pacific Ocean is so large that it is difficult to discuss it as a single ocean. Various parts of it are substantially different from other parts. For purposes of discussion, the Pacific is usually thought of as comprising three distinct regions.

The western region begins in Alaska's western Aleutian Trench, extends through the Kuril and Japanese Trenches, runs south to the Tonga and Kermadec Trenches, then continues on to an area northeast of New Zealand's North Island. This region is characterized by large strings of islands, the largest constituting Japan, New Guinea, New Zealand and the Philippines. Some of these strings, notably Japan, New Guinea and New Zealand, were, through time, sheered off from the continent by tectonic movements. The entire western Pacific is also dotted with volcanic islands, which are the peaks of high underwater mountains. These were formed as molten rock from successive volcanic eruptions cooled and solidified.

The ocean floor of the central area of the Pacific is the largest underwater expanse on Earth. It is the most geologically stable area on the Pacific floor, called the "Abyssal Plain." It contains sprawling underwater plains, most at a depth of about 14,764 ft (4,500 m), that are essentially flat, although they contain some irregularities and some geological formations, called "guyots" that resemble the mesas found in above-water plains.

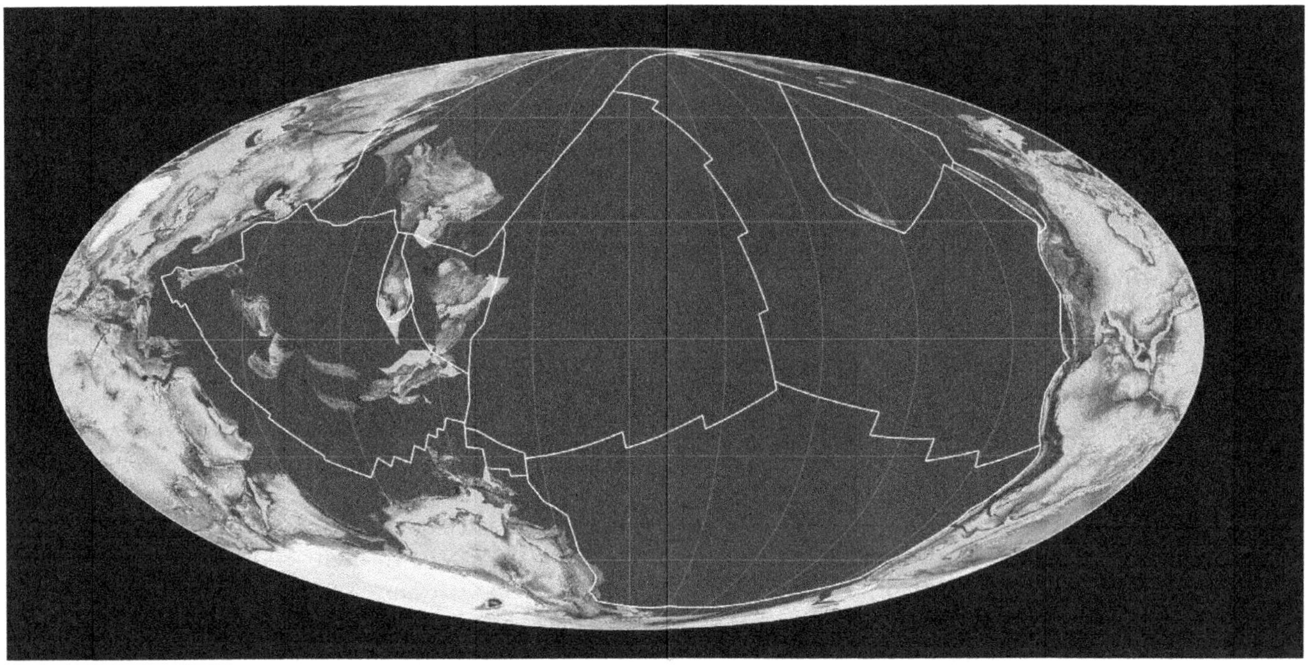

The Panthalassa superocean 250 million years ago (Fama Clamosa via Wikimedia Commons)

The eastern part of the Pacific, which abuts the United States' west coast, has a narrow continental shelf and a steep continental slope so that close to shore there are deep areas, such as the Monterey Canyon, which plunges more than 1.86 mi (3 km) below the water's surface. The two most significant trenches in this part of ocean are the Middle America Trench in the North Pacific and the Peru-Chile Trench in the South Pacific.

Volcanic activity has historically been more pronounced along the western edge of the Pacific than along its eastern edge, although the East Pacific Rise a chain of underwater mountains that runs from Southern California almost to the tip of South America, is also quite volcanically active. In the central and western Pacific, however, there is a profusion of islands that are the tops of underwater volcanic mountains.

PACIFIC CURRENTS

The Pacific, like the Atlantic, has two large systems of currents called "gyres". The Northern Gyre moves in a westerly direction from near the equator and is a survice ocean current that carries warm surface water into the Kuroshio Current and gives Japan moderately warm temperatures. The Kuroshio Current then moves northeast away from the Kamchatka Current, which brings cold water from the area around the Bering Strait. Part of it divides, moving north into the North Pacific Drift. The other part heads southeast.

Off the west coast of North America, part of the North American Drift moves north and becomes the Alaska Current. It brings to the Alaskan coastline waters warm enough to keep its shoreline and harbors from freezing in winter. Another part of the North American Drift veers south and forms the California Current, whose waters are cooled as they flow south to join the North Equatorial Current which again flows west, completing its clockwise motion.

The South Equatorial Current also moves west toward Australia and New Guinea, but it then veers south to begin its counterclockwise course. The East Australian Current skirts Australia's east coast and passes between Australia and New Zealand, south of which it feeds into the West Wind Drift, which moves east toward the coast of South America. Just north of South America's tip, it feeds into the Peru or Humboldt Current, which runs north along South America until it connects with the South Equatorial Current, which moves west to complete the South Pacific Gyre.

In both the northern and southern areas of the Pacific, winds from the west move the water in an easterly direction. Closer to the equator, however, trade winds move the water toward the west. This balance permitted early sailors to head east by riding the westerly winds and to return to the east by riding the trade winds. There are virtually no winds on the equator itself. Sailors can founder there, in the so-called doldrums, for days with none of the propulsive forces that nature provides both north and south of central Earth's dividing line.

The currents of the Pacific are important both for transportation and for bringing warm and cold waters to parts of the ocean, thereby giving them and the land adjoining them more moderate temperatures than might be expected at their latitudes. Occasionally, nature runs afoul of this balance as the southern trade winds diminish and fail to push the cool Peru Current north. Instead, the Peru Current is replaced by the Pacific Equatorial Countercurrent. The Peruvians first dubbed this phenomenon El Niño which is taken to mean "the Christ child," because it usually comes at the Christmas season. In some years, El Niño is followed by its opposite sibling La Niña, resulting in two successive years of hardship. As waters reach unaccustomed temperatures, much sea life fails to appear where it would otherwise be, bringing catastrophe to people dependent upon fish and seaweed for their diets or for income.

FORMATION OF PACIFIC ISLANDS

The Pacific Ocean has more islands than any other ocean. In the west, most of these islands are volcanic, although some, such as Japan, New Guinea and New Zealand, were formed when the corresponding chunks of land became separated from the continental landmass in prehistoric times through tectonic action. The volcanic islands in the western Pacific often rise just barely above sea level. Any change in the ocean's level or tidal range can inundate them. Global warming threatens to melt ice in the polar regions, which will in turn cause sea levels to rise and, in some cases, cause islands to disappear. Indeed, some small, politically independent island nations face total inundation. It is estimated that by the year 2050, the oceans of the world might rise as much as 20 inches

Coral in the Great Barrier Reef (Shwollo via Wikimedia Commons)

Kwajalein Atoll (Matt Kieffer via Wikimedia Commons)

(50 centimeters), which could wipe out currently habitable regions occupied by millions of people in the South Pacific.

Another major structure in the South Pacific is the atoll, a tropical island on which a massive coral reef, often ringlike, generally rests on a volcanic base. Kwajalein is the largest atoll in the world, with a circumference of nearly 201 mi (324 km).

Other coral formations, such as ridges or reefs composed of coral, can grow to enormous lengths, such as the Great Barrier Reef east of Australia. Coral formations are composed of living organisms that require light and a stable environment in order to survive. They are, therefore, always close to the surface of the ocean and often protrude above the surface as tidal sea levels change. Most of them teem with aquatic life. In some cases, when they form around low-lying islands, the island disappears as the ocean surrounding it rises. The coral formation, however, often remains, filled with entrapped sand and rising above the water's surface as an atoll.

Such island chains as the Hawaiian, Pitcairn-Tuamotu and Tubai Islands are the result of great plumes of extraordinarily hot magma that have risen through Earth's mantle over millions of years and erupted through fissures in the oceanic plate above. The resulting formation that occurs when the molten rock cools is an island composed largely of basaltic rock. Over great time spans, the island is slowly carried away to the northwest as the Pacific Ocean tectonic plate continues its normal motion, clearing the way for another such island to be formed above the plume at the same relative location in the oceanic plate. Hundreds of islands, lined up almost like pearls in a necklace, have resulted from this ongoing volcanic activity, though many have not survived to breach the ocean surface due to erosion by waves and currents in the ocean water.

PACIFIC RIM

The areas around the Pacific have many volcanic mountains, large ranges of which line the coasts and still larger ranges of which are submerged. The highest mountains on Earth are the Himalayas, with Mount Everest towering to 28,855 ft (8,795 m); many of the mountains submerged in the Pacific, however, are higher. From the ocean's floor at the Mariana Trench, undersea mountains, or seamounts rise more than 34,777 ft (10,600 m) as they approach the ocean's surface.

Many of the mountains on the Pacific Rim arose as, over time, the oceanic crust is subducted beneath the continental crust. Intense heat is generated by friction in the subduction zone beneath a continental mass, while the pressure exerted by the inexorable motion of the crustal plates causes fractures to form in the overlying rock, permitting magma to erupt through the continental crust as volcanoes. This process occurs all around the so-called Ring of Fire from the North Island of New Zealand, through the Philippines, Malaysia Japan and Siberia, to the Aleutian Islands and on down the west coast of North and South America to Peru and Chile.

The coast along the Pacific Rim is narrow, giving way to the mountain ranges that run close to the shoreline. An exception to this is in eastern China where the Yellow and the Yangtze Rivers have, for many centuries, carried silt toward the ocean, creating fertile coastal plains.

Geological activity beneath the Pacific results in thousands of small earthquakes every year along the Pacific Rim as oceanic plates move relative to each other. These earthquakes, most of which are too insignificant to be felt by humans, result in little or no damage. They relieve pressure that builds up between tectonic plates as they collide. When such pressure builds up without relief over a period of many years, however, a major earthquake, usually causing considerable property damage and loss of life, may occur. Severe earthquakes in the seafloor may generate extraordinarily large waves called "tsunamis." Tsunamis, known to have attained heights of more than 98 ft (30 m), can engulf large stretches of shoreline, as happened in Alaska in 1964. The worst tsunamis are usually caused by sudden displacements of crustal material in the subduction process, in turn causing a massive displacement of seawater that translates to the surface as an energy wave. Such waves travel through the ocean at speeds of up to 497 mi (800 km) per hour, and can cross the entire Pacific Ocean in a matter of hours. The tsunami that devastated the coast of Japan in 2011, producing worldwide economic consequences, was produced by a subduction zone earthquake.

FOOD RESOURCES

The feeding hierarchy that exists among the creatures of the sea is usually referred to as a food chain. One can envision this hierarchy as a triangle, at whose broad base are phytoplankton, microscopic plant organisms that require sunlight for their survival and are plentiful near the surfaces of oceans.

Zooplankton, microscopic animal organisms, feed on the phytoplankton. They, in turn, constitute the diet of numerous species of small fish called "anchoveta." These fish in turn are food for such larger fish as tuna and dolphin, as well as such birds as cormorants and pelicans, which dive into the water to catch fish. Animal waste, decaying plants, and dead fish and birds sink to the ocean floor, where scavengers consume them. Bacteria at those levels cause such droppings to decay, in the process releasing nutrients into the seawater that provide the phytoplankton with nourishment, thereby completing the food chain.

The food chain triangle becomes smaller as it approaches the top. Small fish survive by eating plankton. Larger fish eat the smaller fish, then sharks and whales eat the larger fish. A food chain that begins with millions of plankton may end up producing hundreds of thousands of anchoveta; thousands of cod, hake, and mackerel; and one-half dozen sharks for every whale that is part of this intricate hierarchy.

Although fish are the major food source harvested from the world's oceans, the Pacific also yields a great deal of kelp and seaweed, which, particularly in Asian countries, constitute a significant part of human diets.

Seaweed is often harvested and laid in strips beside the ocean, where the sun dries it, thereby preserving it and making it easy to transport and to store. Even though they are not consumed as readily in the eastern parts of the Pacific Rim as in Asia, kelp and other sea plants are used in many food and pharmaceutical preparations. They have valuable pharmacological properties that are in great demand by drug manufacturers throughout the world.

MINERAL RESOURCES

The mineral wealth beneath the Pacific Ocean has barely been tapped. Among the minerals available near the shore are chromite, gold, iron, monazite, phosphorus, tin, titanium, and zircon. The greatest exploitation of the ocean's treasure trove has been by the petroleum industry. Offshore drilling takes place worldwide for the recovery of crude oil and natural gas. In addition, large quantities of sand and gravel are harvested every year for use in construction and manufacturing.

The deep sea has remained a mysterious place. The historic descent in 1960 of the bathyscaphe *Trieste* into the Mariana Trench, the deepest area of ocean anywhere in the world, unlocked many mysteries. Subsequent exploration of the deepest areas of the ocean has challenged many long-held beliefs and revolutionized deep-sea research. It was long thought impossible for life to exist under the enormous pressure of the ocean below 2,493 ft (760 m), but the *Trieste* discovered life at the very bottom of the Mariana Trench much to the surprise of oceanographers. In early 2012, James Cameron undertook a solo dive mission to the floor of the Mariana Trench in a specially-built one-man submarine, returning with a priceless video record of that environment.

The bathyscaphe Trieste *(US Navy via Wikimedia Commons)*

The deep ocean is richest in its deposits of cobalt, copper, manganese, and nickel. These deposits remain largely unharvested because of the difficulty of getting to them, but such problems will undoubtedly be overcome as deep-sea mining technology is developed.

—*R. Baird Shuman*

Further Reading

American Museum of Natural History. *Ocean*. New Dorling Kindersley Limited, 2006.

Bamforth, Tom. *The Rising Tide: Among the Islands and Atolls of the Pacific Ocean.* Hardie Grant Publishing, 2019.

Benson, Keith Rodney, and Philip F. Rhebock. *Oceanographic History: The Pacific and Beyond*. U of Washington P, 2002.

Clarke, Allan J. *An Introduction to the Dynamics of El Niño and the Southern Oscillation*. Academic Press/Elsevier, 2008.

Duarte, Jono, and Wouter P. Schellart, editors. *Plate Boundaries and Natural Hazards*. Wiley, 2016.

Fache, Elodie, and Simonne Pauwels. *Fisheries in the Pacific: The Challenges of Governance and Stability*. pacific-credo Publications, 2018.

Henderson, Bonnie. *Strand: An Odyssey of Pacific Ocean Debris*. Oregon State UP, 2008.

Kowahata, Hodaka, and Yoshio Awaya. *Global Climate Change and Response of the Carbon Cycle in the Equatorial Pacific and Indian Oceans and Adjacent Landmasses*. Elsevier, 2006.

Nunn, Patrick D. *Climate, Environment and Society in the Pacific during the Last Millennium*. Elsevier, 2007.

Ogawa, Yujiro, Ryo Anma, and Yildirim Dilek. *Accretionary Prisms and Convergent Margin Tectonics in the Northwest Pacific Basin*. Springer Science+Business Media, 2011.

PATTERN RECOGNITION

Fields of Study: computer science; oceanography; resource management; marine biology

ABSTRACT

Pattern recognition is a branch of science concerned with identifying patterns within any type of data, from mathematical models to visual and auditory information. Applied pattern recognition aims to create machines capable of independently identifying patterns and using patterns to perform tasks or make decisions. The field involves cooperation between statistical analysis, mechanical and electrical engineering, and applied mathematics. Pattern recognition technology emerged in the 1970s from work in advanced theoretical mathematics. The development of the first pattern recognition systems for computers occurred in the 1980s and early 1990s. Pattern recognition technology is found both in complex analyses of systems such as ocean currents, aquatic migration, economic markets and physiology and in everyday electronics such as personal computers.

KEY CONCEPTS

datum (pl: data): a datum is a fact or a true condition with respect to a particular subject; the term usually refers to something that is measurable

facial recognition: a computer function in which optical analysis and data manipulation algorithms enable a device to recognize and identify human faces

neural network: an artificial construct whose component pieces function in imitation of the interconnected neurons of the human brain

pattern: a collection of individual items that are related by a specific function or property

speech recognition: a computer function in which sound analysis and data manipulation algorithms enable a device to be activated and controlled by the sound of one's voice

DEFINITION AND BASIC PRINCIPLES

Pattern recognition is a field of science concerned with creating machines and programs used to categorize objects or bits of data into various classes. A pattern is broadly defined as a set of objects or parts that have a relationship. Recognizing patterns therefore requires the ability to distinguish individual parts, identify the relationship between the parts, and remember the pattern for future applications. Pattern recognition is closely linked to machine learning and artificial intelligence (AI), which are fields concerned with creating machines capable of learning new information and using it to make decisions. Pattern recognition is one of the tools used by learning machines to solve problems.

One example of a pattern recognition machine is a computer or smartphone capable of facial recognition. The computer first evaluates the face using a visual sensor and then divides the face into parts, including the eyes, lips, and nose. Next, the system assesses the relationship between individual parts, such as the distance between the eyes, and notes features such as the length of the lips. Once the machine has evaluated an individual's face, it can store the data in memory and later compare the information against other facial scans. Facial recognition is often used to confirm identity in security applications or as a password to log into computers and cellphones.

BACKGROUND AND HISTORY

The earliest work on pattern recognition came from theoretical statistics. Early research concentrated on creating algorithms that would later be used to control pattern recognition machines. Pattern recognition became a distinct branch of mathematics and statistics in the early 1970s. Engineer Jean-Claude Simon, one of the pioneers of the field, began publishing papers on optical pattern recognition in the early 1970s. His work was followed by a number of researchers, both in the United States and abroad. The first international conference on pattern recognition was held in 1974, followed by the creation of the International Association for Pattern Recognition in 1978.

In 1985, the United States spent $80 million on the development of visual and speech recognition systems. During the 1990s, pattern recognition systems became common in household electronics and were also used for industrial, military, and economic applications. By the twenty-first century, pattern recognition had become a robust field with applications ranging from consumer electronics to neuroscience and marine biology. The field continues to evolve along with developments in AI and computer engineering.

HOW IT WORKS

Pattern recognition systems are based on algorithms, or sets of equations that govern the way a machine performs certain tasks. There are two branches of pattern recognition research: developing algorithms for pattern recognition programs and designing machines that use pattern recognition to perform a function.

Obtaining Data. The first step in pattern recognition is to obtain data from the environment. Data can be provided by an operator or obtained through a variety of sensory systems. Some machines use optical sensors to evaluate visual data, and others use chemical receptors to detect molecules or auditory sensors to evaluate sound waves.

Most pattern recognition computers focus on evaluating data according to a few simple rules. For example, a visual computer may be programmed to recognize only red objects and to ignore objects of any other color, or it may be programmed to look only at objects larger than a target length.

Translating Data. Once a machine has intercepted data, the information must be translated into a digital format so that it can be manipulated by the computer's processor. Any type of data can be encoded as digital information, from spatial relationships and geometric patterns to musical notes. A character recognition program is programmed to recognize symbols according to their spatial geometry. In other words, such a program can distinguish the letter *A* from the letter *F* based on the unique organization of lines and spaces. As the machine identifies characters, these characters are encoded as digital signals. A user can then manipulate the digital signals to create new patterns, using a computer interface such as a keyboard.

A character recognition system that is familiar to many computer users is the spelling assistant found on most word-processing programs. The spelling assistant recognizes patterns of letters as words and can therefore compare each word against a preprogrammed list of words. If a word is not recognized, the program looks for a word that is similar to the one typed by the user and suggests a replacement.

Memory and Repeated Patterns. In addition to recognizing patterns, machines must also be able to use patterns in problem solving. For example, the spelling assistant program allows the computer to compare patterns programmed into its memory against input given by a user. Word processors can also learn to identify new words, which become part of the machine's permanent memory.

Advanced pattern recognition machines must be able to learn without direct input from a user or engineer. Certain learning robots, for instance, are programmed with the capability to change their own programming according to experience. With repeated exposure to similar patterns, the machines can become faster and more accurate.

APPLICATIONS AND PRODUCTS

Marine Research. Although primarily developed for terrestrial uses, pattern recognition technology also has many real and potential applications in marine science and marine research. Optical recognition systems are capable of being adapted to automatic identification of fish and other marine species, while pattern recognition software can parse observational data to identify patterns of behavior. These functionalities are useful in fisheries management and marine ecological studies. Similarly, the technology can be applied to monitoring of shoreline erosion, water movement and other features that are dynamic in nature over time. Identification of such features enables more accurate prediction of future trends and the ability to make sound policy decisions with regard to marine resources.

Computer-Aided Diagnosis. Pattern recognition technology is used by hospitals around the world in the development of computer-aided diagnosis (CAD) systems. CAD is a field of research concerned with using computer analysis to more accurately diagnose disease. CAD research is usually conducted by radiologists and also involves participation from computer and electrical engineers. CAD systems can be used to evaluate the results taken from a variety of imaging techniques, including radiography, computed tomography (CT), magnetic resonance imaging (MRI), and ultrasound. Radiologists can use CAD systems to evaluate disorders affecting any body system, including the pulmonary, cardiac, neurologic, and gastrointestinal systems.

At the University of Chicago Medical Center, the radiology department has obtained more than seventy patents for CAD systems and related technology. Among other projects, specialists use CAD systems to evaluate potential tumors and to detect breast cancer.

Speech Recognition Technology. Some pattern recognition systems allow machines to recognize and respond to speech patterns. Speech recognition computers function by recording and analyzing sound waves; individual speech patterns are unique, and the computer can use a recorded speech pattern for comparison.

Speech recognition technology can be used to create security systems in which an individual's speech pattern is used as a passkey to gain access to private information. Speech recognition programs are also used to create dictation machines that translate speech into written documents. This makes programs such as "Dragon" especially useful as assistive devices for users who have difficulty manipulating a standard keyboard and require accommodations. In the 2010s, speech recognition programs were integrated into virtual assistants like Apple's Siri and Amazon's Echo, which play music, search the internet, a can control various home electronics through voice command.

Neural Networks. Many animals, including humans, use pattern recognition to navigate their environments. By examining the way that the brain behaves when confronted with pattern recognition problems, engineers are attempting to design artificial neural networks, which are machines that emulate the behavior of the brain.

Biological neural networks have a complex, nonlinear structure, which makes them unpredictable and adaptable to new problems. Artificial neural networks are designed to mimic this nonlinear function by using sets of algorithms organized into artificial neurons that imitate biological neurons. The artificial network is designed to be adaptive, so that repeated exposure to similar problems creates strong connections among the artificial neurons—similar to memory in the human brain.

Artificial neural networks can be used for any pattern recognition application from economic analysis to fingerprint identification. They are also commonly used in machine learning. Many engineers have come to believe that artificial neural networks are the future of pattern recognition technology and will eventually replace the linear algorithms that have been used most frequently.

Military Applications. Pattern recognition is one of the most powerful tools in the development of military technology, including both surveillance and offensive equipment. For example, the Tomahawk cruise missile, sometimes called a smart bomb, is an application of pattern recognition used for offensive military applications. The missile uses a digital scene area matching correlation (DSAMC) system to guide the missile toward a specific target identified by the pilot. The missile is equipped with sensors, an onboard computer, and flight fins that can be used to adjust its trajectory. After the missile is fired, the DSAMC adjusts the missile's flight pattern by matching images from its visual sensors with the target image.

CAREERS AND COURSEWORK

The most direct route to achieve a career in pattern recognition would be to receive advanced training in electrical engineering. Professional statisticians, mathematicians, neurobiologists, and physicists also participate in pattern research. Alternatively, some medical professionals work with pattern recognition, most notably radiologists who participate in CAD research and development.

Texts and training materials in pattern recognition generally require a strong background in mathematics and statistics. Basic knowledge of statistical analysis is a prerequisite for the most basic college courses in statistical engineering. Those hoping to work at the forefront of pattern recognition research will also need experience with machine learning, AI, artificial neural network design, and related areas.

SOCIAL CONTEXT AND FUTURE PROSPECTS

Pattern recognition technology has become familiar to many consumers. Voice-activated telephones, fingerprint security and facial recognition for personal computers and cellphones, and character analysis in word-processing software are just a few of the many applications that affect daily life. The use of pattern recognition technologies in social media, email sorting, and online banking grew exponentially in the 2010s. Examples range from the social media app Facebook using facial recognition to tag users in pictures to smartphone banking applications using fingerprint security to access finances. Advances in medicine, military technology, economic analysis, criminal justice, and advertising are further examples of how pattern recognition has come to shape the development of society.

Projects that may represent the future of pattern recognition technology include the development of autonomous robots, space exploration, and evaluating complex dynamics. In the field of robotics, pattern recognition is being used to create robots with the ability to locate objects in their environment. It has also begun to be integrated into robots that assist people with disabilities during rehabilitation and into some prosthetic devices. The National Aeronautics and Space Administration (NASA) has begun research to create probes capable of using pattern recognition to find objects or sites of interest on alien landscapes. Combined with research on artificial neural networks, automated systems have begun to make complex decisions based on the recognition of patterns.

—*Micah L. Issitt*

Further Reading

Brighton, Henry, and Howard Selina. *Introducing Artificial Intelligence*. Edited by Richard Appignanesi. 2003. Reprint. Totem Books, 2007.

Chen, C. H., editor. *Emerging Topics in Computer Vision and Its Applications*. World Scientific, 2012.

Cracknell, Arthur P., editor. *Remote Sensing Applications in Marine Science and Technology*. Springer Science + Business Media, 2012.

Frenay, Robert. *Pulse: The Coming Age of Systems and Machines Inspired by Living Things*. 2006. Reprint. U of Nebraska P, 2008.

Jaso, Danel. *Pattern Recognition Techniques, Technology and Applications*. Scitus Academics LLC, 2016.

Marques de Sá, J. P. *Pattern Recognition: Concepts, Methods and Applications*. Springer Books, 2001.

McCorduck, Pamela. *Machines Who Think: A Personal Inquiry into the History and Prospects of Artificial Intelligence*. 2nd ed.,a A. K. Peters, 2004.

Singer, Peter Warren. *Wired for War: The Robotics Revolution and Conflict in the Twenty-first Century*. Penguin Press, 2009.

Theodoridis, Sergio, and Konstantinos Koutroumbas. *Pattern Recognition*. 4th ed., Academic Press, 2009.

PERSIAN GULF ECOSYSTEM

Fields of Study: oceanography; marine biology; marine geology; marine biogeochemistry

ABSTRACT

The Persian Gulf is a shallow marginal sea located in the northwestern Indian Ocean between southwestern Iran to the east; the Arabian Peninsula to the west and south; and Iran, Iraq, and Kuwait to the north. It is sometimes also referred to as the Arabian Gulf.

KEY CONCEPTS

cyclonic: atmospheric and oceanic circulation that is naturally counterclockwise in direction

megafauna: various species of large animals

pneumophores: plant organs that do not enter the ground, instead drawing in water and nutrients from the air

practical salinity: salinity determined by electrical conductivity related to the total concentration of dissolved salts

THE SEA OF THE ARABIAN NIGHTS OF SCHEHEREZADE

The Persian Gulf is a shallow marginal sea located in the northwestern Indian Ocean between southwestern Iran to the east; the Arabian Peninsula to the west and south; and Iran, Iraq, and Kuwait to the north. The number of countries in total that surround the Persian Gulf are eight, and include: Iran, Iraq, Kuwait, Saudi Arabia, Bahrain, Qatar, United Arab Emirates, and the Sultanate of Oman. It is sometimes also referred to as the Arabian Gulf by people of the Arab nations here. The Persian Gulf has a maximum length of 614 miles (989 kilometers), a maximum width of 224 mi (362 km), a surface area of 155,964 square miles (251,000 square kilometers), average depth of 164 feet (50 meters), and maximum depth of 295 ft (90 m).

The Indian Ocean connects to the southeast of the Persian Gulf through the Gulf of Oman via the Straits of Hormuz. The dominant river discharge into the Persian Gulf is from the north, by way of the Shatt al Arab estuary that delineates the border between Iran and Iraq. The Shatt al Arab receives freshwater from the Tigris and Euphrates River system that flow through Iraq, as well as the Karun River that flows from Iran. Other rivers that discharge freshwater into the gulf are exclusively on the Iranian side; these flow down from the Zagros Mountains and include the Mand, Hilleh, and Hendijan Rivers.

Due to the arid climate, the Persian Gulf experiences evaporation rates of some 5 to 7 ft (1.5 to 2 m) per year off of the total surface area, an extreme situation that causes sharp water density and salinity variations in the Gulf. Surface salinity varies between 36 and 41 practical salinity units (PSU) in the summer, with a range of 38 to 43 PSU in the winter. Surface temperatures here are in the range of 75 to 93 degrees Fahrenheit (24 to 34 degrees Celsius) in the summer, and a range of 59 to 68 degrees Fahrenheit (15 to 20 degrees Celsius) in the winter. Circulation in the gulf is cyclonic, or counter-clockwise, and deflects freshly entering water from the Indian Ocean initially toward the Iranian coast.

BIODIVERSITY

The arid environment of the Persian Gulf region creates a unique biome. Mangroves here are an ecologically vital marine coastal habitat, one that has been under threat because of human activities and development. The dominant mangrove species of the Persian Gulf is *Avicennia marina* or grey mangrove, which is distributed along both Iranian and Arabian Peninsula coasts. The grey mangroves inhabit the salt flats, or *sabkha,* adjacent to the coast. The grey mangrove typically is not submerged underwater; the root system consists of pneumatophores or aerial roots.

The less-common Asiatic mangrove, *Rhizophora muronata*, is more restricted in its distribution along both coasts; it occasionally fringes embayments on either side of the gulf. The Asiatic mangrove has a similar habitat to other species of the genus *Rhizophora*, which grow submerged roots in seawater. There have been efforts of restoring the Asiatic mangrove around the Persian Gulf, such as the UAE Agency for the Environment sponsored program in 2008.

The Persian Gulf biome is home to several charismatic megafauna that are threatened by extinction due to region-wide conflicts, unregulated coastal development and expansion, and accelerated climate change. One of the most widely known is the dugong (*Dugong dugon*). The dugong is an herbivorous marine mammal belonging to the order Sirenia; its closest relative of the same order is the Atlantic Ocean manatee (*Trichechus* spp.), whose distribution includes both sides of the Atlantic.

The geographic distribution of the dugong is throughout the Indo-Pacific region spanning from the Persian Gulf to the coasts of Australia. In the Persian Gulf, the dugong can be found off the coasts of Saudi Arabia, Iran, Bahrain, Qatar, and the United Arab Emirates. All these countries recognize that the International Union for Conservation of Nature (IUCN) Red List has the dugong listed as vulnerable, and discourage hunting of the species.

The dugong lifespan is similar to humans, and they can reach an age of up to seventy years. The adults can reach a length of 9 ft (3 m) and approximately 881 pounds (400 kilograms). The young reach sexual maturity between the ages of eight and eighteen; the female dugong typically bears a single calf once every three years, and rears the calf for eighteen months.

The dugong inhabits shallow seagrass beds, which are also their main source of food. When seagrasses are scarce, dugongs may feed on deep-water algae and small invertebrates. Dugongs can hold their breath underwater for about eight minutes, similar to dolphins, but short compared to other marine mammals such as whales (at least twenty minutes and up to two hours), seals (seventy minutes), or walruses (thirty minutes).

Species of whales, dolphins, and porpoise have been recorded in the Persian Gulf. The whale species include Bryde's whale (*Balaenoptera edeni*), humpback whale (*Magaptera indica*), and false killer whale (*Pseudorca crassidens*). Whales generally have a global distribution, and may have traversed the gulf but gone largely unnoticed due to the lack of a whaling industry here. The same is true in identifying dolphin and porpoise species in the Persian Gulf, which are also a globally cosmopolitan group of mammals.

The speckled dolphin (*Sotalia lentiginosa*), Indo-Pacific bottle nosed dolphin (*Tursiops aduncus*), pantropical spotted dolphin (*Stenella attenuate*), and Asiatic black finless porpoise (*Neophocaena phocaenoides*) have been documented in the Persian Gulf. The warm and highly productive waters of the Persian Gulf are likely to be appealing as a feeding ground for cetaceans of all sizes. The great abundance of fishes and shrimp in the Persian Gulf appears to be able to support and sustain large populations of whales, dolphins and porpoises. However, the activities associated with the oil industry and decades-long armed conflicts may have rendered these species endangered in these waters.

The other charismatic megafauna group of the Persian Gulf is the sea turtle. There are five species of sea turtles identified in the Persian Gulf: loggerhead turtle (*Caretta caretta*), green sea turtle (*Chelonia mydas*), hawksbill (*Eretmochelys imbricata*), olive ridley (*Lepidochelys olivacea*), and the leatherback sea turtle

(*Dermochelys coriacea*). The IUCN Red List has the green turtles listed as endangered, the hawksbill turtles listed as critically endangered, and the olive ridley, loggerhead, and leatherback turtles listed as vulnerable as of 2018. Females of all these species are known to lay their eggs on the shores of islands around the Persian Gulf; many of the known egg-laying sites are protected by the respective governments.

The Persian Gulf hosts over seventy species of migratory birds that use the Gulf as a flight path, rest stop, and feeding grounds as they fly south to winter in Africa, or fly north to summer in northern Eurasia. These include various species of pelicans, cranes, flamingos, and warblers. Most habitats for migratory and endemic birds are in a grave state of degradation due to development. One rare species of kingfisher that is in danger of extinction is the collared kingfisher (*Todiramphus chloris kalbaensis*).

Coral reefs are the hot spots of diversity in the Persian Gulf biome. Coral reefs and coral species are distributed throughout the shallow waters here, except for the northernmost reaches that are dominated by alluvial mud flats associated with the Shatt al Arab outflow. In general, a Persian Gulf coral reef consists of several coral species and can go up to thirty-four species. The dominant hard (scleractinian) coral genus in the Gulf is *Acropora* spp., which is endemic (found nowhere else); the most common endemic species in the gulf is *Acropora arabensis*. Other genera include *Porites*, *Siderastrea*, *Pseudosiderastrea*, *Favia*, and *Platygyra*. Anthropogenic activities and climatic factors have led to serious degradation of the coral reef habitat—jeopardizing many species of marine reef fishes and invertebrates.

Fishes of the Persian Gulf number over 700 species, and over 500 of these are associated with coral reefs. Among important commercial crustacean species of the Persian Gulf are the green tiger prawn

(*Penaeus semisulcatus*) and the penaeid shrimp (*Metapenaeus affinis*), which uses the southern Iraqi marshes as a nursery ground by traveling up the Shatt al Arab as juveniles and returning to the Persian Gulf upon maturity. There are other brackish water shrimp that are part of the northern Persian Gulf biome that have no commercial value; however, they are important players in the aquatic ecosystem of this extreme environment.

THREATS

Threats to the flora and fauna in this region come from the potential for oil spills and other human-made disasters. As was seen from the BP oil disaster in the Gulf of Mexico, such spills can do great harm to the local environment and wildlife, causing untold damage.

Climate change threats to the Persian Gulf region include the potential for food shortages, as acidification of the waters may decrease fish populations. Many people in the region rely upon fish as both a food and income source, and any decreases in the available amount and types of fish can create economic hardship and habitat disruption.

In the second decade of the twenty-first century, researchers were studying the Persian Gulf's coral reefs to determine how the coral is able to withstand such high-water temperatures. As the temperatures of the world's bodies of water have increasingly risen over the years due to climate change, damaging marine ecosystems through events such as extended and widespread coral bleaching, interest in this adaptation ability has also increased. In 2015, it was discovered that the majority of the Persian Gulf coral was covered in a new species of algae that was likely enabling the coral to survive in the hot conditions. As a result of such studies, the debate was ignited over whether humans should intervene to plant such adapted coral into other bodies of water around the world in an effort to help other reefs resist the increasing temperatures as well.

—Nasseer Idrisi

Further Reading

Abuzinada, Abdulaziz H., Hans-Jürg Barth, Friedheim Krupp, et al. *Protecting the Gulf's Marine Ecosystems from Pollution*. Birkhauser, 2008.

Al-Azab, M., W. El-Shorbagy, and S. Al-Ghais. *Oil Pollution and Its Environmental Impact in the Arabian Gulf Region*. Elsevier, 2005.

Alosairi, Y., et al. "Mixing and Flushing in the Persian Gulf (Arabian Gulf)." *Journal of Geophysical Research*, vol. 116, 2011.

Idrisi, N., and S. D. Salman. "Distribution, Development, and Metabolism of Larval Stages of the Warm Water Shrimp, Caridina Babaulti Basrensis (Decapoda, Atyidae)." *Marine and Freshwater Behavior Physiology*, vol. 38, 2005.

Lazarus, Sarah. "Heat-Resistant Corals in the Middle East Could Save the World's Dying Reefs." *CNN*, www.cnn.com/2018/06/26/middleeast/middle-east-corals/index.html. Accessed 30 July 2018.

Riegl, B., and Sam Purkis. *Coral Reefs of the Gulf: Adaptation to Climatic Extremes*. Springer, 2012.

Ulrichsen, Kristian Coates. *Insecure Gulf: The End of Certainty and the Transition to the Post-Oil Era*. Oxford UP, 2015.

Petroleum Extraction and Processing

Fields of Study: geology; engineering; oceanography

ABSTRACT

Petroleum extraction and processing is the human exploitation of fossil fuels consisting of hydrocarbons in the form of crude oil and natural gas. The primary use of processed petroleum products is as powerful fuels, particularly for transportation in the form of gasoline, diesel fuel, or jet fuel, and for heating purposes, which account for 84 percent of petroleum use. The remainder of processed petroleum products is used in the petrochemical industry as fuel additives and to create applications such as plastics, specialty chemicals, solvents, fertilizer, pesticides, and pharmaceuticals. Worldwide, about 37 billion barrels of crude oil and natural gas are extracted each year.

KEY CONCEPTS

alkylation: the addition of an alkyl group (a hydrocarbon segment) to an existing molecule

distillation: a process for separating components from a mixture such as crude oil by vaporizing and collecting the portions, or "fractions," according to their individual boiling points

isomerization: changing the structure of a molecule to that of another molecule with the same chemical formula

seal rock: an impermeable rock layer that serves to trap petroleum or natural gas within a lower substrate rock

visbreaking: a mild thermal cracking process applied in processing of crude oil to reduce the viscosity of the material and produce fuel oil and lighter fractions and increase the distillate yield in a refinery

DEFINITION AND BASIC PRINCIPLES

Petroleum extraction and processing encompass the activities through which crude oil and natural gas are taken from their natural reservoirs below the surface of Earth on land and sea and treated so they can be used as fuels and materials for the petrochemical industry. Reservoirs can be detected by applying the results of petroleum geology, the science describing under what circumstances oil and gas reservoirs were created during ancient times and how they can be found. Paleoceanography is an integral part of this practice of geology.

Once a likely reservoir is identified, exploratory drilling through the ground begins, either on land or on sea. When a well strikes a reservoir, the size of its area is estimated. Depending on whether crude oil or natural gas is dominant, the reservoir is called an "oil or gas field." Many wells are drilled to extract oil and gas from fields that can be quite vast, extending for hundreds or thousands of miles (or square kilometers) under the surface. As of 2010, there were about 40,000 producing oil and gas fields in the world. In 2016, Rystad Energy estimated the world's proved reserves contained 381 billion barrels of oil and, together with probable and undiscovered fields, may contain up to 2,092 billion barrels.

Usually, natural pressure forces gas and oil to the surface. In mature wells, however, pressure has to be added by technical means. After gas and oil are extracted, both are sent by pipelines for further processing.

Processing natural gas separates almost all other components from its key ingredient, methane. Processing crude oil occurs in a refinery. Because the different components of crude oil vaporize at different temperatures, they can be separated by distillation. To obtain the most desirable end products, the components are further treated through chemical processes called "cracking and reforming."

BACKGROUND AND HISTORY

Modern petroleum use was based on demand for kerosene for lighting. Polish pharmacist Ignacy Łukasiewicz discovered how to distill kerosene from petroleum in 1853. He dug the first modern oil well in 1846 and the world's first refinery in 1856.

The first commercial oil well in North America was founded in Oil Springs, Ontario, in September, 1858, by James Miller Williams, after what began as a water well excavation struck oil instead of water. Almost directly across Lake Erie from Oil Springs, retired railway conductor Edwin Drake dug the first oil well in the United States at Titusville, Pennsylvania, on August 28, 1859. His drilling method, using a pipe to cover a drill suspended from a rig and leading the pipe into the earth to prevent collapse of the borehole, was widely copied as he did not patent it. It is still the basic drilling mechanism. Horizontal rather than vertical drilling, which allows easier access to the reservoir of an oil field, revolutionized crude oil extraction in the 1990s.

The appearance and spread of the internal combustion engine by the late nineteenth century led to crude oil being refined to produce gasoline. The quest to extract more gasoline led to the application of thermal cracking in 1913, and by 1975, many subsequent new cracking techniques had culminated in residual hydrocracking. Beginning in 1916, processes were invented to decrease the unwanted sulfur content of crude oil. Since 1939, visbreaking has been used to make smoother flowing products, and isomerization and alkylation have increased gasoline yield and quality since 1940. Modern refineries try to make these processes as efficient and clean as possible.

HOW IT WORKS

Exploration. Almost all exploration of new oil and gas fields is done by geophysical means such as seismic reflection, or by other seismic, gravity, or magnetic surveys. These surveys use the different densities of subterranean rock formations to identify source rock formations likely to hold hydrocarbons. Most, if not all, oil-bearing formations represent the signature of the living paleoenvironment of marshes, lakes and oceans, and as such are in the purvey of geologists who are also paleoceanographers. These specialists seek to identify the structures that result in the present time from the environmental structures that existed long ago. Based on this information, three-dimensional (3D) computer models are generated and analyzed. Once petroleum engineers and geologists have decided on a promising prospect, exploratory drilling begins.

Extraction. Oil and gas deposits are accessed by drilling a well from a rig, either on land or in the sea. Rigs are adapted to their particular environment but share some common elements. Suspended from a metal tower called a "derrick," the drilling bit hangs from its drilling string over the well bore, the borehole where the well is dug. As the drilling bit goes deeper, the borehole is fitted with a pipe to prevent its collapse. The pipe is placed in a larger steel casing, leaving space between the pipe and casing. Drilling mud descends through the pipe to cool the drilling bit as it bores into the rock and percolates back to the surface between the pipe and casing. On the sea, the rig is placed on an offshore oil platform. Modern drills can dig thousands of feet (or meters) deep.

Drilling is successful when the well strikes petroleum deposits below their seal rock. Natural pressure ejects gas and oil through the pipe. A wellhead is placed on the top of the well, and natural gas is separated from crude oil in a separator unit. Both products are collected and transported by intermediary pipelines connecting the many wells of a single field.

As long as natural pressure ejects petroleum, primary recovery continues. Once this pressure decreases, water or gas is pumped into the reservoir to create artificial pressure during secondary recovery. When maintaining artificial pressure is no longer economical, the well is considered exhausted.

Processing. Natural gas has to be cleaned in gas processing plants typically built close to the point of extraction. It is separated from its water and natural gas condensate (also called "natural gasoline"), and the latter is sent to a refinery. Then it is stripped of its acid gases (carbon dioxide, hydrogen sulfide, and other gases with sulfur content), which are desulfurized and often burned as tail gases. To gain as much pure methane, its main component, as is possible, natural gas undergoes a further series of processes to remove contaminants (such as mercury) and nitrogen. Natural liquid gases (NLGs) are removed for use as feedstock in petrochemical plants. Then, consumer-grade natural gas is ready for sale.

Crude oil is processed at a refinery. Refineries need not be close to oil fields as crude oil can be transported by pipeline or ocean tanker. A refinery, which is a highly integrated chemical plant, seeks to create the most desirable products out of crude oil. These products are hydrocarbon chains with few rather than many carbon atoms (called "light" instead of heavy).

First, crude oil is desalted, then it is separated into its various components, called "fractions," by atmospheric distillation. As crude oil is heated, its different fractions vaporize at different temperatures and are removed separately. All fractions are processed further. The lightest fraction, distilled gas, is subjected to sulfur extraction. The heaviest fraction, called "atmospheric bottom," undergoes vacuum distillation. Most distilled fractions undergo hydrotreatment, which is infusion of hydrogen to remove sulfur.

At the heart of a refinery are its fluid catalytic cracker and hydrocracker, which break down heavy (long) hydrocarbon chains into lighter (shorter) ones. A catalytic reformer is used to create higher octane reformate, a more valuable naphtha distillate. Isomerization and alkylation are two chemical processes designed to boost the distillate's octane rating. Refineries send their products to end users or the petrochemical industry.

APPLICATIONS AND PRODUCTS

Gasoline. The most common and valuable petroleum product is gasoline. Typically, about 46.5 percent of US crude oil is processed into gasoline of various qualities. As fuel for the internal combustion engine, gasoline is essential for any industrial society. It is used by most cars and many light trucks. Aviation gasoline fuels the piston-engines airplanes.

Refineries seek to optimize their gasoline output and strive to create blends that minimize damaging spontaneous combustion in engines (known as "knocking"). An organometallic compound called "tetraethyl lead" was used until the 1970s, but its use has been prohibited in most industrialized nations because of the harmful effects of lead salts that are ejected in the engine exhaust after combustion of leaded fuel. In its stead, chemists have developed a variety of alternative additives.

Diesel and Fuel Oils. The second most important petroleum products, which account for about 28 percent of US crude oil, are diesel and other fuel oils. These fuel oils have longer hydrocarbon chains than gasoline and thus boil at higher temperatures. Diesel is used as fuel for trucks, buses, ships, locomotives, and automobiles. In Europe, diesel engines are very popular for automobiles as they are more fuel efficient. Environmentalists have expressed concern about burning these fuels because the process creates the pollutant sulfur dioxide.

Heavy fuel oils account for another 2.4 percent of petroleum products processed from crude oil. Among them, bunker oil is the heaviest. It is used to fuel large ship engines.

Natural Gas. Natural gas, which may be found alongside oil and accounts for less than 4 percent of US refinery yields, is used for generation of electric power, domestic and industrial heating, and increasingly for transportation by bus or automobile. Europe has a dense network of pipelines bringing natural gas, primarily from Russia, to industrial and domestic consumers.

Natural gas can be liquefied (as liquefied natural gas, LNG) for transport by gas tankers across oceans. This is the preferred method to bring natural gas from gas fields in the Middle East, North Africa, and the Caribbean to Europe and the United States. After reaching the destination, LNG is gasified and pumped into a pipeline at the port terminal.

Jet Fuel. Jet fuel powers gas-turbine engines of airplanes (jets). Its production amounts to about 9.5 percent of processed crude oil in the US. Commercial jet fuel is produced according to international standards. It is either Jet A or Jet A-1, with Jet A freezing at -40 degrees Fahrenheit (-40 degrees Celsius) and A-1 at -52.6 degrees Fahrenheit (-47 degrees Celsius). Jet B is a lighter, more flammable jet fuel for cold regions, such as northern Canada. Military jet fuel is produced according to individual national standards.

Liquefied Petroleum Gas. The result of distilling the remaining gas from crude oil at a refinery, liquefied petroleum gas (LPG) is widely used for domestic heating and cooking in Asia, South America, and Eastern Europe, as well as a global fuel alternative for automobiles.

Petrochemicals. Petrochemicals are often byproducts of gasoline production, such as naphthas. They are grouped as olefins or aromatics. The value of petrochemicals has been rising consistently as new chemical processes have found additional applications.

After they leave a refinery, most petrochemicals are used as feedstock for petrochemical plants. Among olefins, ethylene and propylene are building blocks for plastics and specialty chemicals. The olefin butadiene is processed to eventually form synthetic rubber. Among aromatics, benzene is used in the production of synthetic dyes, detergents, and such innovative products as polycarbonates that form light hard plastic shells in many electronic items such as mobile phones. Toluene and xylene are used for solvents or as building blocks for other chemicals, creating, for example, polyester fibers.

Petrochemicals are also used to create adhesives, food ingredients, and pharmaceuticals. They have become a source of much innovative chemical research.

Lubricating Oils. Although their volume accounts for only 1 percent of petroleum products leaving a US refinery, lubricating oils and greases are important for the machines and engines of industrialized nations. Refineries create a great variety of lubricating oils through blending with different additives.

Asphalt. The heaviest product from crude oil, asphalt makes up about 2 percent of US refinery products. Concrete asphalt for durable roads and runways accounts for the majority of North American asphalt use; asphalt roof shingles are the next most common use.

CAREERS AND COURSEWORK

Students interested in the petroleum industry should take science and mathematics courses, particularly chemistry and physics, in high school. In college, the aim should be a bachelor of science. The choice of science to study depends on whether the student wishes to pursue a career in exploration and production or processing and petrochemicals.

For the former, the geosciences, particularly geology, geophysics, mining, hydrology, and hydrogeology, as well as chemistry, are useful. Computer literacy and ability to work with specific applications in geology, particularly 3D sensing and modeling, is important. Oil companies will welcome successful graduates who hold a bachelor of science in geology, hydrogeology and engineering, chemical engineering, geology and mining, or even physics. A master's or engineering degree in these fields will provide additional opportunities. For example, a master's of science in petroleum engineering or in environmental engineering geology can lead to an advanced entrance position as reservoir geologist. One of the most important criteria for top positions, however, is years of practical work experience. Because increasing responsibilities lead toward management, it would not be uncommon for a senior geologist to hold a master of business administration degree in addition to science degrees.

For a career in petroleum processing at a refinery or petrochemical plant, a bachelor of science in chemistry or a related field is a good foundation. A master's or doctoral degree in chemistry or chemical engineering, as well as in mechanical, environmental, or control engineering, will serve for an advanced position.

The oil and gas industry also employs many laboratory scientists and technicians. Corporate research and development departments are typically headed by scientists with doctorates in chemistry and related sciences.

Because of the cyclical nature of the oil and gas industry, there are periods in which skilled employees such as drilling engineers are in high demand, but these workers also experience layoffs during downturns. Work can be at very remote locations, and global mobility is often expected.

SOCIAL CONTEXT AND FUTURE PROSPECTS

The late-nineteenth-century discovery that crude oil and natural gas could be turned into powerful fuels for industrialized society meant that those nations where oil and gas were found possessed a very valuable natural resource. However, use and distribution of wealth from petroleum is a huge social challenge.

The industrialized nations' dependence on petroleum products has significantly influenced global politics since the early twentieth century. Since the late 1980s, there has been increasing concern that the massive burning of hydrocarbon fuel causes global warming because of the accumulation of carbon dioxide and other greenhouse gases in the atmosphere. Over the ensuing decades, various methods were employed around the world to address the problem—from increasing the efficiency of hydrocarbon-reliant technologies, such as automobiles and power plants, to creating alternative devices and energy sources to creating emissions caps and carbon pricing schemes for polluters. However, as of 2018, petroleum-related carbon emissions continued to rise.

Oil spills, whether from wells or when crude oil tankers run aground, can be major environmental disasters. On April 20, 2010, an explosion destroyed the Deepwater Horizon oil drilling platform in the Gulf of Mexico. The ensuing spill of gas and crude oil from a well on the ocean floor, about 5,000 ft (1,500 m) below the water, became the world's second-biggest spill and the United States' biggest spill to date, releasing between 35,000 to 60,000 barrels per day and proving very hard to contain.

There is ongoing scientific debate over when Earth's petroleum resources will be depleted. In 1956, American geoscientist Marion King Hubbert published his theory of peak oil, in which he tried to calculate when the volume of petroleum extracted would exceed the volume of remaining petroleum reserves. A concerted effort involving new exploration and extraction technologies seeks to discover even more remote reserves, particularly deep under the sea and in the Arctic, and to develop oil shale and oil sands more efficiently and as economically as possible. Another area of ongoing interest is the recycling of petroleum products, such as asphalt, lubricants, and petrochemicals.

Scientists have also predicted that future alternative fuels, including those derived from hydrogen or renewable resources, will take the role of petroleum in coming years.

—R. C. Lutz

Further Reading

Burdick, Donald. *Petrochemicals in Nontechnical Language*. 3rd ed., Penn Well, 2010.

Duffield, John. *Over a Barrel: The Costs of U.S. Foreign Oil Dependence*. Stanford UP, 2008.

Gary, James, et al. *Petroleum Refining: Technology and Economics*. 5th ed., CRC Press, 2007.

Hyne, Norman. *Nontechnical Guide to Petroleum Geology, Exploration, Drilling, and Production*. 2nd ed., Penn Well, 2001.

Katz, Barry J., editor. *Petroleum Source Rocks*. Springer, 2012.

Matthews, Chris. "America Now Leads the World in This Surprising Category." *Fortune*, 5 July 2016, fortune.com/2016/07/05/oil-reserves-us.

Palmer, Andrew. *Introduction to Petroleum Exploration and Engineering*. World Scientific, 2017.

Raymond, Martin, and William Leffler. *Oil and Gas Production in Nontechnical Language*. Penn Well, 2005.

Saraswati, Pratul Kumar, and M. S. Srinavasan. *Micropaleontology: Principles and Applications*. Springer, 2016.

"U.S. Refinery Yield." *EIA*, US Energy Information Administration, US Department of Energy, 31 Aug. 2018, www.eia.gov/dnav/pet/PET_PNP_PCT_DC_NUS_PCT_A.htm.

Yeomans, Matthew. *Oil: Anatomy of an Industry*. The New Press, 2004.

PHILIPPINE SEA

Fields of Study: oceanography; marine biology; marine history; marine biogeochemistry; marine geology

ABSTRACT

The Philippine Sea stretches across some 2 million square miles (5 million square kilometers) of the North Pacific Ocean. It is surrounded by Japan, Taiwan and the Sea of China, the Philippine Archipelago, Papua New Guinea; Guam and the Northern Mariana Islands.

KEY CONCEPTS

archipelago: a series of relatively small islands that are related to each other geologically

Coral Triangle: an approximately triangular region of the Pacific Ocean in which coral reefs are the primary ecosystem

dynamite: a form of powerful explosive once based on nitroglycerine but now based on trinitrotoluene (TNT) and other compounds that are not at shock-sensitive as nitroglycerine and therefore safer to handle

ecosystem: an identifiable environmental region, including the plants, animals and mineral base that it encompasses and the manner in which they interact

DESCRIPTION

Lying to the east and north of the Philippine Islands, the Philippine Sea stretches across some 2 million sq mi (5 million sq km) of the North Pacific Ocean. North of the sea is Japan; to the northwest is Taiwan and the Sea of China; due west and extending southward is the Philippine Archipelago. To the south lies Papua New Guinea; Guam and the Northern Mariana Islands are to the southeast. Several seas—including the Celebes Sea, Sulu-Sulawesi Sea, South China Sea, and the East China Sea—are adjacent to the Philippine Sea but separated by various islands.

Because of geological formations that have resulted from plate tectonic activity, the floor of the sea is marked by faults and fractures that lie within the Philippine Sea Plate. One of the most significant features is the deep-sea trench, the deepest point on the entire Earth, that has been labeled the Philippine or Mariana Trench.

Because the Philippine Sea is located in the tropical climate zone, it is hot and humid all year, averaging around 80 degrees Fahrenheit (27 degrees Celsius). It also has some of the warmest sea surface temperatures in the world. This part of Earth is an area of intense tropical convection, a dominant heat source for global atmospheric circulation. It contains complex ocean current systems that link the Pacific and Indian Oceans, and is intermediate between the Asian monsoon regime to the north and the Australian monsoon to the south.

BIODIVERSITY

The Philippine Sea is home to a vast aquatic ecosystem. More than 500 species of hard and soft coral lie within its waters. The Philippine Sea includes part of the Coral Triangle, which consists of waters off Indonesia, Malaysia, Papua New Guinea, the Philippines, the Solomon Islands, and East Timor. Near the apex of the triangle lies the Sulu-Sulawesi Sea, containing an ecosystem that is home to some 2,500 species of fish.

About one-fifth of all species of shellfish in the world live in the Philippine Sea, which is also home to sharks, moray eels, octopi, and sea snakes. The Japanese eel, tuna, and various species of whales regularly spawn or give birth in the sea waters.

While much is still unknown about the marine diversity here, it is estimated that at least 5,000 species of clams, snails, and mollusks live in these waters, along with more than 900 species of bottom-living algae and thousands of other organisms of various kinds. A 2004 survey revealed the presence of 1,200 species of crabs and shrimp, 600 mollusk species, and hundreds of species that have yet to be identified.

Of the seven known species of sea turtles, six are found in the Philippine Sea. At least 2,824 marine fish species have been recorded in the Philippine Sea, 33 of which are endemic to the Philippine Sea, and are found nowhere else. More than 1,700 of these are reef-associated fish species, with 169 pelagic and 336 deep-water species.

In addition to live species, there is fossilized evidence of extinct snails. The area around the Philippine Sea is also home to 86 species of birds and 895 species of butterflies. Some 352 of the latter are endemic to the Philippines.

HUMAN IMPACT

The Philippine Sea is recognized as a center of marine biodiversity because of its wealth of marine ecosystems. However, environmentalists around the world have also identified the Philippines as a region of environmental concern, and in 2000, the International Union for the Conservation of Nature and Natural Resources (IUCN) labeled the area as a biodiversity hot spot, noting that, at a minimum, 418 species here are threatened. Current levels and methods of harvesting fish and other marine life are not sustainable, among other problems.

The Asian Development Bank has estimated that the quantity of marine organisms had declined by as much as 90 percent in some areas of the Philippine Sea. Major concerns have been raised at both the national and international levels about the lack of responsible management of marine resources here. For much of its history, the Philippines was exploited by colonial powers, and the local population was unable to make decisions about protecting the environment. Even in the twenty-first century, the situation is complicated by the fact that China controls the adjacent Western Philippine Sea (South China Sea), and Chinese interests do not always run in tandem with those of the Philippines. There is little regulation of either commercial or local fishing in the area, and Chinese fishermen are among those who have been known to use dynamite, which wipes out fragile ecosystems.

The waters and the coast of the Philippine Sea are vulnerable to a host of environmental practices that combine to damage ecosystems in the area. Only about one-tenth of sewage in the entire country of the Philippines is adequately treated, for example. This means that 90 percent is deposited elsewhere, and most of that waste goes directly into the sea. Furthermore, the coastal mangrove swamps that are important fish breeding grounds are being stripped to provide wood for development. Abandoned mines, meanwhile, have been responsible for highly toxic mercury seeping into surrounding waters.

Habitat loss and degradation, increasing pollution, destructive fishing on both commercial and local scales, and overexploitation as a result of expanding markets for marine products are all damaging this complex ecosystem. Compared with several other Asian nations, including Vietnam, Thailand, and Indonesia, the Philippines was ranked the highest in reef degradation and the lowest on taking action to protect its fragile reefs. Environmentalists accuse the government of being swayed by the fact that the fishing industry employs more than 1 million people and comprises a major portion of the Philippine economy. Another environmental concern for marine life here is the red tide algal blooms that first came to the attention of scientists in 1983. Believed to occur naturally, but also influenced by environmental changes including rising sea-surface temperatures, red tide is a high concentration of algae that depletes oxygen levels in water, and releases toxins that are poisonous to some fish.

Climate change is already having a big impact on marine and coastal ecosystems in the Philippine Sea by warming, acidifying, and raising sea levels. It is believed that rising global temperatures severely challenge, and could even eliminate, many coral-dominated reef systems. Global warming also means there will likely be more variability of monsoon events, and more intense cyclones and typhoons. There is an urgent need for improved management and cooperation between countries that surround this sea in conserving and protecting their resources against this swarm of threats.

—Elizabeth Rholetter Purdy

Further Reading

Adraneda, Katherine. "Scientists: The Philippines Is the World's 'Center of Marine Biodiversity.' Announce Alarm Over Threats." *Underwater Times*, 8 June 2006, www.underwatertimes.com/news.php?article_id=64017239105.

Batongbacal, Jay L., and Porfirio M. Aliño. "More Than Just Oil in the West Philippine Sea." *Philippine Daily Inquirer*, 25 June 2011.

Castilla, Juan Carlos, and Tim R. McClanahan, editors. *Fisheries Management: Progress Toward Sustainability*. Blackwell, 2007.

Goldoftas, Barbara. *The Green Tiger: The Costs of Ecological Decline in the Philippines*. Oxford UP, 2006.

R

Red Sea Ecosystem

Fields of Study: oceanography; marine biology; marine geology; tectonics

ABSTRACT
Formed by Arabia splitting from Africa, the Red Sea is a long, narrow inlet of the Indian Ocean between Africa and Asia. The gulfs of Aqaba and Suez to its north are sometimes included informally. The International Hydrographic Organization defined the precise limits of the Red Sea as the southern limits of the Gulfs of Suez, the south point of Shadwan Island, the coast of Africa and Aqaba through Tiran Island, and the coast of the Sinai Peninsula.

KEY CONCEPTS
archaea: the class of bacteria that includes the methane producers, the thermohalophytes, and other extremophilic bacteria

benthic: refers to algae that inhabit the seafloor under mangrove forests; generally, refers to the seafloor environment

epiphyte, epiphytic: refers to algae that live on the surface of mangrove trees in mangrove forests; generally, refers to any plant life that lives on trees, using them for support

fringing reef: reef structures that develop and grow parallel to a shoreline

sea sawdust: the red-colored *Trichodesmium* algae

OVERVIEW
The world's northernmost tropical sea, the Red Sea is home to numerous species of coral, 1,200 species of fish, and more than 1,000 species of invertebrates. The name Red Sea, used since ancient times, may refer to blooms of "sea sawdust" or to the arcane use of colors to refer to cardinal directions (in this case, south).

The sea's coastline is about 1,400 miles (2,253 kilometers) long and it is an average of 174 mi (280 km) wide. The average depth is 1,600 feet (488 meters), but the water may exceed 8,000 ft (2,438 m) at its center point. Most of the sea is shallow, and a quarter of it is less than 200 ft (61 m) deep.

The Red Sea is surrounded by arid land masses, and due to low rainfall and the high rate of evaporation in the north, it has become one of the most saline bodies of water in the world. Salinity ranges from 3.6 percent (36 parts per thousand [ppt]) in the south, where water is exchanged with the Gulf of Aden, to 4.1 percent (41 ppt in the north), with an average salinity of 4 percent (40 ppt). By contrast, worldwide average seawater salinity is 3.5 percent (35 ppt). Some of the salinity is contributed by highly salty brines emanating from sub-seafloor hydrothermal vents at temperatures approaching 140 degrees Fahrenheit (60 degrees Celsius), a result of the continued widening of the sea due to the tectonic movement of the Red Sea Rift.

Rainfall over the sea and its coasts is low, at about 2.3 inches (58.4 millimeters) per year, usually in the form of brief thunderstorms. Wind is dominated by persistent northwest winds in the north. The rest of the Red Sea is subject to winds that are seasonally variable in both speed and direction, with speed generally increasing northward. Wind-induced currents shift much of the sediment in the sea, and affect the erosion and accretion of coastal rock exposure. The climate results from the northeasterly and southwesterly monsoons, and the surface water temperature is among the hottest in the world year-round, averaging 72 degrees Fahrenheit (22 degrees Celsius), with summer highs around 90 degrees Fahrenheit (32 degrees Celsius).

The health of the coral reefs here is maintained in part by the great depths of the sea and the efficiency of its water circulation. The water mass of the sea exchanges its water with the Arabian Sea via the Gulf of Aden to the south. The Red Sea is rich in minerals with many sediment constituents. These include: nanofossils, foraminifera, pteropods, siliceous fossils, tuffites, volcanic ash, montmorillionite, cristobalite, zeolites, quartz, feldspar, mica, clay minerals, sulfide minerals, aragonite, calcite, dolomite, chalcedony, magnesite, gypsum, anhydrite, halite, hematite, siderite, pyrite, and rhodochrosite. Despite this diverse sediment, the water is exceptionally clear because of the lack of river discharge and the low levels of precipitation.

BIODIVERSITY

The harsh conditions of the hydrothermal vents are inhospitable to most forms of life. The water is hot and highly saline, with high concentrations of heavy metals, and there is little light or oxygen at the bottom of the rift. For many years, the vents were believed to be lifeless, but studies have discovered that like many other inhospitable parts of the world, they are home to extremophile bacteria and archaea—microbes adapted to extreme conditions. Some such species in the Red Sea include *Halorhabdus tiamatea*, which is related to *H. utahensis* of the Great Salt Lake in Utah, *H. contractile*, which forms tentacle-like protrusions of unknown purpose and *Salisphaera shabanensis*, which lives at the interface between the sludgy brine and the seawater.

The sea is surrounded by salt flats characterized by evaporite-carbonate deposits, which form shoals similar to the *sabkha* of the Persian Gulf, long flat salt pans covered in fine-grained alkali salt sediment and salt marshes that are home to halophytic, or salt-tolerant, plant life and migratory birds.

The coastal zones include stands of mangrove (*Avicennia marina* and *Rhyzophora mucronata*), trees that are specially adapted to low oxygen and high salinity. *A. marina* has developed salt glands to rid itself of excess salt by excreting it through the leaves. *R. mucronata* is found only in the mangrove stands of the Farasan Archipelago, in the southern part of the sea, which is richer in nutrients due to the proximity of the Indian Ocean. As scientists have discovered that mangrove trees are some of the best types of trees for absorbing carbon dioxide, it has become especially important to conserve such stands in the global effort to reduce the negative effects of climate change.

Nearly 100 species of algae are found in mangrove stands, including the sea's characteristic red algaes, the most common species of which are *Bostrychia tenella, Spyridia filamentosa,* and *Laurencia papillosa.* These types dominate in shallow mangrove pools. Blue-green algae is the least diverse algae group, but also the most widespread, growing in both benthic and epiphytic forms throughout mangrove ecosystems.

Local wildlife in the mangrove stands includes mangrove crab species such as the very abundant *Uca inversa inversa, Metapograpsus messor, Portunus pelagicus, Ocypode saraten,* and *Macrophthalamus telescopius,* as well as a species of mudskipper fish (*Oxudercinae*).

The Red Sea is the most northern coral-reef ecosystem. The most common reef type in the Red Sea is fringing reef, which (unlike atolls and barrier reefs) grows close to and often directly from the shoreline. The back-reef area has the least species diversity and largely consists of seagrass meadows and the fish that feed from them.

Seagrasses are the only flowering plants that can live completely submerged, and seagrass ecosystems are home to numerous rays and echinoderms: sea cucumbers, sea urchins, and starfish. The reefs are home to reef sharks, dolphins, and numerous species of mollusks. About 10 to 20 percent of the fish found around the coral reefs are endemic, or found only here.

The Red Sea coast also is home to four species of seaturtles: green, hawksbill, loggerhead, and leatherback. Greens and hawksbills usually forage around the reefs; the greens eat algae and seagrass, while the hawksbills feed on sponges and invertebrates. Leatherbacks, which feed on jellyfish, are usually found farther from the reefs. However, all of the species of sea turtles were considered threatened or endangered as of 2018.

Migratory birds in the Red Sea coastal zones include swallows, songbirds, stocks, cranes, and birds of prey.

Fish, phytoplankton, and zooplankton populations in the Red Sea closely resemble those of the Indian Ocean. Common species include the shagreen ray (*Raja fullonica*) and the black-striped pipefish (*Syngnathus algeriensis*), which anchors itself to seagrass by wrapping its tail around it. Endemic species include the tiger bass (*Terapon jarbua*), which feeds on insects and plant matter; the leopard torpedo (*Torpedo panthera*), a ray capable of generating electricity as a defense and feeding mechanism; and the great hammerhead shark (*Sphyrna mokarran*), the largest hammerhead species, with a maximum length of 20 feet (6 meters).

HUMAN AND ENVIRONMENTAL THREATS

The tidal amplitude of the sea is very low, and in the mangrove stands, it is often only about 1 ft (0.3 m), which helps preserve the stability of the mangrove ecosystems. The roots of the mangroves provide oyster habitats and increase sediment deposits, resulting in sinks for heavy metals.

Disrupting the mangrove stands could release the trapped heavy metals and recontaminate the sea with these and other pollutants. Mangroves on the Egyptian coast are protected for this reason, and because they help protect the coast from erosion.

Because the land surrounding the Red Sea is arid, there is elevated demand for freshwater to meet both drinking and industrial demands. As of early 2018, there were around eighteen desalination plants along the Saudi Arabian coast working to harvest seawater and strip away the salts. With such high-powered efforts, Saudi Arabia has become one of the world leaders in water desalination. Later that year, it was announced that the country planned to build nine more of these facilities on the coast of the Red Sea. This results in discharges of warm, highly saline water and treatment chemicals, which puts local fish stocks in jeopardy.

The use of the Red Sea water as a coolant by factories and refineries has a similar effect. However, these effects seem to be localized to the coastal zone immediately surrounding the individual plant, refinery, or factory. Commercial fishing in the sea focuses on lobster, tuna, bonito, herring, sardine, and anchovy.

Although warm water temperatures in other seas are a major cause of coral bleaching, that is apparently not the case here. The corals of the southern Red Sea have been under study to determine how they are able to dodge this threat.

The greatest danger to the Red Sea's coral seems to be contaminants introduced into the water, particularly because most of the coral is so close to the coast, where there is not only industrial runoff, but also waste from hotels and offshore oil spillage. Hotel waste has been a considerable problem since at least the 1990s, and as tourism has increasingly become a bigger industry for countries such as Jordan, even greater amounts of waste have entered the sea. Egyptian governments have banned the use of plastic bags in supermarkets in an attempt to reduce the prevalence of the bags in the sea's waters and reef systems.

—*Bill Kte'pi*

Further Reading

Agius, Dionisius A., Emad Khalil, Eleanor M.I. Scerri, and Alun Williams, editors. *Human Interaction with the Environment in the Red Sea.* Brill, 2017.

Almahasheer, Hanan. "Red Sea Mangroves Fight Back in the Face of Global Decline." *HuffPost,* 28 Feb. 2017, www.huffingtonpost.com/entry/red-sea-mangroves-fight-back-in-the-face-of-global_us_58b5b36de4b02f3f81e44ce2.

Chandrasekharam, D, Aref Lashin, Nassir Al Arifi, and Abdulaziz M. Al-Bassam. *Red Sea Geothermal Provinces.* CRC Press, 2019.

Marshall, N. A., P. A. Marshall, A. Abdulla, et al. *Preparing for Climate Change in the Red Sea: Recognizing Early Impacts Through Perceptions of Dive Tourists and Dive Operators in Egypt.* IUCN, 2009.

Rasul, Najeeb M. A., and Ian C. F. Stewart, editors. *The Red Sea: The Formation, Morphology, Oceanography and Environment of a Young Ocean Basin.* Springer, 2015.

Tesfamichael, Dawit, and Daniel Pauly, editors. *The Red Sea Ecosystem and Fisheries.* Springer, 2016.

Voolstra, Christian R., and Michael L. Berumen, editors. *Coral Reefs of the Red Sea.* Springer, 2019.

Zakai, David, and Nanette E. Chadwick-Furman. "Impacts of Intensive Recreational Diving on Reef Corals at Eilat, Northern Red Sea." *Biological Conservation,* vol. 105, no. 2, 2002.

Reefs

Fields of Study: oceanography; marine geology; marine biology; marine chemistry

ABSTRACT

Reefs are among the oldest known communities, having existed for at least 2 billion years. They exert considerable control on the surrounding physical environment, influencing turbulence levels and patterns of sedimentation. Ancient reefs are often important hydrocarbon reservoirs.

KEY CONCEPTS

carbon-14 dating: analysis of the ratios of carbon isotopes in a once-living and carbon-containing materials to determine the age of the material according to the mathematics governing radioactive decay processes and the half-life of carbon-14

geomorphic: term for the three-dimensional (3D) physical shape of a geologic feature

petrographic microscope: an optical microscope that uses a plane-polarized light source to illuminate very thin slices of mineral material, which normally enables identification of the mineral

terrigenous: refers to marine particulate material that originated on land, such as dust and sand particles

uranium-series radiometric dating: analysis of the ratios of elemental isotopes that occur in the radioactive decay of uranium atoms in order to determine the age of the material according to the mathematics governing radioactive decay processes and the half-lives of radioactive uranium isotopes

"TRUE" REEFS VERSUS REEFLIKE STRUCTURES

Reefs or reeflike structures are among the oldest known communities, extending more than 2 billion years into Earth's history. These earliest reefs were vastly different in their biotic composition and physical structure from modern reefs, which are among the most diverse of biotic communities and display amazingly high rates of biotic productivity (carbon fixation) and calcium carbonate deposition, despite their existence in a virtual nutrient "desert." Reefs are among the few communities to rival the power of humankind as a shaper of the planet. The Great Barrier Reef of Australia for example, forms a structure some 1,242 miles (2,000 kilometers) in length and up to 93 mi (150 km) in width.

It is necessary to distinguish between "true," or structural, reefs and reeflike structures or banks. Reefs are carbonate structures that possess an internal framework. The framework traps sediment and provides resistance to wave action. Thus, reefs can exist in very shallow water and may grow to the surface of the oceans. Banks are also biogenically produced but lack an internal framework. Thus, banks are often restricted to low-energy, deep-water settings. "Bioherm" refers to moundlike carbonate buildups, either reefs or banks, and "biostrome" to low, lens-shaped buildups.

REEF CLASSIFICATION

Modern reefs are classified into several geomorphic types: atoll, barrier, fringing, and patch. Many of these may be further subdivided into reef crest or flat, back-reef or lagoon, and fore-reef zones. Atoll reefs are circular structures with a central lagoon, thought to form on subsiding volcanic islands. Barrier reefs are elongate structures that parallel coastlines and possess a significant lagoon between the exposed reef crest and shore. These often occur on the edges of shelves that are uplifted by faulting. Fringing reefs are elongate structures paralleling and extending seaward from the coastline that lack a lagoon between shore and exposed reef crest. Patch reefs are typically small, moundlike structures, occurring isolated on shelves or in lagoons. The majority of fossil reefs would be classified as patch reefs, although many examples of extensive, linear, shelf-edge trends are also known from the geologic record.

Reefs form one of the most distinctive and easily recognized sedimentary facies (or environments). In addition to possessing a characteristic fauna consisting of corals, various algae, and stromatoporoids, they are distinguished by a massive (nonlayered) core that has abrupt contacts with adjacent facies. Associated facies include flat-lying lagoon and steeply inclined fore-reef talus, the latter often consisting of large angular blocks derived from the core. The reef core is typically a thick unit relative to adjacent deposits. The core also consists of relatively pure calcium carbonate with little contained terrigenous material.

REEF ENVIRONMENTS

Modern reefs are restricted to certain environments. They occur abundantly only between 23 degrees north and south latitudes and tend to be restricted to the western side of ocean basins which lack upwelling of cold bottom waters. This restriction is based on temperature, as reefs do not flourish where temperatures frequently fall below 64.4 degrees Fahrenheit (18 degrees Celsius). Reef growth is largely restricted to depths less than 197 ft (60 m), as there is insufficient penetration of sunlight below this depth for symbiont-bearing corals to flourish. Reefs also require clear waters lacking suspended terrigenous materials, as these interfere with the feeding activity of many reef organisms and also reduce the penetration of sunlight. Finally, most reef organisms require salinities that are in the normal oceanic range. It appears that many fossil reefs were similarly limited in their environmental requirements.

Some of the most striking features of modern reefs include their pronounced zonation, great diversity, and high productivity and growth rates. Reefs demonstrate a strong bathymetric (depth-related) zonation. This zonation is largely mediated through depth-related changes in turbulence intensity and in the quantity and spectral characteristics (reds are absorbed first, blues last) of available light. Shallow (3- to 16-feet [1- to 5-meter]) fore-reef environments are characterized by strong turbulence and high light intensity and possess low-diversity assemblages of wave-resistant corals, such as the elk-horn coral *Acropora palmata*, and crustose red algae.

With increasing depth (33 to 66 ft [10 to 20 m]), turbulence levels decrease and coral species diversity increases, with the occurrence of mound and delicate branching colonies. At greater depths (98 to 197 ft [30 to 60 m]), corals assume a flattened, platelike form in an attempt to maximize surface area for exposure to ambient light. Sponges and many green algae are also very important over this range. Finally, corals possessing zooxanthellae, which live in the coral tissues and provide food for the coral host, are rare or absent below 197 ft (60 m) because of insufficient light. Surprisingly, green and red calcareous algae extend to much greater depths (328 to 656 ft [100 to 200 m]), despite the very low light intensity (much less than 1 percent of surface irradiance). Sponges are also important members of these deep reef communities.

REEF COMMUNITIES

Coral reefs are among the most diverse communities on Earth; however, there is no consensus on the mechanisms behind the maintenance of this great diversity. At one time, it was believed that reefs existed in a low-disturbance, highly stable environment, which allowed very fine subdivision of food and habitat resources and thus permitted the coexistence of a great number of different species. Upon closer inspection, however, many reef organisms appear to overlap greatly in food and habitat requirements. Also, it has become increasingly apparent that disturbance, in the form of disease extreme temperatures, and hurricanes, is no stranger to reef communities.

Coral reefs exhibit very high rates of productivity (carbon fixation), which is a result of extremely tight recycling of existing nutrients. This is necessary, as coral reefs exist in virtual nutrient "deserts." Modern corals exhibit high skeletal growth rates, up to 4 inches (10 centimeters) per year for some branching species. Such high rates of skeletal production are intimately related to the symbiosis existing between the hermatypic or reef-building scleractinian corals (also gorgonians and many sponges) and unicellular algae or zooxanthellae. Corals that, for some reason, have lost their zooxanthellae or that are kept in dark rooms exhibit greatly reduced rates of skeleton production.

In addition to high individual growth rates for component taxa, the carbonate mass of the reefs may grow at a rate of some 7 ft (2 m) per 1,000 years, a rate that is much higher than that of most other sedimentary deposits. This reflects the high productivity or growth rates of the component organisms and the efficient trapping of derived sediment by the reef frame. Although the framework organisms, most notably corals, are perhaps the most striking components of the reef system, the framework represents only 10 to 20 percent of most fossil reef masses. The remainder of the reef mass consists of sedimentary fill derived from the reef community through a combination of biosynthesis (secretion) and bioerosion (breaking down) of calcium carbonate. An example of the relative contributions of reef organisms to sediment can be found in Jamaica where shallow-water, back-reef sediment consists of 41 percent coral, 24 percent green calcareous algae, 13 percent red calcareous algae, 6 percent foraminifera, 4 percent mollusks, and 12 percent other grains. The most important bioeroders are boring sponges, bivalves, and various "worms," which excavate living spaces within reef rock or skeletons, and parrot fish and sea urchins, which remove calcium carbonate as they feed upon surface films of algae.

TYPES OF REEF COMMUNITIES

A diversity of organisms has produced reef and reeflike structures throughout Earth's history. Several distinct reef community types have been noted, as well as four major "collapses" of reef communities. The oldest reefs or reeflike structures existed more than 2 billion years ago during the Precambrian age. These consisted of low-diversity communities dominated by soft, blue-green algae which trapped sediment to produce layered, often columnar structures known as stromatolites that are similar, if not identical, to those being formed today. During the Early Cambrian period, blue-green algae were joined by calcareous, conical, spongelike organisms known as archaeocyathids, which persisted until the end of the Middle Cambrian. Following the extinction of the archaeocyathids, reefs again consisted only of blue-green algae until the advent of more modern reef communities in the Middle Ordovician period. These reefs consisted of corals (predominantly tabulate and, to a much lesser extent, rugose corals), red calcareous algae, bryozoans (moss animals), and the spongelike stromatoporoids. This community type persisted through the Devonian period, at which time a global collapse of reef communities occurred.

The succeeding Carboniferous period largely lacked reefs, although algal and crinoidal (sea lily) mounds were common. Reefs again occurred in the Permian period, consisting mainly of red and green calcareous algae, stromatolites, bryozoans, and chambered calcareous sponges known as sphinctozoans, which resembled strings of beads. These reefs were very different from those of the earlier Paleozoic era; in particular, the tabulates and stromatoporoids no longer played an important role. The famous El Capitan reef complex of West Texas formed during this interval. The Paleozoic era ended with a sweeping extinction event that involved not only reef inhabitants but also other marine organisms.

After the Paleozoic extinctions, reefs were largely absent during the early part of the Mesozoic era. The advent of modern-type reefs consisting of scleractinian corals and red and green algae occurred in the Late Triassic period. Stromatoporoids once again occurred abundantly on reefs during this interval; however, the role of the previously ubiquitous blue-green algal stromatolites in reefs declined. Late Cretaceous reefs were often dominated by conical, rudistid bivalves that developed the ability to form frameworks and may have possessed symbiotic relationships with algae, as do many modern corals. Rudists, however, became extinct during the sweeping extinctions that occurred at the end of the Cretaceous period. The reefs that were reestablished in the Cenozoic era lacked stromatoporoids and rudists and consisted of scleractinian corals and red and green calcareous algae. This reef type has persisted, with fluctuations, until the present.

STUDY OF MODERN REEFS

Modern reefs are typically studied by divers, which enables observation and sampling to a comfortably accessible depth of approximately 164 ft (50 m). Deeper environments have been made accessible through the availability of manned submersibles and unmanned, remotely operated vehicles that carry mechanical samplers as well as still and video cameras. The biological compositions of reef communities are determined by census (counting) methods commonly employed by plant ecologists. Studies of symbioses, such as that between corals and their zooxanthellae, employ radioactive tracers to determine the transfer of products between symbiont and host. Growth rates are measured by staining the calcareous skeletons of living organisms with a dye, such as Alizarin red, and then later collecting and sectioning the specimen and measuring the amount of skeleton added since the time of staining. Another method for determining growth is to X-ray a thin slice of skeleton and then measure and count the yearly growth bands that are revealed on the radiograph. Variations in growth banding reflect, among other factors, fluctuations in ocean temperature.

Reef sediments, which will potentially be transformed into reef limestones, are examined through sieving, X-ray diffraction and epoxy impregnation and thin-sectioning. Sieving enables the determination of sediment texture, the relationships of grain sizes and abundance (which will reflect environmental energy and the production), and erosion of grains through biotic processes. X-ray diffraction produces a pattern that is determined by the internal crystalline structure of the sediment grains. As each mineral possesses a unique structure, the mineralogical identity of the sediment may be determined. Thin sections of embedded sediment or lithified rock are examined with petrographic microscopes which reveal the characteristic microstructures of the individual grains. Thus, even highly abraded fragments of coral or algae may be identified and their contributions to the reef sediment determined.

STUDY OF FOSSIL REEFS

Because of their typically massive nature, fossil reefs are usually studied by thin-sectioning of lithified rock samples collected either from surface exposures or well cores. Reef limestones that have not undergone extensive alteration may be dated through carbon-14 dating if relatively young, or through uranium-series radiometric dating methods.

Diver studying a coral reef (para827 via iStock)

NATURAL LABORATORIES AND ECONOMIC RESOURCES

Modern reefs serve as natural laboratories, enabling the geoscientist to witness and study phenomena, such as carbonate sediment production, bioerosion, and early cementation, that have been responsible for forming major carbonate rock bodies in the past. The study of cores extracted from centuries-old coral colonies shows promise for deciphering past climates and perhaps predicting future trends. This is made possible by the fact that the coral skeleton records variations in growth that are related to ocean temperature fluctuations. The highly diverse modern reefs also serve as ecological laboratories for testing models on the control of community structure. For example, the relative importance of stability versus disturbance and recruitment versus predation in determining community structure is being studied within the reef setting.

Modern reefs are economically significant resources, particularly for many developing nations in the tropics. Reefs and the associated lagoonal seagrass beds serve as important nurseries and habitats for many fish and invertebrates. The standing crop of fish immediately over reefs is much higher than that of adjacent open shelf areas. Reef organisms may one day provide an important source of pharmaceutical compounds, such as prostaglandins, which may be extracted from gorgonians (octocorals). In addition, research has focused upon the antifouling properties exhibited by certain reef encrusters. Reefs also provide recreational opportunities for snorkelers and for scuba divers, a fact that many developing countries are utilizing to promote their tourist industries. Finally, reefs serve to protect shorelines from wave erosion.

Because of the highly restrictive environmental tolerances of reef organisms, the occurrence of reefs in ancient strata enables fairly confident estimation of paleolatitude, temperature, depth, salinity, and water clarity. In addition, depth- or turbulence-related variation in growth form (mounds in very shallow water, branches at intermediate depths, and plates at greater depths) enables even more precise estimation of paleobathymetry or turbulence levels. Finally, buried ancient reefs are often important reservoir rock structures containing hydrocarbons and thus represent an important economic resource.

—*W. David Liddell*

Further Reading

Birkeland, Charles. *Coral Reefs in the Anthropocene.* Springer Netherlands, 2016.

Dubinsky, Zvy, and Noga Stambler, editors. *Coral Reefs: An Ecosystem in Transition*. Springer, 2011.

Goldberg, Walter M. *The Biology of Reefs and Reef Organisms.* U of Chicago P, 2013.

Harris, Peter T., and Elaine K. Baker, editors. *Seafloor Geomorphology as Benthic Habitat*. Elsevier, 2011.

Hopley, David. *Encyclopedia of Modern Coral Reefs: Structure, Form and Process*. Springer, 2011.

Hovland, Martin. *Deep-Water Coral Reefs: Unique Biodiversity Hot-Spots*. Praxis, 2008.

Phinney, Jonathan T., Ove Hoegh-Guldberg, Joanie Kleypas, William Skirving, and Al Strong, editors. *Coral Reefs and Climate Change: Science and Management.* American Geophysical Union, 2006.

Rohwer, Forest, Merry Youle, and Derek Vosten. *Coral Reefs in the Microbial Seas.* Plaid Press, 2010.

Schroeder, J. H., and B. H. Purser. *Reef Diagenesis.* Springer Science & Business Media, 2012.

Sheppard, Charles R. C., Simon K. Davy, and Graham M. Pilling. *The Biology of Coral Reefs*. Oxford UP, 2009.

Vacher, Leonard, and Terrence M. Quinn. *Geology and Hydrogeology of Carbonate Islands*. Elsevier, 2004.

Wolanski, Eric. *Oceanographic Processes of Coral Reefs: Physical and Biological Links in the Great Barrier Reef.* CRC Press, 2001.

Remote Sensing of the Oceans

Fields of Study: electronics; computer simulation; oceanography; climatology

ABSTRACT

Remote sensory systems are essential pieces of technology in oceanic studies. Because the oceans are whole systems, satellite-based remote sensors are rapidly becoming the tool of choice (although airborne sensors and shipboard systems also are used). Remote sensors are effective in detecting temperature changes, currents, sea levels, and even topography. Such technologies continue to evolve, providing clearer and more detailed images (and greater volumes of data). In light of concerns about global climate change and about environmental protection, wide-ranging remote sensors are integral for oceanic research.

KEY CONCEPTS

active sensors: devices that emit a signal or stimulus to a target in order to determine specific characteristics of that target

coastal plume: a stream of colder water that flows into and beneath warmer water, carrying with it terrogenic materials and nutrients

contact sensors: devices that collect ocean and other data by being in physical contact with the water or another appropriate target

passive sensors: devices that monitor signals originating from a source rather than sending a signal and measuring the response of a target

remote sensors: devices that collect ocean and other data without being in physical contact with the water or another appropriate target

STUDYING THE OCEANS

Oceans cover about 70 percent of Earth, spanning about 129 million square miles (334 square kilometers), with a volume of 329 million cu mi (1,370 million cubic km). About 97 percent of Earth's water is found in oceans, which reach an average depth of about 2.4 mi (3.8 km).

Oceans play a critical role in many natural processes, including generating precipitation and weather patterns, transferring sediment, supporting countless marine ecosystems and delivering essential chemical elements and compounds (such as carbon and methane).

Scientists have long studied the ocean. In the twenty-first century, as evidence of global climate change continues to surface, scientists are paying increased attention to changes in the extensive series of processes and systems under the ocean surface. In this arena, scientists are focusing on key indicators of shifts in oceanic systems and processes. Among these indicators are coastline erosion, temperature changes, changes in ocean currents, and the presence of organic materials (such as ice and algae blooms). Researchers also are carefully monitoring the discharge of human-made compounds, such as oil and waste materials, into the oceans.

In addition to researching isolated geographic areas, scientists are increasingly focusing on the system of oceans as a whole. Such pursuits require the application of satellite-based and airborne research technologies, which can compile data and images of broad oceanic regions.

REMOTE SENSORS AND THE OCEANS

Remote sensors are systems that, from a distance, collect data and imagery of a given object or phenomenon. In the study of oceans, such sensors are usually placed aboard aircraft or satellites. Remote sensors may detect thermal conditions, wave height, current speeds, water vaporization, and debris movements (including the movements of icebergs and sediment).

There exists a wide range of remote sensor technologies, most of which rely on the object in question emitting some form of detectable energy (such as microwaves or infrared radiation). For example, radar emits radio waves at its target, then receives the waves that reflect back from the target. Lidar uses concentrated beams of light (lasers) to assess a target.

Remote sensors usually fall into one of two general categories: passive and active. Passive sensors target a specific area and seek naturally occurring energy emissions, such as the scattering of sunlight. To function, most passive sensors therefore rely on daylight hours. Active sensors emit a form of radiation of their own, targeting a geographic area on the ocean. When the light, microwave, or other form of emission is absorbed or reflected, the sensor reacquires the data.

DATA AND IMAGE TYPES

Remote sensors may be used to glean information about the spread of pollutants and erosion. For example, scientists use such technologies to study a key indicator of ocean pollution and erosion: the coastal plume, which appears along many coasts.

In a coastal plume, denser water pushes lighter, unpolluted water upward and, along with it, plankton and other materials. To detect coastal plumes, scientists frequently use thermal radiometers, which use infrared to capture the heat signatures of the dense substances caught in the plume.

mAnother highly useful tool for scientists studying the oceanic system is a scatterometer. Scatterometers are active radar systems that usually are mounted on a satellite. The device transmits high-frequency microwave pulses at the ocean surface. The pulses that bounce back to the satellite are then measured. The different return waves provide information about the winds at the ocean surface, creating a detailed composite of wave sizes and other aspects of the ocean's surface. Knowledge of these surface winds, combined with data from other instrumentation (such as the aforementioned thermal radiometers, radar, lidar, and other sensor systems), helps scientists understand the speed at which coastal plumes, algae blooms, and oceanic storm fronts move and provide information on surface water temperatures and wave sizes.

Microwave sensors also have a number of other highly useful applications to oceanic studies, as every object on the surface emits some type of microwave energy (although that energy is usually at a low level of radiation). Sea ice is one type of surface object that emits a detectable microwave signature that can be scanned from orbit, even through cloud cover and at any time of day. Passive microwave sensors (which were introduced in the early 1970s and have since evolved considerably) are found on many different satellites and are used to track sea ice, revealing clues about surface temperatures, wind conditions, and currents. Passive microwave sensors cannot assess the temperature of a piece of ice. Still, they can be used to detect the physical and chemical properties of a block of sea ice, including its crystalline structure.

SYNTHETIC APERTURE RADAR

Another useful remote sensor in this arena is synthetic aperture radar (SAR). Like other radar systems, SAR emits radio waves at a target, reading the echo to create an image of the target. However, SAR, which is usually attached to an aircraft or a satellite, continuously sends such waves at the target as it moves overhead. The constant echoes create a more comprehensive, multidimensional image of the target. SARs are useful in the study of coastal plumes, as they generate images of the plumes themselves and track their growth and movements.

SAR systems are useful in gathering comprehensive data about oceanic conditions. Wind velocities and wave height are essential factors for scientists seeking to monitor changes in ocean currents, sedimentation transfer, and evaporation. Many SAR systems can even penetrate the ocean surface and provide detailed images of the topography of the ocean floor. Such detailed information can help scientists generate computer models and more accurately predict changes.

INNOVATIVE REMOTE SENSORS

In addition to using well-established remote sensors, scientists are exploring the use of innovative sensor technologies. When added to a series of other sensors, these sensors provide even greater clarity to the oceanic target at hand. For example, polarimetric passive radiometers can detect electromagnetic waves emanating from the source target. Used in concert with radiometers and other remote sensors, polarimetric passive radiometers can help scientists gain readings on ocean salinity (the volume of salt in ocean water), adding a new dimension to the study of oceanic conditions.

Earth's oceans are dynamic and are changing in a wide range of areas almost constantly. Knowing this, scientists must consider certain environmental conditions when studying oceanic phenomena and characteristics. For example, in the study of circulation (that is, the water's flow patterns), scientists must account for rising and falling water levels. Ideally, such studies are conducted when the ocean is at its lowest tidal level so that elements such as high waves can be avoided (such conditions can disguise or muddle examinations).

SAR antenna of the SAOCOM satellites at the Teófilo Tabanera Space Center (CONAE via Wikimedia Commons)

An innovative tool in the calculation of oceanic elevation is the interferometric radar altimeter. This type of remote sensor observes two sets of waves from the target (such as the energy between high waves). An examination of this wave interference provides data on an ocean's elevation and provides greater resolution to scientific measurements.

Remote sensors have proved effective at analyzing shallow-depth and surface conditions. A vexing pursuit, however, in the study of oceans (particularly in Arctic and Antarctic regions) is deep currents. Remote sensors have had problems penetrating the dense underwater environment, creating only low-resolution images that are largely unusable.

Scientists have applied another innovative remote sensory approach to this problem. In many areas of the Northern and Southern oceans, subsurface density can be used to provide clues about deep currents. In one study, researchers analyzed the relationship between subsurface density and surface elevation (a theoretical concept known as the gravest empirical mode [GEM]). The study's authors argued that these sizable areas of density can be used to calculate salinity and temperatures (two key elements that play a role in ocean currents). Because of these findings, researchers may begin to use satellite-based and aircraft-mounted remote sensors known as altimeters (which measure elevation) to isolate these columns of dense surface and subsurface water.

From the data collected through this GEM-altimeter approach, researchers may find another useful tool in studying the oceans' dynamic systems. Furthermore, scientists are considering new techniques in the study of sea ice through the use of microwave sensors. As stated earlier, passive microwave sensors do not detect the temperature of a piece of ice. However, some researchers have proposed utilizing the weather to assist in this capacity. This approach entails analysis of an ice block's microwave emissions and its brightness (as created by local weather conditions). Using a two-step mathematical algorithm, scientists are generating a catalog of global sea ice that may be used to track temperature changes on the ocean's surface.

IMPLICATIONS AND FUTURE PROSPECTS

In many ways the evolution of remote sensors has helped scientists better understand some of the most challenging aspects of the oceanic system. Researchers have long used contact sensors (such as those mounted on buoys or dragged behind moving ships) to study ocean elevations, wave height, floor topography, and temperature changes. Remote sensors used today add many more dimensions to these studies. Satellite-based radars, altimeters, and radiometers, for example, generate data on these conditions but on a much larger scale and with much greater resolution.

Adding to the benefits of remote sensors in the study of the oceans is that the ever-evolving technologies can be used in locations in which previous systems would acquire unreliable data only. Also, remote sensors can now be used to speed up thorough studies that used to take much longer. The remote and harsh environments of the Arctic and Antarctica (the Northern and Southern ocean regions) provide excellent examples of this fact.

Today, remote sensors can detect thermal changes, the formation of icebergs, density shifts, currents, and other important phenomena from orbit. Such developments in research of the Northern and Southern oceans are critical, as scientists who are concerned with global climate change tend to focus on ocean elevation and debris in and around these geographic areas.

Additionally, the remote sensors found particularly on satellites are easily accessed by participating scientists around the world. There exists a wide range of internationally managed satellites with onboard remote sensors that are trained on Earth's oceans. Among these satellites are GOES, RapidScat, MODIS, and MERIS, which are part of a joint program of the National Aeronautics and Space Administration and the National Oceanic and Atmospheric Administration (NOAA). Government agencies such as NOAA and university-based researchers and graduate students can access such networks with relative ease.

Through such easy access and global communications networks, scientists can share data and develop large-scale international studies. With the continued evolution of SAR, scatterometers, altimeters, and other remote sensors (and with the application of such innovative remote-sensing techniques as GEM), scientists will likely continue to generate voluminous and detailed data and images of Earth's oceanic system.

—Michael P. Auerbach

Further Reading

Alberotanza, Luigi. "Active and Passive Remote Sensors as a New Technology for Coastal and Lagoon Monitoring." *Aquatic Conservation*, vol. 11, no. 4, 2002, pp. 267–72.

Martin, Seelye. *An Introduction to Ocean Remote Sensing*. 2nd ed., Cambridge UP, 2014.

Mityagina, M. I., O. Y. Larova, and S. S. Karimova. "Multi-sensor Survey of Seasonal Variability in Coastal Eddy and Internal Wave Signatures in the North-Eastern Black Sea." *International Journal of Remote Sensing*, vol. 31, nos. 17/18, 2010, pp. 4779–90.

Raizer, Victor. *Advances in Passive Microwave Remote Sensing of Oceans*. CRC Press, 2017.

———. *Optical Remote Sensing of Ocean Hydrodynamics*. CRC Press, 2019.

Rees, W. G. *Physical Principles of Remote Sensing*. 2nd ed., Cambridge UP, 2001.

Robinson, Ian S. *Discovering the Ocean from Space: The Unique Applications of Satellite Oceanography*. Springer, 2010.

Sikora, Todd D., et al. "A Synthetic Aperture Radar-Based Climatology of Open-Cell Convection Over the Northeast Pacific Ocean." *Journal of Applied Meteorology and Climatology*, vol. 50, no. 3, 2011, pp. 594–603.

Tang, DanLing, editor. *Remote Sensing of the Changing Oceans*. Springer, 2011.

"What Is Remote Sensing?" *National Ocean Service*, National Oceanic and Atmospheric Administration, oceanservice.noaa.gov/facts/remotesensing.html. Accessed 7 Feb. 2019.

Yang, Xiaofeng, Xiaofeng Li, and Ferdinando Nunziata. *Ocean Remote Sensing with Synthetic Aperture Radar*. MDPI, 2018.

ROBOTICS

Fields of Study: electronics; mechanical engineering; oceanography

ABSTRACT

Robotics is an interdisciplinary scientific field concerned with the design, development, operation, and assessment of electromechanical devices used to perform tasks that would otherwise require human action. Robotics applications can be found in almost every arena of modern life. Robots, for example, are essential for undersea exploration and recovery of materials. They are widely used in industrial assembly lines to perform repetitive tasks. They have also been developed to help physicians perform difficult surgeries and are essential to the operation of many advanced military vehicles. Among the most promising robot technologies are those that draw on biological models to solve problems, such as robots whose limbs and joints are designed to mimic those of insects and other animals.

KEY CONCEPTS

actuator: a device that when activated causes a specific movement to occur

machine learning: computer programs for robotic control that contain algorithms designed to register situations encountered by the robot that prevent or interfere with carrying out an instruction; learning "by experience"

proprioception: in humans, the innate and automatic location and position of all parts of the body; in robots, the same function carried out through active self-monitoring

sensor: an electrical, magnetic, mechanical, or combination device that detects an environmental condition and provides that as input to the robotic control mechanism

teleoperation: operation of a device from a distance, by remote control

DEFINITION AND BASIC PRINCIPLES

Robotics is the science of robots—machines that can be programmed to carry out a variety of tasks independently, without direct human intervention, or semi-independently, with human input and guidance. Although robots in science fiction tend to be androids or humanoids (robots with recognizable human form), most real-life robots, especially those designed for industrial use, do not resemble humans. Robots typically consist of at least four parts: a control module in which its instruction program is stored and carried out; a mechanical structure (most commonly a robotic arm) that enables the robot to physically affect either itself or its task environment; sensors that gather information about physical properties such as sound, temperature, motion, and pressure; and some kind of processing system that transforms data from the robot's sensors into instructions about the actions it is to perform. Some devices, such as the search-engine bots that mine the internet daily for data about links and online content, lack mechanical components. However, they are nevertheless often considered robotic because they can perform repeated tasks without supervision.

Many robotics applications also involve the use of artificial intelligence, or AI. This is a complex concept with a shifting definition, but in its most basic sense, a robot with AI possesses features or capabilities that mimic human thought or behavior. For example, one aspect of AI involves creating parallels to the human senses of vision, hearing, or touch. The friendly voices at the other ends of customer service lines, for example, are increasingly likely to be robotic speech-recognition devices capable not merely of hearing callers' words but also of interpreting their meanings and directing the customers' calls intelligently.

More advanced AI applications give robots the ability to assess their environmental conditions, make decisions, and independently develop efficient plans of action for their situations—and then modify these plans as circumstances change. Chess-playing robots do this each time they assess the state of the chessboard and make a new move. The ultimate goal of AI research is to create machines whose responses to questions or problems are so humanlike as to be indistinguishable from those of human operators. This standard is the so-called Turing test, named after the British mathematician and computing pioneer Alan Turing.

BACKGROUND AND HISTORY

The word "robot" comes from a Czech word for "forced labor" that the Czech writer Karel Èapek (1890–1938) used in his 1921 play *R.U.R.* about a man who invents a humanlike automatic machine to do his work. During the 1940s, as computing power began to grow, the influential science-fiction writer Isaac Asimov (1920–1992) began applying the term "robotics" to the technology behind robots. The 1950s saw the development of the first machines that could properly be called robots. These prototypes took advantage of such new technologies as transistors (compact, solid-state devices that control electrical flow in electronic equipment) and integrated circuits (complex systems of electronic connections etched onto single chips) to enable more complicated mechanical actions. In 1959, an industrial robot was designed that could churn out ashtrays automatically. Over the ensuing decades, public fascination with robots expanded far beyond their actual capabilities. It was becoming clear that creating robots that could accomplish seemingly simple tasks—such as avoiding obstacles while walking—was a surprisingly complex problem.

During the late twentieth century, advances in computing, electronics, and mechanical engineering led to rapid progress in the science of robotics. These included the invention of microprocessors, single integrated circuits that perform all the functions of computers' central processing units; production of better sensors and actuators; and developments in AI and machine learning, such as a more widespread use of neural networks. (Machine learning is the study of computer programs that improves their performance through experience.)

Cutting-edge robotics applications are being developed by an interdisciplinary research cohort of computer scientists, electrical engineers, neuroscientists, psychologists, and others, and combine a greater mechanical complexity with more subtle information processing systems than were once possible. Homes may not be populated with humanoid robots with whom one can hold conversations, although Google's Alexa and similar devices are somewhat capable of this, but mechanical robots have become ubiquitous in industry. Also, unmanned robotic vehicles and planes are essential in warfare, search-engine robots crawl the World Wide Web every day collecting and analyzing data about internet links and content, and robotic surgical tools are indispensable in healthcare. All this is evidence of the extraordinarily broad range of problems robotics addresses.

In marine science, research at hazardous depths, in complex undersea structures or in dangerous environmental situations would be impossible without robotic devices. Every aspect of oceanography has some reliance on submersibles that either operate independently or by remote control. These may be guided through complex structures such as shipwrecks and coral reefs, or to extremely dangerous locations such as the "black smokers" of the mid-ocean ridges, undersea fault lines and undersea volcanic eruptions. Robotic manipulators on the submersibles, typically controlled by a human either inside the submersible or from a remote location, enable the manipulation and recovery of material and biological samples for study.

HOW IT WORKS

Sensing. To move within and react to the conditions of task environments, robots must gather as much information as possible about the physical features of their environments. They do so through a large array of sensors designed to monitor different physical properties. Simple touch sensors consist of electric circuits that are completed when levers receive enough pressure to press down on switches. Robotic dogs designed as toys, for example, may have touch sensors in their backs or heads to detect when they are being petted and signal them to respond accordingly. More complex tactile sensors can detect properties such as torque (rotation), texture or proximity. Such sensors may be used, for example, to help an assembly-line robot's end effector control the grip and force it uses to turn an object it is screwing into place, or to detect if a component has been positioned correctly.

Light sensors consist of one or more photocells that react to visible light with decreases in electrical resistance. They may serve as primitive eyes, allowing unmanned robotic vehicles, for example, to detect the bright white lines that demarcate parking spaces and maneuver between them. Reflectance sensors emit beams of infrared light, measuring the amounts of that light that reflect back from nearby surfaces. They can detect the presence of objects in front of robots and calculate the distances between the robots and the objects—allowing the robots either to follow or to avoid the objects. Temperature sensors rely on internal thermistors (resistors that react to high temperatures with decreases in electrical resistance). Robots used to rescue human beings trapped in fires may use temperature sensors to navigate away from areas of extreme heat. Similarly, altimeter sensors can detect changes in elevation, allowing robots to determine whether they are moving up or down slopes.

Other sensor types include magnetic sensors, sound sensors, accelerometers, and proprioceptive sensors that monitor the robots' internal systems and tell them where their own parts are located in space. After robots have collected information through their sensors, algorithms (mathematical processes based on predefined sets of rules) help them process that information intelligently and act on it. For example, a robot may use algorithms to help it determine its location, map its surroundings, and plan its next movements.

Motion and Manipulation. Robots can be made to move around spaces and manipulate objects in many different ways. At the most basic level, a moving robot needs to have one or more mechanisms consisting of connected moving parts, known as links. Links can be connected by prismatic or sliding joints, in which one part slides along the other, or by rotary or articulated joints, in which both parts rotate around the same fixed axis. Combinations of prismatic and rotary joints enable robotic manipulators to perform a host of complex actions, including lifting, turning, sliding, squeezing, pushing, and grasping. Actuators are required to move jointed segments or robot wheels. Actuators may be electric or electromagnetic motors, hydraulic gears or pumps (powered by compressed liquid), or pneumatic gears or pumps powered by pressurized gas. To coordinate the robots' movements, the actuators are controlled by electric circuits.

Motion-description languages are a type of computer programming language designed to formalize robot motions. They consist of sets of symbols that can be combined and manipulated in different ways to identify whole series of predefined motions in which robots of specified types can engage. Motion-description languages were developed to simplify the process of manipulating robot movements by allowing different engineers to reuse common sets of symbols to describe actions or groups of actions, rather than having to formulate new algorithms to describe every individual task they want robots to perform.

Control and Operation. A continuum of robotic control systems ranges from fully manual operation to fully autonomous operation. On the one hand, a human operator may be required to direct every movement a robot makes. For example, some bomb disposal robots are controlled by human operators working only a few feet away, using levers and buttons to guide the robots as they pick up and remove the bombs. On the other side of the spectrum are robots that operate with no human intervention at all, such as the KANTARO—a fully autonomous robot that slinks through sewer pipes inspecting them for damage and obstructions. Many robots have control mechanisms lying somewhere between these two extremes.

Robots can also be controlled from a distance. Teleoperated systems can be controlled by human operators situated either a few centimeters away, as in robotic surgeries, or millions of miles away, as in outer space applications. "Supervisory control" is a term given to teleoperation in which the robots themselves are capable of performing the vast majority of their tasks independently; human operators are present merely to monitor the robots' behavior and occasionally offer high-level instructions.

Artificial Intelligence. Three commonly accepted paradigms, or patterns, are used in AI robotics: hierarchical, reactive, and hybrid. The hierarchical paradigm, also known as a top-down approach, organizes robotic tasks in sequence. For example, a robot takes stock of its task environment, creates a detailed model of the world, uses that model to plan a list of tasks it must carry out to achieve a goal, and proceeds to act on each task in turn. The performance of hierarchical robots tends to be slow and disjointed since every time a change occurs in the environment, the robot pauses to reformulate its plan. For example, if such a robot is moving forward to reach a destination and an obstacle is placed in its way, it must pause, rebuild its model of the world, and begin lurching around the object. Familiar examples of this type of robot are the automatic floor sweepers called iRobot, Roomba, and others like them.

In the reactive (or behavioral) paradigm, also known as a bottom-up approach, no planning occurs. Instead, robotic tasks are carried out spontaneously in reaction to a changing environment. If an obstacle is placed in front of such a robot, sensors can quickly incorporate information about the obstacle into the robot's actions and alter its path, causing it to swerve momentarily.

The hybrid paradigm is the one most commonly used in AI applications being developed during the twenty-first century. It combines elements of both the reactive and the hierarchical models.

APPLICATIONS AND PRODUCTS

Industrial Robots. In the twenty-first century, almost no factory operates without at least one robot—more likely several—playing some part in its manufacturing processes. Welding robots, for example, consist of mechanical arms with several degrees of movement and end effectors in the shape of welding guns or grippers. They are used to join metal surfaces together, and produce faster, more reliable, and more uniform results than human welders. They are also much less vulnerable to injury than human workers. Another common industrial application of robotics is silicon-wafer manufacturing, which must be performed within meticulously clean rooms so as not to contaminate the semiconductors with dirt or oil. Humans are far more prone than robots to carry contaminants on them. The silicon alloy from which the "wafers" will be cut for etching is slowly drawn out as a single, large crystal from its molten source while also being rotated slowly, a function that can only be carried out consistently by a robot.

Six major types of industrial robots are defined by their different mechanical designs. Articulated robots are those whose manipulators (arms) have at least three rotary joints. They are often used for vehicle assembly, die casting (pouring molten metal into molds), welding, and spray painting. Cartesian robots, also known as gantry robots, have manipulators with three prismatic joints. They are often used for picking objects up and placing them in different locations, or for manipulating machine tools. Cylindrical robots have manipulators that rotate in a cylindrical shape around a central vertical axis. Parallel robots have both prismatic and rotary joints on their manipulators. Spherical robots have manipulators that can move in three-dimensional (3D) spaces shaped like spheres. SCARA (Selective Compliant Assembly Robot Arm) robots have two arms connected to vertical axes with rotary joints. One of their arms has another joint that serves as a wrist. SCARA robots are frequently used for palletizing (stacking goods on platforms for transportation or loading).

Service Robots. Unlike industrial robots, service robots are designed to cater to the needs of individual people. Robopets, such as animatronic dogs, provide companionship and entertainment for their human owners. The Sony Corporation's AIBO (Artificial Intelligence roBOt) robopets use complex systems of sensors to detect human touch on their heads, backs, chins, and paws, and can recognize the faces and voices of their owners. They can also maintain their balance while walking and running in response to human commands. AIBOs also function as home-security devices, as they can be set to sound alarms when their motion or sound detectors are triggered. Consumer appliances, such as iRobot Corporation's robotic vacuum cleaner, the Roomba, and the robotic lawn mower, the RoboMower, developed by Friendly Robotics, use AI approaches to safely and effectively maneuver around their task environments while performing repetitive tasks to save their human users time.

Factory automation with industrial robots for palletizing food products like bread and toast at a bakery in Germany (KUKA Roboter GmbH, Bachmann via Wikimedia Commons)

Even appliances that do not much resemble public notions of what robots should look like often contain robotic components. For example, digital video recorders (DVRs) such as TiVos, contain sensors, microprocessors, and a basic form of AI that enable them to seek out and record programs that conform to their owners' personal tastes. Increasingly, cars and trucks can assist their owners with driving tasks such as parallel parking. Other cars have robotic seats that can lift elderly or disabled passengers inside. Many companies now actively explore technologies for autonomous vehicles (self-driving vehicles), with the goal of making mass-produced models available to the public that would function flawlessly on existing roads. Several prototypes have been introduced, although much controversy surrounds the potential hazards and liabilities of such technology. Other companies that deliver goods, such as Amazon and some food delivery companies, have also looked to the use of drones to deliver goods, and foresee the combination of autonomous vehicles and delivery of goods on demand.

Many companies or organizations rely on humanoid robots to provide services to the public. The Smithsonian National Museum of American History, for example, has used an interactive robot named Minerva to guide visitors around the museum's exhibits, answering questions and providing information about individual exhibits. Other professional roles filled by robots include those of receptionists, floor cleaners, librarians, bartenders, and secretaries. At least one primary school in Japan even experimented with a robotic teacher developed by a scientist at the Tokyo University of Science. However, an important pitfall of humanoid robots is their susceptibility to the uncanny valley phenomenon. This is the theory that as a robot's appearance and behavior becomes more humanlike, people will respond to it more positively—but only up to a point. On a line graph plotting positive response against degree of human likeness, the response dips (the "uncanny valley") as the likeness approaches total realism but does not perfectly mimic it. In other words, while people will prefer a somewhat anthropomorphic robot to an industrial-looking one, a highly humanlike robot that is still identifiably a machine will cause people to feel revulsion and fear rather than empathy.

Medical Uses. Robotic surgery has become an increasingly important area of medical technology. In most robotic surgeries, a system known as a master-slave manipulator is used to control robot movements. Surgeons look down into electronic displays showing their patients' bodies and the robots' tool tips. The surgeons use controls attached to consoles to precisely guide the robots' manipulators within the patients' bodies. A major benefit of robotic surgeries is that they are less invasive—smaller incisions need to be made because robotic manipulators can be extremely narrow. These surgeries are also safer because robotic end effectors can compensate for small tremors or shakes in the surgeons' movements that could seriously damage their patients' tissues if the surgeons were making the incisions themselves. Teleoperated surgical robots can even allow surgeons to perform operations remotely, without the need to transport patients over long distances. Surgical robots such as the da Vinci system are used to conduct operations such as prostatectomy, cardiac surgery, bariatric surgery, and various forms of neurosurgery.

Humanoid robots are also widely used as artificial patients to help train medical students in diagnosis and procedures. These robots have changing vital signs such as heart rates, blood pressure, and pupil dilation. Many are designed to breathe realistically, express pain, urinate, and even speak about their conditions. With their help, physicians-in-training can practice drawing blood, performing cardiopulmonary resuscitation (CPR), and delivering babies without the risk of harm to real patients.

Other medical robots include robotic nurses that can monitor patients' vital signs and alert physicians to crises and smart wheelchairs that can automatically maneuver around obstacles. Scientists are also working on developing nanorobots the size of bacteria that can be swallowed and sent to perform various tasks within human bodies, such as removing plaque from the insides of clogged arteries.

Robot Exploration and Rescue. One of the most intuitive applications of robotic technology is the concept of sending robots to places too remote or too dangerous for human beings to work in—such as outer space, great ocean depths, and disaster zones. The six successful manned Moon landings of the Apollo program carried out during the late 1960s and early 1970s are dwarfed in number by the unmanned robot missions that have set foot not only on the Moon but also on other celestial bodies, such as planets in the solar system. The wheeled robots Spirit and Opportunity, for example, began analyzing material samples on Mars and sending photographs back to Earth in 2004. Roboticists have also designed biomimetic robots inspired by frogs that take advantage of lower gravitational fields, such as those found on smaller planets, to hop nimbly over rocks and other obstacles.

Robots are also used to explore the ocean floor. The Benthic Rover, for example, drags itself along the seabed at depths up to 2.5 miles (4 kilometers) below the surface. It measures oxygen and food levels, takes soil samples, and sends live streaming video up to the scientists above. The rover is operated by supervisory control and requires very little intervention on the part of its human operators.

Rescue robots seek out, pick up, and safely carry injured humans trapped in fires, under rubble, or in dangerous battle zones. For example, the US Army's Bear (Battlefield Extraction-Assist Robot) is a bipedal robot that can climb stairs, wedge itself through narrow spaces, and clamber over bumpy terrain while carrying weights of up to three hundred pounds.

Military Robots. Militaries have often been among the leading organizations pioneering robotics, as robots have the potential to complete many military tasks that might otherwise prove dangerous to humans, including armed conflict. While many projects, such as bomb-removal robots, have proven highly useful and have been widely accepted, other military applications have proven more controversial. For example, many observers and activists have expressed concern over the proliferation of drones, particularly unmanned aerial vehicles (UAVs) capable of enacting military strikes such as rocket or missile launches while going virtually undetected by radar. The US government has used such technology (including the Predator drone) to successfully destroy terrorist positions, including in remote territory that would otherwise be difficult and dangerous to access, but there have also been notable examples of misidentified targets and civilian collateral damage caused by drone strikes. Opponents of drones argue that the potential for mistakes or abuse of their capabilities is a dangerous possibility that would be all too easily exploited.

CAREERS AND COURSEWORK

Courses in advanced mathematics, physics, computer science, electrical engineering, and mechanical engineering make up the foundational requirements for students interested in pursuing careers as robotics engineers. Earning a bachelor of science degree in any of these fields would serve as an appropriate preparation for graduate work in a similar area. In most circumstances, either a master's degree or a doctorate is a necessary qualification for the most advanced future career opportunities in both academia and industry. However, an advanced degree is not generally required for a career as a robotics technician—someone who maintains and repairs robots rather than designs them.

Students should take substantial coursework in more than one of the primary fields of study related to robotics (physics, mathematics, computer science, and engineering) because designing and testing robots requires skills drawn from multiple disciplines. In addition, anyone desiring to work in robotics should possess skills that go beyond the academic, including good physical coordination, excellent manual dexterity, and an aptitude for mechanical details. A collaborative mind-set is also an asset, as robotics work tends to be done by teams.

Careers in the field of robotics can take several different shapes. The manufacturing industry is the biggest employer of robotics engineers and technicians. Within this industry, robotics professionals might focus on developing, maintaining, or repairing production-line robots used in factory assembly. Other industries in which robotics engineers and technicians often find work include aviation, agriculture, nuclear energy, telecommunications, electronics, mining, health care and medicine, and education.

Many roboticists prefer employment within academic settings. Such professionals divide their time between teaching university classes on robotics and conducting their own research projects. Others find work in government agencies such as the National Aeronautics and Space Administration (NASA) and Defense Advanced Research Project Agency (DARPA), focusing on large-scale robotics applications in such areas as space exploration, warfare, and disaster management.

SOCIAL CONTEXT AND FUTURE PROSPECTS

In the twenty-first century, the presence of robots in factories all over the world is taken for granted. Meanwhile, robots are also increasingly entering daily life in the form of automated self-service kiosks at supermarkets, electronic lifeguards that detect when swimmers in pools are struggling, and cars whose robotic speech-recognition software enables them to respond to verbal commands. A science-fiction future in which ubiquitous robotic assistants perform domestic tasks such as cooking and cleaning may not be far away, but many technological limitations must be overcome for that to become a reality. For example, it can be difficult for robots to process multipart spoken commands that have not been preprogrammed—a problem for researchers in the AI field of natural language processing. However, the voice-activated personal assistant software in smartphone and computer operating systems, such as Apple's Siri, shows how such technology is rapidly evolving.

Robots that provide nursing care or companionship to the infirm are not merely becoming important parts of the healthcare industry but may also provide a solution to the problem increasingly faced by countries in the developed world—a growing aging population who need more caretakers than can be found among younger adults. Another area where robots can be particularly useful is in performing dangerous tasks that would otherwise put a human's life at risk. The use of robotics in the military remains a growing field—not only in the controversial use of combat drones, but also for tasks such as minesweeping. Robots are also more and more heavily used in space exploration. In the future, robots may be used in other dangerous fields as well; during the 2014 ebola outbreak, roboticists discussed the possibility of using robots in future medical emergencies of the type for purposes such as disinfecting contaminated areas.

There are also concerns about the growing use of robots to perform tasks previously performed by humans, however. As robotics technology improves and becomes less expensive, companies may well turn to cheap, efficient robots to do jobs that are typically performed by immigrant human labor, particularly in such areas as agriculture and manufacturing. Meanwhile, some observers are concerned that the rise of industrial and professional service robots is already eliminating too many jobs held by humans. Many of the jobs lost in the United States during the 2008–2009 recession within the struggling automotive industry, for example, are gone for good, because costly human workers were replaced by cheaper robotic arms. However, the issue is more complicated than that. In certain situations, the addition of robots to a factory's workforce can actually create more jobs for humans. Some companies, for example, have been able to increase production and hire additional workers with the help of robot palletizers that make stacking and loading their products much faster.

Safety concerns can sometimes hinder the acceptance of new robotic technologies, even when they have proven to be less likely than humans to make dangerous mistakes. Robotic sheep shearers in Australia, for example, have met with great resistance from farmers because of the small risk that the machines may nick a major artery as they work, causing the accidental death of a sheep. And while fully autonomous vehicles, including self-driving cars, are seen by many automotive and technology companies alike as a major area of innovation, crashes and other accidents by several prototypes in the 2010s drew significant concerns from regulators and the general public. It is critical in the field of robotics to not only develop the technology necessary to make a design a reality, but to understand the cultural and economic landscape that will determine whether a robot is a success or failure.

—*M. Lee*

Further Reading

Antonelli, Gianluca. *Underwater Robots*. 3rd ed., Springer, 2014.

Chin, Cheng Siong, and Rongxin Cui, editors. *Intelligent Marine Robotics: Modelling, Simulation and Applications*. MDPI, 2020.

Dinwiddie, Keith. *Basic Robotics*. Cengage Learning, 2016.

Faust, Russell A., editor. *Robotics in Surgery: History, Current and Future Applications*. Nova Science, 2007.

Floreano, Dario, and Claudio Mattuissi. *Bio-Inspired Artificial Intelligence: Theories, Methods, and Technologies*. MIT Press, 2008.

Gutkind, Lee. *Almost Human: Making Robots Think*. 2006. Reprint. Norton, 2009.

Jaulin, Luc et al., editors. *Marine Robotics and Applications*. Springer, 2018.

Jones, Joseph L. *Robot Programming: A Practical Guide to Behavior-Based Robotics*. McGraw-Hill, 2004.

Moore, Steven W., Harry Bohm, and Vickie Jensen. *Underwater Robotics: Science, Design and Fabrication*. Marine Advanced Technology Education Center, 2010

"Our Friends Electric." *Economist*, 7 Sept. 2013.

Popovic, Marko B. *Biomechanics and Robotics*. CRC Press, 2013.

Samani, Hooman. *Cognitive Robotics*. CRC Press, 2016.

Siciliano, Bruno, et al. *Robotics: Modeling, Planning, and Control*. Springer, 2009.

Siciliano, Bruno, and Oussama Khatib, editors. *The Springer Handbook of Robotics*. Springer, 2016.

Springer, Paul J. *Military Robots and Drones: A Reference Handbook*. ABC-CLIO, 2013.

S

Salinity and Desalination

Fields of Study: oceanography; limnology; marine engineering; marine chemistry

ABSTRACT

Salinity is the total amount of salts dissolved per unit of water; the average salinity of seawater, for example, is 35 grams of salts per kilogram of water. Utilization of saline water for domestic purposes thus requires removal of the dissolved materials. The desalination of saline or brackish water involves costly operations and is typically performed only in coastal or arid regions where alternative water resources are limited or nonexistent.

KEY CONCEPTS

absolute salinity: concentration of salts in seawater as determined by complete evaporation of a specific volume of seawater

chlorinity: calculation of salinity by determination of dissolved halide ions (fluoride, chloride, bromide, iodide)

halocline: the differentiation point between a layer of less dense freshwater and an underlying layer of saline water having higher density

practical salinity: an indirect measurement of salinity by the electrical conductivity of seawater

saline wedge: a layer of salt water that intrudes into a freshwater outlet, forming a saline layer beneath the freshwater

tonne: also called a "metric ton"; 1,000 kilograms, the equivalent of 2,200 pounds

DISSOLVED SOLIDS

Salinity is one of the most important physical properties of seawater. It is defined as the total amount of dissolved salts found in a unit of water, measured in grams of dissolved matter per kilogram of seawater. The average salinity of seawater is approximately 35 grams per kilogram, corresponding to a concentration of 35 parts of salt per thousand parts of seawater (3.5 percent, or 35 parts per thousand). That is to say, evaporation of all water from 2.2 lbs (1 kg) of seawater will, on average, leave 1.23 oz (35 g) of dry solids, primarily sodium chloride and other metal salts.

The main ionic components of dissolved matter in seawater and the percentages in which they are found in a typical sample are as follows: chloride ions (55.0 percent), sodium ions (30.6 percent), sulfate ions (7.7 percent), magnesium ions (3.7 percent), calcium ions (1.2 percent), and potassium ions (1.1 percent). Accordingly, 0.24 cubic mile (1 cubic kilometer) of typical seawater contains approximately 83 million tons (75 million tonnes) of sodium chloride, 12 million tons (11 million tonnes) of magnesium chloride, 5.5 tons (5 million tonnes) of magnesium sulfate, 4.4 tons (4 million tonnes) of calcium sulfate, 2.75 tons (2.5 million tonnes) of potassium sulfate, and 379 tons (344,000 tonnes) of calcium carbonate.

Depending on the prevailing physicochemical conditions, salinity can vary significantly, from brackish water (0.2 to 1.5 percent, or 2 to 15 ppt) in estuaries and riverine deltas to highly saline water such as in the Mediterranean Sea (3.85 percent, or 38.5 ppt) or Red Sea (4.15 percent, or 42.5 ppt). Even higher levels of salinity prevail in landlocked bodies of water such as the Dead Sea. Atmospheric precipitation or freshwater discharges from inland waters (surface water or groundwater) reduce salinity, while evaporation acts to increase salt content. In spite of the fact that salinity levels vary from one water body to another, the relative abundance of the main dissolved components remains almost unchanged.

The presence of dissolved salts increases the density of seawater. A salinity of 3.5 percent (35 ppt) amounts to a density difference between fresh and salt waters of approximately 2 percent. Thus, whenever there is a confluence of fresh and salt waters, some interesting phenomena occur. Depending on the degree of energy input by wind currents, and tidal action, the fresh and salt waters can be thoroughly or partially mixed, or they may remain stratified. Under quiescent conditions, if all other variables (such as temperature and pressure) are the same for both the fresh and salt waters, then the heavier saltwater will sink. In many estuaries with high riverine discharge and low tidal range (such as the Mississippi River), seawater can penetrate many miles (or kilometers) in the upstream direction, moving along the bottom under the freshwater. This phenomenon is known as a "saline wedge," and the separating interface between fresh and salt waters is called the "halocline." Mixing conditions between fresh and salt waters are important from an ecological point of view because salinity levels affect the diversity and population of aquatic flora and fauna.

Sea salt is also a major contributor to atmospheric aerosol particles. During wave breaking, characterized by the formation of "white caps," small droplets are carried upward by air currents into the atmosphere. When the droplets evaporate, sea-salt particles of very small sizes (0.5 to 20 micrometers) are transported by the wind over the continents. About 10 percent of the total annual amount of salt that is generated from the oceans (about 1.8 billion tons [1.64 billion tonnes]) is deposited as airborne sea-salt particles on the continents.

The total volume of oceanic waters is about 0.33 billion cu mi (1.37 billion cu km). This volume constitutes approximately 96.5 percent of the total water on Earth. The other major water resources are groundwater, of which 2.52 million cu mi (10.53 million cu km) (0.76 percent) is freshwater and 3.09 million cu mi (12.87 million cu km) (0.93 percent) is saltwater; and the polar ice, which contains 5.8 million cu mi (24.02 million cu km (1.7 percent) of water. The remaining amount of water is in lakes, rivers, marshes, soil moisture, atmosphere, and biota. Freshwater thus

Sunset in Badwater, Death Valley National Park (Oliclimb via Wikimedia Commons)

makes up only 2.5 percent of the total amount of Earth's water. Most freshwater supplies are in the polar ice (68.6 percent) and in groundwater (30.1 percent). The percentage of freshwater found in lakes is only 0.26 percent, while the amount found in rivers is even smaller (0.006 percent).

INLAND WATER AND SALINITY MEASUREMENT

Not all inland water is fresh. The amount of the dissolved salts occurring in inland waters, surface or underground, depends on the composition of the soils through which the water passes. Streams and rivers flowing over rocks containing chloride and sodium compounds contribute significantly to the generation of salt. Therefore, some inland waters have a high salt content and are not suitable for any use unless properly pretreated.

The most extreme examples of inland brine waters are the Great Salt Lake in Utah (salt concentration of about 12 percent, or 120 ppt) and the Dead Sea in the Middle East (salt concentration of about 27 precent, or 270 ppt). Salt is also found as surface crust or layers in swamps and dry lake bottoms, particularly in arid climate areas. The famous Bonneville Salt Flats, in Utah, and Death Valley, in California both reflect the accumulation of various salts over time. The water that periodically inundates these areas transports relatively small amounts of dissolved salts from the surrounding area and forms a landlocked body of water as the runoff collects and pools. When the water evaporates, those salts remain, and have been slowly built up over time to produce the massive deposits of salts for which those regions are known.

In coastal regions, excessive pumping of groundwater can lead to seawater intrusion into an aquifer. This can have a long-term negative effect on the water resources of a region, with subsequent devastating impact on the regional economy. The average amount of freshwater that falls daily on the United States is about 4,160 billion gallons (15,750 billion liters). This water feeds the various surface water bodies (lakes and rivers), recharges aquifers, evaporates back to the atmosphere, or flows into the oceans. Therefore, only 0.66 trillion gal (2.5 trillion L) of precipitated water can be used for beneficial purposes. From this amount, 0.33 trillion gal (1.26 trillion L) is used by industry, 140 billion gals (530 billion L) by agriculture, and 25 billion gal (94 billion L) by domestic and rural consumers. The daily per-capita water use in the United States is about 1,849 gal (7,000 L). The average domestic water consumption ranges from 20 to 79 gal (75 to 300 L) per day. Thus, the average amount of water used by a person during a seventy-year life span is estimated at 1.48 million gal (5.6 million L).

Direct measurement of seawater salinity by evaporation or chemical analysis is too complicated to be used on a routine basis. In the past, salinity (known as "absolute salinity") was estimated in terms of chlorinity. Chlorinity is the amount of chlorine ions plus the chlorine equivalent of bromine and iodine ions. From chlorinity, salinity was estimated by multiplying the value of chlorinity by a factor of 1.80655.

Salinity (known as "practical salinity") is estimated indirectly by measuring the electrical conductivity of the seawater. However, because electrical conductivity is strongly affected by both salinity and temperature, the conductivity readings are properly corrected to compensate for temperature effects.

WATER TREATMENT

Freshwater is a precious commodity that is used for a variety of domestic, rural, industrial, and agricultural purposes. Water is not distributed evenly throughout the world and is subject to short-term and long-term temporal variations. In many regions, the available freshwater resources cannot meet the water demands. After the Industrial Revolution accelerated anthropogenic pollution added to the water resource problem. In situations of limited water resources, the only alternative solutions are either cleanup and reuse of domestic and agricultural wastewater or, for islands and coastal regions, desalination of seawater. Both of these operations are costly and require construction of appropriate water-treatment facilities. Several countries in the Middle East, including Israel and Saudi Arabia depend heavily on water desalination for their drinking-water supplies. Under emergency conditions, freshwater is obtained via the use of portable water-purification equipment.

Freshwater can contain a number of impurities that have to be removed or treated before it can be used. These impurities include calcium, magnesium, iron, lead, copper, chloride, sulfate, nitrates, fluorides, sodium, different organic compounds, and suspended solids. Impurities can be hazardous to human health or can give water a disagreeable taste, smell or appearance. In addition, they can create scaling or corrosion problems in pipes or in machinery that use the water. Because saltwater contains a large amount of dissolved salts, it is always subjected to desalination before use. Generally, the acceptable quality standards for drinking water are 0.5 microgram per kilogram of total dissolved solids and 0.2 microgram per kilogram of chloride.

Water treatment for production of domestic water involves a number of operations such as filtration, softening, distillation deionization, chemical disinfection, exposure to ultraviolet radiation and reverse osmosis. The number and the type of operations required depend on the quality and properties of the water supplies. The problem of high salt content can be treated by a variety of methods. Desalination is the process by which dissolved salts are removed from the water; there are a number of desalination methods. Because no one method is applicable in all situations, selection of the most appropriate desalination method is based on such variables as the amount and type of dissolved salts in water, the degree of purification of the water to be produced, and the associated costs. More than one hundred cities worldwide use desalination plants to provide freshwater for their needs.

THERMAL PROCESSING

There are two general methods of desalination: thermal processing (distillation) and membrane separation. The main principle behind thermal processing is as follows: Saltwater is heated until it boils, then the released steam condenses as it cools, forming essentially pure water. Membrane separation is achieved by using reverse osmotic pressure, so that the water passes through a membrane while the salt ions are retained by the membrane.

Distillation, the earliest desalination method, was used in steamships as early as 1884. There are several different thermal processing methods for desalination, such as the thin-film multiple effect distillation (TFMED), multistage flash desalination (MSFD), mechanical vapor compression desalination (MVCD), and thermal vapor compression desalination (TVCD). At standard atmospheric pressure (measured at sea level), water boils at 212 degrees Fahrenheit (100 degrees Celsius). However, boiling temperature decreases with decreasing pressure. Also, if the water is heated under high pressure at 212 degrees Fahrenheit (100 degrees Celsius) and is suddenly released into a vacuum chamber, it flashes into vapor. This technique is used in both the TFMED method and the MSFD method, in which the pressure is continuously reduced in sequential stages. The number of sequential stages can range from fifteen to twenty-five. The MVCD and the TVCD methods use compression to increase the pressure and thus increase the temperature of a constant volume of steam. Heating of the steam as it flows through the process facilitates the desalination process. Water obtained through thermal processing can easily have a purity of less than 0.1 microgram per kilogram of dissolved matter. The brine wastewater resulting from desalination of seawater has a salinity of approximately 70 grams per kilogram.

MEMBRANE SEPARATION

Membrane separation includes two major desalination methods: pressure membrane processes (PMP) and electrodialysis reversal (EDR). In PMP, only pure water is able to pass through the membrane, while the majority of the dissolved material is detained by the membrane. The EDR method utilizes ion-specific membranes placed between anodic (positively charged) and cathodic (negatively charged) electrodes. Dissolved material is then collected as the salt ions move under the electric current. Although the EDR method is relatively common, the most widespread desalination methods involve PMP.

There are four pressure-membrane desalination processes: microfiltration (MF), ultrafiltration (UF), nanofiltration (NF), and reverse osmosis (RO). The common principle of these methods is the forced passing (under high pressure) of saline water through a membrane. The main difference among the PMP methods is the size of the particles that are removed from the saline water. For example, MF removes particles larger than 10 microns (1 micron equals 0.000001 meter), UF removes particles of sizes from 0.001 to 10 microns, NF removes particles greater than 0.001 micron, and RO removes particles ranging from 0.0001 to 0.001 micron. Another difference among the PMP methods is the pressure needed for forcing the water through the membrane. MF operates at a pressure of less than 10 pounds per square inch (psi), UF at a range of 15 to 75 psi, NF at a range of 75 to 250 psi, and RO at a range of 200 to 1,200 psi.

Reverse osmosis is an energy-consuming method that can be used very effectively, particularly for production of water for domestic use, whenever good-quality saline water is available. Reverse osmosis can be accomplished using different design modules such as the tubular, the plate-and-frame, the spiral-wound, or the hollow fiber. In all these designs, the water that is allowed to pass through the membrane, the "permeate," is collected as the product water, while the water retained by the membrane forms the so-called concentrate or reject.

There are various types of membranes used for reverse osmosis, such as the cellulose acetate group, which includes cellulose acetate (CA), cellulose acetate butyrate (CAB), and cellulose triacetate (CTA), or polyamide (PA). Production of CA membranes involves four stages: casting, evaporation, gelation, and shrinkage. During the first stage, a solution of cellulose acetate in acetone containing certain additives is cast into flat or tubular thin surfaces. The acetone evaporates, leaving a porous surface. During the gelation stage, the cast is immersed in cold water, forming a gel, while at the same time the additives dissolve in the water. In the last stage, the film shrinks, forcing reduction of the pore sizes. High temperatures result in smaller pore openings.

The El Prat Desalination Plant, a seawater reverse osmosis facility near Barcelona, Catalonia (James Grellier via Wikimedia Commons)

Electrodialysis is also an effective desalination process whereby ions are separated from the water by being forced through selective ion-permeable membranes under the action of an electric current. The ion-permeable membranes alternate between those allowing only the passage of cations (such as potassium, K^+ and those allowing only the passage of anions (such as chloride, Cl^-.

Another potential methodology for desalination involves freezing of the water. Since ice theoretically excludes any dissolved material as it forms, various techniques have been proposed for application of freezing for desalination. However, this method does not have any widespread applicability.

Both the thermal and membrane methods require some form of energy to accomplish desalination. The energy efficiency of the desalination methods is expressed either as the gained output ratio (GOR) or the performance ratio (PR). The GOR is defined as the ratio between the mass of distillate over the mass of the steam. The PR, also known as economy, is estimated as pounds of distillate per 1,000 British thermal units (BTUs) or as kilograms of distillate per 2,326 kilojoules (kJ). Criteria for the selection of a desalination method include energy consumption, process efficiency, operational and maintenance costs, auxiliary services, and growth of demand. Energy is provided mostly by electrically driven pumps, but diesel engines can also be utilized.

SIGNIFICANCE

Domestic (potable) water is a very valuable commodity. Population growth and high demand for water for industrial and agricultural practices have created severe water shortages in many parts of the world. In addition, increased water contamination and extreme hydrologic conditions, such as prolonged drought can adversely affect the socioeconomic and human health conditions of a region.

In addition, there are other applications that require high-quality water. For example, high-purity water is required by pharmaceutical companies for processing drugs and medications, by hospitals for kidney dialysis, by power and other energy-intensive plants in the form of low-scaling water, and by semiconductor manufacturing for production of high-performance chips.

Since most of the conventional water resources are already under stress, the only alternative solution to the water-resource problem is desalination of the vast oceanic water masses or reuse of wastewater. The various existing thermal or membrane methods can effectively provide high-quality freshwater. However, the high costs associated with these methodologies limit their applicability only to communities that can afford the financial burden. There is ongoing research into improving the efficiency and reducing the cost of existing desalination methods and developing new methodologies. What began as a small distillation process used to provide freshwater to a handful of ships' crew members has evolved into a complex water-treatment operation that supplies freshwater to large populations.

—Panagiotis D. Scarlatos

Further Reading

Abdelly, Chedly, et al. *Biosaline Agriculture and High Salinity Tolerance*. Birkhauser Basel, 2008.

American Water Works Association. *Desalination of Seawater: Manual of Water Supply Practices. M61*. Author, 2011.

Belessiotis, Vassilis, Soteris Kalogirou, and Emmy Delyannis. *Thermal Solar Desalination: Methods and Systems*. Elsevier, 2016.

Eyvaz, Murat, and Ebubekir Yüksel, editors. *Desalination and Water Treatment*. InTechOpen, 2018.

Kress, Nurit. *Marine Environmental Impact of Seawater Desalination*. Elsevier, 2019.

Metcalf & Eddy, Inc. *Wastewater Engineering: Treatment and Reuse*. 4th ed., McGraw-Hill, 2003.

Micale, Giorgio, Andrea Cipollina, and Lucio Rizutti, editors. *Seawater Desalination: Conventional and Renewable Energy Processes*. Springer-Verlag, 2009.

National Research Council. *Desalination: A National Perspective*. National Academies Press, 2008.

Ratnayaka, Don D., Malcolm J. Brandt, and K. Michael Johnson. *Twort's Water Supply*. 6th ed., Butterworth-Heinemann, 2009.

Rizutti, Lucio, Hisham Mohamed Ettouney, and Andrea Cipollina, editors. *Solar Desalination for the 21st Century: A Review of Modern Technologies and Researches on Desalination Coupled to Renewable Energies*. Springer, 2007.

Sarp, Sarper and Nidal Hilal, editors. *Membrane-Based Salinity Gradient Processes for Water Treatment and Power Generation*. Elsevier, 2018.

Segar, Douglas. *Introduction to Ocean Sciences*. 2nd ed., W. W. Norton, 2007.

Talley, Lynne D., et al. *Descriptive Physical Oceanography: An Introduction*. 6th ed., Academic Press, 2011.

Viessman, W., Jr., and M. J. Hammer. *Water Supply and Pollution Control*. 8th ed., Prentice-Hall, 2008.

Warren, John K. *Evaporites: Sediment, Resources and Hydrocarbons*. Springer-Verlag, 2006.

Sargasso Sea Ecology

Fields of Study: oceanography; marine biogeochemistry; marine biology

ABSTRACT

The Sargasso Sea is a part of North Atlantic Ocean that overlaps the northern part of the Bermuda Triangle. The name is from the abundant Sargassum seaweeds (dominated by S. natans and S. fluitans) floating on the surface water in the region. The Sargasso Sea is a slow-moving gyre area separated from the rest of the Atlantic Ocean by several surrounding strong currents: the Gulf Stream, North Atlantic, Canary, and the North Atlantic Equatorial Currents. Found here is a large expanse covered with yellow Sargassum, plastic wastes, and occasional derelict shipwrecks.

KEY CONCEPTS

bioremediation: the use of specific bacteria and other organisms with specific abilities to "clean up" polluted areas by consumption of environmental contaminants

carbon cycle: the path of carbon in the environment involving carbon dioxide, organic plant and animal matter and carbonate rock, ultimately returning to carbon dioxide to complete the cycle

epiphytes: plant organisms that use other plants for physical support

gyre: essentially a circular eddy-like flow produced as the mass of water enclosed within a series of connecting currents rotates sympathetically with the various current flows

polychlorinated biphenyls (PCBs): a family of synthetic organic compounds that have been produced industrially in mass quantities to make use of their high thermal stability, lack of electrical conductivity and inert chemical reactivity; molecular structures of PCBs consist of two connected benzene rings bearing chlorine atoms instead of hydrogen atoms

A SEA OF MANY LEGENDS

The Sargasso Sea is a slow-moving gyre area separated from the rest of the Atlantic Ocean by several surrounding strong currents: the Gulf Stream, North Atlantic, Canary, and the North Atlantic Equatorial. Found here is a large expanse covered with yellow *Sargassum*, plastic wastes, and occasional shipwreck derelicts (the source of many legends). The Sargasso Sea is a part of North Atlantic Ocean that overlaps the northern part of the Bermuda Triangle. The name is from the abundant *Sargassum* seaweeds (dominated by *S. natans* and *S. fluitans*) floating on the surface water in the region. They were reported by Christopher Columbus and his crew during their expedition to the New World. The Sargasso Sea is a slow-moving gyre area separated from the rest of the Atlantic Ocean by several surrounding strong currents: the Gulf Stream, North Atlantic, Canary, and the North Atlantic Equatorial. Found here is a large expanse covered with yellow *Sargassum*, plastic wastes, and occasional shipwreck derelicts (the source of many legends).

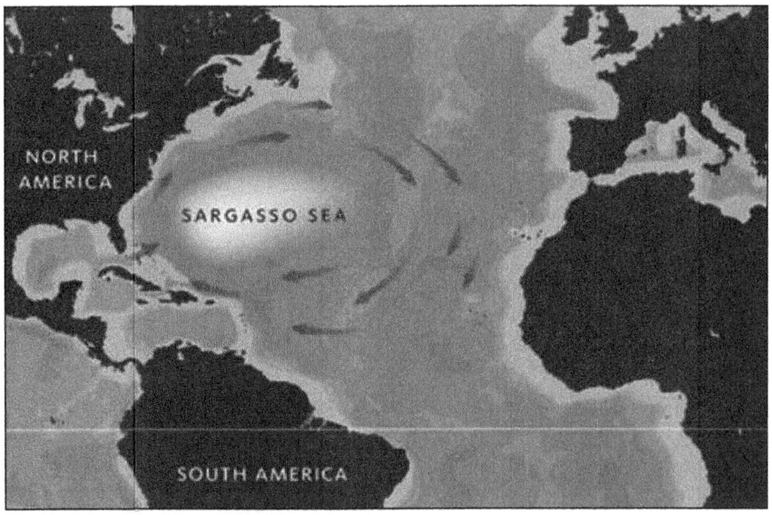

Map of the Sargasso Sea (Wikimedia Commons)

The Sargasso Sea is one of the few areas of the world lacking coastlines but designated as a sea; it is approximately 700 miles (1,100 kilometers) wide and 2,000 mi (3,200 km) long, covering an area of at least 1.35 million square miles (3.5 million square kilometers) of water.

The water in the Sargasso Sea is warm, maintaining a high salinity concentration of around 3.6 percent (36 parts per thousand). The temperature in the euphotic (sunlight-penetrated) zone averages up to 72 degrees Fahrenheit (22 degrees Celsius). For these reasons, even though there are several species of plankton and massive amounts of seaweed floating on the water surface, the Sargasso Sea is still not nutritious enough to attract large communities of fish. The factors of low wind, low nutrients, and high salinity are chief reasons why the Sargasso Sea is sometimes considered a "desert among oceans."

FLORA AND FAUNA

The *Sargassum* weed is known to harbor epiphytes (micro and macro), diverse fungi, over 100 invertebrate species, dozens of fish species, and about four seaturtle species—one of which is the threatened green seaturtle (*Chelonia mydas*). The abundant floating vegetation of *Sargassum* provides a habitat for several endemic (found nowhere else) species of marine animals, including

Sargassum pipefish (*Syngnathus pelagicus*), Sargassum snail (*Litiopa melanostoma*), slender Sargassum shrimp (*Latreutes fucorum*), and the Sargassum crab (*Planes minutes*). In addition, the deeper reaches that feature drift algae provide critical spawning sites for major eel species including American eel (*Anguilla rostrata*) and European eel (*A. Anguilla*).

This generally nutrient-poor, or oligotrophic, sea has been one of the best-characterized oceans in its physical and biogeochemical properties, especially at the Bermuda Atlantic Time-series Study (BATS) site. Since the late 1950s, the Sargasso Sea has been an important research location for marine biogeochemistry. More recently, the first major meta-genome research was carried out in the Sargasso Sea by J. Craig Venter and his colleagues.

They have found many new lineages for the clade (or family) of SAR11 bacteria, which later were shown to constitute a major proportion of all marine prokaryotes (both bacteria and *Archaea*) from the major oceans. This bacterium can surprisingly replicate efficiently in a low-nutrient environment, and is one of the smallest self-replicating cells found.

Other microbes sustained here include cyanobacteria, luminous bacteria, *Berkholderia*, and *Shewanella*. One of the cyanobacteria, *Prochlorococcus marinus*, though among the smallest photosynthetic organisms known, is a major phototroph in the ocean and greatly impacts the carbon cycle. *Berkholderia* was originally thought to be only a terrestrial bacteria. When this new, similar species was first discovered in the ocean, it was found to contain similar deoxyribonucleic acid (DNA) and genes, suggesting genomic transfer. Terrestrial *Berkholderia* bacteria are known to be able to biodegrade a very toxic class of industrial product, the polychlorinated biphenyls (PCBs). Because of this ability, the bacteria have an important ecological and commercial potential for bioremediation, and sequencing this genome may play a major role in environmental protection.

HUMAN INTERACTION

Beginning with the establishment of Hydrostation S in 1957, the time-series biogeochemical sampling between November 1957 to April 1960 by D. M. Menzel and J. H. Ryther effectively set the foundation for the modern science of marine biogeochemistry. They defined the seasonal cycle of primary productivity, and proposed many profound topics related to global biogeochemistry and global climate change. More recently, genomics research has allowed scientists to find evidence of genes that form the basis of the key chemical processes of ocean-dwelling microbes.

This genetic research has shed light on many new, and as yet unknown, microorganisms. The Sargasso Sea, once the source for many legends and science fiction, has been a self-supporting environment for unique marine biota including some endemic fish and shells. It has also now been identified as a global hot spot of microbial diversity.

The projected effects of global warming on the Sargasso Sea biome are difficult to predict. It is known that sea surface temperature here has been more or less constantly increasing over the last 300 years; researchers tend to see this as a return to the historic norm, however. Global sea-level rise will affect this oceanic area, but consequences to its biota are unclear. More crucial will be any changes that precipitate from alterations in the complex suite of temperature, density, salinity, and velocity variables of the surrounding currents that form the gyre that sustains the Sargasso Sea.

—Sanghoon Kang

Further Reading

Giovannoni, Stephen J., et al. "Proteorhodopsin in the Ubiquitous Marine Bacterium SAR11." *Nature*, vol. 438, 2005.

Lipschultz, Fredric, et al. "New Production in the Sargasso Sea: History and Current Status." *Global Biogeochemical Cycles*, vol. 16, no. 1001, 2002.

Morris, Robert M., et al. "SAR11 Clade Dominates Ocean Surface Bacterioplankton Communities." *Nature*, vol. 420, Dec. 2002.

van den Thillart, Guido, Sylvie Dufour, and J. Cliff Rankin, editors. *Spawning Migration of the European Eel: Reproduction Index, A Useful Tool for Conservation Management*. Springer, 2009.

Venter, J. Craig, et al. "Environmental Genome Shotgun Sequencing of the Sargasso Sea." *Science*, vol. 304, 2004.

SEA LEVEL

Fields of Study: oceanography; marine geology; paleogeology; hydrodynamics

ABSTRACT

Sea level is the average position of the surface of the oceans relative to the land, providing a frame of reference for land elevations and ocean depths. Sea levels change through time, affected by a variety of factors, and these changes have a major influence on the geology of Earth.

KEY CONCEPTS

continental shelf: the outer edge of a continental land mass, extending outward from land below sea level at relatively shallow depths and ending where it becomes a steep drop-off into the abyssal depths

eustatic sea level change: a global increase or decrease in mean sea level

hydrated crystals: minerals that incorporate water molecules either as an intimate part of their crystal structure or as entrained water

isostasy: the vertical raising or lowering of areas of Earth's crust

local sea level change: any change of mean sea level, usually temporary, that occurs in a limited area rather than globally

subsidence: the lowering of land area when material is removed from some supporting subsurface structure such as an aquifer, causing the land above it to sink to a lower position relative to sea level

CHANGING SEA LEVEL

Sea level is an increasingly important aspect of the modern world. The distinction between land and sea depends on the level of the sea. A rise in sea level means the flooding of adjacent low-lying land; a drop in sea level means exposure of some of the seafloor. Much of the world's commerce, food supply, and recreation is associated with shallow coastal waters. The position and rate of change of sea level is important to all these activities, especially as construction has encroached right up to the water's edge in many places. Sea level is most accurately defined as mean (average) sea level, which is the average height of the sea surface measured over an extended period of time for all conditions of tides, seasons and storms. Land elevations are measured by reference to sea level. Nautical charts, however, use mean low water, the average position of low tide, to measure ocean depths, while mean high water, the average position of high tide, is used for marking the exposure of adjacent land areas. The reason for use of low- and high-tide values, for depth and land, respectively, is purely practical, as successful navigation requires accurate knowledge of water depths and land exposures. By referring depths to mean low tide, ship captains can be more assured that, regardless of tide position, they know the minimum water depth available for their vessels. The same principle applies to land areas. Using a high-tide value to show land allows land users to be sure that the low-lying areas will not flood with the high-tide cycle.

The most fascinating aspect of sea level is that it is not constant and that through time it can change. The time of change can be hours, days, years, or centuries. When sea level rises and floods the land, it is called a "marine transgression." The flooded land becomes part of the ocean environment, and marine sediments are deposited on what was once dry land. Marine regression, by contrast, occurs when sea level drops and the shallow seafloor is exposed. The exposed seafloor becomes part of the land environment and is subject to the same type of erosion processes that function on land. Almost every continental coastal area includes a continental shelf, or a flat-lying or gently sloping portion of the continent that extends under the sea to depths up to about 492 feet (150 meters) over a width of several tens of miles (or kilometers) to over 62 miles (100 kilometers). The gentle slope of the continental shelf means that a minor rise in sea level results in a major amount of the shelf being drowned; a minor drop in sea level produces a major exposure of the continental shelf seafloor. The rocks of the continents show that over Earth's history, sea level has transgressed onto, and regressed from, the continents many times and to many different levels.

Sea level can be changed in six basic ways: by oscillating the ocean surface. by changing the force of gravity, by moving the land up or down, by changing the characteristics of the ocean water, by changing the amount of water in the oceans and by changing the volume of the ocean basins. The change in sea level brought about by these six ways results in two major types of sea-level change: local and eustatic. Local sea-level change means that only a specific area of coastline is involved and that coastlines relatively far away or on other continents are not changed. Oscillation of the water surface, changing gravity, and land moving up and down also produce local sea-level change. Eustatic sea-level change means that coastlines around the planet all experience a sea-level change of the same magnitude at the same time. Changing water characteristics, changing the amount of water on the planet, or changing the volume of the ocean basins produces eustatic sea-level change.

LOCAL SEA-LEVEL CHANGE

Sea-level change produced by oscillation of the ocean surface refers to waves, storm surges, and tides. A wave breaking on the beach runs up the beach, then slides back down. Each wave event can be considered a short-term, low-magnitude sea-level change. In extreme cases of large waves, erosion and coastline modification can occur during each sea-level "microevent." Storm surges are changes in sea level brought about by the movement of surface waters under the influence of strong winds and low atmospheric pressures, such as occurs with hurricanes. Low atmospheric pressure appears to pull sea level up a small amount, though in actuality this is the result of higher air pressure surrounding the system that pushes down on the surrounding water while lower air pressure inside the system allows the interior water to rise. As for the direct effect of wind on water levels, high onshore winds can pile water up on coastal areas by 9 to 13 ft (3 to 4 m) or more, while offshore winds can lower sea level in the same way but by lesser amounts.

Tides are produced in the oceans by the force of lunar gravity, solar gravity, and by the rotation of the Earth-moon system about its common center of mass. Coastline configuration, latitude, time of the lunar month, and many other factors control the timing and magnitude of ocean tides. Tide characteristics are extremely variable from place to place, but in a given area, the changes in sea level are quite predictable, with magnitudes from a fraction of a meter to more than 33 ft (10 m) possible. Because ocean surface oscillations are common, low in magnitude, and regu-

Sandsend Beach, Whitby, North Yorkshire, England, UK (travellinglight via iStock)

lar, they are often not recognized as sea-level changes. They are different for every coastline, producing local sea-level change.

The force of gravity is not perfectly uniform over Earth. It depends on the amount of mass beneath the surface and thus, because of tectonics and other activities, may vary by a small amount from location to location. This variation is most marked between the continental landmasses, where there is a great deal more matter to exert gravitational force, and the ocean floors, where there is much less crustal mass. The ocean, as a fluid, responds to the force of gravity. If gravity is slightly weaker in one area than in another, the sea will rise slightly higher, but if it is slightly stronger, the sea will sink slightly lower. This sea-level variation occurs only in time frames of millions of years, but it is used to explain local sea levels that are different in certain areas from predicted values.

In many areas, the land itself is moving up or down. This movement has three sources: tectonics, isostasy, and subsidence. Tectonics is the movement and distortion of Earth's crust by convective forces generated within the magma of the mantle, influenced by the thermal activity in the planetary core. Tectonic action is responsible for the formation of volcanoes, mountains, and earthquakes. In an area where tectonic processes are active, the land may be forced up or down, causing it to rise or sink with respect to sea level. In tectonically active areas such as Southern California or in the Mediterranean basin, this type of land movement is common, and sea-level changes of many meters up and down have been historically documented. Isostasy is the vertical movement of the crust downward when a load is applied—as when a glacial mass of ice forms over the land—or its rebound upward when the load is removed. When isostasy occurs in a coastal area, the land will rise or sink, with a subsequent change in sea level.

Glaciers are examples of how a load can be applied to the crust, causing isostatic subsidence. When the glacier melts, the load is removed, and the crust can isostatically rebound. This is the process responsible for the gradual rise of the land west of the Niagara Escarpment in southwestern Ontario, Canada, and of the rising east coast of Britain such that thousand-year-old coastal installations now stand as much as 1,312 ft (400 m) above sea level. Coral atolls and volcanic islands are other examples of a load being placed on the crust, which then isostatically sinks. Land can sink beneath sea level by a process called "subsidence." Subsidence is often caused by compaction of the land material, which commonly happens when oil or water is withdrawn from the ground. The loss of the fluid allows the rock to compact, and the overlying land sinks. All three of these methods of moving land up or down occur in localized areas and so result in local sea-level change.

EUSTATIC SEA-LEVEL CHANGE

Changing sea level eustatically, or worldwide, requires changes that affect the oceans as opposed to the land. The ocean basins generally have a fixed volume. If the nature of the water in it is changed, if the amount of water is changed, or if the shape of the container (ocean basin) is changed, sea level will change. The two characteristics of water that control sea level are its temperature and its salinity. For each degree Fahrenheit (or Celsius) that the oceans warm, thermal expansion will raise sea level 7 ft (2 m) due to the lesser density (mass per unit volume) of the water. Increasing the salinity makes the water solution denser and causes sea level to fall. Sea-level variations caused by changes in salinity measure approximately 3 to 7 ft (1 to 2 m).

The amount of water available to the oceans is changed in three ways: by the growth and melting of glaciers, by steam released from volcanoes and by water lost to the formation of hydrated crystals in sediments. Glaciation is one of the most important sea-level controls. As ice sheets grow, they are fed by evaporated seawater falling as snow; this seawater is then "trapped" as ice on the continents, and sea level falls by the corresponding amount. This is a negligibly small amount per snowflake, but over thousands and millions of years it has produced glacial coverage up to 2.5 mi (4 km) in thickness in some places such as Greenland and Antarctica. If the ice melts and the water flows back to the ocean, sea level rises. During the last two million years, glaciations have come and gone at least four times, and sea level has risen and fallen over a range of approximately 410 ft (125 m). This range is enough to almost totally expose continental shelves during maximum ice advance on the continents. Only eighteen thousand years ago, sea level was 410 ft (125 m) below its present level, yet by three thousand years ago, it was essentially at today's level. If the remaining ice on Antarctica were to melt entirely, sea level would rise an additional 197 ft (60 or more m), drowning coastal cities worldwide.

The oceans are thought to have originated by the outgassing and release of water as steam from early volcanic activity. Volcanoes are still actively adding water to the oceans. When sediments are deposited in the ocean, they normally contain some entrapped water. Tectonic activity can return these water-bearing sediments back into the crust by the process of subduction. There appears to be a rough balance or equilibrium between water escaping the crust from volcanoes and water returning to the crust through sediment deposition, with no overall sea-level change caused by these processes in the modern world.

Sea level is also affected by changes in the volume of the ocean basins. Continents are continually losing sediments to the sea through processes of erosion. This addition of sediment is slowly filling in the ocean basins, pushing sea level up. At the same time, tectonic activity is distorting the edges of continents, often folding them, which increases the volume of the ocean basins. Tectonic activity in the ocean floor can force the seafloor upward, limiting ocean basin volume. Tectonic activities tend to occur episodically, in fits and starts, rather than as a smooth, continuous process. When tectonics is active, it may increase or decrease ocean volume, causing a drop or rise in sea level. Sediments deposited on the ocean floor may be plastered back onto the continents by tectonic activity, first decreasing and then increasing ocean basin volume. Sea-level changes, in response to sediment balance and tectonic location and magnitude, occur slowly over millions of years.

STUDY OF SEA LEVEL

When sea level changes, it leaves an imprint on the land. Transgressing ocean waters erode the land surface with waves and then, as waters deepen again, deposit marine sediments. When regression occurs, the shallowing water leads to wave erosion of the seafloor. Further regression exposes the seafloor to the air and to erosion by wind, rain, and running water from precipitation. Casual examination of the continents reveals clear evidence of past marine transgression and regression. Limestones, marine shales, and marine sandstones are common around the world on the dry land of the continents. They range in age from billions of years old to only a few hundred years old. The sea has transgressed and regressed many times in Earth's history. The timing and exact nature of the sea-level change may be more difficult to determine.

In the Bahamas, fossil corals exist several feet (or meters) above sea level. By sampling the corals and dating them by means of trace amounts of radioactive elements in the samples, scientists can determine their age. In this case, their age is about 125,000 years. Therefore, more than 100,000 years ago, the land to which the fossil coral is attached was under water. Did the land rise, or did the water drop? Measurement of tectonic activity shows that the islands of the Bahamas are not rising or sinking. Examination of the history of glaciation shows that 125,000 years ago, ice sheets had melted back a little more than they are presently. From this information, scientists concluded that the coral grew when sea level was eustatically higher than it is today.

Further evidence of sea-level change can be obtained by examining the tectonic, deposition, and erosion history of the planet. Examination of major episodes of tectonic activity, sediment deposition, and erosion can allow the construction of graphs showing how sea level rose and fell, in general terms, over much of Earth's history. Most of the sea-level changes preserved in the rock record were tectonically generated. Glaciation has occurred only a few times in the past, and the sea-level changes caused by glaciation have left a unique and distinctive record in rock structures.

If sea level never changed, the geology of all surface rocks would be very simple. There would be no exposed marine rocks at all. If there were no tectonics to help drive sea-level change, the continents would eventually erode to sea level. Earth would eventually become a flat, relatively featureless place.

The study of changing sea levels has become critical. Worldwide industrialization has greatly escalated climate change and global warming. As such, polar ice caps and other glaciers have been melting at an alarming rate, causing global sea levels to rise. Studies show that they are rising faster than they have for 2,800 years, and indeed, the rise may be accelerating over time. The National Oceanic and Atmospheric Administration (NOAA) reports that the rate of sea level rise held steady at about 0.067 in (1.7 millimeters) annually for much of the twentieth century, but from 1993 to 2017 it sped up to 0.133 in (3.4 mm) per year.

In 2013, the Intergovernmental Panel on Climate Change projected that global sea levels could rise 0.85 to 2.7 ft (26 to 82 centimeters) by 2100; however, several studies and reports published in 2016 and 2017 suggested those figures might be too low. An October 2017 *Environmental Research Letters* article suggested sea levels could reach as much as 6.6 ft (200 cm) by 2100, under a high fossil fuel-burning scenario, but 1.8 ft (55 cm) if emissions are drastically reduced.

In short, coastal cities across the globe are under threat, and small island countries, such as those in Micronesia, could disappear completely under the ocean, unless decisive action is taken.

—John E. Mylroie

Further Reading

Coe, Angela L., editor. *The Sedimentary Record of Sea-Level Change*. Cambridge UP, 2003.

Davis, Richard A., Jr. *Sea-Level Change in the Gulf of Mexico.* Everbest Printing, 2011.

Dawson, Alastair. *Introducing Sea Level Change.* Dunedin Academic Press, 2018.

De Angelis, Hernan, and Pedro Skvarca. "Glacier Surge After Ice Shelf Collapse." *Science*, vol, 299, 2003, pp. 1560–62.

Janin, Hunt, and Scott A. Mandia. *Rising Sea Levels. An Introduction to Cause and Impact.* McFarland & Company, 2012.

Kelley, Joseph T., Orrin H. Pilkey, and J. A. G. Cooper. *America's Most Vulnerable Coastal Communities*. Geological Society of America, 2009.

Lindsey, Rebecca. "Climate Change: Global Sea Level." *Climate.gov*, NOAA, 11 Sept. 2017, www.climate.gov/news-features/understanding-climate/climate-change-global-sea-level.

Montgomery, C. W. *Environmental Geology*. 9th ed., McGraw-Hill, 2010.

Mooney, Chris. "New Science Suggests the Ocean Could Rise More—And Faster—Than We Thought." *Chicago Tribune*, 26 Oct. 2017, www.chicagotribune.com/news/nationworld/science/ct-climate-change-sea-level-rise-20171026-story.html.

Nauels, Alexander, et al. "Linking Sea Level Rise and Socioeconomic Indicators under the Shared Socioeconomic Pathways." *Environmental Research Letters*, vol. 12, no. 11, 26 Oct. 2017. *IOP Science* iopscience.iop.org/article/10.1088/1748-9326/aa92b6#erlaa92b6f2.

Pilkey, Orrin H., and Rob Young. *The Rising Sea.* Island Press, 2009.

Shennan, Ian, Antony J. Long, and Benjamin P. Horton, editors. *Handbook of Sea-Level Research.* Wiley, 2015.

Wagreich, M., M. B. Hart, B. Sames, and I. O. Yilmaz, editors. *Cretaceous Climate Events and Short-Term Sea-Level Changes.* The Geological Society, 2020.

SEA OF JAPAN ECOSYSTEM

Fields of Study: oceanography; marine biogeochemistry; marine biology

ABSTRACT

The Sea of Japan is a marginal sea in the Pacific Ocean. It has several unique characteristics due to its near-complete enclosure by land and resulting isolation from the rest of the Pacific. It has its own unique ecosystem, is significantly less saline than the surrounding ocean, and its tidal activity is very low. The Sea of Japan lacks any major bays, capes, or large islands.

KEY CONCEPTS

landlocked sea: a sea entirely surrounded by land, with no accessible link to the ocean

mercury: a silvery, liquid metal at room temperature, one of the "heavy metals" that is neurotoxic to animals and bioaccumulates up the food chain; mercury poisoning fell under the spotlight in the 1960s as "Minamata disease" due to the high number of cases observed among the fish-consuming populace of the Minamata Bay region of Japan

radioactive pollution: the presence of materials such as uranium and plutonium that emit dangerous X-rays and decay particles; such materials are believed to have been disposed of in the Russian area of the Sea of Japan as a byproduct of nuclear research efforts

river discharge: the freshwater entering a saline body as outflow from lakes, rivers and streams

typhoon: the name for weather events in the Pacific Ocean that are called "hurricanes" in the Atlantic Ocean

OVERVIEW

A marginal sea in the Pacific Ocean between the archipelago of Japan, the Russian and Korean mainland, and the Russian island of Sakhalin, the Sea of Japan has several unique characteristics due to its near-complete enclosure by land and resulting isolation from the rest of the Pacific. Just as islands are home to species found nowhere else, the Sea of Japan has its own unique ecosystem. It is significantly less saline than the surrounding ocean, and, like the Mediterranean Sea, its tidal activity is very low. The Sea of Japan lacks any major bays, capes, or large islands.

HYDROLOGY AND CLIMATE

The contribution of river discharge to the water exchange is low—about 1 percent overall. Most of the rivers flowing into the sea are from the Japanese archipelago. Japan's four largest rivers—the Shinano, Ishikari, Agano, and Mogami—all empty into the Sea of Japan. The largest river from mainland Asia flowing into the sea is the Tumen River, which begins in Mount Baekdu and forms part of the boundary separating China, North Korea, and Russia.

The water balance of the Sea of Japan is maintained by the inflow and outflow through its five straits: the Strait of Tartary, between the Asian mainland and Sakhalin; La Perouse Strait, between Sakhalin and Hokkaido; the Tsugaru Strait, between Hokkaido and Honshu; the Kanmon Straits (also known as the Straits of Shimonoseki), between Honshu and Kyushu; and the Korea Strait, between South Korea and Kyushu.

Once a landlocked sea, the Sea of Japan became joined to the Pacific Ocean over the Miocene period, as the land fragmented sufficiently for Japan to detach from mainland Asia. The Tsugaru and Tsushima straits formed first, about 2.6 million years ago; La Perouse is the youngest, forming as recently as 11,000 years ago and disrupting the migration of mammoths to and from Hokkaido. Even today, all the straits are shallow, limiting the exchange of water with the Pacific and the migration of marine life to and from the sea.

The isolation of the sea leads to clearly separated layers in the water, except during winter in the northern reaches. Winter temperatures in the central and southern sea hover close to 32 degrees Fahrenheit (0 degrees Celsius) at the bottom of the sea; 34 to 35 degrees Fahrenheit (1.1 to 1.6 degrees Celsius) at depths of 1,310 to 1,640 feet (400 to 500 m); 36 to 40 degrees Fahrenheit (2.2 to 4.4 degrees Celsius) at 656 to 1,310 ft (200 to 400 m); and as warm as 50 degrees Fahrenheit (10 degrees Celsius) at a depth of 328 ft (100 m) below the surface. Because of the counterclockwise circulation of the sea currents, the eastern waters are consistently warmer than the western waters. The difference is greatest in the winter.

The waters of the Sea of Japan thus are predominantly warm, and when the northwestern monsoon wind brings cold, dry air over the sea from the Asian mainland (from October to March), powerful storms and typhoons can result. The western coasts of Japan are regularly pummeled with waves in excess of 25 ft (8 m). The monsoon also carries seawater evaporation south to the Japanese mountains, where it precipitates as snow and adds to the convection of surface waters.

In the northern area of the sea, particularly in the Strait of Tartary, where the water is especially shallow, the surface freezes for months, beginning as early as October and lasting as late as June. The domination of evaporation over precipitation also leads to higher levels of salinity in the southern sea in the winter; this more-saline water is gradually carried to the north, balancing out the introduction of less-saline water as winter ice thaws in the spring. Summer winds blow in the opposite direction, as warm, humid air from the Pacific is carried over the Sea of Japan to the Asian mainland.

Though tidal activity is limited, it is also complex. The tidal flow of the Pacific pushes through the Korea Strait and the northern portion of the Strait of Tartary twice a day, and pushes through the Tsugaru Strait once a day, causing a daily tide on the Koreas' eastern shores, the Far East of Russia, and Honshu and Hokkaido in the Japanese archipelago. The amplitude of these tides is low but varies considerably across the Sea of Japan. Near the Korea Strait, for example, it is 10 ft (3 m); at the southern tip of the Korean peninsula, it is only half that.

BIODIVERSITY

Along with the nutrient richness that is aided by its unique combination of temperature and salinity gradients and tidal cycles, the sea's waters are extremely rich in dissolved oxygen—as high as 95 percent of the saturation point just below the surface and still as high as 70 percent at depths of 9,840 ft (3,000 m). This leads to a thriving environment for some 3,500 species of fauna and 800 of marine flora, including nearly 1,000 species of crustaceans.

While lacking the deep-water fauna richness of the greater Pacific Ocean, the Sea of Japan biome does support cod, Atka mackerel, sardines in great abundance, and the anadromous fish such as salmon and trout. Sea lions and cetaceans are here in great populations. Squid spawn in the East China Sea before migrating north into the Sea of Japan via the Korea Strait.

Coastal areas of the northern reaches of the Sea of Japan provide a wintering haven for Steller's sea-eagle (*Haliaeetus pelagicus*), and vital nesting areas for the endangered spotted greenshank (*Tringa guttifer*). Cranes and shorebirds use Furugelm Island, off the Russian coast, as a stopover in migrations.

Along the southern waters of the Sea of Japan, the wetlands and rice paddies of the Lower Maruyama River (in Japan's Hyogo prefecture) were designated a Ramsar Wetland of International Importance in 2012. Authorities say several thousand species live in or depend on the 1,380-acre (560-hectare) area. It is vital, for instance, to such threatened fish species as chum salmon (*Oncorhynchus keta*), three-spined stickleback (*Gsterosteus aculeatus*), and Japanese rice fish (*Oryzias latipes latipes*). It is here that the previously extirpated oriental white stork (*Ciconia boyciana*) was successfully reintroduced in the 1950s.

Common fish include saury, mackerel, sardines, anchovies, herrings, sea bream, salmon, trout, squid, cod, pollock, and Atka mackerel. The sardine is the most plentiful pelagic fish, constituting more than 70% of the total catch. Squid spawn in the East China Sea before migrating to the Sea of Japan. Seals and whales are common mammals. Because the straits are so shallow, there is no characteristic oceanic deep-water fauna.

THREATS AND CONTROVERSY

A significant ecological problem in the Sea of Japan is oil pollution, particularly along major shipping routes, where the number of accidents of seagoing vessels has increased in the twenty-first century. There are inconclusive reports of radioactive pollution from Soviet-era reactor equipment and materials disposal at sea. Mercury contamination is a recurring problem, particularly in the Agano River, which feeds into the Sea of Japan at Minamata Bay, near Niigata, site of a mass poisoning event in 1965.

Climate change is projected to warm the average seawater temperatures here, a process that seems to be already underway. Major outbreaks of giant jellyfish (*Nemopilema nomurai*) have become far more common than they were historically in the Sea of Japan, to the point where commercial fisheries are registering their repeated swarms as a serious threat.

Coral reefs in the Sea of Japan are extending their range further to the north, but this buildup is counterbalanced by the fact that reef-dwelling fish are migrating here from more tropical waters, likely at the expense of at least some of the cold-water species native to the Sea of Japan. Additionally, even the new reefs are threatened by the crown-of-thorns starfish (*Acanthaster planci*), the classic foe of coral reef-building organisms. This starfish only thrives in warm-water conditions.

A SEA BY ANY OTHER NAME

While the name *Sea of Japan* is the most commonly used, other nations have pushed for different nomenclature. South Korea prefers the name *East Sea*, while North Korea refers to the sea as *East Sea of Korea*. The South Korean government specifically argues that the name *Sea of Japan* serves as a reminder of the period when Korea was occupied by imperial Japan. Although *Sea of Japan* has been the most common name in international use for more than two centuries (in large part because of Japan's prominence in international trade), the term *East Sea* was found on Western maps as early as the 1700s. *Sea of Korea* was actually the most common Western use in that century, likely because of Japan's strict isolationist policies from 1633 to 1853, which made Western interaction with the country almost negligible. *Oriental Sea* was about as common in Western usage as *Sea of Japan* until the opening of Japan to international trade in the mid-nineteenth century.

—Bill Kte'pi

Further Reading

Chang, Kyung-Il, Chang-Ik Zhang, Chul Park et al., editors. *Oceanography of the East Sea (Japan Sea)*. Springer, 2016.

Gershwin, Lisa-Ann. *Stung! On Jellyfish Blooms and the Future of the Ocean*. U of Chicago P, 2013.

Ishimure, Michiko. *Paradise in the Sea of Sorrow: Our Minamata Disease*. Center for Japanese Studies, U of Michigan, 2003.

Seamounts

Fields of Study: oceanography; marine geology; marine biology

ABSTRACT

Seamounts, formed by undersea volcanoes, are far more numerous and much larger than their volcanic counterparts on land. Some may become large enough to rise above sea level as islands for a brief period, but almost all eventually resubmerge as guyots or atolls. Radiometric dating of seamounts has helped to verify the theory of continental drift.

KEY CONCEPTS

asthenosphere: the layer of mobile liquid rock in the mantle directly beneath the lithosphere

atoll: a generally circular coral structure typically surrounding the submerged top of a tropical oceanic seamount

bank: designates the top of a guyot from below the surface to a depth of 656 feet (200 meters)

continental drift: term describing the drifting movement of the continents of Earth's crust driven by convection currents in the underlying magma of Earth's mantle

guyot: a subsurface oceanic mountain having a flat top like that of a terrestrial mesa

hot spot: a region of Earth's mantle where a column of extraordinarily hot magma rises toward the overlying crust at the same relative location, bringing about the formation of volcanic islands

ORIGIN AND GROWTH

Seamounts are volcanoes that rise from the seafloor but either lack sufficient height to rise above the surface of the ocean, or have been quickly eroded to a level below the surface. Like their volcanic counterparts on land, seamounts are roughly conical or domelike in shape, and many have craters at the top. Compared to volcanoes on the continents, seamounts are much larger and higher, and may rise to as much as 32,808 ft (10,000 m) above the ocean floor. A seamount that has a flat top is called a "guyot." Almost all guyots were islands at one time, and they acquired their flat tops through the erosive action of waves, the growth of corals, or a combination of both processes. The islands also may have undergone resubmergence caused by the subsidence or sinking of the seafloor or by eustatic rise in sea level. The tops of some guyots can be as much as 9,843 ft (3,000 m) below present sea level, but most are about 3,280 ft (1,000 m) deep.

A guyot that is less than 656 ft (200 m) deep is called a "bank." Seamounts and guyots are a prominent feature of the ocean floor, particularly in the southwestern Pacific Ocean although a great many also exist in the Atlantic Ocean. Some, including the Grand Banks found some distance off the coast of Newfoundland, Canada, are associated with vast midocean ridges, but many rise from the deep, flat sections of the ocean floor known as the abyssal plains. Some seamounts are isolated, others occur in clusters and form a large volcanic plateau, and still others form a long chain.

Little is known regarding the origins of seamounts. It is known, however, that magma (molten rock) rising through the seafloor squeezes its way beneath the saturated but relatively buoyant bottom sediments and then crystallizes. After many of these intrusions, a solid base is built up, and cone development can begin. The magma that makes up seamounts is not of the explosive variety, so seafloor eruptions are relatively quiet affairs. In fact, the water pressure in the deep ocean is so great that the gas and steam normally associated with volcanoes on the continents cannot escape. Instead, the lava (magma that has escaped the interior) oozes out quietly in submarine flows. Because deep seawater is quite cold, the lava is chilled very rapidly; often, the lava cools into wrinkled, glassy crusts that enclose round blobs. These structures, common on the seafloor, are known as lava pillows, or pillow lava. The rock that makes up the pillows is a dark, fine-grained rock known as basalt, which is rich in iron and magnesium-bearing minerals.

As the seamount grows, its increasing size begins to bring it nearer the sea's surface and consequently into areas of lower water pressure. At a depth of about 6,562 ft (2,000 m), the gases associated with the lava and the heated seawater itself can begin to expand and explode violently as the lava cools, resulting in a glassy, fragmented structure known as hyaloclastite. It

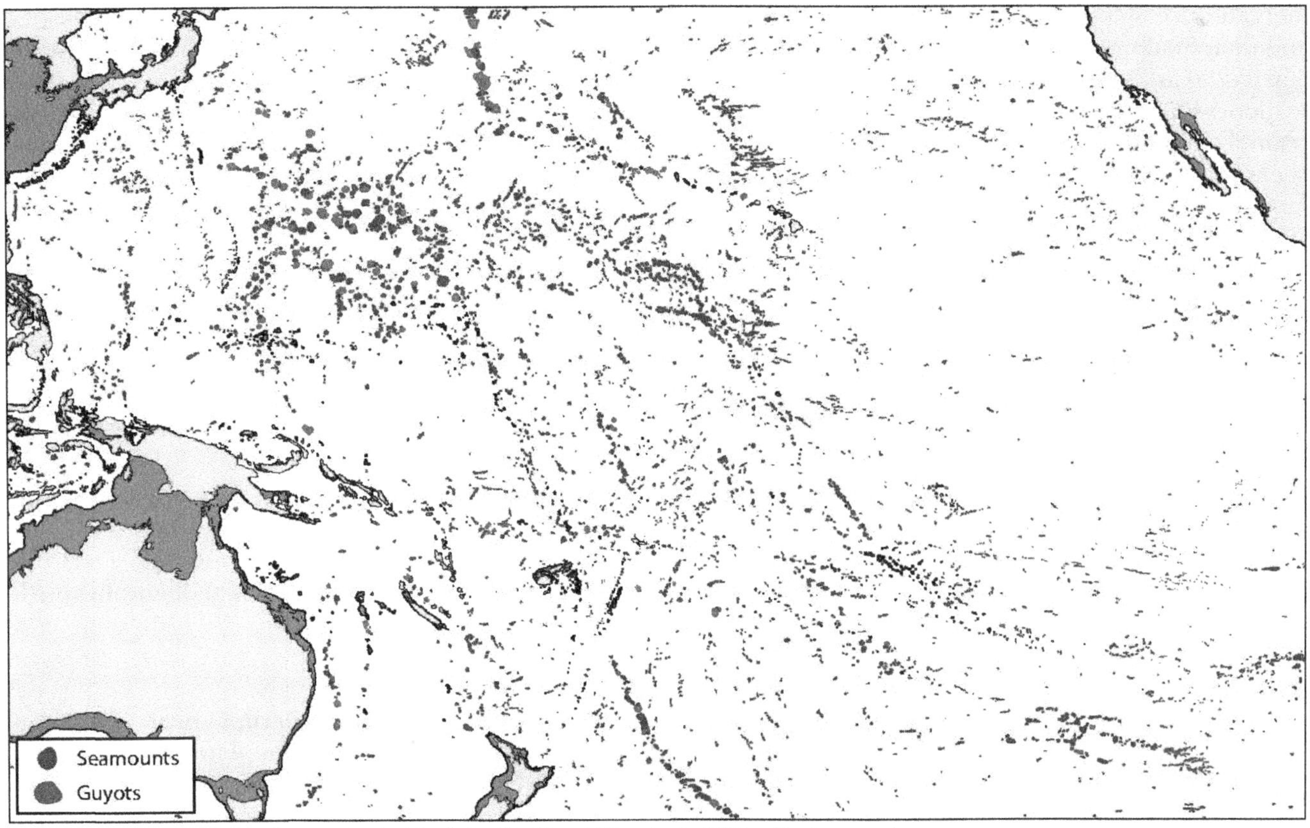

Distribution of seamounts and guyots in the North Pacific (PeterTHarris via Wikimedia Commons)

is also at these or shallower depths that the submarine eruption may be noticed by passing ships as an area of dark, turbulent water, or floating, fragmented and bubbly rocks called "pumice," and gas and ash clouds at the surface. During both of these stages, magma may continue to be injected along cracks and layers inside the growing seamount and crystallize internally. If the magma finds a place where it can escape along the sides or base of the mountain, a flank eruption may build a parasitic cone.

The next stage in the growth of a seamount, when the eruption is close to sea level, is critical. The effects of the exploding gases, combined with the effects of wave erosion on the loose, fragmented debris, tend to prevent the seamount from emerging above the water to become an island. If the volume of lava erupted is sufficient to keep pace with these destructive forces, however, subaerial lava flows may begin to cover the hyaloclastites with an armor plating that is resistant to wave erosion, and a volcanic island is born.

EROSION AND SUBSIDENCE

During its life as an island (which is likely to be only a few million years, a relatively brief period compared to its life as a seamount and guyot), the mountain is attacked by wind, running water, and glaciers (depending on its latitude). All these factors tend to reduce its elevation even as subsequent eruptions may build it up. If the conditions are tropical, coral reef growth may encrust the wave-battered shores. Every island eventually succumbs, however, to the combined effects of erosion and subsidence (sinking) of the seafloor that begin to submerge it once again. Coral growth may be able to keep pace with the effects of subsidence for a time. Sometimes, this extensive coral growth results in a completely coral-capped volcano known as an atoll. Many atolls in the South Pacific have more than 3,280 ft (1,000 m) of massive coral growth above their volcanic bases.

If coral growth cannot keep pace with the rate of seafloor subsidence, the island or atoll becomes submerged once again, this time in the form of a bank and then finally as a guyot once it reaches a depth of 656 ft (200 m). The mountain erodes gradually on the seafloor: The sides are subject to erosion by currents, crumbling, and slumping, and the rocks undergo chemical reactions with the seawater. The entire mountain is also gradually buried in the ubiquitous "rain" of dust and the tiny remains of microscopic plants and animals, which cap the top and drape the sides in thick layers of fine sediments.

CONTINENTAL DRIFT

No guyot lasts for a very long period of geologic time however, because all are eventually carried by continental drift to subduction zones along the edges of continents, where the seafloor is effectively recycled as it moves downward into the mantle below the continental mass. The study of seamounts and guyots has helped to verify the idea of continental drift (plate tectonics) and to show how fast plate movement occurs. The seafloor consists of great slabs (plates) of quasi-rigid rock 31 to 50 miles (50 to 80 kilometers) thick. The "oceanic lithosphere," as it is called, is created at the great midocean ridges, where two plates are being driven apart from each other by convection currents in the underlying magma. Magma wells up to fill the gap thus created. As the magma cools and crystallizes, it is added to the slab and becomes part of the plates. Thus, new seafloor lithosphere is added to both of the diverging plates at midocean ridges. Sometimes, a seamount will grow at the midocean ridge (Iceland is an example of a volcanic island in this location) and actually become split in half as the two plates pull apart.

The slabs of oceanic lithosphere are able to slide about on a slippery zone in the upper mantle, known as the asthenosphere, just as slabs of ice on a winter pond can drift on the water. Where stresses push two plates together, one is forced to override the other. The plate that is forced underneath (subducted) plunges deep into the asthenosphere and eventually melts. Because oceanic lithosphere is thin and dense compared to continental lithosphere, it is the seafloor that gets subducted, or recycled, most often. Thus, the seafloor, carrying its seamounts, is geologically much younger than the continents.

HOT SPOTS

Although some seamounts erupt near midocean ridges, many that erupt far from plate boundaries are associated with hot spots. Hot spots or plumes, are believed to be long, narrow fountains of magma that rise

"Kahalewai" seamount mapped during the NOAA's Mountains in the Deep: Exploring the Central Pacific Basin expedition (National Oceanic and Atmospheric Administration)

from deep within the mantle, well below the asthenosphere. These magmas have a chemical composition that is quite different from that of the magma of the midocean ridges. Seamounts are generally made up of alkaline basalts (basalts enriched in the elements sodium and potassium compared to midocean-ridge basalts).

The upwelling plume of magma at the hot spot has a dramatic effect on the seafloor lithosphere passing over it. The rising plume pushes the lithosphere upward as much as 3,280 to 4,921 ft (1,000 to 1,500 m) into a broad arch or swell and heats the lithosphere from below, making it thin and stretched. Finally, if magma can break through the weakened lithosphere, a seamount is born. Eventually, however, the oceanic lithosphere is carried by plate motion away from the hot spot, which, unlike the lithosphere, has a relatively fixed location. The seamount or volcanic island begins to move, along with the lithosphere in which it is embedded, down the side of the bulge caused by the upwelling magma. In addition, the heavy burden of a large volcano on the thin oceanic lithosphere causes it to sag downward into the soft asthenosphere below. Both of these effects combine to pull a seamount downward or to submerge an island into a guyot.

At the same time that the old volcano is moving off the hot spot and subsiding, magma may continue to erupt through the new lithosphere over the hot spot, and a new volcano will emerge next to the old one. Over time, repeated eruptions as the plate slowly moves over a stationary hot spot create a type of submarine volcanic chain called an "island arc." The shape of the arc reflects the direction of movement of the plate over the plume below. In this way, over millions of years, a whole chain of seamounts, some of which may have been islands at one time, extend from the location of the hot spot in the direction of plate motion for thousands of miles (or kilometers). It was in this fashion that the Hawaiian Islands and the closely related Emperor Seamounts were formed as the Pacific plate was pulled over a hot spot in a northwesterly direction toward the subduction zone near the Aleutian Islands of Alaska.

—*Sara A. Heller*

Further Reading

Bardintzeff, Jacques-Marie, and Alexander R. McBirney. *Volcanology*. 2nd ed., Jones & Bartlett Publishers, 2000.

Bull, William B. *Tectonic Geomorphology of Mountains: A New Approach to Paleoseismology*. Wiley-Blackwell, 2007.

Edwards, John. *Plate Tectonics and Continental Drift*. Smart Apple Media, 2006.

Floyd, P. A., editor. *Oceanic Basalts*. Springer Science & Business Media, 2012.

Garrison, Tom S. *Essentials of Oceanography*. 6th ed., Brooks/Cole Cengage Learning, 2012.

———. *Oceanography: An Invitation to Marine Science*. Brooks/Cole, Cengage Learning, 2010.

Harris, Peter T., and Elaine K. Baker, editors. *Seafloor Geomorphology as Benthic Habitat*. Elsevier, 2011.

Kennish, Michael J. *Practical Handbook of Marine Science*. 3rd ed., CRC Press, 2001.

Pitcher, Tony J., et al. *Seamounts: Ecology, Fisheries and Conservation*. John Wiley & Sons, 2008.

Seibold, E., and W. H. Berger. *The Sea Floor: An Introduction to Marine Geology*. 3rd ed., Springer-Verlag, 2010.

Yuen, David A., Shigenori Maruyama, Shun-ichiro Karato, and Brian F. Windley, editors. *Superplumes: Beyond Plates Tectonics*. Springer, 2007.

Seawater Composition

Fields of Study: oceanography; marine chemistry; geology; paleogeology

ABSTRACT

The properties of seawater are determined primarily by the properties of pure water, and secondarily by its nature as a solution. Because water is a liquid and has great capacity as a solvent, seawater is well mixed and salty due to the ions that have been dissolved from the rocks of the continental crust. Seawater is a source of mineral wealth for humankind.

KEY CONCEPTS

chemical weathering: a mode of erosion in which materials are broken down by chemical reaction processes occurring with water solutions such as acidic rain, or by simple extraction of water-soluble components

corrosion: degradation of the material integrity of a pure metal surface by chemical interaction with some component of seawater, such as chloride ion and dissolved oxygen, resulting in the formation of compounds that were not previously present on the metal surface

hydrologic cycle: also called the "water cycle," in which Earth's water changes state from liquid and solid to atmospheric gas, and condenses back to liquid or solid and falls as precipitation

messenger weight: an object that slides down the adjacent wire of a water-sampling line to trip the latches retaining the sample bottles at different depths, allowing them to capture water samples and be recovered

salinometer: a device that measures the salinity of a water sample by determining its electrical conductivity, which is then related to its saline composition

THE FORMATION OF EARTH

The hydrosphere consists of the water areas of Earth, which include ponds, lakes, rivers, groundwater and the oceans. The oceans form the largest portion of the hydrosphere, covering 71 percent of Earth's surface. The composition of Earth's water derives from the circumstances surrounding the formation of the solar system and the planet Earth.

In the beginning, the simple hypothesis is that a mass of gases and space dust came together and gravitational eddies eventually formed separate clouds of aggregate materials, which presumably later consolidated into planets. When the young planetary mass started to cool, it formed a crust, which being thin, allowed heat, molten material, and gaseous material to escape from the interior through numerous cracks. The gaseous material formed the first atmosphere, probably made up primarily of hydrogen molecules and helium atoms. These materials, being extremely light in weight and highly energized by atmospheric temperatures and incoming solar radiation were probably quickly lost to space. Eventually, they were replaced by other gases derived from the interior by the ongoing volcanic activity as the planet's surface continued to cool. This subsequent atmosphere was composed mainly of carbon dioxide and water vapor, with higher percentages of other gases released from entrapment in rock structures than are found today. All the water on the planet is believed to have been released from within the mineral structures of the material from which the planet formed.

Clouds formed from condensing water vapor in the ancient atmosphere, shielding Earth's surface and allowing less than 60 percent of the energy coming to Earth from the Sun to penetrate. As the surface continued to cool, the water vapor condensed into liquid and began to fall and accumulate in depressions. With widespread volcanic activity continuing, water vapor was continuously being released, along with smaller amounts of carbon dioxide, chlorine, nitrogen, and hydrogen gas, which then underwent chemical reactions driven by sunlight to produce methane and ammonia. As the surface continued to cool, vast amounts of condensation eventually formed the oceans. Geologic evidence shows that the oceans have existed for at least 3 billion years. This evidence comes in the form of algal fossils presumed to have grown in a marine environment.

Earth's crust was formed by primary crystalline rocks, the first molten material that solidified into rock as the planet cooled. These rocks were weathered and eroded into the particles that became deposited and slowly dissolved in water, carried into the ocean basins and accumulated. Also carried into the ocean were great loads of particulate material weathered from the primary crystalline rocks and deposited as sediments. Thus the components of the primary rocks were freed by chemical weathering and dissolved in ocean water or chemically bonded with sediments and carried into the ocean by rivers. The chemical weathering took place because of the different compounds in the atmosphere, some of which combined with water vapor to form acidic rainwater ranging in strength from only slightly acidic solutions of carbonic acid to solutions of much stronger nitric and sulfuric acids. Each of these various acidic solutions had the ability to leach out metal ions, such as sodium, potassium, magnesium, iron, and others, from the materials that they contact. Seawater gained its characteristic saltiness through these processes.

Comparing the discharge of gases by hot springs in the United States to the average rate throughout the 3 billion years of the oceans' existence shows that enough water vapor is produced to fill the oceans to one hundred times their present volume. Water therefore must be recycled and not all newly formed; the excess amount does not represent new water released from the original crystallization of magma. Only 1 percent of this water requires such an origin to account for the present volume of the oceans.

Water on land comes daily from the sea, as part of the hydrologic cycle, or water cycle. The seas hold about 155 billion cubic feet (4.4 billion cubic meters)

of saltwater. Of this amount, about 424 million cu ft (12 million cu m) enters the atmosphere each year through evaporation and is returned by rainfall and the flow of rivers, and about 106 million cu ft (3 million cu m) descends each year over continents, replenishing ponds, lakes, and rivers.

GASES AND SOLIDS IN SEAWATER

Ocean water is a mixture of gases and solids dissolved in pure water (96 percent pure water and 4 percent dissolved elements, by weight). Nearly every natural element has been found or is expected to be found in seawater, although some occur only in miniscule amounts. The most abundant mineral found in ocean water is sodium chloride (familiar as common table salt), which makes up 85 percent of the dissolved minerals. It is interesting to note that the composition of human blood is very similar to that of seawater; in it are all the elements of the sea, dispensed in different proportions.

The seven most abundant minerals in seawater include sodium chloride (2.72 percent, or 27.2 parts per thousand), magnesium chloride (0.38 percent, or 3.8 ppt), magnesium sulfate (0.17 percent, or 1.7 ppt), calcium sulfate (0.13 percent, or 1.3 ppt), potassium sulfate (0.09 percent, or 0.9 ppt), calcium carbonate (0.01 percent, or 0.1 ppt), and magnesium bromide (0.01 percent, or 0.1 ppt). It is important to note that these "minerals" do not occur in the form suggested by their molecular formulas, but as the component ions dissolved in water. Six ionic species actually make up 99.3 percent of the total mass of the dissolved material: chloride (55.2 percent), sodium (30.4 percent), sulfate (7.7 percent), magnesium (3.7 percent), calcium (1.2 percent), potassium (1.1 percent), and others (0.7 percent).

The salinity of seawater is fairly uniform across the oceans at different latitudes and at various depths, mainly because of winds, waves, and currents. Local variations in mineral content are attributed to freshwater streams entering oceans, glacial melt, and human activity, but the variations overall are small. The salinity averages about 3.5 percent (35 ppt). This has been verified by the *Glomar Challenger* expedition, using Nansen bottles to take samples of seawater at different depths around the world and salinometers to measure salinity, as well as many other types of tests. There is more variation of salinity at the surface than at depths because of freshwater (rivers, glacial melt) entering the ocean at a given location, plus the biological activity and climate at different latitudes.

The Glomar Challenger *(Wikimedia Commons)*

In areas where freshwater is entering the ocean, the salinity will decrease. Biological activity changes the salinity according to which species and how many marine plants and animals reside in the area. In climates that are hot and dry, the rate of evaporation is high and rainfall is low, so the ratio of dissolved salts to water is higher, making the water more saline (saltier). Similarly, in the polar regions in winter, the water freezes but most of the minerals do not, which changes the water-to-mineral ratio and thus increases salinity. In most climates, rainfall is greater than evaporation, thus diluting seawater and decreasing salinity.

The most abundant dissolved gases in the ocean are nitrogen, carbon dioxide, and oxygen. The amount varies with depth, and as oxygen and carbon dioxide are vital to life, most living plants and animals are found in the top 328 feet (100 meters) of the ocean, or the sunlight penetration layer. The amount of dissolved gases will also vary with temperature. Warm water holds less dissolved gas than cold water because cold water is heavier and sinks, carrying oxygen-rich water to the ocean depths, which, in turn, allows fish and marine life to live in the deepest parts of the ocean. At the surface, gases (oxygen and carbon dioxide) are exchanged between the ocean and the atmosphere as well as between the plants and animals that live in the top layers of the ocean, where the most abundant life is found. Marine plants take in carbon dioxide along with water and then, with the aid of sunlight, produce the sugar glucose and oxygen in the process of photosynthesis. Marine animals and photosynthetic plants take in oxygen and exhale or release carbon dioxide in the process of respiration.

Most precipitation that falls finds its way back to the ocean largely by rivers, bearing salts from soils and rock in solution. Many things affect the salinity of the ocean, including the exchange of water between the ocean and the atmosphere, which is determined by climate, and the absorption of salts by plants and animals. Considering the history of ocean salinity, one might ask if the oceans have possessed a relatively uniform salinity throughout their history or if they are becoming more saline. By far the most important component of salinity is the chloride ion, which is produced in the same manner as water vapor. The ratio of chloride ions to water vapor has not fluctuated throughout geologic time so scientists conclude, on the basis of present evidence, that ocean water salinity has been relatively constant over the lifetime of the oceans.

It is now evident that the oceans became salty early in their history because of weathering by acidic rainfall and the erosion of primary crystalline rock. Also, through continual volcanic activity, water vapor and gases were amply supplied to the atmosphere to aid in this weathering and eroding process. These particles were carried to the ocean and dissolved in seawater, making the water salty. This process has been going on since the formation of the oceans, so it is safe to assume that the ocean has been salty since it was formed approximately 3 billion years ago.

Comparing the salinity of oceans both past and present by studying vents and hot springs, the salinity appears to maintain a certain balance. Salinity ranges between 3.3 and 3.8 percent (33 and 38 ppt), by weight. This variation is caused by atmospheric effects at different latitudes, by freshwater entering into the oceans from rivers and glacial melt, and by biological activity in the ocean itself as plants and animals remove minerals needed for their growth and development (photosynthesis, respiration, shell building, and so on) from the seawater and put into it other forms (waste products, skeletal structures, shells casts, and so on). In local areas, such as bays, coves, and estuaries, the salinity of seawater is further affected by human activities such as industry and agriculture, and by the pollution from these activities.

STUDY OF SEAWATER COMPOSITION

The investigation of seawater composition is accomplished primarily by collecting samples of seawater at different depths around the world and then measur-

Nansen bottle on a dropline (Donald Emmerich/U.S. National Archives and Records Administration via Wikimedia Commons)

ing the salinity of the samples. A variety of instruments are used to collect samples. Nansen bottles are special metal cylinders fastened at a measuring point on a strong wire, which is then lowered into the sea to the desired depth. A messenger weight is dropped down the wire; when it strikes the bottle, it releases a catch. The bottle then turns upside down, and its valves close, trapping the water at that depth inside. The Nansen bottle commonly used does not seal completely, however; a better apparatus is the Fjorlie sampler or Niskin bottle. The Fjorlie sampler or Niskin bottle is attached to a line at both ends with spring-closing hinged ends. A messenger closes the bottle with a good seal.

Corrosion of metal-lined samplers may cause changes in the water composition in an hour or so. Copper, zinc, lead, and iron in metal linings often contaminate seawater samples. Plastics have solved this problem for both collection and storage of seawater. It is also necessary to filter out any organic matter that could alter the seawater composition. For maximum accuracy of testing, samples should be tested as soon as possible and not stored.

A salinometer is used to electronically measure salinity. It is easy to use and gives immediate readings. Because the ions dissolved in seawater affect its properties as a conductor of electricity, the more ionic mineral matter in the sample, the better it conducts. The results are then compared to a table of standard measurements to obtain the sample's salinity.

—Joyce Gawell

Further Reading

Braitsch, O. *Salt Deposits: Their Origin and Composition.* Springer Science & Business Media, 2012.

Emerson, Steven, and John Hedges. *Chemical Oceanography and Marine Carbon Cycle.* Cambridge UP, 2008.

Grasshoff, Klaus, Klaus Kremling, and Manfred Ehrhardt, editors. *Methods of Seawater Analysis.* 3rd Completely Revised and Extended ed., John Wiley & Sons, 2009.

Garrison, Tom S. *Essentials of Oceanography.* 6th ed., Brooks/Cole Cengage Learning, 2012.

Harrison, Roy M. *Principles of Environmental Chemistry.* Royal Society of Chemistry, 2007.

Hester, R. E., and R. M. Harrison. *Chemistry in the Marine Environment.* Royal Society of Chemistry, 2000.

Marshall, John, and R. Alan Plumb. *Atmosphere, Ocean and Climate Dynamics: An Introductory Text.* Academic Press/Elsevier, 2008.

Spaulding, Nancy E., and Samuel N. Namowitz. *Earth Science.* Houghton Mifflin Harcourt, 2004.

Steele, John H., Steve A. Thorpe, and Karl K. Turekian, editors. *Marine Chemistry and Geochemistry.* Academic Press, 2010.

Trujillo, Alan P., and Harold V. Thurman. *Essentials of Oceanography.* 10th ed., Prentice-Hall, 2010.

Williams, Richard G., and Michael J. Follows. *Ocean Dynamics and the Carbon Cycle: Principles and Mechanisms.* Cambridge UP, 2011.

South Atlantic Ocean

Fields of Study: oceanography; marine biology; marine chemistry; marine biogeochemistry

ABSTRACT

The South Atlantic Ocean, the portion of the Atlantic that lies south of the equator, has historically been relatively untouched in terms of economic exploitation, but humanity's search for natural resources is increasingly focused on its waters.

KEY CONCEPTS

endemic species: species that have adapted to and evolved within a specific region

indigenous species: species that have always existed and evolved in a particular region rather than having adapted to that region

krill: small shrimp-like organisms that swarm by the billions after hatching from the eggs of their respective parent species; krill form the base of the marine food web for many larger species

ocean acidification: a reduction of the pH of ocean water due to the formation of carbonic acid and other acidic compounds as carbon dioxide and other gases are absorbed from the atmosphere by ocean water

DESCRIPTION

The generally accepted boundaries of the South Atlantic are rather fluid, changing several times over the course of the twentieth century as experts debated the existence and boundaries of an ocean further south variously named the Southern, Austral, or Antarctic Ocean. Early in the twentieth century, the Antarctic

Ocean was defined as lying below the Antarctic Convergence, an area where cold waters from Antarctica sink below warmer subtropical waters from the north. For most of the latter part of the twentieth century, the International Hydrographic Office failed to recognize such an ocean, but in 2000 proposed resurrecting the Antarctic Ocean in the form of the Southern Ocean, whose northern boundary would be 60 degrees south latitude. The proposal had not been ratified as of 2012.

Southern boundaries aside, the western boundary of the South Atlantic is defined as the east coast of South America and the great Rio de La Plata estuary. On the north, it is bounded by the equator and the Gulf of Guinea. On the east, it is bounded by the west coast of Africa. With these as the boundary, then the South Atlantic is slightly smaller in area than the North Atlantic, but it is deeper and has a greater volume than its northern counterpart.

The South Atlantic Ocean, like other large ocean basins, has a gyre—the South Atlantic Gyre—at its heart. Here, it circulates in counterclockwise manner, bringing warm water poleward and cold water toward the equator. Its western arm, the Brazil Current, is a weaker, southern counterpart to the North Atlantic's Gulf Stream, flowing south toward Antarctica. Off the tip of South America, it collides with the Malvinas (Falklands) Current, itself a north-flowing offshoot of the eastward-flowing Antarctic Circumpolar Current. Part of the Brazil Current breaks away in the vicinity of this confluence to form the eastward-flowing South Atlantic Current, which, upon reaching Africa, splits, with a northward-flowing branch forming the Benguela Current.

The Benguela Current follows the west coastline of Africa. Ekman transport (a wind-driven force) triggers upwelling of cold, nutrient-rich waters inshore of this current, creating conditions for traditionally rich fisheries in the upwelling zone; these are found off the southwestern coast of Africa and the Cape of Good Hope. As the Benguela Current flows northward, its waters warm. It turns westward near the equator, becoming part of the South Atlantic Equatorial Current—the northern part of the South Atlantic Gyre. In the vicinity of Brazil's Cabo de Sao Roque, the South Atlantic Equatorial Current splits, one portion flowing north as the North Brazil Current, the other portion flowing south, becoming the Brazil Current and completing the circulation. The North Brazil Current is much the stronger; most of it flows northwestward into the Northern Hemisphere, but part of it splits off, flowing eastward as the North Equatorial Countercurrent.

MARINE AND COASTAL LIFE

The upwelling zone of the Benguela has created conditions for rich fisheries of species such as rock lobster, pilchard, anchovy, Cape horse mackerel, and hake. Overharvesting, however, has depleted the stocks of rock lobster and pilchard, among other species. Other notable species include the jackass penguin and a number of sea mammals, such as the Cape fur seal, southern right whale, and Heaviside's dolphin. Kelp forests, similar to those that occur off the coast of California, are locally important.

The cold, stormy waters of the southwest Atlantic have been havens for a wide variety of marine life, offering breeding areas for marine birds and mammals —including Magellanic penguin, imperial cormorant, pale-faced sheathbill, southern right whale, humpback whale, Commerson's dolphin, southern elephant seal, southern sea lion, and South American fur seal. The whaling and sealing industries began taking a heavy toll on many marine mammal species in the nineteenth century. Currently, the region supports economically important fisheries for hake, common squid, shortfin squid, and other species.

In subtropical and tropical waters farther north, collisions of warm currents with upwelling zones support a diverse environment, including mangroves and coral reefs. The beaches are important nesting areas for communities of leatherback, green, hawksbill, olive ridley, and loggerhead turtles. The waters are important tropical breeding areas for humpback whales and the smalleye hammerhead.

The South Atlantic features a number of isolated islands or island groups that encompass a significant amount of environmental and biological diversity. South Georgia and the South Sandwich Islands feature mountain ranges covered by snow and ice; their coastlines are breeding grounds and havens for marine birds and mammals.

Near the equator, Ascension Island—roughly midway between Africa and South America—appears to be the definitive desert island, both on land and at sea. The large expanses of rust-brown volcanic cinder cones and grayish-white lava flows were once breeding grounds for a number of seabirds, including the wideawake tern and Ascension frigatebird. The island's beaches are important nesting sites for green and hawksbill turtles. The low-nutrient marine environment has no coral reefs and lacks the diversity of reef ecosystems found at similar latitudes. Ascension's nearshore waters are noted for abundance of black durgon and ocean triggerfish; they also host a number of endemic (adapted and native uniquely to a biome) species such as resplendent angelfish and Lubbock's yellowtail damselfish.

IMPACT OF MANKIND

Many of the ecosystems of the South Atlantic had long been relatively sheltered from human activity, but in the twentieth century, a burgeoning human population began putting greater and greater demands on the environment. Petroleum exploration and development, for example, is bringing much-needed cash to developing nations of Africa and South America, but at the cost of environmental disruption because of construction and accidents, including oil spills. Additionally, introduced species ravage once-isolated regions, and coastal construction-such as for resorts in tropical areas—leads to increasing loss of natural coastal areas and seascapes. Newly discovered resources, such as deep-sea fisheries, are depleted before there is sufficient scientific understanding of the resource to guide the development of sound management policies. In addition to commercial fishing, whaling activity in the region has been a concern for many years. Since 1998, the International Whaling Commission has discussed on many occasions a proposal to establish a sanctuary for whales in the South Atlantic, but as of 2018 had not yet established such a sanctuary. The government of Brazil, which has spearheaded the effort to create a whale sanctuary, expanded marine conservation areas in its own waters in 2018. However, the election of Jair Bolsonaro in 2019 threatens to undo this and many other environmental policies of his predecessors.

Ocean pollution has also been a major issue. Like the North Atlantic and the Pacific, the South Atlantic has a "garbage patch" made up of litter, mainly plastic, that has entered the ocean and been carried to the South Atlantic Gyre. The patch, smaller than its Pacific and North Atlantic counterparts, was first detected in 2014 and at that time had a maximum density of 100 pieces of litter per square mile (or kilometer).

Much work remains to be done to identify and understand the ecosystems of the South Atlantic. A further overriding concern, of course, is the oncoming suite of changes being wrought by global climate change. A heavier atmospheric carbon dioxide load migrating into seawater causes its acidification—a process with often striking negative effects on corals, crustaceans, mollusks, and other organisms whose ability to create their calcium structures is blocked by the unbalanced seawater chemistry.

Ocean warming (and further south, melting sea ice) has been shown to rather dramatically reduce the supply of Antarctic krill in the South Atlantic; the depletion of this bottom-of-the-pyramid food source directly affects the population stability of such higher animals as penguins and baleen whales. Increased temperatures in the major surface currents along the southern African coast, several studies suggest, will likely cause increased rainfall in coastal areas, which in turn would alter the mix and augment the amount of land-based pollutants and nutrients flowing into the South Atlantic.

—David M. Lawrence

Further Reading

"Brazil to Expand Marine Conservation Areas in the South Atlantic." *Global Environment Facility*, 22 Mar. 2018, www.thegef.org/news/brazil-expand-marine-conservation-areas-south-atlantic.

Britt-Gallagher, Susan, and Tricia Hayne. *St. Helena: Ascension. Tristan da Cunha.* 3rd ed., Bradt Travel Guides, 2015.

Duarte, Érico, and Manuel Correia de Barros, editors. *Maritime Security Challenges in the South Atlantic.* Palgrave Macmillan, 2019.

Earle, Sylvia A., and Linda K. Glover. *Ocean: An Illustrated Atlas.* National Geographic Books, 2008.

Hutchinson, Stephen, Johann R. E. Lutjeharms, Beverly McMillan, John Musick, Bernard Stonehouse, and Matthias Tomczak. *The Illustrated Atlas of the Sea.* Weldon Owen, 2010.

Ryan, Peter G. "Litter Survey Detects the South Atlantic 'Garbage Patch.'" *Marine Pollution Bulletin*, vol. 72, nos. 1–2, 2014, pp. 220–24.

Spalding, Mark D., et al. "Marine Ecoregions of the World: A Bioregionalization of Coastal and Shelf Areas." *BioScience*, vol. 57, no. 7, 2007.
Thurman, Harold V., and Alan P. Trujillo. *Introductory Oceanography*. 10th ed., Prentice-Hall, 2004.
van Aken, Hendrik M. *The Oceanic Thermohaline Circulation: An Introduction*, Springer, 2007.
Wefer, Gerold, Wolfgang H. Berger, Gerold Siedler, and David J. Webb, editors. *The South Atlantic: Present and Past Circulation*. Springer Science & Business Media, 2012.

SOUTH CHINA SEA ECOSYSTEM

Fields of Study: oceanography; marine geology; marine biology; marine biogeochemistry; international affairs

ABSTRACT
The South China Sea contains up to one-third of the world's marine biodiversity, which gives the region a substantial renewable resource base. Below its floor is an unusually large amount of crude oil and natural gas, which has sparked international conflict over extraction rights. Conservation and preservation of the sea's unique and diverse ecosystems are essential.

KEY CONCEPTS

reef raiding: illegal removal of reef fish and other reef organisms, whether for food, the coral market or the aquarium trade

strait: a relatively narrow passage between two landmasses, typically with a current flow of water passing through; straits may be between two islands or groups of islands, between an island and a continental mass, or between two continental land masses

tablemount: a large, flat-topped undersea rock structure similar to guyots and banks

GEOLOGY AND ORIGINS

Awash in the waters of the Indian and Pacific oceans, the South China Sea contains up to one-third of the world's marine biodiversity, which gives the region a substantial renewable resource base—but one that by some accounts is slowly crumbling. Sealed below its floor is an unusually large amount of crude oil and natural gas, which has sparked international conflict over extraction rights. Conservation and preservation of the sea's unique and diverse ecosystems may prove to be the key that opens the way for cooperation among the people of the region.

The South China Basin is the deepest part of the sea in this region. It is surrounded by three significant geomorphologic features: a segment of Eurasia's continental shelf in the west; the Reed Tablemount and Manila Trench to the east; and the Sunda Shelf to the south. Around 30 million years ago, a vast block of rock called the "Reed Tablemount" slowly separated from the Eurasian continent. In between, upwelling magma cooled and hardened to compose the basin floor, a process that ended 17 million years ago. The Spratly Islands sitting atop the Reed Tablemount are mostly uninhabited, but are the most politically contested archipelago in the world. In 2018, tensions regarding the islands increased further after China had a system of defensive missiles installed at some of its outposts on the islands.

To the east, this submarine plateau drops into the Manila Trench, an abyssal habitat where marine life remains mostly a mystery. As the Eurasian tectonic plate slowly moves below the Philippine plate, the upper layer of Earth's crust cracks and fissures, giving magma from below an opportunity to flow to the surface and create volcanic islands.

In the warm tropical and subtropical climates, corals colonize these mountains to form coral reefs. When the forces of erosion cause rock to disappear below the water's surface, the remaining living rings of coral create an atoll surrounding a lagoon. The coral reefs and atolls stretching from the Philippines island of Palawan down to northern Borneo form the base of southeast Asia's Coral Triangle, which consists of a network of protected areas that tapers toward Australia. Environments like these are mainsprings of marine biodiversity—and extremely rich hatcheries for both the Indian and Pacific Oceans.

BIODIVERSITY

Much of the South China Sea's biodiversity springs from the continental and Sunda shelves here, which, along with the Tablemount, endow it with general shallowness. Along the coastal rim, terrestrial biomes such as tropical rainforests and savannas interact with inland aquatic biomes such as the Mekong, Red, and

Pearl rivers, with marine estuaries creating life zones bristling with myriad species of plants and animals.

Among the most animated transition zones along the coast of the South China Sea are mangrove forests, in which gentle warm-water sea currents bring salt to the mouths of freshwater rivers and streams. Such areas produce exceptionally high levels of biomass that in various ways become dynamic natural resources. Mangroves protect coastlines from erosion and the added destructive effects of recurrent typhoons, which usually form between May and September. The region supports about 30 species of mangrove, and has extensive swaths of seagrass beds in the intertidal zones around its perimeter—yet another rich area providing shelter, breeding and feeding grounds for myriad mollusk, crustacean, fish, and marine mammal species. However, progressively more such habitat is being lost to industrial pollution, agricultural runoff, coastal development, aquaculture, and overfishing. In 2015, a report released by the Paulson Institute, which worked alongside the Chinese government, found that the speed of the country's land reclamation in efforts to support economic development had led to the loss of 73 percent of its coastal mangroves, threatening the habitat of several migratory birds and decreasing the natural level of protection against rising seas.

The Sunda Shelf and Coral Triangle have some of the world's most spectacular coral reefs, teeming with species of mollusks, crustaceans, sea snakes and turtles, but they, too, are succumbing to environmental threats. Reef raiding, blasting, and poisoning have depleted these creatures; subsistence and commercial fishers compete for depleted fish stocks, while demand for seafood is at an all-time high. The seagrass and soft bottom beds suffer similar damage, as trawling ships scrub the floor of all that can and cannot be eaten by human beings. International bodies have made some headway in outlawing such practices and introducing more sustainable fishing regimes, but the struggle is by no means won.

HUMAN FACTORS

The territories of Brunei, Cambodia, China, Indonesia, Malaysia, Philippines, Thailand, and Vietnam extend into the South China Sea. Approximately 270 million people live in or near the coast of the South China Sea, a population expected to double by 2040. The economies of most of these countries depend heavily on international trade in manufactured and agricultural goods. Because shipping is the least expensive form of transport, the sea is central to intraregional trade.

Further, it is also one of the world's major global shipping lanes, because its waters funnel through the Straits of Malacca, which for hundreds of years has been a significant passageway for people and goods traveling east and west. With the opening of the Suez Canal in 1869, and the more recent rise of China and Japan as first-rate economies, the interaction among Europe, southwest Asia, and Africa with the countries of the South China Sea has never been greater.

Climate change may impact fishing in the area; as the temperatures of the waters change, and species of fish move in and out of the waters, the nations that rely upon commercial fishing in the South China Sea may become increasingly protective of their areas, vying for whatever fish remain. Not only can climate change impact the marine life living in the South China Sea, but it also may have social and political impacts upon the area, depending on the physical outcomes upon the natural systems here.

To stanch these threats, the United Nations Environment Programme (UNEP) and the Global Environment Facility, along with seven regional nations, have been involved in the South China Sea Project since 2006. Comprehensive scientific surveys on several transboundary ecosystems and subsystems have been completed; these will provide a baseline from which to gauge future changes to this large marine ecosystem.

National governments along with a consortium of international universities are working together to give local people the skills to survey and manage ecosystems; promote public awareness of environmental issues; and understand legal and regulatory issues concerning restoration, preservation, and conservation of local, transboundary environments.

The hope is that increasing cooperation among the people of the South China Sea will lead to sustainable development policies and practices, so that certain geopolitical tensions will no longer threaten to snap the string of cultural and natural pearls adorning the sea. However, as of 2018, such tensions had prevailed, leading to concerns regarding overfishing and increasingly depleted fish stocks as fisheries in countries in the region face greater competition without any real regulation of practices. In 2016, the Chinese government rejected the decision by a Hague tribunal stating that its claims of historic rights to much of the territory in the South China Sea were not legal.

—*Ken Whalen*

Further Reading

Bale, Rachael. "One of the World's Biggest Fisheries Is on the Verge of Collapse." *National Geographic*, 29 Aug. 2016, news.nationalgeographic.com/2016/08/wildlife-south-china-sea-overfishing-threatens-collapse/.

Blueprint of Coastal Wetland Conservation and Management in China. Paulson Institute, 2016, www.paulsoninstitute.org/wp-content/uploads/2016/08/Wetland-Report-EN-Final.pdf. Accessed 30 July 2018.

Dang, Vu Hai. *Marine Protected Areas Network in the South China Sea: Charting a Course for Future Cooperation*. Martinus Nijhof, 2014.

Ho, Chung-Ro, Lingling Xia Jianyu Hu, and Quanan Zheng, editors. *Regional Oceanography of the South China Sea*. World Scientific Publishing, 2020.

Perlez, Jane. "Tribunal Rejects Beijing's Claims in South China Sea." *The New York Times*, 12 July 2016, www.nytimes.com/2016/07/13/world/asia/south-china-sea-hague-ruling-philippines.html.

Robles. Alfredo C. *Endangered Species and Fragile Ecosystems in the South China Sea: Philippines vs. China Arbitration*. Palgrave Macmillan, 2020.

Rogers, Will. "Climate Change and the South China Sea." *Center for New American Security*, www.cnas.org/blogs/naturalsecurity/2011/09/climate-change-fisheries-and-south-china-sea.html.

South China Sea Project. "Reversing Environmental Degradation Trends in the South China Sea and Gulf of Thailand." *United Nations Environment Programme*, www.unepscs.org.

Weightman, Barbara A. *Dragons and Tigers: A Geography of South, East, and Southeast Asia*. 3rd ed., John Wiley & Sons, 2011.

SOUTH PACIFIC OCEAN

Fields of Study: oceanography; international relations; marine biology; marine geology

ABSTRACT

The Pacific Ocean is the largest geographic feature on Earth. The South Pacific is dotted with islands, including Hawaii, the eastern Pacific volcanic island chains of Pitcairn, French Polynesia, the Cook Islands, and the Samoa group. The Pacific tectonic plate forms the Pacific Ocean floor, slowly moving west from the East Pacific Rise until it is subducted beneath the Australian plate.

KEY CONCEPTS

geomorphology: study and analysis (*-ology*) of the shapes or forms (*-morph-*) of various features of the solid Earth (*geo-*)

Gondwanan fragment: a fragment of continental rock remaining from the Gondwana supercontinent that existed about 190 million years ago

subduction: the movement of seafloor tectonic plates against continental rock masses causing the denser basaltic rock of the seafloor plate to be forced downward and under the less dense silicate rock of the continental mass

volcanic hot spot: a position on Earth's crust that is directly above a superheated plume of magma in Earth's mantle layer, generally giving rise to constant emission of lava, as occurs in Hawaii; the Hawaiian Islands are a string of volcanic islands formed by emission of lava as the tectonic plate moved across the location

ISLANDS AND OCEAN

The Pacific Ocean is the largest geographic feature on Earth's surface, an area greater than all the continents combined. The South Pacific is dotted with islands because of its geological structure; the divergent plate boundary of the East Pacific Rise is the formative source of the Pacific tectonic plate. In its westward movement, the plate passes over a number of volcanic hot spots; these have formed the eastern Pacific volcanic island chains of Pitcairn, French Polynesia, the Cook Islands, and the Samoa group. The Pacific plate subducts beneath the Australian plate to the east of the main islands of Tonga, Fiji, and Vanuatu, and north of the Solomons and Papua New Guinea, causing islands to the west to be larger and less isolated.

The South Pacific region includes sixteen countries and territories comprising over 8.6 million square miles (22.3 million square kilometers) of ocean area, and less than 232,000 million sq mi (600,000 sq km)

of land, with 84 percent of this land in Papua New Guinea. Most of the remaining land area is in the other larger Melanesian islands of the Solomons, Vanuatu, Fiji, and New Caledonia, again with the islands overall declining in size from west to east. There are approximately 1,735 islands in total. Land is a primary control on ecosystem diversity, and physical conditions of geomorphology and tropical climates exert controls on habitats and ecosystems.

South Pacific ecosystems are biogeographically fragmented by physical separation of habitats across these islands, and controlled by island type: whether the island groups are characterized by high topography, by limestone formation, or by coral atoll structure. Of these major island groups, the high type is found mainly in Papua New Guinea, Solomon Islands, New Caledonia, Vanuatu, Fiji, Wallis and Futuna, Samoa and American Samoa, and French Polynesia. The atoll types include Tuvalu, Kiribati, and Tokelau. Nauru and Niue are limestone types; Tonga is mixed high and limestone type; and the Cook Islands are mixed high and atoll type.

Ecosystem diversity is at a maximum in the west at Papua New Guinea, and decreases toward the east as island sizes become smaller, and distances between islands become larger. Endemic species (those found nowhere else) occur particularly in terrestrial ecosystems, such as freshwater habitats, where the greatest threats are from introduced weed species. In coastal ecosystems, the greatest threats are also from direct human impacts, but increasingly from climate change and sea-level rise.

Papua New Guinea, Solomons, New Caledonia, Vanuatu, Fiji, Wallis and Futuna, and Tonga are all west of the tectonic margin line, which influences the biogeography of island ecosystems, as larger islands to the west are older continental crust. The three geologically old, large islands of New Guinea, New Caledonia, and Viti Levu (largest of the Fiji Islands) form three diverse biogeographic sub-regions in the Pacific Islands, from which species have migrated to other islands. Southern New Guinea and New Caledonia are Gondwanan fragments, while the Fijian large islands were formed from uplifted crust over 40 million years ago. Other islands here were uplifted coral or formed volcanically in association with the subduction margin.

Island groups to the east include Nauru, Tuvalu, Kiribati, Tokelau, the Samoas, Niue, Cook Islands, and French Polynesia; these are "true islands," mainly atoll types. This has a great influence on ecosystem biodiversity, with these being low in native species, having only secured those adapted to long-distance oceanic migration.

The ecosystems of this region are biogeographically fragmented by physical separation of habitats within the coastal biomes of coral reefs, mangroves, seagrasses, and even more so within the freshwater wetland types. Ecosystems of the South Pacific have many analogous environmental settings, but due to the filter effect of oceanic separation, the species present in ecosystems all differ between countries. This results from two factors: first, the center for species diversity within all groups is on the margins of the western edge of the region; and second, island size tends to decrease and island isolation increase toward the east and north.

Ecosystem habitats on islands are dependent on island types. While volcanic islands can host crater lakes and river catchments, over time, as they start to erode and submerge, coastal plains, mangroves, and freshwater swamps are likely to develop. Fringing reefs become barrier reefs with further island submergence, increasing area and habitat variety in reef environments and introducing lagoons and seagrass habitats. Further subsidence leads to atolls, losing all freshwater wetlands—but increasing reefs and lagoons.

FOUR BIOLOGICAL GROUPS

The biological diversity of corals, reef fish, mangroves, and seagrasses of the Pacific region result from their biogeographic history. The Indo-Malayan archipelago to the west of the region is the center of greatest species diversity for all four groups. It is where they first appeared in the ancient Tethys Sea at equatorial latitudes. Species richness declines from this center toward the east.

Coral reefs are the most extensive ecosystem type in the Pacific island region, occurring offshore of nearly all coastlines that lack turbid river discharge. New Caledonia has the second-longest barrier reef in the world at a length of about 930 m (1,497 km), which surrounds the main island and encloses a lagoon of about 9,300 sq mi (24,000 sq km). Papua New Guinea has about 15,500 sq mi (40,000 sq km) of coral reef, and some countries in Micronesia and Polynesia are almost entirely composed of coral atolls.

There are 5,000 sq mi (13,000 sq km) of coral reefs in just the countries east of Samoa, compared with 2,300 sq mi (6,000 sq km) of land. Coral reefs and freshwater mollusks are the best-studied wetland invertebrate groups in the Pacific island region; research on other groups has pointed to the need for baseline surveys and inventories.

Fiji's marine environment is better studied than most other Pacific Island countries, facilitated by a marine research and training capability at the University of the South Pacific. In 2000, Fiji's reefs suffered an extensive mass bleaching event, causing loss of 40 to 80 percent of stony corals. As a result, reef monitoring commenced to allow ongoing surveillance of coral reef biodiversity and condition, and provide better understanding of natural and human impacts. In some coastal areas, poor inshore water quality with nutrients, suspended sediment, and trace metals have been found to be affecting inshore reefs. American Samoan reefs are well studied and a detailed inventory is available, thanks to United States government research. The reefs of Tonga and Vanuatu are little studied, compared with New Caledonia or Fiji, and other islands, such as Niue, even less.

The diversity of reef-building corals in the equatorial western Pacific at Papua New Guinea is close to the highest in the world. Diversity declines to the north and the south with cooler temperatures, and from west to east across the Pacific with increasing distance.

Similar regional patterns occur in the shallow water finfish, and nearshore shallow water conches. The finfish families, which include 461 genera and approximately 1,312 species, are important sources of protein for the traditional village peoples of the South Pacific region. Reef and shore mollusks also decline in diversity from west to east across the region; species present are however dependent on habitat variety. Different molluscan assemblages live on reef slopes, outer lagoons, and inner lagoons, for example. Coastal sponges, polychaetes, and crustaceans also decline in diversity from west to east; however, the records of all these groups are patchy and more research is needed.

Six of the seven marine turtles are found in the South Pacific region; most common are the green turtle, hawksbill, and leatherback. Of these, the hawksbill is, however, critically endangered; the green and leatherback are endangered. Less common are the loggerhead, flatback, and olive ridley, of which the olive ridley is endangered, and the loggerhead and flatback turtle vulnerable.

Marine turtles are slow to reach maturity, and migrate long distances between nesting and feeding areas. Individuals tracked from a nest at Rose Atoll, American Samoa, moved to the rich seagrass and algae beds of Fiji, spending 90 percent of their time there. Impacts to turtles include bycatch from long-line fisheries, over-exploitation by local communities, and human hunting when nesting on beaches.

Mangroves occur on intertidal habitats of sheltered shorelines, which are most extensive at sedimentary estuaries. In the Pacific Islands region, the total mangrove area is about 2,196 sq mi (5,687 sq km), with the largest areas in Papua New Guinea, Solomon Islands, Fiji, and New Caledonia. The species mixture is unique in each country or territory, and the mangroves provide useful ecosystem services. There are thirty-one species of mangroves of the Indo-Malayan assemblage, with highest global biodiversity in southern Papua New Guinea; diversity declines from west to east across the Pacific, reaching a present limit at Samoa, with a possibly introduced outlier in French Polynesia.

Seagrasses occur mostly in shallow lagoons between coral reefs and shoreline mangroves in the South Pacific, with a close proximity and connectivity of these wetland types typical of island shorelines. Seagrasses evolved from terrestrial grass in the Tethys Sea during the Cretaceous Period, and spread into the Pacific. The center of biodiversity for seagrass is New Guinea, with up to thirteen species; two families and seven genera extend across the South Pacific, declining in diversity with distance, except *H. ovalis sp. bullosa*, which is endemic to Fiji, Tonga, and Samoa. In the eastern Pacific, seagrass is absent from the Cook Islands, and only one species is recorded from French Polynesia. The fruits of most seagrass species are not buoyant, hence reducing their ability to disperse over long distances.

LAND-BASED FAUNA

The tropical Pacific terrestrial mammal fauna is much less in number and types compared with elsewhere in the world, due to island isolation and lack of landbridges, and is dominated by bats. No mammals at all are present in French Polynesia, Tuvalu, Kiribati, the Marshalls, or Tokelau, and one bat species is the only native mammal species present in the Cook Islands and Niue. This most widespread species is the Pacific flying fox *Pteropus tonganus*, which roosts in tall coastal trees.

Unusual species remain, however, on remote islands; for instance, on the volcanic island of Niuafo'ou in Tonga is found the endangered endemic Niuafo'ou megapode, which incubates its eggs by burying them in warm sands near volcanic ducts. Recently chicks and eggs have been released on the predator-free Tongan islands of Late and Fonualei.

The estuarine crocodile occurs in the South Pacific, being fairly common in Papua New Guinea and extending to the Solomon Islands and Vanuatu, which is the limit of its Pacific range. The Solomon Islands once had substantial populations, but these have been decimated by human impact and typhoon damage. The crocodile is currently vulnerable in the South Pacific.

Numbers of amphibian and reptile species decrease rapidly with distance to the east of New Guinea, also due to decrease in island size, proximity of other islands, and island age. Amphibians are intolerant of salt water and therefore have difficulty migrating in this region, a factor also influencing the prawns, shrimps, and freshwater fish. As of 2015, studies of freshwater fish species on the large islands in Fiji found a total of sixty-four fish species from nineteen families.

VULNERABILITY

Pacific island ecosystems are vulnerable because of their relatively small size, poor state of protection or available knowledge, and their proximity to rapidly increasing human populations that put heavy demands on these natural resources to support their economic development. Weeds and introduced species are a great threat to these special biodiversity groups, particularly freshwater ecosystems. Climate change, sea-level rise, and typhoon damage are increasing vulnerability, particularly of coastal ecosystems, and some South Pacific island systems are so close to sea level that even a relatively small rise in sea level will leave them entirely submerged.

Gaps in ecosystem management in the region include weak or vague legislation for their protection in many countries, and the need for more knowledge and capacity to manage.

—Joanna C. Ellison

Further Reading

Ellison, Joanna. "Wetlands of the Pacific Island Region." *Wetlands Ecology and Management*, vol. 17, 2009.

Ellison, Joanna, and Monifa Fiu. *Vulnerability of Fiji's Mangroves and Associated Coral Reefs to Climate Change*. World Wildlife Fund South Pacific Programme, 2010.

"List of Freshwater Fishes Reported from Fiji (Republic Of)." *Fishbase*, 2015, www.fishbase.de/Country/CountryChecklist.php?resultPage=2&c_code=242&vhabitat=fresh. Accessed 17 Aug. 2018.

Pavia, Frank. *Biogeochemical Studies of the South Pacific Ocean Using Thorium and Protactinium Isotopes.* Columbia UP, 2020.

Paterson, Donald Edgar, and Susan Farran. *South Pacific Land Systems.* U of the South Pacific P, 2013.

Randall, John E. *Reef and Shore Fishes of the South Pacific: New Caledonia to Tahiti and the Pitcairn Islands.* U of Hawaii P, 2005.

Steadman, David. *Extinction and Biogeography of Tropical Pacific Birds.* U of Chicago P, 2006.

Wilkinson, Clive. *Status of Coral Reefs of the World: 2004: Volume 2.* Australian Institute of Marine Science, 2004.

SOUTHERN OCEAN ECOSYSTEM

Fields of Study: oceanography; marine ecology; marine biology; hydrodynamics

ABSTRACT

The Antarctic Ocean—or the Southern Ocean, as it is usually known—is approximately 13.7 million square miles (35.5 million square kilometers) in area, or twice the size of the United States. It is the world's fourth-largest ocean. Because of its unique position and characteristics, the Southern Ocean plays a unique role in regulating the global climate.

KEY CONCEPTS

biodiversity: the variety of plant and animal organisms found within an ecosystem

Chilean sea bass: a commercial name used to hide the identity of the fish as the protected, and therefore illegally caught, Antarctic cod

krill: small, shrimp-like crustaceans that swarm the Antarctic waters and other cold-water regimes in vast numbers and form the base of the food web for larger creatures ranging from tiny fish to blue whales

long-line fishing: the commercial fishing practice of running out miles- (or kilometers-) long lines carrying several baited hooks for the purpose of harvesting ocean fish

DESCRIPTION

Encircling the continent of Antarctica, the Southern Ocean is formed by the convergence of the Pacific, Atlantic, and Indian Oceans as they flow together around Antarctica—the coldest, windiest, and driest of all the world's continents. The Southern Ocean encompasses the Amundsen Sea, the Bellinghausen Sea, part of the Drake Passage, the Ross Sea, a small section of the Scotia Sea, and the Weddell Sea. The deepest point of the Southern Ocean is 27,737 feet (8,454 meters) below sea level, at the South Sandwich Trench.

The waters of the Southern Ocean are characterized by the Antarctic Circumpolar Current (ACC), also known as the Wind Drift Current. As it travels eastward, this current transports from 4,556 to 5,120 cubic feet (129 to 145 million cubic meters) of seawater per second, a higher volume of water than any other current in the world, amounting to about 100 times that transported by all the world's rivers. The ACC is the only current that encircles the globe. The Antarctic Convergence, the globe-circling region where the ACC waters blend with the somewhat warmer waters of the other oceans, is a band about 20 to 30 mi (32 to 48 km) wide, ranging between 48 degrees south and 61 degrees south.

FISH SPECIES

Even though the land area of the Antarctic is covered with snow and ice, the biome of the Southern Ocean supports a wealth of plant and invertebrate animal life. The most abundant groups of fish in the Southern waters are the Antarctic cod and the ice fish, and research tends to concentrate on these two species.

Within the Southern Ocean, upwelling currents cause rich nutrients to be pulled up from the seabed to the surface to provide nourishment to the microscopic algae (plankton) that live near the surface water layers. Krill, small shrimplike crustaceans, feed on the plankton and in turn are eaten by the fish, whales, seals, and birds that live in the Southern Ocean. These cold waters are considered to be as much as four times more productive per hectare than that of any other ocean in the world. The abundant krill that teem in the surface waters are responsible for the red color sometimes observed in the waters here. Squid and octopus are also important links in the ocean's food chain because they serve as nourishment for sperm whales, seals, penguins, seabirds, and fish. Estimates place the number of squid consumed by whales alone at some 61 million tons (55 million tonnes) annually.

Because krill are so plentiful and high in protein, efforts have been made to use them to solve some food shortages in some of the poorer nations of the world. The idea has proved to be largely impractical, however, because of the need for rapid processing and distribution. However, both Japan and Russia have begun to perfect methods of using krill for other commercial purposes.

BIODIVERSITY

Marine mammals that live in the Southern Ocean are no strangers to human activity. During the eighteenth and nineteenth centuries, the seals of the Southern Ocean were exploited for their furs, resulting in their ultimately being classified as protected. By the twenty-first century, they have begun a comeback. These seals cannot breed in the water; thus, they head toward land or pack ice during mating seasons. Antarctic seals are larger than those that live in the Arctic, in part due to the easy availability of food and the presence of fewer predators in the area (polar bears are a major predator of Arctic seals, but there are no such predators in the Antarctic).

The Antarctic polar winter generally lasts from November to March and is a period in which violent storms occur. Daylight is absent due to the angle of Earth's axial tilt, and the coastal waters become icebound. When the polar winter ends, more than 100 million seabirds arrive at the Southern Ocean to breed on the mainland or on offshore islands. The albatross, largest of all flying seabirds, is one of the most distinctive. Other species include fulmers, prions, petrels, and shearwaters; more widespread species include shore birds, skuas, gulls, terns, and penguins.

Because ice-free land is scant along the shores of the Southern Ocean, these birds tend to breed in large concentrations. When they sense the return of polar winter, the seabirds move to the open sea, sometimes temporarily populating the pack ice.

The penguin may be the most emblematic animal in the Southern Ocean. Much celebrated popularly, the various species are under considerable environmental pressures, however. The Emperor penguin colony of Terre Adélie in east Antarctica, for example, shrank by 50 percent in the 1970s and has yet to recover from that loss. The direct cause was the diminishing acreage of sea ice—as these birds breed and nurture their young on the sea ice; the problem is exacerbated because much of their diet is also dependent on the ice. The fish, squid, and krill that Emperor penguins consume in turn eat the specialized plankton that grows on the underside of sea ice floes. An incremental rise in sea temperature has been enough to upset the balance of this food web.

Seven species of penguins are classified as Antarctic penguins, and more penguins than any other bird species live on this continent. Unlike most birds, penguins do not fly; they use their feet as propellers to move themselves through the waters in a motion that is torpedo-like yet much like flying. The most common species is the Adélie, which may be 24 to 28 inches (60 to 70 centimeters) tall; and the Emperor, which grows to 4 ft (1.2 m) in height and can weigh up to 95 pounds (43 kilograms). Some concentrations of penguins may include as many as 180,000 birds. The nests of the Adélie are built of stone; just one or two eggs are incubated in these nests at a time.

ENVIRONMENTAL STRESSES

Despite its isolation, the Southern Ocean is not immune to some of the same environmental stresses that are threatening ecosystems around the world. The Southern Ocean serves as a massive storehouse for carbon dioxide (CO_2), with up to one-third of all CO_2 stored by the world's oceans present here, according to some estimates. However, scientists are warning that the Southern Ocean is losing its capability to serve as a buffer for the rest of the world because global warming has accelerated the production and accumulation of CO_2 while simultaneously reducing the ability of the Southern Ocean to absorb it, due to altered temperature and salinity regimes.

Human industrial activity has also contributed to the hole in the atmospheric ozone layer over the Antarctic, which tends to add to a warming in its regional climate. This depleted-ozone area allows more solar ultraviolet radiation to penetrate to the surface waters, which some research suggests has reduced the amount of plankton here by as much as 15 percent, and likely has altered the deoxyribonucleic acid (DNA) of some fish species.

Exploitation of fishing grounds has become a major problem. Australian scientists have sent out an international call for help because of the increased levels of fishing piracy in the Southern Ocean. One such casualty is the highly prized Patagonian toothfish (*Dissostichus eleginoides*) and its relative the Antarctic cod (*Dissostichus mawsoni*), which are considered a delicacy in both Japan and the United States —where they are marketed as "Chilean sea bass." These fish, known to commercial fishers as the "white gold of Antarctica," have both been placed on the seafood red list of fish to avoid, posted by Greenpeace International. Armed surveillance ships are in some cases being used to protect Patagonian toothfish, which are an integral part of the diets of sperm whales and elephant seals.

A "collateral" casualty of the practice of using long-line baited hooks that stretch out for miles over the surface of the Southern Ocean are some 100,000 seabirds a year. Attracted to the bait, the birds go after the hooks and are trapped as they are pulled below the surface. The great albatross and petrels, according to some sources, may have lately joined the ranks of the most endangered birds on the planet as a result of such practices.

—Elizabeth Rholetter Purdy

Further Reading

David, Bruno, and Thomas SaucPde. *Biodiversity in the Southern Ocean.* ISTE Press, 2015.

Everson, Inigo, editor. *Krill: Biology, Ecology and Fisheries.* John Wiley & Sons, 2008.

Forcada, Jaume, and Philip N. Trathana. "Penguin Responses to Climate Change in the Southern Ocean." *Global Change Biology*, vol. 15, no. 7, 2009.

Lowen, James. *Antarctic Wildlife: A Visitor's Guide.* Princeton UP, 2011.

McCann, Joy. *Wild Sea: A History of the Southern Ocean.* U of Chicago P, 2018.

Mikhalev, Yuri. *Whales of the Southern Ocean: Biology, Whaling and Perspectives of Recovery.* Springer, 2019.

Nicol, Stephen I. *The Curious Life of Krill: A Conservation Story from the Bottom of the World.* Island Press, 2018.
Roberts, Leslie Carol. *The Entire Earth and Sky: Views on Antarctica.* U of Nebraska P, 2008.
Siegfried, W. R., P. R. Condy, and R. M. Laws, editors. *Antarctic Nutrient Cycles and Food Webs.* Springer Science & Business Media, 2013.
Turner, John. *Climate Change in the Polar Regions.* Cambridge UP, 2011.
Vacchi, Marino, Eva Pisano, and Laura Ghigliotti, editors. *The Antarctic Silverfish: A Keysone Species in a Changing Ecosystem.* Springer, 2017.
Zarembo, Alan. "Southern Ocean Study Sounds More Global-Warming Alarms." *Los Angeles Times,* May 17, 2007.

Surface Ocean Currents

Fields of Study: oceanography; fluid dynamics; climatology

ABSTRACT
Ocean currents represent a dynamic system that, along with atmospheric circulation, helps to distribute heat evenly across the planet. Responding to the seasons, ocean currents play important roles in climate, marine life, and ocean transportation.

KEY CONCEPTS

Coriolis effect: a perceived force due to Earth's rotation that seems to cause a moving air mass to change its north-south direction of movement to an east-west direction in the northern and southern hemispheres

core rings: areas of still ocean water trapped between two branches of a surface ocean current

gyres: slow circular surface currents produced in oceans by the action of wind and the Coriolis effect

heat budget: the balance between the incoming energy from the Sun and the outgoing energy radiating back into space from Earth

thermohaline circulation: circulation driven by differences in temperature and salt content

OVERVIEW

The "heat budget" of Earth results in a temperature range that makes life on the planet possible. Ocean currents play a vital role in the heat budget. These currents are major determinants of climates and strongly influence the distribution of marine life. Ocean currents must be studied in relation to other aspects of the environment with which they interact. The currents are, for example, closely associated with atmospheric circulation because the planetary winds are the prime movers of the currents. The friction of the wind blowing over the ocean surface began the slow, shallow movement of surface waters that eventually became a global circulation of immense volumes of seawater. There are also deeper ocean currents that are much slower moving and are difficult to monitor, whose significance, therefore, is less well understood. In part, the deep currents derive from the physical fact that water is continuous in structure, such that water moved from one location must be replaced by water from a different location, and so on down the line. The deep currents, however, are primarily caused by thermohaline circulation driven by slight differences in seawater densities resulting from differences in temperature and salinity. The shallow, wind-driven currents—the ones affecting the surface waters—although they may be hundreds of feet (or meters) deep, are effectively independent of such deep ocean currents.

The most significant features of ocean currents are their geographic locations and their directions of flow. It is helpful to recognize overall patterns. There are large-circulation gyres in each of the major ocean basins discernible as an apparent overall circular movement of surface waters. These gyres, or gyrals, move clockwise in the Northern Hemisphere and counterclockwise in the Southern Hemisphere. The North Central Atlantic Gyre, for example, located east of the United States, is one of the best known and most studied. The Florida Current (part of the Gulf Stream system) is on the west side of the gyre and is a warm current flowing generally northward. The Canaries Current, on the east side of the gyre, is a cold current that flows generally southward. The North Atlantic Drift and the North Equatorial Current form the eastward and westward components of the gyre, respectively. The result of circulation in the gyre is that warm water from the equatorial region is transported poleward to heat-deficient areas. Simultaneously, the Canaries Current transports colder water back toward the equator. The ocean currents thus help to distribute surface and atmospheric heat more evenly throughout the world.

In the Atlantic Ocean south of the equator, a large gyre moves counterclockwise. The warm Brazil Current on the west side of the gyre flows southward, transporting heat away from the equator. The Benguela Current on the east side of the gyre moves colder water toward the equator.

In the Pacific Ocean similar patterns of clockwise and counterclockwise gyres are apparent. North of the equator, the Japan Current (also known as the Kuroshio) transports warm water toward the pole, and the California Current moves colder water toward the equator. In the Pacific Ocean south of the equator, the cold, nutrient-rich Humboldt Current flows northward off the west coast of South America and is renowned historically as one of the most fertile commercial marine fishery areas in the world. The Indian Ocean possesses similar gyres, although the attenuated portion north of the equator presents some special features.

WIND AND SOLAR ENERGY

The forces that drive the oceanic current circulations are the planetary winds. The planetary winds are in turn driven by solar energy. The ocean currents, therefore, are sun-driven as sunshine energizes the Gulf Stream and the other currents. Some general principles about Earth's heat budget can be stated. The sun heats Earth, its atmosphere, oceans, and land, but each portion captures heat differently. The atmosphere, the most fluid and most responsive of the three, has developed large bands of alternating pressure belts and wind belts, such as the Northeast Trade Winds and the Prevailing Westerlies. The Northeast Trade Winds lie between 5 and 25 degrees north latitude. The winds flow predominantly from the northeast and form one of the most constant of the wind belts.

The friction of the wind moving over the ocean surface causes the surface waters to move with the wind, but because of the Coriolis effect, caused by Earth's rotation, the movement of the water current in the Northern Hemisphere tends to be about 45 degrees to the right of the winds that cause the current. The resultant current, the North Equatorial Current, is fragmented into different oceans because of the intervening continents. Largely as a result of the Coriolis effect, the current deflects to its right and, in the Atlantic Ocean, eventually becomes the Gulf Stream. In the Pacific Ocean, the comparable current is the Japan Current. One can thus see the origins of the clockwise gyrals in the Northern Hemisphere.

Another wind belt in the Northern Hemisphere, the prevailing westerlies, is located between 35 and 55 degrees north latitude. These winds are not as constant as the northeast trade winds, and they flow prevalently from the west. The correlation of the latitudes of the westerlies and the west-to-east-moving currents of the gyres is apparent. The currents slow down and become more widespread, shallower, and less distinguishable but are urged on toward the east by the Westerlies. The North Atlantic Drift and the North Pacific Current result from this relationship.

NORTHERN HEMISPHERE

Again, analyzing the North Central Atlantic Gyre, the blocking position of the Iberian Peninsula causes the North Atlantic Drift to split, part moving southward toward the equator with the cold Canaries Current and part moving poleward into the Arctic Ocean as the warm Norwegian Current. The Canaries Current merges into the North Equatorial Current to complete the gyre. The temperature characterizations of currents and drifts as "warm" or "cold" are relative. There are no absolute temperature divisions. Some warm currents are actually lower in temperature than some cold currents. For example, the Norwegian Current is considered a warm current only because it is warmer than the Arctic water into which it is entering. Only a few degrees above freezing in winter, the Norwegian Current nevertheless transfers significant amounts of heat into these high latitudes and moderates the winter temperatures in Western and Northern Europe. A compensating movement of cold water out of the Arctic is accomplished by the southward flowing Labrador Current between Greenland and North America.

The Gulf Stream is the world's greatest ocean current. There is, however, some confusion about what constitutes the Gulf Stream. The Gulf Stream is generally taken to include the entire warm-water transport system from Florida to the point at which the warm water is lost by diffusion into the Arctic Ocean. It would thus include both the North Atlantic Drift and the Norwegian Current. Technically, the Gulf

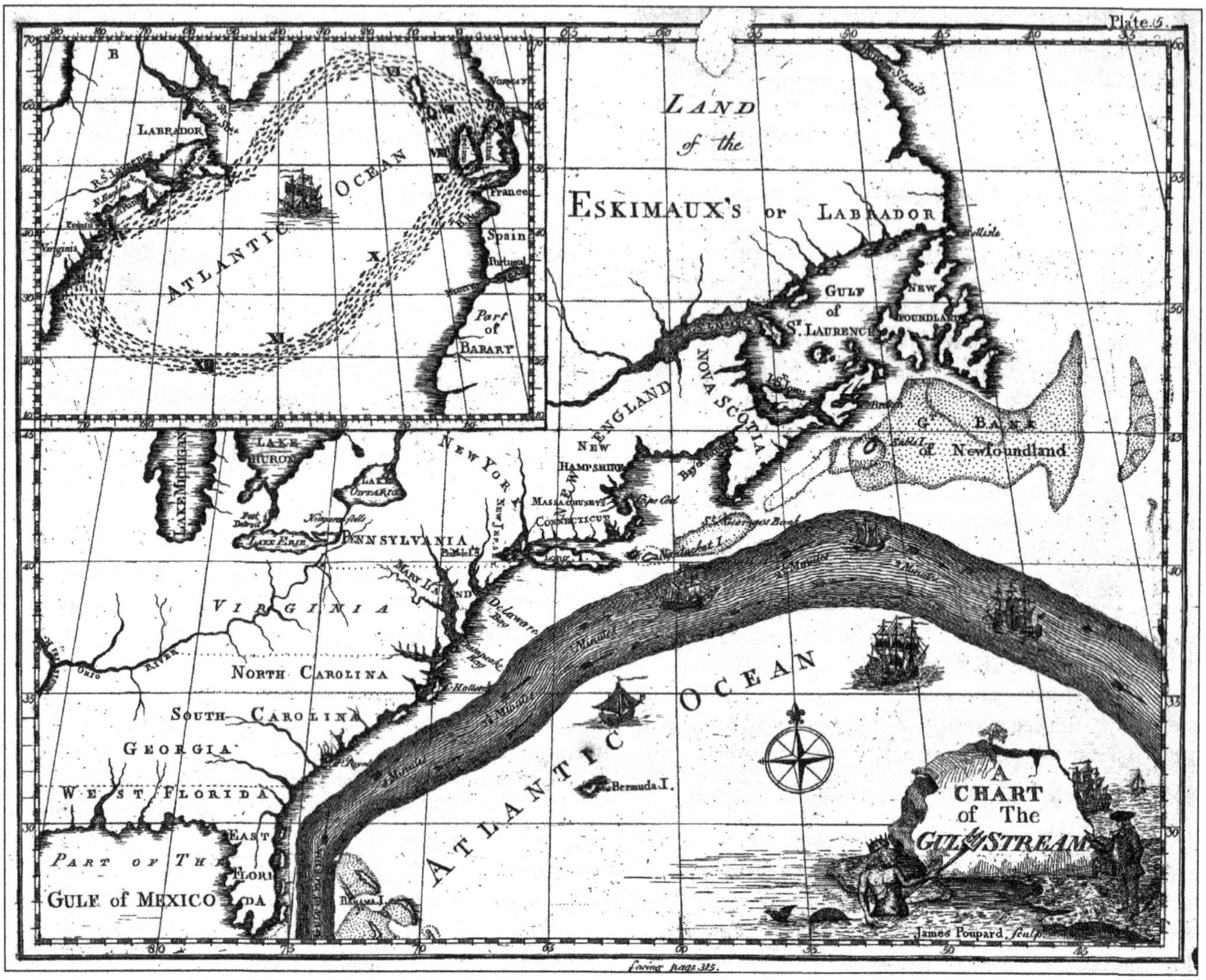

Earliest known map of the Gulf Stream (Benjamin Franklin/Library of Congress via Wikimedia Commons)

Stream is a smaller segment of that transport system—the portion off the northeast coast of the United States. The Gulf Stream system thus includes the Florida Current, the Gulf Stream, the North Atlantic Drift, and the Norwegian Current.

SOUTHERN HEMISPHERE

Ocean current patterns in the Southern Hemi-sphere are almost a mirror image of those in the Northern Hemisphere, adjusted for differences in the configuration of the continents. The southeast trade winds drive the South Equatorial Current, while the Coriolis effect causes it to deflect to the left. Resultant gyres are counterclockwise, but again, the poleward-moving currents transfer the heat away from the equator, and the currents moving toward the equator return colder water. In general, cold currents are richer in nutrients, have a higher oxygen content, and support a greater amount of life than warm currents.

Most products of the world's commercial fisheries are yielded by cold waters. In contrast, cold currents offshore are associated with desert climates onshore. The atmospheric circulations that drive the ocean currents also create conditions that are not conducive to precipitation in the latitudes of these cold currents. Examples are the Sahara Desert adjacent to the Canaries Current, the Atacama Desert adjacent to the Humboldt Current, the Sonoran Desert adjacent to the California Current, and the Kalahari Desert adjacent to the Benguela Current.

There are other currents that are sporadic in occurrence, such as the warm El Niño current that periodically develops off the west coast of northwestern South America for reasons that are not well understood. Numerous small local currents are also caused by tides, storms, and local weather conditions.

STUDY OF OCEAN CURRENTS

The study of ocean currents has acquired new significance as scientists have discerned the effects of the currents on climate and marine life. Information comes from many sources. One of the earliest attempts to identify and chart an ocean current was made by Benjamin Franklin (1706–1790) when he was postmaster general of colonial America. His map of the Gulf Stream was published in 1770 and has proved to be remarkably accurate when one considers his sources of information. Franklin noted that vessels sailing westward from England to America in the midlatitudes of the Atlantic Ocean were taking longer than ships moving eastward and longer than ships moving westward but in lower latitudes. He correctly concluded that the vessels were moving against a slow, eastward-moving current.

Since that time, vast amounts of data have been acquired to detect, measure, and chart the currents. One of the early methods still employed is the use of drift bottles. Sealed bottles are introduced into the sea at various locations and dates and are allowed to float with the currents. Finders are requested to note the date and location of the bottle-find and to return the data to the address in the bottle. Ocean current data have also been obtained serendipitously through the loss of floating goods in midocean shipping accidents. The points of landfall of such materials, as well as of materials washed out to sea by some natural disaster such as a tsunami, provide important information about the surface movement of ocean waters.

Various types of more technologically advanced current meters are also used. Some are moored to the sea bottom and can transmit results by radio. It is difficult for a ship at sea to measure currents because the ship itself is drifting with the current. Currents are generally very slow and difficult to measure. A few currents may be measured at 3.7 to 5 miles (6 to 8 kilometers) per hour, but much more common are those less than 0.6 mi (1 km) per hour. The average surface velocity of the North Atlantic Drift is about 0.8 mi (1.3 km) per hour. The currents also vary in width and depth. The Florida Current off Miami is about 20 mi (32 km) wide, 984 feet (300 meters) deep, and moving at about 3 to 5 mi (5 to 8 km) per hour. It transports more than 4 billion tons (3.6 billion tonnes) of water per minute. The volume of flow is more than one hundred times that of the Mississippi River. As the flow proceeds north and then east as the North Atlantic Drift, it spreads, thins, slows, and splits into individual meandering flows that are difficult to follow. Spin-off eddies, or "core rings," occur that can persist for months.

One useful method of tracking ocean currents is to be able to identify water of slight temperature variations and salinity differences. When the flow movement is so slow as to be practically undetectable with current meters, the slight temperature and salinity differences can be used as "tracers" to identify current movements. This method is also used in identifying the even slower-moving deep-ocean currents. More recently, advanced satellite imagery and high-altitude aerial photography have become extremely important in monitoring ocean currents. Using sensors that detect radiation at selected bands of the electromagnetic spectrum, satellites collect data on broad patterns of seawater temperatures and thus help scientists to understand the movements and extent of the currents. This type of sea monitoring is also useful in detecting any changes that might occur in the oceans in the future.

SIGNIFICANCE

Ocean currents play a vital role in the environment. Along with atmospheric circulation, ocean currents serve to distribute the heat absorbed from the sun to different parts of the world. Immense volumes of relatively warm seawater are slowly moving poleward, transporting heat from the heat-surplus equatorial regions to the heat-deficient regions nearer the poles. Cold-water currents in turn move colder water back toward the equator. Although neither solar energy nor rainfall is evenly distributed over the planet, the mixing actions of the ocean currents function to keep the global environment in a steady state. These moderating effects of the ocean currents affect the climates of coastal areas in the middle and high latitudes, especially in Europe. The densely populated nations of northwestern Europe experience much milder winters than would otherwise be expected for such high latitudes. Northwestern North America similarly benefits.

Life in the sea is also aided by this current-driven ocean water mixing. In addition to heat, ocean currents distribute oxygen and nutrients, the result being the formation of certain areas in the oceans where very favorable life-supporting conditions occur. These fertile areas of mixing are concentrated sources of commercial marine fishery products. Where mixing is limited, nutrient-poor regions arise in the ocean such as the Sargasso Sea located in the center of the North Central Atlantic Gyre. Global warming may alter ocean currents, creating a further need to study these currents and their effects on climate and marine life.

—*John H. Corbet*

Further Reading

Joseph, Antony. *Measuring Ocean Currents: Tools, Technologies, and Data.* Elsevier, 2014.

Broecker, Wally. *The Great Ocean Conveyor*. Princeton UP, 2010.

Brown, Joan. *Ocean Circulation: Prepared by an Open University Course Team.* Elsevier, 2016.

Clarke, Allan J. *An Introduction to the Dynamics of El Niño and the Southern Oscillation. Academic Press, 2008.*

Hohn, Donovan. *Moby Duck: The True Story of 28,800 Bath Toys Lost at Sea & of the Beachcombers, Oceanographers, Environmentalists & Fools Including the Author Who Went in Search of Them.* Penguin, 2011.

Nagai, Takeyoshi, Hiroaki Saito, Koji Suzuki, and Motomitsu Takahashi, editors. *Kuroshio Current: Physical, Biogeochemical and Ecosystem Dynamics.* Wiley, 2019.

Schwartz, M. *Encyclopedia of Coastal Science*. Springer, 2005.

Sverdrup, Keith A., Alyn C. Duxbury, and Alison B. Duxbury. *An Introduction to the World's Oceans*. 8th ed., McGraw-Hill, 2004.

Talley, Lynne D., et al. *Descriptive Physical Oceanography: An Introduction*. 6th ed., Academic Press, 2011.

Thurman, Harold V., and Alan P. Trujillo. *Introductory Oceanography*. 10th ed., Prentice-Hall, 2003.

Trujillo, Alan P., and Harold V. Thurman. *Essentials of Oceanography.* 10th ed., Prentice-Hall, 2010.

Ulanski, Stan. *The Gulf Stream: Tiny Plankton, Giant Bluefin, and the Amazing Story of the Powerful River in the Atlantic*. U of North Carolina P, 2008.

Voituriez, Bruno. *The Gulf Stream.* UNESCO, 2006.

T

Tasman Sea Ecology

Fields of Study: oceanography; marine biogeochemistry; marine biology; marine conservation

ABSTRACT

The Tasman Sea is located in the southwest Pacific Ocean, running between Australia and New Zealand. It is named for the Dutch explorer Abel Tasman (1603-1659).

KEY CONCEPTS

chlorophyll biomass: the quantity of chlorophyll-bearing algae and plankton that appear in a body of water

coralivore: a creature that consumes coral polyps for its primary sustenance

ocean acidification: atmospheric carbon dioxide that is absorbed by ocean water combines with water molecules to produce carbonic acid, a weak acid that is nonetheless capable of increasing the acid content of seawater

DESCRIPTION AND HISTORY

Nestled in the southwestern portion of the Pacific Ocean, the Tasman Sea is a saline body spanning the distance between the landmasses of Australia to the west and New Zealand to the east. The Tasman Sea is approximately 1,200 miles (1,931 kilometers) from west to east, and 1,700 mi (2,736 km) north to south.

The Maori name for the Tasman Sea is *Te Tai-o-Rehua,* or *the great sea of Rehua,* the divine being who is the son of the sky god and earth goddess associated with immortality and various stars. The European name pays homage to Abel Tasman (1603–1659), the Dutch explorer who was the first European to discover New Zealand and the island of Tasmania, formerly called "Van Diemen's Land." Tasman's voyage, in 1642–43, mapped substantial portions of the region, and in the following century, the first voyage of British explorer Captain James Cook (1728–1779) extensively navigated the Tasman Sea, producing a complete and substantially correct map of the New Zealand coastline.

Australians and New Zealanders commonly refer to the Tasman Sea as "the Ditch," especially in the phrase 'crossing the Ditch', referring to passage across the sea in either direction.

Like other Australian waters, the Tasman Sea is generally nutrient-poor, and its ecosystem relies on processes such as upwelling to bring nutrients to the food chain. The mid-ocean ridge of the Tasman Sea was formed about 85 million years ago. Much later, Australia became drier, and Antarctica cooler when the island of Tasmania separated from Antarctica and changed the ocean currents around the major landmasses. As the rift valley sank, sediment accumulated, forming the Otway and Strzelecki ranges in southeastern Australia. The mid-ocean ridge is considerably closer to Australia than New Zealand, having formed essentially at the midway point between their continental margins.

Where the Otway ranges meet the Tasman Sea is Dinosaur Cove, a minor bay of seafront cliffs with fossil-bearing strata dating back over 100 million years. Many of the fossils show the kinds of dinosaurs that lived in the Tasman Sea region, including the taxa called the "polar dinosaurs of Australia," which some paleontologists believe to have been warm-blooded.

BIODIVERSITY

The seamounts of the Tasman Sea are extremely biodiverse, home to hundreds of species of fish, including the orange roughy, which can live for over a century. In 2004, the deep-sea research ship *Tangaroa* found more than 100 previously undiscovered species of fish during a four-week expedition of the Tasman Sea, as well as fossilized teeth of the megalodon, the prehistoric shark that was three times the size of the modern-day great white shark.

Seamounts, undersea mountains which are isolated from many disturbances, are good homes for coral reefs, which are both long-lived and slow-growing. One of the southernmost platform reefs in the world, Middleton Reef is home to the protected species *Epinephelus daemeli,* one of the fish known as black cod. Declines were noted in the early 1900s, and began to decline severely no later than the 1950s. The cod is vulnerable because of commercial interest in the fish and resulting overfishing, as well as its slow speed, large size, and territorial use of inshore habitats, making it an easy catch for spear fishers. Furthermore, some of the black cod populations in the Tasman Sea are believed to be nonbreeding populations as a result of drifting larvae, slowing the replenishment of populations.

On the other side of a pass some 25 mi (40 km) across is another platform coral reef, Elizabeth Reef. At low tide, most of the reef flat is above water; at high tide only one cay (Elizabeth Island) remains, just barely above sea level. Australia's National Heritage Trust manages the Elizabeth and Middleton Reefs Marine National Park. In addition to black cod and a variety of other species, Elizabeth Reef is home to a Galapagos shark (*Carcharhinus galapagensis*) nursery, large numbers of sea cucumbers (*Holothuria whitmaei*), and a recently growing number of crown-of-thorns starfish (*Acanthaster planci*).

The crown-of-thorns starfish is the second-largest starfish in the world, a solitary animal covered in venomous spines. Though it sometimes preys on brittle stars, the crown-of-thorns is principally a coralivore, feeding on the polyps of reef coral by climbing the reef, extruding its stomach onto the structure, and liquefying coral polyps in order to absorb nutrients from them. Its spines, which exude a strong neurotoxin, protect it from shrimp, mollusks, and large reef fish who attempt to prey on it.

The crown-of-thorns is best known as a threat to the coral reef ecosystem of the Great Barrier Reef elsewhere in Australian waters. A single crown-of-thorns consumes 65 square feet (6 square meters) of living coral reef surface each year, and the other factors that jeopardize a reef ecosystem exacerbate the effects of the crown-of-thorns's predation. It may also promote the spread of coral diseases. So far, the crown-of-thorns is not seen in Elizabeth Reef in the same numbers as in the Great Barrier Reef, but the population has shown noticeable growth since 2003.

Middleton and Elizabeth Reef are part of the same underwater plateau, the Lord Howe Rise, consisting of about 0.6 million square miles (1.5 million square kilometers) of the Tasman Sea. The central-east area of the rise is known as the Lord Howe platform, and includes a seamount capped by Lord Howe Island and Ball's Pyramid. Lord Howe Island is a crescent-shaped volcanic remnant about 6 mi (10 km) long and 1 mi (1.6 km) wide at its widest point, with a population of 348 in the 2006 census.

The island is a distinct terrestrial ecoregion and more than half of its plant species are endemic (found only here) to the island; many locals are involved in the Kentia palm industry, revolving around the endemic genus of *Arecaceae Howea*. The critically endangered creeping vine *Calystegia affinis* is native only to Lord Howe Island and nearby Norfolk Island. The island is also known for its banyans, pandanus trees, the spring-flowering bush orchid, and the glowing mushrooms (*Mycena chlorophanos* and *Omphalotus nidiformis,* both endemic to the island) that crop up in the palm forests after heavy rain.

Ball's Pyramid is an eroded remnant of a shield volcano, consisting of an uninhabited volcanic stack. The most notable life on the Pyramid is a population of Lord Howe Island stick insects living among the growth of *Melaleuca howeana* shrubs that grow out of crevices in the rocks.

The central and southern Tasman Sea experiences a spring (September to December) bloom, with the maximum chlorophyll biomass occurring in early October. The surface of the sea includes a rich assortment of microscopic flora and fauna, including crustacea, pteropoda, heteropoda, radiolarians, dinoflagellates, foraminifers, polycaeta, chaeotgnatha, urochorda, vorticellids, and polychlads. Seabirds include the endangered kiwi species *Apteryx haastii, A. rowi, A. owenii,* and *A. australis* along the New Zealand coast, as well as the brown booby, brown tern, black-naped tern, bridled tern, crested tern, roseate tern, sooty tern, black noddy, red-tailed tropicbird, silver gull, wedge-tailed shearwater, and eastern reef egret.

Marine macrofauna include the humpback whale, the spinner dolphin, the dumbo octopus (which navigates with the help of two flaps, giving it an appearance reminiscent of the cartoon character Dumbo the Flying Elephant), and the Pacific spookfish, which uses a long snout to detect the electrical impulses of hiding prey. Along the Fjordland coast, the Fjordland Crested Penguin nests.

WARMING THREATS

Climate change could affect the coral reefs with some intensity. As sea temperatures rise, the coral reefs of the Tasman Sea could be put at risk; those of Lord Howe Island are considered to be the most vulnerable. Furthermore, changes to ocean chemistry as the waters become acidified by absorption of increased greenhouse gas would hurt the shellfish of the Tasman Sea, making it more difficult for them to grow their shells, as well as the coral reefs.

Changes in water temperature have already been observed, as the East Australian Current reaches further south than before, bringing subtropical species into temperate waters. The long-spined sea urchin, previously found only off the shores of New South Wales, has recently been introduced to the western Tasman Sea. The European shore crab, introduced to Tasmania in the 1990s, has spread to the entire eastern coast, and other new species arrivals have impacted the ecosystem of Tasmania's eastern coast.

Over the next 50 years, the temperature of the Tasman Sea is expected to keep rising incrementally. Phytoplankton are expected to push further south, and algal blooms may increase, while the phytoplankton populations of the temperate region will shrink. Temperature changes can also affect the sex of the embryos of turtles and other species, biasing the sex ratio toward females, which may lead to long-term fertility problems in some populations and overpopulation in others.

—Bill Kte'pi

Further Reading

Longhurst, Alan R. *Ecological Geography of the Sea*. Elsevier Science and Technology Books, 2006.

Rich, Thomas H., and Patricia Vickers-Rich. *Dinosaurs of Darkness: In Search of the Lost Polar World.* 2nd ed., Indiana UP, 2020.

Sole, Tony. *Nga ti Ruanui: A History*. HUIA, 2005.

van Oppen, Madeleine J.H., and Janice M. Lough, editors. *Coral Bleaching: Patterns, Processes, Causes and Consequences*. Springer, 2018.

TROPICAL AND SUBTROPICAL OCEANIC REGIONS

Fields of Study: oceanography; climatology; geography; marine biology; historical geography

ABSTRACT

The area of Earth between the Tropics of Cancer and Capricorn (23.5 degrees north and south latitude) makes up the region called the "tropics." The tropics are bound in the north and in the south by the subtropical regions, where much of the water that enters the atmosphere from the tropics falls as rain. Both the tropical and subtropical regions are symmetrically divided by the equator. Both regions are areas where high biodiversity can be found, including most human life.

KEY CONCEPTS

benthos: the lowest depths of the oceans, the seafloor

fauna: the animal life found in a particular region, whether terrestrial or marine

flora: the plant life found in a particular region, whether terrestrial or marine

solar plane: the plane with the Sun at its center in which the orbits of the major planets are found

THE TROPICAL AND SUBTROPICAL REGIONS

Earth rotates on its axis, which is is tilted by 23.5 degrees relative to the solar plane. This results in seasons because of the exposed sun angle. Some regions remain less affected and are exposed to the sunlight in a more constant fashion. The regions between the Tropics of Cancer and Capricorn (23.5 degrees north and south latitude) make up the region called the "tropics." The center of the tropical region is the equator, with the Intertropical Convergence Zone (ITCZ) at its core. In concert with the polar regions, it is a global weather engine. Many ocean and wind currents are also found in this region. The tropics are bound in the north and in the south by the subtropical regions, where much of the rain falls that is created by the ITCZ. Both regions, tropical and subtropical, make for areas where high biodiversity can be found, including human life.

Humans fared very well in this ocean region because of the warm temperatures and the superabundance of food and resources, including plants and fish. Approximately 20 percent of the tropical and subtropical region is covered by land including much of Africa, South America, Asia, as well as Polynesia, Greece, Papua New Guinea, Melanesia, and Japan. The historic use of ocean vessels was commonly found in these regions, for example, off Chile, in Polynesia, Portugal, Spain, and in ancient Greece.

Coastal fisheries have been a mainstay of human life in the tropical and subtropical ocean regions. The coastal wetlands of this region, including estuaries and mangroves, play a major role for global well-being. The mangroves of this region are a major contributor when it comes to global carbon cycling. Mangroves are also nurseries for fish, and host to conches (shells that were harvested) and marine mammals like dugongs and manatees (many of these species are now endangered, or even extinct). Further, the Amazon, Nile, Congo basin, Ganges, Jangtsekiang, and Mekong Rivers all drain into tropical and subtropical oceans, and are major ecological hot spots because of sediment inflow, mixing, and fish nurseries. Marine mammals can be found there (indicative of rich food chains), as well as many human population centers (now often expressed as modern port locations).

The tropical seas are the 'hot spots' of mankind. Most people live in the tropics, and there, most of them actually live near the coast. Coastal fisheries play a major role. Famous interfaces between islands and oceans are found, for instance, in the southeast Pacific with its millions of islands, the Florida Keys, the Caribbean islands, and the Seychelles. Such ocean-land interfaces include entire cultures and nations like the Philippines, Indonesia, Papua New Guinea, Jamaica, Madagascar, Japan, and northern Australia. Subtropical examples are found in New Zealand, the Azores, the Baleares, and Easter Island.

Examples of tropical and subtropical oceans can be found in the Yellow Sea, Indian Ocean, Caribbean Sea, Sargasso Sea, Central Atlantic, and the Central Pacific Ocean. More landlocked seas of such water bodies are represented by the Mediterranean, Black Sea, Red Sea, Gulf of Mexico, and Sea of Japan. These ocean regions include such famous and diverse ecosystems as the coral reefs (e.g., the Great Barrier Reef, Caribbean Reefs, and Coral Reef Triangle), the deepest ocean floor (Mariana Trench), seagrass (Sargasso Sea), the Ring of Fire (a Pacific chain of tectonic plates and volcanoes), and some of the most remote islands of the world (e.g., Easter Island and Hawaii).

The tropical and subtropical oceans are a large resource of biodiversity, and many sea mammal and seabird species can be found there. Much of the marine benthos still awaits its description, as do many plankton and fish species. However, the loss of biodiversity and habitat is already a major problem.

Invasive species and diseases are on the rise. Marine protected areas (MPA) are planned, but these initiatives will hardly protect relevant areas. The well-being of these oceans and their sophisticated and evolved land- and seascapes, ecosystems, and human cultures have declined rapidly in less than 50 years. Overfishing is occurring on a global level. Illegal fishing and piracy are also big unresolved issues. The freedom of the international seas cannot be upheld.

The approach of single-species management fisheries has been criticized for decades. Top predators disappear (e.g., marine mammals such as sharks and tuna) in the study area. Dynamite fishing is still practiced, destroying entire ecosystems.

Coral reefs are bleaching on a massive scale. Hypoxias occur, and so does ocean acidification. Plastic pollution has become another major issue. Shipping, ports, marinas, oil pollution, and offshore gas and oil development are further unresolved problems. Oceanic destructions are directly linked with the promotion of economic growth, the increase in poverty, political heritages, and the rise of neocolonialism schemes. Policies set by the Rio Convention or the Convention on International Trade in Endangered Species of Wild Fauna and Flora (CITES) will not be able to stop these destructions alone, judged by the results thus far. The situation has become worse in most areas, including missed biodiversity targets, widening poverty gaps, economic crises, ideological conflicts and seaborne conflict, and increased human populations. The tropic and subtropic ocean regions and their web of life will not disappear any time soon, but with the increase of climate change, sea-level rise, a population of more than 9 billion people projected by 2050, and a corresponding increase in the consumption of products and goods, the outlook for the tropical and subtropical ocean regions, and thus the future for the world, seem rather grim.

—Falk Huettmann

Further Reading

Talley, Lynne D., George L. Pickard, William J. Emery, and James H. Swift. *Descriptive Physical Oceanography: An Introduction.* 6th ed., Elsevier, 2011.

Wang, Bin. *The Asian Monsoon.* Springer/Praxis, 2006.

Webster, Peter. *Dynamics of the Tropical Atmosphere and Oceans.* Wiley Blackwell, 2020.

Tsunamis

Fields of Study: oceanography; hydraulics: wave physics; computer modeling; seismology

ABSTRACT

A tsunami is a series of potentially catastrophic ocean waves most commonly caused by movement of the seafloor associated with large earthquakes. Once known as tidal waves, tsunamis have resulted in hundreds of thousands of deaths, primarily around the Pacific Ocean.

KEY CONCEPTS

bore: a distinct wave produced by rapid increase in water level in a tapering channel

constructive interference: a phenomenon of wave mechanics by which the coincidence of two or more waves produces a resultant wave with the magnitude of the incident waves combined

inundation: the depth to which a tsunami or other water flood covers the area it has flooded

run-up: the distance to which a tsunami or other water flood encroaches beyond the normal shoreline

seiche: a resonance wave, which can be very large, that forms in an enclosed or partially enclosed body of water

seismometer: a device that detects vibrations in Earth's crust due to earthquakes and other seismic activity

CAUSES OF TSUNAMIS

Giant ocean waves that were once, incorrectly, called "tidal waves" or "seismic sea waves" have resulted in hundreds of thousands, if not millions, of deaths in countries around the Pacific Ocean. These destructive waves are now officially known by their Japanese name as tsunamis. In the simplest terms, a tsunami is a series of waves usually caused by either a violent movement of the seafloor, a volcanic eruption or a landslide, any of which results in a massive displacement of water.

The movement at the seafloor that causes the tsunami can be produced by three different types of violent geologic or seismic activity. By far the most common and, therefore, the most important of these is submarine faulting, which occurs when a section of the ocean floor is thrust upward or suddenly drops at a subduction zone fault. Such fault movements are accompanied by earthquakes.

Probably the second most common cause of tsunamis is sudden landslide by which a mass of rock and earth falls suddenly into a body of water. A tsunami may be generated by a landslide starting above sea level and then plunging into the sea, or by a submarine landslide. The highest tsunami waves ever officially reported were produced by a landslide at Lituya Bay, a confined fjord in Alaska on July 9, 1958. A massive rockslide at the head of the bay produced a tsunami wave that attained a high-water mark more than 1,640 feet (500 meters) above the shoreline.

The third major cause of tsunamis is nearshore or submarine volcanic activity. In most cases, the flank of a volcano is suddenly uplifted or depressed, producing a tsunami in much the same way as faulting activity does. However, tsunamis have also been produced by the actual explosion of submarine or shoreline volcanoes. In 1883, the violent explosion of the famous island volcano Krakatau sent tsunami waves as high as 131 ft (40 m) crashing ashore in Java and Sumatra, killing more than 36,000 people.

Although tsunamis caused by landslides or volcanic activity may become very large near their sources and may cause great damage there, they have relatively little energy. They decrease rapidly in size in the deep waters of the open ocean, becoming small or even unnoticeable at any great distance. As they approach land and enter shallower waters again, their size increases according to the kinetic energy that they have retained. The giant tsunami waves that cross entire oceans are almost all caused by submarine faulting associated with large subduction earthquakes. Most tsunamis occur in the Pacific Ocean because the Pacific Ocean basin is surrounded by a zone of very tectonically active features, such as deep ocean subduction trenches, explosive volcanic islands, and dynamic mountain ranges.

CHARACTERISTICS OF TSUNAMIS

Tsunami waves are very different from ordinary ocean waves, most of which are caused by the wind blowing over the water. Wind-generated waves rarely have a wavelength (distance from crest to crest) greater than 984 ft (300 m). Tsunami waves may have a wavelength of as much as 99 miles (160 kilometers). Wind-generated ocean surface waves never travel at more than about 62 mi (100 km) per hour and are usually much slower. In the deep water of an ocean basin, tsunami waves may travel at 497 mi (800 km) per hour, but are often only 20 inches (50 centimeters) high, thus passing unnoticed by ships at sea.

A tsunami is not a single giant wave but rather consists of a series of several waves, perhaps ten or more, that form what is called a "tsunami wave train." These individual waves follow one behind the other, between five and ninety minutes apart. When tsunami waves move into shallower water and approach shore, they start to change. The shape of the nearshore seafloor has an effect on how tsunami waves behave. The waves tend to be smaller near small, isolated islands, where the bottom drops away quickly into deep water. Near large islands, such as the main Hawaiian Islands the waves are strongly influenced by the bottom; they bend around the land and may be reflected from the shoreline. The reflected waves may augment other waves and create extremely large wave heights in unexpected places. As video records of the 2011 tsunami that struck Japan indicate, the tsunami wave does not appear like a typical wind-driven wave breaking on the shore, but more like an abrupt, massive increase in sea level as the water rushes inland.

Though it is not yet possible to accurately predict the production, size, or effects of tsunami waves as they arrive in coastal areas, prediction of the velocity of tsunamis is made possible by their great wavelength. To understand how the arrival time of a tsunami is determined, it is necessary to look at ocean waves in general. Oceanographers divide waves into various categories based on the relationship between wavelength and the depth of water through which they are passing. When the water depth is less than one-twentieth of the wavelength, the waves are known as shallow-water waves, and their speed is determined solely by the depth of water. Knowing the water depth, one can calculate the velocity of any shallow- water wave.

Tsunami waves may have a wavelength of more than 99 mi (160 km), and if the water depth is less than one-twentieth of the 99-mi (160-km) wavelength, or 5 mi (8 km), then the tsunami waves would be shallow-water waves. Most of the deep Pacific

Aerial view of tsunami damage in an area north of Sendai, Japan, 2011 (US Navy via Wikimedia Commons)

A photograph of the 2004 tsunami in Ao Nang, Krabi Province, Thailand (David Rydevik via Wikimedia Commons)

Ocean basin is less than 3 mi (5 km) deep, so most tsunamis there are shallow-water waves. To determine the speed of such tsunamis, all that is necessary to know is the depth of the water through which they pass. In most of the deep Pacific Ocean, this speed works out to be around 450 mi (725 km) per hour, but it will vary depending on the exact water depth along the path of the tsunami.

EFFECTS OF TSUNAMIS

As the waves approach the shore, frictional effects of contact with the seafloor begin to affect the tsunami waves, and they travel progressively more slowly, with the forward speed dropping to around 40 mi (65 km) per hour. At this point, the wave height usually begins to increase dramatically. A tsunami wave that was 20 in (50 cm) high at sea may reach a height of 33 ft (10 m) or more at the shoreline.

Like storm waves, tsunami waves are often more severe on headlands, where the wave energy is concentrated. The presence of a well-developed coral reef off a shoreline also appears to have a strong effect on tsunami waves. The reef may serve to absorb a significant amount of the wave's energy, reducing its height and the intensity of its impact on the shoreline. Unlike ordinary waves, however, tsunamis are often quite large in bays. In this way, tsunami waves somewhat resemble tides.

Another wave phenomenon may also be produced in bays when a tsunami strikes. The water in any basin, be it a small bay or a large sea, will tend to oscillate back and forth with a fixed period is determined by the size and shape of the basin. This oscillation is known as a seiche. A tsunami wave may initiate a seiche, and if the following tsunami wave arrives in conjunction with the next natural oscillation of the seiche, the water may reach even greater heights than the tsunami waves alone, by the wave property of constructive interference. Much of the great height of tsunami waves in bays may be explained by this constructive combination of a seiche wave and a tsunami wave arriving simultaneously.

The popular image of a tsunami wave approaching shore is a nearly vertical wall of water, similar to the front of a breaking wave in the surf. Actually, most tsunamis do not form such wave fronts; the water surface instead is very close to horizontal, and the surface itself moves up and down. The waves arrive like a very rapidly rising tide. Under certain circumstances, however, an arriving tsunami wave can develop an abrupt, steep front that will move inland at high speeds. This phenomenon, generally encountered only as a tidal phenomenon, is known as a bore.

Bores produced by tides occasionally occur in the mouths of rivers. Well-known examples occur on the Solway Firth and the Severn River in Great Britain, on the Petitcodiac River in Maine, near the mouth of the Amazon River in Brazil, and, most strikingly, on the Chientang River in China, where the bore may attain a height of nearly 16 ft (5 m). In place of the usually gradual rise of the tide, the onset of high tide is delayed by the outflow of the estuary and augmented by a funneling effect as the stream width becomes narrower. When the high tide does arrive, it comes in very quickly, and the rapidly moving wall of water is followed by a less steep, but still quite dramatic, rise in the water level, accompanied by swift upstream currents.

Because the height of tsunami waves is strongly influenced by the submarine topography and shape of the shoreline and by reflected waves, and because they may be further modified by seiches, tides, and wind waves, the actual inundation and flooding produced by a tsunami may vary greatly from place to place over even a short distance. Though the image of a bore is the most dramatic—and such a wall of water can destroy nearly everything in its path—it is the flooding and backwash effect of a tsunami that typically cause the most damage. Two different terms are often used to describe the extent of tsunami flooding: "inundation" and "run-up." Inundation is the depth of water above the normal level and is usually measured from sea level at average low tide. Inundation may be measured at any location reached by the tsunami waves. Run-up is the inundation at the maximum distance inland from the shoreline reached by the tsunami waters.

The withdrawal of the tsunami waves can cause significant damage. As the water rapidly withdraws toward the sea, the force of its movement scours out bottom sediments, undermines the foundations of buildings, and carries almost everything in its path out to sea. Entire beaches have been known to disappear as the sand is carried out to sea by the withdrawing tsunami waves, as have thousands of people, cars,

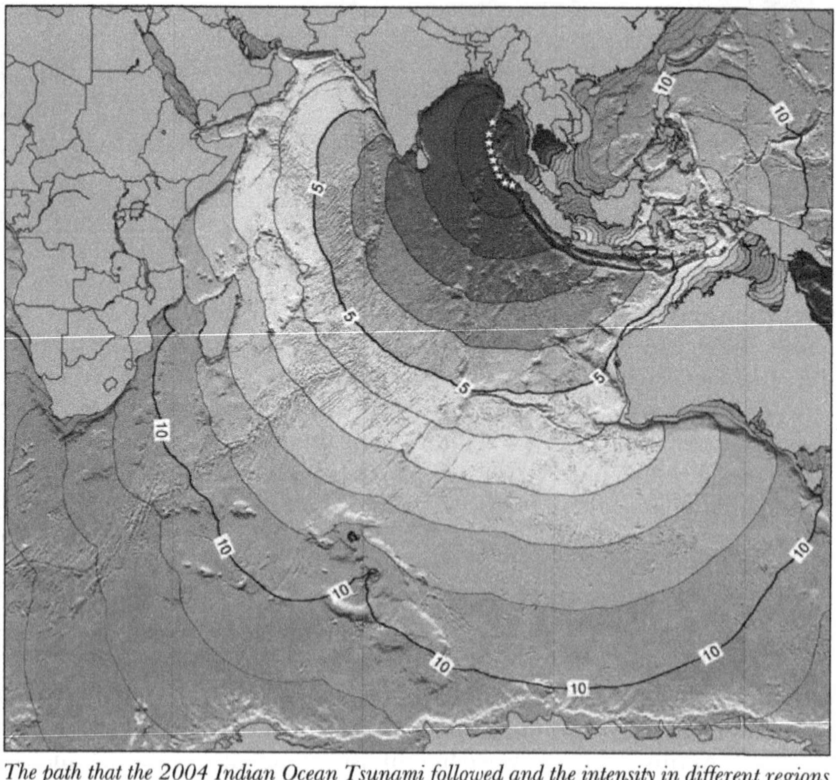

The path that the 2004 Indian Ocean Tsunami followed and the intensity in different regions (National Oceanic and Atmospheric Administration via Wikimedia Commons)

trucks, houses, boats, and any other material caught in the retreating backwash.

STUDY OF TSUNAMIS

The best way to learn about tsunamis is to study the tsunami waves themselves. Much of this work is carried out in the Hawaiian Islands because of their susceptibility to tsunamis from all parts of the Pacific Ocean basin and because Hawaii is the headquarters of the Pacific Tsunami Warning System. Because tsunamis are, fortunately, not everyday events, it is imperative to collect the most information possible from each occurrence. When a tsunami is en route to the Hawaiian Islands, trained observers head toward preselected shoreline vantage points, where time-lapse surveillance cameras are set up to film the waves. Portable tsunami gauges are deployed from piers and in designated shoreline areas. These gauges sense the change in water pressure as the waves pass over them and record their measurements.

Patrol aircraft take to the air before the arrival of the first waves. The aircraft fly at an altitude of 984 ft (300 m) in a racetrack pattern over critical shoreline areas, covering the same spot every fifteen minutes. Special cameras mounted in the belly of each plane document the arrival of each tsunami wave on the shores of the islands.

After a tsunami, a ground survey and damage assessment are conducted by a team from the US Army Corps of Engineers, which is joined by volunteers from the American Society of Civil Engineers. A posttsunami aerial photographic survey is under-

A village near the coast of Sumatra lays in ruin after the 2005 Tsunami that struck South East Asia (US Navy via Wikimedia Commons)

Aerial photo of Hilo Bay, Hawaii (Kanoa Withington via Wikimedia Commons)

taken jointly by the Navy, the Coast Guard and the Civil Air Patrol, using National Weather Service surveillance aircraft.

Laboratory techniques have also been employed to study tsunami waves. Much of the damage caused by tsunamis results from the run-up of the giant waves on shore. Various modeling techniques have been used to try to simulate the run-up phase. One such technique, called "hydraulic modeling," uses a physical scale model. Hilo Bay on the island of Hawaii is particularly sensitive to tsunami waves. After the disastrous tsunami of 1960, a hydraulic model of the bay was constructed. The model measured 82 by 62 ft (25 by 19 m) and represented the triangular shape of Hilo Bay. Model tsunami waves were produced by releasing water from large tanks according to a program. Experiments with this hydraulic model have shown that almost any wave that enters the bay either hits downtown Hilo directly or is bounced off the northeast coast into the town. Sometimes the direct and reflected waves interact constructively to produce especially large waves in the center of the bay.

Another type of model used to study tsunamis is the numerical model, which uses high-speed digital computers to calculate mathematical simulations of tsunami waves. The models, however, are only as accurate as the data on which they are based, and these data can only come from measurements and observations of actual tsunamis. An improvement in data collection has been the use of satellites to transmit data from remote tide stations.

Satellite-telemetered tide stations have been installed at some twenty-five sites across the Pacific. These stations operate on their own independent power sources, secure from electricity outages that may result from earthquakes. Sea-level measurements are made every two seconds and averaged over a three- or four-minute interval, and the data are routinely transmitted by satellite to the Pacific Tsunami Warning Center every three to four hours. In the event of a tsunami wave, however, an "event detector" almost instantaneously sends a message to the warning center over a special emergency satellite channel.

Beginning in the 1990s, even more sophisticated measuring devices were developed, including devices that do not have to be installed on land. A warning system was proposed that would be based on the detection of tsunamis by highly sensitive bottom-pressure gauges located on the midocean seafloor. To study how tsunami waves change upon entering shallow water, an observational program was conducted off the Galápagos Islands, with instruments at depths of 9,842 ft (3,000 m), 33 ft (10 m), and 3 ft (1 m). Such testing eventually led to the creation of Deep-ocean Assessment and Reporting of Tsunami (DART) technology placed on buoy stations. The 2004 Indian Ocean earthquake and tsunami led to the expansion of the DART system, with the United States completing its buoy array in 2008 and other countries establishing their own programs.

In addition to seafloor tsunami detectors, seismometers have been deployed on the ocean bottom. Japanese scientists successfully operated a permanent ocean-bottom seismometric system off the southern coast of central Honshu. Attached to their seismometers is a tsunami gauge. The most advanced tsunami warning systems use both shore-based and ocean-bottom earthquake and tsunami sensors that transmit real-time data via satellite to the warning center, where computers provide scientists with the information they need to issue confirmed tsunami alerts. These efforts have increased knowledge of tsunami wave generation, propagation, and run-up, though the waves are still not fully understood and research continues.

—Walter C. Dudley, Jr.

Further Reading

Bernard, Eddie N., and Allan R. Robinson. *The Sea, Volume 15: Tsunamis.* Harvard UP, 2009.

Chester, Roy. *Furnace of Creation, Cradle of Destruction.* AMACOM Books, 2008.

Clague, John, Chris Yorath, and Richard Franklin. *At Risk: Earthquakes and Tsunamis on the West Coast.* Tricouni Press. 2006.

Corrigan, Eamonn. *Tsunamis: Causes and Effects.* 12-Story Library, 2020.

Engel, Max, Jessica Pilarczyk, Simon Matthias May, Dominik Brill, and Ed Garrett, editors. *Geological Records of Tsunamis and Other Extreme Waves.* Elsevier, 2020.

Guidoboni, Emanuela, and John E. Ebel. *Earthquakes and Tsunamis in the Past: A Guide to Techniques in Historical Seismology.* Cambridge UP, 2009.

Joseph, Antony. *Tsunamis: Detection, Monitoring and Early Warning Technologies.* Academic Press/Elsevier, 2011.

O'Laughlin, Karen Fay, and James F. Lander. *Caribbean Tsunamis: A 500-Year History from 1498–1998.* Kluwer Academic Publishers, 2003.

Prothero, Donald R. *Catastrophes!: Earthquakes, Tsunamis, Tornadoes, and Other Earth-Shattering Disasters.* Johns Hopkins UP, 2011.

Satake, Kenji, editor. *Tsunamis: Case Studies and Recent Developments.* Springer, 2005.

Schneider, Bonnie. *Extreme Weather: A Guide to Surviving Flash Floods, Tornadoes, Hurricanes, Heat Waves, Snowstorms, Tsunamis and Other Natural Disasters.* Palgrave Macmillan, 2012.

Sutton, Gerard K., and Joseph A. Cassalli, editors. *Catastrophe in Japan: The Earthquake and Tsunami of 2011.* Nova Science Publishers, 2011.

Tsuchiya, Yoshito, and Nobuo Shuto, editors. *Tsunami: Progress in Prediction, Disaster Prevention, and Warning.* Kluwer Academic Publishers, 2010.

"Tsunamis." *National Oceanic and Atmospheric Administration,* US Department of Commerce, June 2011, www.noaa.gov/resource-collections/tsunamis.

"What Are Tsunamis, and What Causes Them?" *International Tsunami Information Center,* UNESCO, 2018, itic.ioc-unesco.org/index.php?option=com_content&view=article&id=1017:what-are-tsunamis-and-what-causes-them&catid=1004&Itemid=1004.

"What Is a Tsunami?" *National Ocean Service,* National Oceanic and Atmospheric Administration, 10 Oct. 2017, oceanservice.noaa.gov/facts/tsunami.html. Accessed 20 Mar. 2018.

TURBIDITY CURRENTS AND SUBMARINE FANS

Fields of Study: oceanography; marine geology; fluid dynamics; limnology

ABSTRACT
Turbidity currents are a major mechanism whereby sediment from nearshore areas is transported to deeper parts of the oceans. This sediment often accumulates as part of large depositional systems called "submarine fans," which could well be exploited as a major source of petroleum at some time in the future.

KEY CONCEPTS

continental rise: the region of gradually rising seafloor from the abyssal plain in the direction of the continental shelf

density currents: submarine flows of dense mixtures of water and sediment particles, another name for turbidity currents

depositional lobes: surface formations in sea fan topography, formed when smaller branch channels direct the flow of turbidite material onto an existing sea fan

inclined flume: a kind of trough in which the nature of material flows such as turbidity currents can be studied, adjustable to different slope angles

oversteepening: the buildup of sediment beyond its natural angle of stability, typically resulting in a sudden collapse of the too-steep side

turbidite: a sedimentary deposit on the ocean floor, in which particle size graduates continuously from the largest and densest particles at the bottom of the bed to the smallest and least dense at the top of the bed

DEVELOPMENT OF THE CONCEPT

A turbidity current is a dense mass of water and sediment that flows downhill along the bottom of an ocean or any other standing body of water such as a lake. Turbidity currents may reach high speeds and are capable of carrying large quantities of sediment, and of eroding or scouring the ocean floor as they move.

The concept of the turbidity current has a long history. In the late nineteenth century, geologists observing the waters of the Rhone River entering Lake Geneva noted that instead of mixing with the lake water, the river water moved along the bottom of the lake in a channel. This behavior was interpreted to reflect the elevated density of the cold, sediment-laden Rhone River water. These dense, bottom-hugging currents were called "density currents." It was suggested in the late 1930s that dense currents composed of sediment and water could be produced by wave activity on the continental shelves, the flat or gently sloping submerged edges of the continents, during periods when global sea level is lower than it is today. Furthermore, it was postulated that these dense, turbulent mixtures of suspended sediment and water would flow across the continental shelves and down large canyon-like features into deeper parts of the ocean.

From the 1930s to the 1950s, researchers conducted numerous laboratory studies to test the hypothesis that turbidity currents could actually erode the ocean floor and produce such features as submarine canyons on the continental shelf. This work was done by pouring muddy water into the end of an inclined flume (a long, straight trough in which the hydrodynamic properties of moving fluids can be studied) in order to produce artificial turbidity currents. Results of these investigations indicated that sediment in a turbidity current moves in a very chaotic or random pattern, much like the movement of snow and other materials in an avalanche. Additionally, geologists recognized that turbidity currents can move rapidly and that their velocity is dependent upon the slope of the sea bottom along which they move as well as the density of the sediment-water mixture.

These experimental studies also yielded insight into the nature of the sedimentary deposits that accumulate from turbidity currents. These deposits, referred to as turbidites, form as graded beds, sedimentary layers in which the largest or most coarse-grained sediment particles are concentrated at the bottom of the bed and grade gradually upward to the smallest or most fine-grained sediment at the top. The graded nature of the bedding was deduced correctly to be a consequence of a reduction in the flow velocity of the turbidity current upon reaching the very gentle slopes that are typical of the bottom of the ocean, so that the largest sediment particles settle out of the current to the bottom of the ocean first, followed by progressively finer particles. Thus, each turbidity current produces a single-graded bed or turbidite.

Geologists working in the field began to realize that the graded beds such as those produced in laboratory studies could be observed in sequences of sedimentary rocks exposed on land in various mountain belts throughout the world (the Appalachians, Apennines, and Carpathians). Moreover, oceanographic research demonstrated the presence of turbidite sands in some of the deepest areas of the modern oceans. It is now apparent that turbidity currents are and were a dominant, if not the dominant, mechanism of sediment transport from shallow-water environments on the continental shelf to the deeper parts of the ocean. Evidence of this includes the recovery of shallow-water organic remains from very deep parts of the oceans far removed from land. The laboratory studies and subsequent oceanographic research indicated that turbidites differ from most other sedimentary deposits in that they are nearly instantaneous deposits that may accumulate over a period of just a few hours to a few days. Moreover, turbidity currents are quite capable of eroding the great submarine canyons that cut into many continental shelves.

CAUSES OF TURBIDITY CURRENTS

Downhill movement of turbidity currents can be triggered by various causes. Commonly, earthquakes affecting the continental shelves will cause sediment on the continental slope—the part of the continental margin characterized by an increase in gradient immediately seaward of the continental shelf—to destabilize and slide downslope, thereby mixing with seawater to form a turbidity current. A rapid sequence of successive breaks in transatlantic cables on the continental margin bordering southern Newfoundland was apparently caused by the downslope movement of a dense, turbulent mixture of seawater and sediment generated from the continental shelf and slope in response to the Grand Banks earthquake of November 8, 1929. Those cables closest to the point on the ocean floor directly above the earthquake, the epicenter, broke first, whereas cables farther from this point broke later. The breaking of the cables indicated the erosive capability of turbidity currents. Subsequent drilling of that part of the ocean floor traversed by the turbulent flow recovered a 3-foot- (1-meter)-thick graded bed containing shallow-water organic remains; this discovery strengthened the argument that the flow was indeed a turbidity current. The earthquake-induced turbidity current averaged 17 miles (27 kilometers) per hour, although it reached velocities in excess of 43 mi (70 km) per hour in steeper parts of the continental margin, and it covered an area of more than 75,290 square miles (195,000 square kilometers).

Turbidity currents can also be generated by wave activity on continental shelves, an idea, as already

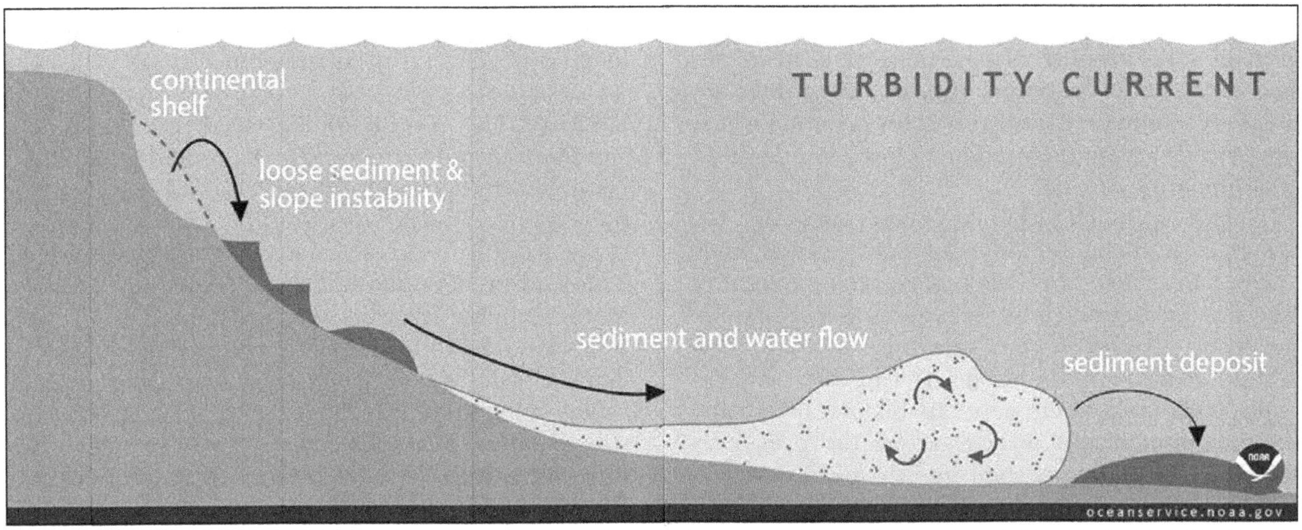

A diagram showing the process of formation and propagation of a turbidity current (National Oceanic and Atmospheric Administration via Wikimedia Commons)

A prominent submarine dune with large ripples, possible result of a turbidity current (National Oceanic and Atmospheric Administration via Wikimedia Commons)

noted, that was postulated in the 1930s. More specifically, large waves produced during great storms such as hurricanes are capable of creating the turbulence required to mix sediment and seawater, thereby creating a turbidity current. An equally plausible mechanism of turbidity-current generation involves oversteepening of the continental slope by the sudden addition of sediment. This is especially common where large rivers deposit their sediment load near the head of a submarine canyon.

The rate at which turbidity currents are generated was greatest during periods when sea level was much lower than it is today. This was especially typical of glacial periods, when more seawater was locked up in the enlarged polar ice caps, resulting in the lowering of the global level of the oceans. During these times of lowered sea level, rivers were capable of transporting their sediment loads directly across the previously submerged continental shelf into the head of a submarine canyon. There, the sediment mixed with seawater and was fed, via the submarine canyon, directly into the deeper parts of the ocean as turbidity currents. When sea level rose and the shoreline retreated landward as the exposed continental shelf was once again submerged, the river was cut off from the canyon head, thereby precluding the infusion of sediment directly to the canyon and reducing the likelihood of turbidity-current generation.

CHARACTERISTICS OF SUBMARINE FANS

Turbidity currents triggered on continental shelves or continental slopes move downhill until they reach a point at which the reduced gradient of the ocean floor causes a reduction in the velocity of the sediment flow. This leads to deposition of a graded bed or turbidite. Many turbidite deposits that are funneled through submarine canyons ultimately accumulate as part of large fan-shaped or cone-shaped sediment bodies called "submarine fans." The submarine fans, which spread outward from the mouths of the submarine canyons, merge with the bottom of the continental slope and comprise the continental rise, the broad, gently sloping feature that rises from the abyssal plain of the ocean floor and merges with the base of the continental slope. Where submarine canyons are close together along the continental shelf and continental slope, the attached submarine fans may coalesce to form a wide, laterally extensive continental rise.

In general, submarine fans are subdivided into three major morphologic elements or parts: the upper fan, midfan, and lower fan. The upper fan, also referred to as the inner fan, is typically characterized by a single submarine channel that is connected to the submarine canyon. Upper fan channels range from 1.24 to 11 mi (2 to 18 km) wide and may be as deep as 2,953 ft (900 m). The single channel is commonly flanked on both sides by levees, low ridges that run along the length of the channel.

Most of the sediment transported through the upper fan channel via the submarine canyon is deposited into the midfan, that area of the submarine fan composed of raised, lobelike sequences of turbidites called "depositional lobes." The depositional lobes are fed by numerous shallow and unstable distributive channels that branch off the main upper fan channel. Because these channels are generally relatively shallow (several tens of feet [or meters] deep), they are more likely to be filled in during passage of extraordinarily dense turbidity currents. When that happens, the channels are abandoned and subsequent turbidity currents erode or scour out new channels in adjacent areas of the midfan portion of the submarine fan.

The lower fan is characterized by a smooth ocean floor that passes imperceptibly seaward into the abyssal plain. The low gradient of the lower fan relative to that of the midfan leads to a much-reduced abundance of turbidites in the lower fan. Turbidite deposits are not common on the abyssal plain because ocean floor gradients in this area are typically too low to sustain movement of the dense sediment cloud.

Submarine fans display great variations in size. The Bengal Fan, in the Bay of Bengal in the northeast Indian Ocean is the largest submarine fan. Its total length exceeds 1,864 mi (3,000 km). The main or upper fan channel ranges from 8 to 11 mi (13 to 18 km) wide and from 492 to 2,953 ft (150 to 900 m) deep. Most submarine fans, however, are much smaller than the Bengal Fan.

The growth of submarine fans, because they are major sites of turbidite sedimentation is controlled by variations in sea level. As noted previously, turbidity currents are generated more frequently during periods of low sea level. Accordingly, submarine fans are likely to grow fastest, as manifested by increased rates of turbidite sedimentation, during periods of low global sea level. Indeed, geologists have demonstrated that the Mississippi submarine fan, which is fed by the Mississippi River was the site of abundant turbidite sedimentation at the end of the most recent ice age, approximately 15,000 to 20,000 years ago, when sea level was as much as 394 to 427 ft (120 to 130 m) lower than it is today. Since then, the rate of turbidite sedimentation has dropped off greatly as sea level gradually rose, and the shoreline retreated landward to its present position.

—*Gary G. Lash*

Further Reading

Brown, Joan. *Ocean Circulation: Prepared by an Open University Course Team.* Elsevier, 2016.

Huneke, Heiko, and Thierry Mulder. *Deep Sea Sediments.* Elsevier, 2011.

Leeder, Mike R. *Sedimentology and Sedimentary Basins: From Turbulence to Tectonics.* Wiley-Blackwell, 2011.

McKinney, Frank. *The Northern Adriatic Ecosystem: Deep Time in a Shallow Sea.* Columbia UP, 2007.

Pickering, Kevin T., and Richard N. Hiscott. *Deep Marine Systems: Processes, Deposits, Environments, Tectonics and Sedimentation.* Wiley, 2016.

Schulz, Horst D., and Matthias Zabel. *Marine Geochemistry.* 2nd ed., Springer-Verlag, 2006.

Schwartz, M., editor. *Encyclopedia of Coastal Science.* Springer, 2005.

Shanmugam, G. *Deep-Water Processes and Facies Models: Implications for Sandstone Petroleum Reservoirs.* Elsevier, 2006.

Talley, Lynne D., George L. Pickard, William J. Emery, and James H. Swift. *Descriptive Physical Oceanography: An Introduction.* 6th ed., Elsevier, 2011.

Thurman, H. V., and Alan P. Trujillo. *Introductory Oceanography.* 10th ed., Prentice-Hall, 2003.

U

Upwelling Ecosystems

Fields of Study: oceanography; marine biology; fisheries management

ABSTRACT
Upwelling ecosystems occur where cold, often nutrient-rich waters from the ocean depths rise to the surface. They can also occur in freshwater bodies such as lakes. Upwelling ecosystems are considered some of the most productive ecosystems in the world. Upwelling is the process in which surface currents move away from one another, or when winds push surface waters away from the shoreline. This draws deeper water upward to replace the surface water. This process of the rising of cold nutrient-rich waters is essential for productivity in marine and freshwater ecosystems.

KEY CONCEPTS
benthic: referring to the lowest depths of the ocean, the seafloor

endemism: the occurrence of species that are or have become endemic to a specific area

epilimnion: the upper level of a body of water, where the water becomes warmed by sunlight

hypolimnion: the lower level of a body of water, where the water is not warmed by sunlight

thermocline: the transition region between the epilimnion and the hypolimnion, where a steep change of water temperature takes place

trophic levels: the various stages in the structure of a food chain or food web

upwelling cell: a site of cold water upwelling that is not associated with a current rising up as it encounters a continental mass, appearing as an isolated area of upwelling either due to atmospheric disturbance at that location or by diversion of the flow of a deep ocean current encountering an undersea obstacle such as a seamount or guyot

UPWELLING AND THE FOOD CHAIN
Upwelling ecosystems occur where cold, often nutrient-rich waters from the ocean depths rise to the surface. They can also occur in freshwater bodies such as lakes. Upwelling ecosystems are considered some of the most productive ecosystems in the world. Upwelling is the process in which surface currents move away from one another, or when winds push surface waters away from the shoreline. This draws cooler water upward from lower depths to replace the surface water. The warmer surface water cools and eventually sinks to the lower depths to complete the convection cycle.

This process of the rising of cold nutrient-rich waters is essential for productivity in marine and freshwater ecosystems. The importance of the nutrient-rich waters has a direct tie to the food web. This is because most primary productivity in the oceans occurs in surface waters, but most of the organic nutrient matter can be found at the bottom of the sea. The reason for this is because almost all food chains begin with photosynthesis, and photosynthesis occurs in sunlight. Marine plants, therefore, will only live near the surface where light can penetrate, keeping herbivores and their predators up near the surface. But once these organisms die or shed scales, eggs, leaves, shells, or feces, that detritus sinks down toward the seafloor and decomposes.

Upwelling thus brings all these organic nutrients back up to the surface water. With these nutrients, the phytoplankton and seaweed population grow, forming the basis of many food webs in marine and freshwater ecosystems. These blooms form the ultimate energy base for large animal populations higher in the food chain, including marine and freshwater fish, mammals, and seabirds. The raising of benthic nutrients to the surface waters occurs in regions where the flow of water brings currents of differing temperatures together. There are at least five types of upwellings: coastal upwelling, large-scale wind-driven

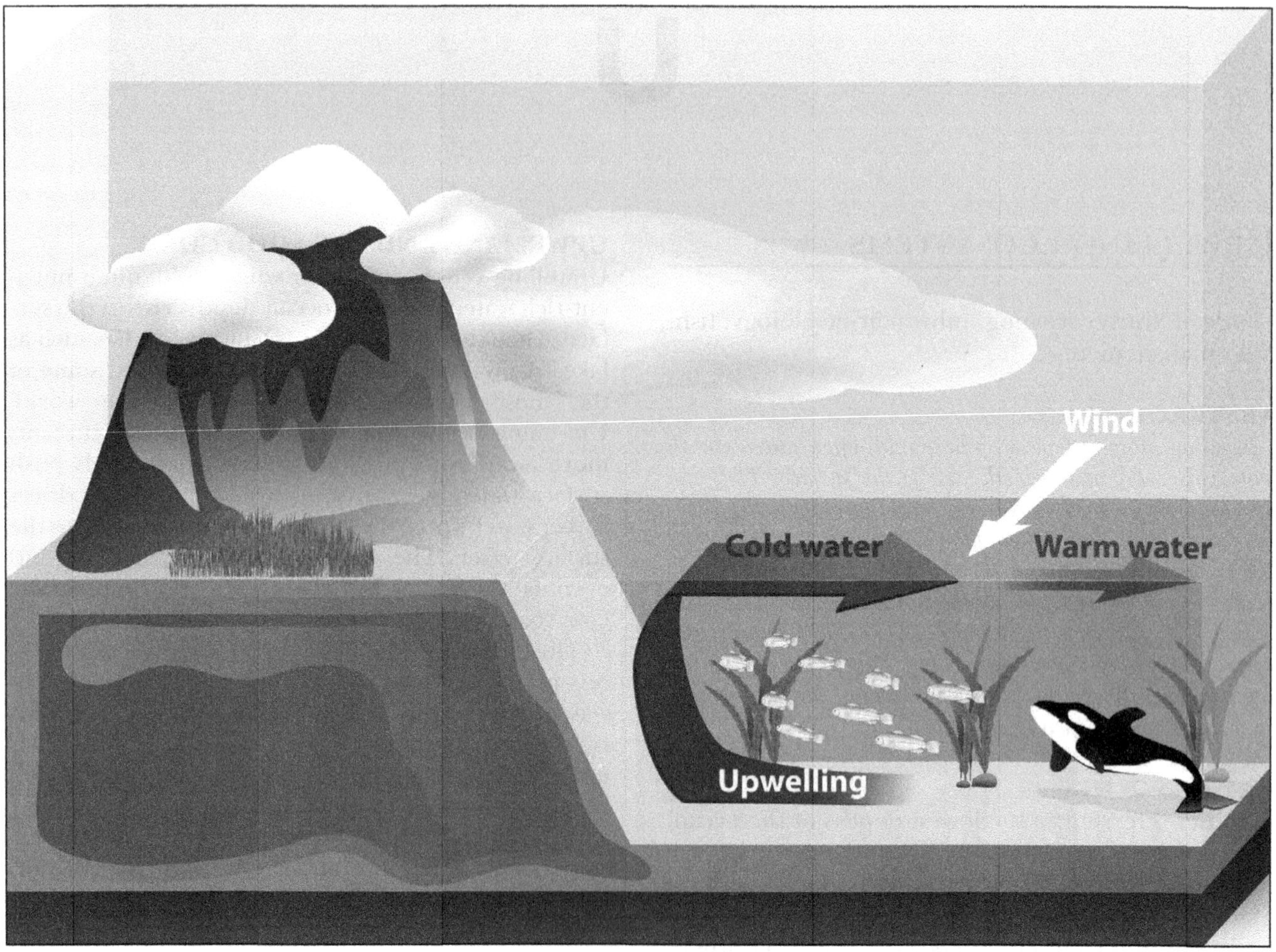

Upwelling currents bring cold, nutrient-rich water up from the ocean floor to the surface, and support the growth of plankton and seaweed (ttsz via iStock)

upwelling in the ocean interior, upwelling associated with eddies, topographically associated upwelling, and broad-diffusive upwelling in the ocean interior.

COASTAL UPWELLING

Winds affected by the rotation of the Earth create a phenomenon known as the Coriolis effect. Along a coastline oriented north-south, like much of the west coast of the United States, winds that blow from the north tend to drive ocean surface currents to the right of the wind direction, thus pushing surface waters offshore. As surface waters are pushed offshore, water is drawn from below to replace them; these are coastal upwellings. Coastal upwelling usually occurs in the subtropics along the continental coasts, where prevailing trade winds drive the surface water away from shore, drawing deeper water upward to take its place.

Because of the abundance of krill and other nutrients in the colder waters, these regions are rich feeding grounds for a variety of marine species, including whales, sharks, and many other fish species, as well as avian species (coastal birds). Some of the most well-known examples of this type of upwelling include: the Canary Current near Northern Africa, the Benguela Current near Southern Africa, the California Current off of the west coast of the United States, the Humboldt Current near Chile and Peru in South America, and the Somali Current off the western side of India.

All of these currents support major fisheries. Along the west coast of southern Africa, the Benguela upwelling distinguishes itself from other upwelling systems by the intrusion of warm waters at its two boundaries: to the north, via the Angola Benguela front, and to the south, via the Agulhas Current that terminates the western boundary current of the Indian Ocean.

To spawn, sardines (*Sardinops sagax*) and anchovies (*Engraulis encrasicolus*) migrate toward the Agulhas Bank, where the warm waters transported by the Agulhas Current create a highly stratified environment. Eggs and larvae are then rapidly transported northward by a coastal current. After several days, they reach the upwelling region of the east coast of Africa, where certain numbers are transported out to the open sea by wind-driven currents, a source of mortality. Other sources of early life cycle mortality must not be overlooked, in particular, those linked to starving larvae and predation of eggs and larvae. A week after hatching, larvae must feed. As their swimming abilities are limited, they need not only a high density of suitably sized prey (mainly microzooplankton), but also optimal turbulence conditions. Excessive turbulence disperses plankton aggregations and reduces capture. Using more precise modeling of currents and production should shed new light on these issues.

Although primary productivity of the Benguela system seems globally in excess, deficits may occur in certain spatiotemporal strata, such as that of the Agulhas Bank or the offshore feeding area during the reproductive season. In these conditions, feeding oases may be found within the large offshore eddies caused by the meeting of different currents and water masses. The low productivity of the Agulhas Bank creates intense food competition among the millions of tons (or tonnes) of spawners that congregate there during the spawning season. This encourages parental cannibalism and the predation of eggs and larvae by other species. Because of the absence of monthly spatialized data on fish distribution, modeling of this phenomenon remains difficult but efforts at quantification are possible.

The Benguela ecosystem can be divided into two subsystems, the northern Benguela (southern Angola and Namibia) and the southern Benguela (South Africa and Namibia), separated by the permanent upwelling cell of Luderitz, the strongest in the world. In South Africa, historical fisheries data suggest a stock collapse of sardines in the late 1960s, followed by slow recovery, whereas the first level of collapse was noted in Namibia during the same period, but worsened in the late 1970s. The biomass of the Namibian stock has remained at a very low level. Scientists are not sure if these collapses are because of an environmental process (upwelling) or overfishing.

The structure and ecosystem dynamics of the southern Benguela have been undergoing progressive changes for over 30 years, whereas the northern Benguela ecosystem has been completely "reorganized" following a "regime shift." Today, the northern Benguela is dominated by jellyfish and fish feeding on detritus (gobies). Pelagic fishes have become low in abundance, and fisheries for most commercial species (hake and horse mackerel) are threatened.

UPWELLING IN THE OCEAN INTERIOR

Upwelling can also occur in the middle of the oceans where cyclonic circulation is relatively permanent or where southern trade winds cross the equator. The churning of a cyclone eventually draws up cooler water from lower layers of the ocean, which, in turn, causes the cyclone to weaken. Another very important example of upwellings that occur in very special habitats are upwellings that occur off the shore of islands, ridges, or seamounts because of a deflection of deeper currents. Just like other upwelling habitats, this provides a nutrient-rich area in otherwise low productivity ocean areas.

Upwelling over seamounts has made them a very unique habitat with much endemism, and can be considered a biodiversity hot spot. A specific example includes upwellings around the Galapagos Islands with the Cromwell Current, and the Seychelles Islands, both which have major pelagic fisheries. In the Galapagos, the Cromwell Current is a very important feeding ground for the endemic Galapagos penguin, which heavily depends on the smaller fish that feed on the plankton caused by the upwelling. This upwelling is also very important for the marine iguana, also endemic to the Galapagos Islands. When this upwelling fails to occur during an El Niño year, there is great devastation to the populations of these endemic animals that depend so heavily on it.

Another type of powerful upwelling occurs around Antarctica. The deep water in the global conveyer belt flows across the seafloor all the way from the North Atlantic before it reaches Antarctica, so it has plenty of time to fill up with nutrients. Antarctica has many surface currents flowing around it, which stir up the waters, including the southern parts of the Pacific, Atlantic, and Indian Ocean gyres and the Circumpolar and Subpolar Currents. Also causing upwelling in this region are the strong polar winds that usually blow offshore.

FRESHWATER UPWELLING

Upwelling in freshwater ecosystems tends to be seasonal. It is most likely to occur in lakes in temperate regions. During the summer, the lake surface is heated by solar radiation, causing a thermal stratification (also known as thermocline) in the lake. It is because of this stratification that upwelling does not occur. By the time fall comes around, the stratification breaks down, and the temperature of the epilimnion (surface layers) becomes the same as the temperature at the hypolimnion (deeper layers). This means that the thermocline shifts and allows upwelling to occur; also, stronger seasonal winds aid the process of upwelling. When winter comes, the surface water of the lake freezes and upwelling or mixing does not occur. As the ice begins to melt during the spring, lakes become isothermal, which means that the epilimnion and the hypolimnion have the same temperature because of the wind action that brings deeper waters to the surface as a result of mixing.

CANARIES CURRENT

In the Canaries Current, there are three major ecosystems: the northern Moroccan coast with seasonal upwellings in the summer; the south Moroccan and north Mauritanian coast with permanent upwellings (Sahara Desert); and the south Mauritanian and Senegalese coast, with boreal upwellings. The southern part of this ecosystem is characterized by high seasonal variability, alternating between an ecosystem under tropical influence in the summer and a coastal upwelling ecosystem in the winter. This alternation is accompanied by a migration of certain tropical species (tuna and tuna-like fishes) up to 20 degrees north during summer, and by a southward extension of the habitat of temperate species such as the sardine *S. pilchardus* during the winter. The Canaries Current system has wide continental shelves in the south, whereas the eastern regions are generally characterized by narrow continental shelves because of their volcanic origin and young geological age. Principal spawning areas have been linked to regions with wide shelves. This association would be the result of physical processes developing over wide and shallow continental shelves, which would limit exchanges between the coast and pelagic areas.

The Canaries Current ecosystems were dominated by large demersal fish, which were rapidly overexploited. In the southern part of the area, the adaptability of artisanal fisheries took advantage of the seasonal migration of numerous species (a balance between seasonal upwelling and the warm season dominated by tropical influences). Over a longer period of time, West African fisheries have been spared the sudden collapses observed in other systems. A lower fishing exploitation rate until the 1980s and high seasonal fluctuations could have contributed to this relative resilience.

Three countries experienced rapid and unexpected growth of octopus stocks in the 1970s. Because of its commercial value, octopus fishing has become one of the essential components of West African fisheries. This shift in species was interpreted as the result of the absence of top-down control, following the overexploitation of demersal species that favored the development of short-lived prey species, such as the octopus (as well as shrimp and pelagic fishes). Further to the north, the reasons that have lead to the southern displacement of seiner fleets exploiting sardines remain largely unknown.

HUMBOLDT CURRENT

The Humboldt Current system, with its permanent upwellings off the coast of Peru and its seasonal upwelling along the coast of Chile, is by far the most productive in fish landings. With less than 1 percent of the world's ocean surface, it provides 15 to 20 percent of world marine catches (up to nearly 20 million tons (18 million tonnes) per year for Peru and Chile combined). A second particularity lies in the presence of a very intense, extensive shallow zone that is very low in oxygen.

A final particularity is its position under the direct influence of the El Niño Southern Oscillation (ENSO) mechanism. Alternate upwellings of nutrient-poor and nutrient-rich waters off the coast of Ecuador and

Peru are associated with El Niño and La Niña episodes in the tropical Pacific. During El Niño, the pycnocline is so deep that the upwelled waters come from the nutrient-poor waters above the pycnocline. In extreme cases, nutrient-deficient waters coupled with overfishing cause fisheries to collapse, bringing about severe, extended economic impacts.

Coastal upwelling ecosystems, like the west coast of the United States, are some of the most productive ecosystems in the world and support many of the world's most important fisheries. Although coastal upwelling regions account for only 1 percent of the ocean surface, they contribute roughly 40 to 50 percent of the world's fisheries landings (this refers to the part of the fish catch that is put ashore and does not include fish that are lost at sea or discarded as undesirable catch. Frequently, landings provide the only record and do not accurately reflect the total catch).

In upwelling systems, large variations in fish recruitment in fisheries appear to be from fluctuations in mortality during the early life-history stages of the fish, and thereby must be essentially associated to climatic variability. Survival through these stages is essentially linked to hydrodynamic structures which, in certain spatiotemporal strata, favor the retention, enriching, and concentration of plankton and ichthyoplankton. Fluctuations in abundance of pelagic fish stocks reflect significant changes in the structure and functioning of upwelling ecosystems.

High mortality rates have been observed at upper trophic levels (birds, marine mammals, and large predatory fish) in response to the diminished abundance of prey. Effects at lower trophic levels may also bring to light evidence of reduced predation by pelagic fishes on planktonic species, bringing about in turn changes throughout the entire trophic web. An example is the change in relative abundance of two species of anchovy associated with the El Niño phenomenon. Long-term fluctuations in dominant species have been observed in most upwelling ecosystems, such as the alternating dominance of sardines and anchovies in the Humbolt, Benguela, and Kuroshio (offshore Japan) Currents.

CLIMATE AND CLIMATE CHANGE

Coastal upwellings also influence weather and climate. Along the northern and central California coast, upwellings lower sea-surface temperatures and increase the frequency of summer fogs. Relatively cold surface waters chill the overlying humid marine air to saturation, so that thick fog develops. Upwelling cold water inhibits formation of tropical cyclones (e.g., hurricanes) because tropical cyclones derive their energy from warm surface waters. During El Niño and La Niña, changes in sea-surface temperature patterns associated with warm- and cold-water upwelling off the northwest coast of South America and along the equator in the tropical Pacific affect the interannual distribution of precipitation around the globe. A great impact on upwelling is posed by the new installation of large wind farms.

Some upwellings are wind-driven; these new large wind farms exert a significant disturbance on the wind speed in the vicinity of the installations. A recent study shows that the size of the wind wake is an important factor for the oceanic response to the wind farm. At certain wind speeds, sufficient upwelling may be generated that the local ecosystem will most likely be strongly influenced by the presence of a wind farm.

Another factor that is currently affecting upwelling is climate change. In any one region, upwelling is intermittent; it can be strong in some years, and weak in others. The success of fishermen is greatly affected by this, since a weakening of an upwelling system can bring economic disaster. Upwelling brings nutrient-rich waters up to the surface, where algae can thrive in the sunlight, feeding the fish. Without nutrients, there will be no fish (as happens during El Niño conditions).

Scientists are trying to determine how climate change will affect upwelling and coastal productivity. The strong dependency of upwelling processes on the strength of trade winds contains one hint. Trade winds are zonal winds, which feed off the latitudinal temperature gradient; as this gradient weakens (computer simulations show that high latitudes become warmer than low ones), the zonal flow will weaken. Thus, the upwelling that derives its energy from trade winds will weaken. Off California, upwelling of cold water has become less common since 1975, and the productivity of the California Current has diminished accordingly. A similar decrease in productivity may be expected elsewhere in the eastern boundary currents, and in the eastern equatorial high production regions. Upwelling that depends on monsoon activity (as in the Arabian Sea) should be much less affected, or may even benefit from the change.

According to some scientists, wind-driven upwellings along the California coast have increased over the past 30 years, and some scientists postulate that the increase in wind-driven upwellings is largely because of increased greenhouse gas, but such an association has been speculative. A study on the effects of future climate change and upwellings in California showed negative results from the intensification of wind-driven upwelling events in that specific area.

According to a model scientists have made, the intensification of upwelling will likely have impacts on terrestrial and marine ecosystems. With intensified upwelling, enrichment can be increased that would be beneficial to organisms; however, concentration may decrease because of increased mixing, and retention may also be decreased by increased seaward transport of surface water. Overall, this could have a negative effect on marine ecosystems, as the current balance of these three factors will change with changes in upwelling.

—Alexandra M. Avila

Further Reading

Boje, R., and M. Tomczak, editors. *Upwelling Ecosystems.* Springer Science & Business Media, 2012.

Brusca, Richard C., editor. *The Gulf of California: Biodiversity and Conservation.* U of Arizona P, 2010.

Kämpf, Johan, and Piers Chapman. *Upwelling Systems of the World: A Scientific Journey to the Most Productive Marine Ecosystems*. Springer, 2016.

Mann, K. H., and J. R. N. Lazier. *Dynamics of Marine Ecosystems: Biological-Physical Interactions in the Oceans.* 3rd ed., John Wiley & Sons, 2013.

Monaco, André, and Patrick Prouzet, editors. *Marine Ecosystems, Diversity and Functions*. ISTE/John Wiley & Sons, 2015.

Sinclair, M., and G. Valdimarsson, editors. *Responsible Fisheries in the Marine Ecosystem.* CABI, 2014.

V

VOLCANISM

Fields of Study: geology; marine geology; volcanology; marine volcanology; marine biology; marine biogeochemistry

ABSTRACT

Volcanology is the investigation of volcanism, Earth's volcanic activity in a number of environments. Volcanoes comprise not only the well-known cone-shaped mountains, but can also be found surrounded by dense vegetation, embedded in glaciers, and at the bottom of oceans. Some volcanic activity is sudden and violent, while other volcanic activity is more incremental and subtle. Terrestrial volcanoes often cause severe damage on land and may affect the marine ecology of the affected area. However, submarine volcanoes seldom, if ever, affect the terrestrial environment directly.

KEY CONCEPTS

caldera: the residual crater remaining after the top of a magma chamber has collapsed following an eruption

effusive eruption: a volcanic eruption in which dense magma is not forcefully ejected but simply flows or oozes out of fissures

magma buoyancy: the level of a mass of magma relative to the larger body of magma according to its density

magma chamber: a void within the crustal rock at the site of a volcano, where rising magma collects and pools

plinian eruption: an explosive eruption that ejects a large, towering column of smoke, ash, and other ejecta that suddenly collapses and falls to ground as a pyroclastic flow; so named for the description of the eruption of Mount Vesuvius recorded firsthand by Roman historian Pliny in 79 BCE

volatiles: liquids and gases that are entrained in or released from magma when it reaches the surface

BASIC PRINCIPLES

Volcanism is the eruption of materials from Earth's inner layers. Terrestrial volcanism may cause great damage on land and affect the local marine environment. However, submarine volcanoes seldom, if ever, affect the terrestrial environment directly. But in the seas, a submarine eruption may have devestating effects on the local marine ecology. Eruptions from the seafloor often produce undersea mountains called "guyots," and on rare occasions may produce a permanent island, such as Surtsey, near Iceland.

The planet features three general layers: the crust (the outermost layer, on which life exists), the mantle, and the core. High pressures and heat located in the core and mantle push molten rock (magma) outward into the cooler rock layers under the outer crust. Magma, according to the theory of plate tectonics (a model that argues that the various plates of the mantle, known collectively as the lithosphere, drift across a sea of superheated rock), is pushed through the boundaries of those plates and outward through openings in the crust. These openings are called "volcanoes."

A volcano releases material in many ways. A key factor in the magnitude of a volcanic eruption is gas pressure. High degrees of gas pressure can eject magma, steam, ash, and other materials explosively, while low degrees of pressures can lead to a more subdued eruption or no eruption at all. Explosive eruptions, however, send magma, ash, rock, and smoke high into Earth's atmosphere. Effusive eruptions, in contrast, are extremely slow and sometimes move no faster than human walking speed. In some cases, volcanism involves no eruption, as is the case when magma pushes slowly through the crust, cooling as it moves upward to form a volcano.

BACKGROUND AND HISTORY

Volcanoes have played a role throughout Earth's history. Although they are noted for their destructiveness, volcanoes have also been instrumental in the creation of much of the planet's landscape and composition.

Mountains, oceans, and even the atmosphere were formed as volcanoes pushed gas and material through the crust. As a result, volcanoes are largely responsible for the conditions that support life on Earth.

While this creative element has piqued scientific interest, it may be said that most research on volcanism has been inspired by the destructive history of volcanoes. Some scientists believe that volcanoes may be responsible for ending the reign of the dinosaurs, linking their extinction to a series of tremendous volcanic explosions 65 million years ago. This show of volcanism would have sent enormous amounts of soot and ash into the atmosphere, choking the air and preventing plants from growing. In 1650 BCE, what is believed to be the largest volcanic eruption in the last 10,000 years took place under what is now the Greek island of Santorini. A tremendous eruption killed a great many people in the Mediterranean region. In 79 CE, Mount Vesuvius erupted, sending a pyroclastic cloud of hot gas and ash down the mountainside, instantly killing everyone in the Italian cities of Pompeii and Herculaneum (modern day Ircolano).

In the nineteenth century, the massive eruptions of two Indonesian volcanoes, Mount Tambora and another under the island of Krakatoa, affected weather around the world for years. In 1980, Washington State's Mount St. Helens exploded, sending ash around the world within two weeks. In 2010, the Icelandic volcano Eyjafjallajökull sent an ash cloud into the sky that disrupted airline travel in Europe for weeks. In October and November of that same year, Mount Merapi erupted in Indonesia, prompting the evacuation of hundreds of thousands of people from the affected area. Despite the evacuations, more than 350 people died. These high-profile events are but a sample of the volcanism that has affected life on Earth.

LOCATIONS AND TYPES OF VOLCANISM

Volcanic eruptions are some of the most violent natural occurrences on Earth. They endanger lives, threaten property, and even cause major changes in the planet's topography. Although scientists are unable to predict major eruptions, they have an extensive knowledge of how volcanism occurs and can interpret indicators of the likelihood of eruption.

Volcanoes are typically found along the edges of Earth's tectonic plates. The majority of the world's volcanoes (about 90 percent) are found along the Ring of Fire, a seismically and volcanically active region along the "edges" of the Pacific Ocean. Iceland, in the northern Atlantic region, also has a significant number of active volcanoes due to its position as part of the Mid-Atlantic Ridge.

Volcanoes erupt in three general ways. The first of these eruptions centers on magma buoyancy. The mass of rock remains the same as it melts, even when its volume increases. When the rock melts, however, its density becomes lighter than that of the surrounding rock, causing the lighter magma to flow up through Earth's crust. The rate of magma flow increases as the magma's density decreases. The buoyancy element relies on the notion that magma is pure (that is, not polluted by compounds that can affect its structure).

In the second scenario, magma flow is aided by the presence of volatile substances (or volatiles). Some examples of volatiles are water, sulfur dioxide, and carbon dioxide. Volatiles also include certain forms of crystallized rock, which increase the density of the magma, causing greater pressure as it moves outward. When rock melts into magma, volatiles start to bubble, causing the magma to flow upward. As the magma approaches the surface, the bubbles intensify, releasing gas as the flow accelerates and causing a violent eruption. Scientists believe that the presence of these volatiles leads to explosive volcanism, whereas buoyancy alone leads to magma pooling and, possibly, cooling at the crust level unless the magma assumes an increased degree of buoyancy.

The third scenario involves the injection of new magma into a highly pressurized magma chamber (an underground reservoir of molten rock located beneath Earth's crust). As magma flows upward toward the crust, it may pool in a magma chamber. As more magma is pushed into the chamber, however, pressure increases steadily. If a conduit presents itself under these conditions, magma will flow upward using the opening. However, if there is no conduit, the magma will continue to push until the crust fractures and allows the magma to escape. Volcanoes above magma chambers are known to erupt repeatedly over time as new magma is injected. This ongoing volcanism can lead to significant changes in a volcano's structure.

DIFFERENT DEGREES OF INTENSITY

Volcanic eruptions range from simple magma emissions to violent explosions, such as when Japan's

Mount Ontake erupted without warning in September 2014, killing dozens of people including mountain climbers who were surprised by the blast. Some eruptions, as was the case with Mount St. Helens and Mount Etna, take place within the conical shape of mountain volcanoes. Others, like the Kilauea volcano in Hawaii, simply burst from a fissure in the ground. The Eyjafjallajökull event in Iceland started out as an effusive eruption but later sent a large plume of steam and ash several miles (or kilometers) into the sky.

The presence of volatiles within the magma is a major determinant in an eruption's degree of intensity. Volatiles may also determine the shape of the eruption. For example, the Vesuvius eruption was considered a plinian eruption, forcing a column of ash and other material high into the sky. Plinian eruptions are some of the most dangerous types, capable of producing pyroclastic flows of fluidized and superheated gas that can slide down the side of a volcano at hurricane- like speeds.

Sometimes an eruption completely depletes the magma chamber beneath a volcano. The volcano then collapses into the empty chamber, creating a bowl-shaped indentation called a "caldera." A volcano's collapse into a caldera after an eruption does not mean the volcano will become extinct or dormant. The Mount Aso volcano in Japan, for example, collapsed into a tremendous caldera more than 100,000 years ago, but its caldera remains the site of frequent volcanic activity.

METHODS OF STUDY

In the study of volcanism, the data that are collected at a given eruption site are both voluminous and complex. Adding to this challenge is the pursuit of general theories based on more than one eruption and, therefore, comparing the data sets from each event. It is here that mathematics proves an invaluable tool.

Scientists seeking to understand the eruption process at Mount Etna, for example, sifted through data from sixty-one earthquakes that took place before a 2001 eruption of the volcano. Researchers used a tensor (a mathematical concept that assigns nonlinear vectors to various geographic sites—in this case, points in and around the volcano at which the seismic activity was recorded) and compared the data input to other vector data. The tensor approach helped provide scientists with a more comprehensive profile of the processes and mechanisms that powered the volcano's eruption.

To analyze the conditions of a volcano, scientists deploy a large array of sensor equipment at a given site. For example, seismographs are placed in key locations to detect any tremors and seismic waves radiating from deep beneath the site. Tiltmeters are used to measure any changes in the horizontal level of the ground. Furthermore, pressure sensors and water detectors are deployed at and near water sources; a change in pressure or the appearance of new streams can be a useful indicator of volcanic activity. Scientists also take gas samples using specialized bottles; as magma approaches the surface, the increased presence of sulfur dioxide and carbon dioxide (among other gases) emanating from that substance is a strong indicator of volcanic activity.

In addition to ground-based sensor equipment, aerial sensors are deployed near the volcanic site. Aircraft fly over the volcano using infrared scanners, thermal imaging cameras, and radar to detect any changes in the volcano's environment. Satellite technology is utilized as well, focusing on heat sources, cloud plumes, and any other characteristics of the site. The use of such technology helps scientists maintain vigilance over an active volcanic site, generating scientific data and enabling greater awareness of the risks of potential eruption.

One of the most challenging aspects of studying volcanism is the impossibility of directly studying the mantle and the superheated rock beneath the lithosphere. In some cases, even sensors are rendered unusable because of the risk of eruption. Scientists are still able to study volcanism through chemical analysis. Studying lava that flowed from a fissure, for example, can reveal the type of magma that was released and the gases that were contained in the magma (based on the types of crystals found in the composition). Scientists also can study the chemical properties of the ash and gas released after an eruption. These airborne elements, upon returning to Earth, can reveal a great deal about the volcanic activity.

An example of the value of chemical analysis may be found in a study of the Piton de la Fournaise eruption. This volcano, located on an island near Mauritius in the Indian Ocean, experienced its largest eruption in April 2007. The remoteness of the island prevented scientists from installing sensor equipment before the eruption. However, volcanologists were able to study the chemical composition of steam, ash, and gas ejected above the volcano. Using the Ozone Monitoring Instrument's special aerosol-sensing technologies (this device is onboard a National Aeronautics and Space Administration [NASA] satellite), scientists gathered data about the volume and a mixture of sulfur dioxide that was part of the plume. The data collected during the eruption helped scientists gain vital information about the processes that created Piton de la Fournaise's eruption.

RESEARCH

There are several US government agencies that have departments dedicated to the study of volcanic activity. Within the Department of the Interior, for example, is the US Geological Survey (USGS). The USGS operates the Volcano Hazards Program, which monitors volcanism within the United States. The USGS also focuses on active volcanoes whose activity has the potential to affect US interests. In addition, the agency maintains several volcano observatories. Meanwhile, the National Oceanic and Atmospheric Administration (through the US Department of Commerce) also works in volcanism, particularly concerning eruption plumes and how they affect air quality and transportation routes.

Universities and their faculties and research staff play a key role in studying volcanism. The University of Hawaii, for example, operates the Hawaii Center for Volcanology, an institution comprising leading researchers. The University of Utah, Oregon State University, and the University of California, Berkeley, all have leading volcano researchers on their faculties.

Because information can be shared so quickly online, the need for global volcano research networks remains high. The Smithsonian Institution, for example, operates a global network of volcano researchers, whose data and theories are shared within the group through the Smithsonian's central repository. Meanwhile, the European Union formed the Network for Observation of Volcanic and Atmospheric Change, which allows experts from all over the world to collaborate on research on such topics as gas emission monitoring and volcanic risk assessment.

IMPLICATIONS AND FUTURE PROSPECTS

There is no evidence to suggest that there is either less or more volcanic activity in the twenty-first century than in previous eras. However, human civilization has advanced significantly in terms of its transportation and telecommunications abilities. When a massive eruption occurs, the potential for regional and even global economic disruption goes hand in hand with risks to public safety. For example, the massive eruption of Eyjafjallajökull in Iceland caused a major disruption in world travel patterns. For several days, planes to and from Europe were either diverted or grounded, affecting travel around the world.

The development of computer modeling, cutting-edge chemical analysis, and photographic technologies (including satellite technologies) has helped volcanologists study these phenomena with greater clarity than ever before. One of the most important technological developments, the internet, has brought the study of volcanic activity to a whole new level. Now, scientists from around the world can almost instantly share data from active volcanic sites and laboratory experiments. In addition to the ability to transfer large data sets and documents, scientists can share photographic images and videos. Furthermore, scientists can use the internet to monitor active volcanic sites from around the world first-hand, simply by logging into an observatory's network. Although the need for field research in the arena of volcanism is still high, a great deal of time can be saved through the application of modern technologies.

—*Michael P. Auerbach*

Further Reading

Castro, Jonathan M., and Donald B. Dingwell. "Rapid Ascent of Rhyolitic Magma at Chaiten Volcano, Chile." *Nature*, vol. 461, 2009, pp. 780–83.

Gottsman, Joachim, and Joan Marti, editors. *Analysis, Modeling and Response*. Vol. 10 in *Caldera Volcanism*. Elsevier Science, 2008.

Lockwood, John P., and Richard W. Hazlett. *Volcanoes: Global Perspectives*. Wiley-Blackwell, 2010.

Paone, Angelo. "The Geochemical Evolution of the Mt. Somma-Vesuvius Volcano." *Mineralogy and Petrology*, vol. 87, nos. 1–2, 2006, pp. 53–80.

Schminke, Hans-Ulrich. *Volcanism*. Springer, 2005.

Taylor, Alan. "The Eruption of Japan's Mount Ontake." *The Atlantic*, 30 Sept. 2014, www.theatlantic.com/photo/2014/09/the-eruption-of-japans-mount-ontake/100823/.

Walter, Thomas R. "Structural Architecture of the 1980 Mount St. Helens Collapse: An Analysis of the Rosenquist Photo Sequence Using Digital Image Correlation." *Geology*, vol. 39, no. 8, 2011, pp. 767–70.

W

Wadden Sea

Fields of Study: oceanography; marine biology; marine ecology; environmental studies

ABSTRACT

Situated in the southeastern North Sea, the Wadden Sea is one of the largest and most naturally thriving intertidal zones in the world. It ranges from the banks of the Netherlands, Germany, and Denmark to the northwestern Frisian Islands, covering a total area of approximately 3,900 square miles (10,100 square kilometers), with about 310 mi (500 km) of coastline.

KEY CONCEPTS

diking: construction of impervious walls to separate normally water-covered stretches of land from the ocean, enabling water to be pumped away from the sequestered area, rendering it suitable for land uses such as agriculture or building development

halligen: a low island in the Wadden Sea

intertidal zone: the area of a shore that is under water at high tide and exposed at low tide

terpen: an artificially raised area of ground on a halligen, created as a stable base for construction of various buildings and to maintain those buildings above water level at high tides

DESCRIPTION

Situated in the southeastern North Sea, the Wadden Sea is one of the largest and most naturally thriving intertidal zones in the world. The word 'wad' in Dutch translates to "mud flat," the defining characteristic of this area. This intertidal zone ranges from the banks of the Netherlands, Germany, and Denmark to the northwestern Frisian Islands (also known as the Wadden Islands), an archipelago of barrier islands that separates the Wadden from the North Sea. The Wadden Sea covers a total area of approximately 3,900 sq mi (10,100 sq km), and has about 310 mi (500 km) of coastline.

The chain of Frisian Islands, of which there are approximately 50, was formed some 7,000 years ago, making this isolated sea one of the youngest bodies of water in the world. Even as recently as 1,000 years ago, the barrier islands underwent significant erosion that detached them from the mainland, truly making them a group of islands and defining the Wadden Sea from the North Sea.

Furthermore, the landscape of the sea is constantly changing. The turbid sea and wind of both the North and Wadden Seas actively remove and deposit sediment from sandbars and islands over many years, causing the layout of the islands and water channels to vary over short time scales. The general movement of the islands is from west to east.

The climate in the region is mainly temperate and marine, with cool summers and mild winters. The average temperature ranges from 34 to 41 degrees Fahrenheit (1 to 5 degrees Celsius) in January and from 55 to 72 degrees Fahrenheit (13 to 22 degrees Fahrenheit) in July. It rains throughout the year, and extreme cold is rare.

BIODIVERSITY

Because of daily tide cycles and easy access to its mudflats, the Wadden Sea is an important research site for studying organismal adaptation to changes in salinity, populations of benthic micro- and macroorganisms, and shorebird behavior. It is estimated that more than 10,000 species rely on this ecoregion for all or part of their life cycles. At the center of the ecosystem are the microorganisms, which are critical to nutrient cycling and largely dominated by mudflat bacteria. Despite a troubled history of aquaculture, mollusks continue to play an important commercial and ecological role in the sea; more than 27,000 tons (24,500 tonnes) of blue mussels (*Mytilus edulis*) are harvested annually for commercial purposes. However, harvests are closely monitored because of the reliance of multiple shorebird species on the blue mussel.

The Wadden Sea has been named one of the top 100 birding sites in the world. In addition to the permanent shorebird populations here, the mudflats are a critical habitat for more than 50 species of migratory birds. This makes the area one of the largest migratory bird habitats in western Europe. The region is a crucial element of the East Atlantic Flyway; an estimated 12 million birds migrate to the Wadden Sea every year.

An incredible array of bird species can be found on the mudflats and dunes of the coast and islands, including the pied avocet (*Recurvirostra avosetta*), osprey (*Pandion haliaetus*), barnacle goose (*Branta leucopsis*), sandwich tern (*Thalasseus sandvicensis*), and bar-tailed godwit (*Limosa lapponica*). Other birds of note include the spoonbill, eider, ringed plover, oystercatcher, and the peregrine falcon.

The rich diversity of the islands and marshes also makes the Wadden Sea an attractive migratory location. More than 900 species of plants, 300 mosses, 650 fungi, and 350 lichens are found in and around the sea and the islands. Some areas of the sea boast an abundance of 2.2 pounds (1 kilogram) of shorebird food per 11 square feet (1 square meter).

This ecosystem is also home to large marine mammals, including the harbor seal, the gray seal, and the harbor porpoise. These animals can avoid predators in the shallow, warm water of the sea, and the seals use the sandbanks for nursing their young. Additionally, the strong water currents and daily tides in the region provide the Wadden Sea approximately 20 times more nutrients than the adjacent North Sea, allowing this ecoregion to sustain its incredible diversity.

HUMAN IMPACT

Like many waterways, the Wadden Sea has been significantly affected by chemical and agricultural runoff from the mainlands and increased turbidity from mainland diking. In the 1960s, shorebirds that relied on the shallow waters of this area for fishing experienced population declines when the water became too polluted from chemical runoff. In the mid-twenty-first century, the sea also became used for aquaculture, which resulted in a continuous disturbance of the seafloor and increased turbidity. Besides pollution, recent studies suggest that climate change has influenced and will continue to influence the migration of birds to the Wadden Sea. Some scholars suggest that the shoreline is one of the most human-modified environments in the world.

The Wadden Sea became a tourist destination for many Europeans in the nineteenth century, when several of the islands opened resorts. Since then, more than 1.8 million visitors per year have had the opportunity to hike on the mudflats during low tide to observe shorebirds and invertebrates. Because of the increased boating and mudflat hiking in the region, the Learning Center of the Wadden Sea (*Schutzstation Wattenmeer*), operated by the German government, has established various conservation areas (*Schutzzonen*) throughout the zone and has initiated research to ensure the protection of critical mudflat species.

Many of the islands continue to be occupied by residents today. Hallig Hooge, the second-largest halligen (German island), has a population of 83 residents living on 10 terpen, which are artificially formed mounds where homes and businesses have been built. These mounds are similar to compact neighborhoods. Some mounds host smaller structures, such as the local school and church, whereas the hallig nature center, movie theater, gift shop, grocery store, and various cafes are located on a larger mound.

These terpen are essential for hallig living, because the smaller islands experience annual floods that would otherwise destroy homes and businesses. Additionally, some of the homes and businesses on the terpen are remarkably sustainable, using geothermal and solar energy for their main sources of heat and electricity. Because the low-lying areas of the island are used for research, tourism, and cattle rearing, residents do have to visit the mainland for specialty items or bulk food.

The Wadden Sea is a remarkably diverse habitat, and is considered to be one of the last pristine intertidal zones in the world. The International Union for Conservation of Nature and Natural Resources (IUCN), an international conservation organization, has characterized the sea as a Category II national park. This recognizes the immense value of this area for its contribution to ecosystem processes and educational, scientific, and cultural enrichment.

In June 2009, the Dutch and German portions of the sea were placed on the World Heritage Site List by the United Nations Educational, Scientific, and Cultural Organization (UNESCO). The list denotes places or objects of cultural or natural significance around the world, furthering protection for the region. Since its induction in 2009, the area's perceived threat for ecological harm has remained at zero.

In 2005, an annual Wadden Sea Day was established as an opportunity for researchers, educators, and conservationists to meet and discuss important ecological trends in the region. One recent Wadden Sea Day focused on the food web of the sea, particularly the importance of fish, a currently understudied group of animals in the area. In 1997, Germany, Denmark, and the Netherlands signed the Trilateral Wadden Sea Plan to ensure shared management and protection of the sea by the three countries. Following this plan in 2002, they developed the Wadden Sea Forum to oversee the conservation of the sea and monitor the protection status of the region.

—Catherine G. Fontana

Further Reading

Dittmann, Sabine, editor. *The Wadden Sea Ecosystem: Stability, Properties and Mechanisms.* Springer Science & Business Media, 2012.

Gehrman, Sven. *Fire! In the Wadden Sea...Or What the Fauna of the North Sea Urgently Wants to Tell Us.* BoD, 2020.

United Nations Educational, Scientific, and Cultural Organization (UNESCO). "The Wadden Sea." World Heritage Centre, whc.unesco.org/en/list/1314.

Van Beusekom, Ruud, Nick Langley, and Manon Tentij, editors. *The Wadden Sea: A Vision for the Conservation of a Natural Heritage.* BirdLife International, 2009.

Vollmer, M., M. Guldberg, M. Maluck, D. Marrewijk, and G. Schlicksbier. "Landscape and Cultural Heritage in the Wadden Sea Region: Project Report." *Wadden Sea Ecosystem,* vol. 12, 2001.

WATER AND ICE

Fields of Study: physical chemistry; oceanography; engineering; environmental science

ABSTRACT

Water is a molecule formed from two hydrogen atoms covalently bonded to a single oxygen atom. The chemical properties of water are unique among molecular substances on Earth and have had a dominant effect on the evolution of life. Water covers some 70 percent of Earth's surface, forming marine environments that harbor innumerable life forms and maintain the habitability of the planet.

KEY CONCEPTS

biosphere: a term encompassing all life of Earth, whether plant or animal in nature

density: generally defined as mass per unit volume, the standard being the density of pure water at 62.4 pounds per cubic foot (1 gram per cubic centimeter)

electric dipole: an electric dipole in a molecule whenever there is charge separation within the molecule, resulting in an identifiable region of relatively positive charge at one end and a region of relatively negative charge at the other end of the dipole

hydrologic cycle: the cyclic system of evaporation, condensation and precipitation of water on Earth

phase: the state of a substance, as either a solid, a liquid or a gas

polarity: the existence and direction of a dipole in a molecule

CHEMICAL PROPERTIES OF WATER

Water (H_2O), identified chemically as dihydrogen monoxide, is a naturally occurring molecule that exists on Earth in solid, liquid, and gaseous phases. Water is the most abundant molecule on Earth, covering approximately 70 percent of Earth's surface.

The hydrosphere, which is composed of the collective waters of the seas, oceans, rivers, lakes, and the lower atmosphere, consists of more than 312 million cubic miles (1.3 billion cubic kilometers) of water. The chemical properties of the water molecule are in large part responsible for the nature of life on Earth. The bodies of most living organisms are more than 60 percent water, and some living cells contain more than 90 percent water by volume.

The molecular structure of water consists of a single oxygen atom joined with two hydrogen atoms through covalent bonding, which involves the sharing of electrons between two atoms. Water molecules are often described as having a bent shape, due to the fixed orientation at an angle of 109.5 degrees of the two hydrogen atoms relative to the oxygen atom. Because the "pull" of oxygen on its shared electrons is stronger than the "pull" of the hydrogen atoms on their electrons, the electrons from the hydrogen atoms are generally situated in the bonds with the oxygen atom rather than in their proper atomic orbitals. This leaves the positiviely charged hydrogen nuclei exposed. Accordingly, the oxygen end of the molecule has exposed negative charge due to the rest of the oxygen atom's electrons, but the hydrogen end has exposed positive charge. This produces an electric dipole that allows water molecules to coordinate to other water molecules, and to other molecules and ions that have a positive or negative charge.

Because of their slight positive and negative charges, water molecules bond to one another through hydrogen bonding, which is the momentary attraction of the negative end of one water molecule with the positive end of a nearby molecule. In liquid water, molecules are constantly moving, and the hydrogen bonds between them are frequently broken and re-formed. The average hydrogen bond lasts for less than one trillionth of a second before being broken. However, the millions of hydrogen bonds forming across a body of water provide a collective strength that gives water unique structural and adhesive properties.

The polarity of water molecules makes water an excellent solvent, meaning it can cause the dissolution of other molecules and substances, as water molecules coordinate with other polar molecules they encounter. Any compound with a negative or positive net charge will dissolve in liquid water. Water's function as a solvent is important in a variety of biological processes and also allows water to dissolve minerals and other elements from Earth's lithosphere, returning them to the hydrosphere and biosphere and thereby acting as the most important catalyst in many geochemical cycles.

Hydrogen bonding gives water high surface tension, which is the capacity of the layer on the surface of a liquid to resist external pressure. Therefore, certain substances and materials will float on top of a body of water rather than penetrate the surface. In addition, hydrogen bonds cause water to tend to gather into drops, rather than spreading evenly, which is important to plants that use water's natural adhesion to help distribute water throughout the plant's tissues.

The collective energy of hydrogen bonding also means that water has a high specific heat, which is the heat at which the liquid form will give rise to a gas. Water therefore must absorb a large amount of heat energy before it will change temperature, and it also tends to release heat energy slowly back into the environment. These properties mean that water is a natural regulator of climate, tending to absorb excess heat when available and to release heat later when the ambient heat has reduced.

PROPERTIES OF ICE

Most substances become denser as they cool because heat energy is lost, causing collisions between molecules to decrease and thereby allowing molecules to settle closer to one another. As water cools, its density increases until the temperature reaches 39 degrees Fahrenheit (4 degrees Celsius), reaching the maximum density for liquid water. If water continues to cool beyond this point, the vibrational energy of individual water molecules decreases to such a point that the hydrogen bonds between molecules become fixed, rather than alternate between bonds. At this point, water begins to decrease in density as the molecules align into a matrix, simultaneously increasing in volume.

At approximately 32 degrees Fahrenheit (0 degrees Celsius) liquid water becomes solid ice. The organized arrangement of water molecules in ice allow for more space between molecules, thereby increasing the volume by more than 9 percent. This property is unusual among liquids, but it also occurs in the common mineral silica (SiO_2) as it transitions from liquid magma to solid silica rock. Because ice is less dense than water, it does not sink and will rise to the surface of the liquid, further buoyed by the surface tension of the liquid. In areas where freezing occurs, this creates a layer of frozen ice that insulates the liquid water beneath. This insulation prevents further water from reaching the freezing point and thereby allows organisms to survive under the surface of frozen bodies of water. Similarly, the preservation of this liquid water means that chemical reactions can continue under the surface of the ice.

THE HYDROLOGIC CYCLE

The hydrologic or water cycle is a geochemical cycle that moves water molecules between the environmental spheres of Earth. The lithosphere is the combination of rocky minerals, sediment, and molten rock that make up Earth's crust and part of Earth's mantle.

The atmosphere is an envelope of gaseous elements that surround Earth in layers and differentiate the planetary environment from that of outer space. The biosphere is the sum of all living organisms on the planet and their interactions with the other environmental spheres. Geochemical cycles are the processes that move various elements through the environmental spheres through interconnected sets of chemical and physical reactions.

The hydrologic cycle begins with evaporation, which is the process by which liquid water transitions into a gaseous form and enters Earth's atmosphere. Approximately 80 percent of evaporation occurs over the ocean, with the remainder occurring over lakes, rivers, and water temporarily deposited on the lithosphere. Water rises into the atmosphere as water vapor and, as it gains in altitude, the temperature of the atmosphere decreases, causing water vapor to condense. Hydrogen bonds cause water vapor to form into clouds, which are pockets of water droplets or ice crystals that move according to the development of thermal currents in the atmosphere.

Water vapor in clouds and in more diffuse gaseous form throughout the atmosphere is transported around Earth on thermal currents and eventually returns to the surface as precipitation, which is the return of condensed water to the hydrosphere and lithosphere in the form of rain and snow. Precipitation occurs when the density of water vapor in the atmosphere reaches a certain critical level, often triggered by temperature changes in the atmosphere that are related to windstorms that develop over the ocean.

A portion of the water delivered to the lithosphere as precipitation returns to the atmosphere through evaporation, because the temperature of the lithosphere is generally warmer than the temperature of the hydrosphere; this causes water to more easily vaporize into gas. Some of the precipitation over the lithosphere penetrates the surface and becomes groundwater, which is water that saturates and remains within the mineral layers below the surface. Flooding occurs when groundwater levels rise to a point that the soil is oversaturated and can no longer absorb additional water.

Groundwater eventually returns to the hydrosphere as it filters into streams and lakes that are connected to the ocean. Additional groundwater enters the ocean as seepage in coastal areas, where the lithosphere and the ocean meet. Water that filters back into the hydrosphere through the lithosphere carries a variety of minerals and other trace elements because of the solvent properties of water. These minerals enrich water sources, thereby contributing to the nutrient cycle in aquatic environments.

The organisms of the biosphere also absorb water into their tissues, using it to fuel and to carry the products of biochemical reactions. Water from within organisms is returned to the atmosphere during respiration and as a component of biological waste. Water returned to the hydrosphere through the biosphere also carries a variety of minerals and other elements that are thereafter transferred through the hydrosphere into the other environmental spheres.

HYDROELECTRICITY

Hydroelectricity is the process of using water to generate electricity for human consumption. The basic method is to harness the kinetic energy of moving water and then to transition this energy into electric currents that can be stored and used to power electric devices.

Generally, the generation of hydroelectricity involves using running water to power turbine engines, devices with spinning axles attached to blades that move in response to the flow of a fluid through the blades. As the blades turn, they power an electric generator, which forces a flow of electrons through a circuit and generates electric currents.

Hydroelectric generators can utilize the gravitational flow of liquid water, which is the tendency for water to flow downward with the force of gravity from an area of higher elevation or depth. To harness this energy, engineers control the flow of water and direct the water through channels containing turbines.

A number of different methods generate hydroelectricity, depending on environmental conditions. Hydroelectric dams function by blocking the flow of water along the path of a river or other waterway. This creates an artificially generated imbalance in water depth on either side of the dam. The higher water on one side of the dam can then be directed to flow through turbines located at the bottom of the dam. Another way of artificially generating water flow is to utilize a series of pumps to control or enhance the flow of water in a certain area. In this case, hydroelec-

Abénaquis hydroelectric dam on the Magog River, Sherbrooke, Québec (DXR via Wikimedia Commons)

tricity can be generated continuously, rather than relying on naturally occurring gravitational imbalances to generate flow. There is an unfortunate trade-off in this, however, in terms of the energy required to pump the water versus the energy recovered in electrical power generation; there is no such thing as an eternal motion machine.

Turbines can also be installed in rivers to capture energy from the flow of the river current. Similarly, turbines can be installed in oceanic areas, where they derive energy from the tidal movement and currents of the ocean waters.

Tidal power and wave power are two of the most recent forms of hydroelectricity generation and have developed into major areas of research. Energy present in the kinetic movement of tides provides a consistent and predictable source of energy generation but is utilized in relatively few areas.

All forms of hydroelectric energy offer the advantage of being infinitely renewable, as they are not based on the supply of a limited ingredient. However, this type of energy usage often carries environmental consequences because the environment must be altered significantly to install hydroelectric generators and other equipment.

—*Micah L. Issitt*

Further Reading

Barry, Roger, and Thian Yew Gan. *The Global Cryosphere: Past, Present, and Future*. Cambridge UP, 2011.

Brutsaert, Wilfried. *Hydrology: An Introduction*. Cambridge UP, 2005.

Gosnell, Mariana. *Ice: The Nature, the History, and the Uses of an Astonishing Substance*. Alfred A. Knopf, 2005.

Ismail, Basel I., editor. *Renewable Hydropower Technologies*. Intech Open, 2017.

Jacobson, Michael C., et al. *Earth System Science: From Biogeochemical Cycles to Global Change*. Academic Press, 2006.

Laybourn-Parry, Johanna, Martyn Tranter, and Andrew J. Hodson. *The Ecology of Snow and Ice Environments*. Oxford UP, 2012.

Macdougall, Douglas. *Frozen Earth*. U of California P, 2006.

Pounder, Elton R. *The Physics of Ice*. Elsevier, 2013.

Tagare, Digamber M. *Electricity Power Generation: The Changing Dimensions*. Wiley-Blackwell, 2011.

Wallace, John M., and Peter V. Hobbs. *Atmospheric Science: An Introductory Survey*. 2nd ed., Academic Press, 2006.

WATER MANAGEMENT

Fields of Study: fluid dynamics; civil engineering; agricultural management; limnology

ABSTRACT

Water management is the process of developing, distributing, and managing water resources. Water resources are essential to agricultural, industrial, household, recreational, and environmental activities. Almost all these water uses by humans require sources of freshwater. Freshwater is defined as having little dissolved salts or other dissolved solids. Though freshwater is a renewable resource, the world's supply of readily available, clean freshwater is decreasing and is being put under growing pressure as populations increase globally.

KEY CONCEPTS

consumptive use: any use or application of water that removes the water from the immediately available supply

nonconsumptive use: any use or application of water that permits immediate reuse of the water in the same or another application, or that retains the water in the immediately available supply

sewage: untreated water containing human and animal wastes (urine and feces) as well as other organic solid and liquid wastes

vacuum distillation: removal of a liquid component of a solution or mixture by boiling or evaporating it under reduced pressure and collecting the distillate by condensation

wastewater: all of the liquid waste, including the solid waste within it, that is produced by a community

THE GOAL OF WATER MANAGEMENT

The goal of water management is optimizing the use of water resources and reducing negative effects on the environment. In an ideal world, water management balances the competing water demands of human populations for sanitation, drinking, manufacturing, leisure, and agriculture.

WATER AND ECOSYSTEMS

Water resources are an important component of ecosystems. Many biomes are defined by their rainfall and water levels. Tropical rainforest ecology, for example, is highly dependent on its namesake year-round high levels of precipitation and humidity. Water also plays an integral role in ecosystem services. Ecosystem services are benefits that humans derive from natural systems, including a wide range of material resources and processes.

These services fall into four main categories: provisioning, which includes the production of food and water; regulating, which includes the control of climate and disease; supporting, which includes nutrient cycling and crop pollination; and cultural, which includes spiritual and recreational benefits. Water resource use can affect each of these categories of ecosystem services.

As human populations grow, so do pressures on ecosystems and water resources. More than half the world's wetlands were lost during the twentieth century as a result of human activity and improper water resource management. Surface and groundwater systems are being heavily drawn upon and contaminated. Ocean ecosystems are also being contaminated and overfished. Due to improper water resource management and climate change, deserts are the only ecosystems increasing worldwide, and desertification is a significant global environmental problem.

WATER USE

Uses of water can be consumptive, where water afterward is not immediately available for other uses, or nonconsumptive (also called "renewable"), where water can be put to additional use after it is used. Consumptive uses of water include activities that allow seepage into the ground, evaporation into the air, or incorporation into a product (such as agricultural crops). Nonconsumptive uses include treated sewage that is returned as surface water. The main categories of human water use are agricultural, urban, industrial, household, recreational, and environmental.

Agriculture is the biggest consumer of freshwater resources in the world, using about 70 percent of available water resources, according to data from the World Bank in 2017; a significant percent of that water is used for irrigation in an unsustainable manner, drawing more water than an area can provide in the long term. An estimated 793 gallons (3,000 liters) of water is used to produce enough food to feed one person daily, compared with the 0.5 to 1.3 gal (2 to 5 L) that one person drinks daily.

The amount of water required to produce food for each person is increasing as people tend toward eating more meat and produce, which are more water-intensive to produce than grains. Agriculture can cause water pollution through use of fertilizers, pesticides, and livestock manure that is then washed into rivers by rainfall.

As of 2020, more than half of the world's people live in towns and cities, and this proportion is expected to reach two-thirds by 2050, according to the United Nations. Agricultural areas surrounding urban centers must compete for water from the same local supply as the city. Traditional water sources, both ground and surface, are being polluted by urban wastewater. Cities offer the best choice to sell produce, so many farmers are located near urban areas and use polluted waters to irrigate their crops. These polluted waters, depending on the city treatment processes, can pose health hazards when consumed in produce. Urban wastewater is a mixture of pollutants from kitchens, toilets, and rainwater runoff, and contains excessive nutrients and salts, pathogens, heavy metals, trace antibiotics, endocrine-disrupting chemicals, and estrogens.

Developing countries have the least amount of water treatment; according to the World Health Organization (WHO), as of 2017, approximately 780 million people do not have access to improved drinking water. Pathogens of greatest concern include bacteria, viruses, and parasitic worms in the water supply. These pathogens can directly affect the health of farmers who come into contact with them, as well as indirectly affect consumers who ingest affected produce.

Contaminated water and produce can carry a variety of pathogens, such as salmonella, cholera, and diarrhea. According to the WHO, as of 2017, diarrhea kills about 525,000 children under the age of five annually and is the second-most-common cause of death in this age range.

According to the Food and Agriculture Organization (FAO) of the United Nations, in 2019, industry used about 19 percent of available water resources for such functions as running hydroelectric dams, thermoelectric power plants, refineries, and manufacturing plants. While industry is not as water-consumptive as agriculture, it can cause a large amount of pollution by discharging wastewater that is tainted with manufacturing chemicals. In addition, plants that use water

Water fountain found in a small Swiss village. They are used as a drinking water source for people and cattle. Almost every Alpine village has such a water source. (Juhanson via Wikimedia Commons)

from rivers and lakes as coolant and return the warmed water directly to the water body can cause damage to ecosystems through temperature changes.

According to the FAO, households were estimated to use 12 percent of the world's available water. This includes water for drinking, cooking, sanitation, bathing, and gardening. It is estimated that—excluding gardening—one person requires around 13 gal (5 L) of water per day for household uses.

Drinking water, also called "potable water," must be of high quality such that it can be consumed without causing immediate or long-term harm. In most developed countries, water supplies piped into houses are all drinking-water quality, despite the fact that only a small percentage of that water is used for drinking.

Recreational water use makes up perhaps the smallest percentage of water resource use. An example of this type of water management is the aquarium trade. Another is when reservoirs are used for recreation. Often, water levels are kept higher in reservoirs to allow for recreational activities such as boating, which, while nonconsumptive in nature, might prevent that water from being used by agriculture downstream. Sometimes, the release of water from reservoirs is timed to enhance recreation, such as whitewater rafting downstream of the reservoir. Golf courses use large amounts of water to irrigate their lawns, and also apply large amounts of fertilizers and pesticides that can contaminate water supplies through runoff.

Human use of water resources to augment the environment makes up a very small percentage of overall available water use. Such activities include restoring natural or artificial wetlands and lakes, creating fish ladders to allow the movement of fish around dams, and releasing water from reservoirs to help fish spawn or restore river flow regimes. Again, these uses are mostly nonconsumptive, but may reduce water availability for other needs.

THREATS TO THE WATER SUPPLY

There are many difficulties with proper water management, including the fact that sources of water often cross national boundaries. However, water that spans international boundaries has been shown by the International Union for the Conservation of Nature (IUCN) to be a source for cooperation over water issues rather than a reason for conflict. In addition, many noncomsumptive water uses are hard to give financial value to, and are generally difficult to manage. These nonconsumptive uses could include protecting water bodies as habitats for rare species or for ecosystem services, as well as protecting ancient groundwater reserves. Climate change may also affect the availability of freshwater by changing precipitation cycles, melting ice caps, and raising ocean levels.

Other challenges facing proper water management include water scarcity and water contamination. To improve water availability, there is a need to improve data, treasure the environment, reform governance, revitalize agricultural water use so that farmers can increase productivity to meet growing needs for food, control urban and industrial water use, and empower underrepresented people in water management.

The highly publicized water crisis that began occurring in Cape Town, South Africa, in 2016 increased concerns about links between climate change and water management and conservation. Scientists argued that after the second-largest South African city started to experience a prolonged drought in 2015, a combination of climate change, poor water management, and unchecked population growth led to a severe water shortage in which a deadline was set for when the city's reservoirs would reach such a critically low point that residents' water taps would be shut off. A series of water restrictions eventually forced residents to find ways to recycle water and cut down on use by early 2018, and by mid-2018, reduced use combined with strong rains ended the worst of the crisis, although the long-term problem persisted.

WATER SCARCITY

Water scarcity is becoming an increasing concern as human populations grow exponentially, and as water use for industry, development, and agriculture for food and emerging biofuels increase. A total of 1.2 billion (20 percent) of the world's people live in areas of physical water scarcity where there is not enough water to meet demands. A total of 1.6 billion people live in areas of economic water scarcity where there is a lack of investment in water infrastructure or insufficient human capacity to supply the demand for water.

Rainwater harvesting is one potential solution for water scarcity. Rainwater is collected with rooftop harvesting systems, water barrels, or water gardens. Harvested rainwater can then be stored for later use. Rainwater can be used as it is for activities like gardening, or can be treated for consumption as drinking water.

Water recycling, using treated wastewater for additional uses, is another solution to water scarcity. Recycled water is used in a wide range of activities, such as agriculture, landscaping, public parks, golf-course irrigation, cooling water for power plants and oil refineries, processing water for mills, watering plants, toilet flushing, dust control, construction activities, concrete mixing, artificial lakes, and replenishing or recharging groundwater basins. Recycling water is especially useful in water-scarcity areas, but is also used to save cost and energy.

Another type of recycled water is called "gray water." Gray water is reusable wastewater from residential, commercial, and industrial bathroom sinks, bathtub shower drains, and laundry-equipment drains. Gray water is generally reused on-site and requires the use of nontoxic, low-sodium soaps to protect vegetation.

Water crisis sign in Cape Town, South Africa, 2018 (D Tankersley via Wikimedia Commons)

WATER TREATMENT

Water treatment is the process of making water more acceptable for its desired use. Water is often treated for use in industry, drinking water, and medicine, as well as to return it to the environment without adverse effect. Treating water generally involves either removing contaminants or reducing them to a satisfactory level.

As stated earlier, water contamination can be harmful to humans through exposure to pathogens, toxins, and chemicals, directly and through uptake into produce. In addition, contaminated water can harm ecosystems and wildlife health. Actions to reduce or remove contaminants in water have the potential to save lives and to improve ecosystems and livelihoods in affected areas.

The multiple-barrier approach to preventing produce contamination involves analyzing the food production process from growth to consumption, looking for where a barrier to contamination might be possible. Such barriers could be safer irrigation practices, on-farm wastewater treatment, actions to kill pathogens, or the implementation of crop washing after harvest in markets and restaurants before sale and consumption.

Treating water to make it safe for drinking can solve both the problems of water scarcity and contamination by making otherwise unusable water safe and available. There are many processes involved in treating drinking water. Solids are separated out using physical means such as settling and filtration, chemical means such as disinfection and coagulation, and biological means such as aerated lagoons, activated sludge, and slow sand filters.

Water is generally prechlorinated to control algae and bacteria growth; aerated to remove dissolved iron and manganese; coagulated, sedimented, and filtered to remove solids and particles; and disinfected to remove bacteria. This process varies from location to location based on local water quality, needs, and regulations. To ensure quality for drinking water, its treatment, transport, and distribution are generally highly regulated.

Taking water samples from a treatment plant in Wellington, New Zealand (CORE via Wikimedia Commons)

Another type of water treatment involves desalination, also called "desalinization," which is the removal of dissolved salts and other minerals from water to make it safe for drinking or irrigation. Desalination is expensive both in terms of energy and infrastructure, and therefore is considered only in areas where freshwater sources are otherwise limited. Australia, for example, receives little rainfall and relies on desalination to augment its recycled water supplies. Saudi Arabia as well operates numerous desalination plants to convert seawater to freshwater for use in that very arid land.

There are two main types of desalination: membrane and thermal. Membrane desalination uses membranes to filter salt out of saline water through reverse osmosis. This uses less energy than the alternative but more widely used process of thermo or vacuum distillation. Vacuum distillation boils the water to separate it from the salt. To reduce the energy needed to boil the water, vacuum distillation is performed at below-atmospheric pressures, which lowers the boiling point of the water, saving energy and, therefore, money.

Desalination can have negative environmental effects by killing marine creatures caught in ocean inflow water, and increased temperature and salinity in outflow water returned to the ocean. On the positive side, desalination sometimes produces table salt as a byproduct.

Sewage treatment has the goal of returning water that was used for household or industrial purposes back to the natural environment without causing adverse effects to the ecosystem. This generally involves removing human waste and chemicals from water that is piped to treatment plants. The process of removing these contaminants is similar to the one used to treat drinking water and involves multiple physical, chemical, and biological steps.

Sewage treatment generally results in the production of a liquid effluent that is safe to dispose into the environment, as well as a solid waste sludge. This sludge is generally treated further and discarded, or is sometimes reused as an agricultural fertilizer or a combustible solid fuel when dried. With new advanced technology, it is even possible to reuse wastewater safely as drinking water.

In areas where sewage is not properly treated, or when sewage pipes leak and sewage is released before it reaches treatment, there can be negative effects on the environment. Increased nutrient inputs from organic waste like feces can cause eutrophication in water bodies, leading to algal blooms, low oxygen conditions, and fish die-offs. Contaminants from wastewater can have a diverse range of effects on the ecosystem. Endocrine-disrupting chemicals, such as those found in the form of synthetic estrogens in many pharmaceuticals, can cause reproductive deformities in vertebrates and have even been found to cause intersex individuals among fish and frogs.

—Hannah Bement

Further Reading

"Aquastat: Overview: Water Use." *Food and Agriculture Organization of the United Nations*, www.fao.org/aquastat/en/overview/methodology/water-use/. Accessed 21 Nov. 2019.

Chartres, C., and S. Varma. *Out of Water: From Abundance to Scarcity and How to Solve the World's Water Problems*. FT Press, 2010.

Chen, Daniel H., editor. *Sustainable Water Management and Technologies: Sustainable Water Management Volume 1.* CRC Press, 2017.

David, Matej, and Stephan Gollasch, editors. *Global Maritime Transport and Ballast Water Management: Issues and Solutions.* Springer, 2015.

"Diarrhoeal Disease." *World Health Organization*, 2 May 2017, www.who.int/news-room/fact-sheets/detail/diarrhoeal-disease.

Earle, Anton, Anders Jägerskog, and Joakim Öjendal, editors. *Transboundary Water Management Principles and Practice.* Earthscan, 2013.

Grigg, Neil S. *Total Water Management: Practices for a Sustainable Future.* American Water Works Association, 2011.

Khokhar, Tariq. "Chart: Globally, 70% of Freshwater Is Used for Agriculture." *The World Bank*, 22 Mar. 2017, blogs.worldbank.org/opendata/latinamerica/chart-globally-70-freshwater-used-agriculture.

Molden, D. *Water for Food, Water for Life: A Comprehensive Assessment of Water Management in Agriculture.* Earthscan/IWMI, 2007.

Vickers, Amy. *Water Use and Conservation.* Water Plow Press, 2002.

Welch, Craig. "Why Cape Town Is Running Out of Water, and Who's Next." *National Geographic*, 5 Mar. 2018, news.nationalgeographic.com/2018/02/cape-town-running-out-of-water-drought-taps-shutoff-other-cities/.

Water-Pollution Control

Fields of Study: ecology; chemistry; biology; oceanography; limnology

ABSTRACT

Water-pollution control is a multidisciplinary field that endeavors to reduce or eliminate discharge of chemical and biological compounds into natural waterways. Legal standards set a level playing field, so that the economic burdens and opportunities fall with approximate equality on all enterprises, industries, and municipalities.

KEY CONCEPTS

biological oxidation: a method of treating organic wastes by reliance upon living aerobic bacteria and microorganisms to consume the material

flocculant: a chemical or combination of chemicals that when added to water form a substance that will aggregate pollutant materials and so facilitate their removal

nonpoint source: a pollution source in for which no specific source can be identified, such as surface water runoff from fields over a broad area

point source: a pollution source that can be identified as the specific source of pollution, such as an effluent discharge pipe from a specific factory

pollutant: essentially, any material that is not normally found in a natural space, whether by identity or quantity

DEFINITION AND BASIC PRINCIPLES

Water-pollution control involves preventing contaminants from reaching any usable watercourse. Understanding all the various contaminants that may enter a body of water is an essential part of this subject. It is also necessary to understand the extent of the reactions of receiving waters to wastewater discharges.

Water pollution itself has such a commonly understood everyday meaning that definition can appear almost unnecessary. However, scientific measurement and legal enforcement require precise definition. The Clean Water Act defines "pollutant" as "dredged spoil, solid waste, incinerator residue, sewage, garbage, sewage sludge, munitions, chemical wastes, biological materials, radioactive materials, heat, wrecked or discarded equipment, rock, sand, cellar dirt and industrial, municipal, and agricultural waste discharged into water" (33 U.S.C. Sec. 1362). A toxic pollutant is separately defined to mean those that "after discharge and upon exposure, ingestion, inhalation or assimilation into any organism, either directly from the environment or indirectly by ingestion through food chains, will... cause death, disease, behavioral abnormalities, cancer, genetic mutations, physiological malfunctions" in organisms or their offspring. The act then references a table of toxic pollutants assembled by the Committee on Public Works of the House of Representatives and authorizes the administrator of the Environmental Protection Agency (EPA) to add additional pollutants to the list.

Organic wastes include fecal material from humans and animals, either from municipal sewage systems or concentrated livestock agriculture. Most water naturally has some minerals dissolved in it, which vary by

Hazardous chemicals flow into a river (AykKokcu via iStock)

locality. When higher concentrations of iron, copper, chromium, platinum, mercury, nickel, zinc, and tin are discharged into waterways because of mining or industrial operations, the concentration can be toxic to aquatic life and cause metabolic diseases or cancer in humans who drink or bathe in the water. Discharges of petroleum and a tremendous number of oil-derived hydrocarbon compounds are toxic or carcinogenic. Common fertilizers, based on phosphorous and nitrogen compounds, cause overgrowth of some aquatic plants, particularly algae.

BACKGROUND AND HISTORY

Cities have had sewers and drains since ancient times, but only to flush raw sewage into nearby rivers to be carried away by the current. This certainly contributed to many plagues and epidemics, including the spread of cholera in the nineteenth century. From the dawn of the industrial revolution, manufacturing and mining have routinely been located near sources of water, because water is used in industrial processes and water sources are convenient places to dump waste products. In the United States, concern that refuse and pollutants were creating obstacles to safe navigation inspired the federal Rivers and Harbors Appropriation Act of 1899. Not until the Federal Water Pollution Control Act of 1948 did the first federal legislation tentatively address water pollution as a hazard to human life and safety.

The Federal Water Pollution Control Act Amendments of 1972 established the first comprehensive regulatory framework for water-pollution control. Similar measures emerged in North America and in Europe, Australia, and Japan, during the 1960s and 1970s. There are some nineteenth-century precedents, including attempts in Britain to regulate major manufacturing sources of pollution and secure clean water for household use, German legislation allowing local authorities to place conditions on methods of manufacturing, and Dutch laws aimed at controlling hazards and nuisances from industrial sources. Still, the German Civil Code of 1873 was typical of the era: It set forth that individuals should tolerate pollution as it was essential to economic development.

HOW IT WORKS

Water-pollution laws in the United States, since adoption of the Clean Water Act of 1972, begin with the baseline standard that the discharge of any pollutant by any person into any navigable waterway, or tributary flowing into a navigable waterway, is prohibited. Since immediate adherence to this standard could prove physically and economically infeasible, the law then provides that any exception must be authorized by a permit, issued under the National Pollutant Discharge Elimination System (NPDES), either by the EPA or an authorized state agency. The type of pollutants allowed, concentration, and total amount, are specified in the license. The standard for issuing permits requires use of the best available technology to reduce or eliminate water pollution. The ruling agencies have also weighed economically practical cost.

Europe, Japan, and Developing Nations. Similar standards and enforcement processes are used in Western Europe and are mandated by the European Union (EU)—however, many Eastern and central European countries have been slow to adopt best available technology standards, for fear that their economies cannot afford the cost. British pollution-control laws mandate best available technologies not entailing excessive cost (BATNEEC). Japan intensified pollution control from 1965, when industry was investing less than ¥50 billion per year, to 1976, when investment per year rose to ¥1 trillion, about 0.9 percent of the increase in gross domestic product (GDP) during this period. In Italy, at the close of the twentieth century, half of municipal sewage was discharged without treatment into receiving waters. In Latin America as a whole, only about 11 percent of wastewater is properly treated. Developing nations generally lack comprehensive regulatory laws, and up until the turn of the twenty-first century developing economies such as Brazil, China, India, and South Africa had not placed great emphasis on pollution control. These countries are beginning to make significant new investments.

Treatment Methods. Three common types of treatment are known as primary treatment, secondary treatment, and tertiary treatment. Primary treatment extracts pollutants that will settle out by grit removal, screening, grinding, and sedimentation. Secondary treatment utilizes the natural degradation of organic wastes by biological oxidation, using activated sludge, trickling filters, and oxidation ponds. Tertiary treatment involves a wide range of chemicals to remove substances that remain in water after primary and secondary treatments. Sludge removed in either of the first two stages must be disposed of, sometimes by incineration, sometimes by landfill. Tertiary treatment uses chemical and electrical processes to remove more elusive nonorganic pollutants.

Point and Nonpoint Source Pollution. Point source pollution comes from a discrete, easily identifiable source, such as a pipe. Responsibility can be established, chemical content and volume measured, and controls put in place. Nonpoint source pollution enters a watershed through runoff from large areas, such as agricultural land, paved urban surfaces draining into storm drains, even lawns and golf courses, or in precipitation, as in the case of acid rain. While most federal law addresses point source pollution, Section 319 of the Clean Water Act authorized the EPA's Nonpoint Source Management Program (NSMP), which provides grants to states, territories, and tribes, and calls for state-level plans to implement measures for nonpoint source pollution control.

APPLICATIONS AND PRODUCTS

The two sources of water pollutants most effectively targeted by NPDES are manufacturing and domestic waste. Most of the latter enter municipal sewage systems, but so does some manufacturing waste. Industries are sometimes required to pretreat their effluent before discharging it into municipal sewers. This is more difficult to enforce than a specific factory with its own discharge system into a river or lake.

Physical Technology. One of the first steps in primary treatment is a screening system to remove large objects. Removing grit often requires a conveyer system, which allows sand and dirt to settle out of the water stream. The activated sludge process is one of the most visually familiar to anyone who has viewed a common wastewater-treatment plant: large concrete tanks with walkways between them, continually agitated by rotary aerators to mix oxygen into the sludge, which hastens biological processes to break down organic compounds. Thermal conditioning units, a kind of pressure cooker, digest sludge and separate the majority of water from the sludge. Sludge is further dried by vacuum-filtration equipment—a rotating roll spread with sludge with a vacuum applied through the roll to extract more water, leaving a semidried cake that falls off. This can be disposed of in a landfill but is often incinerated.

Chemical Processes. Tertiary treatment uses many methods to remove pollutants that do not settle out in primary or secondary treatment. These include coagulation, flocculation, nitrification, denitrification, adsorption, ion exchange, and electrodialysis. These have high capital investment and operation costs and require highly skilled personnel for operation and maintenance. Flocculation removes solids suspended in water-but it is not effective in removing dissolved compounds. The name derives from "floc," a kind of clumping of different precipitated materials. Generally, one or more chemicals are added to the water being treated, which form a gel or other precipitate that attracts other suspended materials. One common combination is potassium aluminum sulfate and ammonia, which react to form aluminum hydroxide. Adsorption introduces powdered activated carbon to remove even small traces of pollutants. Since there is no chemical bonding, recovered pollutants can be collected and cycled back into industrial processes, offsetting costs or even generating a net profit. Adsorption technology is appropriate for removal of heavy metals and complex organic compounds, including pesticides, silicates, aluminum compounds, wood- and petroleum-based carbon compounds, and resins.

CAREERS AND COURSEWORK

A wide variety of careers are available in the public, private for-profit, and private not-for-profit sectors. These include design, manufacture, and operation of wastewater-treatment facilities, both by municipal government and industrial sources. Most government jobs relate to enforcement of water-pollution control laws, preparation of long-term strategies, or the review and funding of research projects. Water is one of four program divisions within the EPA, which also has a research and development program concerned in part with water pollution. EPA includes ten regional offices and laboratories and a network of laboratories and research centers. A large portion of careers in this field involve measuring existing pollution, establishing what is and is not a hazardous substance and what concentrations are tolerable, advocating for new measures, and litigation.

Undergraduate degrees in environmental science typically require courses in biology, natural history, soil science, geology, chemistry, physics, calculus, instrumental methods of analysis, scientific presentations, and interdisciplinary courses such as principles of life sciences, understanding the earth, and ecology. Some colleges offer specific courses on pollution-control technology, toxicology, water-quality analysis and management, solid-waste management, industrial hygiene, and water-pollution biology. In addition to graduate programs in environmental science, many law schools offer degree programs in environmental law, including courses on water-pollution control, natural resources law, regulation of toxic-substance risk, and environmental issues in business transactions.

Environmental law is a considerable specialty, which requires a law degree for attorneys, and for paralegals, coursework with specific emphasis on environmental law, preferably with some relevant science courses. Understanding the facts to be litigated requires at least a basic undergraduate familiarity with chemistry and biology. Legal practice options include working for government enforcement agencies at the state or federal level, defending alleged polluters against administrative or judicial actions, advising businesses on compliance, and representing advocacy organizations in civil actions.

Monitoring pollutants requires processing and analyzing huge quantities of data, which requires complex computer software. The EPA uses software called BASINS (Better Assessment Science Integrating Point and Nonpoint Sources) to perform assessments of water quality and of specific watersheds. Compliance with a variety of detailed regulations requires software both for enforcement and helping businesses integrate compliance into their daily operation.

SOCIAL CONTEXT AND FUTURE PROSPECTS

There are literally hundreds of companies in the business of supplying instrumentation for water-quality monitoring, sanitation systems, water-storage systems, filtration systems, chemicals to neutralize pollutants (adding a basic compound to neutralize an acid or vice versa), piping and pumping equipment, and a variety of components and wastewater-treatment products. The market is not going to decline in North America, Europe, or Japan. It is likely to expand exponentially in developing nations such as China, India, Brazil, Nigeria, Indonesia, and South Africa. These countries, particularly China, will develop their own research and production. Finding new methods that would make reclaiming and recycling into profitable operations is an open area of research that has barely been touched. More efficient methods of recovery, which would bring the cost of reclaiming metals in particular below the cost of purchasing newly mined quantities, has yet to be thoroughly explored.

—Charles Rosenberg

Further Reading

Abel, P. D. *Water Pollution Biology*. 2nd ed., Taylor, 1996.

Arias, Andres Hugo, and Sandra Elizabeth Botté. *Coastal and Deep Ocean Pollution*. CRC Press, 2020.

Beiras, Ricardo. *Marine Pollution: Sources, Fate and Effects of Pollutants in Coastal Ecosystems*. Elsevier, 2018.

Dodson, Roy D. *Storm Water Pollution Control: Municipal, Industrial and Construction NPDES Compliance*. McGraw, 1999.

Heath, Alan G. *Water Pollution and Fish Physiology*. 2nd ed., CRC Press, 2018.

Helmer, Richard, and Ivanildo Hespanhol, editors. *Water Pollution: A Guide to the Use of Water Quality Management Principles*. Spon, 1997.

McBeath, Jenifer Huang, and Jerry McBeath. *Environmental Change and Food Security in China*. Springer, 2010.

Pei, De-Sheng, and Muhammad Junaid, editors. *Marine Ecology: Current and Future Developments (Volume 1). Marine Pollution: Current Status, Impacts and Remedies*. Bentham Books, 2019.

Sell, Nancy J. *Industrial Pollution Control: Issues and Techniques*. 2nd ed., Wiley, 1992.

Walker, Colin H., and David R. Livingstone, editors. *Persistent Pollutants in Marine Ecosystems*. Elsevier, 2013.

Zeng, Eddy Y., editor. *Microplastic Contamination in Aquatic Environments: An Emerging Matter of Environmental Urgency*. Elsevier, 2018.

Water-Rock Interactions

Fields of Study: geochemistry; marine geochemistry; hydrogeology

ABSTRACT

Water-rock interactions occur as fluids circulate through the rock structures of Earth's crust. Isotopes of common elements are exchanged between a fluid and its host rock. As a result of these reactions, a rock may preserve a record of the fluids that

have passed through its pore spaces. Studies of water-rock interactions provide information on the nature of fluid movement through rocks and on the origin of economic ore deposits.

KEY CONCEPTS

contact exchange of atoms: as different materials contact each other, atoms from one material may change places with atoms from the other material; in isotopic studies the exchange of atoms of the same element occurs, which may exchange one isotope for another in the process of isotopic fractionation

isotopes: atoms of a single element containing the same number of protons but different numbers of neutrons in the nucleus

isotopic fractionation: the relative enrichment and depletion of isotopes of the same element in two different materials through contact exchange of atoms

juvenile water: water that has been newly released from magma and has not yet become part of the different aspects of the water cycle; very rarely observed

stable isotopes: isotopes of an element that do not decay over time

stable isotope ratios: the relative concentrations of stable isotopes of a particular element in a sample; the ratio of stable isotopes can be used to determine different time-related properties such as the age of the material and the amount of water that has been in contact with the material over time

STABLE ISOTOPES

Fluids that circulate within Earth's crust take part in chemical reactions involving an exchange of elements between the fluid and the host rock. Such reactions, referred to as fluid-rock interactions, are an important mechanism in the concentration of minerals in economically valuable ore deposits. A useful means of studying fluid-rock interactions is by measurement of the stable isotopic composition of rocks that have undergone such an exchange history.

Rocks are exposed to fluids that are diverse in their origin and composition. The most important volatile constituents of natural fluids are water, carbon dioxide, carbon monoxide, hydrogen fluoride, sulfur compounds, and light hydrocarbons such as methane. In addition, fluids contain dissolved solids derived from the crustal rocks through which they pass. As the list of volatiles makes clear, the dominant elements present in fluids are oxygen, hydrogen, carbon, and sulfur.

An element may be characterized by its atomic number and atomic weight. Atomic number refers to the number of protons present in the nucleus of an atom and is a constant value for each element. Atomic weight is determined by adding together the number of protons and neutrons contained in an atom. Because the number of neutrons present often varies within a limited range, atoms of a particular element may have several different atomic weights. These atoms are referred to as isotopes. Oxygen, with an atomic number of 8, occurs most commonly with eight neutrons but may have nine or ten. Therefore, three isotopes of oxygen occur in nature: ^{16}O or oxygen-16, ^{17}O or oxygen-17, and ^{18}O or oxygen-18. Similarly, carbon occurs as ^{12}C or carbon-12 or ^{13}C or carbon-13. (^{14}C or Carbon-14 is a radioactive isotope that forms in Earth's upper atmosphere and will not be considered here.) Hydrogen contains only one proton; however, a small fraction of hydrogen atoms also contains a neutron, which doubles the mass of the atom. This heavy hydrogen isotope ^{2}H is called "deuterium." Sulfur has four stable isotopes: ^{32}S, ^{33}S, ^{34}S and ^{36}S or sulfur-32, -33, -34, and -36.

STABLE ISOTOPE RATIOS

For any given mineral or fluid, the relative concentrations of the isotopes of a particular element may be expressed as a stable isotopic ratio, or the ratio of the second most abundant isotopic species over the most abundant isotope. Any physical process that results in the enrichment or depletion of the concentration of a heavy isotope is referred to as isotopic fractionation. A common fractionation process is evaporation of water. Water molecules containing the lighter isotope of oxygen (oxygen-16) will preferentially evaporate so that the remaining liquid will be enriched in the heavier oxygen isotope (oxygen-18). Fractionation also occurs during the growth of minerals in either a magma or a water-rich solution. Some minerals, because of the nature of their chemical bonds and their crystal structure, tend to contain a greater number of heavy isotopes than do other minerals. Quartz, dolomite, and calcite are common examples of minerals that contain high concentrations of heavy oxygen, while oxides such as ilmenite and magnetite have very little of the heavy isotope.

The effectiveness of fractionation during mineral growth is dependent upon temperature. Low temperatures permit minerals to be more selective in choosing atoms for growing crystal sites, resulting in large differences in isotopic ratios between different minerals. At high temperatures, the selection of atoms is a more random process, and differences between isotopic ratios become progressively smaller.

Because the heavier isotopes of elements naturally occur in such small concentrations, isotopic ratios of oxygen and carbon, for example, have numerical values that are very small and difficult to measure accurately. For this reason, isotopic ratios for a particular sample are presented as a relative enrichment or depletion of the heavy isotope of an element as compared with a defined standard. The difference between the sample and the standard is measured in parts per thousand or per million and is expressed by the Greek letter delta (ä). For carbon isotopic values, for example, the accepted standard is called PDB (Pee Dee Belemnite) and is obtained from a belemnite fossil of the Cretaceous-age Pee Dee formation of North Carolina.

EXCHANGE OF STABLE ISOTOPES

As fluids migrate through rocks, reactions occur that involve the exchange of stable isotopes between the fluid and the solid. The exchange process may be pervasive, where fluid movement is diffusive and affects the entire rock mass, or localized along specific fluid channel ways, such as fractures, where only the wall of the rock is altered along the route of water movement. The degree to which the isotopic composition of the rock is altered depends on the initial composition of the rock and the fluid, the temperature at which isotopic exchange is occurring, and the amount of fluid present. Typically, the oxygen and hydrogen isotopic values for crustal water are light compared to those for most rocks. Therefore, as isotopic exchange proceeds, the isotopic composition of the rock becomes progressively lighter, while the water becomes increasingly enriched in the heavier isotopes. If the fluid-rock interaction has occurred under constant temperature conditions, the final isotopic value of the rock is proportional to the volume of water that has passed through the rock.

RESERVOIRS OF CRUSTAL FLUIDS

There are four principal sources, or reservoirs, of crustal fluids. Each of these fluid reservoirs contains stable isotopic ratios that reflect the fractionation mechanisms at work and the result of chemical reactions between the fluids and their host rocks.

"Meteoric water" is a term applied to fluids that take part in the surface hydrologic cycle. Water that undergoes evaporation, precipitation, and runoff to lakes and to the ocean is capable of penetrating Earth's crust to a depth of several miles (or kilometers). This penetration is usually accomplished by fluid migration along weaknesses in the crust, such as faults or fracture systems. Natural hot springs are an example of meteoric water that has been heated deep in the crust and then reemerges at the surface. The oxygen and hydrogen isotopic ratios associated with meteoric water are controlled principally by the distillation effect of evaporation and precipitation. As a result of the general transport of air masses from the equator toward the poles, meteoric water isotopic values vary in a systematic manner, with heavier ratios found near the equator and progressively lighter isotopic values occurring toward the poles.

Meteoric water that is trapped in the pore spaces of accumulating sediments is referred to as connate water. When loose sediments have lithified to form hard sedimentary rock, the enclosed pore fluids may become isolated for very long periods. Connate water reveals an isotopic trend similar to that of meteoric water, except that the oxygen values tend to be heavier because of the capacity of the lighter oxygen (oxygen-16) contained in the trapped pore fluids to be exchanged for some of the heavier oxygen (oxygen-18) of the host sedimentary rock. Isotopic exchange between connate water and the host sediments continues until equilibrium is achieved.

As rocks undergo increases in temperature and pressure associated with metamorphism, water is frequently released in a process called "dehydration." The escaping water may be from pore spaces in the rock or from hydrous minerals such as micas or amphiboles. Because of their origin, these dehydration fluids are also referred to as metamorphic fluids. Dehydration water has a very wide range of isotopic compositions, which reflects the diversity of the original sediments.

Juvenile water, which originates in the upper mantle, escapes from ascending magma and represents the fourth important fluid source. Not all water derived from magma should be considered juvenile, as meteoric or connate water will frequently be present in sediments that undergo melting deep in the crust. True juvenile water has a very narrow range of isotopic compositions. Because of mixing of crustal fluids and exchange reactions with igneous rocks, samples of unaltered juvenile water are found very rarely.

ISOTOPIC COMPOSITIONS OF ROCKS

The isotopic compositions of rocks reflect the formation history of the particular rock type. Igneous rocks are controlled by the composition of their magmatic source area, which is usually in the lower crust or upper mantle. Other factors include the temperature of crystallization, the type of minerals, and the degree to which the magma remains isolated or mixes with other constituents. Unaltered igneous rocks typically have a narrow range of isotopic values as compared with natural fluids. Deep mantle rocks contain minerals that are low in oxygen-18 and therefore fall in a narrow range of isotopically light compositions. Crustal rocks, with a higher proportion of silicate minerals, which concentrate oxygen-18, are isotopically heavy. Sedimentary rocks have two distinct modes of formation.

Clastic sedimentary rocks are made of transported particles of weathered material and therefore have isotopic ratios that reflect the individual components. Chemical sedimentary rocks precipitate directly and often involve biological activity. The oxygen isotopic values of these rocks are usually much heavier than those of fluids. Because of fractionation factors associated with organisms, carbon isotope ratios are highly variable. Rocks that have undergone metamorphism contain the widest range of isotopic compositions, as a result of chemical reactions in the presence of fluids that may be derived from any of the reservoirs previously described.

MASS SPECTROMETRY

The most important analytical tool in the study of fluid-rock interactions is the stable isotope mass spectrometer. This instrument separates isotopes according to their mass differences, as determined by the deflection of charged ions moving within a magnetic field. Elements of interest are extracted from minerals through appropriate chemical reactions and then converted to a gas, which is entered into the mass spectrometer. The sample gas is bombarded by a stream of electrons, converting the gas molecules to positively charged ions. The ions are accelerated along a tube into a chamber maintained under very high vacuum, where a powerful magnet deflects the charged molecules into curved pathways. Lighter particles are deflected more than heavier ones, so several streams of ions result, each with a particular mass. The relative proportions of each isotope are measured by comparing the induced current produced by each ion stream at a collector.

RESEARCH INVOLVING IGNEOUS ROCKS

Studies of oxygen and hydrogen isotopes have been useful in research on water-rock interactions involving igneous rocks. Hydrous minerals, such as biotite and hornblende, are separated out of granitic rocks in order to extract hydrogen isotopic values. Oxygen isotopic compositions are usually measured from feldspars and quartz. Two areas of water-rock study have been of particular interest. The first concerns the origin and quantity of water responsible for the isotopic alteration of large igneous intrusions within the continental crust. The second area of investigation addresses the interaction of ocean water with ocean-crust basalt and the formation of associated sulfide ore deposits.

The initial isotopic composition of igneous rocks is predictable according to their mineral content and temperature of crystallization. Therefore, it is possible to recognize when igneous rocks have been influenced by exchange with a fluid. Many examples have been found of shallow igneous intrusions that have been depleted in oxygen-18 through interaction with large volumes of isotopically light meteoric water. Hydrogen isotope exchange is even more sensitive than with oxygen, because igneous rocks have so little hydrogen relative to the amount in water. A very small quantity of water may produce a large isotopic shift in a rock's hydrogen value, while the oxygen is largely unaffected.

Isotopes associated with the ocean crust have been studied through cores obtained from seabed drilling and through the analysis of ophiolites, which represent portions of ocean crust exposed on land. In the vicinity of the mid-ocean ridge, where temperatures exceed 572 degrees Fahrenheit (300 degrees Celsius), ocean water circulates to a depth of 1.87 to 2.49 miles (3 or 4 kilometers) and is responsible for depleting oxygen-18 by 1 or 2 parts per thousand.

STUDY OF METAMORPHISM AND PRECIPITATED ROCK

Fluids associated with metamorphism have also been intensively studied by stable isotopic methods. Increasing metamorphic grade is associated with progressively lighter isotopic values for oxygen and hydrogen. Areas of regional metamorphism may show two end-member types of fluid behavior. Consistent depletion of oxygen and hydrogen isotopic values throughout a large terrain points to exchange with an external source of water that flows pervasively through the region. Alternatively, only the trapped connate water may be involved in exchange reactions, leading to higher isotopic composition values and to enrichment of deuterium in hydrous minerals. Because of the large difference in isotopic composition between magma and sedimentary rocks, contact metamorphism is particularly appropriate for study. Samples from intrusions indicate that the margin of the igneous rock is enriched in oxygen-18 through exchange with the country rock, while the interior of the body remains unaltered. The early stages of fluid-rock interaction are dominated by magmatic water, while meteoric water becomes increasingly important as cooling proceeds.

Sedimentary rocks that form by precipitation, such as limestone and chert, have heavy isotopic compositions as a result of the large fractionation between water and either calcite or quartz at low temperature. Clastic sandstones, which contain transported quartz grains, are characterized by the lighter isotopic values of the component particles. Sedimentary rocks selected from a large sample area frequently show a progressive change in the amount of isotopic depletion that has resulted from water-rock exchange. As water continues the exchange process, the isotopic composition of the fluid also shifts. During the next increment of water-rock exchange, the potential amount of depletion of the rock will not be as great. By plotting isotopic values of sedimentary rocks, it is possible to determine directions of fluid motion.

SEARCH FOR ECONOMIC RESOURCES

Water-rock interactions are important primarily for their role in the formation of economically vital ore deposits. These are localized regions where metals such as gold, silver, lead, copper, zinc, and tungsten occur in unusually high concentrations and can be extracted. Water, circulating within Earth's crust, plays an important role in the formation of most ore deposits by leaching elements from rocks and concentrating them in zones of new mineral growth. During this process, the isotopic compositions of the host rock and fluid are progressively changed. Analysis of the resulting stable isotope ratios of the ore rocks allows geologists to understand the sources and quantity of the mineralizing fluids. Better understanding of the formation of ore deposits has led to greater success in the discovery of new mineral resources.

Stable isotope research into water-rock interaction is not limited to studying water-rich fluids associated with precious mineral deposits. Petroleum and natural gas flow from source rocks rich in organic material to porous reservoir rocks, where they may become trapped. Rocks through which hydrocarbons have migrated frequently show well-depleted carbon isotope values and thus preserve a record of fluid movement. Oil companies have used carbon data to track the migration history of hydrocarbons and to locate regions where petroleum leaks to the surface. The use of carbon isotopes has also proved successful in identifying source rocks associated with producing oil fields. This technology will become even more important as resources become increasingly scarce.

The source of fresh drinking water for almost half the population in the United States is subsurface groundwater. This crucial resource is jeopardized by contamination with common pollutants, such as pesticides, and by depletion through the withdrawing of water from aquifers faster than it is replenished at recharge zones. Research involving stable isotope studies has become important in hydrology to track fluid movement within aquifers and identify sources of recharge water.

HAZARDOUS WASTE STORAGE

Another area of concern associated with water-rock interaction is the safe, long-term storage of toxic and nuclear waste. One of the most important criteria for the isolation of dangerous wastes is that the enclosing rocks be relatively dry and impermeable to water movement so that hazardous material is not transported into water supply aquifers. The record of water flow recorded by stable isotopes is sensitive to even very small fluid volumes and provides one of the means of assessing the risks associated with a toxic disposal site.

—Grant R. Woodwell

Further Reading

Albarede, Francis. *Geochemistry: An Introduction*. 2nd ed., Cambridge UP, 2009.

Birkle, Peter, and Ignacio S. Torres-Alvarado, editors. *Water-Rock Interaction: Proceedings of the 13th International Conference on Water-Rock Interaction WRI-13, Guanajuato, Mexico, 16–20 August, 2010*. CRC Press, 2010.

Blatt, Harvey, and Robert J. Tracy. *Petrology: Igneous, Sedimentary, and Metamorphic*. 3rd ed., W. H. Freeman, 2005.

Brantley, Susan, James Kubicki, and Art White, editors. *Kinetics of Rock-Water Interactions*. Springer, 2007.

Faure, Gunter. *Isotopes: Principles and Applications*. 3rd ed., John Wiley & Sons, 2004. (Originally titled *Principles of Isotope Geology*.)

Hoefs, Jochen. *Stable Isotope Geochemistry*. 6th ed., Springer-Verlag, 2009.

Hurai, Vratislav, Monika Huraiová, Marek Slobodnik, and Rainer Thomas. *Geofluids: Developments in Microthermometry, Spectroscopy, Thermodynamics and Stable Isotopes*. Elsevier, 2015.

Krauskopf, Konrad B. *Introduction to Geochemistry*. 3rd ed., McGraw-Hill, 2003.

Lebedev, Albert T., editor. *Comprehensive Environmental Mass Spectrometry*. ILM Publications, 2012.

Oelkers, Eric H., editor. *Thermodynamics and Kinetics of Water-Rock Interaction: Reviews in Mineralogy and Geochemistry*. Mineralogical Society of America, 2009.

Stober, Ingrid, and Kurt Bucher, editors. *Water-Rock Interaction*. Springer Science & Business Media, 2012.

Stober, Ingrid, and Kurt Bucher, editors. *Hydrogeology of Crystalline Rocks*. Springer Science & Business Media, 2013.

White Sea Ecosystem

Fields of Study: oceanography; marine biogeochemistry; marine biology

ABSTRACT

The White Sea is a nearly-landlocked body of water in northwest Russia. It is home to four major bays or gulfs.

KEY CONCEPTS

eider duck: a species of duck (aquatic fowl) renowned for the softness and insulating properties of its down feathers; the down feathers are used to line the nests in large quantities and are easily harvested for use for use in insulated dry goods

peninsula: a long, relatively narrow protrusion of land into a body of water

GEOGRAPHY AND CLIMATE

The White Sea is located at the northwest corner of Russia, just touching the Arctic Circle parallel of latitude; it is a southern inlet of the Barents Sea. The White Sea is almost entirely surrounded by land, with the Kola Peninsula to its north, Karelia to the west, and Arkhangelsk Oblast to the south and east. The strait connecting to the Barents Sea runs along the Kanin Peninsula to the northeast.

There are four main bays or gulfs in the sea. In the south is Onega Bay; in the southeast is Dvina Bay, which holds the major port of Arkhangelsk. Kandalaksha Gulf is in the western part of the White Sea and is the deepest part of the sea, reaching 1,115 feet (340 m). Opposite the Kola Peninsula, on the east side of the strait, is Mezen Bay. The main rivers draining into these bays or gulfs are the Onega, Northern Dvina, Mezen, and Kuloy. Other significant rivers that flow into the White Sea are the Vyg, Niva, Umba, Varzuga, and Ponoy.

The White Sea climate varies between polar and moderate continental. The winter is long, severe, and often unpredictable. The mean temperature in February is about 5 degrees Fahrenheit (-15 degrees Celsius) and may fall as low as -22 degrees Fahrenheit (-40 degrees Celsius). The ice on the surface of the sea may be up to 5 ft (1.5 m) thick. At times, warm air from the Atlantic raises the air temperature to a "comfortable" 43 degrees Fahrenheit (6 degrees Celsius). The temperature of the water in winter is about 28 to 30 degrees Fahrenheit (-1 to -2 degrees Celsius). The salinity of the White Sea is 2.75 to 2.8 percent (27.5–28 parts per thousand), lower than the mean salinity of the Arctic Ocean.

The White Sea is affected somewhat by the Atlantic Ocean, as its hydrology reflects processes linked with the warm Gulf Current. Together with its generally Arctic character, the Atlantic and freshwater links define many of the species in the White Sea. Most recently, nutrients, pollutants, sedimentation, and invasive species issues have recurred, because of many disturbances across the inland watersheds.

BIODIVERSITY

Since 1930, a large part of the coastline of the Kandalaksha Bay and about 100 small islands in the White Sea have been declared protected areas. This was mainly to protect the eider ducks during nesting season and while the ducklings were growing up. During this time, hunting, fishing, and even visiting of these areas has been forbidden or strongly regulated.

The White Sea provides marine habitat for more than 700 species of invertebrates, about 60 fish species and five marine mammal species. The fishing industry is relatively small, mostly targeting harp and ringed seals (hunted mainly by indigenous peoples), herring, saffron cod, European smelt, Atlantic cod, and Atlantic salmon. There is also a small but developing seaweed industry.

HUMAN IMPACT

The White Sea has been used by the indigenous population for thousands of years, and has played a role in western civilizations for more than 1,000 years. The sea was known to the Novgorod peoples since at least the eleventh century, and probably to the Vikings before them. It was used to link Russian trading cities with northern and central European cities.

One of the earliest settlements near the sea was established in the late fourteenth century in Kholmogory on the Northern Dvina River. Trade in furs, grains, and fish continued through the centuries between Russia, Denmark, England, and Netherlands. Of note is the White Sea Canal that opened in 1933, which—along with the existing natural waterways, Lake Vygozero, Lake Onega, Lake Ladoga, Svir River, and the Neva River—connects the White Sea to the Baltic Sea. The canal was built almost entirely with forced manual labor.

Global climate change impacts all of the Arctic regions, with habitat-altering changes recorded in sea-ice cover, sea level, air and water temperatures, and precipitation patterns. Strategic military and extraction-industry interests present other concerns in the White Sea, such as waste dumping, ongoing for over seventy years. Arctic shipping routes, oil and gas production, commercial fisheries, and inland watershed misuse have left strong traces of pollution in this otherwise robust and somewhat intact environment.

There is a small amount of tourism here in the form of deep-sea diving, but little threat to the environment stems from this. A Russian management regime has dominated the White Sea for centuries; more recently, a Norwegian and European Union agreement has been provisionally established. In either case, an efficient and sustainable management regime has not yet been implemented in a manner or scale sufficient to aid full recovery of the White Sea biome.

Global relevance of conservation practices in the White Sea has not been realized on an international level yet. There is some controversy whether, and what type of, human-made changes have already occurred in the White Sea region. Some Russian sources report little or no changes for local regions in the White Sea. It is noteworthy that diseases and invasive species have not entered this discussion yet, but there are ongoing observations to monitor any environmental changes.

Further risky economic growth schemes involving oil and mineral extraction are still planned for the White Sea. Yet, there are few relevant climate change policies in place for the overall preservation of this unique ecoregion.

—*Falk Huettmann*

Further Reading

Berger V. and S. Dahle, editors. *White Sea—Ecology and Environment.* Derzhavets Publishers, 2001.

Filatov, Nikolai, Dmitry Pozdnyakov, Ola M. Johannessen, et al. *White Sea. Its Marine Environment and Ecosystem Dynamics Influenced by Global Change.* Springer, 2005.

Lisitsyn, Alexander P., and Viacheslav V. Gordieev, editors. *Biogeochemistry of the Atmosphere, Ice and Water of the White Sea: The White Sea Environment, Part 1.* Springer, 2019.

Lisitsyn Alexander P., and Liudmila L. Demina, editors. *Sedimentation Processes in the White Sea: The White Sea Environment, Part 2.* Springer, 2019.

Pascual-Fernández, José J., Cristina Pita, and Maarten Bavinck, editors. *Small-Scale Fisheries in Europe: Status, Resilience and Governance.* Springer, 2020.

Pertsova N. M., and K. N. Kosobokova. "Interannual and Seasonal Variation of the Population Structure, Abundance, and Biomass of the Arctic Copepod *Calanus Glacialis* in the White Sea." *Oceanology*, vol. 50, no. 4, 2010.

Stiansen, J. E., and A. A. Filin, editors. "Joint Polar Research Institute of Marine Fisheries and Oceanography (PINRO)/Institute of Marine Research (IMR) Report on the State of the Barents Sea Ecosystem 2006, with Expected Situation and Considerations for Management." *IMR/PINRO Joint Report Series*, vol. 2, 2007.

Solyanko, K., V. Spiridonov, and A. Naumov. "Biomass: Commonly Occurring and Dominant Species of Macrobenthos in Onega Bay (White Sea, Russia): Data from Three Different Decades." *Marine Ecology*, vol. 32, no. 1, 2011.

Wilson Rowe, E., editor. *Russia and the North.* U of Ottawa P, 2009.

World Ocean Circulation Experiment

Fields of Study: oceanography; computer modeling; climatology; fluid dynamics

ABSTRACT

The World Ocean Circulation Experiment (WOCE) was an ambitious international project designed to increase knowledge about the movement of water, heat, and various substances in the sea. The data obtained are of major importance in predicting future long-term changes in climate.

KEY CONCEPTS

bathythermograph: a shipboard device that measures and records water temperatures at depths, and does not require the ships to remain stationary

Coriolis effect: a perceived force due to Earth's rotation that seems to cause a moving air mass to change its north-south direction of movement to an east-west direction in the northern and southern hemispheres

Ekman layer: roughly the uppermost 328 feet (100 meters) of ocean, where the water is affected by air movement

friction: the physical interaction between two surfaces or boundary layers, that transfers energy from one to the other in both directions and may generate heat in the process

wind stress: air moving across the surface of water, acting to push the surface water in a direction and creating a surface flow

MOVEMENT OF SEAWATER

The WOCE, which began in 1990, brought together scientists from around the globe to study the way in which water moves in all parts of the sea. The WOCE was also intended to obtain information on the movement of heat in the ocean, as well as the movement of substances such as salt and oxygen.

The movement of seawater, known as ocean circulation, consists of horizontal and vertical motion. Horizontal movements are known as currents. Vertical motions are known as upwellings and downwellings. Currents vary widely in speed, ranging from a few inches (or centimeters) per second to as much as 13 ft (4 m) per second. Surface currents typically move between 1.97 and 19.7 inches (5 and 50 centimeters) per second, with deeper currents generally moving more slowly. Vertical movement of seawater is much slower, with a typical speed of only a few feet (or meters) per month.

Ocean circulation is primarily caused by two major factors. Wind-driven circulation is caused by air moving across the surface of the sea. This induces friction, known as wind stress, between the water and the air, thus applying a directional force and setting the water in motion. Thermohaline circulation is caused by differences in temperature and salt concentration. These differences cause variations in the density of seawater, leading to differences in pressure and resulting in motion. Surface currents are mostly caused by wind-driven circulation. Deeper currents and vertical movements are mostly the result of thermohaline circulation.

Several factors are involved in determining the size, shape, and speed of ocean circulation patterns. An important influence is the Coriolis effect, named for the French scientist Gustave-Gaspard de Coriolis (1792–1843). This effect, caused by the rotation of the planet, causes the actual path of a moving object to be curved away from the straight-line path that it would otherwise follow. The Coriolis effect causes ocean currents to bend to the right in the Northern Hemisphere and to bend to the left in the Southern Hemisphere.

Friction also influences the nature of ocean circulation. Layers of water moving at different speeds produce friction where they meet, forming an intermediate zone of turbulent eddy currents as the two layers interact. This transfers energy from one layer to the other, and causes the faster layer to move more slowly and the slower layer to move more quickly. Friction also occurs between moving water and the continents, and between currents at the bottom of the sea and the ocean floor. This friction, with the attendant resulting turbulence zones, tends to slow the motion of seawater.

Important effects on ocean circulation are seen in the region of the sea known as the Ekman layer named for the Swedish scientist Vagn Walfrid Ekman (1874–1954). The Ekman layer extends from the surface of the ocean to a depth of about 328 ft (100 m). In this layer, the wind has a direct effect on water movement. Wind stress, the Coriolis effect, and friction between layers of water combine to move the Ekman layer in complex ways.

Vagn Walfrid Ekman (Wikimedia Commons)

Gaspard-Gustave de Coriolis (Wikimedia Commons)

Water at the surface of the ocean tends to move at an angle of about 45 degrees to the direction of the wind because of the Coriolis effect. This angle is bent to the right in the Northern Hemisphere and to the left in the Southern Hemisphere. With increasing depth, the water moves more slowly and the angle increases. At a depth where the speed of the water is about 4.3 percent of the surface speed, the water moves in the opposite direction to the wind. The overall effect is that the average movement of the water is at about 90 degrees to the wind.

This movement in the Ekman layer, combined with differences in wind stress, creates areas on the surface of the ocean where water converges or diverges. Where it converges, water sinks in downwellings. Where it diverges, water rises in upwellings. Downwellings and upwellings also occur where the wind blows parallel to the coast of a landmass. They also occur because of differences in temperature and salt concentration, both of which alter the density of the seawater. Cold water and salty water tend to sink, while warm water and less salty water tend to rise.

WORLDWIDE PATTERNS OF OCEAN CIRCULATION

The numerous factors involved in ocean circulation, combined with the irregular shapes of the continents, result in complex patterns of water movement. Although major surface currents have been known since the earliest days of ocean travel, much less is known about deeper currents. Similarly, less is also known about the Southern Hemisphere than the Northern Hemisphere. The WOCE was designed to fill these gaps in scientific knowledge.

Before the WOCE, the basic pattern of surface currents was fairly well understood. In the Northern Hemisphere, strong currents tend to move northward along the eastern coasts of the continents. These include the Gulf Stream-North Atlantic-Norway Current, along North America, and the Kuroshio-North Pacific Current, along Asia. In the Southern Hemisphere, strong currents tend to move northward along the western coasts of the continents. These include the Peru Current along South America; the Benguela Current, along Africa; and the Western Australia Current, along Australia.

In the regions north and south of the equator, major surface currents move westward. These are the Pacific North Equatorial Current and the Pacific South Equatorial Current, between South America and Asia; the Atlantic North Equatorial Current and the Atlantic South Equatorial Current, between Africa and South America, and the Indian South Equatorial Current, between Australia and Africa.

At the equator, narrow eastward currents are found between the wider westward currents. These are the Pacific Equatorial Countercurrent, between Asia and South America; the Atlantic Equatorial Countercurrent, between South America and Africa; and the Indian Equatorial Countercurrent, between Africa and Asia. Another major surface current is the Antarctic Current, moving eastward around Antarctica.

Although less is known about deep currents, certain broad patterns of movement are now understood from WOCE data. Cold water in the northern part of the Atlantic Ocean sinks in downwellings. This deep, cold water tends to move southward along the eastern coasts of North and South America to join the deep, cold water that sinks in downwellings near Antarctica. This water then tends to flow eastward in a deep current around Antarctica. Some of this water then moves northward along the coasts of Asia and Africa, rising in upwellings as it warms. Overall, the deep ocean currents function something like a continuous conveyor belt system, moving thermal energy in ocean water and distributing it throughout the oceans. Cold water upwelling in the Indian and north Pacific Oceans circulates westward between Australia and Asia, and around the southern tip of Africa to make its way to the North Atlantic. There it sinks again to depths and enters the deep current that takes it back eastward, where it splits and rises again in the Indian and northern Pacific Ocean.

STUDYING OCEAN CIRCULATION

Scientific studies of surface currents began in the eighteenth century in order to aid navigation. Later studies concentrated on the effect of changes in ocean circulation on the weather. The need for a major effort to increase the amount of information known about the movement of seawater became clear in the 1980s. The best models of ocean circulation, based on the available data, failed to describe the conditions actually observed in the sea with complete accuracy.

A major determining factor in the establishment of the WOCE was the development of new techniques for studying ocean circulation. Temperature measurements could be made from a ship without stopping the movement of the vessel using an instrument known as a bathythermograph. Devices designed to drift in the ocean, both on the surface and at specific depths, were developed that could be tracked for months or years. Advanced methods of accurately measuring the concentration of substances present in seawater in very low concentrations were also developed. In addition, computers able to handle the enormous amount of data that the WOCE would generate were made possible through advances in electronic manufacturing methods.

Perhaps the most important new instruments available for the WOCE were satellites capable of obtaining data on ocean circulation. In 1979, the Seasat satellite mission, lasting one hundred days demonstrated that detection of radar echoes and microwave radiation from the sea could be used to produce detailed information. After years of planning, the WOCE project was ready to begin collecting data in 1990.

Scientists from more than thirty nations participated in the many studies involved in the WOCE. In the United States, the headquarters of the WOCE is located at the Department of Oceanography at Texas Agricultural and Mechanical University. Data collection ended in 1998, but analysis of the information was expected to last until at least 2002, and many scientists expected processing of the data to last until at least 2005.

The first WOCE study began with the launching of the German research ships *Polarstern* and *Meteor* in 1990. These ships collected data in the southern part of the Atlantic Ocean between Antarctica and South Africa. Other early WOCE studies also concentrated on the Southern Hemisphere because this area had been studied in less detail prior to the WOCE than the Northern Hemisphere. Later WOCE studies moved into the Indian Ocean the North Atlantic Ocean, and the Pacific Ocean.

Satellites used by the WOCE included the ERS series launched by the European Space Agency, the TOPEX/POSEIDON, a joint project of France and the United States, and the Japanese ADEOS. More than one thousand drifting instruments, designed to remain at specific depths far below the surface of the sea, were also used. The movement of these instruments was measured by satellites or by sonic equipment. Tens of thousands of measurements were made at the surface of the ocean as well.

SIGNIFICANCE

The WOCE project is likely to be one of the most important sources of oceanographic data from the early twenty-first century. The official goals of the WOCE included providing a complete description of the general circulation of the ocean, creating a numerical model of ocean circulation for use in advanced computers, accounting for seasonal changes in ocean circulation, obtaining data on the exchange of substances between layers of water in the ocean, providing detailed information on the interaction between the ocean and the atmosphere, and obtaining data on the movement of heat within the ocean.

The most important application of WOCE data is in the study of the effect of ocean circulation on climate changes. This information aided scientists in predicting the effect on long-term weather patterns of various human activities, such as the increase in carbon dioxide in the atmosphere. Such data is also useful in predicting natural changes in climate that take place over years or decades.

Several examples of the interaction between ocean conditions and changes in weather are well documented. The Sahel, a region of Africa along the southern fringe of the Sahara Desert, experienced severe droughts in the 1970s and 1980s after having experienced much wetter conditions in the 1950s. These droughts were associated with higher-than-normal surface temperatures in the South Atlantic Ocean, Indian Ocean, and southeast Pacific Ocean, and with lower-than-normal temperatures in the North Atlantic Ocean and most of the Pacific Ocean. Ocean temperature is also an important factor in the formation of tropical cyclones and other powerful storms.

Currents have a powerful effect on weather patterns. The Gulf Stream-North Atlantic-Norway Current brings relatively warm tropical water northward, moderating the climates of eastern North America, Ireland, the British Isles, and the coast of Norway. The Kuroshio-North Pacific Current does the same for Japan and western North America. These warm currents also encourage increased water evaporation, resulting in increased rainfall in these areas. The Peru Current brings cold polar water northward along the western coast of South America, decreasing water evaporation and maintaining deserts in Peru and Chile. The Benguela Current, running northward along the western coast of Africa, has the same effect in Namibia.

Perhaps the best-known example of the effect of changes in ocean circulation on weather is the El NiZo phenomenon. This situation occurs at irregular intervals in the eastern Pacific Ocean. Increased water temperatures, typically 35.6 to 46.4 degrees Fahrenheit (2 to 8 degrees Celsius) higher than normal, are associated with changes in climate. Typical effects seen during an El NiZo condition are droughts in Australia, northeastern Brazil and southern Peru; excessive summer rainfall in Ecuador and northern Peru; severe winter storms in Chile; and warm winter conditions in North America.

The El Niño effect is also associated with large reductions in fish populations along the western coast of South America. During normal conditions, the water near the coast consists of a thin layer of warm, nutrient-poor water above a thick layer of cooler, nutrient-rich water. The top layer is thin enough to allow coastal upwellings to bring nutrients to the surface, supporting marine life. During El Niño conditions, the top layer of warm water is much thicker, preventing nutrients from reaching the surface. Data from the WOCE project are expected to aid in the prediction of climate changes such as El Niño, with the possibility of having a major impact on human activities.

—*Rose Secrest*

Further Reading

Broecker, Wally. *The Great Ocean Conveyor*. Princeton UP, 2010.

Brown, Joan. *Ocean Circulation: Prepared by an Open University Course Team*. Elsevier, 2016.

Field, J. G., Gotthilf Hempel, and C. P. Summerhayes. *Oceans 2020: Science, Trends and the Challenge of Sustainability*. Island Press, 2002.

Jacques, Peter. *Globalization and the World Ocean*. Rowman and Littlefield Publishing Group, 2006.

Jochum, Markus, and Raghi Murtugudde, editors. *Physical Oceanography: Developments Since 1950*. Springer Science + Business Media, 2006.

Joseph, Antony. *Measuring Ocean Currents: Tools, Technologies, and Data*. Elsevier, 2014.

Klinger, Barry A., and Thomas W. N. Haine. *Ocean Circulation in Three Dimensions*. Cambridge UP, 2019.

Miller, Robert N. *Numerical Modeling of Ocean Circulation*. Cambridge UP, 2007.

Siedler, Gerold, Stephen M. Griffies, John Gould, and John A. Church, editors. *Ocean Circulation and Climate: A 21st Century Perspective*. Elsevier, 2013.

Pedlosky, Joseph. *Ocean Circulation Theory*. Rev. ed. Springer, 2010.

Schmittner, Andreas, John C. H. Chjiang, and Sidney R. Hemming. *Ocean Circulation: Mechanisms and Impacts-Past and Future Changes of Meridional Overturning*. John Wiley & Sons, 2013.

Vallis, Geoffrey K. *Atmospheric and Oceanic Fluid Dynamics: Fundamentals and Large-Scale Circulation*. Cambridge UP, 2006.

van Aken, Hendrik Mattheus. *The Oceanic Thermohaline Circulation: An Introduction*. Springer Science + Business Media, 2007.

Y

Yellow Sea Ecosystem

Fields of Study: oceanography; marine biology; marine biogeochemistry

ABSTRACT

The Yellow Sea, located in the center of northeastern Asia, is one of the 25 major semi-enclosed seas in the world. It is surrounded by the East China Sea to the south, the Chinese landmass to the west and north, and the Korean peninsula to the east. It features two main bays, the Bohai Sea and Korea Bay.

KEY CONCEPTS

demersal fish: fish that normally inhabit deep waters
domestic sewage: sewage waste water from residential and convenience sanitary facilities
industrial wastewater: waste water from industrial processes and facilities

OVERVIEW

The Yellow Sea, located in the center of northeastern Asia, is one of the 25 major semi-enclosed seas in the world. It is surrounded by the East China Sea to the south, the Chinese landmass to the west and north, and the Korean peninsula to the east. It features two main bays, the Bohai Sea and Korea Bay.

Total surface area of the sea is 154,441 square miles (400,000 square kilometers) with a maximum length of 621 mi (1,000 km) and maximum width of 435 mi (700 km). The average depth of the Yellow Sea is 180 to 394 feet (55 to 120 meters), rather shallow compared to other seas in this part of the world.

The Yellow Sea contains more than 200 fish species, 140 phytoplankton species, a wide variety of zooplankton, some seabirds, and a few types of large marine mammals. In addition, aquaculture has been developed along the coasts of the sea. Yet, this unique marine system faces serious problems in maintaining sustainable development. Issues such as overfishing and pollution are continuously addressed by the three nations surrounding the sea.

BIODIVERSITY

The Yellow Sea is rich in marine resources. As the home of more than 100 species possessing commercial value, the Yellow Sea has been intensively exploited over the years. Among 200 fish species that have been found here, 45 percent are warm-water species, 46 percent are warm-temperate forms, and 9 percent are cold-temperature species.

In addition, the long-tailed crustaceans comprise an amazing 54 species, which include 65 and 35 percent of warmwater and boreal forms, respectively. The cephalopod group contains 14 species, which consist of nine warm-water forms and five warm-temperate forms.

The habitats of marine populations in the Yellow Sea can be divided into two groups, near-shore and migratory. Near-shore species include those that are mainly found in bays, estuaries, and around islands, such as skate, greenline, black snapper, scaled sardine, and spotted sardine. During winter, with colder water temperatures, these species generally move to deeper waters.

The migratory species such as small yellow croaker, hairtail, and Pacific herring respond to water temperature very actively and have distinct seasonal movements. During winter, some species, such as chub mackerel, Spanish mackerel, and filefish, even migrate out of the Yellow Sea and enter the warmer East China Sea.

Reflecting the complicated oceanographic conditions of the Yellow Sea, microbial communities in the sea are fairly complex with regard to species composition, spatial distribution, and community structure. The total number of phytoplankton species is 140, including 91 diatoms, 37 dinoflagellates, seven euglenoids, four silicoflagellates, and one of cryptomonads. Composition of these species in the Yellow

Sea has a distinct seasonal shift. In different seasons, different species dominate.

The overall biomass of zooplankton in the Yellow Sea is lower than that of adjacent seas; the species composition also varies with season and location. The most abundant group, copepods, accounts for 60 percent of the total species throughout the year. Zooplankton is a vital food for pelagic and demersal fish and invertebrates here.

Of marine mammals, the whales that have been observed in the Yellow Sea include the fin, humpback, and grey whale. They represent the remnant of far greater pods that used to migrate and breed here. Some marine mammals here are endangered species, including the black right whale, whitefin dolphin, Kurile harbor seal, and Japanese sea lion.

The Yellow Sea ecosystem also hosts seabirds. Two key types that breed off the eastern coast of China are the streaked shearwater (*Calonectris leucomelas*) and Bulwer's petrel (*Bulweria bulwerii*). Among the endangered seabirds of the China coast, the relict gull (*Larus relictus*) and the Chinese crested tern (*Sterna bernsteini*) live in the Yellow Sea region. On the Korea side, of 370 bird species identified in South Korea, 112 breeds in the Yellow Sea region, and 17 locations have been designated as protected breeding grounds.

The yellow croaker is one of the most commercially important deepwater species to Yellow Sea fisheries (Creative Life via iStock)

EFFECTS OF HUMAN ACTIVITY

Fisheries in the Yellow Sea represent a multinational business involving China, the Koreas, and Japan. The total catch increased to approximately 5.5 million tons (5 million tonnes) in 1984 from about 3 million tons (2.7 million tonnes) in 1970. Of this amount, only one-third to one-half is considered sustainable. The number of species that are commercially harvested is about 100, including crustaceans and cephalopods.

Demersal, or deepwater, species are the major component of the fishery resources and account for 65 to 90 percent of the annual total catch. The two most commercially important demersal species are small yellow croaker and hairtail. Because of increases in fishing effort and intensive fishing of spawning stock and young fish, catches of both species have been declining.

Overfishing has also caused serious declines in stock abundance of Pacific cod, flatfish, sea robin, red seabream, and white croaker. Surprisingly, the abundance of species such as cephalpods, skates, and daggertooth pike-congers seems to be unaffected by fishing pressure. Two possible reasons could be their scattered distribution or their strong adaptive nature.

Besides abundant natural resources, the Yellow Sea has tremendous value in terms of beautiful coastlines, aquaculture, and beaches for recreation. All coastal provinces of China along the Yellow Sea have marine aquaculture, which uses seawater to culture different species. Shangdong and Liaoning are the two provinces where the aquaculture enterprises are most advanced. Similarly, South Korea has developed its own aquaculture. In 1977, the total invertebrate mariculture produced 52 percent of the fish yield, or 280,865 tons (254,796 tonnes).

ENVIRONMENTAL THREATS

Pollution of the Yellow Sea, especially near shore, has been a serious problem for both China and South Korea. Before strict environmental laws were put into effect, both countries dumped industrial and domestic wastes directly into the sea. In addition, oil discharged from vessels, ports, and oil exploration was released into the sea.

In the western part of the Yellow Sea, where China is, more than 100 million tons (91 million tonnes) of domestic sewage and about 500 million tons (454 million tonnes) of industrial wastewater are discharged into the sea each year. The major pollutants are organic chemicals that absorb quantities of oxygen; heavy metals such as mercury, lead, and cadmium; oils; and inorganic nitrogen. On the eastern side of the Yellow Sea, the pollution level has been similar. However, due to the dynamic nature of the sea, which leads to strong and rapid mixing and biodegradation, the marine environmental quality has been measured as roughly normal. Over the past several decades, climate change—particularly the slowly rising average seawater temperature—seems to have played a role in various algal and jellyfish blooms, as well as species shifts in the overall population.

The Yellow Sea is believed to have great potential for future oil and gas exploitation in the seabed; this will produce a new challenge in maintaining the biologic health of the ecosystem. Overfishing and pollution have been the two major issues concerning the sustainable use of this precious sea up until now. Numerous agreements between nations and a great number of environmental laws and rules have been put into action. Ideally, human beings will support efforts to enforce those mandates and will add new elements to the campaign to protect the natural habitats of the Yellow Sea.

—*Yanna Liang*

Further Reading

Liang, X. San. *The Slow Coastal-Trapped Waves Off Subei Bank in the Yellow Sea and Their Climatic Change in the Past Decades.* IntechOpen, 2018.

Organisation for Economic Development and Cooperation. *OECD Territorial Reviews: Trans-border Urban Co-operation in the Pan Yellow Sea Region.* OECD, 2009.

Park, Chul. *Yellow Sea and East China Sea Reported by PICES and Korean Monitoring Program.* Chungnam National U, 2012.

Wang, Pinxian, Qianyu Li, and Chun-Feng Li. *Geology of the China Seas.* Elsevier, 2014.

Glossary

abyssal plain: the ocean floor between the edges of the continental shelves

active sensors: devices that emit a signal or stimulus to a target in order to determine specific characteristics of that target

actuator: a device that when activated causes a specific movement to occur

aerobic: requiring or in the presence of oxygen

aerobic digestion: breaking down of organic materials by bacteria that require the presence of oxygen to function

albedo: the percentage of incoming light that is reflected from Earth's atmosphere and surface

alewife: a non-native species of fish found in the Great Lakes

algorithm: a computing routine designed to identify and coordinate data points according to user-defined criteria

aliphatics: hydrocarbon compounds in which the molecular structures consist solely of single carbon-carbon and carbon-hydrogen bonds

alkylation: the addition of an alkyl group (a hydrocarbon segment) to an existing molecule

alluvial: describes a river or watercourse that carries and deposits its silt load in its associated floodplain

anadromous fish: fish species that split their life cycle between freshwater and salt water, typically spawning in freshwater then developing to maturity in salt water before returning to freshwater to spawn

anaerobic: requiring or in the absence of oxygen

anaerobic digestion: breaking down of organic materials by bacteria that do not function if oxygen is present or that do not require oxygen to function

angular momentum: the principle of physics by which a mass rotating about an axis tends to maintain its rotation in the same angular direction (*cf.* Newton's law of inertia, which states that an object in motion tends to stay in motion unless acted upon by an external force)

anoxic: containing no dissolved oxygen

anthropogenic: having been caused directly by or occurring as a result of human activities

aphotic zone: the lower part of the ocean water below the maximum depth to which any sunlight can reach

aquaculture: farming of plants and animals that inhabit marine environments, as opposed to agriculture, the farming of plants and animals that inhabit terrestrial environments

aquifer: a mass of water contained within an underground soil or rock structure, maintained from a recharge point and flowing toward a discharge point

archaea: the class of bacteria that includes the methane producers, the thermohalophytes, and other extremophilic bacteria

archipelago: a series of relatively small islands that are related to each other geologically

aromatics: hydrocarbon compounds in which the molecular structures are cyclic and contain a series of alternating single and double carbon-carbon bonds, which imparts greater stability and chemical reactivity to them relative to their aliphatic and non-aromatic counterparts

asthenosphere: the layer of mobile liquid rock in the mantle directly beneath the lithosphere

atmosphere: the gaseous layer that envelopes the planet

atoll: a generally circular coral structure typically surrounding the submerged top of a tropical oceanic seamount

atoll reef: a circular coral reef structure that typically builds around the edges of a sunken volcanic crater, although some atoll reefs form naturally without the presence of a volcanic substrate on which to grow

autochthonous: a synonym for 'native'

average salinity: the mathematical average of salinities found in different parts of the sea

bank: an extensive plateau-like region of seafloor that may be an extension of a continental shelf or a segment of continental rock that lies at some depth below the surface

barnacles: types of molluscs that attach themselves to submerged surfaces such as pier foundations, rocks, and the hulls of ships

barrage: a structure constructed across an river mouth or estuary with openings to house generators to be driven by the influx and eflux of water as the rising tide drives water to flow into the estuary and the falling tide allows the water to flow out again

basalt: the rock type that is formed when molten magma breaks through the ocean floor and solidifies to form new ocean floor

base flow: the minimum amount of water that normally flows in a watercourse or aquifer

basement rock: the most fundamental rock layer of Earth's crust, typically ocean floor basalt

basin: an area of deep sea bottom defined by rough boundaries presented by continental shelves and island structures

bathysphere: a spherical steel chamber devised by William Beebe and Otis Barton, capable of descending to a depth of 900 meters

bathythermograph: a ship-board device that measures and records water temperatures at depths, and does not require the ships to remain stationary

benthic: referring to the lowest depths of the ocean, the sea floor

benthic organisms: organisms that live on or very near to the sea bottom

benthos: the lowest depths of the oceans, the seafloor

bioaccumulation: increasing amounts of toxic materials stored in body fats and organs progressing upwards in a food chain

biodegradation: the breaking down of out-of-place materials by living organisms and metabolic processes

biodiversity: the variety of plant and animal organisms found within an ecosystem

bioerosion: erosion or structural breakdown caused by the activity of living organisms

biofouling: fouling of marine devices by the build-up of algae, barnacles and other biological entities, leading to disruption of the optimum functioning of the particular device

biogenic: produced by or as the result of activity by living organisms

biological oxidation: a method of treating organic wastes by reliance upon living aerobic bacteria and microorganisms to consume the material

biome: the flora, fauna and mineral structures existing within a particular defined environment

bioremediation: the use of specific bacteria and other organisms with specific abilities to 'clean up' polluted areas by consumption of environmental contaminants

biosphere: a term encompassing all life of Earth, whether plant or animal in nature

biostratigraphy: the assignment of geological period and age to geologic strata through identification of their fossil content

biota: the flora and fauna of a particular ecological niche or environment

biotic zones: areas within a particular environment where life is supported

black smoker/white smoker: seafloor hydrothermal vents; those spewing hotter water quickly produce clouds, like smoke, of inky black sulfide minerals, those spewing less hot water produce white clouds of mineral particles as their ejection contacts the cold water of the ocean deep

bleaching: the loss of colored symbiotes from coral polyps, leaving the innate white appearance of the coral structures

bolide: a stony object from space, whether meteor or comet, that strikes Earth's surface

bora: cold, dry winds that blow into Europe from the mountains ranges to the east

bore: a distinct wave produced by rapid increase in water level in a tapering channel

bottom trawling: a method of capturing and removing large numbers of fish by indiscriminately dragging nets over the sea bottom

boundary currents: currents that border the eastern and western ranges of a gyre

brine channels: small tunnel-like openings found in sea ice, caused by the incomplete formation and erosion of ice as it forms from saline sea water

buffer zone: the area and structure of a shoreline that tends to absorb and mitigate the energy of incoming waves

by-catch: the undesired aquatic species that are collected along with the target species in fishing operations

caldera: the residual crater remaining after the top of a magma chamber has collapsed following an eruption

Calvin cycle: a metabolic pathway of photosynthesis occurring in the stroma of chlorophyll-containing chloroplasts, during which carbon dioxide becomes fixed to an enzyme and is reduced to form carbohydrates

capillary force: the effect of the intermolecular forces between water molecules that are in narrow channels and interparticle pores

capillary waves: the first and smallest identifiable waves to form when wind blows across still surface water

capture fisheries: any fishery in which fish are captured by some means and removed from their aquatic environment for use as food and other purposes

carbon cycle: the path of carbon in the environment involving carbon dioxide, organic plant and animal matter and carbonate rock, ultimately returning to carbon dioxide to complete the cycle

carbon-14 dating: analysis of the ratios of carbon isotopes in a once-living and carbon-containing materials to determine the age of the material according to the mathematics governing radioactive decay processes and the half-life of carbon-14

carbonate compensation depth: the depth in the ocean at which carbonate becomes highly soluble due to the increased pressure and cold temperature

carbonic acid: H_2CO_3, a weak acid produced by the reaction of carbon dioxide (CO_2) and water (H_2O)

carnivorous fish: fish that eat other fish and aquatic animal species

cathodoluminescence: the emission of light following irradiation by a beam of electrons

caviar: preserved and processed sturgeon roe (eggs) considered a great delicacy; having very high commercial value, this has driven Caspian Sea sturgeon almost to extinction due to poaching and overfishing

cementation: the binding of relatively large particles into a solid whole by the formation of crystalline structures in the spaces between the particles

chemical weathering: a mode of erosion in which materials are broken down by chemical reaction processes occurring with water solutions such as acidic rain, or by simple extraction of water-soluble components

chemosynthesis: a natural process of producing carbohydrates and other compounds from carbon dioxide and water that uses energy derived from chemical reduction reactions rather than sunlight

chemosynthetic communities: communities of organisms that derive energy from chemosynthesis rather than from photosynthesis

Chilean sea bass: a commercial name used to hide the identity of the fish as the protected, and therefore illegally caught, Antarctic cod

chlorophyll biomass: the quantity of chlorophyll-bearing algae and plankton that appear in a body of water

chromatography: a means of physically separating components of a solution by using the differences in their abilities to stick to interact temporarily with a suitable solid phase while being eluted through it by a suitable fluid phase

cladistics: also called phylogenetic analysis, a method that recognizes and orders characters in an objective fashion to show the most likely evolution of organisms through time

clastic: produced from mineral sources rather than by living organisms, the opposite of biogenic

clathrate: a non-bonded compound that forms when water molecules freeze into a cage-like str4ucture about a gas molecule such as methane

climate change: the observed principle that Earth's average temperature is increasing over time and the global climate is adjusting to the increase in average temperature

climate forcing: the difference between energy received from sunlight and energy radiated back to space, the imbalance forcing the temperature of the atmosphere to rise or fall depending on whether the difference is positive or negative

coastal plume: a stream of colder water that flows into and beneath warmer water, carrying with it terrogenic materials and nutrients

coelenteron: the central tube-like structure found in many invertebrate species, which is essentially their digestive tract

cold zone: a thermally isolated band of water at a depth below the surface that is considerably colder than the water both above and below it

column stratification: denotes the various temperature and salinity layers in a body of water; the water above any particular point of the seafloor is called the 'water column'

compressible fluids: fluids that can change volume by the application of external pressure, particularly gases

condensate: a mixture of light hydrocarbon compounds that separate or 'condense out' during the processing of natural gas

condensed matter: matter that has definite quantifiable properties as a solid, liquid or gas

conjunction: when the sun and the moon are on the same side of Earth and the three bodies are positioned in a straight line with each other

constructive interference: a phenomenon of wave physics in which the coincidence of two or more waves adds together both their peak height and energy, resulting in the formation of a much larger wave; applies to waves of all types

consumptive use: any use or application of water that removes the water from the immediately available supply

contact exchange of atoms: as different materials contact each other, atoms from one material may change places with atoms from the other material; in isotopic studies the exchange of atoms of the same element occurs, which may exchange one isotope for another in the process of isotopic fractionation

contact sensors: devices that collect ocean and other data by being in physical contact with the water or another appropriate target

continental drift: term describing the drifting movement of the continents of Earth's crust driven by convection currents in the underlying magma of Earth's mantle

continental rise: the region of gradually rising seafloor from the abyssal plain in the direction of the continental shelf

continental shelf: the extension underwater of the continental mass

convectional precipitation: precipitation resulting from the elevation of humid air in an atmospheric convection cycle

convective overturning: an event in which colder, denser water forms above less dense, less cold water, resulting in an inversion of relative position as the colder water sinks and the warmer water rises

convergent boundary: the boundary where the edges of tectonic plates are pushed towards each other by underlying magma currents

coordinate system: a system of designating the position of a point in space relative to a fixed standard origin; e.g.: latitude, relative to the equator, and longitude, relative to the meridian of Greenwich

coral bleaching: a stress response of corals that causes symbiotic zooxanthellae to be expelled from within the coral polyps, leaving them essentially colorless against the white calcium carbonate structure of the reef

Coral Triangle: an approximately triangular region of the Pacific Ocean in which coral reefs are the primary ecosystem

coralivore: a creature that consumes coral polyps for its primary sustenance

core barrel: the hollow cylindrical opening within the first section of a drill string, which also functions as the drill bit; typically contains a plastic liner that encases the core sample as it is produced

core rings: areas of still ocean water trapped between two branches of a surface ocean current

core sample: a cylindrical rod of material recovered either from a corer, a piston corer or a circular drill corer; the sample exhibits the strata of the material within the sediments or rock that has been sampled

Corexit: a chemical dispersant used to break up oil from oil spills and cause it to disperse into the water

Coriolis effect: the apparent deflection of the linear motion of air and water currents due to Earth's rotation

corrosion: the degradation of metal through oxidation processes that are chemical or biochemical in nature

covalent bond: a connection between two atoms in which each atom 'shares' an electron from the other atom in order to form a more stable electron configuration around its nucleus

coy: a thermal gradient that develops in response to the different rates at which the land and the water absorb energy from the Sun

creep: gradual downslope movement of soil under the force of gravity, typically identified along gullies and other depressions by the presence of trees that lean toward the downslope side or that have grown with a distinct upward bend from the downslope direction

crown-of-thorns starfish: an aggressive, voracious species of starfish that feeds exclusively on coral polyps; attempts to kill the starfish by catching it and chopping it to pieces have historically failed because each of the pieces subsequently grows and develops into a new starfish if they are simply tossed back into the sea

crude oil: as the name implies, the cruse oil material, petroleum, that is taken from oil wells before any processing

cryosphere: the portion of Earth consisting of ice and snow

cryptocrystalline: consisting of submicroscopic crystals

crystallization: the formation of a solid material from a solution, either by combining dissolved ions to form an insoluble material, or by exceeding the solubility of the material in that solvent under the existing conditions

current: a directional flow of water within a body of water such as a sea or ocean, driven by wind or as thermohaline circulation

cyclonic: atmospheric and oceanic circulation that is naturally counterclockwise in direction

datum (pl: data): a datum is a fact or a true condition with respect to a particular subject; the term usually refers to something that is measurable

daughter isotope: in the process of radioactive decay, an unstable 'parent' isotope releases part of its nucleus, producing a daughter isotope of a lower atomic number

dead zone: a region within an ocean or other body of water where life is not supported and so does not exist

decadal: refers to phenomena that recur on an essentially regular ten-year cycle

deep: a relatively broad expanse like a deep valley in the ocean floor

definite integral: the value of the parent function of a derivative determined as the difference between two specific values of the parent function: eg. The definite integral of $y = 2x$ between $x = 4$ and $x = 2$ is $4^2 - 2^2 = 16 - 4 = 12$. The integral notation specifies the starting and ending values of the variable of integration.

demersal fish: fish that normally inhabit deep waters

dendrochronology: a system of determining the absolute date of an artifact by relating it to the identifiable dates of growth rings in trees

density: generally defined as mass per unit volume, the standard being the density of pure water at 1 gram per cubic centimeter

density currents: submarine flows of dense mixtures of water and sediment particles, another name for turbidity currents

density stratification: the formation of identifiable layers in ocean water according to density

depositional lobes: surface formations in sea fan topography, formed when smaller branch channels direct the flow of turbidite material onto an existing sea fan

derivative function: a function that describes the change of another 'parent' function with respect to a particular variable: eg. The derivative of $y = x^2$ is $y' = 2x$

desertification: the conversion of normally fertile and arable land into desert, occurring by the expansion of an existing desert, or by prolonged dryness and lack of precipitation resulting in desert formation

diagenesis: the process whereby sediment particles are transformed into rock through pressure and the formation of opaline microcrystalline matter in the pores between the sediment particles

diking: construction of impervious walls to separate normally water-covered stretches of land from the ocean, enabling water to be pumped away from the sequestered area, rendering it suitable for land uses such as agriculture or building development

directional drilling: a method of well-drilling that allows control of the direction in which the drill head will cut

disarticulation: the separation of skeletal parts from each other by various means, typically due to predation and environmental activity such as flooding and landslides

discharge: the amount of water flowing past a particular reference point in a specific amount of time

disphotic zone: the region between the euphotic and aphotic zones, where sunlight can still be detected but is insufficient to support photosynthesis

distillation: a process for separating components from a mixture such as crude oil by vaporizing and collecting the portions, or 'fractions', according to their individual boiling points

diurnal: high and low tides occurring once in a period of one lunar day

diurnal tide: high and low tides that occur once per lunar day

divergent boundary: the boundary between two tectonic plates that are being driven away from each other by underlying magma currents

dodecahedron: a geometric shape having twenty (ie: *dodeca*) individual faces, as a soccer ball does

domestic sewage: sewage waste water from residential and convenience sanitary facilities

doughnut hole: a relatively small area of international waters between the exclusive economic zones of the nations bordering the Bering Sea

downcurrent: an abbreviation of 'downstream current', refers to locations farther downstream of a current from any arbitrary reference point

drill string: the long line of attached pipes that connects the drill head to the engine that drives the boring apparatus

drilling mud: an engineered mixture used in drilling of oil wells, designed to stabilize the piping in the bore hole against subterranean pressures while also cooling and lubricating the drillhead

dynamic positioning: the process of actively monitoring and continually adjusting the position of a drilling platform ship above the bore hole by using stationary reference points, made much easier in the present day through computer-controlled thrusters and GPS

dynamite: a form of powerful explosive once based on nitroglycerine but now based on trinitrotoluene (TNT) and other compounds that are not at shock-sensitive as nitroglycerine and therefore safer to handle

Earth system science: an approach to the many disciplines of science, in which Earth is viewed as a single entity composed of many interrelated systems

echolocation: the identification of size and distance by the generation of sound waves and measurement of the time lag and frequency of their reflection from an object

echo-sounding: the use of sonography, or sonar, for the specific purpose of measuring the vertical depths in a body of water to produce a topographical map of the bottom

ecosystem: an identifiable environmental region, including the plants, animals and mineral base that it encompasses and the manner in which they interact

ecotope: a particular ecological zone

effusive eruption: a volcanic eruption in which dense magma is not forcefully ejected but simply flows or oozes out of fissures

eider duck: a species of duck (aquatic fowl) renowned for the softness and insulating properties of its down feathers; the down feathers are used to line the nests in large quantities and are easily harvested for use for use in insulated dry goods

Ekman layer: roughly the uppermost 100 meters of ocean, where the water is affected by air movement

Ekman spiral: a downward spiral movement of water from the ocean surface to depths under the influence of trade winds and the Coriolis effect

electric dipole: a condition that exists when the distribution of negative charge (electrons) in a molecule is not symmetrical

electrolysis: the process by which ions are extracted from solution by an electric field to be deposited on a collector as solid metals. Because atoms have different electrical properties, different voltages can extract different materials

electronegativity: term for the propensity of a particular atom to gain or surrender electrons relative to another kind of atom

elevation isobar: a line marking a specific elevation about a feature on a topographical map

El Niño: a quasi-regular phenomenon that brings abnormal quantities of warm surface water across the Pacific Ocean toward South America and subsequently affects climate globally

emersion: removal of the aqueous medium that normally surrounds an aquatic species, as when the tide goes out; the opposite of submersion

empirical: describes any system based on direct observation

endemic: refers to species that have evolved and developed in adaptation to a specific area

endemic species: species that have adapted to and evolved within a specific region

endemism: the occurrence of endemic species unique to a particular environment or ecosystem

epilimnion: the upper level of a body of water, where the water becomes warmed by sunlight

epiphytes: plant organisms that use other plants for physical support

Equatorial countercurrent: a current that exists between the northern and southern equatorial currents and moves in the opposite direction, returning surface water that has been moved by the equatorial currents

equatorial currents: currents flowing from east to west across oceans on either side of the equator, as part of their respective gyres

erosion: the movement or displacement of soil and larger particles by the action of a fluid in motion, such as by wind or water in motion

estuary: locations where a stream or river of freshwater meets and flows into an ocean or other large body of saline water

eukaryotic cells: cells in which the genetic material DNA and RNA is carried within a distinct nucleus

euphotic zone: the upper layer of ocean water to the maximum depth to which there is sufficient sunlight to support photosynthesis

eustatic sea level change: a global increase or decrease in mean sea level

eutrophic: oxygen-poor water that contributes to a reduced ability to support life, such as cold water upwelling into a body of warmer water

eutrophication: stagnation of a body of water by the intrusion of large quantities of organic matter, especially algae, which consumes the dissolved oxygen in the water, thereby rendering it incapable of supporting a healthy ecosystem

exclusive economic zone (EEZ): the area of ocean waters extending from the shore of a nation to a distance of 322 kilometers (200 miles), which can only be harvested by fishers and other industries from the corresponding nation

extensometer: a device that measures the strength of an Earth tide by measuring the tension it causes in a taut wire

facial recognition: a computer function in which optical analysis and data manipulation algorithms enable a device to recognize and identify human faces

fauna: the animal life found in a particular region, whether terrestrial or marine

fetch: the extent of open water the wind pushes a surface wave before striking a beach

fish emulsion: a fertilizer produced from the fluid remainder of fish already processed for fish oil and fish meal

fish ladder: a structure built in association with a dam that redirects a small amount of the river water to cascade down over a series of stages, allows fish to ascend to the higher level of the dam by leaping up stage by stage to the top

fish meal: a high-protein by-product of rendering and processing fish for fish oil, used as a food supplement in aquaculture

fish oil: natural oils obtained from fish by various means of processing

fishing down the food web: the almost stepwise shift of capture fishing from species at the top of the food web to species at increasingly lower trophic levels

fissure flow: typically refers to lava that flows relatively smoothly from a long fissure vent rather than being ejected from a volcano

fjord: a deep, narrow canyon-like coastal inlet with steep sides that was typically formed by glacial action

fjordic: resembling high, steep-walled ravines like the fjords of Norway

flocculant: a chemical or combination of chemicals that when added to water form a substance that will aggregate pollutant materials and so facilitate their removal

flocculation: the process of binding suspended particles in a colloid

flooding: the condition in which water has surpassed its normal environmental containment features, ranging from stream banks to the ocean's verge

floodplain: the area bordering a river or watercourse that will become inundated by flood waters

flora: the plant life found in a particular region, whether terrestrial or marine

foundation species: species that the major food web of a particular ecosystem is dependent upon for its maintenance

frazil ice: a loose formation of small, plate-like ice crystals that forms on the surface of freshwater under suitable conditions; in Arctic sea waters becomes compacted into pack ice, but can also be seen in terrestrial areas like Yosemite National Park where it can flow haphazardly over flooded ground like thick slush

frequency: the characteristic of a sound, defined as the number of complete cycles of the sound frequency in one second, stated in hertz

friction: the physical interaction between two surfaces or boundary layers, that transfers energy from one to the other in both directions and may generate heat in the process

fringing corals: corals that build structures that develop around the perimeter of a small rocky island

fringing reef: a coral reef that builds parallel to a shore rather than as an isolated structure such as an atoll

fringing reef: reef structures that develop and grow parallel to a shoreline

frontal precipitation: precipitation resulting from the interaction of warm and cold fronts

fumarole: a volcanic vent that emits steam and other gases, but not lava

gabion: a large, usually rectangular, metal cage filled with rocks, used to construct erosion control structures

geodesy: measurement of the shortest distance between two points on Earth's surface

geomorphic: term for the three-dimensional physical shape of a geologic feature

geomorphology: the study of how and why the shapes of geologic features change over time

geostrophic current: an oceanic current in which the force of downslope movement of an elevated mass of surface water is balanced by the Coriolis force

geothermal: literally, 'Earth heat', describing phenomena that are driven by the heat of Earth's interior

Gondwonaland: one of two supercontinents formed by the sundering of the original supercontinent Pangaea about 200 million years ago

Gondwanan fragment: a fragment of continental rock remaining from the Gondwana supercontinent that existed about 190 million years ago

goniometer: a device used in the study of crystal structures for the measurement of angles between the faces of a crystal

GPS: the Global Positioning System, an array of satellites in geosynchronous orbit for the purpose of identifying the position of any point on Earth's surface by triangulation

gravimeter: a device that measures the strength of Earth tides by measuring the small change in Earth's gravity

gravity corer: a device, essentially a length of pipe, that is dropped vertically to the bottom of a body of water in order to recover a vertical sample of bottom sediment

greenhouse effect: increase of the average temperature of the global atmosphere due to the action of so-called 'greenhouse gases'

greenhouse gas: an atmospheric gas capable of absorbing infrared radiation (heat) coming from Earth's surface and reemitting a significant portion of it back into the atmosphere rather than out into space

groin: solid structures that are perpendicular to the shoreline and extend into the water below normal wave heights as a means of controlling longshore drift

guyot: an underwater volcanic mountain whose top has been sheared off by the action of waves, leaving a submerged mesa

gyre: the slow, roughly circular movement of ocean water over the expanse of the particular ocean

halligen: a low island in the Wadden Sea

halocline: the intermediate layer or interface between a lower layer of water of higher salinity and density, and an upper layer of water of lower salinity and density

hard coastal engineering: engineering and construction of solid, man-made structures such as docks and ports, walkways, fences, etc., structures that are not part of the natural structure of coastlines

head: the vertical difference between low and high tides at a certain location

headwater: the collective term for the area that begins the watershed of a particular watercourse, typically the farthest points from the exit point of the watercourse

heat budget: the balance between the incoming energy from the Sun and the outgoing energy radiating back into space from Earth

heave compensator: a system that functions to prevent the drillhead and piping from rising and falling in the borehole as waves raise and lower the drilling platform ship

heavy metals: metals, such as cadmium, lead, mercury and others from the same region of the periodic table, that are readily absorbed and bioaccumulated within a trophic chain where it can cause neurological and other health issues

hertz: the basic frequency unit, defined as one complete cycle of the frequency signal

hot spot: a region of Earth's mantle where a column of extraordinarily hot magma rises toward the overlying crust at the same relative location, bringing about the formation of volcanic islands

hundred-year flood: the magnitude of a flood that has a one-in-one-hundred chance of occurring in any given year

hydrate: any molecule or compound that has the component elements of water in its structure

hydrated minerals: minerals that contain water as part of their crystal structure or that is entrained within the overall structure of the mineral without being part of the crystal structure

hydrated crystals: minerals that incorporate water molecules either as an intimate part of their crystal structure or as entrained water

hydrocarbons: organic compounds consisting only of carbon and hydrogen atoms, the products of the petroleum industry

hydrogen bonding: not a true bond, but a pronounced attractive force between the unshielded positive charge on the nucleus of a covalently-bonded hydrogen atom of one molecule and the exposed negative charge of the outermost electron pair(s) on a covalently-bonded atom (typically a nitrogen or oxygen atom) of another molecule

hydrogen sulfide: having the molecular formula H_2S; as a gas, familiar as the compound responsible for the odor of rotten eggs; dissolved in water, acts as a weak acid, but can also react with dissolved oxygen to form the much stronger sulfuric acid molecule H_2SO_4; this reaction acts to deplete the amount of dissolved oxygen in the water

hydrologic cycle: the cyclic system of evaporation, condensation and precipitation of water on Earth

hydrology: the study of natural water movement in the environment

hydrosphere: the portion of Earth consisting of liquid water

hydrothermal mineralization: the deposition of mineral deposits from hot water solutions from natural sources, such as hot springs and undersea themal vents

hydrothermal plume: a column of relatively warm water rising through the cold water column above a hydrothermal vent or other source of heated water

hydrothermal vents: seafloor structures that emit volcanically superheated mineral-laden water, found at mid-seafloor ridges where two tectonic plates are being slowly driven apart

hypolimnion: the lower level of a body of water, where the water is not warmed by sunlight

hypoxic: containing low quantities of dissolved oxygen relative to normal oxygenation levels

igneous rock: rock such as granite and basalt that is formed by crystallization of molten rock

immiscible liquids: two liquids that do not dissolve in each other due to their different chemical natures, for example oil and water

inclined flume: a kind of trough in which the nature of material flows such as turbidity currents can be studied, adjustable to different slope angles

incompressible fluids: fluids that do not change volume with the application of external pressure, particularly liquids

indigenous species: species that have always existed and evolved in a particular region rather than having adapted to that region

individual transferable quota: an allotment of a portion of the allowable annual catch from the Bering Sea, which can be purchased by individual fishers who exceed their allotted quota from those who have not fulfilled their quota

industrial wastewater: waste water from industrial processes and facilities

infalling: objects from space that are captured by Earth's gravity and eventually fall to the surface of the planet

infiltration capacity: the ability of a soil to absorb water rather than allow it to flow away as surface runoff

infrared energy: electromagnetic radiation having wavelengths longer than red in the visible portion of the electromagnetic spectrum, experienced as 'heat'

infrasonic: the designation of sound frequencies that are below the lower limit of human hearing

integral function: generally, the parent function of a derivative function: eg the integral function of $y = 2x$ is $y = x^2 + c$, where c is a constant value with no derivative function

interbedding: formation of a layer, or bed, of one kind of sedimentary rock between two beds of sedimentary rock of a different nature

interception capacity: the ability of different types of foliage to capture rainfall

intertidal area: the strip of terrain that becomes flooded during a high tide and then bared during a low tide

intertidal flat: a broad, flat area of land that is under water at high tide and exposed at low tide

intertidal zone: the area of a shore that is under water at high tide and exposed at low tide

ion: an atom or molecule that contains either more or less electrons than it contains as a neutral atom or molecule and thus carries either a negative or a positive electric charge respectively

ionic bond: a strong attraction between two ions carrying opposite charges, forming a rigid connection that does not involve sharing of electrons

inundation: the depth to which a tsunami or other water flood covers the area it has flooded

isomerization: changing the structure of a molecule to that of another molecule with the same chemical formula

isostasy: the vertical raising or lowering of areas of Earth's crust

isostatic rebound: the gradual uplifting of an area of Earth's crust that has been depressed by the weight of glacial ice that has been on top of it

isostructural: describes the unit cell structures of two crystalline materials having the same angular orientation of crystal faces in the unit cells

isotopes: atoms of a single element containing the same number of protons but different numbers of neutrons in the nucleus

isotopic fractionation: the relative enrichment and depletion of isotopes of the same element in two different materials through contact exchange of atoms

isotopic ratio: the relative amounts of different isotopes of an element in a sample; the ratio may be used to determine the age of the material in the sample according to the mathematics of radioactive decay processes

isotopic tracing: using the relative abundances of specific isotopes in rocks to determine the timeline of their formation

juvenile water: water that has been newly released from magma and has not yet become part of the different aspects of the water cycle; very rarely observed

Kara-Bogaz-Gol: one of the two bays in the Caspian Sea

karst: a landscape of limestone, characterized by surface structures and the presence of cave formations that have been deeply eroded into the landscape by the action of mildly acidic water

karstification: the formation of karst landscape through erosion of limestone beds by acidic rainfall and wind

krill: small, shrimp-like crustaceans that swarm the Antarctic waters and other cold water regimes in vast numbers and form the base of the food web for larger creatures ranging from tiny fish to blue whales

laminar flow: flow in which there is no turbulence in the fluid as it flows, as all particles making up the fluid flow uniformly in parallel paths, or 'lamina'

landlocked: any body of water or land region that does not have an ocean shoreline or shipping ports

lava: molten rock that has been ejected from a volcano or other volcanic source

leachate: the solution of soluble materials that is produced as water percolates through a solid layer such as soil or landfill

linear: describing the manner in which one property, such as pressure, changes with respect to another property, such as depth or altitude

liquefaction: an effect of seismic activity that causes groundwater to separate from between the pores of soil, turning the solid soil into a mobile slurry

lithosphere: generally termed the crust, consisting of the solid outermost rock layer and the very viscous molten layer between the crust and the asthenosphere

littoral zone: the area of shoreline between high and low tide levels

live-fish trade: capture of desirable and colorful reef fish in the wild for sale to the aquarium market

local sea level change: any change of mean sea level, usually temporary, that occurs in a limited area rather than globally

long-line fishing: the commercial fishing practice of running out kilometers-long lines carrying several baited hooks for the purpose of harvesting ocean fish

longshore current: a current that is produced by wave action on a shoreline and that flows parallel to that shoreline

lotic: refers to water that is moving, that has a current flow

luminescence: the property of emitting light

luminophors: particles that have been treated artificially so that they can be detected in the environment

lunar day: a period of 24 hours and 50 minutes, the length of time for the moon to complete one orbit around Earth

lysocline: the physical region of ocean water at the carbonate compensation depth in which carbonate is actually dissolved

machine learning: computer programs for robotic control that contain algorithms designed to register situations encountered by the robot that prevent or interfere with carrying out an instruction; learning 'by experience'

magma: molten rock contained within Earth's crust and that has not been ejected onto the surface

magma buoyancy: the level of a mass of magma relative to the larger body of magma according to its density

magma chamber: a void within the crustal rock at the site of a volcano, where rising magma collects and pools

magmatic differentiation: decomposition of mineral rock through melting in Earth's mantle layer, resulting in the release of water molecules that have been incorporated into the molecular structures of the minerals or that have been entrained and trapped within the rock as the mineral formed

magnetometer: a device that detects and measures the embedded magnetic 'signature' embedded in underlying rock structures

manometer: a device used to measure differences in pressure between two systems

map: a symbolic representation of anything that may be described as a terrain, such as the surface of the planet, a range of connected ideas, etc.

mariculture: fish farming and other pursuits in which 'crops' are raised in the sea rather than on land

marine snow: the remains of dead planktonic and other creatures that constantly falls through the water to the seafloor

marker fossils: fossils of specific types and species that can be coordinated to identifiable geologic times and terrains

marsh accretion: the building up of sediments over time that gradually rtaises the bottom of the marsh to higher levels

mass spectrometer: an apparatus used to identify the mass of atoms introduced into it, by providing them with an electrical charge and measuring their detection as they pass through a magnetic field

mass wasting: the downslope loss of material mass due to creep, slip, slump or landslide

meander: a current that branches away from the flow of the main current and may or may not rejoin the main current to form an ox-bow if the current is terrestrial or an eddy circle if the current is oceanic

measurement: determination of the magnitude of a physical property relative to a fixed unit standard for that property

megafauna: various species of large animals

mercury: a silvery, liquid metal at room temperature, one of the 'heavy metals' that is neurotoxic to animals and bioaccumulates up the food chain; mercury poisoning fell under the spotlight in the 1960s as "Minamata disease" due to the high number of cases observed among the fish-consuming populace of the Minamata Bay region of Japan

messenger weight: an object that slides down the adjacent wire of a water-sampling line to trip the latches retaining the sample bottles at different depths, allowing them to capture water samples and be recovered

methane: the simplest of the hydrocarbon molecules, having the molecular formula CH_4, about sixty times more potent than carbon dioxide as a greenhouse gas

methane hydrate: a type of compound in which methane molecules have become associated with water molecules to form a solid complex under the influence of cold water temperatures and high pressure on the sea floor

metrology: the study of methods, practices and applications of measurement

microplastics: tiny fragments of plastics that find their way into the environment from normal human activities such as washing clothes made with synthetic polymer fibers

microtidal: characterized by very low tide heights

mid-ocean ridge: an undersea mountain range situated approximately down the middle of the Atlantic Ocean where tectonic activity causes molten magma to ooze up as the seafloor slowly spreads away to east and west

millimole: simply, one one-thousandth of a mole, in which the mole is defined as the weight of a pure element or compound that corresponds to the number of identical particles equal to the Avogadro constant of 6.02259×10^{23}; accordingly, 1 mole of ^{12}C atoms weighs exactly 12 grams

mineralization: the deposition of mineral deposits from natural aqueous solutions, whether hot or cold

modeling: the use of computer programs that employ data inputs, calculations and algorithms to describe conditions and predict the behaviors of a particular phenomenon that changes with respect to one or more variables

mole: defined as a quantity of individual particles such as atoms, molecules, subatomic particles or ions equal in number to the Avogadro constant

mousse: a stable mixture of oil and water as a rough emulsion

mud volcano: volcanic vents that emit mud rather than lava

multiscalar: involving study and analysis at scales ranging from the minutiae to the broad physical area in which the minutiae are found

nanoparticulates: solid particles less than 1 micron in size, such as particles of lead oxide produced in the combustion of leaded gasoline and soot from burning wood or coal

navigation: basically, the practice of finding one's way from point A to point B by relating direction to fixed points of reference

nekton: term for all creatures that move through the water by their own self-propulsion, as opposed to merely floating

neural network: an artificial construct whose component pieces function in imitation of the interconnected neurons of the human brain

nomograph: typically a one-page document that shows a graphical representation of the relationship between a cause and effect, such as rate of surface evaporation and ambient temperature

non-consumptive use: any use or application of water that permits immediate re-use of the water in the same or another application, or that retains the water in the immediately available supply

non-point source: a pollution source in for which no specific source can be identified, such as surface water run-off from fields over a broad area

nutrient loading: the introduction of higher-than-normal quantities of nutrients such as fertilizer or organic waste into a body of water

ocean acidification: a reduction of the pH of ocean water due to the formation of carbonic acid and other acidic compounds as carbon dioxide and other gases are absorbed from the atmosphere by ocean water

ocean conveyor belt: a global circulation system that ties all individual ocean currents into one continuous current system

ocean warming: increase in the average temperature of ocean waters, locally or globally, as a reflection of the corresponding increase in average atmospheric temperature due to climate change or local environmental influences

one atmosphere: the standard pressure experienced at Earth's surface

opposition: when the sun and moon are on opposite sides of Earth and the three bodies are positioned in a straight line with each other

organophosphorus compounds: organic molecules that include a phosphorus-based component in their molecular structure; highly toxic, and typically interact with biochemical systems such as the central nervous system

orographic precipitation: precipitation caused by the rising of humid air to regions of cooler temperatures as it approaches and surmounts a mountain range

oscillation: movement of a property or a structure between two values or positions that are determined by the the parameters of the physical system

osmosis: the passage of water molecules through a semipermeable membrane from the region of lower salinity to the region of higher salinity

outgassing: the release of trapped or entrained gases, either directly or by vaporization

overfishing: harvesting species of fish in quantities that exceed the ability of the population of the target species to maintain its numbers

oversteepening: the buildup of sediment beyond its natural angle of stability, typically resulting in a sudden collapse of the too-steep side

oxygen: O_2, a biatomic molecule that aerobic biological systems use for production of biochemical energy

oxygen catastrophe: a mass die-off of anaerobic life forms, essentially the only life on Earth at the time, due to poisoning by the rapidly forming abundance of oxygen in the atmosphere as photosynthesis became prominent

oyster bed, oyster reef: an area of shallow sea bottom that supports colonies of oysters

ozone: a highly reactive molecular form of oxygen consisting of three oxygen atoms rather the more stable diatomic form

pack ice: the uniform layer of solid ice that covers the Arctic waters in the cold season

paleoceanography: the study of oceanography of Earth's oceans in the distant past through the records left in geological formations and how that relates to Earth's oceans in the present

paleomagnetism: the magnetic 'signature' of Earth's magnetic field that is acquired by seafloor basaltic rock as it cools from its molten state and solidifies

Pangaea: a supercontinent that existed some 250 million years ago, when all of the current continental masses were conjoined in a single, giant continent

parabolic: any two-dimensional motion or shape that can be described mathematically by an equation having the form $y = ax^2 + b$

passive sensors: devices that monitor signals originating from a source rather than sending a signal and measuring the response of a target

pattern: a collection of individual items that are related by a specific function or property

pedosphere: the layer of fertile earth or non-rock overlying the lithosphere

pelagic: refers to the open water of the oceans above the abyssal plains

peninsula: a long, relatively narrow protrusion of land into a body of water

permafrost: a contraction of permanent frost, the name given to soils that remain well frozen throughout the entire year

petrographic microscope: an optical microscope that uses a plane-polarized light source to illuminate very thin slices of mineral material, which normally enables identification of the mineral

pH: a relative measure of the acidity or alkalinity of a water solution, defined as the negative logarithmic value of the hydrogen ion concentration in the solution

photic zone: the upper part of the ocean that can be penetrated by sunlight

photogrammetry: the practice of reproducing a three-dimensional representation or mapping of an area from two-dimensional photographic images

photosynthetic compensation point: the depth to which sunlight reaches in ocean water where the rate of formation of products of photosynthesis is balanced by the rate of consumption of those products in respiration

phytoplankton: small marine organisms that rely on photosynthesis for life; essentially green plants composed of as little as one cell

pibal: a pilot weather balloon that is typically followed by direct observation to determine wind speeds and directions at altitude

piezoelectric: crystalline material that produces an electrical output in response to a physical stimulus

piracy: 'seaway robbery', in which a group of heavily-armed individuals use high-speed boats to accost a commercial vessel in open ocean waters, then board the vessel by force to take control of the ship and its crew; crew quarters and cargo are typically ransacked for valuables, and at times the cargo, crewmembers and even the ship itself may be held for ransom

piston corer: a modified type of gravity corer that can recover bottom sediment samples of much greater depth

placer: a deposit of sand or gravel containing particles of gold or other valuable minerals that can be washed out readily

plane-polarized light: light for which the normal radially-symmetric electromagnetic vibration perpendicular to the direction of motion has been restricted to a single plane by reflection from a surface or by passing through a polarizing filter

plankton: term for organisms that float in the water and are incapable of self-propulsion

platform reefs: reefs forming a relatively long, broad structure that is of a consistent height throughout its extent

plinian eruption: an explosive eruption that ejects a large, towering column of smoke, ash and other ejecta that suddenly collapses and falls to ground as a pyroclastic flow; so named for the description of the eruption of Mt. Vesuvius recorded first-hand by Roman historian Pliny in 79 BCE

plume: a rising column of superheated magma within Earth's mantle

pneumophores: plant organs that do not enter the ground, instead drawing in water and nutrients from the air

point source: a pollution source that can be identified as the specific source of pollution, such as an effluent discharge pipe from a specific factory

polarity: the existence and direction of a dipole in a molecule

polarizer: a filter that polarizes light that passes through it

pollutant: essentially, any material that is not normally found in a natural space, whether by identity or quantity

polychlorinated biphenyls (PCBs): a family of synthetic organic compounds that have been produced industrially in mass quantities to make use of their high thermal stability, lack of electrical conductivity and inert chemical reactivity; molecular structures of PCBs consist of two connected benzene rings bearing chlorine atoms instead of hydrogen atoms

polymorphism: a phenomenon in which a single compound or material may have different intrinsic forms in nature

polynya: a limited area of open water that does not freeze over within an unbroken ice pack

positive feedback loop: any process in which the presence of its product acts to increase the effect or output of the process

practical salinity: the salt content of water as determined by electrical conductivity, as opposed to absolute salinity which is determined as the weight of dry salts residue obtained from a specific volume of ocean water

predation: the exploitation of 'prey' species for food by 'predator' species

pressure: the force exerted uniformly by a fluid over all parts of a surface, commonly due to the weight of the overlying air or water column

proprioception: in humans, the innate and automatic location and position of all parts of the body; in robots, the same function carried out through active self-monitoring

public trust: the principle by which natural resources are deemed to belong to the people and are administered as a trust by governments

quantify: to assign a standardized numerical value to a property in order to enable comparison to other values of that property and relation of the particular property to other properties

radar: an electronic method of detection at a distance by measuring the elapsed time between transmitting a radio frequency signal and receiving the reflected signal at a detector

radioactive pollution: the presence of materials such as uranium and plutonium that emit dangerous X-rays and decay particles; such materials are believed to have been disposed of in the Russian area of the Sea of Japan as a by-product of nuclear research efforts

radiocarbon dating: a technique that uses the ratio of the carbon isotope ^{14}C (carbon-14) in a non-living material of organic origin to the ratio of ^{14}C that is maintained within the material while it was living

radionuclides: atoms that are radioactive isotopes of various elements that are used in different nuclear industries

raster mapping: portraying map data as as an array of rectangular sections, with the sections having equal dimensions but different data values, like an image made by using different background colors in the cells of a spreadsheet

rawin: a weather balloon carrying a radar target to allow its progress to be monitored by radar rather than by direct observation

red tide: a type of algal bloom consisting of dynoflagellate organisms that produce a toxic red pigment rather than the green chlorophyll of other algae

reef raiding: illegal removal of reef fish and other reef organisms, whether for food, the coral market or the aquarium trade

remote sensors: devices that collect ocean and other data without being in physical contact with the water or another appropriate target

revetment: a solid structure consisting of stones or other materials that acts to prevent erosion of beach sands as waves break on and recede from the shoreline

Ring of Fire: the name given to the coastal regime of the Pacific Ocean due to its volcanic character and many active volcanoes

riparian lands: the lands beside freshwater water-courses

river discharge: the freshwater entering a saline body as outflow from lakes, rivers and streams

riverine: indicates that water and other materials that flow into the oceans are carried there by freshwater rivers

rogue wave: also called a 'monster wave', a wave of unusual size formed as a result of constructive interference

run-up: the distance to which a tsunami or other water flood encroaches beyond the normal shoreline

salinity: the content of dissolved salt(s) in water

salinity count: the concentration of dissolved salt stated relative to the amount of water, as for example 17 parts per thousand, corresponding to a salt solution of 0.017%

saltation: the motion of particles under the influence of wind or water, by which individual particles bump into particles that lie ahead of them and essentially 'leapfrog' over them

salinometer: a device that measures the salinity of a water sample by determining its electrical conductivity, which is then related to its saline composition

scale: the ratio of the depiction of a feature to the actual physical size of the feature

scirocco: warm, moist winds that originate in northern Africa and blow across the Mediterranean Sea into Europe

scuba: acronym for Self-Contained Underwater Breathing Apparatus, typically refers to the use of the equipment rather than just the equipment itself

sea ice: ice that has formed from saline sea water, typically refers to the broken remnants of the pack ice in the warm season

sea sawdust: the red-colored *Trichodesmium* algae

seafans: an undersea avalanche, and the manner in which the debris is scattered over the ocean floor with the heaviest, densest particles traveling the furthest from the site of collapse and the fine silt being spread out in the water before settling; the name comes from the appearance of the materials after they have settled

seafloor spreading: the movement of the seafloor to the east and west of a mid-ocean ridge driven by convective currents in the underlying magma of Earth's mantle layer

seagrass: collective term for a variety of plant species that grow on shallow sea bottoms in large colonies

seal rock: an impermeable rock layer that serves to trap petroleum or natural gas within a lower substrate rock

seamount: the mesa-like subsurface remnant of an undersea volcanic mountain

seasonality of light: the variation in hours of daylight received each day throughout the year

seawater stratification: the formation of discernible layers in the ocean water as characterized by differences of density due to salinity or temperature

Secchi disk: a round, white disk of standard size that is used to measure the depth to which it can still be seen from the surface

seiche: a resonance wave, which can be very large, that forms in an enclosed or partially enclosed body of water

seismic reflection: a system that uses sound impulses and their reflection from underlying rock structures to identify the stratigraphy of an area

seismometer: a device that detects vibrations in Earth's crust due to earthquakes and other seismic activity

semidiurnal: high and low tides occurring twice in a period of one lunar day

semi-enclosed body of water: one that is bordered on three 'sides' by land, thus connecting to the ocean by a single relatively narrow opening

semipermeable membrane: a porous membrane in which the pores are large enough to allow water molecules to pass through, but not large enough to allow dissolved salts to pass through

sensor: an electrical, magnetic, mechanical or combination device that detects an environmental condition and provides that as input to the robotic control mechanism

sewage: untreated water containing human and animal wastes (urine and feces) as well as other organic solid and liquid wastes

sinkholes: essentially a vertical cave formation produced when acidified water seeps through an underlying limestone bed into an underground void, causing the roof of the void to collapse

slip: the sudden downslope movement of material that has sheared off from the face of a structure or that has otherwise become unstable, the typical landslide or avalanche

slump: a failure of slope integrity in which a large body of soil shears under gravitational stress and moves suddenly downslope with a rotational motion, characterized by an arc-shaped shear face

sofar channel: an oceanic layer between two salinity layers that have the effect of trapping and propagating sound such that it travels long distances without significant loss of energy

soft coastal engineering: engineering and construction that seeks to protect or restore the natural structures of a coastline, such as beach 'nourishment' and sand dune stabilization

solar plane: the plane with the Sun at its center in which the orbits of the major planets are found

solifaction: melting of water in frozen soils such as permafrost, turning them from hard solids to muddy liquid slurries

sonar: similar to radar, but uses a sound pulse traveling through water rather than a radio frequency signal

speech recognition: a computer function in which sound analysis and data manipulation algorithms enable a device to be activated and controlled by the sound of one's voice

spring constant: a unique, constant value, inherent to each particular spring, that determines the input force needed to deflect the spring; formulated in Hooke's Law equation as $F = -kx$ where F is the force required, x is the extent of deflection and k is the spring constant

stable isotopes: isotopes of an element that do not decay over time

stable isotope ratios: the relative concentrations of stable isotopes of a particular element in a sample; the ratio of stable isotopes can be used to determine different time-related properties such as the age of the material and the amount of water that has been in contact with the material over time

stoichiometry: determination of the proportions of the materials with which complete chemical reactions take place

strait: a relatively narrow passage between two landmasses, typically with a current flow of water passing through; straits may be between two islands or groups of islands, between an island and a continental mass, or between two continental land masses

stratigraphy: study of the rock types and ordering of different layers, or strata, of underlying rock structures

subduction: a consequence of seafloor spreading, in which seafloor being displaced from the mid-ocean ridge eventually encounters a less dense continental mass and is forced beneath it and back into the mantle layer

subduction zone: the area at which a segment of the dense basaltic ocean floor is driven against a less dense continental mass by the action of seafloor spreading and is forced downward into the mantle; characterized by seafloor trenches, and capable of producing massive earthquakes and volcanic activity

subgyre: a smaller gyre adjacent to a main gyre

submarine topography: the shape and changes of seafloor depth in a body of water

submersion: the return of the aqueous medium that normally surrounds an aquatic organism, as when the tide comes in; the opposite of emersion

subsidence: the lowering of land area when material is removed from some supporting subsurface structure such as an aquifer, causing the land above it to sink to a lower position relative to sea level

surficial: in relation to Earth's surface

sympagic: indicates a close relationship with ice

synchronized rotation-revolution: the situation in which a body in orbit rotates about is axis at the same rate as its orbital rotation about its orbital center, as the Moon rotates on its axis at the same rate that it orbits Earth

tablemount: a large, flat-topped undersea rock structure similar to guyots and banks

taiga: the evergreen forests typical of northern regions

taphonomy: the science of the processes that lead to fossilization

tectonic plates: segments of Earth's crust somewhat analogous to the shell of an egg that has cracked in several places yet maintains the egg's shape

teleoperation: operation of a device from a distance, by remote control

terpen: an artificially raised area of ground on a halligen, created as a stable base for construction of various buildings and to maintain those buildings above water level at high tides

terrigenous: refers to marine particulate material that originated on land, such as dust and sand particles

thermal maturation: essentially, cooking; the acquisition of certain characteristics according to the time and temperature to which something has been exposed

thermocline: a virtual surface between warmer and colder water, where there is an abrupt change from one temperature to the other

thermohaline circulation: vertical circulation of an ocean current due to differences in temperature and salinity of the water

tidal wave: the long waveform of the tides following the gravitational 'pull' of the moon around Earth along the equator (NB: this is NOT a tsunami)

tide pools: low spots or catchments in the intertidal zone that are refreshed and replenished with water during high tide and then retain that water during low tide

tidal zonation: also called vertical zonation, identifiable zones of characteristic organisms corresponding to different water levels in the intertidal zone

tide: the periodic rising and falling of ocean water under the influence of the gravitational force of the Moon and the Sun

tide mill: a mill constructed at the shore of the ocean and powered by the rise and fall of the tides

topographical map: a two-dimensional map that shows the relative elevations of three-dimensional features, typically as elevation isobars

topography: the three-dimensional features of the surface of Earth's crust

tracers: the chemical materials used to treat particles to create luminophors

trade winds: winds to the north and south of the equator that typically blow in an east-to-west direction, so named because sailing ships on trade missions used these winds to sail to the west

trench: a relatively narrow and more sharply defined deep region of the ocean floor where the edge of a tectonic plate of the basaltic ocean floor meets a less dense continental mass and is forced downward into the mantle layer, often the site of massive earthquakes and volcanic activity

triangulation: a method of locating a specific location based on the relationships of angles and the fact that the interior angles of a triangle always add up to a total of 180 degrees

trigonometry: the mathematics of angles and their measurement, particularly of triangles

trophic cascade: the result of a single change in the trophic chain (the food web and species distribution) of an environment, often as the result of a single change to one small part of the trophic chain

trophic level: a step in the ordering of a food chain or food web; at the highest level are the top predator species, and at the lowest level are the 'bottom dweller' prey species, while in between are various species that are both predator and prey species

trough: a long, deep depression in the sea bottom, formed as sea floor tectonic plate sections move away from each other over time as is the case for the Cayman Trough which also has a developing mid-trough ridge

tsunami: a large, dangerous wave produced by sudden, severe displacement of water caused by earthquake, landslide or volcanic eruption; often erroneously called a tidal wave

tundra: treeless expanses that extend to the north from the taiga regions, typically overlying permafrost

turbidite: a sedimentary deposit on the ocean floor, in which particle size graduates continuously from the largest and densest particles at the bottom of the bed to the smallest and least dense at the top of the bed

turbulence: non-laminar flow, characterized by chaotic motion within the fluid, as the flowing motion of particles making up the fluid is disrupted and forced to flow in non-parallel paths, or 'lamina'

turnover: an exchange of level as surface water that has been cooled by cold winter air becomes denser and drops to a level below the warmer and less dense water beneath it

typhoon: the name for weather events in the Pacific Ocean that are called hurricanes in the Atlantic Ocean

ultrasonic: the designation of sound frequencies that are beyond the upper limit of human hearing

upwelling: the rise to the surface of relatively cold, nutrient-rich water from depths

upwelling cell: a site of cold water upwelling that is not associated with a current rising up as it encounters a continental mass, appearing as an isolated area of upwelling either due to atmospheric disturbance at that location or by diversion of the flow of a deep ocean current encountering an undersea obstacle such as a seamount or guyot

uranium-series radiometric dating: analysis of the ratios of elemental isotopes that occur in the radioactive decay of uranium atoms in order to determine the age of the material according to the mathematics governing radioactive decay processes and the half-lives of radioactive uranium isotopes

urban heat island effect: an elevated average temperature over the area covered by a city due to the generation and enhanced absorption of heat in that area

vacuum distillation: removal of a liquid component of a solution or mixture by boiling or evaporating it under reduced pressure and collecting the distillate by condensation

vector mapping: portraying map data as an array of points, each having a unique value and directon derived from the data itself, such as elevation relative to a set base level

Venturi principle: the phenomenon by which a fast-moving stream of a fluid draws more fluid along with it as it flows, as a flow of water under pressure will draw surrounding water along with it

Venturi tube: tubing used in undersea mining and dredging in which an injected stream of water under pressure causes more water to flow into the open intake end of the tube, so that the tube acts to 'suck up' material from the sea floor that is then ejected from the other end of the tube for collection

visbreaking: a mild thermal cracking process applied in processing of crude oil to reduce the viscosity of the material and produce fuel oil and lighter fractions and increase the distillate yield in a refinery

viscosity: the temperature-dependent property relating the ability of different fluids to flow

volatile: able to enter the gas phase by evaporation or sublimation

volcanic hot spot: a position on Earth's crust that is directly above a superheated plume of magma in Earth's mantle layer, generally giving rise to constant emission of lava, as occurs in Hawaii; the Hawaiian Islands are a string of volcanic islands formed by emission of lava as the tectonic plate moved across the location

wastewater: all of the liquid waste, including the solid waste within it, that is produced by a community

water cycle: the cycle of phase changes that water undergoes in the natural environment, from solid and liquid to gas and back to liquid and solid (snow, ice and liquid water to water vapour and back to condensation, rain and snow)

waterlogged: indicates that the soil contains more water than it is capable of absorbing

wave orbit: the circular motion of water in a wave

wave-powered pump: a pump system in which the power to drive the pump mechanism is provided by the up-and-down motion of waves

wave refraction: an effect of bottom topography that waves running parallel to the shoreline to break on an angle against the shore

wind farm: an array of large windmills that 'harvest' the energy of wind to drive the blades of the mill, causing them as they turn to drive a generator that converts the energy of the wind to electrical energy

wind stress: directional surface pressure exerted against surface water by wind causing the surface water to move in the same direction

x-ray diffraction: the refraction of a beam of x-rays from the atoms within a crystal to produce a characteristic pattern of refracted beams that is then used to determine the relative orientations of the atoms making up that crystal

x-ray fluorescence: when x-rays of a specific energy impinge upon a mineral, the atoms of the mineral absorb the energy and then re-emit it as x-ray energies that are characteristic of the particular atoms

zonation: the graduation of plant species from the most water-tolerant in the wettest part of a marsh to dryland plants outside of the marsh environs, according to the nature of the terrains in the intervening area

zooplankton: small marine organisms that rely on consumption of phytoplankton and other zooplankton for life; essentially animals composed of as little as one cell

zooxanthellae: colored symbiont organisms that exist within the interstitial spaces between the cells of coral polyps; expulsion of these organisms due to stress results in coral bleaching

FURTHER READING

Abaimov, S. G., et al. "Earthquakes: Recurrence and Interoccurrence Times." *Pure and Applied Geophysics*, vol. 165, 2008, pp. 777-95.

Abdelly, Chedly, et al. *Biosaline Agriculture and High Salinity Tolerance*. Birkhauser Basel, 2008.

Abel, P. D. *Water Pollution Biology*. 2nd ed., Taylor, 1996.

"About Our Great Lakes: Ecology." *Great Lakes Environmental Research Laboratory*. NOAA Great Lakes Environmental Research Laboratory, n.d. Accessed 22 Aug. 2016.

Abuzinada, Abdulaziz H., Hans-Jürg Barth, Friedheim Krupp, et al. *Protecting the Gulf's Marine Ecosystems from Pollution*. Birkhauser, 2008.

Ackerman, Steven A., and John A. Knox. *Meteorology: Understanding the Atmosphere*. 4th ed., Jones, 2015.

Adams, Thomas E., and Thomas C. Pagano, editors. *Flood Forecasting; A Global Perspective*. Academic Press, 2016.

Adraneda, Katherine. "Scientists: The Philippines Is the World's 'Center of Marine Biodiversity.' Announce Alarm Over Threats." *Underwater Times*, 8 June 2006, www.underwatertimes.com/news.php?article_id=64017239105.

Agius, Dionisius A., Emad Khalil, Eleanor M.I. Scerri, and Alun Williams, editors. *Human Interaction with the Environment in the Red Sea*. Brill, 2017.

Ahluwalia, V. K. *Advanced Environmental Chemistry*. TERI, 2017.

Ahr, Wayne M. *Geology of Carbonate Reservoirs*. John Wiley & Sons, 2008.

Ahrens, C. Donald. *Meteorology Today: An Introduction to Weather, Climate and the Environment*. 8th ed., Thomson Brooks-Cole, 2007.

Ahrens, C. Donald. *Meteorology Today: An Introduction to Weather, Climate, and the Environment*. 10th ed., Brooks, 2013.

Ahuja, Satinder, editor. *Chemistry and Water: The Science Behind Sustaining the World's Most Crucial Resource*. Elsevier, 2017.

Akulichev, Viktor Anatol'evich. *Far Eastern Seas of Russia*. Nauka Press, 2007.

Al-Azab, M., W. El-Shorbagy, and S. Al-Ghais. *Oil Pollution and Its Environmental Impact in the Arabian Gulf Region*. Elsevier, 2005.

Albarede, Francis. *Geochemistry: An Introduction*. 2nd ed., Cambridge UP, 2009.

Alberotanza, Luigi. "Active and Passive Remote Sensors as a New Technology for Coastal and Lagoon Monitoring." *Aquatic Conservation*, vol. 11, no. 4, 2002, pp. 267-72.

Aldrete, Gregory S. *Floods of the Tiber in Ancient Rome*. Johns Hopkins UP, 2007.

Aleshire, Peter. *Ocean Ridges and Trenches*. Chelsea House, 2007.

Alexandrov, Vitaly Y., Thomas Martin, Josef Kolatschek, et al. "Sea Ice Circulation in the Laptev Sea and Ice Export to the Arctic Ocean: Results from Satellite Remote Sensing and Numerical Modeling." *Journal of Geophysical Research*, vol. 105, no. C7, July 2000.

Almahasheer, Hanan. "Red Sea Mangroves Fight Back in the Face of Global Decline." *HuffPost*, 28 Feb. 2017, www.huffingtonpost.com/entry/red-sea-mangroves-fight-back-in-the-face-of-global_us_58b5b36de4b02f3f81e44ce2.

Alosairi, Y., et al. "Mixing and Flushing in the Persian Gulf (Arabian Gulf)." *Journal of Geophysical Research*, vol. 116, 2011.

American Geosciences Institute, "What Is Atmospheric Pressure?" www.americangeosciences.org/education/k5geosource/content/weather/what-is-atmospheric-pressure. Accessed 19 Mar. 2018.

American Museum of Natural History. *Ocean*. Dorling Kindersley Limited, 2006.

American Society of Civil Engineers, Hurricane Katrina External Review Panel. *The New Orleans Hurricane Protection System: What Went Wrong and Why*. ASCE, 2007. *ASCE Library*, doi:10.1061/9780784408933. Accessed 6 Sept. 2017.

American Society of Mechanical Engineers. *Pressure Measurement*. American Society of Mechanical Engineers, 2010.

American Water Works Association. *Desalination of Seawater: Manual of Water Supply Practices. M61*. Author, 2011.

Amirahmadi, Hooshang, editor. *The Caspian Region at a Crossroad: Challenges of a New Frontier of Energy and Development*. St. Martin's Press, 2000.

Anderson, John D., Jr. *A History of Aerodynamics and Its Impact on Flying Machines*. Cambridge UP, 1997.

Anderson, William C. "A History of Environmental Engineering in the United States." *Environmental and Water Resources History*, edited by Jerry R. Rogers and Augustine J. Fredrich, American Society of Civil Engineers, 2003.

Andrews, John, and Nick Jelley. *Energy Science: Principles, Technologies, and Impact*. Oxford UP, 2007.

Annin, Peter. *The Great Lakes Water Wars*. Island Press, 2018.

Annual Report 2017. Ocean Energy Systems, International Energy Agency, report2017.ocean-energy-systems.org. Accessed 29 Mar. 2018.

Anthony, Abigail, et. al. "Coastal Lagoons and Climate Change: Ecological and Social Ramifications in US Atlantic and Gulf Coast Ecosystems." *Ecology and Society*, vol. 14, no. 1, 2009.

Antonelli, Gianluca. *Underwater Robots*. 3rd ed., Springer, 2014.

Aquado, Edward, and James E. Burt. *Understanding Weather and Climate*. 7th ed., Prentice, 2015.

"Aquastat: Overview: Water Use." *Food and Agriculture Organization of the United Nations*, www.fao.org/aquastat/en/overview/methodology/water-use/. Accessed 21 Nov. 2019.

Archer, David. *Global Warming: Understanding the Forecast*. 2nd ed., Wiley, 2012.

Archer, David. *The Global Carbon Cycle*. Princeton UP, 2010.

Arias, Andres Hugo, and Sandra Elizabeth Botté. *Coastal and Deep Ocean Pollution. CRC Press, 2020.*

Arup, Tom. "Great Barrier Reef Coral Bleaching More Widespread Than First Thought, Surveys Reveal." *Sydney Morning Herald*, 31 Mar. 2016, www.smh.com.au/environment/great-barrier-reef-coral-bleaching-more-widespread-than-first-thought-surveys-reveal-20160331-gnux7x.html.

Askeland, Donald R. *The Science and Engineering of Materials*. 3rd ed., Chapman & Hall, 1998.

"Atmospheric Scientists, Including Meteorologists." *Occupational Outlook Handbook*. Bureau of Labor Statistics, 17 Dec. 2015, www.bls.gov/ooh/life- physical-and-social-science/atmospheric-scientists-including-meteorologists.htm.

Audubon Alaska. "Chukchi Sea." ak.audubon.org/chukchi-sea.

Austin, Jay, and Steve Colman. "A Century of Temperature Variability in Lake Superior." *American Society of Limnology and Oceanography*, vol. 53, no. 6, 2008, pp. 2724-30.

Australian Government. "About the Reef." *Great Barrier Reef Marine Park Authority*, 2016, www.gbrmpa.gov.au/about-the-reef. Accessed 22 Aug. 2016.

Avallone, Eugene A., Theodore Baumeister III, and Ali M. Sadegh. *Marks' Standard Handbook for Mechanical Engineers*. 11th ed., McGraw-Hill, 2006.

Ayers, Jonathan. *Introduction to Environmental Chemistry*. Callisto Reference, 2019.

Bailey, Geoff, Nina Galanidou, Hans Peeters, Hauke Jöns, and Moritz Mennenga. *The Archaeology of Europe's Drowned Landscapes*. Springer International Publishing, 2020.

Baird, Colin, and Michael Cann. *Environmental Chemistry*. W. H. Freeman, 2008.

Baker, Victor R., editor. *Catastrophic Flooding: The Origin of the Channeled Scabland*. Dowden, Hutchinson & Ross, 1981.

Bakun, A. "The California Current, Benguela Current, and Southwestern Atlantic Shelf Ecosystems: A Comparative Approach to Identifying Factors Regulating Biomass Yields." In Sherman, K., Alexander, L.M. and Gold, B.D. eds. *Large Marine Ecosystems: Stress, Mitigation, and Sustainability. Washington, DC: Am. Ass. Advancement of Science, pp. 199-224.*

Bale, Rachael. "One of the World's Biggest Fisheries Is on the Verge of Collapse." *National Geographic*, 29 Aug. 2016, news.nationalgeographic.com/2016/08/wildlife-south-china-sea-overfishing-threatens-collapse/.

Ballard, Robert D., and Malcolm McConnell. *Explorations: My Quest for Adventure and Discovery Under the Sea*. Hyperion, 1995.

Ballard, Robert D., editor. *Archaeological Oceanography*. Princeton UP, 2008.

Bamforth, Tom. *The Rising Tide: Among the Islands and Atolls of the Pacific Ocean*. Hardie Grant Publishing, 2019.

Banner, Adrian. *The Calculus Lifesaver: All the Tools You Need to Excel at Calculus*. Princeton UP, 2007.

Banyopadhyay, Jayanta. *Water, Ecosystems, and Society*. Sage Publishing, 2009.

Barange, M., et al., editors. *Marine Ecosystems and Global Change*. Oxford UP, 2010.

Bar-Cohen, Yoseph, and Kris Zacny, editors. *Drilling in Extreme Environments: Penetration and Sampling on Earth and Other Planets*. Weinheim, Wiley-VCH. 2009.

Bardi, Jason Socrates. *The Calculus Wars: Newton, Leibniz, and the Greatest Mathematical Clash of All Time*. Thunder's Mouth Press, 2006.

Bardi, Ugo. *Extracted. How the Quest for Mineral Wealth Is Plundering the Planet: The Past, Present and Future of Global Mineral Depletion. A Report to the Club of Rome*. Chelsea Green Publishing, 2014.

Bardintzeff, Jacques-Marie, and Alexander R. McBirney. *Volcanology*. 2nd ed., Jones & Bartlett Publishers, 2000.

Barendse, R. J. *The Arabian Seas: The Indian Ocean World of the Seventeenth Century*. Routledge, 2016.

Barr, Brian R., Marni E. Kooperman, Cindy Deacon Williams, et al. *Preparing for Climate Change in the Klamath Basin*. National Center for Conservation Science & Policy and the Climate Leadership Initiative, 2010.

Barry, Roger G., and Eileen A. Hall-McKim. *Polar Environments and Global Change*. Cambridge UP, 2018.

Barry, Roger, and Thian Yew Gan. *The Global Cryosphere: Past, Present, and Future*. Cambridge UP, 2011.

Bartlett, Darius, and Louis Celliers. *Geoinformatics for Marine and Coastal Management*. Taylor & Francis, 2019.

Bashkin, V. N., and Robert W. Howarth. *Modern Biogeochemistry*. Kluwer Academic, 2002.

Bass, Henry E., and William J. Cavanaugh, editors. *ASA at Seventy-Five*. Acoustical Society of America, 2004.

Batongbacal, Jay L., and Porfirio M. Aliño. "More Than Just Oil in the West Philippine Sea." *Philippine Daily Inquirer*, 25 June 2011.

Baturin, G. N. *The Geochemistry of Manganese and Manganese Nodules in the Ocean*. Springer Science & Business Media, 2012.

Batzer, Darold P., and Rebecca R. Sharitz, editors. *Ecology of Freshwater and Estuarine Wetlands*. 2nd ed., U of California P, 2014.

Baxter, Roberta. "Maintaining Vertical: Techniques for Slope Stabilization." *Erosion Control*, vol. 11, no. 2, Mar.-Apr. 2004.

Beamish, Richard J. *Impacts of Climate and Climate Change on the Key Species in the Fisheries in the North Pacific*. North Pacific Marine Science Organization, 2008.

Beamish, Richard J., and Brian J. Rothschild, editors. *The Future of Fisheries Science in North America*. Springer, 2009.

Beard, James M. *Environmental Chemistry in Society*. Taylor & Frances, 2017.

"Beaufort Sea Integrated Management Area." *Fisheries and Oceans Canada*, Government of Canada, 28 Oct. 2017, www.dfo-mpo.gc.ca/oceans/management-gestion/beaufort-eng.html.

Beckman, Daniel W. *Marine Environmental Biology and Conservation*. Jones & Bartlett, 2013.

Bedard, Roger, et al. "An Overview of Ocean Renewable Energy Technologies." *Oceanography*, vol. 23, no. 2, June 2010, pp. 22-31.

Bedient, Philip B., Wayne C. Huber, and Baxter E. Vieux. *Hydrology and Floodplain Analysis*. 5th ed., Prentice-Hall, 2013.

Beets, J., L. Muehlstein, K. Haught, and H. Schmitges. "Habitat Connectivity in Coastal Environments: Patterns and Movements of Caribbean Coral Reef Fishes with Emphasis on Bluestriped Grunt, *Haemulon sciurus*." *Gulf and Caribbean Research*, vol. 14, no. 2, 2003.

Beiras, Ricardo. *Marine Pollution: Sources, Fate and Effects of Pollutants in Coastal Ecosystems*. Elsevier, 2018.

Belessiotis, Vassilis, Soteris Kalogirou, and Emmy Delyannis. *Thermal Solar Desalination: Methods and Systems*. Elsevier, 2016.

Bell, F. G. *Basic Environmental and Engineering Geology*. Whittles Publishing, 2007.

Bellwood, Peter, editor. *The Global Prehistory of Human Migration*. Wiley Blackwell, 2013.

Beltran, Carmenza Duque, and Edisson Tello Camacho, editors. *Corals in a Changing World*. IntechOpen, 2018.

"Beluga Whale." *NOAA Fisheries*, www.fisheries.noaa.gov/species/beluga-whale. Accessed 27 July 2018.

Bender, Oliver, et al., editors. *Geoinformation Technologies for Geocultural Landscapes: European Perspectives*. CRC Press, 2009.

Benedict, Robert P. *Fundamentals of Temperature, Pressure, and Flow Measurements*. 3rd ed., John Wiley & Sons, 1984.

Benson, Keith Rodney, and Philip F. Rhebock. *Oceanographic History: The Pacific and Beyond*. U of Washington P, 2002.

Benton, Michael J. *Vertebrate Palaeontology*. 4th ed., Wiley, 2015.

Benton, Michael J., and David A. T. Harper. *Introduction to Paleobiology and the Fossil Record*. Wiley, 2009.

Berger V. and S. Dahle, editors. *White Sea-Ecology and Environment*. Derzhavets Publishers, 2001.

Berger, Joel. *Extreme Conservation: Life at the Edges of the World*. U of Chicago P, 2018.

Bergmann, Melanie, Lars Gutow, and Michael Klages, editors. *Marine Anthropogenic Litter*. Springer Open, 2015.

Berkman, Paul Arthur, and Alexander N. Vylegzhanin, editors. *Environmental Security in the Arctic Ocean*. Springer, 2012.

Bernard, Eddie N., and Allan R. Robinson. *The Sea, Volume 15: Tsunamis*. Harvard UP, 2009.

Bertness, Mark D. *Atlantic Shorelines: Natural History and Ecology*. Princeton UP, 2006.

Beyer, Robert T. *Sounds of Our Times: Two Hundred Years of Acoustics*. Springer, 1999

Bigg, Grant R. *The Oceans and Climate*. 2nd ed., Cambridge UP, 2003.

Birkeland, Charles. *Coral Reefs in the Anthropocene*. Springer Netherlands, 2016.

Birkle, Peter, and Ignacio S. Torres-Alvarado, editors. *Water-Rock Interaction: Proceedings of the 13th International Conference on Water-Rock Interaction WRI-13, Guanajuato, Mexico, 16-20 August, 2010*. CRC Press, 2010.

Bischoff, James L., and David Z. Piper. *Marine Geology and Oceanography of the Pacific Manganese Nodule Province*. Springer Science & Business Media, 2013.

Bittinger, Marvin L., et al. *Calculus and Its Applications*. 11th ed., Pearson Education, 2016.

Bjorndal, Trond, and Gordon R. Munro. *The Economics and Management of World Fisheries*. Oxford UP, 2012.

Bjorno, Leif. *Applied Underwater Acoustics*. Elsevier, 2017.

Black, Brian, and Gary J. Weisel. *Global Warming*. Greenwood, 2011.

Bland, Alistair. "Ending the Ecologically Harmful Capture of Tropical Fish." *Pacific Standard*, 9 Jan. 2018, psmag.com/environment/tropical-fish-collection-kills-coral-reefs.

Blatt, Harvey, and Robert J. Tracy. *Petrology: Igneous, Sedimentary, and Metamorphic*. 3rd ed., W. H. Freeman, 2005.

Blatt, Harvey, Robert J. Tracy, and Brent Owens. *Petrology: Igneous, Sedimentary, and Metamorphic*. 3rd ed., W. H. Freeman, 2005.

Blomberg, Lindsey E. "Dead in the Gulf." *The Environmental Magazine*, vol. 22, no. 5, 2011.

Blomberg, Lindsey E. "The Great Pacific Garbage Patch." *Environmental Magazine*, vol. 22, no. 3, 2011.

Blueprint of Coastal Wetland Conservation and Management in China. Paulson Institute, 2016, www.paulson institute.org/wp-content/uploads/2016/08/Wetland-Report-EN-Final.pdf. Accessed 30 July 2018.

Blumberg, Alan F., and Michael S. Bruno. *The Urban Ocean: The Interaction of Cities with Water*. Cambridge UP, 2018.

Bockheim, J. G., and A. N. Gannadiyev. "Soil-Factorial Models and Earth-System Science: A Review." *Geoderma*, vol. 159, nos. 3-4, 2010, pp. 243-51.

Boggs, Sam, Jr. *Petrology of Sedimentary Rocks*. Cambridge UP, 2009.

Boje, R., and M. Tomczak, editors. *Upwelling Ecosystems*. Springer Science & Business Media, 2012.

Bolius, David. *Paleo Climate Reconstructions Based on Ice Cores: Results from the Andes and the Alps*. SVH, 2010.

Bolstad, Paul. *GIS Fundamentals: A First Text on Geographic Information Systems*. 5th ed., XanEdu, 2016.

Bonan, Gordon. *Climate Change and Terrestrial Ecosystem Modeling*. Cambridge UP, 2019.

Boon, John D. *Secrets of the Tide: Tide and Tidal Current Application and Prediction, Storm Surges and Sea Level Trends*. Horwood, 2004.

Boorstein, M. and Richard M. Renneboog, editors. *Earth Science: Earth's Weather, Water and Atmosphere*. Salem Press, 2013

Bottjer, David J. *Paleoecology: Past, Present and Future*. Wiley-Blackwell, 2016.

BouDagher-Fadel, Marcelle K. *Biostratigraphic and Geological Significance of Planktonic Foraminifera*. UCL Press, 2015.

Boulding, Russell J., editor. *EPA Environmental Engineering Sourcebook*. Routledge, 2019.

Bowen, James, and Mavourna Collits. *The Coral Reef Era: From Discovery to Decline, a History of Scientific Investigation from 1600 to the Anthropocene Epoch*. Springer, 2015.

Boyle, Godfrey, editor. *Renewable Energy*. 2nd ed., Oxford UP, 2004.

Braitenberg, Carla, Giuliana Rossi et al., editors. *Geodynamics and Earth Tides Observations from Global to Micro Scale*. Birkhauser, 2019.

Braitsch, O. *Salt Deposits: Their Origin and Composition*. Springer Science & Business Media, 2012.

Brant, Cory. *Great Lakes Sea Lamprey: The 70 Year War on a Biological Invader*. U of Michigan P, 2019.

Brantley, Susan, James Kubicki, and Art White, editors. *Kinetics of Rock-Water Interactions*. Springer, 2007.

"Brazil to Expand Marine Conservation Areas in the South Atlantic." *Global Environment Facility*, 22 Mar. 2018, www.thegef.org/news/brazil-expand-marine-conservation-areas-south-atlantic.

Brewer, Cynthia A. *Designing Better Maps: A Guide for GIS Users*. 2nd ed., ESRI Press, 2015.

Bridgman, Howard A., and John E. Oliver. *The Global Climate System. Patterns, Processes and Teleconnections*. Cambridge UP, 2014.

Brighton, Henry, and Howard Selina. *Introducing Artificial Intelligence*. Edited by Richard Appignanesi. 2003. Reprint. Totem Books, 2007.

The Britannica Guide to the One Hundred Most Influential Scientists. Constable & Robinson, 2008.

Britt-Gallagher, Susan, and Tricia Hayne. *St. Helena: Ascension. Tristan da Cunha*. 3rd ed., Bradt Travel Guides, 2015.

Broad, William J. *The Universe Below: Discovering the Secrets of the Deep Sea*. Simon & Schuster, 1997.

Broecker, Wally. *The Great Ocean Conveyor: Discovering the Trigger for Abrupt Climate Change*. Princeton UP, 2010.

Bromhead, Edward N., and Vernon R. Schaefer. *Stability of Slopes and Landslide Engineering*. 3rd ed., CRC Press, 2021.

Brosche, P., and Jürgen Sünderman. *Tidal Friction and the Earth's Rotation*. Springer Science & Business Media, 2012.

Brown, Joan. *Ocean Circulation: Prepared by an Open University Course Team*. Elsevier, 2016.

Brunet, M.-F., M. Wilmsen, and J. W. Granath, editors. *South Caspian to Central Iran Basins*. Special Publication 312. The Geological Society of London, 2009.

Brusca, Richard C., editor. *The Gulf of California: Biodiversity and Conservation*. U of Arizona P, 2010.

Brutsaert, Wilfried. *Hydrology: An Introduction*. Cambridge UP, 2005.

Bull, William B. *Tectonic Geomorphology of Mountains: A New Approach to Paleoseismology*. Wiley-Blackwell, 2007.

Burch, David. *Emergency Navigation: Find Your Position and Shape Your Course at Sea Even if Your Instruments Fail*. 2nd ed., McGraw-Hill, 2008.

Burch, David. *The Barometer Handbook: A Modern Look at Barometers and Applications of Barometric Pressure*. Starpath Publications, 2009.

Burdick, Donald. *Petrochemicals in Nontechnical Language*. 3rd ed., Penn Well, 2010.

Burdige, David J. *Geochemistry of Marine Sediments*. Princeton UP, 2020.

Burgan, Michael. *Exxon Valdez. How a Massive Oil Spill Triggered an Environmental Catastrophe*. Compass Point, 2018.

Burk, C. A., and C. A. Drake, editors. *The Geology of Continental Margins*. Springer, 2013.

Burns, Bob, and Mike Burns. *Wilderness Navigation: Finding Your Way Using Map, Compass, Altimeter and GPS*. 2nd ed., Mountaineers, 2004.

Buster, Noreen A., and Charles W. Holmes, editors. *Gulf of Mexico: Origin, Waters, and Biota*. Texas A&M UP, 2011.

Butcher, Kenneth S., Linda D. Crown, and Elizabeth J. Gentry, editors. *The International System of Units (SI): Conversion Factors for General Use*. National Institute of Standards and Technology, 2006.

Butler, James H., and Stephen A. Montzka. "The NOAA Annual Greenhouse Gas Index (AGGI)." *Earth System Research Laboratory Global Monitoring Division*, National Oceanic & Atmospheric Administration, Spring 2019, www.esrl.noaa.gov/ gmd/aggi/aggi.html. Accessed 22 May. 2019.

Butts, Edward. *Henry Hudson: New World Voyageur.* Dundurn Press, 2009.

Calewaert, Jan-Bart, and Frank Maes. *Science and Sustainable Management of the North Sea: Belgian Case Studies.* Academia Press, 2007.

Camp, Emma F., Verena Schoepf, Peter J. Mumby, and David J. Suggett, editors. *The Future of Coral Reefs Subject to Rapid Climate Change: Lessons from Natural Extreme Environments.* Frontiers Media, 2019.

Campbell, Maria Shauna. *Fisheries, Marine Conservation, Marine Renewable Energy and Displacement: A Fresh Approach.* U of Plymouth P, 2017.

Cardarelli, François. *Encyclopaedia of Scientific Units, Weights and Measures: Their SI Equivalences and Origins. Trans. M. J. Shields. Rev. and expanded ed. Springer, 2003.*

Carlisle, Rodney P. *Where the Fleet Begins: A History of the David Taylor Research Center.* Naval Historical Center, 1998.

Carpenter, Kent E., et al. "One-Third of Reef-Building Corals Face Elevated Extinction Risk from Climate Change and Local Impacts." *Science*, vol. 321, no. 1, 2008.

Carson, Rachel L. *The Sea Around Us*. 1951. Oxford UP, 2003.

Carter, R. W. G. *Coastal Environments: An Introduction to the Physical, Ecological, and Cultural Systems of Coastlines*. Academic Press, 1988.

Cartwright, William, Michael P. Peterson, and Georg F. Garner, editors. *Multimedia Cartography*. 2nd ed., Springer, 2007.

Castilla, J. C. "Roles of Experimental Marine Ecology in Coastal Management and Conservation." *Journal of Experimental Marine Biology and Ecology*, vol. 250, nos. 1-2, 2000.

Castilla, Juan Carlos, and Tim R. McClanahan, editors. *Fisheries Management: Progress Toward Sustainability.* Blackwell, 2007.

Castro, Jonathan M., and Donald B. Dingwell. "Rapid Ascent of Rhyolitic Magma at Chaiten Volcano, Chile." *Nature*, vol. 461, 2009, pp. 780-83.

"The Causes of Climate Change." *Global Climate Change: Vital Signs of the Planet*, NASA, 1 July 2019, climate.nasa.gov/causes/.

Cave, Damien, and Justin Gillis. "Large Sections of Australia's Great Reef Are Now Dead, Scientists Find." *The New York Times*, 15 Mar. 2017, www.nytimes.com/ 2017/03/15/science/great-barrier- reef-coral-climate-change-dieoff.html.

Cavnar, Bob. *Disaster on the Horizon. High Stakes, High Risks, and the Story Behind the Deepwater Well Blowout*. Chelsea Green Publishing, 2010.

Cazes, Jack. *Ewing's Analytical Instrumentation Handbook*. 3rd ed., Marcel Dekker, 2005.

Cech, Thomas V. *Principles of Water Resources: History, Development, Management, and Policy*. 3rd ed., Wiley, 2010.

Çengel, Yunus A., and John M. Cimbala. *Fluid Mechanics: Fundamentals and Applications.* McGraw-Hill, 2010.

Chandra, A. M. *Geoinformatics*. New Academic Science, 2017.

Chandrasekharam, D, Aref Lashin, Nassir Al Arifi, and Abdulaziz M. Al-Bassam. *Red Sea Geothermal Provinces.* CRC Press, 2019.

Chang, Fi-John, Kuolin Hsu, and Li-Chiu Chang, editors. *Flood Forecasting Using Machine Learning Methods*. MDPI, 2019.

Chang, Kyung-Il, Chang-Ik Zhang, Chul Park et al., editors. *Oceanography of the East Sea (Japan Sea).* Springer, 2016.

Chang, Raymond. *Chemistry*. McGraw-Hill Higher Education, 2007.

Charabe, Yassine, and Salim al-Hatrushi. *Indian Ocean Tropical Cyclones and Climate Change*. Springer, 2010.

"Characterising Meso-Marine Ecosystems of the North Pacific," S. D. Batten, et al. *Deep-Sea Research II*, vol. 53, 2006.

Charlier, Roger H., and Charles W. Finkl. *Ocean Energy. Tide and Tidal Power.* Springer, 2009.

Chartres, C., and S. Varma. *Out of Water: From Abundance to Scarcity and How to Solve the World's Water Problems*. FT Press, 2010.

Chassignet, Eric P., and Jacques Verron, editors. *Ocean Weather Forecasting: An Integrated View of Oceanography*. Springer, 2006.

Chatterjee, Piu, and Dona Ganguly. *India and Indian Ocean: Past, Present and Future.* Mittal Publications, 2018.

Chen, C. H., editor. *Emerging Topics in Computer Vision and Its Applications*. World Scientific, 2012.

Chen, Daniel H., editor. *Sustainable Water Management and Technologies: Sustainable Water Management Volume 1*. CRC Press, 2017.
Chen, Falin. *The Kuroshio Power Plant*. Springer Science & Business Media, 2013.
Chen, J. L., et al. "Long-Term Caspian Sea Level Change." *Geophysical Research Letters*, vol. 44, no. 13, 2017, doi:10.1002/2017GL073958.
Chernicoff, Stanley, and Donna Whitney. *Geology: An Introduction to Physical Geology*. Prentice-Hall, 2007.
Chernicoff, Stanley. *Geology: An Introduction to Physical Geology*. 4th ed., Prentice-Hall, 2006.
Chesapeake Bay Program, 2018, www.chesapeakebay.net/. Accessed 29 June 2018.
Chester, Roy. *Furnace of Creation, Cradle of Destruction*. AMACOM Books, 2008.
Chester, Roy. *Marine Geochemistry*. Springer Science and Business Media, 2012.
Chin, Cheng Siong, and Rongxin Cui, editors. *Intelligent Marine Robotics: Modelling, Simulation and Applications*. MDPI, 2020.
Chiocci, F. L., and A. R. Chivas, editors. *Continental Shelves of the World: Their Evolution During the Last Glacioeustatic Cycle*. Geological Society, 2014.
Chiras, Daniel D., and John P. Reganold. *Natural Resource Conservation: Management for a Sustainable Future*. 10th ed., Addison Wesley, 2009.
Christopherson, Robert W. *Geosystems: An Introduction to Physical Geography*. 8th ed., Prentice-Hall, 2012.
Christopherson, Robert W., and Mary-Louise Byrne. *Geosystems: An Introduction to Physical Geography*. Pearson Prentice Hall, 2006.
Ciavola, Paolo, and Giovanni Coco. *Coastal Storms: Processes and Impacts*. Wiley-Blackwell, 2017.
CIESM (International Commission for Scientific Exploration of the Mediterranean Sea). "The Mediterranean Science Commission." www.ciesm.org.
Clague, John J., and Douglas Stead, editors. *Landslides: Types, Mechanisms and Modeling*. Cambridge UP, 2012.
Clague, John, Chris Yorath, and Richard Franklin. *At Risk: Earthquakes and Tsunamis on the West Coast*. Tricouni Press. 2006.
Clark, John R. *Coastal Seas: The Conservation Challenge*. John Wiley & Sons, 2009.
Clarke, Allan J. *An Introduction to the Dynamics of El Niño and the Southern Oscillation*. Academic Press, 2008.
Claudino-Sales, Vanda. *Coastal World Heritage Sites*. Springer, 2019.
Clift, Peter D., and R. Alan Plumb. *The Asian Monsoon*. Cambridge UP, 2008.
Clover, Charles. *The End of the Line: How Overfishing Is Changing the World and What We Eat*. U of California P, 2008.
Coad, Bryan W., and James D. Reist. *Marine Fishes of Arctic Canada*. U of Toronto P, 2018.
Coe, Angela L., editor. *The Sedimentary Record of Sea-Level Change*. Cambridge UP, 2003.
Coetzee, Ken. *Caring for Natural Rangelands*. U of KwaZulu-Natal P, 2005.
Colasimone, Luisa, editor. *MedWaves: Focus on Biodiversity*. United Nations Environment Programme, Mediterranean Action Plan, 2010.
Coley, David A. *Energy and Climate Change: Creating a Sustainable Future*. John Wiley & Sons, 2008.
Coll, Marta, et al. "Biodiversity of the Mediterranean Sea: Estimates, Patterns, and Threats." *PLoS ONE*, vol. 5, no. 8, 2010.
Colling, Angela. *Ocean Circulation*. 2nd ed., Butterworth-Heinemann, 2001.
Comette, James L., and Ralph A. Ackerman. *Calculus for the Life Sciences: A Modeling Approach*. MAA Press, 2015.
Committee on Nonnative Oysters in the Chesapeake Bay. *Nonnative Oysters in the Chesapeake Bay*. National Academies Press, 2004.
Committee to Review Individual Fishing Quotas, et al. *Sharing the Fish: Toward a National Policy on Individual Fishing Quotas*. National Academy Press, 1999.
"Conservation International." Coral Triangle Initiative." www.conservation.org/global/marine/initiatives/oceanscapes/cti/pages/overview.aspx.
Cooper, C. David. *Introduction to Environmental Engineering*. Waveland Press, 2015.
"Coral Reefs." *WWF Global*. WWF-World Wide Fund for Nature, 2016. Accessed 8 Aug. 2016.
Cornforth, Derek H. *Landslides in Practice*. John Wiley & Sons, 2005.
Corrigan, Eamonn. *Tsunamis: Causes and Effects*. 12-Story Library, 2020.
Crabbe, M. James C., et al. "The Impact of Weather and Climate Extremes on Coral Growth." *Climate Extremes and Society*, edited by Henry F. Diaz and Richard J. Murnane, Cambridge UP, 2011.
Cracknell, Arthur P., editor. *Remote Sensing Applications in Marine Science and Technology*. Springer Science + Business Media, 2012.
Craig, James R., David J. Vaughan, and Brian Skinner. *Earth Resources and the Environment*. Prentice, 2010.
Craig, John F., editor. *Freshwater Fisheries Ecology*. Wiley Blackwell, 2016.
Crawford, Christopher Blair, and Brian Quinn. *Microplastic Pollutants*. Elsevier, 2016.
Crittenden, John C. *Water Treatment Principles and Design*. Wiley, 2005.
Crocker, Malcolm J., editor. *The Encyclopedia of Acoustics*. 4 vols. Wiley, 1997.

Cronan, D. S. *Handbook of Marine Mineral Deposits*. CRC Press, 2000.
Cropper, Thomas. "Did El Niño Drive the Record Heat of 2015?" *Niskanen Center*. Niskanen Center, 25 Jan. 2016.
Cruz, Joan. *Ocean Wave Energy: Current Status and Future Perspectives*. Springer-Verlag, 2010.
Cullen, Robert. "The Rise and Fall of the Caspian Sea." *National Geographic*, vol. 195, May 1999, pp. 2-35.
Curley, Robert. *New Thinking About Pollution*. Britannica Educational Publishing, 2011.
Cushman-Roisan, Benoit, Miroslav Gačić, Pierre-Marie Poulain, and Antonio Artegiani. *Physical Oceanography of the Adriatic Sea. Past, Present and Future*. Springer Science & Business Media, 2013.
Cutler, Thomas J. *Dutton's Nautical Navigation*. 15th ed., Naval Institute, 2004.
Cyronak, Tyler et al. *Spatial and Temporal Variability of Seawater Chemistry in Coastal Ecosystems in the Context of Global Change*. Frontiers Media, 2019.
Dale, Virginia H., et al. *Hypoxia in the Northern Gulf of Mexico*. Springer, 2010.
Dang, Vu Hai. *Marine Protected Areas Network in the South China Sea: Charting a Course for Future Cooperation*. Martinus Nijhof, 2014.
Danovaro, Roberto, Antonio Dell'Anno, Mauro Fabiano, Antonio Pusceddu, and Anastasios Tselepides. "Deep-Sea Ecosystem Response to Climate Changes: The Eastern Mediterranean Case Study." *Trends in Ecology & Evolution*, vol. 16, no. 9, 2001.
Danson, Ted, and Michael Orso *Oceana: Our Endangered Oceans and What We Can Do to Save Them*. Potter/Ten Speed/Harmony/Rodale, 2011.
Darrigol, Olivier. *Worlds of Flow: A History of Hydrodynamics from the Bernoullis to Prandtl*. Oxford UP, 2005.
David, Bruno, and Thomas SaucPde. *Biodiversity in the Southern Ocean*. ISTE Press, 2015.
David, Matej, and Stephan Gollasch, editors. *Global Maritime Transport and Ballast Water Management: Issues and Solutions*. Springer, 2015.
Davidson, Helen. "Coral Reef Bleaching 'the New Normal' and a Fatal Threat to Ecosystems." *The Guardian*, 4 Jan. 2018, www.theguardian.com/ environment/2018/jan/04/ coral-reef-bleaching-the- new-normal-and-a-fatal-threat-to-ecosystems.
Davidson, Jon P., Walter E. Reed, and Paul M. Davis. *Exploring Earth: An Introduction to Physical Geology*. 2nd ed., Prentice-Hall, 2001.
Davidson-Arnott, Robin, Bernard Bauer, and Chris Hauser. *Introduction to Coastal Processes and Geomorphology*. 2nd ed., Cambridge UP, 2019.
Davis, Mackenzie L., and Susan J. Masten. *Principles of Environmental Engineering and Science*. 3rd ed., McGraw, 2013.
Davis, Richard A., and Duncan M. FitzGerald. *Beaches and Coasts*. 2nd ed., Wiley, 2020.
Davis, Richard A., and Robert W. Dalrymple, editors. *Principles of Tidal Sedimentology*. Springer, 2012.
Davis, Richard A., Jr. *Sea-Level Change in the Gulf of Mexico*. Everbest Printing, 2011.
Davis, Richard Jr., and Duncan Fitzgerald. *Beaches and Coasts*. 2nd ed., John Wiley & Sons, 2020.
Dawson, Alastair. *Introducing Sea Level Change*. Dunedin Academic Press, 2018.
De Angelis, Hernan, and Pedro Skvarca. "Glacier Surge After Ice Shelf Collapse." *Science*, vol, 299, 2003, pp. 1560-62.
Dean, Robert G., and Robert A. Dalrymple. *Coastal Processes with Engineering Applications*. 2002. Reprint. Cambridge UP, 2004.
Decker, Robert, and Barbara Decker. *Volcanoes*. 4th ed., W. H. Freeman, 2005.
Delamare, Francois, and Bernard Guineau. *Colors: The Story of Dyes and Pigments*. H. N. Abrams, 2000.
Delino-Musgrave, Virginia E. *Marine Archaeology: A Handbook*. Council for British Archaeology, 2012.
DeMers, Michael N. *Fundamentals of Geographic Information Systems*. 4th ed., John Wiley & Sons, 2009.
Demina, Liudmila L., and Sergey V. Galkin, editors. *Trace Metal Biogeochemistry and Ecology of Deep-Sea Hydrothermal Vent Systems*. Springer, 2016.
Demirbas, Ayhan. *Methane Gas Hydrates*. Springer, 2010.
Demuth, Bathsheba. *The Floating Coast: An Environmental History of the Bering Strait*. W. W. Norton and Company, 2019.
Denny, Mark. *How the Ocean Works*. Princeton UP, 2008.
Dera, Jerzy. *Marine Physics*. Elsevier, 1992.
Derocher, Andrew E. *Polar Bears: A Complete Guide to Their Biology and Behavior*. Johns Hopkins UP, 2012.
"Diarrhoeal Disease." *World Health Organization*, 2 May 2017, www.who.int/news-room/fact-sheets/detail/ diarrhoeal-disease.
Ding, H., et al. "An Overview of Spatial Management and Marine Protected Areas in the East China Sea." *Coastal Management*, vol. 36, no. 5, 2008.
Dinwiddie, Keith. *Basic Robotics*. Cengage Learning, 2016.
Dittmann, Sabine, editor. *The Wadden Sea Ecosystem: Stability, Properties and Mechanisms*. Springer Science & Business Media, 2012.
Dodds, Walter K., and Matt R. Whiles. *Freshwater Ecology: Concepts and Environmental Applications of Limnology*. 2nd ed., Academic Press, 2010.
Dodge, Martin, Rob Kitchin, and Chris Perkins, editors. *Rethinking Maps: New Frontiers in Cartographic Theory*. Routledge, 2009.

Dodson, Roy D. *Storm Water Pollution Control: Municipal, Industrial and Construction NPDES Compliance*. McGraw, 1999.

Donn, Jeff. "AP Exclusive: Gulf Oil Wells Lingering with Temporary Seals." *AP News*, 18 July 2015, apnews.com/e68c094dbe7d4d1b91e3657c2314d364/ap-exclusive-share-aging-temporarily-sealed-wells-grows.

Doody, J. Patrick. *Saltmarsh Conservation, Management and Restoration*. Springer, 2008.

Doremus, Holly, and A. Dan Tarlock. *Water War in the Klamath Basin: Macho Law, Combat Biology and Dirty Politics*. Island Press, 2008.

Dorton, Peter, Saman Alavi, and T. K. Woo. "Free Energies of Carbon Dioxide Sequestration and Methane Recovery in Clathrate Hydrates." *Journal of Chemical Physics*, vol. 127, no. 12, 2007.

Drew, Lisa W. "The Promise and Peril of Seafloor Mining: Can Minerals Be Extracted from the Seafloor without Environmental Impact?" *Oceanus*, vol. 47, no. 3, 2009, pp. 8-14.

Drew, Lisa W. "Who Regulates Mining on the Seafloor?" *Oceanus*, vol. 47, no. 3, 2009, p. 15.

Drumond, Anita, Belen Rodrigues-Fonseca, Chris Reason, and Silvina A. Solman, editors. *Wet and Dry Periods in Regions Surrounding the Atlantic Ocean Basin*. Frontiers Media SA, 2016.

Duarte, Érico, and Manuel Correia de Barros, editors. *Maritime Security Challenges in the South Atlantic. Palgrave Macmillan, 2019.*

Duarte, Jono, and Wouter P. Schellart, editors. *Plate Boundaries and Natural Hazards*. Wiley, 2016.

Dubinsky, Zvy, and Noga Stambler, editors. *Coral Reefs: An Ecosystem in Transition*. Springer, 2011.

Duffield, John. *Over a Barrel: The Costs of U.S. Foreign Oil Dependence*. Stanford UP, 2008.

Dumont, Henri J. *The Nile: Origin, Environments, Limnology and Human Use*. Springer Science & Business Media, 2009.

Dumont, Henri, Tamara A. Shiganova, and Ulrich Niermann, editors. *Aquatic Invasions in the Black, Caspian, and Mediterranean Seas*. Kluwer Academic Publishers, 2004.

Duncan, J. Michael, Stephen G. Wright, and Thomas L. Brandon. *Soil Strength and Slope Stability*. 2nd ed., Wiley, 2014.

Dunham, William. *The Calculus Gallery: Masterpieces from Newton to Lebesgue*. Princeton UP, 2005.

Dunne, Thomas, and Luna B. Leopold. *Water in Environmental Planning*. Freeman, 1978.

Dyke, Phil. *Modeling Coastal and Offshore Processes*. Imperial College Press, 2007.

Dzurik, Andrew A. *Water Resources Planning*. 3rd ed., Rowman, 2003.

Earle, Anton, Anders Jägerskog, and Joakim Öjendal, editors. *Transboundary Water Management Principles and Practice*. Earthscan, 2013.

Earle, Steven. *Physical Geology*. BC Open Textbook Project, 2019.

Earle, Sylvia A. *The World Is Blue: How Our Fate and the Ocean's Are One*. National Geographic Society, 2009.

Earle, Sylvia A., and Linda K. Glover. *Ocean: An Illustrated Atlas*. National Geographic Books, 2008.

Eckert, Michael. *The Dawn of Fluid Dynamics: A Discipline Between Science and Technology*. Wiley-VCH, 2006.

"Ecosystems of the Hudson Bay Coast." *Churchill Wild*, 27 Apr. 2017, www.churchillwild.com/ecosystems-of-the-hudson-bay-coast/.

Edwards, Charles T. T., and Dorothy J. Dankel, editors. *Management Science in Fisheries: An Introduction to Simulation-based Methods*. Routledge, 2016.

Edwards, John. *Plate Tectonics and Continental Drift*. Smart Apple Media, 2006.

Edzwald, James K. *Water Quality and Treatment: A Handbook on Drinking Water.* McGraw-Hill, 2011.

Egan, Dan. "Commercial Fishing, Once a Great Lakes Way of Life, Slips Away." *LA Times*, 30 Aug. 2011.

Egan, Dan. *The Death and Life of the Great Lakes*. W. W. Norton & Company, 2017.

Ellis, Erle C., and Peter K. Haff. "Earth Science in the Anthropocene: New Epoch, New Paradigm, New Responsibilities." *EOS, Transactions, American Geophysical Union*, vol. 90, no. 49, 2009, p. 473.

Ellison, Joanna, and Monifa Fiu. *Vulnerability of Fiji's Mangroves and Associated Coral Reefs to Climate Change*. World Wildlife Fund South Pacific Programme, 2010.

Ellison, Joanna. "Wetlands of the Pacific Island Region." *Wetlands Ecology and Management*, vol. 17, 2009.

Emelyanov, Emelyan M. *The Barrier Zones in the Ocean*. Springer-Verlag, 2005.

Emergency Management BC. *A Simple Explanation of Earthquake Magnitude and Intensity*. Ministry of Public Safety and Solicitor General, Provincial Emergency Program. 2007.

Emerson, Steven, and John Hedges. *Chemical Oceanography and Marine Carbon Cycle*. Cambridge UP, 2008.

Engel, Max, Jessica Pilarczyk, Simon Matthias May, Dominik Brill, and Ed Garrett, editors. *Geological Records of Tsunamis and Other Extreme Waves*. Elsevier, 2020.

England, Gordon. "Implementing LID for New Development." *Stormwater*, vol. 11, no. 5, July-Aug. 2010.

"EPA Outlines Final Stage of Chesapeake Bay Restoration Effort." *Storm Water Solutions*, 28 June 2018, www.estormwater.com/epa-outlines-final-stage-chesapeake-bay-restoration-effort.

Ettenger, Kreg, and Samantha Hogenson. *Marine Tourism, Climate Change and Resilience in the Caribbean. Volume 1: Ocean Health, Fisheries and Marine Protected Areas.* Business Expert Press, 2017.

Etter, Paul C. *Underwater Acoustic Modeling and Simulation* 4th ed., CRC Press, 2017.

European Environment Agency. "Natura 2000 Data-The European Network of Protected Sites." www.eea.europa.eu/data-and-maps/data/natura-2000.

Everest, F. Alton, and Ken C. Pohlmann. *Master Handbook of Acoustics*. 6th ed., McGraw, 2015.

Everhart, W. Harry, and William D. Youngs. *Principles of Fisheries Science*. 2nd ed., Cornell UP, 1981.

Everson, Inigo, editor. *Krill: Biology, Ecology and Fisheries.* John Wiley & Sons, 2008.

Ewen, Dale, Neill Schurter, and P. Erik Gundersen. *Applied Physics*. 11th ed., Pearson, 2017.

Exon, Neville. "Scientific Drilling Beneath the Oceans Solves Earthly Problems." *Australian Journal of Maritime and Ocean Affairs*, vol. 2, 2010, pp. 37-47.

Eyvaz, Murat, and Ebubekir Yüksel, editors. *Desalination and Water Treatment. InTechOpen, 2018.*

Fache, Elodie, and Simmone Pauwels, editors. *Fisheries in the Pacific: Challenges of Governance and Sustainability.* Pacific-Credo Publications, 2018.

Fagan, Brian M., editor. *The Complete Ice Age: How Climate Change Shapes the World*. Thames, 2009.

Fagan, Brian. *Before California: An Archaeologist Looks at Our Earliest Inhabitants*. Rowman, 2004.

Fall, J. A., et al. *Long-Term Consequences of the Exxon Valdez Oil Spill for Coastal Communities of South-central Alaska*. Alaska Department of Fish and Game, Division of Subsistence, Anchorage, Alaska. Technical Report 163, 2001.

Fang, Huacan, and Menglan Duan. *Offshore Operation Facilities: Equipment and Procedures*. Gulf Professional Publishing, 2014.

FAO-UN. *The State of Mediterranean and Black Sea Fisheries 2018*. Food and Agriculture Organization of the United Nations, 2019.

Fariña, José Miguel, and Andrés Camaño, editors. *The Ecology and Natural History of Chilean Salt Marshes.* Springer, 2017.

Faulkner, Edwin B., and Russell J. Schwartz. *High Performance Pigments*. Wiley, 2009.

Faure, Gunter. *Isotopes: Principles and Applications*. 3rd ed., John Wiley & Sons, 2004. (Originally titled *Principles of Isotope Geology*.)

———. *Origin of Igneous Rocks: The Isotopic Evidence*. Springer-Verlag, 2010.

Faust, Russell A., editor. *Robotics in Surgery: History, Current and Future Applications*. Nova Science, 2007.

Fears, Darryl. "How We Are All Contributing to the Destruction of Coral Reefs: Sunscreen." *The Washington Post*, 20 Oct. 2015.

Feidi, Izzat H. "Mechanism for Fisheries Management in the Red Sea and Gulf of Aden." *Arab Agricultural Statistics Yearbook 2012*, no. 1, 2012.

Felder, Darryl L., and David K. Camp, editors. *Gulf of Mexico-Origin, Waters, and Biota: Vol. 1, Biodiversity.* Texas A&M UP, 2009.

Ferguson, Steve H., Lisa L. Loseto, and Mark L. Mallory, editors. *A Little Less Arctic: Top Predators in the World's Largest Northern Inland Sea, Hudson Bay.* Springer, 2010.

Ferreiro, Larrie D. *Ships and Science: The Birth of Naval Architecture in the Scientific Revolution, 1600-1800*. MIT Press, 2007.

Field, J. G., Gotthilf Hempel, and C. P. Summerhayes. *Oceans 2020: Science, Trends and the Challenge of Sustainability.* Island Press, 2002.

Field, Kenneth. *Cartography: A Compendium of Design Thinking for Mapmakers*. ESRU, Press, 2018.

Filatov, Nikolai, Dmitry Pozdnyakov, Ola M. Johannessen, et al. *White Sea. Its Marine Environment and Ecosystem Dynamics Influenced by Global Change.* Springer, 2005.

Fingas, Merv, editor. *Handbook of Oil Spill Science and Technology.* Wiley, 2015.

Finkl, Charles W., and Christopher Makowski, editors. *Coastal Wetlands: Alteration and Remediation.* Springer, 2017.

Finlayson, C. Max, Angela H. Arthington, and Jamie Pittock, editors. *Freshwater Ecosystems in Protected Areas: Conservation and Management.* Routledge, 2018.

Finley, Fred N., Younkeong Nam, and John Oughton. "Earth Systems Science: An Analytic Framework." *Science Education*, vol. 95, no. 6, 2011, pp. 1066-85.

Finucane, Edward W. *Concise Guide to Environmental Definitions, Conversions, and Formulae*. Lewis, 1999.

"Fishery Stock Status Updates." *NOAA Fisheries*, 2018, www.fisheries.noaa.gov/national/population-assessments/fishery-stock-status-updates. Accessed 26 July 2018.

Flemming, Nicholas C. et al., editors. *Submerged Landscapes of the European Continental Shelf: Quaternary Paleoenvironments.* Wiley Blackwell, 2017.

Floreano, Dario, and Claudio Mattuissi. *Bio-Inspired Artificial Intelligence: Theories, Methods, and Technologies*. MIT Press, 2008.

Floyd, Peter A., editor. *Oceanic Basalts*. Springer Science + Business Media, 2012.

Food and Agriculture Organization (FAO) of the United Nations. *Trends in Oceanic Captures and Clustering of Large Marine Ecosystems-Two Studies Based on the FAO Capture Database*. FAO, 2003.

Food and Agriculture Organization of the United Nations. "Fisheries and Aquaculture Department Website." 2012. www.fao.org/fishery/en.

Foote, Michael, and Arnold L. Miller. *Principles of Paleontology*. 3rd ed., Freeman, 2007.

Forcada, Jaume, and Philip N. Trathana. "Penguin Responses to Climate Change in the Southern Ocean." *Global Change Biology*, vol. 15, no. 7, 2009.

Fotheringham, Stewart, and Peter Rogerson. *Spatial Analysis and GIS*. CRC Press, 2013.

Freitag, Bob, et al. *Floodplain Management: A New Approach for a New Era*. Island Press, 2009.

Frenay, Robert. *Pulse: The Coming Age of Systems and Machines Inspired by Living Things*. 2006. Reprint. U of Nebraska P, 2008.

Freuchen, Peter. *Peter Freuchen's Book of the Seven Seas*. The Lyons Press, 2003.

Freudenburg, William R., and Robert Gramling. *Blowout in the Gulf: The BP Oil Spill Disaster and the Future of Energy in America*. MIT Press, 2011.

Fuerst-Bjelis, Borna, editor. *Mediterranean Identities: Environment, Society, Culture*. IntechOpen, 2017.

Fusetani, Nobuhiro, and Anthony S. Clare, editors. *Antifouling Compounds*. Springer, 2006.

Galati, Steven R. *Geographic Information Systems Demystified*. Artech House, 2006.

Galgani, Luisa, Ricardo Beiras, Francois Galgani, et al. *Impacts of Marine Litter*. Frontiers Media, 2019.

Gamble, Clive. *Archaeology: The Basics*. 2nd ed., Routledge, 2008.

Garcia, Carmelo Consesa, and Mario Aristide Lenzi. *Check Dams, Morphological Adjustments and Erosion Control in Torrential Streams*. Nova Science, 2013.

Garrison, Tom S. *Essentials of Oceanography*. 6th ed., Brooks/Cole Cengage Learning, 2012.

———. *Oceanography: An Invitation to Marine Science*. Thomson, 2007.

———. *Oceanography: An Invitation to Marine Science*. Brooks/Cole, Cengage Learning, 2010.

———. *Oceanography: An Invitation to Marine Science*. 8th ed., Brooks, 2013.

Gary, James, et al. *Petroleum Refining: Technology and Economics*. 5th ed., CRC Press, 2007.

Gat, Joel R. *Isotope Hydrology: A Study of the Water Cycle*. Imperial College Press, 2010.

Gates, Nancy. *The Alaska Almanac: Facts About Alaska*. 30th anniversary ed., Alaska Northwest Books, 2006.

Gautier, Catherine. *Oil, Water, and Climate: An Introduction*. Cambridge UP, 2008.

Geggel, Laura. "78-Foot Wave Is the Largest Ever Recorded in Southern Hemisphere." *LiveScience*, 11 May 2018, www.livescience.com/62543-largest-wave-hits-southern-hemisphere.html.

Gehrman, Sven. *Fire! In the Wadden Sea...Or What the Fauna of the North Sea Urgently Wants to Tell Us*. BoD, 2020.

Gershwin, Lisa-Ann. *Stung! On Jellyfish Blooms and the Future of the Ocean*. U of Chicago P, 2013.

Gharibreza, Mohammadreza, and Muhammad Aqeel Ashraf. *Applied Limnology: Comprehensive View from Watershed to Lake*. Springer, 2014.

Giancoli, Douglas C. *Physics for Scientists and Engineers with Modern Physics*. Pearson Prentice Hall, 2008.

Giavarini, Carlo, and Keith Hester. *Gas Hydrates: Immense Energy Potential and Environmental Challenges*. Springer, 2011.

Gill, Adrian E. *Atmosphere-Ocean Dynamics*. Elsevier, 2016.

Gill, Robin. *Igneous Rocks and Processes: A Practical Guide*. Wiley-Blackwell, 2011.

Gillum, Donald R. *Industrial Pressure, Level, and Density Measurement*. 2nd ed., International Society for Automation, 2009.

Giovannoni, Stephen J., et al. "Proteorhodopsin in the Ubiquitous Marine Bacterium SAR11." *Nature*, vol. 438, 2005.

Girard, James E. *Principles of Environmental Chemistry*. 2nd ed., Jones and Bartlett, 2010.

Global Warming of 1.5 °C. UN Intergovernmental Panel on Climate Change, 2018.

"Global Wind Statistics 2017." Global Wind Energy Council. *GWEC*, 14 Feb. 2018, gwec.net/wp-content/uploads/vip/GWEC_PRstats2017_EN-003_FINAL.pdf.

Glade, Thomas, Paola Albini, and Felix Francés. *The Use of Historical Data in Natural Hazard Assessments*. Springer Science & Business Media, 2013.

Glikson, Andrew Y. *Evolution of the Atmosphere, Fire and the Anthropocene Climate Event Horizon*. Springer, 2014.

Glover, David M., William J. Jenkins, and Scott C. Doney. *Modeling Methods for Marine Science*. Cambridge UP, 2011.

Gobas, Frank A. P. C. *Chemical Dynamics in Freshwater Ecosystems*. CRC Press, 2018.

Göbel, Ernst O., and Uwe Seigner. *The New International System of Units (SI): Quantum Metrology and Quantum Standards*. Wiley-VCH, 2019.

Goes, Joaquim I., Helga Gomes, Prasad Thoppil, Prabhu Matondkar, and Adnan Al Azri. *Eurasian Warming-Hydrography and Biological Productivity in the Arabian Sea*. Bigelow Laboratory for Ocean Sciences, 2007.

Gohn, Gregory S., Christian Koebert, Kenneth G. Miller, and Wolf Uwe Reimold, editors. *The ICDP-USGS Deep Drilling Project in the Chesapeake Bay Impact Structure: Results from the Eyreville Core Holes*. Geological Society of America, 2009.

Gökçe, Didem, editor. *Limnology: Some New Aspects of Inland Water Ecology*. IntechOpen, 2019.

Goldberg, Walter M. *The Biology of Reefs and Reef Organisms.* U of Chicago P, 2013.

Goldoftas, Barbara. *The Green Tiger: The Costs of Ecological Decline in the Philippines.* Oxford UP, 2006.

Goni, G. J., and Paola Malanotte-Rizzoli, editors. *Interhemispheric Water Exchange in the Atlantic Ocean.* Elsevier, 2003.

Goniea, Tom. "The Story of State-Licensed Commercial Fishing History on the Great Lakes." *Michigan Department of Natural Resources.* State of Michigan, n.d. Accessed 22 Aug. 2016.

Gönülal, Onur. *Deep-sea Biodiversity in the Aegean Sea.* IntechOpen, 2017.

Goreau, Thomas J., and Robert Kent Trench, editors. *Innovative Methods of Marine Ecosystem Restoration.* CRC Press, 2013.

Gosnell, Mariana. *Ice: The Nature, the History, and the Uses of an Astonishing Substance.* Alfred A. Knopf, 2005.

Gottsman, Joachim, and Joan Marti, editors. *Analysis, Modeling and Response.* Vol. 10 in *Caldera Volcanism.* Elsevier Science, 2008.

Graham, Bob, et al. *Deep Water: The Gulf Oil Disaster and the Future of Offshore Drilling.* National Commission on the BP Deepwater Horizon Oil Spill and Offshore Drilling, 2011.

Grasshoff, Klaus, Klaus Kremling, and Manfred Ehrhardt, editors. *Methods of Seawater Analysis.* 3rd Completely Revised and Extended ed., John Wiley & Sons, 2009.

Great Barrier Reef Marine Park Authority. *Great Barrier Reef Outlook Report 2014: In Brief,* 2014, www.gbr.qld.gov.au/documents/gbr-outlook-report-2014-brief.pdf. Accessed 6 Apr. 2017.

Greaves, Deborah, and Gregorio Iglesias, editors. *Wave and Tidal Energy.* Wiley, 2018.

Grebmeier, Jacqueline M., and Wieslaw Maslowski, editors. *The Pacific Arctic Region: Ecosystem Status and Trends in a Rapidly Changing Environment.* Springer, 2014.

Green, Don W., and Robert H. Perry. *Perry's Chemical Engineers' Handbook.* 8th ed., McGraw-Hill, 2008.

Greenberg, Arnold, et al., editors. *Standard Methods for the Examination of Water and Wastewater.* 21st ed., American Public Health Association, 2005.

Greenwood, Gregory B., and J. F. Shroder Jr., editors. *Mountain Ice and Water: Investigations of the Hydrologic Cycle in Alpine Environments.* Elsevier, 2016.

Grigg, Neil S. *Total Water Management: Practices for a Sustainable Future.* American Water Works Association, 2011.

Grimsditch, Gabriel D., and Rodney V. Salm. *Coral Reef Resilience and Resistance to Bleaching.* International Union for Conservation of Nature, 2006.

Gritti, E. S., B. Smith, and M. T. Sykes. "Vulnerability of Mediterranean Basin Ecosystems to Climate Change and Invasion by Exotic Plant Species." *Journal of Biogeography,* vol. 33, no. 1, 2006.

Grossman, David. "Scientists Confirm the Largest Wave Ever Recorded." *Popular Mechanics,* 15 Dec. 2016, www.popularmechanics.com/science/environment/a24326/worlds-largest-wave-recorded-near-scotland.

Grotzinger, John, et al. *Understanding Earth.* 5th ed., W. H. Freeman, 2006.

Guastoni, Alessandro, and Roberto Appiani. *Minerals.* Firefly, 2005.

Guidoboni, Emanuela, and John E. Ebel. *Earthquakes and Tsunamis in the Past: A Guide to Techniques in Historical Seismology.* Cambridge UP, 2009.

"Gulf of Carpenteria Marine Park." *Australian Marine Parks,* Parks Australia, parksaustralia.gov.au/marine/parks/north/gulf-of-carpentaria/. Accessed 9 July 2018.

Gupta, Rabin Sen, and Ehrlich Desa, editors. *The Indian Ocean-A Perspective. Volumes 1 & 2.* A. A. Balkema Publishers, 2001.

Gurvich, Evgeny G. *Metalliferous Sediments of the World Ocean.* Springer-Verlag, 2010.

Gutkind, Lee. *Almost Human: Making Robots Think.* 2006. Reprint. Norton, 2009.

Halfar, Jochen, and Rodney M. Fujita. *Science,* vol. 316, no. 5827, 2007, p. 987.

Hamblin, Kenneth W., and Eric H. Christiansen. *Earth's Dynamic Systems.* 10th ed., Prentice-Hall, 2003.

Hamilton, Kevin, and Wataru Ohfuchi, editors. *High Resolution Numerical Modeling of the Atmosphere and Ocean.* Springer Science+Business Media, 2008.

Hancock, James G., Hennie Haworth, Stuart Hill, and Sarah King. *The Art of Map Illustration.* Quarto Publishing, 2018.

Hand, Carol. *Dead Zones: Why Earth's Waters Are Losing Oxygen.* Twenty-First Century Books, 2016.

Hansen, Kathryn. "Historic Low Sea Ice in the Bering Sea." *NASA Earth Observatory,* earthobservatory.nasa.gov/images/92084/historic-low-sea-ice-in-the-bering-sea. Accessed 27 July 2018.

Haq, Bilal U., and Anne Boersma, editors. *Introduction to Marine Micropaleontology.* Elsevier, 1998.

Harder, Christian, and Clint Brown. *The ArcGIS Book: 10 Big Ideas About Applying the Science of Where.* ESRI Press, 2017.

Harris, Peter T., and Elaine K. Baker, editors. *Seafloor Geomorphology as Benthic Habitat.* Elsevier, 2011.

Harrison, Roy M. *Principles of Environmental Chemistry.* Royal Society of Chemistry, 2007.

Harvey, Francis. *A Primer of GIS: Fundamental Geographic and Cartographic Concepts.* Guildford Press, 2008.

Haslett, Simon K. *Coastal Systems*. 2nd ed., Routledge, 2009.

Hatfield, Jerry L., et al., editors. *Impacts of El Niño and Climate Variability on Agriculture.* ASA Special Publication Number 63. American Society of Agronomy, 2001.

Hawksworth, David L., and Alan T. Bull, editors. *Marine, Freshwater, and Wetlands Biodiversity Conversation.* Springer, 2006.

Heath, Alan G. *Water Pollution and Fish Physiology*. 2nd ed., CRC Press, 2018.

Heileman, S., R. Guevara, F. Chavez, A. Bertrand, and H. Soldi. "XVII-56 Humboldt Current: Large Marine Ecosystems #13." 2009, www.lme.noaa.gov/ lmeweb/ LME_Report/lme_13.pdf.

Helmer, Richard, and Ivanildo Hespanhol, editors. *Water Pollution: A Guide to the Use of Water Quality Management Principles*. Spon, 1997.

Hem, J. D. *Study and Interpretation of the Chemical Characteristics of Natural Water.* US Geological Survey Water Supply Paper 2254. UP of the Pacific, 2005.

Henderson, Bonnie. *Strand: An Odyssey of Pacific Ocean Debris*. Oregon State UP, 2008.

Herman, Yvonne, editor. *The Arctic Seas. Climatology, Oceanography, Geology and Biology.* Springer Science & Business Media, 2012.

Herman, Yvonne. *Marine Geology and Oceanography of the Arctic Seas.* Springer Science & Business Media, 2012.

Hester, R. E., and R. M. Harrison. *Chemistry in the Marine Environment*. Royal Society of Chemistry, 2000.

Hilborn, Ray, and Ulrike Hilborn. *Overfishing: What Everyone Needs to Know.* Oxford UP, 2012.

Hilgenkamp, Kathryn. *Environmental Health: Ecological Perspectives*. Jones and Bartlett Publishers, 2005.

Himbert, M. E. "A Brief History of Measurement." *European Physical Journal Special Topics*, vol. 172, 2009, pp. 25-35.

"History." *International Ocean Discovery Program: Exploring the Earth Under the Sea*, 2017, www.iodp.org/about-iodp/ history. Accessed 11 Sept. 2017.

Hithin, N. K., V. Sanil Kumar, and P. R. Shanas. *Trends of Wave Height and Period in the Central Arabian Sea from 1996 to 2012: A Study Based on Satellite Altimeter Data.* Elsevier, 2015.

Ho, Chung-Ro, Lingling Xia Jianyu Hu, and Quanan Zheng, editors. *Regional Oceanography of the South China Sea.* World Scientific Publishing, 2020.

Hobbs, Carl H. *The Beach Book: Science of the Shore*. Columbia UP, 2012.

Hobbs, John E. *Applied Climatology.* Elsevier, 2016.

Hoefs, J. *Stable Isotope Geochemistry*. 6th ed., Springer-Verlag, 2009.

Hoefs, Jochen. *Stable Isotope Geochemistry*. 6th ed., Springer-Verlag, 2009.

Hogarth, Peter J. *The Biology of Mangroves and Seagrass*. 3rd ed., Oxford UP, 2015.

Hohn, Donovan. *Moby Duck: The True Story of 28,800 Bath Toys Lost at Sea & of the Beachcombers, Oceanographers, Environmentalists & Fools Including the Author Who Went in Search of Them.* Penguin, 2011.

Hopkins, Rae. *Hydrothermal Vents: Treasures of the Deep.* Heinemann, 2020.

Hopley, David. *Encyclopedia of Modern Coral Reefs: Structure, Form and Process*. Springer, 2011.

Horner, Rita A., editor. *Sea Ice Biota.* CRC Press, 2018.

Hough, Susan. *Predicting the Unpredictable: The Tumultuous Science of Earthquake Prediction*. Princeton UP, 2010.

Hovland, Martin. *Deep-Water Coral Reefs: Unique Biodiversity Hot-Spots*. Praxis, 2008.

"How Much Oil Is Consumed in the United States?" *US Energy Information Administration*, 27 Sept. 2017, www.eia.gov/tools/faqs/faq.php?id=33&t=6.

Hoyt, Erich. *Marine Protected Areas for Whales, Dolphins and Porpoises: A World Handbook for Cetacean Habitat Conservation and Planning*. Earthscan, 2011.

Hu, Xiumian, Chengshan Wang, Robert W. Scott, Michael Wagreich, and Luba Jansa, editors. *Cretaceous Oceanic Red Beds: Stratigraphy, Composition, Origins and Paleoceanographic and Paleoclimatic Significance*. SEPM, 2009.

Hughes, Terrence B. *Mediterranean Sea: Ecosystems, Economic Importance and Environmental Threats.* NOVA Publishers, 2013.

"Human Disturbances to Estuaries." *NOAA Ocean Service Education*, 25 Mar. 2008. http://oceanservice. noaa.gov/education/kits/estuaries/estuaries09_ humandisturb.html.

Huneke, H., and T. Mulder. *Deep-Sea Sediments*. Elsevier, 2011.

Huneke, Heiko, and Thierry Mulder. *Deep Sea Sediments.* Elsevier, 2011.

Hunt, Constance Elizabeth. *Thirsty Planet: Strategies for Sustainable Water Management*. Room 400, 2004.

Huo, Hu, et al. "Mechanical and Thermal Properties of Methane Clathrate Hydrates as an Alternative Energy Resource." *Journal of Renewable and Sustainable Energy*, vol. 3, no. 6, 2011.

Hurai, Vratislav, Monika Huraiová, Marek Slobodnik, and Rainer Thomas. *Geofluids: Developments in Microthermometry, Spectroscopy, Thermodynamics and Stable Isotopes. Elsevier, 2015.*

Hurd, Catriona, Paul J. Harrison, Kai Bischof, and Christopher S. Lobban. *Seaweed Ecology and Physiology.* 2nd ed., Cambridge UP, 2014.

Hutchings, P. A., et al., editors. *The Great Barrier Reef: Biology, Environment, and Management*. Springer, 2008.

Hutchings, Pat, Mike Kingsford, and Ove Hoegh-Guldberg, editors. *The Great Barrier Reef: Biology, Environment and Management.* 2nd ed., CRC Press, 2019.

Hutchinson, S., J. R. E. Lutjeharms, B. McMillan, J. Musick, B. Stonehouse, and M. Tomczak. *Illustrated Atlas of the Sea.* Weldon Owen, 2010.

Hutchinson, Stephen, Johann R. E. Lutjeharms, Beverly McMillan, John Musick, Bernard Stonehouse, and Matthias Tomczak. *The Illustrated Atlas of the Sea.* Weldon Owen, 2010.

Hyne, Norman J. *Nontechnical Guide to Petroleum Geology, Exploration, Drilling, and Production.* 2nd ed., PennWell Books, 2001.

Idrisi, N., and S. D. Salman. "Distribution, Development, and Metabolism of Larval Stages of the Warm Water Shrimp, Caridina Babaulti Basrensis (Decapoda, Atyidae)." *Marine and Freshwater Behavior Physiology,* vol. 38, 2005.

Ingebritsen, Steve, Ward Sanford, and Chris Neuzil. *Groundwater in Geologic Processes.* 2nd ed., Cambridge UP, 2006.

International Council for the Exploration of the Sea. "ICES Advice Book 6: North Sea." www.ices.dk/committe/acom/comwork/report/2008/2008/6.1-6.2 North Sea ecosystem overview.pdf.

International Tsunami Information Center, UNESCO, 2018, "What Are Tsunamis, and What Causes Them?," itic.ioc-unesco.org/index.php?option=com_content&view=article&id=1017:what-are-tsunamis-and-what-causes-them&catid=1004&Itemid=1004.

Ishibashi, Jun-ichiro, Kyoko Okino, and Michinari Sunamura, editors. *Subseafloor Biosphere Linked to Hydrothermal Vents: TAIGA Concept.* Springer Open 2015.

Ishimatsu, Atsushi, and Heung-Jae Lie, editors. *Coastal Environmental and Ecosystem Issues of the East China Sea.* Terrapub and Nagasaki U, 2010.

Ishimure, Michiko. *Paradise in the Sea of Sorrow: Our Minamata Disease.* Center for Japanese Studies, U of Michigan, 2003.

Ismail, Basel I., editor. *Renewable Hydropower Technologies.* Intech Open, 2017.

Iudicello, Suzanne, Michael Weber, and Robert Wieland. *Fish, Markets, and Fishermen: The Economics of Overfishing.* Island Press, 1999.

Izdar, E., and James W. Murray. *Black Sea Oceanography.* Springer Science & Business Media, 2012.

Jackson, B. C., et al., editors. *Shifting Baselines: The Past and the Future of Ocean Fisheries.* Island Press, 2011.

Jackson, Derek, and Andrew D. Short, editors. *Sandy Beach Morphodynamics.* Elsevier, 2020.

Jackson, Jeremy B. C. "Ecological Extinction and Evolution in the Brave New Ocean." *PNAS,* vol. 105, supp. 1, pp. 11458-65, doi:10.1073/pnas.0802812105. Accessed 2 July 2018.

Jacobson, Michael C., et al. *Earth System Science: From Biogeochemical Cycles to Global Change.* Academic Press, 2006.

Jacobson, Michael, et al. *From Biogeochemical Cycles to Global Changes.* Vol. 7 in *Earth System Science.* Academic Press, 2000.

Jacoby, Wolfgang, and Peter L. Smilde. *Gravity Interpretation: Fundamentals and Application of Gravity Inversion and Geological Interpretation.* Springer-Verlag, 2009.

Jacques, Peter. *Globalization and the World Ocean.* Rowman and Littlefield Publishing Group, 2006.

James, Noel P., and Brian Jones. *Origin of Carbonate Sedimentary Rocks.* John Wiley & Sons, 2015.

Janin, Hunt, and Scott A. Mandia. *Rising Sea Levels. An Introduction to Cause and Impact.* McFarland & Company, 2012.

Jaso, Danel. *Pattern Recognition Techniques, Technology and Applications.* Scitus Academics LLC, 2016.

Jaulin, Luc et al., editors. *Marine Robotics and Applications.* Springer, 2018.

Jawad, Laith A. *The Arabian Seas: Biodiversity, Environmental Challenges and Conservation Measures.* Springer International Publishing, 2020.

Jeffreys, Elizabeth, and Ruth Gertwagen. *Shipping, Trade and Crusade in the Medieval Mediterranean.* Ashgate, 2012.

Jerram, Dougal, and Nick Petford. *The Field Description of Igneous Rocks.* Wiley-Blackwell, 2011.

Jochum, Markus, and Raghi Murtugudde, editors. *Physical Oceanography: Developments Since 1950.* Springer Science + Business Media, 2006.

Johannessen, Ola M., Leonid P. Bobylev, Elena V. Shalina, and Stein Sandven, editors. *Sea Ice in the Arctic: Past, Present and Future.* Springer, 2019.

Johnson, Craig Richard. *Seaweed Invasions: A Synthesis of Ecological, Economic and Legal Imperatives.* Walter de Gruyter, 2007.

Johnson, Richard W., editor. *The Handbook of Fluid Dynamics.* CRC Press, 1998.

Jones, Joseph L. *Robot Programming: A Practical Guide to Behavior-Based Robotics.* McGraw-Hill, 2004.

Joseph, Antony. *Measuring Ocean Currents. Tools, Technologies, and Data.* Elsevier, 2014.

———. *Tsunamis: Detection, Monitoring and Early Warning Technologies.* Academic Press/Elsevier, 2011.

Joseph, P. *Deep Water Sedimentation in the Alpine Basin of SE France.* Special Publication 221. Geological Society of London, 2004.

Juuti, Petri S., Tapio S. Katko, and Heikki Vuorinen. *Environmental History of Water: Global Views on Community Water Supply and Sanitation*. IWA, 2007.

Kagan, Boris A. *Ocean-Atmosphere Interaction and Climate Modeling*. Cambridge UP, 2006.

Kagan, Yan Y. *Earthquakes: Models, Statistics, Testable Forecasts*. John Wiley & Sons, 2013.

Kaligh, Alireza, and Omer C. Onar. *Energy Harvesting: Solar, Wind, and Ocean Energy Conversion Systems*. CRC Press, 2017.

Kallmeyer, Jens, editor. *Life at Vents and Seeps*. Walter de Gruyter, 2017.

Kämpf, Johan, and Piers Chapman. *Upwelling Systems of the World: A Scientific Journey to the Most Productive Marine Ecosystems*. Springer, 2016.

Kamphuis, J. William. *Introduction to Coastal Engineering and Management*. 3rd ed., World Scientific Publishing, 2020.

Kandel, Robert. "Understanding and Measuring Earth's Energy Budget: From Fourier, Humboldt, and Tyndall to CERES and Beyond." Surveys in Geophysics, Jan. 2012, pp. 1-12.

Kandel, Robert. *Water from Heaven: The Story of Water from the Big Bang to the Rise of Civilization and Beyond*. Columbia UP, 2003.

Kappelle, Maarten. *Costa Rican Ecosystems*. U of Chicago P, 2016.

Karapanagioti, Hrissi K., and Ioannis K. Kalavrouziotis. *Microplastics in Water and Wastewater*. IWA Publishing, 2019.

Karimi, Hassan A. *Big Data: Techniques and Technologies in Geoinformatics*. CRC Press, 2017.

Karson, Jeffrey A., Deborah S. Kelley, Daniel J. Farnari, et al. *Discovering the Deep: A Photographic Atlas of the Seafloor and Ocean Crust*. Cambridge UP, 2015.

Kassem, Ken, and Eric Madeja. *The Coral Triangle*. John Beaufoy Publishing, 2014.

Kassens, Heidemarie, Henning A. Bauch, Igor A. Dmitrenko, et al. *Land-Ocean Systems in the Siberian Arctic: Dynamics and History*. Springer Science & Business Media, 2012.

Katz, Barry J., editor. *Petroleum Source Rocks*. Springer, 2012.

Kaufman, Wallace, and Orrin Pilkey. *The Beaches Are Moving: The Drowning of America's Shoreline*. Duke UP, 1983.

Kaye, Catheryn Berger, and Philippe Cousteau. *Going Blue*. Free Spirit Publishing, 2009.

Kearny, Philip, editor. *The Encyclopedia of the Solid Earth Sciences*. Blackwell, 1994.

Keller, E. A. *Environmental Geology*. 9th ed., Prentice-Hall, 2010.

Keller, Edward A. *Environmental Geology*. 9th ed., Prentice-Hall 2010.

Keller, Edward A. *Introduction to Environmental Geology*. Prentice-Hall, 2012.

Kelley, Joseph T., Orrin H. Pilkey, and J. A. G. Cooper. *America's Most Vulnerable Coastal Communities*. Geological Society of America, 2009.

Kelly, Robert L., and David Hurst Thomas. *Archaeology*. 6th ed., Wadsworth, 2013.

Kenchington, Richard, Laura Stocker, and David Wood, editors. *Sustainable Coastal Management and Climate Adaptation: Global Lessons from Regional Approaches in Australia*. CRC Press, 2012.

Kennedy, Merrit. "Great Barrier Reef Hit By Bleaching for the Second Year in a Row." *NPR*, 10 Apr. 2017, www.npr.org/sections/thetwo-way/2017/04/10/523254085/great-barrier-reef-hit-by-bleaching-for-the-second-year-in-a-row.

Kennedy, Victor S. *Shifting Baselines in the Chesapeake Bay: An Environmental History*. Johns Hopkins UP, 2018.

Kennish, Michael J. *Practical Handbook of Marine Science*. 3rd ed., CRC Press, 2001.

Kent, Alexander, and Peter Vujakovic, editors. *The Routledge Handbook of Mapping and Cartography*. Routledge, 2017.

Kernan, Martin, Richard W. Battarbee, and Brian Moss, editors. *Climate Change Impacts of Freshwater Ecosystems*. Wiley Blackwell, 2010.

Khare, Vikas, Cheshta Khare, Savita Nema, and Prashant Baredar. *Tidal Energy Systems: Design, Optimization and Control*. Elsevier, 2019.

Khare, Vikas, Savita Nema, and Prashant Baredar. *Ocean Energy Modeling and Simulation with Big Data*. Butterworth-Heinemann, 2020.

Khodorevskaya, R. P., G. I. Ruban, and D. S. Pavlov. *Behaviour, Migrations, Distribution and Stocks of Sturgeons in the Volga-Caspian Basin*. World Sturgeon Conservation Society Special Publication No. 3, 2009.

Khokhar, Tariq. "Chart: Globally, 70% of Freshwater Is Used for Agriculture." *The World Bank*, 22 Mar. 2017, blogs.worldbank.org/opendata/latinamerica/chart-globally-70-freshwater-used-agriculture.

King, Charles. *The Black Sea: A History*. Oxford UP, 2005.

Kirby, Richard R. *Ocean Drifters: A Secret World Beneath the Waves*. Studio Cactus, 2010.

Kituyi, Mukhisa. "90% of Fish Stocks Are Used Up-Fisheries Subsidies Must Stop." *United Nations Conference on Trade and Development*, 13 July 2018, unctad.org/en/pages/newsdetails.aspx?Original VersionID=1812.

Klein, Cornelis, and Barbara Dutrow. *Manual of Mineral Science*. Wiley, 2008.

Klein, Cornelis, and C. S. Hurlbut, Jr. *Manual of Mineralogy*. 23rd ed., John Wiley & Sons, 2008.

Klinger, Barry A., and Thomas W. N. Haine. *Ocean Circulation in Three Dimensions*. Cambridge UP, 2019.

Knauss, John A., and Newell Garfield. *Introduction to Physical Oceanography.* 3rd ed., Waveland Press, 2017.

Knowles, Anne Kelly, and Amy Hillier, editors. *Placing History. How Maps, Spatial Data and GIS Are Changing Historical Scholarship.* ESRI Press, 2008.

Kolata, Gina Bari. "Geodesy: Dealing with an Enormous Computer Task." *Science*, vol. 200, no. 4340, 28 Apr. 1978, pp. 421-66.

Komar, Paul D., and J. Robert Moore, editors. *Handbook of Coastal Processes and Erosion.* CRC Press, 2018.

Kondratyev, K. Y., V. F. Krapivin, and Costas A. Varotsos. *Global Carbon Cycle and Climate Change.* Springer, 2003.

Konecny, Gottfried. *Geoinformation: Remote Sensing, Photogrammetry and Geographic Information Systems.* Taylor & Francis, 2003.

Koppel, Tom. *Ebb and Flow: Tides and Life on Our Once and Future Planet.* Dundern Press, 2007.

Koslow, J. A. *The Silent Deep: The Discovery, Ecology, and Conservation of the Deep Sea.* U of Chicago P, 2007.

Kostianoy, Andrey G., and A. Kosarev. *The Caspian Sea Environment.* Springer-Verlag, 2005.

Kostianoy, Andrey G., and Alexey N. Kosarev. *The Black Sea Environment.* Springer, 2007.

Kostigen, Thomas M. "The World's Largest Dump: The Great Pacific Garbage Patch." *Discover*, July 2008, discovermagazine.com/2008/jul/10-the-worlds-largest-dump.

Kosyan, Ruben, editor. *The Diversity of Russian Estuaries and Lagoons Exposed to Human Influence.* Springer, 2017.

Kowahata, Hodaka, and Yoshio Awaya. *Global Climate Change and Response of the Carbon Cycle in the Equatorial Pacific and Indian Oceans and Adjacent Landmasses.* Elsevier, 2006.

Kraak, Menno-Jan, and Ferjan Ormeling. *Cartography: Visualization of Spatial Data.* 3rd ed., Prentice-Hall, 2010.

Kraus, Nicholas, editor. *History and Heritage of Coastal Engineering: A Collection of Papers on the History of Coastal Engineering in Countries Hosting the International Coastal Engineering Conference 1950-1996.* American Society of Civil Engineers, 1996.

Krauskopf, Konrad B. *Introduction to Geochemistry.* 3rd ed., McGraw-Hill, 2003.

Krebs, Robert E. *The History and Use of Our Earth's Chemical Elements: A Reference Guide.* 2nd ed., Greenwood Press, 2006.

Kress, Nurit. *Marine Environmental Impact of Seawater Desalination.* Elsevier, 2019.

Kurlansky, Mark. *Cod: A Biography of the Fish That Changed the World.* Walker, 2007.

Kurten, T., et al. "Large Methane Releases Lead to Strong Aerosol Forcing and Reduced Cloudiness." *Atmospheric Chemistry and Physics Discussions*, vol. 11, no. 3, 2011, pp. 9057-81.

Laik, Sukumar. *Offshore Petroleum Drilling and Production.* CRC Press, 2018.

Laist, David W. *North Atlantic Right Whales: From Hunted Leviathan to Conservation Icon.* Johns Hopkins UP, 2017.

Lalic, Branislava, et al. *Agricultural Meteorology and Climatology.* Firenze UP, 2018.

Lam, Buuan, et al. "Major Structural Components in Freshwater Dissolved Organic Matter." *Environmental Science & Technology*, vol. 41, 2007, pp. 8240-47.

Larkum, Anthony W. D., Robert J. Orth, and Carlos M. Duarte, editors. *Seagrasses: Biology, Ecology and Conservation.* Springer, 2006.

Launer, Donald. *Navigation Through the Ages.* Sheridan House, 2009.

Law, Kara Lavender, Skye Morét-Ferguson, Nikolai A. Maximenko, et al. "Plastic Accumulation in the North Atlantic Subtropical Gyre." *Science*, vol. 329, no. 5996, 2010.

Law, Michael, and Amy Collins. *Getting to Know ArcGIS.* 4th ed., ESRI Press, 2015.

Laybourn-Parry, Johanna, Martyn Tranter, and Andrew J. Hodson. *The Ecology of Snow and Ice Environments.* Oxford UP, 2012.

Lazarus, Sarah. "Heat-Resistant Corals in the Middle East Could Save the World's Dying Reefs." *CNN*, www.cnn.com/2018/06/26/middleeast/middle-east-corals/index.html. Accessed 30 July 2018.

Le Quéré, Corinne, and Eric S. Saltzman, editors. *Surface Ocean-Lower Atmosphere Processes.* John Wiley & Sons, 2013.

Le'vêque, C. *Ecology from Ecosystem to Biosphere.* Science, 2003.

Lebedev, Albert T., editor. *Comprehensive Environmental Mass Spectrometry.* ILM Publications, 2012.

Leeder, Mike R. *Sedimentology and Sedimentary Basins: From Turbulence to Tectonics.* Wiley-Blackwell, 2011.

Leidy, Chris. *The Coral Triangle.* Assouline, 2020.

Lele, Ajey. *Strategic Technologies for the Military: Breaking New Frontiers.* Sage, 2009.

Lenton, Tim. *Earth System Science: A Very Short Introduction.* Oxford UP, 2016.

Leonard, Barry, editor. *Floods: The Awesome Power.* DIANE Publishing, 2009.

Leonard, Kathleen M. "Brief History of Environmental Engineering: 'The World's Second Oldest Profession.'" *ASCE Conference Proceedings*, vol. 265, no. 47, 2001, pp. 389-93.

Leposky, Rosalie E. "Retaining Walls: What You See and What You Don't." *Erosion Control*, vol. 11, no. 1, Jan.-Feb. 2004.

Lesser, Michael P. "Coral Bleaching: Causes and Mechanisms." *Coral Reefs: An Ecosystem in Transition*, edited by Zvy Dubinsky and Noga Stambler, Springer, 2011.

———. "Coral Reef Bleaching and Global Climate Change: Can Corals Survive the Next Century?" *PNAS*, vol. 104, no. 13, 2007, pp. 5259-60.

Letcher, Trevor M., editor. *Future Energy: Improved, Sustainable, and Clean*. Elsevier, 2008.

Letcher, Trevor, and Daniel Vallero, editors. *Waste: A Handbook for Management*. Academic Press/Elsevier, 2011.

Liang, X. San. *The Slow Coastal-Trapped Waves Off Subei Bank in the Yellow Sea and Their Climatic Change in the Past Decades*. IntechOpen, 2018.

Libes, Susan M. *Introduction to Marine Biogeochemistry*. Academic Press, 2009.

Lighthill, James, and Robert Pearce, editors. *Monsoon Dynamics*. Cambridge UP, 2009.

Lim, Wee Ho, and Michael L. Roderick. *An Atlas of the Global Water Cycle: Based on the IPCC AR4 Climate Models*. ANU E Press, 2009.

Lindsey, Rebecca. "Climate Change: Global Sea Level." *Climate.gov*, NOAA, 11 Sept. 2017, www.climate.gov/news-features/understanding-climate/climate-change-global-sea-level.

Lipps, Jere H. "What, if Anything, Is Micropaleontology?" *Paleobiology*, vol. 7, no. 2, 1981, pp. 167-99.

Lipschultz, Fredric, et al. "New Production in the Sargasso Sea: History and Current Status." *Global Biogeochemical Cycles*, vol. 16, no. 1001, 2002.

Lisitsyn Alexander P., and Liudmila L. Demina, editors. *Sedimentation Processes in the White Sea: The White Sea Environment, Part 2*. Springer, 2019.

Lisitsyn, Alexander P., and Viacheslav V. Gordieev, editors. *Biogeochemistry of the Atmosphere, Ice and Water of the White Sea: The White Sea Environment, Part 1*. Springer, 2019.

"List of Freshwater Fishes Reported from Fiji (Republic Of)." *Fishbase*, 2015, www.fishbase.de/Country/CountryChecklist.php?resultPage=2&c_code=242&vhabitat=fresh. Accessed 17 Aug. 2018.

Livingston, Robert J. *Climate Change and Coastal Ecosystems*. CRC Press, 2015.

Lockwood, John P., and Richard W. Hazlett. *Volcanoes: Global Perspectives*. Wiley-Blackwell, 2010.

Longhurst, Alan R. *Ecological Geography of the Sea*. Elsevier Science and Technology Books, 2006.

Longhurst, Alan. *Mismanagement of Marine Fisheries*. Cambridge UP, 2010.

Longley, Paul A., et al. *Geographic Information Systems and Science*. 4th ed., Wiley, 2015.

Lotze, Heike K., Marta Coll, and Jennifer A. Dunne. "Historical Changes in Marine Resources, Food-Web Structure and Ecosystem Functioning in the Adriatic Sea, Mediterranean." *Ecosystems*, vol. 14, no. 2, 2011.

Lowell, Robert P., Jeffrey S. Seewald, Anna Metaxas, and Michael R. Perfit, editors. *Magma to Microbe: Modeling Hydrothermal Processes at Oceanic Spreading Centers*. John Wiley & Sons, 2013.

Lowen, James. *Antarctic Wildlife: A Visitor's Guide*. Princeton UP, 2011.

Lucero, Olivia C., and Carter B. Keyes. *New Developments in Global Warming Research*. Nova Science, 2013.

Ludewig, Elke. *On the Effect of Offshore Wind Farms on the Atmosphere and Ocean Dynamics*. Springer, 2015.

Lundgren, Lawrence W. *Environmental Geology*. 2nd ed., Prentice-Hall, 2010.

Lund-Hansen, et al. *Arctic Sea Ice Ecology: Seasonal Dynamics in Algal and Bacterial Productivity*. Springer, 2020.

Lutgens, Frederick K., and Edward J. Tarbuck. *Essentials of Geology*. Prentice-Hall, 2012.

———. *Essentials of Geology*. Illus. Dennis G. Tasa. 12th ed., Prentice, 2015.

———. *The Atmosphere: An Introduction to Meteorology*. 11th ed., Prentice-Hall, 2010.

———. *The Atmosphere: An Introduction to Meteorology*. 12th ed., Prentice, 2013.

Lutgens, Frederick K., Edward J. Tarbuck, and Dennis Tasa. *Earth: An Introduction to Physical Geology*. 10th ed., Prentice-Hall, 2010.

———. *Essentials of Geology*. 11th ed., Prentice-Hall, 2011.

Lymbery, Philip. *Dead Zone: Where the Wild Things Were*. Bloomsbury Publishing, 2017.

Macdougall, Douglas. *Frozen Earth*. U of California P, 2006.

Mackelworth, Peter, editor. *Marine Transboundary Conservation and Protected Areas*. Routledge, 2016.

MacLeish, Summer. *Seven Words for Wind: Essays and Field Notes from Alaska's Pribilof Islands*. U of Alaska P, 2008.

Madrigal, Alexis C. "The World Has Never Seen an Oil Spill Like This." *The Atlantic*, 19 Jan. 2018, www.theatlantic.com/technology/archive/2018/01/the-oil-spill-that-wasnt/550820/.

Mahan, A. T. *The Influence of Sea Power upon History, 1660-1783*. 1890. Reprint. Barnes & Noble Books, 2004.

Mani, J. S. *Coastal Engineering*. 2nd ed., PHI Learning, 2018.

Mann, K. H., and J. R. N. Lazier. *Dynamics of Marine Ecosystems: Biological-Physical Interactions in the Oceans*. 3rd ed., John Wiley & Sons, 2013.

March, James H. "Hudson Bay." *The Canadian Encyclopedia*, Historica Canada, 4 Mar. 2015, www.thecanadianencyclopedia.ca/en/article/hudson-bay/.

Margariti, Roxani Eleni. *Aden & the Indian Ocean Trade: 150 Years in the Life of a Medieval Arabian Port*. U of North Carolina P, 2007.
Marques de Sá, J. P. *Pattern Recognition: Concepts, Methods and Applications. Springer Books, 2001.*
Marsh, William M., and Martin M. Kaufman. "Physical Geography" Great Systems and Global Environments. Cambridge UP, 2012.
Marshal, John, and R. Alan Plumb. *Atmosphere, Ocean and Climate Dynamics: An Introductory Text*. Elsevier Academic Press, 2008.
Marshall, N. A., P. A. Marshall, A. Abdulla, et al. *Preparing for Climate Change in the Red Sea: Recognizing Early Impacts Through Perceptions of Dive Tourists and Dive Operators in Egypt*. IUCN, 2009.
Martin, James E., and David C. Parris, editors. *The Geology and Paleontology of the Late Cretaceous Marine Deposits of the Dakotas*. Geological Society of America, 2009.
Martin, Karen L. M. *Beach-Spawning Fishes. Reproduction in an Endangered Ecosystem*. CRC Press, 2015.
Martin, Seelye. *An Introduction to Ocean Remote Sensing*. 2nd ed., Cambridge UP, 2014.
Martini, I. Peter, Victor R. Baker, and Guillermina Garzón, editors. Flood and Megaflood Processes and Deposits: Recent and Ancient Examples. John Wiley & Sons, 2009.
Maser, Chris. *Interactions of Land, Ocean and Humans: A Global Perspective*. CRC Press, 2015.
Maslin, Mark. *Global Warming: Causes, Effects and the Future*. MBI, 2007.
Masud, Muhammad Mehedi. *Conservation of Marine Resources and Sustainable Coastal Development in Malaysia*. Palgrave Macmillan, 2019.
Mathiesen, Karl. "World's Oceans Facing Biggest Coral Die-Off in History, Scientists Warn." *Guardian*, 8 Oct. 2015, www.theguardian.com/environment/2015/oct/08/worlds-oceans-facing-biggest-coral-die-off-in-history-scientists-warn.
Matthews, Chris. "America Now Leads the World in This Surprising Category." *Fortune*, 5 July 2016, fortune.com/2016/07/05/oil-reserves-us.
Max, Michael D., editor. *Natural Gas Hydrate in Oceanic and Permafrost Environments*. Springer Science & Business Media, 2012.
Maxlow, James. *On the Origin of Continents and Oceans: A Paradigm Shift in Understanding*. Terrella Press, 2014.
Mays, Larry W. *Ground and Surface Water Hydrology*. Wiley, 2011.
Mays, Larry W. *Water Resources Engineering*. 2nd ed., Wiley, 2010.
McBeath, Jenifer Huang, and Jerry McBeath. *Environmental Change and Food Security in China*. Springer, 2010.
McCabe, Declan. "Rivers and Streams: Life in Flowing Water." *Nature Education*, 2011. http://www.nature.com/scitable/knowledge/library/rivers-and-streams-life-in-flowing-water-23587918. Accessed on 2 Nov. 2016.
McCann, Joy. *Wild Sea: A History of the Southern Ocean*. U of Chicago P, 2018.
McCarthy, John K., Jonathan Benjamin, Trevor Winton, and Wendy van Duivenvoorde, editors. *3D Recording and Interpretation for Maritime Archaeology*. Springer, 2019.
McCarthy, Kevin T., Thomas Pichler, and Roy E. Price. "Geochemistry of Champagne Hot Springs Shallow Hydrothermal Vent Field and Associated Sediments, Dominica, Lesser Antilles." *Chemical Geology*, vol. 224, 2005, pp. 55-68.
McCorduck, Pamela. *Machines Who Think: A Personal Inquiry into the History and Prospects of Artificial Intelligence*. 2nd ed.,a A. K. Peters, 2004.
McCully, James Greig. *Beyond the Moon: A Conversational, Common Sense Guide to Understanding the Tides*. World Scientific Publishing, 2006.
McDonnell, Michael. *Masters of Empire: Great Lakes Indians and the Making of America*. Hill and Wang, 2015.
McGoodwin, James R. *Crisis in the World's Fisheries: People, Problems, and Policies*. Stanford UP, 1990.
McKeown, Alice. "Coral Reefs Under Threat." *World Watch*, vol. 23, no. 1, 2010.
McKinnell, S. M., and M. J. Dagg, editors. *Marine Ecosystems of the North Pacific Ocean 2003-2008*. PICES, 2010.
McKinney, Frank K. *The Northern Adriatic Ecosystem, Deep Time in a Shallow Sea*. Columbia UP, 2007.
McKinney, Michael L., et al. *Environmental Science: Systems and Solutions*. 4th ed., Jones and Bartlett Publishers, 2007.
McLachlan, Anton, and Alec Brown. *The Ecology of Sandy Shores*. Elsevier, 2006.
McLellan, Hugh J. *Elements of Physical Oceanography*. Elsevier, 2016.
Merritts, Dorothy, Andrew De Wet, and Kirsten Menking. *Environmental Geology: An Earth System Science Approach*. W. H. Freeman, 1998.
Metcalf & Eddy, Inc. *Wastewater Engineering: Treatment and Reuse*. 4th ed., McGraw-Hill, 2003.
Micale, Giorgio, Andrea Cipollina, and Lucio Rizutti, editors. *Seawater Desalination: Conventional and Renewable Energy Processes*. Springer-Verlag, 2009.
Micallef, Anton, and Allan Williams. *Beach Management: Principles and Practice*. Earthscan, 2009.
Middelburg, Jack J. *Marine Carbon Biogeochemistry. A Primer for Earth System Scientists*. Springer Open, 2019.
Middleton, Gerard V., editor. *Encyclopedia of Sediments and Sedimentary Rocks*. Springer, 2003.

Miglorie, Catherine. *Vermont's Marble Industry*. Arcadia, 2013.

Migon, Christophe, Paul Nival, and Antoine Sciandra, editors. *The Mediterranean Sea in the Era of Global Change 2: 30 Years of Interdisciplinary Study of the Ligurian Sea*. Wiley, 2020.

Mikhalev, Yuri. *Whales of the Southern Ocean: Biology, Whaling and Perspectives of Recovery*. Springer, 2019.

Miles, Kathryn. "Can Anyone Save the North Atlantic Right Whale?" *The Boston Globe*, 11 Apr. 2018, www.bostonglobe.com/magazine/2018/04/11/can-anyone-save-north-atlantic-right-whale/iS76fBOUn9PdnL9LySgmhN/story.html.

Miller, Robert N. *Numerical Modeling of Ocean Circulation*. Cambridge UP, 2007.

Millero, Frank J. *Chemical Oceanography*. 4th ed., CRC Press, 2013.

Milliman, J. D. *Recent Sedimentary Carbonates. Part 1: Marine Carbonates*. Springer Science & Business Media, 2012.

Milman, Oliver. "Deepwater Horizon Disaster Altered Building Blocks of Ocean Life." *The Guardian*, 28 June 2018, www.theguardian.com/environment/ 2018/jun/28/bp-deepwater-horizon-oil-spill-report.

Miloslavich, Patricia, et al., editors. *Caribbean Marine Biodiversity: The Known and Unknown*. DEStech Publications, 2005.

Mines Jr., Richard O. *Environmental Engineering Principles and Practices*. Wiley Blackwell, 2014.

Mityagina, M. I., O. Y. Larova, and S. S. Karimova. "Multi-sensor Survey of Seasonal Variability in Coastal Eddy and Internal Wave Signatures in the North-Eastern Black Sea." *International Journal of Remote Sensing*, vol. 31, nos. 17/18, 2010, pp. 4779-90.

Molden, D. *Water for Food, Water for Life: A Comprehensive Assessment of Water Management in Agriculture*. Earthscan/IWMI, 2007.

Molnar, Peter. *Plate Tectonics: A Very Short Introduction*. Oxford UP, 2015.

Molyneaux, Paul. *Swimming in Circles: Aquaculture and the End of Wild Oceans*. Thunder's Mouth Press, 2007.

Monaco, André, and Patrick Prouzet, editors. *Marine Ecosystems, Diversity and Functions. ISTE/John Wiley & Sons, 2015.*

Monroe, James S., Reed Wicander, and Richard Hazlett. *Physical Geology: Exploring the Earth*. 6th ed., Thomson Higher Education, 2007.

Montgomery, C. W. *Environmental Geology*. 9th ed., McGraw-Hill, 2010.

Mooney, Chris. "New Science Suggests the Ocean Could Rise More-And Faster-Than We Thought." *Chicago Tribune*, 26 Oct. 2017, www.chicagotribune.com/news/nationworld/science/ct-climate-change-sea-level-rise-20171026-story.html.

Mooney, Chris. *Storm World: Hurricanes, Politics and the Battle Over Global Warming*. Harcourt, 2007.

Moore, Charles, and Cassandra Phillips. *Plastic Ocean: How a Sea Captain's Chance Discovery Launched a Determined Quest to Save the Oceans*. Avery, 2011.

Moore, John E. *Field Hydrogeology: A Guide for Site Investigations and Report Preparation*. 2nd ed., CRC, 2012.

Moore, Steven W., Harry Bohm, and Vickie Jensen. *Underwater Robotics: Science, Design and Fabrication*. Marine Advanced Technology Education Center, 2010

Moore, Sue E., Douglas P. DeMaster, and Paul K. Dayton. "Cetacean Habitat Selection in the Alaskan Arctic During Summer and Autumn." *Arctic*, vol. 53, no. 4, 2000.

Moores, Eldridge M., and Robert J. Twiss. *Tectonics*. Waveland, 2014.

Mora, Camilo, editor. *Ecology of Fishes on Coral Reefs*. Cambridge UP, 2015.

Morewood, Steve. "Suez: The Canal before the Crisis." *History Today*, vol. 56, no. 11, 2006, p. 38.

Morgan, Lisa A. *Integrated Geoscience Studies in the Greater Yellowstone Area: Volcanic, Tectonic, and Hydrothermal Processes in the Yellowstone Geoecosystem*. U.S. Geological Survey, 2007.

Morosov, Eugene G. *Oceanic Internal Tides: Observations, Analysis and Modeling. A Global View*. Springer, 2018.

Morozov, Eugene G., Alexander N. Demidov, Roman Y. Tarakanov, and Walter Zenk. *Abyssal Channels in the Atlantic Ocean: Water Structure and Flows*. Springer, 2010.

Morris, Robert M., et al. "SAR11 Clade Dominates Ocean Surface Bacterioplankton Communities." *Nature*, vol. 420, Dec. 2002.

Mossop, Joanna. *The Continental Shelf Beyond 200 Nautical Miles: Rights and Responsibilities*. Oxford UP, 2016.

Mukhopadhyay, Ranadhir, Anil K. Ghosh, and Sridhar D. Iyer. *The Indian Ocean Nodule Field: Geology and Resource Potential,* edited by M. Hale, Elsevier, 2012.

Mukhopadhyay, Ranadhir, Anil K. Ghosh, and Sridhar D. Iyer. *The Indian Ocean Nodule Field: Geology and Resource Potential*. 2nd ed., Elsevier, 2018.

Murawski, Steven A., Cameron H. Ainsworth, Sherryl Gilbert, et al, editors. *Deep Oil Spills. Facts, Fate and Effects*. Springer, 2020.

Murdy, Edward O., and John A. Musick. *Field Guide to Fishes of the Chesapeake Bay*. Johns Hopkins UP, 2013.

Murray, John, and Alphonse J. Renard. *Deep Sea Deposits*. Neill, 1891.

Murray, Steven N., Richard F. Ambrose, and Megan N. Dethier. *Monitoring Rocky Shores*. U of California P, 2006.

Murray, Tim. *Milestones in Archaeology: A Chronological Encyclopedia*. ABC-CLIO, 2007.

Nagai, Takeyoshi, Hiroaki Saito, Koji Suzuki, and Motomitsu Takahashi, editors. *Kuroshio Current. Physical, Biogeochemical, and Ecosystem Dynamics.* Wiley, 2019.

Nagarkar, Sanjay. *The Ecology of Intertidal Epilithic Biofilms with Special Reference to Cyanobacteria.* BiblioBazaar, 2017.

Nairn, Alan E. M., and Francis G. Stehli. *The Ocean Basins and Margins Volume 6: The Indian Ocean.* Springer Science & Business Media, 2013.

"NASA, NOAA Analyses Reveal Record-Shattering Global Warm Temperatures in 2015." *Goddard Institute for Space Studies.* NASA, 20 Jan. 2016.

Nash, J. Madeleine. *El Niño: Unlocking the Secrets of the Master Weather-Maker.* Warner Books, 2002.

National Academy of Sciences, Engineering and Medicine, 2019. *A Research Review of Interventions to Increase the Persistence and Resilience of Coral Reefs.* National Academies Press, 2019.

National Ocean Service, National Oceanic and Atmospheric Administration, "What Is a Tsunami?" oceanservice.noaa.gov/facts/tsunami.html. Accessed 20 Mar. 2018.

National Ocean Service, National Oceanic and Atmospheric Administration, "What Is Remote Sensing?" oceanservice.noaa.gov/facts/remote sensing.html. Accessed 7 Feb. 2019.

National Oceanic and Atmospheric Administration. "Chesapeake Bay Office." chesapeakebay.noaa.gov.

National Research Council. *Achieving Nutrient and Sediment Reduction Goals in the Chesapeake Bay: An Evaluation of Program Strategies and Implementation.* National Academies Press, 2011.

National Research Council. *Desalination: A National Perspective.* National Academies Press, 2008.

National Research Council. *Hydrology, Ecology and Fishes of the Klamath River.* National Academies Press, 2008.

National Research Council. *Macondo Well* Deepwater Horizon *Blowout: Lessons for Improving Offshore Drilling Safety.* National Academies Press, 2012.

National Research Council. *Restoring and Protecting Marine Habitat: The Role of Engineering and Technology.* National Academy Press, 1994.

National Research Council. *Scientific Ocean Drilling. Accomplishments and Challenges.* National Academies Press, 2011.

National Snow and Ice Data Center. "All About Sea Ice." nsidc.org/seaice/characteristics/difference.html.

Nauels, Alexander, et al. "Linking Sea Level Rise and Socioeconomic Indicators under the Shared Socioeconomic Pathways." *Environmental Research Letters,* vol. 12, no. 11, 26 Oct. 2017. *IOP Science* iopscience.iop.org/article/10.1088/1748-9326/aa92b6#erlaa92b6f2.

Navarra, Antonio, editor. *Beyond El Niño: Decadal and Interdecadal Climate Variability.* Springer Science & Business Media, 2012.

Nazarenko, Sergey. *Fluid Dynamics via Examples and Solutions.* CRC Press, 2015.

Neill, Simon P., and M. Reza Hashemi. *Fundamentals of Ocean Renewable Energy. Generating Electricity from the Sea.* Academic Press, 2018.

Nentwig, Wolfgang, editor. *Biological Invasions.* Springer, 2008.

Nichols, C. Reid, et al. *Recent Advances in Oceanography.* Greenwood, 2003.

Nichols, Gary. *Sedimentology and Stratigraphy.* John Wiley & Sons, 2013.

Nicol, Stephen I. *The Curious Life of Krill: A Conservation Story from the Bottom of the World.* Island Press, 2018.

Nicolas, Adolphe. *The Mid-Oceanic Ridges: Mountains Below Sea Level.* Springer Science & Business Media, 2013.

Nihoul, Jacques C. J., Peter O. Zavialov, and Philip P. Micklin, NATO Scientific Affairs Division. *Dying and Dead Seas: Climatic versus Anthropic Causes.* Kluwer Academic Publishing, 2004.

Normile, Dennis, and Richard A. Kerr. "A Sea Change in Ocean Drilling." *Science,* vol. 300, 2003, p. 410.

Norris, Joanne E., et al., editors. *Slope Stability and Erosion Control: Ecotechnological Solutions.* Springer, 2010.

Norton, Robert. L. *An Inventory of Sea Birds of the Caribbean.* UP of Florida, 2009.

Novotny, Vladimir. *Water Quality: Diffuse Pollution and Watershed Management.* 2nd ed., John Wiley & Sons, 2003.

Nunn, Patrick D. *Climate, Environment and Society in the Pacific during the Last Millennium.* Elsevier, 2007.

O'Laughlin, Karen Fay, and James F. Lander. *Caribbean Tsunamis: A 500-Year History from 1498-1998.* Kluwer Academic Publishers, 2003.

Oceanography Course Team. *Ocean Circulation.* 2nd ed., Butterworth-Heinemann, 2001.

Oelkers, Eric H., editor. *Thermodynamics and Kinetics of Water-Rock Interaction: Reviews in Mineralogy and Geochemistry.* Mineralogical Society of America, 2009.

Ogata, Kei, Andrea Festa, and Gian Andrea Pini, editors. *Submarine Landslides: Subaqueous Mass Transport Deposits from Outcrops to Seismic Profiles.* Wiley, 2020.

Ogawa, Yujiro, Ryo Anma, and Yildirim Dilek. *Accretionary Prisms and Convergent Margin Tectonics in the Northwest Pacific Basin.* Springer Science+ Business Media, 2011.

Okada, Alexandra, Simon J. Buckingham Shum, and Tony Sherborne, editors. *Knowledge Cartography: Software Tools and Mapping Techniques.* Springer, 2008.

Oke, Timothy R., Gerald Mills, Andreas Christen, and James A. Voogt. *Urban Climates.* Cambridge UP, 2017.

Oreskes, Naomi. *Plate Tectonics: An Insider's History of the Modern Theory.* CRC Press, 2018.

Organisation for Economic Development and Cooperation. *OECD Territorial Reviews: Trans-border Urban Co-operation in the Pan Yellow Sea Region.* OECD, 2009.

"Our Friends Electric." *Economist*, 7 Sept. 2013.

Palmer, Andrew. *Introduction to Petroleum Exploration and Engineering.* World Scientific, 2017.

Paone, Angelo. "The Geochemical Evolution of the Mt. Somma-Vesuvius Volcano." *Mineralogy and Petrology*, vol. 87, nos. 1-2, 2006, pp. 53-80.

Park, Chul. *Yellow Sea and East China Sea Reported by PICES and Korean Monitoring Program.* Chungnam National U, 2012.

Parsons, Timothy R., et al. *Biological Oceanographic Processes.* Elsevier, 2013.

Partain, Roy Andrew. *Environmental Hazards from Offshore Methane Hydrate Operations.* Wolters Kluwer, 2017.

Pascual-Fernández, José J., Cristina Pita, and Maarten Bavinck, editors. *Small-Scale Fisheries in Europe: Status, Resilience and Governance.* Springer, 2020.

Pastero, Linda, editor. *Carbonates.* MDPI, 2019.

Paterson, Donald Edgar, and Susan Farran. *South Pacific Land Systems.* U of the South Pacific P, 2013.

Patterson, William D. *Klamath River Estuary Wetlands Restoration Prioritization Plan v1.0.* Yurok Tribe Environmental Program, 2009.

Pauly, D. "Global Fisheries: A Brief Review." *Journal of Biological Research-Thessaloniki*, vol. 9, no. 3-9, 2008.

Pavia, Frank. *Biogeochemical Studies of the South Pacific Ocean Using Thorium and Protactinium Isotopes.* Columbia UP, 2020.

Pearson, Michael Naylor. *The Indian Ocean.* Routledge, 2003.

Pecher, Arthur, and Jens Peter Kofoed, editors. *Handbook of Ocean Wave Energy.* Springer Open, 2017.

Pedersen, Soren Anker, et al. "Natura 2000 Sites and Fisheries in German Offshore Waters." *ICES Journal of Marine Science*, vol. 6, 2009.

Pedlosky, Joseph. *Ocean Circulation Theory.* Rev. ed. Springer, 2010.

Pei, De-Sheng, and Muhammad Junaid, editors. *Marine Ecology: Current and Future Developments (Volume 1). Marine Pollution: Current Status, Impacts and Remedies.* Bentham Books, 2019.

Pelinovsky, Efim, and Christian Kharif, editors. *Extreme Ocean Waves.* 2nd ed., Springer, 2016.

Pellant, Chris. *Smithsonian Handbook: Rocks and Minerals.* Dorling, 2002.

Perillo, Gerardo M. E., Eric Wolanski, Donald R. Cahoon, and Mark M. Brinson, editors. *Coastal Wetlands: An Integrated Ecosystem Approach.* Elsevier, 2009.

Perkins, Dexter. *Mineralogy.* Prentice-Hall, 2011.

Perkins, Sid. "Oceans Yield Huge Haul of Plastic." *Science News*, vol. 177, no. 7, 2010.

Perlez, Jane. "Tribunal Rejects Beijing's Claims in South China Sea." *The New York Times*, 12 July 2016, www.nytimes.com/2016/07/13/world/asia/south-china-sea-hague-ruling-philippines.html.

Perry, Sarah J., Shayne McGregor, Alex Sen Gupta, and Matthew H. England. *Future Changes to El Niño-Southern Oscillation Temperature and Precipitation Teleconnections.* John Wiley & Sons, 2017.

Pertsova N. M., and K. N. Kosobokova. "Interannual and Seasonal Variation of the Population Structure, Abundance, and Biomass of the Arctic Copepod *Calanus Glacialis* in the White Sea." *Oceanology*, vol. 50, no. 4, 2010.

Philpotts, Anthony, and Jay Aque. *Principles of Igneous and Metamorphic Petrology.* 2nd ed., Cambridge UP, 2009.

Phinney, Jonathan T., Ove Hoegh-Guldberg, Joanie Kleypas, William Skirving, and Al Strong, editors. *Coral Reefs and Climate Change: Science and Management.* American Geophysical Union, 2006.

Pick, James B. "Geographic Information Systems: A Tutorial and Introduction." *Communications of AIS*, vol. 14, 2004, pp. 307-31.

Pickering, Kevin T., and Richard N. Hiscott. *Deep Marine Systems: Processes, Deposits, Environments, Tectonics and Sedimentation.* Wiley, 2016.

Pierrehumbert, Raymond T. *Principles of Planetary Climate.* Cambridge UP, 2011.

Pilkey, Orrin H., and Rob Young. *The Rising Sea.* Island Press, 2009.

Pinet, Paul R. *Invitation to Oceanography.* 5th ed., Jones and Bartlett Publishers, 2009.

Pitcher, Tony J., et al. *Seamounts: Ecology, Fisheries and Conservation.* John Wiley & Sons, 2008.

Plummer, Charles C., Diane H. Carlson, and Lisa Hammersley. *Physical Geology.* 13th ed., McGraw-Hill, 2009.

Polovina J. J., et al. "The Ocean's Least Productive Waters Are Expanding." *Geophysical Research Letters*, vol. 35, 2008.

Pongrácz, Eva, Victor Pavlov, and Niko Hänninen, editors. *Arctic Marine Sustainability: Arctic Maritime Businesses and the Resilience of the Marine Environment.* Springer, 2020.

Popovic, Marko B. *Biomechanics and Robotics.* CRC Press, 2013.

Pounder, Elton R. *The Physics of Ice.* Elsevier, 2013.

Powderham, Alan J., and Sancia Van Der Meij. *At the Heart of the Coral Triangle: Celebrating Biodiversity.* Taylor & Frances, 2020.

Prinet, Dominique F. *Celestial Navigation.* Friesen Press, 2015.

Probert, P. Keith. *Marine Conservation.* Cambridge UP, 2017.

Prothero, Donald R. *Bringing Fossils to Life: An Introduction to Paleobiology*. 3rd ed., Columbia UP, 2013.

———. *Catastrophes! Earthquakes, Tsunamis, Tornadoes, and Other Earth-Shattering Disasters*. Johns Hopkins UP, 2011.

Qu, J., Z. Xu, Q. Long, L. Wang, X. Shen, J. Zhang, and Y. Cai. *East China Sea-Global International Waters Assessment*. Linnaeus U and United Nations Environment Programme, 2005.

Quante, Marcus, and Franciscus Colijn, editors. *North Sea Region Climate Change Assessment.* Springer Open, 2016.

Raffaelli, David, and Stephen Hawkins, editors. *Intertidal Ecology.* Springer Science & Business Media, 2012.

Railkin, Alexander I. *Marine Biofouling: Colonization Processes and Defenses.* CRC Press, 2004.

Raizer, Victor. *Advances in Passive Microwave Remote Sensing of Oceans.* CRC Press, 2017.

———. *Optical Remote Sensing of Ocean Hydrodynamics.* CRC Press, 2019.

Rajput, Sanjeev, and Naresh Kumar Thakur. *Geological Controls for Gas Hydrates and Unconventionals*. Elsevier, 2016.

Rana, Sanjay, and Jayant Sharma, editors. *Frontiers of Geographic Information Technology*. Springer, 2006.

Randall, John E. *Reef and Shore Fishes of the South Pacific: New Caledonia to Tahiti and the Pitcairn Islands.* U of Hawaii P, 2005.

Rapai, William. *Lake Invaders: Invasive Species and the Battle for the Future of the Great Lakes*. Wayne State UP, 2016.

Rasul, Najeeb M. A. and Ian C. F. Stewart, editors. *The Red Sea: The Formation, Morphology, Oceanography and Environment of a Young Ocean Basin.* Springer, 2015.

Ratnayaka, Don D., Malcolm J. Brandt, and K. Michael Johnson. *Twort's Water Supply*. 6th ed., Butterworth-Heinemann, 2009.

Ray, G. Carleton, and Jerry McCormick-Ray. *Marine Conservation Science Policy and Management*. Wiley Blackwell, 2014.

Raymond, Martin, and William Leffler. *Oil and Gas Production in Nontechnical Language*. Penn Well, 2005.

Reddy, K. Ramesh, and Ronald D. DeLaune. *Biogeochemistry of Wetlands: Science and Applications* CRC Press, 2008.

"Redefining the SI Units." *National Physical Laboratory*, www.npl.co.uk/si-units/redefining-the-si-units/. Accessed 12 Dec. 2018.

Rees, W. G. *Physical Principles of Remote Sensing*. 2nd ed., Cambridge UP, 2001.

Reeve, Dominic, Andrew Chadwick, and Christopher Fleming. *Coastal Engineering: Processes, Theory and Design Practice*. 2nd ed., CRC Press, 2015.

Renfrew, Colin, and Paul G. Bahn, editors. *Archaeology: The Key Concepts*. Routledge, 2005.

Reynolds, Pamela L. "Seagrass and Seagrass Beds." *Ocean Portal*. Smithsonian Institution, 2015. Accessed 8 Aug. 2016.

Rice, Tony. *Deep Ocean*. Smithsonian Institution, 2009.

Rich, Thomas H., and Patricia Vickers-Rich. *Dinosaurs of Darkness: In Search of the Lost Polar World.* 2nd ed., Indiana UP, 2020.

Richardson, S. M., H. Y. McSween, Jr., and Maria Uhle. *Geochemistry Pathways and Processes*. 2nd ed., Prentice-Hall, 2003.

Riding, Robert E., and Stanley M. Awramik. *Microbial Sediments*. Springer-Verlag, 2010.

Riegl, B., and Sam Purkis. *Coral Reefs of the Gulf: Adaptation to Climatic Extremes*. Springer, 2012.

Rieutord, Michel. *Fluid Dynamics. An Introduction*. Springer, 2015.

Rigo, M., et al. "A Rise in the Carbonate Compensation Depth of Western Tethys in the Carnian (Late Triassic): Deep-Water Evidence for the Carnian Pluvial Event." *Palaeogeography, Palaeoclimatology, Palaeoecology*, vol. 246, 2007, pp. 188-205.

Riley, John L. *The Once and Future Great Lakes Country: An Ecological History.* McGill-Queens UP, 2013.

———. *Wetlands of the Hudson Bay Lowland: An Ontario Overview.* Nature Conservancy of Canada, 2011.

Rilov, Gil, and Jeffrey A. Crooks, editors. *Biological Invasions in Marine Ecosystems.* Springer, 2006.

Rizutti, Lucio, Hisham Mohamed Ettouney, and Andrea Cipollina, editors. *Solar Desalination for the 21st Century: A Review of Modern Technologies and Researches on Desalination Coupled to Renewable Energies*. Springer, 2007.

Roberts, Callum. *The Unnatural History of the Sea.* Island Press, 2007.

Roberts, Leslie Carol. *The Entire Earth and Sky: Views on Antarctica.* U of Nebraska P, 2008.

Robinson, A. R. *Eddies in Marine Science.* Springer Science & Business Media, 2012.

Robinson, Allan R., and Kenneth H. Brink. *The Global Coastal Ocean: Regional Studies and Syntheses.* Harvard UP, 2005.

Robinson, Arthur H., et al. *Elements of Cartography*. 6th ed., Wiley, 1995.

Robinson, Ian S. *Discovering the Ocean from Space: The Unique Applications of Satellite Oceanography*. Springer, 2010.

———. *Measuring the Oceans from Space: The Principles and Methods of Satellite Oceanography*. Springer/Praxis, 2004.

Robles. Alfredo C. *Endangered Species and Fragile Ecosystems in the South China Sea: Philippines vs. China Arbitration.* Palgrave Macmillan, 2020.

Roff, John, and Mark Zacharias. *Marine Conservation Ecology*. Earthscan, 2011.
Rogers, Rudy. *Offshore Gas Hydrates: Origins, Development and Production*. Elsevier, 2015.
Rogers, Will. "Climate Change and the South China Sea." *Center for New American Security*, www.cnas.org/blogs/naturalsecurity/2011/09/climate-change-fisheries-and-south-china-sea.html.
Rohling, Eelco. *The Oceans: A Deep History*. Princeton UP, 2017.
Rohwer, Forest, Merry Youle, and Derek Vosten. *Coral Reefs in the Microbial Seas*. Plaid Press, 2010.
Rona, Peter A. "Resources of the Seafloor." *Science*, vol. 299, no. 5607, 2003, pp. 673-74.
Rona, Peter A., Colin W. Devey, Jérôme Dyment and Bramley J. Murton, eds. *Diversity of Hydrothermal Systems on Slow Spreading Ocean Ridges. John Wiley & Sons, 2013.*
Rona, Peter A., Kurt Boström, Lucien Laubier, and Kenneth L. Smith Jr., editors. *Hydrothermal Processes at Seafloor Spreading Centers*. Springer Science & Business Media, 2013.
Rose, George A. *Cod: The Ecological History of the North Atlantic Fisheries*. Breakwater Books, 2007.
Rosenberg, Eugene, and Yossi Loya, editors. *Coral Health and Disease*. Springer, 2004.
Ross, Pierre-Simon, et al. "Basaltic to Andesitic Volcaniclastic Rocks in the Blake River Group, Abitibi Greenstone Belt: 2. Origin, Geochemistry, and Geochronology." *Canadian Journal of Earth Sciences*, vol. 48, 2011, pp. 757-77.
Rossing, Thomas, and Neville Fletcher. *Principles of Vibration and Sound*. 2nd ed., Springer, 2004.
Rothwell, Donald R. *Arctic Ocean Shipping. Navigation, Security and Sovereignty in the North American Arctic*. Brill, 2008.
Rumsey, Francis, and Tim McCormick. *Sound and Recording: An Introduction*. 7th ed., Elsevier, 2014.
Rusu, Eugen, and Vengatesan Venugopal. *Offshore Renewable Energy: Ocean Waves, Tides and Offshore Winds*. MDPI, 2019.
Ryan, Peter G. "Litter Survey Detects the South Atlantic 'Garbage Patch.'" *Marine Pollution Bulletin*, vol. 72, nos. 1-2, 2014, pp. 220-24.
Ryann, Amy L., and Nathan J. Perkins. *The Black Sea: Dynamics, Ecology and Conservation*. NOVA Science Publishers, 2011.
Ryans, J. L. "Pressure Measurement." *Kirk-Othmer Encyclopedia of Chemical Technology*. 5th ed., John Wiley & Sons, 2000.
Sala, Enric. "The Past and Present Topology and Structure of Mediterranean Subtidal Rocky-Shore Food Webs." *Ecosystems*, vol. 7, no. 4, 2004.
Samani, Hooman. *Cognitive Robotics*. CRC Press, 2016.
Santamaria-Fernandez, Rebeca. "Precise and Traceable Carbon Isotope Ratio Measurements by Multicollector ICP-Ms: What Next?" *Analytical & Bioanalytical Chemistry*, vol. 397, 2010, pp. 973-78.
Santhanam, R. *Fisheries Science*. Astral International, 2013.
Sarachik, Edward S., and Mark A. Cane. *The El Niño-Southern Oscillation Phenomenon*. Cambridge UP, 2010.
Saraswati, Pratul Kumar, and M. S. Srinavasan. *Micropaleontology: Principles and Applications*. Springer, 2016.
Sarp, Sarper and Nidal Hilal, editors. *Membrane-Based Salinity Gradient Processes for Water Treatment and Power Generation*. Elsevier, 2018.
Sassa, Kyoji, and Paolo Canuti, editors. *Landslides: Disaster Risk Reduction*. Springer, 2009.
Satake, Kenji, editor. *Tsunamis: Case Studies and Recent Developments*. Springer, 2005.
Schlager, Wolfgang. "Earth's Layers, Their Cycles, and Earth System Science." *Austrian Journal of Earth Sciences*, vol. 101, 2008, pp. 4-16.
Schminke, Hans-Ulrich. *Volcanism*. Springer, 2005.
Schmittner, Andreas, John C .H. Chiang, and Sidney R. Hemmings, editors. *Ocean Circulation: Mechanisms and Impacts-Past and Future Changes of Meridional Overturning*. John Wiley & Sons, 2013.
Schneider, Bonnie. *Extreme Weather: A Guide to Surviving Flash Floods, Tornadoes, Hurricanes, Heat Waves, Snowstorms, Tsunamis and Other Natural Disasters*. Palgrave Macmillan, 2012.
Schoff, Wilfred Harvey, and Aman Aman. *The Periplus of the Erythraean Sea: Travel and Trade in the Indian Ocean*, Creative Media Partners, 2018.
Scholten, Henk J., Rob van de Velde, and Niels van Manen, editors. *Geospatial Technology and the Role of Location in Science*. Springer, 2009.
Schopf, Thomas, and J. M. Schopf. *Paleoceanography*. Harvard UP, 2013.
Schroeder, J. H., and B. H. Purser. *Reef Diagenesis*. Springer Science & Business Media, 2012.
Schrope, Mark. "Conservation: Providential Outcome." *Nature*, vol. 451, no. 1, 2008.
Schulz, Horst D., and Matthias Zabel. *Marine Geochemistry*. 2nd ed., Springer-Verlag, 2006.
Schwartz, Maurice L. *Encyclopedia of Coastal Science*. Springer, 2005.
Schwartzstein, Peter. "The Black Sea Is Dying, and War Might Push It over the Edge." *Smithsonian*, 11 May 2016, www.smithsonianmag.com/science-nature/ black-sea-dying-and-war-might-push-it-over-edge- 180959053.
Scudder, Thayer. *The Future of Large Dams: Dealing with Social, Environmental, Institutional and Political Costs*. Earthscan, 2005.

Segar, D. A. *Introduction to Ocean Sciences*. 2nd ed., W. W. Norton & Company, 2007.
Segar, Douglas A. *Introduction to Ocean Sciences*. 2nd ed., W. W. Norton, 2007.
Seibold, E., and W. H. Berger. *The Sea Floor: An Introduction to Marine Geology*. 3rd ed., Springer-Verlag, 2010.
Seibold, Eugen, and Wolfgang Berger. *The Sea Floor: An Introduction to Marine Geology*. 4th ed., Springer, 2017.
Sei-Ichi, Saitoh, et al. "Some Operational Uses of Satellite Remote Sensing and Marine GIS for Sustainable Fisheries and Aquaculture." *ICES Journal of Marine Science/Journal du Conseil*, vol. 68, no. 4, 2011, pp. 687-95.
Seinfeld, J. H. *Atmospheric Chemistry and Physics of Air Pollution*. John Wiley & Sons, 1986.
Şekercioglu, Çagan Hakki, Sean Anderson, Erol Akçay, and Raşit Bilgin. "Turkey's Rich Natural Heritage Under Assault." *Science*, vol. 23, no. 12, 2011.
Sell, Nancy J. *Industrial Pollution Control: Issues and Techniques*. 2nd ed., Wiley, 1992.
Sen, P. C. *Principles of Electric Machines and Power Electronics*. 2nd ed., John Wiley & Sons, 1997.
Sene, Kevin. *Flash Floods: Forecasting and Warning*. Springer, 2013.
———. *Flood Warning, Forecasting and Emergency Response*. Springer, 2008.
Shanmugam, G. *Deep-Water Processes and Facies Models: Implications for Sandstone Petroleum Reservoirs*. Elsevier, 2006.
Shanov, Stefan, and Konstantin Kostov. *Dynamic Tectonics and Karst*. Springer, 2015.
Sharma, Rahul, editor. *Deep Sea Mining: Resource Potential, Technical and Environmental Considerations*. Springer, 2017.
———. *Environmental Issues of Deep Sea Mining: Impacts, Consequences and Policy Perspectives*. Springer, 2019.
Sharma, S. R. *Thermo-Physical Properties of Rocks: Special Reference to Deccan Trap Basalts*. Allied Publishers, 2015.
Shaw, George H. *Earth's Early Atmosphere and Oceans, and the Origin of Life*. Springer, 2016.
Shen, Samuel S. P., and Richard C. J. Somerville. *Climate Mathematics Theory and Applications*. Cambridge UP, 2019.
Shennan, Ian, Antony J. Long, and Benjamin P. Horton, editors. *Handbook of Sea-Level Research*. Wiley, 2015.
Sheppard, Charles R. C., Simon K. Davy, and Graham M. Pilling. *The Biology of Coral Reefs*. Oxford UP, 2009.
Sheppard, Charles. *Coral Reefs: A Very Short Introduction*. Oxford UP, 2014.
Shibayama, Tomoya. *Coastal Processes: Concepts in Coastal Engineering and Their Application to Multifarious Environment*. World Scientific Publishing, 2009.
Shoemaker, Robert W. *Metric for Me! A Layperson's Guide to the Metric System for Everyday Use with Exercises, Problems and Estimations*. 2nd ed., Blackhawk, 1998.
"SI Redefinition." *National Institute of Standards and Technology*, US Department of Commerce, 2018, www.nist.gov/si-redefinition. Accessed 12 Dec. 2018.
Siciliano, Bruno, and Oussama Khatib, editors. *The Springer Handbook of Robotics*. Springer, 2016.
Siciliano, Bruno, et al. *Robotics: Modeling, Planning, and Control*. Springer, 2009.
Siedler, Gerold, John Church, and John Gould. *Ocean Circulation and Climate: Observing and Modelling the Global Ocean*. Academic, 2001.
Siedler, Gerold, Stephen M. Griffies, John Gould, and John A. Church, editors. *Ocean Circulation and Climate: A 21st Century Perspective*. Elsevier, 2013.
Siegfried, W. R., P. R. Condy, and R. M. Laws, editors. *Antarctic Nutrient Cycles and Food Webs*. Springer Science & Business Media, 2013.
Siegfried, W. Roy. *Rocky Shores Exploitation in Chile and South Africa*. Springer Science & Business Media, 2013.
Sikora, Todd D., et al. "A Synthetic Aperture Radar-Based Climatology of Open-Cell Convection Over the Northeast Pacific Ocean." *Journal of Applied Meteorology and Climatology*, vol. 50, no. 3, 2011, pp. 594-603.
Simmons, George F. *Calculus Gems: Brief Lives and Memorable Mathematics*. 1992. Mathematical Association of America, 2007.
Simon, Thomas P., and Paul M. Stewart. *Coastal Wetlands of the Laurentian Great Lakes: Health, Habitat and Indicators*. AuthorHouse, 2007.
Sinclair, M., and G. Valdimarsson, editors. *Responsible Fisheries in the Marine Ecosystem*. CABI, 2014.
Sinding-Larsen, Richard, and Friedrich-W. Wellmer, editors. *Non-renewable Resource Issues: Geoscientific and Societal Challenges*. Springer, 2012.
Singer, Peter Warren. *Wired for War: The Robotics Revolution and Conflict in the Twenty-first Century*. Penguin Press, 2009.
Skinner, Brian J., and Barbara W. Murck. *The Blue Planet: An Introduction to Earth System Science*. 3rd ed., Wiley, 2011.
Slezak, Michael. "World's Largest Marine Park Created in Ross Sea in Antarctica in Landmark Deal." *The Guardian*, 27 Oct. 2016, www.theguardian.com/world/2016/oct/28/worlds-largest-marine-park-created-in-ross-sea-in-antarctica-in-landmark-deal.
Sloan, E. Dendy, Jr., and Carolyn Koh. *Clathrate Hydrates of Natural Gases*. 3rd ed., CRC Press, 2007.
Smith, F. G. Walton. "Measuring Trujillo, Alan P., and Harold V. Thurman. *Essentials of Oceanography*. 11th ed., Prentice, 2014.

Smithson, Peter, Kenneth Addison, and Kenneth Atkinson. "Energy and Earth." Fundamentals of the Physical Environment. 3rd ed., Routledge, 2002.

———. "Energy Flows and Nutrient Cycles in Ecosystems." Fundamentals of the Physical Environment. 3rd ed., Routledge, 2002.

———. "Heat and Energy in the Atmosphere." Fundamentals of the Physical Environment. 3rd ed., Routledge, 2002.

Snedden, John W, and William E. Galloway. *The Gulf of Mexico Sedimentary Basin: Depositional Evolution and Petroleum Applications.* Cambridge UP, 2019.

Sobolev, A. V., et al. "The Amount of Recycled Crust in Sources of Mantle-Derived Melts." *Science*, vol. 316, 2007, pp. 412-17.

Soennichsen, John. *Bretz's Flood: The Remarkable Story of a Rebel Geologist and the World's Greatest Flood.* Sasquatch Books, 2010.

Sokolow, Susanne Helene. *Ecology of Caribbean Coral Disease.* U of California P, 2007.

Sole, Tony. *Nga ti Ruanui: A History.* HUIA, 2005.

Soloviev, Alexander, and Roger Lukas. *The Near-Surface Layer of the Ocean: Structure, Dynamics and Applications.* 2nd ed., Springer, 2014.

Solyanko, K., V. Spiridonov, and A. Naumov. "Biomass: Commonly Occurring and Dominant Species of Macrobenthos in Onega Bay (White Sea, Russia): Data from Three Different Decades." *Marine Ecology*, vol. 32, no. 1, 2011.

Sommerville, Richard C. J. *The Forgiving Air: Understanding Environmental Change.* 2nd ed., American Meteorological Society, 2008.

Sonnenberg, G. J. *Radar and Electronic Navigation.* 6th ed., Elsevier, 2013.

Sorensen, Robert. *Basic Coastal Engineering.* 3rd ed., Springer, 2006.

Sorokin, I. O. U. *The Black Sea: Ecology and Oceanography.* Backhuys Publishing, 2002.

Sosson, M, et al., editors. *Sedimentary Basin Tectonics from the Black Sea and Caucasus to the Arabian Platform.* Special Publication 340. Geological Society of London, 2010.

Souchay, Jean, Stéphane Mathis, and Tadashi Tokieda, editors. *Tides in Astronomy and Astrophysics.* Springer, 2013.

South China Sea Project. "Reversing Environmental Degradation Trends in the South China Sea and Gulf of Thailand." *United Nations Environment Programme*, www.unepscs.org.

Spalding, Mark D., et al. "Marine Ecoregions of the World: A Bioregionalization of Coastal and Shelf Areas." *BioScience*, vol. 57, no. 7, 2007.

Spalding, Mark. *A Guide to the Coral Reefs of the Caribbean.* U of California P, 2004.

Spaulding, Nancy E., and Samuel N. Namowitz. *Earth Science.* Houghton Mifflin Harcourt, 2004.

Speight, James G. *Handbook of Offshore Oil and Gas Operations.* Elsevier, 2015.

Spellman, Frank R., and Nancy E. Whiting. *Environmental Science and Technology: Concepts and Applications.* 2nd ed., Government Institutes, 2006.

Sposito, Garrison, editor. *The Environmental Chemistry of Aluminum.* 2nd ed., CRC Press, 2020.

Springer, Paul J. *Military Robots and Drones: A Reference Handbook.* ABC-CLIO, 2013.

Stambler, Noga. *Life in the Mediterranean Sea: A Look at Habitat Changes.* NOVA Science, 2012.

The State of World Fisheries and Aquaculture. Food and Agriculture Organization of the United Nations, 2016, www.fao.org/3/a-i5555e.pdf. Accessed 26 July 2018.

Steadman, David. *Extinction and Biogeography of Tropical Pacific Birds.* U of Chicago P, 2006.

Steele, John H., Steve A. Thorpe, and Karl K. Turekian, editors. *Ocean Currents.* Academic Press, 2010.

Steele, John H., Steve A. Thorpe, and Karl K. Turekian, editors. *Marine Chemistry and Geochemistry.* Academic Press, 2010.

Stein, Ruediger, Donna K. Blackman, Fumio Inagaki, and Hans-Christian Larsen, editors. *Earth and Life Processes Discovered from Subseafloor Environments: A Decade of Science Achieved by the Integrated Ocean Drilling Program (IODP).* Elsevier, 2014.

Stein, Ruediger. *Arctic Ocean Sediments: Processes, Proxies and Paleoenvironment.* Elsevier, 2008.

Stephenson, F. Richard. *Historical Eclipses and Earth's Rotation.* Cambridge UP, 2008.

Stewart, D. B., and W. L. Lockhart. "An Overview of the Hudson Bay Marine Ecosystem." Fisheries and Oceans Canada, 2005. www.dfo-mpo.gc.ca/ libraries-biblio theques/toc-tdm/314704-eng.htm.

Stiansen, J. E., and A. A. Filin, editors. "Joint Polar Research Institute of Marine Fisheries and Oceanography (PINRO)/Institute of Marine Research (IMR) Report on the State of the Barents Sea Ecosystem 2006, with Expected Situation and Considerations for Management." *IMR/PINRO Joint Report Series*, vol. 2, 2007.

Stober, Ingrid, and Kurt Bucher, editors. *Hydrogeology of Crystalline Rocks.* Springer Science & Business Media, 2013.

———. *Water-Rock Interaction.* Springer Science & Business Media, 2012.

Stolberg, F., United Nations Environment Program. *Caspian Sea.* United Nations, 2006.

Stow, Dorrik A. V. *Sedimentary Rocks in the Field: A Color Guide.* Academic Press, 2006.

Strahler, Alan. *Introducing Physical Geography*. 5th ed., John Wiley & Sons, 2011.
Strong, William J., and George R. Plitnik. *Music, Speech, Audio*. 3rd ed., BYU Academic Press 2007.
Sundar, V. *Ocean Wave Mechanics: Applications in Marine Structures*. John Wiley & Sons, 2016.
Sutton, Gerard K., and Joseph A. Cassalli, editors. *Catastrophe in Japan: The Earthquake and Tsunami of 2011*. Nova Science Publishers, 2011.
Sverdrup, Keith A., Alyn C. Duxbury, and Alison B. Duxbury. *An Introduction to the World's Oceans*. 8th ed. McGraw-Hill Science, 2004.
Sverdrup, Keith A., and E. Virginia Armbrust. *An Introduction to the World's Oceans*. 10th ed., McGraw-Hill, 2009.
Swart, Peter K., Gregor Eberli, and Judith A. McKenzie, editors. *Perspectives in Carbonate Geology*. Wiley-Blackwell, 2009.
Swift, Gregory. "Thermoacoustic Engines and Refrigerators." *Physics Today*, vol. 48, no. 7, pp. 22-28.
Szalay, Jessie. "Caspian Sea: Largest Inland Body of Water." *LiveScience*, 24 Feb. 2017, www.livescience.com/57999-caspian-sea-facts.html.
Szczepanek, MalGorzata, et al. "Hydrogen, Carbon, and Oxygen Isotopes in Pine and Oak Tree Rings from Southern Poland as Climatic Indicators in Years 1900-2003." *Geochronometria: Journal on Methods & Applications of Absolute Chronology*, vol. 25, 2006, pp. 67-76.
Tagare, Digamber M. *Electricity Power Generation: The Changing Dimensions*. Wiley-Blackwell, 2011.
Talley, Lynne D., et al. *Descriptive Physical Oceanography: An Introduction*. 6th ed., Academic Press, 2011.
Talley, Lynne D., George L. Pickard, William J. Emery, and James H. Swift. *Descriptive Physical Oceanography: An Introduction*. 6th ed., Elsevier, 2011.
Talwani, Pradeep, editor. *Intraplate Earthquakes*. Cambridge UP, 2014.
Tang, DanLing, editor. *Remote Sensing of the Changing Oceans*. Springer, 2011.
Tang, Q., and K. Sherman, editors. *Large Marine Ecosystems of the Pacific Rim*. Blackwell Science, 1999.
Tang, Quihong, Qi Youcum, Zhihui Wang, and Yun Pan, editors. *Remote Sensing of the Terrestrial Hydrologhic Cycle*. MDPI, 2020.
TAO Project Office of NOAA/Pacific Marine Environmental Laboratory. *Upper Ocean Heat Content and ENSO*. Seattle: National Oceanic and Atmospheric Administration (NOAA) Website, www.pmel.noaa.gov/tao.
Tarbuck, Edward J., and Frederick K. Lutgens. *Earth: An Introduction to Physical Geology*. Illus. Dennis G. Tasa. 11th ed., Prentice, 2014.
Tarbuck, Edward J., Frederick K. Lutgens, and Dennis Tasa. *Earth: An Introduction to Physical Geology*. 10th ed., Prentice-Hall, 2010.
Tartar, Luc. *An Introduction to Navier-Stokes Equation and Oceanography*. Springer, 2006.
Taylor, Alan. "The Eruption of Japan's Mount Ontake." *The Atlantic*, 30 Sept. 2014, www.theatlantic.com/ photo/ 2014/09/the-eruption-of-japans-mount- ontake/100823/.
Taylor, George Frederic. *Elementary Meteorology*. Prentice-Hall, 1954.
Tedesco, M., editor. *Remote Sensing of the Cryosphere*. John Wiley & Sons, 2015.
Terence, Erica, et al. "Restoring the Shasta for Coho Salmon." *Klamath River News*, vol. 8, Summer 2012.
Tesfamichael, Dawit, and Daniel Pauly, editors. *The Red Sea Ecosystem and Fisheries*. Springer, 2016.
Theodoridis, Sergio, and Konstantinos Koutroumbas. *Pattern Recognition*. 4th ed., Academic Press, 2009.
Thomas, D. N., G. E. Fogg, and P. Convey. *The Biology of Polar Regions*. Oxford UP, 2008.
Thompson, J. William, and Kim Sorvig. *Sustainable Landscape Construction: A Guide to Green Building Outdoors*. Island Press, 2008.
Thomson, William. *Tides and the Ocean: Water's Movement Around the World from Waves to Whirlpools*. Running Press, 2018.
Thrall, Grant Ian. "MapInfo Professional 8.5: Internet-Enabled GIS." *Geospatial Solutions*, vol. 16, no. 9, 2006, pp. 36-39.
Thresher, R., and W. Musial. "Ocean Renewable Energy's Potential Role in Supplying Future Electrical Needs." *Oceanography*, vol. 23, no. 2, June 2010, pp. 16-21.
Thrower, Norman J. W. *Maps and Civilization: Cartography in Culture and Society*. 3rd ed., U of Chicago P, 2007.
Thurman, Harold V., and Alan P. Trujillo. *Essentials of Oceanography*. 10th ed., Pearson Education, 2009.
———. *Introductory Oceanography*. 10th ed., Prentice-Hall, 2003.
Tomczak, Matthias, and J. Stuart Godfrey. *Regional Oceanography: An Introduction*. Elsevier, 2013.
Trenberth, K. E. "An Imperative for Climate Change Planning: Tracking Earth's Global Energy." Current Opinion in Environmental Sustainability, vol. 1, 2009, pp. 19-27.
Trenberth, K. E., and John T. Fasullo. "Tracking Earth's Energy." Science, vol. 328, Apr. 2010, pp. 316-17.
Trimm, Harold H., and William Hunter, III. *Environmental Chemistry: New Techniques and Data*. Apple Academic Press, 2011.
Trujillo, Alan P., and Harold V. Thurman. *Essentials of Oceanography*. 10th ed., Prentice-Hall, 2010.

Tsuchiya, Yoshito, and Nobuo Shuto, editors. *Tsunami: Progress in Prediction, Disaster Prevention, and Warning*. Kluwer Academic Publishers, 2010.

"Tsunamis." *National Oceanic and Atmospheric Administration*, US Department of Commerce, June 2011, www.noaa.gov/resource-collections/tsunamis.

Tucker, Maurice E. *Sedimentary Rocks in the Field*. 4th ed., John Wiley & Sons, 2011.

Tucker, Maurice E., and Robin G. C. Bathurst, editors. *Carbonate Diagenesis*. John Wiley & Sons, 2009.

Tundisi, José Galizia, and Takako Matsumura-Tundisi. *Limnology*. CRC Press, 2011.

Tunnell, John W. Jr., et al., editors. *Coral Reefs of the Southern Gulf of Mexico*. Texas A&M UP, 2007.

Turner, David P. *The Green Marble. Earth System Science and Global Sustainability*. Columbia UP, 2018.

Turner, John. *Climate Change in the Polar Regions*. Cambridge UP, 2011.

Tyler, Carrie L., and Chris L. Schneider, editors. *Marine Conservation Paleobiology*. Springer, 2018.

Udías, Agustin, Raúl Madariaga, and Elisa Buforn. Source Mechanisms of Earthquakes: Theory and Practice. Cambridge UP, 2014.

Ulanski, Stan. *Gulf Stream: Tiny Plankton, Giant Bluefin, and the Amazing Story of the Powerful River in the Atlantic*. U of North Carolina P, 2010.

Ulrichsen, Kristian Coates. *Insecure Gulf: The End of Certainty and the Transition to the Post-Oil Era*. Oxford UP, 2015.

United Nations Educational, Scientific, and Cultural Organization (UNESCO). "The Wadden Sea." World Heritage Centre, whc.unesco.org/en/list/1314.

United Nations Environment Programme Staff. *One Planet, Many People: Atlas of Our Changing Environment*. UNEP/Earthprint, 2005.

United Nations Environmental Programme (UNEP). *Marine Plastic Debris and Microplastics: Global Lessons and Research to Inspire Action and Guide Policy Change*. United Nations, 2016.

United States Geologic Survey. "The Chesapeake Bay Bolide: Modern Consequences of an Ancient Cataclysm." woodshole.er.usgs.gov/epubs/bolide.

Unsworth, Richard K.F., Julian Clifton, and Adrian J. Smith, editors. *Marine Research and Conservation in the Coral Triangle: The Wakatobi National Park*. Nova Science, 2010.

"U.S. Refinery Yield." *EIA*, US Energy Information Administration, US Department of Energy, 31 Aug. 2018, www.eia.gov/dnav/pet/PET_PNP_PCT_ DC_NUS_ PCT_A.htm.

Vacchi, Marino, Eva Pisano, and Laura Ghigliotti, editors. *The Antarctic Silverfish: A Keysone Species in a Changing Ecosystem*. Springer, 2017.

Vacher, Leonard, and Terrence M. Quinn. *Geology and Hydrogeology of Carbonate Islands*. Elsevier, 2004.

Valencia, Raymond P., editor. *Applied Physics in the Twenty-First Century*. Nova Science, 2010.

Vallero, Daniel A. *Environmental Biotechnology: A Biosystems Approach*. Academic Press, 2010.

Vallis, Geoffrey K. *Atmospheric and Oceanic Fluid Dynamics: Fundamentals and Large-Scale Circulation*. Cambridge UP, 2006.

"The Value of Coastal Ecosystems." *UN Environment*, United Nations Environment Programme, 1 Feb. 2016, web.unep.org/coastal-eba/it/node/63.

van Aken, Hendrik Mattheus. *The Oceanic Thermohaline Circulation: An Introduction*. Springer Science + Business Media, 2007.

Van Beusekom, Ruud, Nick Langley, and Manon Tentij, editors. *The Wadden Sea: A Vision for the Conservation of a Natural Heritage*. BirdLife International, 2009.

van den Thillart, Guido, Sylvie Dufour, and J. Cliff Rankin, editors. *Spawning Migration of the European Eel: Reproduction Index, A Useful Tool for Conservation Management*. Springer, 2009.

Van der Voo, Rob. *Paleomagnetism of the Atlantic, Tethys and Iapetus Oceans*. Cambridge UP, 2004.

Van Loon, Gary W., and Stephen J. Duffy. *Environmental Chemistry: A Global Perspective*. Oxford UP, 2010.

Van Oppen, Madeleine J. H., and J. M. Lough, editors. *Coral Bleaching: Patterns, Processes, Causes, and Consequences*. Springer, 2009.

Vaughn, Jacqueline. *Environmental Politics: Domestic and Global Dimensions*. 6th ed., Wadsworth, 2011.

Venkataraman, K., C. Sivareuman, and C. Raghunathan, editors. *Ecology and Conservation of Tropical Marine Faunal Communities*. Springer, 2013.

Venter, J. Craig, et al. "Environmental Genome Shotgun Sequencing of the Sargasso Sea." *Science*, vol. 304, 2004.

Veron, J. E. N. *A Reef in Time: The Great Barrier Reef from Beginning to End*. Belknap Press/Harvard UP, 2008.

Veron, J., et al. "Delineating the Coral Triangle." *Galaxea, Journal of Coral Reef Studies*, vol. 11, no. 1, 2009.

Vespremeanu, Emil, and Mariana Golumbeanu. *The Black Sea: Physical, Environmental and Historical Perspectives*. Springer, 2018.

Vickers, Amy. *Water Use and Conservation*. Water Plow Press, 2002.

Viessman, W., Jr., and M. J. Hammer. *Water Supply and Pollution Control*. 8th ed., Prentice-Hall, 2008.

Viessman, Warren, Jr., and Gary L. Lewis. *Introduction to Hydrology*. 5th ed., Prentice-Hall, 2003.

Vincent, Warwick F., and Johanna Leybourn-Parry, editors. *Polar Lakes and Rivers: Limnology of Arctic and Antarctic Aquatic Ecosystems*. Oxford UP, 2008.

Vogt, H. "Coral Reefs and Coral Bleaching in the Region, Reefs in the Red Sea and Gulf of Aden." *Al Sanbouk*, vol. 2, no. 1, 2000.

Voituriez, Bruno. *The Gulf Stream*. UNESCO, 2006.

Vollmer, M., M. Guldberg, M. Maluck, D. Marrewijk, and G. Schlicksbier. "Landscape and Cultural Heritage in the Wadden Sea Region: Project Report." *Wadden Sea Ecosystem*, vol. 12, 2001.

Voolstra, Christian R., and Michael L. Berumen, editors. *Coral Reefs of the Red Sea*. Springer, 2019.

Wade, Timothy G., et al. "A Comparison of Vector and Raster GIS Methods for Calculating Landscape Metrics Used in Environmental Assessments." *Photogrammetric Engineering and Remote Sensing*, vol. 69, no. 12, pp. 1399-405.

Wagner, Martin, and Scott Lambert, editors. *Freshwater Microplastics: Emerging Environmental Contaminants?* Springer Open, 2017.

———. *Freshwater Microplastics: Emerging Environmental Contaminants?* Springer Open, 2018.

Wagreich, M., M. B. Hart, B. Sames, and I. O. Yilmaz, editors. *Cretaceous Climate Events and Short-Term Sea-Level Changes*. The Geological Society, 2020.

Walker, Colin H., and David R. Livingstone, editors. *Persistent Pollutants in Marine Ecosystems*. Elsevier, 2013.

Walker, Sharon, and David McMahon. *Biochemistry Demystified*. McGraw-Hill, 2008.

Wallace, John M., and Peter V. Hobbs. *Atmospheric Science: An Introductory Survey*. 2nd ed., Academic Press, 2006.

Walsh, John J. *On the Nature of Continental Shelves*. Elsevier, 2013.

Walsh, John J., editor. *Wind-Borne Illness from Coastal Seas: Present and Future Consequences of Marine Aerosols*. Academic Press, 2019.

Walter, Thomas R. "Structural Architecture of the 1980 Mount St. Helens Collapse: An Analysis of the Rosenquist Photo Sequence Using Digital Image Correlation." *Geology*, vol. 39, no. 8, 2011, pp. 767-70.

Walters, Carl J., and Steven J. D. Martell. *Fisheries Ecology and Management*. Princeton UP, 2004.

Wang, Bin. *The Asian Monsoon*. Springer/Praxis, 2006.

Wang, Pinxian, Qianyu Li, and Chun-Feng Li. *Geology of the China Seas*. Elsevier, 2014.

Wang, Qian-Wen, Gui-Peng Yang, Ze-Ming Zhang, and Shan Jian. *Perfluoralkyl Acids in Surface Sediments of the East China Sea*. Elsevier Science, 2017.

Wang, S.-Y. Simon, Jin-Ho Yoon, Christopher C. Funk, and Robert R. Gillies, editors. *Climate Extremes: Patterns and Mechanisms*. Wiley, 2017.

Ward, C. Herb, editor. *Habitats and Biota of the Gulf of Mexico: Before the Deepwater Horizon Oil Spill. Volume 1*. Springer Open, 2017.

Warner, Jeroen Frank, et al., editors. *Making Space for the River: Governance Experiences with Multifunctional River Flood Management in the US and Europe*. IWA Publishing, 2013.

Warren, John K. *Evaporites: Sediment, Resources and Hydrocarbons*. Springer-Verlag, 2006.

Waterlow, Julia. *The Atlantic Ocean*. Raintree Steck-Vaughn, 1997.

Webster, Peter. *Dynamics of the Tropical Atmosphere and Oceans*. Wiley Blackwell, 2020.

Wefer, Gerold, Wolfgang H. Berger, Gerold Siedler, and David J. Webb, editors. *The South Atlantic: Present and Past Circulation*. Springer Science & Business Media, 2012.

Weightman, Barbara A. *Dragons and Tigers: A Geography of South, East, and Southeast Asia*. 3rd ed., John Wiley & Sons, 2011.

Weiner, Eugene. *Applications of Environmental Aquatic Chemistry: A Practical Guide*. 2nd ed., CRC Press, 2008.

Weintrit, Adam, editor. *Navigational Systems and Simulators*. CRC Press, 2011.

Weis, Judith S. *Marine Pollution: What Everyone Needs to Know*. Oxford UP, 2015.

Welch, Craig. "Why Cape Town Is Running Out of Water, and Who's Next." *National Geographic*, 5 Mar. 2018, news.nationalgeographic.com/2018/02/cape-town-running-out-of-water-drought-taps-shutoff-other-cities/.

Welland, Michael. *Sand: The Never-Ending Story*. U of California P, 2009.

Wells, Neil C. *The Atmosphere and Ocean: A Physical Introduction*. 3rd ed., John Wiley &Sons, 2012.

Wiens, John A. *Oil in the Environment. Legacies and Lessons of the Exxon Valdez Oil Spill*. Cambridge UP, 2015.

Wiggert, Jerry D. *Indian Ocean Biogeochemical Processes and Ecological Variability*. American Geophysical Union, 2009.

Wilford, John N. *The Mapmakers: The Story of the Great Pioneers in Cartography from Antiquity to the Space Age*. Pimlico, 2002.

Wilkinson, Clive. *Status of Coral Reefs of the World: 2004: Volume 2*. Australian Institute of Marine Science, 2004.

Williams, Richard G., and Michael J. Follows. *Ocean Dynamics and the Carbon Cycle: Principles and Mechanisms*. Cambridge UP, 2011.

Williams, Rick. *A Future for the Fishery: Crisis and Renewal in Canada's Neglected Fishing Industry*. Nimbus Publishing, 2020.

Wilson Rowe, E., editor. *Russia and the North*. U of Ottawa P, 2009.

Wilson, Samuel T., and Matthew J. Church, editors. *Microbial Ecology in the North Pacific Subtropical Gyre*. Frontiers Media, 2018.

Windham-Myers, Lisamarie, Stephen Crooks, and Tiffany G. Troxler, editors. *A Blue Carbon Primer: The State of Coastal Wetland Carbon Science, Practice and Policy.* CRC Press, 2019.

Wohl, Ellen. *Sustaining River Ecosystems and Water Resources.* Springer, 2018.

Wolanski, Eric, John W. Day, Michael Elliott, and Ramesh Ramachandran, editors. *Coasts and Estuaries: The Future.* Elsevier, 2019.

Wolanski, Eric. *Oceanographic Processes of Coral Reefs: Physical and Biological Links in the Great Barrier Reef.* CRC Press, 2001.

Wolfson, Richard. *Energy, Environment, and Climate.* W. W. Norton, 2008.

Woods Hole Oceanographic Institution (WHOI), www.whoi.edu/index.html.

Worm, B., et al. "Impacts of Biodiversity Loss on Ocean Ecosystem Services." *Science*, vol. 314, no. 1, 2006.

Wright, Dawn J., editor. *Undersea with GIS.* ESRI Press, 2002.

Wright, Glen, Sandy A. Kerr, and Kate Johnson. *Ocean Energy: Governance Challenges for Wave and Tidal Stream Technologies.* Taylor & Frances, 2018.

Würtz, Maurizio. *Meditarranean Pelagic Habitat: Oceanographic and Biological Processes: An Overview.* IUCN, 2010.

Wylson, Anthony. *Aquatecture: Architecture and Water.* Elsevier, 2013.

Xu, Guochang, editor. *Sciences of Geodesy-I. Advances and Future Directions.* Springer, 2010.

Yalçiner Ahmet C., Efim N. Pelinovsky, Emile Okal, and Costas E. Synolakis, editors. *Submarine Landslides and Tsunamis. Springer Science & Business Media, 2012.*

Yang, Xiaofeng, Xiaofeng Li, and Ferdinando Nunziata. *Ocean Remote Sensing with Synthetic Aperture Radar.* MDPI, 2018.

Yanko-Hombach, Valentina, et al., editors. *The Black Sea Flood Question: Changes in Coastline, Climate, and Human Settlement.* Springer, 2007.

Yeomans, Matthew. *Oil: Anatomy of an Industry.* The New Press, 2004.

Young, Ian, and Alexander Babanin. *Ocean Wave Dynamics.* World Scientific, 2020.

Young, Kim, editor. *Handbook of Coastal and Ocean Engineering.* World Scientific Publishing, 2010.

Yuen, David A., Shigenori Maruyama, Shun-ichiro Karato, and Brian F. Windley, editors. *Superplumes: Beyond Plates Tectonics.* Springer, 2007.

Zalasiewicz, Jan, and Mark Williams. *Ocean Worlds. The Story of Seas on Earth and Other Planets.* Oxford UP, 2014.

Zarembo, Alan. "Southern Ocean Study Sounds More Global-Warming Alarms." *Los Angeles Times,* May 17, 2007.

Zektser, Igor S., et al., editors. *Geology and Ecosystems.* Springer, 2010.

Zeng, Eddy Y., editor. *Microplastic Contamination in Aquatic Environments: An Emerging Matter of Environmental Urgency.* Elsevier, 2018.

Zhu, Qian. *Biological Assessment of the Ecologically Important Areas for Marine Mammals in the Yellow Sea Ecoregion.* Shandong U, 2008.

Zonn, Igor S., et al. *The Caspian Sea Encyclopedia.* Springer-Verlag, 2010.

Index

F

G

P

Q

R

S

X

Y

Z